T0265310

Introduction to Strings and Branes

Supersymmetry, strings and branes are believed to be essential ingredients in a single unified consistent theory of physics. This book gives a detailed, step-by-step introduction to the theoretical foundations required for research in strings and branes.

After a study of the different formulations of the bosonic and supersymmetric point particles, the classical and quantum bosonic and supersymmetric string theories are presented. This book contains accounts of brane dynamics including D-branes and the M5-brane as well as the duality symmetries of string theory. Several different accounts of interacting strings are presented; these include the sum over world-sheets approach and the original S-matrix approach. More advanced topics include string field theory and Kac–Moody symmetries of string theory.

The book contains pedagogical accounts of conformal quantum field theory, supergravity theories, Clifford algebras and spinors, and Lie algebras. It is essential reading for graduate students and researchers wanting to learn strings and branes.

Peter West is a Professor at King's College London and a Fellow of the Royal Society. He is a pioneer in the development of supersymmetry and its application to strings and branes.

Introduction to Strings and Branes

PETER WEST

King's College London

Shaftesbury Road, Cambridge CB2 8EA, United Kingdom

One Liberty Plaza, 20th Floor, New York, NY 10006, USA

477 Williamstown Road, Port Melbourne, VIC 3207, Australia

314–321, 3rd Floor, Plot 3, Splendor Forum, Jasola District Centre, New Delhi – 110025, India

103 Penang Road, #05–06/07, Visioncrest Commercial, Singapore 238467

Cambridge University Press is part of Cambridge University Press & Assessment, a department of the University of Cambridge.

We share the University's mission to contribute to society through the pursuit of education, learning and research at the highest international levels of excellence.

www.cambridge.org
Information on this title: www.cambridge.org/9781009434096

© P. West 2012

This publication is in copyright. Subject to statutory exception and to the provisions of relevant collective licensing agreements, no reproduction of any part may take place without the written permission of Cambridge University Press & Assessment.

First published 2012
First paperback edition 2024

A catalogue record for this publication is available from the British Library

ISBN   978-0-521-81747-9   Hardback
ISBN   978-1-009-43409-6   Paperback

Cambridge University Press & Assessment has no responsibility for the persistence or accuracy of URLs for external or third-party internet websites referred to in this publication and does not guarantee that any content on such websites is, or will remain, accurate or appropriate.

Contents

v

Preface

If we have told lies you have told half lies. A man who tells lies merely hides the truth, but a man who tells half truths has forgotten where he put it.

> The British consul to Lawrence of Arabia before he arrived with the Arab army in Damascus.

In the late 1960s a small group of theorists concluded that quantum field theory could not provide a suitable description for the main problem of the time, that is, to account for hadronic physics. As a result, they began a quest to find an S-matrix that had certain preordained properties. The search culminated in the discovery of such an S-matrix for four, and then any number of, spin-0 particles. By using physical principles and mathematical consistency it was found that these S-matrix elements were part of a larger theory that possessed an infinite number of particles. Remarkably, the early pioneers found the scattering amplitudes for any number, and any type, of these particles; they even found these results at any loop order. It was subsequently realised that this was the theory of string scattering and that the theory was more suited to describe fundamental, rather than hadronic, physics.

Supersymmetry was unearthed from the world-sheet action for the ten-dimensional string and also found by independent quantum field theory considerations in Russia. Supersymmetry is entwined with string theory, but it is an independent subject. Hopefully, it will be found at the Large Hadron Collider at CERN, but even if it is, this is unlikely to be direct evidence for string theory. Supersymmetry and string theory are believed to be essential ingredients in a unified consistent theory of all physics. It was thought initially that this theory would just be the theory of ten-dimensional strings, but we now realise that it must also include branes on an equal footing. We are quite far from having a systematic understanding of the quantum properties of branes and what the underlying theory is remains unclear. Indeed, even the concepts on which it is based may be quite different to those we know now. Ironically when string theory was first discovered it was not called string theory, but the dual model, as researchers were unaware of its stringy origins, where as nowadays all discoveries on fundamental physics involving supersymmetry and supergravity are also packaged up in the term string theory. As the subject has developed sometimes string theory, and sometimes supersymmetry, has provided the dominant insights, but it remains to be seen what the mix of ideas will be in the final theory. In this book I have tried to reflect this.

The aim of this book is to provide a systematic and, hopefully, pedagogical account of the essential topics in the subject known as string theory. Almost all of the computations are carried out explicitly. The book also contains some more advanced topics; these have been selected on the basis that I know something about them and I have a wish to explain them. There are also some pedagogical chapters such as those on Clifford algebras and Lie algebras that students should know and which could well play an even more important role in future developments. There are several very important topics that are missing: Calabi–Yau compactifications, string based black hole entropy computations and the AdS–CFT duality. However, these are rapidly developing and perhaps not yet ready for a systematic, or complete, treatment. There is also a long chapter on supergravity theories reflecting the important role they have played in the subject; this includes the methods used to construct them, their symmetries and the properties of these theories in ten and eleven dimensions. Many aspects of supersymmetric theories which are not discussed in this book can be found in my book *Introduction to Supersymmetry and Supergravity* [1.11].

This book has evolved over more than 25 years and some of the calculations were performed many years ago. Although almost certainly correct when first derived they may have developed transcription errors since then. As such, if you find a factor of 2, or a minus sign out, or some other defect in the occasional place you could be correct. Hopefully, these can be corrected in a second edition.

I have tried to reference the original papers in order to give the reader a better guide to the literature and in particular access to some of the best accounts of the material presented. I have studied quite a number of the papers that I had not read before, but I may well have missed some references. For this I apologise, and I hope to put such mistakes right in the future.

The reader who wants to get to grips with the basics of string theory in the quickest possible time could take the following path: first sections 1.1–1.2.3, then chapters 2, 3, 4, 5, 7, then sections 8.1–8.3, followed by the chapters 9 and 10, then sections 13.3–13.8.4, chapter 14, and finally sections 18.1 and 18.2.

I wish to thank Paul Cook for designing the cover, which shows the projection of the roots of E_{11} onto the Coxeter plane using the SimpLie computer programme of Teake Nutma, and Pascal Anastasopoulos for drawing and helping to construct the figures. I also wish to thank Andreas Braun, Lars Brink, Lisa Carbone, Paul Cook, Finn Gubay, Arthur Greenspoon, Joanna Knapp, Neil Lambert, Andrew Pressley, Sakura Schafer-Nameki, Duncan Steele and Arkardy Tseytlin for help with proof reading sections and references. My thanks also goes to the staff and students of the Department of Mathematics, King's College London and the Technical University of Vienna for many useful comments on my lectures which were taken from this book.

1   The point particle

"We must proceed very slowly, for we are in a great hurry."
Statement at the beginning of the peace talks with the Palestinians

A number of the features of string theory are shared by the point particle. This is not too surprising as the point particle can be obtained in the limit as the string collapses to a point. Although one might think that the relativistic free point particle is a rather trivial system, it is a system with constraints and must be quantised with corresponding care. In this chapter we give the classical description of the point particle and then quantise it using first the Dirac method and then Becchi–Rouet–Stora–Tutin (BRST) quantisation techniques.

These steps are then repeated for the superparticle. There are two ways to incorporate supersymmetry into the point particle and these lead to different formulations that have, after quantisation, different physical states.

1.1   The bosonic point particle

1.1.1   The classical point particle and its Dirac quantisation

As the point particle moves through a Minkowski space-time of dimension D with coordinates x^μ, $\mu = 0, 1, \ldots, D - 1$, it sweeps out a one-dimensional curve called the world line which we choose to parameterise by τ. We may write the world line as $x^\mu(\tau)$. The motion of the point particle is taken to be so as to be an extremum of the action

$$ A = -m \int d\tau \sqrt{-\dot{x}^\mu \dot{x}^\nu \eta_{\mu\nu}}, \tag{1.1.1} $$

where $\dot{x}^\mu \equiv dx^\mu/d\tau$ and $\eta_{\mu\nu}$ is the Minkowski metric, which in our conventions is given by $\eta_{\mu\nu} = \mathrm{diag}(-1, +1, +1, \ldots, +1)$. For a time-like particle, that is, one moving at less than the speed of light, $-ds^2 = -\eta_{\mu\nu}dx^\mu dx^\nu = dt^2 - \sum_{i=1}^{D-1} dx^i dx^i > 0$, since in our units the speed of light is set to 1. For such a motion the quantity under the square root is positive. This is the reason for the minus sign under the square root. The proper time u of the particle is defined by $du^2 = -ds^2$ and we recognise that $A = -m \int du$. As a result, we conclude that the point particle moves so as to extremise its proper time.

The choice of parameterisation of the world line is of no physical significance and indeed the action of equation (1.1.1) is invariant under the reparameterisations $\tau \to \tau'(\tau)$. Such an infinitesimal transformation can be written as $\tau' = \tau - f(\tau)$, where f is a small quantity. The field x^μ is taken to be a scalar under reparameterisations, namely $x'^\mu(\tau') = x^\mu(\tau)$.

1

The infinitesimal variation of any field ϕ which is defined on a space-time labelled by ξ, all indices being suppressed, is defined to be $\delta\phi(\xi) = \phi'(\xi) - \phi(\xi)$. Therefore, an infinitesimal transformation acts on the field x^μ as

$$\delta x^\mu = x^{\mu\prime}(\tau) - x^\mu(\tau) = x^\mu(\tau + f) - x^\mu(\tau) = f(\tau)\dot{x}^\mu. \tag{1.1.2}$$

We may write the above action in an alternative, but classically equivalent way, namely

$$A = \tfrac{1}{2} \int d\tau \{ e^{-1} \dot{x}^\mu \dot{x}^\nu \eta_{\mu\nu} - m^2 e \}, \tag{1.1.3}$$

where x^μ and e are independent fields. This action is also reparameterisation invariant under the transformation $\tau' = \tau - f(\tau)$, which acts on the above fields as the infinitesimal transformations

$$\delta x^\mu = f\dot{x}^\mu, \quad \delta e = f\dot{e} + \dot{f}e. \tag{1.1.4}$$

We recognise the first term in equation (1.1.3) as D scalar fields coupled to one-dimensional gravity; e is the einbein on the one-dimensional world line and the metric is given by $g_{\tau\tau} = -e^2$.

It is often useful to introduce explicitly the momentum p^μ and write the action of equation (1.1.3) in a first order form, namely

$$A = \int d\tau \left\{ \dot{x}^\mu p^\nu \eta_{\mu\nu} - \frac{e}{2}(p^\mu p^\nu \eta_{\mu\nu} + m^2) \right\}. \tag{1.1.5}$$

Eliminating p^μ by its algebraic equation of motion in this action we recover the action of equation (1.1.3), demonstrating that these two actions are equivalent. The equations of motion of the action of equation (1.1.3) are

$$\frac{d}{d\tau}(e^{-1}\dot{x}^\mu) = 0, \quad e^2 m^2 + \dot{x}^\mu \dot{x}^\nu \eta_{\mu\nu} = 0. \tag{1.1.6}$$

Substituting e from its equation of motion into the action of equation (1.1.3), we recover the action of equation (1.1.1), while the first equation becomes the equation of motion of the action of equation (1.1.1).

All the above actions are Poincaré invariant, but the second two actions have the advantage that they can be used even in the massless case, that is, $m = 0$. In this case, the actions of equation (1.1.3) and (1.1.5) are also invariant under space-time conformal transformations. Indeed, it is straightforward to verify that the action of equation (1.1.3), with $m = 0$, is invariant under $\delta x^\mu = w^\mu$, $\delta e = (2/D)e\partial_\nu w^\nu$ provided $\partial_\mu w_\nu + \partial_\nu w_\mu = (2/D)\eta_{\mu\nu}\partial_\rho w^\rho$. We recognise the latter condition as that required for a reparameterisation to preserve the Minkowski space line element up to an arbitrary scale factor, in other words a conformal transformation. We refer the reader to chapter 8 for further details of conformal transformations.

To analyse the point particle, we may start from any of the above actions. Let us first take the action of equation (1.1.1) Taking τ as our time evolution parameter, the corresponding canonical momentum is given by

$$p^\mu = \frac{\partial L}{\partial \dot{x}_\mu(\tau)} = \frac{m\dot{x}^\mu}{\sqrt{-\dot{x}^\mu \dot{x}^\nu \eta_{\mu\nu}}}, \tag{1.1.7}$$

where L is the Lagrangian and is given by $L = -m\sqrt{-\dot{x}^\mu \dot{x}^\nu \eta_{\mu\nu}}$.

We find by inspection that the momenta automatically satisfy the constraint

$$\phi \equiv p^{\mu} p_{\mu} + m^2 = 0 \tag{1.1.8}$$

and so there are fewer momenta than coordinates. This can be viewed as a consequence of the reparameterisation invariance of the action. The Hamiltonian

$$H = p^{\mu} \dot{x}_{\mu} - L \tag{1.1.9}$$

vanishes once we substitute for p^{μ}. As the Hamiltonian is the generator of time translations and so the motion of the system, we might appear, at first sight, to have no dynamics.

The method of dealing with such a constrained system was given by Dirac in [1.1] and we encourage the reader to consult this reference. However, we give a summary of the Dirac method in appendix A that will be sufficient to completely understand the following sections. The reader who is unfamiliar with this method may wish to read appendix A before going further. However, in the discussion below the steps are rather natural and for those in a hurry it can be followed without an additional reading.

We now apply the Dirac method of quantisation to the point particle. The Poisson brackets that involve x^{μ} and p^{μ} are given by

$$\{x^{\mu}, x^{\nu}\} = 0 = \{p^{\mu}, p^{\nu}\}, \quad \{x^{\mu}, p^{\nu}\} = \eta^{\mu\nu}. \tag{1.1.10}$$

We take the Hamiltonian, which by usual methods vanishes, to be proportional to the constraint ϕ multiplied by $v(\tau)$, which is an arbitrary function of τ. It is given by

$$H = v(\tau)(p^{\mu} p_{\mu} + m^2). \tag{1.1.11}$$

Continuing the Dirac procedure we must demand that the constraint is preserved in time. However, in this case, we find that $(d/d\tau)\phi = \{\phi, H\} = 0$ and so there is no new constraint. Hence we have a system which has only one constraint which obviously obeys $\{\phi, \phi\} = 0$, and so it is a first class constraint in the language of Dirac.

The Hamiltonian H generates time translations and so the equations of motion are

$$\frac{dx^{\mu}}{d\tau} = \{x^{\mu}, H\} = 2v(\tau)p^{\mu}, \quad \frac{dp^{\mu}}{d\tau} = \{p^{\mu}, H\} = 0. \tag{1.1.12}$$

We note that reparameterisations change the dependence of the coordinates x^{μ} on time while time evolution shifts the time dependence. As such, reparameterisations and time evolution have much in common and in particular $\tau \to \tau +$ constant is a reparameterisation which can also be viewed as a time evolution. This is why the right-hand side of the above equations of motion resembles a reparameterisation when multiplied by an appropriate parameter.

To quantise the theory we make the usual transition, according to the Dirac rule, from Poisson brackets to commutators, with an appropriate factor of $i\hbar$. The commutators for the coordinates and momenta are then given by $[x^{\mu}, x^{\nu}] = 0 = [p^{\mu}, p^{\nu}]$ and $[x^{\mu}, p^{\nu}] = i\hbar\eta^{\mu\nu}$. These commutators are represented by the replacements

$$x^{\mu} \to x^{\mu}; \quad p^{\mu} \to -i\hbar\frac{\partial}{\partial x_{\mu}}. \tag{1.1.13}$$

Setting $\hbar = 1$, the constraint becomes

$$\hat{\phi} = (-\partial^2 + m^2). \tag{1.1.14}$$

This is no longer an algebraic condition, but a differential operator. To proceed further, we consider the particle to be described by a wavefunction, or field, $\psi(x^\mu, \tau)$ and we impose the constraint

$$\hat{\phi}\psi = (-\partial^2 + m^2)\psi = 0. \tag{1.1.15}$$

We also impose the Schrödinger equation

$$i\hbar\frac{\partial\psi}{\partial\tau} = H \cdot \psi. \tag{1.1.16}$$

However, the right-hand side of this equation vanishes once the constraint ϕ is imposed and we find that ψ is independent of τ. We recognise the usual formulation of a second-quantised spin-0 particle namely, the τ dependence has disappeared and we are left with the Klein–Gordon equation.

Let us now briefly consider starting from the alternative action of equation (1.1.3). The momentum conjugate to x^μ is $p^\mu = e^{-1}\dot{x}^\mu$, but the momentum p_e conjugate to e vanishes. We take this latter condition as a constraint: $r \equiv p_e = 0$. The Hamiltonian is found to be $H = (e/2)(p^\mu p_\mu + m^2) + s(\tau)p_e$, where s is an arbitrary function of τ. The non-vanishing fundamental Poisson brackets are $\{x^\mu, p^\nu\} = \eta^{\mu\nu}$, $\{e, p_e\} = 1$. Insisting that the time development of the constraint $r = 0$ should vanish implies that $\dot{r} = \{r, H\} = \frac{1}{2}(p^\mu p_\mu + m^2) = 0$. Hence, we recover the constraint of equation (1.1.7) and the Hamiltonian vanishes as it is proportional to the constraints. It is easy to verify that there are no further constraints.

To quantise the system we proceed much as before. We turn the Poisson brackets into commutators with an $i\hbar$ factor and adopt a Schrödinger representation. The constraints are then imposed on the wavefunction, which depends on the coordinates x^μ and e. The constraint $p^\mu p_\mu + m^2 = 0$ becomes the Klein–Gordon equation, while the other constraint states that the wavefunction does not depend on e. The Schrödinger equation simply states that the wavefunction does not depend on τ. Hence, we arrive at the same quantum system.

Although we may not implement a gauge choice naively in the action, we may use it on the equations of motion of any system. We note that the equations of motion of the action of equation (1.1.3) given in equation (1.1.6) can be simplified by a suitable gauge choice. Indeed, we may use the reparameterisation invariance of equation (1.1.4) to choose $e = 1$, whereupon they become

$$\ddot{x}^\mu = 0, \quad \dot{x}^\mu\dot{x}^\nu\eta_{\mu\nu} + m^2 = 0. \tag{1.1.17}$$

We note that the second equation, when expressed in terms of momenta, is just the constraint found above. The constraint $\dot{x}^\mu\dot{x}^\nu\eta_{\mu\nu} + m^2 = 0$ is none other than the condition that the energy-momentum tensor of the one-dimensional system in the absence of gravity should vanish. We may read off the energy-momentum tensor by substituting $e = 1 + h$ into the action of equation (1.1.3) expanding in terms of h and taking the coefficient of the term linear in h. Another possible gauge choice is to take $\dot{x}^0 = 1$. This is called the static gauge and an analogue of it will be used extensively when we come to discuss branes.

1.1.2   The BRST quantization of the point particle

We now wish to apply the BRST approach to the point particle. The BRST transformations were found in [1.2], they were further developed in [1.3] and the relation between the BRST charge and the physical states were found in [1.4]. This approach will be particularly important for the string. The reader may wish to first read appendix A, where the BRST formulation of Yang–Mills theory is given and the general procedure is explained. We begin with the action of equation (1.1.3), which is world-line reparameterisation invariant under the transformations of equation (1.1.2). Corresponding to the one local invariance with parameter $f(\tau)$, we introduce the ghost $c(\tau)$ and anti-ghost $b(\tau)$, which are both Grassmann odd.

The BRST transformations of the original fields are found by the substitution $f(\tau) \to \Lambda c(\tau)$, where Λ is a Grassmann odd BRST parameter, into the original local transformations of equation (1.1.4). The result is

$$\delta x^\mu = (\Lambda c)\dot{x}^\mu, \quad \delta e = \frac{d}{d\tau}((\Lambda c)e). \tag{1.1.18}$$

We choose c to be Hermitian, but Λ is taken to be anti-Hermitian in order that Λc be real. The standard rule for taking the complex conjugate of two Grassmann odd variables is to reverse their order and then take their individual complex conjugates; thus

$$(\Lambda c)^* = c^*\Lambda^* = -\Lambda^* c^* = \Lambda c. \tag{1.1.19}$$

The transformation law of the ghost is given by

$$\delta c = (\Lambda c)\dot{c}. \tag{1.1.20}$$

Under an infinitesimal reparameterisation x^μ transforms as $\delta x^\mu = f\dot{x}^\mu$; carrying out the commutation of two infinitesimal reparameterisations f_1 and f_2 yields a third reparameterisation with parameter

$$f_{12} = (-f_1\dot{f}_2 + f_2\dot{f}_1). \tag{1.1.21}$$

Following the prescription in appendix A, we find that $\tilde{f}_{12} = 2\Lambda c\dot{c}$ and so equation (1.1.20).

To the fields x^μ, e, c, b we add the Lagrange multiplier λ (called B in appendix A). The anti-ghost b transforms into the multiplier λ, namely,

$$\delta\lambda = 0, \quad \delta b = \Lambda\lambda. \tag{1.1.22}$$

Although the above may seem like a cookery book recipe, we have arrived at one of the desired results, namely a set of nilpotent transformations. For example, two transformations on c are given by

$$\delta_{\Lambda_1}\delta_{\Lambda_2}c = \delta_{\Lambda_1}\{(\Lambda_2 c)\dot{c}\}$$
$$= \Lambda_2((\Lambda_1 c)\dot{c})\dot{c} + \Lambda_2 c\frac{d}{d\tau}\{(\Lambda_1 c)\dot{c}\} = 0. \tag{1.1.23}$$

We now choose the gauge fixing function to be

$$G = \ln e. \tag{1.1.24}$$

Setting $G = 0$ sets $e = 1$. Consequently, we should add to the original action of equation
(1.1.3) the gauge fixing term

$$A^{gf} = \int d\tau \lambda \, \ln e. \tag{1.1.25}$$

A BRST invariant action is given by

$$
\begin{aligned}
A^{BRST} &= A^{orig} + A^{gf} + A^{gh} \\
&= \tfrac{1}{2} \int d\tau \{ e^{-1} \dot{x}^\mu \dot{x}^\nu \eta_{\mu\nu} - m^2 e \} + \int d\tau \lambda \, \ln e - \int d\tau b D_\tau c,
\end{aligned} \tag{1.1.26}
$$

where A^{orig} is the action of equation (1.1.3) and A^{gf} is that of equation (1.1.25). Due to its
original reparameterisation invariance A^{orig} is automatically BRST invariant. We find by
cancelling the variations of A^{gf} that

$$A^{gh} = - \int d\tau b D_\tau c, \tag{1.1.27}$$

where $D_\tau c = \dot{c} + (d \ln e / d\tau)c$. An alternative way of arriving at the above result is to note
that under a BRST transformation

$$\delta_\Lambda \{ b(\ln e) \} = \Lambda (\lambda (\ln e) - b D_\tau c), \tag{1.1.28}$$

and use the nilpotency of δ_Λ to establish the BRST invariance of $A^{gf} + A^{gh}$.

The quantum theory is then given by the functional integral

$$\int De Dx^\mu Dc Db D\lambda \, \exp(iA^{BRST}).$$

In this functional integral, we can carry out the λ integration which sets $e = 1$, whereupon
the BRST action becomes

$$A^{final} = \int d\tau \left(\tfrac{1}{2} \dot{x}^\mu \dot{x}^\nu \eta_{\mu\nu} - \tfrac{1}{2} m^2 - b\dot{c} \right). \tag{1.1.29}$$

This result is still BRST invariant; however, for δb we must substitute the value of λ given
by the e equation of motion with $e = 1$, that is,

$$\delta b = \Lambda \left(\frac{\dot{x}^\mu \dot{x}_\mu}{2} + \frac{m^2}{2} - \frac{d}{d\tau}(bc) \right), \tag{1.1.30}$$

the other variations being unchanged.

As the action of equation (1.1.29) is BRST invariant, we can in the standard way deduce
the associated Noether current Q which, in this one-dimensional case, is also the BRST
charge. We find that it is given by

$$Q = c(p^\mu p_\mu + m^2), \tag{1.1.31}$$

where $p^\mu = \dot{x}^\mu$ is the momentum conjugate to x^μ. We take the definition of momenta for
Grassmann odd variables to be left differentiation of the action by the coordinate. Hence,
the momentum for the coordinate c is given by

$$\frac{\vec{\partial}}{\partial \dot{c}} A = b. \tag{1.1.32}$$

We could also take b as our coordinate and then c would be the corresponding momentum.

We now give a Poisson bracket suitable for a general system which contains Grassmann even and odd variables. Given a system with coordinates q_A and corresponding momenta p_A, some of which may be Grassmann odd, we define the Poisson bracket of two functions f and g of q_A and p_A as

$$\{f, g\}_{PB} = \sum_A \left\{ f \frac{\overleftarrow{\partial}}{\partial q_A} \frac{\overrightarrow{\partial}}{\partial p_A} g - (-1)^{fg} g \frac{\overleftarrow{\partial}}{\partial q_A} \frac{\overrightarrow{\partial}}{\partial p_A} f \right\}, \tag{1.1.33}$$

where

$$(-1)^{fg} = \begin{cases} -1 & \text{if } f \text{ and } g \text{ are Grassmann odd,} \\ 1 & \text{otherwise.} \end{cases}$$

It satisfies the relations

$$\{f, g\}_{PB} = -(-1)^{fg}\{g, f\}_{PB}, \quad \{f, gk\}_{PB} = (-1)^{fg}g\{f, k\}_{PB} + \{f, g\}_{PB}k,$$
$$\{f_1 + f_2, g\}_{PB} = \{f_1, g\}_{PB} + \{f_2, g\}_{PB} \tag{1.1.34}$$

as well as a generalised Jacobi identity.

Returning to the point particle and the action of equation (1.1.29). We find that the resulting non-zero Poisson brackets for the coordinates and momenta are

$$\{x^\mu, p^\nu\}_{PB} = \eta^{\mu\nu}, \quad \{c, b\}_{PB} = 1. \tag{1.1.35}$$

The Hamiltonian associated to the action of equation (1.1.29) is given by

$$H = p^\mu p_\mu + m^2. \tag{1.1.36}$$

The BRST charge is the generator of transformations in the usual sense that

$$\delta \bullet = \{\bullet, \Lambda Q\}_{PB}, \tag{1.1.37}$$

where $\bullet$ is any field. The reader may verify that, on-shell, these transformations agree with those previously given and are an invariance of the Hamiltonian equations of motion. We note that the Q satisfies the equation $\{Q, Q\}_{PB} = 0$.

To quantise the system we apply the Dirac rule

$$\{,\}_{PB} \rightarrow \begin{cases} \dfrac{1}{i\hbar} & \{,\} \text{ for two odd quantities,} \\[2mm] \dfrac{1}{i\hbar} & [,] \text{ otherwise} \end{cases} \tag{1.1.38}$$

to the Poisson brackets. In the above, $\{A, B\} \equiv AB + BA$ and $[A, B] \equiv AB - BA$. The use of the symbol $\{,\}$ in the quantum theory should not be confused with the classical Poisson bracket used in the other sections in this book where it is not generally given the subscript *PB*. Consequently, we must demand

$$[x^\mu, p^\nu] = i\hbar\eta^{\mu\nu}, \quad \{c, b\} = i\hbar. \tag{1.1.39}$$

We may use the generalisation of the Schrödinger representation:

$$x^\mu \rightarrow x^\mu, \quad c \rightarrow c \quad p^\mu \rightarrow -i\hbar\frac{\partial}{\partial x^\mu} \quad b \rightarrow i\hbar\frac{\partial}{\partial c}. \tag{1.1.40}$$

In checking the appearance of *i*s, it is important to remember that *b* is anti-Hermitian.

The BRST charge now becomes the operator

$$Q = c(-\partial^2 + m^2) \tag{1.1.41}$$

and it is obviously nilpotent, that is, $Q^2 = 0$ as $c^2 = 0$.

We now consider wavefunctions of the coordinates, that is, $\Psi(x^\mu, c)$. Taylor expanding in c, the wavefunction has the form

$$\Psi(x^\mu, c) = \psi(x^\mu) + c\phi(x^\mu) \tag{1.1.42}$$

as c is Grassmann odd. In the BRST formalism, physical states are taken to satisfy the condition

$$Q\Psi = 0. \tag{1.1.43}$$

This condition is equivalent to

$$(-\partial^2 + m^2)\psi = 0, \tag{1.1.44}$$

which is the usual Klein–Gordon result. Two physical states are equivalent if they differ by a term of the form $Q\Omega$. We may take $\Omega(x^\mu, c)$ to be of the form $\Omega(x^\mu, c) = a(x^\mu) + cb(x^\mu)$ and so $Q\Omega(x^\mu, c) = c(-\partial^2 + m^2)a(x)$. Thus two fields ϕ are equivalent if they differ by a term of the form $(-\partial^2 + m^2)a(x)$ for any a and so any ϕ is equivalent to the trivial field 0. Thus the physical states are described by only the field ψ subject to equation (1.1.44), in other words the well-known result.

We observe that the results of the usual quantisation carried out in section (1.1.1) can be rewritten in terms of the BRST formalism. For example, equation (1.1.15) can be written as $Q\Psi = 0$ if the field Ψ is subject to $b\Psi = 0$. The latter condition sets ϕ to zero. We also note that the equation $Q\Psi = 0$ has a local invariance under $\delta\Psi = Q\Lambda$, where now Λ is a arbitrary function of space-time. At first sight, these observations could be regarded as a cumbersome way of describing the point particle. However, with hindsight, and after the discovery of an analogous equation for the string, it was realised that what had looked like an artificial manipulation of the BRST formalism had a deeper significance. The meaning of the above statements will become clearer when we study gauge covariant string theory in chapter 12.

The method of BRST quantization originated [1.2] in the context of Yang–Mills theory, which is still the prototype example of how to proceed. The systematic use of the BRST charge, for general systems with first class constraints, was carried out in [1.3]. For some reviews of this procedure, see [1.4]. It must be stated, however, that the BRST method, as with any quantisation method, is more like an art than a science. Its justification is that the final result, namely a nilpotent set of transformations and an invariant action, usually defines a quantum theory which is unitary and whose physical observables are independent of how the gauge was fixed.

1.2   The super point particle

The key to finding systems whose quantisation leads to particles with spin is the introduction of Grassmann odd degrees of freedom [1.5, 1.6]. There are, however, two different ways to proceed: we may extend the ordinary point particle by encoding either a world-line supersymmetry or a space-time supersymmetry. These two formulations, called the spinning particle [1.7, 1.8] and the Brink–Schwarz particle [1.9], respectively, look very different and indeed lead to different physics. While the former may be quantised straightforwardly, the covariant quantisation of the latter presents formidable problems. In what follows we explain both formulations beginning with the spinning particle. We also explain how twistors can be used to give alternative formulations of the super point particle that avoid the quantisation problems found in one of the formulations. It is interesting to compare the discovery of the supersymmetric formulation of the spinning particle with the corresponding development, as discussed in chapter 6, of the superstring with world-sheet supersymmetry whose supersymmetric action was found very shortly afterwards.

1.2.1   The spinning particle

The massless bosonic particle is described by D fields x^μ, $\mu = 0, 1, \ldots, D - 1$ coupled to one-dimensional gravity with einbein e. To extend this to a formulation with world-line supersymmetry we begin with a theory of D fields x^μ and their superpartners χ^μ which is invariant under rigid supersymmetry. We will then couple this system to one-dimensional, or world-line, supergravity which is described by the graviton and its superpartner, the gravitino ψ.

Let us therefore consider the action

$$A^F = \tfrac{1}{2} \int d\tau (\dot{x}^\mu \dot{x}^\nu - i\chi^\mu \dot{\chi}^\nu) \eta_{\mu\nu}. \tag{1.2.1}$$

We take x^μ and χ^μ to be Grassmann even and odd, respectively. We recall that in the classical theory, Grassmann even objects commute with Grassmann even and Grassmann odd objects, but that Grassmann odd objects anti-commute with Grassmann odd objects, that is, $\chi^\mu \chi^\nu = -\chi^\nu \chi^\mu$. The action of equation (1.2.1) is invariant under rigid time translations

$$\delta x^\mu = a\dot{x}^\mu, \ \delta \chi^\mu = a\dot{\chi}^\mu \tag{1.2.2}$$

and rigid supersymmetry, whose transformations are given by

$$\delta x^\mu = i\epsilon \chi^\mu, \quad \delta \chi^\mu = \dot{x}^\mu \epsilon, \tag{1.2.3}$$

where ϵ is Grassmann odd. We choose ϵ and χ^μ to be real, that is, $\epsilon^* = \epsilon$, $\chi^{\mu*} = \chi^\mu$. The commutator of two such transformations is found to be

$$[\delta_1, \delta_2]\bullet = 2i\epsilon_2\epsilon_1 \frac{d}{d\tau} \bullet \tag{1.2.4}$$

acting on either x^μ or χ^μ, which are denoted by $\bullet$ in the above equation. We recognise the result as a time translation of magnitude $2i\epsilon_2\epsilon_1$.

The supersymmetry current j is given by

$$j = \dot{x}^\mu \chi^\nu \eta_{\mu\nu}. \tag{1.2.5}$$

One standard method of finding the current corresponding to any rigid symmetry is to let the parameter of the symmetry become local, that is, space-time dependent, and then compute the variation of the action. Since the action is invariant when the parameter is constant, the variation of the action must contain the space-time derivative of the parameter times a quantity that is just the current. This identification follows from the fact that any variation of the action is given by the equation of motion multiplied by the field variation and so it must vanish when the equations of motion are enforced. As a result, in the case of the above variation we conclude that the object identified as the current is indeed conserved if the equations of motion hold. In the case under consideration here we let $\epsilon \to \epsilon(\tau)$ and write the variation of the action as $\delta A = i \int d\tau ((d/d\tau)\epsilon) j$. Carrying out this calculation we find the supersymmetry current of equation (1.2.5).

The coupling of the action of equation (1.2.1) to world-line supergravity (e, ψ) is given by

$$A = \tfrac{1}{2} \int d\tau \left(e^{-1}\dot{x}^{\mu}\dot{x}^{\nu} - i\chi^{\mu}\dot{\chi}^{\nu} - \kappa e^{-1}i\psi\chi^{\mu}\dot{x}^{\nu}\right)\eta_{\mu\nu} \tag{1.2.6}$$

and the local supersymmetry transformations which leave it invariant are

$$\delta x^{\mu} = i\epsilon\chi^{\mu}, \quad \delta\chi^{\mu} = \left(\dot{x}^{\mu} - \frac{\kappa}{2}i\psi\chi^{\mu}\right)\epsilon e^{-1},$$

$$\delta e = i\kappa\epsilon\psi, \quad \delta\psi = \frac{2}{\kappa}\dot{\epsilon}, \tag{1.2.7}$$

where the supersymmetry parameter ϵ is an arbitrary function of τ. It is also invariant under reparameterisation symmetry $\delta x^{\mu} = k\dot{x}^{\mu}$, $\delta\chi^{\mu} = k\dot{\chi}^{\mu}$, $\delta e = d(ke)/d\tau$ and $\delta\psi = d(k\psi)/d\tau$, where k is an arbitrary function of τ.

We now explain how this result is found using the Noether method since it provides a simple example which illustrates most of the points required to construct supergravity theories using this method. An explanation of the Noether technique is given in chapter 13. The Noether technique is not required again for the super point particle and the reader who is not interested in this derivation may skip the following discussion and resume at equation (1.2.24).

We start with the action of equation (1.2.1) with the rigid, that is, constant, supersymmetry transformations of equation (1.2.3) and the linearized supergravity fields h and ψ which have the rigid supersymmetry transformations

$$\delta h = i\epsilon\psi, \quad \delta\psi = 0. \tag{1.2.8}$$

The supergravity fields have the Abelian local transformations

$$\delta h = \dot{g}, \quad \delta\psi = \dot{\eta}, \tag{1.2.9}$$

where g and η are arbitrary functions of τ, which are Grassmann even and odd, respectively. Despite the unusual appearance of the transformations of equations (1.2.8) it is trivial to verify that they close provided one allows for the occurrence of the transformations of equations (1.2.9). Of course, h and ψ can be gauged away using these latter transformations corresponding to the fact there is no gravity, or supergravity, in one dimension. In what follows we will suppress the μ, ν indices until we have found the local result.

We now let the previously constant supersymmetry parameter ϵ become τ-dependent. The action A^F is no longer invariant, but

$$A^1 = A^F - i \int d\tau \frac{\kappa}{2} \psi j \tag{1.2.10}$$

is invariant to order κ^0 if we identify $\eta = (2/\kappa)\epsilon$; that is, link the η invariance to the now local supersymmetry. The variation of A^1 is readily found to be

$$\delta A^{(1)} = -i \int d\tau \frac{\kappa}{2} (\psi \dot{x} \ddot{x} \epsilon + i \psi \epsilon \dot{\chi} \chi). \qquad (1.2.11)$$

We now gain invariance order by order in κ by adding terms either to the action or to the transformation rules. The last term above is cancelled by adding $-i(\kappa/2)\psi \chi \epsilon$ to $\delta \chi$ since to the order to which we are working this new addition to $\delta \chi$ only contributes to the variation of the kinetic term of χ^μ. The first term above is cancelled by adding the term

$$\int d\tau \left(-\frac{\kappa}{2} h \dot{x} \ddot{x} \right) \qquad (1.2.12)$$

to the action. However, this variation also creates two new terms, namely

$$\int d\tau \{ -i\kappa h \dot{\epsilon} \chi \dot{x} - i\kappa h \epsilon \dot{\chi} \dot{x} \}, \qquad (1.2.13)$$

which we cancel by adding the term

$$\int d\tau i \frac{\kappa^2}{2} h \psi \dot{x} \chi \qquad (1.2.14)$$

to the action and the term $-\kappa h \dot{x} \epsilon$ to $\delta \chi$. Consequently, we find that the action

$$A^{(2)} = \int d\tau \left\{ \frac{1}{2} \dot{x} \ddot{x} (1 - \kappa h) - \frac{i}{2} \chi \dot{\chi} - i \frac{\kappa}{2} \psi \dot{x} \chi (1 - \kappa h) \right\} \qquad (1.2.15)$$

is invariant to order κ^1 under the transformations

$$\delta x = i\epsilon \chi, \qquad \delta \chi = \left(\dot{x}(1 - \kappa h) - i \frac{\kappa}{2} \psi \chi \right) \epsilon,$$

$$\delta h = i\epsilon \psi, \qquad \delta \psi = \frac{2}{\kappa} \dot{\epsilon}. \qquad (1.2.16)$$

We can now proceed to the next stage; varying $A^{(2)}$ we find the κ^2 terms

$$\delta A^{(2)} = \int d\tau \left(-\frac{\kappa^2}{2} \psi \epsilon \dot{\chi} \chi h - \frac{1}{2} i \kappa^2 \psi \dot{x} \ddot{x} \epsilon h \right). \qquad (1.2.17)$$

The first term is cancelled by adding $+i(\kappa^2/2)\psi \chi \epsilon h$ to $\delta \chi$ while adding $\int d\tau \frac{1}{2} \dot{x} \ddot{x} \kappa^2 h^2$ cancels the last term but creates two more terms which we in turn cancel by adding $+\kappa^2 \dot{x} \epsilon h^2$ to $\delta \chi$ and $\int d\tau (-i(\kappa^3/2) \psi \dot{x} \chi h^2)$ to the action. Hence, we find the action

$$A^{(3)} = \int d\tau \frac{1}{2} \dot{x} \ddot{x} (1 - \kappa h + \kappa^2 h^2) - \frac{i}{2} \chi \dot{\chi} - \frac{\kappa}{2} i \psi \dot{x} \chi (1 - \kappa h + \kappa^2 h^2) \qquad (1.2.18)$$

is invariant to order κ^2 under the transformations

$$\delta x = i\epsilon \chi, \qquad \delta \chi = \left[\dot{x}(1 - \kappa h + \kappa^2 h^2) - i \frac{\kappa}{2} \psi \chi (1 - \kappa h) \right] \epsilon,$$

$$\delta h = i\epsilon \psi, \qquad \delta \psi = \frac{2}{\kappa} \dot{\epsilon}. \qquad (1.2.19)$$

We could carry on step by step, but the general pattern now emerges. If we define $e = 1 + \kappa h$, then the action [1.7] invariant to all orders under local supersymmetry takes the form

$$A = \int d\tau \left\{ \frac{e^{-1}}{2} \dot{x}\dot{x} - \frac{i}{2}\chi\dot{\chi} - i\frac{\kappa}{2}\psi\dot{x}\chi e^{-1} \right\} \tag{1.2.20}$$

and the transformation laws become

$$\delta x = i\epsilon\chi, \qquad \delta\chi = e^{-1}\left(\dot{x} - i\frac{\kappa}{2}\psi\chi\right)\epsilon,$$

$$\delta e = i\kappa\epsilon\psi, \qquad \delta\psi = \frac{2}{\kappa}\dot{\epsilon}. \tag{1.2.21}$$

They agree with the above results to the relevant orders of κ. It is straightforward to verify that the transformations of equation (1.2.21) leave the action of equation (1.2.20) invariant to all orders in κ. Thus we have derived the action of equation (1.2.20) once we restore the μ, ν indices.

We must also check that the new transformations form a closed algebra. We find that

$$[\delta_1, \delta_2]x = 2i\epsilon_2\epsilon_1 e^{-1}\dot{x} + \kappa\epsilon_2\epsilon_1\psi\chi e^{-1}$$

$$= k\dot{x} + i\epsilon_c\chi, \tag{1.2.22}$$

where $k = 2i\epsilon_2\epsilon_1 e^{-1}$ and $\epsilon_c = -i\kappa\epsilon_2\epsilon_1\psi e^{-1}$ are the parameters of a reparameterisation and supersymmetry transformation. The same closure relation holds for χ. For the fields h and ψ we find that

$$[\delta_1, \delta_2]e = 2i\frac{d}{d\tau}(\epsilon_2\epsilon_1)$$

$$= \frac{d}{d\tau}(ke) = i\kappa\epsilon_c\psi + \frac{d}{d\tau}(ke), \tag{1.2.23}$$

$$[\delta_1, \delta_2]\psi = 0 = \frac{d}{d\tau}(k\psi) + \frac{2}{\kappa}\frac{d}{d\tau}\epsilon_c.$$

We recognise these results as consistent with the fact that h and ψ carry the indices $h_{\mu\nu}$ and $\psi_{\mu\alpha}$, but as these indices only take one value in one dimension they are not shown.

We observe that in the process of constructing a locally supersymmetric theory we have automatically found one that is also reparameterisation invariant. This is to be expected as the commutator of two supersymmetry transformations is a translation and so if the supersymmetries become local we expect invariance under local translations or reparameterisations. In the above process therefore, the rigid translation a of equation (1.2.2) has become local and knitted together with the local Abelian transformation of equation (1.2.9) using the identification $a/\kappa = g$ so that at the lowest order $\delta h = a\dot{h} + \dot{a}/\kappa$. The corrections at higher orders then give the result $\delta e = a\dot{e} + \dot{a}e$.

When carrying out the Noether procedure we can often save considerable work by anticipating the local reparameterisation invariance of the theory and encoding it at the end of each step in the process. Carrying this out for the super point particle we would stop after only the first step since then we would have arrived at the final result. This completes the derivation of equation (1.2.6) using the Noether method.

We can write the superparticle action in a first order form as follows:

$$\int d\tau \left(\dot{x}^\mu p_\mu - \frac{1}{2}ep^\mu p_\mu - \frac{i}{2}\chi_\mu\dot{\chi}^\mu - \frac{i\kappa}{2}\psi\chi^\mu p_\mu \right), \tag{1.2.24}$$

which is invariant under the local supersymmetry transformations

$$\delta x^\mu = i\epsilon\chi^\mu, \qquad \delta\chi^\mu = p^\mu\epsilon, \ \delta p^\mu = 0,$$
$$\delta e = i\kappa\epsilon\psi, \qquad \delta\psi = \frac{2}{\kappa}\dot\epsilon. \tag{1.2.25}$$

The equations of motion of equation (1.2.24) are

$$\dot p^\mu = 0,$$
$$\dot\chi^\mu - \frac{\kappa}{2}p^\mu\psi = 0, \tag{1.2.26}$$
$$\dot x^\mu - ep^\mu - i\frac{\kappa}{2}\psi\chi^\mu = 0$$

as well as those for the fields e and ψ:

$$\phi \equiv p_\mu p^\mu = 0,$$
$$\varphi \equiv p_\mu\chi^\mu = 0. \tag{1.2.27}$$

Eliminating p^μ, using its equation of motion, we recover the action and transformations of equations (1.2.6) and (1.2.7). The constraints of equation (1.2.27) also arise from the action of equation (1.2.20) by following the Dirac method. The analogous derivation was carried out below equation (1.1.16) for the point particle in the formulation of the above action with the Grassmann odd fields vanishing. In particular the constraints of equation (1.2.27) arise by demanding that the constraints that momenta for e and ψ vanish are preserved in time.

Using the reparameterisation and local supersymmetry invariance we can choose gauges where

$$e = 1, \qquad \psi = 0. \tag{1.2.28}$$

Implementing these choices, the equations of motion become

$$\dot p^\mu = 0, \quad \dot\chi^\mu = 0, \quad p^\mu = \dot x^\mu, \tag{1.2.29}$$

which are those from the free action of equation (1.2.1), after eliminating p_μ. However, we also have the two constraints of equation (1.2.27).

Let us now consider a Hamiltonian approach using the action of equation (1.2.24). We now *adopt* the Poisson brackets for the fundamental variables

$$\{x^\mu, x^\nu\} = 0 = \{p^\mu, p^\nu\}, \quad \{x^\mu, p^\nu\} = \eta^{\mu\nu}, \quad \{\chi^\mu, \chi^\nu\} = i\eta^{\mu\nu}, \tag{1.2.30}$$

with the Poisson brackets between these Grassmann even and odd variables vanishing.

The Hamiltonian is read off from the above action to be

$$H = \frac{1}{2}ep_\mu p^\mu + i\frac{\kappa}{2}\psi p_\mu\chi^\mu, \tag{1.2.31}$$

which is just proportional to the constraints. The Hamiltonian H generates the equations of motion of equation (1.2.26) by the formula

$$\frac{d\bullet}{d\tau} = \{\bullet, H\}, \tag{1.2.32}$$

where • denotes any field. The reader may verify that one does indeed find the correct equations of motion, that is, those of equation (1.2.26). This calculation provides a justification for the Poisson bracket of equations (1.2.30). Using the Poisson brackets we can show that the two constraints ϕ and φ satisfy the closed algebra

$$\{\phi, \varphi\} = 0, \quad \{\varphi, \varphi\} = i\phi, \tag{1.2.33}$$

which is a reflection of the underlying supersymmetry algebra. Since the Poisson brackets of the constraints again give constraints we conclude that the constraints are by definition first class.

In the Dirac method we do not use the definition of the Poisson bracket of say equation (1.1.30), but we adopt from the beginning the standard Poisson brackets between the coordinates and their momenta. The Poisson brackets are taken to obey the relations of equation (1.1.34), namely $\{f, g\} = -(-1)^{fg}\{g, f\}$, $\{f, gk\} = (-1)^{fg}g\{f, k\} + \{f, g\}k$, $\{f_1 + f_2, g\} = \{f_1, g\} + \{f_2, g\}$, where $(-1)^{fg}$ is -1 if f and g are Grassmann odd and $+1$ otherwise. In particular, we note that $\{\chi^\mu, \chi^\nu\} = \{\chi^\nu, \chi^\mu\}$ as these are two Grassmann odd quantities. In this way we can evaluate the Poisson bracket of any two functions of coordinates and momenta. As usual, only after the Poisson brackets are evaluated are the constraints of equation (1.2.27) implemented.

When following the Dirac method we encounter a subtlety. The momentum conjugate to x^μ is clearly p_μ. However, the momentum conjugate to χ^μ is $\frac{i}{2}\chi^\mu$. To correctly take account of the fact that χ^μ and its momentum p_χ^μ are equal we must adopt the constraint $s^\mu \equiv p_\chi^\mu - (i/2)\chi^\mu = 0$. Starting from the standard Poisson bracket $\{\chi^\mu, p_\chi^\nu\} = \eta^{\mu\nu}$, we discover that the constraint $s^\mu = 0$ is second class, that is, its Poisson bracket with the other constraints, and in particular itself, is not a constraint. The correct procedure is to calculate the modified Poisson bracket, sometimes called the Dirac bracket. The net effect of this is the Poisson bracket for χ^μ of equation (1.2.30) instead of the naive result. This explains from first principles why the Poisson bracket for χ^μ of equation (1.2.30) is out by a factor of 2 from what one might naively expect.

To quantise the superparticle [1.8] we apply the Dirac rule of equation (1.1.38) to arrive at the relations

$$[x^\mu, x^\nu] = 0 = [p^\mu, p^\nu], \ [x^\mu, p^\nu] = i\hbar\eta^{\mu\nu}, \quad \{\chi^\mu, \chi^\nu\} = -\hbar\eta^{\mu\nu} \tag{1.2.34}$$

all others vanishing. As explained in the previous subsection the Grassmann odd variables obey an anti-commutator in quantum theory. Hence, in above equation $\{A, B\} = AB + BA$. We use the same symbol as for the classical Poisson bracket, but the correct meaning will be clear from the context, that is, whether it is a classical or quantum discussion. Setting $\hbar = 1$, the first relations are represented, as usual, by

$$x^\mu \to x^\mu, \quad p^\mu \to -i\frac{\partial}{\partial x^\mu}. \tag{1.2.35}$$

We recognise the final relation as none other than the well-known γ-matrix relation up to a factor of $-\frac{1}{2}$. Consequently, we can adopt the choice

$$\chi^\mu = i\frac{\gamma^\mu}{\sqrt{2}}, \tag{1.2.36}$$

where by definition the γ^μ obey the Clifford algebra relations $\gamma^\mu \gamma^\nu + \gamma^\nu \gamma^\mu = 2\eta^{\mu\nu} I$, where I is the identity matrix.

The constraints now become the differential operators

$$\phi = -\partial^2, \quad \varphi = \frac{\slashed{\partial}}{\sqrt{2}}, \tag{1.2.37}$$

where $\slashed{\partial} = \gamma^\mu \partial_\mu$. We impose these on a wavefunction $\zeta_\alpha(x^\mu)$, which must carry a spinor index in order to carry a realisation of the constraints and in particular the γ^μ matrices. Since the fermionic constraint squares to give the bosonic constraint, it suffices to impose only the former, namely

$$\slashed{\partial}\zeta = 0. \tag{1.2.38}$$

We recognise this equation as the Dirac equation for a spin-$\frac{1}{2}$ particle. Hence, a particle with local $N = 1$ world-line supersymmetry describes a spin-$\frac{1}{2}$ particle upon quantisation. A detailed account of the properties of Clifford algebras can be found in chapter 5.

Let us now carry out a BRST quantization (see appendix A) of the spinning particle starting this time from the action of equation (1.2.6). In the BRST method, we introduce for each local symmetry a ghost. In this case, this means introducing the ghosts c and γ for reparameterisations and local supersymmetry, respectively. Since local supersymmetry has a Grassmann odd parameter, the corresponding ghost, γ, has opposite Grassmann parity and so is Grassmann even. Making the replacement $\epsilon \to \Lambda\gamma$ we must choose γ to be anti-Hermitian so that ϵ is real. The BRST transformations of the original fields are then of the form

$$\delta x^\mu = i\Lambda\gamma\chi^\mu + \Lambda c\dot{x}^\mu, \qquad \delta\chi^\mu = \left(\dot{x}^\mu - \frac{i\kappa}{2}\psi\chi^\mu\right)\Lambda\gamma e^{-1} + \Lambda c\dot{\chi}^\mu,$$

$$\delta e = i\kappa\Lambda\gamma\psi + \frac{d}{d\tau}(e\Lambda c), \quad \delta\psi = \frac{2}{\kappa}\frac{d}{d\tau}(\Lambda\gamma) + \frac{d}{d\tau}(\Lambda c\psi). \tag{1.2.39}$$

By considering the composite parameter of two supersymmetries given in equation (1.2.22) and the recipe in appendix A, we deduce that the ghosts transform as

$$\delta c = \Lambda c\dot{c} - i\Lambda\gamma^2 e^{-1}, \quad \delta\gamma = +\frac{i\kappa}{2}\Lambda\gamma^2\psi e^{-1} + \Lambda c\dot{\gamma}. \tag{1.2.40}$$

The reader can verify that the BRST transformations are nilpotent:

$$\delta_{\Lambda_1}\delta_{\Lambda_2}\bullet = 0. \tag{1.2.41}$$

The next step in the BRST procedure is to introduce the anti-ghosts b and β which are Grassmann odd and even, respectively, and transform as

$$\delta b = \Lambda\lambda, \quad \delta\beta = \Lambda\rho,$$

$$\delta\lambda = 0, \quad \delta\rho = 0. \tag{1.2.42}$$

To fix the gauge we add to the original action the terms

$$A^{gf} = \int d\tau \left(\lambda \ln e + \frac{\kappa}{2}\rho\psi\right) \tag{1.2.43}$$

in order to impose the gauge conditions of equation (1.2.28).

To construct an invariant action we note that

$$\int d\tau \delta_\Lambda \left(b \ln e - \frac{\kappa}{2} \beta \psi \right)$$

$$= \int d\tau \left\{ \Lambda [\lambda \ln e + \rho \psi] + \Lambda \left[-b \nabla_\tau c - i\kappa b\gamma \psi + \beta \dot{\gamma} + \frac{\kappa}{2} \beta \frac{d}{d\tau} (c\psi) \right] \right\}$$

$$\equiv \Lambda (A^{gf} + A^{gh}) \qquad (1.2.44)$$

and consequently

$$A^{BRST} = A^{orig} + A^{gf} + A^{gh} \qquad (1.2.45)$$

is BRST invariant.

Eliminating λ and ρ enforces the gauge conditions, whereupon our final action is given by

$$A^{tot} = \int d\tau \{ \tfrac{1}{2} (\dot{x}^\mu \dot{x}^\nu - i\chi^\mu \dot{\chi}^\nu) \eta_{\mu\nu} - b\dot{c} - \beta\dot{\gamma} \}. \qquad (1.2.46)$$

The conjugate momenta to c and γ are b and β, respectively, and consequently we adopt the Poisson brackets

$$\{c, b\} = 1, \quad \{\gamma, \beta\} = 1. \qquad (1.2.47)$$

Finally, we will construct the BRST charge Q, which in this case is just the current associated with the BRST invariance. After a calculation one finds that the charge associated with the action of equation (1.2.46) is given by

$$Q = cp_\mu p^\mu + \gamma p_\mu \chi^\mu + i\frac{b}{2}\gamma^2. \qquad (1.2.48)$$

It is straightforward to check that

$$Q^2 = 0 \qquad (1.2.49)$$

and that it generates a set of BRST transformations by $\delta \bullet = \{\bullet, \Lambda Q\}$, which leaves the action of equation (1.2.46) invariant.

To quantise the system we replace, according to the Dirac rule, the Poisson brackets by commutators or anti-commutators depending on the statistics of the objects involved. In addition to $[x^\mu, p^\nu] = i\hbar \eta^{\mu\nu}$ and equation (1.1.13), we find the relations

$$\{c, b\} = i\hbar, \quad [\gamma, \beta] = i\hbar \qquad (1.2.50)$$

and so, after setting $\hbar = 1$,

$$b = +i\frac{\partial}{\partial c}, \quad \beta = -i\frac{\partial}{\partial \gamma}. \qquad (1.2.51)$$

The quantum BRST charge becomes

$$Q = -c\partial^2 - \frac{i}{\sqrt{2}}\gamma \slashed{\partial} - \frac{1}{2}\gamma^2 \frac{\partial}{\partial c}. \qquad (1.2.52)$$

One might be tempted to state that the physical states should satisfy

$$Q\Psi = 0, \qquad (1.2.53)$$

where the wavefunction Ψ is a function of the coordinates x^μ, c and γ. However, γ commutes with Q and so Ψ, even when subject to the above equation, is an arbitrary function of γ. The coefficients of γ include an arbitrary number of space-time fields which in turn lead to an infinite number of on-shell states. To overcome this problem we assume that Ψ has the specific form

$$\Psi = \psi + \gamma\phi - \sqrt{2}i\slashed{\partial}\phi c. \tag{1.2.54}$$

Applying Q one finds that

$$Q\Psi = -i\frac{1}{\sqrt{2}}\slashed{\partial}\psi\gamma - \partial^2\psi c. \tag{1.2.55}$$

In carrying out this step we took c to anti-commute with $\slashed{\partial}$ since c and ψ^μ are Grassmann odd. We observe that the form of $Q\Psi$ is compatible with the original form of Ψ. Hence, provided we take Ψ to have the form of equation (1.2.54) we can adopt $Q\Psi = 0$ as our physical state condition since we then recover the required result, namely the Dirac equation. The reader will have noticed that in order to recover the correct result we have had to supplement the description of physical states in the BRST formalism in a rather non-trivial way.

Under the usual equivalence relation $\Psi_1 \sim \Psi_2$ only if $\Psi_1 = \psi_2 + Q\Lambda$, we find that $\psi_1 = \psi_2$, $\phi_1 = \phi_2 - \frac{1}{\sqrt{2}}\slashed{\partial}\lambda_1$ provided that we take Λ to have the same form as Ψ, namely $\Lambda = \lambda_1 + \gamma\lambda_2 - \sqrt{2}i\slashed{\partial}\lambda_2 c$. This explains why ϕ is absent from the physical states.

1.2.2   The Brink–Schwarz superparticle

In the previous section we discussed a particle that was supersymmetric in the sense that it possessed a supersymmetry on its one-dimensional world-line. In this section, we construct an alternative superparticle, called the Brink–Schwarz superparticle [1.9], which has space-time supersymmetry instead of world-line supersymmetry. It is constructed from the fields x^μ, $\mu = 0, 1, \ldots, D-1$ and θ_A. The latter is a space-time Majorana spinor which as explained in chapter 5 satisfies the equation $\bar{\theta}^M = \theta^T C = \bar{\theta}^D$. As the two definitions of the conjugate spinor are equal we can simply denote it by $\bar{\theta}$. The range of A runs over the dimension of the corresponding Clifford algebra which is $2^{D/2}$ if it is just Majorana and a factor of 2 less if it is Majorana–Weyl. Their supersymmetry transformations are

$$\delta x^\mu = i\bar{\epsilon}\gamma^\mu\theta, \;\; \delta\theta_A = \epsilon_A, \tag{1.2.56}$$

where ϵ_A is a constant space-time Majorana spinor. These transformations are those corresponding to x^μ and θ_A being the coordinates of superspace [1.10] (see reference [1.11] for an account). It is obvious that the commutator of two supersymmetry transformations is a space-time translation. As $\bar{\lambda}\gamma^\mu\rho = -\bar{\rho}\gamma^\mu\lambda$ for any two Majorana spinors λ and ρ. The quantity

$$\Pi^\mu \equiv \dot{x}^\mu - i\bar{\theta}\gamma^\mu\dot{\theta} \tag{1.2.57}$$

is invariant under the above supersymmetry transformations. The coordinates transform in the usual way under a Lorentz transformation, or more precisely a Spin(1, $D-1$)

transformation,

$$\delta x^{\mu} = w^{\mu}_{\ \nu} x^{\nu}, \quad \delta \theta = \tfrac{1}{4} w_{\mu\nu} \gamma^{\mu\nu} \theta, \tag{1.2.58}$$

and consequently Π^{μ} transforms as a vector.

It is clear how to extend the action of equation (1.1.3) with $m = 0$ for the bosonic point particle to be space-time supersymmetric: we simply replace x^{μ} by Π^{μ} to find the action [1.8]

$$A = \tfrac{1}{2} \int d\tau \, e^{-1} \Pi^{\mu} \Pi^{\nu} \eta_{\mu\nu}. \tag{1.2.59}$$

The field e is inert under supersymmetry transformations. The equations of motion for this action are

$$\frac{d}{d\tau} (e^{-1} \Pi^{\mu}) = 0, \quad \slashed{\Pi} \dot{\theta} = 0, \quad \Pi^{\mu} \Pi^{\nu} \eta_{\mu\nu} = 0, \tag{1.2.60}$$

where $\slashed{\Pi} = \gamma^{\mu} \Pi_{\mu}$. In finding these equations, we have used the Majorana property of θ which implies that $\delta \bar{\theta} \gamma^{\mu} \dot{\theta} = -\dot{\bar{\theta}} \gamma^{\mu} \delta \theta$ and in finding the θ_A equation of motion we have used the equation of motion for x^{μ}.

The corresponding first order action is given by

$$A = \int d\tau \big(\Pi^{\mu} p^{\nu} - \tfrac{1}{2} e p^{\mu} p^{\nu} \big) \eta_{\mu\nu}. \tag{1.2.61}$$

This action is equivalent to the action of equation (1.2.59) upon elimination of p^{μ}. It is invariant under the supersymmetry transformations if we take those of equation (1.2.56) as well as $\delta e = 0$ and $\delta p^{\mu} = 0$.

The matrix $\slashed{\Pi}$ is not invertible on-shell, as is evident from the equation $\slashed{\Pi} \slashed{\Pi} = \Pi^{\mu} \Pi_{\mu} = 0$. Consequently, one cannot conclude from the equations of motion that $\dot{\theta}_A = 0$. This observation is related to the fact that the action possesses an additional local symmetry [1.12], called κ symmetry, which has the transformations

$$\delta x^{\mu} = i \bar{\theta} \gamma^{\mu} \delta \theta, \quad \delta \theta = i \slashed{\Pi} \kappa, \quad \delta e = -4 e \dot{\bar{\theta}} \kappa, \tag{1.2.62}$$

where κ_A is a Majorana spinor that is an arbitrary function of τ. Using these variations, we find that $\delta \Pi^{\mu} = -2 \dot{\bar{\theta}} \gamma^{\mu} \slashed{\Pi} \kappa$ and so $\delta (\Pi^{\mu} \Pi_{\mu}) = -4 \Pi^{\mu} \Pi_{\mu} \dot{\bar{\theta}} \kappa$. It is then straightforward to verify the invariance of the action of equation (1.2.59). The first order action of equation (1.2.61) is invariant under κ symmetry if we use the variations for x^{μ} and e of equation (1.2.62), but take $\delta \theta = i \slashed{p} \kappa$ and $\delta p^{\mu} = 0$.

We leave it to the reader to verify that the κ transformations do form a closed algebra, but only if one uses the equations of motion and introduces the transformations $\delta x^{\mu} = i \bar{\theta} \gamma^{\mu} \delta \theta$, $\delta \theta = l \dot{\theta}$ and $\delta e = 0$, where l is an arbitrary function of τ. One can verify that these transformations are indeed an additional symmetry of the action. In fact, they leave Π^{μ} inert due to the identity $\bar{\theta} \dot{\gamma}^{\mu} \theta = 0$.

The Brink–Schwarz particle does not allow a Lorentz covariant quantization. This problem becomes apparent when one carries out the Dirac method of quantization as described in appendix A. The first step in this process is to calculate the canonical momenta and identify the constraints. The bosonic constraint $p^{\mu} p_{\mu} = 0$ is dealt with as for the bosonic point particle and does not pose a problem. As such, we focus on the fermionic variables.

The momentum $\bar{\lambda}^A$ corresponding to θ^A is by definition given by $\bar{\lambda}^A \equiv \partial L/\partial \dot{\theta}_A$. Using the first order action and taking into account the Grassmann odd nature of θ_A, we find from the first order action of equation (1.2.61) that

$$\bar{\lambda}^A = i(\bar{\theta} \not{p})^A. \qquad (1.2.63)$$

This equation tells us that the momentum $\bar{\lambda}^A$ and the coordinate θ_A are related and as such we must impose the above relation as a constraint:

$$\bar{\chi}^A \equiv \bar{\lambda}^A - i(\bar{\theta} \not{p})^A = 0. \qquad (1.2.64)$$

Although it is typical of fermionic systems that the momenta and coordinates are related, it is the presence of the $\not{p}$ factor that causes the difficulties. In particular, this factor is not invertible when the equations of motion hold, while for conventional fermions one finds γ_0, which is invertible. We now adopt the usual Poisson bracket relations between coordinates and their momenta, namely

$$\{\theta_A, \bar{\lambda}^B\} = \delta_A^B, \qquad (1.2.65)$$

with the other Poisson brackets vanishing. Using this result, we find that

$$\{\bar{\chi}^A, \bar{\chi}^B\} = -2i(C\not{p})^{AB} \text{ or } \{\chi_A, \bar{\chi}^B\} = -2i(\not{p})_A^B. \qquad (1.2.66)$$

To quantise the system, we must apply the rest of the Dirac method outlined in appendix A. This entails separating the constraints into first and second class constraints. We recall that first class constraints are such that their Poisson brackets are proportional to the constraints themselves. Second class constraints are those for which this does not happen and for these one finds that their Poisson brackets equal quantities that do not vanish when one imposes the constraints.

We will not enter here into the details of the Dirac procedure, but it is perhaps not too surprising to learn that one must treat the two types of constraints separately in order to quantise the system. In particular, we must separate the constraints of equation (1.2.59) into second class constraints and first class constraints. This is achieved by carrying out the rearrangement $\bar{\chi}'^A = E^A_B \bar{\chi}^B$, where E is an invertible matrix (that is, $\det E \neq 0$) such that the matrix $\{\bar{\chi}'^A, \bar{\chi}'^B\}$ is made up of only two diagonal blocks, one of which is the zero matrix if we use the constraints whereas the other block has non-zero determinant even if we use of the constraints. The constraints that are associated with the zero matrix are the first class constraints, while those associated with the other block are the second class constraints. The number of constraints that are second class is therefore the rank of the matrix $\{\bar{\chi}'^A, \bar{\chi}'^B\}$ when we impose the constraints. We recall that the rank of a matrix is the dimension of the image of the linear operator that has this matrix as its representative in the chosen basis. This latter definition makes it clear that the rank of a matrix is unchanged by a similarity transformation such as E. The equation $\not{p}\not{p} = p^\mu p_\mu = 0$ implies that $\det \not{p} = 0$ and consequently, $\det EC\not{p}E^T = \det\{\bar{\chi}'^A, \bar{\chi}'^B\}\frac{i}{2} = 0$. Hence, some of the constraints of equation (1.2.59) are first class, but some must also be second class as $\not{p} \neq 0$ even if $p^2 = 0$. In the above we set $e = 1$ as it plays no role.

In chapter 5 we will find all the irreducible representations of a Clifford algebra in a space of dimension D. It will be shown that there is a unique irreducible representation of dimension greater than 1. This is the representation of the corresponding spin group, the covering group of the Lorentz group, and as it is irreducible there is no smaller representation. The constraints $\bar{\chi}^A$, like θ_A, belong to this representation, but in order to separate out the second class constraints we must select a subset of the constraints and in this process we must break spin and so Lorentz symmetry.

We will now analyse in more detail how many first and second class constraints there are and take into account the possibility of Weyl spinors. The rank of the matrix that appears on the right-hand side of the Poisson bracket can be determined by introducing the projectors

$$P_1 = p\bar{p}, \ P_2 = \bar{p}p, \tag{1.2.67}$$

where $\bar{p}^\mu$ is any massless vector such that $p \cdot \bar{p} = \frac{1}{2}$. The properties of the Clifford algebra imply the relations $P_1^2 = P_1, P_2^2 = P_2, P_1 P_2 = 0, P_1 + P_2 = I$, which mean that P_1 and P_2 form a complete set of projectors. Given any spinor u we can decompose it into $u = u_1 + u_2 \equiv p\bar{p}u + \bar{p}pu$; clearly $P_1 u_1 = u_1, P_2 u_1 = 0$ and similarly for u_2. Consequently, we may split the space of spinors, that is, the space of constraints, V, into the direct sum of two spaces V_1 and V_2 corresponding to the above projectors in the obvious way. The symmetry between p^μ and $\bar{p}^\mu$ implies that V_1 and V_2 have equal dimensions. The relations $pP_1 = 0$, $pP_2 = p$ and similar relations for $\bar{p}$ imply that p annihilates states in V_1 and maps V_2 into V_1, while $\bar{p}$ has the opposite effect. On the other hand, P_1 is the product of these two operators and it is clearly a one-to-one onto map from V_1 to itself. We may conclude that p is a one-to-one map from V_2 onto V_1 and similarly for $\bar{p}$. It then follows that the image of p is V_2 and so it has a rank that is the dimension of V_2, which is half the dimension of V. The kernel of p is V_1 and any element of this space can be written as pv. We have therefore shown that the rank of $\{\bar{\chi}^{\prime A}, \bar{\chi}^{\prime B}\}$ is half that of the number of constraints.

Thus half the constraints are first class and half are second class. Since p anti-commutes with γ^{D+1}, the constraints of opposite chirality commute and so we may treat the spinors of each chirality separately when evaluating the number of second class constraints. Consequently, the number of second class constraints for a given chirality is half the total number of constraints of a given chirality. As noted above, a Clifford algebra possesses only one irreducible representation of dimension greater than 1, which is just the usual spinor representation which can be Majorana, Weyl or Majorana–Weyl. Hence, we can never find a representation of the Clifford algebra or Spin group which is smaller. However, the second class constraints of the super point particle considered in this section have half the number of components of the smallest representation of the relevant Spin group. Clearly, the second class constraints cannot carry a representation of the the Spin group in the normal way. For example, in ten dimensions an unrestricted spinor has 32 components; if it is a Majorana spinor these 32 components are real, while if it is Majorana–Weyl it has 16 real components. There are then only eight second class constraints for a given chirality Majorana spinor, however, there is no representation of the Spin group with only eight components. Hence any split into first and second class constraints must be non-covariant with respect to the Spin or Lorentz group and any quantization must be non-covariant [1.13].

It is instructive to work through the above discussion for a particular choice of ten-dimensional Dirac matrices in the Lorentz frame in which $p^0 = p^1 = 1$ and all other components vanish.

One can, however, quantise the Brink–Schwarz particle non-covariantly. This is most easily carried out by choosing a special gauge called the light-cone gauge. This procedure is similar to the light-cone treatment of the bosonic string given in chapter 4. One finds that the super point particle considered in this section has a massless vector and space-time spinor as its physical states. We note that this is not the same content as the super point particle that had world-line supersymmetry, studied in section 1.2.1, which had only a spinor when quantised. This had to be the case as the Brink–Schwarz particle possesses space-time supersymmetry and so its field content when quantised must be a multiplet of this space-time supersymmetry algebra. As such it must possess equal numbers of bosons and fermions on-shell.

1.2.3   Superspace formulation of the point particle

Studying the superspace formulation of the super point particle considered in section 1.2.1 gives us the opportunity to explain some features of the simplest possible superspace, that is, one involving one bosonic coordinate τ and one Grassmann odd coordinate θ. This superspace is the appropriate one for the point particle that possesses world-line supersymmetry. Despite its simplicity, this superspace illustrates many of the features of superspaces found in higher-dimensional theories.

The supercharges of the supersymmetry algebra carried by the point particle were given in equation (1.2.4) and the commutator of two supersymmetry transformations was found to be given by

$$[\delta_1, \delta_2] = [\epsilon_1 Q, \epsilon_2 Q] = -\epsilon_1 \epsilon_2 \{Q, Q\} = 2i\epsilon_2 \epsilon_1 P, \tag{1.2.68}$$

where Q and $P = d/d\tau$ are the generators of supersymmetry and τ translations. From this equation, taking into account the Grassmann odd nature of the supersymmetry parameter, we can deduce the anti-commutator of two supercharges. The results for this and the other transformation are

$$\{Q, Q\} = +2iP, \quad [Q, P] = 0, \tag{1.2.69}$$

which are those of the simplest possible supersymmetry algebra.

Let us then consider the supergroup generated by Q and P. In the exponential representation, the group elements may be represented by

$$g(\theta, t) = e^{\theta Q + \tau P}, \tag{1.2.70}$$

where θ and τ are Grassmann odd and Grassmann even coordinates, respectively. Acting with a group element $g(a, \epsilon) = e^{-\epsilon Q - aP}$, the point (τ, θ) is transformed to (τ', θ') according to the group composition rule:

$$g(a, \epsilon)g(\theta, \tau) = g(\theta', \tau') = \exp((-a + \tau)P + (\theta - \epsilon)Q + \tfrac{1}{2}[\epsilon Q, \theta Q])$$
$$= \exp((\tau - a)P + (\theta - \epsilon)Q + i\epsilon\theta P). \tag{1.2.71}$$

We have used the fact that Q commutes with P and expanded the exponentials of anti-commuting parameters. We find that

$$\tau' = \tau - a + i\epsilon\theta, \quad \theta' = \theta - \epsilon. \tag{1.2.72}$$

We now define a superfield to be a function φ on the group which transforms under supersymmetry as

$$\varphi'(\tau', \theta') = \varphi(\tau, \theta). \tag{1.2.73}$$

An infinitesimal transformation is given by

$$\delta\varphi = \varphi'(\tau, \theta) - \varphi(\tau, \theta) = (aP + \epsilon Q)\varphi, \tag{1.2.74}$$

where the generators P and Q are read off to be

$$P = \frac{d}{d\tau} \equiv d_\tau, \quad Q = \frac{d}{d\theta} - id_\tau\theta, \tag{1.2.75}$$

consistent with our previous considerations.

The covariant derivatives

$$d_\tau \quad \text{and} \quad D = \frac{d}{d\theta} + id_\tau\theta \tag{1.2.76}$$

commute and anti-commute with P and Q, respectively, that is, $\{D, Q\} = 0$, $[d_\tau, Q] = 0$. Consequently, $D\varphi$ is also a superfield, that is, it obeys equation (1.2.73), or equivalently equation (1.2.74). The anti-commutator of two covariant derivatives is $\{D, D\} = id_\tau$.

Let us apply the above discussion to formulate the super point particle in superspace. We consider a scalar superfield $\mathbf{X}$, which we may Taylor expand in θ to find

$$\mathbf{X}(\tau, \theta) = x(\tau) + i\theta\chi(\tau). \tag{1.2.77}$$

The component fields x and χ depend on τ and using equation (1.2.73) we find δx and $\delta\chi$ in agreement with equation (1.2.3). We can also define the component fields of $\mathbf{X}$ by taking $x = \mathbf{X}\,|_{\theta=0}$, $i\chi = D\mathbf{X}\,|_{\theta=0}$. Using these definitions allows us to find the supersymmetry variations in a neater way by making use the algebra of covariant derivatives and the fact that $\delta\bullet = \epsilon Q\bullet = \epsilon D\bullet$ at $\theta = 0$, where $\bullet$ is any field. We find that their supervariations are given by

$$\delta x = \epsilon Q\mathbf{X}\,|_{\theta=0} = \epsilon D\mathbf{X}\,|_{\theta=0} = i\epsilon\chi,$$
$$\delta\chi = -i\epsilon QD\mathbf{X}\,|_{\theta=0} = -i\epsilon DD\mathbf{X}\,|_{\theta=0} = \dot{x}\epsilon. \tag{1.2.78}$$

A more detailed discussion of this technique can be found in chapter 14 or in [1.11]. The alert reader will note that carrying out the anti-commutation $\{Q, Q\}$ using the expressions for the supercharge of equation (1.2.75) results in $-2iP$, which is out by a minus sign compared with the result found in equation (1.2.69). This sign difference is due to the difference between active and passive viewpoints.

Grassmann integration is defined by

$$\int d\theta\,\theta = 1, \qquad \int d\theta\,1 = 0 \tag{1.2.79}$$

and so

$$\int d\theta\,\mathbf{X} = i\chi. \tag{1.2.80}$$

We note that the second expression in equation (1.2.79) is Grassmann odd and so there is nothing it can equal except 0. In fact, Grassmann integration is just equivalent to differentiation and we may write full superspace integrals as

$$\int d\tau \, d\theta = \int d\tau \, \frac{d}{d\theta} = \int d\tau \, D \qquad (1.2.81)$$

acting on any superfield. Here we have discarded a total τ derivative. For any scalar superfield φ, $\int d\tau \, d\theta \varphi$ is invariant under supersymmetry transformations using equation (1.2.74) and the fact that $\int d\tau \, d\theta Q\varphi = 0$.

The action of equation (1.2.1) can be written in superspace as

$$-\frac{i}{2} \int d\tau d\theta \dot{\mathbf{X}} D \mathbf{X}, \qquad (1.2.82)$$

which is clearly supersymmetric since $\dot{\mathbf{X}}$ and $D\mathbf{X}$ are superfields. This action can be expressed in component fields as follows:

$$-\frac{i}{2} \int d\tau \, D(\dot{\mathbf{X}} D\mathbf{X}) = -\frac{i}{2} \int d\tau \, (D\dot{\mathbf{X}} D\mathbf{X} + \dot{\mathbf{X}} D^2 \mathbf{X}) = \frac{1}{2} \int d\tau \{\dot{x}\dot{x} - i\chi \dot{\chi}\} \qquad (1.2.83)$$

as it should be. One could also have evaluated this integral by simply substituting the expressions for $\mathbf{X}$ and using equation (1.2.79).

The supergravity fields can be encoded in the superfield

$$\mathbf{E} = e(1 + i\kappa\theta\psi) \qquad (1.2.84)$$

and the action of equation (1.2.20) for the spinning particle is given by

$$-\frac{i}{2} \int d\tau d\theta \mathbf{E}^{-1} \dot{\mathbf{X}} D \mathbf{X}, \qquad (1.2.85)$$

provided one makes the rescaling $\psi \to e^{-\frac{1}{2}}\psi, \chi \to e^{\frac{1}{2}}\chi$. In fact, one can introduce a supereinbein in the superspace and then one finds that $\mathbf{E}$ is the determinant of the supereinbein in a particular gauge.

We will now show that the action of equation (1.2.85) is superconformally invariant, for which we will require the superconformal transformations written in superspace. A conformal transformation is one which preserves the line element $ds^2 = -d\tau^2$ up to scale that is, $ds^2 \to \Omega(t)ds^2$. Clearly, in one dimension conformal transformations are just arbitrary reparameterisations in τ. Superconformal transformations can be defined in several ways, see chapter 25 of [1.11]. We take here superconformal transformations to be those that preserve the spinor covariant derivative D up to scale, that is, $D \to \Omega(\tau, \theta)D$. The most general supercoordinate transformation can be written as

$$\tau' = f(\tau) + \theta\alpha_1(\tau), \quad \theta' = \theta g(\tau) + \alpha(\tau), \qquad (1.2.86)$$

where $f(\tau), g(\tau)$ are Grassmann even functions of τ and $\alpha_1(\tau), \alpha_2(\tau)$ are Grassmann odd functions of τ. Under this transformation

$$D = g\frac{d}{d\theta'} + \alpha_1\frac{d}{d\tau'} + i\theta \left(\frac{df}{d\tau}\frac{d}{d\tau'} + \frac{d\alpha}{d\tau}\frac{d}{d\theta'} \right). \qquad (1.2.87)$$

This can be written as

$$D = \left(g + i\theta\frac{d\alpha}{d\tau}\right)\left(\frac{d}{d\theta'} + i\theta'\frac{d}{d\tau'}\right),$$ (1.2.88)

provided

$$\alpha_1 - ig\alpha = 0; \quad g^2 = \frac{df}{d\tau} - i\frac{d\alpha}{d\tau}\alpha.$$ (1.2.89)

Consequently, a superconformal transformation is of the form

$$\tau' = f(\tau) + i\theta\alpha(\tau)\sqrt{\frac{df}{d\tau}}, \quad \theta' = \alpha(\tau) + \theta\sqrt{\frac{df}{d\tau} - i\frac{d\alpha}{d\tau}\alpha}.$$ (1.2.90)

The super-Jacobian of this transformation is

$$\mathrm{sdet}\begin{pmatrix}\dfrac{d\tau'}{d\tau} & \dfrac{d\tau'}{d\theta} \\[2mm] \dfrac{d\theta'}{d\tau} & \dfrac{d\theta'}{d\theta}\end{pmatrix} = \left(g + i\theta\frac{d\alpha}{d\tau}\right) = D\theta'.$$ (1.2.91)

We observe that on the coordinates of superspace when $f(\tau) = \tau$, an infinitesimal superconformal transformation with parameter $\alpha(\tau)$ is obtained from an ordinary supersymmetry transformation by the replacement $\epsilon \to \alpha(\tau)$.

If we denote $D\theta' = \Omega$, then we have seen that under a superconformal transformation $D' = \Omega^{-1}D$ and $d\tau'd\theta' = \Omega d\tau d\theta$. As a result of the latter relationship $d'_\tau = \Omega^{-2}d_\tau - (D\Omega/\Omega^3)D$. We take $\mathbf{X}$ to be a scalar under superconformal transformations, that is, $\mathbf{X}'(\tau', \theta') = \mathbf{X}(\tau, \theta)$, while $\mathbf{E}'(\tau', \theta') = \Omega^2\mathbf{E}(\tau, \theta)$. It is then straightforward to see that the action of equation (1.2.85) is invariant under superconformal transformations provided one uses the relation $(D\mathbf{X})^2 = 0$, due to its Grassmann odd character. Taking into account the field redefinitions $\psi \to e^{-\frac{1}{2}}\psi$, $\chi \to e^{\frac{1}{2}}\chi$ and $\alpha \to e^{-\frac{1}{2}}\alpha$ it is straightforward to compute the superconformal transformations of x, χ and e, ψ. Clearly, the $f(\tau)$ transformation is just a reparameterisation, however, the $\alpha(\tau)$ transformation is in agreement with that of equation (1.2.21) once we identify the transformed α with ϵ. Consequently, world-line superconformal transformations are just reparameterisations and local supersymmetry transformations. We note that this is a peculiarity of one dimension and in two or more dimensions the superconformal group is finite-dimensional.

1.3   The twistor approach to the massless point particle

In the last section, we gave two different formulations of the superparticle: one possessed world-line supersymmetry while the other had space-time supersymmetry. However, the latter formulation could not be quantised so as to maintain Lorentz invariance in a manifest manner. This stemmed from the inability to separate the constraints into first and second class constraints in a Lorentz covariant way.

It was discovered that this difficulty could be overcome if one used an alternative formulation [1.14] that involved twistor-like variables. Twistors [1.15], introduced by Penrose, were invented as an alternative to our usual Minkowski space description of space-time. Rather than space-time points being the most important objects, twistors stress the role played by light-like lines and perhaps not surprisingly they are most suited to describing massless particles. One substantial benefit of these twistor-like formulations of the point particle is that they possess both world-line and space-time supersymmetry. We begin by giving a review of those aspects of twistor theory that are most relevant to the point particle.

1.3.1   Twistors in four and three dimensions

The twistor description in four dimensions makes essential use of the isomorphism between the part of $SO(3, 1)$ connected to its identity element and $SL(2, \mathbf{C})/Z_2$. It will be instructive to demonstrate this isomorphism before giving the twistors themselves. The isomorphism is established through the identification

$$x_\mu (\sigma^\mu)^{A\dot{B}} \equiv x^{A\dot{B}} = \begin{pmatrix} -x^0 + x^3 & x^1 - ix^2 \\ x^1 + ix^2 & -x^0 - x^3 \end{pmatrix}, \tag{1.3.1}$$

where $\sigma^\mu = (I, \sigma^1, \sigma^2, \sigma^3)$ and $\sigma^i, i = 1, 2, 3$ are the Pauli matrices. The self-adjoint nature of the Pauli matrices and the reality of x^μ imply that $x^{A\dot{B}}$ is a 2×2 self-adjoint matrix. Conversely, any 2×2 self-adjoint matrix can be expressed in terms of a sum of the matrices σ^μ with four real coefficients which we may identify as belonging to Minkowski space. Hence, there is a one-to-one correspondence between points of Minkowski space and 2×2 self-adjoint matrices. Consider the transformation

$$x \to x' = UxU^\dagger \tag{1.3.2}$$

induced by any matrix U of $SL(2, \mathbf{C})$ and where x is a 2×2 self-adjoint matrix. Since x' is also a self-adjoint matrix, it follows that x and x' correspond to two points in Minkowski space, denoted by x^μ and $x^{\mu'}$ respectively. Being a linear transformation, the above transformation relates x^μ to $x^{\mu'}$ by $x^{\mu'} = (R_U)^\mu_{\ \nu} x^\nu$, where $(R_U)^\mu_{\ \nu}$ is a 4×4 matrix. Since $\det U = 1$, the transformation above preserves the determinant of $x^{A\dot{B}}$. The negative of this determinant is equal to $-(x^0)^2 + (x^1)^2 + (x^2)^2 + (x^3)^2 = x^\mu x^\nu \eta_{\mu\nu} \equiv x \cdot x$, consequently $x^\mu x^\nu \eta_{\mu\nu} = x'^\mu x'^\nu \eta_{\mu\nu}$ and so the matrix R_U preserves the Minkowski scalar product. That is, it satisfies the relation $R_U^T \eta R_U = \eta$ and so must be an element of $O(3, 1)$. Taking the determinant of this last relation, we conclude that $\det R_\mu = \pm 1$ and similarly one can show that $|R^0_{\ 0}| \geq 1$. In fact, it belongs to the part, denoted $\mathbf{L}^\uparrow_+$, of $O(3, 1)$ which is connected to the identity element, that is, the part with $\det R = 1$ and $R^0_{\ 0} \geq 1$. This follows from the fact, that the group $SL(2, \mathbf{C})$ is connected, that is, any point in the group can be reached by a continuous path from the identity element, and that the map U to $\det R_U$ from $SL(2, \mathbf{C})$ to R is continuous. As a result quantities such as $\det R_U$ cannot change their values by discrete jumps.

The map between these two groups is not one to one, since it is clear that $\pm U$ lead to the same element of $SO(3, 1)$. This, however, is the only ambiguity, since if two elements U_1 and U_2 of $SL(2, \mathbf{C})$ lead to the same element R_U of the Lorentz group, then $U_1 x U_1^\dagger = U_2 x U_2^\dagger$ for all x^μ, which implies that $\sigma^\mu (U_1^{-1} U_2)^\dagger = U_1^{-1} U_2 \sigma^\mu$, $\mu = 0, 1, 2, 3$. This is only true if $U_1^{-1} U_2 = aI$, but since U_1 and U_2 belong to $SL(2, \mathbf{C})$ then $a = \pm 1$. As a result, $U_1 = \pm U_2$. Thus the map from $SL(2, \mathbf{C})$ to $\mathbf{L}^\uparrow_+$ has kernel $\pm I$ which generates the group Z_2. The map is easily seen to satisfy $R_{U_1 U_2} = R_{U_1} R_{U_2}$ and so is a homomorphism. Finally, one can show that the map is from SL(2, **C**) onto $\mathbf{L}_+^\uparrow$. This is usually achieved by an explicit examination of the map itself. Thus we have a homomorphism from SL(2, **C**) onto $\mathbf{L}_+^\uparrow$ whose kernel is Z_2. It follows from a well-known theorem in group theory that $\mathbf{L}_+^\uparrow$ is isomorphic to SL(2, **C**)$/Z_2$.

The above isomorphism and the principles of quantum mechanics allow the existence of spinors that transform under the group SL(2, **C**). In fact, there exists a direct connection between null vectors and *commuting* spinors. Strictly speaking we should say Grassmann even rather than commuting, but the latter is the expression which is often used. Such spinors are also often called pure spinors. It follows from the discussion above that there is a one-to-one correspondence between null vectors and 2×2 self-adjoint matrices with determinant 0. A Majorana spinor in four dimensions has four components which in two-component notation take the form $u^A, u^{\dot{B}}, A, \dot{B} = 1, 2$, subject to the reality condition $(u^A)^\star = -u^{\dot{A}}$, $(u_A)^\star = u_{\dot{A}}$. We raise and lower indices using the epsilon symbols $\epsilon^{AB} = -\epsilon^{BA} = \epsilon_{AB} = \epsilon^{\dot{B}\dot{A}} = -\epsilon^{\dot{A}\dot{B}} = -\epsilon_{\dot{A}\dot{B}} = \epsilon^{\dot{B}\dot{A}}$, $A, B, \dot{A}, \dot{B} = 1, 2$, $\epsilon_{12} = 1$ as follows:

$$\epsilon^{AB}u_B = u^A, \quad u^A\epsilon_{AB} = u_B,$$
$$\epsilon^{\dot{A}\dot{B}}u_{\dot{B}} = u^{\dot{A}}, \quad u^{\dot{A}}\epsilon_{\dot{A}\dot{B}} = u_{\dot{B}}. \tag{1.3.3}$$

The reader may consult appendix A of [1.11] for more details.

From this spinor we can form the matrix $u^A u^{\dot{B}}$, which the above reality conditions imply is a self-adjoint matrix. Using the above discussion we can identify from this matrix a vector p^μ by

$$p_\mu(\sigma^\mu)^{A\dot{B}} \equiv p^{A\dot{B}} = u^A u^{\dot{B}}. \tag{1.3.4}$$

It is useful to introduce the matrix $(\bar{\sigma}^\mu)_{\dot{B}A} \equiv (\sigma^\mu)_{A\dot{B}}$, where we have lowered the indices with the ϵ tensor as described above. In terms of matrix multiplication this equation becomes $\bar{\sigma}^\mu = \epsilon\sigma^{\mu T}\epsilon$, where the T indicates to the transpose and the matrix ϵ is identified with ϵ_{AB}. As a result, $\bar{\sigma}_\mu = (-I, \sigma^1, \sigma^2, \sigma^3)$. Consequently, we find that $u_{\dot{B}}u_A = (\bar{\sigma}_\mu)_{\dot{B}A}p^\mu = p_{A\dot{B}}$. Since the spinors are commuting, it follows that

$$u^A u_A = u^A u^B\epsilon_{BA} = u^B u^A\epsilon_{BA} = -u^B u^A\epsilon_{AB} = -u^B u_B = 0, \tag{1.3.5}$$

as a result of which

$$p^{A\dot{B}}p_{C\dot{B}} = \delta^A_C p^\mu p_\mu = 0. \tag{1.3.6}$$

Consequently, given any real commuting Majorana spinor we can identify a null four-vector.

Conversely, given any null vector we can form a 2×2 self-adjoint matrix $p^{A\dot{B}}$ with determinant 0 and try to identify a spinor using equation (1.3.4). By explicitly examining this equation we find that $|u^1| = \sqrt{p^{1\dot{1}}}$, $|u^2| = \sqrt{p^{2\dot{2}}}$ and $\arg(u^1 - u^2) = \arg(p^{1\dot{2}})$, however, $\arg(u^1 + u^2)$ is left undetermined. We regard u_1^A and u_2^A as subject to an equivalence relation $u_1 \sim u_2$ if

$$u_2^A = \exp(i\alpha)u_1^A \tag{1.3.7}$$

for any α. Then null vectors are in one-to-one correspondence with commuting Majorana spinors subject to the above equivalence relation.

A particle is usually described by its position x^μ and its momentum p^μ. In terms of these variables a light-like line of a massless particle is described by

$$x^\mu(\tau) = x_0^\mu + p^\mu g(\tau), \tag{1.3.8}$$

where x_0^μ is a constant, $p_\mu p^\mu = 0$ and g is any monotonic function of time. Indeed, this is just the solution to the point particle equations of motion of the action of equation (1.1.5) with $m = 0$, provided $\dot{g}e^{-1} = $ constant and $\dot{p}^\mu = 0$. Clearly, we can change g by $g \to g + s$, where s is an arbitrary function of τ, but this does not change the light-like line only the way it is parameterised and it corresponds to the reparameterisation of τ discussed at the beginning of this chapter. In particular, it leads to the change $x^\mu \to x^\mu + sp^\mu$, which is a reparameterisation on x^μ. We may replace the null momentum p^μ by the commuting Majorana spinor u^A together with its conjugate $u^{\dot{A}}$, subject to the equivalence relation discussed above which can be implemented by the gauge transformation $u^A \to \exp(i\alpha)u^A$. As such we may describe the light-like world-line by x^μ and u^A.

However, we can go further and introduce another *commuting* Majorana spinor v^A with conjugate $v^{\dot{A}}$. We will refer to the pair of commuting Majorana spinors u^A, $u^{\dot{A}}$ and v^A, $v^{\dot{A}}$ as a twistor and as we will now show they can be used to define a light-like line. The twistor defines a matrix $x^{A\dot{B}}$ by the equation

$$v^A = x^{A\dot{B}} u_{\dot{B}}, \tag{1.3.9}$$

together with the conjugate equation $v^{\dot{A}} = -(x^{A\dot{B}})^* u_{\dot{B}}$. By examining the equation $\eta^A \zeta_A = 0$ for any two commuting Majorana spinors in components, we find it implies that $\eta^A = a\zeta^A$, where a is an arbitrary constant, as well as a similar result for the conjugate spinor. Consequently, the most general solution to the above equation is

$$x^{A\dot{B}} = x_0^{A\dot{B}} + u^A u^{\dot{B}} f^{-1}, \tag{1.3.10}$$

where $x_0^{A\dot{B}}$ is the particular solution and f is an undetermined function. We recognise this equation as that for the light-like line of equation (1.3.8), provided we identify f with e. Consequently, points in twistor space, that is, u^A, v^B and their conjugates, describe light-like lines in Minkowski space. In fact, we take the twistors to be subject to the constraint

$$(v^{\dot{A}})^* u_A = (u_A)^* v^{\dot{A}} \quad \text{or} \quad v^A u_A = -u_{\dot{A}} v^{\dot{A}}. \tag{1.3.11}$$

Substituting equation (1.3.9) into this condition we find that it implies the reality of x^μ which in turns means that the matrix $x^{A\cdot B}$ is self-adjoint. In fact, we used the self-adjoint nature of x when deriving equation (1.3.10) in the knowledge that we would adopt equation (1.3.11).

The gauge symmetry $u^A \to e^{i\alpha} u^A$ we used for the theory before we introduced v^A becomes extended to be $u^A \to e^{i\alpha} u^A$, $v^A \to e^{-i\alpha} v^A$ in the presence of the new field. We note that it leaves x^μ and p^μ inert and also preserves the constraint of equation (1.3.11). Let us denote u and v to be the column vectors with components $u_{\dot{A}}$ and v^A, respectively, then equation (1.3.9) can be written as $v = xu$. A transformation U of the spin group SL(2, **C**) is realised on the twistor as $v' = Uv$, $u' = (U^\dagger)^{-1}u$; taking these, together with, equation (1.3.2), one finds that equation (1.3.9) is preserved. The reader may verify if we denote the row vector $\tilde{u}$ and the column vector $\tilde{v}$ to have the components u_A and $v_{\dot{A}}$ respectively, then they transform as $\tilde{u}' = \tilde{u}U^{-1}$ and $\tilde{v}' = (U^\dagger)^{-1}\tilde{v}$, as expected. In fact, the twistor belongs to the vector representation of the four-dimensional conformal group which is isomorphic to Sp(4, **C**).

To summarise, we have replaced the usual phase space variables x^μ, p^μ for a massless particle by the twistor u^A, $u^{\dot A}$, v^A, $v^{\dot A}$. The degrees of freedom match since the twistors are subject to the reality constraint of equation (1.3.11) and the gauge invariance of equation (1.3.7) making six degrees of freedom in all, while the usual variables of phase space have the null condition $p_\mu p^\mu = 0$ and reparameterisation symmetry also making six degrees of freedom.

The twistor formulation of the three-dimensional massless point particle is simpler than that in four dimensions. We begin by defining our three-dimensional spinor conventions. We take as our Dirac matrices $\gamma^\mu = (i\sigma^2, \sigma^1, \sigma^3)$, $\mu = 0, 1, 2$, where σ^i are the Pauli matrices and the corresponding metric is $\eta^{\mu\nu} = \text{diag}(-1, 1, 1)$. A three-dimensional spinor has two components, that is, ψ_α, $\alpha = 1, 2$. Its Dirac conjugate is defined to be $\bar\psi^D = \psi^\dagger \gamma_0$, while its Majorana conjugate is given by $\bar\psi^M = -\psi^T(\gamma_0)$. With our choice of Dirac matrices, we observe that $\gamma^0 = \varepsilon$, where ε is the matrix whose components are fixed by the conditions $\varepsilon_{\alpha\beta} = -\varepsilon_{\beta\alpha}$ together with $\varepsilon_{12} = 1$. We may use $\varepsilon_{\alpha\beta}$ to raise and lower spinor indices; given a spinor with a lower index χ_α we may define $\chi^\alpha = \varepsilon^{\alpha\beta}\chi_\beta$ and conversely $\chi^\beta \varepsilon_{\beta\alpha} = \chi_\alpha$ where $\varepsilon_{\alpha\beta} = \varepsilon^{\alpha\beta}$. We then note that $\bar\psi^{M\alpha} = \varepsilon^{\alpha\beta}\psi_\beta = \psi^\alpha$. Also we find that $(\psi_\alpha)^* = -\bar\psi_\alpha$ for a Majorana spinor, which by definition obeys the relation $\bar\psi^D = \bar\psi^M$. We may also use the ε tensor to raise indices on the γ matrices, that is, $(\gamma^\mu)^{\alpha\beta} = \varepsilon^{\alpha\delta}(\gamma^\mu)_\delta^{\ \beta} = (\varepsilon\gamma^\mu)^{\alpha\beta}$, which are three real symmetric matrices. Given any vector x^μ we can construct a real symmetric matrix by the equation

$$(x)^{\alpha\beta} = (\gamma^\mu)^{\alpha\beta} x_\mu = \begin{pmatrix} -x_0 + x_1 & -x_2 \\ -x_2 & -x_0 - x_1 \end{pmatrix}. \tag{1.3.12}$$

Conversely, given any 2×2 real symmetric matrix, we can, by expanding it in terms of the $(\gamma^\mu)^{\alpha\beta}$ matrices, construct a unique real three-vector which we can identify as a point in three-dimensional Minkowski space. This correspondence can be used to show the isomorphism between the part of SO(2, 1) connected to the identity and SL(2, **R**)/Z_2. The derivation of this isomorphism proceeds much as its four-dimensional analogue. The map between $U \in$ SL(2, **R**) and $R_U \in$ SO(2, 1) is given by

$$U x_\mu (\gamma^\mu)^{\alpha\beta} U^T = (R_U x)_\mu (\gamma^\mu)^{\alpha\beta}. \tag{1.3.13}$$

Clearly, $(R_U x)^{\alpha\beta}$ is a symmetric matrix and represents the transformed point in Minkowski space. The group SL(2, **R**) is isomorphic to SU(1, 1).

In three dimensions, there is a one-to-one correspondence between null vector, p^μ ($p_\mu p^\mu = 0$) and Majorana spinors u_α up to the equivalence relation, $u_1 \sim u_2$ if $u_1 = \pm u_2$. We first note that the number of degrees of freedom match: a null vector, p^μ, has only two degrees of freedom, once we subtract one for the null condition, while u_α also has two degrees of freedom. Given a Majorana spinor u_α we can construct the symmetric real matrix $u^\alpha u^\beta$ provided u_α is a *commuting* spinor. From this matrix we can define a 3-vector p^μ by

$$u^\alpha u^\beta = (\gamma^\mu)^{\alpha\beta} p_\mu = p^{\alpha\beta}, \tag{1.3.14}$$

which, using the result $u^\alpha u_\alpha = -u^\beta u_\beta = 0$, is null since

$$p_\mu p^\mu = -p^{\alpha\beta} p_{\beta\alpha} = -u^\alpha u^\beta u_\beta u_\alpha = 0. \tag{1.3.15}$$

Conversely, given a massless vector p^μ we can construct a Majorana spinor; we define $u^1 = \sqrt{-p_0 + p_1}$, $u^2 = -\sqrt{-p_0 - p_1}$ and then, using the fact that p^μ is null we find $(\gamma^\mu p_\mu)^{\alpha\beta} = p^{\alpha\beta} = u^\alpha u^\beta$. As noted above, the change $u^\alpha \to -u^\alpha$ gives the same $p^{\alpha\beta}$ and so we can also take minus the above choice of u^α.

To summarise, we can parameterise phase space by x^μ and u^α rather than x^μ and p^μ. However, we can also eliminate x^μ by considering the twistor u^α, v^α where the latter is also a commuting Majorana spinor. The twistors can be used to define the matrix $x^{\alpha\beta}$ by the Penrose equation, $v^\alpha = x^{\alpha\beta} u_\beta$. The most general solution to this equation is $x^{\alpha\beta} = x_0^{\alpha\beta} + u^\alpha u^\beta e^{-1}$, which we recognise as a light-like line. Associated with the ambiguity in parameterising the line we have a corresponding reparameterisation invariance. The number of degrees of freedom matches since the x^μ, p^μ have four degrees of freedom taking into account the null nature of p^μ and the reparameterisation symmetry, while the unconstrained twistors u^α, v^α have the same number of degrees of freedom.

Twistors cannot naturally be constructed in any dimension, since we require a correspondence between null vectors and spinors. Apart from the cases of three and four dimensions considered above, they also exist in six and ten dimensions. The pattern in these higher dimensions is similar to that in four dimensions, where the spinors are subject to gauge symmetries and reality constraints. Of course, ten dimensions is particularly important since this is the dimension in which the superstring lives.

1.3.2   The twistor point particle actions

Having introduced new variables to describe phase space in the above we now give actions which also describe the bosonic particle, but that are built out of these new variables. We then generalise these results to the superparticle using superspace techniques. Since the actions for the three-dimensional particle are simpler than those in four dimensions we consider these first. An action built out of the variables $x^{\alpha\beta}$, $p^{\alpha\beta}$ and u_α is given by [1.14, 1.16]

$$\int d\tau \, p^{\alpha\beta} (\dot{x}_{\alpha\beta} - u_\alpha u_\beta). \tag{1.3.16}$$

This action is invariant under the reparameterisation

$$\delta x_{\alpha\beta} = f(\tau)\dot{x}_{\alpha\beta}, \quad \delta p^{\alpha\beta} = f(\tau)\dot{p}^{\alpha\beta}, \quad \delta u_\alpha = f(\tau)\dot{u}_\alpha + \tfrac{1}{2}\dot{f}(\tau)u_\alpha. \tag{1.3.17}$$

The equations of motion are

$$\dot{p}^{\alpha\beta} = 0, \quad \dot{x}_{\alpha\beta} = u_\alpha u_\beta, \tag{1.3.18}$$

$$p^{\alpha\beta} u_\beta = 0. \tag{1.3.19}$$

We may solve the latter equation as

$$p^{\alpha\beta} = u^\alpha u^\beta (e(\tau))^{-1}, \tag{1.3.20}$$

where $e(t)$ is an arbitrary function of t. This is the most general solution since, without loss of generality, we can take $p^{11} = (u^1)^2 e^{-1}$ and then writing equation (1.3.19) out in components solve for p^{12} and p^{22} to recover the above equation. Invariance of equation (1.3.20) under reparameterisation invariance implies

$$\delta e = \frac{d}{dt}(ef) \qquad (1.3.21)$$

and so we can identify e as the einbein. The equations of motion can then be rewritten as

$$\dot{p}_\mu = 0, \quad ep^\mu = \dot{x}^\mu, \quad p_\mu p^\mu = 0, \qquad (1.3.22)$$

which are those of the standard point particle.

There exists an alternative action that uses only the variables $x^{\alpha\beta}$ and u_α which is given by

$$-\int d\tau \dot{x}^{\alpha\beta} u_\alpha u_\beta, \qquad (1.3.23)$$

which is reparameterisation invariant under the transformations

$$\delta x^{\alpha\beta} = f\dot{x}^{\alpha\beta}, \quad \delta u_\alpha = f\dot{u}_\alpha. \qquad (1.3.24)$$

It is also invariant under

$$\delta x^{\alpha\beta} = s(\tau) u^\alpha u^\beta, \quad \delta u_\alpha = 0, \qquad (1.3.25)$$

where $s(\tau)$ is an arbitrary function of τ. The equations of motion are

$$\dot{x}^{\alpha\beta} u_\beta = 0, \quad u_\alpha \dot{u}_\beta + \dot{u}_\alpha u_\beta = 0, \qquad (1.3.26)$$

which in turn imply

$$\dot{x}^{\alpha\beta} = e(t) u^\alpha u^\beta, \quad \dot{u}_\alpha = 0. \qquad (1.3.27)$$

The latter follows trivially by writing out the equation of motion in components and assuming u_α never vanishes. Defining $p^{\alpha\beta} = u^\alpha u^\beta$, we recover the usual equations of motion, namely $\dot{x}^\mu = ep^\mu$, $\dot{p}^\mu = 0$ and $p^\mu p_\mu = 0$.

Yet another action for the bosonic point particle which is built out of a twistor u_α, v_α is given by

$$\int d\tau \quad (v^\alpha \dot{u}_\alpha - u_\alpha \dot{v}^\alpha). \qquad (1.3.28)$$

It is invariant under the transformations

$$\delta v^\alpha = fv^\alpha, \quad \delta u_\alpha = fu_\alpha. \qquad (1.3.29)$$

The equations of motion are

$$\dot{v}_\alpha = 0, \quad \dot{u}_\alpha = 0. \qquad (1.3.30)$$

We recover the more familiar variables by defining a null vector as $p^{\alpha\beta} = u^\alpha u^\beta$ and taking x^μ defined by the equation

$$v^\alpha = (x)^{\alpha\beta} u_\beta, \qquad (1.3.31)$$

where $x^{\alpha\beta} = x_\mu (\gamma^\mu)^{\alpha\beta}$. We recall that this equation does not define $x^{\alpha\beta}$ uniquely since v^α is inert when $x^{\alpha\beta}$ transforms as in equation (1.3.25). In terms of $p^{\alpha\beta}$ and $x^{\alpha\beta}$ the equations of motion become $\dot{p}^{\alpha\beta} = 0 = \dot{x}^{\alpha\beta} u_\beta$. Hence, we recover the equations of motion of the action of equation (1.3.26).

We note that the equations of motion of the actions of equations (1.3.16), (1.3.23) and (1.3.28) all possess a reparameterisation symmetry either explicitly or, in the last case, when x^μ is introduced. However, none of these actions involves explicitly an einbein. Hence, we have the interesting phenomenon of theories which possess general coordinate invariance without any gravity fields!

We now consider the supersymmetric analogues of the point particle actions considered above. The superextension of twistors was given in [1.16] and their use to construct actions for the super point particle was given in [1.14, 1.16, 1.18–1.28]. The simplest way to find the supersymmetric extensions of the above actions is to use superspace techniques. We begin by finding the supersymmetric extension of the action of equation (1.3.16). For each of the bosonic fields $x^{\alpha\beta}$, $p^{\alpha\beta}$ and u_α we introduce a corresponding superfield:

$$\mathbf{P}^{\alpha\beta} = p^{\alpha\beta} + i\theta\varrho^{\alpha\beta}, \quad \mathbf{X}_{\alpha\beta} = x_{\alpha\beta} + i\theta\chi_{\alpha\beta}, \quad \Theta_\alpha = \eta_\alpha + \theta u_\alpha; \tag{1.3.32}$$

the superfields $\mathbf{P}^{\alpha\beta}$ and $\mathbf{X}_{\alpha\beta}$ are Grassmann even while Θ_α is Grassmann odd. A supersymmetric action is given by

$$-i \int d\tau d\theta \, \mathbf{P}^{\alpha\beta} \left(D\mathbf{X}_{\alpha\beta} - \frac{i}{2} D\Theta_\alpha \Theta_\beta - \frac{i}{2}\Theta_\alpha D\Theta_\beta \right). \tag{1.3.33}$$

Using similar steps to those in equation (1.2.83) we find its component form to be

$$\int d\tau \left(\varrho^{\alpha\beta} \left(i\chi_{\alpha\beta} - \frac{i}{2}(u_\alpha\eta_\beta + \eta_\alpha u_\beta) \right) + p^{\alpha\beta} \left(\dot{x}_{\alpha\beta} - \frac{i}{2}(\dot{\eta}_\alpha\eta_\beta - \eta_\alpha\dot{\eta}_\beta) - u_\alpha u_\beta \right) \right). \tag{1.3.34}$$

The rigid supersymmetry transformations of the component fields are

$$\delta p^{\alpha\beta} = i\varepsilon\varrho^{\alpha\beta}, \quad \delta\varrho^{\alpha\beta} = \dot{p}^{\alpha\beta}\varepsilon, \quad \delta x_{\alpha\beta} = i\varepsilon\chi_{\alpha\beta}, \tag{1.3.35}$$

$$\delta\chi_{\alpha\beta} = \dot{x}_{\alpha\beta}\varepsilon, \quad \delta\eta_\alpha = \varepsilon u_\alpha, \quad \delta u_\alpha = -i\dot{\eta}_\alpha\varepsilon. \tag{1.3.36}$$

We observe that the $\varrho^{\alpha\beta}$ equation of motion just sets $\chi_{\alpha\beta} = \frac{1}{2}(u_\alpha\eta_\beta + \eta_\alpha u_\beta)$. Since these are algebraic equations we may implement them in the action, which then becomes

$$\int d\tau \, p^{\alpha\beta} \left(\dot{x}_{\alpha\beta} - \frac{i}{2}(\dot{\eta}_\alpha\eta_\beta - \eta_\alpha\dot{\eta}_\beta) - u_\alpha u_\beta \right) \tag{1.3.37}$$

with the rigid supersymmetry transformations

$$\delta p^{\alpha\beta} = 0, \quad \delta x_{\alpha\beta} = i\varepsilon(u_\alpha\eta_\beta + u_\beta\eta_\alpha), \tag{1.3.38}$$

$$\delta\eta_\alpha = \varepsilon u_\alpha, \quad \delta u_\alpha = -\frac{i}{2}\dot{\eta}_\alpha\varepsilon. \tag{1.3.39}$$

The equations of motion of this action are

$$\dot{p}^{\alpha\beta} = 0 = p^{\alpha\beta}\dot{\eta}_\beta, \quad \dot{x}_{\alpha\beta} - \frac{i}{2}(\dot{\eta}_\alpha\eta_\beta - \eta_\alpha\dot{\eta}_\beta) - u_\alpha u_\beta = 0, \quad p^{\alpha\beta}u_\beta = 0. \tag{1.3.40}$$

The last of these equations implies that $p_{\alpha\beta} = e^{-1}u_\alpha u_\beta$, whereupon we find that $ep_{\alpha\beta} = \dot{x}_{\alpha\beta} - \frac{1}{2}i(\dot{\eta}_\alpha\eta_\beta - \eta_\alpha\dot{\eta}_\beta)$.

In fact, the above action is invariant under the world-line superconformal transformations of equation (1.2.90) since the measure $d\tau d\theta$ and D scale by (1.2.91) and (1.2.88),
respectively. As explained in the discussion around these equations, the superconformal
transformations are just the reparamterisation and local supersymmetry transformations.
The latter transformations on the component fields are essentially found by replacing ε
by $\alpha(t)$ in the transformations of rigid supersymmetry, that is, those of either equations
(1.3.35) and (1.3.37) or equations (1.3.38) and (1.3.39) for the first and second actions,
respectively, taking into account any required redefinitions of fields and parameters.

Another very important property of the actions of equations (1.3.34) and (1.3.37) is that
they are not only invariant under world-line supersymmetry, but also invariant under rigid
space-time supersymmetry, that is, under

$$\delta \mathbf{P}^{\alpha\beta} = 0, \quad \delta \mathbf{X}_{\alpha\beta} = \frac{i}{2}(\Theta_\alpha \varepsilon_\beta + \Theta_\beta \varepsilon_\alpha), \quad \delta \Theta_\beta = \varepsilon_\beta. \tag{1.3.41}$$

In fact, they are also invariant under space-time superconformal transformations. Hence the
above supertwistor actions possess space-time and world-sheet superconformal invariance.

We have shown above how the actions of equations (1.3.34) and (1.3.37) lead to the
usual formulations of the super point particle which possess world-sheet supersymmetry
and were given in section 1.2. However, they are parent actions in that one can *classically*
recover the Brink–Schwarz formulation that possesses space-time supersymmetry. Indeed,
substituting the expressions for $p_{\alpha\beta} = (\dot{x}_{\alpha\beta} - \frac{1}{2}i(\dot{\eta}_\alpha \eta_\beta - \eta_\alpha \dot{\eta}_\beta))e^{-1}$, given below equation
(1.3.40), into the field equations (1.3.40) and comparing the result with those from the
Brink–Schwarz action of equation (1.2.61) we find they are the same. To recover the
κ transformations of the Brink–Schwarz actions of equation (1.2.62), we must set $\alpha =
-ie^{-1}u^\beta \kappa_\beta$ into the transformations of $\delta x^{\alpha\beta}$ and $\delta \eta$ of equations (1.3.38) and (1.3.39) and
identify $u^\alpha u^\beta = ep^{\alpha\beta}$. Consequently, the κ transformations of the Brink–Schwarz theory
are part of the superconformal invariance of the supertwistor theory of equation (1.3.37).

The other forms of the bosonic point particle actions given previously also possess supersymmetric extensions. The supersymmetric extension of the action of equation (1.3.23)
becomes

$$\int d\tau \, u^\alpha u^\beta \left(\dot{x}_{\alpha\beta} - \frac{i}{2}(\dot{\eta}_\alpha \eta_\beta - \eta_\alpha \dot{\eta}_\beta) \right). \tag{1.3.42}$$

It is invariant under the rigid supersymmetry transformations

$$\delta x_{\alpha\beta} = \frac{i}{2}\epsilon(\eta_\alpha \eta_\beta + \eta_\alpha \eta_\beta), \quad \delta u_\alpha = 0, \quad \delta \eta_\alpha = \epsilon u_\alpha, \tag{1.3.43}$$

while the action of equation (1.3.28) becomes

$$\int d\tau \, (v^\alpha \dot{u}_\alpha - u_\alpha \dot{v}^\alpha - \tfrac{1}{2}u_\alpha u_\beta (\dot{\eta}^\alpha \eta^\beta - \eta^\alpha \dot{\eta}^\beta)) \tag{1.3.44}$$

with the rigid supersymmetry invariance

$$\delta u_\alpha = 0, \quad \delta \eta_\alpha = \epsilon u_\alpha, \quad \delta v^\alpha = \tfrac{1}{2}\epsilon u^\alpha \eta^\beta u_\beta. \tag{1.3.45}$$

The latter is constructed from the supertwistor $u^\alpha, v^\alpha, \eta^\alpha$ [1.17].

We now turn to the case of the four-dimensional point particle beginning with the bosonic particle. In terms of the variables x^μ, p^μ, or their equivalent matrix forms, and the commuting Majorana spinor u_A, $u_{\dot{A}}$ we have the action

$$\int d\tau \, p_{\dot{B}A}(\dot{x}^{A\dot{B}} - u^A u^{\dot{B}}). \qquad (1.3.46)$$

This action is reparameterisation invariant under

$$\delta x^{A\dot{B}} = f(\tau)\dot{x}^{A\dot{B}}, \; \delta p^{A\dot{B}} = f(\tau)\dot{p}^{A\dot{B}}, \; \delta u_A = f\dot{u}_A + \tfrac{1}{2}\dot{f}u_A, \; \delta u_{\dot{A}} = f\dot{u}_{\dot{A}} + \tfrac{1}{2}\dot{f}u_{\dot{A}}. \qquad (1.3.47)$$

An alternative action is given by

$$\int d\tau \, \dot{x}^{A\dot{B}} u_A u_{\dot{B}}. \qquad (1.3.48)$$

The reparameterisation invariance becomes

$$\delta x^{A\dot{B}} = f(\tau)\dot{x}^{A\dot{B}} , \delta u_A = f\dot{u}_A, \; \delta u_{\dot{A}} = f\dot{u}_{\dot{A}}, \qquad (1.3.49)$$

where x denotes the matrix $x^{A\dot{B}}$.

We may also work entirely with the twistor variables and use the action

$$\int d\tau \, (v^A \dot{u}_A + u_{\dot{A}} \dot{v}^{\dot{A}}). \qquad (1.3.50)$$

All the actions are invariant under the gauge invariance $u^A \to e^{i\alpha} u^A$, $v^A \to e^{-i\alpha} v^A$. We leave it to the reader to verify that they do indeed describe the standard dynamics of a massless point particle; the demonstration is similar to the three-dimensional case.

The description of the four-dimensional supersymmetric point particle follows the three-dimensional case closely. The superspace action extending the bosonic action of equation (1.3.46) is given by

$$-i \int d\tau \, d\theta \, \mathbf{P}^{A\dot{B}} \left(D\mathbf{X}_{\dot{B}A} - \frac{i}{2}(D\Theta_A \Theta_{\dot{B}} + \Theta_A D\Theta_{\dot{B}}) \right), \qquad (1.3.51)$$

where the superfields have the component expansions

$$\mathbf{P}^{A\dot{B}} = p^{A\dot{B}} + i\theta \varrho^{A\dot{B}}, \; \mathbf{X}^{A\dot{B}} = x^{A\dot{B}} + i\theta \chi^{A\dot{B}}, \; \Theta_A = \eta_A + \theta u_A, \; \Theta_{\dot{A}} = \eta_{\dot{A}} + \theta u_{\dot{A}}. \qquad (1.3.52)$$

The discussion of this action proceeds as for the three-dimensional case. From this one action one can *classically* derive the equations of motion of the spinning particle which has world-line supersymmetry and also those of the Brink–Schwarz particle, which possesses only space-time supersymmetry. Indeed, the action of equation (1.3.51) is invariant under space-time supersymmetry transformations and invariant under world-line superconformal transformations, which are just the same as reparameterisations and local supersymmetry transformations.

The supersymmetric extension of the action of equation (1.3.48) which involved just $x^{A\dot{B}}$, u^A and $u^{\dot{A}}$ is given by

$$-\int d\tau \left\{ u_A u_{\dot{B}} \left(\dot{x}^{A\dot{B}} - \frac{i}{2}(\dot{\eta}^A \eta^{\dot{B}} - \eta^A \dot{\eta}^{\dot{B}}) \right) \right\}, \qquad (1.3.53)$$

whereas the supersymmetric extension of the action of equation (1.3.50) which was constructed from twistors is built out of a supertwistor which contains the fields $u_{\dot{A}}$, $v^{\dot{A}}$, η^B and their conjugates is given by

$$\int d\tau \left\{ v^{\dot{A}} \dot{u}_{\dot{A}} + u_{\dot{A}} \dot{v}^{\dot{A}} - \frac{1}{2} u_{\dot{A}} u_B (\eta^{\dot{A}} \dot{\eta}^B - \dot{\eta}^{\dot{A}} \eta^B) \right\}. \tag{1.3.54}$$

One remarkable feature of the above actions is that they are invariant under local supersymmetry, but have no einbein or gravitino. These supergravity fields only appear when one solves the equations of motion.

Unlike the Brink–Schwarz theory the twistor-like formulations of the supersymmetric point particle can be first quantised in a Lorentz covariant manner. We illustrate this for the three-dimensional action of equation (1.3.37). Calculating the momenta we find the constraints

$$p_u = 0, \ d^{\alpha} \equiv P_{\eta}^{\alpha} + i p^{\alpha\beta} \eta_{\beta} = 0. \tag{1.3.55}$$

The Hamiltonian is $H = p^{\alpha\beta} u_{\alpha} u_{\beta}$ and one finds that the time development of the first constraint in the above equation implies the further constraint $c^{\alpha} \equiv p^{\alpha\beta} u_{\beta} = 0$. As should be the case, the d^{α} constraints satisfy the supersymmetry algebra: $\{d^{\alpha}, d^{\beta}\} = 2i p^{\alpha\beta}$. It was explained in section 1.2.2 that due to the null nature of p^{μ}, that is, $p_{\mu} p^{\mu} = 0$, these two constraints are a mixture of first and second class constraints. However, for this theory we can split them into the constraints [1.14]

$$u_{\alpha} d^{\alpha}, \ v_{\alpha} d^{\alpha}, \tag{1.3.56}$$

where $v^{\alpha} = (x^{\alpha\beta} - i\eta^{\alpha}\eta^{\beta}) u_{\beta}$. The former constraint is first class as a result of the constraint $c^{\alpha} = 0$, whereas, the latter constraint has the Poisson bracket

$$\{v_{\alpha} d^{\alpha}, v_{\beta} d^{\beta}\} = 2i v_{\alpha} v_{\beta} p^{\alpha\beta}. \tag{1.3.57}$$

Since the left-hand side is non-vanishing for non-zero v_{α}, it is second class.

The actions given in this sections can be derived from a non-linear realisation of the superconformal group [1.28]. Twistors can, in fact, be viewed as a coset of the (super)conformal group divided by an appropriate subgroup. In this way of viewing things, the various parameterisations of phase space emerge by considering the different parameterisations of the coset and the Penrose equation and reality constraints emerge automatically.

The ten-dimensional twistor-like actions for the bosonic point particle are much like the three- and four-dimensional analogues given above. The same cannot be said for the ten-dimensional superparticle. Indeed, the existence of twistors is related to the properties of the conformal group, and it is interesting to note that the conformal group in ten dimensions possesses a minimal supersymmetric extension that involves central charges in a non-trivial way.

One of the motivations to quantise the Brink–Schwarz particle covariantly was the desire to covariantly quantise its string analogue, the Green–Schwarz string, in such a way as to preserve its space-time supersymmetry. In fact, this hope has largely been achieved in a formulation of the string that possesses the type of twistor-like variables discussed above [1.29]. It is intriguing that twistors have played an important role in the solution to this and other problems in string theory.

Although the gauge choice of equation (1.1.24) was used in this chapter and is almost universally adopted in the literature, it is not quite correct. This is due to the fact that the reparameterisations in the world-line are fixed for the intial and final configurations [1.30]. If one makes a correct choice and then carries out the corresponding BRST quantisation of the bosonic point particle, one finds a formulation of the point particle which possesses an ortho-symplectic symmetry [1.31, 1.32]. The true significance of this result has yet to be understood.

2 The classical bosonic string

> Men wanted for hazardous journey, small wages, bitter cold, long months of complete
> darkness, constant danger, safe return doubtful, honour and recognition in case of success.
> Ernest Shackleton advertising for crew for the journey to the South Pole

In this chapter, we will discuss the classical bosonic string. As the string moves through space-time, it sweeps out a two-dimensional world sheet. By analogy with the point particle, we will take the string action to be the area of the world sheet swept out [2.1, 2.2]. From this action we will derive the equations of motion for the string. In preparation for its quantisation in the next chapter we will describe the string in terms of new variables.

As explained in chapter 18, string theory was discovered in a very unexpected way. After most of the scattering amplitudes had been found the early pioneers were still unaware that they were dealing with the scattering of strings. We refer the reader to section 18.1 for an account of this fascinating history.

2.1 The dynamics

Strings may be open or closed as shown in figure 2.1.1. We take the length along the string to be parameterised by σ and its passage in time to be parameterised by τ. Hence the world sheet has coordinates ξ^α, $\alpha = 0, 1$ with $\xi^\alpha = (\tau, \sigma)$ and is embedded in D-dimensional Minkowski space-time according to $x^\mu(\xi^\alpha)$, $\mu = 0, 1, \ldots, D - 1$. For the open string we take the range of σ to be $0 \leq \sigma \leq \pi$, while for the closed string we take $-\pi < \sigma \leq \pi$. It is natural for the closed string to take the boundary condition $x^\mu(\tau, -\pi) = x^\mu(\tau, \pi)$ as $\sigma = -\pi$ and $\sigma = \pi$ are the same point on the string. We will discuss the boundary condition for the open string below.

We take the action, first discussed in this context by Nambu and Gotto [2.1, 2.2], for both the open and the closed string, to be given by

$$A = -\frac{1}{2\pi\alpha'} \int d^2\xi \sqrt{-\det\{\partial_\alpha x^\mu \partial_\beta x^\nu \eta_{\mu\nu}\}}, \tag{2.1.1}$$

where $\partial_\alpha x^\mu = \partial x^\mu / \partial \xi^\alpha$. The dimension of x^μ is [mass]$^{-1}$ and as the dimension of ξ^α cancels out from the dimension of the action we must take the constant α' to have the dimension of [mass]$^{-2}$ in order that A be dimensionless. We note that $\det\{\partial_\alpha x^\mu \partial_\beta x^\nu \eta_{\mu\nu}\} = (\partial_0 x^\mu \partial_0 x^\nu \eta_{\mu\nu})(\partial_1 x^\rho \partial_1 x^\kappa \eta_{\rho\kappa}) - (\partial_0 x^\mu \partial_1 x^\nu \eta_{\mu\nu})^2$.

If we move from ξ^α to $\xi^\alpha + d\xi^\alpha$ on the world sheet, the corresponding change induced in the flat Minkowski space-time through which the string moves is $x^\mu \to x^\mu + \partial_\alpha x^\mu d\xi^\alpha$.

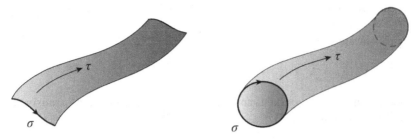

Figure 2.1.1   The world sheets of the free open and closed strings.

Consequently, the (distance)^2 between the two points ξ^α and $\xi^\alpha + d\xi^\alpha$ on the world sheet is given by $d\xi^\alpha \partial_\alpha x^\mu d\xi^\beta \partial_\beta x^\nu \eta_{\mu\nu}$ and so $g^I_{\alpha\beta} \equiv \partial_\alpha x^\mu \partial_\beta x^\nu \eta_{\mu\nu}$ is the metric induced on the world sheet from the Minkowski metric in space-time. The index I indicates the induced metric and should not be confused with the independent metric we will encounter shortly. We may therefore write the integral above as $\int d^2\xi \sqrt{-\det g^I_{\alpha\beta}}$ and the integrand as $\sqrt{-\det g^I_{\alpha\beta}} = \det e^a_\alpha$, where e^a_α is the two-dimensional vielbein on the world sheet defined by $e^a_\alpha \eta_{ab} e^b_\beta = g^I_{\alpha\beta}$ and $\eta_{ab} = \mathrm{diag}(-1, +1)$. Since the area of the infinitesimal element defined by the world sheet vielbein tangent vectors $e^a_\alpha \delta\xi^\alpha$, $a = 0, 1$, is just $\det e^a_\alpha \delta\xi^0 \delta\xi^1$, we recognise the above action as just the area swept out by the string times the constant $-1/2\pi\alpha'$.

Remarkably, the action, as the volume swept out, for what we now call a 2-brane, that is, an extended object whose spatial surface is two-dimensional, was written down and analysed by Dirac [2.3]. The idea was that the muon could be some excited state of the electron and both could arise as the dynamics of some extended object. The analysis of the dynamics of the action of equation (2.1.1) was given in [2.4] at both the classical and the quantum level. This chapter contains the classical analysis.

As for the point particle considered in section 1.1, the minus sign under the square root in equation (2.1.1) is required in order that the quantity under the square root be positive for a string that propagates causally.

The analogue of the mass of the point particle is the quantity $T \equiv (2\pi\alpha')^{-1}$, which is called the string tension. For a static string the string tension multiplies the length of the string. In this case the kinetic term vanishes and so the Hamiltonian equals minus the Lagrangian, and therefore we can interpret the string tension as the energy, or mass per unit length, of the string.

The factors of $(2\pi\alpha')^{-1}$ outside the action of equation (2.1.1) lead to messy factors in the subsequent equations. To simplify these expressions one can, as in the old days, take the scale of energy to be such that $2\alpha' = 1$ for the open string and $\alpha' = 2$ for the closed string or alternatively scale $x^\mu \to \sqrt{2\alpha'} x^\mu$ in all expressions.

The action of equation (2.1.1) is invariant under the following symmetries.

(i) Reparameterisation invariance:

$$\xi^\alpha \to \xi^{\alpha\prime}(\xi) \tag{2.1.2}$$

under which the fields $x^\mu(\xi)$ are scalars, that is,

$$x^{\mu\prime}(\xi') = x^\mu(\xi). \tag{2.1.3}$$

For such an infinitesimal transformation $\delta\xi^\alpha \equiv \xi^{\alpha\prime} - \xi^\alpha$ is small and $\delta x^\mu \equiv x^{\mu\prime}(\xi) - x^\mu(\xi) = -\delta\xi^\alpha \partial_\alpha x^\mu$.

(ii) Rigid Poincaré symmetry:

$$x^{\mu\prime}(\xi) = \Lambda^\mu{}_\nu x^\nu(\xi) + a^\mu, \tag{2.1.4}$$

where a^μ and $\Lambda^\mu{}_\nu$ are constants and the Lorentz transformations $\Lambda^\mu{}_\nu$ satisfy their usual defining condition

$$\eta_{\mu\nu}\Lambda^\mu{}_\rho \Lambda^\nu{}_\tau = \eta_{\rho\tau}. \tag{2.1.5}$$

Reparameterisation invariance means that the particular coordinate system used to parameterise the world-sheet is of no physical significance. From the two-dimensional world-sheet viewpoint, the Poincaré invariance is an internal symmetry. We shall see that the quantum bosonic string is consistent only in 26 dimensions. Although it is not apparent why one should assume that Poincaré invariance is true in all dimensions, in this first treatment, we take this to be the case.

To verify the reparameterisation invariance of the Nambu action of equation (2.1.1) we use the chain rule for differentiation to obtain the formula

$$\frac{\partial x^{\mu\prime}}{\partial\xi^{\alpha\prime}}\frac{\partial x^{\nu\prime}}{\partial\xi^{\beta\prime}} = \frac{\partial\xi^\gamma}{\partial\xi^{\alpha\prime}}\frac{\partial\xi^\delta}{\partial\xi^{\beta\prime}}\partial_\gamma x^\mu \partial_\delta x^\nu. \tag{2.1.6}$$

Writing this equation in matrix form, and using the relations $\det AB = \det A \det B$ and $\det A^T = \det A$ for any two matrices A and B, we find that

$$\sqrt{-\det\frac{\partial x^{\mu\prime}}{\partial\xi^{\alpha\prime}}\frac{\partial x^{\nu\prime}}{\partial\xi^{\beta\prime}}\eta_{\mu\nu}} = J\sqrt{-\det\frac{\partial x^\mu}{\partial\xi^\alpha}\frac{\partial x^\nu}{\partial\xi^\beta}\eta_{\mu\nu}}, \tag{2.1.7}$$

where

$$J = \det\frac{\partial\xi^\alpha}{\partial\xi^{\beta\prime}}. \tag{2.1.8}$$

The final step is to realise that the transformation of the measure is given by

$$d^2\xi' = J^{-1}d^2\xi. \tag{2.1.9}$$

As a result, the reparameterisation invariance of the action of equation (2.1.1) is apparent.

An action with a square root is not particularly easy to handle especially when one comes to quantisation. Fortunately, by introducing the independent field $g_{\alpha\beta}$, there exists another reparameterisation invariant action [2.5, 2.6]:

$$A = -\frac{1}{4\pi\alpha'}\int d^2\xi\sqrt{-g}g^{\alpha\beta}\partial_\alpha x^\mu \partial_\beta x^\nu \eta_{\mu\nu}, \tag{2.1.10}$$

where $g_{\alpha\beta}$ can be identified as the two-dimensional metric on the world sheet. As usual $g^{\alpha\beta}$ is its inverse and $g \equiv \det g_{\alpha\beta}$. In this form the action is a set of D scalar fields x^μ coupled to two-dimensional gravity $g_{\alpha\beta}$. We will show after equation (2.1.23), that this action is equivalent, at the classical level, to the previous action of equation (2.1.1). It is invariant under reparameterisations, the transformations being those of equation (2.1.3) and

$$x^{\mu\prime}(\xi') = x^\mu(\xi), \quad g'_{\alpha\beta}(\xi') = \frac{\partial\xi^\gamma}{\partial\xi^{\alpha\prime}}\frac{\partial\xi^\delta}{\partial\xi^{\beta\prime}}g_{\gamma\delta}(\xi). \tag{2.1.11}$$

The Poincaré transformations, which now take the form

$$x^{\mu\prime}(\xi) = \Lambda^\mu{}_\nu x^\nu(\xi) + a^\mu, \quad g'_{\alpha\beta}(\xi) = g_{\alpha\beta}(\xi) \tag{2.1.12}$$

also lead to an invariance of the action of equation (2.1.10). However, the action of equation (2.1.10) also has a Weyl symmetry which is realised by the transformations

$$g'_{\alpha\beta}(\xi) = \Lambda(\xi)g_{\alpha\beta}(\xi), \qquad x^{\mu\prime}(\xi) = x^\mu(\xi), \tag{2.1.13}$$

where $\Lambda(\xi)$ is the local parameter. The action of equation (2.1.1) is also Weyl invariant in the sense that all the fields (that is, x^μ) are inert.

To show that the action of equation (2.1.10) is invariant under general coordinate transformations is a textbook exercise in general relativity. In essence, it is obvious using the usual rules of the tensor calculus. The metric is a (0, 2) tensor and so the inverse metric is a (2, 0) tensor; however, $\partial_\alpha x^\mu$ is a (0, 1) tensor and so $g^{\alpha\beta}\partial_\alpha x^\mu \partial_\beta x^\nu \eta_{\mu\nu}$ is a scalar. Given that $\int d^2\xi \sqrt{-g}$ is also invariant we find the desired result. When verifying this statement explicitly for infinitesimal variations, it is useful to use the equations

$$\delta g = g g^{\alpha\beta}\delta g_{\alpha\beta} = -g\delta g^{\alpha\beta}g_{\alpha\beta}, \quad \delta g^{\alpha\beta} = -g^{\alpha\delta}\delta g_{\delta\gamma}g^{\gamma\beta} \tag{2.1.14}$$

for the variations $\delta g_{\alpha\beta} = -\delta\xi^\gamma \partial_\gamma g_{\alpha\beta} - \partial_\alpha \delta\xi^\gamma g_{\gamma\beta} - \partial_\beta \delta\xi^\gamma g_{\alpha\gamma}$ and $\delta x^\mu = -\delta\xi^\gamma \partial_\gamma x^\mu$.

We now have several choices in deriving the equations of motion. We can adopt a Lagrangian or a Hamiltonian viewpoint and we can work with the actions of either equation (2.1.1) or equation (2.1.10). The Hamiltonian treatment of the action of equation (2.1.1) is given in chapter 4 in the context of the light-cone formalism. Now we will use the action of equation (2.1.10) whose Lagrangian is given by $L = (-1/4\pi\alpha')\sqrt{-g}g^{\alpha\beta}\partial_\alpha x^\mu \partial_\beta x^\nu \eta_{\mu\nu}$. The results are, of course, the same, no matter which formalism one uses.

We recall that the variational principle states that a system moves from a fixed configuration at the initial time τ_1 to a fixed configuration at a final time τ_2 in such a way as to be an extremum of the action. In our case this means that

$$0 = \delta A = \int_a^\pi d\sigma \int_{\tau_1}^{\tau_2} d\tau \left\{ \frac{\partial L}{\partial g_{\alpha\beta}}\delta g_{\alpha\beta} + \frac{\partial L}{\partial(\partial_\sigma x^\mu)}\delta(\partial_\sigma x^\mu) + \frac{\partial L}{\partial(\partial_\tau x^\mu)}\delta(\partial_\tau x^\mu) \right\} \tag{2.1.15}$$

for arbitrary $\delta g_{\alpha\beta}$ and δx^μ provided only that these quantities vanish at τ_1 and τ_2. In accord with our discussion on the parameterisation of the open and closed strings above equation (2.1.1), we take $a = 0$ and $a = -\pi$ for the open and closed strings, respectively, in equation (2.1.15). Integrating the last two terms by parts, according to the above instruction, and using the fact that $\delta\partial_\alpha x^\mu = \partial_\alpha \delta x^\mu$, we find that

$$0 = \int_a^\pi d\sigma \int_{\tau_1}^{\tau_2} d\tau \left\{ \frac{\partial L}{\partial g_{\alpha\beta}}\delta g_{\alpha\beta} + \left[-\partial_\sigma \frac{\partial L}{\partial(\partial_\sigma x^\mu)} - \partial_\tau \frac{\partial L}{\partial(\partial_\tau x^\mu)} \right] \delta x^\mu \right\}$$

$$+ \int_{\tau_1}^{\tau_2} d\tau \frac{\partial L}{\partial(\partial_\sigma x^\mu)}\delta x^\mu \Big|_{\sigma=a}^{\sigma=\pi}. \tag{2.1.16}$$

We find no integral over σ in the integration by parts as the variational principle requires fixed initial and final configurations. Consequently, for arbitrary $\delta g_{\alpha\beta}$, but $\delta x^\mu = 0$, we find the equation

$$\frac{\partial L}{\partial g_{\alpha\beta}} = 0. \tag{2.1.17}$$

Similarly, for $\delta g_{\alpha\beta} = 0$ and δx^μ arbitrary but vanishing at $\sigma = \pi$ and $\sigma = a$, we find that

$$\partial_\alpha \Pi^{\alpha\mu} = 0, \tag{2.1.18}$$

where

$$\Pi^{\alpha\mu} = \frac{\partial L}{\partial(\partial_\alpha x_\mu)}. \tag{2.1.19}$$

Evaluating equations (2.1.17) and (2.1.18) we find, respectively,

$$0 = \partial_\alpha x^\mu \partial_\beta x^\nu \eta_{\mu\nu} - \tfrac{1}{2} g_{\alpha\beta} g^{\gamma\delta} \partial_\gamma x^\mu \partial_\delta x^\nu \eta_{\mu\nu} \tag{2.1.20}$$

and

$$\partial_\alpha \left(\sqrt{-g} \, g^{\alpha\beta} \partial_\beta x^\mu \right) = 0. \tag{2.1.21}$$

In fact, it is slightly easier to compute the equations of motion by explicitly varying the action under $\delta g^{\alpha\beta}$ and using equation (2.1.14) to find equation (2.1.20) and δx^μ to find equation (2.1.21).

For the closed string the boundary term of equation (2.1.16) vanishes automatically as $x^\mu(\tau, -\pi) = x^\mu(\tau, \pi)$. However, for the open string we find it leads to non-trivial condition. By continuity we may also take equation (2.1.18) and therefore equation (2.1.21) to hold at the boundary $\sigma = 0$ and at $\sigma = \pi$. Since the σ dependence of δx^μ is not restricted at the boundaries $\sigma = 0$ and $\sigma = \pi$ we find that setting the variation of the action to vanish also implies, corresponding to the last term in equation (2.1.16), the boundary condition

$$\Pi^{1\mu} \equiv \frac{\partial L}{\partial(\partial_\sigma x_\mu)} = 0 \text{ at } \sigma = 0 \text{ and } \sigma = \pi. \tag{2.1.22}$$

Hence, for the open string we conclude that

$$\sqrt{-g} \, g^{1\beta} \partial_\beta x^\mu = 0, \text{ at } \sigma = 0 \text{ and } \sigma = \pi. \tag{2.1.23}$$

We must impose this condition at the two ends of the string separately as the variations δx^μ are not related at the two boundaries. This is just as well, as the two ends may not be causally connected at all times. In fact, there is another possible boundary condition that we will discuss at length in chapter 15.

Using equation (2.1.20) we can express $g_{\alpha\beta}$ by $g_{\alpha\beta} = 2f \partial_\alpha x^\mu \partial_\beta x^\nu \eta_{\mu\nu}$, where $f = \left(g^{\gamma\delta} \partial_\gamma x^\mu \partial_\delta x^\nu \eta_{\mu\nu} \right)^{-1}$. Taking the determinant we find that

$$g = \det g_{\alpha\beta} = 4f^2 \det \partial_\alpha x^\mu \partial_\beta x^\nu \eta_{\mu\nu} \tag{2.1.24}$$

and substituting for $\sqrt{-g}$ gives the relation

$$\sqrt{-g} \, g^{\gamma\delta} \partial_\gamma x^\mu \partial_\delta x^\nu \eta_{\mu\nu} = 2\sqrt{-\det \partial_\alpha x^\mu \partial_\beta x^\nu \eta_{\mu\nu}}. \tag{2.1.25}$$

This formula allows us to algebraically eliminate $g_{\alpha\beta}$ in the action of equation (2.1.10) and so obtain the action of equation (2.1.1), thus demonstrating the equivalence of the two actions at the classical level. At the quantum level the equivalence is more delicate.

The above equations are rather complicated, but simplify considerably when we make a convenient gauge choice which we will now discuss. In the mathematical literature a Riemann surface is taken to be any two-dimensional surface with a Euclidean metric. The string world sheet is a two-dimensional surface, but with a Minkowski metric. In most discussions of string theory this difference is ignored and many theorems are just 'borrowed' from the Euclidean case. Of course, it could turn out that for some subtle points this difference is important.

It can be shown that on any oriented Riemann surface, we may locally choose a coordinate system such that the metric takes the form

$$g_{\alpha\beta}(\xi) = \eta_{\alpha\beta} e^{\phi(\xi)} \tag{2.1.26}$$

or, in terms of the line element,

$$ds^2 = e^{\phi} d\xi^{\alpha} d\xi^{\beta} \eta_{\alpha\beta}, \tag{2.1.27}$$

where $\eta_{\alpha\beta}$ is the flat metric on the world sheet, that is, $\eta_{\alpha\beta} = \text{diag}(-1, +1)$.

On a degree of freedom count one may expect such a theorem. The most general metric $g_{\alpha\beta}$ has three degrees of freedom while the restricted metric of equation (2.1.26) has only one degree of freedom corresponding to the field $\phi(\xi)$. However, we can expect to remove two degrees of freedom by using the two ξ^{α} reparameterisations. The proof of the theorem, which is constructive by nature, can be found in many places. A readable account is contained in [2.7]. Essentially one starts with the most general metric and then changes coordinates such that the metric becomes that of equation (2.1.27). This implies that the new coordinates as a function of the old coordinates satisfy certain partial differential equations depending upon the old metric. One can then show that such solutions to these differential equations always exist.

Consequently, we may choose the metric $g_{\alpha\beta}$ that appears in the above formulation to be of the form of equation (2.1.26). It is illegal to impose a gauge choice in the action before we carry out the variation as it would mean that we would not find the equations of motion for the gauge degrees of freedom being fixed. However, we can impose a gauge choice on the equations of motion. Using the gauge choice of equation (2.1.26), equation (2.1.21) becomes the wave equation, namely

$$\partial_{\alpha} \partial^{\alpha} x^{\mu} = 0, \tag{2.1.28}$$

while equation (2.1.20) can be written as

$$T_{\alpha\beta} = 0, \tag{2.1.29}$$

where

$$T_{\alpha\beta} = \partial_{\alpha} x^{\mu} \partial_{\beta} x^{\nu} \eta_{\mu\nu} - \tfrac{1}{2} \eta_{\alpha\beta} \partial_{\gamma} x^{\mu} \partial^{\gamma} x^{\nu} \eta_{\mu\nu}. \tag{2.1.30}$$

The use of the symbol $T_{\alpha\beta}$ is deliberate as we recognise it as the energy-momentum tensor of the theory of D free scalar fields x^{μ} with action

$$\int d^2\xi \{ -\tfrac{1}{2} \partial_{\alpha} x^{\mu} \partial_{\beta} x^{\nu} \eta_{\mu\nu} \eta^{\alpha\beta} \}. \tag{2.1.31}$$

As a consequence of Weyl invariance, under which $\sqrt{-g}g^{\alpha\beta}$ is invariant, the field ϕ does not appear in the equations of motion. This invariance also ensures that $T_{\alpha\beta}$ is traceless. The two non-zero components are

$$T_{00} = T_{11} = \tfrac{1}{2}(\dot{x}^\mu\dot{x}^\nu + x^{\mu\prime}x^{\nu\prime})\eta_{\mu\nu} \tag{2.1.32}$$

and

$$T_{01} = T_{10} = \dot{x}^\mu x^{\nu\prime}\eta_{\mu\nu}, \tag{2.1.33}$$

where $\dot{x}^\mu = \partial x^\mu/\partial\tau$, $x^{\mu\prime} = \partial x^\mu/\partial\sigma$. The prime here should not be confused with the notation used earlier for field transformations. For this choice of metric the boundary condition for the open string also takes on a particularly simple form:

$$x^{\mu\prime}(\tau, \sigma) = 0 \quad \text{at } \sigma = 0 \text{ and at } \sigma = \pi. \tag{2.1.34}$$

The momentum $P^\mu(\tau, \sigma)$ is given by

$$P^\mu = \frac{\partial L}{\partial(\partial_\tau x_\mu)} = \frac{\dot{x}^\mu}{2\pi\alpha'}, \tag{2.1.35}$$

where in the last line we have used the the choice of metric of equation (2.1.26) after differentiating the Lagrangian. Using the standard definition of the Poisson brackets, the Poisson brackets for the coordinates x^μ and momenta P^μ are

$$\{x^\mu(\sigma), x^\nu(\sigma')\} = 0 = \{P^\mu(\sigma), P^\nu(\sigma')\},$$
$$\{x^\mu(\sigma), P^\nu(\sigma')\} = \eta^{\mu\nu}\delta(\sigma - \sigma'). \tag{2.1.36}$$

In these formulae we have not shown the τ dependence as the Poisson brackets relations are at the same τ.

The different boundary conditions of the open and closed strings require us to treat them separately for the rest of this chapter. We begin with the closed string.

2.1.1   The closed string

Rather than describe the closed string by $x^\mu(\tau, \sigma)$, it will prove useful to introduce normal modes much as one does for a violin string. Corresponding to the range of σ of $-\pi \le \sigma < \pi$ and the boundary condition $x^\mu(\tau, \sigma) = x^\mu(\tau, \sigma + 2\pi)$ of the closed string, we may write

$$x^\mu(\tau, \sigma) \equiv \sum_{n=-\infty}^{\infty} e^{in\sigma}x_n^\mu(\tau), \tag{2.1.37}$$

where $x_{-n}^\mu = (x_n^\mu)^*$ to ensure that $x^\mu(\tau, \sigma)$ is real. In terms of the normal modes, equation (2.1.28), which was $\ddot{x}^\mu = x^{\mu\prime\prime}$, becomes $\ddot{x}_n^\mu = -n^2 x_n^\mu$. Hence, the closed string is described by centre of mass coordinates x_0^μ, which move with uniform velocity, and an infinite set of harmonic oscillators. However, these modes are subject to the constraint $T_{\alpha\beta} = 0$.

Every student recalls the quantisation of the usual harmonic oscillator. The Hamiltonian is given by $H = \tfrac{1}{2}(p^2 + x^2)$, where we have set the frequency to 1, and introduced the well-known harmonic oscillators. In carrying out this step first at the classical level one replaces the parameterisation of phase space given by $x(\tau)$ and $p(\tau)$ with the new variables $a(\tau) =$

$\frac{1}{\sqrt{2}}(x+ip)$ and $(a(\tau))^\dagger = \frac{1}{\sqrt{2}}(x-ip)$. The usual Poisson bracket $\{x,p\}=1$ becomes $\{a,a^\dagger\} = -i$, all other relations being zero, and $H = a^\dagger a + \frac{1}{2}$. The quantisation is now straightforward: the Poisson bracket becomes the commutator $[a,a^\dagger] = 1$ and the energy eigenstates are given by $(a^\dagger)^n|0\rangle$, where the vacuum satisfies $a|0\rangle = 0$. We will now essentially follow this path for the string. In doing so we will carry out the transformation in phase space carefully and resist the temptation to solve the equations of motion and quantise the constants that appear. This recipe is much shorter but it does not follow from any physical principles and will not work for physical systems in general.

For the closed string, we will introduce an infinite number of creation and annihilation operators. We can do this first at the classical level as indeed we just did for the usual harmonic oscillator. This step changes the parameterisation of phase space of the string by introducing the quantities

$$\wp^\mu(\tau,\sigma) \equiv \frac{1}{\sqrt{2\alpha'}}(\dot{x}^\mu(\tau,\sigma) + x^{\mu\prime}(\tau,\sigma)),$$

$$\bar{\wp}^\mu(\tau,\sigma) \equiv \frac{1}{\sqrt{2\alpha'}}(\dot{x}^\mu(\tau,\sigma) - x^{\mu\prime}(\tau,\sigma)). \tag{2.1.38}$$

The reader should not confuse $\wp^\mu(\tau,\sigma)$ with the momentum of the string, which is given by P^μ. The classical analogues of the infinite set of creation and annihilation operators are given by α_n^μ and $\bar{\alpha}_n^\mu$, which are defined by

$$\wp^\mu(\tau,\sigma) \equiv \sum_{n=-\infty}^{\infty} \alpha_n^\mu(\tau)e^{-in\sigma}, \qquad \bar{\wp}^\mu(\tau,\sigma) \equiv \sum_{n=-\infty}^{\infty} \bar{\alpha}_n^\mu(\tau)e^{in\sigma}. \tag{2.1.39}$$

The total momentum p^μ of the closed string is by definition given by

$$p^\mu = \int_{-\pi}^{\pi} d\sigma\, P^\mu = \frac{\dot{x}_0^\mu}{\alpha'}, \tag{2.1.40}$$

using equation (2.1.35). Taking the integral over σ of $\wp^\mu$ and $\bar{\wp}^\mu$, and using equation (2.1.40) and equations (2.1.38) and (2.1.39) we find they both give, up to the same factor, the total momentum

$$p^\mu = \frac{\dot{x}_0^\mu}{\alpha'} = \frac{1}{\sqrt{2\alpha'}\pi}\int_{-\pi}^{\pi} d\sigma\,\wp^\mu = \frac{1}{\sqrt{2\alpha'}\pi}\int_{-\pi}^{\pi} d\sigma\,\bar{\wp}^\mu = \sqrt{\frac{2}{\alpha'}}\alpha_0^\mu = \sqrt{\frac{2}{\alpha'}}\bar{\alpha}_0^\mu. \tag{2.1.41}$$

We note that, unlike for the non-zero modes, $\bar{\alpha}_0^\mu = \alpha_0^\mu$.

The momentum of the string was given by $P^\mu = \dot{x}^\mu/2\pi\alpha'$ and so $\wp^\mu$ and $\bar{\wp}^\mu$ can be expressed as

$$\wp^\mu(\sigma) = \sqrt{2\alpha'}\pi P^\mu + \frac{x^{\mu\prime}}{\sqrt{2\alpha'}}, \quad \bar{\wp}^\mu(\sigma) = \sqrt{2\alpha'}\pi P^\mu - \frac{x^{\mu\prime}}{\sqrt{2\alpha'}}. \tag{2.1.42}$$

Using equation (2.1.36) we find the Poisson brackets

$$\{\wp^\mu(\sigma),\wp^\nu(\sigma')\} = 2\pi\eta^{\mu\nu}\delta'(\sigma-\sigma'), \{\bar{\wp}^\mu(\sigma),\bar{\wp}^\nu(\sigma')\} = -2\pi\eta^{\mu\nu}\delta'(\sigma-\sigma'),$$

$$\tag{2.1.43}$$

where $\delta'(\sigma - \sigma') = (d/d\sigma)\delta(\sigma - \sigma')$. The Poisson bracket between $\wp^\mu$ and $\bar{\wp}^\mu$ vanishes. Taking the Fourier modes we find that the harmonic oscillators obey the Poisson brackets

$$\{\alpha_n^\mu, \alpha_m^\nu\} = -in\delta_{n+m,0}\eta^{\mu\nu}, \quad \{\bar{\alpha}_n^\mu, \bar{\alpha}_m^\nu\} = -in\delta_{n+m,0}\eta^{\mu\nu}, \quad \{\alpha_n^\mu, \bar{\alpha}_m^\nu\} = 0. \quad (2.1.44)$$

Hence, we have introduced the variables $x_0^\mu(\tau)$, $\alpha_0^\mu(\tau)$, $\alpha_n^\mu(\tau)$ and $\bar{\alpha}_n^\mu(\tau)$, $n \neq 0$, to describe the phase space of the string which was originally given by $x^\mu(\tau, \sigma)$ and $P^\mu(\tau, \sigma)$.

The next step is to find the equations of motion in terms of these new variables. Taking the τ derivative of $\wp^\mu$, and using the equation of motion of x^μ, namely $\ddot{x}^\mu = x^{\mu\prime\prime}$, we note that $\dot{\wp}^\mu = \wp^{\mu\prime}$. This, and a similar manipulation for $\bar{\wp}^\mu$, implies that

$$\dot{\alpha}_n^\mu(\tau) = -in\alpha_n^\mu(\tau); \quad \dot{\bar{\alpha}}_n^\mu(\tau) = -in\bar{\alpha}_n^\mu(\tau), \quad (2.1.45)$$

and hence

$$\alpha_n^\mu(\tau) = e^{-in\tau}\alpha_n^\mu(0), \quad \bar{\alpha}_n^\mu = \bar{\alpha}_n^\mu(0)e^{-in\tau}. \quad (2.1.46)$$

For the zero mode x_0^μ, we find, using equation (2.1.41), that $x_0^\mu(\tau) = q^\mu + \sqrt{2\alpha'}\alpha_0^\mu\tau$, where $q^\mu \equiv x_0^\mu(0)$ and α_0^μ are independent of τ.

We now must express $x^\mu(\tau, \sigma)$ in terms of the new variables. Using equation (2.1.38) we may solve for $x^{\mu\prime}$ to find that

$$x^{\mu\prime}(\tau, \sigma) = \sqrt{\frac{\alpha'}{2}}(\wp^\mu(\tau, \sigma) - \bar{\wp}^\mu(\tau, \sigma)) = \sqrt{\frac{\alpha'}{2}} \sum_{n=-\infty}^{\infty} \left(\alpha_n^\mu(\tau)e^{-in\sigma} - \bar{\alpha}_n^\mu(\tau)e^{in\sigma}\right). $$

$$(2.1.47)$$

Integrating with respect to σ, we find that

$$x^\mu(\tau, \sigma) = x_0^\mu(\tau) + i\sqrt{\frac{\alpha'}{2}} \sum_{\substack{n=-\infty \\ n\neq 0}}^{\infty} \frac{1}{n}\left(\alpha_n^\mu(\tau)e^{-in\sigma} + \bar{\alpha}_n^\mu(\tau)e^{in\sigma}\right). \quad (2.1.48)$$

Substituting equations (2.1.46) into equation (2.1.48) we find that

$$x^\mu(\tau, \sigma) = q^\mu + \alpha' p^\mu\tau + i\sqrt{\frac{\alpha'}{2}} \sum_{\substack{n=-\infty \\ n\neq 0}}^{\infty} \frac{1}{n}\left(\alpha_n^\mu(0)e^{-in(\tau+\sigma)} + \bar{\alpha}_n^\mu(0)e^{-in(\tau-\sigma)}\right),$$

$$(2.1.49)$$

where we recall that $\sqrt{\frac{1}{2}\alpha'}p^\mu = \alpha_0^\mu = \bar{\alpha}_0^\mu$. We may therefore write x^μ in the form

$$x^\mu(\tau, \sigma) = x_L^\mu(\tau + \sigma) + x_R^\mu(\tau - \sigma), \quad (2.1.50)$$

where

$$x_L^\mu(\tau + \sigma) = \frac{q^\mu}{2} + \alpha'\frac{p^\mu}{2}(\tau + \sigma) + i\sqrt{\frac{\alpha'}{2}} \sum_{\substack{n=-\infty \\ n\neq 0}}^{\infty} \frac{\alpha_n^\mu(0)}{n}e^{-in(\tau+\sigma)}, \quad (2.1.51)$$

$$x_R^\mu(\tau - \sigma) = \frac{q^\mu}{2} + \alpha'\frac{p^\mu}{2}(\tau - \sigma) + i\sqrt{\frac{\alpha'}{2}} \sum_{\substack{n=-\infty \\ n\neq 0}}^{\infty} \frac{\bar{\alpha}_n^\mu(0)}{n}e^{-in(\tau-\sigma)}. \quad (2.1.52)$$

In the gauge in which we are working the equation of motion of x^μ is the wave equation $\partial^\alpha\partial_\alpha x^\mu = 0$ and clearly the above expression for x^μ is a solution to the equation of motion, indeed x_R^μ and x_L^μ are functions of $\xi^\pm = \tau \pm \sigma$. The subscripts L and R correspond to the left and right moving characters, respectively, of the objects to which they refer. Closed strings moving on a torus can wind around the torus; this leads to some important modifications to the above formulae that are discussed in chapter 10. We also observe that

$$\wp^\mu(\tau, \sigma) = \sum_{n=-\infty}^{\infty} \alpha_n^\mu(0) e^{-in(\tau+\sigma)}, \qquad \bar{\wp}^\mu(\tau, \sigma) = \sum_{n=-\infty}^{\infty} \bar{\alpha}_n^\mu(0) e^{-in(\tau-\sigma)}.$$

$$(2.1.53)$$

In the next chapter we will quantise the closed string and this will be much easier when the phase space of the string is described by q^μ, α_n^μ and $\bar{\alpha}_n^\mu$ rather than the original description of $x^\mu(\tau, \sigma)$ and $P^\mu(\tau, \sigma)$. As such, it is instructive to carefully carry out the change of variables, and compute the Poisson brackets, from $x^\mu(\tau, \sigma)$ and $P^\mu(\tau, \sigma)$ to these new phase space variables as we have done above. We could have found equations (2.1.51) and (2.1.52) by simply solving the wave equation and then adopted the constants as variables. However, as mentioned above, we wish to pursue a conventional path for quantising a system. In general, one cannot correctly quantise a system by first solving the classical equations of motion and then using the constants so obtained as variables in the quantum theory.

Finally, we consider the constraints of equation (2.1.29), which state that the energy-momentum tensor vanishes. The latter is related to $\wp^\mu$ and $\bar{\wp}^\mu$ by

$$(\wp)^2 = \frac{1}{\alpha'}\left(T_{00} + T_{01}\right) \equiv \frac{2}{\alpha'} T_{++} \quad \text{and} \quad (\bar{\wp})^2 = \frac{1}{\alpha'}\left(T_{00} - T_{01}\right) \equiv \frac{2}{\alpha'} T_{--}.$$

$$(2.1.54)$$

The constraints of equation (2.1.29) can be written as

$$L_n = \bar{L}_n = 0 \qquad \forall\, n, \qquad\qquad (2.1.55)$$

where

$$L_n \equiv \frac{1}{2\pi\alpha'} \int_{-\pi}^{\pi} d\sigma\, e^{in\sigma}\, T_{++} = \frac{1}{4\pi} \int_{-\pi}^{\pi} d\sigma\, e^{in\sigma}\, \wp^2 = \frac{1}{2} \sum_{m=-\infty}^{+\infty} \alpha_m^\mu \alpha_{n-m}^\nu \eta_{\mu\nu},$$

$$(2.1.56)$$

$$\bar{L}_n \equiv \frac{1}{2\pi\alpha'} \int_{-\pi}^{\pi} d\sigma\, e^{-in\sigma}\, T_{--} = \frac{1}{4\pi} \int_{-\pi}^{\pi} d\sigma\, e^{-in\sigma}\, \bar{\wp}^2 = \frac{1}{2} \sum_{m=-\infty}^{+\infty} \bar{\alpha}_m^\mu \bar{\alpha}_{n-m}^\nu \eta_{\mu\nu}.$$

It is easy to verify that the constraints obey the algebra $\{L_n, L_m\} = -i(n - m)L_{n+m}$, $\{\bar{L}_n, \bar{L}_m\} = -i(n - m)\bar{L}_{n+m}$ and $\{L_n, \bar{L}_m\} = 0$. Hence, the constraints are first class.

In chapters 8 and 9 we will see that two-dimensional conformal symmetry plays an important role in the formulation of string theories. In order to make contact with the discussion there we now introduce light-cone coordinates

$$\xi^+ = \xi^0 + \xi^1 = \tau + \sigma, \quad \xi^- = \xi^0 - \xi^1 = \tau - \sigma, \qquad (2.1.57)$$

which are discussed in detail in appendix B. The conformal symmetry is particularly apparent in these coordinates and we will observe here that many expressions take a particularly simple form in these coordinates. For example, the two-dimensional wave equation (2.1.28) can be written as $\partial_+\partial_- x^\mu = 0$ and

$$\wp^\mu = \sqrt{\frac{2}{\alpha'}}\partial_+ x^\mu \quad \text{and} \quad \bar{\wp}^\mu = \sqrt{\frac{2}{\alpha'}}\partial_- x^\mu. \tag{2.1.58}$$

The objects T_{++} and T_{--} that appear above are just the energy-momentum tensor in light-cone coordinates.

2.1.2 The open string

The treatment of the open string is much like that for the closed string; however, important differences arise from the different boundary condition. It is convenient in the case of the open string to extend the range of σ from $0 < \sigma < \pi$ to $-\pi < \sigma \le \pi$. This is achieved by setting

$$x^\mu(\tau, \sigma) = \begin{cases} x^\mu(\tau, \sigma) & \text{for } 0 \le \sigma \le \pi, \\ x^\mu(\tau, -\sigma) & \text{for } -\pi \le \sigma \le 0. \end{cases} \tag{2.1.59}$$

In this equation the x^μ that appears on the right-hand side is always in the range $0 \le \sigma \le \pi$ and so the equation is an identity in the region $0 \le \sigma \le \pi$ but it is equivalent to the reflection condition $x^\mu(\tau, \sigma) = x^\mu(\tau, -\sigma)$ for $-\pi < \sigma \le \pi$. The boundary condition at $\sigma = 0$ of equation (2.1.34) is automatically incorporated into this identification if $x^\mu(\tau, \sigma)$ is a smooth function of σ. We can also consider further extending the range of σ to be from $-\infty$ to ∞ by taking $x^\mu(\tau, \sigma) = x^\mu(\tau, \sigma + 2\pi)$. Such an extension is allowed as the boundary conditions $x^\mu(\tau, \pi) = x^\mu(\tau, -\pi)$ and $x^{\mu\prime}(\tau, \pi) = x^{\mu\prime}(\tau, -\pi) = 0$ ensure that $x^\mu(\tau, \sigma)$ is continuous and possesses a first derivative at $\sigma = \pm\pi$. The existence of higher derivatives is ensured by virtue of the equation of motion. Conversely, taking $x^\mu(\tau, \sigma) = x^\mu(\tau, \sigma + 2\pi)$ and equation (2.1.59) ensures that all the boundary conditions are obeyed. The mathematical extension of equation (2.1.59) allows us to rewrite the two constraints of equations (2.1.32) and (2.1.33) as one equation, namely

$$\wp^\mu(\tau, \sigma)\wp_\mu(\tau, \sigma) = 0, \qquad -\pi < \sigma \le \pi, \tag{2.1.60}$$

where

$$\wp^\mu(\tau, \sigma) = \frac{1}{\sqrt{2\alpha'}}(\dot{x}^\mu(\tau, \sigma) + x^{\mu\prime}(\tau, \sigma)), \qquad 0 \le \sigma \le \pi. \tag{2.1.61}$$

The value of $\wp^\mu(\tau, \sigma)$ for $-\pi < \sigma \le 0$ is found by using equation (2.1.59). Equations (2.1.32) and (2.1.33) are the parts of equation (2.1.60) which are even and odd under $\sigma \to -\sigma$. This is readily apparent once one realises that $\dot{x}^\mu$ and $x^{\mu\prime}$ are even and odd under this exchange.

As for almost any string, subject to boundary conditions, we will find it convenient to use normal modes or Fourier transformed quantities. Corresponding to the open string boundary conditions of equation (2.1.34) we can write

$$x^\mu(\tau, \sigma) = \sum_{n=0}^{\infty} x_n^\mu(\tau)\cos n\sigma = x_0^\mu(\tau) + \frac{1}{2}\sum_{\substack{n=-\infty \\ n\neq 0}}^{\infty} x_n^\mu(\tau)e^{in\sigma}$$

or

$$x_n^\mu(\tau) = \frac{2}{\pi} \int_0^\pi d\sigma' \cos n\sigma' x^\mu(\tau, \sigma'), n \neq 0, \quad x_0^\mu(\tau) = \frac{1}{\pi} \int_0^\pi d\sigma' x^\mu(\tau, \sigma'),$$

(2.1.62)

where $x_{-n}^\mu = x_n^\mu$ due to the fact that x^μ is even under $\sigma \to -\sigma$.

In terms of the normal modes the equation of motion of equation (2.1.28) becomes

$$\ddot{x}_n^\mu + n^2 x_n^\mu = 0,$$

(2.1.63)

which we recognize for $n \geq 1$ as the equations of motion of an infinite set of harmonic oscillators as well as a centre of mass coordinate x_0^ν. However, we also have the constraints of equation (2.1.29). As for the closed bosonic string, in preparations for quantisation it will also be useful to introduce the classical analogue of creation and annihilation operators for the infinite number of harmonic oscillators using $\wp^\mu(\tau, \sigma)$ of equation (2.1.61). To this end we define

$$\wp^\mu(\tau, \sigma) \equiv \sum_{n=-\infty}^{\infty} \alpha_n^\mu(\tau) e^{-in\sigma}$$

(2.1.64)

or

$$\alpha_n^\mu(\tau) = \frac{1}{2\pi} \int_{-\pi}^\pi d\sigma \, e^{in\sigma} \wp^\mu(\tau, \sigma).$$

(2.1.65)

To express x^μ in terms of α_n^μ and x_0^μ, we integrate the equation

$$\frac{2x^{\mu\prime}(\tau, \sigma)}{\sqrt{2\alpha'}} = \wp^\mu(\tau, \sigma) - \wp^\mu(\tau, -\sigma)$$

(2.1.66)

to find

$$x^\mu(\tau, \sigma) = x_0^\mu(\tau) + i\sqrt{\frac{\alpha'}{2}} \sum_{\substack{n=-\infty \\ n \neq 0}}^{\infty} \frac{1}{n}\left(\alpha_n^\mu(\tau) - \alpha_{-n}^\mu(\tau)\right) e^{-in\sigma},$$

(2.1.67)

where we have recognised $x_0^\mu(\tau)$ as the constant of integration. The equations of motion, when expressed in terms of the $\alpha_n^\mu(\tau)$, take the simple form

$$\dot{\alpha}_n^\mu(\tau) = -in\alpha_n^\mu(\tau) \quad \Rightarrow \quad \alpha_n^\mu(\tau) = e^{-in\tau}\alpha_n^\mu(0).$$

(2.1.68)

To derive these results, we make use of the field equation (2.1.28), $\ddot{x}^\mu = x^{\mu\prime\prime}$, directly in equation (2.1.67) or to take the τ derivative of equation (2.1.64) and so derive the relation

$$\dot{\wp}^\mu = \left(\ddot{x}^\mu + \dot{x}^{\mu\prime}\right)\frac{1}{\sqrt{2\alpha'}} = \left(x^{\mu\prime\prime} + \dot{x}^{\mu\prime}\right)\frac{1}{\sqrt{2\alpha'}} = \wp^{\mu\prime}.$$

(2.1.69)

Using equation (2.1.64) and considering the zero mode term, that is, $n = 0$, we conclude that

$$\dot{x}_0^\mu(\tau) = \frac{\sqrt{2\alpha'}}{2}\frac{1}{2\pi}\int_{-\pi}^\pi d\sigma \left(\wp^\mu(\tau, \sigma) + \wp^\mu(\tau, -\sigma)\right)$$

$$= \sqrt{2\alpha'}\alpha_0^\mu \quad \Rightarrow \quad x_0^\mu(\tau) = q^\mu + \sqrt{2\alpha'}\alpha_0^\mu\tau,$$

(2.1.70)

where $q^\mu \equiv x_0^\mu(0)$ and α_0^μ is independent of τ. Thus $x^\mu(\tau, \sigma)$ is expressed in terms of the variables $\alpha_n^\mu(0)$ and q^μ by

$$x^\mu(\tau, \sigma) = q^\mu + \sqrt{2\alpha'}\alpha_0^\mu \tau + i\sqrt{\frac{\alpha'}{2}} \sum_{\substack{n=-\infty \\ n\neq 0}}^{\infty} \frac{1}{n}\left(\alpha_n^\mu(0)e^{-in(\tau+\sigma)} + \alpha_n^\mu(0)e^{-in(\tau-\sigma)}\right).$$

(2.1.71)

From the fundamental Poisson brackets of equation (2.1.36) we deduce that

$$\{\wp^\mu(\sigma), \wp^\nu(\sigma')\} = 2\pi\eta^{\mu\nu}\delta'(\sigma - \sigma') \equiv 2\pi\frac{d}{d\sigma}\delta(\sigma - \sigma')\eta^{\mu\nu}.$$

(2.1.72)

To derive this equation, we used the relation

$$\wp^\mu(\sigma) = \sqrt{2\alpha'}\pi P^\mu + \frac{x^{\mu\prime}}{\sqrt{2\alpha'}},$$

(2.1.73)

where P^μ is the momentum of the string which is given by $\dot{x}^\mu/2\pi\alpha'$.

Taking the Fourier components of equation (2.1.36) implies the Poisson brackets

$$\{\alpha_n^\mu, \alpha_m^\nu\} = -in\delta_{n+m,0}\eta^{\mu\nu}.$$

(2.1.74)

The total momentum of the open string, denoted by p^μ, is given by

$$p^\mu \equiv \int_0^\pi d\sigma\, P^\mu = \frac{\dot{x}_0^\mu}{2\alpha'} = \frac{\alpha_0^\mu}{\sqrt{2\alpha'}}.$$

(2.1.75)

It follows from equation (2.1.62), and using equation (2.1.36), that q^μ and p^μ obey the Poissson bracket $\{q^\mu, p^\nu\} = \eta^{\mu\nu}$. Thus we have expressed the phase space of the string in terms of q^μ and α_n^μ instead of the original $x^\mu(\tau, \sigma)$, $P^\mu(\tau, \sigma)$ variables.

Taking the Fourier transform of the constraints of equation (2.1.60) of the open string, we find they become

$$L_n = 0 \qquad \forall\, n,$$

(2.1.76)

where

$$L_n \equiv \frac{1}{4\pi} \int_{-\pi}^{\pi} d\sigma\, e^{in\sigma} \wp^\mu \wp_\mu = \frac{1}{2} \sum_{m=-\infty}^{+\infty} \alpha_m^\mu \alpha_{n-m}^\nu \eta_{\mu\nu}.$$

(2.1.77)

It is straightforward, using equation (2.1.72) to show that the Poisson bracket of the constraints is given by

$$\{L_n, L_m\} = -i(n - m)L_{n+m}.$$

(2.1.78)

As a result, we find that the constraints are first class.

To make contact with the later discussion on conformal symmetry we display the dependence of the above quantities on $\xi^\pm$. Using equation (B.1.3), we may rewrite equations (2.1.61) and (2.1.64) as

$$\wp^\mu(\tau, \sigma) = \frac{2}{\sqrt{2\alpha'}}\partial_+ x^\mu(\tau, \sigma) \equiv \sum_{n=-\infty}^{\infty} \alpha_n^\mu(0)e^{-in\xi^+} \text{ for } -\pi \leq \sigma < \pi, \quad (2.1.79)$$

while equation (2.1.71) can be written as

$$x^{\mu}(\tau,\sigma) = q^{\mu} + \frac{1}{2}\sqrt{2\alpha'}\alpha_0^{\mu}(\xi^+ + \xi^-) + i\sqrt{\frac{\alpha'}{2}}\sum_{\substack{n=-\infty \\ n\neq 0}}^{\infty}\frac{1}{n}$$

$$\times \left(\alpha_n^{\mu}(0)e^{-in\xi^+} + \alpha_n^{\mu}(0)e^{-in\xi^-}\right). \tag{2.1.80}$$

Such a form can be obtained in a quicker manner by realising that $x^{\mu}(\tau,\sigma)$ satisfies the wave equation, whose solution is of the form $f(\xi^+) + g(\xi^-)$, and imposing this form on equation (2.1.67). However, in the next chapter we will quantise the string and the comments below equation (2.1.53) in the context of the closed string apply just as well here.

The reader may be puzzled by the preference for ξ^+ in equation (2.1.79), but the symmetry of x^{μ} under $\sigma \to -\sigma$ changes ξ^+ to ξ^- and so when expressed in terms of the x^{μ} defined in the region $0 \leq \sigma < \pi$, $\wp^{\mu}$ takes the form

$$\wp^{\mu} = \begin{cases} \dfrac{2}{\sqrt{2\alpha'}}\partial_+ x^{\mu} & \text{for } 0 \leq \sigma \leq \pi, \\[2mm] \dfrac{2}{\sqrt{2\alpha'}}\partial_- x^{\mu} & \text{for } -\pi < \sigma < 0. \end{cases} \tag{2.1.81}$$

2.2   The energy-momentum and angular momentum of the string

Let us first explain a general technique for finding the conserved current associated with any rigid symmetry. We consider an action that has a symmetry under a rigid symmetry, that is, the parameter T of the symmetry is a constant. This will, in general, no longer be invariant when the parameter becomes a function of space-time, but the variation of the action must be of the generic form $\int d^d x \partial_{\mu} T j^{\mu}$, as it is a symmetry when T is a constant. However, the variation of any action is given by the field equation for each field times the variations of the corresponding field summed over all fields. Clearly, when the field equations hold it must vanish. This holds for any variation including the case when T is any function of space-time. As such, we find the $\partial_{\mu} j^{\mu} = 0$, that is, j^{μ} is the conserved current associated with the symmetry, often called the Noether current.

From the two-dimensional viewpoint, the Poincaré symmetry is an internal symmetry and so has corresponding conserved two-dimensional currents. If we let the parameter of the associated transformation become dependent on the world-sheet coordinates, then the current is the coefficient of the derivative of the now local parameter in the variation of the action. One finds that the current for translations is given by

$$-\Pi^{\mu\alpha} = -\frac{\partial L}{\partial(\partial_{\alpha} x_{\mu})}, \tag{2.2.1}$$

while that for Lorentz rotations is

$$M^{\mu\nu\alpha} = x^{\mu}\frac{\partial L}{\partial(\partial_{\alpha} x_{\nu})} - (\mu \leftrightarrow \nu) = x^{\mu}\Pi^{\nu\alpha} - x^{\nu}\Pi^{\mu\alpha}. \tag{2.2.2}$$

Using the equations of motion, we find these are conserved, that is,

$$\partial^{\alpha}\Pi_{\alpha}^{\mu} = 0; \quad \partial^{\alpha}M_{\alpha}^{\mu\nu} = 0. \tag{2.2.3}$$

Taking the metric of equation (2.1.26), we find that

$$\Pi^\mu_\alpha = \frac{1}{2\pi\alpha'}\partial_\alpha x^\mu, \quad M^{\mu\nu}_\alpha = \frac{1}{2\pi\alpha'}(x^\mu\partial_\alpha x^\nu - x^\nu\partial_\alpha x^\mu). \tag{2.2.4}$$

For any closed curve C on the world sheet with surface S within, we have, using Stokes' theorem, the relation

$$\int_C (\Pi^\mu_0 d\sigma + \Pi^\mu_1 d\tau) = -\int_S d\tau d\sigma\, \partial^\alpha \Pi^\mu_\alpha = 0. \tag{2.2.5}$$

We recall that Stokes' theorem is proved by first establishing it for any infinitesimal closed curve and the area it encloses and then summing over all such areas enclosed in the curve C. In terms of differential forms the above relation becomes $\int_C \hat\Pi^\mu = \int_S d\hat\Pi^\mu$, where $\hat\Pi^\mu = \epsilon_{\alpha\beta}\Pi^{\mu\alpha}d\xi^\beta$.

It follows from equation (2.2.5) that the flow of energy and momentum into any segment of the curve C is balanced by the flow out of the remaining segment. Consequently, we may interpret

$$\int_C (\Pi^\mu_0 d\sigma + \Pi^\mu_1 d\tau) \tag{2.2.6}$$

as the flow of energy-momentum through any open curve C on the world sheet. In particular, the flow across an infinitesimal curve from (τ, σ) to $(\tau + d\tau, \sigma + d\sigma)$ is $\Pi^\mu_0 d\sigma + \Pi^\mu_1 d\tau$. For the open string, the boundary condition of equation (2.1.22) implies that the flow of energy and momentum through the ends of the string is zero.

The total energy-momentum of the string is, in the case of the open string, given by integrating along any curve C from one end of the string to the other, while for the closed string we integrate around a loop. By the above, the result is independent of the curve chosen. One natural choice is the curve $\tau = $ constant, for which

$$p^\mu = -\int_a^\pi \Pi^\mu_0 d\sigma, \tag{2.2.7}$$

where $a = 0$ and $a = -\pi$ for the open and closed strings, respectively.

A similar analysis applies to the Lorentz current and one finds that the total angular momentum of the string is given by

$$J^{\mu\nu} = \int_a^\pi M^{\mu\nu}_0 d\sigma = -\int_a^\pi (x^\mu P^\nu - x^\nu P^\mu)d\sigma. \tag{2.2.8}$$

2.3   A classical solution of the open string

Although the open string in the gauge of equation (2.1.26) satisfies the wave equation (2.1.28), whose solution is well known, it is also subject to the constraints of equation (2.1.29). To get a feeling for this system, it is very educational to find an explicit solution to equations (2.1.28) and (2.1.29). To be specific, let us consider the open string. We can choose to correlate the time on the string τ with the time x^0 of space-time by making the identification

$$t \equiv x^0 = \tau d, \tag{2.3.1}$$

where d is a constant. We then find that the constraint of equation (2.1.29) for T_{00} implies that

$$-1 + \sum_{i=1}^{D-1} \frac{\partial x^i}{\partial t} \frac{\partial x^i}{\partial t} + \frac{1}{d^2} x^{\mu\prime} x'_\mu = 0. \tag{2.3.2}$$

However, for the open string the end points satisfy $x^{\mu\prime} = 0$. As a result, the above equation states that the ends of the strings move with the speed of light, which we have set to 1.

Clearly, a static open string cannot be a solution. The next simplest possibility is to have a rigid straight string rotate about its mid-point. We can take this point to be the origin of the space-time and the rotation to be in the 1–2 plane. Let us choose the string to be of length L, then the motion is given by [2.8]

$$x^0 = \tau d, \quad x^1 = \frac{L}{2} \sin \tau \cos \sigma, \quad x^2 = \frac{L}{2} \cos \tau \cos \sigma,$$
$$x^3 = x^4 = \cdots = x^{D-1} = 0 \tag{2.3.3}$$

or $x^1 + ix^2 = \frac{1}{2} iLe^{-i\tau} \cos\sigma$. This is a solution as it solves both the constraints of equation (2.1.29) and the wave equation (2.1.28), provided $d = \frac{1}{2}L$.

The total energy E and angular momentum $J^{\mu\nu}$ can be found by substituting the above solution into equations (2.2.3) and (2.2.4), respectively. For E, the result is

$$E \equiv \Pi^0 = \frac{L}{4\alpha'}, \tag{2.3.4}$$

while the only non-zero component of $J^{\mu\nu}$ is given by

$$J^{12} = -\frac{L}{4} \frac{L}{4\alpha'}. \tag{2.3.5}$$

We observe that the total angular momentum $J = |J^{12}|$ is related to the energy by

$$J = \alpha' E^2. \tag{2.3.6}$$

This equation survives with only a minor modification in quantum theory. For historical reasons, the parameter α' was called the Regge slope. In chapter 18 we will discuss the historical development of string theory and this includes a discussion of the Regge slope.

One could envisage more complicated solutions in which the string not only rotates but also vibrates along its length. These solutions lead to a more complicated relation between J and E than that given above.

3 The quantum bosonic string

Don't worry, I know you'll never understand it.

> Wittgenstein to Bertrand Russell, his Ph.D supervisor,
> after he had defended his Ph.D thesis

The initial reaction of many people when told that the fundamental entities of Nature may be strings is the response: what is so special about strings? This attitude is a reflection of the unexciting or, perhaps, all too familiar nature of string-like objects that we encounter in everyday life. The string we are interested in here is different in two important respects to the more familiar violin string as unlike the latter it embodies the principles of relativity and quantum mechanics. In the previous chapter on the classical relativistic string we did see some interesting properties, but we did not, at first sight, encounter any properties that might lead one to expect that strings might provide an important element in a unified theory of physics.

In fact, it is only when one quantises the relativistic string that it becomes apparent that such strings are magical objects. In some sense, they take to quantum mechanics like a duck to water. We shall see that just as classically the string can be viewed as an infinite collection of point particles, when quantised it can be thought of as an infinite set of quantum particles, each of which corresponds to an irreducible representation of the Poincaré group and so to a given spin. The range of spins is from zero to infinity, there being, in general, more than one particle of a given spin.

Furthermore, we will see that some of these particles are massless. For the open bosonic string there is only one massless particle which carries a vector index. This is the same type of gauge particle that leads to the electromagnetic, the nuclear weak and the nuclear strong forces. The closed bosonic string can be of two types, unoriented or oriented; the only massless particles for the unoriented closed bosonic string are a graviton and a scalar, while for the oriented closed bosonic string we have in addition a second rank anti-symmetric gauge field. These bosonic strings, which have only space-time bosons, contain the bosonic massless particles found so far in nature, namely the graviton and gauge bosons of Yang–Mills theory. In fact, in the limit in which the string shrinks to a point, only the massless states survive and one can show that for the interacting open and closed strings the spin-2 and spin-1 particles are described by Einstein's gravity and Yang–Mills theories respectively [3.1, 3.2]. This limit is given by taking $\alpha' \to 0$. The spectrum for the closed superstring theories is even more exciting in the sense that some of these theories have massless particles with spins 2, 3/2, 1 and 1/2 and 0. Hence these theories can potentially also contain the known fermions, that is, the quarks and leptons and even the Higgs boson.

The interacting theories one finds at low energy are just the supergravity theories, as these theories possess the required local space-time supersymmetry.

Einstein's theory of gravity is known not to be a renormalizable theory and so it is not likely be provide a consistent theory of gravity and quantum mechanics. With the discovery of supergravity the first non-trivial extension of Einstein's theory was found. It was achieved by encoding supersymmetry in such a way that the graviton had a spin-3/2 particle as its superpartner. Furthermore, in this theory the spin-3/2 particle propagates in a causal way, a feature that had not been achieved before except for a free spin-3/2 particle. It was hoped for some years that supergravity theories would be consistent with quantum mechanics, but it was eventually realised that this was not likely to be the case. We now believe that string theory when combined with supersymmetry leads to a theory that includes gravity, is consistent with quantum mechanics, and propagates all the higher spin particles that it contains in a causal manner.

The early history of string theory is explained in chapter 18, which recounts the fascinating story of how string theory emerged from the scattering amplitude for four spin-0 particles. From this amplitude, and that for any number of spin-0 particles, a theory for the scattering of an infinite number of particles was developed, called the dual model. Indeed, the early pioneers were able to find the scattering amplitudes at tree level for any external states and the loop amplitudes were also found, up to some problems with negative norm states. However, in this development what was not known for quite some time was that this theory was due to string scattering. The development of the theory from the string perspective began with the quantisation of the Nambu action [2.1, 2.2] of equation (2.1.1) and its quantization [2.4]. The major obstacle to quantising this action is that one must take into account the constraints, discussed in chapter 2, which result from the reparameterisation invariance of the string. One procedure involves a clever choice of gauge which allows one to solve the constraints and leads to what is called the light-cone theory in which one is left with an unconstrained set of variables which may be straightforwardly quantised [2.4]. This approach is given in chapter 4. The disadvantage of this approach is that one loses manifest Lorentz invariance, and if one is not rather careful, one loses Lorentz invariance altogether. The other method of quantising the string, called the old covariant method, works with the constraints and is automatically Lorentz covariant [2.4]. The authors of [2.4] were able to recover many features found in the original approach to string theory, that is, the dual model: these included the constraints on physical states and the spectrum of the quantum theory.

The old covariant approach was developed before the discovery of BRST symmetries [1.2]. The BRST method has subsequently also been used for string theory and it seems particularly powerful in this context. The gauge invariance upon which the formalism is applied is the two-dimensional reparameterisation symmetry which is as usual fixed and has the ghosts and anti-ghosts introduced in [3.3]. Despite the power of this formalism, it is sometimes difficult to see the wood for the trees and even familiar manoeuvres may appear unfamiliar in this context. We will therefore first give the old covariant method.

Before carrying out the quantisation using these methods, however, let us give a heuristic argument which determines the particle spectrum of the string. Consider the classical solution given in equation (2.3.3) which was the rotating string of length L.

The string in this configuration has an energy E and total angular momentum $J = |J^{12}|$ given by

$$E = \frac{1}{4\alpha'}L, \quad J = \frac{L^2}{16\alpha'}. \tag{3.0.1}$$

We observe that the energy E and angular momentum J are related by

$$J = \alpha' E^2. \tag{3.0.2}$$

When we quantise the system, the angular momentum J is quantised, that is, $J = n\hbar$ and as a result we have the relation

$$\alpha' E^2 = J = n\hbar. \tag{3.0.3}$$

We may regard E as the energy retained in the configuration and so interpret it as the rest mass. As a result, we find that we have a series of rest masses m_n related to their spins by

$$\alpha' m_n^2 = J = n\hbar. \tag{3.0.4}$$

Thus the quantum string has an infinite number of particles whose $(\text{mass})^2$ is proportional to their spin J.

Corresponding to the string vibrating as it rotates, we can find further particles whose masses and spins have a more complicated relation than that of equation (3.0.4). Despite the heuristic nature of the above argument, we will find that the results are essentially correct.

3.1   The old covariant method [2.4]

To quantise the classical string, we follow Dirac and demand that the Poisson brackets be rewritten as commutators according to the following rule:

$$\{A, B\} = C \tag{3.1.1}$$

becomes

$$[A, B] = i\hbar C, \tag{3.1.2}$$

where in equation (3.1.1) A, B and C are classical variables while in equation (3.1.2) they are quantum operators. As is well known, to apply this rule to all operators leads to inconsistencies, but it can always be applied to the coordinates and momenta.

For the string the momentum is given by

$$P^\mu = \frac{\partial L}{\partial (\partial_\tau x_\mu)}, \tag{3.1.3}$$

which in the gauge of equation (2.1.26) becomes $P^\mu = (1/2\pi\alpha')\dot{x}^\mu$. The classical Poisson brackets for the string are given by

$$\{x^\mu(\sigma), x^\nu(\sigma')\} = 0 = \{P^\mu(\sigma), P^\nu(\sigma')\}, \quad \{x^\mu(\sigma), P^\nu(\sigma')\} = \eta^{\mu\nu}\delta(\sigma - \sigma'). \tag{3.1.4}$$

We therefore impose on the quantum theory the equal time commutation relations

$$\left[x^\mu(\sigma), x^\nu(\sigma')\right] = 0 = \left[P^\mu(\sigma), P^\nu(\sigma')\right], \left[x^\mu(\sigma), P^\nu(\sigma')\right] = i\hbar\delta(\sigma - \sigma')\eta^{\mu\nu}. \tag{3.1.5}$$

The Schrödinger representation of these commutators corresponds to taking

$$x^\mu(\sigma) \to x^\mu(\sigma); \quad P^\mu(\sigma) \to -i\hbar \frac{\delta}{\delta x_\mu(\sigma)}. \tag{3.1.6}$$

In this equation $\delta/\delta x^\mu(\sigma)$ is the functional derivative and by definition it satisfies $\delta x^\nu(\sigma)/\delta x^\mu(\sigma') = \delta^\nu_\mu \delta(\sigma - \sigma')$.

We now find, as for the point particle of chapter 1, that the constraints of the classical theory given in chapter 2, which were $T_{\alpha\beta} = 0$, or $L_n = 0$ for the open string and $L_n = 0$ and $\bar{L}_n = 0$ for the closed string, become differential conditions which we must implement on a wave functional, $\psi[x^\mu(\tau, \sigma)] = \langle x^\mu(\tau, \sigma)|\psi\rangle$. However, as we will see, unlike for the point particle, the quantum operators L_n do not commute and we must be careful only to implement a compatible set of constraints on the wave functional.

3.1.1   The open string

For the open bosonic string we found that by extending the range of σ we could combine the two constraints into one condition whose Fourier transform is $L_n = 0 \; \forall \; n$. The correct choice of constraints to impose on the wavefunction is given by [3.4]

$$(L_0 - 1)\psi = 0, \quad L_n\psi = 0 \quad n \geq 1. \tag{3.1.7}$$

These equations are called the physical state conditions. Since $L_n^\dagger = L_{-n}$, the expectation value of L_n, vanishes for all $n \neq 0$, that is, swapping to Dirac Bra-ket notation $\langle\psi|L_n|\psi\rangle = 0$, $n \neq 0$. We will give a justification of the above choice of constraints, including the choice of the 1, later in this chapter. In particular, we will explain that once we have given a precise meaning to the operators L_n we would set ψ to be zero if we imposed all the Virasoro constraints to vanish on it, that is, take $L_n\psi = 0 \; \forall n$. The situation here is like that which occurs in the Gupta–Bleuler formulation of electrodynamics where only half of the classical constraint $\partial^\mu A_\mu = 0$ is implemented on the wavefunction, that is, $(\partial^\mu A_\mu^+)\psi = 0$, where A_μ^+ is the positive frequency part of A_μ. To impose $(\partial^\mu A_\mu)\psi = 0$ at the quantum level would imply ψ itself was zero.

In chapter 2 we began by describing the phase space of the open bosonic string by $x^\mu(\tau, \sigma)$ and $P^\mu(\tau, \sigma)$. However, we then carried out a change of variables to describe the phase space by $q^\mu, \alpha_0^\mu = \sqrt{2\alpha'}p^\mu$ and α_n^μ for $n \neq 0$. One of the main advantages of these new variables is that they allow us to develop a precise formalism which takes care of the required normal ordering. As we will see this is essential for explaining the meaning of the physical state conditions of equation (3.1.7). The functional representation of equation (3.1.6) does not have this level of precision.

If we begin with the variables q^μ, p^μ and α_n^μ we can quantise the string by applying the usual Dirac rule of equations (3.1.1) and (3.1.2) to equation (2.1.74) to find that, after setting $\hbar = 1$, the α_n^μs satisfy the commutation relations

$$[\alpha_n^\mu, \alpha_m^\nu] = n \, \delta_{n+m,0} \, \eta^{\mu\nu} \tag{3.1.8}$$

and $[q^\mu, p^\nu] = i\eta^{\mu\nu}$. The last relation can be represented in the usual way by taking $\alpha_0^\mu = \sqrt{2\alpha'}p^\mu = -i\sqrt{2\alpha'}\partial/\partial q_\mu$ as $x_0^\mu = q^\mu$. We note that from the self-adjoint nature of $\wp^\mu(\sigma)$ we conclude that

$$\alpha_n^{\mu\dagger} = \alpha_{-n}^\mu. \tag{3.1.9}$$

This leads to the so-called oscillator formalism first found by factorising the Veneziano amplitude. The oscillators first arose in the dual model in [3.5–3.9].

The reader in a hurry may skip this paragraph, but it is instructive to rederive the above equations by starting from the more familiar description of phase space given by $x^\mu(\tau, \sigma)$ and $P^\mu(\tau, \sigma)$ and make the change to the oscillator formalism after the quantisation. We define the oscillators α_n^μ, as for the classical string, by taking the Fourier components of the operator $\wp^\mu$ of equation (2.1.64):

$$\wp^\mu(\sigma) = -i\sqrt{2\alpha'}\,\pi\,\frac{\delta}{\delta x_\mu(\sigma)} + \frac{x^{\mu\prime}(\sigma)}{\sqrt{2\alpha'}} = \sum_{n=-\infty}^{\infty} \alpha_n^\mu\, e^{-in\sigma}. \tag{3.1.10}$$

Using equation (2.1.62) to carry out the functional differentiation in terms of the variables x_n^ν, we find that

$$\frac{\delta}{\delta x^\mu(\sigma)} = \sum_{n=0}^{\infty} \frac{\delta x_n^\nu}{\delta x^\mu(\sigma)}\frac{\partial}{\partial x_n^\nu} = \frac{1}{\pi}\sum_{n=-\infty}^{\infty} e^{in\sigma}\frac{\partial}{\partial x_n^\mu}. \tag{3.1.11}$$

Substituting this result into equation (3.1.10) and, after setting $\hbar = 1$, we may make the identification

$$\alpha_n^\mu = -i\left(\sqrt{2\alpha'}\,\frac{\partial}{\partial x_{n\mu}} + \frac{n}{\sqrt{2\alpha'}}\frac{x_n^\mu}{2}\right). \tag{3.1.12}$$

From these explicit expressions it is easy to see that the oscillators obey equation (3.1.8).

The Virasoro operators L_n of equation (2.1.77) can be expressed in terms of α_n^μ, by substituting equation (3.1.10), to yield the result

$$L_n = \tfrac{1}{2}: \sum_{m=-\infty}^{\infty} \alpha_m^\mu\, \alpha_{n-m}^\nu\, \eta_{\mu\nu} :. \tag{3.1.13}$$

As we now explain, in common with the quantum formulation of many classical systems, with the introduction of the quantum operators L_n we encounter an operator ordering problem. This is solved by adopting a particular ordering of the operators, called normal ordering, which is denoted by ': ... :'. This notation is used to indicate that α_{-n}^μ is to be put to the left of α_n^μ for $n > 0$. In other words, we set

$$: \alpha_n^\mu \alpha_m^\nu := \begin{cases} \alpha_n^\mu \alpha_m^\nu & \text{if } m > n, \\ \alpha_m^\nu \alpha_n^\mu & \text{if } n > m. \end{cases}$$

Clearly, this reordering only makes a difference if one finds a term containing $\alpha_n^\mu \alpha_{-n}^\mu$ and if $n < 0$. Normal ordering ensures that the L_n have a finite action on any state consisting of a finite number of oscillators acting on the vacuum $|0, p\rangle$ which satisfies

$$\alpha_n^\mu |0, p\rangle = 0, \ n \geq 1, \ \alpha_0^\mu |0, p\rangle = \sqrt{2\alpha'}\,p^\mu |0, p\rangle. \tag{3.1.14}$$

In fact, the only L_n which requires normal ordering is L_0.

The classical Poisson brackets for the L_n were given in equation (2.1.78); however, the above normal ordering of L_0 results in an additional modification to the quantum analogue of equation (2.1.78) over and above the implementation of the rule of equations (3.1.1) and

(3.1.2). In fact, as we will discuss below, the Virasoro operators of the quantum theory obey the relations

$$[L_n, L_m] = (n - m)L_{n+m} + \frac{D}{12}n(n^2 - 1)\delta_{n+m,0}, \tag{3.1.15}$$

where D is the dimension of the space-time. This algebra was introduced in the dual model in [3.10, 3.11] without the last term which is called the central term and is found in paper [3.12]. The algebra emerged from the dual model after the discovery of the physical state conditions of equation (3.1.7).

As we will explain in chapter 8, one can show on general grounds that the central term can only have the above form. Although it may look like a rather insignificant term it has a very important role in string theory. This is in accord with the well-known fact that the Dirac rule cannot be consistently applied to all operators due to operator ordering problems. Equation (3.1.15) can be evaluated by using equation (3.1.13), but it is most easily found by carrying out the calculations between the vacuum state. For $n \geq 1$, we have that

$$\langle 0, 0|[L_n, L_{-n}]|0, 0\rangle = \langle 0, 0|L_n L_{-n}|0, 0\rangle$$

$$= \frac{1}{4} \sum_{m=1}^{n-1} \sum_{p=1}^{n-1} \langle 0, 0| \alpha_m \cdot \alpha_{n-m} \, \alpha_{n-p}^\dagger \cdot \alpha_p^\dagger|0, 0\rangle$$

$$= \frac{D}{2} \sum_{m=1}^{n-1} m(n - m)\langle 0, 0|0, 0\rangle$$

$$= \frac{D}{12} n(n^2 - 1)\langle 0, 0|0, 0\rangle, \tag{3.1.16}$$

where $|0, 0\rangle$ is the vacuum state defined above in equation (3.1.14), but with zero momentum p^μ. To go from the first and to the second line we have used the expression of equation (3.1.13) for L_n. For $n = \pm 1, 0$, the central term vanishes.

As mentioned above, imposing $L_n\psi = 0$ for all n implies, using equation (3.1.15) and the presence of the central term, that ψ itself is zero. Indeed $L_n\psi = 0$ and $L_{-n}\psi = 0$ imply that $(2nL_0 + \frac{1}{12}Dn(n^2 - 1))\psi = 0$ and so imposing such constraints for two values of n, one of which is not -1, 0 or $+1$, implies that $\psi = 0$. However, we can get a non-zero value of ψ by imposing $L_n\psi = 0$, $n \geq 1$, and in addition either $L_{-1}\psi = 0$, which implies $L_0\psi = 0$, or $(L_0 + a)\psi = 0$. We will see that only the latter condition leads to acceptable physical states and that the value of the constant a must be 1. Only the L_0 constraint of equation (3.1.7) has a constant term, which can be viewed as a consequence of the fact that only this operator requires normal ordering.

It will prove useful to write L_0 and $L_{\pm 1}$ in the more explicit forms

$$L_0 = \frac{1}{2}\alpha_0^\mu\alpha_{0\mu} + \sum_{m=1}^{\infty} \alpha_m^{\mu\dagger}\alpha_{m\mu} \equiv \frac{1}{2}\alpha_0^\mu\alpha_{0\mu} + N, \tag{3.1.17}$$

$$L_1 = L_{-1}^\dagger = \alpha_0^\mu\alpha_{1\mu} + \sum_{m=1}^{\infty} \alpha_m^{\mu\dagger}\alpha_{m+1\mu}, \tag{3.1.18}$$

where

$$\alpha_0^\mu = -i\sqrt{2\alpha'}\frac{\partial}{\partial x_0^\mu} \equiv \sqrt{2\alpha'}p^\mu, \tag{3.1.19}$$

that is, α_0^μ is proportional to the momenta associated with the coordinates of Minkowski space-time.

We can now analyse the content of the physical state conditions. Given any functional $\psi[x^\mu(\sigma)] \equiv \langle x^\mu(\sigma)|\psi\rangle$ of $x^\mu(\sigma)$, we may express it in terms of the occupation number basis by using the creation operators $\alpha_n^{\mu\dagger}$, $n \geq 1$. The first few terms in this expansion are

$$\psi[x^\nu(\sigma)] = \left\{\phi(x^\nu) + iA_\mu^1(x^\nu)\alpha_1^{\mu\dagger} + iA_\mu^2(x^\nu)\alpha_2^{\mu\dagger} + h_{\mu\nu}(x^\nu)\alpha_1^{\mu\dagger}\alpha_1^{\nu\dagger} + \cdots\right\}\psi_0,$$

(3.1.20)

where $x^\mu = x_0^\mu$. In the above expression we are using the coordinate x_0^μ of the zero mode and so we are treating the zero and non-zero modes differently. The vacuum ψ_0 refers to the non-zero modes and satisfies the equation

$$\alpha_n^\mu\psi_0 = 0, \quad n \geq 1. \tag{3.1.21}$$

The vacuum of equation (3.1.21) is of the form

$$\langle x^\mu(\sigma)|0, p\rangle = c\prod_{n=1}^\infty \exp\left(-\frac{n}{4\alpha'}x_n^\mu x_{n\mu}\right), \tag{3.1.22}$$

where c is a constant. The reader will find it useful to recall the simple quantum harmonic oscillator. In this exposition one uses the theorem that states that any function of position x can be expressed in terms of the complete set of Hermite polynomials. This expansion is just the sum over all terms consisting of arbitrary powers of the creation operator acting on the vacuum with constant coefficients. The above expansion for an arbitrary $\psi[x^\mu(\tau, \sigma)]$ relies on the same theorem, except that now we have an infinite number of creation operators. The dependence of $\psi[x^\mu(\tau, \sigma)]$ on the coordinates x^μ of Minkowski space-time is not altered in any way and goes along for the ride. As such, the coefficients in the expansion just depend on the coordinates of space-time. The level of a term appearing in $\psi[x^\mu(\tau, \sigma)]$ is defined to be the sum of all the n indices on the $\alpha_n^{\mu\dagger}$ oscillators occurring in the term being considered. This is none other than the value of $N = L_0 - \frac{1}{2}\alpha_0 \cdot \alpha_0$.

In terms of component fields, we find that the physical state conditions imply that

$$\left(\alpha'\partial^2 + 1\right)\phi = 0 = \left(\alpha'\partial^2 - (s-1)\right)A_\mu^s = \left(\alpha'\partial^2 - 1\right)h_{\mu\nu} = \cdots \tag{3.1.23}$$

as well as

$$\partial^\mu A_\mu^1 = 0 = -\sqrt{2\alpha'}\partial^\mu h_{\mu\nu} + A_\nu^2 = 2\sqrt{2\alpha'}\partial^\mu A_\mu^2 + h_\nu^\nu = \cdots. \tag{3.1.24}$$

As a result, the string contains a tachyon ϕ, a massless 'spin-1', or vector, A_μ^1, a massive 'spin-2' $h_{\mu\nu}$ and a further infinite tower of states of ever increasing mass. The reader should not confuse the symbol $h_{\mu\nu}$ with that used for the graviton when we later treat the closed string. A tachyon is a state whose mass squared is negative. States at level n have a mass $m_n^2 = (1/\alpha')(n-1)$ and the particle of highest spin at level n is contained in the contribution $\alpha_{-1}^{\mu_1}\cdots\alpha_{-1}^{\mu_n}t_{\mu_1\cdots\mu_n}$ in ψ and has spin $J = n$. It is these states which correspond to those of equation (3.0.4). At first sight, the presence of a tachyon may seem to be a problem, but it has been suspected for a long time that its presence is the signal that the bosonic string as formulated above is not in the correct vacuum state. We recall that a tachyon, that is, the Higgs field, occurs in the Standard Model of particle physics and its shift to the true vacuum of the theory gives masses to most of the particles in the Standard Model including the part of the Higgs fields which acquires a positive mass.

Imposing the alternative set of conditions, $L_n \psi = 0$, $n \geq -1$, mentioned above, one finds that the component fields of equation (3.1.20) obey equation (3.1.23) with the 1s removed and equation (3.1.24). The constraint $L_{-1} \psi = 0$ implies $\partial_\mu \phi = 0 = \partial_\mu A_\nu + \partial_\nu A_\mu$ etc. The only solution is that ϕ is a constant and the remaining fields vanish; as such, the solution is $\psi[x^\mu(\sigma)] = \langle x^\mu(\sigma)|p\rangle$. Clearly, this choice does not lead to physical particles.

A general functional of $x^\mu(\sigma)$ contains an infinite number of possible negative norm states associated with the oscillators α_n^0 which obey the relation $[\alpha_n^0, \alpha_{-n}^0] = -n$. One such example is the state $\alpha_{-n}^0|0, p\rangle$ which has norm $-n\langle 0, p|0, p\rangle$. Physical states must have positive norms as their norms are interpreted as probabilities. Hence, unless the physical state conditions eliminate all the negative norm states the bosonic string will be a physically unacceptable theory. One of the miracles of string theory is contained in the following theorem which will be proved in chapter 11.

The no-ghost theorem [3.13, 3.14] For space-time dimension $D \leq 26$ states ψ which satisfy the equations $L_n \psi = 0$, $n \geq 1$, $(L_0 - 1)\psi = 0$ have positive norm.

It is this theorem which provides the most important justification for the physical state conditions adopted earlier. In fact, for $D < 26$ the bosonic string is inconsistent for another reason associated with the singularity structure in one-loop graphs, that is, it has a cut rather than the required pole structure [3.15]. In fact, this was how it was first realised that the string had special properties in 26 dimensions. The reader may find it a good exercise to show, for $D > 26$, that the lowest level negative norm state occurs among the level 2 massive states. The proof is given in chapter 11. The no-ghost theorem was one of the results that convinced the early pioneers working in string theory that they did indeed have a very special theory.

Even for $D \leq 26$ the norms of the physical states are not all positive definite. Consider, for example, the physical state

$$|s\rangle = L_{-1}|\Omega\rangle, \tag{3.1.25}$$

where $|\Omega\rangle$ satisfies $L_0|\Omega\rangle = 0$, $L_n|\Omega\rangle = 0$, $n \geq 1$. The reader will have little difficulty showing that $|s\rangle$ satisfies the physical state conditions of equation (3.1.7). If $|\varphi\rangle$ is any other physical state, that is, it satisfies equation (3.1.7), then

$$\langle \varphi|s\rangle = \langle \varphi|L_{-1}|\Omega\rangle = (\langle \Omega|L_1|\varphi\rangle)^* = 0. \tag{3.1.26}$$

Clearly, this result also holds if $|\varphi\rangle$ is taken to be $|s\rangle$ and hence $|s\rangle$ itself has zero norm. It suffices to verify the physical state conditions for L_1 and L_2 as these imply all the others, except for the L_0 condition which is readily checked.

For $D = 26$ another zero-norm physical state can be found at the next level by considering the state

$$|s'\rangle = (L_{-2} + bL_{-1}^2)|\Omega'\rangle, \tag{3.1.27}$$

where $L_n|\Omega'\rangle = 0$, $n > 1$, $(L_0 + 1)|\Omega'\rangle = 0$. Demanding that $|s'\rangle$ be a physical state, we find that

$$L_1|s'\rangle = 0 = (3L_{-1} - 2bL_{-1})|\Omega'\rangle, \tag{3.1.28}$$

which implies that $b = +3/2$. A short calculation shows that $L_2|s'\rangle = 0$ implies that $D = 26$. Consequently, if $D = 26$ and $b = 3/2$ we find that $|s'\rangle$ is a physical state. As before, if $|\varphi\rangle$ is any physical state, then

$$\langle s'|\varphi\rangle = 0. \tag{3.1.29}$$

The existence of zero-norm physical states of the form

$$L_{-n_1} \cdots L_{-n_p}|\Omega''\rangle, \tag{3.1.30}$$

where $L_n|\Omega''\rangle = 0, n \geq 1, L_0|\Omega''\rangle = (\sum_{q=1}^{p} n_q + 1)|\Omega''\rangle$, can be read off from the Kac determinant (see chapters 8 and 9). The reader may verify by explicit computation that at the next level such a state only exists if $D = 28$.

The existence of a zero-norm physical state at level 1 is consistent with the presence of a massless 'spin-1' particle at this level and we now give a detailed analysis of the states of such a field as they appear in quantum field theory. Let us consider a vector particle A^μ which is subject to Maxwell's equation $\partial^2 A_\mu - \partial_\mu \partial^\nu A_\nu = 0$. If we choose the gauge $\partial_\mu A^\mu = 0$ the field equation becomes $\partial^2 A^\mu = 0$. The electromagnetic field has the creation and annihilation operators $a^{\mu\dagger}(k)$ and $a^\mu(k)$, respectively, and the corresponding states are of the form $\epsilon_\mu a^{\mu\dagger}(k)|0\rangle$, where k^μ is the momentum of the particle and ϵ_μ is the polarisation tensor. This is precisely what we find at level 1 in the expansion of a functional $\psi[x^\mu(\tau, \sigma)]$, as given in equation (3.1.20), if we identify $\alpha^\mu_{-1} \exp(ik_\mu q^\mu) = a^{\mu\dagger}(k)$ and $\alpha^\mu_1 \exp(-ik_\mu q^\mu) = a^\mu(k)$, where q^μ is the position operator, and also take $A^1_\mu(x) = \epsilon_\mu \exp(-ik_\mu x^\mu)$, which is a momentum eigenstate. The additional factor $\exp(-ik_\mu q^\mu)$ is required to inject the momentum k^μ into the vacuum $|0\rangle$, that is, $\exp(ik_\mu q^\mu)|0\rangle = |k\rangle$. The equation of motion and the above gauge choice imply that the physical states satisfy $k^2 = 0$ and $k^\mu \epsilon_\mu = 0$. These are the same conditions as are implied by the physical state conditions in equations (3.1.23) and (3.1.24). In the Lorentz frame, where $k^\mu = (k, 0, \ldots, 0, k)$, the D arbitrary states of the form $\epsilon_\mu (\alpha^\mu_1)^\dagger |0, k\rangle$ can be divided into two sets:

$$(\alpha^i_1)^\dagger |0, k\rangle, i = 1, 2, \ldots, D-2, \quad (-(\alpha^0_1)^\dagger + (\alpha^{D-1}_1)^\dagger)|0, k\rangle \tag{3.1.31}$$

and

$$((\alpha^0_1)^\dagger + (\alpha_1{}^{D-1})^\dagger)|0, k\rangle. \tag{3.1.32}$$

The last state is unphysical as $k^\mu \epsilon_\mu \neq 0$. The first $D-1$ states are physical states, but only the first $D-2$ of these have positive definite norm. The $(D-1)$th state has polarisation vector k^μ and so has zero norm and zero scalar product with the other physical states. As such, we can discard it as it never enters in any physical process. In fact, even though we have fixed the gauge by $k^\mu \epsilon_\mu = 0$, there is a residual gauge symmetry, namely $\epsilon_\mu \to \epsilon_\mu + k_\mu \lambda$, which can be used to remove this zero-norm state. The effect of this is to leave us with $D-2$ physical positive definite norm states in accordance with the vector representation of the little group $SO(D-2)$ which occurs in the application of the theory of induced representations to find the irreducible representations of the Poincaré group for a massless spin-1 particle. The zero-norm physical state, that is, the last state in (3.1.31), is of the form of $|s\rangle$ in equation (3.1.25) if we take $|\Omega\rangle = |0, k\rangle$ with $k^2 = 0$.

The scattering amplitude in quantum electrodynamics (QED) for fermions and photons, which are physical states, vanishes if one of the photons is a zero-norm physical state. This can be seen to be a consequence of the $U(1)$ gauge invariance and in particular its Ward identity. Were gauge invariance to fail, then unitarity would be violated. In string theory we must also verify that such decoupling of null states does indeed take place. This requirement has important consequences for interacting string theory. In point particle field theory, we are used to the idea that zero-norm physical states are associated with massless particles, but for the string we have found such states even at levels which only contain massive particles. As such, we expect gauge symmetries for the string to occur at levels greater than 1. This is the subject of chapter 12.

For $D = 26$, the number of physical states with positive definite norm, $d(n)$, at level n which have mass $\alpha' m^2 = n - 1$ is given by the formula (a derivation is given in chapters 4 or 11)

$$\prod_{n=1}^{\infty} d(n)x^n = \prod_{n=1}^{\infty} \frac{1}{(1 - x^n)^{24}} = 1 + 24x + \cdots . \tag{3.1.33}$$

This expansion is in agreement with the explicit computations above; we have one tachyon at level 0 and one 'spin-1' at level 1 which has 24 on-shell degrees of freedom. The reader may analyse the field content at level 2 and check that the result agrees with the above formula. The occurrence of the 24 is most easily understood in the light-cone formulation, given in chapter 4, where the constraints have been solved leaving 26 centre of mass coordinates x^μ and p^μ as well as the additional coordinates x_n^i, $i = 1, \ldots, 24$. Any functional of these which satisfies the mass condition is a physical state and such a state may be built up from oscillators $\alpha_n^{i\dagger}$, $i = 1, \ldots, 24$, $n \geq 1$, acting on a suitable vacuum with coefficients which are functions of $x_0^\mu = x^\mu$.

Point particles in D dimensions are labelled by the irreducible representations of the Poincaré group, which are themselves labelled by irreducible representations of $SO(D-1)$ if massive and $SO(D-2)$ if massless. In equation (3.1.33) the numbers that arise for massive states must be those for the representations of $SO(25)$. The reader may verify that at the first massive level one finds the number appropriate for the 'massive spin-2' particle mentioned above which should be taken to mean the symmetric traceless representation of $SO(D-1)$.

The above no-ghost theorem is proved in chapter 11, where a method of explicitly constructing the physical states to all orders is given. We now give a brief flavour of the derivation. In this construction it is required to introduce a vector k^μ; if the momentum of the tachyon is labelled by p^μ, $p^2 = 2$, we specify a vector k^μ such that $k^\mu k_\mu = 0$ and $k_\mu p^\mu = 1$. We will then show that one can decompose any physical state $|\psi\rangle$ as $|\psi\rangle = |D\rangle + |S\rangle$, where $|D\rangle$ and $|S\rangle$ are each separately physical states, but in addition $K_n|D\rangle = 0, n > 0$, where $K_n = \alpha_n^\mu k_\mu$. In this decomposition, we may write the state $|S\rangle = a|s\rangle + b|s'\rangle$, where $|s\rangle$ and $|s'\rangle$ are given in equations (3.1.27) and (3.1.28), respectively, and a and b are constants. The states $|D\rangle$ have positive definite norm and they will be shown to be a certain set of states that can be explicitly constructed called Del Giudice–DiVecchia–Fubini (DDF) states. As such, all zero-norm physical states are of the form $a|s\rangle + b|s'\rangle$. Since these states are orthogonal to all other physical states, including themselves, they decouple from all physical processes.

It remains to express the total momentum and angular momentum in terms of oscillators. Evaluating the total momentum of equation (2.2.7) we find that

$$p^\mu = \int_0^\pi \frac{1}{2\pi\alpha'} \partial_\tau x^\mu d\sigma = \frac{1}{4\pi\alpha'} \int_{-\pi}^\pi \partial_\tau x^\mu d\sigma = \frac{1}{2\pi\sqrt{2\alpha'}} \int_{-\pi}^\pi \wp^\mu d\sigma$$

$$= \frac{1}{2\pi\sqrt{2\alpha'}} \int_{-\pi}^\pi \sum_n \alpha_n^\mu e^{in\sigma} d\sigma = \frac{\alpha_0^\mu}{\sqrt{2\alpha'}} = -i\frac{\partial}{\partial x_0^\mu} . \tag{3.1.34}$$

We have used equations (2.1.59), (2.1.61) and (3.1.10) and recalled that we have previously denoted the total momentum by p^μ.

To compute the total angular momentum we require the expression for x^μ in terms of oscillators given in the previous section. Evaluating equation (2.2.8) we find

$$J^{\mu\nu} = \frac{1}{2\pi\alpha'} : \int_0^\pi (x^\mu \partial_\tau x^\nu - x^\nu \partial_\tau x^\mu) d\sigma := \frac{1}{2\pi\alpha'} \sqrt{\frac{\alpha'}{2}} : \int_{-\pi}^\pi (x^\mu \wp^\nu - x^\nu \wp^\mu) d\sigma :$$

$$= \frac{1}{\sqrt{2\alpha'}} (x_0^\mu \alpha_0^\nu - x_0^\nu \alpha_0^\mu) - i \sum_{n=1}^\infty \frac{(\alpha_{-n}^\mu \alpha_n^\nu - \alpha_{-n}^\nu \alpha_n^\mu)}{n}, \tag{3.1.35}$$

where we have used the relation

$$\dot{x}^\mu(\tau,\sigma) = \sqrt{\frac{\alpha'}{2}} (\wp^\mu(\tau,\sigma) + \wp^\mu(\tau,-\sigma)). \tag{3.1.36}$$

The first term is the usual orbital expression for the angular momentum, while the second term takes account of the Lorentz indices carried by the α_n^μ. This completes the quantisation of the free open bosonic string.

3.1.2   The closed string

In chapter 2 we began by describing the phase space of the closed bosonic string by $x^\mu(\tau,\sigma)$ and $P^\mu(\tau,\sigma)$, but then transformed, using equations (2.1.38) and (2.1.39), to the variables α_n^μ, $\bar{\alpha}_n^\mu$, $n \neq 0$ together with the centre-of-mass coordinates $x_0^\mu = q^\mu$ and p^μ. We recall that

$$\alpha_0^\mu = \bar{\alpha}_0^\mu = \sqrt{\frac{\alpha'}{2}} p^\mu. \tag{3.1.37}$$

As for the open string the oscillators provide a much better set of variables with which to carry out the quantisation as they allow us to resolve the normal ordering problems. Taking these as the fundamental variables we quantise the closed string by taking them to be operators subject to the Dirac quantisation condition. As such, the Poisson brackets of equations (2.1.44) become the commutation relations

$$[\alpha_n^\mu, \alpha_m^\nu] = n\eta^{\mu\nu}\delta_{n+m,0}, \ [\bar{\alpha}_n^\mu, \bar{\alpha}_m^\nu] = n\eta^{\mu\nu}\delta_{n+m,0}, \ [\alpha_n^\mu, \bar{\alpha}_m^\nu] = 0, \ [q^\mu, p^\nu] = i\eta^{\mu\nu}. \tag{3.1.38}$$

The reality of $\wp$ and $\bar{\wp}$ implies that

$$\alpha_n^{\mu\dagger} = \alpha_{-n}^\mu, \qquad \bar{\alpha}_{-n}^\mu = \bar{\alpha}_n^{\mu\dagger}, \tag{3.1.39}$$

while q^μ and p^μ are self-adjoint.

We now recover the commutators of equations (3.1.38) by taking $x^\mu(\tau,\sigma)$ and $P^\mu(\tau,\sigma)$ as our fundamental variables as we did for the open string, but the reader who is in a hurry may proceed directly to equation (3.1.43). We adopt the commutators of equation (3.1.5) with the representation of equation (3.1.6). Introducing the momentum P^μ we find, using equation (2.1.38), that the annihilation and creation operators α_n^μ and $\bar{\alpha}_n^\mu$ are given by

$$\wp^\mu = \sqrt{2\alpha'}\pi P^\mu + \frac{x^{\mu\prime}}{\sqrt{2\alpha'}} = \sum_{n=-\infty}^{\infty} \alpha_n^\mu(\tau) e^{-in\sigma}, \tag{3.1.40}$$

$$\bar{\wp}^\mu = \sqrt{2\alpha'}\pi P^\mu - \frac{x^{\mu\prime}}{\sqrt{2\alpha'}} = \sum_{n=-\infty}^{\infty} \bar{\alpha}_n^\mu(\tau) e^{in\sigma}. \tag{3.1.41}$$

Substituting for P^μ using equation (3.1.6) and using equation (2.1.37) to change variables, we find that we may express α_n^μ and $\bar{\alpha}_n^\mu$ in terms of the normal coordinates x_n^μ as follows:

$$\alpha_n^\mu = -i\sqrt{\frac{\alpha'}{2}}\frac{\partial}{\partial x_n^\mu} - \frac{inx_{-n}^\mu}{\sqrt{2\alpha'}}, \quad \bar{\alpha}_n^\mu = -i\sqrt{\frac{\alpha'}{2}}\frac{\partial}{\partial x_{-n}^\mu} - \frac{inx_n^\mu}{\sqrt{2\alpha'}}. \tag{3.1.42}$$

Calculating their commutators we recover equation (3.1.38).

The Virasoro operators of equation (2.1.56) are expressed in terms of the oscillators by

$$L_n = \tfrac{1}{2}: \sum_{m=-\infty}^{\infty} \alpha_m^\mu \alpha_{n-m}^\nu : \eta_{\mu\nu}, \quad \bar{L}_n = \tfrac{1}{2}: \sum_{m=-\infty}^{\infty} \bar{\alpha}_m^\mu \bar{\alpha}_{n-m}^\nu : \eta_{\mu\nu}, \tag{3.1.43}$$

where we adopt the normal ordering prescription we used in the open string for the α_m^μ and an identical prescription for the barred oscillators. Since the α_m^μ obey the same relations as in the open string case and they commute with the barred oscillators, the Virasoro operators obey the algebra

$$[L_n, L_m] = (n-m)L_{n+m} + \frac{D}{12}n(n^2-1)\delta_{n+m,0},$$

$$[L_n, \bar{L}_m] = 0, \tag{3.1.44}$$

$$[\bar{L}_n, \bar{L}_m] = (n-m)\bar{L}_{n+m} + \frac{D}{12}n(n^2-1)\delta_{n+m,0}.$$

The physical state conditions for the closed string are given by

$$L_n\psi = \bar{L}_n\psi = 0, \quad n \geq 1, \quad (L_0 + \bar{L}_0 - 2)\psi = 0, \tag{3.1.45}$$

$$(L_0 - \bar{L}_0)\psi = 0. \tag{3.1.46}$$

The last constraint has a simple interpretation. From equations (2.2.3) and (2.1.37) we find $L_0 - \bar{L}_0$ in functional form to be

$$L_0 - \bar{L}_0 = \frac{1}{4\pi}\int_{-\pi}^{\pi} d\sigma\left((\wp)^2 - (\bar{\wp})^2\right) = \frac{1}{4\pi}\int_{-\pi}^{\pi} d\sigma\, 2x^\mu(\sigma)'\frac{\delta}{\delta x^\mu(\sigma)}. \tag{3.1.47}$$

Consequently, $2\pi\varepsilon(L_0 - \bar{L}_0)$ implements the change $\sigma \to \sigma + \varepsilon$ and so the constraint $(L_0 - \bar{L}_0)\psi = 0$ tells us, as expected, that there is no preferred point along the length of the closed string. Just like for the open string, the physical states of the closed bosonic string do not contain any negative norm states if $D \leq 26$.

We now solve the physical state conditions at low levels. Expanding the most general functional $\psi[x^\mu(\sigma)] \equiv \langle x^\mu(\sigma)|\psi\rangle$ in terms of Hermite polynomials, or equivalently creation operators acting on the vacuum, we find the expression

$$\psi = (\phi(x) + h_\mu(x)\alpha_{-1}^\mu + k_\mu(x)\bar{\alpha}_{-1}^\mu + h_{\mu\nu}(x)\alpha_{-1}^\mu\bar{\alpha}_{-1}^\nu + h_\mu^{(2)}(x)\alpha_{-2}^\mu$$

$$+ k_\mu^{(2)}(x)\bar{\alpha}_{-2}^\mu + k_{\mu\nu}^{(2)}(x)\alpha_{-1}^\mu\alpha_{-1}^\nu + l_{\mu\nu}^{(2)}(x)\bar{\alpha}_{-1}^\mu\bar{\alpha}_{-1}^\nu + \cdots)\psi_0, \tag{3.1.48}$$

where the non-zero mode vacuum satisfies

$$\alpha_n^\mu \psi_0 = \bar{\alpha}_n^\mu \psi_0 = 0, \quad n \geq 1. \tag{3.1.49}$$

The fields depend on x_0^μ, which is identified to be the coordinate of Minkowski space-time and so is also denoted simply by x^μ. The constraint $(L_0 - \bar{L}_0)\psi = 0$ does not contain any space-time derivatives and just counts the level of the αs minus the $\bar{\alpha}$s. It implies that $h_\mu = k_\mu = h_\mu^{(2)} = k_\mu^{(2)} = k_{\mu\nu}^{(2)} = l_{\mu\nu}^{(2)} = 0$, etc. For the remaining level 0 and 1 states, the other physical state conditions imply the equations $(\alpha'\partial^2 + 4)\phi = 0$, $\partial^2 h_{\mu\nu} = 0$; $\partial^\mu h_{\mu\nu} = 0 = \partial^\nu h_{\mu\nu}$. This, in turn, implies that we have a tachyon ϕ at level 0. At level 1 we have some massless states and above this we have an infinite number of massive states.

There are two types of closed strings, which are called oriented and unoriented. An *unoriented* string has the symmetry $\sigma \to -\sigma$, which implies that for every wave moving around the closed string which depends on $\tau + \sigma$ we have a wave of the same amplitude, but which depends on $\tau - \sigma$. That is, we have waves of equal amplitudes moving in opposite directions around the closed string. Examining equation (2.1.49) we find that the symmetry $\sigma \leftrightarrow -\sigma$ implies that we must impose the symmetry $\alpha_n^\mu \leftrightarrow \bar{\alpha}_n^\mu$ on any state of the unoriented closed string. An *oriented* string has no such symmetry and so the amplitudes for left and right moving waves are unrelated. The oriented and unoriented closed strings are sometimes referred to as the extended Shapiro–Virasoro model and the restricted Shapiro–Virasoro model, respectively.

For an *oriented* string we therefore have a tachyon ϕ and massless states that comprise the graviton, a scalar and a particle corresponding to a second rank anti-symmetric tensor gauge field as well as an infinite number of massive states. We note that on-shell the graviton is described by the second rank symmetric tensor gauge field $\tilde{h}_{\mu\nu} = h_{(\mu\nu)} - \frac{1}{D}\eta_{\mu\nu}h^\lambda{}_\lambda$ that obeys the on-shell conditions $\partial^2 \tilde{h}_{\mu\nu} = 0$; $\partial^\mu \tilde{h}_{\mu\nu} = 0$ as well as $\tilde{h}^\mu{}_\mu = 0$. The massless scalar is contained in the trace of $h_{\mu\nu}$ and the anti-symmetric second rank tensor is just $h_{[\mu\nu]}$. In terms of the representations of $SO(D-2)$, these particles are just given by a symmetric traceless symmetric second rank tensor, a scalar and an anti-symmetric second rank tensor, respectively. To recover these states from $h_{\mu\nu}$ one must fix a momentum k_μ, such that $k^2 = 0$, and then find the most general state in $h_{\mu\nu}$ subject to $k^\mu h_{\mu\nu} = 0$ in the frame of reference of the chosen momentum and use the residual gauge symmetry $h_{\mu\nu} \to h_{\mu\nu} + \lambda_\mu k_\nu + \bar{\lambda}_\nu k_\mu$ with $k^\mu \lambda_\mu = 0 = k^\mu \bar{\lambda}_\mu$ to eliminate further states. The calculation is along the lines of that for the photon given around equation (3.1.32).

For an *unoriented* string we must also impose $\alpha_n^\mu \leftrightarrow \bar{\alpha}_n^\mu$ and hence $h_{\mu\nu}$ becomes a symmetric field. This string contains a tachyon ϕ, a graviton and a massless scalar as well as an infinite number of massive states.

3.2   The BRST approach

3.2.1   The BRST action

We begin with the string action of equation (2.1.10), which we rewrite here for convenience

$$A^{\text{orig}} = -\frac{1}{4\pi\alpha'} \int d^2\xi \sqrt{-g}\, g^{\alpha\beta}\, \partial_\alpha x^\mu \partial_\beta x^\nu \eta_{\mu\nu}. \tag{3.2.1}$$

As explained in chapter 2, this action is invariant under the two-dimensional diffeomorphisms given by

$$\delta x^{\mu} = f^{\alpha}\, \partial_{\alpha}\, x^{\mu}, \quad \delta g_{\alpha\beta} = f^{\gamma}\, \partial_{\gamma} g_{\alpha\beta} + \partial_{\alpha}\, f^{\gamma} g_{\gamma\beta} + \partial_{\beta} f^{\gamma} g_{\alpha\gamma}. \tag{3.2.2}$$

It will be useful to rewrite the variation of fields in terms of the covariant derivative. For completeness, we recall that the covariant derivative ∇_{α} acts on a tensor $T_{\gamma\cdots}^{\beta\cdots}$ according to the rule

$$\nabla_{\alpha} T_{\gamma\cdots}^{\beta\cdots} = \partial_{\alpha} T_{\gamma\cdots}^{\beta\cdots} - \Gamma_{\alpha\gamma}^{\delta} T_{\delta\cdots}^{\beta\cdots} + \Gamma_{\delta\alpha}^{\beta} T_{\gamma\cdots}^{\delta\cdots} + \cdots. \tag{3.2.3}$$

In the absence of torsion, and for $\nabla_{\delta} g_{\alpha\beta} = 0$, the Christoffel symbol is given by

$$\Gamma_{\alpha\beta}^{\gamma} = \tfrac{1}{2} g^{\gamma\delta}(\partial_{\alpha} g_{\beta\delta} + \partial_{\beta} g_{\alpha\delta} - \partial_{\delta} g_{\alpha\beta}). \tag{3.2.4}$$

One easily verifies that we may write equation (3.2.2) as

$$\delta x^{\mu} = f^{\alpha} \nabla_{\alpha} x^{\mu}, \quad \delta g_{\alpha\beta} = \nabla_{\alpha} f_{\beta} + \nabla_{\beta} f_{\alpha}. \tag{3.2.5}$$

We recall that the μ index is inert under two-dimensional general coordinate transformations. The action contains the inverse of $g_{\alpha\beta}$ and we may write its variation as

$$\delta g^{\alpha\beta} = -g^{\alpha\gamma} \delta g_{\gamma\delta} g^{\delta\beta}, \quad g^{\alpha\beta} = -(\nabla^{\alpha} f^{\beta} + \nabla^{\beta} f^{\alpha}), \tag{3.2.6}$$

where $\nabla^{\alpha} \equiv g^{\alpha\delta} \nabla_{\delta}$. The variation of $g \equiv \det g_{\alpha\beta}$ is

$$\delta g = g g^{\gamma\delta} \delta g_{\gamma\delta} = 2g \nabla_{\alpha} f^{\alpha}. \tag{3.2.7}$$

We now gauge fix the above action and introduce ghosts and anti-ghosts according to the 'standard' BRST method. The final goal is to gauge fix the symmetry so as to yield an invertible propagator and introduce ghosts and anti-ghosts in such a way that the BRST transformations are nilpotent and leave the action invariant. The reader unfamiliar with the procedure will find it useful to carry out the analogous steps for Yang–Mills theory at every step (see appendix A). The original BRST treatment of the string was in [3.3]. We first adopt a gauge-fixing term. According to the theorem given in equation (2.1.36), we may locally bring the metric to the form $g_{\alpha\beta} = e^{\phi(\xi)} \eta_{\alpha\beta}$, using the two-dimensional general coordinate invariance. We could therefore choose this as our gauge-fixing condition; however, in order to avoid the appearance of the field ϕ, which did not occur in the original action, we choose to gauge fix the ϕ independent (that is, Weyl invariant) combination $\sqrt{-g}\, g^{\alpha\beta}$. We therefore adopt the gauge-fixing condition

$$\sqrt{-g} g^{\alpha\beta} = \sqrt{-\widehat{g}} \widehat{g}^{\alpha\beta}, \tag{3.2.8}$$

where $\widehat{g}_{\alpha\beta}$ is any two-dimensional metric. One choice is $\sqrt{-\widehat{g}} \widehat{g}_{\alpha\beta} = \eta_{\alpha\beta}$, but we can use any $\widehat{g}_{\alpha\beta}$. We therefore introduce the gauge-fixing term

$$A^{gf} = -\frac{1}{\pi} \int d^2\xi\, \lambda_{\alpha\beta}(\sqrt{-g} g^{\alpha\beta} - \sqrt{-\widehat{g}} \widehat{g}^{\alpha\beta}), \tag{3.2.9}$$

where $\lambda_{\alpha\beta}$ is often called the Lautrup field. The reason for the choice of the factor $-1/\pi$ will become apparent later.

The constraint $\sqrt{-g} g^{\alpha\beta} = \sqrt{-\widehat{g}} \widehat{g}^{\alpha\beta}$ involves only two conditions and so only two components of $\lambda_{\alpha\beta}$ contribute in the above action. We may take $\lambda_{\alpha\beta}$ symmetric (that is, $\lambda_{\alpha\beta} = \lambda_{\beta\alpha}$)

and traceless with respect to either $g_{\alpha\beta}$ or $\hat{g}_{\alpha\beta}$. Although the latter condition does not correspond precisely with the part of $\lambda_{\alpha\beta}$ which drops out of the action, this is immaterial as we have retained in $\lambda_{\alpha\beta}$ the two components that do remain.

As usual, the BRST variations of the original fields are obtained by the substitution

$$f^{\alpha} \to \Lambda c^{\alpha} \tag{3.2.10}$$

into equation (3.2.2), where Λ is a rigid parameter, (that is, ξ^{α} independent) and c^{α} are the ghost fields. Both c^{α} and Λ are Grassmann odd. Carrying out this procedure, the BRST variations of the original fields are found to be

$$\delta x^{\mu} = \Lambda c^{\alpha} \partial_{\alpha} x^{\mu}, \quad \delta g^{\alpha\beta} = -\Lambda(\nabla^{\alpha} c^{\beta} + \nabla^{\beta} c^{\alpha}). \tag{3.2.11}$$

Introducing anti-ghosts $b_{\alpha\beta}$, we take them to have the variations

$$\delta \lambda_{\alpha\beta} = 0, \quad \delta b_{\alpha\beta} = \Lambda \lambda_{\alpha\beta}. \tag{3.2.12}$$

In accord with the latter equation, the tensor character of the anti-ghosts is chosen to match that of $\lambda_{\alpha\beta}$ and hence we take $b_{\alpha\beta}$ to be also symmetric and traceless. Finally, we take the variation of the ghosts to be

$$\delta c^{\alpha} = \Lambda c^{\gamma} \nabla_{\gamma} c^{\alpha} = \Lambda c^{\gamma} \partial_{\gamma} c^{\alpha}. \tag{3.2.13}$$

As explained in appendix A, for the Yang–Mills theory the variation of the ghosts is given by the 'structure' constant with two of its indices contracted with those of the ghost fields. In the case we are dealing with here, the commutator of two general coordinate transformations with parameters ζ_1^{α} and ζ_2^{β} on x^{μ}, say, yields a general coordinate transformation with parameter

$$\zeta_1^{\alpha} \partial_{\alpha} \zeta_2^{\beta} - (1 \leftrightarrow 2) = \zeta_1^{\alpha} \nabla_{\alpha} \zeta_2^{\beta} - (1 \leftrightarrow 2). \tag{3.2.14}$$

Making the replacement of one of the parameters ζ^{α} by Λc^{α} and the other by c^{β} we find equation (3.2.13).

The most important property of BRST transformations is that they should be nilpotent, that is, $\delta_{\Lambda_1} \delta_{\Lambda_2}$ on any field vanishes. For example, on x^{μ} we have

$$\delta_1 \delta_2 x^{\mu} = \Lambda_1 c^{\alpha} \partial_{\alpha}(\Lambda_2 c^{\beta} \partial_{\beta} x^{\mu}) + \Lambda_1 (\Lambda_2 c^{\delta} \partial_{\delta} c^{\alpha}) \partial_{\alpha} x^{\mu} = 0. \tag{3.2.15}$$

The reader may verify the same result for the remaining fields.

The final step is to find an action A^{gh} involving $b_{\alpha\beta}$ and c^{γ} such that

$$A^{BRST} = A^{\text{orig}} + A^{gf} + A^{gh} \tag{3.2.16}$$

is BRST invariant. The result is given by

$$A^{gh} = -\frac{1}{\pi} \int d^2\xi \sqrt{-g} b_{\alpha\beta} (\nabla^{\alpha} c^{\beta} + \nabla^{\beta} c^{\alpha} - g^{\alpha\beta} \nabla_{\delta} c^{\delta}). \tag{3.2.17}$$

One can find A^{gh} by explicitly varying A^{gf} and verifying that the full action is BRST invariant. However, a quicker method is to observe that

$$\Lambda(A^{gf} + A^{gh}) = -\frac{1}{\pi} \delta \int d^2\xi b_{\alpha\beta} \left(\sqrt{-g} g^{\alpha\beta} - \sqrt{-\hat{g}} \hat{g}^{\alpha\beta} \right) \tag{3.2.18}$$

and since $\delta^2 = 0$, it follows that $\delta(A^{gf} + A^{gh}) = 0$. It is appropriate to comment on the Hermitian character of the fields. Let us take c^{α} to be Hermitian; then the reality of the action implies that $b_{\alpha\beta}$ is anti-Hermitian. We recall that for two Grassmann odd variables n and p, complex conjugation is implemented on the product by $(np)^* = p^*n^* = -n^*p^*$. Clearly, if they are Grassmann even, changing the order upon complex conjugation makes no difference to the result. Examining the transformations of the fields, we conclude that Λ is anti-Hermitian. Should the above procedure remind the reader of a cookery book recipe, that is because the BRST procedure is rather like one. We refer the reader to the discussions in chapter 1 and appendix A.

We now give an alternative way to arrive at the gauge-fixed action of equation (3.2.16). We will insert 1 into the vacuum to vacuum functional integral

$$\int \mathbf{D}x^\mu \mathbf{D}g_{\alpha\beta} e^{iA^{orig}}.$$

However, we will write 1 in the form

$$1 = \Delta_{FP} \int d^2\zeta \delta(\sqrt{-g_\zeta}\, g_\zeta^{\alpha\beta} - \sqrt{-\widehat{g}}\widehat{g}^{\alpha\beta}), \tag{3.2.19}$$

where $g_\zeta^{\alpha\beta}$ is a diffeomorphism of $g^{\alpha\beta}$ with parameter ζ^α and Δ_{FP} is defined by this equation. Using standard techniques (see, for example, equation (2.17) in [3.16]), we find that Δ_{FP} is reparameterisation independent and can be written as

$$\begin{aligned}
\Delta_{FP} &\equiv \det\left\{\frac{\delta}{\delta\zeta^\gamma}\left(\sqrt{-g_\zeta}\, g_\zeta^{\alpha\beta}\right)\right\}_{\zeta^\gamma=0} \\
&= \det\sqrt{-g}\frac{\delta}{\delta\zeta^\gamma}(\nabla^\alpha \zeta^\beta + \nabla^\beta \zeta^\alpha - g^{\alpha\beta}\nabla_\delta \zeta^\delta)_{\zeta^\gamma=0} \\
&= \det\{\sqrt{-g}(\nabla^\alpha\delta_\gamma^\beta + \nabla^\beta \delta_\gamma^\alpha - g^{\alpha\beta}\nabla_\gamma)\delta^2\}. \tag{3.2.20}
\end{aligned}$$

In going between the first and second lines we have used equations (3.2.6) and (3.2.7) and to get to the third line we used that $\delta\zeta^\beta(\xi)/\delta\zeta^\gamma(\xi') = \delta_\gamma^\beta\delta^2(\xi - \xi')$. Here δ^2 is the two-dimensional delta function whose argument is $\xi - \xi'$. We may therefore write Δ_{FP} as $\Delta_{FP} = |\det P| = (\det PP^\dagger)^{\frac{1}{2}}$, where P is the operator defined by the equation $(P\zeta)^{\alpha\beta} = \nabla^\alpha \zeta^\beta + \nabla^\beta\zeta^\alpha - g^{\alpha\beta}\nabla_\gamma\zeta^\gamma$, which maps vectors into traceless symmetric tensors. Its adjoint $P^\dagger$ acts on symmetric traceless tensors $t^{\alpha\beta}$, taking them to vectors by the relation $(P^\dagger t)^\alpha = -2\nabla_\beta t^{\alpha\beta}$.

We may rewrite Δ_{FP} as an integral over Faddeev–Popov ghosts with the action of equation (3.2.17). Since nothing now depends on ζ, the integral over ζ may be discarded. The delta function of equation (3.2.19) can be implemented by an integral over $\lambda_{\alpha\beta}$ using the action of equation (3.2.9). Thus we recover our previous result of equation (3.2.16), to which we now return. In the above we have glossed over a number of points which will be treated carefully in section 18.2.

Integrating over $\lambda_{\alpha\beta}$ in the functional integral of equation (3.2.16) enforces the constraint $\sqrt{-g}\, g^{\alpha\beta} = \sqrt{-\widehat{g}}\widehat{g}^{\alpha\beta}$ and we find the action

$$\begin{aligned}
A^T = &-\frac{1}{4\pi\alpha'} \int d^2\xi\sqrt{-\widehat{g}}\widehat{g}^{\alpha\beta}\partial_\alpha x^\mu \partial_\beta x^\nu \eta_{\mu\nu} \\
&-\frac{1}{\pi} \int d^2\xi\sqrt{-\widehat{g}}\, b_{\alpha\beta}(\widehat{\nabla}^\alpha c^\beta + \widehat{\nabla}^\beta c^\alpha - \widehat{g}^{\alpha\beta}\widehat{\nabla}_\delta c^\delta). \tag{3.2.21}
\end{aligned}$$

In this action, the $\Gamma^\delta_{\alpha\beta}$ in $\widehat{\nabla}^\alpha$ is given by equation (3.2.4) but with $g_{\alpha\beta}$ replaced by $\hat{g}_{\alpha\beta}$. A particularly simple choice of gauge is given by $\hat{g}_{\alpha\beta} = \eta_{\alpha\beta}\, e^\phi$, whereupon we find the action becomes

$$A^F = -\frac{1}{4\pi\alpha'} \int d^2\xi\, \eta^{\alpha\beta}\partial_\alpha x^\mu \partial_\beta x^\nu \eta_{\mu\nu} - \frac{1}{\pi}\int d^2\xi\, b_{\alpha\beta}\big(\partial^\alpha c^\beta + \partial^\beta c^\alpha - \eta^{\alpha\beta}\partial_\delta c^\delta\big).$$

(3.2.22)

To obtain this action, which we denote by A^F, we substituted the Christoffel symbol for the metric $\hat{g}_{\alpha\beta} = \eta_{\alpha\beta}e^\phi$, which is given by

$$\Gamma^\gamma_{\alpha\beta} = \tfrac{1}{2}(\partial_\alpha\phi\delta^\gamma_\beta + \partial_\beta\phi\delta^\gamma_\alpha - \partial^\gamma\phi\eta_{\alpha\beta}).$$

(3.2.23)

The result is independent of ϕ, which is to be expected if Weyl invariance is maintained. In fact, this is only true in 26 dimensions. In less than 26 dimensions, the Weyl symmetry is anomalous and one develops an additional scalar degree of freedom.

A much more extensive discussion of the sum over world-sheet approach including the role of the ghosts is given in section 18.2.

The manoeuvres used to arrive at the action of equation (3.2.22), like those for the point particle, can be criticised from several points of view. One problem centres around whether the gauge fixing of equation (3.2.8) is actually a good one. The invariance of the action places restrictions on the parameter ζ^α at the initial and final points of the string and these missing degrees of freedom in ζ^α do not actually allow one to reach the gauge of equation (3.2.8). The good gauge-fixing condition involves derivatives on $g_{\alpha\beta}$ [3.17]. The resulting differences are essential for one of the formulations of gauge covariant string theory where those parts of $g_{\alpha\beta}$ which remain after gauge fixing acquire derivatives in the BRST procedure and so become dynamical fields which play the role of moduli in the string field theory [3.17].

We now quantise the action of equation (3.2.22), and then use the result to find an alternative description of the on-shell states of the quantised string. We proceed in the same way as in the 'old covariant method' given in the previous chapter, but now we have to include the ghost coordinates. The equations of motion of the action of equation (3.2.22) are found by varying x^μ, c^γ and $b_{\alpha\beta}$; they are given by

$$\partial_\alpha\partial^\alpha x^\mu = 0, \quad \partial_\beta b^{\alpha\beta} = 0, \quad \partial^\alpha c^\beta + \partial^\beta c^\alpha - \eta^{\alpha\beta}\partial_\delta c^\delta = 0.$$

(3.2.24)

We must, however, apply the analogous discussion on boundary terms given in chapter 2 for the string without any ghosts. For the *closed string*, the boundary term vanishes due to the boundary condition

$$x^\mu(-\pi) = x^\mu(\pi), \quad b_{\alpha\beta}(-\pi) = b_{\alpha\beta}(\pi), \quad c^\beta(-\pi) = c^\beta(\pi).$$

(3.2.25)

For the *open string*, we must require $x^{\mu\prime} = 0$ at $\sigma = 0$ and $\sigma = \pi$, as before, as well as

$$\delta c^\beta \frac{\delta A}{\delta(\partial_\sigma c^\beta)} = -2\big(\delta c^1 b_{11} + \delta c^0\, b_{10}\big) = 0$$

(3.2.26)

at $\sigma = 0$ and $\sigma = \pi$. Equation (3.2.26) will be satisfied if both

$$b_{01} = 0 \quad \text{and} \quad c^1 = 0$$

(3.2.27)

at $\sigma = 0$ and $\sigma = \pi$. We note that the variation of a field must have the same boundary conditions as the field itself. Clearly, there exist other choices of boundary condition which satisfy equation (3.2.26). However, one should choose to restrict the variables so as to ensure that conjugate variables, discussed below, have the same number of degrees of freedom. Equation (3.2.27) is one such choice. Equation (3.2.27) implies, using the equations of motion of equation (3.2.24), that

$$\partial_1 c^0 = 0 \quad \text{and} \quad \partial_1 b_{00} = 0 \tag{3.2.28}$$

at $\sigma = 0$ and $\sigma = \pi$.

Writing the ghost action out explicitly, we find it is given by

$$-\frac{2}{\pi} \int d^2\xi \left\{ b_{01}\left(-\partial_0 c^1 + \partial_1 c^0\right) + b_{00}\left(-\partial_0 c^0 + \partial_1 c^1\right) \right\}. \tag{3.2.29}$$

Defining the conjugate variable by differentiation of the action with respect to the variable from the left, and taking into account the Grassmann odd nature of the ghost, we find that c^1 is conjugate to $-(2/\pi)b_{01}$ and c^0 is conjugate to $-(2/\pi)b_{00}$. Changing the corresponding Poisson brackets of the classical theory (see chapter 1 for a discussion of Poisson brackets for anti-commuting variables) to anti-commutators according to the Dirac rule we have

$$\left\{ b_{01}(\sigma), c^1(\sigma') \right\} = -\frac{i\pi}{2}\delta(\sigma - \sigma'),$$

$$\left\{ c^0(\sigma), b_{00}(\sigma') \right\} = -\frac{i\pi}{2}\,\delta(\sigma - \sigma'), \tag{3.2.30}$$

with the rest vanishing. Due to the symmetric nature of the anti-commutator it does not matter which variables we choose as coordinates and which as momenta.

One possible Schrödinger representation extending that of equation (3.1.6), which we will *not* use, is given by

$$b_{00} = -\frac{i\pi}{2}\frac{\delta}{\delta c^0}, \qquad c^1 = -\frac{i\pi}{2}\frac{\delta}{\delta b_{01}}. \tag{3.2.31}$$

For the *open string*, it is advantageous to extend the range of σ from $0 < \sigma < \pi$ to $\pi \le \sigma < \pi$ by taking

$$c^0(\sigma) = c^0(-\sigma), \qquad c^1(\sigma) = -c^1(-\sigma), \tag{3.2.32}$$

$$b_{00}(\sigma) = b_{00}(-\sigma), \qquad b_{01}(\sigma) = -b_{01}(-\sigma) \tag{3.2.33}$$

for $0 < \sigma < \pi$. This choice encodes the boundary conditions of equations (3.2.27) and (3.2.28) at $\sigma = 0$ and is consistent with the tensor character of the fields under $\sigma \to -\sigma$. On the range $-\pi < \sigma < \pi$ we can then define the fields

$$c(\sigma) \equiv c^0(\sigma) + c^1(\sigma) = c^+(\sigma) \tag{3.2.34}$$

and

$$b(\sigma) \equiv -2(b_{00}(\sigma) + b_{01}(\sigma)) = -4b_{++}(\sigma), \tag{3.2.35}$$

which have no particular symmetry as $\sigma \to -\sigma$. However, we do still have the boundary conditions of equations (3.2.27) and (3.2.28) at $\sigma = \pi$ which may be expressed as $c(\pi) = -c(-\pi)$ and $(\partial/\partial\sigma)c(\pi) = (\partial/\partial\sigma)c(-\pi)$ and similarly for b. We may further extend the range of σ to all σ since these boundary conditions are consistent with the equations

$$c(\sigma) = c(\sigma + 2\pi), \qquad b(\sigma) = b(\sigma + 2\pi). \tag{3.2.36}$$

Conversely, taking the above conditions and demanding that c and b be continuous and have first derivatives ensures all the above boundary conditions. In terms of the fields $c(\sigma)$ and $b(\sigma)$, equation (3.2.30) becomes

$$\{b(\sigma), b(\sigma')\} = \{c(\sigma), c(\sigma')\} = 0,$$

$$\{b(\sigma), c(\sigma')\} = 2i\pi\delta(\sigma - \sigma') \tag{3.2.37}$$

for $-\pi < \sigma < \pi$, $-\pi < \sigma' < \pi$. We recognise $b(\sigma)$ and $c(\sigma)$ as conjugate variables and we adopt this choice in all future discussions.

Corresponding to the boundary conditions of equation (3.2.36), the normal mode expansions of the ghosts are given by

$$c(\sigma) = \sum_{n=-\infty}^{\infty} c_n e^{-in\sigma} \tag{3.2.38}$$

and

$$b(\sigma) = i \sum_{n=-\infty}^{\infty} b_n e^{-in\sigma}, \tag{3.2.39}$$

where c_n and b_n are independent variables subject only to the Hermiticity conditions

$$c_n^\dagger = c_{-n}, \qquad b_n^\dagger = b_{-n}. \tag{3.2.40}$$

We recall that $c(\sigma)$ and $b(\sigma)$ are Hermitian and anti-Hermitian respectively. Using the relations

$$c_n = \frac{1}{2\pi} \int_{-\pi}^{\pi} d\sigma \, e^{in\sigma} c(\sigma), \quad b_n = -\frac{i}{2\pi} \int_{-\pi}^{\pi} d\sigma \, e^{in\sigma} b(\sigma), \tag{3.2.41}$$

we find that equation (3.2.37) becomes

$$\{c_n, c_m\} = 0 = \{b_n, b_m\}, \quad \{c_n, b_m\} = \delta_{n+m,0}. \tag{3.2.42}$$

A representation of equation (3.2.37) is provided by

$$b(\sigma) = 2i\pi \frac{\delta}{\delta c(\sigma)}, \quad \text{or equivalently}, \; b_n = \frac{\partial}{\partial c_{-n}}. \tag{3.2.43}$$

Note that, unlike for $\delta/\delta x^\mu(\sigma)$, we do not require an i for Hermiticity since we are dealing with anti-commuting quantities and consequently

$$\left(\frac{\delta}{\delta c(\sigma)}\right)^\dagger = \frac{\delta}{\delta c(\sigma)}. \tag{3.2.44}$$

The two zero modes c_0 and b_0 are Hermitian and obey $\{b_0, c_0\} = 1$. They will play an important role in what follows. The action of these zero modes on the vacuum requires careful consideration [3.3]. We can define a $|+\rangle$ vacuum by the condition

$$c_0|+\rangle = 0. \tag{3.2.45}$$

Under the action of b_0 we find an alternative vacuum, denoted by $|-\rangle$ and given by

$$b_0|+\rangle = |-\rangle. \tag{3.2.46}$$

Since $(b_0)^2 = 0$, we find that $b_0|-\rangle = 0$. While from the relation $\{c_0, b_0\} = 1$, we conclude that $c_0|-\rangle = |+\rangle$. We note that

$$\langle +|+\rangle = \langle -|c_0 c_0|-\rangle = 0 \tag{3.2.47}$$

and similarly $\langle -|-\rangle = 0$. We also have the relations

$$\langle +|-\rangle = \langle -|c_0 b_0|+\rangle = \langle -|+\rangle. \tag{3.2.48}$$

We choose $\langle +|-\rangle = 1$ and take the $|-\rangle$ vacuum to be Grassmann odd. Consequently, the $|+\rangle$ vacuum is Grassmann even since c_0 is Grassmann odd. The action of the non-zero modes on the vacuum is taken to be

$$c_n|\pm\rangle = 0 = b_n|\pm\rangle, \qquad n \geq 1. \tag{3.2.49}$$

Let us consider the most general functional χ of $x^\mu(\sigma)$, and the ghost coordinates $c(\sigma)$ and $b(\sigma)$. In terms of the ghost oscillator basis χ may be written as

$$\begin{aligned}
|\chi\rangle &= \psi|-\rangle + \varphi|+\rangle \\
&\equiv \sum_{\{n\}\{m\}} c^\dagger_{n_1} \cdots c^\dagger_{n_b} b^\dagger_{m_1} \cdots b^\dagger_{m_a} \psi^{m_1 \cdots m_a}_{n_1 \cdots n_b} [x^\mu(\sigma)]|-\rangle \\
&\quad + \sum_{\{n\}\{m\}} c^\dagger_{n_1} \cdots c^\dagger_{n_{b'}} b^\dagger_{m_1} \cdots b^\dagger_{m_{a'}} \varphi^{m_1 \cdots m_{a'}}_{n_1 \cdots n_{b'}} [x^\mu(\sigma)]|+\rangle.
\end{aligned} \tag{3.2.50}$$

The Grassmann odd character of the ghosts implies the anti-symmetry $\psi^{m_1 \cdots m_a}_{n_1 \cdots n_b} = \psi^{[m_1 \cdots m_a]}_{[n_1 \cdots n_b]}$ $\equiv \psi^a_b$. If $a + b$ is an odd integer, then ψ^a_b is a Grassmann odd field. The field $\varphi^{m_1 \cdots m_{a'}}_{n_1 \cdots n_{b'}} \equiv \varphi^{a'}_{b'}$ is also anti-symmetric in its upper and lower indices separately. This field is Grassmann even if $a' + b'$ is an odd integer and Grassmann odd if $a' + b'$ is an even integer. The difference comes about as a result of the Grassmann odd nature of $|+\rangle$. The above assignments assume that $|\chi\rangle$ is Grassmann even; they should be reversed if $|\chi\rangle$ is Grassmann odd.

We can also expand the functionals of $x^\mu(\sigma)$ in the above equation in terms of α^μ_n and the centre-of-mass coordinate x^μ to obtain

$$\begin{aligned}
|\chi\rangle &= (\varphi(x) + iA_\mu(x)\alpha^\mu_{-1} + eb_{-1} + \bar{e}c_{-1} + e^\mu\alpha^\mu_{-1}b_{-1} + fb_{-2} + \cdots)\psi_{0-} \\
&\quad + (\varphi'(x) + iA'_\mu\alpha^\mu_{-1} + \cdots)\psi_{0+},
\end{aligned} \tag{3.2.51}$$

where

$$\psi_{0\pm} = |\pm\rangle \psi_0. \tag{3.2.52}$$

Let us repeat the above discussion for the *closed string*, the main difference from that for the open string being in the boundary conditions of the fields which are all periodic with period 2π for the closed string. Let us choose, as our coordinates $c \equiv c^0 + c^1 = c^+$ and $\bar{c} \equiv c^0 - c^1 = c^-$. The corresponding momenta are $b/2\pi \equiv -(1/\pi)(b_{00} + b_{01}) = -(2/\pi)b_{++}$ and $\bar{b}/2\pi = -(1/\pi)(b_{00} - b_{01}) = -(2/\pi)b_{--}$, and consequently, the anti-commutators are given by

$$\{c(\sigma), b(\sigma')\} = 2i\pi\, \delta(\sigma - \sigma'), \tag{3.2.53}$$

$$\{\bar{c}(\sigma), \bar{b}(\sigma')\} = 2i\pi\, \delta(\sigma - \sigma'). \tag{3.2.54}$$

The remaining anti-commutators vanish. The normal mode expansions appropriate to their periodic character are given by

$$c(\sigma) = \sum_n e^{-in\sigma} c_n, \quad \bar{c}_n = \sum_n e^{in\sigma} \bar{c}(\sigma), \tag{3.2.55}$$

$$b(\sigma) = i \sum_n e^{-in\sigma} b_n, \quad \bar{b}(\sigma) = i \sum_n e^{in\sigma} \bar{b}_n \tag{3.2.56}$$

and the corresponding Hermiticity properties are

$$c_n^\dagger = c_{-n}, \quad \bar{c}_n^\dagger = \bar{c}_{-n}, \quad b_n^\dagger = b_{-n}, \quad \bar{b}_n^\dagger = \bar{b}_{-n}. \tag{3.2.57}$$

They obey the relations

$$\{c_n, b_m\} = \delta_{n+m,0}; \quad \{\bar{c}_n, \bar{b}_m\} = \delta_{n+m,0}, \tag{3.2.58}$$

corresponding to equations (3.2.53) and (3.2.54). The remaining anti-commutators vanish.

For the closed string, we have twice as many zero modes as for the open string, namely $b_0, \bar{b}_0, c_0$ and $\bar{c}_0$. Corresponding to the existence of the four zero modes above, we have four types of vacuum

$$|+, +\rangle, \quad |+, -\rangle, \quad |-, +\rangle \quad \text{and} \quad |-, -, \rangle. \tag{3.2.59}$$

The first entry refers to the vacuum conditions of equations (3.2.45) and (3.2.46) under c_0, b_0 and the second to analogous conditions under $\bar{c}_0, \bar{b}_0$. These vacua are annihilated by $c_n, b_n, \bar{c}_n$ and $\bar{b}_n$ for $n \geq 1$. It is straightforward to write the most general functional of $x^\mu(\sigma)$ and ghost coordinates in terms of the oscillators $\alpha_n^\mu, \bar{\alpha}_n^\mu, c_n, b_n, \bar{c}_n$ and $\bar{b}_n$ acting on the vacua. The most general state can also be written by acting with $c_{-n}, b_{-n}, \bar{c}_{-n}$ and $\bar{b}_{-n}$ for $n \geq 1$ on the vacua, the coefficients being functionals of $x^\mu(\sigma)$.

3.2.2   The world-sheet energy-momentum tensor and BRST charge

The action of equation (3.2.22) is invariant under rigid two-dimensional Poincaré and BRST transformations. Correspondingly, we may compute the associated conserved currents and charges. Given any action A invariant under a rigid symmetry with a, necessarily constant, parameter Λ (any indices on Λ are not explicitly shown), then the variation of the action, once the parameter Λ is made space-time dependent, must be of the form

$$\delta A = \int d^2\xi (\partial_\alpha \Lambda) j^\alpha. \tag{3.2.60}$$

We identify j^α as the Noether current associated with the rigid symmetry Λ. The variation δA vanishes for any field variations when the equations of motion are used, including those with an arbitrary parameter $\Lambda(x)$. Consequently, it follows that j^α is conserved on-shell. The first step in finding a locally Λ invariant theory is to introduce the gauge field h_α (again suppressing any additional indices) and coupling constant g such that $\delta h_\alpha = (1/g)\partial_\alpha \Lambda + O(g^0)$ terms. The locally invariant action to order g^0, or first order in h_α, is then given by

$$A + g \int d^2\xi h_\alpha j^\alpha. \tag{3.2.61}$$

This procedure is the first step in the Noether method, which also works for determining the self-coupling of h_α, although in this case j^α involves h_α and in general a factor of $\frac{1}{2}$ is required in the above equation. Hence, one can find the current either from equation (3.2.60) given the rigid theory, or as the coefficient of h_α in equation (3.2.61) given the local theory.

The energy-momentum tensor is the conserved current corresponding to the invariance of the theory under space-time translations. We can therefore deduce it by coupling the theory to gravity and reading off the lowest order coupling to $h_{\mu\nu} = g_{\mu\nu} - \eta_{\mu\nu}$, or by making the translations depend on space-time in the theory and using equation (3.2.60). In fact we will use the latter method. Using the variations

$$\delta x^\mu = \zeta^\gamma \partial_\gamma x^\mu, \quad \delta c^\alpha = \zeta^\gamma \partial_\gamma c^\alpha, \quad \delta b_{\alpha\beta} = \zeta^\gamma \partial_\gamma b_{\alpha\beta} \tag{3.2.62}$$

but with the previously rigid parameter ζ^γ now a function of ξ^γ, we find by carrying out the variation of the action of equation (3.2.22) that

$$T_{\alpha\beta} = \partial_\alpha x^\mu \partial_\beta x^\nu \eta_{\mu\nu} - \tfrac{1}{2}\eta_{\alpha\beta}(\partial_\gamma x^\nu \partial^\gamma x^\mu \eta_{\mu\nu}) + 4\alpha'(b_{\alpha\delta}\partial_\beta\, c^\delta - \tfrac{1}{2}\eta_{\alpha\beta}b_{\gamma\delta}\partial^\gamma c^\delta). \tag{3.2.63}$$

We have normalised $T_{\alpha\beta}$ so that the x^μ part agrees with that given previously in equation (2.1.30). The above $T_{\alpha\beta}$ is traceless and conserved on-shell; however, it is not symmetric. We may always add to $T_{\alpha\beta}$ a term of the form

$$a\partial^\gamma(b_{\gamma\beta}c_\alpha - b_{\alpha\beta}c_\gamma), \tag{3.2.64}$$

which is automatically conserved and, being a total derivative, does not contribute to the total energy and momentum if the fields die off sufficiently fast at infinity. Using the equations of motion (3.2.24), we may write this term as

$$-a(b_{\gamma\beta}\partial_\alpha\, c^\gamma + \partial^\gamma\, b_{\alpha\beta}c_\gamma) \tag{3.2.65}$$

and, taking $a = -4\alpha'$, we find that

$$T_{\alpha\beta} = \partial_\alpha x^\mu \partial_\beta x^\nu \eta_{\mu\nu} - \tfrac{1}{2}\eta_{\alpha\beta}\partial_\gamma x^\mu \partial^\gamma x^\nu \eta_{\mu\nu}$$
$$+ 4\alpha'\left\{(b_{\alpha\delta}\partial_\beta c^\delta + b_{\beta\delta}\partial_\alpha c^\delta) + (\partial^\gamma b_{\alpha\beta})c_\gamma\right\}. \tag{3.2.66}$$

The final step is to realise that the equation of motion for $b_{\alpha\beta}$ can be written as

$$\partial_\delta b_{\alpha\beta} - \partial_\alpha b_{\delta\beta} = 0. \tag{3.2.67}$$

Using this repeatedly on the last term, we find the desired symmetric energy-momentum tensor

$$T_{\alpha\beta} = T_{\alpha\beta}^x + T_{\alpha\beta}^{gh}, \tag{3.2.68}$$

where

$$T_{\alpha\beta}^x = \partial_\alpha x^\mu \partial_\beta x^\nu \eta_{\mu\nu} - \tfrac{1}{2}\eta_{\alpha\beta}(\partial_\gamma x^\mu \partial^\gamma x^\nu \eta_{\mu\nu}),$$
$$T_{\alpha\beta}^{gh} = 4\alpha'\left\{(b_{\alpha\delta}\partial_\beta c^\delta + b_{\beta\delta}\partial_\alpha c^\delta) + \tfrac{1}{2}\left\{(\partial_\alpha b_{\beta\delta})\, c^\delta + (\partial_\beta\, b_{\alpha\delta})c^\delta\right\}\right\} - \text{trace}. \tag{3.2.69}$$

In fact, the trace vanishes on-shell.

We can now define the Virasoro generators L_n for the full system, which consists of the x^μ and ghosts. For the *open string* we find, using equation (2.1.80), that $(\wp)^2$ of equation (2.1.60) is given in terms of the energy-momentum tensor by

$$\left(\wp(\sigma)\right)^2 = \frac{2}{\alpha'}T_{++}^x, \quad \text{for } -\pi \le \sigma < \pi. \tag{3.2.70}$$

Using the full $T_{\alpha\beta}$ of equation (3.2.68), we define, by analogy with equation (2.1.78), the total Virasoro operators, now denoted L_n, by

$$L_n \equiv \frac{1}{2\pi\alpha'} \int_{-\pi}^{\pi} d\sigma \, e^{in\sigma} T_{++}. \tag{3.2.71}$$

In quantum theory, we find the expression for L_n by substituting the expressions for the oscillators and normal ordering. For the open string, the result is

$$L_n = L_n^x + L_n^{gh}, \tag{3.2.72}$$

where

$$L_n^x = \tfrac{1}{2} : \sum_m \alpha_m^\mu \alpha_{n-m}^\nu \eta_{\mu\nu} :, \tag{3.2.73}$$

$$L_n^{gh} =: \sum_m (n-m) b_{n+m} \, c_{-m} : -\delta_{n,0}. \tag{3.2.74}$$

The operator L_0^{gh} has a normal ordering ambiguity and so has a corresponding constant. The reason for the constant being $-1\delta_{n,0}$ will become apparent shortly. For the ghost oscillators, we normal order with respect to the $|\pm\rangle$ vacuum. To be precise, we place b_{-n} to the left of c_n and c_{-n} to the left of b_n for $n \ge 1$ and assign a minus sign for a change of order due to their Grassmann odd character. This is equivalent to the rule

$$:c_n b_{-m}: = -b_{-m}c_n \quad \text{for } m > 0, n > 0. \tag{3.2.75}$$

Since a $b_0 c_0$ term does not occur in any of the L_n we do not need to specify how to normal order b_0 with respect to c_0 at this stage.

For the L_n commutator, we find

$$[L_n, L_m] = (n-m) L_{n+m} + \frac{(D-26)}{12} n(n^2-1)\delta_{n+m,0}. \tag{3.2.76}$$

This result used the fact that for the Virasoro operator with only ghost oscillators we have the relation

$$\left[L_n^{gh}, L_m^{gh}\right] = (n-m)L_{n+m}^{gh} - \frac{26}{12}n(n^2-1)\delta_{n+m,0}. \tag{3.2.77}$$

The normal ordering constant in equation (3.2.74) was adjusted so as to have no central term in $[L_1^{gh}, L_{-1}^{gh}]$. One may verify the central term in equation (3.2.77) by the same method used for L_n^x in equation (3.1.16). The ghost Virasoro operators have central charge -26. Consequently, for $D = 26$, there is no central term. As we will see in chapter 9, a central term is equivalent to an anomaly in the conformal symmetry. As a result, demanding no conformal anomaly implies that $D = 26$. In fact, theories with local symmetries are inconsistent if that local symmetry has an anomaly. In a string theory the conformal symmetry is a local symmetry and so we require $D = 26$ for the consistency of the open bosonic string.

For the *closed string* to define L_n and $\bar{L}_n$ we use equation (2.1.37), but substitute $T_{\alpha\beta}$, now given by equation (3.2.68), into equation (2.2.3). We find that

$$L_n = L_n^x + L_n^{gh}; \quad \bar{L}_n = \bar{L}_n^x + \bar{L}_n^{gh}, \tag{3.2.78}$$

where

$$L_n^x = \,: \tfrac{1}{2} \sum_m \alpha_m^\mu \alpha_{n-m}^\nu \eta_{\mu\nu} :, \quad \bar{L}_n^x = \tfrac{1}{2} : \sum_m \bar{\alpha}_m^\mu \bar{\alpha}_{n-m}^\nu \eta_{\mu\nu} :,$$

$$L_n^{gh} = \,: \sum_m b_{n+m}\, c_{-m}(n-m): -\delta_{n,0}, \tag{3.2.79}$$

$$\bar{L}_n^{gh} = \,: \sum_m \bar{b}_{n+m}\bar{c}_{-m}(n-m): -\delta_{n,0}.$$

Let us now turn our attention to the BRST current. We may compute it by the same method using the BRST variations of equations (3.2.21), (3.2.22) and (3.2.23). However, as we wish to examine the action with $\lambda_{\alpha\beta}$ eliminated, we must substitute for $\lambda_{\alpha\beta}$ in $\delta b_{\alpha\beta}$ using the equation of motion of $g_{\alpha\beta}$. The reader may verify that the BRST current is given by

$$J_\alpha^{BRST} = c^\gamma (T_{\alpha\gamma}^x + \tfrac{1}{2} T_{\alpha\gamma}^{gh}). \tag{3.2.80}$$

The first term arises as a consequence of the fact that the BRST variation of the original field x^μ is just a translation with parameter Λc^α. The factor of $\tfrac{1}{2}$ in the second term is typical of a self-coupling problem. It is conserved as $T_{\alpha\gamma}^x$ and $T_{\alpha\gamma}^{gh}$ are separately conserved and $\partial^\alpha c^\gamma N_{\alpha\gamma} = 0$ by the equation of motion of c^γ if $N_{\alpha\gamma}$ is any symmetric and traceless tensor.

The BRST charge Q is given by

$$Q = \frac{1}{2\pi\alpha'} \int_a^\pi d\sigma J_0^{BRST}, \tag{3.2.81}$$

where $a = 0$ for the open string and $a = -\pi$ for the closed string. We may rewrite Q for the *open string* as

$$Q = \frac{1}{2\pi\alpha'} \int_{-\pi}^\pi d\sigma c(\sigma)(T_{++}^x + \tfrac{1}{2} T_{++}^{gh}). \tag{3.2.82}$$

Substituting the oscillator expressions, we find that

$$Q =: \sum_{n=-\infty}^\infty c_{-n}\big(L_n^x + \tfrac{1}{2} L_n^{gh} - \tfrac{1}{2}\delta_{n,0}\big): . \tag{3.2.83}$$

In the quantum theory the operator Q has a normal ordering ambiguity and we have made a choice for the corresponding constant that will be justified below. This constant and the one in L_0^{gh} of equation (3.2.74) will be the subject of a more systematic discussion in chapter 9. Using the above expressions for L_n^x and L_n^{gh} we may write Q as

$$Q =: \sum_{n=-\infty}^\infty c_{-n}L_n^x - \tfrac{1}{2} \sum_{\substack{n,m,p \\ =-\infty}}^\infty f_{nm}{}^p b_p c_{-m}c_{-n} - c_0: , \tag{3.2.84}$$

where f_{nm}^p are the structure constants of the conformal algebra, which is just the Virasoro algebra in the absence of the central charge:

$$[L_n, \; L_m] = (n - m)L_{n+m} \equiv f_{nm}{}^p L_p. \tag{3.2.85}$$

For the *closed string*, we find

$$Q =: \sum_{n=-\infty}^{\infty} c_{-n}\left(L_n^x + \tfrac{1}{2}L_n^{gh}\right) + \sum_{n=-\infty}^{\infty} \bar{c}_{-n}\left(\bar{L}_n^x + \tfrac{1}{2}\bar{L}_n^{gh}\right): . \tag{3.2.86}$$

The work of [1.3] generalised the BRST prescription of appendix A and gave a formula for the BRST charge for any system with first class constraints. Let us consider such a system whose first class constraints, denoted by φ_i, generate an algebra in the quantum theory with structure constants $f_{ij}{}^k$:

$$[\varphi_i, \; \varphi_j] = f_{ij}{}^k \varphi_k. \tag{3.2.87}$$

The authors of [1.3] introduced for each such constraint a ghost c^i and an anti-ghost b^i such that

$$\{c^i, b_j\} = \delta_j^i. \tag{3.2.88}$$

The BRST charge is then given by

$$Q = c^i \varphi_i - \tfrac{1}{2} \sum_{i,j,k} f_{ij}{}^k b_k c^i c^j. \tag{3.2.89}$$

It is straightforward to verify that

$$Q^2 = 0 \tag{3.2.90}$$

using equations (3.2.87) and (3.2.88) and the Jacobi identity for $f_{ij}{}^k$. Indeed, given the first term of Q, one finds that demanding that $Q^2 = 0$ leads one to the second term.

As Yang–Mills theories have just first class constraints, their BRST charge should be of the above form. Indeed, it is true that the BRST charge constructed according to the above procedure, that is, equation (3.2.89), agrees with that given in our BRST recipe of appendix A. In particular, every first class constraint implies a corresponding local invariance for the system and so in the above procedure we also end up by introducing a ghost and anti-ghost for each local symmetry.

For the open string, the procedure is as follows. The first class constraints are $L_n^x = 0$, $n \in \mathbf{Z}$ and so we introduce ghosts c_n and anti-ghosts b_n, $n \in \mathbf{Z}$. Neglecting the central term, the L_n obey the algebra $[L_n^x, L_m^x] = (n - m)L_{n+m}^x = f_{nm}{}^p L_p^x$, where $f_{nm}{}^p = (n - m)\delta_{p,n+m}$ and applying equation (3.2.89) leads to the BRST charge

$$Q =: \sum_{n=-\infty}^{\infty} c_{-n}L_n^x - \tfrac{1}{2} \sum_{n,m=-\infty}^{\infty} (n - m)b_{n+m}c_{-n}c_{-m} - ac_0: , \tag{3.2.91}$$

where a is a constant. We note that the last term was not included in the above procedure, but as we have seen it is a consequence of the normal ordering of Q which is required as a result of the infinite number of degrees of freedom of the string.

One finds that [3.3]

$$Q^2 = 0 \tag{3.2.92}$$

if and only if $D = 26$ and $a = 1$, which was the value of this constant adopted above in equation (3.2.84). The derivation of the expression for Q of the *closed string* follows the same pattern and is given in chapter 12.

The action with the ghosts of equation (3.2.22) also has the rigid ghost number invariance under the transformations $c^\alpha \to e^{i\lambda} c^\alpha$, $b_{\alpha\beta} \to e^{-i\lambda} b_{\alpha\beta}$ and we find that the corresponding current is given by

$$j_\alpha = b_{\alpha\beta} c^\beta. \tag{3.2.93}$$

The corresponding charge, called the ghost number operator, is given by

$$N = \frac{1}{\pi} \int_a^\pi d\sigma \, j_0, \tag{3.2.94}$$

where $a = 0$ for the *open string* and $a = -\pi$ for the closed string. For the *open string* we may extend the range of σ and rewrite N as

$$N = \frac{1}{2\pi} \int_{-\pi}^\pi d\sigma \, c(\sigma) b(\sigma), \tag{3.2.95}$$

while for the *closed string*

$$N = \frac{1}{2\pi} \int_{-\pi}^\pi d\sigma \, (c(\sigma) b(\sigma) + \bar{c}(\sigma) \bar{b}(\sigma)). \tag{3.2.96}$$

For the open string, we find that the suitably normal ordered oscillator expression is

$$N = \sum_{n=1}^\infty (c_{-n} b_n - b_{-n} c_n) + \tfrac{1}{2}(c_0 b_0 - b_0 c_0). \tag{3.2.97}$$

The choice of ordering in the last term is such that N is Hermitian. For the closed string N is given by (3.2.97) plus a similar term with $c_n \to \bar{c}_n$ and $b_n \to \bar{b}_n$. The BRST charge has ghost number 1 in accordance with the equation

$$Q = [N, Q]. \tag{3.2.98}$$

An important property of Q with respect to the Virasoro algebra is given by

$$\{b_n, Q\} = L_n \equiv L_n^x + L_n^{gh}, \tag{3.2.99}$$

which in turn implies that

$$[L_n, Q] = 0 \tag{3.2.100}$$

as

$$[L_n, Q] = \{b_n, Q\}Q - Q\{b_n, Q\} = b_n Q^2 - Q^2 b_n = 0. \tag{3.2.101}$$

3.2.3 The physical state condition

We must now find what the physical state conditions of equation (3.1.7) look like in the BRST formalism. At first sight, it would seem that we have gained very little since we now have a very much larger Fock space, as it now also contains states generated by the ghost oscillators. However, as we shall see the physical state condition is particularly simple in the

BRST language. For Yang–Mills theories, and many other theories with local symmetries, the physical states in the BRST formalism are given by

$$Q|\chi\rangle = 0, \tag{3.2.102}$$

where Q is the BRST charge. This constraint leads to the physical states for most theories, but, as we will see, not for all theories. As $Q^2 = 0$, if a state $|\chi\rangle$ is a solution, so is $|\chi\rangle + Q|\Lambda\rangle$ for any state $|\Lambda\rangle$. A state of the form $Q|\Lambda\rangle$ has zero scalar product with all physical states including itself, since

$$\langle \chi|Q|\Lambda\rangle = (\langle\Lambda|Q|\chi\rangle)^* = 0 \tag{3.2.103}$$

for any physical state $|\chi\rangle$. Consequently, the physical states $|\chi\rangle$ and $|\chi\rangle + Q|\Lambda\rangle$ have the same scalar product with all physical states and so lead to identical physical results. Since from the physical viewpoint $|\chi\rangle$ and $|\chi\rangle + Q|\Lambda\rangle$ are indistinguishable, it is useful to remove this redundancy by setting up the equivalence relation $|\chi_1\rangle \sim |\chi_2\rangle$ if

$$|\chi_1\rangle = |\chi_2\rangle + Q|\Lambda\rangle \tag{3.2.104}$$

for some state $|\Lambda\rangle$.

Let us consider all states $|\chi\rangle$ subject to $Q|\chi\rangle = 0$ and then consider the equivalence classes according to the above equivalence relation. We call the set of all such equivalence classes the cohomology of Q. One can show in Yang–Mills theories that the cohomology of Q is the set of all physical states with positive norm. The reader will recognise many similarities with de Rham cohomology.

One can make a similar discussion with operators. A physical operator S is one which commutes with Q, that is, $[S, Q] = 0$. Consequently, a physical operator S maps all physical states into physical states. Clearly, if S is of the form $[Q, U]$, then it automatically commutes with Q and its action on any physical state is to map it to the cohomology class containing the zero state. Consequently, S and $S + [Q, U]$ are the same operator from the physical point of view and we define the corresponding equivalence relation.

In QED we may choose our representatives of the equivalence classes to not depend on ghost oscillators and then the physical state condition of equation (3.2.102) enforces, in effect, the condition $(\partial^\mu A_\mu^{(+)})|\psi\rangle = 0$, where $+$ denotes the positive frequency part of A_μ, that is, the Gupta–Bleuler condition.

In string theory, the situation is similar, but there is an important difference. The physical state conditions for the *open string* are given by the two equations

$$Q|\chi\rangle = 0, \tag{3.2.105}$$

$$b_0|\chi\rangle = 0, \tag{3.2.106}$$

where $|\chi\rangle$ is, in the coordinate representation, a general functional of $x^\mu(\sigma)$ and $c(\sigma)$. In the oscillator representation this was given in equations (3.2.50) and (3.2.51). We can still consider the equivalence relation of equation (3.2.104), but we now subject all states to equation (3.2.106). Clearly, all states satisfying equations (3.2.105) and (3.2.106) also satisfy $L_0|\chi\rangle = \{Q, b_0\}|\chi\rangle = 0$.

The reason for the two conditions can be illustrated by considering a state $|\chi\rangle$ which is independent of the ghosts non-zero mode oscillators. Such a state has the form

$$|\chi\rangle = \psi|-\rangle + \phi|+\rangle, \tag{3.2.107}$$

where ψ and ϕ are functionals of $x^\mu(\sigma)$ alone. We find that equation (3.2.105) implies that

$$Q|\chi\rangle = \sum_{n=0}^{\infty} c_{-n}(L_n^x - \delta_{n,0})\psi|-\rangle + \sum_{n=1}^{\infty} c_{-n}L^x\phi|+\rangle, \qquad (3.2.108)$$

and so

$$(L_n^x\psi - \delta_{n,0})\psi = 0, \quad n \geq 0, \ L_n^x\phi = 0, \ n \geq 1. \qquad (3.2.109)$$

We recognise that ψ obeys the physical state conditions of equation (3.1.7) and so it contains all the physical states of the string. However, we also have the state ϕ which obeys all the physical state conditions except that involving L_0. By considering the state $Qb_0\varphi|+\rangle = (L_0^x - 1)\varphi|+\rangle$, where φ is a functional of $x^\mu(\sigma)$ alone, we can see that any parts of ϕ which do not satisfy $(L_0 - 1)\phi = 0$ are of the form $Q|\Lambda\rangle$ and so can be removed. Hence ϕ, in effect, also satisfies the physical state conditions of equation (3.1.7) and so also carries a set of the physical states. Since we wish to have only one set of physical states we impose the condition of equation (3.2.106), which, due to the equation $b_0|+\rangle = |-\rangle \neq 0$, eliminates all the states built on the $|+\rangle$ vacuum and so sets the field ϕ to zero.

What the above also shows is that a state $|\chi\rangle$ subject to equations (3.2.105) and (3.2.106) does contain the physical states of equation (3.1.7) found in the old covariant approach. The question of what other states are solutions of equations (3.2.105) and (3.2.106) remains. We will show in chapter 11 that all states which satisfy equations (3.2.105) and (3.2.106) and have ghost excitations (that is, any b_{-n} and c_{-n} factors acting on the vacua $|\pm\rangle$) are of the form $Q|\Lambda\rangle$ with one exception. This one exception is the state

$$b_{-1}|0,0\rangle|-\rangle, \qquad (3.2.110)$$

where the state $|0,0\rangle$ is annihilated by $\alpha_n, n \geq 1$ and it carries no momentum. The reader can easily verify that this state is annihilated by Q.

Hence, apart from the state of equation (3.2.110), we may conclude that all solutions of equations (3.2.105) and (3.2.106) which are not of the form $Q|\Lambda\rangle$ are of the form $\psi|-\rangle$, where ψ is a functional of $x^\mu(\sigma)$ alone and satisfies the Virasoro conditions, $L_n\psi = 0$, $n \geq 1$, $(L_0 - 1)\psi = 0$. Since some of these latter states have zero norm we may suspect that some of them are also of the form $Q|\Lambda\rangle$. In particular, let us consider how the zero-norm physical states of equations (3.1.25) and (3.1.27) occur in the new formalism. Counting the ghost number, they must be of the form $Q|\Lambda\rangle$, where $|\Lambda\rangle$ must contain one b ghost oscillator acting on the vacuum. Their precise expression is given by

$$Q\{b_{-1}|\Omega\rangle|-\rangle\} \text{ and } Q\{(b_{-2} + \tfrac{3}{2}b_{-1}L_{-1})|\Omega'\rangle|-\rangle\}, \qquad (3.2.111)$$

where $|\Omega\rangle$ and $|\Omega'\rangle$ satisfy the conditions below equations (3.1.25) and (3.1.27). Using the relation $\{Q, b_n\} = L_n$ we find that the first state becomes

$$\begin{aligned} Q\{b_{-1}|\Omega\rangle|-\rangle\} &= L_{-1}|\Omega\rangle|-\rangle - b_{-1}Q\{|\Omega\rangle|-\rangle\} \\ &= L_{-1}^x|\Omega\rangle|-\rangle + L_{-1}^{gh}|\Omega\rangle|-\rangle - b_{-1}(-c_o)|\Omega\rangle|-\rangle \\ &= L_{-1}^x|\Omega\rangle|-\rangle, \end{aligned} \qquad (3.2.112)$$

which is our previous state of equation (3.1.25) multiplied by a $|-\rangle$ ghost vacuum. A similar analysis applies to the second state of equation (3.2.111). Thus our previous zero-norm physical states are indeed of the form $Q|\Lambda\rangle$. In chapter 11 we will also show that all zero-norm physical states with no ghosts are of the form of equations (3.1.25) and (3.1.27), and so it follows that all zero-norm physical states with no ghosts are of the form $Q|\Lambda\rangle$.

We may therefore conclude that the positive norm solutions of equation (3.1.7) and the solutions of equations (3.2.105) and (3.2.106) when subject to the above equivalence relation are in one-to-one correspondence. Put another way, for $D = 26$, the cohomology classes of Q which are also subject to $b_0|\chi\rangle = 0$ are in one-to-one correspondence with the physical states of equation (3.1.7) with positive definite norm. The precise correspondence being

$$\psi|-\rangle \leftrightarrow \psi, \tag{3.2.113}$$

where ψ is a positive definite norm physical state. Consequently, at level n there are $d^{24}(n)$ cohomology classes, where $d^{24}(n)$ is given in equation (3.1.33). Of course, these statements are modulo the one additional state of equation (3.2.110).

The reader may wonder what are the cohomology classes of Q when not subject to equation (3.2.106). We will see in chapter 11 that these classes can be represented by the states mentioned above which satisfy equation (3.2.106) as well as the states

$$\hat{\phi}|+\rangle \text{ and } c_{-1}|+\rangle, \tag{3.2.114}$$

where $\hat{\phi}$ satisfies the Virasoro conditions of equation (3.1.7) and as a consequence has positive norm. Thus we gain exactly twice as many additional solutions, which are built on states with ghost numbers $3/2$ and $1/2$. We can think of these as being a kind of reflection of the original cohomology classes built on states with ghost number $-3/2$ and $-1/2$, respectively. This reflection property of the cohomology of Q is also shared by the cohomology of the de Rham operator d.

In chapter 12 we will discuss the gauge covariant string theory. It turns out that this makes extensive use of the BRST formalism. Since the *classical* formulation of any gauge invariant theory considered before string theory used the BRST formalism for its construction, the use of this formalism for string theory came as a considerable surprise. Of course, gauge covariant string theory, like any other gauge covariant theory, does not contain any space-time ghosts and so it must contain space-time fields which are Grassmann even. Hence, even though we will use the BRST formalism we will place additional restrictions on the fields to eliminate all the Grassmann odd fields.

4 The light-cone approach

> Taniyama was not very careful as a mathematician, he made a lot of mistakes, but he made mistakes in a good direction, eventually he got the right answers. I tried to imitate him, but I found that it was very difficult to make good mistakes.
>
> Gora Shimura speaking about the Shimura–Taniyama conjecture

Given any constrained system such as the string we can either work with the constraints or we can solve them. The former course of action was pursued for the string under the name of the 'old covariant quantization' in chapter 3. The latter approach has the advantage that having solved the classical constraints in terms of independent variables, it is then straightforward to quantise the theory. The disadvantage is that when solving the constraints by expressing some variables in terms of others one is left with independent variables which no longer transform in a simple way under the Lorentz group. The Lorentz transformations of the remaining variables become very non-linear and the Lorentz symmetry is no longer manifest. As a result, one must verify explicitly that Lorentz invariance is not broken in the quantisation procedure. Indeed, one finds that this leads to non-trivial conditions even for the free quantum string. For the interacting string, the verification of Lorentz invariance in the quantum light-cone theory is a very non-trivial calculation.

At first sight, solving the constraints of the string looks like a very non-trivial task. However, we are free to use the reparameterisation invariance to choose a gauge and we will find that there exists a particularly useful gauge which reduces the problem of solving the constraints to an almost trivial task. The light-cone approach to the free string was worked out in [2.4].

4.1 The classical string in the light-cone

In chapter 2 we used the action of equation (2.1.10), which involved the two-dimensional metric $g_{\alpha\beta}$ and x^μ, to find the equations of motion and constraints of the bosonic string. In this chapter we start from the Nambu formulation of the string of equation (2.1.1) and find the equations of motion and constraints. The results are, of course, the same, but since the derivation is slightly different and we want to solve the constraints from this starting point we now give the steps in detail. We recall that the Nambu action of equation (2.1.1) is given by

$$A = -\frac{1}{2\pi\alpha'}\int d^2\xi\sqrt{-g'},$$

81

where

$$g^I = \det g^I_{\alpha\beta}, \text{ and } g^I_{\alpha\beta} = \partial_\alpha x^\mu \partial_\beta x^\nu \eta_{\mu\nu}.$$

The latter is the metric induced on the world sheet from the Minkowski metric in space-time. More explicitly we find that

$$g^I = (\dot{x} \cdot \dot{x})(x' \cdot x') - (\dot{x} \cdot x')^2.$$

The canonical momentum, taking τ as the evolution parameter, is given by

$$P_\mu = \frac{\delta A}{\delta(\partial_\tau x^\mu)} = \frac{1}{2\pi\alpha'} \frac{\dot{x}_\mu (x' \cdot x') - x'_\mu (\dot{x} \cdot x')}{\sqrt{-\det(\partial_\alpha x^\mu \partial_\beta x^\nu \eta_{\mu\nu})}}. \tag{4.1.1}$$

It is straightforward to verify that the momentum satisfies the constraints

$$P^2 + \frac{1}{(2\pi\alpha')^2}(x')^2 = 0 \tag{4.1.2}$$

and

$$x^{\mu\prime} P_\mu = 0. \tag{4.1.3}$$

The existence of two constraints can be traced to the invariance of the action under the local two-dimensional reparameterisation invariance. As discussed in chapter 2 (see equation (2.1.26)), any two-dimensional metric can locally be brought to be proportional to the two-dimensional Minkowski metric $\eta_{\alpha\beta}$ by using the two-dimensional reparameterisations. Applying this theorem to the induced world-sheet metric we may choose $g^I_{\alpha\beta} = e^{\phi(\xi)} \eta_{\alpha\beta}$, where $\phi(\xi)$ is an arbitrary function. Choosing such a set of coordinates on the world sheet for the metric $g^I_{\alpha\beta}$ we find the constraints become

$$\dot{x} \cdot \dot{x} + x' \cdot x' = 0 \tag{4.1.4}$$

and

$$\dot{x} \cdot x' = 0. \tag{4.1.5}$$

in agreement with equation (2.1.29). In carrying out these last steps we have used the fact that for this choice of metric the momentum density becomes

$$P^\mu = \frac{1}{2\pi\alpha'} \dot{x}^\mu, \tag{4.1.6}$$

while

$$\Pi^\mu \equiv \frac{\delta A}{\delta x'_\mu} = -\frac{1}{2\pi\alpha'} x^{\mu\prime}. \tag{4.1.7}$$

The equation of motion resulting from varying x^μ is

$$\partial_\tau P^\mu + \partial_\sigma \Pi^\mu = 0 \tag{4.1.8}$$

and so with the above choice of metric it takes the simple form

$$\ddot{x}^\mu - x^{\mu\prime\prime} = 0. \tag{4.1.9}$$

Following our discussion in chapter 2, we also find that, for the case of the open string, the boundary conditions

$$\Pi^\mu(0) = \Pi^\mu(\pi) = 0 \tag{4.1.10}$$

take the form

$$x^{\mu\prime}(0) = x^{\mu\prime}(\pi) = 0. \tag{4.1.11}$$

The choice of metric resulting in the constraints of equations (4.1.4) and (4.1.5) and the equation of motion of equation (4.1.9) does not, however, fix all the reparameterisation invariance. This is most easily seen by writing the equations in terms of the variables

$$\xi^+ = \xi^o + \xi^1 = \tau + \sigma \text{ and } \xi^- = \xi^o - \xi^1 = \tau - \sigma, \tag{4.1.12}$$

whereupon, the derivatives become

$$\frac{\partial}{\partial \tau} = \frac{\partial}{\partial \xi^+} + \frac{\partial}{\partial \xi^-}, \quad \frac{\partial}{\partial \sigma} = \frac{\partial}{\partial \xi^+} - \frac{\partial}{\partial \xi^-}. \tag{4.1.13}$$

The constraints of equations (4.1.4) and (4.1.5) then become

$$\frac{\partial x^\mu}{\partial \xi^+} \frac{\partial x_\mu}{\partial \xi^+} = 0, \quad \frac{\partial x^\mu}{\partial \xi^-} \frac{\partial x_\mu}{\partial \xi^-} = 0. \tag{4.1.14}$$

It is now apparent that these conditions are invariant under the coordinate changes

$$\tilde{\xi}^+ = f(\xi^+), \ \tilde{\xi}^- = g(\xi^-), \tag{4.1.15}$$

where f and g are arbitrary functions. Under a conformal transformation we take x^μ to be a scalar, that is, it transforms as $\tilde{x}^\mu(\tilde{\xi}^+, \tilde{\xi}^-) = x^\mu(\xi^+, \xi^-)$. The equation of motion is also invariant under such a change as it can be written as $(\partial^2 x_\mu/\partial\xi^+\partial\xi^-) = 0$. For the open string, we must also preserve the boundary condition $x^{\mu\prime}(\sigma) = 0$ at $\sigma = 0$ and $\sigma = \pi$ under this transformation. We may rewrite the boundary condition as $\partial x^\mu/\partial\xi^+ = \partial x^\mu/\partial\xi^-$ at $\xi^+ - \xi^- = 0$ and $\xi^+ - \xi^- = 2\pi$. Hence, we demand that $\partial\tilde{\xi}^+/\partial\xi^+ = \partial\tilde{\xi}^-/\partial\xi^-$ at $\xi^+ - \xi^- = 0$ and $\xi^+ - \xi^- = 2\pi$. Thus f and g are related for the open string.

The transformations of equation (4.1.15) are those of the conformal group (see chapter 8). This is to be expected once we realise that the constraints and equation of motion do not involve the function ϕ that occurs in the special gauge choice for the induced two-dimensional metric $g^I_{\alpha\beta} = \partial_\alpha x^\mu \partial_\beta x^\nu \eta_{\mu\nu} = e^{\phi(\xi)}\eta_{\alpha\beta}$. As a result, any transformation whose effect is just to change ϕ will preserve the constraints of equations (4.1.4) and (4.1.5) and equations of motion of equation (4.1.9). However, by definition conformal transformations are those that leave the metric invariant up to a scale factor, that is, $g^I_{\alpha\beta} \rightarrow e^{\psi(\xi)} g^I_{\alpha\beta}$. Consequently, it follows that conformal transformations will leave the constraints invariant.

In order to fix this remaining invariance in the most useful way, we introduce light-cone coordinates for the Minkowski space-time in addition to those we have introduced already for the string world-sheet coordinates. We replace the x^μ, $\mu = 0, 1, 2, \ldots, D - 1$ of Minkowski space-time by the coordinates

$$x^\pm = \frac{1}{\sqrt{2}}(x^{D-1} \pm x^0), \ x^i, i = 1, \ldots, D - 2. \tag{4.1.16}$$

This coordinate change implies that for vectors of the type U^μ we use the components

$$U^\pm = \frac{1}{\sqrt{2}}(U^{D-1} \pm U^0), \quad U^i, i = 1, \ldots, D-2, \tag{4.1.17}$$

while for vectors of the type V_μ we use

$$V_\pm = \frac{1}{\sqrt{2}}(V_{D-1} \pm V_0), \quad V_i, i = 1, \ldots, D-2. \tag{4.1.18}$$

The scalar product of U^μ and S^μ is given in terms of the light-cone components by

$$U^\mu S^\nu \eta_{\mu\nu} = -U^0 S^0 + U^{D-1} S^{D-1} + \sum_{i=1}^{D-2} U^i S^i = U^+ S^- + U^- S^+ + \sum_{i=1}^{D-2} U^i S^i. \tag{4.1.19}$$

By examining the above equation, or by carrying out the coordinate change associated with equation (4.1.16), we find that the metric in the new coordinates has the non-zero components

$$\eta_{+-} = \eta_{-+} = 1, \quad \eta_{ij} = \delta_{ij}. \tag{4.1.20}$$

We caution the reader that although the use of the indices, $\pm$ as used in this chapter, is that traditionally employed in the light-cone formalism for Minkowski space-time, it differs with the use of these indices for the coordinates of the string world sheet.

As we will now explain, using the transformations of equation (4.1.15) one can choose

$$x^+(\tau, \sigma) = x_0^+(0) + c\tau, \tag{4.1.21}$$

where c and $x_0^+(0)$ are constants, that is, they do not depend on τ and σ. The constant $x_0^+(0)$ can be identified with the centre-of-mass position in the $+$ direction at $\tau = 0$ and can be set to zero by a shift of the coordinates along the $+$ direction. Such a choice is natural as it allows us to identify the time evolution of the world sheet as proportional with the time evolution in Minkowski space. To show that such a choice is possible we first note that since x^μ satisfies the wave equation it can be written as $x^\mu = l^\mu(\xi^+) + k^\mu(\xi^-)$. However, we have at our disposal precisely two such functions in our residual conformal symmetry and so we might hope to be able to choose one of the components of x^μ to be of a particular form. We also note that, for the open string, the boundary conditions imply that the functions l^μ and k^μ satisfy $\partial l^\mu / \partial \xi^+ = \partial k^\mu / \partial \xi^-$ at $\sigma = 0$ and $\sigma = \pi$ and these correspond to the boundary conditions on f and g mentioned above for the open string. Let us write x^+ in the form $x^+ = x_0^+ + c\tau + K^+(\xi^+, \xi^-)$. Under an infinitesimal conformal transformation of equation (4.1.15), x^+ transforms as $x^+ \rightarrow \tilde{x}^+ = x_0^+(0) + f(\xi^+)(\partial x^+/\partial \xi^+) + g(\xi^-)\partial x^+/\partial \xi^-$, which to lowest order in f, g and K^+ implies that

$$\tilde{x}^+ = x_0^+(0) + c\tau + K^+(\xi^+, \xi^-) + \frac{c}{2}f(\xi^+) + \frac{c}{2}g(\xi^-) + \cdots. \tag{4.1.22}$$

The coordinates x^μ obey the wave equation (4.1.9) and as a result so does K^+, which is therefore of the form $(l^+(\xi^+) - \frac{1}{2}c\xi^+) + (k^+(\xi^-) - \frac{1}{2}c\xi^-)$, where l^+ and k^+ are arbitrary functions of ξ^+ and ξ^-, respectively, which we introduced just above. Clearly, we can use f and g to choose $K^+ = 0$, leaving the choice of equation (4.1.21) after we drop the tilde on x^+. What this demonstrates is that, for any x^+, (τ, σ) in the infinitesimal neighbourhood of

$x^+ = x_0^+ + c\tau$ can be chosen to be given by $x^+ = x_0^+ + c\tau$ using a conformal transformation. To be complete one should show this choice is possible for any x^+.

With the choice of equation (4.1.21) the momentum density in the + direction becomes

$$P^+ = \frac{1}{2\pi\alpha'}c, \qquad (4.1.23)$$

which is independent of σ and τ. The total momentum p^μ, which is of course conserved, is given by

$$p^\mu = \int_a^\pi d\sigma\, P^\mu(\sigma),$$

where $a = 0$ for the open string and $\sigma = -\pi$ for the closed string. Using equation (4.1.23) we find

$$p^+ = \int_a^\pi d\sigma\, P^+ = \frac{(\pi - a)}{2\pi\alpha'}c.$$

Consequently, for the *open string*

$$c = 2\alpha' p^+ \text{ implying } P^+ = \frac{p^+}{\pi}, \qquad (4.1.24)$$

while for the *closed string*

$$c = \alpha' p^+, \text{ implying } P^+ = \frac{p^+}{2\pi}. \qquad (4.1.25)$$

Thus the constant c is determined in terms of p^+. This is consistent as the total momentum is conserved and so it is also a constant. However, the value of c also depends on how one chooses to parameterise the string. Although taking the parameterisation from 0 to π for the open string would seem to be the most obvious choice, when one is dealing with interacting *open* strings it is natural to let σ take the range $0 < \sigma < \pi\alpha' p^+$. Following the above arguments, we find that in this case $c = 2$. The underlying reason for this choice is that strings can interact by joining at their end points. The third string so formed inherits the parameterisation $0 < \sigma < \pi\alpha'(p_1^+ + p_2^+)$ from the initial two strings of + momenta p_1^+ and p_2^+. Due to momentum conservation in the + direction we may write the range of σ as $0 < \sigma < \pi\alpha' p_3^+$, where p_3^+ is the momentum of the outgoing string in the + direction. A similar discussion applies for the closed string. In this chapter, which deals only with the free string, we will always take $a < \sigma < \pi$ to parameterise the string; however, it is straightforward to change to the parameterisation mentioned just above.

We now solve the constraints for the *open string*. In terms of light-cone coordinates, and using equations (4.1.21) and (4.1.24), equation (4.1.2) becomes

$$2\frac{p^+}{\pi}P^- + \sum_{i=1}^{D-2}\left\{P_i^2 + \frac{1}{(2\pi\alpha')^2}x_i'^2\right\} = 0. \qquad (4.1.26)$$

Clearly, we can solve for P^- to find that

$$P^- = -\frac{\pi}{2p^+}\sum_{i=1}^{D-2}\left\{P_i^2 + \frac{1}{(2\pi\alpha')^2}x_i'^2\right\}. \qquad (4.1.27)$$

The other constraint of equation (4.1.3) can be written as

$$\frac{p^+}{\pi}x^{-\prime} + \sum_{i=1}^{D-2} P^i x_i' = 0, \tag{4.1.28}$$

since $x^{+\prime} = 0$. We can solve this equation for $x^{-\prime}$ and then integrate with respect to σ incurring one integration constant denoted A. For the open string, we find the result

$$x^-(\tau, \sigma) = -\frac{\pi}{p^+} \int_0^\sigma d\sigma' \sum_{i=1}^{D-2} x_i'(\tau, \sigma') P^i(\tau, \sigma') + A. \tag{4.1.29}$$

We will determine the integration constant in terms of the centre of mass of the string in the $-$ direction, which is defined by

$$q^-(\tau) = \frac{1}{\pi} \int_0^\pi d\sigma' x^-(\tau, \sigma'). \tag{4.1.30}$$

We now write $A = \tilde{A} + q^-$ and determine the constant $\tilde{A}$ by integrating x^- with respect to σ from 0 to π and then dividing by π. Substituting (4.1.29) into (4.1.30) we find that

$$q^-(\tau) = q^-(\tau) - \frac{1}{p^+} \int_0^\pi d\sigma \int_0^\sigma d\sigma' \sum_{i=1}^{D-2} x_i'(\tau, \sigma') P^i(\tau, \sigma') + \tilde{A}. \tag{4.1.31}$$

Integrating the second term by parts we find $\tilde{A}$ and so

$$x^-(\tau, \sigma) = q^-(\tau) - \frac{\pi}{p^+} \int_0^\sigma d\sigma' \sum_{i=1}^{D-2} x_i'(\tau, \sigma') P^i(\tau, \sigma')$$

$$- \frac{1}{p^+} \int_0^\pi d\sigma'(\sigma' - \pi) \sum_{i=1}^{D-2} x_i'(\tau, \sigma') \dot{P}^i(\tau, \sigma').$$

Hence, we conclude that

$$x^-(\tau, \sigma) = q^-(\tau) + \frac{\pi}{p^+} \int_\sigma^\pi d\sigma' \sum_{i=1}^{D-2} x^i(\tau, \sigma') P^i(\tau, \sigma')$$

$$- \frac{1}{p^+} \int_0^\pi d\sigma'\sigma' \sum_{i=1}^{D-2} x_i(\tau, \sigma') P^i(\tau, \sigma'). \tag{4.1.32}$$

For the *closed string* the analogous steps yield the corresponding result:

$$x^-(\tau, \sigma) = q^-(\tau) - \frac{\pi}{p^+} \int_{-\pi}^\sigma d\sigma' \sum_{i=1}^{D-2} x_i(\tau, \sigma') P_i(\tau, \sigma')$$

$$- \frac{1}{2p^+} \int_{-\pi}^\pi d\sigma'(\sigma' - \pi) \sum_{i=1}^{D-2} x_i'(\tau, \sigma') P^i(\tau, \sigma'), \tag{4.1.33}$$

where now

$$q^-(\tau) = \frac{1}{2\pi} \int_{-\pi}^\pi x^-(\tau, \sigma') d\sigma'. \tag{4.1.34}$$

Let us summarise: we started with $x^\mu(\tau, \sigma)$ and $P^\mu(\tau, \sigma)$, and used the residual conformal symmetry to choose $x^+(\tau, \sigma)$ to be given by equation (4.1.21). This implied that P^+ was independent of σ and we could then solve the constraints of equations (4.1.4) and (4.1.5) to find $P^-(\tau, \sigma)$ and $x^{-\prime}(\tau, \sigma)$ in terms of the remaining variables. Identifying the integration constant in terms of $q^-(\tau)$ and swapping P^+ for the total momentum p^+ in the $+$ direction we have solved all the constraints and so we are left with the independent variables

$$x^i(\tau, \sigma),\ P^i(\tau, \sigma),\ \text{and}\ q^-(\tau),\ p^+(\tau). \tag{4.1.35}$$

The next step is to identify the Hamiltonian, H. By definition the Hamiltonian is the generator of time translations. Due to our gauge choice of equation (4.1.21), translations in τ are directly related to translations in x^+, indeed $\delta\tau = (1/2\alpha' p^+)\delta x^+$. Hence, the generator of x^+ translations is proportional to the zero mode of its conjugate momentum p^- and so we take

$$H = -2\alpha' p^+ p^- = -2\alpha' p^+ \int_0^\pi d\sigma P^-(\sigma). \tag{4.1.36}$$

Using equation (4.1.27), which is solved for $P^-(\sigma)$, we find that

$$H = \pi\alpha' \int_0^\pi d\sigma \sum_{i=1}^{D-2} \left\{ P_i^2(\sigma) + \left(\frac{x_i'(\sigma)}{2\pi\alpha'}\right)^2 \right\}. \tag{4.1.37}$$

The great advantage of the light-cone approach is that we are left with the independent variables, given above, and no constraints to solve. It only remains to give the Hamiltonian formalism in terms of these independent variables and then quantise them. The Poisson brackets for the independent variables of equation (4.1.35) are those of the original theory, which are given by

$$\{x^i(\sigma), P^j(\sigma')\} = \delta^{ij}\delta(\sigma - \sigma'),$$
$$\{q^-, p^+\} = 1, \tag{4.1.38}$$

as well as Poisson brackets that vanish. It is instructive to verify that the Hamiltonian of equation (4.1.37) does indeed generate the correct relations between coordinates and momenta and the equations of motion. We find that

$$\dot{x}^i(\sigma) = \{x^i(\sigma), H\} = \pi\alpha' \int_0^\pi d\sigma' \left\{ x^i(\sigma), \sum_j P_j^2(\sigma') \right\} = 2\pi\alpha' P^i(\sigma), \tag{4.1.39}$$

while

$$\dot{P}^i(\sigma) = \{P^i(\sigma), H\} = \frac{\pi\alpha'}{(2\pi\alpha')^2} \int_0^\pi d\sigma' \left\{ P^i(\sigma), \sum_j (x_j'(\sigma'))^2 \right\} = \frac{1}{2\pi\alpha'} x^{i\prime\prime}(\sigma) \tag{4.1.40}$$

and consequently $\ddot{x}^i - x^{i\prime\prime} = 0$, as we expect. We can regard the above calculation as a confirmation for the Hamiltonian of equation (4.1.36). From the definition of equation

(4.1.30) it follows that

$$\dot{q}^- = \frac{1}{\pi} \int_0^\pi d\sigma' \dot{x}^-(\sigma') = 2\alpha' \int_0^\pi d\sigma' P^-(\sigma') = 2\alpha' p^- = -\frac{H}{p^+}. \tag{4.1.41}$$

Since H and p^+ are constants of the motion, $\ddot{q}^- = 0$ and so the time evolution of q^- is given by $q^- = q^-(0) - (H/p^+)\tau$.

In preparation for quantisation we introduce the Fourier modes of the independent variables, which are the same those given in chapter 2. For the *open string* we define

$$x^i(\tau, \sigma) = x_0^i(\tau) + \frac{1}{2} \sum_{n=-\infty}^{\infty} x_n^i(\tau) e^{in\sigma}, \tag{4.1.42}$$

where $x_n^i = x_{-n}^i = (x_n^i)^*$, and

$$P^i(\tau, \sigma) = \frac{1}{\pi} \sum_{n=\infty}^{-\infty} P_n^i(\tau) e^{in\sigma}, \tag{4.1.43}$$

where $P_n^i = P_{-n}^i = (P_n^i)^*$. The time evolution of x_0^i is given by $x_0^i(\tau) = x_0^i(0) + 2\alpha' p^i \tau$ since $P_0^i = p^i$ is a constant of the motion. Substituting equations (4.1.42) and (4.1.43) into the Hamiltonian it becomes

$$H = \alpha' \sum_{n=-\infty}^{n=\infty} \sum_{i=1}^{D-2} \left\{ (P_n^i)^2 + \frac{n^2}{16\alpha'^2} (x_n^i)^2 \right\}. \tag{4.1.44}$$

We immediately recognise this Hamiltonian as being an infinite number of decoupled harmonic oscillators. One of these (that for $n = 0$) has zero frequency and corresponds to a free point particle. The quantisation is now reduced to an exercise in undergraduate quantum mechanics. Extending the range of σ from $0 < \sigma < \pi$ to $-\pi < \sigma < \pi$ by $x^i(\sigma) = x^i(-\sigma)$ as explained in chapter 2, we introduce the oscillators

$$\wp^i(\tau, \sigma) \equiv \frac{\dot{x}^i(\tau, \sigma) + x^{i\prime}(\tau, \sigma)}{\sqrt{2\alpha'}} = \pi\sqrt{2\alpha'} P^i(\tau, \sigma) + \frac{x^{i\prime}(\tau, \sigma)}{\sqrt{2\alpha'}}$$

$$\equiv \sum_{n=-\infty}^{n=\infty} \alpha_n^i(\tau) e^{-in\sigma}. \tag{4.1.45}$$

It is straightforward to verify the Poisson brackets

$$\{\alpha_n^i, \alpha_m^j\} = -in\delta^{ij}\delta_{n+m,0}, \text{ or equivalently } \{x_n^i, p_m^j\} = \delta^{ij}\delta_{n+m,0}. \tag{4.1.46}$$

Writing the Hamiltonian in terms of $\wp^i$ we can express it in terms of the oscillators as follows:

$$H = \frac{1}{4\pi} \int_{-\pi}^{\pi} d\sigma \sum_{i=1}^{D-2} (\wp^i(\tau, \sigma))^2 = \frac{1}{2} \sum_{i=1}^{D-2} \sum_{n=-\infty}^{\infty} \alpha_{-n}^i \alpha_n^i, \tag{4.1.47}$$

where $\alpha_0^i = \sqrt{2\alpha'} p^i$. Recalling equations (4.1.19) and (4.1.36) we may write

$$\alpha' p^\mu p_\mu = 2\alpha' p^+ p^- + 2\alpha' \sum_{i=1}^{D-2} p^i p^i = -\sum_{i=1}^{D-2} \sum_{n=1}^{\infty} \alpha_{-n}^i \alpha_n^i. \tag{4.1.48}$$

4.2   The quantum string in the light-cone

To quantise the system we make the Dirac replacement of the Poisson brackets for the independent variables by their commutators divided by $i\hbar$. The non-vanishing commutators are then given by

$$[x^i(\sigma), P^j(\sigma')] = i\hbar\delta(\sigma - \sigma')\delta^{ij}, \quad [q^-, p^+] = i\hbar. \qquad (4.2.1)$$

One finds, using the same arguments as in chapter 2, that the oscillators obey the relations

$$[\alpha_n^i, \alpha_m^j] = n\delta^{ij}\delta_{n+m,0}. \qquad (4.2.2)$$

The most general functional Ψ of the coordinates $x^i(\sigma), q^-$ and τ, or equivalently x^+, is most conveniently expressed using a Hermite polynomial basis, that is, using the α_n^i oscillators, since these yield eigenfunctions of the Hamiltonian. We write Ψ as

$$\Psi = (\phi(x^\mu) + i\alpha_1^{i\dagger}A_i(x^\mu) + \alpha_1^{i\dagger}\alpha_1^{j\dagger}B_{ij}(x^\mu) + i\alpha_2^{i\dagger}C_i(x^\mu) + \cdots)\Psi_0, \qquad (4.2.3)$$

where x^μ stands for x_0^i, q^- and τ and

$$\alpha_n^i\Psi_0 = 0, \quad n \geq 1. \qquad (4.2.4)$$

When defining the quantum Hamiltonian we add a constant α_0 to take into account any normal ordering that may be required:

$$H = \alpha'\sum_{i=1}^{D-2}p^{i2} + \sum_{i=1}^{D-2}\sum_{n=1}^{\infty}\alpha_n^{i\dagger}\alpha_n^i - \alpha_0. \qquad (4.2.5)$$

Equation (4.1.48) then becomes a differential equation which we impose on the wavefunction:

$$\left(-\alpha'\partial_\mu\partial^\mu - \alpha_0 + \sum_{i=1}^{D-2}\sum_{n=1}^{\infty}\alpha_n^{i\dagger}\alpha_n^i\right)\Psi = 0. \qquad (4.2.6)$$

In fact, this is none other than the Schrödinger equation $i\hbar\partial\Psi/\partial\tau = H\Psi$ once we have traded the differential with respect to τ for that with respect to x^+ using the discussion above equation (4.1.36).

In the light-cone formalism there are no constraints and so all the degrees of freedom in Ψ are physical states. Equation (4.2.6) just tells us the mass of a given state. Let us now examine the mass spectrum when $\alpha_0 = 1$. The lowest mass state ϕ has (mass)$^2 = -1/\alpha'$ and so is a tachyon. The $D - 2$ states A_i at the first level are massless while those states at the next level, namely B_{ij} and C_i, have (mass)$^2 = 1/\alpha'$ etc. Massless states are classified by representations of the little group SO$(D - 2)$, while the massive states are classified by the little group SO$(D - 1)$. We have $D - 2$ states A_i which can belong to the vector representation of SO$(D - 2)$, but cannot belong to any representations of SO$(D - 1)$ other than the trivial representation, which is ruled out as these states must transform in some representation as a result of their i index. As such, these $D - 2$ states at the first level must be massless if we are to have Lorentz symmetry in the theory. We note that had we not chosen $\alpha_0 = 1$, then we would not have had a Lorentz invariant theory. As we shall see $\alpha_n^i, i = 1, \ldots, D - 2$, transforms, as one might expect, by a straightforward rotation under SO$(D - 2)$. Consequently, the states which are coefficients of the products of $\alpha_n^{i\dagger}$

naturally belong to tensor product representations of SO($D - 2$). These representations are, in general, not irreducible representations as they are not traceless. At a given mass level, except for the tachyon and the massless states, one finds more than one of the above sets of SO($D - 2$) representations corresponding to the different possible combinations of $\alpha_n^{i\dagger}$ which result in the same mass level. However, as we noted above, all the states except for those at level 1 discussed above are massive and so must be classified by SO($D - 1$). Hence, assuming Lorentz invariance is not broken, all these SO($D - 2$) representations must assemble into representations of SO($D - 1$).

Let us illustrate this procedure at level 2. The states at this level are C_i, $i = 1, \ldots, D - 2$ and B_{ij}, $i, j = 1, \ldots, D - 2$. This makes

$$D - 2 + \frac{(D-1)(D-2)}{2} = \frac{(D+1)(D-2)}{2} \tag{4.2.7}$$

states in all. The symmetric traceless representation of SO($D - 1$), t_{ij}, $i, j = 1, \ldots, D - 1$ with $t_i^i = 0$ has

$$\frac{D(D-1)}{2} - 1 = \frac{(D+1)(D-2)}{2} \tag{4.2.8}$$

states. We may indeed identify

$$t_{ij} = \begin{cases} B_{ij}, & i, j = 1, 2, \ldots, D - 2, \\ C_i, & i = 1, 2, \ldots, D - 2, j = D - 1, \end{cases} \tag{4.2.9}$$

while the remaining component $t_{D-1,D-1}$ is equal to $\sum_i B_{ii}$ as t_{ij} is traceless. In four dimensions this would be called a massive Spin-2 state. The reader is encouraged to analyse the states at the next levels.

The general state at level n is of the form

$$\alpha_{m_1}^{i_1\dagger} \cdots \alpha_{m_p}^{i_p\dagger} S_{i_1 \ldots i_p} \Psi_0, \tag{4.2.10}$$

provided $\sum_i m_i = n$. If $c(n)$ is the number of such states at level n, then, as we will shortly show, $c(n)$ is given by the formula

$$\sum_n c(n) x^n = \prod_{n=1}^{\infty} \frac{1}{(1 - x^n)^{D-2}} = 1 + (D - 2)x$$

$$+ \left[\frac{(D-2)(D-1)}{2} + (D-2) \right] x^2 + \cdots. \tag{4.2.11}$$

One is to regard the right-hand side as a formal power series whose coefficients are $c(n)$. The first few terms that have been computed above are in agreement with our previous considerations.

To establish this formula at all levels let us first digress to the case of one harmonic oscillator with creation operator $a^{\dagger}$ and destruction operator a which obey $[a, a^{\dagger}] = 1$. This case is none other than the usual harmonic oscillator. It has only one energy, or number, eigenstate at each energy level which is given by $|n\rangle = ((a^{\dagger})^n / \sqrt{(n!)})|0\rangle$, where $a|0\rangle = 0$. The number of energy states at level n is in agreement with the formula

$$\sum_n \tilde{c}(n) x^n = \frac{1}{(1 - x)}, \tag{4.2.12}$$

where $\tilde{c}(n) = 1$. For future discussion it will be useful to note that we may rewrite this equation in terms of a trace over the Hilbert space, for which the energy eigenstates form a basis, as follows:

$$\sum_n x^n = \frac{1}{(1-x)} = \mathrm{Tr} x^{a^\dagger a} = \mathrm{Tr} \exp(a^\dagger a \ln x) = \sum_{n=0}^{\infty} \langle n|\exp(a^\dagger a \ln x)|n\rangle.$$

$$(4.2.13)$$

A little thought shows that for the case of one dimension, where we have the oscillators $\alpha_n \equiv \alpha_n^1$, $n = 1, 2, \ldots$ and their Hermitian conjugates, the number of states $\tilde{c}(n)$ at level n is given by the formula

$$\sum_n \tilde{c}(n)x^n = \prod_{n=1}^{\infty} \frac{1}{(1-x^n)} = \frac{1}{(1-x)}\frac{1}{(1-x^2)}\cdots.$$

$$(4.2.14)$$

The states at level n are of the form

$$\sum_p \sum_{n_1,\ldots,n_p} \alpha_{n_1}^\dagger \cdots \alpha_{n_p}^\dagger t_{n_1\cdots n_p} \delta(n_1 + \cdots + n_p - n)\Psi_0.$$

$$(4.2.15)$$

The number of ways of selecting $\alpha_n^\dagger$ such that $\sum_p n_p = n$ is precisely the same as the number of ways of selecting the power x^n from the right-hand side of equation (4.2.14). For example, the choice $\alpha_{n_1}^\dagger \ldots \alpha_{n_p}^\dagger$ with $n_i \neq n_j$ corresponds to selecting, for each i, an x^{n_i} from the first term in the expansion of the term

$$\frac{1}{(1-x^{n_i})}$$

$$(4.2.16)$$

in the right-hand side of equation (4.2.14).

A more formal derivation of equation (4.2.14) begins by writing

$$\sum_n c(n)x^n = \mathrm{Tr} x^N,$$

$$(4.2.17)$$

where $N \equiv \sum_n^{\infty} \alpha_{-n}\alpha_n$ has eigenvalue n on the state of level n and Tr means the trace of the operator x^N in the Hilbert space of states. It is most convenient to take the space of states to have a basis that consists of the eigenstates of the Hermitian operator N, however, the trace is independent of the particular basis used. If we label the eigenstates of the operator N by $|n, i\rangle$, where n is the eigenvalue of N and i is the degeneracy label, then the latter takes $c(n)$ values and equation (4.2.17) is obviously true. For the case of one dimension, we have the oscillators α_n, $n = 1, 2, \ldots$, and their conjugates and we may write

$$\sum_n c_n x^n = \mathrm{Tr} x^{\sum_{n=1}^\infty \alpha_{-n}\alpha_n} = \mathrm{Tr}\left(\exp\sum_{n=1}^{\infty}\alpha_{-n}\alpha_n \ln x\right) = \mathrm{Tr}\prod_{n=1}^{\infty} \exp\{\alpha_{-n}\alpha_n \ln x\}$$

$$= \prod_{n=1}^{\infty}(\mathrm{Tr}^{(n)} \exp(\alpha_{-n}\alpha_n \ln x)).$$

$$(4.2.18)$$

In the last line we have used the fact that the Hilbert space of states can be written as a tensor product of Hilbert spaces $H^{(n)}$ for each species of oscillator α_n and $\mathrm{Tr}^{(n)}$ is the trace with respect to $H^{(n)}$. The α_n are related to the usual oscillators a_n which satisfy $[a_n, a_m^\dagger] = \delta_{n+m,0}$, by $\alpha_n \equiv \sqrt{n}a_n$, $\alpha_{-n} \equiv \sqrt{n}a_{-n}$ for $n > 1$. Utilising this result equation (4.2.18) becomes

$$\sum_n c_n x^n = \prod_{n=1}^{\infty} \mathrm{Tr}^{(n)} \exp(na_{-n}a_n \ln x) = \prod_{n=1}^{\infty} \mathrm{Tr}^{(n)} \exp(a_{-n}a_n \ln x^n)$$

$$= \prod_{n=1}^{\infty} \frac{1}{(1 - x^n)}, \tag{4.2.19}$$

which is the same result as that of equation (4.2.14).

Following the same argument and using an obvious notation, we find that for $D - 2$ oscillators α_n^i, $i = 1, 2 \ldots, D - 2$,

$$\sum_n c(n) x^n = \mathrm{Tr} x^{\sum_{n=1}^{\infty} \sum_i \alpha_{-n}^i \alpha_n^i} = \mathrm{Tr} \exp\left(\sum_{n=1}^{\infty} \sum_i \alpha_{-n}^i \alpha_n^i \ln x\right)$$

$$= \mathrm{Tr} \prod_{n=1}^{\infty} \prod_{i=1}^{D-2} \exp(\alpha_{-n}^i \alpha_n^i \ln x) = \mathrm{Tr} \prod_{n=1}^{\infty} \prod_{i=1}^{D-2} \exp(a_{-n}^i a_n^i \ln x^n)$$

$$= \prod_{n=1}^{\infty} \prod_{i=1}^{D-2} \mathrm{Tr}^{(n,i)} \exp(a_{-n}^i a_n^i \ln x^n) = \prod_{n=1}^{\infty} \frac{1}{(1 - x^n)^{D-2}} \tag{4.2.20}$$

and we recover equation (4.2.11).

One important observation is that in contrast to the covariant case all the light-cone states must have positive definite norm since they are created with $\alpha_n^{i\dagger}$, $n = 1, 2, \ldots$, and they obey the relation $[\alpha_n^i, \alpha_m^{j\dagger}] = n\delta^{ij}\delta_{n,m}$ for $n, m > 0$. Thus in the light-cone approach there are no negative norm states and the no-ghost theorem automatically holds. At first sight, this is surprising as within the context of the covariant approach it was noted in chapter 3, and will be shown in chapter 11, that the no-ghost theorem requires the dimension of space-time to be 26 or less. As a formalism cannot change the fundamental results, it must be that something also goes wrong in the light-cone approach if $D > 26$. This is the subject of the next section, where we will show that Lorentz symmetry is only valid in the light-cone approach if $D = 26$ and $\alpha_0 = 1$.

4.3   Lorentz symmetry

Although the quantisation in the light-cone formalism is straightforward, Lorentz invariance is not manifest and we must verify explicitly whether it is maintained or not in the quantisation procedure. In the previous section, we found that Lorentz symmetry enforced an intercept of $\alpha_0 = -1$. This resulted from the fact that there are $D - 2$ states at level 1 which only formed a representation of $SO(D - 2)$ and not $SO(D - 1)$ and so must be massless. Clearly, further conditions are possible at higher levels.

As a prelude to this calculation, let us consider Lorentz invariance in the classical theory. The Lorentz generators $J_{\mu\nu}$ of equation (2.2.8) are given by

$$J^{\mu\nu} = \int_0^{\pi} d\sigma \, (x^{\mu}P^{\nu} - x^{\nu}P^{\mu}). \tag{4.3.1}$$

The generators J^{ij}, $i,j = 1,\ldots,D-2$ have a straightforward form involving the fundamental variables x^i and P^j, namely

$$J^{ij} = \int_0^\pi d\sigma\,(x^i P^j - x^j P^i). \tag{4.3.2}$$

In terms of the oscillators J^{ij} is given by

$$J^{ij} = \frac{1}{\sqrt{2\alpha'}}(x_0^i \alpha_0^j - x_0^j \alpha_0^i) + \sum_{n=1}^\infty \frac{(\alpha_{-n}^i \alpha_n^j - \alpha_{-n}^j \alpha_n^i)}{in}, \tag{4.3.3}$$

the derivation being identical to that given in chapter 2.

The generators J^{i+} are of the form

$$J^{i+} = \int_0^\pi d\sigma\,(x^i P^+ - x^+ P^i) = p^+ x_0^i(\tau) - 2\alpha'\tau p_0^i = p^+ x_0^i(0), \tag{4.3.4}$$

while the generators J^{+-} have the simple form

$$J^{+-} = \int_0^\pi d\sigma\,(x^+ P^- - x^- P^+) = 2\alpha'\tau p^+ \int_0^\pi d\sigma P^- - p^+ \int_0^\pi d\sigma x^-$$

$$= -\tau H - p^+ q^-(\tau) = -p^+ q^-(0). \tag{4.3.5}$$

The generators J^{i-} are of the form

$$J^{i-} = \int_0^\pi d\sigma\,(x^i P^- - x^- P^i). \tag{4.3.6}$$

They involve x^- and P^- which we solved for and so generator J^{i-} is not of such a simple form. They are most simply calculated in terms of oscillators by introducing the oscillators α_n^-, which are defined by

$$\wp^-(\tau,\sigma) \equiv \frac{1}{\sqrt{2\alpha'}}(\dot{x}^-(\tau,\sigma) + x^{-\prime}(\tau,\sigma)) = \pi\sqrt{2\alpha'}P^-(\tau,\sigma) + \frac{x^{-\prime}(\tau,\sigma)}{\sqrt{2\alpha'}}$$

$$= \sum_n \alpha_n^-(\tau)e^{-in\sigma}. \tag{4.3.7}$$

These oscillators are not independent of α_n^i. Their dependence can be found by using the expressions for P^- and x^- in terms of these oscillators. However, we can also recall the constraints of equations (4.1.4) and (4.1.3), which can be written as

$$0 = \wp^\mu \wp_\mu = 2\wp^- \wp^+ + \sum_{i=1}^{D-2} \wp^i \wp^i. \tag{4.3.8}$$

We note that

$$\wp^+ = \frac{1}{\sqrt{2\alpha'}}(\dot{x}^+ + x^{+\prime}) = \pi\sqrt{2\alpha'}P^+ = \sqrt{2\alpha'}p^+ \tag{4.3.9}$$

and then equation (4.3.8) becomes

$$\wp^- = -\frac{1}{2p^+\sqrt{2\alpha'}}\sum_{i=1}^{D-2}\wp_i^2. \tag{4.3.10}$$

Taking the Fourier transform with $e^{in\sigma}$, and substituting equation (4.1.45), we find that

$$\alpha_n^- = -\frac{1}{2p^+\sqrt{2\alpha'}}\sum_{i=1}^{D-2}\sum_m \alpha_{n-m}^i \alpha_m^i. \tag{4.3.11}$$

The evaluation of J^{i-} is then straightforward: we substitute for x^-, x^i and P^-, P^i in terms of oscillators to find the result

$$J^{i-} = \frac{1}{\sqrt{2\alpha'}}(x_0^i \alpha_0^- - x_0^- \alpha_0^i) + \sum_{n=1}^{\infty}\frac{(\alpha_{-n}^i \alpha_n^- - \alpha_{-n}^- \alpha_n^i)}{in}. \tag{4.3.12}$$

One can verify that the Poisson brackets $\{J^{\mu\nu}, J^{\rho\kappa}\}$ do indeed give the required result corresponding to the Lorentz algebra. Although a complicated calculation, this result in the classical theory is guaranteed, because we have algebraically solved a set of Lorentz covariant constraints and chosen a gauge, but none of these steps can violate Lorentz invariance. It is instructive to compute the variation of $x^i(\tau, \sigma)$ under a Lorentz transformation. In the classical theory this is given by

$$\delta x^i(\tau, \sigma) = \left\{x^i(\tau, \sigma), \frac{\omega_{\mu\nu}}{2}J^{\mu\nu}\right\}. \tag{4.3.13}$$

It is obvious that under ω_{ij} the fields $x^i(\tau, \sigma)$ transform as

$$\delta x^i(\tau, \sigma) = -\omega^i{}_j x^j(\tau, \sigma). \tag{4.3.14}$$

This is as expected since the $SO(D-2)$ sub-group generated by J^{ij} is manifest in the light-cone formalism.

To evaluate the variation under w_{k-} we require the Poisson brackets

$$\{x^i(\tau, \sigma), P^-(\tau, \sigma')\} = -\frac{\pi}{p^+}\delta(\sigma - \sigma')P^i(\tau, \sigma') \tag{4.3.15}$$

and

$$\{x^i(\tau, \sigma), x^-(\tau, \sigma')\} = \frac{1}{p^+}x^{i\prime}(\tau, \sigma)[\pi\theta(\sigma - \sigma') - \sigma], \tag{4.3.16}$$

where we have first substituted for P^- and x^- and then evaluated the Poisson bracket, and $\theta(\sigma) = 1$ if $\sigma > 0$ and is zero otherwise. As a result,

$$\delta x^i(\tau, \sigma) = \{x^i(\tau, \sigma), \omega_{k-}J^{k-}\}$$

$$= -\omega_{k-}\left[\delta^{ik}x^-(\tau, \sigma) + \frac{\pi}{p^+}x^k(\tau, \sigma)P^i(\tau, \sigma)\right.$$

$$\left. + \frac{x^{i\prime}(\tau, \sigma)}{p^+}\int_0^{\pi}d\sigma' P^k(\tau, \sigma')[\pi\theta(\sigma - \sigma') - \sigma]\right]. \tag{4.3.17}$$

Only the first term is what we might naively expect. However, we have to recall that a ω_{k-} rotation will also change x^+ by $\delta x^+ = -w^+{}_k x^k = +w_{k-}x^k$, which we have gauge fixed to be $x^+ = x_0^+(0) + c\tau$. In what follows we will choose our x^+ coordinate such that $x_0^+(0) = 0$ for simplicity. To restore the gauge choice in the new Lorentz frame we must make a compensating conformal transformation. The need to compensate is often the case when one fixes a local symmetry by using fields which carry a representation of a rigid symmetry. Another example is the Wess–Zumino gauge in super-Yang–Mills theories, where the local symmetry is gauge invariance and the rigid symmetry is supersymmetry. Recalling that $P^i(\tau, \sigma) = (1/2\pi\alpha')\dot{x}^i(\tau, \sigma)$ we may write

$$\delta x^i(\tau, \sigma) = -\omega_{i-}x^-(\tau, \sigma) + \eta^0\dot{x}^i(\tau, \sigma) + \eta^1 x^{i\prime}(\tau, \sigma), \tag{4.3.18}$$

where

$$\eta^0 = -\frac{1}{2\alpha' p^+}\omega_{k-}(x^k - 2\alpha' p^k\tau),$$

$$\eta^1 = -\frac{1}{p^+}\omega_{k-}\int_0^\pi d\sigma' P^k(\sigma')[\pi\theta(\sigma - \sigma') - \sigma]. \tag{4.3.19}$$

Thus we have to make the compensating transformation

$$\tau' = \tau + \eta^0, \quad \sigma' = \sigma + \eta^1.$$

We can check that these are indeed conformal transformations; indeed, one readily verifies that

$$\frac{\partial\eta^0}{\partial\tau} = -\frac{1}{2\alpha' p^+}\omega_{k-}(\dot{x}^k - 2\alpha' p^k) = \frac{\partial\eta^1}{\partial\sigma} \tag{4.3.20}$$

and

$$\frac{\partial\eta^1}{\partial\tau} = -\frac{\omega_{k-}x^{k\prime}}{2\alpha' p^+} = \frac{\partial\eta^0}{\partial\sigma}, \tag{4.3.21}$$

where in the last step we used $P^k = \dot{x}^k/2\pi\alpha'$ and that $\ddot{x}^k = x^{k\prime\prime}$. In terms of the coordinates ξ^+ and ξ^- of equation (4.1.12) the above result is equivalent to

$$\frac{\partial}{\partial\xi^-}(\eta^0 + \eta^1) = 0, \quad \frac{\partial}{\partial\xi^+}(\eta^0 - \eta^1) = 0, \tag{4.3.22}$$

which we recognise as the conformal transformations of equation (4.1.15).

The compensating transformations can also be computed directly from the variation of the gauge choice $x^+ = 2\alpha' p^+\tau$. Under Lorentz and conformal transformations

$$\delta x^\mu = -\omega^\mu{}_\nu x^\nu + \eta^\alpha\partial_\alpha x^\mu \tag{4.3.23}$$

and in particular

$$\delta x^+ = -\omega^+{}_\nu x^\nu + \eta^0 2\alpha' p^+. \tag{4.3.24}$$

On the other hand,

$$\delta(2\alpha' p^+\tau) = -2\alpha' p^\nu\tau\omega^+{}_\nu. \tag{4.3.25}$$

Equating these two results we find that

$$\eta^0 = \frac{\omega^+{}_\nu}{p^+}(x^\nu - 2\alpha' p^\nu\tau) = \frac{\omega^+{}_k}{p^+}(x^k - 2\alpha' p^k\tau). \tag{4.3.26}$$

The restriction to a sum over spatial indices in the last step follows as $w^+{}_- = w_{--} = 0$ and $w^+{}_+$ multiplies $x^+ - 2\alpha' p^+\tau = 0$. We can obtain η^1 from η^0 by demanding that it be a conformal transformation, that is, by integrating the equations

$$\frac{\partial\eta^1}{\partial\tau} = \frac{\partial\eta^0}{\partial\sigma} \quad \text{and} \quad \frac{\partial\eta^0}{\partial\tau} = \frac{\partial\eta^1}{\partial\sigma}. \tag{4.3.27}$$

Thus, we recover the result of equation (4.3.19). Only the Lorentz transformation $\omega^+{}_k$ requires a compensating conformal transformation. This is to be expected, since J^{i-} is the only non-linear generator.

In quantum theory we must normal order all our operators including the Lorentz generators. Since these are not manifestly Lorentz covariant, it is when performing this step that we may lose Lorentz invariance. The generators must be normal ordered in such a way as to preserve their Hermitian character. The standard way to achieve this is to symmetrise in x^μ and p^ν, that is, $x^\mu p^\nu \rightarrow \frac{1}{2}(x^\mu p^\nu + p^\nu x^\mu)$, and then normal order the resulting oscillator expression. Carrying out this step we find that the simplest Lorentz generators are given by

$$J^{ij} = \frac{1}{\sqrt{2\alpha'}}(x_0^i \alpha_0^j - x_0^j \alpha_0^i) + \sum_{n=1}^{\infty} \frac{(\alpha_{-n}^i \alpha_n^j - \alpha_{-n}^j \alpha_n^i)}{in}, \tag{4.3.28}$$

$$J^{i+} = p^+ x_0^i(0) = -J^{+i}, \tag{4.3.29}$$

$$J^{+-} = -\frac{1}{2}(p^+ q^-(0) + q^-(0)p^+) = -J^{+-}. \tag{4.3.30}$$

For J^{i-} we must also take into account any normal ordering constants which could occur in α_n^-. In fact, only α_0^- has a normal ordering ambiguity and, as a result, we only introduce a constant α_0 in its expression. We can then write the quantum version of equation (4.3.11) as

$$\alpha_n^- = -\frac{1}{\sqrt{2\alpha'}p^+}l_n, \; n \neq 1, \; \alpha_0^- = -\frac{1}{\sqrt{2\alpha'}p^+}(l_0 - \alpha_0), \tag{4.3.31}$$

where

$$l_n = \frac{1}{2}\sum_{i=1}^{D-2} : \sum_{m=-\infty}^{\infty} \alpha_{n-m}^i \alpha_m^i : . \tag{4.3.32}$$

We note that

$$H = l_0 - \alpha_0 \text{ and } x_0^- = q^-. \tag{4.3.33}$$

We see from the above equation for H that the constant α_0 is the same as the one we introduced in equation (4.2.5). Symmetrising and normal ordering, J^{i-} becomes

$$\begin{aligned} J^{i-} = \; &: \left\{ -\frac{1}{\sqrt{2\alpha'}}x_0^- \alpha_0^i - \frac{1}{4\alpha'p^+}[x_0^i(l_0 - \alpha_0) + (l_0 - \alpha_0)x_0^i] \right. \\ &\left. - \sum_{n=1}^{\infty} \frac{(\alpha_{-n}^i l_n + l_n \alpha_n^i - l_{-n}\alpha_n^i - \alpha_n^i l_{-n})}{2in\sqrt{2\alpha'}p^+} \right\} : \end{aligned}$$

$$\begin{aligned} = \; &: \left\{ -\frac{1}{\sqrt{2\alpha'}}x_0^- \alpha_0^i - \frac{1}{4\alpha'p^+}[x_0^i(l_0 - \alpha_0) + (l_0 - \alpha_0)x_0^i] \right. \\ &\left. - \sum_{n=1}^{\infty} \frac{(\alpha_{-n}^i l_n - l_{-n}\alpha_n^i)}{in\sqrt{2\alpha'}p^+} \right\} : \end{aligned} \tag{4.3.34}$$

We may simultaneously replace x_0^- and x_0^i by $x_0^-(0)$ and $x_0^i(0)$, respectively, as their τ dependent pieces cancel in J^{i-}. The objects l_n are very similar to the Virasoro generators L_n of covariant theory except that the former only contains the $D-2$ oscillators α_n^i. It follows, by the same arguments as in chapter 3, that the l_n obey the relations

$$[l_n, l_m] = (n - m)l_{n+m} + \frac{(D-2)}{12} n(n^2 - 1)\delta_{n+m,0},$$

$$[l_n, \alpha_m^j] = -m\alpha_{n+m}^j, \quad [x_0^-, p^+] = i, \quad [x_0^i, l_n] = i\sqrt{2\alpha'}\alpha_n^i.$$

(4.3.35)

All the commutators of the Lorentz generators are straightforward and yield the expected result except for the $[J^{i-}, J^{j-}]$ commutator. We will now spend some time evaluating this commutator as it has been a little neglected in some reviews.

The first step is to split J^{i-} into pieces

$$J^{i-} = S_1^i + S_2^i + S_3^i + S_4^i,$$

(4.3.36)

where

$$S_1^i = -\frac{1}{\sqrt{2\alpha'}}x_0^- \alpha_0^i, \quad S_2^i = -\frac{1}{4\alpha' p^+}[x_0^i(l_0 - \alpha_0) + (l_0 - \alpha_0)x_0^i],$$

$$S_3^i = -\sum_{n=1}^{\infty} \frac{(\alpha_{-n}^i \tilde{l}_n - \tilde{l}_{-n}\alpha_n^i)}{in\sqrt{2\alpha'}p^+}, \quad S_4^i = -\sum_{n=1}^{\infty} \frac{(\alpha_{-n}^i \alpha_n^k - \alpha_{-n}^k \alpha_n^i)\alpha_0^k}{in\sqrt{2\alpha'}p^+},$$

(4.3.37)

with

$$\tilde{l}_n = l_n - \sum_{i=1}^{D-2} \alpha_0^i \alpha_n^i, \quad \tilde{l}_0 = l_0 - \tfrac{1}{2}\sum_{i=1}^{D-2}(\alpha_0^i)^2.$$

(4.3.38)

In $[J^{i-}, J^{j-}]$ we find all the commutators of $[S_p^i, S_q^j]$, which we now evaluate one by one. Using the relation

$$\left[x_0^-, \frac{1}{p^+}\right] = -\frac{1}{p^{+2}},$$

(4.3.39)

the non-vanishing commutators involving S_1^i are

$$[S_1^i, S_2^j] - (i \leftrightarrow j) = -\frac{i}{4\alpha' p^{+2}\sqrt{2\alpha'}}[(x_0^j \alpha_0^i - x_0^i \alpha_0^j)(l_0 - \alpha_0)$$

$$+ (l_0 - \alpha_0)(x_0^j \alpha_0^i - x_0^i \alpha_0^j)],$$

(4.3.40)

$$[S_1^i, S_3^j] - (i \leftrightarrow j) = -\frac{1}{2\alpha' p^{+2}}\left[\sum_{n=1}^{\infty}(\alpha_{-n}^j \tilde{l}_n - \tilde{l}_{-n}\alpha_n^j)\alpha_0^i - (i \leftrightarrow j)\right],$$

(4.3.41)

$$[S_1^i, S_4^j] - (i \leftrightarrow j) = -\frac{1}{2\alpha' p^{+2}}\left[\sum_{n=1}^{\infty}\frac{(\alpha_{-n}^j \alpha_n^k - \alpha_{-n}^k \alpha_n^j)\alpha_0^i \alpha_0^k}{n} - (i \leftrightarrow j)\right].$$

(4.3.42)

It is straightforward to evaluate $[S_2^i, S_2^j]$, which we find cancels against $[S_1^i, S_2^j] - (i \leftrightarrow j)$. The commutator $[S_2^i, S_3^j] - (i \leftrightarrow j)$ vanishes, but

$$[S_2^i, S_4^j] = \frac{1}{\alpha' p^{+2}}\sum_{n=1}^{\infty}\frac{(\alpha_{-n}^i \alpha_n^j - \alpha_{-n}^j \alpha_n^i)(l_0 - \alpha_0)}{n}.$$

(4.3.43)

When evaluating commutators containing S_4^i it is useful to use the fact that this object is closely related to the non-zero mode part of J^{ij}. One finds that

$$[S_3^i, S_4^j] - (i \leftrightarrow j) = -\{[S_1^i, S_3^j] - (i \leftrightarrow j)\}.$$

(4.3.44)

Also one readily discovers that

$$[S_4^i, S_4^j] = -\{[S_1^i, S_4^j] - (i \leftrightarrow j)\} + \frac{1}{2\alpha'p^{+2}} \sum_{n=1}^{\infty} \frac{(\alpha_{-n}^i \alpha_n^j - \alpha_{-n}^j \alpha_n^i)(\alpha_0^k \alpha_{k0})}{n}.$$

(4.3.45)

When evaluating the only remaining commutator, $[S_3^i, S_3^j]$, care must be taken with the normal ordering. One finds that it is given by

$$[S_3^i, S_3^j] = -\frac{1}{\alpha'p^{+2}} \sum_{n=1}^{\infty} \frac{(\alpha_{-n}^i \alpha_n^j - \alpha_{-n}^j \alpha_n^i)(\tilde{l}_0 - \alpha_0)}{n}$$

$$\times \left[\left(\tilde{l}_0 - \frac{(D-2)}{24} \right) + n^2 \left(\frac{(D-2)}{24} - 1 \right) \right].$$

(4.3.46)

In fact, the first three terms of this expression arise from the use of the relation

$$[\tilde{l}_n, \tilde{l}_{-n}] = 2n\tilde{l}_0 + \frac{(D-2)}{12} n(n^2 - 1)$$

(4.3.47)

in the relevant terms. Collecting together all the above results we find that

$$[J^{i-}, J^{j-}] = -\frac{1}{\alpha'p^{+2}} \sum_{n=1}^{\infty} \frac{(\alpha_{-n}^i \alpha_n^j - \alpha_{-n}^j \alpha_n^i)}{n}$$

$$\times \left[\left(\alpha_0 - \frac{(D-2)}{24} \right) + n^2 \left(\frac{(D-2)}{24} - 1 \right) \right],$$

(4.3.48)

where we recall that α_0 is the normal ordering constant for l_0. Lorentz invariance implies that this commutator must vanish, which it does provided

$$D = 26 \text{ and } \alpha_0 = +1.$$

(4.3.49)

A result of this type is to be expected, since in the covariant approach we found $D \le 26$ and $\alpha_0 = 1$, but for that formalism the result occurred from demanding the physical states have positive norm, while in the light-cone formalism the problem shows up in the only place it can, namely, in the failure of Lorentz invariance.

One could argue for the result of equation (4.3.48) in a heuristic manner as follows. The failure of Lorentz invariance can only arise from our normal ordering procedures and some thought shows that this can only result in terms bilinear or less in α_n^i. However, the Jacobi identity $[J^{ki}, [J^{i-}, J^{j-}]] + \text{cyclic} = 0$ implies that $[J^{i-}, J^{j-}]$ must transform under $SO(D-2)$ as its indices suggest. This, taken with the fact that it commutes with the number operator $\tilde{l}_0$, means it can only be of the form

$$\sum_{n=1}^{\infty} (\alpha_{-n}^i \alpha_n^j - \alpha_{-n}^j \alpha_n^i) F,$$

(4.3.50)

where F is a function of n, α' and p^+ only. To find F, one can study

$$\langle 0|[J^{i-}, J^{j-}]|0\rangle \tag{4.3.51}$$

in a way similar to that which we used to locate the central term for $[L_{-n}, L_n]$ in chapter 2.

4.4   Light-cone string field theory

We now construct the free second quantised string field theory [4.1]] in the light-cone formalism. The lack of constraints makes this a particularly simple task. As explained in section 4.2 the coordinates are $x^i(\sigma)$, q^- and they, together with their momenta, satisfy the commutation relations of equations (4.2.1). We therefore consider the functional Ψ of equation (4.2.3) which is a functional of $x^i(\sigma)$, q^- and the time τ. We may rewrite equation (4.2.6) in an illuminating way. We take the Fourier transform of Ψ with respect to q^- to define

$$\chi(p^+, x^i(\sigma), \tau) = \int dq^- e^{iq^- p^+} \Psi(q^+, x^i(\sigma), \tau), \tag{4.4.1}$$

where $\chi(p^+, x^i(\sigma), \tau)^* = \chi(-p^+, x^i(\sigma), \tau)$ due to the reality of Ψ. Rewriting equation (4.2.6) in terms of x^+ and q^- and using equation (4.1.24) it becomes

$$\left\{ \frac{1}{p^+} \frac{\partial}{\partial\tau} \frac{\partial}{\partial q^-} - H \right\} \Psi = 0. \tag{4.4.2}$$

Substituting the Fourier transform gives the equation

$$i\frac{\partial}{\partial\tau}\chi(p^+, x^i(\sigma), \tau) = H\chi(p^+, x^i(\sigma), \tau), \tag{4.4.3}$$

which we recognise as the Schrödinger equation.

This equation for the free theory is derivable from the action

$$A = \frac{1}{2} \int_0^\infty dp^+ \int Dx^i(\sigma) \int d\tau \chi(p^+, x^i(\sigma), \tau)^* \left(i\frac{\partial}{\partial\tau} - H \right) \chi(p^+, x^i(\sigma), \tau). \tag{4.4.4}$$

The interacting second quantised field theory [4.1] requires the addition of interaction terms which are cubic and quartic in χ for the open string, but remarkably only cubic for the closed string. This is the subject of section 18.4.

5   Clifford algebras and spinors

So I asked her the nearest way to Learning's home.... Ask for the direct road, she said,
from here to Suffer-both-weal-and-woe-if you are willing to learn that lesson. Then ride
past Riches, and dont stop there, for if you become attached to them you will never reach
Learning. And avoid the lecherous meadow that is called Lust; leave it a good mile or
more to the left, and continue till you come to a mansion called Keep-your-toungue-from-
lying-and-slander-and-your-mount-from-spicy-drinks. There you will meet Sobriety and
Simplicity of-Speech, and while they are with you every man will be glad to show you his
wisdom. So you will come to Learning, who knows most of the answers.
                    Taken from the 14th century text Piers the Ploughman by William Langland

In this chapter we define a Clifford algebra in a flat space-time of arbitrary dimension and
find its irreducible representations and their properties. This enables us to find which types
of spinors are allowed in a flat space-time of a given dimension and signature. We will
find that in a space-time of even dimension D, a generic, that is, Dirac, spinor has $2^{D/2}$
complex components; however, for certain dimensions and signatures this representation is
not irreducible. In these cases we can place restrictions on the Dirac spinor and find spinors
which are called Majorana, Weyl or Majorana–Weyl spinors.

The starting point for the construction of a supersymmetric theory is the supersymmetry
algebra which underlies it. Supersymmetry algebras contain supercharges which transform
as spinors under the appropriate Lorentz group. Hence, even to construct the supersymmetry
algebras, we must first find out what types of spinors are possible in a given dimension and
what are their properties. We will find in section 5.5 and chapter 10 that supersymmetric
algebras and the supersymmetric theories on which they are based rely for their existence
in an essential way on the detailed properties of Clifford algebras, which we will derive in
this chapter.

As far as I am aware the first physics oriented discussion of spinors in arbitrary dimensions
was given in [5.1] and many of the steps in this section are taken from this paper. Use has
also been made of the reviews in [5.2], [5.2] and [5.3].

5.1   Clifford algebras

A Clifford algebra in D dimensions is defined as a set containing D elements γ_m and an
identity element I which satisfy the relation

$$\{\gamma_m, \gamma_n\} \equiv \gamma_m\gamma_n + \gamma_n\gamma_m = 2\eta_{mn}I, \tag{5.1.1}$$

where the labels $m, n, \ldots$ take D values and η_{mn} is the flat metric in $R^{s,t}$ ($s + t = D$); that is, the metric η_{mn} is a diagonal matrix whose first t entries down the diagonal are -1 and whose last s entries are $+1$. We can raise and lower the $m, n, \ldots$ indices using the metric $\eta_{mn} = \eta^{mn}$ in the usual way. The identity element I commutes with γ_m, that is, $I\gamma_n = \gamma_n I$, and $I^2 = I$.

Under multiplication, the D elements γ_n and the identity element I of the Clifford algebra generate a finite group denoted C_D which consists of the elements

$$C_D = \{\pm I, \pm \gamma_m, \pm \gamma_{m_1 m_2}, \ldots, \pm \gamma_{m_1 \cdots m_D}\}, \qquad (5.1.2)$$

where $\gamma_{m_1 \cdots m_p} = \gamma_{[m_1} \cdots \gamma_{m_p]}$. Hence, $\gamma_{m_1 m_2 \cdots}$ is non-vanishing only if all indices $m_1, m_2, \ldots$ are different, in which case it equals

$$\gamma_{m_1 m_2 \cdots} = \gamma_{m_1} \gamma_{m_2} \cdots. \qquad (5.1.3)$$

This is sufficient to define what is meant by the anti-symmetrisation of the indices on $\gamma_{[m_1} \cdots \gamma_{m_p]}$. One can write an explicit expression for $\gamma_{[m_1} \cdots \gamma_{m_p]}$ for all possible values of $m_1, m_2, \ldots, m_p$, for example, $\gamma_{[m_1 m_2]} = \frac{1}{2}(\gamma_{m_1} \gamma_{m_2} - \gamma_{m_2} \gamma_{m_1})$. While the general expression is just $1/p!$ times a sum over all possible permutations of the indices with a plus sign if it is an even permutations and a minus sign if it is odd permutation. In fact, this most useful definition is that of equation (5.1.3). The set of elements $\gamma_{m_1 m_2 \cdots m_p}$ for all possible different values of the ms contains

$$\frac{D!}{(D-p)!p!} = \binom{D}{p}$$

different elements. As a result, the group C_D generated by the γ_m has

$$2 \sum_{p=0}^{D} \binom{D}{p} = 2(1+1)^D = 2^{D+1} \qquad (5.1.4)$$

elements. The number of elements in a group is called its order.

5.2   Clifford algebras in even dimensions

To find the representations of C_D is a standard exercise in the representation theory of finite groups; see, for example, [5.4]. We will use without proof a number of the standard theorems found in [5.4]. We recall that an irreducible representation of a group G is one such that no subspace of it is also a representation of G. We will first consider the case of even D.

One well-known theorem is that the number of irreducible representations of any finite dimensional group G equals the number of its conjugacy classes. We recall that the conjugacy class $[a]$ of $a \in G$ is given by

$$[a] = \{gag^{-1} \quad \forall \, g \in G\}. \qquad (5.2.1)$$

For even D it is straightforward to show, using equation (5.1.1), that the conjugacy classes of C_D are given by

$$[+1], [-1], [\gamma_m], [\gamma_{m_1 m_2}], \ldots, [\gamma_{m_1 \cdots m_D}]. \qquad (5.2.2)$$

Hence for even D there are $2^D + 1$ inequivalent irreducible representations of C_D.

Next we use the theorem that the number of inequivalent one-dimensional representations of any finite group G is equal to the order of G divided by the order of the commutator group of G. We denote the commutator group of G by $\text{Com}(G)$. It is defined to be the group $\text{Com}(G) = aba^{-1}b^{-1} \; \forall \, a, b \in G$. For even D the commutant of C_D is just the elements ± 1 and so has order 2. As a result, the number of inequivalent irreducible one-dimensional representations of C_D is 2^D. Since the total number of irreducible representations is $2^D + 1$, we conclude that there is only one irreducible representation of C_D whose dimension is greater than 1.

Finally, we make use of the classic theorem which states that if we denote the order of any finite group by ord G and it has p irreducible inequivalent representations of dimension n_p, then

$$\text{ord}G = \sum_p (n_p)^2. \tag{5.2.3}$$

Applying this theorem to C_D we find that

$$2^{D+1} = 1^2 2^D + n^2, \tag{5.2.4}$$

where n is the dimension of the only irreducible representation whose dimension is greater than 1. We therefore conclude that $n = 2^{D/2}$. These results are summarised in the following theorem.

Theorem   For D even the group C_D has $2^D + 1$ inequivalent irreducible representations. Of these irreducible representations 2^D are one-dimensional and the remaining representation has dimension $2^{D/2}$.

This means that we can represent the γ_m as $2^{D/2} \times 2^{D/2}$ matrices for the irreducible representation with dimension greater than 1 and where I is the $2^{D/2} \times 2^{D/2}$ unit matrix. Our next task is to find the properties of this representation under complex conjugation and transpose.

That the above are irreducible representations of the group C_D means that they provide a representation of the group which consists of the elements given in equation (5.1.2) together with a group composition law which is derived from the Clifford algebra relations using only the operation of multiplication. In particular, the group operations do not include the operations of addition and subtraction, which, however, do occur in the Clifford algebra defining the condition of equation (5.1.1). Hence, the irreducible representations of C_D are not necessarily irreducible representations of the Clifford algebra itself. In fact, all the one-dimensional irreducible representations of C_D do not extend to be also representations of the Clifford algebra as they do not obey the rules for addition and subtraction. As such, the only representation of C_D and the Clifford algebra is the unique irreducible representation of dimension greater than 1 described above. It follows that the Clifford algebra itself has only one irreducible representation and this has dimension $2^{D/2}$. It is, of course, the well-known representation that acts on spinors. In fact, the one-dimensional representations are not faithful representations of C_D and we shall not consider them in what follows.

Given an irreducible representation of the Clifford algebra, also denoted γ_m, with dimension greater than 1 we can take its complex conjugate. Denoting the complex conjugate of the representation by γ_m^*, it is obvious that γ_m^* also satisfies equation (5.1.1) and so forms a representation of the same Clifford algebra. It follows that it also forms a representation of

C_D. However, there is only one irreducible representation of C_D of dimension greater than 1 and as a result, the complex conjugate representation and the original representation must be equivalent. Consequently, there must exist a matrix B such that

$$\gamma_m^* = B\gamma_m B^{-1}. \tag{5.2.5}$$

We can choose the scale of B such that $|\det B| = 1$ as it does not enter in the above equation. Taking the complex conjugate of equation (5.2.5) we find that

$$\gamma_m = (\gamma_m^*)^* = +B^* B\gamma_m B^{-1} B^{-1*}. \tag{5.2.6}$$

Hence, we conclude that B^*B commutes with the irreducible representation and by Schur's lemma it must be a constant times the identity matrix, that is,

$$B^*B = \epsilon I. \tag{5.2.7}$$

Taking the complex conjugate of the above relation we find that $BB^* = \epsilon^* I$ and so $BB^*BB^{-1} = \epsilon^* I = \epsilon I$, thus $\epsilon = \epsilon^*$, that is, ϵ is real. Since we have chosen $|\det B| = 1$ we conclude that $|\epsilon| = 1$ and so $\epsilon = \pm 1$.

We can also consider the transpose of the irreducible representation γ_m. Denoting the transpose of γ_m by γ_m^T, we find, using a very similar argument, that γ_m and γ_m^T are equivalent representations and so there exists a matrix C, called the charge conjugation matrix, such that

$$\gamma_m^T = -C\gamma_m C^{-1}. \tag{5.2.8}$$

The reader may find the minus sign in the above equation puzzling. Clearly, as γ_n satisfies the Clifford algebra so does $-\gamma_n$, hence one can take either sign. However, different signs lead to different possibilities for C with different consequences for spinors. It turns out that the most interesting possibilities are generically for the above minus sign. In section 5.6 we will carry out the same analysis for an arbitrary signature and we will allow both possible signs.

We denote the Hermitian conjugate of γ_m by $\gamma_m^\dagger = \gamma_m^{*T}$. We can relate C to B if we know the Hermiticity properties of the γ_m. From now until section 5.6, where we discuss the analogous results for a space-time of general signature, we will assume we are dealing with Minkowski space-time whose metric η_{mn} is given by $\eta = \text{diag}(-1, +1, +1, \ldots, +1)$. Any finite-dimensional irreducible representation of a finite group G can be chosen to be unitary. Making this choice for our group C_D, we have $\gamma_m \gamma_m^\dagger = I$. Taking into account the relationship $\gamma_n \gamma_n = \eta_{nn}$ we conclude that

$$\gamma_0^\dagger = -\gamma_0, \quad \gamma_m^\dagger = \gamma_m; \quad m = 1, \ldots, D-1. \tag{5.2.9}$$

We could, as some texts do, regard this equation as part of the definition of the Clifford algebra. We may rewrite equation (5.2.9) as

$$\gamma_m^\dagger = \gamma_0 \gamma_m \gamma_0, \quad m = 0, 1, \ldots, D-1. \tag{5.2.10}$$

We may take C to be given by $C = -B^T \gamma_0$ as then

$$C\gamma_m C^{-1} = -B^T \gamma_0 \gamma_m (\gamma_0 ((B)^T)^{-1}) = -B^T \gamma_m^\dagger (B)^{T-1} = -(B^{-1} \gamma_m^* B)^T = -\gamma_m^T \tag{5.2.11}$$

as required.

Further restrictions on B can be found by computing γ_m^T in two ways: we see that $\gamma_m^T = (\gamma_m^*)^\dagger = (\gamma_m^\dagger)^*$ implies

$$(B^{-1})^\dagger \gamma_0 \gamma_m \gamma_0 B^\dagger = B\gamma_0 \gamma_m \gamma_0 B^{-1}. \tag{5.2.12}$$

Using Schur's lemma we deduce that $-\gamma_0 B^\dagger B\gamma_0$ is proportional to the unit matrix and as a result so is $B^\dagger B$, that is, $B^\dagger B = \mu I$. Since $|\det B| = 1$ we find that $|\mu| = 1$, but taking the matrix element of $B^\dagger B = \mu I$ with any vector we conclude that μ is real and positive. Hence $\mu = 1$ and consequently B is unitary, that is, $B^\dagger B = I$. This result and the previously derived equation $BB^* = \epsilon I$ imply that

$$B^T = \epsilon B, \quad C^T = -\epsilon C. \tag{5.2.13}$$

We now wish to determine ϵ in terms of the space-time dimension D. Consider the set of matrices

$$I, \gamma_m, \gamma_{m_1 m_2}, \gamma_{m_1 m_2 m_3}, \cdots, \gamma_{m_1 \cdots m_D}. \tag{5.2.14}$$

There are

$$2^D = \sum_p \binom{D}{p} = (1+1)^D$$

such matrices and as they are linearly independent they form a basis for the space of all $2^{D/2} \times 2^{D/2}$ matrices. Using equation (5.1.1) we can relate $\gamma_{m_1 \cdots m_p}$ to $\gamma_{m_p \cdots m_1}$, to find that the sign change required to reverse the order of the indices is given by

$$\gamma_{m_1 \cdots m_p} = (-1)^{p(p-1)/2} \gamma_{m_p \cdots m_1}. \tag{5.2.15}$$

This equation, together with equation (5.2.8), implies that

$$C\gamma_{m_1 \cdots m_p} C^{-1} = (-1)^p (-1)^{p(p-1)/2} \gamma_{m_1 \cdots m_p}{}^T \tag{5.2.16}$$

or equivalently

$$C\gamma_{m_1 \cdots m_p} = \epsilon(-1)^{(p-1)(p-2)/2} (C\gamma_{m_1 \cdots m_p})^T$$

or

$$\gamma_{m_1 \cdots m_p} C^{-1} = \epsilon(-1)^{(p-1)(p-2)/2} (\gamma_{m_1 \cdots m_p} C^{-1})^T. \tag{5.2.17}$$

To derive this result we proceed as follows:

$$C\gamma_{m_1 \cdots m_p} C^{-1} = (-1)^p \gamma_{m_1}^T \cdots \gamma_{m_p}^T = (-1)^p (\gamma_{m_p \cdots m_1})^T$$

$$= (-1)^p (-1)^{p(p-1)/2} (\gamma_{m_1 \cdots m_p})^T$$

and hence the result. We have used that $(-1)^{2p} = 1$ and so $(-1)^p = (-1)^{-p}$.

Using this result, we can calculate the number of anti-symmetric matrices in the complete set of equation (5.2.14) when multiplied from the left by C; it is given by

$$\sum_{p=0}^{D} \frac{1}{2} \left(1 - \epsilon(-1)^{(p-1)(p-2)/2} \right) \binom{D}{p} \tag{5.2.18}$$

as, using equation (1.2.17), the last expression in the bracket is -1 when the matrix is anti-symmetric.

Using the relationship

$$(-1)^{(p-1)(p-2)/2} = -\tfrac{1}{2}[(1+i)i^p + (1-i)(-i)^p],$$

(5.2.19)

we can carry out the summation in equation (5.2.18). Equation (5.2.19) can be proved by observing that under $p \to p + 2$ both the left- and right-hand sides of equation (5.2.19) change by a factor of -1 and so it suffices to test the equation for $p = 0, 1$. We know, however, that the number of anti-symmetric $2^{D/2} \times 2^{D/2}$ matrices is $\tfrac{1}{2}2^{D/2}(2^{D/2} - 1)$. Equating these two methods of evaluating the number of anti-symmetric matrices we find that

$$\epsilon = -\sqrt{2}\cos\left\{\frac{\pi}{4}(D+1)\right\}.$$

(5.2.20)

Put another way $\epsilon = +1$ for $D = 2, 4$ mod 8 and $\epsilon = -1$ for $D = 6, 8$ mod 8. In deriving this result we have used the identity

$$\cos\left\{\frac{\pi}{4}(D+1)\right\} = \frac{1}{2}\left(e^{i\pi(D+1)/4} + e^{-i\pi(D+1)/4}\right) = \frac{1}{2}\left(\left(e^{i\pi/4}\right)^{D+1} + \left(e^{-i\pi/4}\right)^{D+1}\right)$$

$$= \frac{2^{-D/2}}{2\sqrt{2}}\left((1+i)^{D+1} + (1-i)^{D+1}\right).$$

It follows from equation (5.2.13) that for $D = 2, 4$ mod 8, B is a symmetric unitary matrix. Writing B in terms of its real and imaginary parts $B = B_1 + iB_2$, where B_1 and B_2 are symmetric and real, the unitarity condition becomes $B_1^2 + B_2^2 = 1$ and $[B_1, B_2] = 0$. Under a change of basis of the γ_m matrices, $\gamma^{m'} = A\gamma^n A^{-1}$, we find that the matrix B changes as $B' = A^*BA^{-1}$. In fact, we can use A to diagonalise B_1 and B_2 which, still being unitary, must be of the form $B = \text{diag}(e^{i\alpha_1}, \ldots, e^{i\alpha_D})$. Carrying out another A transformation of the form $A = \text{diag}(e^{i\alpha_1/2}, \ldots, e^{i\alpha_D/2})$ we find the new B equals the identity matrix. Hence if $D = 2, 4$ mod 8 the γ_m matrices can be chosen to be real and $C = \gamma^0$.

We close this section by giving a short exercise in how to manipulate γ matrices. Let us consider the product $\gamma^p \gamma^{mn}$; this can involve either three different γ matrices or two the same and another one which may be the same or not. In the first case we get the product of the three γ matrices and in the second case we just get the remaining γ matrix after using that the square of a γ matrix is ± 1 times the identity matrix. Hence we may write

$$\gamma_p \gamma_{mn} = b\gamma_{pmn} + a(\eta_{pm}\gamma_n - \eta_{pn}\gamma_m),$$

(5.2.21)

where a and b are constants. We can determine their value by taking specific cases. Let us consider $\gamma_1\gamma_{23}$, which obviously equals γ_{123} and so $b = 1$. We then try $\gamma_1\gamma_{12}$, which obviously equals $\gamma_1\gamma_1\gamma_2 = \gamma_2$ and so $a = 1$. Using the same line of argument it is easy to see that

$$\gamma_p \gamma_{m_1 \cdots m_q} = \gamma_{pm_1 \cdots m_q} + q\eta_{p[m_1}\gamma_{m_2 \cdots m_q]}.$$

(5.2.22)

Taking the γ matrices in equation (5.1.21) in the other order and taking the commutator we find that

$$\left[\tfrac{1}{2}\gamma_{mn}, \gamma_p\right] = \eta_{pn}\gamma_m - \eta_{pm}\gamma_n.$$

(5.2.23)

The reader may wish to also verify, by taking cases, the more sophisticated relation

$$\gamma^{a_1 \cdots a_p} \gamma_{b_1 \cdots b_q} = \sum_r (-1)^{r(r+1)/2} (-1)^{pr} \frac{p!q!}{r!(p-r)!(q-r)!} \delta^{[a_1 \cdots a_r}_{[b_1 \cdots b_r} \gamma^{a_{r+1} \cdots a_p]}_{b_{r+1} \cdots a_q]},$$

(5.2.24)

where

$$\delta^{a_1 \cdots a_r}_{b_1 \cdots b_r} = \delta^{a_1}_{[b_1} \cdots \delta^{a_r}_{b_r]}$$

(5.2.25)

and the sum is from $r = 0$ to $r = p$, or $r = q$, whichever is the least. We take our anti-symmetry to have strength 1, which means that

$$T_{[a_1 \cdots a_r]} = \frac{1}{r!} \sum_{perms, p} (-1)^p T_{p(a_1) \cdots p(a_r)},$$

(5.2.26)

where $(-1)^p$ is $+1$ for an even permutation and -1 for an odd permutation. As a result if $A_{a_1 \cdots a_r}$ is totally anti-symmetric in all its indices, then $A_{[a_1 \cdots a_r]} = A_{a_1 \cdots a_r}$. Let us give two examples of the anti-symmetry operation

$$T_{[a_1 a_2]} = \frac{1}{2} (T_{a_1 a_2} - T_{a_2 a_1})$$

(5.2.27)

and

$$T_{[a_1 a_2 a_3]} = \frac{1}{6} (T_{a_1 a_2 a_3} + T_{a_2 a_3 a_1} + T_{a_3 a_1 a_2} - T_{a_2 a_1 a_3} - T_{a_1 a_3 a_2} - T_{a_3 a_2 a_1}).$$

(5.2.28)

We note that for an object of the form $S_{a_1, a_2 \cdots a_n}$ which is totally anti-symmetric in its second set of indices, that is, $S_{a_1, [a_2 \cdots a_n]} = S_{a_1, a_2 \cdots a_n}$, but has no other symmetry, then

$$S_{[a_1, a_2 \cdots a_n]} = \frac{1}{n} (S_{a_1, a_2 \cdots a_n} \pm \text{cyclic permuations}),$$

(5.2.29)

where $\pm$ is $+$ for an even and $-$ for an odd permutation; for example,

$$S_{[a_1, a_2 a_3]} = \frac{1}{3} (S_{a_1, a_2 a_3} + S_{a_2, a_3 a_1} + S_{a_3, a_2 a_1}).$$

(5.2.30)

Equation (5.2.24) implies the commutator

$$[\gamma^{a_1 \cdots a_n}, \gamma_{b_1 \cdots b_m}] = \sum_{r=0}^{n} \frac{n!m!}{r!(n-r)!(m-r)!} (-1)^{r(r+1)/2} (-1)^{rn}$$

$$\times (1 - (-1)^{(nm+r)}) \delta^{[a_1 \cdots a_r}_{[b_1 \cdots b_r} \gamma^{a_{r+1} \cdots a_n]}_{b_{r+1} \cdots a_q]},$$

(5.2.31)

valid for $m > n$.

5.3   Spinors in even dimensions

The group Spin$(1, D - 1)$ is by definition the group generated by $\frac{1}{2}\gamma_{mn}$. In other words, it has infinitesimal elements of the form $I + \frac{1}{4} w^{mn} \gamma_{mn}$, where $w^{mn} = -w^{nm}$ are the parameters and, in a suitable region containing the identity, the group elements can be written as $A = \exp \frac{1}{4} w^{mn} \gamma_{mn}$. The group Spin$(1, D - 1)$ is assumed to be connected to its identity element, that is, any element can be reached by a continuous path from the identity element. We note that $\det A = \exp(\text{Tr} \ln A) = \exp(\frac{1}{4} w^{mn} \text{Tr} \, \gamma_{mn}) = 1$ as $\text{Tr} \, \gamma_{mn} = 0$.

It is straightforward to check that for an infinitesimal spin transformation an element of the Clifford algebra of the form $v = v^m \gamma_m$ for arbitrary v^m is preserved in form under $v \to v' = AvA^{-1}$ for any spin transformation A. In particular, using equation (5.2.23) we find that if $A = I + \frac{1}{4} w^{mn} \gamma_{mn}$, then $v'^n = v^n + w^n{}_m v^m$, which we recognise as an infinitesimal Lorentz transformation. Using the relation

$$e^B C e^{-B} = I + [B, C] + \tfrac{1}{2}[B, [B, C]] + \cdots$$

for any two matrices B and C, we conclude that $v \to v' = AvA^{-1}$ preserves the form of v for any spin transformation A and induces a finite Lorentz transformation on v^n. In fact, one may verify that an alternative definition of the group $\mathrm{Spin}(1, D - 1)$ is to take the set of all elements in the enveloping algebra of the Clifford algebra, that is, $\sum_{n_1 \cdots n_p} c_{n_1 \cdots n_p} \gamma^{n_1 \cdots n_p}$, that preserve the form of v.

We now prove the following crucial *theorem*: $\mathrm{Spin}(1, D - 1)/Z_2$ is isomorphic to the part of the Lorentz group connected to its identity.

Clearly, for any $A \in \mathrm{Spin}\,(1, D - 1)$ we can use the above map to induce a linear map on $\mathbf{R}^{(1,D-1)}$, which we may write as $v'^m = \Lambda^m{}_n v^n$. However, since

$$2^{D/2} v'^m v'_m = \mathrm{Tr}(v'v') = \mathrm{Tr}(vv) = 2^{D/2} v^m v_m$$

for arbitrary v^m, we conclude, as we did above, that $\Lambda^m{}_n$ is a Lorentz transformation. Hence, we have constructed a map from $\mathrm{Spin}(1, D - 1)$ to the Lorentz group. We recall that the Lorentz group is by definition the set of matrices Λ^m_n which obey $\Lambda \eta \Lambda^T = \eta$. We can denote the map by $A \to \Lambda_A$. If $v_1 = A_1 v A_1^{-1}$ and $v_2 = A_2 v A_2^{-1}$, then

$$v_{12} = A_2 A_1 v (A_2 A_1)^{-1} = A_2 A_1 v A_1^{-1} A_2^{-1} = A_2 v_1 A_2^{-1}$$

and so $v_{12}^m = \Lambda^m_{A_2 n} \Lambda^n_{A_1 p} v^p$. In other words, $A_1 A_2 \to \Lambda_{A_1} \Lambda_{A_2}$ and we conclude that the map is a homomorphism.

The kernel of the map is A, such that $A \to I$, where I is the identity matrix of the Lorentz group. In this case, $A\gamma_n = \gamma_n A$ and so A commutes with the entire group C_D. It follows that $A = cI$, but $\det A = 1$ and, as c is real, we find that $c = \pm 1$. Hence, the kernel of the map is $\pm I$, or the subgroup Z_2. A Lorentz transformation has $\det \Lambda_A = \pm 1$. We note that the map $A \to \det \Lambda_A$ is a map from $\mathrm{Spin}(1, D - 1)$ to the real numbers. However, this map is continuous and so for any path starting from the identity of $\mathrm{Spin}(1, D - 1)$ the value of this map cannot change discontinuously in value. As a result, the map from $\mathrm{Spin}(1, D - 1)$ to the Lorentz group must be into the part of the Lorentz group which has $\det \Lambda_A = 1$ as all elements of $\mathrm{Spin}\,(1, D - 1)$ can be reached by a connected path from the identity element. From the definition of the Lorentz group, it is also straightforward to show that $(\Lambda^0{}_0)^2 = 1 + (\sum_i \Lambda^i{}_0)^2 \geq 1$. Applying the same argument to $\mathrm{sign} \Lambda^0{}_0$, we conclude that the map is also into the part of the Lorentz group with $\Lambda^0{}_0 \geq 1$. Hence the map must be into the part of the Lorentz group that has $\det \Lambda = 1$ and $\Lambda^0{}_0 \geq 1$, which is the part connected to the identity. It can be shown that the map is *onto* this subgroup. Finally, we having established a homomorphism from $\mathrm{Spin}(1, D - 1)$ onto the part of $SO(1, D - 1)$ connected to the identity, and the kernel of the map is Z_2. From one of the classic theorems of group theory, we may conclude that $\mathrm{Spin}(1, D - 1)/Z_2$ is isomorphic to the part of $SO(1, D - 1)$ connected to the identity. Hence given any $\mathrm{Spin}(1, D - 1)$ we can use the formula $x'^n \gamma_n = Ax^n \gamma_n A^{-1}$ to induce a Lorentz transformation on the coordinates of Minkowski space.

By definition a spinor λ transforms under $\mathrm{Spin}(1, D-1)$ as

$$\delta\lambda = \tfrac{1}{4}w^{mn}\gamma_{mn}\lambda, \tag{5.3.1}$$

where $w^{mn} = -w^{nm}$ are the parameters of the infinitesimal spin transformation. Under a finite transformation $\lambda' = A\lambda$. As γ_n are $2^{D/2} \times 2^{D/2}$ matrices it follows that λ must have $2^{D/2}$ components modulo any further restrictions we may place on it. We now show that the free Dirac equation, that is, $\gamma^n\partial_n\lambda = 0$, is invariant under infinitesimal, and so also finite, $\mathrm{Spin}(1, D-1)$ transformations. As discussed above, an infinitesimal $\mathrm{Spin}(1, D-1)$ induces the Lorentz transformation $x'^n = x^n + w^n{}_m x^m$ on Minkowski space and as a result $\partial'_n = \partial/\partial x^{n'} = (\partial/\partial x^n) + w_n{}^m\partial_m$. Under such transformations

$$\gamma^n\partial_n\lambda \to \gamma^n\partial'_n\lambda' = \gamma^n\partial_n\lambda + \gamma^n\partial_n(\tfrac{1}{4}w^{pq}\gamma_{pq}\lambda) + \gamma^n w_n{}^m\partial_m\lambda$$

$$\gamma^n\partial_n\lambda = +\tfrac{1}{4}w^{pq}\gamma_{pq}\gamma^n\partial_n\lambda$$

and so we obtain the desired result.

The Dirac conjugate, denoted $\bar{\lambda}^D$, must transform such that $\bar{\lambda}^D\lambda \equiv \bar{\lambda}^{D\alpha}\lambda_\alpha$ is invariant and so transforms under a spin transformation as

$$\delta\bar{\lambda}^D = \bar{\lambda}^D(-\tfrac{1}{4}w^{mn}\gamma_{mn}). \tag{5.3.2}$$

Using the relation $\gamma^\dagger_{mn} = -\gamma_0\gamma_{nm}\gamma_0 = \gamma_0\gamma_{mn}\gamma_0$ we find that

$$\delta(\lambda^\dagger\gamma^0) = (\lambda^\dagger\gamma^0)(-\tfrac{1}{4}w^{mn}\gamma_{mn}). \tag{5.3.3}$$

Consequently, we can take the Dirac conjugate to be defined by

$$\bar{\lambda}^D \equiv \lambda^\dagger\gamma^0. \tag{5.3.4}$$

The Majorana conjugate, denoted $\bar{\lambda}^M$, is defined by

$$\bar{\lambda}^M = \lambda^T C. \tag{5.3.5}$$

Using the relationship $\gamma^T_{mn} = C\gamma_{nm}C^{-1} = -C\gamma_{mn}C^{-1}$ we find that

$$\delta\bar{\lambda}^M = \lambda^T\tfrac{1}{4}w^{mn}\gamma^T_{mn}C = -\lambda^T C(\tfrac{1}{4}w^{mn}\gamma_{mn}) = \bar{\lambda}^M(-\tfrac{1}{4}w^{mn}\gamma_{mn}). \tag{5.3.6}$$

Hence the Majorana conjugate transforms like the Dirac conjugate under $\mathrm{Spin}(1, D-1)$ transformations and as a result it is consistent, at least from the $\mathrm{Spin}(1, D-1)$ viewpoint, to define a Majorana spinor to be one whose Dirac and Majorana conjugates are equal:

$$\bar{\lambda}^D = \bar{\lambda}^M. \tag{5.3.7}$$

The above condition can be rewritten as $\lambda^* = -(\gamma^0)^T C^T\lambda$ and, using the relation $C = B^T\gamma^0$, it becomes

$$\lambda^* = B\lambda. \tag{5.3.8}$$

We could have directly verified that $B^{-1}\lambda^*$ transforms under $\mathrm{Spin}(1, D-1)$ in the same way as λ by using the equation $B\gamma_{mn}B^{-1} = (\gamma_{mn})^*$ and as a result have imposed this Majorana condition without any mention of the Dirac conjugate.

We are finally in a position to discover the dimensions in which Majorana spinors exist. If we impose the relationship $\lambda^* = B\lambda$, then taking the complex conjugate we find that it implies the relationship $\lambda = B^*\lambda^*$. Substituting this condition into the first relation we find that

$$\lambda = B^*B\lambda = \epsilon\lambda, \qquad (5.3.9)$$

since $B^*B = \epsilon I$. Consequently, Majorana spinors can only exist if $\epsilon = +1$, which is the case only in the dimensions $D = 2, 4 \mod 8$, that is, $D = 2, 4, 10, 12, \ldots$.

In an even-dimensional space-time we can construct the matrix

$$\gamma^{D+1} = \gamma_0\gamma_1\cdots\gamma_{D-1} = \gamma_{01\cdots D-1}. \qquad (5.3.10)$$

This matrix anticommutes with γ_m and so commutes with the generators $(\frac{1}{4}\gamma_{mn}w^{mn})$ of $\text{Spin}(1, D-1)$, the covering group of $\text{SO}(1, D-1)$. Hence $\gamma^{D+1}\chi$ transforms like a spinor if χ does. A straightforward calculation shows that

$$(\gamma^{D+1})^2 = (-1)^{D(D-1)/2}(-1)I = (-1)^{(D/2)-1}I. \qquad (5.3.11)$$

Hence $(\gamma^{D+1})^2 = I$ for $D = 2 \mod 4$, while $(\gamma^{D+1})^2 = -I$ for $D = 4 \mod 4$. In either case we can define Weyl spinors

$$\gamma^{D+1}\chi = \pm\chi \qquad \text{if } D = 2 \mod 4 \qquad (5.3.12)$$

and

$$i\gamma^{D+1}\chi = \pm\chi \qquad \text{if } D = 4 \mod 4. \qquad (5.3.13)$$

We say a Weyl spinor has plus or minus chirality according to the signs in the above equation. We can now consider when Majorana–Weyl spinors exist. We found that Majorana spinors (that is, $\chi^* = B\chi$) exist if $\epsilon = 1$, that is, when $D = 2, 4 \mod 8$. Taking the complex conjugate of the above Weyl conditions, and using the relationship $(\gamma^{D+1})^* = B\gamma^{D+1}B^{-1}$, we find we get a non-vanishing solution only if $D = 2 \mod 4$. Hence Majorana–Weyl spinors only exist if $D = 2 \mod 8$ that is, $D = 2, 10, 18, 26, \ldots$. The factor of i is necessary for $D = 4 \mod 4$ as the chirality condition must have an operator that squares to 1; however, it is this same factor of i that gets a minus sign under complex conjugation and so rules out the possibility of having Majorana–Weyl spinors in these dimensions. We note that these are the dimensions in which self-dual Lorentzian lattices exist and, except for 18 dimensions, these are the dimensions in which critical strings exist.

We take our spinors to have a lowered index, that is, λ_α with $\alpha = 1, 2, \ldots, 2^{D/2}$ and the Dirac, or Majorana, conjugate spinor has a raised index, that is, $\bar{\lambda}^\alpha$. As such, the gamma matrices have the indices $(\gamma_n)_\alpha{}^\beta$ and the charge conjugation matrix has $C^{\alpha\beta}$ while its inverse has $(C^{-1})_{\alpha\beta}$. Given two Majorana spinors χ_α and λ_α, we find that

$$\bar{\chi}\gamma_{m_1\cdots m_p}\lambda = \chi^T(C\gamma_{m_1\cdots m_p})\lambda = \chi_\alpha(C\gamma_{m_1\cdots m_p})^{\alpha\beta}\lambda_\beta = -\lambda_\beta(C\gamma_{m_1\cdots m_p})^{\alpha\beta}\chi_\alpha$$

$$= -(-1)^{(p-1)(p-2)/2}\bar{\lambda}\gamma_{m_1\cdots m_p}\chi$$

using equation (5.2.17), taking $\epsilon = 1$ and that the two spinors are Grassmann odd. Hence, we can at the cost of a sign swap around the two spinors in this expression. For example, $\bar{\chi}\gamma_n\lambda = -\bar{\lambda}\gamma_n\chi$ and so $\bar{\lambda}\gamma_n\lambda = 0$.

Corresponding to the above chiral conditions we can define projectors onto the spaces of positive and negative chiral spinors. These projectors are given by $P_\pm = \frac{1}{2}(1 \pm a\gamma^{D+1})$, where $a = 1$ if $D = 2 \mod 4$ and $a = i$ if $D = 4 \mod 4$. It is easy to verify that they are indeed projectors: that is, $P_+P_\mp = 0$, $P_\pm^2 = P_\pm$, $P_\mp^2 = P_\mp$ and $P_\pm + P_\mp = I$. Under complex conjugation the projectors transform as

$$P_\pm^* = \begin{cases} BP_\pm B^{-1}, & \text{if } D = 2 \text{ mod } 4, \\ BP_\mp B^{-1}, & \text{if } D = 4 \text{ mod } 4. \end{cases} \tag{5.3.14}$$

This equation places restrictions on the form that the matrices B and the chiral projectors can take. For example, let us write the $2^{D/2} \times 2^{D/2}$ γ-matrices in terms of $2^{(D/2)-1} \times 2^{(D/2)-1}$ blocks. We also choose our basis of spinors such that the projection operators are diagonal and such that P_+ has only its upper diagonal block non-vanishing and equal to the identity matrix and P_- has only its lower diagonal block non-zero and equal to the identity matrix. Applying equation (5.3.14), we find that if $D = 2$ mod 4, the matrix B has only its two diagonal blocks non-zero and if $D = 4$ mod 4, only its off-diagonal blocks non-zero.

Under complex conjugation the chiral spinors transform as

$$B^{-1}(P_\pm \lambda)^* = \begin{cases} P_\pm B^{-1}\lambda^*, & \text{if } D = 2 \text{ mod } 4, \\ P_\mp B^{-1}\lambda^*, & \text{if } D = 4 \text{ mod } 4. \end{cases} \tag{5.3.15}$$

Hence, complex conjugation and multiplication by B^{-1} relate the same chirality spinors if $D = 2$ mod 4 and opposite chirality spinors if $D = 4$ mod 4. As such, in the dimensions $D = 2, 4$ mod 8 where we can define Majorana spinors, the Majorana condition relates same chirality spinors if $D = 2$ mod 4 and opposite chirality spinors if $D = 4$ mod 4.

We now investigate how a matrix transformation on λ acts on its chiral components. Under the matrix transformation $\lambda \to A\lambda$ we find that $B^{-1}\lambda^* \to (B^{-1}A^*B)B^{-1}\lambda^*$ and so the equivalent transformation on $B^{-1}\lambda^*$ is $B^{-1}A^*B$. Clearly, if A is a polynomial in the γ matrices with real coefficients, then this transformation is the same on λ and $B^{-1}\lambda^*$. As we have already discussed this is the case with Lorentz transformations which are generated by $J^{mn} = \frac{1}{2}\gamma^{mn}$. However, if $A = EP_\pm$, where E is a polynomial in the γ-matrices with real coefficients, then the transformation becomes

$$B^{-1}(EP_\pm)^*B = \begin{cases} EP_\pm, & \text{if } D = 2 \text{ mod } 4, \\ EP_\mp, & \text{if } D = 4 \text{ mod } 4. \end{cases} \tag{5.3.16}$$

As an example of the latter, let us consider the chiral projections of Lorentz transformations which are given by $J_\pm^{mn} \equiv \frac{1}{2}\gamma^{mn}P_\pm$. In this case the above equation becomes $B^{-1}J_\pm^{mn*}B = J_\pm^{mn}$ for $D = 2$ mod 4 and $B^{-1}J_\pm^{mn*}B = J_\mp^{mn}$ for $D = 4$ mod 4. Hence for $D = 2$ mod 4 we find that the representation generated by $J_\pm^{mn}$ is a representation which is conjugate to the complex conjugate representation of the same chirality. On the other hand, for $D = 4$ mod 4 we find that the complex conjugate of the chiral Lorentz transformations is conjugate to that for the opposite chirality. As we noted in section 5.2, for $D = 2, 4$ mod 8 we can choose $B = I$ and so, for example, if $D = 2$ mod 8, then $(J_\pm^{mn})^* = J_\pm^{mn}$ and the representation they generate is contained in the group $SL(2^{(D/2)-1}, \mathbf{R})$.

However, for $D = 6, 8$ mod 8 the matrix B is anti-symmetric and so cannot be chosen to be the identity matrix. In particular, for $D = 6$ mod 8 we find that the chiral Lorentz transformations are contained in the group $SU^*(2^{(D/2)-1})$. The group $SU^*(N)$ is the group of $N \times N$ complex matrices of determinant 1 that commute with the operation of complex conjugation and multiplication by an anti-symmetric matrix B which obeys $B^\dagger B = 1$; see, for example, [5.5]. That is, if $A \in SU^*(N)$, then $BAB^{-1} = A^*$. Taking such an infinitesimal transformation $A = I + K$ we find that $SU^*(N)$ has real dimension $N^2 - 1$. The first such case is $D = 6$, since $SU^*(4)$ and $\text{Spin}(1, 5)$ both have dimension 15 we must conclude that the group of six-dimensional chiral Lorentz transformations generated by $J^{mn}_{\pm}$ is isomorphic to the group $SU^*(4)$.

If the spinors carry internal spinor indices that transform under a pseudo-real representation of an internal group, that is, a representation which acquires a minus sign under complex conjugation, then we can also define a kind of Majorana spinor when $D = 6, 8$ mod 8 by the condition

$$(\lambda_i)^* = \Omega^{ij} B \lambda_j. \tag{5.3.17}$$

In this equation Ω^{ij} is a real anti-symmetric matrix which also obeys the relation $\Omega^{ij} \Omega_{jk} = -\delta^i_k$, where $\Omega^{ij} = \Omega_{ij}$. We call spinors that satisfy this new type of Majorana condition symplectic Majorana spinors. Taking the complex conjugate of this symplectic Majorana condition, using the above relations and the fact that $B^* B = -1$ in these dimensions we find that it is indeed a consistent condition.

The symplectic Majorana condition should also be such that the internal group, which acts on the internal indices $i, j, \ldots$, acts in the same way on the left- and right-hand sides of the symplectic Majorana condition. This requires λ_i to carry a pseudo-real representation of the internal group. The vector representation of the group $USp(N)$ provides one of the most important examples of a pseudo-real representation. The group $USp(N)$ consists of unitary matrices that in addition preserve Ω^{ij}, that is, matrices A which satisfy $A^\dagger A = 1$ and $A^T \Omega A = \Omega$. Taking such an infinitesimal transformation we find that this group has dimension $\frac{1}{2}N(N + 1)$. Under $\lambda \to A\lambda$, we find that $\lambda^* \to A^*\lambda^*$, but $\Omega\lambda \to \Omega A\lambda = -\Omega A \Omega \Omega \lambda$. However, using the defining conditions of $USp(N)$ we can show that

$$- \Omega A \Omega = -(A^T)^{-1} \Omega \Omega = (A^T)^{-1} = (A^\dagger)^T = A^*, \tag{5.3.18}$$

which means that the symplectic Majorana condition preserves $USp(N)$. We note that $USp(2) = SU(2)$.

In addition to the symplectic Majorana condition of equation (5.3.17) which requires $D = 6, 8$ mod 8 we can also impose a Weyl constraint if $D = 2$ mod 4. That is, in the dimensions $D = 6$ mod 8 symplectic Majorana–Weyl spinors exist.

An important example of a symplectic Majorana–Weyl spinor is found in six dimensions which transform under $USp(4)$. These spinors naturally arise when we reduce the eleven-dimensional Majorana spinors to six dimensions. The eleven-dimensional spinors transform under $\text{Spin}(1, 10)$, which under the reduction to six dimensions becomes $\text{Spin}(1, 5) \times \text{Spin}(5)$. The group $\text{Spin}(5)$, which is isomorphic to $USp(4)$, becomes the internal group in six dimensions. Such a symplectic Majorana–Weyl spinor has $4 \times 4 = 16$ components and plays an important role in the construction of the M theory 5-brane.

5.4   Clifford algebras in odd dimensions

We now take the dimension of space-time D to be odd. The group C_D of equation (5.1.2) is generated by the γ_n and has order 2^{D+1}. The irreducible representations can be found using the same arguments as we used for the case of a space-time of even dimensions. However, there are some differences, which are a consequence of the fact that the conjugacy classes are not given by the obvious generalisation of those for the even-dimensional case which were listed in equation (5.2.2). From all the γ matrices, γ_m, $m = 0, 1, \ldots, D-1$ we can form the matrix $\gamma_D \equiv \gamma_0 \gamma_1 \cdots \gamma_{D-1}$. This matrix commutes with all the γ_m, $m = 0, 1, \ldots, D-1$ and so all products of the γ_m. As such, $\pm \gamma_D$ form conjugacy classes by themselves and as a result the full list of conjugacy classes is given by

$$[1], [-1], [\gamma_m], [\gamma_{m_1 m_2}], \ldots, [\gamma_{m_1 \cdots m_D}], [-\gamma_{m_1 \cdots m_D}]. \tag{5.4.1}$$

There are $2^D + 2$ conjugacy classes and so $2^D + 2$ inequivalent irreducible representations of C_D.

The commutator group of C_D is given by $\{\pm 1\}$ and so has order 2. As such, the number of inequivalent irreducible one-dimensional representations of C_D is 2^D. Hence, in an odd-dimensional space-time we have two inequivalent irreducible representations of C_D of dimension greater than 1. In either of these two irreducible representations, the matrix γ_D commutes with the entire representation and so by Schur's lemma must be a multiple of the identity, that is, $\gamma_D = a^{-1}I$, where a is a constant. Multiplying both sides by γ_{D-1} we find the result

$$\gamma_{D-1} = a\gamma_0\gamma_1 \cdots \gamma_{D-2} = a\gamma_{01\cdots D-2}. \tag{5.4.2}$$

Using equation (5.3.11), we conclude that $(\gamma_{01\cdots D-2})^2 = -(-1)^{(D-1)/2}$ as the matrix $\gamma_{01\cdots D-2}$ is the same as that denoted by γ^{D+1} for the even-dimensional space-time with one dimension less. However, as $\gamma_{D-1}^2 = +1$ we must conclude that for $D = 3$ mod 4 $a = \pm 1$, while for for $D = 5$ mod 4 $a = \pm i$. The γ_m, $m = 0, 1, \ldots, D-2$, generate an even-dimensional Clifford algebra and we recall that the corresponding subgroup C_{D-1} has a unique irreducible representation of dimension greater than 1, the dimension being $2^{(D-1)/2}$. It follows that the two irreducible representations for D odd which have dimension greater than 1 must coincide with this irreducible representation when restricted to C_{D-1}. Hence, the two inequivalent irreducible representations for D odd are generated by the unique irreducible representation for the γ_m, $m = 0, 1, \ldots, D-2$, with the remaining γ matrix being given by $\gamma_{D-1} = a\gamma_0\gamma_1 \cdots \gamma_{D-2}$. The two possible choices of a, that is, $a = \pm$, correspond to the two inequivalent irreducible representations. Clearly, these two inequivalent irreducible representations both have dimension $2^{(D-1)/2}$ as this is the dimension of the unique irreducible representation with dimension greater than 1 in the space-time with one dimension less. We can check that this is consistent with the relationship between the order of the group, 2^{D+1}, and the sum of the dimensions squared of all irreducible representations. The latter is given by $1^2 \cdot 2^D + (2^{(D-1)/2})^2 + (2^{(D-1)/2})^2 = 2^{D+1}$ as required.

We now extend the complex conjugation and transpose properties discussed previously for even-dimensional space-time to the case of an odd-dimensional space-time. Clearly, for the matrices γ_m, $m = 0, 1, \ldots, D-2$, these properties are the same and are given in equations (1.1.5) and (1.1.8). It only remains to consider $\gamma_{D-1} = a\gamma_0\gamma_1 \cdots \gamma_{D-2} \equiv a\gamma_{01\cdots D-2}$. It follows from the previous section that

$$\gamma_{01\cdots D-2}^* = B\gamma_{01\cdots D-2}B^{-1} \tag{5.4.3}$$

and

$$C\gamma_{01\cdots D-2}C^{-1} = (-1)^{(D-1)/2}\gamma_{01\cdots D-2}^T. \tag{5.4.4}$$

We may also write this last equation as

$$C\gamma_{01\cdots D-2} = -\epsilon(-1)^{(D-1)/2}(C\gamma_{01\cdots D-2})^T \qquad (5.4.5)$$

as $C^T = -\epsilon C$. Taking into account the different possible values of a discussed above we conclude that

$$\gamma^*_{D-1} = -(-1)^{(D-1)/2}B\gamma_{D-1}B^{-1} \qquad (5.4.6)$$

and

$$\gamma^T_{D-1} = (-1)^{(D-1)/2}C\gamma_{D-1}C^{-1}. \qquad (5.4.7)$$

As we did for the even-dimensional case we can adopt the choice $C = B^T\gamma^0$, whereupon we find that $\gamma^\dagger_{D-1} = \gamma^0\gamma_{D-1}\gamma^0$. The representation is automatically unitary as a consequence of being unitary for the C_{D-1} subgroup.

For $D = 3 \bmod 4$ γ_{D-1} has the same relationships under complex conjugation and transpose as do the γ_m, $m = 0, 1, \ldots, D-2$. As a result, for $D = 3 \bmod 4$

$$(C\gamma_{m_1\cdots m_p})^T = \epsilon(-1)^{\frac{(p-1)(p-2)}{2}}(C\gamma_{m_1\cdots m_p}) \quad m_1, \ldots, m_p = 0, \ldots, D-1. \quad (5.4.8)$$

For $D = 5 \bmod 4$, we get an additional minus sign in this relationship if one of the $m_1, \ldots, m_p$ takes the value $D - 1$.

Let us now consider which types of spinors can exist in odd-dimensional space-times. Clearly in odd-dimensional space-times the Weyl condition is not a Lorentz invariant condition and so one cannot define such spinors. However, we can ask which odd-dimensional space-times have Majorana spinors, which we take to be defined by $\chi^* = B\chi$. Since either of the two inequivalent irreducible representations coincides with the unique irreducible representation of dimension greater than 1 when restricted to the subgroup C_{D-1}, the matrix B is the same as in the even-dimensional case. It follows that for a Majorana spinor to exist we require $\epsilon = +1$, which is the case for $D = 3, 5 \bmod 8$. We must, however, verify that the Majorana condition is preserved by all Lorentz transformations. Those that are generated by $\frac{1}{2}\gamma_{mn}$, $m, n = 0, 1 \ldots, D-2$ are guaranteed to work; however, carrying out the Lorentz transformation $\delta\chi = \frac{1}{4}\gamma_{mD-1}\chi$, $m = 0, 1 \ldots, D-2$ we find it preserves the Majorana constraint only if $D = 3 \bmod 4$. Hence, Majorana spinors exist in odd D-dimensional space-time if $D = 3 \bmod 8$. We note that these odd dimensions are precisely one dimension higher than those where Majorana–Weyl spinors exist. This is not a coincidence as the reduction of a Majorana spinor in $D = 3 \bmod 8$ dimensions leads to two Majorana–Weyl spinors of opposite chirality and, since the resulting matrix γ_{D+1} in the even dimensions $D = 2 \bmod 8$ is real, we may Weyl project to find a Majorana–Weyl spinor. We close this section by listing for future reference some of our conventions in odd and even Minkowski space-time of dimension D. We define $\epsilon^{01\cdots D-1} = +1$ and as a result $\epsilon_{01\cdots D-1} = -1$. The identity

$$\epsilon^{a_1\cdots a_n b_1\cdots b_m}\epsilon_{c_1\cdots c_n b_1\cdots b_m} = -m!n!\delta^{a_1\cdots a_n}_{c_1\cdots c_n} \qquad (5.4.9)$$

can easily be proved by taking values, say $a_1 = 1, a_2 = 2, \ldots$ and $c_1 = 1, c_2 = 2, \ldots$, where $m + n = D$. We note that $\delta^{a_1\cdots a_n}_{c_1\cdots c_n} = \delta^{[a_1}_{c_1}\cdots\delta^{a_n]}_{c_n}$, where, as usual, anti-symmetry means take all permutations with the appropriate signs and divide by $n!$. For example, $\delta^{a_1 a_2}_{c_1 c_2} = \frac{1}{2}(\delta^{a_1}_{c_1} \delta^{a_2}_{c_2} - \delta^{a_2}_{c_1} \delta^{a_1}_{c_2})$. For even dimensions we define $\gamma^{D+1} = \gamma_0 \gamma_1 \cdots \gamma_{D-1}$.

5.5   Central charges

One important application of the above theory is to find what central charges can appear in a supersymmetry algebra, that is, what generators can appear in the anti-commutator $\{Q_\alpha, Q_\beta\}$, where Q_α is the generator of supersymmetry transformations. As we shall see, the result depends on the dimension of space-time and on whether the spinor Q_α is Majorana, Weyl or Majorana–Weyl. The existence of central terms was first noticed in four dimensions [5.6] The occurrence of central charges is closely related to the existence of branes that preserve some of the supersymmetry of the theory.

To begin with we take the dimension of space-time to be even. The right-hand side of the anti-commutator of the supercharges takes the form [5.7]

$$\{Q_\alpha, Q_\beta\} = (\gamma_m C^{-1})_{\alpha\beta} P^m + \sum_{p,\; p\neq 1} (\gamma_{m_1 \cdots m_p} C^{-1})_{\alpha\beta} Z^{m_1 \cdots m_p}, \tag{5.5.1}$$

where $Z^{m_1 \cdots m_p}$ are the central charges and P^m is the generator of translations. The sum is over all possible central terms, however, these can only occur if the matrix $(\gamma_{m_1 \cdots m_p} C^{-1})_{\alpha\beta}$ is symmetric in α, β. Examining equation (5.2.17) we find that this is the case if

$$\epsilon (-1)^{(p-1)(p-2)/2} = 1. \tag{5.5.2}$$

For $D = 2, 4 \bmod 8$, $\epsilon = 1$ and so we find central charges for $p = 1, 2 \bmod 4$. In these dimensions we can define Majorana spinors and adopting this constraint still allows these central charges although the Majorana condition will place reality conditions on them. For $D = 6, 8 \bmod 8$ $\epsilon = -1$ and so we find central charges of rank p for $p = 3, 4 \bmod 4$.

Let us now consider the case when the spinors are Weyl, that is, satisfy $(P_\pm Q)_\alpha = 0$. In this case we must modify the terms on the right-hand side of the anti-commutator to be given by

$$\{Q_\alpha, Q_\beta\} = (P_\pm \gamma_m C^{-1})_{\alpha\beta} P^m + \sum_p (P_\pm \gamma_{m_1 \cdots m_p} C^{-1})_{\alpha\beta} Z^{m_1 \cdots m_p}. \tag{5.5.3}$$

In this case $(\gamma_{m_1 \cdots m_p} C^{-1})_{\alpha\beta}$ and $(\gamma^{D+1} \gamma_{m_1 \cdots m_p} C^{-1})_{\alpha\beta}$ must be symmetric. However, the latter matrix is equal to ϵ times $(\gamma_{m_1 \cdots m_{D-p}} C^{-1})_{\alpha\beta}$. Hence, central charges of rank p are possible if $p = 1, 2 \bmod 4$ and $D - p = 1, 2 \bmod 4$, assuming $\epsilon = 1$.

Finally, we can consider Majorana–Weyl spinors, which only exist in $D = 2 \bmod 8$. In this case, we have the above condition on p resulting from the Weyl constraint. Writing $D = 2 + 8n$, for $n \in \mathbf{Z}$ one finds that it only allows central charges of rank $p = 1 \bmod 4$. An example of this case is provided by the $N = 1$ $D = 10$ supersymmetry algebra constructed from a supercharge, that is, a Majorana–Weyl spinor. This algebra underlies the $N = 1$ Yang–Mills theory as well as the type I supergravity theory in ten dimensions. The algebra is then given by

$$\{Q_\alpha, Q_\beta\} = (P_\pm \gamma_m C^{-1})_{\alpha\beta} P^m + (P_\pm \gamma_{m_1 \cdots m_5} C^{-1})_{\alpha\beta} Z^{m_1 \cdots m_5}$$
$$+ (P_\pm \gamma_{m_1 \cdots m_9} C^{-1})_{\alpha\beta} Z^{m_1 \cdots m_9}.$$

In fact, the last term on the right-hand side can be rearranged to be precisely of the same form as the the first term using the identity $\epsilon^{nm_1 \cdots m_9} \gamma_{m_1 \cdots m_9} \propto \gamma^n \gamma_{11}$. Carrying out this we find the algebra becomes

$$\{Q_\alpha, Q_\beta\} = (P_\pm \gamma_m C^{-1})_{\alpha\beta} P^m + (P_\pm \gamma_{m_1 \cdots m_5} C^{-1})_{\alpha\beta} Z^{m_1 \cdots m_5}, \tag{5.5.4}$$

where we have absorbed a term proportional to $\epsilon_{nm_1 \cdots m_9} Z^{m_1 \cdots m_9}$ in P_n. For similar reasons $P_\pm \gamma_{m_1 \cdots m_5}$ is either self-dual or anti-self-dual, depending on the sign taken in $P_\pm$ and therefore so is $Z^{m_1 \cdots m_5}$. The number of central charges is $10 + \frac{1}{2}(\frac{10!}{5! \times 5!}) = 136$. A Majorana–Weyl spinor in ten dimensions has 16 components and so the left-hand side of the supersymmetry algebra is an arbitrary symmetric matrix, which has $16 \times 17/2 = 136$ components as it should.

The above discussion can be generalised to the case of odd-dimensional space-times although in this case we do not have Weyl spinors. For the case of $D = 3 \bmod 4$, γ_{D-1} behaves exactly like the other γ matrices under complex conjugation and transpose and as a result we find from equation (5.4.8) precisely the same condition for the existence of central charges that is, equation (5.5.2) with the value of ϵ being the same as that in one dimension less. The case of $D = 5 \bmod 4$ can be deduced in a similar way by taking into account the discussion below equation (5.4.8).

A very important example is the supersymmetry algebra in 11 dimensions for Majorana supercharges as this algebra underlies the supergravity theory in this dimension. For this algebra we can have central charges of rank p with $p = 1, 2 \bmod 4$ and so the supersymmetry algebra takes the form [5.7].

$$\{Q_\alpha, Q_\beta\} = (\gamma^m C^{-1})_{\alpha\beta} P_m + (\Gamma^{mn} C^{-1})_{\alpha\beta} Z_{mn} + (\Gamma^{mnpqr} C^{-1})_{\alpha\beta} Z_{mnpqr}. \tag{5.5.5}$$

We need only go up to rank 5 thanks to the identity

$$\gamma^{m_1 \cdots m_p} = \frac{(-1)^{p(q-1)}(-1)^{q(q+1)/2}}{q!} \epsilon^{m_1 \cdots m_p n_1 \cdots n_q} \gamma_{n_1 \cdots n_q}$$

$$= -\frac{(-1)^{p(p+3)/2}}{(11-p)!} \epsilon^{m_1 \cdots m_p n_1 \cdots n_q} \gamma_{n_1 \cdots n_q}, \tag{5.5.6}$$

where $p + q = 11$, $\epsilon^{01 \cdots 10} = 1$ and $\gamma_{10} = \gamma_0 \gamma_1 \cdots \gamma_9$, using equation (5.4.2) with $a = +1$. Similar identities are true in all odd-dimensional spaces. As must be the case, the left-hand side is the most general symmetric 32×32 matrix, which has 528 components, and one can verify that this is also the number of central charges together with the translation.

The ten-dimensional supersymmetry algebra with one Majorana supercharge, which underlies the IIA supergravity theory, is found using the techniques above to be given by

$$\{Q_\alpha, Q_\beta\} = (\gamma_m C^{-1})_{\alpha\beta} P^m + (\gamma_{m_1 m_2} C^{-1})_{\alpha\beta} Z^{m_1 m_2} + (\gamma_{m_1 \cdots m_5} C^{-1})_{\alpha\beta} Z^{m_1 \cdots m_5}$$

$$+ (\gamma_{m_1 \cdots m_6} C^{-1})_{\alpha\beta} Z^{m_1 \cdots m_6} + (\gamma_{m_1 \cdots m_9} C^{-1})_{\alpha\beta} Z^{m_1 \cdots m_9}$$

$$+ (\gamma_{m_1 \cdots m_{10}} C^{-1})_{\alpha\beta} Z^{m_1 \cdots m_{10}}. \tag{5.5.7}$$

Again we find 528 central charges. If we decomposed the supercharges into Majorana–Weyl spinors we would find two such spinors of opposite chirality. Indeed, we can find the above IIA algebra by dimensionally reducing the 11-dimensional algebra of equation (5.5.5).

It is straightforward to extend the discussion to supersymmetry algebras which contain supersymmetry generators Q_α^i that carry a representation of an internal algebra. The anti-commutator $\{Q_\alpha^i, Q_\beta^j\}$ is now symmetric under interchange in α, i and β, j. If the central charges are symmetric in i, j we must have a corresponding matrix which is also symmetric in α, β and we can use the criterion above. However, we can also have central charges which are anti-symmetric in i, j, in which case the corresponding matrix is also anti-symmetric in α, β. This can occur when $p = 3, 4 \mod 4$ and if the supercharges are also Weyl we must also have $D - p = 3, 4 \mod 4$.

One of the most important examples is the ten-dimensional supersymmetry algebra constructed from two Majorana–Weyl supercharges Q_α^i, $i = 1, 2$ of the same chirality. The supercharges transform on their internal indices under SO(2) in the obvious way. The resulting supersymmetry algebra is

$$\{Q_\alpha^i, Q_\beta^j\} = (P_\pm \gamma_m C^{-1})_{\alpha\beta} \delta^{ij} P^m + (P_\pm \gamma_m C^{-1})_{\alpha\beta} Z_m^{ij}$$

$$+ (P_\pm \gamma_{m_1 m_2 m_3} C^{-1})_{\alpha\beta} \epsilon^{ij} Z^{m_1 m_2 m_3}$$

$$+ (P_\pm \gamma_{m_1 \cdots m_5} C^{-1})_{\alpha\beta} Z^{m_1 \cdots m_5 ij}, \tag{5.5.8}$$

where Z_m^{ij} is symmetric and traceless in its ij indices, $Z^{m_1 \cdots m_5 ij}$ is only symmetric in its ij indices, but self-dual, or anti-self-dual, in its $m_1 \cdots m_5$ indices. One can verify that there are 528 central charges as there should be, corresponding to the left-hand side of the algebra. This algebra underlies the IIB supergravity theory.

A final example concerns the algebra that underlies the M theory 5-brane. It contains a supercharge which is a USp(4) Majorana–Weyl spinor Q_α^i, $i = 1, \ldots, 4$, which belongs to a six-dimensional space-time. As such, it has $4 \times 4 = 16$ components. The supersymmetry algebra is given by [5.8]

$$\{Q_\alpha^i, Q_\beta^j\} = (P_\pm \gamma_m C^{-1})_{\alpha\beta} \Omega^{ij} P^m + (P_\pm \gamma_m C^{-1})_{\alpha\beta} Z_m^{ij}$$

$$+ (P_\pm \gamma_{m_1 m_2 m_3} C^{-1})_{\alpha\beta} Z^{m_1 m_2 m_3 ij}. \tag{5.5.9}$$

In this equation Ω^{ij} is the anti-symmetric USp(4) invariant tensor and Z_m^{ij} is anti-symmetric and Ω-traceless in i, j, while $Z^{m_1 m_2 m_3 ij}$ is symmetric in i, j and anti-symmetric and self-dual, or anti-self-dual, in m_1, m_2, m_3. In principle, one could write down a rank 5 central charge, but this can be absorbed in the rank 1 charges. The number of central charges is $6 + 5 \times 6 + 10 \times 10 = 136$, which is the number of components of an arbitrary symmetric 16×16 matrix. In carrying out the above construction it is important to note that in six dimensions $\epsilon = -1$ and, as explained above, the symmetry of the γ matrices is different to all the other examples studied above which had $\epsilon = 1$.

5.6 Clifford algebras in space-times of arbitrary signature

An account of Clifford algebras and spinors in non-Lorentzian space-times aimed at physicists was given in [5.3]. We now consider a Clifford algebra in a space-time with metric

$$\eta_{mn} = \text{diag}\{\underbrace{-1, -1, \ldots, -1}_{t}, \underbrace{+1, \ldots, +1}_{s}\} \tag{5.6.1}$$

and we will find the dimensions and signatures for which Majorana and Majorana–Weyl spinors exist. The arguments are very similar to those for the case of Minkowski space-time considered in section 5.3 and hence we will not always spell them out in detail. The dimension of space-time $D = t + s$ is taken to be even. The indices m, n, p take the values $1, \ldots, D$, since there is now in general more than one time-like direction. As we discussed in section 5.1, in any such space-time the Clifford algebra has only one non-trivial representation of dimension $2^{D/2}$. If γ_m is in this representation, so is γ_m^* and so these must be related to each other by

$$\gamma_m^* = \rho B \gamma_m B^{-1}, \tag{5.6.2}$$

where $\rho = \pm 1$. This equation generalizes equation (5.2.5) to admit the possible occurrence of the factor ± 1. Following the discussion around equation (5.2.7) we can, as before, choose $|\det B| = 1$ and we also find that

$$B^* B = \varepsilon I \quad \text{and} \quad \varepsilon = \pm 1. \tag{5.6.3}$$

Choosing a unitary representation implies that $\gamma_m \gamma_m^\dagger = 1, m = 1, \ldots, D$. However, since $\gamma_m \gamma_m = \eta_{mm}$ we conclude that

$$\gamma_m^\dagger = \begin{cases} -\gamma_m, & m = 1, \ldots, t, \\ \gamma_m, & m = t + 1, \ldots, D. \end{cases} \tag{5.6.4}$$

The Hermitian properties of the γ_m can also be written as

$$\gamma_m^\dagger = A \gamma_m A^{-1} (-1)^t, \quad m = 1, \ldots, D, \tag{5.6.5}$$

where

$$A = \gamma_1 \gamma_2 \cdots \gamma_t = \gamma_{12\cdots t}. \tag{5.6.6}$$

From the above formulae relating the adjoint and the complex conjugate of γ_m we can deduce a formula relating the transpose. It is straightforward to verify that if we take

$$C = -B^T A, \tag{5.6.7}$$

then

$$\gamma_m^T = -\rho(-1)^{t+1} C \gamma_m C^{-1}. \tag{5.6.8}$$

Evaluating $(\gamma_m^*)^\dagger = (\gamma^\dagger)^*$ using equations (5.6.2) and (5.6.4) we find that

$$\gamma_m A^{-1} B^\dagger A^* B = A^{-1} B^\dagger A^* B \gamma_m. \tag{5.6.9}$$

Schur's lemma then tells us that

$$A^{-1} B^\dagger A^* B = \mu I. \tag{5.6.10}$$

However, using equation (5.6.2)

$$A^* = \gamma_1^* \cdots \gamma_t^* = \rho^t B A B^{-1} \tag{5.6.11}$$

and equation (5.6.10) then becomes

$$B^\dagger B = \mu \rho^t I. \tag{5.6.12}$$

For any $2^{D/2}$-vector v we then have $v^\dagger B^\dagger B v = \mu \rho' v^\dagger v$ and so $\mu \rho'$ is real and positive definite. Recalling that $|\det B| = 1$, we must set $\mu \rho' = 1$ and so

$$B^\dagger B = I. \tag{5.6.13}$$

Comparing $B^* B = \varepsilon I$ with equation (5.6.13) we find that

$$B^T = \varepsilon B. \tag{5.6.14}$$

Using equations (5.2.15) and (5.6.8) we find that

$$A^T = CAC^{-1}(-1)^{t(t-1)/2}(-1)^{t^2}\rho^t$$

and as a result

$$C^T = \varepsilon \rho^t (-1)^{t(t+1)/2}C. \tag{5.6.15}$$

It is straightforward to verify that $C^\dagger C = I$.

The analogue of equation (5.2.17) is therefore given by

$$(C\gamma_{m_1}\cdots\gamma_{m_p})^T = \varepsilon \rho^{t+p}(-1)^{pt}(-1)^{t(t+1)/2}(-1)^{\frac{p(p-1)}{2}}C\gamma_{m_1\cdots m_p}. \tag{5.6.16}$$

Consequently, the number of anti-symmetric matrices is

$$\sum_{p=0}^{D} \frac{1}{2}\left(1 + \varepsilon(-1)^{t(t+1)/2}(-1)^{p(t+1)}\rho^{p+t}(-1)^{(p-1)(p-2)/2}\right)\binom{D}{p}$$

$$= 2^{D-1} - 2^{(D+1)/2}\varepsilon(-1)^{t(t+1)/2}\rho^t \cos\frac{\pi}{4}(1 + (-1)^{t+1}\rho D). \tag{5.6.17}$$

However, the number of anti-symmetric matrices of dimension $2^{D/2} \times 2^{D/2}$ is obviously $\frac{1}{2}2^{D/2}(2^{D/2} - 1)$ and so we conclude that

$$1 = \sqrt{2}\varepsilon(-1)^{t(t+1)/2}\rho^t \cos\frac{\pi}{4}(1 + (-1)^{t+1}\rho D). \tag{5.6.18}$$

Since for q an odd integer

$$2\left(\cos\frac{\pi}{4}q\right)^2 = \left(1 + \cos\frac{\pi}{2}q\right) = 1, \tag{5.6.19}$$

we can express ε as

$$\varepsilon = \sqrt{2}(-1)^{t(t+1)/2}\rho^t \cos\frac{\pi}{4}(1 + (-1)^{t+1}\rho D). \tag{5.6.20}$$

Since $(-1)^{\frac{1}{2}t(t+1)} = e^{\pm\frac{1}{2}i\pi t(t+1)}$ and $\rho = e^{\pm\frac{1}{2}i\pi(\rho-1)}$ we can rewrite ε as

$$\varepsilon = \sqrt{2}\rho^t \cos\frac{\pi}{4}(1 - \rho(s - t)). \tag{5.6.21}$$

In deriving this result we have used the fact that if t is an even integer $(-1)^{t^2} = 1$, while if t is an odd integer $(-1)^{(t+1)^2} = 1$. Finally, we may write ε as

$$\varepsilon = \cos\frac{\pi}{4}(s - t) + \rho \sin\frac{\pi}{4}(s - t). \tag{5.6.22}$$

Clearly, ε depends only on $s - t$ mod 8 and its values can be written for $\rho = +1$ as

$$\begin{aligned}\varepsilon &= +1 \quad s - t = 0, 2 \text{ mod } 8, \\ \varepsilon &= -1 \quad s - t = 4, 6 \text{ mod } 8\end{aligned} \tag{5.6.23}$$

with corresponding values for $\rho = -1$.

A spinor transforms under Spin(t, s) as $\delta\lambda = \frac{1}{4}\omega^{mn}\gamma_{mn}\lambda$, the generators of the Lie algebra of Spin(t, s) being $\frac{1}{2}\gamma_{mn}$. As for the case of Minkowski space, we can introduce a Majorana spinor which by definition satisfies $\lambda^* = B\lambda$. Following the same argument as in that case, we find that Majorana spinors only exist if $\varepsilon = +1$. The Dirac conjugate is defined by $\bar{\lambda}^D \equiv -\lambda^\dagger A$ and the Majorana conjugate by $\bar{\lambda}^M = \lambda^T C$. Setting $\bar{\lambda}^D = \bar{\lambda}^M$ we recover the above condition for a Majorana spinor. Hence, Majorana spinors only exit in dimensions for which $s - t = 0, 2$ mod 8.

The matrix $\gamma^{D+1} = \gamma_1 \cdots \gamma_D$ commutes with the γ_m, $m = 1, \ldots, D$ and has

$$(\gamma^{D+1})^2 = (-1)^{D(D-1)/2}(-1)^t = (-1)^{(s-t)/2}, \tag{5.6.24}$$

since $s - t$ is even. As a consequence, the matrix

$$(-1)^{(s-t)/4}\gamma^{D+1} \tag{5.6.25}$$

squares to 1 and can be used to define a representation of a Clifford algebra in $D + 1$ dimensions with t time-like directions and $s + 1$ space-like directions.

We define Weyl spinors to satisfy by

$$(-1)^{(s-t)/4}\gamma^{D+1}\chi = \pm\chi. \tag{5.6.26}$$

The factor $(-1)^{(s-t)/4}$ is real only if $s - t = 0$ mod 4 and imaginary if $s - t = 2$ mod 4. By examining the reality of the Weyl condition we find that Majorana–Weyl spinors only exist if $s - t = 0$ mod 8.

6   The classical superstring

"At university we can lecture, we can give seminars, we can invent didactical tricks, but we can never prevent that now and then a youngster comes along understands things in depth and pushes physics forward."

Julius Wess as narrated by Klaus Sibold

The superstring is an extended one-dimensional object which moves through superspace. Although it has many features in common with the bosonic string it has some important differences. The most striking of these is that it can describe not only space-time bosons, but also space-time fermions. The precise particle content of the superstring depends on whether it is open or closed and the type of world-sheet supersymmetry it possesses. Certain superstrings possess massless particles of spins 2, $\frac{3}{2}$, 1, $\frac{1}{2}$ and even 0. Furthermore, superstrings can be formulated in such a way as not to possess a tachyon, although the presence of a tachyon is not necessarily a bad thing since it could just be a manifestation of an instability of the vacuum, as is the case for the Higgs field in the Standard Model before it is shifted to the true vacuum. However, this mechanism has not been shown to exist in string theory.

Although there exist many superstring theories in four dimensions, in this chapter we confine our attention to the original superstrings. These theories are only consistent in ten space-time dimensions and possess $(1,1)$ world-sheet supersymmetry. They were discovered by Ramond [6.1] and Neveu and Schwarz [6.2], and can describe either open or closed strings. In [6.1] Ramond presented a string theory that describes space-time fermions. Subsequently, Neveu and Schwarz [6.2] realised that in theory Ramond's there was in fact a choice of boundary conditions and taking the other choice from that chosen by Ramond leads to a theory that describes space-time bosons. They were also able to show [6.3] that there exist interaction vertices that lead to interactions between the bosons and fermions of the two theories. Subsequently, it was found that one could make a projection, called the Gliozzi–Scherk–Olive (GSO) projection [5.1], which had the effect of eliminating some of the physical states in the original string theories in such a way that the resulting string theories were found to possess ten-dimensional space-time supersymmetry. In this book the name superstring refers to any string theory that possesses world-sheet supersymmetry whether or not this results in a theory with space-time supersymmetry. This differs from the usage in some other places. The supersymmetry transformations of the world-sheet fields were written down in reference [6.4], which was important for the development of the four-dimensional Wess–Zumino model.

The GSO projected theories were found to have an alternative, equivalent, formulation [6.5] which has manifest space-time supersymmetry, but which does not possess manifest world-sheet supersymmetry. This formulation is called the Green–Schwarz formulation and will be discussed at the end of this chapter. While this formulation has all the advantages of having space-time supersymmetry the price to be paid is that its manifestly Lorentz covariant quantization is very difficult.

One unattractive feature of superstrings is their complexity as compared with bosonic strings. Of course, one would expect a certain amount of complexity when dealing with ten-dimensional space-time spinors. However, the calculation of multi-loop superstring amplitudes involving fermions is particularly difficult.

6.1   The Neveu–Schwarz–Ramond (NS–R) formulation

This string [6.1, 6.2] moves through a kind of background superspace which we may parameterise by $\{x^\mu, \chi^\mu_A\}$, $\mu = 0, 1, \ldots, D - 1$, where x^μ are the usual coordinates of Minkowski space and χ^μ_A, $A = 1, 2$ is a Grassmann odd, two-dimensional Majorana spinor which also has space-time vector index μ. This superspace is not the usual one associated with the discussion of supersymmetric theories in D dimensions as χ^μ_A is not a spinor of the background space-time, but is a D-dimensional vector. However, it is a spinor of the two-dimensional world-sheet and Spin(2) transformations act on its A index. Theories with this field content can possess (1,1) supersymmetry on the world sheet and we may add to the physical fields the real auxiliary field N^μ. The reader is referred to the review in [1.11, chapter 21] for a detailed and essentially self-contained discussion of the action and symmetries of this theory which we now summarise.

From the supermultiplet $\{x^\mu, \chi^\mu_A, N^\mu\}$ we can construct the action

$$\int d^2\xi \left(-\frac{1}{2} \partial_\alpha x^\mu \partial^\alpha x^\nu - \frac{i}{2} \bar{\chi}^\mu \slashed{\partial} \chi^\nu + \frac{1}{2} N^\mu N^\nu \right) \eta_{\mu\nu}, \tag{6.1.1}$$

which admits (1,1) rigid supersymmetry. The action for the superstring results from coupling this system to (1,1) supergravity, which comprises the vierbien e^a_α and gravitino $\psi_{\alpha A}$. The final result, which is invariant under local (1,1) supersymmetry, is given by [6.6, 6.7]

$$A = \frac{1}{2\pi\alpha'} \int d^2\xi\, e \left\{ -\frac{1}{2} e_a{}^\alpha \partial_\alpha x^\mu e^{a\beta} \partial_\beta x^\nu - \frac{i}{2} \bar{\chi}^\mu \slashed{\partial} \chi^\nu + \frac{1}{2} N^\mu N^\nu \right.$$

$$\left. + \frac{i\kappa}{2} \bar{\psi}_\alpha \slashed{\partial} x^\mu \gamma^\alpha \chi^\nu - \frac{\kappa^2}{16} \bar{\psi}_\alpha \gamma^\beta \gamma^\alpha \psi_\beta \bar{\chi}^\mu \chi^\nu \right\} \eta_{\mu\nu} \equiv \int d^2\xi\, \mathbf{L}, \tag{6.1.2}$$

where

$$e = \det e_\alpha{}^a, \quad \slashed{\partial} = \gamma^a e_a{}^\alpha \partial_\alpha. \tag{6.1.3}$$

The symmetries of the action are as follows:

(a) Two-dimensional reparameterisation invariance:

$$\delta e_\alpha{}^a = f^\beta \partial_\beta e_\alpha{}^a + \partial_\alpha f^\beta e_\beta{}^a, \quad \delta \psi_{\alpha A} = f^\beta \partial_\beta \psi_{\alpha A} + \partial_\alpha f^\beta \psi_{\beta A},$$

$$\delta x^\mu = f^\beta \partial_\beta x^\mu, \quad \delta \chi^\mu_A = f^\beta \partial_\beta \chi^\mu_A, \quad \delta N^\mu = f^\beta \partial_\beta N^\mu, \tag{6.1.4}$$

where f^α is the local, world-sheet dependent, parameter.

(b) Local two-dimensional Lorentz invariance:

$$\delta e_\alpha{}^a = -w^a{}_b e_\alpha{}^b, \quad \delta \chi_A^\mu = \tfrac{1}{4}(w^{ab}\gamma_{ab})_A{}^B \chi_B^\mu, \quad \delta \psi_{\alpha A} = \tfrac{1}{4}(w^{ab}\gamma_{ab})_A{}^B \psi_{\alpha B}, \quad (6.1.5)$$

while x^μ and N^μ are inert.

(c) Local two-dimensional supersymmetry:

$$\delta x^\mu = i\bar{\epsilon}\chi^\mu, \quad \delta \chi^\mu = (\hat{D}x^\mu + N^\mu)\epsilon, \quad \delta N^\mu = i\bar{\epsilon}\hat{D}\chi^\mu \tag{6.1.6}$$

and

$$\delta e_\alpha{}^a = i\kappa \bar{\epsilon}\gamma^a \psi_\alpha, \quad \delta \psi_\alpha = \frac{2}{\kappa} D_\alpha \epsilon, \tag{6.1.7}$$

where

$$\hat{D}_\alpha x^\mu = \partial_\alpha x^\mu - \frac{i\kappa}{2}\bar{\psi}_\alpha \chi^\mu,$$

$$\hat{D}_\beta \chi^\mu = (D_\beta \chi^\mu - \frac{\kappa}{2}\hat{\rlap{D}x}^\mu + N)\psi_\beta, \tag{6.1.8}$$

$$D_\beta \chi^\mu = (\partial_\beta + \tfrac{1}{4}w_{\beta ab}\gamma^{ab})\chi^\mu, \quad D_\beta \epsilon = (\partial_\beta + \tfrac{1}{4}w_{\beta ab}\gamma^{ab})\epsilon.$$

(d) Weyl symmetry:

$$\delta e_\alpha{}^a = \Lambda e_\alpha{}^a, \quad \delta \psi_\alpha = \tfrac{1}{2}\Lambda \psi_\alpha,$$
$$\delta x^\mu = 0, \quad \delta \chi^\mu = -\tfrac{1}{2}\Lambda \chi^\mu, \quad \delta N^\mu = -\Lambda N^\mu. \tag{6.1.9}$$

(e) Local 'S-supersymmetry':

$$\delta \psi_\alpha = \gamma_\alpha \zeta, \tag{6.1.10}$$

where ζ_A is a Majorana spinor which depends on ξ^α in an arbitrary way. All the other fields are inert under S-supersymmetry.

(f) The Poincaré group:

$$\delta x^\mu = -\Lambda^\mu{}_\nu x^\nu + a^\mu, \quad \delta \chi^\mu = -\Lambda^\mu{}_\nu \chi^\nu \text{ etc.} \tag{6.1.11}$$

In contrast to all the other symmetries this final symmetry is a rigid symmetry. We note that we do not have invariance under the super-Poincaré group, only under the Poincaré subgroup. In fact, the superstring theory only possesses space-time supersymmetry if we make certain projections which are discussed in the next chapter.

The analysis of the action follows the same path as in the bosonic case given in chapter 2. The equations of motion for x^μ and χ^μ are

$$\partial_\alpha (ee_a{}^\alpha e^{a\beta}\partial_\beta x^\mu) - i\frac{\kappa}{2}\partial_\beta (ee^{\beta a}\bar{\psi}_\alpha \gamma_a \gamma^\alpha \chi^\mu) = 0 \tag{6.1.12}$$

and

$$\rlap{\,/}\partial \chi^\mu - \frac{\kappa}{2}\gamma^\alpha \rlap{\,/}\partial x^\mu \psi_\alpha - i\frac{\kappa^2}{8}\chi^\mu \bar{\psi}_\alpha \gamma^\beta \gamma^\alpha \psi_\beta = 0. \tag{6.1.13}$$

The equation for the auxiliary field N^μ is $N^\mu = 0$ and we now implement this in all future equations. In fact, we could have eliminated N^μ from the action and transformation laws from the beginning.

The constraints are the equations of motion for e_a^α and $\psi_{\alpha A}$, which are given by

$$\left\{ +\partial_\alpha x^\mu e^{\alpha\beta} \partial_\beta x^\nu + \frac{i}{2} \bar{\chi}^\mu \gamma^a \partial_\alpha \chi^\nu - \frac{i\kappa}{2} \bar{\psi}_\beta \gamma^a \partial_\alpha x^\mu \gamma^\beta \chi^\nu - \frac{i\kappa}{2} \bar{\psi}_\alpha \partial x^\mu \gamma^\alpha \chi^\nu \right.$$

$$\left. + \frac{\kappa^2}{8} \bar{\psi}_\alpha \gamma^\beta \gamma^a \psi_\beta \bar{\chi}^\mu \chi^\nu \right\} \eta_{\mu\nu} + 2\pi\alpha' e_\alpha{}^a \mathbf{L} = 0, \tag{6.1.14}$$

where $\mathbf{L}$ is the Lagrangian, and

$$\left\{ \partial x^\mu \gamma^\alpha \chi^\nu + \frac{i\kappa}{4} \gamma^\beta \gamma^\alpha \psi_\beta \bar{\chi}^\mu \chi^\nu \right\} \eta_{\mu\nu} = 0. \tag{6.1.15}$$

As the superstring is an extended object, we will also have boundary conditions in addition to the above equations of motion as a consequence of varying the fields of the action of equation (6.1.2). The derivation of these boundary conditions follows closely that for the bosonic string given in chapter 2. From the variation of x^μ we find the equation of motion for this field and the boundary condition

$$\Pi^\mu = \frac{\delta A}{\delta \partial_\sigma x^\mu} \delta x^\mu \tag{6.1.16}$$

must vanish at a and π, while the variation of χ_A^μ implies its equation of motion and that

$$\frac{\delta A}{\delta \partial_\sigma \chi_A^\mu} \delta \chi_A^\mu \tag{6.1.17}$$

must vanish at a and π, where $a = 0$ and $a = -\pi$ for the open and closed string, respectively.

We now choose gauges which simplify the above equations considerably. The vielbein is subject to four local transformations: one local Lorentz transformation, two reparameterisations and one Weyl transformation. Using our previous theorem of chapter 2 we may use the two reparameterisations to choose the metric $g_{\alpha\beta} = e_\alpha{}^a \eta_{ab} e_\beta{}^b$ to be of the form $g_{\alpha\beta} = \eta_{\alpha\beta} e^\phi$. Consequently, $e_\alpha{}^a$ must satisfy $e_\alpha{}^a \eta_{ab} e_\beta{}^b = e^\phi \eta_{\alpha\beta}$. We can rescale $e_\alpha{}^a$ by $e_\alpha{}^a \to e^{\frac{\phi}{2}} e_\alpha{}^a$, whereupon $e_\alpha{}^a$ obeys the condition corresponding to flat space. This condition tells us that $e_\alpha{}^a$ belongs to SO(1, 1) and as such it can be set to $e_\alpha{}^a = \delta_\alpha^a$ using a Lorentz transformation whose action is unaffected by the above rescaling. Consequently, undoing the rescaling we find that $e_\alpha{}^a$ is

$$e_\alpha{}^a = \delta_\alpha^a e^{\phi/2} \text{ or } e^\alpha{}_a = \delta_a^\alpha e^{-\phi/2}. \tag{6.1.18}$$

Finally, using the Weyl symmetry we may choose $e_a^\alpha = \delta_a^\alpha$. We observe that all the above equations of motion are independent of ϕ and so this last step is not really necessary in the classical theory. In the quantum theory, except in the critical dimension $d = 10$ where no Weyl anomaly is present, one finds that the theory develops an additional degree of freedom.

The gravitino has four components, but is also subject to four local transformations: two local supersymmetries and two local 'S-supersymmetries'. As one might expect, we may choose it to vanish. To establish this fact we first use the freedom to write ψ_α in the form

$$\psi_\alpha = \gamma_\alpha \phi + \varphi_\alpha, \tag{6.1.19}$$

where the spinor φ_α is taken to be γ-traceless, namely $(\gamma^\alpha)_A^{\ B}\varphi_{\alpha B} = 0$. Consequently, φ_α has only $2 \times 2 - 2 = 2$ independent components. We may describe the spinor φ_α in terms of a spinor η without a vector index by the equation

$$\varphi_\alpha = \gamma^\beta \gamma_\alpha D_\beta \eta, \tag{6.1.20}$$

which is automatically traceless by virtue of the identity $\gamma^\alpha \gamma_\beta \gamma_\alpha = 0$. This result is apparent once we write equation (6.1.20) in terms of the light-cone and spinor notation of appendix B, whereupon it becomes

$$\varphi_+ = \begin{pmatrix} 2D_+\eta_+ \\ 0 \end{pmatrix}, \quad \varphi_- = \begin{pmatrix} 0 \\ 2D_-\eta_- \end{pmatrix}. \tag{6.1.21}$$

In this chapter we have two-dimensional spinors and vectors and so, as explained in appendix B, the meaning of $+$ depends on the object that carries it. The above zeros can be thought of as a consequence of φ_α being γ-traceless. It is clear that we can integrate to find $\eta_\pm$s which satisfy these conditions for any φ_+ and φ_-. The decomposition of equation (6.1.19) is also obvious in this notation. Inserting (6.1.20) in (6.1.19) and rewriting it we find ψ_α takes the form

$$\psi_\alpha = 2D_\alpha \eta + \gamma_\alpha(\phi - \slashed{D}\eta).$$

Using the local supersymmetry transformation of equation (6.1.7) we may choose $\psi_\alpha = \gamma_\alpha(\phi - \slashed{D}\eta)$. We may also further choose $\psi_\alpha = 0$ using the S-supersymmetry transformation of equation (6.1.10). In fact, this last step is redundant since if ψ_α has the form $\psi_\alpha = \gamma_\alpha(\phi - \slashed{D}\eta)$ it drops out of the action, at least classically.

With the above choices for $e_\alpha^{\ a}$, that is, $e_\alpha^{\ a} = \delta_\alpha^a e^{\phi/2}$ and $\psi_\alpha = \gamma_\alpha \zeta$, the above equations of motion simplify considerably. The equations of motion for x^μ and χ^μ read

$$\partial_\alpha \partial^\alpha x^\mu = 0; \quad \slashed{\partial}\chi^\mu = 0, \tag{6.1.22}$$

while the constraints become

$$T'_{\alpha\beta} \equiv \partial_\alpha x^\mu \partial_\beta x^\nu \eta_{\mu\nu} + \frac{i}{2}\bar{\chi}^\mu \gamma_\beta \partial_\alpha \chi^\nu \eta_{\mu\nu}$$
$$- \frac{\eta_{\alpha\beta}}{2}(\partial_\gamma x^\mu \partial^\gamma x^\nu + i\bar{\chi}^\mu \slashed{\partial}\chi^\nu)\eta_{\mu\nu} = 0 \tag{6.1.23}$$

and

$$J^\alpha \equiv \slashed{\partial}x^\mu \gamma^\alpha \chi^\nu \eta_{\mu\nu} = 0. \tag{6.1.24}$$

We can identify the first and second constraints with the energy-momentum tensor, $T'_{\alpha\beta}$, and supercurrent, $J_{\alpha A}$, of the action for the free fields x^μ, χ_A^μ of equation (6.1.1). This identification is an inevitable result of the Noether coupling of e_a^α and $\psi_{\alpha A}$. The energy-momentum tensor $T'_{\alpha\beta}$ is not symmetric; however, this defect can be remedied by the addition of the anti-symmetric quantity

$$\frac{i}{4}\{-(\bar{\chi}^\mu \gamma_\beta \partial_\alpha \chi^\nu) + (\bar{\chi}^\mu \gamma_\alpha \partial_\beta \chi^\nu)\}\eta_{\mu\nu}, \tag{6.1.25}$$

which vanishes on-shell as it may be rewritten as

$$+\frac{i}{4}\bar{\chi}^\mu(\gamma_\alpha\partial_\beta - \gamma_\beta\partial_\alpha)\chi^\nu\eta_{\mu\nu} = +\frac{i}{4}\bar{\chi}^\mu\epsilon_{\alpha\beta}\epsilon^{\gamma\delta}\gamma_\gamma\partial_\delta\chi^\nu\eta_{\mu\nu}$$

$$= -\frac{i}{4}\bar{\chi}^\mu\epsilon_{\alpha\beta}\gamma_5\partial\!\!\!/\chi^\nu\eta_{\mu\nu} \tag{6.1.26}$$

due to the identity $\epsilon^{\alpha\beta}\gamma_\beta = \gamma_5\gamma^\alpha$. The χ^μ_A field equation also allows us to adjust the coefficient of the final term in equation (6.1.23). The result of these steps is the energy-momentum tensor

$$T_{\alpha\beta} = \partial_\alpha x^\mu\partial_\beta x^\nu\eta_{\mu\nu} + \frac{i}{4}(\bar{\chi}^\mu\gamma_\beta\partial_\alpha\chi^\nu + \bar{\chi}^\mu\gamma_\alpha\partial_\beta\chi^\nu)\eta_{\mu\nu}$$

$$- \eta_{\mu\nu}\frac{1}{2}\left(\partial_\gamma x^\mu\partial^\gamma x^\nu + \frac{i}{2}\bar{\chi}^\mu\partial\!\!\!/\chi^\nu\right)\eta_{\alpha\beta}, \tag{6.1.27}$$

which is conserved, symmetric and traceless. As a result of the identity $\gamma^\alpha\gamma_\beta\gamma_\alpha = 0$, the conserved supercurrent also satisfies the condition

$$(\gamma_\delta)_A{}^B J^\delta_B = 0. \tag{6.1.28}$$

The tracelessness of $T_{\alpha\beta}$ and the γ-tracelessness of J^δ_B are the result of the superconformal invariance. We will now deal separately with the open and the closed superstring.

6.1.1   The open superstring

With the above gauge choices of equations (6.1.18) and (6.1.19) our boundary condition of equation (6.1.17) becomes

$$\bar{\chi}^\mu\gamma^1\delta\chi_\mu\,|^\pi_0 = 0. \tag{6.1.29}$$

We will use the notation of appendix B, which is also used in chapters 21 and 22 of [1.11], that is,

$$\chi^\mu \equiv \begin{pmatrix} \chi^\mu_+ \\ \chi^\mu_- \end{pmatrix};\ \gamma^\beta \equiv (i\sigma^2, \sigma^1),\ \gamma_5 = \gamma^0\gamma^1,\ C = -\gamma^0;$$

$$\bar{\chi}^D \equiv -\chi^{*T}\gamma^0,\ \bar{\chi}^M \equiv \chi^T C, \tag{6.1.30}$$

where $\sigma^i,\ i = 1, 2, 3$ are the Pauli matrices. Since χ^μ is a Majorana spinor the condition of equation (6.1.29) becomes

$$(\chi^\mu_-\delta\chi_{-\mu} - \chi^\mu_+\delta\chi_{\mu+})|^\pi_0 = 0. \tag{6.1.31}$$

Since the ends of the string at $\sigma = 0$ and $\sigma = \pi$ are, in general, not causally connected, we must demand that $\chi^\mu_-\delta\chi_{\mu-} = \chi^\mu_+\delta\chi_{\mu+}$ separately at $\sigma = 0$ and $\sigma = \pi$. One might be tempted, as for the x^μ coordinate to set $\chi^\mu_- = 0 = \chi^\mu_+$ at $\sigma = 0$ and $\sigma = \pi$. However, χ^μ_A obeys a first order equation and if it was subject to four boundary conditions it would not possess interesting solutions. As $\delta\chi^\mu_A$ obeys the same conditions as χ^μ_A, the constraint vanishes if we set

$$\chi^\mu_+(0) = \tau(0)\chi^\mu_-(0),\ \chi^\mu_+(\pi) = \tau(\pi)\chi^\mu_-(\pi), \tag{6.1.32}$$

provided the constants $\tau(0)$ and $\tau(\pi)$ obey $(\tau(0))^2 = 1 = (\tau(\pi))^2$. Since the field χ_+^μ is a spinor, we may always choose its phase so that at $\sigma = 0$

$$\chi_+^\mu(0) = \chi_-^\mu(0) \text{ or } \tau(0) = 1. \tag{6.1.33}$$

As a result, at $\sigma = \pi$ we are left with two choices:

$$\chi_+^\mu(\pi) = \chi_-^\mu(\pi) \qquad (\text{R}) \tag{6.1.34}$$

or

$$\chi_+^\mu(\pi) = -\chi_-^\mu(\pi) \qquad (\text{NS}). \tag{6.1.35}$$

We will refer to these choices as the Ramond (R) and Neveu–Schwarz (NS) sectors, since it is this choice of boundary condition that distinguishes the two models of [6.1] and [6.2], respectively. The above apparently innocuous choice takes on an important significance upon quantisation: one choice (R) leads to space-time fermions while the other (NS) leads to space-time bosons. The formulation will simplify considerably if we use the light-cone notation spelt out in detail in appendix B. In particular, we define $\xi^+ = \tau + \sigma$, $\xi^- = \tau - \sigma$ and so for the tensors S^μ and L_μ we have $S^+ = S^0 + S^1$, $S^- = S^0 - S^1$ and $L_+ = \frac{1}{2}(L_0 + L_1)$, $L_- = \frac{1}{2}(L_0 - L_1)$. The reader may verify that

$$\partial_+ \equiv \frac{\partial}{\partial\xi^+} = \frac{1}{2}\left(\frac{\partial}{\partial\tau} + \frac{\partial}{\partial\sigma}\right), \quad \partial_- \equiv \frac{\partial}{\partial\xi^-} = \frac{1}{2}\left(\frac{\partial}{\partial\tau} - \frac{\partial}{\partial\sigma}\right).$$

We find that the constraints of equation (6.1.27) become

$$T_{++} = \partial_+ x^\mu \partial_+ x^\nu \eta_{\mu\nu} + \frac{i}{2}\bar{\chi}^\mu \gamma_+ \partial_+ \chi^\nu \eta_{\mu\nu}, \tag{6.1.36}$$

$$T_{--} = \partial_- x^\mu \partial_- x^\nu \eta_{\mu\nu} + \frac{i}{2}\bar{\chi}^\mu \gamma_- \partial_- \chi^\nu \eta_{\mu\nu}, \tag{6.1.37}$$

while T_{+-} vanishes automatically as $T_{\alpha\beta}$ is traceless. Recalling from appendix B the definitions

$$\gamma_+ \equiv \frac{1}{2}(\gamma_0 + \gamma_1) = \begin{pmatrix} 0 & 0 \\ 1 & 0 \end{pmatrix}, \quad \gamma_- \equiv \frac{1}{2}(\gamma_0 - \gamma_1) = -\begin{pmatrix} 0 & 1 \\ 0 & 0 \end{pmatrix}, \tag{6.1.38}$$

we find that

$$T_{++} = \partial_+ x^\mu \partial_+ x^\nu \eta_{\mu\nu} - \frac{i}{2}\chi_+^\mu \partial_+ \chi_+^\nu \eta_{\mu\nu}, \tag{6.1.39}$$

$$T_{--} = \partial_- x^\mu \partial_- x^\nu \eta_{\mu\nu} - \frac{i}{2}\chi_-^\mu \partial_- \chi_-^\nu \eta_{\mu\nu}. \tag{6.1.40}$$

The supercurrent of equation (6.1.24) becomes

$$J^+ = \partial\!\!\!/ x^\mu \gamma^+ \chi^\nu \eta_{\mu\nu}, \quad J^- = \partial\!\!\!/ x^\mu \gamma^- \chi^\nu \eta_{\mu\nu}.$$

Using appendix B, aided if need be by the discussion around equation (22.37) of [1.11], we find that

$$J_+^+ = 0 = J_-^-; \quad J_-^+ = -4\partial_- x^\mu \chi_-^\nu \eta_{\mu\nu}, \quad J_+^- = -4\partial_+ x^\mu \chi_+^\nu \eta_{\mu\nu}. \tag{6.1.41}$$

As for the bosonic open string, for the open superstring we may rewrite the constraints in simplified form by extending the range of σ from 0–π to $-\pi$–π by .

$$x^\mu(\sigma) = \begin{cases} x^\mu(\sigma) & 0 \le \sigma < \pi, \\ x^\mu(-\sigma) & -\pi \le \sigma < 0, \end{cases} \qquad (6.1.42)$$

$$\Psi^\mu(\sigma) = \frac{1}{\alpha'} \begin{cases} \chi_+^\mu(\sigma) & 0 \le \sigma < \pi, \\ \chi_-^\mu(-\sigma) & -\pi \le \sigma < 0. \end{cases} \qquad (6.1.43)$$

These conditions, as discussed in chapter 2, automatically place the correct boundary conditions on $x^\mu(\sigma)$ at $\sigma = 0$. Continuity of Ψ^μ at $\sigma = 0$ then automatically encodes the boundary condition at $\sigma = 0$ of equation (6.1.33). At $\sigma = \pi$ the conditions of equations (6.1.34) and (6.1.35) become

$$\Psi^\mu(\pi) = \Psi^\mu(-\pi) \qquad (\text{R}), \qquad (6.1.44)$$

$$\Psi^\mu(\pi) = -\Psi^\mu(-\pi) \qquad (\text{NS}). \qquad (6.1.45)$$

Next we introduce the quantity

$$\wp^\mu(\sigma) = \frac{1}{\sqrt{2\alpha'}}(\dot{x}^\mu + x^{\mu'}). \qquad (6.1.46)$$

From equation (6.1.42) we find that

$$\wp^\mu(\sigma) = \frac{2}{\sqrt{2\alpha'}} \begin{cases} \partial_+ x^\mu & 0 \le \sigma < \pi \\ \partial_- x^\mu & -\pi \le \sigma < 0. \end{cases} \equiv \frac{2}{\sqrt{2\alpha'}} \Delta x^\mu. \qquad (6.1.47)$$

In terms of all the above definitions the energy-momentum constraint of equations (6.1.36) and (6.1.37) are equivalent to the condition

$$\wp^\mu \wp_\mu - i\Psi^\mu \Delta\Psi^\nu \eta_{\mu\nu} = 0 \text{ for } -\pi < \sigma < \pi, \qquad (6.1.48)$$

while the supercurrent constraint of equation (6.1.41) is

$$\wp^\mu \Psi_\mu = 0. \qquad (6.1.49)$$

Although we have reduced the number of constraint equations, the two ranges $0 \le \sigma < \pi$ and $-\pi \le \sigma < 0$ yield two equations as given in the earlier formulation.

In preparation for the quantum theory we introduce the Fourier modes. As usual we write

$$x^\mu = x_0^\mu + \frac{1}{2} \sum_{n=-\infty, \, n\neq 0}^{\infty} e^{in\sigma} x_n^\mu, \qquad (6.1.50)$$

where $x_{-n}^\mu = x_n^\mu$ and

$$\wp^\mu = \sum_{n=-\infty}^{\infty} e^{-in\sigma} \alpha_n^\mu. \qquad (6.1.51)$$

For the fermion field we must introduce modes, as dictated by the boundary conditions of equations (6.1.44) and (6.1.45). Consequently, we take

$$\Psi^\mu = \sum_{n\in\mathbf{Z}} d_n^\mu(\tau)e^{-in\sigma} \qquad \text{(R)}, \tag{6.1.52}$$

$$\Psi^\mu = \sum_{r\in\mathbf{Z}+\frac{1}{2}} b_r^\mu(\tau)e^{-ir\sigma} \qquad \text{(NS)}. \tag{6.1.53}$$

In general the indices $r, s, t, \ldots$ will be half-integer-valued, while $m, n, p, \ldots$ will be integer-valued. The spinor χ_A^μ obeys the wave equation $\slashed{\partial}\chi = 0$ or $\partial_-\chi_+ = 0 = \partial_+\chi_-$ and as a result we may write

$$\Psi^\mu = \sum_{n\in\mathbf{Z}} d_n^\mu(0)e^{-in\xi^+} \qquad \text{(R)}, \tag{6.1.54}$$

$$\Psi^\mu = \sum_{r\in\mathbf{Z}+\frac{1}{2}} b_r^\mu(0)e^{-ir\xi^+} \qquad \text{(NS)}. \tag{6.1.55}$$

The analogous discussion for $\wp^\mu$ was given in chapter 2 and we recall the equation

$$\wp^\mu = \sum_n \alpha_n^\mu(0)e^{-in\xi^+}. \tag{6.1.56}$$

For the constraints we introduce the Fourier modes

$$L_n = \frac{1}{4\pi}\int_{-\pi}^{\pi} e^{+in\sigma}\{\wp^\mu\wp_\mu - i\Psi^\mu\Delta\Psi^\nu\eta_{\mu\nu}\}d\sigma \tag{6.1.57}$$

and

$$F_n = \frac{1}{2\pi}\int_{-\pi}^{\pi} e^{+in\sigma}\wp^\mu\Psi_\mu d\sigma; \quad n = 0, \pm1, \pm3, \pm4, \ldots \qquad \text{(R)}, \tag{6.1.58}$$

$$G_r = \frac{1}{2\pi}\int_{-\pi}^{\pi} e^{+ir\sigma}\wp^\mu\Psi_\mu d\sigma; \quad r = \pm\frac{1}{2}, \pm\frac{3}{2}, \ldots \qquad \text{(NS)}. \tag{6.1.59}$$

In the gauge we are using the canonical momentum for x^μ, denoted P^μ, is given by $P^\mu = \dot{x}^\mu/2\pi\alpha'$. The only non-vanishing Poisson bracket between the coordinates and momenta is

$$\{x^\mu(\sigma), P^\nu(\sigma')\} = \delta(\sigma - \sigma')\eta^{\mu\nu} \tag{6.1.60}$$

as discussed in chapter 2. The analogous procedure for the fermion fields $\chi_\pm^\mu$ is more complicated. Firstly, we must use the Poisson bracket defined for Grassmann odd objects in chapter 1 and, secondly, as this fermion is a Majorana fermion, the canonical momentum of $\chi_\pm^\mu$ is $(i/4\pi\alpha')\chi_\pm^\mu$. In other words, $\chi_\pm^\mu$ is its own momentum. Put another way, if we denote the momentum of $\chi_\pm^\mu$ by $\zeta_\pm^\mu$ we have the constraints

$$\zeta_\pm^\mu - \frac{1}{2\pi\alpha'}\frac{i}{2}\chi_\pm^\mu = 0.$$

These constraints are, however, second class, meaning that if we compute their Poisson bracket the right-hand side is not another constraint. Once we find ourselves in such a situation, we must follow the procedure due to Dirac and replace the Poisson brackets by Dirac brackets [1.1]; the net result is

$$\{\chi_\pm^\mu(\sigma), \chi_\pm^\nu(\sigma')\} = -i\eta^{\mu\nu}2\pi\alpha'\delta(\sigma - \sigma'), \tag{6.1.61}$$

while the other brackets vanish. We note that the right-hand side is a factor of 2 smaller than had we proceeded naively and neglected the Dirac procedure.

There is another way to find equation (6.1.61); whatever the right-hand side is, it must generate the correct equation of motion through the usual Poisson bracket relation. To illustrate this method we make a small detour and consider the bosonic fields $\varphi_i(t)$, $i = 1, \ldots, n$ whose action is

$$A = \int dt \left\{ \sum_{i=1}^{n} a_i(\varphi_j) \frac{d\varphi_i}{dt} - H(\varphi_i) \right\}. \tag{6.1.62}$$

Like a free Majorana fermion, this system is subject to second class constraints associated with the fact that the momentum of φ_i is $a_i(\varphi)$. The equation of motion of φ_i is

$$F_{ij} \frac{d\varphi^j}{dt} = \frac{\partial H}{\partial \varphi_i}, \tag{6.1.63}$$

where $F_{ij} = \partial_i a_j - \partial_j a_i$ and $\partial_i a_j \equiv \partial a_j / \partial \varphi_i$. Defining F^{ij} by $F^{ij} F_{jk} = \delta_k^i$, we may write the equation of motion as

$$\frac{d\varphi^i}{dt} = F^{ij} \partial_j H. \tag{6.1.64}$$

We must compare this equation with the relation

$$\frac{d\varphi^i}{dt} = \{\varphi^i, H\} = \{\varphi^i, \varphi^j\} \frac{\partial H}{\partial \varphi^j} \tag{6.1.65}$$

and as a consequence we identify

$$\{\varphi^i, \varphi^j\} = F^{ij}. \tag{6.1.66}$$

The same strategy can be applied to the free Majorana fermion whose field equation is $\not{\partial}\chi^\mu = 0$ and whose Hamiltonian is

$$H = \frac{i}{4\pi\alpha'} \int d\sigma' \bar{\chi}^\nu(\sigma')\gamma^1 \partial_1 \chi_\nu(\sigma')$$

$$= -\frac{i}{4\pi\alpha'} \int d\sigma' \chi^\nu(\sigma')\gamma^0 \gamma^1 \partial_1 \chi_\nu(\sigma'). \tag{6.1.67}$$

As such, the time derivative of χ_A^μ is given by

$$\dot{\chi}_A^\mu = \{\chi_A^\mu, H\} = -\frac{2i}{4\pi\alpha'} \int d\sigma' \{\chi_A^\mu, \chi^\nu(\sigma')\}\gamma^0 \gamma^1 \partial_1 \chi_\nu(\sigma'). \tag{6.1.68}$$

The factor of 2 comes from the two χ_B^ν factors, which occur anti-symmetrically in H and give rise to two Poisson brackets with χ_A^μ. We must compare this result with the equation of motion $\dot{\chi}^\mu = -\gamma^0 \gamma^1 \partial_1 \chi^\mu$ from which we deduce equation (6.1.61).

For the field Ψ^μ equation (6.1.61) implies the relation

$$\{\Psi^\mu(\sigma), \Psi^\nu(\sigma')\} = -i\eta^{\mu\nu}2\pi\delta(\sigma - \sigma'). \tag{6.1.69}$$

In the above we have used the same symbol, that is, $\{,\}$, for Poisson brackets for all objects even though the precise form for two Grassmann odd quantities is different to that for other quantities. In terms of the oscillators equations (6.1.60) and (6.1.69) translate into the relations

$$\{\alpha_n^\mu, \alpha_m^\nu\} = -in\eta^{\mu\nu}\delta_{n+m,0},$$ (6.1.70)

$$\{d_n^\mu, d_m^\nu\} = -i\eta^{\mu\nu}\delta_{n+m,0}$$ (6.1.71)

and

$$\{b_r^\mu, b_s^\nu\} = -i\eta^{\mu\nu}\delta_{r+s,0}.$$ (6.1.72)

It will prove useful to evaluate the constraints in terms of the oscillators. In the NS sector equations (6.1.59) and (6.1.57) become, respectively,

$$G_r = \sum_{n=-\infty}^{\infty} \alpha_{-n} \cdot b_{r+n},$$ (6.1.73)

$$L_n = \frac{1}{2} \sum_{m=-\infty}^{\infty} \alpha_{n-m}^\mu \cdot \alpha_m^\mu + \frac{1}{2} \sum_{r=-\infty}^{\infty} b_{-r} \cdot b_{n+r}\left(r + \frac{n}{2}\right).$$ (6.1.74)

The reader may not arrive immediately at equation (6.1.74) in the above form, which is the conventional expression; however use can be made of the result

$$\sum_r b_{-r} \cdot b_{n+r} = -\sum_r b_{n+r} \cdot b_{-r} = -\sum_r b_r \cdot b_{-r+n} = 0.$$ (6.1.75)

It is straightforward to compute the algebra of the constraints:

$$\{l_n, l_m\} = -i(n-m)L_{n+m}, \quad \{L_n, G_r\} = -i\left(\frac{n}{2} - r\right)G_{n+r},$$

$$\{G_r, G_s\} = -2iL_{r+s}.$$ (6.1.76)

The analogous results in the R sector are

$$F_n = \sum_{m=-\infty}^{\infty} \alpha_{-m} \cdot d_{n+m},$$

$$L_n = \frac{1}{2} \sum_{m=-\infty}^{\infty} \alpha_{-m} \cdot \alpha_{n+m} + \frac{1}{4} \sum_{m=-\infty}^{\infty} (2m+n)d_{-m} \cdot d_{m+n},$$ (6.1.77)

with the corresponding algebra

$$\{L_n, L_m\} = -i(n-m)L_{n+m}, \quad \{L_n, F_m\} = -i\left(-m + \frac{n}{2}\right)F_{n+m},$$

$$\{F_n, F_m\} = -2iL_{n+m}.$$ (6.1.78)

6.1.2   The closed superstring

This construction is similar to that for the open superstring given earlier; however, one finds that the closed nature of the superstring leads to some significant differences from the open case when one considers the boundary conditions, particularly in the fermionic sector. For the closed superstring the range of σ is from the outset from $-\pi$ to π and as a consequence we get twice as many modes as for the open superstring with a corresponding doubling of the constraints. The string coordinate $x^\mu(\tau, \sigma)$ obeys the boundary condition $x^\mu(\tau, \sigma) = x^\mu(\tau, \sigma + 2\pi)$ and so the boundary conditions involving only this field automatically vanish, as they did for the closed bosonic string studied in chapter 2.

The closed superstring is parameterised by σ in the range $-\pi < \sigma < \pi$. The boundary condition of equation (6.1.31) now becomes

$$\chi^\mu_- \delta \chi_{\mu-} \big|^\pi_{-\pi} = \chi^\mu_+ \delta \chi_{\mu+} \big|^\pi_{-\pi}. \tag{6.1.79}$$

We impose the boundary condition on χ^μ_- and χ^μ_+ separately and demand that they be either periodic or anti-periodic. The difference from the open superstring is that $\chi^\mu_\pm$ are now independent and $\sigma = \pi$ and $\sigma = -\pi$ represent the same point on the string. We can have

$$\chi^\mu_+(\pi) = \chi^\mu_+(-\pi) \qquad \text{(R)} \qquad \text{or}$$
$$\tag{6.1.80}$$
$$\chi^\mu_+(\pi) = -\chi^\mu_+(-\pi) \qquad \text{(NS)}$$

and

$$\chi^\mu_-(\pi) = \chi^\mu_-(-\pi) \qquad \text{(R)} \qquad \text{or}$$
$$\tag{6.1.81}$$
$$\chi^\mu_-(\pi) = -\chi^\mu_-(-\pi) \qquad \text{(NS)}.$$

This gives us four possibilities which we may write as (R, R), (NS, R), (R, NS) and (NS, NS) where the first entry corresponds to the boundary condition in the χ_+ sector and the second entry to that in the χ_- sector.

The constraints for the closed string are those of equations (6.1.27) and (6.1.28). In light-cone notation these become equations (6.1.39) and (6.1.40) which we may write as

$$T_{++} = \partial_+ x^\mu \partial_+ x_\mu - \frac{i}{2} \chi^\mu_+ \partial_+ \chi_{\mu+} = 0,$$
$$\tag{6.1.82}$$
$$T_{--} = \partial_- x^\mu \partial_- x_\mu - \frac{i}{2} \chi^\mu_- \partial_- \chi_{\mu-} = 0$$

as well as those of equation (6.1.41), which become

$$J_{++} = 2\partial_+ x^\mu \chi_{+\mu} = 0, \quad J_{--} = 2\partial_- x^\mu \chi_{-\mu} = 0. \tag{6.1.83}$$

The normal modes for x^μ are, as in chapter 2,

$$x^\mu = \sum_{n=-\infty}^{\infty} e^{in\sigma} x^\mu_n \tag{6.1.84}$$

and the corresponding oscillators are defined by

$$\wp^\mu = \frac{1}{\sqrt{2\alpha'}} (\dot{x}^\mu + x^{\mu'}) = \frac{2}{\sqrt{2\alpha'}} \partial_+ x^\mu = \sum_n \alpha^\mu_n e^{-in\sigma}$$
$$\tag{6.1.85}$$
$$\bar\wp^\mu = \frac{1}{\sqrt{2\alpha'}} (\dot{x}^\mu - x^{\mu'}) = \frac{2}{\sqrt{2\alpha'}} \partial_- x^\mu = \sum_n \bar\alpha^\mu_n e^{in\sigma},$$

whose non-zero Poisson brackets are

$$\{\alpha_n^\mu, \alpha_m^\nu\} = -in\delta_{n+m,0}\eta^{\mu\nu},$$

$$\{\bar\alpha_n^\mu, \bar\alpha_m^\nu\} = -in\delta_{n+m,0}\eta^{\mu\nu}.$$

(6.1.86)

For $\chi_\pm^\mu$, and corresponding to the above boundary conditions, we introduce the oscillators

$$\chi_+^\mu = \sqrt{\alpha'}\sum_{n\in\mathbf{Z}} d_n^\mu(\tau)e^{-in\sigma} \qquad (R)$$

(6.1.87)

or

$$\chi_+^\mu = \sqrt{\alpha'}\sum_{r\in\mathbf{Z}+\frac{1}{2}} b_r(\tau)e^{-ir\sigma} \qquad (NS)$$

(6.1.88)

and

$$\chi_-^\mu = \sqrt{\alpha'}\sum_{n\in\mathbf{Z}} \bar d_n^\mu(\tau)e^{+in\sigma} \qquad (R)$$

(6.1.89)

or

$$\chi_-^\mu = \sqrt{\alpha'}\sum_{r\in\mathbf{Z}+\frac{1}{2}} \bar b_r^\mu(\tau)e^{+ir\sigma} \qquad (NS).$$

(6.1.90)

As χ^μ satisfies the wave equation the τ dependence of d_n^μ, b_r^μ, $\bar d_n^\mu$ and $\bar b_r^\mu$ is found to be

$$\chi_+^\mu = \sqrt{\alpha'}\sum_n d_n^\mu(0)e^{-in\xi^+} \qquad (R)$$

(6.1.91)

or

$$\chi_+^\mu = \sqrt{\alpha'}\sum_r b_r(0)e^{-ir\xi^+} \qquad (NS)$$

(6.1.92)

and

$$\chi_-^\mu = \sqrt{\alpha'}\sum_n \bar d_n^\mu(0)e^{-in\xi^-} \qquad (R)$$

(6.1.93)

or

$$\chi_-^\mu = \sqrt{\alpha'}\sum_r \bar b_r^\mu(0)e^{-ir\xi^-} \qquad (NS).$$

(6.1.94)

The discussion concerning the Poisson brackets of a free Majorana fermion and second class constraints which is above equation (6.1.61) is equally valid for the closed superstring and consequently

$$\{\chi_\pm^\mu(\sigma),\ \chi_\pm^\nu(\sigma')\} = -i\eta^{\mu\nu}2\pi\alpha'\delta(\sigma-\sigma').$$

(6.1.95)

This translates into the following relations for the oscillators of equations (6.1.87)–(6.1.90):

$$\{d_n^\mu, d_m^\nu\} = -i\eta^{\mu\nu}\delta_{n+m,0}, \quad \{b_r^\mu, b_s^\nu\} = -i\eta^{\mu\nu}\delta_{r+s,0}$$

(6.1.96)

and

$$\{\bar d_n^\mu, \bar d_m^\nu\} = -i\eta^{\mu\nu}\delta_{n+m,0}, \quad \{\bar b_r^\mu, \bar b_s^\nu\} = -i\eta^{\mu\nu}\delta_{r+s,0}.$$

(6.1.97)

We also introduce the Fourier transforms of the constraints of equations (6.1.82) and (6.1.83):

$$L_n = \frac{2}{\alpha'} \cdot \frac{1}{4\pi} \int_{-\pi}^{\pi} T_{++} e^{in\sigma} \, d\sigma, \quad \bar{L}_n = \frac{2}{\alpha'} \cdot \frac{1}{4\pi} \int_{-\pi}^{\pi} T_{--} e^{-in\sigma} \, d\sigma, \qquad (6.1.98)$$

but

$$F_n = \frac{1}{\sqrt{2\alpha'}} \cdot \frac{1}{2\pi} \int_{-\pi}^{\pi} e^{in\sigma} J_{++} d\sigma \qquad (R),$$

$$G_r = \frac{1}{\sqrt{2\alpha'}} \cdot \frac{1}{2\pi} \int_{-\pi}^{\pi} e^{ir\sigma} J_{++} d\sigma \qquad (NS) \tag{6.1.99}$$

and

$$\bar{F}_n = \frac{1}{\sqrt{2\alpha'}} \cdot \frac{1}{2\pi} \int_{-\pi}^{\pi} e^{-in\sigma} J_{--} d\sigma \qquad (R),$$

$$\bar{G}_r = \frac{1}{\sqrt{2\alpha'}} \cdot \frac{1}{2\pi} \int_{-\pi}^{\pi} e^{-ir\sigma} J_{--} d\sigma \qquad (NS). \tag{6.1.100}$$

In terms of oscillators they become

$$L_n = \frac{1}{2} \sum_{m=-\infty}^{\infty} \alpha_{n+m} \cdot \alpha_{-m} + \frac{1}{2} \sum_{r=-\infty}^{\infty} \left(r + \frac{n}{2} \right) b_{-r} \cdot b_{n+r},$$

$$G_r = \sum_{m=-\infty}^{\infty} \alpha_m \cdot b_{r-m} \tag{6.1.101}$$

in the NS sector and

$$L_n = \frac{1}{2} \sum_{m=-\infty}^{\infty} \alpha_{n+m} \cdot \alpha_{-m} + \frac{1}{2} \sum_{m=-\infty}^{\infty} \left(m + \frac{n}{2} \right) d_{-m} \cdot d_{m+n},$$

$$F_n = \sum_{m=-\infty}^{\infty} \alpha_m \cdot d_{n-m} \tag{6.1.102}$$

in the R sector. The corresponding expressions for $\bar{L}_n$, $\bar{F}_n$ and $\bar{G}_r$ are obtained by placing bars on all the oscillators. The Poisson bracket algebra of their constraints is as in equations (6.1.76) and (6.1.78) with the barred operators obeying the same algebra. The bracket between the barred and unbarred generators vanishes.

6.2   The Green–Schwarz formulation

The treatment of the ten-dimensional string given earlier in this chapter started from an action which possessed two-dimensional (that is, world-sheet) supersymmetry. We discovered that there was a choice of boundary conditions for the world-sheet fermions which lead to a number of sectors. In the next chapter, where we will quantise this theory, we will find that for the open string this leads to space-time fermions in the R sector, while the NS sector contains space-time bosons. We will also show that when both of these two possibilities are included and the theory is subjected to a certain GSO projection, which eliminates some physical states, the remaining physical states are compatible with the theory possessing a space-time supersymmetry. Indeed, the full theory does indeed possess a space-time super-symmetry. A similar result also holds for the closed string. This procedure 'seems' to be a little ad hoc in that it uses two distinct sectors, one for fermions and one for bosons and then requires a projection. It might seem better if one could have a formulation of the superstring which had fermions and bosons unified in such a way that space-time supersymmetry was manifest. It was this desire that lead Green and Schwarz to a formulation of the superstring [6.5] which does indeed possess a manifest space-time supersymmetry as a symmetry of the two-dimensional action. As we shall see, however, we must pay a high price: we will lose, as a manifest symmetry of the action, the infinite-dimensional superconformal group in two dimensions which played such an important role in our previous considerations. The Green–Schwarz formulation is difficult to quantise in a manifestly Lorentz invariant man-ner. Nonetheless, it is this form of the string that provides the most natural generalisation to branes.

To incorporate space-time supersymmetry we follow the pattern of the Brink–Schwarz point particle given in chapter 1. The action consists of the field $x^\mu(\tau, \sigma)$, $\mu = 0, 1, \ldots, D-1$, and the Grassmann odd field $\theta^{Ai}(\tau, \sigma)$, which is a spinor of the back-ground space-time, with corresponding index A. It also possesses an internal index $i = 1, 2$. The spinor index A runs over the dimension of the Clifford algebra, which varies depending on the space-time dimension D and the type of spinor that is, Majorana, Weyl or Majorana–Weyl. The Green–Schwarz string moves through a background superspace which is the same as the superspace used to construct supersymmetric theories in the back-ground space-time. Indeed, it can be viewed as the coset of the background super Poincaré group with sub-group the Lorentz group. This is in contrast to the R–NS string discussed in the previous section.

The supersymmetry transformations on these fields are just those of the background superspace:

$$\delta\theta^i = \epsilon^i, \quad \delta x^\mu = i\bar\epsilon^j \gamma^\mu \theta^j, \tag{6.2.1}$$

where ϵ^i, $i = 1, 2$ is a constant spinor of the same type as θ^i. We realise that the quantity

$$\Pi^\mu_\alpha = \partial_\alpha x^\mu - i\bar\theta^j \gamma^\mu \partial_\alpha \theta^j \tag{6.2.2}$$

is invariant under supersymmetry and therefore it is natural to consider the action

$$S_1 = -\frac{1}{2\pi} \int d^2\xi \sqrt{-g} g^{\alpha\beta} \Pi^\mu_\alpha \Pi^\nu_\beta \eta_{\mu\nu}, \tag{6.2.3}$$

which is found by substituting $\partial_\alpha x^\mu \to \Pi^\mu_\alpha$ in the usual expression for the bosonic string action given in chapter 1. This term, however, is not sufficient, for it does not possess the additional κ symmetry present for the Brink–Schwarz particle, which is required in order to get the correct on-shell degrees of freedom for θ^i. To recover this symmetry we consider the action

$$S = S_1 + S_2, \tag{6.2.4}$$

where

$$S_2 = \frac{1}{\pi} \int d^2\xi \, \epsilon^{\alpha\beta} \{-i\partial_\alpha x^\mu (\bar{\theta}^1 \gamma^\nu \partial_\beta \theta^1 - \bar{\theta}^2 \gamma^\nu \partial_\beta \theta^2) + \bar{\theta}^1 \gamma^\mu \partial_\alpha \theta^1 \bar{\theta}^2 \gamma^\nu \partial_\beta \theta^2\} \eta_{\mu\nu}.$$

(6.2.5)

Clearly, S is manifestly Lorentz invariant and, since $\epsilon^{\alpha\beta}$ is a tensor density, it is also reparameterisation invariant. However, the second piece, S_2, is not obviously space-time supersymmetric. Its variation is given by

$$\delta S_2 = \frac{1}{\pi} \int d^2\xi \, \epsilon^{\alpha\beta} \{\bar{\epsilon}^1 \gamma^\mu \partial_\alpha \theta^1 \theta^1 \gamma_\mu \partial_\beta \theta^1 - \bar{\epsilon}^2 \gamma^\mu \partial_\alpha \theta^2 \bar{\theta}^2 \gamma_\mu \partial_\beta \theta^2\},$$

(6.2.6)

where we have discarded total derivatives such as

$$\frac{1}{\pi} \int d^2\xi \, \epsilon^{\alpha\beta} \partial_\alpha x^\mu \bar{\epsilon}^1 \gamma_\mu \partial_\beta \theta^1.$$

(6.2.7)

We may consider each of the terms in δS_2 separately. Dropping the index on θ and adding a total derivative the first term can be reexpressed as

$$\frac{1}{\pi} \int d^2\xi \, \epsilon^{\alpha\beta} \{\bar{\epsilon} \gamma^\mu \partial_\alpha \theta \bar{\theta} \gamma_\mu \partial_\beta \theta - \tfrac{1}{3} \partial_\alpha (\bar{\epsilon} \gamma^\mu \theta \cdot \bar{\theta} \gamma_\mu \partial_\beta \theta)\}$$

$$= \frac{1}{\pi} \int d^2\xi \, \epsilon^{\alpha\beta} \{\tfrac{2}{3} \bar{\epsilon} \gamma^\mu \partial_\alpha \theta \bar{\theta} \gamma_\mu \partial_\beta \theta - \tfrac{1}{3} \bar{\epsilon} \gamma^\mu \theta \partial_\alpha \bar{\theta} \gamma_\mu \partial_\beta \theta\}$$

$$= \frac{1}{3\pi} \int d^2\xi \, \{\bar{\epsilon}\} \{2\gamma^\mu \dot{\theta} \bar{\theta} \gamma_\mu \theta' - 2\gamma^\mu \theta' \bar{\theta} \gamma_\mu \dot{\theta} - \gamma^\mu \theta \dot{\bar{\theta}} \gamma_\mu \theta' + \gamma^\mu \theta \bar{\theta}' \gamma_\mu \dot{\theta}\}$$

$$= \frac{1}{3\pi} \int d^2\xi \, \{\bar{\epsilon}\} \{2(\gamma^\mu \dot{\theta} \bar{\theta} \gamma_\mu \theta' + \gamma^\mu \theta' \bar{\theta} \gamma_\mu \dot{\theta} + \gamma^\mu \theta \bar{\theta}' \gamma_\mu \dot{\theta})$$

$$-2\gamma^\mu \theta' \{\bar{\theta} \gamma_\mu \dot{\theta} + \dot{\bar{\theta}} \gamma_\mu \theta\} - \gamma^\mu \theta \{\dot{\bar{\theta}} \gamma_\mu \theta' + \bar{\theta}' \gamma_\mu \dot{\theta}\}\}$$

(6.2.8)

where

$$\theta' = \frac{\partial \theta}{\partial \sigma} \text{ and } \dot{\theta} = \frac{\partial \theta}{\partial \tau}.$$

(6.2.9)

Every term in the above is of the form

$$\bar{\epsilon} \gamma^\mu \lambda \bar{\psi} \gamma_\mu \chi,$$

(6.2.10)

where λ, ψ and χ are distinct spinors. If θ is a Majorana spinor (see chapter 5), then the last two terms in brackets vanish since $\bar{\chi} \gamma_\mu \lambda = -\bar{\lambda} \gamma_\mu \chi$ for two Majorana spinors χ and λ. The remaining three terms are then obtained by cycling the spinors of the first term to yield two more terms. To show that three such cycled terms vanish we must use the Fierz rearrangement. In any dimension this has the generic form

$$\delta_A^B \delta_C^D = \frac{1}{D_F} \sum_R c_R (\gamma_R)_A{}^D (\gamma_R)_C{}^B,$$

(6.2.11)

where $D_F = 2^{D/2}$, γ_R are a complete set of matrices and c_R are numbers. A Dirac spinor has $2^{D/2}$ components in D dimensions if D is even, corresponding to the dimension of the Clifford algebra generated by γ^μ, $\mu = 0, 1, \ldots, D - 1$. As discussed in chapter 5 a complete set of matrices for the space of $2^{D/2} \times 2^{D/2}$ matrices is provided by

$$\gamma_{\mu_1 \cdots \mu_n} \equiv \gamma_{[\mu_1 \cdots \mu_n]}, \ n = 0, \ldots, D. \tag{6.2.12}$$

The square bracket denotes anti-symmetry of the indices and the normalisation is such that $\gamma_{\mu_1 \cdots \mu_n} = \gamma_{\mu_1} \gamma_{\mu_2} \cdots \gamma_{\mu_n}$ if $\mu_1, \ldots, \mu_n$ are distinct. By taking traces it is straight-forward to verify that this set is linearly independent. There are

$$\sum_{n=0}^{D} \frac{D!}{(D-n)(n!)} = (1+1)^D = 2^D \tag{6.2.13}$$

such matrices, demonstrating that they do indeed form a basis of all $2^{D/2} \times 2^{D/2}$ matrices.

If we define

$$\gamma_{D+1} = \gamma^{[0,1,\ldots,D-1]} = \gamma^0 \gamma^1 \cdots \gamma^{D-1}, \tag{6.2.14}$$

then, for D even, half of the above matrices may be expressed in terms of γ_{D+1} and the other half using the relation

$$\gamma_{\mu_1 \cdots \mu_n} = \frac{s}{(D-n)!} \epsilon_{\mu_1 \cdots \mu_n \mu_{n+1} \cdots \mu_D} \gamma^{\mu_{n+1} \cdots \mu_D} \gamma_{D+1}, \tag{6.2.15}$$

where $\epsilon^{01 \cdots D-1} = +1 = -\epsilon_{01 \cdots D-1}$ and s is a constant. To verify this relation we take the trace of the right-hand side with $\gamma_{\nu_1 \cdots \nu_p}$; $p = 0, 1, \ldots, D$ from which it is clear that the left-hand side must only be proportional to $\gamma_{\mu_1 \cdots \mu_n}$ as written above. The constant of proportionality s is most easily found by taking specific values. Let us choose $\mu_i = i - 1$ for $i = 1, \ldots, n$. On the left-hand side we find

$$\gamma_{0 \cdots n-1} = \gamma_0 \cdots \gamma_{n-1}, \tag{6.2.16}$$

while on the right-hand side we find

$$\frac{s}{(D-n)!} \epsilon_{0 \cdots n-1 \mu_{n+1} \cdots \mu_D} \gamma^{\mu_{n+1} \cdots \mu_D} \gamma_{D+1} = -s \gamma^n \gamma^{n+1} \cdots \gamma^{D-1} \gamma^0 \gamma^1 \cdots \gamma^{D-1}$$

$$= -s(-1)^{n(D-n)} (-\gamma_0) \gamma_1 \cdots \gamma_{n-1} \gamma^n \gamma^{n+1} \cdots \gamma^{D-1} \gamma^n \cdots \gamma^{D-1}$$

$$= s(-1)^{n(D-n)} \prod_{p=1}^{D-n} (-1)^{D-n-p} \gamma_0 \gamma_1 \cdots \gamma_{n-1}. \tag{6.2.17}$$

Consequently, we conclude that

$$s = (-1)^{n(D-n)} \prod_{p=1}^{D-n} (-1)^{D-n-p} = (-1)^{(D-n)/2(D+n-1)}. \tag{6.2.18}$$

Since the γ_R matrices form a complete set, we may always express $\delta_A^B \delta_C^D$ as

$$\delta_A^B \delta_C^D = \sum_{R,S} (\gamma^R)_A{}^D (\gamma^S)_C{}^B c_{R,S}.$$

The coefficients $c_{R,S}$ in the Fierz formula (6.2.11) are then found by taking traces with $(\gamma_S)_B{}^C$ to give

$$(\gamma_T)_A{}^D = \frac{1}{D_F} \sum_{R,S} c_{R,S}(\gamma_R)_A{}^D \mathrm{Tr}(\gamma_T \gamma_S). \tag{6.2.19}$$

For the above basis one finds that $\mathrm{Tr}(\gamma_R \gamma_S) = e_R \delta_{RS}$, where e_R are a set of numbers which are easily calculated by taking specific values for the indices. From the linear independence of this basis it follows that $c_{R,S} = c_R \delta_{R,S}$ and that

$$c_R = \frac{1}{D_F e_R}. \tag{6.2.20}$$

Using this formula equation (6.2.10) becomes

$$\sum_R \left\{ -\frac{c_R}{D_F} \bar{\epsilon} \gamma^\mu \gamma_R \gamma_\mu \chi \, \bar{\psi} \gamma_R \lambda \right\}. \tag{6.2.21}$$

Whether this vanishes depends on the dimension of space-time and the nature of the spinors ψ, χ and λ.

We now consider the case of four dimensions in detail. The complete set of γ matrices can be written as

$$\gamma_R = (1, \ \gamma_5, \ \gamma_\mu, \ \gamma_\mu \gamma_5, \ \gamma_{\mu\nu}) \tag{6.2.22}$$

and the corresponding c_R are

$$c_R = (1, \ 1, \ \eta_{\mu\nu}, \ -\eta_{\mu\nu}, \ -\tfrac{1}{4}(\eta_{\mu\rho}\eta_{\nu\kappa} - \eta_{\nu\rho}\eta_{\mu\kappa})). \tag{6.2.23}$$

From equation (6.1.21) we observe that

$$\bar{\epsilon} \gamma^\mu \lambda \bar{\psi} \gamma_\mu \chi = -\bar{\epsilon} \chi \bar{\psi} \lambda + \bar{\epsilon} \gamma_5 \chi \bar{\psi} \gamma_5 \lambda + \tfrac{1}{2} \bar{\epsilon} \gamma_\nu \chi \bar{\psi} \gamma^\nu \lambda + \tfrac{1}{2} \bar{\epsilon} \gamma_\nu \gamma_5 \chi \bar{\psi} \gamma^\nu \gamma_5 \lambda. \tag{6.2.24}$$

If θ is a Majorana spinor, then so are θ, $\dot{\theta}$ and θ' and hence so are λ, ψ and χ. As noted above, the term we are interested in is cycled. Correspondingly, cycling equation (6.2.24) we find, after using the following relation, which is valid for any two Majorana spinors:

$$\bar{\chi} \gamma_R \lambda = -d_R \bar{\lambda} \gamma_R \chi,$$

where $d_R = (1, 1, -1, 1, -1)$, that

$$\tfrac{3}{2} \bar{\epsilon} \gamma^\mu \lambda \bar{\psi} \gamma_\mu \chi + \mathrm{cyclic} = (-\bar{\epsilon} \chi \bar{\psi} \lambda + \bar{\epsilon} \gamma_5 \chi \bar{\psi} \gamma_5 \lambda + \tfrac{1}{2} \bar{\epsilon} \gamma_\nu \gamma_5 \chi \bar{\psi} \gamma^\nu \gamma_5 \lambda) + \mathrm{cyclic}. \tag{6.2.25}$$

However, while the left-hand side of this equation is anti-symmetric under the exchange of any two spinors, the right-hand side is symmetric. For example, the first term on the left-hand side can be written as

$$-\bar{\epsilon} \chi \bar{\psi} \lambda + \mathrm{cyclic} = -\bar{\epsilon} \chi \bar{\psi} \lambda - \bar{\epsilon} \lambda \bar{\chi} \psi - \bar{\epsilon} \psi \bar{\lambda} \chi = -\bar{\epsilon} \chi \bar{\lambda} \psi - \bar{\epsilon} \lambda \bar{\psi} \chi - \bar{\epsilon} \psi \bar{\chi} \lambda, \tag{6.2.26}$$

from which the symmetry is manifest once one uses the fact that $\bar{\chi}\lambda = \bar{\lambda}\chi$. Taking symmetric and anti-symmetric combinations we conclude that each side of equation (6.2.25) vanishes identically, that is,

$$\bar{\epsilon}\gamma_\mu\lambda\bar{\psi}\gamma_\mu\chi + \text{cyclic} = 0. \tag{6.2.27}$$

Hence, we may conclude, in four dimensions, that if θ is a Majorana spinor, $\delta S_2 = 0$, so that S is invariant under space-time supersymmetry.

Let us now suppose that θ is a Weyl spinor; equation (6.2.24) then becomes

$$\bar{\epsilon}\gamma^\mu\lambda\bar{\psi}\gamma_\mu\chi = \bar{\epsilon}\gamma_\mu\chi\bar{\psi}\gamma^\mu\lambda \tag{6.2.28}$$

since $\pm\chi = i\gamma_5\chi$ and similarly for λ and ψ. Examining equation (6.2.8) without the total derivative term we find that

$$\frac{1}{\pi}\int d^2\xi\epsilon^{\alpha\beta}\{\bar{\epsilon}\gamma_\mu\dot{\theta}\bar{\theta}\gamma_\mu\theta' - \bar{\epsilon}\gamma_\mu\theta'\bar{\theta}\gamma_\mu\dot{\theta}\} = 0. \tag{6.2.29}$$

Consequently, S_2 also possesses space-time supersymmetry in four dimensions if θ^A is a Weyl spinor.

In fact, S_2 has space-time supersymmetry in three dimensions if θ is Majorana, in four dimensions if θ is Majorana or Weyl, in six dimensions if θ is Weyl and in ten dimensions if θ is Majorana–Weyl.

We will find in the next chapter that the superstring is only consistent as a quantum theory in ten dimensions, so it is this case which is of most interest to us and which we now investigate in a manner similar to the four-dimensional case. We now assume θ is both Majorana and Weyl. As discussed in chapter 5 such spinors only exist in a Minkowski space-time that has dimension $8n + 2$, $n = 0, 1, 2, 3 \ldots$. As such, we must show that

$$\bar{\epsilon}\gamma^\mu\lambda\bar{\psi}\gamma_\mu\chi + \text{cyclic} = 0, \tag{6.2.30}$$

where λ, ψ and χ are any Majorana–Weyl spinors of the same Weyl chirality. The complete set of matrices in ten dimensions can be taken to be $\gamma_{\mu_1\cdots\mu_n}$, $n = 0, 1, \ldots, 5$, $\gamma_{\mu_1\cdots\mu_n}\gamma_{11}$, $n = 1, \ldots, 4$ and γ_{11}. Since the spinors are Weyl, $\gamma_{\mu_1\cdots\mu_n}\gamma_{11} = \pm\gamma_{\mu_1\cdots\mu_n}$ when between the spinors, the sign depending on their chirality. We also note that $\bar{\psi}\gamma_{\mu_1\cdots\mu_n}\chi$ vanishes if n is even due to the chirality of $\bar{\psi}$ and χ. This is easily seen by inserting γ_{11} and using that $\bar{\psi}\gamma_{11} = \mp\bar{\psi}$ if $\gamma_{11}\psi = \pm\psi$ which follows from equation (5.3.5) and equation (5.2.16) taking $p = 10$. Consequently, when carrying out the step given in equation (6.2.125) we only obtain a sum with $\gamma_{\mu_1\cdots\mu_n}$, $n = 1, 3, 5$ with suitably modified c_R. As such,

$$\bar{\epsilon}\gamma^\mu\lambda\bar{\psi}\gamma_\mu\chi = -\sum_R \frac{c_R}{D_F}\bar{\epsilon}\gamma^\mu\gamma_R\gamma_\mu\chi\bar{\psi}\gamma_R\lambda$$

$$= -c^1\bar{\epsilon}\gamma^\mu\gamma^\lambda\gamma_\mu\chi\bar{\psi}\gamma_\lambda\lambda - d^1\bar{\epsilon}\gamma^\mu\gamma_{\lambda_1\lambda_2\lambda_3}\gamma_\mu\chi\bar{\psi}\gamma^{\lambda_1\lambda_2\lambda_3}\lambda \tag{6.2.31}$$

for suitable constants c^1 and d^1, using the identity $\gamma^\mu\gamma_{\lambda_1\cdots\lambda_p}\gamma_\mu = (-1)^p(D - 2p)\gamma_{\lambda_1\cdots\lambda_p}$, which is valid in any dimension D. As such, the term with $\gamma_{\mu_1\cdots\mu_5}$ is absent in ten dimensions as $\gamma_\mu\gamma_{\lambda_1\cdots\lambda_5}\gamma^\mu = 0$ and the other two terms may be easily processed. Taking the cyclic combination of equation (6.2.31) in the three spinors we therefore find that

$$\bar{\epsilon}\gamma^\mu\lambda\bar{\psi}\gamma_\mu\chi + \text{cyclic} = -8c^1\bar{\epsilon}\gamma^\lambda\chi\bar{\psi}\gamma_\lambda\lambda - 4d^1\bar{\epsilon}\gamma^\mu\gamma_{\lambda_1\lambda_2\lambda_3}\gamma_\mu\chi\bar{\psi}\gamma^{\lambda_1\lambda_2\lambda_3}\lambda$$

$$+ \text{cyclic}. \tag{6.2.32}$$

However, for any two Majorana spinors $\bar{\psi}\lambda = \bar{\lambda}\psi$, $\bar{\psi}\gamma_\mu\lambda = -\bar{\lambda}\gamma_\mu\psi$, but $\bar{\psi}\gamma_{\mu_1\mu_2\mu_3}\lambda = \bar{\lambda}\gamma_{\mu_1\mu_2\mu_3}\psi$. Taking symmetry and anti-symmetry with respect to any two of the spinors implies the desired result, namely equation (6.1.32). Thus we have shown that S is space-time supersymmetric in ten dimensions for Majorana–Weyl spinors.

We now discuss the κ invariance of the action of equation (6.2.4). For the point particle of chapter 1 this invariance was $\delta x^\mu = i\bar{\theta}^j\gamma^\mu\delta\theta^j$, $\delta\theta^j = i\gamma^\mu\Pi_\mu\kappa^j$. When extending to the case of the string we can expect in $\delta\theta^j$ to replace Π^μ by Π_α^μ and so the local parameter should have the three indices $\kappa^{A\alpha i}$: a two-dimensional world index α; a space-time spinor index A and an internal index $i = 1, 2$. The resulting transformations are

$$\delta x^\mu = i\bar{\theta}^j\gamma^\mu\delta\theta^j,$$
$$\delta\theta^i = 2i\gamma^\mu\Pi_{\mu\alpha}\kappa^{i\alpha}. \tag{6.2.33}$$

The parameter $\kappa^{\alpha i}$ (the A index being suppressed) is, however, subject to the projection conditions

$$\kappa^{1\alpha} = P^\alpha_{-\beta}\kappa^{1\beta}, \quad \kappa^{2\alpha} = P^\alpha_{+\beta}\kappa^{2\beta},$$

where

$$P^\alpha_{\pm\beta} = \frac{1}{2}\left(\delta^\alpha_\beta \pm \frac{\epsilon^{\alpha\delta}g_{\delta\beta}}{\sqrt{-g}}\right). \tag{6.2.34}$$

The variation of the two-dimensional metric $g_{\alpha\beta}$ will be given later.

This condition on $\kappa^{\alpha i}$ is the local analogue of the condition in flat two-dimensional space-time that decomposes a vector into its two irreducible parts. From a group theory point of view, the two-dimensional Lorentz group is $SO(1, 1)$ and as is well known all irreducible representations of a commutative group are one-dimensional. Technically, the decomposition relies on the existence of the $\epsilon_{\alpha\beta}$ tensor. Locally we note that

$$-\epsilon_{\gamma\delta}g = \epsilon^{\alpha\beta}g_{\alpha\gamma}g_{\beta\delta}, \tag{6.2.35}$$

where $g = \det g_{\alpha\beta}$. Using this relation we find that $P^\beta_{\pm\alpha}$ satisfy the conditions of a projector, that is,

$$P^\beta_{\pm\alpha}P^\gamma_{\pm\beta} = P^\gamma_{\pm\alpha},$$
$$P^\beta_{\pm\alpha}P^\gamma_{\mp\beta} = 0. \tag{6.2.36}$$

The variation of Π_α^μ under the κ transformations of equations (6.2.33) is

$$\delta\Pi_\alpha^\mu = 2i\partial_\alpha\bar{\theta}^j\gamma^\mu\delta\theta^j, \tag{6.2.37}$$

since θ^j is a Majorana spinor. The variation of S_1 is then given by

$$\delta S_1 = -\frac{1}{\pi}\int d^2\xi\{\sqrt{-g}g^{\alpha\beta}2i\partial_\alpha\bar{\theta}^j\gamma^\mu\delta\theta^j \cdot \Pi^\nu_\beta\eta_{\mu\nu} + \tfrac{1}{2}\delta(\sqrt{-g}g^{\alpha\beta})\Pi^\mu_\alpha\Pi^\nu_\beta\eta_{\mu\nu}\}.$$

$$\tag{6.2.38}$$

To determine the variation of $\sqrt{-g}g^{\alpha\beta}$ we must isolate the $\Pi^2\partial\theta\kappa$ terms. To this end we write S_2 as

$$S_2 = \frac{1}{\pi}\int d^2\xi\,\epsilon^{\alpha\beta}\{(-i\Pi^\mu_\alpha)(\bar\theta^1\gamma_\mu\partial_\beta\theta^1 - \bar\theta^2\gamma_\mu\partial_\beta\theta^2) + \bar\theta^1\gamma^\mu\partial_\alpha\theta^1\bar\theta^2\gamma_\mu\partial_\beta\theta^2$$
$$+ \bar\theta^j\gamma^\mu\partial_\alpha\theta^j(\bar\theta^1\gamma_\mu\partial_\beta\theta^1 - \bar\theta^2\gamma_\mu\partial_\beta\theta^2)\}. \tag{6.2.39}$$

We can carry out the calculation as an order by order calculation in θ. The lowest such term is

$$\delta S = +\frac{1}{\pi}\int d^2\xi\left\{-\frac{1}{2}\Pi^\mu_\alpha\Pi_{\beta\mu}\delta(\sqrt{-g}g^{\alpha\beta}) - 2i\sqrt{-g}g^{\alpha\beta}\partial_\alpha\bar\theta^j\gamma^\mu\Pi_{\mu\beta}\delta\theta^j\right.$$
$$+ i\epsilon^{\alpha\beta}\{(2\partial_\beta\theta^1\gamma^\mu\Pi_{\mu\alpha}\delta\theta^1 - (1\leftrightarrow 2))$$
$$\left. + \partial_\beta\Pi^\mu_\alpha((\bar\theta^1\gamma\mu\delta\theta^1) - (1\leftrightarrow 2))\} + O(\theta^3)\right\}, \tag{6.2.40}$$

where we have integrated the term with a factor $\partial_\beta\delta\theta$ by parts. The very last term in the final bracket can be discarded, since at this order

$$\epsilon^{\alpha\beta}\partial_\beta\Pi^\mu_\alpha = -i\epsilon^{\alpha\beta}(\partial_\beta\bar\theta^i)\gamma^\mu(\partial_\alpha\theta^i). \tag{6.2.41}$$

We write $\gamma^\mu\gamma^\nu = \eta^{\mu\nu} + \gamma^{\mu\nu}$. The resulting $\gamma^{\mu\nu}$ term involving $\kappa^{1\delta}$ is of the form

$$2\partial_\beta\bar\theta^1\gamma^{\mu\nu}\kappa^{1\delta}\{\Pi_{\mu\alpha}\Pi_{\nu\delta}\}\{g^{\alpha\beta}\sqrt{-g} - \epsilon^{\alpha\beta}\} = -\partial_\beta\bar\theta^1\gamma^{\mu\nu}\kappa^{1\delta}\{\Pi_{\mu\lambda}\Pi_{\nu\tau}\epsilon^{\lambda\tau}\}$$
$$\epsilon_{\alpha\delta}\{g^{\alpha\beta}\sqrt{-g} - \epsilon^{\alpha\beta}\} = +\partial_\beta\bar\theta^1\gamma^{\mu\nu}\kappa^{1\delta}\{\Pi_{\mu\lambda}\Pi_{\nu\tau}\epsilon^{\lambda\tau}\}\{\delta^\beta_\delta - \epsilon_{\alpha\delta}\epsilon^{\alpha\beta}\sqrt{-g}\} \tag{6.2.42}$$

Examining this expression we find that it contains a factor $P^\beta_{+\delta}\kappa^{1\delta}$ which vanishes due to equations (6.2.34) and (6.2.36). The term involving $\kappa^{2\delta}$ vanishes for a similar reason. This explains the use of the projection conditions on $\kappa^{i\delta}$ we adopted earlier.

The other terms arising from $\delta\theta^i$ are then exactly cancelled by taking

$$\delta\sqrt{-g}g^{\alpha\beta} = -16\sqrt{-g}(P^\alpha_{-\delta}g^{\delta\gamma}\bar\kappa^{1\beta}\partial_\gamma\theta^1 + P^\alpha_{+\delta}g^{\delta\gamma}\bar\kappa^{2\beta}\partial_\gamma\theta^2). \tag{6.2.43}$$

One may verify that thus S is indeed invariant at order θ^1. We could have deduced S_2 from S_1 by varying S_1.

The remaining terms in the variation of S are of order θ^3 and are of the form

$$\frac{1}{\pi}\int d^2\xi\,\epsilon^{\alpha\beta}\{\delta(-\bar\theta^1\gamma^\mu\partial_\alpha\theta^1\bar\theta^2\gamma_\mu\partial_\beta\theta^2) + 2\partial_\alpha\bar\theta^j\gamma^\mu\delta\theta^j(\bar\theta^1\gamma_\mu\partial_\beta\theta^1 - \bar\theta^2\gamma_\mu\partial_\beta\theta^2)$$
$$+ \partial_\beta\bar\theta^j\gamma^\mu\partial_\alpha\theta^j(\bar\theta^1\gamma_\mu\delta\theta^1 - \bar\theta^2\gamma_\mu\delta\theta^2)\}. \tag{6.2.44}$$

Carrying out the variation of the first term we find all terms involving $(\theta^1)^2$ and $(\theta^2)^2$ drop out leaving

$$\frac{1}{\pi}\int d^2\xi\,\epsilon^{\alpha\beta}\{\partial_\beta\bar\theta^1\gamma^\mu\partial_\alpha\theta^1\bar\theta^1\gamma_\mu\delta\theta^1 + 2\partial_\alpha\bar\theta^1\gamma^\mu\delta\theta^1\cdot\bar\theta^1\gamma_\mu\partial_\beta\theta^1 - (1\leftrightarrow 2)\}. \tag{6.2.45}$$

The terms involving θ^1 can be written, dropping the 1 index, in the form

$$-\frac{2}{\pi}\int d^2\xi\,\delta\bar\theta\{\gamma_\mu\theta\bar\theta'\gamma_\mu\dot\theta + \gamma_\mu\dot\theta\bar\theta\gamma_\mu\theta' + \gamma_\mu\theta'\dot{\bar\theta}\gamma_\mu\theta\} \tag{6.2.46}$$

using the Majorana properties of θ. This, however, is of the same form as we found to vanish previously, provided θ is a Majorana–Weyl spinor in ten dimensions or the other possibilities we mentioned above.

In fact, S has another local symmetry whose transformations are

$$\delta\theta^1 = \sqrt{-g}P^{\alpha}_{-\beta}g^{\beta\delta}\partial_\delta\theta^1\phi_\alpha,$$

$$\delta\theta^2 = \sqrt{-g}P^{\alpha}_{+\beta}g^{\beta\delta}\partial_\delta\theta^2\phi_\alpha,$$

$$\delta x^\mu = i\bar\theta^j\gamma^\mu\delta\theta^j,$$

$$\delta(\sqrt{-g}g^{\alpha\beta}) = 0,$$

(6.2.47)

where ϕ_α is a local bosonic parameter. We leave the demonstration that these are an invariance of the action as an exercise for the reader. In fact, the closure of two $\kappa^{i\alpha}$ transformations requires such a transformation.

The equations of motion for the action are easily found to be

$$\Pi_\alpha\Pi_\beta - \tfrac{1}{2}g_{\alpha\beta}g^{\gamma\delta}\Pi_\gamma\Pi_\delta = 0,$$

$$\gamma^\mu\Pi_{\mu\alpha}P^{\alpha}_{-\beta}g^{\beta\delta}\partial_\delta\theta^1 = 0,$$

$$\gamma^\mu\Pi_{\mu\alpha}P^{\alpha}_{+\beta}g^{\beta\delta}\partial_\delta\theta^2 = 0,$$

(6.2.48)

$$\partial_\alpha\{\sqrt{-g}g^{\alpha\beta}\partial_\beta x^\mu - 2iP^{\alpha}_{-\beta}g^{\beta\delta}\bar\theta^1\gamma^\mu\partial_\delta\theta^1 - 2iP^{\alpha}_{+\beta}g^{\beta\delta}\bar\theta^2\gamma^\mu\partial_\delta\theta^2\} = 0.$$

The equations of motion simplify considerably if we make some suitable gauge choices. Since the theory is invariant under world surface reparameterisations we may, as for the bosonic string, choose the gauge $g_{\alpha\beta} = \eta_{\alpha\beta}e^\phi$. Then $P^{\alpha\beta}_\pm = \tfrac{1}{2}(\eta^{\alpha\beta} \pm \epsilon^{\alpha\beta})$ and the resulting equations of motion are easily found. As discussed in chapter 4, we may also use the residual conformal symmetry to choose the gauge of equation (4.1.21), that is,

$$x^+ = x_0^+ + c\tau,$$

(6.2.49)

where c is given by equation (4.1.24) or equation (4.1.25) and x_0^+ is a constant.

We may make one further gauge choice: using the κ symmetry of equation (6.2.33) we can choose

$$\gamma^+\theta^i = 0.$$

(6.2.50)

This choice implies that $\bar\theta^i\gamma^+ = 0$, from which it follows that

$$\bar\theta^i\gamma^\mu\partial_\alpha\theta^j = 0, \ i, j = 1, 2; \alpha = 0, 1 \ \mu = +, 1, \ldots, D-2.$$

(6.2.51)

The case of $\mu = +$ is obvious, while the result for $\mu = i$ is found by inserting $\gamma^+\gamma^- + \gamma^-\gamma^+ = 2$.

As explained in chapter 4 the first of the equations (6.2.48) can be used to solve for x^- and P^-; the only difference is that we must take into account the presence of the θ^i. The equations of motion for the remaining variable take on the very simple forms

$$\partial_\alpha\partial^\alpha x^i = 0,$$

(6.2.52)

$$(\partial_0 + \partial_1)\theta^1 = 0,$$

(6.2.53)

$$(\partial_0 - \partial_1)\theta^2 = 0.$$

(6.2.54)

The derivation of the first of these equations of motion is straightforward, while the latter two are found by using the above results and then multiplying by γ^+. One can also verify that the equations of motion contain no further information.

The reader will by now be well aware of the much more technical nature of the Green–Schwarz formulation of the superstring as compared to the NS–R formulation given earlier. We have in the former gained space-time supersymmetry as an invariance of the action. As we shall see, this symmetry will also emerge in the NS–R formulation only after combining bosonic and fermionic sectors and then making the GSO projection. However, in the Green–Schwarz formulation we have lost, at first sight, the two-dimensional superconformal group which played such a vital role in the NS-R formulation.

7 The quantum superstring

But mighty Jove cuts short, with just disdain, the long, long views of poor designing man.

Homer

In this chapter we will quantise the classical superstring given in the previous chapter. As with the bosonic string we can achieve this in different ways. We can use the old covariant method of quantisation, we can solve the constraints in the light-cone formalism and then quantise the independent variables or we can use the BRST approach to quantise the theory. In this chapter we will follow the first approach, given for the R–NS strings in the papers in which they were discovered [6.1, 6.2], and in chapter 10 we will give the BRST approach. For the superstring we find a set of constraints which generate the superconformal group and so the physical state conditions will involve not only the Virasoro generators L_n, but also their superanalogues, the G_rs or F_ns.

One major difference from the bosonic string stems from the choice of boundary conditions for the spinors, which we labelled by the letters R and N–S in the previous chapter. It will turn out that for the open superstring the R sector contains space-time fermions while the NS sector contains space-time bosons. For the closed superstring, on the other hand, the (NS, NS) and (R, R) sectors describe space-time bosons, while the (NS, R) and (R, NS) sectors describe space-time fermions. For the open superstring we will show that one can find projectors in both the R and NS sectors which are consistent with the physical state conditions and which lead to a superstring theory that possesses space-time supersymmetry. One effect of this projector in the NS sector is to remove the tachyon. Consequently, as the R sector possesses no tachyons we find a theory which is tachyon free. We will also discuss a projector for the closed string that has similar properties.

In fact, the R–NS superstrings discussed in the previous chapter possess supersymmetry on the two-dimensional world-sheet, but they do not in general possess space-time supersymmetry. This will be apparent when we quantise them as we will not, in general, find equal numbers of space-time bosons and fermions at a given mass level. As we will explain we can apply the GSO [5.1] projection that eliminates certain states in the theory and the resulting theory does possess space-time supersymmetry and indeed coincides with the Green–Schwarz string studied at the end of the last chapter. The quantisation of the latter string theory, whilst maintaining manifest Lorentz symmetry is notoriously difficult and we will confine our attention to the R–NS superstrings in this chapter.

As we let the string tension go to infinity, that is, $\alpha' \to 0$, the classical string string shrinks to a point. The effect for the quantum string is that all except the massless particles acquire infinite mass and disappear from the theory and, since α' is the only dimensional parameter

143

in the string, the theory contains terms of at most second order in space-time derivatives. The theory resulting from taking this limit in a superstring which possesses space-time supersymmetry must be one of the known ten-dimensional supersymmetric theories. If the theory possesses a spin-2 particle, which is the case for a closed superstrings, and it has a space-time supersymmetry with 32 supercharges, then in the low energy limit we must obtain either the IIA or IIB supergravity theories in ten dimensions. This follows as these are the unique supergravity theories with 32 supercharges in ten dimensions up to a possible cosmological constant for the IIA theory. Which theory one finds depends on the chirality of the supercharges. If the superstring possesses a massless spin-2 particle, but has a supersymmetry with only 16 supercharges, then in the low energy limit we will obtain a corresponding locally supersymmetric theory that can only be type I supergravity theory possibly coupled to the ten-dimensional Yang–Mills theory. These ten-dimensional theories are discussed in detail in chapter 13. Clearly, the latter theories can only appear if the string theory contains the corresponding massless states. If the superstring possesses no spin-2 and spin-$\frac{3}{2}$ particles, which is the case for the open superstring, then we will find in the limit only a rigid supersymmetric theory, namely the ten-dimensional, $N = 1$ super Yang–Mills theory. Once we have found the particle spectrum of the superstrings, we will be able to verify that, after carrying out the relevant projections to obtain a theory with space-time supersymmetry and then taking the limit $\alpha' \to 0$, we do indeed recover the particle content of the the the supersymmetric theories mentioned above.

The early pioneers were encouraged by the proof of a no-ghost theorem for the superstring in the bosonic [7.1–7.3] and fermionic sectors [7.4]. A very important development [7.5–7.10] was the calculation of the fermionic scattering amplitudes in the R–NS superstring which were found to be consistent with the general principles of S-matrix theory.

We have been rather loose with our use of the word superstring. In many quarters, superstring theory is taken to be a string theory that possesses space-time supersymmetry. Examples are the open or closed NS–R string with the GSO projection applied. In this book we will give the term a rather more general meaning and take it to apply to any string that has some world-sheet supersymmetry. However, in practice one always does intend apply the GSO projection to the NS–R string.

One property of the superstring which is a very definite advantage is that certain superstring theories appear to be finite consistent quantum theories involving gravity [7.11].

7.1   The old covariant approach to the open superstring

We now give the old covariant quantisation of the superstring which parallels closely the discussion given for the bosonic string in chapter 3. Applying the Dirac rule to the fundamental Poisson brackets of equations (6.1.60)–(6.1.61), and dropping $\hbar$ we obtain

$$[x^{\mu}(\sigma), P^{\nu}(\sigma')] = i\eta^{\mu\nu}\delta(\sigma - \sigma'), \ 0 < \sigma < \pi \tag{7.1.1}$$

and

$$\{\chi_+^{\mu}(\sigma), \chi_+^{\nu}(\sigma')\} = 2\pi\alpha'\eta^{\mu\nu}\delta(\sigma - \sigma'), \ 0 < \sigma < \pi, \tag{7.1.2}$$

$$\{\chi_-^{\mu}(\sigma), \chi_-^{\nu}(\sigma')\} = 2\pi\alpha'\eta^{\mu\nu}\delta(\sigma - \sigma'), \ 0 < \sigma < \pi, \tag{7.1.3}$$

where $\{,\}$ means the anti-commutator as we are discussing the quantum theory. All other commutators or anti-commutators vanish. The last two relations can be written as

$$\{\Psi^\mu(\sigma), \Psi^\nu(\sigma')\} = 2\pi \eta^{\mu\nu} \delta(\sigma - \sigma'), \quad -\pi < \sigma < \pi, \tag{7.1.4}$$

where Ψ is defined in equation (6.1.34). For the oscillators we then find

$$[\alpha_n^\mu, \alpha_m^\nu] = n\delta_{n+m,0}\eta^{\mu\nu}, \tag{7.1.5}$$

$$\{d_n^\mu, d_m^\nu\} = \delta_{n+m,0}\eta^{\mu\nu} \quad \text{(R)}, \tag{7.1.6}$$

$$\{b_r^\mu, b_s^\nu\} = \delta_{r+s,0}\eta^{\mu\nu} \quad \text{(NS)}. \tag{7.1.7}$$

Substituting the oscillators into the constraints of equations (6.1.57)–(6.1.59) they take the form

$$L_n = \tfrac{1}{2} : \sum_m \alpha_{-m} \cdot \alpha_{n+m} : + \tfrac{1}{4} : \sum_s (2s+n)b_{-s} \cdot b_{s+n} :, \tag{7.1.8}$$

$$G_r =: \sum_m \alpha_m \cdot b_{-m+r} : \tag{7.1.9}$$

in the NS sector, and

$$L_n = \tfrac{1}{2} : \sum_m \alpha_m \cdot \alpha_{n+m} : + \tfrac{1}{4} : \sum_m (2m+n)d_{-m} \cdot d_{m+n} :, \tag{7.1.10}$$

$$F_n =: \sum_m \alpha_m \cdot d_{-m+n} : \tag{7.1.11}$$

in the R sector The dots : imply that the expressions are normal ordered. When normal ordering fermionic oscillators, we must take into account their Grassmann odd character, in particular

$$: b_s^\mu b_{-r}^\nu := -b_{-r}^\nu b_s^\mu; \quad r, s, \in Z + \tfrac{1}{2}, \ r, s > 0 \tag{7.1.12}$$

and similarly for d_n^μ. In fact, the only quantity that has a normal ordering ambiguity is L_0 and we could just as well remove all the : from all the other generators.

Using equations (7.1.8)–(7.1.11) it is straightforward to evaluate the algebra of the generators of the constraints. In the NS sector we find that

$$[L_n, L_m] = (n-m)L_{n+m} + \frac{D}{8}n(n^2 - 1)\delta_{n+m,0},$$

$$[L_n, G_r] = -\left(r - \frac{n}{2}\right)G_{n+r}, \tag{7.1.13}$$

$$\{G_r, G_s\} = 2L_{r+s} + \frac{D}{2}\left(r^2 - \frac{1}{4}\right)\delta_{r+s,0},$$

while in the R sector

$$[L_n, L_m] = (n-m)L_{n-m} + \frac{D}{8}n^3\delta_{n+m,0},$$

$$[L_n, F_m] = -\left(m - \frac{n}{2}\right)F_{n+m}, \tag{7.1.14}$$

$$\{F_n, F_m\} = 2L_{n+m} + \frac{D}{2}n^2\delta_{n+m,0}.$$

The central terms may be computed by using the same methods as for the bosonic theory. Their presence indicates the existence of two-dimensional anomalies in the superconformal symmetry.

The L_0, $L_{\pm 1}$ subalgebra of the Virasoro algebra can be generalised in the NS sector to the subalgebra containing L_0, $L_{\pm 1}$ and $G_{+\frac{1}{2}}$. However, in the R sector, it is not possible to find a subalgebra containing L_0, $L_{\pm 1}$ and one or more F_p. If we add F_0, then its commutators with $L_{\pm 1}$ introduce $F_{\pm 1}$, which in turn through their anti-commutators introduce $L_{\pm 2}$. Using F_0 with $L_{\pm 2}$ we find $F_{\pm 2}$ and then $L_{\pm 3}$. Repeating these steps we recover the whole algebra. If, on the other hand, we add F_p, $p > 0$, then its successive commutators with L_{-1} introduce $F_{p-1}, \ldots, F_1, F_0$ and from the above argument we recover the whole algebra. Similarly, if we add F_p, $p < 0$, then its successive commutators with L_1, introduce $F_{p+1}, \ldots F_1, F_0$ with the same result.

The fact that the SU(1,1) subalgebra of the Virasoro algebra has no natural extension in the R sector is one of the factors that results in the considerable complexity encountered in string scattering calculations in this sector. The above algebras, once we drop their central terms, are identical to the usual superconformal algebra, to be precise, the (1,1) superconformal algebra, which is reviewed in chapter 25 of [1.11]. In particular, there the reader will find the transformations on two-dimensional superspace which this algebra generates.

The next step in the procedure is to give the physical state conditions and find the corresponding spectrum in space-time. We must do this in each sector separately.

7.1.1 The NS sector

The on-shell constraints, or physical state conditions, are

$$G_r \mid \Phi\rangle = 0, \ r > 0; \ (L_0 - \tfrac{1}{2}) \mid \Phi\rangle = 0; \ L_n \mid \Phi\rangle = 0, \ n \geq 1. \tag{7.1.15}$$

Due to the central terms in their algebra we cannot impose all the constraints without setting Φ itself to zero. The set above is a maximal set which implies all the constraints in the classical limit. The discussion is similar to that for the bosonic string given in chapter 3. Corresponding to the normal ordering problem of L_0 we have included the constant $\tfrac{1}{2}$, the precise reason for this particular value will be given shortly.

The most general functional Φ can be written as

$$\Phi = (\phi(x) + i\alpha_1^{\mu\dagger} B_\mu(x) + i b_{\frac{1}{2}}^{\mu\dagger} A_\mu(x) + \tfrac{1}{2} b_{\frac{1}{2}}^{\mu\dagger} b_{\frac{1}{2}}^{\nu\dagger} S_{\mu\nu}(x) + \cdots) \mid 0\rangle, \tag{7.1.16}$$

where the vacuum $\mid 0\rangle$ is a vacuum state in the non-zero mode sector that satisfies

$$\alpha_n^\mu \mid 0; p\rangle = 0, \ n \geq 1, \ b_r^\mu \mid 0; p\rangle = 0, \ r \geq \tfrac{1}{2}. \tag{7.1.17}$$

For the lowest components the $(L_0 - \tfrac{1}{2}) \mid \Phi\rangle = 0$ constraint implies that

$$(\alpha' \partial^2 + \tfrac{1}{2})\phi = 0, \ \partial^2 A_\mu = 0,$$
$$(\alpha' \partial^2 - \tfrac{1}{2})S_{\mu\nu} = (\alpha' \partial^2 - \tfrac{1}{2})B_\mu = 0. \tag{7.1.18}$$

The $G_{\frac{1}{2}} \mid \Phi\rangle = 0$ constraint implies that

$$\partial^\mu A_\mu = 0 = \sqrt{2\alpha'}\partial^\mu S_{\mu\nu} - B_\nu. \tag{7.1.19}$$

The other constraints do not imply any further conditions on the above lowest level fields and we conclude that in the NS sector we have one tachyon ϕ, one massless 'spin-1' A_μ, meaning it corresponds to the vector representation of the little group $SO(D-2)$, as well as an infinite number of massive states. We note that in the NS sector the physical states go up in mass squared in units of $1/2\alpha'$.

The above physical states satisfy a no-ghost theorem; it can be shown that the on-shell states which satisfy equation (7.1.15) have positive norm provided the dimension of space-time is less than or equal to 10 [7.1–7.3]. Among these states we find some zero-norm states, one of which is the state

$$G_{-\frac{1}{2}} \mid \Omega\rangle, \tag{7.1.20}$$

where

$$G_r \mid \Omega\rangle = 0, \ r > 0; \ L_n \mid \Omega\rangle = 0, \ n \geq 0. \tag{7.1.21}$$

It is of zero norm as

$$\langle\Omega \mid G_{\frac{1}{2}}, G_{-\frac{1}{2}} \mid \Omega\rangle = \langle\Omega \mid \{G_{\frac{1}{2}} G_{-\frac{1}{2}}\} \mid \Omega\rangle = 2\langle\Omega \mid L_0 \mid \Omega\rangle = 0. \tag{7.1.22}$$

It is also straightforward to show that it is a physical state, in other words it obeys all the conditions of the equation (7.1.15). It is obvious that such a state has zero scalar product with all physical states.

Another potential zero-norm physical state is given by

$$(G_{-\frac{3}{2}} + aL_{-1}G_{-\frac{1}{2}}) \mid \Omega'\rangle, \tag{7.1.23}$$

where

$$(L_0 + 1) \mid \Omega'\rangle = 0, \ L_n \mid \Omega'\rangle, \ n \geq 1, \ G_r \mid \Omega'\rangle = 0, \ r \geq \frac{1}{2} \tag{7.1.24}$$

and a is a constant which is to be determined. Acting with $G_{\frac{1}{2}}$ we find the constraint

$$(2L_{-1} + aG_{-\frac{1}{2}}G_{-\frac{1}{2}} - 2aL_{-1}) \mid \Omega'\rangle = 0. \tag{7.1.25}$$

Since $G_{-\frac{1}{2}}G_{-\frac{1}{2}} = L_{-1}$ we conclude that it is a physical state provided $a = 2$. Acting with $G_{\frac{3}{2}}$ we find that

$$((2L_0 + D) + 2a\{G_{\frac{1}{2}}, G_{-\frac{1}{2}}\}) \mid \Omega'\rangle = (-2 + D - 4a) \mid \Omega'\rangle = 0 \tag{7.1.26}$$

or $D = 2 + 4a = 10$. Thus while the state of equation (7.1.20) is a zero-norm physical state in any dimension, the state of equation (7.1.23) is such a state only in the critical dimension of the string, that is, $D = 10$. The reader can readily verify that the latter is a zero-norm state and obeys all the other physical state conditions.

7.1.2   The R sector

In this sector we encounter the relation

$$\{d_0^\mu, d_0^\nu\} = \eta^{\mu\nu} \tag{7.1.27}$$

by taking $m = n = 0$ in equation (7.1.6). This relation states that the d_0^μ by themselves satisfy a Clifford algebra. We discussed these algebras in chapter 5 and we recover the normalisation of the above relation if we set $d_0^\mu = \frac{1}{\sqrt{2}}\Gamma^\mu$. We showed in chapter 5 that a Clifford algebra has only one irreducible representation, which has dimension $2^{D/2}$. In quantum theory the constraints will act on a wavefunction which must carry a representation of all the oscillators. In particular, it must be a spinor of the Clifford algebra. We found in chapter 5 that in certain dimensions Dirac spinors can be subject to the Lorentz invariant condition that can make it Majorana, Weyl or possibly Majorana–Weyl. It is up to us to choose what type of spinor we are dealing with. We can therefore give the wavefunction a spinor index such as ϵ, where $\epsilon = 1, 2, \ldots, 2^{D/2}$ and take $d_0^\mu = \frac{1}{\sqrt{2}}\Gamma^\mu$ where the Γ^μ are taken to be the irreducible representation. An alternative, but equivalent way to represent the d_0^μ is to divide them up into creation and annihilation operators and introduce a corresponding vacuum. One such choice, for even D, is to take the creation operators

$$D^1 = d_0^0 + id_0^1, D^2 = d_0^2 + id_0^3, \ldots, D^{\frac{D}{2}} = d_0^{D-2} + id_0^{D-1}, \tag{7.1.28}$$

while the destruction operators are

$$\bar{D}^1 = -d_0^0 + id_0^1, \bar{D}^2 = d_0^2 - id_0^3, \ldots, \bar{D}^{\frac{D}{2}} = d_0^{D-2} - id_0^{D-1}. \tag{7.1.29}$$

Such a set satisfies the relations

$$\{D^i, \bar{D}^j\} = 2\delta^{ij}$$

and

$$\{D^i, D^j\} = 0 = \{\bar{D}^i, \bar{D}^j\}. \tag{7.1.30}$$

Of course, we can write any Clifford algebra in this form. A representation is then found by defining the vacuum to satisfy

$$\bar{D}^j \,|\rangle = 0, \tag{7.1.31}$$

and acting with D^i on the vacuum, namely

$$|\rangle, \; D^j \,|\rangle, \; D^i D^j \,|\rangle, \ldots,$$

to create $(1+1)^{D/2} = 2^{D/2}$ states. Of course, this is just the unique irreducible representation of the Clifford algebra, but written in an explicit manner. The latter construction has the advantage that it treats the d_0^μ in a similar way to the d_n^μ, $n \geq 1$. However, in this second formulation the Lorentz transformation properties of the states are less familiar.

In the R sector the physical states must satisfy

$$F_n \,|\,\psi\rangle = 0, \; n \geq 0, \tag{7.1.32}$$

which imply that

$$L_n \,|\,\psi\rangle = 0, \; n \geq 0. \tag{7.1.33}$$

To maintain the Grassmann even–odd grading in any equation, we wish to express Grassmann odd (even) quantities in terms of Grassmann odd (even) quantities. As a result, we cannot introduce any constant in the $F_0 \mid \psi \rangle = 0$ equation as F_0 is odd. Indeed, F_0 has no normal ordering problem and so we do not expect a constant. Since $F_0^2 = L_0$ we can then have no constant, that is, intercept, in the $L_0 \mid \psi \rangle = 0$ equation.

It can be shown that the states which satisfy equation (7.1.32) have positive norm provided $D \le 10$ [7.4]. As for the case of the bosonic string, there exist among these states some states of zero norm. One such a state is given by

$$(L_{-1} - \tfrac{1}{2}F_{-1}F_0) \mid \Omega \rangle, \tag{7.1.34}$$

where

$$F_n \mid \Omega \rangle = 0, \ n \ge 1, \ L_n \mid \Omega \rangle = 0, \ n \ge 1, \ (L_0 + 1) \mid \Omega \rangle = 0. \tag{7.1.35}$$

In fact, there are two states of this type since if $\mid \Omega \rangle$ satisfies the above constraints so does $F_0 \mid \Omega \rangle$. Applying F_1 to the above state we discover that it is a physical state only if $D = 10$. We leave it to the reader to verify that it is a zero-norm state and that it satisfies all the physical state conditions.

The most general functional $\mid \phi \rangle_\epsilon$ which is spinor valued, has the form

$$\mid \phi \rangle_\epsilon = \{ \lambda_\epsilon(x) + i\alpha_{-1}^\mu \psi_{\mu\epsilon}^1(x) + d_{-1}^\mu \psi_{\mu\epsilon}^2(x) + \cdots \} \mid 0; p \rangle, \tag{7.1.36}$$

where the vacuum satisfies

$$\alpha_n^\mu \mid 0; p \rangle = 0, n \ge 1; \ d_n^\mu \mid 0; p \rangle = 0, n \ge 1, \tag{7.1.37}$$

and x^μ stands for the centre-of-mass coordinate x_0^μ. The most general term in the series is of the form

$$d_{-n_1}^{\mu_1} \cdots d_{-n_i}^{\mu_i} \alpha_{-m_1}^{\nu_1} \cdots \alpha_{-m_j}^{\nu_j} S_{\mu_1 \cdots \mu_i; \nu_1 \cdots \nu_j \epsilon}(x) \mid 0; p \rangle, \tag{7.1.38}$$

where $S_{\mu_1 \cdots \mu_i; \nu_1 \cdots \nu_j \epsilon}(x)$ is anti-symmetric in the indices μ_k, μ_l, but symmetric in the indices ν_k, ν_l.

To lowest order, the generators F_0 and F_1 have the form

$$F_0 = \alpha_0^\mu \frac{\Gamma_\mu}{\sqrt{2}} + \alpha_{-1}^\mu d_{1\mu} + \alpha_1^\mu d_{-1\mu} + \cdots,$$

$$F_1 = \alpha_0^\mu d_{1\mu} + \alpha_1^\mu \frac{\Gamma_\mu}{\sqrt{2}} + \cdots, \tag{7.1.39}$$

where

$$\alpha_0^\mu = -i\sqrt{2\alpha'}\partial^\mu. \tag{7.1.40}$$

Consequently, the constraint $F_0 \mid \phi \rangle_\epsilon = 0$ implies that

$$\not\partial\lambda = 0, \ \sqrt{\alpha'}\not\partial\psi_\mu^1 = \psi_\mu^2, \ -\sqrt{\alpha'}\not\partial\psi_\mu^2 = \psi_\mu^1, \ldots, \tag{7.1.41}$$

while $F_1 \mid \phi \rangle_\epsilon = 0$ implies

$$-\sqrt{2\alpha'}\partial^\mu \psi_\mu^2 + \frac{\Gamma^\mu}{\sqrt{2}}\psi_\mu^1 = 0. \tag{7.1.42}$$

The other physical state conditions place no further constraints on λ and $\psi_\mu^i, i = 1, 2$.

In the R sector we therefore find one massless Majorana spinor λ_ϵ and an infinite tower of massive spinors. We recall that the spinors arise in this sector due to the existence of the d_0^μ, which, in turn, arose from the choice of boundary condition in the R sector. The content at higher levels is most easily seen from a light-cone analysis. The first massive states belong to the $128 + 128$ representation of the little group Spin (9) assuming the theory to be in ten dimensions. We note that the physical states go up in mass squared units of $1/\alpha'$, which is twice the rate found in the NS sector.

As already mentioned, the physical states in both the NS and R sectors given by equations (7.1.15) and (7.1.32) satisfy a no-ghost theorem;

Theorem [7.1–7.4] For $D \leq 10$ the physical states of equations (7.1.15) and (7.1.32) have positive norm.

The proof of this result is analogous to that for the bosonic string discussed in detail in chapter 11.

7.2   The GSO projector for the open string [5.1]

The open superstring discussed above has two sectors, the NS and the R sectors, which consist of bosons and fermions, respectively. We might demand that such a theory possess space-time supersymmetry as well as the world-sheet supersymmetry, which was encoded at the very outset. To admit space-time supersymmetry the theory would have to have an equal number of fermionic and bosonic degrees of freedom on-shell at every mass level in the resulting space-time theory. We now assume the string theory is in ten dimensions. We recall the count of physical states for the two sectors at the lowest levels. In the NS sector we have a tachyonic spin-0 particle, next we have a massless vector, that is, a vector representation of the little group SO(8), and then an $S_{\mu\nu}$ and B_μ at the first massive level whose physical states belong to the $9 \times 4 = 36$-dimensional representation of the little group SO(9), while in the R sector we have no tachyonic states and the lowest level state is a massless spinor, that is, $2^4 =$ states of the little group Spin(8), if it is a Majorana spinor. In ten dimensions one can have Majorana–Weyl spinors and the 16 on-shell states of a Majorana spinor decompose into two states of opposite chirality each with eight states. At the next level we have massive states that belong to the $128 + 128$ representation of Spin(9).

Clearly, adding the two sectors together naively cannot produce a theory with space-time supersymmetry. There is no match for the tachyon of the NS sector in the R sector as there are no tachyons in this sector. Furthermore, the massless states are labelled by representations of SO(8), but in the NS sector the field A_μ is a vector representation of SO(8) of dimension 8, while in the R sector we have the Majorana spinor representation of SO(8) which has dimension 16 on-shell.

To gain equal numbers of fermions and bosons we must begin by removing the tachyon. The simplest possibility that consists of an operation that affects the string as a whole is to project out the tachyon and all states at every second level in the NS sector; this is achieved by the projection condition

$$P_{NS} \mid \phi \rangle = \mid \phi \rangle, \tag{7.2.1}$$

where

$$P_{NS} = (-1)^{\sum_{r=\frac{1}{2}}^{\infty} b_r^\dagger \cdot b_r - 1} \equiv (-1)^F. \tag{7.2.2}$$

Since $\sum_{r=\frac{1}{2}}^{\infty} b_r^\dagger \cdot b_r$ just counts the number of b_r^μ oscillators, equation (7.2.1) just states that $|\phi\rangle$ should have an odd number of b_r^μ in any given term. We note that the states that have been projected out have mass squared of $(2s - 1)/2\alpha'$ for integer s and there are no states in the R sector with these masses. We note that

$$P_{NS}^2 = (-1)^{2 \sum_{r=\frac{1}{2}}^{\infty} b_r^\dagger \cdot b_r}$$

which is 1 on the states of the NS sector and so we do indeed have a projector.

The remaining states in the NS sector are left untouched. In fact, as we have just observed the massless level of the NS sector involves only a vector which belongs to the irreducible vector representation of SO(8). Thus, if we were to carry out a projection condition that affected this level we would have to remove all the massless states in the NS sector completely. By the Fermi–Bose balance we would then would have no massless states which is an undesirable possibility if one wants account for the world as we see it.

However, even if we adopt the above projector in the NS sector, the degree of freedom count still does not work for the massless states. The only solution is to impose a Weyl condition on λ in the R sector reducing it to the matching eight degrees of freedom on-shell in the NS sector. It would be incorrect to implement the constraint to all levels by demanding that all spinors be of a given chirality, since at higher levels, we have an infinite number of massive spinors which by their equation of motion must involve both chiralities. The consistent constraint is to impose

$$P_R \mid \phi\rangle_\epsilon = |\phi\rangle_\epsilon, \tag{7.2.3}$$

where

$$P_R = \Gamma_{11}(-1)^{\sum_{n=1}^{\infty} d_n^\dagger \cdot d_n} \equiv (-1)^F \tag{7.2.4}$$

and

$$\Gamma_{11} = \Gamma^0 \Gamma^1 \cdots \Gamma^9 = (\sqrt{2})^{10} d_0^0 \cdots d_0^9. \tag{7.2.5}$$

The wavefunctions in the two sectors are distinguished by the presence of the spinor index ϵ; however, we use the same symbol F in each sector. The reader can distinguish which it is by the context in which it is used. A standard calculation in counting signs gives

$$(\Gamma_{11})^2 = 1. \tag{7.2.6}$$

Since $\sum_{n=1}^{\infty} d_n^\dagger \cdot d_n$ is the number operator in the d_n^μ sector it has positive integer eigenvalues and so

$$P_R^2 = \Gamma_{11}^2(-1)^{2\sum_{n=1}^{\infty} d_n^\dagger \cdot d_n} = 1. \tag{7.2.7}$$

The above remarks on consistency can be more concretely formulated in terms of the requirement that the projectors in each sector should commute, or anti-commute, with the physical state conditions. In the NS sector the relations

$$\{P_{NS}, G_r\} = 0 = [P_{NS}, L_n], \ \forall\, n \in Z, \ r \in Z + \tfrac{1}{2}, \tag{7.2.8}$$

obviously guarantee this requirement. They are a consequence of the relations

$$P_{NS}\alpha_n^\mu P_{NS} = \alpha_n^\mu, \quad P_{NS}b_r^\mu P_{NS} = -b_r^\mu. \tag{7.2.9}$$

In the R sector, the consistency with the physical state conditions is guaranteed by the equations

$$\{P_R, F_n\} = 0 \,\forall\, n, \quad [P_R, L_n] = 0 \,\forall\, n, \tag{7.2.10}$$

which are a result of the relations

$$P_R\alpha_n^\mu P_R = \alpha_n^\mu, \quad P_R d_n^\mu P_R = -d_n^\mu. \tag{7.2.11}$$

As suggested above, Γ_{11} alone would not have these last properties and as a result would not preserve the physical state conditions. These projector conditions have become known as the GSO projectors [5.1].

The projections above were suggested by our desire to ensure supersymmetry and so on-shell Fermi–Bose balance at the first few levels. The reader may wish to verify that this fact also holds at the next few levels. The simplest way to show it holds at all levels is to use the light-cone formulation. After counting the states in each level in the two sectors one finds they match as a result of an identity due to Jacobi.

Systems with equal numbers of fermions and bosons do not necessarily possess space-time supersymmetry. Examples that do not are easily constructed even at the free level by taking non-adjacent spins, say one massless gravitino in four dimensions and two spin-0 particles. However, it can be shown that the superstring with the above projection does indeed preserve supersymmetry not only at the free, but also at the interacting, level. The R–NS theory with the GSO projection was the first string theory discovered that has space-time supersymmetry.

In the limit of infinite string tension (that is, $\alpha' \to 0$) [3.1] we are left, for the GSO projected open superstring, with one Majorana–Weyl spinor and one spin-1 particle, meaning an eight-dimensional representation of SO(8), the massless states. In fact we find $N = 1, D = 10$ super QED theory, discussed in chapter 13, whose action is given by

$$A = \int d^{10}x\{-\tfrac{1}{4}f_{\mu\nu}f^{\mu\nu} - \tfrac{1}{2}\bar\lambda\slashed{\partial}\lambda\}. \tag{7.2.12}$$

It is invariant under the rigid supersymmetry transformations

$$\delta A_\mu = \bar\epsilon\Gamma_\mu\lambda, \quad \delta\lambda = -\tfrac{1}{2}\Gamma^{\mu\nu}f_{\mu\nu}\epsilon, \tag{7.2.13}$$

where ϵ is a Majorana–Weyl spinor like λ_α. If one considered the interacting superstring with Chan–Paton factors, then one would find the $N = 1, D = 10$ super Yang–Mills theory. This theory is given by equations (7.2.12) and (7.2.13), but with the replacements $f_{\mu\nu} \to F_{\mu\nu}$, where $F_{\mu\nu}$ is the full field strength, and $\partial_\mu\lambda \to D_\mu\lambda$, where $D_\mu\lambda$ is the gauge covariant derivative of λ. On grounds of gauge invariance alone, this action is the only one possible that agrees with the free theory in the limit of zero gauge coupling constant. While space-time supersymmetry possesses a number of desirable features the reader may wonder whether the projections are actually required for consistency of the string theory. The closed superstring possesses in its massless states a gravitino; however, this field at the free level necessarily involves a local Grassmann odd spinor symmetry. There are strong arguments to suppose that such a field is only consistent when the theory possesses local supersymmetry. Thus for the closed superstring the projections would seem necessary for consistency. In the original GSO paper [5.1] the reader will also find arguments for the projectors within the context of the open superstring. It does, of course, remove the tachyon, which is commonly considered to be an advantage.

7.3   The old covariant approach to the closed superstring

For the closed superstring we have not only the α_n^μ, b_r^μ and d_n^μ oscillators, but also their barred analogues $\bar{\alpha}_n^\mu$, $\bar{b}_r^\mu$ and $\bar{d}_n^\mu$. The constraints generate two copies of the superconformal algebra: one copy is generated by L_n and G_r or F_n and the other by $\bar{L}_n$ and $\bar{G}_r$ or $\bar{F}_n$. The former are as in equations (7.1.8)–(7.1.11) and the latter are obtained by putting bars on the former in the obvious places. As explained in the previous section we now have four possible sectors, (NS–NS), (R–NS), (NS–R) and (R–R). Apart from the bosonic zero mode, the Fock space for the closed superstring belongs to a tensor product of two open superstring Fock spaces. The first and second factors correspond to the action of the unbarred and barred oscillators acting on the vacuum respectively. Those sectors with an R sector will contain d_0^μ or $\bar{d}_0^\mu$ oscillators or both and so carry the appropriate spinor indices. Corresponding to the above sectors the wavefunctions will carry the spinor index structure $|\phi\rangle$, $|\phi'\rangle_\epsilon$, $|\phi\rangle_\epsilon$ and $|\phi\rangle_{\epsilon\delta}$, respectively. Consequently it is the (R–NS) and (NS–R) sectors which contain the space-time fermions.

We impose the following physical state conditions in the NS–NS sector:

$$G_r \mid \phi\rangle = \bar{G}_r \mid \phi\rangle = 0, \ r \ge \tfrac{1}{2}, \ L_n \mid \phi\rangle = \bar{L}_n \mid \phi\rangle = 0, \ n \ge 1, \qquad (7.3.1)$$

$$(L_0 - \tfrac{1}{2}) \mid \phi\rangle = 0, \ (\bar{L}_0 - \tfrac{1}{2}) \mid \phi\rangle = 0. \qquad (7.3.2)$$

In the NS–R sector we impose

$$G_r \mid \phi\rangle_\epsilon = \bar{F}_n \mid \phi\rangle_\epsilon = 0, \ r \ge \tfrac{1}{2}, n \ge 0, \ L_n \mid \phi\rangle_\epsilon = \bar{L}_n \mid \phi\rangle_\epsilon = 0, n \ge 1, \quad (7.3.3)$$

$$(L_0 - \tfrac{1}{2}) \mid \phi\rangle_\epsilon = 0 = (\bar{L}_0) \mid \phi\rangle_\epsilon. \qquad (7.3.4)$$

The R–NS sector is given by the same conditions as the NS–R sector, but with bars and unbars interchanged, a prime on ϕ and the ϵ index is then associated with the first part. Finally, in the R–R sector the physical state conditions are given by

$$F_n \mid \phi\rangle_{\epsilon\delta} = \bar{F}_n \mid \phi\rangle_{\epsilon\delta} = 0, \ L_n \mid \phi\rangle_{\epsilon\delta} = \bar{L}_n \mid \phi\rangle_{\epsilon\delta} = 0, \ n \ge 0. \qquad (7.3.5)$$

From these we can deduce conditions that involve L_0 and $\bar{L}_0$, such as $(L_0 - \bar{L}_0) \mid \phi\rangle$ in the first and last sectors. As for the bosonic string, these conditions can be interpreted as the statement that no point of the string is preferred.

We now consider the lowest states in each sector and deduce the on-shell conditions implied by the above physical state conditions. We consider first the NS–NS sector. Writing the wavefunction as

$$|\phi\rangle = \{\phi + b_{-1/2}^\mu \bar{b}_{-1/2}^\nu l_{\mu\nu} + \cdots\} \mid 0\rangle, \qquad (7.3.6)$$

the physical state conditions imply that

$$(\alpha' \partial^2 + 1)\phi = 0 = \partial^\mu l_{\mu\nu} = \partial^2 l_{\mu\nu}, \ldots. \qquad (7.3.7)$$

In the NS–R sector the wavefunction can be written as

$$|\phi\rangle_\epsilon = (b^\mu_{-1/2}\psi_{\mu\epsilon} + \cdots) \,|\,0\rangle \tag{7.3.8}$$

and the physical state conditions imply that

$$\not{\partial}\psi_\mu = \partial^\mu \psi_\mu = 0, \,\ldots. \tag{7.3.9}$$

In the R–NS sector the wavefunction can be written as

$$|\phi\rangle_\epsilon = (\bar{b}^\mu_{-1/2}\psi'_{\mu\epsilon} + \cdots) \,|\,0\rangle \tag{7.3.10}$$

and the physical state conditions imply that

$$\not{\partial}\psi'_\mu = \partial^\mu \psi'_\mu = 0, \,\ldots. \tag{7.3.11}$$

In the R–R sector the wavefunction can be written as

$$|\phi\rangle_{\epsilon\delta} = \{S_{\epsilon\delta} + \cdots\} \,|\,0\rangle \tag{7.3.12}$$

and the resulting on-shell conditions are

$$(\not{\partial})_\epsilon{}^\beta S_{\beta\delta} = 0 = (\not{\partial})_\delta{}^\gamma S_{\epsilon\gamma}, \,\ldots. \tag{7.3.13}$$

The vacua above are just the obvious tensor product of the vacua discussed in the previous section for the NS and R sectors except for the bosonic zero mode associated with x^ν, which is identified in the two sectors. When listing the above fields we have not listed those that are set to zero in a trivial way by the physical state conditions. For example, the lowest state in the NS–R sector does in equation (7.3.8) contain a term $b^\mu_{-\frac{1}{2}}\psi_{\mu\epsilon}$ times the vacuum as a consequence of the condition that $(L_0 - \frac{1}{2} - \bar{L}_0)\,|\,\psi\rangle = 0$, but the term with just the vacuum is eliminated by this condition.

In the NS–NS sector we have a tachyon ϕ at the lowest level and the field $l_{\mu\nu}$ at the next level. The on-shell conditions on the fields are the same as for the closed bosonic string although they arise in a different way. The symmetric part of $l_{\mu\nu}$ leads to a graviton $h_{\mu\nu} = h_{\nu\mu}, h_\mu{}^\mu = 0$, the trace to a spin 0, φ and the anti-symmetric part to the gauge fields $B_{\mu\nu} = -B_{\nu\mu} = l_{[\mu\nu]}$. These fields are all massless and belong to the $(\frac{1}{2}(D-2)(D-1) - 1)$-, 1- and $(\frac{1}{2}(D-2)(D-1))$-dimensional representations of the little group $SO(D-2)$, respectively.

In the NS–R sector the particle states are best displayed by writing $\psi_\mu = \hat{\psi}_\mu + \Gamma_\mu\lambda$, where $\Gamma^\mu\hat{\psi}_\mu = 0$. The on-shell conditions of equation (7.3.11) can easily be seen to be equivalent to the conditions

$$\not{\partial}\hat{\psi}_\mu = \partial^\mu \hat{\psi}_\mu = 0, \quad \not{\partial}\lambda = 0, \tag{7.3.14}$$

from which we recognise the on-shell conditions for a gravitino, $\hat{\psi}_\mu$, and a spinor, λ.

The analysis of the R–R sector is more complicated. It describes bosons in a first order formalism. For simplicity, we will carry this out after imposing the GSO projection. In addition to the above states we have a tower of massive states in each sector.

We demand that the spinors in all the wavefunctions in the R sectors are Majorana spinors. This implies that the spinors that arise in the NS–R and R–NS sectors are Majorana spinors and it places a reality condition on $S_{\epsilon\delta}$ in the R–R sector. A Majorana spinor in D dimensions has $2^{D/2}$ real components while if it is Majorana–Weyl it has $2^{(D/2)-1}$ real components.

These correspond to $2^{(D/2)-1}$ and $2^{(D/2)-2}$ on-shell degrees of freedom respectively. A gravitino which is Majorana–Weyl has $2^{(D/2)-2}(D-3)$ on-shell degrees of freedom. The vector index corresponds to the $D-3$ factor. The -3 emerges from the analysis of gauge symmetries; essentially the condition $\Gamma^\mu \hat{\psi}_\mu = 0$ accounts for -1 and the gauge invariance $\delta \hat{\psi}_\mu = \partial_\mu \epsilon$, acts to fix a gauge and provide an on-shell invariance which together give a further -2.

We now consider the imposition of the GSO projector that will ensure Fermi–Bose balance. The projector on the left moving sector is given by

$$P_{NS} = (-1)^F = (-1)^{\sum_{r=\frac{1}{2}}^{\infty} b_r^\dagger \cdot b_r - 1}$$

in the NS sector and

$$P_R = (-1)^F = \Gamma_{11}(-1)^{\sum_{n=1}^{\infty} d_n^\dagger \cdot d_n}$$

in the R sector. The projectors $\bar{P}_{NS} = (-1)^{\bar{F}}$ and $\bar{P}_R = (-1)^{\bar{F}}$ in the right moving sector have the same expression but in terms of barred oscillators. This implies that the Γ_{11} part now acts on the second spinor index of S of equation (7.3.17).

Clearly, we must remove the tachyon from the NS–NS sector. This is achieved by demanding that $(-1)^F = 1$ in this sector and we also demand that $(-1)^{\bar{F}} = 1$ in this sector. These signs are chosen such that we do not remove the massless states at the next level. They also have the effect of removing the tachyon.

In the NS–R sector we must impose $(-1)^F = 1$ in order to keep any massless spinors. We can also then choose $(-1)^{\bar{F}} = \pm 1$. This has the effect of making the spinors chiral, that is, $\Gamma^{11}\psi_\mu = \pm \psi_\mu$.

In the R–NS sector for the same reason we must choose $(-1)^{\bar{F}} = 1$ and we can take $(-1)^F = \pm 1$, or $(-1)^F = \mp 1$, with corresponding chirality conditions on the spinor ψ'_μ. Although there would appear to be four choices, there are really only two as only the relative chirality between the two sectors is physically relevant. We therefore choose only the signs $(-1)^{\bar{F}} = 1$ in the NS–R sector and $(-1)^F = \pm 1$ in the R–NS sector. These lead to the chiral choices

$$\Gamma^{11}\psi_\mu = \psi_m, \tag{7.3.15}$$

$$\Gamma^{11}\psi'_\mu = \pm \psi'_m. \tag{7.3.16}$$

Hence, we have two Majorana–Weyl gravitons of either the same chirality for the first choice of sign or different chirality for the second choice of sign. We will shortly refer to the resulting theories as the IIB and IIA theories, respectively.

When we come to the R–R sector, we have a choice: we can impose $(-1)^F = \pm 1$ and $(-1)^{\bar{F}} = \pm 1$ or $(-1)^F = \pm 1$ and $(-1)^{\bar{F}} = \mp 1$. In the R–R sector we can expand $S_{\epsilon\delta}$ in terms of the complete set of Γ matrices (see the end of chapter 6):

$$S = C^{-1}F + (\Gamma_\mu)C^{-1}F^\mu + (\Gamma_{\mu\nu}C^{-1})F^{\mu\nu} + (\Gamma_{\mu\nu\rho}C^{-1})F^{\mu\nu\rho} + (\Gamma_{\mu\nu\rho\kappa}C^{-1})F^{\mu\nu\rho\kappa}$$

$$+ (\Gamma_{\mu\nu\rho\kappa\epsilon}C^{-1})F^{\mu\nu\rho\kappa\epsilon} + (\Gamma_{\mu\nu\rho\kappa}\Gamma_{11}C^{-1})G^{\mu\nu\rho\kappa} + (\Gamma_{\mu\nu\rho}\Gamma_{11}C^{-1})G^{\mu\nu\rho}$$

$$+ (\Gamma_{\mu\nu}\Gamma_{11}C^{-1})G^{\mu\nu} + (\Gamma_\mu\Gamma_{11}C^{-1})G^\mu + (\Gamma_{11}C^{-1})G. \tag{7.3.17}$$

Imposing $(-1)^F$ is ± 1 on the first R factor implies that

$$F = \pm G, \quad F^{\mu} = \mp G^{\mu}, \quad F^{\mu\nu} = \pm G^{\mu\nu}, \quad F^{\mu\nu\rho} = \mp G^{\mu\nu\rho}, \quad F^{\mu\nu\rho\kappa} \tag{7.3.18}$$

$$= \pm G^{\mu\nu\rho\kappa}, \quad F^{\mu_1\cdots\mu_5} = \pm \frac{1}{5!} \epsilon^{\mu_1\mu_2\cdots\mu_{10}} F_{\mu_6\cdots\mu_{10}} \equiv \pm {}^{\star} F^{\mu_1\cdots\mu_5}. \tag{7.3.19}$$

Imposing $(-1)^{\bar{F}} = \pm 1$ on the second R factor, and using the relation

$$(\Gamma_{11})_\alpha{}^\delta (\Gamma_{\mu_1\ldots\mu_n} C^{-1})_{\epsilon\delta} = (\Gamma_{\mu_1\ldots\mu_n} C^{-1} \Gamma_{11}^T)_{\epsilon\delta} = -(\Gamma_{\mu_1\ldots\mu_n} \Gamma_{11} C^{-1})_{\epsilon\delta} \tag{7.3.20}$$

implies that

$$F = \mp G, \quad F^{\mu} = \mp G^{\mu}, \quad F^{\mu\nu} = \mp G^{\mu\nu}, \quad F^{\mu\nu\rho} = \mp G^{\mu\nu\rho}, \quad F^{\mu\nu\rho\kappa} = \mp G^{\mu\nu\rho\kappa},$$

$$F^{\mu_1\cdots\mu_5} = \pm {}^{\star} F^{\mu_1\cdots\mu_5}. \tag{7.3.21}$$

We observe that it is only the relative sign between $(-1)^F$ and $(-1)^{\bar{F}}$ which matters in the analysis of the physical states in the R–R sector and so for simplicity we will choose $(-1)^F = 1$. The remaining choice $(-1)^{\bar{F}} = \pm 1$ is up to us and it determines the type of string theory that we construct.

Imposing $(-1)^F = 1 = (-1)^{\bar{F}}$ results in keeping only the fields

$$F^{\mu}, \quad F^{\mu\nu\rho}, \quad F^{\mu_1\cdots\mu_5}, \tag{7.3.22}$$

where $F^{\mu_1\cdots\mu_5} = {}^{\star} F^{\mu_1\cdots\mu_5}$. Analysing the on-shell conditions of equation (7.3.13) for these fields implies for F_{μ} that

$$\partial_\mu F^{\mu} = 0, \quad \partial_{[\mu} F_{\nu]} = 0, \tag{7.3.23}$$

which in turn imply that $F^{\mu} = \partial^{\mu}\varphi$, $\partial^2\varphi = 0$. We recognise these as describing one spin-0 particle. For $F^{\mu\nu\rho}$ we find that

$$\partial_\mu F^{\mu\nu\rho} = \partial_{[\mu} F_{\nu\rho\kappa]} = 0, \tag{7.3.24}$$

which implies that $F_{\mu\nu\rho} = \partial_{[\mu} A_{\nu\rho]}$ and so it describes the propagation of a massless anti-symmetric tensor belonging to the $(\frac{1}{2}(D-2)(D-3))$-dimensional representation of $SO(D-2)$. Finally, we have the self-dual field $F^{\mu_1\cdots\mu_5}$, which satisfies

$$\partial_{\mu_1} F^{\mu_1\cdots\mu_5} = 0 = \partial_{[\mu_1} F_{\mu_2\cdots\mu_6]}, \tag{7.3.25}$$

which implies that $F_{\mu_1\cdots\mu_5} = \partial_{[\mu_1} A_{\mu_2\cdots\mu_5]}$. It describes the propagation of

$$\frac{1}{2} \frac{(D-2)(D-3)(D-4)(D-5)}{4!}$$

degrees of freedom in the fourth rank totally anti-symmetric self-dual tensor representation of the little group $SO(D-2)$. The additional $\frac{1}{2}$ is due to the self-dual nature of the field strength.

On the other hand imposing $(-1)^F = 1 = -(-1)^{\bar{F}}$ implies that we keep only the fields

$$F, \quad F^{\mu\nu}, \quad F^{\mu\nu\rho\kappa}. \tag{7.3.26}$$

Repeating the same analysis as above we find the equations

$$\partial_{\mu_1} F^{\mu_1\cdots\mu_n} = 0 = \partial_{[\nu} F_{\mu_1\cdots\mu_n]}, \quad n = 0, 2, 4, \tag{7.3.27}$$

which implies the propagation of only the potentials A_μ and $A_{\mu\nu\rho}$ belonging to the $D-2$ and $((D-2)(D-3)(D-4)/3!)$-dimensional representations of the little group of $SO(D-2)$, respectively. The field F is a constant and so does not propagate any degrees of freedom.

Although we obtain Fermi–Bose balance for the massless physical states for all of the above adopted choices of signs it turns out that they will not all lead to theories that admit space-time supersymmetry. This can been seen by considering if their field content belongs to one of the supersymmetry multiplets known to exist in ten dimensions and discussed further below. In fact, the relative choices of signs between the NS–R and the R–NS sectors and that in the R–R sector are actually tied together by this requirement and we can only have the choice denoted *type IIB* which is given by

$$((-1)^F, (-1)^{\bar F}) = (1, 1) \text{ in all sectors} \tag{7.3.28}$$

or the choice denoted *type IIA* which is given by

$$((-1)^F, (-1)^{\bar F}) = \begin{cases} (1, 1) & \text{in (NS–NS),} \\ (1, -1) & \text{in (NS–R),} \\ (1, 1) & \text{in (R–NS),} \\ (1, -1) & \text{in (R–R).} \end{cases} \tag{7.3.29}$$

Let us summarise the results of applying the above GSO projections. The type IIB superstring has the massless field content

$$h_{\mu\nu}(\underline{35}), \varphi(\underline{1}), A_{\mu\nu}(\underline{28}); \varphi'(\underline{1}), A'_{\mu\nu}(\underline{28}), A_{\mu_1...\mu_4}(\underline{35});$$

$$\hat\psi_{\mu1}(\underline{56}), \hat\psi_{\mu2}(\underline{56}), \lambda_1(\underline{8}), \lambda_2(\underline{8}), \tag{7.3.30}$$

where the underlined numbers in brackets indicate the $SO(8)$ representation and where $\hat\psi_{\mu1}, \hat\psi_{\mu2}$ are Majorana–Weyl and have the same chirality. The field strength $F_{\mu_1\cdots\mu_5}$ for the 4-form potential is the self-dual. The numbers in parentheses correspond to the dimensions of $SO(8)$ little group representations corresponding to the fields.

The type IIA superstring has the massless field content

$$h_{\mu\nu}(\underline{35}), \varphi(\underline{1}), A_{\mu\nu}(\underline{28}); A_\mu(\underline{8}), A_{\mu\nu\rho}(\underline{56}), \hat\psi_{\mu1}(\underline{56}), \hat\psi_{\mu2}(\underline{56}), \lambda_1(\underline{8}), \lambda_2(\underline{8}),$$

$$\tag{7.3.31}$$

where $\hat\psi_{\mu1}, \hat\psi_{\mu2}$ are Majorana–Weyl with the opposite chirality. For both superstrings the bosonic fields that occur before the semi-colon belong to the NS–NS sector and those after belong to the R–R sector. We note that the fields in the NS–NS sectors of the IIA and IIB string are the same. This is an obvious consequence of the fact that the projections in the NS–NS sector are the same for both strings.

No matter what choice of the above signs we take we have two Majorana–Weyl gravitinos and, for consistency of propagation, such a theory must have a local supersymmetry with a parameter that has in all 32 components, since each Majorana–Weyl spinor has 16 components. The transformation of the gravitino under a local supersymmetry variation is of the form $\delta\psi_\mu = \partial_\mu\epsilon + \cdots$ and so the type of gravitinos one has and the type of local supersymmetry are closely related. As such, we are dealing with a supergravity theory based on a supersymmetry algebra whose supercharge has $16 + 16$ components which are either of the same chirality, or not, according to the IIB or IIA choices, respectively. As discussed in chapter 13, there are only two such supergravity theories called IIB supergravity and IIA supergravity theories. Indeed, these are the only two maximal supergravity theories in ten dimensions. In fact, one cannot have a theory which is supersymmetric and have spins less than or equal to 2 and have more than 32 supercharges. The above set of massless states is in precise agreement with those of the IIA and IIB supergravity theories respectively.

In addition to implementing the GSO projection we can demand that the string be oriented or not. We recall that an unoriented string has the symmetry $\sigma' = 2\pi - \sigma$, $\tau' = \tau$, while an oriented string has no such symmetry. This symmetry implies that the wavefunction is symmetric under the interchange α_n^μ, b_r^μ or $d_n^\mu \leftrightarrow \bar\alpha_n^\mu, \bar b_r^\mu$ or $\bar d_n^\mu$ as well as changing the first and second factors in the tensor product. More precisely we introduce the operator Ω, which obeys $\Omega^2 = 1$, that implements these changes in the NS sector as

$$\Omega \alpha_n^\mu \Omega^{-1} = \bar\alpha_n^\mu, \quad \Omega b_r^\mu \Omega^{-1} = \bar b_r^\mu$$

with an identical set of transformations in the R–R sector. We must also specify its action on the vacua and we take

$$\Omega \,|\, 0\rangle_{NS-NS} = -\Omega\,|\,0\rangle_{NS-NS}, \quad \Omega\,|0\rangle_{NS-R} = -\Omega\,|0\rangle_{R-NS}, \quad \Omega\,|0\rangle_{R-R} = -\Omega\,|0\rangle_{R-R}.$$

An oriented closed string has no such symmetry and so keeps all the fields; this is the case for the strings we discussed above. The symmetry of the unoriented string identifies the states in the NS–R sector with those in the R–NS sector and so we have only one gravitino instead of two, that is, one Majorana–Weyl spinor and only one Majorana–Weyl spinor of opposite chirality. As such, the projection conditions in these two sectors must be the same and so we can only apply the unoriented condition to the IIB string. In the NS–NS sector, taking account of the transformations of the vacuum only the graviton and scalar survive. The reader may like to recall the discussion for the bosonic string at the end of section 3.1.2. In the R sector, the symmetry implies that we must keep only anti-symmetric matrices for an unoriented string and, for the reason just stated, take the IIB GSO projection. This anti-symmetry of the R–R sector is best seen by considering the construction below equation (7.1.31). Terms symmetric under the exchange $D^j \leftrightarrow \bar D^j$ are of the form $D^j \bar D^k + \bar D^j D^k$, etc. These correspond to anti-symmetric matrices. This can be seen by acting with this operator on the states and remembering to raise the lowered spinor index with the anti-symmetric charge conjugation matrix C. As a result, we find that only the field strength $F^{\mu\nu\rho}$ survives. An unoriented superstring is referred to as a type I string, while oriented superstrings are called type II strings. It is usual not to use unoriented and oriented for superstrings, but to use type I and type II respectively.

Finally an unoriented, that is, type I, superstring has the massless field content

$$h_{\mu\nu}(\underline{35}), \; \varphi(\underline{1}); \; A_{\mu\nu}(\underline{28}), \; \hat\psi_{\mu 1}(\underline{56}), \; \lambda_1(\underline{8}), \tag{7.3.32}$$

where $\hat\psi_{\mu 1}$ is Majorana–Weyl. As for the type II string above, none of these theories has a tachyon, but they do have an infinite number of massive states. The type I supergravity theory corresponds to the ten-dimensional supergravity theory whose supersymmetry algebra has one Majorana–Weyl supercharge, that is, $D = 10$, $N = 1$ supergravity which one can verify has the above field content.

In fact, the full interacting type IIA and IIB superstrings possess the corresponding space-time supersymmetry. However, this is not the same requirement as demanding they be consistent string theories. As we will discuss in chapter 9, for the latter we require no superconformal anomaly once we have included all the necessary ghosts (as explained in chapter 9) and we require modular invariance, which is explained in chapter 18. Making these demands one finds that the IIA and IIB superstrings given above are consistent. However, there do exist other assignments of signs of the projectors in the various sectors that lead to strings that are also consistent, but have no spinors and so cannot possess any space-time supersymmetry. These are called type OA and OB theories [7.12].

As we will explain in more detail in chapter 13 the low energy limit of the IIB and IIA superstring theories must possess the corresponding space-time supersymmetry and be at most second order in space-time derivatives. As such the low energy limits can only be the IIA and IIB supergravity theories. This simple observation implies that all the effects, perturbative and non-perturbative, of the IIB and IIA superstring theories are encoded in the IIB and IIA supergravity theories. This makes the study of these theories particularly interesting.

Given the type I closed string, we could consider coupling it to the open superstring considered above, as they both have the same type of underlying supersymmetry. In fact the interacting open superstring leads to closed strings and these must be the closed type I superstrings. However, the open superstring or the type I closed string by itself is not a consistent theory, but the type I closed string coupled to the open string is under some very special conditions consistent. This is discussed in chapter 10.

The quantisation of the Green–Schwarz string while keeping manifest Lorentz symmetry is difficult, but it is straightforward to quantise in the light-cone gauge [7.13]. It is equivalent to the IIA and IIB strings discussed above depending on the corresponding supersymmetry one adopts in the Green–Schwarz action.

8 Conformal symmetry and two-dimensional field theory

He cooks with water doesn't he?

German saying

In this chapter we begin by finding the conformal algebra in a Minkowski space-time, that is, the set of transformations that leave the metric invariant up to a scale factor. In dimensions greater than two the conformal algebra is finite dimensional, but in two dimensions it is infinite-dimensional. We will find that if a field theory possesses conformal symmetry, then its energy-momentum tensor is traceless. For a two-dimensional classical theory conformal invariance implies the theory has an infinite number of conserved quantities, which are moments of the energy-momentum tensor.

In two-dimensional quantum theories one finds that the conformal algebra becomes modified by a central term associated with the required normal ordering. The central term contains a constant, which is called the central charge, upon which the nature of the representations of the conformal algebra crucially depends. As for any symmetry of a quantum field theory, the conformal symmetry of a two-dimensional quantum theory implies certain Ward identities. However, only a finite-dimensional sub-algebra of the infinite-dimensional conformal algebra is globally defined on the appropriate space-time and we will find that the Green's functions are only invariant under these transformations. Nonetheless, for a certain class of theories, called the minimal models, which have special central charges we can use the Ward identities corresponding to all the conformal transformations to place constraints on the Green's functions, which actually determine them. The feature of a minimal model that allows it to be solved is connected to the representation of the conformal algebra it carries. In particular, this highest weight representation of the conformal algebra is generically a reducible representation, and demanding that it be irreducible implies certain states in the representation must be set to zero. This is, in turn, leads to differential equations that determine the Green's functions.

These developments were essentially contained in the seminal paper of Belavin, Polyakov and Zamolodchikov [8.1], which created this subject and it is the primary reference for this chapter. This paper was motivated by the study of critical phenomena. It had been realised, through the study of the universality of critical exponents, that second order phase transitions were dominated by long range fluctuations. Indeed, they correspond to systems at a fixed point of their renormalisation group flows and at this point the system is conformally invariant. The prototype example is the Ising model, which has been solved completely for all values of the coupling and not just at the critical point. Some of the earlier work leading up to this development is given in [8.2].

160

8.1   Conformal transformations

8.1.1   Conformal transformations in D dimensions

Although most of this chapter is concerned with two-dimensional theories, we first consider conformal transformations in D dimensions. We work in a space that has coordinates x^μ and a diagonal metric $\eta_{\mu\nu}$ of the form

$$\eta_{\mu\nu} = \operatorname{diag}(\underbrace{-1, -1, \ldots, -1}_{s}, \underbrace{+1, \ldots, +1}_{r}). \tag{8.1.1}$$

In this space, a conformal transformation is a diffeomorphism $x^\mu \mapsto x'^\mu(x)$, such that the associated line element $ds^2 = dx^\mu dx^\nu \eta_{\mu\nu}$ changes only by a scale factor, namely

$$ds'^2 \equiv dx'^\mu dx'^\nu \eta_{\mu\nu} \equiv \Omega(x) ds^2. \tag{8.1.2}$$

Setting $x'^\mu \equiv f^\mu(x)$, the condition on f^μ of equation (8.1.2) is given by

$$\partial_\rho f^\mu \partial_\kappa f^\nu \eta_{\mu\nu} = \eta_{\rho\kappa} \Omega. \tag{8.1.3}$$

Raising the first index with the inverse metric and taking the trace we find that

$$\Omega = \frac{1}{D} \partial_\rho f^\mu \partial_\kappa f^\nu \eta_{\mu\nu} \eta^{\rho\kappa}. \tag{8.1.4}$$

Let us consider an infinitesimal transformation

$$f^\mu = x^\mu + \varepsilon^\mu(x), \tag{8.1.5}$$

then at lowest order, equation (8.1.4) becomes

$$\partial_\mu \varepsilon_\nu + \partial_\nu \varepsilon_\mu = \frac{2}{D} \eta_{\mu\nu} \partial^\rho \varepsilon_\rho, \tag{8.1.6}$$

where

$$\varepsilon_\nu = \varepsilon^\mu \eta_{\mu\nu}.$$

Taking the ∂^ν of this equation we find that

$$(-D + 2)\partial_\mu \partial^\kappa \varepsilon_\kappa = D\partial^2 \varepsilon_\mu. \tag{8.1.7}$$

Acting on this equation with ∂^μ we conclude that provided $D \neq 1$

$$\partial^2 \partial^\mu \varepsilon_\mu = 0. \tag{8.1.8}$$

For $D \neq 2$ and using equation (8.1.6) we note that

$$\partial_\rho \partial_\lambda \partial^\kappa \varepsilon_\kappa = \frac{D}{(2 - D)} \partial_\rho \partial^2 \varepsilon_\lambda. \tag{8.1.9}$$

However, the left-hand side is ρ, λ symmetric and hence, so must be the right-hand side. Consequently, using (8.1.6) and then (8.1.8) we conclude that

$$\partial_\rho \partial_\lambda \partial^\kappa \varepsilon_\kappa = \frac{1}{(2 - D)} \eta_{\rho\lambda} \partial^2 \partial^\kappa \varepsilon_\kappa = 0, \tag{8.1.10}$$

provided also $D \neq 1$.

As we now show the above equations imply that the third derivative of ε_μ vanishes. We have, using equations (8.1.6) and (8.1.7),

$$\partial_\rho \partial_\lambda \partial_\kappa \varepsilon_\mu = -\partial_\rho \partial_\lambda \partial_\mu \varepsilon_\kappa + \frac{2}{D}\eta_{\kappa\mu}\partial_\rho \partial_\lambda \partial^\tau \varepsilon_\tau$$

$$= -\partial_\rho \partial_\lambda \partial_\mu \varepsilon_\kappa = \partial_\rho \partial_\kappa \partial_\mu \varepsilon_\lambda$$

$$= -\partial_\rho \partial_\kappa \partial_\lambda \varepsilon_\mu = 0. \tag{8.1.11}$$

Hence ε^μ, when expressed as a polynomial in x^μ, has no higher terms than quadratic:

$$x^{\mu\prime} = x^\mu + a^\mu + t^\mu{}_\nu x^\nu + t^\mu{}_{\nu\rho} x^\nu x^\rho. \tag{8.1.12}$$

Applying equation (8.1.6) we find that

$$t_{\mu\nu} + t_{\nu\mu} = \frac{2}{D}\eta_{\mu\nu} t^\rho{}_\rho, \tag{8.1.13}$$

$$t_{\mu\nu\rho} + t_{\nu\mu\rho} = \frac{2}{D}\eta_{\mu\nu} t^\kappa{}_{\kappa\rho}. \tag{8.1.14}$$

We may solve (8.1.13) by writing

$$t_{\mu\nu} = w_{\mu\nu} + \lambda\eta_{\mu\nu}, \tag{8.1.15}$$

where

$$w_{\mu\nu} = -w_{\nu\mu}.$$

To solve equation (8.1.14) we consider

$$(t_{\mu\nu\rho} + t_{\nu\mu\rho}) + (t_{\rho\mu\nu} + t_{\mu\rho\nu}) - (t_{\rho\nu\mu} + t_{\nu\rho\mu})$$

$$= 2t_{\mu\nu\rho} = \frac{2}{D}(\eta_{\mu\nu} t^\kappa{}_{\kappa\rho} + \eta_{\mu\rho} t^\kappa{}_{\kappa\nu} - \eta_{\rho\nu} t^\kappa{}_{\kappa\mu}). \tag{8.1.16}$$

Inserting these solutions in $x^{\mu\prime}$ of equation (8.1.12), it becomes for $D \neq 1$ or 2

$$x^{\mu\prime} = x^\mu + a^\mu + w^\mu{}_\nu x^\nu + \lambda x^\mu + \lambda^\mu x^\kappa x_\kappa - 2\lambda_\kappa x^\kappa x^\mu, \tag{8.1.17}$$

where

$$\lambda^\mu = -\frac{1}{D} t^\kappa{}_\kappa{}^\mu. \tag{8.1.18}$$

Since the Poincaré group, which consists of the translations a^μ and Lorentz rotations $w^\mu{}_\nu$, is by definition the group which preserves the line element we necessarily find these transformations among the above. The rescaling of x^μ with parameter λ is a dilation and the transformation with parameter λ_μ is called a special conformal transformation. Counting the number of generators, we have

$$D + \frac{D}{2}(D-1) + 1 + D = \frac{1}{2}(D+1)(D+2). \tag{8.1.19}$$

The infinitesimal operators can be read off from equation (8.1.17) and one readily finds that they realise the Poincaré algebra plus the relations

$$[D, K_\nu] = K_\nu, \quad [D, P_\mu] = -P_\mu,$$

$$[K_\mu, J_{\rho\kappa}] = \eta_{\mu\rho}K_\kappa - \eta_{\mu\kappa}K_\rho, \quad [K_\mu, K_\nu] = 0, \tag{8.1.20}$$

$$[K_\mu, P_\nu] = -2\eta_{\mu\nu}D + 2J_{\mu\nu}.$$

This algebra is called the conformal algebra. It is isomorphic to the algebra of the group $SO(r+1, s+1)$.

The finite transformations are obtained by exponentiating the above infinitesimal transformations. For dilations this is straightforward, we find

$$x'^\mu = \lambda x^\mu. \tag{8.1.21}$$

For special conformal transformations the finite transformation is most easily found by realising that at the infinitesimal level it implies the relation

$$\frac{x'^\mu}{x'^2} = \frac{x^\mu}{x^2} + \lambda^\mu. \tag{8.1.22}$$

A finite transformation is, of course, of the same form and solving for x'^μ it implies that

$$x'^\mu = \frac{x^\mu + \lambda^\mu x^2}{(1 + 2x \cdot \lambda + \lambda^2 x^2)}. \tag{8.1.23}$$

Our definition of conformal transformations utilised the flat metric. However, once one has performed a conformal transformation the metric is no longer flat. A more aesthetic definition uses the notion of a conformally flat metric. The latter is a metric for which there exists local coordinates such that the line element is of the form

$$ds^2 = \Omega(x)dx^\mu dx^\nu \eta_{\mu\nu}. \tag{8.1.24}$$

A conformal transformation is defined to be one which takes one conformally flat metric to another. Clearly, this leads to the same transformations as our original definition.

One could also not use a conformally flat metric, but instead use any metric and define conformal transformations to be those general coordinate transformations which scale the given metric. The resulting conformal group will then change as one takes different metrics.

8.1.2   Conformal transformations in two dimensions

In the previous section, we constructed the conformal algebra and transformations; however, an exception that we did not cover was when $D = 2$. In this case, we find equation (8.1.5) becomes in Minkowski space-time

$$\partial_0 \varepsilon_1 = -\partial_1 \varepsilon_0 \text{ and } \partial_0 \varepsilon_0 + \partial_1 \varepsilon_1 = 0 \tag{8.1.25}$$

and in Euclidean space

$$\partial_1 \varepsilon_2 = -\partial_2 \varepsilon_1 \text{ and } \partial_1 \varepsilon_1 - \partial_2 \varepsilon_2 = 0. \tag{8.1.26}$$

In *Minkowski space-time*, the first equation implies that

$$\varepsilon_\mu = (\partial_0\epsilon, -\partial_1\epsilon),\tag{8.1.27}$$

while the second equation implies that ϵ satisfies the wave equation, namely $(\partial_0^2 - \partial_1^2)\epsilon = 0$. Consequently, ϵ is the sum of two arbitrary functions, one function of x^+ and one function of x^-, where

$$x^+ \equiv x^0 + x^1, \text{ and } x^- \equiv x^0 - x^1.\tag{8.1.28}$$

Unlike the case $D > 2$, for $D = 2$ we find an infinite number of conformal transformations which are most easily analysed by moving to the light-cone coordinates x^+, x^-. In terms of these coordinates, which are discussed in detail in appendix B, finite conformal transformations can be written as

$$x^+ \to x^{+\prime} = f(x^+), \qquad x^- \to x^{-\prime} = g(x^-),\tag{8.1.29}$$

where f and g are arbitrary functions. This result follows from the discussion above, but it is obvious once we write the line element in the form

$$ds^2 = -(dx^0)^2 + (dx^1)^2 = -dx^+ dx^-.\tag{8.1.30}$$

In *Euclidean space*, equations (8.1.26) are the Cauchy–Riemann equations associated with the complex quantity $\varepsilon = \varepsilon_1 + i\varepsilon_2$. They imply that ε is a function of z and $\bar\varepsilon = \varepsilon_1 - i\varepsilon_2$ is a function of $\bar z$. As such, we adopt the complex coordinates

$$z = x^1 + ix^2, \qquad \bar z = x^1 - ix^2\tag{8.1.31}$$

and the line element becomes

$$ds^2 = (dx^1)^2 + (dx^2)^2 = dz d\bar z.\tag{8.1.32}$$

The derivatives are related by

$$\partial \equiv \frac{\partial}{\partial z} = \frac{1}{2}\left(\frac{\partial}{\partial x^1} - i\frac{\partial}{\partial x^2}\right), \qquad \bar\partial \equiv \frac{\partial}{\partial z} = \frac{1}{2}\left(\frac{\partial}{\partial x^1} + i\frac{\partial}{\partial x^2}\right),\tag{8.1.33}$$

while the integration measure becomes

$$d^2x \equiv dx^1 \wedge dx^2 = i\frac{dz d\bar z}{2} \equiv i\frac{dz \wedge d\bar z}{2}.\tag{8.1.34}$$

We can change from Minkowski space-time to Euclidean space by $x^0 \mapsto -ix^1, x^1 \mapsto x^2$. This is just a Wick rotation. The induced changes are

$$x^+ = x^0 + x^- \to -iz, x^- = x^0 - x^- \to -i\bar z, \partial_+ = \frac{1}{2}\left(\frac{\partial}{\partial x^0} + \frac{\partial}{\partial x^1}\right) \to i\partial, \partial_-$$

$$= \frac{1}{2}\left(\frac{\partial}{\partial x^0} - \frac{\partial}{\partial x^1}\right) \to i\bar\partial.$$

Statistical systems are defined in Euclidean space. The string world sheet has a Minkowskian signature and by rotating to Euclidean signature we can take advantage of the large literature on Riemann surfaces. However, it is fair to say that the deeper significance of this change has not been adequately understood. For the remainder of this chapter we work in Euclidean space, except for a small section where it is explicitly stated that we are working in Minkowski space-time.

Once we write the line element in the form of equation (8.1.32) the conformal transformations found above become obvious. The line element is preserved up to a scale factor (that is, $\Omega = f'(z)g'(\bar{z})$) by any transformation of the form

$$z \mapsto f(z), \qquad \bar{z} \mapsto g(\bar{z}) \tag{8.1.35}$$

for any functions f and g. If we are dealing with real Euclidean two-dimensional space, z and $\bar{z}$ are related by complex conjugation and as a result so are f and g. However, we now assume that we have complexified the original space-time and so these complex conjugation relations are not valid. This complex extension of space-time is sometimes adopted after rotating quantum field theories to Euclidean space in a number of different contexts. The only other transformations which preserve the line element up to scale are

$$z \mapsto h(\bar{z}), \qquad \bar{z} \mapsto k(z). \tag{8.1.36}$$

However, these transformations change the orientation defined on the two-dimensional surface and will not be considered further.

A complete set of infinitesimal transformations of the form of equation (8.1.35) are given by

$$z \mapsto z + a_n z^{n+1}, \qquad \bar{z} \mapsto \bar{z} + \bar{a}_n (\bar{z})^{n+1} \tag{8.1.37}$$

and are generated by

$$L_n = z^{n+1} \frac{\partial}{\partial z}, \qquad \bar{L}_n = \bar{z}^{n+1} \frac{\partial}{\partial \bar{z}}. \tag{8.1.38}$$

We will specify later the precise space on which these transformation are carried out. The Lie algebra these generators satisfy is given by

$$[L_n, L_m] = (n - m)L_{n+m},$$
$$[L_n, \bar{L}_m] = 0, \tag{8.1.39}$$
$$[\bar{L}_n, \bar{L}_m] = (n - m)\bar{L}_{n+m}.$$

Naive substitution of equation (8.1.38) in equation (8.1.39) yields the result above with a minus sign. Such a calculation is performed from an active viewpoint. To arrive at the sign in (8.1.39), one must take the passive viewpoint and carry out the manoeuvre on functions as follows. A scalar function φ under a passive transformation $z \mapsto z' = f(z)$ transforms as

$$\varphi(z) \mapsto \varphi'(z) = \varphi(f(z)). \tag{8.1.40}$$

For an infinitesimal transformation $z' = z + \varepsilon(z)$,

$$\delta\varphi = \varphi'(z) - \varphi(z) = \varepsilon(z)\partial\varphi \tag{8.1.41}$$

and we evaluate the commutator as follows

$$[\delta_{\varepsilon_1}, \delta_{\varepsilon_2}]\varphi = \delta_{\varepsilon_1}\varepsilon_2\partial\varphi - (1 \leftrightarrow 2) = \varepsilon_2\partial\delta_{\varepsilon_1}\varphi - (1 \leftrightarrow 2) = (\varepsilon_2\partial\varepsilon_1 - (1 \leftrightarrow 2))\partial\varphi \tag{8.1.42}$$

to arrive at the above commutation relations.

Thus the conformal algebra in two dimensions is an infinite-dimensional algebra which is the direct sum of two identical algebras V, that is, $V \oplus V$, where V is the Virasoro algebra, which is the one generated by one set of the L_n alone. An important sub-algebra of V is that generated by L_0, $L_{\pm 1}$, which is isomorphic to the algebra of SL(2, **R**).

Even if we begin with a space that has a flat metric, we will, after a conformal transformation, have a space with a metric that has a scale factor. Consequently, we are in general working with a space that has a line element of the form

$$e^{\varphi(z,\bar{z})} dz d\bar{z}. \tag{8.1.43}$$

Since the most general line element is of the form

$$ds^2 = g_{zz} dz dz + g_{z\bar{z}} dz d\bar{z} + g_{\bar{z}z} d\bar{z} dz + g_{\bar{z}\bar{z}} d\bar{z} d\bar{z}, \tag{8.1.44}$$

we can read off the components of the metric

$$g_{zz} = g_{\bar{z}\bar{z}} = 0, \qquad g_{z\bar{z}} = g_{\bar{z}z} = \tfrac{1}{2} e^{\varphi}. \tag{8.1.45}$$

The inverse metric is then given by

$$g^{zz} = g^{\bar{z}\bar{z}} = 0, \qquad g^{z\bar{z}} = g^{\bar{z}z} = 2e^{-\varphi}. \tag{8.1.46}$$

Just as for the tensor calculus of general relativity, we may develop a tensor calculus for conformal transformations.

The most general tensor is of the form

$$A{\overbrace{\scriptstyle z\cdots z}^{n}\overbrace{\scriptstyle \bar{z}\cdots \bar{z}}^{\bar{n}} \atop \underbrace{\scriptstyle z\cdots z}_{m}\underbrace{\scriptstyle \bar{z}\cdots \bar{z}}_{m}} \equiv A(z, \bar{z}) \tag{8.1.47}$$

and it transforms under the coordinate change $z \mapsto w(z), \bar{z} \mapsto \bar{w}(\bar{z})$ as

$$A(z, \bar{z}) \mapsto \left(\frac{dw}{dz}\right)^{h} \left(\frac{d\bar{w}}{d\bar{z}}\right)^{\bar{h}} A(w, \bar{w}), \tag{8.1.48}$$

where $h = m - n$ and $\bar{h} = \bar{m} - \bar{n}$ are called the conformal weights or dimensions of the tensor A. It is instructive to recall the general form for the transformation of a tensor used in general relativity; in particular a 1-form R_{μ} transforms as $R_{\mu}(x) dx^{\mu} = R'_{\mu}(x') dx^{\mu'}$ and so $R_{\nu}(x) = (\partial x^{\mu'}/\partial x^{\nu}) R'_{\mu}(x')$. This is in agreement with the above formula if we drop the $'$ on the transformed field and the arrow means we make the replacement, that is, $A(z, \bar{z}) \mapsto A(z, \bar{z}) = (dw/dz)^{h} (d\bar{w}/d\bar{z})^{\bar{h}} A(w, \bar{w})$. This somewhat unconventional way of writing the field transformation is the one universally used in conformal field theory. For the above conformal transformation $z \to w, \bar{z} \to \bar{w}$ a form, Bdz, changes as $B(z, \bar{z}) = (dw/dz) B(w, \bar{w})$ and so has conformal weight $(1, 0)$.

A translation $x^1 \mapsto x^1 + a$ is induced by $z \mapsto z + a, \bar{z} \mapsto \bar{z} + a$ for a real and so is generated by $L_{-1} + \bar{L}_{-1}$, while the shift $x^2 \mapsto x^2 + b$ is induced by $z \mapsto z + ib, \bar{z} \mapsto \bar{z} - ib$, b real, and is generated by $L_{-1} - \bar{L}_{-1}$. The remaining generator of the two-dimensional ISO(2) algebra is the rotation $\delta x^1 = -\alpha x^2, \delta x^2 = +\alpha x^1$, whose finite action on $z, \bar{z}$ is given by $z \mapsto e^{i\alpha} z, \bar{z} \mapsto e^{-i\alpha} z$ and so generated by $L_0 - \bar{L}_0$. The dilation $z \mapsto \lambda z, \bar{z} \mapsto \lambda \bar{z}$ for λ real is generated by $L_0 + \bar{L}_0$.

We find that under a dilation $A \mapsto \lambda^{h+\bar{h}}A$, while under a rotation $A \mapsto e^{i(h-\bar{h})\alpha}A$. Consequently, we will call $\Delta = h + \bar{h}$ the dilation weight of A and $s = h - \bar{h}$ its spin. It will turn out that we will wish to consider fields for which Δ and s are not integers. We raise and lower indices with the metric and so in particular

$$A_z = g_{z\bar{z}}A^{\bar{z}} = \tfrac{1}{2}e^{\varphi}A^{\bar{z}}, \qquad A_{\bar{z}} = \tfrac{1}{2}e^{\varphi}A^z. \tag{8.1.49}$$

The relation between tensors in the $z, \bar{z}$ coordinate system and the original x^1, x^2 coordinate system is given by the usual transformation formula, that is,

$$A_z = \frac{\partial x^\alpha}{\partial z}A_\alpha = \tfrac{1}{2}(A_1 - iA_2), \qquad A_{\bar{z}} = \tfrac{1}{2}(A_1 + iA_2) \tag{8.1.50}$$

and for a second rank tensor by

$$A_{zz} = \tfrac{1}{4}(A_{11} - iA_{12} - iA_{21} - A_{22}),$$
$$A_{z\bar{z}} = \tfrac{1}{4}(A_{11} + iA_{12} - iA_{21} + A_{22}),$$
$$A_{\bar{z}z} = \tfrac{1}{4}(A_{11} - iA_{12} + iA_{21} + A_{22}),$$
$$A_{\bar{z}\bar{z}} = \tfrac{1}{4}(A_{11} + iA_{12} + iA_{21} - A_{11}). \tag{8.1.51}$$

To find the analogous result in Minkowski space-time, we substitute $A_1 \mapsto A_0$, $A_2 \mapsto iA_1$ and $A_{11} \mapsto A_{00}$, $A_{12} \mapsto iA_{01}$, $A_{21} \mapsto iA_{10}$, $A_{22} \mapsto -A_{11}$.

The Christoffel symbol in the $z, \bar{z}$ coordinates is found from its usual definition to be zero except for the components

$$\Gamma_{zz}{}^z = \partial_z\varphi, \qquad \Gamma_{\bar{z}\bar{z}}{}^{\bar{z}} = \partial_{\bar{z}}\varphi. \tag{8.1.52}$$

Of course, if we work with the metric for which $\varphi = 0$, then all the Christoffel symbols vanish.

We take the conformal transformations to act on the Riemann sphere $\mathbf{CP}^1$, which is the complex plane $\mathbf{C}$ with the point at infinity added. The Riemann sphere and the nature of this addition are the subject of appendix C. We now wish to discover which conformal transformations are globally defined on $\mathbf{CP}^1$. An infinitesimal conformal transformation which changes $z \to z + v(z)$ corresponds to the vector field $v(z)\partial/\partial z$. The conformal transformation is globally defined if the corresponding vector field is well defined on all of the Riemann sphere. Before considering when this is the case, we will consider the corresponding, but simpler problem of when a scalar function $f(z)$ is well defined on the Riemann sphere. As discussed in appendix C, the Riemann sphere requires two coordinate patches to cover it. Following appendix C, we denote the coordinates on the two coordinate patches of the Riemann sphere by z and z' and in the overlap of the patches they are related by $z' = 1/z$. A function f that is well defined on the Riemann sphere must have no pole, or other singularity, at any point on the Riemann sphere. Clearly f is well defined in the z coordinate patch if it is of the form $f(z) = \sum_{n=0}^{\infty}a_n z^n$. This patch does not, however, include the point at infinity which is $z' = 0$ in the z' coordinate patch. The function f will be well defined in the z' coordinate patch if it has the expansion $f(z') = \sum_{n=0}^{\infty}b_n z'^n$. However, since f is a scalar function $f(z) = f(z')$. Applying this to the particular conformal transformation relating the two coordinate patches we can equate the above expansions, set $z = 1/z'$ and compare the two sides of the equation. We find that the only consistent term is the constant term and so the only scalar function which is well defined on the Riemann sphere is a constant.

We now apply a similar analysis to analytic vector fields. If $v(z)$ is the component of the analytic vector field it must transform under a coordinate change such that $v(z)\partial/\partial z$ is invariant. In the z' coordinate patch, the vector field has component $-z'^2 v(1/z')$. The vector field generated by L_n has component $v = z^{n+1}$ in the z coordinate patch and component $-z'^{-n+1}$ in the z' coordinate patch. Consequently, the only well-defined analytic vector fields on the Riemann sphere are those generated by L_0 and $L_{\pm 1}$. Hence the only globally defined conformal transformations generated by the L_n are those generated by L_0 and $L_{\pm 1}$. An identical argument applied to the $\bar{L}_n$ shows that the only globally defined transformations are those generated by $\bar{L}_0$ and $\bar{L}_{\pm 1}$.

We will often consider finite transformations generated by the L_n. Let us first consider L_{-1}, writing $e^{-aL_{-1}}$ as $\lim_{N \to \infty}(1 + (aL_{-1}/n))^N$ and using the fact that for small ε $(1 + \varepsilon L_{-1})z = z + \varepsilon$, we conclude that

$$e^{aL_{-1}} : z \mapsto z' = z + a. \tag{8.1.53}$$

Applying the same procedure to L_0 we find that

$$e^{aL_0} : z \mapsto z' = e^a z.$$

The L_n for $n \neq 0$ induce the transformation $(1 + \varepsilon L_n) : z \mapsto z + \varepsilon z^{n+1}$ for small ε and consequently $z^{-n} \mapsto z^{-n} - n\varepsilon$. We may write this as z'^n. Following the same argument, as for L_{-1} above, but now applied to z^{-n} we find that,

$$e^{aL_n} : z^{-n} \mapsto z^{-n} - na\varepsilon = z'^{-n} \tag{8.1.54}$$

from which we find

$$e^{aL_n} : z \mapsto z' = \left[\frac{z^n}{(1 - az^n/n)} \right]^{\frac{1}{n}}. \tag{8.1.55}$$

For $n = 1$ this result reads

$$e^{aL_1} : z \mapsto z' = \frac{z}{(1 - az)}.$$

From the form of equation (8.1.55) it is clear that for $n \neq 0, \pm 1$ z' is not an analytic function of z in that it is not single valued. This agrees with our previous result that only the conformal transformations generated by $L_0, L_{\pm 1}$ are well defined.

The transformation group generated by $L_0, L_{\pm 1}$ has, at least in the neighbourhood of the identity, the form

$$e^{hL_{-1}} e^{gL_0} e^{fL_1} : z \mapsto z' = \frac{gz}{1 - fz} + h. \tag{8.1.56}$$

By letting $g = (ad - bc)/d$, $f = -c/d$ and $h = b/d$ we may rewrite the above transformation as

$$z' = \frac{az + b}{cz + d}, \tag{8.1.57}$$

where a, b, c, d are real numbers. A rescaling of these parameters by an arbitrary real number will not affect the transformation, and so we can choose $ad - bc$ equal to 1 or -1.

The choice depends on the original values of a, b, c, d, that is, if $ad - bc$ is positive or negative.

The group $SL(2, \mathbf{R})$ consists of all 2×2 real matrices

$$g \in \begin{pmatrix} a & b \\ c & d \end{pmatrix} \qquad a, b, c, d \in \mathbf{R} \tag{8.1.58}$$

with $ad - bc = +1$. We may identify $g \in SL(2, \mathbf{R})$ with the transformation of equation (8.1.57). It is straightforward to verify that this identification, denoted by ϕ, preserves the group multiplication law, that is, $\phi(g_1, g_2) = \phi(g_1)\phi(g_2)$, and it is clear that ϕ is a map from $SL(2, \mathbf{R})$ onto P, where P is the set of all transformations of equation (8.1.58) for which $ad - bc = 1$. The identification ϕ is, however, not one to one since the kernel of ϕ consists of elements $+I$ and $-I$, where I is the 2×2 unit matrix. Consequently, we have a homomorphism ϕ from $SL(2, \mathbf{R})$ onto P with kernel $\{\pm I\} = \mathbf{Z}_2$ and by the well-known homomorphism theorem $SL(2, \mathbf{R})/\mathbf{Z}_2$ is isomorphic to P.

We can carry out an identical analysis on the group generated by $\bar{L}_0, \bar{L}_{\pm 1}$ which leaves z inert, but generates identical transformations on $\bar{z}$.

We can also consider the six-parameter group generated by $L_0, \bar{L}_0, L_{\pm 1}$ and $\bar{L}_{\pm 1}$. Following our previous reasoning this leads to transformations of the form

$$z \mapsto z' = \frac{az + b}{cz + d} \qquad a, b, c, d \in \mathbf{C}. \tag{8.1.59}$$

The difference with the previous formula is that a, b, c, d are complex instead of real.

By rescaling the parameters, we may choose $ad - bc = 1$ without affecting the transformations. The set of all such transformationsis is denoted M, and they are called Möbius transformations or fractional linear transformations. It follows by rerunning the previous argument that M is isomorphic to $SL(2, \mathbf{C})/\mathbf{Z}_2$, where $SL(2, \mathbf{C})$ is the group of 2×2 complex matrices with determinant one.

The Möbius group is, in fact, the group of all globally defined one-to-one transformations of $\mathbf{CP}^1$ onto itself. A globally defined transformation must possess globally defined vector fields corresponding to its infinitesimal action. The result then follows from the observation above that the only globally defined analytic vector fields are those corresponding to L_0, $\bar{L}_0, L_{\pm 1}$ and $\bar{L}_{\pm 1}$ which generate the Möbius group.

The group of globally defined conformal transformations is isomorphic to $SO(3, 1)$ since this latter group is isomorphic to $SL(2, \mathbf{C})/\mathbf{Z}_2$. This agrees with the general result that in D dimensions the conformal group of Euclidean space is isomorphic to $SO(D + 1, 1)$.

The Riemann sphere and Möbius transformations play an important role in string theory and we now describe some of their properties. A Möbius transformation can be used to take any three distinct points $z_1, z_2, z_3 \in \mathbf{CP}^1$ to any other three distinct points $z_1', z_2', z_3' \in \mathbf{CP}^1$. To see this we consider the transformation

$$z \mapsto \frac{(z - z_2)(z_3 - z_1)}{(z - z_1)(z_3 - z_2)} = M(z) \tag{8.1.60}$$

which takes z_1, z_2, z_3 to $\infty, 0, 1$ respectively. Its inverse will take $\infty, 0, 1$, to z_1, z_2, z_3. We can use one Möbius transformation M_1 to go from z_1, z_2, z_3 to $\infty, 0, 1$ and another Möbius transformation $\tilde{M}$ from $\infty, 0, 1$ to z_1', z_2', z_3' and so $\tilde{M}M$ will be a Möbius transformation from z_1, z_2, z_3 to z_1', z_2', z_3' as required.

In fact, if there were two such Möbius transformations M and M', then MM'^{-1} would leave z_1, z_2, z_3 fixed, and the only Möbius transformation with three fixed points is the identity. As such, the Möbius transformation which takes three distinct points to another three distinct points is unique and can be specified by these six points. This is sometimes used to denote M by

$$\begin{pmatrix} z_1 & z_2 & z_3 \\ z_1' & z_2' & z_3' \end{pmatrix}.$$

The above discussion makes it clear that there is no Möbius invariant constructed from only three points. A Möbius invariant f which is a function of n points z_i, $i = 1, \ldots, n$, must be inert under the infinitesimal variations of equation (8.1.37) for $n = 0, \pm 1$, namely

$$\sum_i \left(\frac{\partial}{\partial z_i} \right) f = 0, \tag{8.1.61}$$

$$\sum_i \left(z_i \frac{\partial}{\partial z_i} \right) f = 0, \tag{8.1.62}$$

$$\sum_i \left(z_i^2 \frac{\partial}{\partial z_i} \right) f = 0. \tag{8.1.63}$$

The first equation implies that f is a function of differences, $z_{ij} = z_i - z_j$, and the second equation implies that it is homogeneous of degree 0. The final equation may be written in the form

$$\sum_{i,j} (z_i + z_j) z_{ij} \frac{\partial}{\partial z_{ij}} f = 0. \tag{8.1.64}$$

Unlike f this equation is not only a function of differences and as a result the coefficients of z_i, $i = 1, 2, \ldots, n$, must vanish separately; namely

$$\sum_j z_{ij} \frac{\partial}{\partial z_{ij}} f = 0. \tag{8.1.65}$$

If we write f as a ratio of powers of z_{ij}, then these equations imply that a given index i must appear as many times in the denominator as in the numerator. It is then straightforward to write down Möbius invariants associated with any four points z_1, z_2, z_3, z_4. In fact, there are six possible such invariants, called cross ratios, which are given by

$$x = \frac{z_{12} z_{34}}{z_{13} z_{24}}, \qquad 1 - x = \frac{z_{14} z_{23}}{z_{13} z_{24}},$$

$$\frac{x}{1 - x} = \frac{z_{12} z_{34}}{z_{14} z_{23}} \quad \text{and their three inverses.} \tag{8.1.66}$$

As the above equation states, although there are six invariants and they are all related to one of them, say x.

As we will discuss in more detail in the next chapter the world sheet of the classical closed string can be thought of as the Riemann sphere. In particular, the conformal symmetry of the closed string is realised just as was discussed above. However, for open strings it will turn out that we will wish to identify the upper and lower half-planes of the Riemann sphere, in particular to identify $\bar{z}$ with the complex conjugate of z. As such, z and $\bar{z}$ must transform in such a way that they are still related by complex conjugation after the transformation. This is achieved for a group which has only three real parameters by considering the group generated by $L_0 + \bar{L}_0$ and $L_{\pm 1} + \bar{L}_{\pm 1}$. Such transformations are of the form $z' = (az + b)/(cz + d)$ with $a, b, c, d \in \mathbf{R}$ and are such that they map the upper half-plane, denoted H, onto itself. This follows from the result that $\mathrm{Im}z' = (\mathrm{Im}z)/|cz + d|^2$. In fact, the group P consists of all one-to-one conformal (analytic) transformations of H onto itself. The upper half-plane H and the group $\mathrm{SL}(2, \mathbf{R})$ play important roles in open string theories. Using the group P on the upper half-plane H we may move any three points on the real axis to any other three points. However, any element of P reverses or preserves the cyclic ordering of points on the real axis. Any element of P can be made up of a translation $T(a) : z \mapsto z + a$, a dilation $D(\lambda) \mapsto \lambda z$ and an inversion $I : z \mapsto -1/z$. As one readily verifies

$$T\left(\frac{a}{c}\right) D\left(\frac{ad - bc}{c}\right) I T(d) D(c) z = \frac{az + b}{cz + d}. \tag{8.1.67}$$

Translations and inversions preserve the order of the points, while dilations preserve or reverse the order depending on whether $\lambda > 0$ or $\lambda < 0$. Examining the above equation we find that if $ad - bc > 0$ the order is preserved, and if $ad - bc < 0$ it is reversed.

8.2   Conformally invariant two-dimensional field theories

8.2.1   Conformally invariant two-dimensional classical theories

Before considering two-dimensional theories, let us consider a theory in an arbitrary number of dimensions which is Poincaré invariant. For such a theory there exists an energy-momentum tensor $T_{\alpha\beta}$, which we can choose to be symmetric. This tensor is a conserved current, that is, $\partial^\alpha T_{\alpha\beta} = 0$, whose charge generates translations. The current corresponding to Lorentz rotations is a moment of the energy-momentum tensor, namely $x_\alpha T_{\beta\delta} - x_\beta T_{\alpha\delta}$. It is conserved on its δ index due to the symmetry of $T_{\alpha\beta}$.

The theory is dilatation invariant if it is invariant under under $x^\alpha \mapsto \lambda x^\alpha$. The associate current, j_β, is a moment of the energy-momentum tensor and is given by

$$j_\beta = x^\alpha T_{\alpha\beta}, \tag{8.2.1}$$

as might be anticipated since $T_{\alpha\beta}$ generates translations. It is conserved provided that $T^\alpha{}_\alpha = 0$ and so the theory is dilation invariant if and only if the energy-momentum tensor is traceless. However, this condition also allows us to construct further conserved moments of $T_{\alpha\beta}$. Let us consider moments of the form

$$k^\alpha(x) T_{\alpha\beta}, \tag{8.2.2}$$

where k^α may carry further Lorentz indices that are not contracted with $T_{\alpha\beta}$ and which we have not explicitly shown. This current is conserved provided

$$\partial^\beta k^\alpha + \partial^\alpha k^\beta - \varphi \eta^{\alpha\beta} = 0, \tag{8.2.3}$$

where φ is an arbitrary function of x^α. Another conserved current in addition to equation (8.2.1) is given by

$$2x_\alpha x^\delta T_{\delta\beta} - x^\gamma x_\gamma T_{\alpha\beta}. \tag{8.2.4}$$

By considering the Poisson bracket of x^α with the charges corresponding to the currents of equation (8.2.2), it is clear that these charges generate the transformations $x^\alpha \mapsto x^\alpha + k^\alpha$. Since k^α is subject to the constraint of equation (8.2.3), we find, comparing this equation with equation (8.1.5) and identifying ε with k, that these charges generate the transformations of the conformal group. As a result, these charges will have a Poisson bracket algebra corresponding to the conformal algebra. Indeed, the current of equation (8.2.4) corresponds to that of special conformal transformations.

We have therefore shown that any theory that has $T^\alpha_{\ \alpha} = 0$ is not only invariant under dilations, but also conformally invariant. In fact, it can be shown, using only the conformal algebra that the current for dilations is the above moment of the energy-momentum tensor, that it is traceless and that the other currents have the the form of equation (8.2.2) subject to equation (8.2.3).

From now on we restrict our attention to the two-dimensional case and examine how the above general statements are realised. The energy-momentum tensor if it is symmetric and traceless has only two components which we can take to be T_{11} and T_{12} in Euclidean space and T_{00} and T_{01} in Minkowski space-time. In the z, $\bar{z}$ system, the traceless condition is given by

$$T_{z\bar{z}} = 0 \tag{8.2.5}$$

leaving only $T_{zz} = \frac{1}{2}(T_{11} - iT_{12})$ and $T_{\bar{z}\bar{z}} = \frac{1}{2}(T_{11} + iT_{12})$ in Euclidean space and $T_{zz} = \frac{1}{2}(T_{00} + T_{01})$ and $T_{\bar{z}\bar{z}} = \frac{1}{2}(T_{00} - T_{01})$ in Minkowski space-time. The conservation condition is given by

$$\partial_{\bar{z}} T_{zz} = 0 = \partial_z T_{\bar{z}\bar{z}}, \tag{8.2.6}$$

which implies that T_{zz} and $T_{\bar{z}\bar{z}}$ are functions of only z and $\bar{z}$, respectively. Given any two functions $f(z)$ and $g(\bar{z})$, we find an infinite set of conserved currents given by

$$f(z)T_{zz} \quad \text{and} \quad g(\bar{z})T_{\bar{z}\bar{z}} \tag{8.2.7}$$

as clearly $\partial_{\bar{z}}(f(z)T_{zz}) = 0$ and $\partial_z(f(\bar{z})T_{\bar{z}\bar{z}}) = 0$.

For reasons that will become apparent, we will take our time to be generated by $|z|$ and so equal time surfaces are given by $|z| = $ constant. The functions $f(z)$ and $g(\bar{z})$ can be represented in terms of the complete sets z^n and $\bar{z}^n$, respectively. In this case, the infinite set of charges corresponding to the currents of equation (8.2.7) are given by

$$L_n = \oint \frac{dz}{2\pi i} z^{n+1} T_{zz}, \qquad \bar{L}_n = \oint \frac{d\bar{z}}{2\pi i} \bar{z}^{n+1} T_{\bar{z}\bar{z}}, \tag{8.2.8}$$

where the contours of integration are circles around the origin.

Let us illustrate the above in the context of the free scalar field, whose Euclidean action is given by

$$A = -\frac{i}{8\pi} \int d^2x (\partial_\alpha \varphi \partial^\alpha \varphi). \tag{8.2.9}$$

We note that the path integral in Minkowski space-time contains a phase factor containing the action A that is of the form e^{iA}, but in the Euclidean path integral this factor is of the form e^{-A}. This change is achieved by making the transformation $A \to iA$ when going from the Minkowski space-time action to that in Euclidean space. This latter transformation has been carried out in going from the usual scalar field action in Minkowski space-time to the Euclidean action above. The apparently odd normalisation of the action will turn out to be that universally used in conformal field theory for a scalar field action. Under the transformation $\delta\varphi = k^\alpha \partial_\alpha \varphi$, the variation of A is given by

$$\delta A = -\frac{i}{8\pi} \int d^2 x (\partial_\alpha k_\beta + \partial_\beta k_\alpha - \eta_{\alpha\beta} \partial_\gamma k^\gamma) \partial^\alpha \varphi \partial^\beta \varphi$$

$$= -\frac{i}{4\pi} \int d^2 x \partial^\alpha k^\beta (\partial_\alpha \varphi \partial_\beta \varphi - \tfrac{1}{2} \eta_{\alpha\beta} \partial_\gamma \varphi \partial^\gamma \varphi). \tag{8.2.10}$$

Consequently, for the above variation the action is invariant only if equation (8.2.3) holds, whereupon it is invariant under conformal transformations. From the second line in the above equation we can read off the energy-momentum tensor since this current is the coefficient of $\partial^\alpha k^\beta$, that is, a local translation, under the variation of the action of equation (8.2.10), and we conclude that it is given by

$$T_{\alpha\beta} = -\tfrac{1}{2} (\partial_\alpha \varphi \partial_\beta \varphi - \tfrac{1}{2} \eta_{\alpha\beta} \partial_\gamma \varphi \partial^\gamma \varphi). \tag{8.2.11}$$

It is indeed traceless and has been normalised (that is, the factor of $-\tfrac{1}{2}$) so as to agree with later discussions. In the $z, \bar{z}$ coordinates we find therefore

$$T_{z\bar{z}} = 0 \quad \text{and} \quad T_{zz} = -\tfrac{1}{2}\partial_z \varphi \partial_z \varphi, \qquad T_{\bar{z}\bar{z}} = -\tfrac{1}{2}\partial_{\bar{z}}\varphi \partial_{\bar{z}}\varphi \tag{8.2.12}$$

and, using equation (8.1.34), the action takes the form

$$\frac{1}{4\pi} \int dz d\bar{z} \partial_z \varphi \partial_{\bar{z}} \varphi. \tag{8.2.13}$$

8.2.2   Conformal Ward identities

We now generalise the results of the previous section to the quantum theory. We start with a quantum field theory which is Poincaré invariant and has an energy-momentum tensor with vanishing trace. A symmetry in a quantum field theory implies that the correlation functions satisfy certain Ward identities and we first give a general discussion of Ward identities, which with small changes applies to any quantum field theory in any dimension. We then apply the results to find the consequences of conformal symmetry in two dimensions.

In this section, we will work in Euclidean space, which is the signature that arises naturally in statistical mechanics applications. Statistical mechanical systems often live on a strip or cylinder. These may be taken to have the coordinates $\varsigma = \tau + i\sigma$, $\bar{\varsigma} = \tau - i\sigma$, $-\infty \le \tau \le \infty$ with $-\pi \le \sigma < \pi$ for the cylinder and $0 \le \sigma \le \pi$ for the strip. In the case of the cylinder the lines $\sigma = \pi$ and $\sigma = -\pi$ are to be identified. However, we will often wish to work on the Riemann sphere, with coordinates $z, \bar{z}$. The mapping from the cylinder to the Riemann sphere, $\mathbf{CP}^1$ is given by

$$z = e^\varsigma = e^{\tau + i\sigma}, \qquad \bar{z} = e^{\bar{\varsigma}} = e^{\tau - i\sigma}. \tag{8.2.14}$$

Figure 8.2.1   Map from the cylinder to the Riemann sphere.

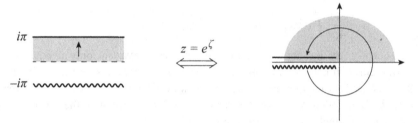

Figure 8.2.2   Map from the strip to the upper half-plane.

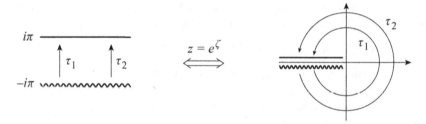

Figure 8.2.3   Lines of constant time on the cylinder mapped to the Riemann sphere.

We use the same map for the strip, but in this case it is not mapped to all of the Riemann sphere, but just the upper half-plane. These mappings are illustrated in figures 8.2.1 and 8.2.2. In the latter figure the strip is the shaded region. Although string world sheets, which for free strings can be represented by a strip for open strings and a cylinder for closed strings, have Minkowski signature we can, as discussed above, perform a rotation to Euclidean space and then the above mapping to $\mathbf{CP}^1$. In what follows we take our theory to be defined on the Riemann sphere. More details on the mapping to the Riemann sphere are given in chapter 9.

Time translations on the strip, or cylinder, $\tau \mapsto \tau + a$ correspond on the Riemann sphere, $\mathbf{CP}^1$, to $z \mapsto \lambda z$, $\bar{z} \mapsto \lambda \bar{z}$, where $\lambda = e^a$. As such they are generated by $L_0 + \bar{L}_0$, which we can think of as the Hamiltonian. Lines of constant time on the strip, that is, $\tau = c$, where c is a constant, correspond on $\mathbf{CP}^1$ to circles of radius e^c; see figure 8.2.3. We note that $\tau \to -\infty$ on the strip is mapped into the point $z = 0$ of the Riemann sphere while $\tau \to \infty$ on the strip is mappped to the point at infinity of the Riemann sphere. A σ translation on the strip, that is, $\sigma \to \sigma + \phi$, corresponds on the Riemann sphere to a rotation by angle ϕ about the origin.

Correlation functions of the form $\langle A_1(\varsigma_1)\cdots A_n(\varsigma_n)\rangle$ are understood to be time ordered with respect to τ on the strip. We may use the Hamiltonian H to rewrite the correlation function as $\langle e^{\tau_1 H}A_1(0,\sigma_1)e^{(\tau_1-\tau_2)H}A_2(0,\sigma_2)e^{(\tau_2-\tau_3)H}A_3(0,\sigma_3)\cdots\rangle$ and if, as is usually the case, H has a real spectrum which is bounded from below, but not from above we require the above time ordering, namely $\tau_p > \tau_{p+1}$, in order for the above correlator to be well defined. The time ordering $\tau_1 > \tau_2 > \cdots > \tau_n$ translates on the Riemann sphere to the ordering $|z_1| > |z_2| > \cdots > |z_n|$. This is referred to as radial ordering and is denoted by R, meaning

$$R\{A(z)B(w)\} = \begin{cases} A(z)B(w), & \text{if } |z| > |w|, \\ B(w)A(z), & \text{if } |w| > |z|, \end{cases} \tag{8.2.15}$$

if A and B are Grassmann even or if A and B are Grassmann odd and even, respectively. The same formula also holds if A and B are both Grassmann odd provided one inserts an appropriate minus sign. We will not always indicate such an ordering, but it will always be understood to be present for all correlation functions.

When one is provided with a theory whose correlation functions can be defined from a Euclidean path integral

$$\langle\varphi(x_1)\varphi(x_2)\cdots\varphi(x_n)\rangle = \int \mathcal{D}\varphi e^{-A}\varphi(x_1)\cdots\varphi(x_n), \tag{8.2.16}$$

then a symmetry of the action A and measure in the path integral leads to a corresponding Ward identity for the correlation functions. In the above equation A is the Euclidean action in the quantum field theory or the Hamiltonian in the partition function of the statistical system and φ denotes the basic field, or fields, of the theory. Now we suppose that A has a rigid (that is, constant) symmetry with infinitesimal parameter ε of the form $\delta\varphi = \varepsilon X$, where X may be a function of φ and its derivatives. Since when ε is a constant $\delta A = 0$, we must conclude that when ε is space-time dependent the variation of the action is of the form

$$\delta A = -\frac{1}{2\pi}\int d^2x\,\partial_\alpha\varepsilon\,j^\alpha. \tag{8.2.17}$$

We recognise j^α as the conserved Noether current associated with the symmetry. Let us now make the change of variables in the path integral from φ' to φ, where $\varphi' = \varphi + \delta\varphi$, with a space-time dependent ε. As the variable of integration in the path integral is a dummy variable we may conclude that

$$\langle\varphi(x_1)\varphi(x_2)\cdots\varphi(x_n)\rangle = \int \mathcal{D}\varphi' e^{-A[\varphi']}\varphi'(x_1)\cdots\varphi'(x_n)$$

$$= \int \mathcal{D}\varphi\{-\delta A(\varphi(x_1)\cdots\varphi(x_n)) + \delta(\varphi(x_1)\cdots\varphi(x_n),)\}e^{-A[\varphi]}$$

$$+ \int \mathcal{D}\varphi e^{-A[\varphi]}\varphi(x_1)\cdots\varphi(x_n), \tag{8.2.18}$$

where

$$\delta(\varphi(x_1)\cdots\varphi(x_n)) = \sum_{k=1}^{n}\varphi(x_1)\cdots\varphi(x_{k-1})\delta\varphi(x_k)\varphi(x_{k+1})\cdots\varphi(x_n) \tag{8.2.19}$$

and we have assumed that the Jacobian from φ' to φ is 1. As a result, we must find that

$$\langle \delta(\varphi(x_1)\cdots\varphi(x_n))\rangle = -\frac{1}{2\pi}\left(\int d^2x' \partial_\alpha \varepsilon(x')\langle j^\alpha(x')\varphi(x_1)\cdots\varphi(x_n)\rangle\right). \quad (8.2.20)$$

If we functionally differentiate with respect to $\varepsilon(x)$ and let $\delta\varphi \equiv \varepsilon\delta_\varepsilon\varphi$, this equation becomes

$$\sum_{i=1}^{n}\delta^2(x-x_i)\langle\varphi(x_1)\cdots\delta_\varepsilon\varphi(x_i)\cdots\varphi(x_n)\rangle = \frac{1}{2\pi}\partial_\alpha\langle j^\alpha(x)\varphi(x_1)\cdots\varphi(x_n)\rangle.$$

This is what is often called the unintegrated form of the Ward identity. We note that the derivative $\partial_\alpha = \partial/\partial x^\alpha$ is outside the time or radial, ordering operation in the Green's function. Integrating the last equation over space, we find that it becomes

$$\sum_{i=1}^{n}\langle(\varphi(x_1)\cdots\delta_\varepsilon\varphi(x_i)\cdots\varphi(x_n)\rangle\delta(x^0-x_i^0)$$

$$= \frac{1}{2\pi}\int dx^1 \frac{\partial}{\partial x^0}\langle j^0(x)\varphi(x_1)\cdots\varphi(x_n)\rangle$$

$$= \sum_{i=1}^{n}\delta(x^0-x_i^0)\langle\varphi(x_1)\cdots[Q,\varphi(x_i)]\cdots\varphi(x_n)\rangle$$

$$+ \sum_i\left\langle\frac{dQ}{dt}\varphi(x_1)\cdots\varphi(x_n)\right\rangle.$$

The last term results from carrying out the time differentiation. This can act either on $j^0(x)$, giving the first term, or on the time ordering to produce the last term. To find the latter term one must implement the time ordering by writing the expression in all possible orders, with each order multiplied by the θ functions appropriate to that ordering and noting that the time derivative of $\theta(x^0-x_i^0)$ produces the factor $\delta(x^0-x_i^0)$. We have defined

$$Q = \frac{1}{2\pi}\int dx^1 j^0(x).$$

Choosing $x^0 \neq x_i^0$ we conclude that dQ/dt vanishes in any correlator and so we are left with

$$\langle\varphi(x_1)\cdots[Q,\varphi(x_i)]\cdots\varphi(x_n)\rangle = \langle\varphi(x_1)\cdots\delta_\varepsilon\varphi(x_i)\cdots\varphi(x_n)\rangle.$$

Namely, Q generates symmetry transformations in correlation functions.

We now give a slightly different treatment of equation (8.2.20) which is more appropriate to our needs. Consider two contours C_1 and C_2 which only contain the point x_i and not the other points x_j, $j \neq i$ and are such are that C_1 is contained entirely in C_2. Let Σ be the region between the two contours. Correspondingly we choose the previously arbitrary function ϵ to be zero on and outside C_2, a constant on and inside C_1 and arbitrary in the region between

C_1 and C_2. Then $\delta\varphi(x_j)$ is non-zero only if $j = i$ and so we may write equation (8.2.20) as

$$\langle\varphi(x_1)\cdots\delta\varphi(x_i)\ldots\varphi(x_n)\rangle = -\frac{1}{2\pi}\int_\Sigma d^2x'\,\partial_\alpha\{\varepsilon(x')\langle j^\alpha(x')\varphi(x_1)\cdots\varphi(x_n)\rangle\}$$

$$+\frac{1}{2\pi}\int_\Sigma d^2x'\,\varepsilon(x')\partial_\alpha\langle j^\alpha(x')\varphi(x_1)\cdots\varphi(x_n)\rangle.$$

We now use Stokes' theorem to rewrite the first term in terms of a line integral over the boundary of Σ. Remembering that ϵ vanishes on C_2 and is a constant on C_1 we find that

$$\langle\varphi(x_1)\cdots\delta\varphi(x_i)\ldots\varphi(x_n)\rangle = +\frac{1}{2\pi}\varepsilon\oint_{C_1}\{dx^2\langle j^1(x')\varphi(x_1)\cdots\varphi(x_n)\rangle$$

$$-dx^1\langle j^2(x')\varphi(x_1)\cdots\varphi(x_n)\rangle\}$$

$$+\frac{1}{2\pi}\int_\Sigma d^2x'\,\varepsilon(x')\partial_\alpha\langle j^\alpha(x')\varphi(x_1)\cdots\varphi(x_n)\rangle.$$

Now the right-hand side of this equation contains an arbitrary function of ϵ, while the left-hand side only contains $\epsilon(x_i)$ and as a result we may conclude that

$$\partial_\alpha\langle j^\alpha(x')\varphi(x_1)\cdots\varphi(x_n)\rangle = 0,$$

provided $x \neq x_1, \ldots, x_n$. Taking this into account, when written in terms of complex coordinates the above equation becomes

$$\langle(\varphi(z_1,\bar z_1)\cdots\delta\varphi(z_i,\bar z_i)\cdots\varphi(z_n,\bar z_n))\rangle$$

$$= \varepsilon\left\{\oint_{C_1}\frac{dw}{2\pi i}\langle j_w(w,\bar w)\varphi(z_1,\bar z_1)\cdots\varphi(z_n,\bar z_n)\rangle\right.$$

$$\left.-\oint_{C_1}\frac{d\bar w}{2\pi i}\langle j_{\bar w}(w,\bar w)\varphi(z_1,\bar z_1)\cdots\varphi(z_n,\bar z_n)\rangle\right\}.$$

Here we have used the equation

$$B^1dx^2 - B^2dx^1 = \frac{1}{i}(B_zdz - B_{\bar z}d\bar z), \tag{8.2.21}$$

which is valid for any vector B^α. As a result we may write the variation of the field as

$$\delta\varphi(z,\bar z) = \varepsilon\oint_{C_1}\left\{\frac{dw}{2\pi i}j_w(w,\bar w)\varphi(z,\bar z) - d\bar w\,j_{\bar w}(w,\bar w)\varphi(z,\bar z)\right\},$$

where this expression is understood to be valid only when used in a correlator.

The above discussion must be modified if the symmetry is anomalous or explicity broken since then $\partial_\mu j^\mu \neq 0$ and the time derivative of the charge is non-vanishing. If the symmetry is spontaneously broken some of the space-time integrals are not finite due to the presence of Goldstone bosons and although equation (8.2.20) and the one below it are valid, the equations below that are not valid as Q does not exist.

In one of the applications we have in mind, namely statistical systems near their critical point, we do not always possess an action formulation and where we do, such as in the Ising model, the critical phenomenon of interest to us is outside the range of usual coupling constant perturbation theory. However, one can use an alternative perturbative expansions,

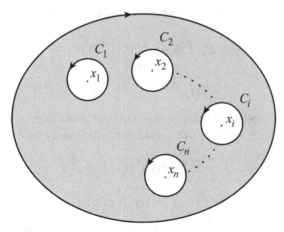

Figure 8.2.4   The contours used in equation (8.2.24).

such as the ϵ expansion, to compute at the critical point. In this chapter, however, we will assume that equation (8.2.22) is valid in all the theories we consider.

Now let us apply the Ward identities to the case of interest to us. In the case of a theory invariant under translations with energy-momentum tensor $T_{\alpha\beta}$ equation (8.2.20) implies that

$$\langle \delta(\varphi(x_1) \cdots \varphi(x_n)) \rangle = -\frac{1}{2\pi} \int d^2x \partial^\alpha \varepsilon^\beta \langle T_{\alpha\beta}(x) \varphi(x_1) \cdots \varphi(x_n) \rangle. \tag{8.2.22}$$

We are free to choose ε. For ε a constant, $\delta\varphi$ is a translation and since the right-hand side of equation (8.2.22) vanishes we find, as expected, that the correlation function is translationally invariant. Due to the symmetry of $T_{\alpha\beta}$, that is, $T_{\alpha\beta} = T_{\beta\alpha}$, we find that for $\varepsilon^\beta = w^\beta{}_\alpha x^\alpha$, where $w^{\alpha\beta} = -w^{\beta\alpha}$, $\delta\varphi$ is a Lorentz rotation and that the right-hand side of equation (8.2.22) vanishes. As a result, we conclude that the correlation functions are invariant under Poincaré symmetry.

As already mentioned, the above discussion from equation (8.2.16) to here is valid in D-dimensional Euclidean space provided one makes the obvious adjustments.

We now suppose that in the quantum theory $T^\alpha{}_\alpha = 0$ is an operator when inserted in correlators. If we choose ε^α to be a conformal transformation and so ε^α satisfies equation (8.1.5), then the right-hand side again vanishes provided ε^α is well defined in the whole region of integration, which is the Riemann sphere. However, this only occurs for the Möbius transformations generated by L_0, $L_{\pm1}$ and $\bar{L}_0$, $\bar{L}_{\pm1}$. Consequently, for these transformations alone

$$\langle \delta(\varphi(x_1) \cdots \varphi(x_n)) \rangle = 0. \tag{8.2.23}$$

To examine the effect of the remaining conformal transformations we must consider other choices of ε^α. In particular, let us consider an ε^α which vanishes on and outside a contour C which encloses all the points x_i and is such that in the neighbourhood of each point x_i enclosed by a contour C_i, ε^α is conformal. With this choice, the region of integration, denoted by Σ, becomes that enclosed by C but exterior to the contours C_i (see figure 8.2.4).

With this choice equation (8.2.22) becomes

$$\langle \delta(\varphi(x_1) \cdots \varphi(x_n)) \rangle = -\frac{1}{2\pi} \left\{ \oint_C - \sum_{i=1}^n \oint_{C_i} \right\} \varepsilon^\beta \langle (dx^2 T_{1\beta} - dx^1 T_{2\beta}) \varphi(x_1) \cdots \varphi(x_n) \rangle$$

$$+ \frac{1}{2\pi} \int_\Sigma d^2 x \varepsilon^\beta \partial^\alpha \langle T_{\alpha\beta}(x) \varphi(x_1) \cdots \varphi(x_n) \rangle, \qquad (8.2.24)$$

after using Stokes' theorem. Due to our choice of ε the contour integration over C vanishes. Using equation (8.2.21) and that $T_{z\bar{z}} = 0$, the above equation may be rewritten in z, $\bar{z}$ coordinates as

$$\langle \delta(\varphi(z_1, \bar{z}_1) \cdots \varphi(z_n, \bar{z}_n)) \rangle$$

$$= \sum_{i=1}^n \oint_{z_i} \frac{dz}{2\pi i} \varepsilon(z) \langle T(z, \bar{z}) \varphi(z_1, \bar{z}_1) \cdots \varphi(z_n, \bar{z}_n) \rangle$$

$$- \sum_{i=1}^n \oint_{\bar{z}_i} \frac{d\bar{z}}{2\pi i} \bar{\varepsilon}(\bar{z}) \langle \bar{T}(z, \bar{z}) (\bar{z}) \varphi(z_1, \bar{z}_1) \cdots \varphi(z_n, \bar{z}_n) \rangle$$

$$+ \frac{1}{4\pi i} \int_\Sigma dz d\bar{z} (\varepsilon(z) \partial_{\bar{z}} \langle T(z, \bar{z}) \varphi(z_1, \bar{z}_1) \cdots \varphi(z_n, \bar{z}_n) \rangle$$

$$+ \bar{\varepsilon}(\bar{z}) \partial_z \langle \bar{T}(z, \bar{z}) (\bar{z}) \varphi(z_1, \bar{z}_1) \cdots \varphi(z_n, \bar{z}_n) \rangle), \qquad (8.2.25)$$

where $T = T_{zz}$, $\bar{T} = T_{\bar{z}\bar{z}}$, $\varepsilon^z = \varepsilon$, $\varepsilon^{\bar{z}} = \bar{\varepsilon}$.

Let us now assume that the variation of φ is local, that is, $\delta\varphi$ depends on only finitely many derivatives of ε. In this case the left-hand side of equation (8.2.25) depends only on the coordinates z_i, $\bar{z}_i$, while the last term of the right-hand side depends on ε, $\bar{\varepsilon}$ at all points in Σ. As a result we must conclude that

$$\bar{\partial} \langle T(z, \bar{z}) \varphi(z_1, \bar{z}_1) \cdots \varphi(z_n, \bar{z}_n) \rangle = 0$$

and

$$\partial \langle \bar{T}(z, \bar{z}) \varphi(z_1, \bar{z}_1) \cdots \varphi(z_n, \bar{z}_n) \rangle = 0, \qquad (8.2.26)$$

except when z, $\bar{z}$ are equal to one of the z_i, $\bar{z}_i$. These equations state that

$$X(z, \bar{z}) \equiv \langle T(z, \bar{z}) \varphi(z_1, \bar{z}_1) \cdots \varphi(z_n, \bar{z}_n) \rangle \text{ and}$$

$$\bar{X}(z, \bar{z}) \equiv \langle \bar{T}(z, \bar{z}) \varphi(z_1, \bar{z}_1) \cdots \varphi(z_n, \bar{z}_n) \rangle$$

obey the Cauchy–Riemann equations away from the points z_i, $\bar{z}_i$, $i = 1, \ldots, n$. The familiar form of the Cauchy–Riemann equations is immediately recognisable from equation (8.2.26) when written in terms of x^α coordinates. As is well known, these equations imply that the derivative of X with respect to z exists and by definition X is analytic everywhere except at the points z_i, $\bar{z}_i$ and similarly for $\bar{X}$.

As a result, equation (8.2.25) becomes

$$\langle \delta(\varphi(z_1, \bar{z}_1) \cdots \varphi(z_n, \bar{z}_n)) \rangle$$

$$= \sum_{i=1}^n \oint_{z_i} \frac{dz}{2\pi i} \varepsilon \langle T(z, \bar{z}) \varphi(z_1, \bar{z}_1) \cdots \varphi(z_n, \bar{z}_n) \rangle$$

$$- \sum_{i=1}^n \oint_{\bar{z}_i} \frac{d\bar{z}}{2\pi i} \bar{\varepsilon} \langle \bar{T}(z, \bar{z}) \varphi(z_1, \bar{z}_1) \cdots \varphi(z_n, \bar{z}_n) \rangle. \qquad (8.2.27)$$

Taking ε and $\bar{\varepsilon}$ to be non-vanishing in the neighbourhood of z_i we may extract the result

$$\delta\varphi(z,\bar{z}) = \oint_z \frac{dw}{2\pi i}\varepsilon(w)T(w)\varphi(z,\bar{z}) - \oint_{\bar{z}} \frac{d\bar{w}}{2\pi i}\bar{\varepsilon}(w)\bar{T}(\bar{w})\varphi(z,\bar{z}), \qquad (8.2.28)$$

which is understood to be valid in any correlation function.

We now consider a special case of local fields called primary fields. These are fields that transform in a particular way under the conformal group; a primary field of conformal weight $(h,\bar{h})$ is one which transforms as

$$\varphi(z,\bar{z}) \mapsto \left(\frac{dw}{dz}\right)^h \left(\frac{d\bar{w}}{d\bar{z}}\right)^{\bar{h}} \varphi(w,\bar{w}) \qquad (8.2.29)$$

under the conformal transformation $z \mapsto w(z)$, $\bar{z} \mapsto \bar{w}(\bar{z})$. Primary fields are operators in the quantum theory that have the same transformation rule as the classical tensors of equation (8.1.48). For such fields we define $\Delta = h + \bar{h}$ as the scaling weight and $s = h - \bar{h}$ as the spin, since under $z \mapsto \lambda z$, $\bar{z} \mapsto \lambda\bar{z}$ and $z \mapsto e^{i\alpha}z$, $\bar{z} \mapsto e^{-i\alpha}\bar{z}$, the primary field φ transforms as $\varphi \mapsto \lambda^{h+\bar{h}}\varphi$ and $\varphi \mapsto e^{i(h-\bar{h})\alpha}\varphi$, respectively.

For the infinitesimal transformation $w = z + \varepsilon(z)$, $\bar{w} = \bar{z} + \bar{\varepsilon}(\bar{z})$ equation (8.2.29) becomes

$$\delta\varphi = \varepsilon\partial\varphi + \bar{\varepsilon}\bar{\partial}\varphi + h\partial\varepsilon\varphi + \bar{h}\bar{\partial}\bar{\varepsilon}\varphi. \qquad (8.2.30)$$

We can then conclude from equation (8.2.28) that a primary field of conformal weight $(h,\bar{h})$ obeys the relation

$$T(w)\varphi(z,\bar{z}) = \frac{h}{(w-z)^2}\varphi(z,\bar{z}) + \frac{\partial\varphi(z,\bar{z})}{(w-z)} + \cdots, \qquad (8.2.31)$$

where $+\cdots$ stands for terms of the form $(w-z)^p a_p$, $p \geq 0$, as well as a similar relation for $\bar{T}$. These are our first examples of relations called operator product expansions which express how two operators behave as their arguments approach each other. Since these equations are only valid inside correlation functions they are equations which are understood to be radially ordered.

A quasi-primary field is one which obeys equation (8.2.30), but only for Möbius transformations which infinitesimally give $\varepsilon = z^n$, $n = 0, 1, 2$. In this case, equation (8.2.30) implies the weaker result

$$T(w)\varphi(z,\bar{z}) = \sum_{p=4}^{\infty} \frac{a_p(z,\bar{z})}{(z-w)^p} + \frac{h\varphi(z,\bar{z})}{(w-z)^2} + \frac{\partial\varphi(z,\bar{z})}{(w-z)} + \cdots, \qquad (8.2.32)$$

where a_p are local operators which are not subject to any restriction due to the quasi-primary nature of φ. In general, the sum in the first term will not contain an infinite number of terms, indeed were this not to be the case we would find operators with arbitrarily low weights. We will also refer to quasi-primary fields as having scaling weight $(h,\bar{h})$.

For fields that are neither primary, nor quasi-primary, but have well-defined transformation properties under the conformal group, we can by an analogous procedure read off their operator product expansions with T and $\bar{T}$.

Given the correlation function of a set of primary fields φ_i we may similarly use $\delta\varphi_i$ of equation (8.2.30) to conclude that equation (8.2.27) can be written as

$$\sum_{i=1}^{n} \left\{ \frac{h_i}{(z-z_i)^2} + \frac{1}{(z-z_i)} \partial_i \right\} \langle \varphi_1(z_1, \bar{z}_1) \cdots \varphi_n(z_n, \bar{z}_n) \rangle$$

$$= \langle T(z)\varphi_1(z_1, \bar{z}_1) \cdots \varphi_n(z_n, \bar{z}_n) \rangle \tag{8.2.33}$$

and

$$\sum_{i=1}^{n} \left\{ \frac{\bar{h}_i}{(\bar{z}-\bar{z}_i)^2} + \frac{1}{(\bar{z}-\bar{z}_i)} \bar{\partial}_i \right\} \langle \varphi_1(z_1, \bar{z}_1) \cdots \varphi_n(z_n, \bar{z}_n) \rangle$$

$$= \langle \bar{T}(\bar{z})\bar{\varphi}_i(z_1, \bar{z}_1) \cdots \varphi_n(z_n, \bar{z}_n) \rangle, \tag{8.2.34}$$

where $\partial_i = \partial/\partial z_i$, $\bar{\partial}_i = \partial/\partial \bar{z}_i$ and $(h_i, \bar{h}_i)$ is the conformal weight of φ_i.

At first sight one might think that one could also add more terms of the form $(z - z_i)^p$, $p \geq 0$, to the left-hand side of the above two equations since such terms would not contribute in equation (8.2.27). However, such additional terms would have to be well-behaved analytic functions on the Riemann sphere, which in the previous section we showed can only be a constants. Furthermore, by examining the behaviour at the point at ∞, the constant can be seen to be zero using equation (8.4.12) which is given later.

Using equation (8.2.29) we find that the version of equation (8.2.23) for finite Möbius transformations takes the form

$$\langle \varphi(z_1, \bar{z}_1) \cdots \varphi(z_n, \bar{z}_n) \rangle = \left(\frac{dw_1}{dz_1} \right)^{h_1} \cdots \left(\frac{dw_n}{dz_n} \right)^{h_n} \left(\frac{d\bar{w}_1}{d\bar{z}_1} \right)^{\bar{h}_1} \cdots \left(\frac{d\bar{w}_n}{d\bar{z}_n} \right)^{\bar{h}_n}$$

$$\times \langle \varphi(w_1, \bar{w}_1) \cdots \varphi(w_n, \bar{w}_n) \rangle \tag{8.2.35}$$

for primary and quasi-primary fields and when written in terms of complex coordinates.

8.3   Constraints due to global conformal transformations

We found in equations (8.2.23) and (8.2.34) that correlation functions were invariant under globally defined conformal transformations, that is, Möbius transformations. Once we specify the way the fields transform under such transformations, we may use this equation to place restrictions on the correlators. For the one-, two- and three-point correlators these constraints are sufficient to determine their dependence on space-time. This is to be expected since a Möbius transformation can be used to take any three distinct points to 0, 1, and ∞. Applying this transformation to such Green's functions, which like all Green's functions are Möbius invariant, we can map all their points to such constant values, and the only function of the coordinates that occurs is that arising from the transformations of the fields which is specified by the transformation character of the fields.

Let us consider correlators of primary or quasi-primary fields φ_i of weights h_i, $i = 1, \dots, N$, whereupon equation (8.2.23) becomes

$$\sum_i \partial_i \langle \varphi_1(z_1, \bar{z}_1) \cdots \varphi_n(z_n, \bar{z}_n) \rangle = 0, \tag{8.3.1}$$

$$\sum_i (z_i \partial_i + h_i) \langle \varphi_1(z_1, \bar{z}_1) \cdots \varphi_n(z_n, \bar{z}_n) \rangle = 0, \tag{8.3.2}$$

$$\sum_i (z_i^2 \partial_i + 2 h_i z_i) \langle \varphi_1(z_1, \bar{z}_1) \cdots \varphi_n(z_n, \bar{z}_n) \rangle = 0 \tag{8.3.3}$$

as a result of translations, dilations and special conformal transformations, respectively. We have an analogous set of equations for the $\bar{z}$ variables. In what follows we shall often just discuss the z dependence, the discussion for the $\bar{z}$ dependence being completely analogous.

Equation (8.3.1) implies that the Green's functions, denoted $G(z_j)$, are functions of differences $z_{ij} \equiv z_i - z_j$ and so can be written as

$$G = \langle \varphi(z_1) \cdots \rangle = e \prod_{i<j} \frac{1}{(z_{ij})^{k_{ij}}}, \tag{8.3.4}$$

where e and k_{ij} are constants. There is, of course, also a $\bar{z}_j$, $j = 1, \dots, n$ dependence which has a similar form.

Assuming $e \neq 0$, equations (8.3.2) and (8.3.3) imply

$$-\sum_{i<j} k_{ij} + \sum_{i=1}^N h_i = 0 \tag{8.3.5}$$

and

$$-\sum_{i<j}(z_i + z_j)k_{ij} + 2\sum_{i=1}^N h_i z_i = 0. \tag{8.3.6}$$

Let us first consider the one-point correlation function $\langle \varphi(z) \rangle$. Equation (8.3.1) implies that it is independent of z and equation (8.3.2) that $h\langle \varphi(z) \rangle = 0$. Consequently, the one-point function of a primary, or quasi-primary, field vanishes unless it has weight 0.

Consider now the two-point function $G = e(z_1 - z_2)^{-k}$; equations (8.3.5) and (8.3.6) imply that

$$-k + h_1 + h_2 = 0; \qquad -(z_1 + z_2)k + 2h_1 z_1 + 2h_2 z_2 = 0. \tag{8.3.7}$$

The unique solution is that G vanishes unless $h_1 = h_2$, $k = 2h_1$. Hence, for any primary, or quasi-primary, fields $\varphi_i(z)$

$$G(z_1, z_2) = \langle \varphi_i(z_1)\varphi_j(z_2) \rangle = \frac{\delta_{ij}}{(z_1 - z_2)^{2h_i}}. \tag{8.3.8}$$

In this equation we have used the freedom to scale our primary fields in such a way that $e = 1$ and also diagonalise any matrix which appears on the right-hand side for primary fields of the same weight by adopting a suitable basis for the fields. Putting in the $\bar{z}$ dependence we find that

$$\langle \varphi_i(z_1, \bar{z}_1) \varphi_j(z_2, \bar{z}_2) \rangle = \frac{\delta_{ij}}{(z_1 - z_2)^{2h_i} (\bar{z}_1 - \bar{z}_2)^{2\bar{h}_i}}$$

$$= \frac{\delta_{ij}}{r^{2\Delta_i}} e^{-2is_i\theta},$$

where $z_{12} = re^{i\theta}$, and Δ_i and s_i are the dilation weight and spin of φ_i, respectively.
   The three-point function has a z dependence of the form

$$\langle \varphi_i(z_1) \varphi_j(z_2) \varphi_k(z_3) \rangle = \frac{C_{ijk}}{(z_{12})^{k_1} (z_{23})^{k_2} (z_{31})^{k_3}}. \tag{8.3.9}$$

Equations (8.3.5) and (8.3.6) for the case of three fields read

$$-(k_1 + k_2 + k_3) + h_1 + h_2 + h_3 = 0 \tag{8.3.10}$$

and

$$-k_1(z_1 + z_2) + 2h_1 z_1 + \text{cyclic} = 0. \tag{8.3.11}$$

Taking the latter equation and reading off the coefficient of z_1 we have $k_1 + k_3 = 2h_1$ and the analogous cycled equations. Solving these equations we find

$$k_1 = \tfrac{1}{2}\{(k_1 + k_3) + (k_2 + k_1) - (k_3 + k_2)\}$$

$$= h_1 + h_2 - h_3 \tag{8.3.12}$$

and similarly for the analogous cycled equations. Hence, the three-point function is given by

$$\langle \varphi_i(z_1) \varphi_j(z_2) \varphi_k(z_3) \rangle = \frac{C_{ijk}}{(z_1 - z_2)^{h_1+h_2-h_3} (z_2 - z_3)^{h_2+h_3-h_1} (z_3 - z_1)^{h_3+h_1-h_2}} \tag{8.3.13}$$

with an analogous dependence on $\bar{z}$. From Möbius invariance alone we cannot determine the constant C_{ijk}, which could vanish; however, the arguments above ensure that the dependence on z is unique.
   For Green's functions with more that three primary fields we cannot, in general, use Möbius, or indeed conformal invariance, to solve for their functional dependence. Equations (8.3.1)–(8.3.3) only provide us with three differential equations, and so we can only hope to determine the space-time dependence of the one-, two- and three-point Green's functions as we have done above. For four points z_i, $i = 1, \ldots, 4$, we can construct the 'cross ratio'

$$u = \frac{(z_1 - z_3)(z_2 - z_4)}{(z_1 - z_4)(z_2 - z_3)}, \tag{8.3.14}$$

which is Möbius invariant as discussed at the end of section (8.1). The Möbius constraints on any N-point Green's function can be automatically solved by writing it in the form

$$G = \prod_{i<j} \left(\frac{1}{z_{ij}}\right)^{\frac{2}{(N-2)}\left[h_i+h_j-\frac{h}{(N-1)}\right]} F(u_i), \tag{8.3.15}$$

where $h = \sum_{i=1}^{N} h_i$ and F is an arbitrary function of the $N - 3$ independent cross ratios u_i, $i = 1, \ldots, N - 3$. There is also a similar dependence on the $\bar{z}_j$ coordinates. The prefactor in front of the invariant function is not unique as the ratio of any two such factors is an invariant. However, any choice of prefactor that solves the Möbius constraints can be changed to the above choice by a change of invariant function F.

8.4   Transformations of the energy-momentum tensor

In a conformally invariant *classical* theory the two non-zero components of the energy-momentum tensor, $T \equiv T_{zz}$ and $\bar{T} \equiv T_{\bar{z}\bar{z}}$, have conformal weights $(2, 0)$ and $(0, 2)$, respectively, and as a result have the transformation law

$$\delta_\varepsilon T = \varepsilon \partial T + 2(\partial \varepsilon)T, \quad \delta_{\bar{\varepsilon}}\bar{T} = \bar{\varepsilon}\bar{\partial}\bar{T} + 2(\bar{\partial}\bar{\varepsilon})\bar{T}, \quad \delta_{\bar{\varepsilon}}T = 0 = \delta_\varepsilon\bar{T} \tag{8.4.1}$$

under $z \mapsto z + \varepsilon, \bar{z} \mapsto \bar{z} + \bar{\varepsilon}$. The last two transformation laws have the above simple form since $\bar{\partial}T = 0 = \partial\bar{T}$. However, in a *quantum* theory, we may find the transformation law is modified due to normal ordering. If we assume that the quantum $T, \bar{T}$ are primary fields of weight $(2, 0)$ and $(0, 2)$, respectively, at least as far as translations and dilations are concerned, then the arguments of the previous section tell us that

$$\langle T(z) \rangle = 0 = \langle \bar{T}(\bar{z}) \rangle, \tag{8.4.2}$$

but

$$\langle T(z_1)T(z_2) \rangle = \frac{c/2}{(z_1 - z_2)^4}, \quad \langle T(z_1)\bar{T}(z_2) \rangle = 0, \quad \langle \bar{T}(\bar{z}_1)\bar{T}(\bar{z}_2) \rangle = \frac{\bar{c}/2}{(\bar{z}_1 - \bar{z}_2)^4}, \tag{8.4.3}$$

where c and $\bar{c}$ are constants. Since $T, \bar{T}$ have already been normalised, say in equation (8.2.21), the values of c and $\bar{c}$ are fixed for a given system. The constants c and $\bar{c}$ are called the central charges, and they are in general non-zero. They will play a crucial role in what follows.

From equation (8.4.3) and equation (8.2.27) applied to T itself we find that

$$\langle \delta_\varepsilon T(z) \rangle = \oint_z \frac{dw}{2\pi i}\varepsilon(w)\langle T(w)T(z) \rangle = \frac{c}{12}\partial^3\varepsilon(z) \tag{8.4.4}$$

as well as a similar result for $\bar{T}$. The transformation of T also has a contribution corresponding to the classical variation of T but these vanish due to equation (8.4.2). Adding these contributions to the one implied by equation (8.4.4) we find that

$$\delta T = \varepsilon \partial T + 2\partial\varepsilon T + \tfrac{1}{12}c\partial^3\varepsilon. \tag{8.4.5}$$

Examining equation (8.2.28) we find that this result in turn implies the operator product expansion

$$T(z)T(w) = \frac{c/2}{(z-w)^4} + \frac{2T(w)}{(z-w)^2} + \frac{\partial T(w)}{(z-w)} + \cdots \tag{8.4.6}$$

as well as an analogous result for $\bar{T}$. We note that this relation does scale correctly under dilations, that is, $T \mapsto \lambda^2 T, z \mapsto \lambda z$ as we assumed it should and that this invariance prevents us from adding further terms without introducing operators other that T. We now adopt this relation which we will discuss in much more detail. Examining equation (8.2.32) we recognise that T is quasi-primary.

We will require the change of T under a finite conformal transformation $z \mapsto w(z)$. The result of the non-trivial procedure of exponentiating equation (8.4.5) is

$$' \quad T(z) \mapsto \left(\frac{dw}{dz}\right)^2 T(w) + \frac{c}{12}\{w, z\}, \tag{8.4.7}$$

where $\{w, z\}$ is the Schwartzian derivative and is given by

$$\{w, z\} = \frac{d^3 w}{dz^3}\left(\frac{dw}{dz}\right)^{-1} - \frac{3}{2}\left(\frac{d^2 w}{dz^2}\right)^2 \left(\frac{dw}{dz}\right)^{-2}. \tag{8.4.8}$$

The finite term must agree with the classical result when $c = 0$ and the addition must vanish for Möbius transformations. Indeed $\{w, z\} = 0$ implies that $w = (az + b)/(cz + d)$ and vice versa. This property follows in a straightforward way from an alternative formulation of the Schwartzian derivative:

$$\{w, z\} = -2\left(\frac{dw}{dz}\right)^{1/2} \frac{d^2}{dz^2}\left[\left(\frac{dw}{dz}\right)^{-1/2}\right]$$

$$= \frac{d^2}{dz^2}\ln\frac{dw}{dz} - \frac{1}{2}\left(\frac{d}{dz}\ln\frac{dw}{dz}\right)^2, \tag{8.4.9}$$

which is itself readily verified. It is easily verified by using equation (8.4.7) that the finite result contains the infinitesimal transformation law of equation (8.4.5).

The Schwartzian derivative also satisfies the relation

$$\left\{\frac{aw + b}{cw + d}, z\right\} = \{w, z\}. \tag{8.4.10}$$

Finally, it obeys the equation

$$\{w, z\} = \left(\frac{d\varsigma}{dz}\right)\{w, \} + \{\varsigma, z\}, \tag{8.4.11}$$

which corresponds to the change in the Schwartzian derivative under the composition rule $z \mapsto \varsigma(z) \mapsto w(\varsigma)$ and can also be verified using equation (8.4.9).

Let us carry out a Möbius transformation on the $\langle T(w)\varphi_1(z_1)\cdots\varphi_n(z_n)\rangle$ correlator, in particular $z \mapsto w = -1/z$. The points $z_k, k = 1, \ldots, N$ are mapped to the points $u_k = -1/z_k$ of the Riemann sphere $\hat{C}$ which are not at infinity. Since T is quasi-primary and the φ_i are also assumed to be to be quasi-primary or primary, the Green's functions are invariant under Möbius transformations once one takes into account the associated field transformations, as are the fields. Using equation (8.4.7), we find therefore that as $w \to \infty$

$$\langle T(w)\varphi_1(z_1)\cdots\varphi_n(z_n)\rangle \propto \frac{1}{w^4}. \tag{8.4.12}$$

The Green's function on the right-hand side of this equation is $\langle T(0)\varphi_1(u_1)\cdots\varphi_n(u_n)\rangle$ which is indeed independent of w.

Let us use equation (8.4.7) to calculate the energy-momentum tensor on an infinitely long cylinder which we regard as the strip which is parameterised by $-\infty < u < \infty, 0 \le v \le l$

and has its edges at $v = 0$ and $v = l$ identified. This strip is related to the Riemann sphere by the conformal mapping

$$z = e^{2\pi \varsigma/l},$$

where $\varsigma = u + iv$. We find that

$$T_{cyl}(\varsigma) = \left(\frac{2\pi}{l}\right)^2 \left(z^2 T_{Riemannsphere}(z) - \frac{c}{24}\right) \tag{8.4.13}$$

and consequently upon using equation (8.4.2)

$$\langle T_{cyl}(\varsigma) \rangle = -\left(\frac{2\pi}{l}\right)^2 \frac{c}{24}. \tag{8.4.14}$$

When considering this geometry previously we took $l = 2\pi$. Thus on the cylinder we find a vacuum energy whose magnitude depends on the geometry and goes like $1/l^2$, where l is the width of the strip. In this respect it is like the Casimir effect of quantum electrodynamics (QED). The magnitude of the effect depends only on the geometry and is independent of the details of a given model apart from being proportional to the central charge c. This observation provides us with a method of calculating the central charge of a system [8.3]. We note that we do not have translation invariance on the strip due to its boundary and so the analogous equation to equation (8.4.2) does not follow.

A non-zero central charge corresponds to the presence of an anomaly in the quantum theory for the conformal transformations which are not globally defined. Let us suppose we couple the theory to gravity $g_{\alpha\beta}$. The lowest order coupling is

$$\int d^2x h_{\alpha\beta} T^{\alpha\beta} = \frac{\kappa}{2i} \int dz d\bar{z} h_{zz} T_{\bar{z}\bar{z}} + \cdots, \tag{8.4.15}$$

where $g_{\alpha\beta} = \eta_{\alpha\beta} + \kappa h_{\alpha\beta}$ to lowest order in $h_{\alpha\beta}$. As a result, a change in the geometry changes an expectation value by inserting an energy-momentum tensor from the action. Hence the change of the expectation value of one energy-momentum tensor is given by the expectation values of two energy-momentum tensors. This two-point function was given in equation (8.4.3) and so the central charge makes its entrance. A proper account of this argument is given in the review mentioned in reference [8.3].

Following the same argument as was used to derive equation (8.2.33) we can, using (8.4.5), show that the correlator for m Ts and n primary fields φ_i obeys the equation

$$\langle T(w)T(w_1)\cdots T(w_m)\varphi_1(z_1, \bar{z}_1)\cdots \varphi_n(z_n, \bar{z}_n)\rangle$$

$$= \left\{\sum_{i=1}^n \left(\frac{h_i}{(w-z_i)^2} + \frac{1}{(w-z_i)}\partial_i\right) + \sum_{j=1}^m \left(\frac{2}{(w-w_j)^2} + \frac{1}{(w-w_j)}\partial_j\right)\right\}$$

$$\times \langle T(w_1)\cdots T(w_m)\varphi_1(z_1, \bar{z}_1)\cdots \varphi_n(z_n, \bar{z}_n)\rangle + \sum_{j=1}^m \frac{\frac{c}{2}}{(w-w_j)^4} \tag{8.4.16}$$

$$\times \langle T(w_1)\cdots T(w_{j-1})T(w_{j+1})\cdots T(w_m)\varphi_1(z_1, \bar{z}_1)\cdots \varphi_n(z_n, \bar{z}_n)\rangle.$$

The analogous equation for $\bar{T}$s is obvious.

8.5   Operator product expansions

Given a correlation function in a D-dimensional field theory of the form

$$\langle A(x)B(y)\varphi_1(x_1)\cdots\varphi_n(x_n)\rangle, \tag{8.5.1}$$

we will be interested in its behaviour as $x^\mu \mapsto y^\mu$, that is, when the operators $A(x)$ and $B(y)$ approach one another. A knowledge of such behaviour is useful for deep inelastic scattering in quantum chromodynamics (QCD) and also as we have seen in section 8.2 in the context of the conformal Ward identities. It was Wilson's idea that we could replace $A(x)B(y)$ in correllators by the equation

$$A(x)B(y) = \sum_l c_l(x^\mu - y^\mu)O^l(y), \tag{8.5.2}$$

where O^l are a set of local operators and c_l are their coefficients which depend on only $x^\mu - y^\mu$ rather than both x^μ and y^μ as a result of translation invariance. Such equations are known as operator product expansions and are valid only within the correlation function and as such the operators which appear in them must be regarded as being normal ordered.

Even if A and B are operators corresponding to the elementary fields of the theory, the operators O^l will, in general, be composite operators. Usually, equation (8.5.2) is given in terms of renormalised operators and so its demonstration involves the often difficult renormalisation of composite operators as well as complicated questions of convergence. However, in certain classes of theories operator product expansions have been shown to hold for certain sets of operators when inserted into certain correlators. The convergence properties of equation (8.5.2), when valid, depend on the operators used and the space-time positions of the other operators in the correlator. In general, it is at best an asymptotic expansion. Using renormalisation group arguments it can be shown that the coefficients c_l are of the form

$$c_l(x) = r^{-d_A - d_B - d_{O_l}}$$

times a polynomial in $\ln r$, where $r = \sqrt{x^\mu x_\mu}$ and d_A, d_B and d_{O_l} are the scaling dimensions of the corresponding operators. We refer the reader to the review of [8.4] for a proof of an operator product expansion in $\lambda\varphi^4$ in four dimensions.

Let us now return to the study of two-dimensional theories, and the set of local operators $\{A_i\}$ which are assumed to have the operator product expansion

$$A_i(z, \bar{z})A_j(w, \bar{w}) = \sum_k c_{ij}{}^k(z - w, \bar{z} - \bar{w})A_k(w, \bar{w}). \tag{8.5.3}$$

We are to regard this as an exact expression and we have tacitly assumed that the set of operators $\{A_i\}$ is complete in the sense that any operator that occurs on the right-hand side is one of them. The argument of the field on the right-hand side is chosen to be w, but we could also use, say $(z + w)/2$, with a corresponding change of the coefficients $c_{ij}{}^k$ as a result of applying Taylor's theorem. In theories with conformal invariance we may use this symmetry to place very strong constraints on the form of $c_{ij}{}^k(z - w, \bar{z} - \bar{w})$, since both sides of equation (8.5.3) must transform under conformal transformations in the same way. In particular, if A_i have scaling weights $(h_i, \bar{h}_i)$, then under $z \mapsto \lambda z$, $\bar{z} \mapsto \lambda\bar{z}$ $A_i(z, \bar{z}) \to$

$UA_i(z, \bar{z})U^{-1} = A_i(z, \bar{z})' = \lambda^{h_i}\bar{\lambda}^{\bar{h}_i}A_i(\lambda z, \lambda\bar{z})$ and the operator product expansion becomes

$$\lambda^{h_i}\bar{\lambda}^{\bar{h}_i}A_i(\lambda z, \lambda\bar{z})\lambda^{h_j}\bar{\lambda}^{\bar{h}_j}A_j(\lambda w, \lambda\bar{w})$$

$$= \sum_k \lambda^{h_i+h_j}\bar{\lambda}^{\bar{h}_i+\bar{h}_j}c_{ij}{}^k(\lambda(z-w), \bar{\lambda}(\bar{z}-\bar{w}))A_k(\lambda w, \lambda\bar{w})$$

$$= \sum_k c_{ij}{}^k(z-w, \bar{z}-\bar{w})\lambda^{h_k}\bar{\lambda}^{\bar{h}_k}A_k(\lambda w, \lambda\bar{w}). \tag{8.5.4}$$

Consequently, we conclude that

$$c_{ij}^k(\lambda(z-w), \bar{\lambda}(\bar{z}-\bar{w})) = c_{ij}^k(z-w, \bar{z}-\bar{w})\lambda^{h_k-h_i-h_j}\bar{\lambda}^{\bar{h}_k-\bar{h}_i-\bar{h}_j} \tag{8.5.5}$$

and so as a result

$$c_{ij}^k(z-w, \bar{z}-\bar{w}) = \frac{c_{ij}^k}{(z-w)^{h_i+h_j-h_k}(\bar{z}-\bar{w})^{\bar{h}_i+\bar{h}_j-\bar{h}_k}}, \tag{8.5.6}$$

where c_{ij}^k are numbers. Hence, unlike in many other theories including most four-dimensional theories, no $\ln|z-w|$ terms occur in the coefficients $c_{ij}{}^k$ unless $h_i+h_j-h_k = 0$. It is the occurrence of these terms that poses particular problems for a demonstration of convergence. Inserting equation (8.5.6) into equation (8.5.3) we have

$$A_i(z, \bar{z})A_j(w, \bar{w}) = \sum_k c_{ij}^k\frac{A_k(w, \bar{w})}{(z-w)^{h_i+h_j-h_k}(\bar{z}-\bar{w})^{\bar{h}_i+\bar{h}_j-\bar{h}_k}}. \tag{8.5.7}$$

To emphasise its importance we reiterate that like the correlator into which it is inserted the operator product expansion is automatically radially ordered. For example, the operator product expansion of $T(z)T(w)$ of equation (8.4.6) should be equal to that for $T(w)T(z)$. The latter is given by

$$T(w)T(z) = \frac{c/2}{(w-z)^4} + \frac{2T(z)}{(w-z)^2} + \frac{\partial T(z)}{(w-z)} + \cdots, \tag{8.5.8}$$

which can be written as

$$T(w)T(z) = \frac{c/2}{(z-w)^4} + \frac{2T(w)}{(z-w)^2} + \frac{\partial T(w)}{(z-w)} + \cdots, \tag{8.5.9}$$

verifying the desired result. We note that a term with $(w-z)^3$ would not be allowed as it does not satisfy this requirement. In general, the knowledge that operator product expansions are radial ordered can be used to place restrictions on them. The above discussion was for Grassmann even operators, but one could also have Grassmann odd operators and all the above would hold provided we introduce the necessary minus signs on exchanging the two operators. In fact, in any well-behaved quantum field theory it can be shown that exchanging the two operators leads either to a plus or minus sign depending on the Grassmann character of the operators. This can be thought of a kind of locality condition. However, we will, in later chapters, discuss the operator product expansion for operators for which this is not true.

Given three operators, A, B and C, we can use operator product expansions to calculate their product in two ways, namely, $(AB)C$ and $A(BC)$. The two results, however, will not be equal for generic operator product expansion. Demanding that they are equal, that is, demanding associativity, can be used to constrain operator product expansions.

8.6   Commutators

There is a close relationship between the commutators and operator product expansions of two operators. Let us consider two operators U and V which have scaling weights $(d_U, 0)$ and $(d_V, 0)$ and are analytic functions of z away from the origin. We can define the modes

$$u_n = \oint_{\Gamma_1} \frac{dz}{2\pi i} U(z) z^{n+d_U-1}, \qquad v_m = \oint_{\Gamma_2} \frac{dz}{2\pi i} V(z) z^{m+d_V-1}. \tag{8.6.1}$$

The contours Γ_1 and Γ_2 must enclose the origin. As a result of the analytic nature of U and V the integral is independent of the precise shape of the contour.

Let us consider the commutator

$$[u_n, v_m] = u_n v_m - v_m u_n$$

$$= \oint_{\Gamma_1} \frac{dz}{2\pi i} z^{n+d_U-1} \oint_{\Gamma_2} \frac{d\zeta}{2\pi i} \zeta^{m+d_V-1} U(z) V(\zeta)$$

$$- \oint_{\Gamma_3} \frac{dz}{2\pi i} z^{n+d_U-1} \oint_{\Gamma_4} \frac{d\zeta}{2\pi i} \zeta^{m+d_V-1} V(\zeta) U(z). \tag{8.6.2}$$

We choose the contours such that Γ_1 encloses Γ_2 and Γ_4 encloses Γ_3. This coincides with the radial operating used for the operator product expansion as $R(U(z)V(\zeta)) = U(z)V(\zeta)$ for $|z| > |\zeta|$ and $R(V(\zeta)U(z)) = R(U(z)V(\zeta)) = V(\zeta)U(z)$ for $|\zeta| > |z|$. As such, we may use the operator product expansion of U and V whose left-hand side we denoted by $\widehat{UV}$. For a wide class of operators the operator product expansion of $U(z)V(\zeta)$ is the same as that for $V(\zeta)U(z)$ when they are analytically continued and so for these operators we may simply use the same expression in the first and second terms. Consequently, when this is the case we have

$$[u_n, v_m] = \left\{ \oint_{\Gamma_1} \frac{dz}{2\pi i} z^{n+d_U-1} \oint_{\Gamma_2} \frac{d\zeta}{2\pi i} \zeta^{m+d_V-1} \right.$$

$$\left. - \oint_{\Gamma_3} \frac{dz}{2\pi i} z^{n+d_U-1} \oint_{\Gamma_4} \frac{d\zeta}{2\pi i} \zeta^{m+d_V-1} \right\} \widehat{U(z)V(\zeta)}. \tag{8.6.3}$$

If we are dealing with Grassmann odd operators the same argument goes through, but one should compute the anti-commutator as the operator product expansions coincide if we introduce a minus sign. However, as we will encounter in section 16.8 there exist operators, such as vertex operators, for which the operator expansions for $U(z)V(\zeta)$ and $V(\zeta)U(z)$ are not the same; however, in such cases one can introduce compensating factors and still carry out the above steps.

Let us choose to identify the contours Γ_2 and Γ_4. Then Γ_1 encloses Γ_2, which, in turn, encloses Γ_3 (see figure 8.6.1) and we consider carrying out the z integration. Since the integrand is analytic everywhere away from the origin except at $z = \zeta$, we have the result

$$[u_n, v_m] = \oint_{\Gamma_2} \frac{d\zeta}{2\pi i} \zeta^{m+d_V-1} \oint_{\Gamma} \frac{dz}{2\pi i} z^{n+d_U-1} \widehat{U(z)V(\zeta)}, \tag{8.6.4}$$

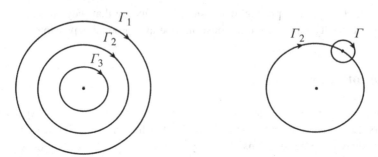

Figure 8.6.1   Contours occurring in the evaluation of the commutator.

where Γ is a contour which surrounds the point $z = \zeta$. Substituting in the operator product expansion, which is of the form $\sum_l (z - \zeta)^{-d_U - d_V - d_l} A_l(\zeta)$, we can evaluate the commutator using Cauchy's theorem.

By a similar argument we may show that

$$[u_n, V(\zeta)] = \oint_\Gamma \frac{dz}{2\pi i} z^{n+d-1} \overbrace{U(z)V(\zeta)}.$$

(8.6.5)

Let us take the case that $U = T(z)$ and $V = \varphi$, a primary field of weight $(h, 0)$, and introduce the Virasoro operators

$$L_n = \oint \frac{dz}{2\pi i} T(z) z^{n+1}, \qquad \bar{L}_n = \oint \frac{d\bar{z}}{2\pi i} \bar{T}(\bar{z}) \bar{z}^{n+1}$$

or

$$T(z) = \sum_n L_n z^{n-2}, \qquad \bar{T}(\bar{z}) = \sum_n \bar{L}_n \bar{z}^{n-2}.$$

(8.6.6)

The commutator can be written as

$$[L_n, \varphi(z)] = \oint_z \frac{dw}{2\pi i} T(w)\varphi(z) w^{n+1},$$

(8.6.7)

which upon inserting equation (8.2.31) becomes

$$[L_n, \varphi(z)] = \left[z^{n+1} \frac{d}{dz} + z^n(n+1)h \right] \varphi(z).$$

(8.6.8)

Thus we recover the standard type of relation between a field variation and its generator.

Similarly, equation (8.4.6) in conjunction with equation (8.6.4) implies that L_n satisfies the Virasoro algebra

$$[L_n, L_m] = (n - m)L_{n+m} + \frac{c}{12} n(n^2 - 1)\delta_{n+m,0},$$

(8.6.9)

with an analogous result for $\bar{L}_n$.

The restricted nature of the central term also becomes apparent from the following argument. Let us consider quantising a two-dimensional conformal field theory. While the L_n and $\bar{L}_n$ classically obey equation (3.1.3), quantum mechanically they become quantum operators which are defined using a normal ordering prescription. Let us assume that the classical Virasoro algebra acquires in the quantum theory a central term $c_{n,m}$ due to normal ordering. As a result, in the quantum theory they can obey

$$[L_n, L_m] = (n - m)L_{n+m} + c_{n,m} \qquad (8.6.10)$$

as well as a similar equation for $\bar{L}_n$. The central term $c_{n,m}$ is assumed to commute with the L_n. Clearly $c_{n,m} = -c_{m,n}$, however, it must also obey other conditions in order that the algebra obey the Jacobi identity. We now consider what is the most general form of $c_{n,m}$ subject to these considerations. The Jacobi identity

$$0 = [[L_n, L_m], L_p] + [[L_p, L_n], L_m] + [[L_m, L_p], L_n] \qquad (8.6.11)$$

implies the result

$$(n - m)c_{n+m,p} + (p - n)c_{n+p,m} + (m - p)c_{m+p,n} = 0. \qquad (8.6.12)$$

Using the redefinitions

$$L_n \mapsto L_n + \frac{1}{n}c_{n,0},$$

$$L_0 \mapsto L_0 + \frac{1}{2}c_{1,-1}, \qquad (8.6.13)$$

we can set $c_{n,0} = 0 = c_{1,-1}$. Taking $p = 0$ in equation (8.6.12) we find

$$(n + m)c_{n,m} = 0 \qquad (8.6.14)$$

and so $c_{n,m} = c_n\delta_{n+m,0}$. Further taking $p = -m - 1, n = 1$, we find the recursion relation

$$c_{m+1} = \frac{m + 2}{m - 1}c_m, \qquad (8.6.15)$$

which implies that

$$c_m = c\frac{m(m^2 - 1)}{12}, \qquad (8.6.16)$$

where c is a constant. As such, we have found that

$$c_{m,n} = c\frac{m(m^2 - 1)}{12}\delta_{m+n,0}. \qquad (8.6.17)$$

The operator product expansions becomes less mysterious when we examine how we might calculate the operator expansion between two operators $U(z)$ and $V(z)$, which can be constructed from a set of free fields. Using Wick normal ordering we may write

$$U(z)V(w) = \sum_l \frac{P_l(w)}{(z - w)^{d_U + d_V - d_{P_l}}} + : U(z)V(w) : \qquad (8.6.18)$$

which is valid only if $|z| > |w|$, where P_l are normal ordered operators and the sum is only over terms that lead to poles in $z - w$. The singularities arise from taking past each other the infinite number of annihilation and creation operators, in order to achieve the normal ordered product of the two operators. Any normal ordered product is finite as $z \mapsto w$ in the sense that it has finite expectation values between any states with finite occupation number. Hence the singular terms as $z \mapsto w$ are found in the first term and as $|z| > |w|$ corresponds to the correct radial ordering inherent in the operator product expansion, we may identify the terms in the operator product expansion with those in the normal ordered expression. Some explicit examples of this proceedure can be found in chapter 9.

8.7   Descendants

We now examine what constitutes a complete set of operators with respect to the operator product expansion. To this end let us examine the operator product expansion between $T(z)$ and a primary field φ of weight $(h, \bar{h})$, which was found in equation (8.2.31) to be of the form

$$T(z)\varphi(w) = \sum_{k=0}^{\infty} \frac{\varphi^{(-k)}(w)}{(z-w)^{2-k}}, \tag{8.7.1}$$

where $\varphi^{(0)} = h\varphi$, $\varphi^{(-1)} = \partial\varphi$ and the remaining $\varphi^{(k)}$, $k = 2, 3, \ldots$ are defined by this equation. For a general field ϕ, primary or not, we can introduce the definition

$$T(z)\phi(w) = \sum_{n=-\infty}^{\infty} \frac{\hat{L}_n\phi(w)}{(z-w)^{2+n}} \tag{8.7.2}$$

or equivalently

$$\phi^{(-n)}(w) = \hat{L}_{-n}\phi(w) = \oint_w \frac{dz}{2\pi i} T(z)\phi(w)(z-w)^{-n+1}. \tag{8.7.3}$$

We can define the action of any number of $\hat{L}_n$ on a field ϕ as follows:

$$\phi^{(-n_1,-n_2,\ldots,-n_m)}(w) = \hat{L}_{-n_1}\hat{L}_{-n_2}\ldots\hat{L}_{-n_m}\phi(w), \tag{8.7.4}$$

where $\hat{L}_{-n_k}$ is defined as in equation (8.7.3), but with ϕ replaced by $\hat{L}_{-n_{k+1}}\cdots\hat{L}_{-n_m}\phi(w)$ and with a contour C_k which encloses w and $z_{k+1}, \cdots z_m$, which are the integration variables of all the $\hat{L}_p$ to the right of $\hat{L}_{-n_k}$. Such fields occur in the repeated operator product expansion of Ts with ϕ, for example, $T(z)\phi^{(k_1)}(w)$ contains $\phi^{(k_1,k_2)}$. More generally, we can also act with $\hat{\bar{L}}_n$ in the same way to create the fields

$$\phi^{(\{p\},\{\bar{p}\})}(w) = \hat{L}_{-p_1}\cdots\hat{L}_{-p_i}\hat{\bar{L}}_{-\bar{p}_1}\cdots\hat{\bar{L}}_{-\bar{p}_j}\phi(w),$$

where

$$\{p\} = (-p_1, \cdots, -p_i), \qquad \{\bar{p}\} = (-\bar{p}_1, \cdots, -\bar{p}_j). \tag{8.7.5}$$

It is straightforward to show that $\hat{L}_n$ and $\hat{\bar{L}}_n$ obey the Virasoro algebra with central charges c and $\bar{c}$, respectively, corresponding to those for T and $\bar{T}$.

By comparing equation (8.7.2) with equation (8.7.3) for a primary field φ of weight $(h, \bar{h})$, we conclude that

$$\hat{L}_n\varphi = 0, \quad n \geq 1 \quad \hat{L}_0\varphi = h\varphi, \quad \hat{L}_{-1}\varphi = \partial\varphi. \tag{8.7.6}$$

Consequently, using the Virasoro property of the $\hat{L}_n$ we have

$$\hat{L}_0\varphi^{(\{n\},\{\bar{n}\})} = \hat{L}_0\varphi^{(-n_1,-n_2,\ldots,-n_p),\{\bar{n}\})} = (h + (n_1 + n_2 + \cdots + n_p))\varphi^{(\{n\},\{\bar{n}\})}. \tag{8.7.7}$$

By applying a conformal transformation $\oint d\tau\varepsilon(\tau)T(\tau)$ to equation (8.7.3) we find that even if the field φ is a primary field, the fields $\varphi^{(\{n\},\{\bar{n}\})}$ are not, in general, primary fields. We call the fields $\varphi^{(\{n\},\{\bar{n}\})}$ descendants of φ and denote the primary field φ and all its descendants by the symbol $[\varphi]$. Clearly, $[\varphi]$ forms a representation of the conformal algebra, which is generated by $L_n, \bar{L}_n$ and the above discussion shows that if a complete set contains the primary field φ it must also contain all the fields contained in $[\varphi]$.

The identity operator I is a primary field of weight $(0,0)$ and its descendants must be polynomials in the energy-momentum tensor. Indeed, using equation (8.7.3) we find $I^{(-1)} = 0$, but

$$I^{(-2)}(z) = \oint \frac{dw}{2\pi i} \frac{T(w)}{(w-z)} = T(z) \qquad (8.7.8)$$

and so $T(z)$ is itself a descendant. This is consistent with our observation that descendants are not, in general, primary fields.

Let us now consider a field φ' which is not necessarily primary or quasi-primary, but has scaling weight $(l, \bar{l})$, that is, $\hat{L}_0\varphi' = l\varphi'$. As a result $\hat{L}_n\varphi'$ satisfies $\hat{L}_0(\hat{L}_n)\varphi' = (h-n)\varphi'$, that is, has scaling weight $h - n$. If acting with $\hat{L}_n$, $n \geq 1$ never leads to a $\psi = \hat{L}_{n_1} \cdots \hat{L}_{n_p}\varphi'$ for which the action of further $\hat{L}_n$'s $n \geq 1$ vanishes, then we have an infinite tower of fields whose scaling weights will eventually become negative. However, the two-point function $\langle\varphi(z_1)\varphi(z_2)\rangle$ for a field with negative scaling weight $(-k, -\bar{k})$ has a space-time dependence $|z_1 - z_2|^{2(k+\bar{k})}$, which rises with space-time separation. In a physically well-defined theory this should never happen. If we exclude this possibility, then by acting with L_n, we can always find a ψ which is annihilated by $\hat{L}_n$, $n \geq 1$, and so is a primary field. By acting with $\hat{L}_{-n}$ on ψ we can write φ' as a descendant of ψ.

Motivated by this observation we shall assume that the complete set of operators is such that all its elements are members of the conformal family of some primary fields, in other words the complete set of operators is made up of primary fields φ_i and their descendants. In general, the complete set will involve an infinite number of primary fields, but in certain theories, called minimal models, which we discuss later, the complete set involves only a finite number of primary fields.

We may therefore write the operator product of two primary fields φ_i and φ_j as

$$\varphi_i(z.\bar{z})\varphi_j(w,\bar{w}) = \sum_k \sum_{\{n\},\{\bar{n}\}} c_{ij}^{k;\{n\},\{\bar{n}\}} \varphi_k^{(\{n\},\{\bar{n}\})}(w,\bar{w})$$

$$\times (z-w)^{-h_i-h_j+h_k+N}(\bar{z}-\bar{w})^{-\bar{h}_i-\bar{h}_j+\bar{h}_k+\bar{N}}, \qquad (8.7.9)$$

where $\varphi^{(0,0)} = \varphi_k$, $N = \sum_i n_i$, $\bar{N} = \sum_i \bar{n}_i$. We will now show that the coefficients $c_{ij}^{k;\{n\},\{\bar{n}\}} \varphi_k^{(\{n\},\{\bar{n}\})}$ associated with the descendants are determined by conformal symmetry in terms of the coefficients $c_{ij}^{k;\{0\};\{0\}}$; we define these latter constants to be just given by c_{ij}^{k} as they specify how the primary field φ_k occurs in the operator product expansion. To this end, we apply the operator $\int_C d\tau T(\tau)(\tau-w)^{m+1}$, where $m \geq 1$ and the contour C contains z and w, to the above form of the operator product expansion. Using equation (8.7.3) the right-hand side of equation (8.7.9) becomes

$$\sum_k \sum_{\{n\},\{\bar{n}\}} c_{ij}^{k;\{n\};\{\bar{n}\}} (z-w)^{-h_i-h_j+h_k+N}(\bar{z}-\bar{w})^{-\bar{h}_i-\bar{h}_j+\bar{h}_k+\bar{N}} \hat{L}_m\varphi_k^{(\{n\};\{\bar{n}\})}. \qquad (8.7.10)$$

After deforming the contour C to encircle the z and w separately, the left-hand side, on the other hand, becomes

$$\oint_C \frac{d\tau}{2\pi i} (\tau - w)^{m+1} \left\{ h_i \frac{\varphi_i(z, \bar{z})}{(\tau - z)^2} + \frac{\partial \varphi(z)}{(\tau - z)} + \cdots \right\} \varphi_j(w)$$

$$= \{ (m + 1) h_i (z - w)^m + (z - w)^{m+1} \partial_z \} \varphi_i(z, \bar{z}) \varphi_j(w, \bar{w})$$

$$= \sum_k \sum_{\{n\},\{\bar{n}\}} c_{ij}^{k;\{n\};\{\bar{n}\}} (-h_i - h_j + h_k + N + (m + 1) h_i)$$

$$\times (z - w)^{-h_i - h_j + h_k + N + m} (\bar{z} - \bar{w})^{-\bar{h}_i - \bar{h}_j + \bar{h}_k + \bar{N}} \varphi_k^{(\{n\},\{\bar{n}\})}(w). \qquad (8.7.11)$$

Since $m \geq 1$ we do not find any poles in $(\tau - w)$ arising from the operator product expansion of $T(\tau)$ with $\varphi_j(w, \bar{w})$ and in the last step we used equation (8.7.9). A useful formula for evaluating contour integrals is

$$\oint_w \frac{dz}{2\pi i} \frac{f(z)}{(z - w)^n} = \frac{1}{(n - 1)!} f^{n-1}(w),$$

where f has no poles in the region enclosed in the contour and we used Taylor's expansion of f and Cauchy's theorem for the case of $n = 1$. Equating coefficients of $(z - w)^{-h_i - h_j + h_k + N}$ we find

$$\hat{L}_m \tilde{\varphi}_k^{(\{n\};\{\bar{n}\})} = (-h_i - h_j + h_k + N - m + (m + 1) h_i) \tilde{\varphi}_k^{(\{p\};\{\bar{n}\})}, \qquad (8.7.12)$$

where

$$\tilde{\varphi}_k^{(\{n\};\{\bar{n}\})} = \sum_{\{n\};\{\bar{n}\}} c_{ij}^{k(\{n\};\{\bar{n}\})} \varphi_k^{(\{n\};\{\bar{n}\})},$$

with the sum over all $(\{n\}; \{\bar{n}\})$, and $\{p\}$ is subject to $\sum_i n_i = N$, $\sum_i \bar{n}_i = \bar{N}$ and $\sum_i p_i = N - m$. The above equation expresses a descendant with an L_0 value of $N - m$ on the left-hand side in terms of a descendant with an L_0 value of N on the right-hand side and it allows us to solve for $\varphi_k^{(\{n\};\{\bar{n}\})}$, $n > 0$, $\bar{n} > 0$, in terms of φ_k. This becomes clear if we carry out this calculation explicitly for $N = 1, 2$ and $\bar{N} = 0$. The lowest level equation which has $N = 1, n = 1$ is

$$\hat{L}_1 \tilde{\varphi}_k^{(1)} = (h_i - h_j + h_k) \varphi_k, \qquad (8.7.13)$$

where $\tilde{\varphi}_k^{(1)} = c_{ij}^{k;(\{1\},\{0\})} \hat{L}_{-1} \varphi_k \equiv c_{ij}^k a \hat{L}_{-1} \varphi_k$ and a is a constant. Evaluating we find the left-hand side is given by

$$c_{ij}^k \hat{L}_1 \hat{L}_{-1} \tilde{\varphi}_k = 2a c_{ij}^k \hat{L}_0 \varphi_k = 2a h_k \varphi_k \qquad (8.7.14)$$

and consequently

$$c_{ij}^{k;\{1\};\{0\}} = \left(\frac{h_1 - h_j + h_k}{2 h_k} \right) c_{ij}^k.$$

At the next level, $N = 2$, we have two equations corresponding to $m = 1, 2$:

$$\hat{L}_1 \tilde{\varphi}_k^{(2)} = (h_i - h_j + h_k + 1) \tilde{\varphi}_k^{(1)},$$

$$\hat{L}_2 \tilde{\varphi}_k^{(2)} = (2 h_i - h_j + h_k) \tilde{\varphi}_k^{(0)}, \qquad (8.7.15)$$

where

$$\tilde{\varphi}_k^{(2)} = \left(c_{ij}^{\ k;(\{2\},\{0\})} \hat{L}_{-2} + c_{ij}^{\ k;(\{1,1\},\{0\})} \hat{L}_{-1}^2 \right) \varphi_k$$

$$\equiv c_{ij}^{\ k} (b\hat{L}_{-2} + c\hat{L}_{-1}^2) \varphi_k. \tag{8.7.16}$$

Substituting this form into the first equation implies

$$[3b + 2c(2h_k + 1)] = (h_i - h_j + h_k + 1) \left(\frac{h_i - h_j + h_k}{2h_k} \right), \tag{8.7.17}$$

while the second equation implies

$$\left(4h_k + \frac{c}{2} \right) + 6ch_k = 2h_i - h_j + h_k. \tag{8.7.18}$$

Clearly, these last two equations can be solved to find the values of b and c.

In general, one finds a solution of the form

$$c_{ij}^{\ k;\{n\};\{\bar{n}\}} = c_{ij}^{\ k} \beta_{ij}^{\ k\{n\}} \bar{\beta}_{ij}^{\ k\{\bar{n}\}}, \tag{8.7.19}$$

where the β and $\bar{\beta}$ factors can be expressed in terms of h_i, h_j, h_k and c and $\bar{h}_i, \bar{h}_j, \bar{h}_k$ and $\bar{c}$. Since $c_{ij}^{\ k;\{0\};\{0\}} = c_{ij}^{\ k}$, we choose $\beta_{ij}^{\ k\{0\}}$ and $\bar{\beta}_{ij}^{\ k\{0\}}$ to be 1.

Thus the operator product expansion is entirely determined by the constants $c_{ij}^{\ k}$ whose values are not determined, in general, by conformal symmetry alone. To express this fact we may write the operator product expansion in the form

$$\varphi_i(z, \bar{z}) \varphi_j(w, \bar{w}) = \sum_k \frac{c_{ij}^{\ k}}{(z - w)^{h_i + h_j - h_k} (\bar{z} - \bar{w})^{\bar{h}_i + \bar{h}_j - \bar{h}_k}} \psi_k(z - w, \bar{z} - \bar{w}), \tag{8.7.20}$$

where

$$\psi_k(w, \bar{w}) = \sum_{\{n\}} \sum_{\{\bar{n}\}} \beta_{ij}^{\ k\{n\}} \bar{\beta}_{ij}^{\ k\{\bar{n}\}} (z - w)^{\sum_i n_i} (\bar{z} - \bar{w})^{\sum_i \bar{n}_i} \varphi_k^{(\{n\},\{\bar{n}\})}. \tag{8.7.21}$$

The correlation functions of descendants are given in terms of the corresponding correlation function of the primary fields. For example, using equations (8.7.3) and (8.2.33) we find that

$$\langle \varphi_1(z_1, \bar{z}_1) \cdots \hat{L}_{-n} \varphi_i(z_i, \bar{z}_i) \cdots (z_n, \bar{z}_n) \rangle$$

$$= \oint_{z_i} \frac{dw}{2\pi i} (w - z_i)^{-n+1} \langle T(w) \varphi_1(z_1, \bar{z}_1) \cdots \varphi_i(z_i, \bar{z}_i) \cdots (z_n, \bar{z}_n) \rangle$$

$$= \sum_{j=1}^{n} \oint_{z_i} \frac{dw}{2\pi i} (w - z_i)^{-n+1} \left\{ \frac{h_j}{(w - z_j)^2} + \frac{1}{(w - z_j)} \frac{\partial}{\partial z_j} \right\}$$

$$\times \langle \varphi_1(z_1, \bar{z}_1) \cdots \varphi_n(z_n, \bar{z}_n) \rangle$$

$$= \sum_{\substack{j=1 \\ j \neq i}}^{n} \left\{ \frac{h_j(n-1)}{(z_j - z_i)^n} - \frac{1}{(z_j - z_i)^{n-1}} \frac{\partial}{\partial z_j} \right\} \langle \varphi_1(z_1, \bar{z}_1) \cdots \varphi_n(z_n, \bar{z}_n) \rangle \tag{8.7.22}$$

for $n \geq 2$. The latter condition ensures that there are no poles when $i = j$ and in going from the second to the third line we used equation (8.2.33). Consequently, all correlation functions are also determined in terms of those for the primary fields φ_i. By systematically using the operator product expansion on two adjacent operators we can determine all correlators in terms of $c_{ij}{}^k$. Thus, we can reconstruct the conformal field theory once we know the weights $(h_i, \bar{h}_i)$ of the primary fields in the complete set of operators, the value of the central charge $(c, \bar{c})$ and the numerical coefficients $c_{ij}{}^k$. It is due to the remarkable power of conformal symmetry that the complete theory can be found from such a small set of data.

Given a Green's function we may evaluate it by using the operator product expansion. However, there are a number of ways to do this corresponding to the different ways of pairing the operators upon which to use the operator product expansion. For a four-point function $\langle \varphi_i(z_1, \bar{z}_1)\varphi_j(z_2, \bar{z}_2)\varphi_k(z_3, \bar{z}_3)\varphi_l(z_4, \bar{z}_4)\rangle$ we can use the operator product expansion on $\varphi_i\varphi_j$ and on $\varphi_k\varphi_l$ which very generically is of the form $\sum_m c_{ij}^m c_{kl}^m$. However, we could equally well use the operator product expansion on $\varphi_i\varphi_k$ and on $\varphi_j\varphi_l$ or on $\varphi_i\varphi_l$ and on $\varphi_j\varphi_k$. Demanding that these three ways of calculating are the same is equivalent to demanding the associativity of the operator product expansion and it places constraints on the central charges $(c, \bar{c})$ the weights of the primary fields $(h_i, \bar{h}_i)$ and the $(c_{ij}^k, \bar{c}_{ij}^k)$. Trying to solve these constraints became known as the conformal bootstrap.

8.8   States, modes and primary fields

Given a primary field $\varphi(z, \bar{z})$ of weight $(h, \bar{h})$ we can associate with it a particular state by letting it act in the vacuum $|0\rangle$ for a particular value of z and $\bar{z}$. Thinking of the state as an in state it is natural to choose $z = \bar{z} = 0$ since this corresponds on the cylinder to taking $\tau \to -\infty$ from equation (8.2.14). As such, we define

$$|\varphi_{in}\rangle = \lim_{z,\bar{z}\to 0} \varphi(z, \bar{z})|0\rangle. \tag{8.8.1}$$

An out state is associated with $\tau \to \infty$ on the cylinder and so the point at infinity on $\hat{C}$. We adopt the definition

$$\langle \varphi_{out}| \equiv \lim_{z,\bar{z}\to\infty} \langle 0|\varphi(z, \bar{z})z^{2h}\bar{z}^{2\bar{h}}. \tag{8.8.2}$$

While the factors of z^{2h}, $\bar{z}^{2\bar{h}}$ may look unusual at first sight, they are consistent with the action of the adjoint operator if φ is self-adjoint. On the strip or cylinder an operator is self-adjoint if

$$(\varphi_{cyl}(\varsigma, \bar{\varsigma}))^\dagger = \varphi_{cyl}(-\bar{\varsigma}, -\varsigma). \tag{8.8.3}$$

In Euclidean space we change $\tau \mapsto -\tau$ under the self-adjoint operation to reflect the change $e^{itH} \mapsto e^{-itH}$ in Minkowski space. The change $\tau \mapsto -\tau$ implies $\varsigma = \tau + i\sigma \mapsto -\tau + i\sigma = -\bar{\varsigma}$. Making the conformal map $z = e^\varsigma$ from the cylinder, or strip with sides identified, to the Riemann sphere, the change $\varsigma \mapsto -\bar{\varsigma}$ corresponds to $z \mapsto (\bar{z})^{-1}$ and as a result on $\hat{C}$ the above equation becomes

$$(\varphi(z, \bar{z}))^\dagger = \varphi\left(\frac{1}{\bar{z}}, \frac{1}{z}\right) \frac{1}{\bar{z}^{2h}} \frac{1}{z^{2\bar{h}}} \tag{8.8.4}$$

upon use of the result

$$\varphi_{\text{cyl}}(\varsigma, \bar{\varsigma}) = \varphi(z, \bar{z}) z^h \bar{z}^{\bar{h}} \tag{8.8.5}$$

and dropping the subscript Riemann sphere on φ. Carrying out the adjoint on $|\varphi_{\text{in}}\rangle$ of equation (8.8.1) we recover equation (8.8.2) by the following steps:

$$|\varphi_{\text{in}}\rangle^\dagger = (\lim_{z,\bar{z}\to 0} \varphi(z, \bar{z})|0\rangle)^\dagger = \lim_{\bar{z},z\to 0} \langle 0|\varphi^\dagger(z, \bar{z}) = \lim_{z,\bar{z}\to 0} \langle 0|\varphi\left(\frac{1}{\bar{z}}, \frac{1}{z}\right) \bar{z}^{-2h} z^{-2\bar{h}}$$

$$= \lim_{w,\bar{w}\to\infty} \langle 0|\varphi(w, \bar{w}) w^{2h} \bar{w}^{2\bar{h}} = \langle \varphi_{\text{out}}|, \tag{8.8.6}$$

where $w = 1/\bar{z}$, $\bar{w} = 1/z$.

The above limits are also natural from the point of view of correlation functions, which are automatically radially ordered. Indeed, equation (8.8.2) leads to finite expectation values. For the two-point function, we find that

$$\langle i|j\rangle = \lim_{z,\bar{z}\to\infty} z^{2h}\bar{z}^{2\bar{h}} \langle 0|\varphi_i(z, \bar{z})\varphi_j(0, 0)|0\rangle = \lim_{z,\bar{z}\to\infty} z^{2h}\bar{z}^{2\bar{h}} \frac{1}{z^{2h}} \frac{1}{\bar{z}^{2\bar{h}}} \delta_{ij} = \delta_{ij} \tag{8.8.7}$$

using $|j\rangle = |(\varphi_j)_{\text{in}}\rangle$, $\langle i| = \langle (\varphi_i)_{\text{out}}|$ and equation (8.3.8) We have assumed the basis for the two-point functions such that the two-point functions are orthogonal. Similarly for the three-point function

$$\lim_{z_1,\bar{z}_1\to\infty} z_1^{2h}\bar{z}_1^{2\bar{h}} \langle 0|\varphi_i(z_1, \bar{z}_1)\varphi_j(z, \bar{z})\varphi_k(0, 0)|0\rangle$$

$$= \langle i|\varphi_j(z, \bar{z})|k\rangle = \frac{\tilde{c}_{ijk}}{z^{h_j+h_k-h_i} \bar{z}^{\bar{h}_j+\bar{h}_k-\bar{h}_i}} \tag{8.8.8}$$

upon use of equation (8.3.13) and where $\tilde{c}_{ijk}$ are constants. We note that as a consequence

$$\langle i|\varphi_j(z, \bar{z})\varphi_k(0)|0\rangle = \langle i|\varphi_j(z, \bar{z})|k\rangle = \frac{\tilde{c}_{ijk}}{z^{h_j+h_k-h_i} \bar{z}^{\bar{h}_j+\bar{h}_k-\bar{h}_i}}.$$

However, we can evaluate this expression in another way by considering the operator product expansion of equation (8.7.20): taking $w = \bar{w} = 0$ and sandwiching it between $\langle i|$ and $|0\rangle$ we find that

$$\sum_l \langle i|c_{jk}{}^l \varphi_l(0)|0\rangle \frac{1}{z^{h_j+h_k-h_l} \bar{z}^{\bar{h}_j+\bar{h}_k-\bar{h}_l}} = \frac{c_{jk}{}^i}{z^{h_j+h_k-h_i} \bar{z}^{\bar{h}_j+\bar{h}_k-\bar{h}_i}}. \tag{8.8.9}$$

In deriving this equation we have used the orthogonality relation of (8.8.7) and the fact that the two-point function of a primary field and a descendant of the same primary field, or any other primary field, vanishes. The two-point function of a primary field and its descendant at must vanish as they have different dilation weights. However, if we have a primary field and the descendant of a different primary field it vanishes using equations (8.7.22) and (8.3.8). As a result, we conclude that

$$\tilde{c}_{ijk} = c_{ij}{}^k. \tag{8.8.10}$$

We used the fact that by virtue of its definition $\tilde{c}_{ijk}$ is symmetric in its indices since the order in which the three fields are written in a Green's function makes no difference as it is radially ordered. Thus the operator product expansions can be deduced from the three-point functions and vice versa.

Given a primary field of weight $(h, 0)$ which is taken to be a holomorphic function of z, it is useful to introduce modes according to the rule

$$\varphi(z) = \sum_n \varphi_n z^{-n-h}. \tag{8.8.11}$$

Mapping back to the cylinder corresponds to the decomposition

$$\varphi_{\text{cyl}}(\varsigma) = \sum_n \varphi_n e^{-n\varsigma}, \tag{8.8.12}$$

which is single valued under $\varsigma \mapsto \varsigma + 2\pi i$ as it should be. We may invert equation (8.8.11) by

$$\varphi_n = \oint_0 \frac{dz}{2\pi i} \varphi(z) z^{n+h-1}. \tag{8.8.13}$$

In terms of the modes the adjoint condition of equation (8.8.4) implies that

$$\varphi_n^\dagger = \varphi_{-n} \tag{8.8.14}$$

and

$$\begin{aligned}
[L_n, \varphi_m] &= \oint \frac{d\varsigma}{2\pi i} \varsigma^{m+h-1} \oint_\varsigma \frac{dz}{2\pi i} z^{n+1} T(z)\varphi(\varsigma) \\
&= \oint \frac{d\varsigma}{2\pi i} \varsigma^{m+h-1} \oint_\varsigma \frac{dz}{2\pi i} z^{n+1} \left\{ \frac{h\varphi(z)}{(z-\varsigma)^2} + \frac{\partial\varphi}{(z-\varsigma)} \right\} \\
&= \oint \frac{d\varsigma}{2\pi i} \varsigma^{m+n+h-1} \{(n+1)h\varphi(\varsigma) + \varsigma\partial\varphi(\varsigma)\} \\
&= (n(h-1) - m)\varphi_{n+m}.
\end{aligned} \tag{8.8.15}$$

We have tacitly assumed that $|\varphi_{\text{in}}\rangle$ is, in fact, well defined. This will be the case if the modes of equation (8.8.11) obey the equations

$$\varphi_n|0\rangle = 0 \qquad n \geq -h+1 \tag{8.8.16}$$

and so $|\varphi_n\rangle = \varphi_{-n}|0\rangle$. Carrying out the same analysis for the bra we find

$$\langle 0|\varphi_n = 0 \qquad n \leq h-1. \tag{8.8.17}$$

For the energy-momentum tensor $T(z) = \sum_n z^{-n-2}$ we find that $L_n^\dagger = L_{-n}$ and

$$L_n|0\rangle = 0 \quad n \geq -1 \quad \langle 0|L_n = 0 \quad n \leq 1. \tag{8.8.18}$$

Consequently, the generators of Möbius transformations $L_n, \bar{L}_n, n = 0, \pm 1$, annihilate both the vacuum $|0\rangle$ and its adjoint $\langle 0|$. It is straightforward to give an alternative derivation of equations (8.3.1)–(8.3.3) starting from Möbius invariance of the vacuum.

One can use the Virasoro algebra to rederive

$$\langle T(z)T(w)\rangle = \frac{c}{2}\frac{1}{(z-w)^4}$$

(that is, equation (8.4.3)) making use of equation (8.8.18). When carrying out this computation it becomes obvious that radial ordering is required for the correlator to be well defined.

The state $|\varphi_{\text{in}}\rangle \equiv |h,\bar{h}\rangle$ of equation (8.8.1) satisfies the conditions

$$L_0|h,\bar{h}\rangle = h|h,\bar{h}\rangle, \quad L_n|h,\bar{h}\rangle = 0, \quad n \geq 1,$$

$$\bar{L}_0|h.\bar{h}\rangle = \bar{h}|h,\bar{h}\rangle, \quad \bar{L}_n|h,\bar{h}\rangle = 0$$

$$(8.8.19)$$

as is evident from applying equation (8.6.8) to $|0\rangle$ and taking $z = 0$. Any state for which an equation of the form of equation (8.8.19) holds is known as a highest weight state. It carries a representation of the conformal algebra of the form

$$L_{-n_1}\cdots L_{-n_p}\bar{L}_{-m_1}\cdots\bar{L}_{-m_q}|h,\bar{h}\rangle; \tag{8.8.20}$$

these are often called descendant states. In fact, the descendant states are closely related to the descendant fields of equation (8.7.3); indeed it is straightforward to show that

$$(\hat{L}_{-n}\varphi)(0)|0\rangle = \oint_0 \frac{dw}{2\pi i}T(w)w^{-n+1}\varphi(0)|0\rangle = L_{-n}|h,\bar{h}\rangle. \tag{8.8.21}$$

The descendant states are eigenstates of L_0 and $\bar{L}_0$ with eigenvalues $h + \sum_i n_i$ and $\bar{h} + \sum_j m_j$, respectively. They are not, however, usually annihilated by L_n and $\bar{L}_n$, $n \geq 1$, but in certain theories this can happen for particular descendant states. When this occurs is the subject of the next section and, as we will see, it can have far reaching consequences for the theory.

The scalar products of descendant states play a crucial role in conformal field theory and will be discussed in the next chapter. We note here that

$$\|L_{-n}|h,\bar{h}\rangle\| = \langle h,\bar{h}|L_n L_{-n}|h,\bar{h}\rangle$$

$$= \langle h,\bar{h}|\left(2nL_0 + \frac{c}{12}n(n^2-1)\right)|h,\bar{h}\rangle$$

$$= \left(2nh + \frac{c}{12}n(n^2-1)\right)\langle h,\bar{h}|h,\bar{h}\rangle. \tag{8.8.22}$$

In a physical theory for which the states have positive norm the above quantity must be positive. For $n = 1$ we conclude that $h \geq 0$, and if $h = 0$, then $L_{-1}|h,\bar{h}\rangle = 0$, while taking $n \to \infty$ implies that $c \geq 0$.

8.9   Representations of the Virasoro algebra and minimal models

Given a space on which the Virasoro generators L_n, $n \in \mathbf{Z}$, act we can form a representation by taking any state $|\varphi\rangle$ and acting on it with the L_n in all possible ways to find the states of the form

$$L_{-n_1}^{i_1} L_{-n_2}^{i_2}\cdots L_{-n_p}^{i_p}|\varphi\rangle. \tag{8.9.1}$$

We may choose $|\varphi\rangle$ to be an eigenstate of L_0, that is, $L_0|\varphi\rangle = \tilde{h}|\varphi\rangle$. Due to the relation $[L_0, L_n] = -nL_n$, it follows that all other states in the representation are eigenstates of L_0. In particular, the state of equation (8.9.1) has eigenvalue $(\tilde{h} + \sum_{m=1}^{p} i_m n_m)$. As noted earlier, on the cylinder L_0 is the generator of time translations and so we adopted it as our Hamiltonian. In a physical system we expect L_0 to be bounded from below; however, this will only happen if after successively acting with L_n, $n \geq 1$, on states of the form of equation (8.9.1) we arrive at a state which is annihilated by L_n, $n \geq 1$. We now consider only representations of this kind and denote by $|h\rangle$ the state for which

$$L_0|h\rangle = h|h\rangle, \quad L_n|h\rangle = 0, \quad n \geq 1. \tag{8.9.2}$$

We call such a representation a highest weight representation and $|h\rangle$ the highest weight state. The other states in the representation are of the form

$$|n_1, i_1; \ldots, n_p, i_p\rangle = L_{-n_1}^{i_1} \ldots L_{-n_p}^{i_p} |h\rangle \tag{8.9.3}$$

and are called descendants. We call $\sum_{q=1}^{p} n_q i_q$ the level and the number of states at level N is $p(N)$, where

$$\prod_{n=1}^{\infty} \frac{1}{(1 - q^n)} = \sum_{n=0}^{\infty} p(N) q^n \tag{8.9.4}$$

for reasons which are virtually identical to that given above equation (4.2.14).

We can now consider when a highest weight representation is irreducible. If the representation is not irreducible it will contain a set of states which transform into themselves under the L_n. This must form a highest weight representation and in particular contain its own highest weight state. Thus for a reducible representation one of the descendent states of equation (8.9.1) must itself be a highest weight state. Of the states in this sub-representation it is clear that such a highest weight state, or null state, will be the first such state to occur in the set of all states when put in order of increasing level.

It is very instructive to find when such highest weight states exist at the lowest levels. At level 1, the only possible state is $L_{-1}|h\rangle$. This will be a highest weight state if

$$L_1 L_{-1}|h\rangle = 2L_0|h\rangle = 2h|h\rangle = 0. \tag{8.9.5}$$

The other L_n, $n \geq 2$, obviously annihilate the candidate state. Hence, we have a highest weight descendant at level 1 if and only if $h = 0$.

At level 2 the possible states are of the form

$$(aL_{-2} + bL_{-1}^2)|h\rangle. \tag{8.9.6}$$

Acting with L_1 we find that it is a highest weight state if $3a + b(2 + 4h) = 0$, while acting with L_2 we find the condition $a(4h + (c/2)) + 6hb = 0$. Solving for a and b we find that a level 2 highest weight state exists if

$$16h^2 + 2h(c - 5) + c = 0. \tag{8.9.7}$$

Hence, at level 2 we have a unique highest weight descendant of the form

$$\left(L_{-2} - \frac{3}{2(2h + 1)} L_{-1}^2\right) |h\rangle \tag{8.9.8}$$

if c and h satisfy equation (8.9.7).

It is clear that the existence of highest weight descendants at a given level depends only on the values of h and c since it relies only on the manipulation of the L_n and their action on $|h\rangle$.

For an arbitrary state at level N we find $p(N)$ simultaneous equations for the $p(N)$ coefficients that occur in an arbitrary state at level N. A non-trivial solution will occur if the determinant of the associated simultaneous equations vanishes. We now study this determinant, called the Kac determinant, directly.

From all descendent states at level N we can form all possible scalar products which define a matrix $M^{(N)}(c, h)$ whose elements are $\langle m_1, j_1; \ldots; m_q, j_q | n_1, i_1; \ldots; n_p, i_p \rangle$, where $\sum_m m_j j_m = N = \sum_n n_m i_m$. This procedure is best illustrated by some examples. At level one we have only 1 state $L_{-1}|h\rangle$ and the corresponding matrix is

$$M^{(1)}(c, h) = \langle h|L_1 L_{-1}|h\rangle = 2h \tag{8.9.9}$$

if we choose to normalise $|h\rangle$, so that $\langle h|h\rangle = 1$. At level 2 we have two possible states which we order as $L_{-2}|h\rangle$, $L_{-1}^2|h\rangle$. The corresponding matrix is

$$M^{(2)}(c, h) = \begin{pmatrix} \langle h|L_2 L_{-2}|h\rangle & \langle h|L_2 L_{-1}^2|h\rangle \\ \langle h|L_1^2 L_{-2}|h\rangle & \langle h|L_1^2 L_{-1}^2|h\rangle \end{pmatrix} = \begin{pmatrix} (c/2) + 4h & 6h \\ 6h & 8h^2 + 4h \end{pmatrix}. \tag{8.9.10}$$

The zeros of the determinant of $M^{(N)}(c, h)$ (called the Kac determinant) are in one-to-one correspondence with the highest weight descendants. In particular, if $|n_1, i_1; \ldots; n_p, i_p\rangle$, or some combination of such states at level N, is a highest weight state for a given c and h, then the matrix will contain a column whose entries are zero and the Kac determinant will vanish. Conversely, if the determinant is zero, we can find an eigenstate of the matrix with vanishing eigenvalue whose coefficients can be used to construct a highest weight state. In particular, M is of the generic form $\langle \chi_n | \chi_m \rangle$ and if a_m are the components of the eigenvector with zero eigenvalue, then $\sum_m a_m |\chi_m\rangle$ has zero scalar product with all $\langle \chi_n |$, since $Ma = 0$, and so it is a highest weight state. At levels 1 and 2 we find the results are

$$\det M^{(1)} = 2h, \quad \det M^{(2)} = 2h(16h^2 + 2h(c - 5) + c), \tag{8.9.11}$$

respectively, which is indeed in accord with our discussion of highest weight states in equations (8.9.7) and (8.9.5).

The Kac determinant is calculated by moving L_n, $n \geq 1$, through L_{-m}, $m \geq 1$, onto $|h\rangle$ and so it depends only on c and h. In fact, the value of the Kac determinant is known at all levels and is given by [8.5]

$$\det M^{(N)}(c, h) = K_N \prod_{\substack{r,s \\ rs \leq N}} (h - h_{r,s})^{p(N-rs)}, \tag{8.9.12}$$

where K_N is a constant,

$$h_{r,s} = \frac{1}{4}(r^2 - 1)t + \frac{1}{4}(s^2 - 1)\frac{1}{t} - \frac{1}{2}(rs - 1)$$

$$= \frac{1}{4}\left(r\sqrt{t} - \frac{s}{\sqrt{t}}\right)^2 - \frac{1}{4}\left(\sqrt{t} - \frac{1}{\sqrt{t}}\right)^2 \tag{8.9.13}$$

and t is related to c by

$$t = \frac{13 - c \pm \sqrt{(1-c)(25-c)}}{12} \quad \text{or} \quad \frac{1}{t} = \frac{13 - c \mp \sqrt{(1-c)(25-c)}}{12}, \quad (8.9.14)$$

which are equivalent to

$$c = 13 - 6t - \frac{6}{t}. \quad (8.9.15)$$

Furthermore, $p(\bullet)$ is given in equation (8.9.4) and we note that the product is over all positive integers r, s such that $rs \leq N$. Clearly t and $1/t$ lead to the same central charge and weights.

For example, at level 2 the Kac formula gives

$$\det M^{(2)}(c, h) = K_2(h - h_{1,1})(h - h_{1,2})(h - h_{2,1}). \quad (8.9.16)$$

Upon using equation (8.9.14) we find that

$$h_{1,1} = 0, \quad h_{1,2} = \frac{3}{4t} - \frac{1}{2}, \quad h_{2,1} = \frac{3t}{4} - \frac{1}{2} \quad (8.9.17)$$

and we recover the result of equation (8.9.11).

If a highest weight descendant $|n_1, i_1; \ldots; n_p, i_p\rangle$ first appears at level N_1, then the states

$$L_{-m_1}^{j_1} \cdots L_{-m_q}^{j_q} |n_1, i_1; \ldots; n_p, i_p\rangle \quad (8.9.18)$$

will also lead to a zero of the Kac determinant since we are taking its scalar product with a state of level $N_1 + \sum_{i=1}^{q} m_i j_i > N_1$. This leads in the scalar product to $N_1 + \sum_{i=1}^{q} m_i j_i$ positively moded L_n acting on the state of equation (8.9.18), which will vanish after the manipulations of the Virasoro operators as we will be left with N_1 positively moded L_n acting on the highest weight state. At level N there will be $p(N - N_1)$ such states based on the level N_1 null state that will lead to zeros of the Kac determinant. This feature was evident in the factor h in $\det M^{(N)}$ of equation (8.9.11) and it explains why the product in $\det M^{(N)}$ of equation (8.9.12) is over all $rs < N$ as well as $rs = N$.

A straightforward proof of equation (8.9.12) can be found in [8.6]. One regards $\det M^{(N)}(c, h)$ as a polynomial in h. The degree of the polynomial is by definition the maximal power of h. Since h can only arise from manipulations of L_n that result in L_0, the maximal power must come from the product of the diagonal elements. The scalar product of $L_{-n_1}^{i_1} \cdots L_{-n_p}^{i_p} |h\rangle$ with itself can yield at most $\sum_p i_p n_p$ powers of h and hence the maximal power of h in $\det M^{(N)}$ is $q^{(N)}$, where $q^{(N)} = \sum i_p$ in which the sum is over all i_p such that $\sum i_p n_p = N$. It follows that

$$q^{(N)} = \sum_{\substack{r,s \\ rs \leq N}} p(N - rs). \quad (8.9.19)$$

The final step in the proof is to show that $h_{r,s}$ for all $rs \leq N$ are zeros of $\det M^{(N)}$. Since these zeros alone lead to a polynomial $\prod_{rs \leq N} h^{p(N-rs)}$ that is of degree $\sum_{rs \leq N} p(N - rs) = q^N$, there can be no more zeros and the Kac determinant must be the form of equation (8.9.12).

Hence, in a theory with central charge c a highest weight representation with $h = h_{(r,s)}$ will lead to a zero in the Kac determinant of equation (8.9.2) and as a consequence one of its descendants, at level rs, will be itself a highest weight state. This state will have the form

$$\sum_{i_1 n_1 + \cdots + i_p n_p = rs} c_{i_1, \ldots, i_p}^{n_1, \ldots, n_p} L_{-n_1}^{i_1} L_{-n_2}^{i_2} \cdots L_{-n_p}^{i_p} |h_{(r,s)}\rangle. \tag{8.9.20}$$

The importance of this fact is that it is consistent to set this highest weight descendant to zero since the action of positive modes of the Virasoro algebra is automatically zero and the scalar product of such states with any other states in the representations is zero. This will have profound effects for the resulting theory allowing us for certain theories to solve for, i.e., compute, all the Greens functions and operator product expansions.

The above discussion of representations of the Virasoro algebra was given in terms of states; however, using section 8.8 we can equally well formulate it in terms of primary fields and their descendants. In particular, we would conclude that a primary field $\varphi_{(r,s)}$ of conformal weight $h_{(r,s)}$ would have a descendant of the form

$$\sum_{i_1 n_1 + \cdots + i_p n_p = rs} c_{i_1, \ldots, i_p}^{n_1, \ldots, n_p} \hat{L}_{n_1}^{i_1} \hat{L}_{n_2}^{i_2} \cdots \hat{L}_{n_p}^{i_p} \varphi_{h_{(r,s)}} \tag{8.9.21}$$

that is itself a primary field.

Let us now investigate the consequences of setting one of the null states, or equivalently null fields, to zero. To demonstrate the ideas we begin with a rather simple example. Let us consider the primary field $\varphi_{(1,1)}$ which is present in all models and has conformal weight $h = 0$. This possesses the null descendant $\hat{L}_{-1}\varphi_{(1,1)}$, which we may set to zero. This implies that $(\partial/\partial z)\varphi_{(1,1)}(z) = 0$ and so for the three-point function of $\varphi_{(1,1)}$ and two primary fields φ_i and φ_j of conformal weights h_i and h_j, respectively, we find that

$$\frac{\partial}{\partial z} \langle \varphi_{(1,1)}(z, \bar{z}) \varphi_i(z_i, \bar{z}_i) \varphi_j(z_j, \bar{z}_j) \rangle = 0. \tag{8.9.22}$$

However, the dependence of the three-point function was found using Möbius invariance in equation (8.3.13) and applying the above constraint one finds that $h_i = h_j$. We showed in equation (8.9.10) that the three-point function of three primary fields and their operator product expansions are related and so we conclude that the operator product expansion of $\varphi_{(1,1)}$ with φ_i has the form $[\varphi_{(1,1)}][\varphi_i] = [\varphi_j]$, where φ_j is any primary field of weight $h_j = h_i$. Here we are using the notation given below equation (8.7.7), where $[\varphi]$ stands for the primary field φ and all its descendants, Putting this into words, we have found that the operator product expansion of $\varphi_{(1,1)}$ with a primary field leads only to primary fields of the same conformal weight and their descendants.

Let us now consider the primary field $\varphi_{(2,1)}$ with conformal weight $h_{(2,1)}$. We find a null field at level 2, which if we set to zero implies the condition

$$\hat{L}_{-2}\varphi_{(2,1)} - \frac{3}{2(2h_{(2,1)} + 1)} \hat{L}_{-1}^2 \varphi_{(2,1)} = 0. \tag{8.9.23}$$

As a result the three-point correlation functions for $\varphi_{(2,1)}$ and two primary fields φ_i and φ_j of conformal weights h_i and h_j, respectively, satisfy the condition

$$\left\langle \left(\hat{L}_{-2}\varphi_{(2,1)} - \frac{3}{2(2h_{(2,1)} + 1)} \hat{L}_{-1}^2 \varphi_{(2,1)} \right)(z, \bar{z}) \varphi_i(z_i, \bar{z}_i) \varphi_j(z_j, \bar{z}_j) \right\rangle = 0. \tag{8.9.24}$$

Using equation (8.7.22) we find that this implies the differential equation

$$
\left\{ \frac{h_i}{(z-z_i)^2} + \frac{1}{(z-z_i)} \frac{\partial}{\partial z_i} + \frac{h_j}{(z-z_j)^2} + \frac{1}{(z-z_j)} \frac{\partial}{\partial z_j} - \frac{3}{2(2h_{(2,1)}+1)} \frac{\partial^2}{\partial z^2} \right\}
$$
$$
\times \langle \varphi_{(2,1)}(z\bar{z}) \varphi_i(z_i, \bar{z}_i) \varphi_j(z_j, \bar{z}_j) \rangle = 0. \tag{8.9.25}
$$

Substituting in the form of the three-point function of equation (8.3.13) we find the condition

$$
h_{(2,1)} - h_i - h_j - \frac{3}{2h_{(2,1)}+1}(h_j - h_i - h_{(2,1)})(h_i - h_j - h_{(2,1)}) = 0. \tag{8.9.26}
$$

In fact, one gets three conditions all of which are the same. Substituting $h_{(2,1)} = \frac{3}{4}t - \frac{1}{2}$ we can rewrite this last equation as

$$
\left(h_i - h_j + \frac{t}{4} \right)^2 = t \left(h_i + \frac{t}{4} - \frac{1}{2} + \frac{1}{4t} \right). \tag{8.9.27}
$$

To solve this equation we write the conformal weights h in terms of the variable α as

$$
h = -\frac{1}{4} \left(\sqrt{t} - \frac{1}{\sqrt{t}} \right)^2 + \frac{1}{4}\alpha^2, \tag{8.9.28}
$$

whereupon the equation becomes

$$
\alpha_j^2 - \alpha_i^2 - t = \pm 2\alpha_i\sqrt{t}, \quad \text{or} \quad \alpha_j = (\alpha_i \pm \sqrt{t}). \tag{8.9.29}
$$

Here α_j corresponds to the conformal weight h_j. Remembering that the primary fields that occur in the operator product expansion of two operators lead to non-trivial three-point functions and vice versa, we conclude that the operator product expansion of $\varphi_{(2,1)}\varphi_i$ contains only two primary fields with conformal weights whose α obeys $\alpha = (\alpha_i + \sqrt{t})$ and $\alpha = (\alpha_i - \sqrt{t})$ and their descendants. In making this observation we used the α_j of third field φ_j in the three-point function. Comparing equation (8.9.13) with equation (8.9.28), we can identify the α corresponding to the primary field of weight $h_{r,s}$ as $\alpha_{r,s} = r\sqrt{t} - (s/\sqrt{t})$. Taking φ_i to be this primary field we conclude that

$$
\varphi_{(2,1)}\varphi_{(r,s)} = [\varphi_{(r+1,s)}] + [\varphi_{(r-1,s)}]. \tag{8.9.30}
$$

Using an almost identical argument we find that

$$
\varphi_{(1,2)}\varphi_{(r,s)} = [\varphi_{(r,s+1)}] + [\varphi_{(r,s-1)}]. \tag{8.9.31}
$$

We now find the form of the operator product expansion of any two primary fields by considering the four-point correlation function

$$
\langle \varphi_{(2,1)}\varphi_{(2,1)}\varphi_{(r,s)}\varphi_{(m,n)} \rangle. \tag{8.9.32}
$$

Using the operator product expansion of the first two fields (that is, $\varphi_{(2,1)}\varphi_{(2,1)} = [\varphi_{(1,1)}] + [\varphi_{(3,1)}]$), we find we get two terms:

$$
\langle \varphi_{(1,1)}\varphi_{(r,s)}\varphi_{(m,n)} \rangle \quad \text{and} \quad \langle \varphi_{(3,1)}\varphi_{(r,s)}\varphi_{(m,n)} \rangle. \tag{8.9.33}
$$

The first term is the same as $\langle \varphi_{(r,s)}\varphi_{(m,n)} \rangle$ since the above operator product expansion of $\varphi_{(1,1)}$ with $\varphi_{(r,s)}$ is just $\varphi_{(r,s)}$ plus descendants. We can also evaluate the four-point function using the operator product expansion of $\varphi_{(2,1)}$ which contains the primary fields $\varphi_{r+1,s}$ and $\varphi_{r-1,s}$ and their descendants and so the four-point functions become

$$\langle \varphi_{(2,1)} \varphi_{(r+1,s)} \varphi_{(m,n)} \rangle \quad \text{and} \quad \langle \varphi_{(2,1)} \varphi_{(r-1,s)} \varphi_{(m,n)} \rangle. \tag{8.9.34}$$

Using the operator product expansion of $\varphi_{(2,1)}$ with the field next to it we find that the four-point function contains the terms

$$\langle \varphi_{(2,1)} \varphi_{(r+2,s)} \varphi_{(m,n)} \rangle, \quad \langle \varphi_{(2,1)} \varphi_{(r,s)} \varphi_{(m,n)} \rangle \quad \text{and} \quad \langle \varphi_{(2,1)} \varphi_{(r-2,s)} \varphi_{(m,n)} \rangle. \tag{8.9.35}$$

We now demand that the four-point function is independent of the way it is evaluated using the operator product expansion. This is equivalent to demanding that operator product expansion is associative. Insisting on this and taking into account that $\varphi_{(m,n)}$ is any field and the two-point functions are orthogonal, and comparing equations (8.9.33) and (8.9.35) we conclude that

$$\varphi_{(3,1)} \varphi_{(r,s)} = [\varphi_{(r+2,s)}] + [\varphi_{(r,s)}] + [\varphi_{(r-2,s)}]. \tag{8.9.36}$$

Using induction one can show that the general operator expansion is given by

$$\varphi_{(n,m)} \varphi_{(r,s)} = \sum_{k=-n+1}^{n-1} \sum_{l=-m+1}^{m-1} [\varphi_{(r+k,s+l)}]. \tag{8.9.37}$$

The sums increase in steps of 2 in the above equation and all other such formulae. We note that there are nm terms consistent with the fact that a null highest weight field at level nm leads to a differential equation of degree nm.

It might seem that the right-hand side of the operator product expansion contains fields with negative indices. However, the above result was derived using the fact that we set to zero the null field of $\varphi_{(m,n)}$ and we can also set to zero the null field for $\varphi_{(r,s)}$. The consequence is found by writing out the corresponding operator expansion for $\varphi_{(r,s)} \varphi_{(m,n)}$, that is using equation (8.9.37) with (r, s) exchanged with (m, n). However, these two operator expansions are the same once we swap their space-time dependence and so they must contain the same set of primary fields and their descendants. The operator expansion for the latter is then given by

$$\varphi_{(r,s)} \varphi_{(m,n)} = \sum_{k=-r+1}^{r-1} \sum_{l=-s+1}^{s-1} [\varphi_{(n+k,m+l)}]. \tag{8.9.38}$$

Comparing the primary fields $\varphi_{i,j}$ which arise in equation (8.9.37) and (8.9.38) we find that $\varphi_{i,j}$ for any negative i, j must be absent. As a result we may write the operator expansion in the form

$$\varphi_{(m,n)} \varphi_{(r,s)} = \sum_{i=|r-n|+l}^{r+n-1} \sum_{j=|s-m|+1}^{s+m-1} [\varphi_{(i,j)}]. \tag{8.9.39}$$

Now all the indices are positive and again we sum in steps of 2.

Let us summarise the development so far. We have studied the descendants of highest weight states, or equivalently primary fields, in the context of the Virasoro algebra with a central charge c. We found that if the weights of the highest weight states are of the form $h_{r,s}$ given in equation (8.9.13), then the descendants also contain highest weight states. The significance of this is that the original representation is not irreducible and one could set these highest weight descendants to zero. Using the conformal Ward identities derived in earlier sections this leads to differential constraints on the Green's function involving the corresponding primary fields of weights $h_{r,s}$ and so also on their operator product expansions. Indeed, the operator product expansion of all such fields form a closed, although infinite, set and one could even derive its form. Clearly, if one has a conformal theory which only has primary fields with weights $h_{r,s}$ one would be able to know much about all its operator product expansions and equivalently Green's functions.

We now suppose that this is the case and assume that the primary fields have the weights $h_{r,s}$ of equation (8.9.13). In a well-behaved quantum field theory we expect all physical states of the theory to have positive norm as the norm of a state is related to a probability which should be positive. However, we note that there are some physically interesting models in statistical models that are not unitary. By analysing the Kac determinant we can see if the descendant states for a given c and h are positive or not. Examining the parameterisation of c and $h_{r,s}$ in terms of t of equation (8.9.15), or equation (8.9.14), we find for $t < 0$ that $c > 25$ and all the $h_{r,s}$ are negative. If t is complex, then $1 < c < 25$, but all the weights $h_{r,s}$ are complex. Such theories may be well defined and then they should not have any degenerate states. We will discard these two cases. This leaves us with $c \leq 1$. We are left with t real and $t > 0$. As t and $1/t$ correspond to the same theory we can choose to take $t > 1$. The exception is $t = 1$, or $c = 1$, which we will not discuss. However, even in the case $t > 1$ there are in general $h_{r,s}$ which are negative. Indeed we may write equation (8.9.13) as $h_{r,s} = (1/4t)((rt - s)^2 - (t - 1)^2)$ and for a subtle choice of s and t we may make the first term smaller than the second. The way forward is to demand that every primary field has two distinct highest weight descendant states.

We therefore suppose that there exist weights such that $h_{r,s} = h_{r',s'}$, which from equation (8.9.13) implies that $(r \mp r')\sqrt{t} = (s \mp s')(1/\sqrt{t})$. This means that we can express t as a rational number that is, $t = p/q$, where p and q are positive coprime integers. In this case we may write equations (8.9.15) and (8.9.13) as

$$c = 13 - 6\frac{(p^2 + q^2)}{pq} \tag{8.9.40}$$

and

$$h_{r,s} = \frac{1}{4pq}\{(rp - sq)^2 - (p - q)^2\}, \tag{8.9.41}$$

respectively. We observe that

$$h_{r,s} = h_{nq\pm r,np\pm s} \tag{8.9.42}$$

for integer n. This just corresponds to the values $r' = nq \pm r$, $s' = np \pm s$ which follow from the above condition for $t = p/q$.

This implies that the state with highest weight $h_{r,s}$ has in particular two highest weight states, amongst its descendants at levels rs and $r's'$, where $r' = n - r$ and $s' = p - s$. Given that we have two highest weight descendants it could happen that one was a descendant of the other. As we will now show this is not the case and they are independent.

We may always assume that $r's' \geq rs$ and consider the possibility that the highest weight state at level $r's'$ occurs also as a highest weight state at level $r's' - rs$ of the representation with highest weight state at level rs. This latter state has a weight $h_{r,s} + rs$. Using equation (8.9.13) we may write this as

$$h_{r,s} + rs = h_{-r,s} = h_{r,-s} = h_{q+r,p-s} = h_{q-r,p+s} \tag{8.9.43}$$

from which we conclude that this state has highest weight descendants at levels $(q + r)(p - s)$ and $(q - r)(p + s)$. However, $r's' - rs = (q - r)(p - s) - rs$, which is obviously less than either of the above possible levels. Hence, we conclude that the two descendants of the highest weight state $|h_{(r,s)}\rangle$ at levels $r's'$ and rs are independent. Thus we find two possible conditions for every primary field instead of just one.

Since the primary fields with $h_{r,s}$ and $h_{q-r,p-s}$ have the same weight we may set them equal, that is, $\varphi_{r,s} = \varphi_{q-r,p-s}$. As a result, the operator product expansion of equation (8.9.39) for $\varphi_{(r,s)}\varphi_{(m,n)}$ must coincide with that for $\varphi_{(q-r,p-s)}\varphi_{(q-n,p-m)}$. Using equation (8.9.39) we find that

$$\varphi_{(q-r,p-s)}\varphi_{(q-n,p-m)} = \sum_{i=|r-n|+1}^{2q-r-n-1}\sum_{j=|s-m|+1}^{2p-s-m-1} [\varphi_{(i,j)}]. \tag{8.9.44}$$

Comparing this with equation (8.9.39), which must have the same set of primary fields on the right-hand side, we see that the only primary fields $\varphi_{(i,j)}$ that can occur are given by

$$|r - n| + 1 \leq i \leq \min\{r + n - 1, 2q - r - n - 1\},$$
$$|s - m| + 1 \leq i \leq \min\{s + m - 1, 2p - s - m - 1\}. \tag{8.9.45}$$

We note that

$$\min\{r + n - 1, 2q - r - n - 1\} = \begin{cases} r + n - 1 & \text{if } r + n \leq q, \\ 2q - r - n - 1 & \text{if } r + n \geq q \end{cases} \tag{8.9.46}$$

and so $\min\{r + n - 1, 2q - r - n - 1\} \leq q - 1$ and by a similar argument $\min\{s + m - 1, 2p - s - m - 1\} \leq p - 1$.

We have shown that the primary fields $\varphi_{r,s}$ for $1 \leq r \leq q - 1$, $1 \leq s \leq p - 1$ form a closed operator product algebra. These primary fields are subject to the identification that $\varphi_{r,s}$ is the same field as $\varphi_{q-r,p-s}$. This was one of the most important results in [8.1].

However, not all the weights of equation (8.9.41) are positive even if we make the above restriction $1 \leq r \leq q - 1$, $1 \leq s \leq p - 1$. It is relatively straightforward to show that the only unitary theories (that is, with $h > 0$) with $0 < c < 1$ have $p = q \pm 1$ [8.7]. The central charges of these theories are given by

$$c = 1 - \frac{6}{m(m + 1)}, \qquad m = 3, 4, \ldots, \tag{8.9.47}$$

and primary fields of weight

$$h_{r,s} = \frac{((m + 1)r - ms)^2 - 1}{4m(m + 1)}, \quad 1 \leq s \leq r \leq m - 1, 1 \leq r \leq m, 1 \leq s \leq m - 1. \tag{8.9.48}$$

As discussed above, they have the symmetry

$$h_{r,s} = h_{m-r,m+1-s}, \tag{8.9.49}$$

Table 8.9.1 The grid of conformal weights for the
Ising model ($m = 3$) and the tricritical Ising model
($m = 4$)

$$\begin{pmatrix} 0 & \\ \frac{1}{2} & \frac{1}{16} \end{pmatrix} \qquad\qquad \begin{pmatrix} 0 & & \\ \frac{7}{16} & \frac{3}{80} & \\ \frac{3}{2} & \frac{3}{5} & \frac{1}{10} \end{pmatrix}$$

which means we have only $\frac{1}{2}m(m - 1)$ values of $h_{r,s}$. These theories are known as the minimal models.

For theories that have $c > 1$ and highest weight states with $h \geq 0$ the Kac determinant has no zeros since for $1 < c < 25$, m and so $h_{r,s}$ is not real, while for $c \geq 25$, $h_{r,s}$ is negative.

For $m = 2$, $c = 0$ and we have only one primary field $\varphi(1, 1) = \varphi(1, 2)$ with conformal weight 1; it is just the identity operator.

For $m = 3$, $c = \frac{1}{2}$ and we have three primary fields: $h_{1,1} = h_{(2,3)} = 0$, $h_{1,2} = h_{(2,2)} = \frac{1}{16}$, $h_{2,1} = h_{(1,3)} = \frac{1}{2}$. The corresponding primary fields are $\varphi(1, 1) \equiv I$, the identity operator, $\varphi(1, 2) \equiv \sigma$, the spin operator, and $\varphi(2, 1) \equiv \epsilon$, the energy operator, respectively. The null states can be read off from equation (8.9.23) to be

$$\left(L_{-2} - \tfrac{3}{4}L_{-1}^2\right)|\tfrac{1}{2}\rangle \text{ and } \left(L_{-2} - \tfrac{4}{3}L_{-1}^2\right)|\tfrac{1}{16}\rangle. \tag{8.9.50}$$

The operator product expansions can be read off from equation (8.9.39) to be

$$II = [I], \quad I\epsilon = [\epsilon], \quad I\sigma = [\sigma], \quad \sigma\sigma = [I] + [\epsilon], \quad \epsilon\sigma = [\sigma], \quad \epsilon\epsilon = [I]. \tag{8.9.51}$$

This model coincides with the Ising model near its critical point and is referred to for simplicity as the Ising model. The above values of the conformal dimensions are the critical exponents in this model.

It is common practice to display these values in a conformal grid, which is just the matrix $h_{r,s}$ plotted for $r \leq s$. The conformal grids for the Ising model, $m = 3$, $c = \frac{1}{2}$, and the model with $m = 4$, $c = \frac{7}{10}$, which is called the tricritical Ising model, are displayed in table 8.9.1.

Although we have concentrated on the part of the Virasoro algebra that contains the operators L_n, and so is related to the z coordinate, a two-dimensional theory will also have a $\bar{z}$ coordinate and so possess the analogous operators $\bar{L}_n$. As the two algebras are identical the discussion of the latter is essentially the same as that for the former. However, the interesting part comes when one puts the two together. In general, for a given $c = \bar{c}$ we can have primary fields with different conformal weights h and $\bar{h}$. How the 'chiral' Green's functions, or chiral blocks, in each sector are put together is restricted by demanding that the theory be invariant under monodromony transformations, $z - a \to e^{2\pi i}(z - z_i)$, and under the interchange of the insertion points, that is, $(z_i, \bar{z}_i) \leftrightarrow (z_j, \bar{z}_j)$. It is also interesting to take the conformal field theory to be on a torus in which case one finds that demanding modular invariance [8.8] determines the way the conformal blocks are put together. The result has a correspondence with the ADE classification of finite-dimensional semi-simple Lie algebras discussed in chapter 16.

Remarkably, the differential equations that result from setting to zero the highest weight descendants, that is, null fields, can be used to find the explicit form of all the Green's functions of any of the minimal model. To illustrate this we consider the example of four $h = \frac{1}{16}$ fields for the Ising model. Using equation (8.9.50) and equation (8.7.22) we conclude that

$$\left\{ \frac{4}{3} \frac{\partial^2}{\partial z_1^2} - \frac{1}{16} \sum_{j=2}^{4} \{ \frac{1}{(z_1 - z_j)^2} - \frac{1}{(z_1 - z_j)} \frac{\partial}{\partial z_j} \right\}$$

$$\times \langle \sigma (z_1, \bar{z}_1) \sigma (z_2, \bar{z}_2) \sigma (z_3, \bar{z}_3) \sigma (z_4, \bar{z}_4) \rangle = 0. \tag{8.9.52}$$

Using equation (8.2.15) we may write the four-point function in the form

$$\langle \sigma (z_1, \bar{z}_1) \sigma (z_2, \bar{z}_2) \sigma (z_3, \bar{z}_3) \sigma (z_4, \bar{z}_4) \rangle$$

$$= \left(\frac{1}{z_{12} z_{13} z_{14} z_{23} z_{24} z_{34}} \right)^{\frac{1}{24}} \left(\frac{1}{\bar{z}_{12} \bar{z}_{13} \bar{z}_{14} \bar{z}_{23} \bar{z}_{24} \bar{z}_{34}} \right)^{\frac{1}{24}} F'(y, \bar{y})$$

$$= \left(\frac{z_{14} z_{23}}{z_{12} z_{13} z_{24} z_{34}} \right)^{\frac{1}{8}} \left(\frac{\bar{z}_{14} \bar{z}_{23}}{\bar{z}_{12} \bar{z}_{13} \bar{z}_{24} \bar{z}_{34}} \right)^{\frac{1}{8}} F(y, \bar{y}), \tag{8.9.53}$$

where

$$y = \frac{x}{(x-1)} = \frac{z_{12} z_{34}}{z_{14} z_{32}}.$$

In going from the first to the second line we have recognised that we may rewrite the prefactor of the first equation in terms of that of the second times a cross ratio that we absorb into F'. Substituting this into the differential equation (8.9.52) we find that

$$\left\{ x(1-x) \frac{d^2}{dx^2} + \left(\frac{1}{2} - x \right) \frac{d}{dx} + \frac{1}{16} \right\} F = 0. \tag{8.9.54}$$

The two solutions to this equation are $\sqrt{1 + \sqrt{1-x}}$ and $\sqrt{1 - \sqrt{1-x}}$ and so we find that

$$F = a \sqrt{1 + \sqrt{1-x}} \sqrt{1 + \sqrt{1-\bar{x}}} + b \sqrt{1 + \sqrt{1-x}} \sqrt{1 - \sqrt{1-\bar{x}}}$$

$$+ c \sqrt{1 - \sqrt{1-x}} \sqrt{1 + \sqrt{1-\bar{x}}} + d \sqrt{1 - \sqrt{1-x}} \sqrt{1 - \sqrt{1-\bar{x}}}, \tag{8.9.55}$$

where a b, c and d are constants. Requiring invariance under monodromy transformations and interchange of the insertion points sets $b = 0 = c$ and fixes $a = b$.

9 Conformal symmetry and string theory

If Ed had thought of this, it would have worked

David Farlie

In this chapter we will illustrate the general methods of conformal field theory given in chapter 8. This will include the theory of a free boson and also that of a free fermion. We then explain how such systems occurred when we quantised the bosonic string and superstring in earlier chapters. This allows us to give a general procedure for constructing string theories using conformal field techniques. Finally, using a very special free field which possesses what is known as a background charge we give a method of explicitly constructing all the correlation functions of any of the minimal models.

9.1   Free field theories

We now illustrate many of the features of conformal field theories discussed in the previous chapter within the context of the simplest of all possible theories, namely certain free field theories. These particular examples will play an important role in string theory. Although the theories of interest that arise in string theory are defined in two-dimensional Minkowski space, as discussed in the previous chapter, we will Wick rotate them to Euclidean space and work on the Riemann sphere $\hat{C}$. We will discuss the relation between the Euclidean and Minkowski space theories in more detail later in this chapter.

9.1.1   The free scalar

We begin with the Euclidean action of equation (8.2.13)

$$\frac{1}{4\pi} \int dz d\bar{z}\, \partial\varphi \bar{\partial}\varphi, \tag{9.1.1}$$

where we use the $z, \bar{z}$ coordinates which parameterise the Riemann sphere $\hat{C}$ and which were discussed in section 8.1. We found there that this action is conformally invariant, that is, invariant under $\delta\varphi = \varepsilon\partial\varphi + \bar{\varepsilon}\bar{\partial}\varphi$, and possesses an energy-momentum tensor whose non-vanishing components were found in equation (8.2.12) to be

$$T = - : \tfrac{1}{2}\partial\varphi\partial\varphi : , \quad \bar{T} = - : \tfrac{1}{2}\bar{\partial}\varphi\bar{\partial}\varphi : . \tag{9.1.2}$$

210

As we will be considering the quantum theory we have added the required normal ordering
to the classical operator that emerges from this calculation. The propagator is defined to be

$$G(z, \bar{z}, w, \bar{w}) \equiv \langle \varphi(z, \bar{z})\varphi(w, \bar{w}) \rangle \equiv \langle 0|R\varphi(z, \bar{z})\varphi(w, \bar{w})|0 \rangle$$

$$= \int D\varphi e^{-S[\varphi]} \varphi(z, \bar{z})\varphi(w, \bar{w}), \tag{9.1.3}$$

where in the last expression we have given the path integral expression for the propagator
and in the next to last expression, the symbol R denotes that propagators, like all Green's
functions, contain radial ordered operators. As we will again discuss later in this section,
this is the Riemann sphere equivalent of time ordering on the cylinder. The propagator
obeys the equation

$$\partial\bar{\partial}G(z, \bar{z}, w, \bar{w}) = -2\pi\delta^2(z - w) \equiv -2\pi\delta(z - w)\delta(\bar{z} - \bar{w}). \tag{9.1.4}$$

The propagator can be deduced from the path integral by introducing sources in the usual
way. However, one of the simplest ways to derive equation (9.1.4) is to use the fact that the
total derivative in the integrand of the path integral vanishes and so

$$\int D\varphi \frac{\delta}{\delta\varphi(z, \bar{z})}\{\exp(-S[\varphi])\varphi(w_1, \bar{w}_1) \cdots \varphi(w_n, \bar{w}_n)\} = 0.$$

Taking only one field $\varphi(w, \bar{w})$ in the integrand we find that a δ function arises from
the functional differentiation of $\varphi(w, \bar{w})$ in the above equation, while the left-hand side
arises from the differentiation of the action. Were we to carry out explicitly the space-time
differentiation of the two-point Green's function we must differentiate the time ordering,
or in our case the radial ordering, step function and evaluate the resulting commutator of
the fields that arises. The solution to equation (9.1.4) is given by

$$G(z, \bar{z}, w, \bar{w}) = -\ln|z - w|^2 = -[\ln(z - w) + \ln(\bar{z} - \bar{w})]. \tag{9.1.5}$$

It satisfies equation (9.1.4) due to the identity

$$\bar{\partial}_z \frac{1}{(z - \zeta)} = 2\pi\delta^2(z - \zeta), \tag{9.1.6}$$

which we verify in the usual way for a δ function by considering its integral with a test
function

$$\frac{1}{2\pi i} \int_D dz d\bar{z} f(z) \left[\bar{\partial} \frac{1}{(z - \zeta)} \right] = \oint_{\partial D} \frac{dz}{2\pi i} f(z) \frac{1}{(z - \zeta)} = f(\zeta), \tag{9.1.7}$$

where D is some domain containing the point ζ and ∂D is its boundary. The δ function
is to be understood as acting with functions which are only analytic, or anti-analytic, in the
domain D, but not with functions that have singularities or depend on both z and $\bar{z}$.

It will be useful to recall Wick's theorem (for a review, see [9.1]), which expresses the
time ordered product of a set of operators in terms of their normal ordered product and the
two-point Green's function. This will allow us to find the operator product expansions of
composite operators of the free scalar field. For any free field $\phi(x)$, Wick's theorem states

that

$$
\begin{aligned}
T\{\phi(x_1)\cdots\phi(x_N)\} = \ &:\phi(x_1)\cdots\phi(x_N): \\
&+\sum_{k<l} :\phi(x_1)\cdots\hat{\phi}(x_k)\cdots\hat{\phi}(x_l)\cdots\phi(x_N): \\
&\times \langle 0|T\phi(x_k)\phi(x_l)|0\rangle + \cdots \\
&+\sum_{k_1<k_2<\ldots<k_{2p}} :\phi(x_1)\cdots\hat{\phi}(x_{k_1})\cdots\hat{\phi}(x_{k_{2p}})\cdots\phi(x_N): \\
&\times \left\{\sum_p \langle 0|T\phi(x_{k_{p_1}})\phi(x_{k_{p_2}})|0\rangle \cdots \langle 0|T\phi(x_{k_{2p-1}})\phi(x_{k_{2p}})|0\rangle\right\} \\
&+\cdots,
\end{aligned}
\tag{9.1.8}
$$

where a caret above a term means it is to be omitted and $\sum_p$ is over all distinct permutations.

We recall that to normal order an operator we express it in terms of annihilation and creation operators and then just place the annihilation operators to the right of the creation operators. The simplest example of such a result is

$$
T\phi(x_1)\phi(x_2) = :\phi(x_1)\phi(x_2): + \langle 0|T\phi(x_1)\phi(x_2)|0\rangle. \tag{9.1.9}
$$

One can apply the formula to expressions which themselves contain normal ordered pieces such as

$$
T\left[\phi(x_1)\cdots :\phi(x_p)\cdots\phi(x_q): \cdots\phi(x_r)\cdots\right], \tag{9.1.10}
$$

the only modification being that one should not include two-point Green's functions of fields which were originally in the same normal ordered piece. In the above we gave Wick's theorem using the more familiar time ordering, but, as we noted, as we are working on the Riemann sphere, time ordering is in fact radial ordering.

Given any two operators $A(z)$, $B(w)$ which are polynomial in φ and its derivatives we may use Wick's theorem to express their radially ordered product in terms of normal ordered pieces and the two-point functions. The normal ordered pieces contain no singularities as $z \to w$, as can be seen by taking their expectation values with any states of finite occupation number. Clearly, in such an expectation value one is left with finite polynomials of z, w and their inverses which are not singular away from zero and infinity. This remark will become obvious when we consider some of the examples below. However, as in equation (9.1.18), singularities do arise from the terms containing the two-point functions. The operator product of $A(z)$ with $B(w)$ can therefore be derived by applying Wick's theorem. As an example, let us consider the operator product of $\varphi(z)$ with $\varphi(w)$. In equation (9.1.9) we may substitute equation (9.1.5) to find

$$
\varphi(z,\bar{z})\varphi(w,\bar{w}) = -\ln(z-w) - \ln(\bar{z}-\bar{w}) + :\varphi(z,\bar{z})\varphi(w,\bar{w}): . \tag{9.1.11}
$$

As a result we find their operator product expansion to be

$$
\varphi(z,\bar{z})\varphi(w,\bar{w}) = -\ln(z-w) - \ln(\bar{z}-\bar{w}) + \cdots, \tag{9.1.12}
$$

where $+\cdots$ stands for non-singular terms as $z \to w$. In equation (9.1.12) the radial ordering is automatically understood to be present. This is in accord with our discussion of section 8.5, where we noted that operator product expansions are only to be used in Green's functions, which are by definition always radially ordered. As a result, operator product expansions are automatically radially ordered and, although we do not explicitly relocate their radial ordering, it is understood to be present. As another example, we consider the operator product expansion

$$T(z)\varphi(w,\bar{w}) = -\tfrac{1}{2} :\partial\varphi(z)\partial\varphi(z): \varphi(w,\bar{w})$$

$$= -\langle\partial\varphi(z)\varphi(w,\bar{w})\rangle\partial\varphi(z) - \tfrac{1}{2} :\partial\varphi(z)\partial\varphi(z)\varphi(w,\bar{w}):$$

$$= \frac{\partial\varphi(z)}{(z-w)} + \cdots. \tag{9.1.13}$$

This equation, together with the analogous result for $\bar{T}$, tells us that φ transforms as a conformal field of weight $(0,0)$. This is consistent with its classical transformation law given at the beginning of this section.

The operator φ obeys the equation of motion $\partial\bar{\partial}\varphi = 0$ even in Green's functions if we neglect δ function terms which occur at the positions of other operators such as those in equation (9.1.4). Hence, it can be written as the sum of a holomorphic and an anti-holomorphic piece:

$$\varphi(z,\bar{z}) = \varphi_L(z) + \varphi_R(\bar{z}), \tag{9.1.14}$$

where

$$\varphi_L(z) = q_L - ip_L \ln z - i \sum_{\substack{n=-\infty\\n\neq 0}}^{\infty} \frac{\alpha_{-n}}{n} z^n \tag{9.1.15}$$

and

$$\varphi_R(\bar{z}) = q_R - ip_R \ln\bar{z} - i \sum_{\substack{n=-\infty\\n\neq 0}}^{\infty} \frac{\bar{\alpha}_{-n}}{n} \bar{z}^n. \tag{9.1.16}$$

We observe that $q_L - q_R$ does not occur in φ and demanding invariance of φ under $z \to ze^{2\pi i}$ enforces the condition $p_L - p_R = 0$. In fact, a more general possibility can occur and will be discussed in chapter 10 on string compactifications, but for the time being we take $p = p_L = p_R$ and $q = q_L/2 = q_R/2$. We note that

$$\partial\varphi = \partial\varphi_L = -i \sum_{n=-\infty}^{\infty} \alpha_{-n} z^{n-1}, \tag{9.1.17}$$

$$\bar{\partial}\varphi = \bar{\partial}\varphi_R = -i \sum_{n=-\infty}^{\infty} \alpha_{-n} \bar{z}^{n-1}, \tag{9.1.18}$$

where $\alpha_0 = \bar{\alpha}_0 = p$. It is straightforward to show by computing their operator product expansions with T and $\bar{T}$ that $\partial\varphi$ and $\bar{\partial}\varphi$ have conformal weights $(1,0)$ and $(0,1)$, respectively. We note, however, that although φ_L transforms as a field of weight $(0,0)$, unlike $\partial\varphi$, it is not holomorphic due to its $\ln z$ term, which is not single valued. Due to this feature one does not usually regard φ_L, or φ_R, as a primary field.

From the Green's function of equation (9.1.5) we conclude that

$$\langle \varphi_L(z)\varphi_L(w)\rangle = -\ln(z-w), \quad \langle \varphi_R(\bar{z})\varphi_R(\bar{w})\rangle = -\ln(\bar{z}-\bar{w}),$$

$$\langle \varphi_L(z)\varphi_R(\bar{w})\rangle = 0. \tag{9.1.19}$$

By differentiating equation (9.1.11) with respect to z and w we find that

$$\partial\varphi(z)\partial\varphi(w) = -\frac{1}{(z-w)^2} + \sum_{n=0}^{\infty}\frac{(z-w)^n}{n!} :\partial^{n+1}\varphi(w)\partial\varphi(w), \tag{9.1.20}$$

where we have Taylor expanded $\partial\varphi(z)$ about w. We then read off the operator product expansion:

$$\partial\varphi(z)\partial\varphi(w) = -\frac{1}{(z-w)^2} + \cdots. \tag{9.1.21}$$

Similarly, we find that

$$\bar{\partial}\varphi(\bar{z})\bar{\partial}\varphi(\bar{w}) = -\frac{1}{(\bar{z}-\bar{w})^2} + \cdots, \quad \partial\varphi(z)\bar{\partial}\varphi(\bar{w}) = 0 + \cdots. \tag{9.1.22}$$

As the reader will have noticed we can also readily calculate the regular terms that occur in these operator product expansions.

We note from (9.1.15) that

$$\alpha_n = \oint_0 \frac{dz}{2\pi i}\, i\partial\varphi(z)z^n \tag{9.1.23}$$

and using equation (8.6.4) we find that

$$[\alpha_n, \alpha_m] = -\oint_0 \frac{d\zeta}{2\pi i}\zeta^m \oint_\zeta \frac{dz}{2\pi i}\, z^n \overline{\partial\varphi(z)\partial\varphi(\zeta)}, \tag{9.1.24}$$

which, upon using equation (9.1.21), becomes

$$[\alpha_n, \alpha_m] = \oint_0 \frac{d\zeta}{2\pi i}\zeta^m \oint_\zeta \frac{dz}{2\pi i}\, z^n \frac{1}{(z-\zeta)^2} = n\oint_0 \frac{d\zeta}{2\pi i}\zeta^{m+n-1} = n\delta_{n+m,0}. \tag{9.1.25}$$

Similarly, one finds

$$[\bar{\alpha}_n, \bar{\alpha}_m] = n\delta_{n+m,0}, \quad [\alpha_n, \bar{\alpha}_m] = 0, \quad [q, p] = i. \tag{9.1.26}$$

The reader may still be wondering why we use radial ordering on the Riemann sphere. However, as we will discuss in section 18.2 and we will discuss in more detail later one can map the Riemann sphere onto a cylinder with coordinates ς, $\bar{\varsigma}$ by $\varsigma = \tau + i\sigma$, where $z = e^\varsigma, \bar{z} = e^{\bar{\varsigma}}$. We can think of the cylinder as a two-dimensional space-time whose spatial coordinate σ is on a circle, $-\pi \leq \sigma \leq \pi$, and whose length is the time τ. We will now show that the usual quantisation on the cylinder involving time ordering, when mapped onto the Riemann sphere, leads to precisely the same results as radial quantisation on the latter. We first note that lines of constant time on the cylinder correspond to circles of fixed radius on the Riemann sphere. Hence, time ordering on the cylinder becomes radial ordering on the Riemann sphere.

Since the scalar action of equation (9.1.1) is conformally invariant, it takes the same form in the new coordinates when mapped to the cylinder, namely

$$\frac{1}{4\pi} \int d\varsigma d\bar{\varsigma} \, \partial\varphi\bar{\partial}\varphi = \frac{1}{8\pi i} \int d\tau d\sigma \, (-\partial_\tau\varphi\partial_\tau\varphi + \partial_\sigma\varphi\partial_\sigma\varphi). \tag{9.1.27}$$

In the last expression we have given the Minkowski space action using the Wick rotation discussed in the previous chapter. We now quantise the latter in the standard way, indeed as we did in chapters 2 and 3. The momentum of φ is $-(1/4\pi i)\partial_\tau\varphi$ and so the equal time commutation relation is of the form

$$\left[\varphi(\tau,\sigma), \partial_\tau\varphi(\tau,\sigma')\right] = 4\pi \delta(\sigma - \sigma'). \tag{9.1.28}$$

Due to the invariance of $\dot{\varphi}$ under $\varsigma \to \varsigma + 2\pi i$, or $\sigma \to \sigma + 2\pi$, we can write φ as

$$\varphi(\tau,\sigma) = q(\tau) + i \sum_{\substack{n=-\infty \\ n\neq 0}}^{\infty} \frac{a_n(\tau)}{n} e^{-in\sigma}. \tag{9.1.29}$$

The equation of motion implies $\ddot{a}_n(\tau) - n^2 a_n(\tau) = 0$ and $\ddot{q}(\tau) = 0$. The solution is given by $a_n(\tau) = \alpha_n e^{-in\tau} + \bar{\alpha}_n e^{-in\tau}$ and $q(\tau) = q + 2p\tau$. Substituting these into $\varphi(\tau,\sigma)$ we find that

$$\varphi(\tau,\sigma) = q + 2p\tau + i \sum_{\substack{n=-\infty \\ n\neq 0}}^{\infty} \left(\frac{\alpha_n}{n} e^{-in(\tau+\sigma)} + \frac{\bar{\alpha}_{-n}}{n} e^{-in(\tau-\sigma)} \right), \tag{9.1.30}$$

where q and p are independent of time. We now make the rotation to the Euclidean space corresponding to the cylinder by identifying $\varsigma = i(\tau + \sigma)$, $\bar{\varsigma} = i(\tau - \sigma)$ as explained in section 8.1.2. Since φ transforms as a conformal field of weight $(0,0)$ we can, using equation (8.2.29), carry out the mapping from the cylinder to the Riemann sphere $\hat{C}$ by $z = e^\varsigma$, $\bar{z} = e^{\bar{\varsigma}}$ to recover the form of φ of equation (9.1.12). Substituting equation (9.1.30) into equation (9.1.28) we also recover the commutators of equation (9.1.28). Consequently, for this example at least, we find radial quantisation and equal time canonical quantisation on the cylinder yield the same result.

Let us return to working on the Riemann sphere. Using equation (8.8.4) and taking into account that $\partial\varphi$ has conformal weight $(1, 0)$ we take the vacuum to satisfy

$$\alpha_n|0\rangle = 0, \quad n \geq 0 \tag{9.1.31}$$

and we correspondingly adopt the normal ordering prescription

$$:\alpha_n\alpha_m: = \begin{cases} \alpha_n\alpha_m & \text{if } m > n, \\ \alpha_m\alpha_n & \text{if } n > m \end{cases} \tag{9.1.32}$$

for all $m, n \in \mathbf{Z}$, with completely analogous formulae for $\bar{\alpha}_n$. Clearly, this reordering only matters if $m + n = 0$. Less obvious is that we also take factors of the coordinates q to the left of the momentum p. We will comment on this further below.

It is instructive to calculate the operator product expansion from first principles by explicitly carrying out the normal ordering rather than using Wick's theorem. As an example, let us consider the naive product of $i\partial\varphi(z)$ and $i\partial\varphi(w)$, which is given by

$$i\partial\varphi(z)i\partial\varphi(w) = \sum_{n=-\infty}^{\infty} \alpha_{-n}z^n \sum_{m=-\infty}^{\infty} \alpha_{-m}w^m .$$

This differs from the corresponding normal ordered expression, which is finite as $z \to w$, by the term

$$i\partial\varphi(z)i\partial\varphi(w) - :i\partial\varphi(z)i\partial\varphi(w):$$

$$= \sum_{n=1}^{\infty} \alpha_n z^{-n-1} \sum_{m=1}^{\infty} \alpha_{-m}w^{m-1} - \sum_{m=1}^{\infty} \alpha_{-m}w^{m-1} \sum_{n=1}^{\infty} \alpha_n z^{-n-1}$$

$$= \sum_{n=1}^{\infty} \frac{n}{zw} \left(\frac{w}{z}\right)^n = \frac{1}{(z-w)^2}. \tag{9.1.33}$$

The last step is only valid if $|w| > |z|$ since the series only converges in this domain. The restriction $|w| > |z|$ corresponds to the radial ordering used inside a correlation function. Consequently, we recover the operator product expansion of equation (9.1.20).

Let us compute the operator product expansion of the energy-momentum tensor with itself using Wick's theorem:

$$T(z)T(w) = \frac{1}{4}:(\partial\varphi)^2(z)::(\partial\varphi)^2(w):$$

$$= :\partial\varphi(z)\partial\varphi(w): \left[-\frac{1}{(z-w)^2}\right] + \frac{1}{2}\left(-\frac{1}{(z-w)^2}\right)^2$$

$$= \frac{1}{2}\frac{1}{(z-w)^4} + \frac{2T(w)}{(z-w)^2} + \frac{\partial T(w)}{z-w} - \sum_{n=2}^{\infty} \frac{(z-w)^{n-2}}{(n)!}\partial^{n+1}\varphi(w)\partial\varphi(w)$$

$$= \frac{1}{2}\frac{1}{(z-w)^4} + \frac{2T(w)}{(z-w)^2} + \frac{\partial T(w)}{z-w} + \cdots. \tag{9.1.34}$$

From the first three terms we recognise equation (8.5.8) and conclude that a free boson has central charge 1 in its left sector and, taking the operator product with $\bar{T}$ central charge, also 1 in its right sector. We may write both central charges as $(c, \bar{c}) = (1, 1)$. Using equation (8.6.6), the moments of the energy momentum tensor are given in terms of oscillators by

$$L_n = \frac{1}{2} : \sum_m \alpha_{n-m}\alpha_m : .$$

We may reverse the above discussion and use the operator product of two operators to define their normal ordered product. Using Wick's theorem of equation (9.1.8) we observed that the radially ordered product is expressed in terms of the normal ordered and contracted terms. The latter are just two-point Green's functions which lead to the singular terms in the operator product of the two operators. Hence, given two operators $A(z)B(w)$ we can define their normal ordered product by subtracting all singular terms in their operator product and then taking the limit $z \to w$. For example, for the two-point Green's function we have

$$: \partial\varphi(z)\partial\varphi(w): = \lim_{z \to w} \left\{ \partial\varphi(z)\partial\varphi(w) - \left(-\frac{1}{(z-w)^2}\right) \right\}. \tag{9.1.35}$$

Clearly, the definition of normal ordering implies certain operator product expansions and vice versa.

We will also make extensive use of operator product expansions involving the operator

$$:e^{ik\varphi_L(z)}: = e^{ik\varphi_{L<}(z)} e^{ik\cdot q} e^{k \cdot p \ln z} e^{ik\varphi_{L>}(z)}, \tag{9.1.36}$$

where $\varphi_L(z) = \varphi_{L<}(z) + \varphi_{L0}(z) + \varphi_{L>}(z)$ and

$$\varphi_{L<}(z) = -i \sum_{n=1}^{\infty} \frac{\alpha_{-n}}{n} z^n, \quad \varphi_{L0<}(z) = q - ip\ln z, \quad \varphi_{L>}(z) = i \sum_{n=1}^{\infty} \frac{\alpha_n}{n} z^{-n}. \tag{9.1.37}$$

We see that $\varphi_{L<}(z)$ is just that defined in equation (9.1.15) but we take the zero modes to obey $[q, p] = i$. Equation (9.1.35) is consistent with the notion of normal ordering given earlier, namely it places destruction operators to the right of creation operators; however, it also places factors of q to the left of factors of p. It turns out that this is the normal ordering implied by the radially ordered operator product expansions. We encourage the reader to deduce the operator product expansion of $\varphi_L(z)\varphi_L(w)$ and verify that the result of this normal ordering agrees with the operator product expansion expected consistent with (9.1.19). One finds that

$$\partial\varphi_L(z):e^{ik\varphi_L(w)}: = \partial\varphi_L(z) \sum_{n=0}^{\infty} :\frac{[ik\varphi_L(w)]^n}{n}: = -\frac{ik}{z-w} \sum_{n=1}^{\infty} :\frac{[ik\varphi_L(w)]^{n-1}}{(n-1)}: + \cdots$$

$$= -\frac{i}{z-w} e^{ik\varphi_L(w)} + \cdots, \tag{9.1.38}$$

while a similar calculation involving a single and a double contraction gives

$$T(z):e^{ik\varphi_L(w)}: = :[-\partial\varphi_L(z)]\left(\frac{-ik}{z-w}\right) e^{ik\varphi_L(w)}: - \frac{1}{2}\left[\frac{-ik}{z-w}\right]^2 :e^{ik\varphi_L(w)}:$$

$$= \frac{k^2/2}{(z-w)^2} e^{ik\varphi_L(w)} + \frac{1}{z-w} \partial e^{ik\varphi_L(w)} + \cdots, \tag{9.1.39}$$

where we take $T = -\frac{1}{2}\partial\varphi_L(z)\partial\varphi_L(z)$. The last calculation implies that $e^{ik\varphi_L(w)}$ is a primary field of conformal weight $k^2/2$.

We will also make extensive use of the relation

$$:e^{ik_1\varphi_L(z)}::e^{ik_2\varphi_L(w)}: = (z-w)^{k_1 k_2} :e^{ik_1\varphi_L(z)+ik_2\varphi_L(w)}:, \tag{9.1.40}$$

which can be proved using Wick's theorem or, more simply, by carrying out the normal ordering by hand using the relation

$$e^A e^B = e^B e^A e^{[A,B]}, \tag{9.1.41}$$

which is valid for any A, B for which $[A, B]$ is a constant. For our case, we take $A = ik_1\varphi_{L>}(z)$, $B = ik_2\varphi_{L<}(w)$, for which $[A, B] = k_1 k_2 \ln(1 - (w/z))$ and we use the same formula to move all ps to the right of all qs.

A generalisation of this result, which we can prove by induction using a straightforward extension of the above argument, is

$$\prod_{i=1}^{n} :e^{ik_i\varphi_L(z_i)}: = \prod_{\substack{i,j=1 \\ i<j}}^{n} (z_i - z_j)^{k_ik_j} : \exp\left[\sum_{i-1}^{n} ik_i\varphi_L(z_i)\right] :. \tag{9.1.42}$$

We now consider a scalar ϕ which is slightly different to the one above [9.2, 9.3]. It has the same operator product expansion, that is,

$$\phi(z, \bar{z})\phi(w, \bar{w}) = -\ln(z - w) - \ln(\bar{z} - \bar{w}) + \cdots. \tag{9.1.43}$$

However, its energy-momentum tensor is different and is given by

$$T = -\tfrac{1}{2}\partial\phi\partial\phi - Q\partial^2\phi, \quad \bar{T} = -\tfrac{1}{2}\bar{\partial}\phi\bar{\partial}\phi - Q\bar{\partial}^2\phi, \tag{9.1.44}$$

where Q is a constant called the background charge. The operator product expansion and energy-momentum tensor given above are sufficient to specify the theory.

Calculating the operator product expansion of $T(z)T(w)$ we find that it is of the form of equation (8.4.6), but the central charge is $c^{\phi} = 1 + 12Q^2$. We also find that

$$T(z)\phi(w) = -\frac{2Q}{(z - w)^3} + \frac{\partial\phi(w)}{(z - w)^2} + \frac{\partial^2\phi(w)}{(z - w)} + \cdots. \tag{9.1.45}$$

Hence, ϕ does not transform under a conformal transformation in quite the same way as φ and it is not a primary field.

Despite this latter result one finds, by computing its operator product expansion with $T(w)$, that $e^{i\beta\phi(z)}$ is a primary field of weight $\tfrac{1}{2}\beta(\beta - 2iQ)$. As the operator product expansion of ϕ is the same as that for φ, equations (9.1.40) and (9.1.42) still hold with φ replaced by ϕ.

We define the oscillators by $\partial\phi = -i\sum_n \alpha_n z^{-n-1}$. The Virasoro operators L_n are given by equation (8.6.6) and using equation (9.1.44) we find that they are given by

$$L_n = \tfrac{1}{2}\sum_m :\alpha_{m-n}\alpha_n: - i(n + 1)Q\alpha_n. \tag{9.1.46}$$

In fact the extra terms in the energy-momentum tensor can be accounted for by taking ϕ to have an action which consists of the usual free action of equation (9.1.1) and an additional term, namely

$$\frac{1}{4\pi}\int dz d\bar{z} \sqrt{g}(\partial\phi\bar{\partial}\phi - 2QR\phi), \tag{9.1.47}$$

where R is the two-dimensional Riemann scalar curvature and g is the determinant of the two-dimensional metric. Expanding in $h_{\alpha\beta} \equiv g_{\alpha\beta} - \eta_{\alpha\beta}$, the coefficient of the term linear in $h_{\alpha\beta}$ is the energy-momentum tensor, which is of the form of equation (9.1.44). One can show that the presence of this extra term in the action implies that momenta in any correlator must sum to $2iQ$ rather than zero. This can be seen by taking $\phi \to \phi + c$, where c is a constant and using the corresponding Ward identity.

9.1.2   The free fermion

As explained in chapter 5 a generic two-dimensional fermion has two components and so a Majorana fermion has two real components which we denote by ψ and $\bar{\psi}$. The Euclidean action derived from the usual Minkowski space-time action for such a free fermion takes the form

$$A = -\frac{1}{4\pi} \int dz d\bar{z} \{ \psi \bar{\partial} \psi + \bar{\psi} \partial \bar{\psi} \}. \tag{9.1.48}$$

For a detailed derivation of this result see chapter 22 of [1.11]. The fields ψ and $\bar{\psi}$ have conformal weights $(\frac{1}{2}, 0)$ and $(0, \frac{1}{2})$, respectively, and so transform as

$$\delta \psi = \epsilon \partial \psi + \tfrac{1}{2} \partial \epsilon \psi + \bar{\epsilon} \bar{\partial} \psi, \ \ \delta \bar{\psi} = \bar{\epsilon} \bar{\partial} \bar{\psi} + \tfrac{1}{2} \bar{\partial} \bar{\epsilon} \bar{\psi} + \epsilon \partial \bar{\psi}. \tag{9.1.49}$$

These conformal transformations leave the above action invariant.

The equations of motion are

$$\bar{\partial} \psi = 0, \ \ \partial \bar{\psi} = 0, \tag{9.1.50}$$

implying that ψ and $\bar{\psi}$ are a mereomorphic functions of z and $\bar{z}$, respectively. Taking ϵ and $\bar{\epsilon}$ to depend on both z and $\bar{z}$ we can read off the energy-momentum tensor from the variation of the action. The non-vanishing components are

$$T \equiv T_{zz} = \tfrac{1}{2} : \psi \partial \psi :, \ \ \bar{T} \equiv \bar{T}_{\bar{z}\bar{z}} = \tfrac{1}{2} : \bar{\psi} \bar{\partial} \bar{\psi} :. \tag{9.1.51}$$

Much of the discussion below follows a similar path to that for the free scalar given in the previous section and so we will give it only in outline. The action of equation (9.1.48) implies that the propagator $\langle \psi(z) \psi(w) \rangle$ satisfies the equation

$$\bar{\partial} \langle \psi(z) \psi(w) \rangle = -2\pi \delta^2(z - w). \tag{9.1.52}$$

Using equation (9.1.6) we conclude that it is given by

$$\langle \psi(z) \psi(w) \rangle = -\frac{1}{(z - w)}. \tag{9.1.53}$$

Similarly, we find that

$$\langle \bar{\psi}(\bar{z}) \bar{\psi}(\bar{w}) \rangle = -\frac{1}{(\bar{z} - \bar{w})}. \tag{9.1.54}$$

A quantum field theory which contains fields that are Grassmann odd as well as Grassmann even also obeys a version of Wick's theorem which takes the form of equation (9.1.8) provided one introduces additional minus signs that are just those required to reorder the Grassmann odd fields from their positions on the left-hand side to positions on the right-hand side of the equation using the naive reordering rules for Grassmann odd and even objects. We therefore find that

$$\psi(z) \psi(w) = \langle \psi(z) \psi(w) \rangle + :\psi(z) \psi(w): = -\frac{1}{(z - w)} + \cdots, \tag{9.1.55}$$

where radial ordering is as always implied. From this last equation we can read off the operator product expansion as the normal ordered expression is finite in the limit of $z \to w$. Similarly, we find that

$$\bar{\psi}(\bar{z})\bar{\psi}(\bar{w}) = -\frac{1}{(\bar{z} - \bar{w})} + \cdots, \quad \bar{\psi}(\bar{z})\psi(w) = 0 + \cdots. \tag{9.1.56}$$

Using the above operator product expansions and the expressions for T and $\bar{T}$ of equation (9.1.49), one can verify that $T(z)\psi(w)$ has an operator product expansion of the form of equation (8.2.31) and so we conclude that ψ is a primary field of conformal weight $(\frac{1}{2}, 0)$. We also find that $\bar{\psi}$ is a primary field of conformal weight $(0, \frac{1}{2})$.

Corresponding to the equation of motion of equation (9.1.50) we can write the fermion fields as

$$\psi = -i \sum_{r \in \mathbb{Z}+\frac{1}{2}} b_r z^{-r-\frac{1}{2}}, \quad \bar{\psi} = -i \sum_{r \in \mathbb{Z}+\frac{1}{2}} \bar{b}_r \bar{z}^{-r-\frac{1}{2}}. \tag{9.1.57}$$

Above we have assumed that the fermion satisfies the boundary condition $\psi(e^{2\pi i}z) = \psi(z)$. However, we could also have chosen $\psi(e^{2\pi i}z) = -\psi(z)$ in which case we would have had an integer mode expansion. We will see in section 9.3 that these correspond to the NS and R sectors of the superstring, respectively. The reader should note that the map between the cylinder and the Riemann sphere inverts these two boundary conditions, that is, takes periodic to anti-periodic and vice vera.

The operator product expansions of equations (9.1.55) and (9.1.56), together with equation (8.6.4), imply that the oscillators b_r and $\bar{b}_r$ obey the relations

$$\{b_r, b_s\} = \delta_{r+s}, \quad \{\bar{b}_r, \bar{b}_s\} = \delta_{r+s}, \quad \{b_r, \bar{b}_s\} = 0. \tag{9.1.58}$$

Although equation (8.6.4) was deduced for Grassmann even quantities, the reader can verify that it also holds for Grassmann odd quantities. It is important in this derivation to use the appropriate radial ordering for Grassmann odd quantities, that is, to introduce the appropriate minus signs when changing the order of operators. The Hermiticity properties of the oscillators are $(b_r)^\dagger = b_{-r}$ and $(\bar{b}_r)^\dagger = \bar{b}_{-r}$.

The vacuum is defined to obey the equations

$$b_r|0\rangle = 0, \quad \bar{b}_r|0\rangle = 0, r > 0. \tag{9.1.59}$$

Normal ordering is defined with respect to this vacuum in the sense that destruction operators are put to the right of creation operators taking into account the signs required for their Grassmann character. Hence we adopt the rule

$$:b_r b_s: = \begin{cases} b_r b_s & \text{if } s > r, \\ -b_s b_r & \text{if } r > s. \end{cases} \tag{9.1.60}$$

In terms of oscillators the moments of the energy-momentum tensor are given by

$$L_n \equiv \oint \frac{dz}{2\pi i} z^{n+1} T(z) = : \sum_r \left(r + \frac{n}{2}\right) b_r b_{n-r} : \tag{9.1.61}$$

In deriving this result we have used the identity $\sum_r : b_r b_{n-r} := 0$ as can be proved by taking $r \to n - r$ in the sum and interchanging the order within the normal ordering. The second term in the bracket does not contribute to L_n, but it is often written in this way.

A remarkable peculiarity of two dimensions is that is possible to express fermions in terms of bosons and vice versa. A free scalar with no background charge has central charge $(1,1)$, while a single Majorana fermion consists of the real components ψ and $\bar{\psi}$, each of which has conformal weights $(\frac{1}{2},0)$ and $(0,\frac{1}{2})$ and which are left and right moving, respectively. A minimal requirement to express two sets of fields in terms of each other is that their energy-momentum tensors have the same left and right central charges. Hence we might hope to express one real scalar φ in terms of two free Majorana fermions ψ_1, ψ_2, or equivalently one complex fermion $\Psi^{\pm} = \frac{1}{\sqrt{2}}(\psi_1 \pm i\psi_2)$ and vice versa.

The left moving part of the action for the two Majorana fermions becomes

$$A = -\frac{1}{4\pi}\int dz d\bar{z}\{\psi_1 \bar{\partial}\psi_1 + \psi_2 \bar{\partial}\psi_2\} = -\frac{1}{4\pi}\int dz d\bar{z}\{\Psi^* \bar{\partial}\Psi + \Psi \bar{\partial}\Psi^*\}, \quad (9.1.62)$$

while the energy-momentum tensor T is given by

$$T = \tfrac{1}{2}: \psi_1 \partial \psi_1: + :\tfrac{1}{2}\psi_2 \partial \psi_2: = \tfrac{1}{2}: \Psi^- \partial \Psi^+: + \tfrac{1}{2}: \Psi^+ \partial \Psi^-: . \quad (9.1.63)$$

In fact, this bosonisation is possible and the relation between the fields is

$$\Psi^+(z) = c_+ : e^{i\varphi_L(z)}:, \qquad \Psi^-(z) = c_- : e^{-i\varphi_L(z)}:, \quad (9.1.64)$$

where $\Psi^{\pm}$ are the left moving components of the fermion with related formulae for the right moving components, which are given in terms of φ_R. We observe using equation (9.1.40) that they have the operator product expansion

$$\Psi(z)^+\Psi^+(w) = (c_+)^2 : e^{i\varphi_L(z)} :: e^{i\varphi_L(w)} := (z-w)e^{i(\varphi_L(z)+\varphi_L(w))} = 0 + \cdots,$$

$$\Psi(z)^-\Psi(w)^- = 0 + \cdots, \quad (9.1.65)$$

$$\Psi(z)^+\Psi^-(w) = c_+c_- : e^{i\varphi_L(z)} :: e^{-i\varphi_L(w)} := \frac{c_+c_-}{z-w}e^{i(\varphi_L(z)-\varphi_L(w))} = \frac{c_+c_-}{z-w} + \cdots.$$

To agree with the operators product for the fermion fields of equation (9.1.53) we require $c_+c_- = -1$. We also observe, using equation (9.1.39), that, since $T = -\frac{1}{2}(\partial\varphi_L)^2$, the $\Psi(z)^{\pm}$ given in equation (9.1.64) both have conformal weight $(\frac{1}{2},0)$ as required for a fermion. Since the Green's functions of $\Psi(z)^{\pm}$ can be evaluated using their operator product expansions, which by this relation are the same as for φ_L, it follows that the two fields also give identical correlation functions.

We can also verify that the energy-momentum tensor of equation (9.1.63) is related that for a free scalar if the fields are related by equation (9.1.64). This best done by computing the required normal ordering by subtracting the singular pieces of the operator product expansion of the two operators, as explained above. As a warm up exercise let us consider the fermion current $:\Psi^+\Psi^-:$. Expressed in terms of φ_L, these two operators have the operator product expansion of equation (9.1.65), which we may write as

$$\Psi^+(z)\Psi^-(w) = -\frac{1}{z-w} + i\partial\varphi_L(w) + O((z-w)^1). \quad (9.1.66)$$

Subtracting the singular term we conclude that

$$:\Psi^+\Psi^-: = i\partial\varphi_L. \quad (9.1.67)$$

Differentiating equation (9.1.65) with respect to w and subtracting the singular terms we conclude that

$$:\Psi^+(z)\partial\Psi^-(z): = -c_+c_- \left(:\frac{(\partial\varphi_L)^2}{2}: + \frac{i}{2}\partial^2\varphi_L\right). \tag{9.1.68}$$

Adding the corresponding term for $:\Psi^-(z)\partial\Psi^+(z):$ we indeed find that $T = -\frac{1}{2}(\partial\varphi_L)^2$. We note that all the checks carried out require only $c_+c_- = -1$ and so there is more than one choice of $c_\pm$ possible. The most symmetric choice is $c_+ = c_- = i$.

9.2   First order systems

Let us consider a system which consists of two Grassmann odd fields b and c of conformal weights $(\lambda, 0)$ and $(1 - \lambda, 0)$, respectively. Their conformal transformations are then given by

$$\delta b = \epsilon\partial b + \lambda(\partial\epsilon)b + \bar{\epsilon}\bar{\partial}b,$$
$$\delta c = \epsilon\partial c + (1 - \lambda)(\partial\epsilon)c + \bar{\epsilon}\bar{\partial}c. \tag{9.2.1}$$

Their Euclidean action is given by

$$A^{b,c} = \frac{1}{2\pi}\int dz d\bar{z} b\bar{\partial}c. \tag{9.2.2}$$

Taking the above transformation for an ϵ which is an arbitrary function of space-time the variation of the action becomes,

$$\delta A^{b,c} = \frac{1}{2\pi}\int dz d\bar{z}(\bar{\partial}\epsilon)[\lambda b\partial c - (1 - \lambda)\partial bc]. \tag{9.2.3}$$

For a conformal transformation $\bar{\partial}\epsilon = 0$ and so $A^{b,c}$ is conformally invariant. The coefficient in the integrand of $-\bar{\partial}\epsilon/2\pi$ is the $T_{zz} = T$ component of the energy-momentum tensor, which we read off to be

$$T^{b,c} = -:\lambda b\partial c+(1-\lambda)\partial bc: =:(\tfrac{1}{2}(\partial bc - b\partial c) + \tfrac{1}{2}(1 - 2\lambda)\partial(bc)):. \tag{9.2.4}$$

The second of the alternative expressions in the above will prove useful later. Since there is no $\bar{\epsilon}$ in equation (9.2.3) we conclude that $\bar{T}^{b,c} = 0$. The expressions for right moving fields with conformal weights $(0, \lambda)$ and $(0, 1 - \lambda)$ are just those above with obvious replacements.

When radially ordering Grassmann odd quantities we must use a minus sign when we interchange their order, for example,

$$Rc(z)b(w) = -b(w)c(z) \quad if \quad |w| > |z|. \tag{9.2.5}$$

The equations of motion are

$$\bar{\partial}b = 0 = \bar{\partial}c. \tag{9.2.6}$$

From the action of equation (9.2.2) we can deduce, in a similar way to that for the scalar of section 9.1.1, that the propagator obeys the equation $\partial_z \langle b(z)c(w) \rangle = 2\pi \delta^2(z - w)$ and so is given by

$$\langle b(z)c(w) \rangle = \langle 0|Rb(z)c(w)|0 \rangle = \frac{1}{z-w}. \tag{9.2.7}$$

Using the Grassmann odd character of the fields we deduce that

$$\langle c(z)b(w) \rangle = -\langle 0|Rb(w)c(z)|0 \rangle = -\frac{1}{w-z} = \frac{1}{z-w}, \tag{9.2.8}$$

where in the last two equations we have demonstrated the consistency of the propagator with the Grassmann odd character of the fields.

As already mentioned, Wick's theorem also applies to Grassmann odd quantities provided we incorporate the appropriate minus signs when we exchange the order of such operators. We find that

$$Rc(z)b(w) =: c(z)b(w): + \langle c(z)b(w) \rangle = \frac{1}{z-w} +: c(z)b(w): . \tag{9.2.9}$$

Hence, we deduce the operator product expansions

$$c(z)b(w) = \frac{1}{z-w} + \cdots, \quad b(z)c(w) = \frac{1}{z-w} + \cdots,$$

$$c(z)c(w) = 0 + \cdots = b(z)b(w). \tag{9.2.10}$$

As before $+ \cdots$ corresponds to finite corrections. Due to their equation of motion, b and c are mereomorphic functions of z which, taking account of their conformal weights and using equation (8.8.11), have the mode expansions

$$c(z) = \sum_{n=-\infty}^{\infty} c_n z^{-n-(1-\lambda)}, \quad b(z) = \sum_{n=-\infty}^{\infty} b_n z^{-n-\lambda}. \tag{9.2.11}$$

In accord with equation (8.8.4), the vacuum $|0\rangle$ is defined to satisfy

$$b_n|0\rangle = 0 \quad n \geq 1 - \lambda; \quad c_n|0\rangle = 0 \quad n \geq \lambda. \tag{9.2.12}$$

The normal ordering is carried out with respect to this definition of the vacuum. In terms of oscillators the moments of the energy-momentum tensor are given by

$$L_n \equiv \oint \frac{dz}{2\pi i} z^{n+1} T^{b,c}(z) =: \sum_p b_{n-p}c_p(p-(1-\lambda)n): . \tag{9.2.13}$$

One can verify that the vacuum is SL(2, **C**) invariant in that L_0 and $L_{\pm 1}$ annihilate the vacuum. Using the operator product expansion of equation (9.2.10) and equation (8.6.4) we find that

$$\{c_n, b_m\} = \delta_{m+n,0}, \quad \{c_n, c_m\} = 0 = \{b_n, b_m\}. \tag{9.2.14}$$

As for the free scalar field, one can find these results by using the usual time ordering on the cylinder and then mapping back to the Riemann sphere.

Using Wick's theorem we find the operator product expansions

$$T^{b,c}(z)b(w) = \frac{1}{(z-w)^2} \lambda b(w) + \frac{1}{z-w} \partial_w b(w) + \cdots \tag{9.2.15}$$

and

$$T^{b,c}(z)c(w) = \frac{(1-\lambda)c(w)}{(z-w)^2} + \frac{1}{z-w}\partial_w c(w) + \cdots. \tag{9.2.16}$$

We also find that for the energy-momentum tensor

$$T^{b,c}(z)T^{b,c}(w) = \frac{1}{2}\frac{c^{b,c}}{(z-w)^4} + \frac{2T^{b,c}(w)}{(z-w)^2} + \frac{1}{z-w}\partial_w T^{b,c}(w) + \cdots. \tag{9.2.17}$$

We leave the reader to verify the first two terms. The third term comes from contracting all the fields. Although there is only one way to perform these two contractions, there are four terms in which the contractions must be carried out. One of them is given by

$$: -\lambda b(z)\partial_z c(z) :: -\lambda b(w)\partial_w c(w) : . \tag{9.2.18}$$

Carrying out both contractions, we have

$$(-\lambda)^2 \partial_w \langle b(z)c(w)\rangle \partial_z \langle c(z)b(w)\rangle = -\frac{(\lambda)^2}{(z-w)^4}. \tag{9.2.19}$$

A similar analysis on the remaining three terms yields

$$c^{b,c} = 2(-1 + 6\lambda - 6\lambda^2). \tag{9.2.20}$$

Equation (9.2.20) does not apply to the real Grassmann fields with $\lambda = 1 - \lambda = \frac{1}{2}$, which we have already found at the end of the last section to have $c = \frac{1}{2}$. This is because we have assumed above that b and c are distinct fields.

We now consider a very similar first order system whose fields β and γ are Grassmann even, but have the same conformal weights as b and c, that is, $(\lambda, 0)$ and $(1 - \lambda, 0)$, respectively. Their conformal transformations are given by equation (9.2.1) with $b \to \beta$ and $c \to \gamma$ and the Euclidean action is

$$A^{\beta,\gamma} = \frac{1}{2\pi}\int dz d\bar{z}\beta\bar{\partial}\gamma. \tag{9.2.21}$$

The energy-momentum tensor is

$$T^{\beta,\gamma} = :-\lambda\beta\partial\gamma + (1-\lambda)\partial\beta\gamma: . \tag{9.2.22}$$

The calculations are much as for the b, c system. The propagators are

$$\langle\gamma(z)\beta(w)\rangle = \frac{1}{z-w}, \quad \langle\beta(z)\gamma(w)\rangle = -\frac{1}{z-w}. \tag{9.2.23}$$

The minus sign in the latter result compared to the b, c system arises due to their Grassmann even character. In particular, it arises by applying the analogous steps to those in equation (9.2.8), or using the fact that the total derivative of the functional integral vanishes. The operator product expansions are

$$\beta(z)\gamma(w) = -\frac{1}{z-w} + \cdots, \quad \gamma(z)\beta(w) = \frac{1}{z-w} + \cdots,$$

$$\gamma(z)\gamma(w) = 0 + \cdots, \quad \beta(z)\beta(w) = 0 + \cdots. \tag{9.2.24}$$

The equations of motion $\bar{\partial}\beta = 0 = \bar{\partial}\gamma$ imply the mode expansion

$$\beta(z) = \sum_n \beta_n z^{-n-\lambda}, \quad \gamma(z) = \sum_n \gamma_n z^{-n-(1-\lambda)} \tag{9.2.25}$$

with the vacua defined by

$$\beta_n|0\rangle = 0 \quad n \geq 1 - \lambda; \quad \gamma_n|0\rangle = 0 \quad n \geq \lambda. \tag{9.2.26}$$

Using the operator product expansion we find that

$$[\gamma_n, \beta_m] = \delta_{n+m,0} \tag{9.2.27}$$

with other commutators vanishing. The moments of the energy-momentum tensor are given by

$$L_n \equiv \oint \frac{dz}{2\pi i} z^{n+1} T^{b,c}(z) = : \sum_p \beta_{n-p} \gamma_p (p - (1-\lambda)n): . \tag{9.2.28}$$

It is straightforward to verify that the energy-momentum tensor of equation (9.2.22) has the correct operator product expansion with $\beta(z)$ and $\gamma(z)$ corresponding to their conformal weights. Finally, we find the operator product expansion for the energy-momentum tensor

$$T^{\beta,\gamma}(z)T^{\beta,\gamma}(w) = \frac{c^{\beta,\gamma}}{(z-w)^4} + \frac{2T^{\beta,\gamma}(w)}{(z-w)^2} + \frac{1}{z-w}\partial_w T^{\beta,\gamma}(w) + \cdots, \tag{9.2.29}$$

but now the central charge is given by

$$c^{\beta,\gamma} = -2(-1 + 6\lambda - 6\lambda^2). \tag{9.2.30}$$

For the b, c system we can define a classically conserved U(1) current $j(w) = -: b(w)c(w):$. Using equation (8.6.4), we find that its charge $j_0 = \oint (dw/2\pi i)j(w)$ induces the transformations $\delta b = -\alpha b$ and $\delta c = \alpha c$ which are a symmetry of the classical action. We find that

$$j(z)j(w) = \frac{1}{(z-w)^2} + \cdots, \tag{9.2.31}$$

while

$$T^{b,c}(z)j(w) = \frac{(1-2\lambda)\epsilon}{(z-w)^3} + \frac{j(w)}{(z-w)^2} + \frac{\partial_w j(w)}{z-w} + \cdots. \tag{9.2.32}$$

In this equation we now that $\epsilon = +1$. However, the presence of the first term implies that $j(w)$ is not a primary field and it also implies that the U(1) current $j(w)$ is anomalous. In terms of oscillators $j_n = \oint (dz/2\pi i)j(z)z^{-n} = \sum_n \epsilon: c_{n-m}b_m:$ and so j_0 counts the number of c minus b oscillators and so is the ghost number operator; again we have taken $\epsilon = +1$.

The β, γ system also has a symmetry but with current $j = -:\beta(w)\gamma(w):$. All the formulae in the previous paragraph hold with $b \to \beta, c \to \gamma$ provided that we now take $\epsilon = -1$.

We can bosonise the b, c system in terms of a scalar, ϕ, with a background charge. The b, c system has a central charge $c^{b,c} = 1 - 3(2\lambda - 1)^2$ and hence we require a scalar field, σ, with a background charge $\frac{1}{2}i(2\lambda - 1)$. However, rather than work with an imaginary background charge we can take our field to be purely imaginary, $\sigma = i\phi$. We take

$$b = :e^{-\sigma}:, \quad c = :e^{\sigma}:, \tag{9.2.33}$$

and our energy-momentum tensor becomes

$$T = : \frac{1}{2}(\partial\sigma)^2 : + \frac{(2\lambda - 1)}{2}\partial^2\sigma. \tag{9.2.34}$$

We note that the first term has the opposite sign to that which is normal for a scalar field. As a consequence, its kinetic energy term also has the opposite sign and the propagator has the opposite sign, that is, $\langle \sigma(z)\sigma(w)\rangle = \ln(z - w)$, and so does the operator product expansion:

$$\sigma(z)\sigma(w) = \ln(z - w) + \cdots. \tag{9.2.35}$$

It follows from equation (9.2.33) that the operator product expansion of the fields b and c defined in this way is given by

$$b(z)c(w) = :e^{-\sigma(z)}::e^{\sigma(w)}: = \frac{1}{z - w} + \cdots \tag{9.2.36}$$

as it should be. We have used equation (9.1.40), but taken into account the unusual sign for the operator product expansion and the fact that the momenta are imaginary. The ghost number current can be calculated in terms of the field σ as follows:

$$j(z) = - : b(z)c(z) := -\ln_{z \to w}\left(b(z)c(w) - \frac{1}{z - w}\right) = \partial\sigma(z). \tag{9.2.37}$$

As a result, we can express σ in terms of the b, c fields by

$$\sigma(z) = \int^z dw\, j(w). \tag{9.2.38}$$

The reader may verify that $\sigma(z)\sigma(w) = \ln(z - w) + \cdots$ is a consequence of $j(z)j(w) = (1/(z - w)^2) + \cdots$. Similarly, one can also recover the energy-momentum tensor of equation (9.2.34) from that of the b, c system of equation (9.2.4).

The β, γ system cannot be written in terms of a scalar field even with a background charge. We require in addition a first order Grassmann odd system η, ξ with conformal weights $(1, 0)$ and $(0, 0)$, respectively, with $T = -\eta\partial\xi$ and a real scalar field ρ, with background charge $-\frac{1}{2}(2\lambda - 1)$, that is, it has an energy-momentum tensor $T = -\frac{1}{2}(\partial\rho)^2 - \frac{1}{2}(2\lambda - 1)\partial^2\rho$. We note its energy-momentum tensor has the conventional sign for its first term and so its propagator is given by $\langle\rho(z)\rho(w)\rangle = -\ln(z - w)$ with its operator product expansion $\rho(z)\rho(w) = -\ln(z - w) + \cdots$. The ξ, η system contributes -2 to the central charge, while the scalar field contributes $1 + 12Q^2 = 1 + 3(2\lambda - 1)^2$, which together do indeed give the correct background charge of equation (9.2.30).

The bosonisation formulae are

$$\beta(z) = :\partial\xi e^{-\phi}:, \quad \gamma = :\eta e^{\phi}: . \tag{9.2.39}$$

We can verify that

$$\beta(z)\gamma(w) = (\partial\xi e^{-\phi})(z)(\eta e^{\phi})(w) = \frac{-1}{(z - w)^2}(z - w) :e^{-\phi(z)+\phi(w)}: = -\frac{1}{z - w}$$

$$\tag{9.2.40}$$

as required.

Finally, we may bosonise the η, ξ system as discussed above. We require a scalar σ_1 with background charge $\frac{1}{2}$ which satisfies equations (9.2.34) and (9.2.35) and we can write $\eta = e^{-\sigma_1}$ and $\xi = e^{\sigma_1}$. Substituting these into the above expressions for β and γ we find that

$$\beta(z) = :\partial\sigma_1 e^{\sigma_1 - \phi}:, \quad \gamma = :e^{-\sigma_1 + \phi}: . \tag{9.2.41}$$

As above, one can relate σ_1 to the current for the system, namely $:\eta\xi: = -\partial\sigma_1$.

9.3   Application to string theory

9.3.1   Mapping the string to the Riemann sphere

In chapters 2 and 3 we considered the bosonic string which was either open or closed. We noted that the algebra of constraints was the conformal algebra in two dimensions and that expressions simplified considerably in the light-cone coordinates $\xi^{\pm} = \tau \pm \sigma$. In this section we will map the open and closed classical string world sheets to the Riemann sphere $\hat{C}$ and fully expose the underlying conformal symmetry of the bosonic string. Let us first consider the free closed string, which is parameterised by τ and σ with $-\infty \leq \tau \leq -\infty$, $-\pi < \sigma \leq \pi$ with $x^{\mu}(\sigma + 2\pi) = x^{\mu}(\sigma)$. As explained in chapter 8 we first make a Wick rotation to Euclidean space by taking $\tau \to -i\tau$ whereupon $\xi^{\pm} \to -i(\tau \pm i\sigma)$. We now define the complex coordinates $\varsigma = \tau + i\sigma$ and $\bar{\varsigma} = \tau - i\sigma$ on the cylinder, or strip, with sides at $\sigma = -\pi$ and $\sigma = \pi$ identified. The map from the familiar string world sheet to the Riemann sphere is given by

$$\tau + i\sigma = \varsigma = \ln z \quad \text{or} \quad z = e^{\varsigma} = e^{\tau + i\sigma} \tag{9.3.1}$$

with the analogous relations for the complexified coordinates i.e. $\bar{z} = e^{\bar{\varsigma}} = e^{\tau - i\sigma}$. The mapping is illustrated in figure 18.2.1.

The open string is parameterised by $-\infty \leq \tau \leq -\infty$, but $0 \leq \sigma \leq \pi$. However, in equation (2.1.59) we extended the range from $-\pi \leq \sigma \leq \pi$ by reflecting in $\sigma \to -\sigma$ (that is, $x^{\mu}(\tau, -\sigma) = x^{\mu}(\tau, \sigma)$). The map from the strip to the Riemann sphere is then given by equation (9.3.1). We note that $\sigma \to -\sigma$ corresponds to $z \to \bar{z}$ and so the upper and lower halves of the Riemann sphere are identified for the open string; see figure 18.2.2.

We note lines of constant τ on the open and closed string world sheets are mapped to circles about $z = 0$ on $\hat{C}$. Hence, as noted previously, time ordering on the cylinder corresponds to radial ordering on the Riemann sphere. The incoming string at $\tau = -\infty$ corresponds to the circle of radius 0 around $z = 0$, that is, the south pole of $\hat{C}$, while the outgoing string at $\tau = \infty$ corresponds to the circle of infinite radius around $z = \infty$, that is, the north pole of $\hat{C}$. A time translation on the world sheet, that is, $\tau \to \tau' = \tau + \lambda$, $\sigma \to \sigma' = \sigma$, induces the change $z \to z' = e^{\lambda}z$, $\bar{z} \to \bar{z}' = e^{\lambda}\bar{z}$ and is a dilaton on $\hat{C}$ which is generated by $L_0 + \bar{L}_0$ for the closed string and L_0 for the open string. This explains why $L_0 + \bar{L}_0$ and L_0 play the roles of Hamiltonians for the open and closed strings, respectively, on $\hat{C}$.

A primary field, R, of weight $(\lambda, \bar{\lambda})$ transforms under a conformal transformation as in equation (8.2.29); applying this to the transformation of equation (9.3.1) we find that the primary field on the cylinder is related to that on the Riemann sphere by

$$R(\varsigma, \bar{\varsigma}) = z^{\lambda} \bar{z}^{\bar{\lambda}} R(\bar{z}, \bar{z}), \tag{9.3.2}$$

where the arguments of R indicate if it is defined on the cylinder or on the Riemann sphere. As a result, if $R(\varsigma, \bar{\varsigma})$ has a mode expansion

$$R(\varsigma) = \sum_{n} R_n e^{-n\varsigma} \tag{9.3.3}$$

on the cylinder, then it has the mode expansion

$$R(z) = \sum_{n} R_n z^{-n-\lambda} \tag{9.3.4}$$

on $\hat{C}$ and vice versa. We note that the modes R_n are the same in both descriptions.

We will now give some of the string theory expressions that we found in chapters 2 and 3 on the cylinder in their Riemann sphere description. We begin with the string field x^{μ} itself, which has conformal weight $(0, 0)$. We find, after making the transformation to Euclidean space $\tau \to -i\tau$, as discussed below equation (8.1.35), in equations (2.1.51) and (2.1.52) for the closed string, that

$$x^{\mu}(z, \bar{z}) = x_L^{\mu}(z) + x_R^{\mu}(\bar{z}),$$

where

$$x_L^{\mu}(z) = q_L^{\mu} - i\frac{\alpha'}{2} p_L^{\mu} \ln z + i\sqrt{\frac{\alpha'}{2}} \sum_{\substack{n=-\infty \\ n \neq 0}}^{\infty} \frac{1}{n} \alpha_n^{\mu} z^{-n}$$

and

$$x_R^{\mu}(z) = q_L^{\mu} - i\frac{\alpha'}{2} p_R^{\mu} \ln \bar{z} + i\sqrt{\frac{\alpha'}{2}} \sum_{\substack{n=-\infty \\ n \neq 0}}^{\infty} \frac{1}{n} \bar{\alpha}_n^{\mu} \bar{z}^{-n},$$

where $z = e^{\varsigma} = e^{\tau + i\sigma}$. For the closed string $q^{\mu} = q_L^{\mu} + q_R^{\mu}$ and $p^{\mu} = \frac{1}{2}(p_L^{\mu} + p_R^{\mu})$ and we recall that, except for some compactifications to be discussed in chapter 10 in which the left moving and right moving states live on different internal spaces, $\alpha_0^{\mu} = \sqrt{(\alpha'/2)} p^{\mu} = \bar{\alpha}_0^{\mu}$, and $p_L^{\mu} = p_R^{\mu}$. It is usual to take $\alpha' = 2$ for the closed string. This is just D copies of the field φ given in section 9.1.1 and so, apart from the index μ, the x^{μ} have the all the properties that the φ field has, namely the same energy-momentum tensor and operator product expansion, for example, equations (9.1.11)–(9.1.13). The states and scattering of the closed string involve the operator $:e^{ik \cdot x(z,\bar{z})}:$, where the normal ordering is just as described below equation (9.1.36). The operator product expansions of this operator with itself and x^{μ} and its derivatives can be found just as they were for φ_L. The results are very similar.

For the field x^μ of the open string we find that the expansion of equation (2.1.71), after the rotation to Euclidean space $\tau \to -i\tau$, as discussed below equation (8.1.35), is given on the boundary, that is, $\sigma = 0$, by

$$Q^\mu(z) = q^\mu - i\sqrt{2\alpha'}\alpha_0^\mu \ln z + i\sqrt{2\alpha'} \sum_{\substack{n=-\infty \\ n \neq 0}}^{\infty} \frac{1}{n}\alpha_n^\mu z^{-n},$$

where $z = e^\tau$. For the open sting the pioneers usually took $2\alpha' = 1$. We also recall that for the open string $\alpha_0^\mu = \sqrt{2\alpha'}p^\mu$. As we will see this operator plays an important role in open string theory, particularly in the construction of the physical states and string scattering. Open string scattering occurs by interactions at the boundary of the string, which is where the above operator is defined. One can extend it into the Riemann sphere by adopting the same form. However, this is not the same as the field x^μ. Indeed, making the rotation to Euclidean space on the string field of equation (2.1.71) we find that it becomes

$$x^\mu(z) = q^\mu - i\sqrt{\frac{\alpha'}{2}}\alpha_0^\mu(\ln z + \ln\bar{z}) + i\sqrt{\frac{\alpha'}{2}} \sum_{\substack{n=-\infty \\ n \neq 0}}^{\infty} \left(\frac{1}{n}\alpha_n^\mu z^{-n} + \frac{1}{n}\alpha_n^\mu \bar{z}^{-n}\right),$$

which clearly depends on z and $\bar{z}$. However, we notice that after the rotation to Euclidean space

$$\mathscr{J}^\mu = \sqrt{\frac{2}{\alpha'}}i\partial x^\mu = \frac{i}{\sqrt{2\alpha'}}\partial Q^\mu = \sum_{n=-\infty}^{n=\infty} \alpha_n^\mu z^{-n}.$$

The energy-momentum tensor for the open string can be written as

$$T = -\frac{1}{2\sqrt{2\alpha'}}\partial Q^\mu \partial Q^\nu \eta_{\mu\nu}.$$

It will prove useful to have the operator product expansion of Q^μ. Using the normal ordering procedure as discussed in section 9.1 we find that

$$Q^\mu(z)Q^\nu(w) = {:}Q^\mu(z)Q^\nu(w){:} - 2\eta^{\mu\nu}\alpha' \ln(z-w) = -2\eta^{\mu\nu}\alpha' \ln(z-w) + \cdots.$$

It follows that

$$\partial Q^\mu(z)\partial Q^\nu(w) = -2\alpha'\eta^{\mu\nu}\frac{1}{(z-w)^2} + \cdots.$$

One can verify that Q^μ has the operator product expansion with the energy-momentum tensor that corresponds to a conformal field of weight 0. In fact, the field q^μ has exactly the same coordinate, momentum and oscillator expressions as the field φ_L introduced in equation (9.1.36). This explains why the above operator product expansions agree and indeed one can take the further operator product expansions given below equation (9.1.36) with the replacement $\varphi_L \to Q$, such as in equations (9.1.39)–(9.1.42).

The energy-momentum tensor is not a primary field but we found in section 8.4 that the energy-momentum tensors on the cylinder and Riemann sphere are related by

$$T_{cyl}(\varsigma) = z^2 T_{\hat{c}}(z) - \frac{c}{24}. \tag{9.3.5}$$

In chapter 3 we quantised the bosonic string using BRST methods. This involved fixing the gauge for reparameterisations and then introducing the ghosts b_{++} and c^+ and b_{--} and c^-.

We have written the ghosts in terms of the light-cone coordinates of appendix B; however, we will now use the alternative notation of chapter 8 and write them as $b_{\varsigma\varsigma}$ and c^ς and $b_{\bar\varsigma\bar\varsigma}$ and $c^{\bar\varsigma}$. We recognise from their indices that these have conformal weights $(2, 0)$, $(-1, 0)$, $(0, 2)$ and $(0, -1)$, respectively. The final action following from the BRST procedure was given in terms of light-cone coordinates in equation (3.2.23). Making the Wick rotation as explained in section 8.1.2, that is, $\tau \to -i\tau$, and then using the coordinates $\varsigma, \bar\varsigma$ we find the action is given by

$$A = -\frac{1}{2\pi} \int d\varsigma d\bar\varsigma \{ -\tfrac{1}{2}\partial_\varsigma x^\mu \partial_{\bar\varsigma} x^\nu \eta_{\mu\nu} - (b_{\varsigma\varsigma}\partial_{\bar\varsigma}c^\varsigma + b_{\bar\varsigma\bar\varsigma}\partial_\varsigma c^{\bar\varsigma}) \}. \qquad (9.3.6)$$

In deriving this result we have used the fact that the Euclidean action A^E and the Minkowski action A^M are related by $A^E = iA^M$, the measure etc. changing as $d\xi^+ d\xi^- \to \tfrac{1}{2}id\varsigma d\bar\varsigma$, $\partial_+ \to i\partial_\varsigma$ and $\partial_- \to i\partial_{\bar\varsigma}$. We have also carried out the rescaling

$$x^\mu \to \tfrac{1}{2}x^\mu, \quad c^\xi \to c^\varsigma, \quad c^{\bar\xi} \to c^{\bar\varsigma}, \quad b_{\xi\xi} \to -\tfrac{1}{4}b_{\xi\xi}, \quad b_{\bar\xi\bar\xi} \to -\tfrac{1}{4}b_{\bar\xi\bar\xi}, \qquad (9.3.7)$$

set $2\alpha' = 1$ and finally taken $b_{\varsigma\varsigma} \to ib_{\varsigma\varsigma}$. The rescalings are required to get the expressions we used for the string theory quantities in chapter 3, which have inherited conventions from the beginnings of string theory, to agree with the conventions used in conformal field theory such as in chapter 8.

The components of the energy-momentum tensor in light-cone coordinates were given in equation (3.2.69). Carrying out the Wick rotation they are given in the $\varsigma, \bar\varsigma$ coordinate system by

$$T_{\varsigma\varsigma} = -\tfrac{1}{2}\partial_\varsigma x^\mu \partial_\varsigma x^\nu \eta_{\mu\nu} - 2b_{\varsigma\varsigma}(\partial_\varsigma c^\varsigma) - (\partial_\varsigma b_{\varsigma\varsigma})c^\varsigma,$$
$$T_{\bar\varsigma\bar\varsigma} = -\tfrac{1}{2}\partial_{\bar\varsigma} x^\mu \partial_{\bar\varsigma} x^\nu \eta_{\mu\nu} - 2b_{\bar\varsigma\bar\varsigma}(\partial_{\bar\varsigma} c^{\bar\varsigma}) - (\partial_{\bar\varsigma} b_{\bar\varsigma\bar\varsigma})c^{\bar\varsigma}. \qquad (9.3.8)$$

In finding this result, we have rescaled the energy-momentum tensor by a factor of -2.

Having expressed all the quantities of interest in terms of the $\varsigma, \bar\varsigma$ coordinates on the cylinder, we can now easily map them to the Riemann sphere $\hat C$. The fields map according to their conformal weights as in equation (8.1.48). For example, the ghost b_{zz} maps as

$$b_{zz}(z) = z^{-2}b_{\varsigma\varsigma}(\varsigma). \qquad (9.3.9)$$

Consequently, if $b_{\varsigma\varsigma}$ has the expansion $b_{\varsigma\varsigma} = \sum_n b_n e^{-n\varsigma}$, then b_{zz} has the expansion $b_{zz} = \sum_n b_n z^{-n-2}$. This is in accord with the general result of equation (8.8.12). It is straightforward to transform the action of equation (9.3.6) to the Riemann sphere, $\hat C$: since it is conformally invariant it just takes the same form in the $z, \bar z$ coordinates

$$A = \frac{1}{2\pi} \int dz d\bar z \{ \tfrac{1}{2}\partial x^\mu \bar\partial x^\nu \eta_{\mu\nu} + (b\bar\partial c + \bar b\partial \bar c) \}, \qquad (9.3.10)$$

while the energy-momentum tensor of equation (9.3.8) becomes

$$T_{zz} = -\tfrac{1}{2}\partial x^\mu \partial x^\nu \eta_{\mu\nu} - 2b(\partial c) - (\partial b)c,$$
$$T_{\bar z\bar z} = -\tfrac{1}{2}\bar\partial x^\mu \bar\partial x^\nu \eta_{\mu\nu} - 2\bar b(\bar\partial \bar c) - (\bar\partial b)\bar c, \qquad (9.3.11)$$

where we have set

$$b = b_{zz}, \quad c = c^z, \quad \bar b = b_{\bar z\bar z}, \quad \bar c = c^{\bar z} \qquad (9.3.12)$$

and $\partial = \partial/\partial z$, $\bar\partial = \partial/\partial\bar z$. For the open string these identifications are for $0 \leq \sigma \leq \pi$.

The reader will immediately recognise from sections 9.1 and 9.2 that equations (9.3.10) and (9.3.11) are precisely the action and energy-momentum tensors for a set of D scalar fields x^μ and a first order Grassmann odd $\lambda = 2$ system (b, c) and their anti-mereomorphic counterparts $(\bar{b}, \bar{c})$.

If we compute the central charge for the holomorphic sector, we have $c = D - 26$ for x^μ and (b, c) and similarly $\bar{c} = D - 26$ in the anti-holomorphic sector. The central charge is an anomaly in the conformal symmetry which is part of two-dimensional general coordinate transformations. However, the bosonic string possesses this symmetry as a local symmetry and, as such, an anomaly would be an inconsistency in the theory. This anomaly can be seen in the correlator $\langle T(z)T(0)\rangle$, which we found in equation (8.4.3) was given by $\langle T(z)T(0)\rangle = (c/2)(1/z^4)$. Differentiating with respect to $\bar{\partial}_z$ corresponds to conservation of conformal transformations, but we see this does not vanish and is proportional to the central charge c.

Demanding no anomaly implies $D = 26$. Thus we recover the fact that the bosonic string is only consistent in 26 dimensions, a result we found earlier by demanding either that the BRST charge vanishes or that the theory in light-cone coordinates possesses Lorentz symmetry.

The BRST charge of equation (3.2.82) when mapped to the Riemann sphere takes the form

$$Q = \oint_0 \frac{dz}{2\pi i} c(z) \left(T^x(z) + \frac{1}{2} T^{b,c}(z) \right), \tag{9.3.13}$$

where the two parts of the energy-momentum tensor are given in equation (9.3.11). The anti-commutator can be evaluated using equation (8.6.4) to find

$$\{Q, Q\} = \oint_{\Gamma_2} \frac{dz}{2\pi i} \oint_\Gamma \frac{dw}{2\pi i} \{c(z)(T^x(z) + \tfrac{1}{2}T^{b,c}(z))\}\{c(w)(T^x(w) + \tfrac{1}{2}T^{b,c}(w))\}. \tag{9.3.14}$$

Substituting in the operator product expansion, using Wick's theorem and carrying out the first contour integration around z we do indeed find it vanishes if $D = 26$.

The discussion in section 8.8 on highest weight states and primary fields can naturally be applied to the physical states of the string. Let us consider the open string whose physical states obey $(L_0 - 1)|\psi\rangle = 0$ and $L_n|\psi\rangle = 0, n \geq 1$. As such they are highest weight states with weight 1 and we can expect that to each such state there is associated a primary field of weight 1, the two being related by equation (8.8.1). Let us consider the tachyon state $|0, p\rangle$ with $p^2 = 2$. The associated primary field is $:e^{ip\cdot Q(z)}:$ with $p^2 = 2$, where the definition of Q^μ and the normal ordering prescription are given in section 9.3. Indeed, one see that

$$\lim_{z \to 0} :e^{ip\cdot Q(z)}: |0, 0\rangle = |0, p\rangle. \tag{9.3.15}$$

This is consistent because $:e^{ip\cdot Q(z)}:$ with $p^2 = 2$ does, indeed, have conformal weight 1 as required. Let us deduce the physical state at the next level but working from the conformal field theory perspective. The candidate primary field at the next level is of the form $:\epsilon_\mu \partial Q^\mu(z) e^{ik\cdot Q(z)}:$. Let us test if this operator is a primary field of weight 1; we find using equations (9.1.38) that

$$T(z) :\!\epsilon_\mu \partial Q^\mu(w)e^{ik\cdot Q(w)}\!: = \frac{:\!\epsilon_\mu \partial Q^\mu(w):e^{ik\cdot Q(w)}\!:}{(z-w)^2} + \frac{\partial(:\!\epsilon_\mu \partial Q^\mu(w)e^{ik\cdot Q(w)}\!:)}{(z-w)}$$

$$+ \frac{k^2}{2} \frac{:\!\epsilon_\mu \partial Q^\mu(w):e^{ik\cdot Q(w)}\!:}{(z-w)^2}$$

$$+ i\frac{:\!\epsilon_\mu k^\mu : e^{ik\cdot Q(w)}\!:}{(z-w)}, \tag{9.3.16}$$

from which we conclude that it is a conformal operator of weight 1 if $k^2/2 = 0 = \epsilon_\mu k^\mu$. The reader will readily find that this does, indeed, lead to the photonic states, that is,

$$\lim_{z\to 0} :\!\epsilon_\mu \partial Q^\mu(z)e^{ik\cdot Q(z)}\!: |0,0\rangle = \epsilon_\mu \alpha^\mu_{-1}|0, K\rangle \tag{9.3.17}$$

and that $k^2/2 = 0 = \epsilon_\mu k^\mu$ are the required physical state conditions.

Te reader may like to verify that the tachyonic 1 states at level 1 for the closed string have associated the weight 1 primary fields

$$:\!e^{ip\cdot x(w,\bar{w})}\!:, \quad p^2 = 2; \epsilon_{\mu\nu} :\partial x^\mu(w)\partial x^\nu(w)e^{ik\cdot x(w,\bar{w})}\!:, \quad k^2 = 0 = k_\mu\epsilon_{\mu\nu}, \tag{9.3.18}$$

respectively. We recall that x^μ has, except for the index μ, all the properties of the φ of section 9.1.1, for example, equation (9.1.10) and the equations that follow it.

9.3.2 Construction of string theories

The construction of the bosonic string given in the previous section illustrates the general procedure to construct string theories. We can summarise the essential steps as follows.

- We took a two-dimensional conformally invariant theory and coupled it to two-dimensional gravity.
- We then demanded that it has no conformal anomaly. This was equivalent to insisting that the central charge was zero in the left and right moving sectors separately in the BRST quantised theory. Hence the central charge of the matter, that is, the original fields, and the ghosts must cancel.

To construct superstrings we carry out a very similar procedure. We start from a two-dimensional superconformally invariant theory which we couple to the corresponding supergravity and demand that there is no superconformal anomaly. The BRST quantisation of this theory is more involved as we must introduce ghosts for all the local symmetries. In the left moving sector this means the b, c ghosts for the diffeomorphisms as before, but now also β, γ ghosts for the local supersymmetry. The latter are commuting as they must be of opposite Grassmann parity to the supersymmetry parameter and they will be a $\lambda = \frac{3}{2}$ system as then β and γ have conformal weights $\frac{3}{2}$ and $-\frac{1}{2}$, respectively. We note that $\gamma\Lambda$, where Λ is the BRST parameter, replaces the local supersymmetry parameter when finding the BRST transformations (see appendix A) and so it must have the same index structure as the supersymmetry parameter. Now the central charge of the original fields, the so-called matter, and all the ghosts must cancel. The same must be true for the right moving sector. Actually, one should also check that the system has no local supersymmetry anomaly.

One very important feature of the bosonic superstring string constructions and, indeed, any string is that the ghost action is independent of the details of the matter fields used and only depends on the symmetry algebra of the matter fields that are BRST quantized. This follows from the BRST procedure explained in appendix A. For the bosonic string we have just diffeomorphism symmetry and the required ghosts and their action follows entirely from the algebra of diffeomorphisms. In this case we must cancel the ghost central charge of -26 and so include matter that has a central charge of 26. Clearly, the simplest possibility is to use 26 free scalar fields x^μ, $\mu = 0, 1, \ldots, 25$. For a superstring the local symmetries are diffeomorphism and supersymmetry. The corresponding b, c and β, γ ghosts have a central charge of $-26 + 11 = 15$ and so we must have a matter content that carries a representation of conformal supersymmetry and have a central charge of 15. Here, we have assumed that in the left moving sector we have a supersymmetry with a Majorana–Weyl parameter. If we had a larger supersymmetry we would have a more complicated ghost structure which we will discuss below.

We can construct more general string theories by considering matter systems with other symmetries and carrying out a similar procedure; we couple the system to the corresponding gravity and after BRST quantising the theory we demand that the total central charge be 0 in the left and right moving sectors separately.

Let us now carry out the construction of the *superstring* in more detail. As discussed above we need a matter system with central charge 15 that carries a representation of super-conformal symmetry. The simplest possibility is to consider the matter to be constructed from free bosons and fermions. In particular, we can take the bosonic fields x^μ and the fermionic fields χ^μ for $\mu = 0, 1, \ldots, D-1$, which do indeed carry a representation of conformal supersymmetry with one Majorana supercharge. In a given left, or right, moving sector the symmetry has Majorana–Weyl parameter of a given chirality depending on the sector. The central charge of this matter is $D + \frac{1}{2}D = \frac{3}{2}D$. Hence, we find that such a superstring is consistent only in ten space-time dimensions. The free bosons have an action that is given by

$$\frac{1}{4\pi} \int dz d\bar{z} \, (\partial x^\mu \bar{\partial} x^\nu - \chi^\mu \bar{\partial} \chi^\nu - \bar{\chi}^\mu \partial \bar{\chi}^\nu) \eta_{\mu\nu}. \tag{9.3.19}$$

The coupling to supergravity is described in chapter 1. The corresponding energy-momentum tensor and supercurrent can be read off from equations (9.1.2) and (9.1.51) and is given by

$$T = -\tfrac{1}{2} : \partial x^\mu \partial x^\nu : \eta_{\mu\nu} + \tfrac{1}{2} : \chi^\mu \partial \chi^\nu : \eta_{\mu\nu}, \tag{9.3.20}$$

$$J = -4 \partial x^\mu \chi^\nu \eta_{\mu\nu}, \tag{9.3.21}$$

with corresponding expressions for the right moving sector. We note that T and J are quasi-primary fields with conformal weights 2 and $\frac{3}{2}$, respectively. As was explained earlier in this chapter we can introduce oscillators for x^μ and ψ^μ and find T and J in terms of them. In the old covariant quantisation we would implement some of these as constraints on the wavefunction.

In the BRST quantisation we introduce the Grassmann odd b, c system with $\lambda = 2$ and the Grassmann even β, γ system with $\lambda = \frac{3}{2}$, whose energy-momentum tensors can be read off from equations (9.2.4) and (9.2.22) to be

$$T^{b,c} = :-2b\partial c - \partial bc: \text{ and } T^{\beta,\gamma} = :-\frac{3}{2}\beta\partial\gamma - \frac{1}{2}\partial\beta\gamma:. \tag{9.3.22}$$

The total energy-momentum tensor in the ghost sector is given by

$$T^{ghost} = T^{b,c} + T^{\beta,\gamma}. \tag{9.3.23}$$

In the above we have just given the expressions for the left moving sector, but the right moving sector has analogous expressions and an equivalent discussion. The reader will recognise all the features of the superstring discussed in chapter 7.

We now construct the *heterotic string* [10.29]. This string is also built from free bosons and fermions, but it treats the left and right sectors differently. The left moving sector is just like the superstring: it has ten free bosonic fields x_L^μ and ten free fermionic field χ^μ and the discussion proceeds just like above. However, the right moving sector has no supersymmetry and so is rather like the bosonic string: it has only the $\lambda = 2$ b, c ghosts and so the right moving matter is made up of 26 free bosons, which we label as $x_R^\mu, \mu = 0, 1, \ldots, 9$ and $x_R^I, I = 1, \ldots 16$. Clearly, once we add the appropriate ghosts in each sector the total central charges are 0 as they should be. The coordinates $x^\mu = x_L^\mu + x_R^\mu$ make up the ten-dimensional space-time of this theory. It is straightforward to write down the energy-momentum tensors in the left and right moving sectors. Thus we arrive at the heterotic string discussed in more detail in chapter 10. As we will find there, the x^I are associated with a Lie algebra which must have rank 16 and be self-dual. The only possible such algebras are $E_8 \otimes E_8$ and $SO(32)/Z_2$.

We might also consider a matter system with $N = 2$ supersymmetry as opposed to the ones with $N = 1$ considered above. Now the underlying supersymmetry algebra has two supercharges which are exchanged by a U(1) symmetry. Hence the corresponding $N = 2$ supergravity has one graviton, two gravitinos and a vector corresponding to the U(1) current. Hence, for this system we must introduce for the ghosts one $\lambda = 2$ b, c system, two $\lambda = \frac{3}{2}$ β, γ ghosts and one $\lambda = 1$ b, c ghost. The last corresponds to the U(1) symmetry. The total central charge of all the ghosts is then given by $c = -26 + 2.11 - 2 = 6$. The simplest free matter which carries a representation of $N = 2$ superconformal symmetry has two scalars A, B and two fermions $\psi_i, i = 1, 2$. The central charge in each sector is $2 + \frac{1}{2} \times 2 = 3$. Hence, we can must have two such multiplets leading to the conclusion that this string lives in only two space-time dimensions.

It will have occurred to the reader that one can construct many more string theories as we are allowed to use matter systems that are not just built out of free fermions and bosons. Indeed, there is no need to incorporate Lorentz symmetry other than in the part that defines our four-dimensional space-time and we can use interacting matter systems as long as they have the required symmetry and central charge. For example, in the bosonic case we may use four x^μs, which will be used to construct space-time and then make up the remaining central charge (that is, 22) out of minimal models. Clearly, this procedure allows a vast number of string theories to be constructed.

We may also construct strings based on more general algebras. One such example is W algebras, the simplest of which contains an energy momentum tensor and a spin three current.

Not all string theories constructed from the above procedure are consistent theories. Indeed, the above considerations concern only the behaviour of the string theories at the free level and they can develop diseases which result from their interactions. One such disease is anomalies in any space-time gauge or diffeomorphism symmetries of the theory. These can occur if the theory has space-time fields that are chiral fermions or bosonic fields that have self-dual field strengths. Such fields do occur in certain superstring theories. It has been shown that these space-time anomalies do not occur in closed strings if the theory is modular invariant [9.6]. We therefore add modular invariance, which is discussed in chapter 18, to our list of requirements for a good string theory.

The current attitude is that any (super)conformally invariant theory which has no (super)conformal anomaly, when coupled to (super)gravity, provides a good string theory if it is modular invariant. This recipe has included a staggering number of four-dimensional theories.

Unitarity, that is, space-time unitarity, is not obviously guaranteed in the above string theories. Nonetheless, it is thought that their physical states have positive norm and that the probabilities sum to 1. The former is easily seen once the theories are formulated in light-cone gauge, while the latter is guaranteed by the existence of a string field theory formulation.

A final comment concerns BRST anomalies. Gauge covariant string theory possesses not only the local symmetries associated with the massless modes, but also local symmetries at all higher levels. One may wonder if anomalies that might arise in these higher level symmetries are automatically absent in any modular and (super)conformally invariant theory. Anomalies of this type are known to arise in the open string as boundary terms in moduli space. One might accordingly think that modular invariance would ensure their absence in closed string theories.

9.4   The free field representation of the minimal models

A formulation of the minimal models discussed in section 8.9 can be given in terms of a real scalar field with background charge Q with an energy-momentum tensor given by $T = -\frac{1}{2}(\partial\phi)^2 - Q\partial^2\phi$, which has $c = 1 + 12Q^2$ [9.2–9.5]. We recall from section 9.1 that the field ϕ is not a primary field. However, $V_\beta = :e^{i\beta\phi}:$ is a primary field with conformal weight $\Delta_\beta = \frac{1}{2}\beta(\beta - 2iQ)$. We note that there are two possible solutions for β for a given weight, that is, if β is a solution, so is β and $2iQ - \beta$. Important for the construction is the observation that $\Delta = 1$ for the two values $\alpha_\pm = iQ \pm \sqrt{2 - Q^2}$. We note that $\alpha_+ + \alpha_- = 2iQ$ and that $\alpha_+\alpha_- = -2$.

The field ϕ has the action of equation (9.1.47). We are working on the Riemann sphere, which has two patches, one around the south pole at $z = 0$ and the other around the north pole at $z = \infty$. We can choose our metric, using conformal transformations, so that the curvature of the Riemann sphere is only at the point at infinity and so the additional term, the action, only contributes at this point. Hence away from this point ϕ behaves just as a free field.

Let us consider the current $j = i\partial\phi$ which is usually associated with translations and so momentum conservation, the corresponding current being $j_0 = \oint_0 (dz/2\pi i)Jz$. In the presence of the background charge Q this current is no longer conserved as from equation (9.1.47) the equation of motion of ϕ is given by $\bar{\partial}(i\partial\phi) = -iQ\sqrt{g}R$. However, the non-conservation only results from the presence of the curvature, and so background charge Q, at infinity. As the current is not conserved we should expect some modification to the usual conservation of momentum, or charge, as it is called in this case.

From the operator product expansion

$$j(z){:}e^{i\beta\phi(w)}{:} = \frac{\beta}{(z-w)}{:}e^{i\beta\phi(w)}{:} + \cdots \tag{9.4.1}$$

we conclude that $[j_0, V_\beta(z)] = \beta V_\beta(z)$. Hence the operator V_β inserts a 'charge' β into the correlator.

If we consider the correlators of the primary fields V_β, the momenta will not, however, in general sum to the required $2iQ$ even if we allow for the possibility of taking either V_β or $V_{2iQ-\beta}$. The way out of this dilemma is to use the two primary fields with conformal weight 1. In particular, we introduce the so-called screening charges

$$Q_\pm = \int_c dz e^{i\alpha_\pm\phi(z)}. \tag{9.4.2}$$

These commute with the Virasoro operators L_n and so will not change the conformal properties of the correlator. We now consider the correlator of primary fields to be of the form

$$\langle V_{\beta_1}(z_1)\cdots V_{\beta_n}(z_n)\int \prod_i^p du_i e^{i\alpha_+\phi(u_i)}\int\prod_j^q dv_j e^{i\alpha_-\phi(v_j)}\rangle. \tag{9.4.3}$$

The condition for momentum conservation is now

$$\sum \beta_i + p\alpha_+ + q\alpha_- = l2iQ. \tag{9.4.4}$$

The appearance of l factors of $2iQ$ allows for the possibility that we can use β_k or $2iQ - \beta_k$. Considering the correlator of primary fields of the same weight we find that momentum or charge conservation implies $s\beta + p\alpha_+ + q\alpha_- = l2iQ$. In general, this will not be possible unless the momenta we are interested in are of the form

$$\beta = \tfrac{1}{2}\{(1-n)\alpha_+ + (1-m)\alpha_-\}. \tag{9.4.5}$$

We note that we can always reduce the number of βs in the momentum sum to one or two, depending on whether the number of primary fields is even or odd, by using $2iQ - \beta$. The conformal weights corresponding to these charges are

$$h_{n,m} = \frac{Q^2}{2} + \frac{1}{8}(n\alpha_+ + n\alpha_-)^2. \tag{9.4.6}$$

Let us now choose the value of Q that corresponds to the central charge of the minimal models, that is, $1 + 12Q^2 = 1 - (6(p-q)^2/pq)$ for positive integers p and q which have no common factor. This implies that

$$Q = \frac{i}{\sqrt{2}}\frac{(p-q)}{\sqrt{pq}} \text{ and so } \alpha_+ = \frac{\sqrt{2q}}{\sqrt{pq}}, \quad \alpha_- = -\frac{\sqrt{2p}}{\sqrt{pq}}. \tag{9.4.7}$$

The weights of equation (9.4.6) are then given by

$$h_{n,m} = \frac{1}{4pq}\{(nq - mp)^2 - (p - q)^2\}. \tag{9.4.8}$$

These are just the weights of the primary fields contained in the minimal models. Hence, a single field ϕ with a background charge provides a method of explicitly calculating the correlators of the minimal models. The screening charges of equation (9.4.2) are conformally invariant and so do not alter the conformal properties of the Green's function, or the singularities when $T(z)$ is inserted in the Green's function. Consequently, the Green's function calculated in this way obeys the differential equations due to the null states discussed in section 8.10 and so are indeed correct. It is then straightforward using equation (9.1.42) to evaluate the product of $e^{i\beta_n\phi(z)}$ factors, although there remains the choice of contour in the integrals of the screening charge.

Let us illustrate how it works for the Ising model. The central charge $c = \frac{1}{2}$ requires a background charge $Q = i/2\sqrt{6}$ and the screening momenta are given by $\alpha_+ = -6iQ$ and $\alpha_- = 8iQ$. The fields of the Ising model ϕ_0, $\phi_{1/2}$ and $\phi_{1/16}$ can be constructed with $\beta_0 = 0$ or $2iQ$, $\beta_{1/2} = -\alpha_-/2$ or $\beta_{1/2} = \alpha_+$ and $\beta_{1/16} = -\alpha_-/8$ or $-\alpha_+/2$, respectively.

The correlator for four $\Delta_n = \frac{1}{16}$ primary fields can be constructed by taking $\beta_{1/16} = -\alpha_+/2$ for three of them, $\beta_{1/16} = -\alpha_-/8$ for one and a screening charge Q_+. We verify that the total background charge is $2iQ$ as it must be. The four-($\Delta_n = \frac{1}{16}$) correlator is given by

$$\langle\phi_{\frac{1}{16}}(z_1)\phi_{\frac{1}{16}}(z_2)\phi_{\frac{1}{16}}(z_3)\phi_{\frac{1}{16}}(z_4)\rangle = \int_C dw \langle e^{i(2iQ-\alpha)\phi(z_1)} \prod_{j=2}^{4} e^{i\alpha\phi(z_j)} e^{i\alpha+\phi(w)}\rangle,$$

$$(9.4.9)$$

where $\alpha = -\alpha_+/2$. Using equation (9.1.42) we may evaluate this to be

$$\langle\phi_{1/16}(z_1)\phi_{1/16}(z_2)\phi_{1/16}(z_3)\phi_{1/16}(z_4)\rangle = \int_C dw (z_1 - w)^{1/4} \prod_{i=2}^{4}(z_1 - z_i)^{-1/8}$$

$$\times \prod_{i,j=2,i<j}^{4}(z_i - z_j)^{3/8} \prod_{i=2}^{4}(z_1 - w)^{-3/4}.$$

$$(9.4.10)$$

We now use Möbius invariance to choose $z_1 = \infty$, $z_2 = 1$, $z_4 = 0$ and label $z_3 = z$, whereupon we find that

$$\langle\phi_{1/16}(z_1)\phi_{1/16}(z_2)\phi_{1/16}(z_3)\phi_{1/16}(z_4)\rangle$$

$$= \int_C dw((1 - z)z)^{3/8}(w(1 - w)(z - w))^{-3/4}. \tag{9.4.11}$$

The question now arises of what is the contour C. The integrand corresponds to a Riemann sphere with two cuts which can be taken to be from 1 to ∞ and from 0 to z. The independent contours which lead to non-trivial results can be taken to be those around the cuts and this is equivalent to taking the integral to be from 1 to ∞ and from 0 to z. These are, indeed, the contours we will take, but first we must accumulate some facts about hypergeometric functions. The hypergeometric function [8.7] is given by

$$\int_1^\infty dw\, w^{a-c}(w-1)^{c-b-1}(w-z)^{-a} = \int_0^1 dt\, t^{b-1}(1-t)^{c-b-1}(1-tz)^{-a}$$

$$= \frac{\Gamma(b)\Gamma(c-b)}{\Gamma(c)}F(a,b;c;z), \qquad (9.4.12)$$

where we have made the substitution $t = 1/w$. The following equation also holds:

$$\int_0^z dw\, w^c (1-w)^b (z-w)^a$$

$$= z^{1+a+c}\frac{\Gamma(a+1)\Gamma(c+1)}{\Gamma(a+c+2)}F(-b,a+1;a+c+2;z). \qquad (9.4.13)$$

The following identities, also found in [18.7], will be useful

$$\cos a\theta = F\left(\frac{a}{2},-\frac{a}{2}:\frac{1}{2};\sin^2\theta\right)$$

$$= \cos\theta F\left(\frac{1}{2}+\frac{a}{2},\frac{1}{2}-\frac{a}{2}:\frac{1}{2};\sin^2\theta\right), \qquad (9.4.14)$$

$$\sin a\theta = a\sin\theta F\left(\frac{1}{2}+\frac{a}{2},\frac{1}{2}-\frac{a}{2}:\frac{3}{2};\sin^2\theta\right)$$

$$= a\sin\theta\cos\theta F\left(1+\frac{a}{2},1-\frac{a}{2}:\frac{3}{2};\sin^2\theta\right) \qquad (9.4.15)$$

and if we take $a = \frac{1}{2}$ they become

$$F\left(\frac{1}{4},-\frac{1}{4}:\frac{1}{2};\sin^2\theta\right) = \cos\frac{\theta}{2}, \quad F\left(\frac{3}{4},\frac{1}{4}:\frac{1}{2};\sin^2\theta\right) = \frac{\cos(\theta/2)}{\cos\theta},$$

$$F\left(\frac{3}{4},\frac{1}{4}:\frac{3}{2};\sin^2\theta\right) = \frac{2\sin(\theta/2)}{\sin\theta},$$

$$F\left(\frac{5}{4},\frac{3}{4}:\frac{3}{2};\sin^2\theta\right) = \frac{2\sin(\theta/2)}{\cos\theta\sin\theta}. \qquad (9.4.16)$$

A useful property is $F(a,b;c;z) = F(b,a;c;z)$.

Taking the above mentioned contours we then find that

$$\langle\phi_{\frac{1}{16}}(z_1)\phi_{\frac{1}{16}}(z_2)\phi_{\frac{1}{16}}(z_3)\phi_{\frac{1}{16}}(z_4)\rangle = ((1-z)z)^{3/8}\left\{a'F\left(\frac{3}{4},\frac{5}{4}:\frac{3}{2};z\right)\right.$$

$$\left. + bz^{-\frac{1}{2}}F\left(\frac{3}{4},\frac{1}{4}:\frac{1}{2};z\right)\right\}$$

$$= ((1-z)z)^{\frac{3}{8}}\left\{a'\frac{\sin(\theta/2)}{\sin\theta\cos\theta} + b\frac{\cos(\theta/2)}{\cos\theta}\right\}$$

$$= ((1-z)z)^{-\frac{1}{8}}\left\{a''\sin\frac{\theta}{2} + b\cos\frac{\theta}{2}\right\}. \qquad (9.4.17)$$

In carrying out these steps we have used the above identities, a', a'' and b are constants and have taken $z = \sin^2 \theta$. We note that $2 \sin \frac{\theta}{2} = \sqrt{1 - \sqrt{1 - z}}$ and $2 \cos \frac{\theta}{2} = \sqrt{1 + \sqrt{1 - z}}$

In the above we have computed only the right moving part of the Green's functions. These are normally called the conformal blocks and for this four-point function we have two of them. The actual Green's function is a product of a left and a right moving part and the way to put them together is determined by requiring that the result is single valued, that is, single valued under monodromy transformations of the form $z - z' \to e^{2\pi i}(z - z')$; where z' is an insertion point that is invariant under the interchange of the insertion points. One finds that the Green's function considered here is given by the modulus squared of the first block plus the modulus squared of the second block with a suitable normalisation between them which is determined by this proceedure. This agrees with the result given at the end of chapter 8.

10 String compactification and the heterotic string

But all these four Muses (Bach, Handel, Gluck, and Haydon) are amalgamated in Mozart. He who knows Mozart also knows what is good in these four, because being the greatest and most potent of all musical creators, he was not adverse, even, to taking them under his wings and saving them from oblivion. They are rays lost in the sun of Mozart.

Tchaikovsky 1886

In this chapter we consider the dimensional reduction of a closed bosonic string on a torus. The string is an extended object and this leads to some important differences in its dimensional reduction compared to that for a point particle. One of the main differences is that the string can wind itself around the torus leading to an enlarged set of states. The resulting compactified string possesses some unexpected symmetries. We also consider constructions for which the left and right moving modes on the world sheet of the string are treated independently and belong to different tori.

These discussions naturally motivate the construction of the heterotic string in the final section.

10.1 Compactification on a circle

Let us consider the closed bosonic string moving in a space-time in which one dimension, with coordinate, say x^{25}, is a circle of radius R. Its dynamics is determined by the same actions as given in chapter 2. As we have a closed string we must consider the possibility that the string can wind itself around the circle [10.1]. If it winds $l \in \mathbf{Z}$ times, then as we change σ by adding 2π (that is, going the length of the string) we must travel around the circle l times. As a result, we find that

$$x^{25}(\tau, \sigma + 2\pi) = x^{25}(\tau, \sigma) + 2\pi R l. \tag{10.1.1}$$

We may write x^{25} as

$$x^{25}(\tau, \sigma) = x_s^{25}(\tau, \sigma) + l R \sigma \tag{10.1.2}$$

from which we conclude that $x_s^{25}(\tau, \sigma + 2\pi) = x_s^{25}(\tau, \sigma)$ and, consequently, x_s^{25} has the Fourier expansion

$$x_s^{25} = \sum_{n=-\infty}^{\infty} x_n^{25}(\tau) e^{in\sigma}, \tag{10.1.3}$$

240

where $x_n^{25} = (x_{-n}^{25})^*$. As before, we introduce the oscilators using equations (2.1.57), (2.1.79), namely

$$\wp^{25} \equiv \frac{1}{\sqrt{2\alpha'}}(\dot{x}^{25} + x^{25'}) \equiv \sum_{n=-\infty}^{\infty} \alpha_n^{25} e^{-in\sigma}, \tag{10.1.4}$$

$$\bar{\wp}^{25} \equiv \frac{1}{\sqrt{2\alpha'}}(\dot{x}^{25} - x^{25'}) \equiv \sum_{n=-\infty}^{\infty} \bar{\alpha}_n^{25} e^{in\sigma}. \tag{10.1.5}$$

Substituting $x^{25}(\tau, \sigma)$, as given in equation (10.1.2) and (10.1.3), into the above we find that α_n^{25} and $\bar{\alpha}_n^{25}$ for $n \neq 0$ have the same expression in terms of x_n^{25} as for the uncompactified directions. The $n = 0$ components are, however, given by

$$\alpha_0^{25} = \frac{1}{\sqrt{2\alpha'}}(\dot{x}_0^{25} + Rl), \quad \bar{\alpha}_0^{25} = \frac{1}{\sqrt{2\alpha'}}(\dot{x}_0^{25} - Rl). \tag{10.1.6}$$

This differs from the uncompactified oscillators which satisfy equation (2.1.41) and so $\alpha_0^\mu = \bar{\alpha}_0^\mu, \mu \neq 25$. The additonal terms follow in an obvious way from the extra term in equation (10.1.2). Clearly, the Poisson brackets remain unchanged, even for the compactified oscillators.

The equation of motion is still $\ddot{x}^\mu - x^{\mu''} = 0$ and consequently, equation (2.1.45) holds with its solution (2.1.46). However, integrating equation (10.1.6) we find that the compactified zero mode is given by

$$x_0^{25}(\tau) = q^{25} + \frac{(\alpha_0^{25} + \bar{\alpha}_0^{25})}{2}\sqrt{2\alpha'}\tau, \tag{10.1.7}$$

where α_0^{25}, $\bar{\alpha}_0^{25}$ and q^{25} are independent of τ. As in chapter 2 we can obtain the expression for $x^{25}(\tau, \sigma)$ in terms of oscillators by integrating the expression $x^{25'} = \frac{1}{2}\sqrt{2\alpha'}(\wp^{25} - \bar{\wp}^{25})$, using the equations (10.1.4) and (10.1.5), to find

$$x^{25}(\tau, \sigma) = q^{25} + \sqrt{2\alpha'}\left(\frac{\alpha_0^{25} + \bar{\alpha}_0^{25}}{2}\right)\tau + Rl\sigma$$

$$+ i\sqrt{\frac{\alpha'}{2}} \sum_{\substack{n=-\infty \\ n \neq 0}}^{\infty} \frac{1}{n}\left[\alpha_n^{25}(0)e^{-in(\tau+\sigma)} - \bar{\alpha}_{-n}^{25}(0)e^{in(\tau-\sigma)}\right]$$

$$= x_L^{25}(\tau + \sigma) + x_R^{25}(\tau - \sigma), \tag{10.1.8}$$

where

$$x_L^{25}(\tau + \sigma) = q_L + \frac{\sqrt{2\alpha'}}{2}\alpha_0^{25}(\tau + \sigma) + i\sqrt{\frac{\alpha'}{2}} \sum_{\substack{n=-\infty \\ n \neq 0}}^{\infty} \frac{1}{n}\alpha_n^{25}e^{-in(\tau+\sigma)} \tag{10.1.9}$$

and

$$x_R^{25}(\tau - \sigma) = q_R + \frac{\sqrt{2\alpha'}}{2}\bar{\alpha}_0^{25}(\tau - \sigma) + i\sqrt{\frac{\alpha'}{2}} \sum_{\substack{n=-\infty \\ n \neq 0}}^{\infty} \frac{1}{n}\bar{\alpha}_n^{25}e^{-in(\tau-\sigma)}, \tag{10.1.10}$$

$$q_L^{25} + q_R^{25} = q^{25} \tag{10.1.11}$$

and the difference $q_L - q_R$ is as yet unspecified.

We now quantise the string wound on the circle. The canonical momentum P^μ is still given by equation (2.1.35) and so equations (2.1.42) still hold with the Poisson brackets of equations (2.1.43) and (2.1.44). While the zero mode, or total momentum, p^μ is given by equation (2.1.40), but one sees that equation (2.1.41) does not hold. As such the commutation relations for the oscillators are the same and are given by equation (3.1.38), and for $n \neq 0$ they are as in equation (3.1.42). The zero modes x_0^{25} and the total momentum $p_0^{25} = p^{25} = \dot{x}_0^{25}/\alpha'$ can be represented in the quantum theory by x_0^{25} and $-i\partial/\partial x_0^{25}$. As such we see that equation (10.1.6) becomes

$$\alpha_0^{25} = \sqrt{\frac{\alpha'}{2}}p^{25} + \frac{1}{\sqrt{2\alpha'}}Rl = -i\sqrt{\frac{\alpha'}{2}}\frac{\partial}{\partial x_0^{25}} + \frac{1}{\sqrt{2\alpha'}}Rl \qquad (10.1.12)$$

and

$$\bar{\alpha}_0^{25} = \sqrt{\frac{\alpha'}{2}}p^{25} - \frac{1}{\sqrt{2\alpha'}}Rl = -i\sqrt{\frac{\alpha'}{2}}\frac{\partial}{\partial x_0^{25}} - \frac{1}{\sqrt{2\alpha'}}Rl. \qquad (10.1.13)$$

Let us define

$$p_L = \frac{2}{\sqrt{2\alpha'}}\alpha_0^{25}, \quad p_R = \frac{2}{\sqrt{2\alpha'}}\bar{\alpha}_0^{25}. \qquad (10.1.14)$$

For the closed string, it is customary to take $\alpha' = 2$, upon which all the above expressions simplify.

It is natural to consider wavefunctions that are single valued on the circle, that is, under $x^{25} \to x^{25} + 2\pi R$. Imposing this requirement on the momentum eigenstates, $\exp(ip^{25}x^{25})$ implies that

$$p^{25} = \frac{h}{R}, \quad h \in \mathbf{Z}. \qquad (10.1.15)$$

These are just the usual Kaluza–Klein states that are found in the reduction of any field theory and so are not special to string theory. Demanding this condition is equivalent to demanding that the vertex operators that create all the physical states be single valued. This is because the vertex operator contains this exponential multiplied by other operators such as ∂x that are automatically single valued. As a result, we find that the zero modes α_0^{25} and $\bar{\alpha}_0^{25}$ are quantised as follows:

$$\alpha_0^{25} = \frac{\sqrt{2\alpha'}}{2}\frac{h}{R} + \frac{1}{\sqrt{2\alpha'}}Rl, \quad \bar{\alpha}_0^{25} = \frac{\sqrt{2\alpha'}}{2}\frac{h}{R} - \frac{1}{\sqrt{2\alpha'}}Rl$$

or

$$p_L = \frac{h}{R} + \frac{1}{\alpha'}Rl, \quad p_R = \frac{h}{R} - \frac{1}{\alpha'}Rl. \qquad (10.1.16)$$

The quantisation in the directions not on the circle, that is, $\mu = 0, 1, \ldots, 24$, is the same as in chapter 3.

The L_n and $\bar{L}_N$ are given in terms of α_n^μ and $\bar{\alpha}_n^\mu$ by equation (3.1.43) and still obey equation (3.1.44). The physical state conditions are equations (3.1.45) and (3.1.46). One of these is the equation

$$(L_0 + \bar{L}_0 - 2)|\psi\rangle = 0, \qquad (10.1.17)$$

which may be rewritten as

$$\left(\frac{\alpha'}{2}p^2 + \frac{\alpha'}{4}(p_L^2 + p_R^2) + N + \bar{N} - 2 \right)|\psi\rangle = 0, \tag{10.1.18}$$

where $p^2 = \sum_{\mu=0}^{24} p_\mu p^\mu$ and $N = \frac{1}{2}\sum_{n=1}^{\infty}\sum_{\mu=0}^{25} \alpha_{-n}^\mu \alpha_{\mu n}$ and similarly for $\bar{N}$. The mass levels are therefore given by

$$\text{mass}^2 = \frac{1}{2}(p_L^2 + p_R^2) + \frac{2}{\alpha'}(N + \bar{N} - 2)$$

$$= \frac{h^2}{R^2} + \frac{1}{(\alpha')^2}R^2 l^2 + \frac{2}{\alpha'}(N + \bar{N} - 2). \tag{10.1.19}$$

The first term comes from the usual contribution from the momentum in the 25 direction; however, the second term is the winding energy of the string on the circle.

Another of the physical state conditions is $(L_0 - \bar{L}_0)|\psi\rangle = 0$, which means that

$$\frac{\alpha'}{4}(p_L^2 - p_R^2 + N - \bar{N})|\psi\rangle = 0 \quad \text{or} \quad hl|\psi\rangle = (-N + \bar{N})|\psi\rangle. \tag{10.1.20}$$

Thus, when we have a non-zero winding number the left and right numbers of $|\psi\rangle$ are no longer equal.

We label the momentum and winding number of the state on the circle by the notation.

$$|h, l\rangle, \tag{10.1.21}$$

where

$$p_L|h, l\rangle = \left(\frac{h}{R} + \frac{1}{\alpha'}Rl \right)|h, l\rangle, \quad p_R|h, l\rangle = \left(\frac{h}{R} - \frac{Rl}{\alpha'} \right)|h, l\rangle. \tag{10.1.22}$$

The state $|h, l\rangle$ is to understood not to carry any oscillator excitations and so is annihilated by $\bar{\alpha}_n^\mu$, $\bar{\alpha}_n^{25}$, $n > 0$. We have also suppressed the dependence on p^μ, $\mu = 0, 1, \ldots, 24$.

Let us now consider the physical states of a string compactified on a circle, or one-dimensional torus, as discussed above. The lowest mass state is the tachyon which has no oscillator excitations, winding or Kaluza–Klein momenta, and is represented by the state

$$|\psi\rangle = |0, 0\rangle \tag{10.1.23}$$

which has $(\text{mass})^2 = -4/\alpha'$.

The masses of the physical states with non-zero h and l depend on the radius of the circle and so we will consider states according to their value of hl. We begin by considering $hl = 0$ for which equation (10.1.20) implies that $N = \bar{N}$. In this case either $l = 0$ or $h = 0$, and the corresponding $(\text{mass})^2$ is given by equation (10.1.19) to be

$$\frac{h^2}{R^2} + \frac{4}{\alpha'}(N - 1) \quad \text{and} \quad \frac{R^2 l^2}{(\alpha'^2)} + \frac{4}{\alpha'}(N - 1),$$

respectively. The simplest possibility is to take $h = 0$ and $l = 0$ whose states therefore have $N = \bar{N}$ and the $(\text{mass})^2 = (4/\alpha')(N - 1)$. In addition to the tachyon state considered above, we have the states for which $N = \bar{N} = 1$, which are as follows:

$$\alpha_{-1}^\mu \bar{\alpha}_{-1}^\nu | 0, 0\rangle, \quad (\alpha_{-1}^\mu \bar{\alpha}_{-1}^{25} + \alpha_{-1}^{25} \bar{\alpha}_{-1}^\mu)| 0, 0\rangle, \quad (\alpha_{-1}^\mu \bar{\alpha}_{-1}^{25} - \alpha_{-1}^{25} \bar{\alpha}_{-1}^\mu)| 0, 0\rangle,$$

$$\alpha_{-1}^{25} \bar{\alpha}_{-1}^{25} | 0, 0\rangle, \quad \mu, \nu = 0, 1, \ldots, 24. \tag{10.1.24}$$

These are the only massless states with $h = 0 = l$. The first of these states describes the graviton, an anti-symmetric tensor and a scalar, while the remainder describe two vectors and one more scalar (the dilaton). These massless states are equivalent to those found in the standard Kaluza–Klein compactification on a circle of a point particle theory consisting of gravity coupled to a second rank anti-symmetric gauge field and a scalar. For higher values of $N = \bar{N}$ we find massive states.

We next consider physical states that have $h = \pm 1$, $l = 0$, $N = 0 = \bar{N}$ with $(\text{mass})^2 = (1/R^2) - (4/\alpha')$ and those with $h = 0$, $l = \pm 1$, $N = 0 = \bar{N}$ with $(\text{mass})^2 = (R^2/(\alpha'^2)) - (4/\alpha')$. Whether these spinless states, which are given by $|\pm 1, 0\rangle$ and $|0, \pm 1\rangle$, respectively, are tachyons, massless or have $(\text{mass})^2 > 0$ depends on the value of R. For generic values of R they are massive, but we note that one of the two states is massless if either $R = 2\sqrt{\alpha'}$ or $R = \frac{1}{2}\sqrt{\alpha'}$. In addition, we have the states $|\pm 2, 0\rangle$ and $|0, \pm 2\rangle$ also with $N = 0 = \bar{N}$ which have $(\text{mass})^2$ given by $4((1/R^2) - (1/\alpha'))$ and $4((R^2/(\alpha')^2) - (1/\alpha'))$, respectively. We observe that for $R = \sqrt{\alpha'}$ these four spin-0 states are massless.

Let us now consider the states with non-zero hl. The state for which $hl = \pm 1$ must have $h = \pm 1$ and $l = \pm 1$ or $h = \pm 1, l = \mp 1$. Equation (10.1.20) implies that $-N + \bar{N} = \pm 1$ and so either $\bar{N} = 1$, $N = 0$ or $N = 1$, $\bar{N} = 0$ depending on the choice of sign and taking only the lowest possible values. These states can only be

$$\bar{\alpha}^{\mu}_{-1}|\pm 1, \pm 1\rangle, \quad \bar{\alpha}^{25}_{-1}|\pm 1, \pm 1\rangle \tag{10.1.25}$$

and

$$\alpha^{\mu}_{-1}|\pm 1, \mp 1\rangle, \quad \alpha^{25}_{-1}|\pm 1, \mp 1\rangle \tag{10.1.26}$$

giving four vectors and four more scalars, all of which have

$$(\text{mass})^2 - \frac{1}{R^2} + \frac{1}{(\alpha')^2}R^2 - \frac{2}{\alpha'} = \frac{1}{R^2\alpha'^2}(R^2 - \alpha')^2. \tag{10.1.27}$$

No matter what choice of R we adopt, the $(\text{mass})^2$ of the above states is always positive as is evident from the inequality $((1/R) - (R/\alpha'))^2 \geq 0$. For $R = \sqrt{\alpha'}$, however, the particles are massless, giving four vectors and four scalars. The next set of states have $hl = \pm 2$. Consequently $\bar{N} - N = \pm 2$ and so either N or $\bar{N}$ is 2 or more and all such states are massive. The same conclusion holds for $|hl| > 2$.

The most important fact that emerges from the above discussion of the mass spectrum is that the number of massless states depends on the value of R. For a generic value of R we have as massless states only the graviton, anti-symmetric tensor and scalar as well as two vectors and one further scalar given in equation (10.1.24). The vectors must generate the gauge group $U(1)_L \times U(1)_R$, the left and right indices referring to the way the states arise as left and right moving objects on the circle.

For the value $R = \sqrt{\alpha'}$, we find the same massless states as for a generic R, that is, the graviton, anti-symmetric tensor, two vectors and a scalar of equation (10.1.24); however, in addition we find a further four scalars in the paragraph above equation (10.1.25) as well as four vectors and another four scalars in equations (10.1.25) and (10.1.26). Thus the massless states are a graviton, an anti-symmetric tensor, the dilaton, six vectors and nine scalars. Only by examining the interacting theory can one tell which six-dimensional gauge group these vectors generate. Assuming it does not contain a series of $U(1)$ factors it must be $SU(2)_L \times SU(2)_R$ and indeed, this is the case. As one might expect, by examining their left–right indices the nine scalars belong to the $(\underline{3},\underline{3})$ representation of $SU(2)_L \times SU(2)_R$. We disuss this further at the end of this section.

These additional symmetries for $R = \sqrt{\alpha'}$ do not correspond to isometries of the circle as the symmetry that arises from the string compactification is non-Abelian while the circle only has Abelian isometries. These additional massless states do not therefore arise in a Kaluza–Klein compactification and their presence arises as a special feature of string theory and in particular the possibility for strings to wind around the circle. Since the number of massless vectors and their gauge symmetries are of vital importance when trying to reconcile string theory with nature, the existence of additional symmetries is of considerable interest.

Examining equation (10.1.19) we note that the masses of the states are invariant under [10.2, 10.3]

$$R \leftrightarrow \frac{\alpha'}{R} \tag{10.1.28}$$

provided we also interchange the momentum and winding number (that is, $h \leftrightarrow l$). Examining equation (10.1.16), we find that this is equivalent to the changes $\alpha_0^{25} \rightarrow \alpha_0^{25}$ and $\bar{\alpha}_0^{25} \rightarrow -\bar{\alpha}_0^{25}$. We can generalise this to all the oscillators by taking $\alpha_n \rightarrow \alpha_n^{25}$ and $\bar{\alpha}_n^{25} \rightarrow -\bar{\alpha}_n^{25}$. All momenta and oscillators in the remaining 24 directions are unchanged. One can easily verify that this preserves the mass shell conditions of equations (10.1.17)–(10.1.20) and so it preserves the spectrum of the string. In fact, this transformation is equivalent to the simple transformation $x_L^{25} \rightarrow x_L^{25}$ and $x_R^{25} \rightarrow -x_R^{25}$. This is our first example of what has become known as a T duality transformation. It has the interesting consequence that the mass spectrum is the same whether the circle has a very large, or a very small radius, provided the radii are related as above. In fact, it can also be shown to be a symmetry of the scattering amplitudes of the compactified string. It is our first hint that a string cannot probe arbitrarily small distances.

As a final comment on the circle compactification we elucidate the ambiguity inherent in equations (10.1.9), (10.1.10) and (10.1.11) for q_L and q_R. Given the existence of winding numbers, it is natural to introduce an operator L, where

$$L = \frac{\sqrt{2\alpha'}}{2}(\alpha_0^{25} - \bar{\alpha}_0^{25}). \tag{10.1.29}$$

Clearly, on the state $|h, l\rangle$ the effect of L is

$$L|h, l\rangle = Rl|h, l\rangle. \tag{10.1.30}$$

Let us now introduce the conjugate operator W, which by definition satisfies $[W, L] = i$, $[W, q^{25}] = 0 = [W, p^{25}]$. Similar to the derivation of the equation

$$e^{ip^{25}q^{25}}|h, l\rangle = |h + p^{25}, l\rangle \tag{10.1.31}$$

we find that

$$e^{il'W}|h, l\rangle = |h, l + l'\rangle. \tag{10.1.32}$$

and so W creates winding states. It is therefore natural to define

$$q_L^{25} = \frac{q^{25}}{2} + \frac{\alpha'}{2}W, \quad q_R^{25} = \frac{q^{25}}{2} - \frac{\alpha'}{2}W \tag{10.1.33}$$

and we then find that the commutators are

$$[q_L^{25}, p_L^{25}] = i, \quad [q_R^{25}, p_R^{25}] = i$$

$$[q_R^{25}, p_L^{25}] = [q_L^{25}, p_R^{25}] = 0, \quad [p_L^{25}, p_R^{25}] = [q_L^{25}, q_R^{25}] = 0. \tag{10.1.34}$$

Although W does not appear in $x^{25}(\tau, \sigma)$, it does appear in the vertex operator which is of the form

$$:e^{ip_L x_L^{25}} e^{ip_R x_R^{25}}: \tag{10.1.35}$$

times polynomials in ∂x^{25} and $\bar{\partial} x^{25}$ together with expressions involving $x^\mu, \mu = 0, 1, \ldots, 24$ and its derivatives. This expression contains the terms required to create momenta and winding numbers, namely it has the sub-factor

$$e^{ip_L q_L} e^{ip_R q_R} = e^{ip^{25} q^{25}} e^{iL.W}. \tag{10.1.36}$$

We note that in contrast the usual vertex operator for an uncompactified direction contains the factor $e^{ip^{24} x^{24}(\tau,\sigma)}$. This operator differs from that given in equation (10.1.35) and makes apparent the different way compactified and ordinary strings are treated.

We now discuss in more detail the symmetry enhancement at the self-dual radius $R = \sqrt{\alpha'}$. The vector particles arising from the metric and anti-symmetric tensor field, that is, those of the usual Kaluza–Klein origin, have, respectively, the forms

$$:A_\mu^{(1)} (\partial x^\mu \bar{\partial} x^{25} + \bar{\partial} x^\mu \partial x^{25}) e^{ik_\nu x^\nu}: \quad \text{and}$$

$$:A_\mu^{(2)} (\partial x^\mu \bar{\partial} x^{25} - \bar{\partial} x^\mu \partial x^{25}) e^{ik_\nu x^\nu}: \mu = 0, 1, \ldots, 24, \tag{10.1.37}$$

where $A_\mu^{(1)}$ and $A_\mu^{(2)}$ are the polarisation tensors. While the vector states of equations (10.1.25) and (10.1.26) have the forms

$$:A_\mu^{(L)} \bar{\partial} x^\mu e^{\pm(2i/\sqrt{\alpha'})x_L^{25}} e^{ik_\nu x^\nu}: \quad \text{and} \quad :A_\mu^{(R)} \partial x^\mu e^{\pm(2i/\sqrt{\alpha'})x_R^{25}} e^{ik_\nu x^\nu}:, \tag{10.1.38}$$

where we have used equation (10.1.22) to find that $p_L = (1/\sqrt{\alpha'})(h + l)$ and $p_R = (1/\sqrt{\alpha'})(h - l)$, we have employed the expression for the vertex operator of equation (10.1.35), and $A_\mu^{(L)}$ and $A_\mu^{(R)}$ are the polarisation tensors.

The vertex operator corresponding to the Kaluz–Klein scalar of equation (10.1.24) is of the form

$$:\phi_{33} \partial x^{25} \bar{\partial} x^{25} e^{ik_\nu x^\nu}:. \tag{10.1.39}$$

This can be written together with the other eight scalars as

$$\phi_{ij} : j_L^i(z) j_R^j(\bar{z}) e^{ik_\nu x^\nu}:, \tag{10.1.40}$$

where

$$j_L^3 =: i\partial x_L^{25}:, \quad j_L^\pm =: e^{\pm(2i/\sqrt{\alpha'})x_L^{25}}: \tag{10.1.41}$$

and the expressions for $j_R^j(\bar{z})$ are identical except for $L \rightarrow R$ and $\partial \rightarrow \bar{\partial}$. We note that for the self-dual radius $R = \sqrt{\alpha'}$ all these vertex operators are single valued as we go around the circle, that is, $x^{25} \rightarrow x^{25} + 2\pi \sqrt{\alpha'}$.

The vertex operators which appear in correlation functions are integrated over the world sheet and so for the vertex operator of equation (10.1.39), with zero momentum in the uncompactified directions, we find the factor $\int d^2z \phi_{33} \partial x^{25} \bar{\partial} x^{25}$. For small ϕ_{33} this is the same as $\exp(\int d^2z \phi_{33} \partial x^{25} \bar{\partial} x^{25} - 1)$ and we can interpret the first term as an addition to the world-sheet action which simply changes the g_{2525} component of the background metric. However, the physical radius of the circle is $2\pi R \sqrt{g_{2525}}$ and so it is equivalent to a change in the radius of the circle. Hence, a shift in the scalar ϕ_{33} corresponds to a change in the physical radius of compactification of the circle.

In fact, one can readily show, using the vertex operator construction of Lie algebras of section 16.8, that the vertex operators $j^i_L(z)$ and $j^j_R(\bar{z})$ lead to the algebra $SU(2)_L \otimes SU(2)_R$. Indeed using the latter parts of this section one can even construct the affine extension of this algebra.

If the radius of the circle R changes from its self-dual value of $\sqrt{\alpha'}$, then four of the vectors become massive and the gauge symmetry is reduced to $U(1)_L \otimes U(I)_R$. Indeed, using equation (10.1.27) we find that for a small deviation $R = \sqrt{\alpha'}(1 + \epsilon)$ they acquire the mass squared $4\epsilon^2/\alpha'$. We observed that this small change is induced by a shift in the scalar ϕ_{33}. Thus a shift in this scalar gives the four vectors a mass and the gauge symmetry changes from $SU(2)_L \otimes SU(2)_R$ to $U(1)_L \otimes U(1)_R$. It is also straightforward to compute the massess of the nine scalars. This is very reminiscent of the Higgs effect. Indeed, one can use the above vertex operators to compute the correlation functions of the vectors and the scalars, as explained in chapter 9, and so deduce the effective action. One finds that the nine scalars have a potential which possesses a flat direction involving shifts in the ϕ_{33} field. As such, one can check that this really does correspond to the Higgs effect.

10.2   Torus compactification [10.4, 10.5]

The compactification on a circle given in the previous section can be generalised in a relatively straightforward manner to a compactification on a higher-dimensional torus. To define the higher-dimensional torus it will be instructive to first give a definition of a circle that is easily generalised. Given $x \in \mathbf{R}^1$, the real numbers, we may define the equivalence relation $x' \sim x$ if

$$x' = x + 2\pi nR, \text{ for some } n \in \mathbf{Z}. \tag{10.2.1}$$

The above equivalence relation means that we are to identify any two points x' and x of $\mathbf{R}^1$ if $x' \sim x$. The set of all equivalence classes of $\mathbf{R}^1$ with respect to the above equivalence relation is a circle of radius R. The set $\mathbf{R}^1$ is a group with addition being the group operation. A subgroup of $\mathbf{R}^1$ is the set

$$\Lambda^1 = \{2\pi nR, n \in \mathbf{Z}\}. \tag{10.2.2}$$

The equivalence relation of equation (10.2.1) is just the standard one for the group $\mathbf{R}^1$ with respect to its subgroup Λ^1. The resulting factor group, which by definition is the set of all equivalence classes, with the inherited group multiplication, is the coset space $S^1 = \mathbf{R}^1/\Lambda^1$, which is just the circle.

To generalise the above we take $\mathbf{R}^D$ to replace $\mathbf{R}^1$ and replace Λ^1 by

$$\Lambda^D = \left\{ \sum_{i=1}^{D} n_i e_i, n_i \in \mathbf{Z} \right\}, \tag{10.2.3}$$

where $e_i = e_i^J = (e_i^1, e_i^2, \ldots, e_i^D)$, $i = 1, \ldots, D$, are D linearly independent vectors in $\mathbf{R}^D$. The set Λ^D defines a lattice and is obviously a sub-group of $\mathbf{R}^D$. The corresponding equivalence relation, throwing in a factor of 2π, is given by

$$x' \sim x \text{ for } x', \ x \in \mathbf{R}^D \text{ if } x' = x + 2\pi \sum_{i=1}^{D} n_i e_i \text{ for some } n_i \in \mathbf{Z}. \tag{10.2.4}$$

That is, x and x' are equivalent if their difference is a member of the lattice Λ^D times 2π. The torus T^D is by definition the coset space

$$T^D = \frac{\mathbf{R}^D}{2\pi \Lambda^D}. \tag{10.2.5}$$

The factor $2\pi \Lambda^D$ means the lattice Λ^D scaled up by 2π. The torus T^D inherits the flat Euclidean space metric of $\mathbf{R}^D$.

It is now straightforward to describe the string compactified on T^D: we essentially replace the index 25 of the previous section by the internal indices I and sum on I where appropriate. As the logic is unchanged we simply summarise the results. The compactified components $x^I(\tau, \sigma)$ of the string are subject to the equivalence relation of equation (10.2.4) and the winding condition around the torus is given by

$$x^I(\tau, \sigma + 2\pi) = x^I(\tau, \sigma) + 2\pi L^I; \ I = 25 - D + 1, \ldots, 25 \tag{10.2.6}$$

for $L^I \in \Lambda^D$. Introducing the oscillators as before by the definition

$$\wp^I \equiv \frac{1}{\sqrt{2\alpha'}}(\dot{x}^I + x'^I) \equiv \sum_{n=-\infty}^{\infty} \alpha_n^I e^{-in\sigma}, \tag{10.2.7}$$

$$\bar{\wp}^I \equiv \frac{1}{\sqrt{2\alpha'}}(\dot{x}^I - x'^I) \equiv \sum_{n=-\infty}^{\infty} \bar{\alpha}_n^I e^{in\sigma}, \tag{10.2.8}$$

we may write $x^I(\tau, \sigma)$ as

$$x^I(\tau, \sigma) = x_L^I(\tau + \sigma) + x_R^I(\tau - \sigma), \tag{10.2.9}$$

where

$$x_L^I = q_L^I + \frac{\sqrt{2\alpha'}}{2}\alpha_0^I(\tau + \sigma) + i\sqrt{\frac{\alpha'}{2}} \sum_{\substack{n=-\infty \\ n\neq 0}}^{\infty} \frac{1}{n}\alpha_n^I \, e^{-in(\tau+\sigma)}, \tag{10.2.10}$$

$$x_R^I = q_R^I + \frac{\sqrt{2\alpha'}}{2}\bar{\alpha}_0^I(\tau - \sigma) + i\sqrt{\frac{\alpha'}{2}} \sum_{\substack{n=-\infty \\ n\neq 0}}^{\infty} \frac{1}{n}\bar{\alpha}_n^I e^{-in(\tau-\sigma)} \tag{10.2.11}$$

and

$$q_L^I = \frac{q^I}{2} + \frac{\alpha'}{2}W^I, q_R^I = \frac{q^I}{2} - \frac{\alpha'}{2}W^I, \ \alpha_0^I + \bar{\alpha}_0^I = \sqrt{2\alpha'}p^I, \alpha_0^I - \bar{\alpha}_0^I = \frac{2}{\sqrt{2\alpha'}}L^I. \tag{10.2.12}$$

We also define

$$p_L^I = \frac{2}{\sqrt{2\alpha'}} \alpha_0^I, \quad p_R^I = \frac{2}{\sqrt{2\alpha'}} \bar{\alpha}_0^I. \tag{10.2.13}$$

These operators obey the commutation relations

$$[\alpha_n^I, \alpha_m^J] = n\delta_{n+m,0}\delta^{IJ}, \quad [\bar{\alpha}_n^I, \bar{\alpha}_m^J] = n\delta_{n+m,0}\delta^{IJ}, \quad [W^I, L^J] = i\delta^{IJ},$$

$$[q^I, p^J] = i\delta^{IJ}, \tag{10.2.14}$$

all other comutators vanishing. Imposing that the wavefunction $\exp(i \sum_{I=1}^D p^I x^I(\tau, \sigma))$ be single valued on T^D, that is, under equation (10.2.6), implies that

$$\sum_{I=1}^D p^I L^I \in \mathbf{Z}. \tag{10.2.15}$$

The physical state condition $(L_0 + \bar{L}_0 - 2)\psi = 0$ implies that the mass levels are given by

$$(\text{mass})^2 = \frac{1}{2} \sum_{I=1}^D \{(p_L^I)^2 + (p_R^I)^2\} + \frac{2}{\alpha'}(N + \bar{N} - 2), \tag{10.2.16}$$

while the $(L_0 - \bar{L}_0)\psi = 0$ condition implies

$$\frac{\alpha'}{4} \sum_{I=1}^D \{(p_L^I)^2 - (p_R^I)^2\} = -(N - \bar{N}). \tag{10.2.17}$$

To interpret the above conditions it will prove advantageous to develop the theory of lattices. The D-dimensional lattice is given by

$$\Lambda = \left\{ \sum_{i=1}^N n_i e_i : n_i \in \mathbf{Z} \right\}, \tag{10.2.18}$$

where e_i, $i = 1, \ldots, N$, form a basis of a vector space V equipped with a scalar product. In our previous consideration $V = \mathbf{R}^D$, but we will also be interested in the case $V = \mathbf{R}^{N-M,M}$. The corresponding lattices have a scalar product which they inherited from the scalar product of the underlying vector space. We take the metric of $V = \mathbf{R}^D$ to be that for flat Euclidean space and so it just has $+1$s down the diagonal, while for the case of $V = \mathbf{R}^{N-M,M}$, the metric is diagonal and has $+1$s in the first $N - M$ entries and -1s in the last M entries. We call the lattice Λ Euclidean or Lorentzian when V is Euclidean or Lorentzian, respectively. We take a vector space whose diagonal metric that has one -1 and otherwise $+1$s on its diagonal to be called Lorentzian. A lattice is said to be integral if $x \cdot y \in \mathbf{Z}$ for all $x, y \in \Lambda$. An even lattice is an integral lattice for which $x \cdot x \in 2\mathbf{Z} \,\forall\, x \in \Lambda$. If an integral lattice is not even it is called odd. A more complete discussion of lattices and their properties can be found in chapter 16.

The matrix $g_{ij} \equiv e_i \cdot e_j$ is the metric of the lattice inherited from the underlying vector space. The unit cell of a lattice is the set of points $\{x = \sum_i a_i e_i : 0 \le a_i < 1\}$; clearly it only contains one lattice point, namely $x = 0$. The volume of the unit cell denoted vol(Λ) is then given by vol$(\Lambda) = \sqrt{|\det g_{ij}|}$. A lattice is unimodular if vol$(\Lambda) = 1$ in which case there is only one point of Λ per unit volume.

The dual lattice Λ^* is the set of points

$$\Lambda^* = \{y : y \cdot x \in \mathbf{Z} \, \forall \, x \in \Lambda\}. \tag{10.2.19}$$

If the inner product is non-singular, Λ^* is a lattice with a basis consisting of the vectors e_i^*, $i = 1, \ldots, N$, where $e_i^* \cdot e_j = \delta_{ij}$. The metric on Λ^* is $g_{ij}^* = e_i^* \cdot e_j^*$ and the volume of the unit cell $\text{vol}(\Lambda^*) = \sqrt{|\det g_{ij}^*|}$. The definition of e_i^* implies that $\sum_{i=1} e_i (e_i^* \cdot e_j) = e_j$, which, in turn, implies the result

$$\sum_{i=1}^{N} e_i^I e_i^{J*} = \delta^{IJ}. \tag{10.2.20}$$

Consequently, the metrics of the lattices Λ and Λ^* are related by

$$g_{ij}^* = (g_{ij})^{-1} \tag{10.2.21}$$

and $\text{vol}(\Lambda^*) = (\text{vol}(\Lambda))^{-1}$. A lattice is self-dual if $\Lambda^* = \Lambda$. Clearly, Λ is integral if and only if $\Lambda \subset \Lambda^*$. However, if $\text{vol}(\Lambda) = 1$, then Λ and Λ^* have the same number of points per unit volume and so $\Lambda = \Lambda^*$ if and only if Λ is unimodular and integral.

To give a simple example, let us consider the two-dimensional lattice $\Lambda_{SU(3)}$ generated by the roots of SU(3). A suitable basis for the roots of SU(3) is provided by the simple roots which can be chosen to be $e_1 = (\sqrt{2}, 0)$, $e_2 = (-\frac{1}{\sqrt{2}}, \sqrt{\frac{3}{2}})$. The dual lattice is generated by $e_1^* = (\frac{1}{\sqrt{2}}, \sqrt{\frac{1}{6}})$, $e_i^* = (0, \sqrt{\frac{2}{3}})$. Since

$$g_{ij} = \begin{pmatrix} 2 & -1 \\ -1 & 2 \end{pmatrix},$$

we conclude that $\Lambda_{SU(3)}$ is an integral and even lattice. However, $\text{vol} \Lambda_{SU(3)} = \sqrt{3}$, implying that $\Lambda_{SU(3)}$ is not unimodular.

The introduction of the lattice and its related concepts allows us to give a natural interpretation to the above conditions. The condition of single valuedness of the wavefunction of equation (10.2.15) can be restated as the $p \in \Lambda^{D*}$ and therefore may be written as

$$p^I = \sum_{i=1}^{D} m_i e_i^{I*}; \quad m_i \in \mathbf{Z}. \tag{10.2.22}$$

Clearly, if $L = \sum_{i=1}^{D} n_i e_i$, then $\sum_{I=1}^{D} p^I L^I = \sum_{i=1}^{D} m_i n_i$.

To interpret equation (10.2.17) we introduce an extended lattice $\Lambda^{D,D}$ which consists of all $2D$ component vectors of the form $W = (w_L, w_R)$, where

$$w_L^I = \sqrt{\frac{\alpha'}{2}} p_L^I = \alpha_0^I = \sqrt{\frac{\alpha'}{2}} p^I + \frac{1}{\sqrt{2\alpha'}} L^I$$

$$= \sqrt{\frac{\alpha'}{2}} \sum_{i=1}^{D} m_i e_i^{I*} + \frac{1}{\sqrt{2\alpha'}} \sum_{i=1}^{D} n_i e_i^I, \tag{10.2.23}$$

$$w_R^I = \sqrt{\frac{\alpha'}{2}} p_R^I = \bar{\alpha}_0^I = \sqrt{\frac{\alpha'}{2}} p^I - \frac{1}{\sqrt{2\alpha'}} L^I$$

$$= \sqrt{\frac{\alpha'}{2}} \sum_{i=1}^{D} m_i e_i^{I*} - \frac{1}{\sqrt{2\alpha'}} \sum_{i=1}^{D} n_i e_i^I. \tag{10.2.24}$$

This lattice has dimension $2D$. It is embedded in the underlying space $R^{D,D}$ that has a non-Euclidian metric which is diagonal and its first D entries along the diagonal are 1 and its last D entries are -1. As a result, the scalar product of two lattice elements is given by

$$W \cdot W' = \sum_{I=1}^{D} w_L^I w_L^{I'} - \sum_{I=1}^{D} w_R^I w_R^{I'}. \tag{10.2.25}$$

Using equation (10.2.24) we conclude that

$$W \cdot W' = \frac{\alpha'}{2}(p_L \cdot p'_L - p_R \cdot p'_R) = \sum_{i=1}^{D}(m_i n'_i + m'_i n_i). \tag{10.2.26}$$

The $2D$ basis elements, or generators, of the lattice $\Lambda^{D,D}$ are given by

$$\sqrt{\frac{\alpha'}{2}}(e_i^{I*}, e_i^{I*}) \quad \text{and} \quad \frac{1}{\sqrt{2\alpha'}}(e_i^I, -e_i^I). \tag{10.2.27}$$

The metric of this lattice is of the form

$$\begin{pmatrix} 0 & 1_{D \times D} \\ 1_{D \times D} & 0 \end{pmatrix}, \tag{10.2.28}$$

where each entry represents a $D \times D$ matrix and $1_{D \times D}$ is the $D \times D$ unit matrix. We note that the determinant of the lattice metric is $(-1)^D$.

Equation (10.2.26) implies that $\Lambda^{D,D}$ is an even lattice as $W \cdot W = 2\sum_{i=1}^{D} m_i n_i$. Further, as the determinant of the lattice metric has modulus 1, $\Lambda^{D,D}$ is self-dual. It is important to stress that $\Lambda^{(D,D)}$ has a signature which has nothing to do with that of space-time, which is Euclidean in the compactified directions.

In terms of the lattice $\Lambda^{D,D}$ the physical state condition of equation (10.2.17) becomes

$$W \cdot W = w_L^2 - w_R^2 = 2\sum_{i=1}^{D} m_i n_i = -2(N - \bar{N}), \tag{10.2.29}$$

while the masses are given by

$$\alpha' m^2 = w_L^2 + w_R^2 + 2(N + \bar{N} - 2)$$

$$= \alpha' \sum_{i,j} m_i g_{ij}^* m_j + \frac{1}{\alpha'} \sum_{i,j} n_i g_{ij} n_j + 2(N + \bar{N} - 2). \tag{10.2.30}$$

Clearly, massless states can only occur if $N + \bar{N} \le 2$.

To find the massless physical states for a given lattice Λ^D we must find sets of integers m_i, n_i such that $m^2 = 0$ in equation (10.2.30) and solve equation (10.2.29). We begin with states with winding numbers $L^I = \sum_{i=1}^{D} n_i e_i = 0$, which implies that $N = \bar{N}$. If the states with $N = \bar{N} = 0$ have no oscillator excitations and hence must be a scalar which will only be massless if $w_L^2 = w_R^2 = 2$. We also consider the case when $N = \bar{N} = 1$. Such states which are massless must have $w_L = w_R = 0$, or equivalently, $n_i = m_i = 0$ and so do not possess any momenta in the compactified directions. Apart from the usual states that correspond to the graviton, the anti-symmetric tensor gauge field and the dilaton, these states are of the forms

$$\alpha_{-1}^I \bar{\alpha}_{-1}^\mu |0, 0\rangle, \, \alpha_{-1}^\mu \bar{\alpha}_{-1}^I |0, 0\rangle \tag{10.2.31}$$

as well as

$$\alpha_{-1}^I \bar{\alpha}_{-1}^J |0, 0\rangle. \tag{10.2.32}$$

Here, the state $|w_L^I, w_R^I\rangle$ has no oscillator excitations and the dependence on $p^\mu, \mu = 0, 1, \ldots, 25 - D$ is suppressed. The states of equations (10.2.31) describe $2D$ massless vectors. One would expect these states to correspond to the gauge interaction $U(1)_L^D \times U(1)_R^D$. The states of equations (10.2.32) describe D^2 scalars. These states just correspond to the massless states one would find if one carried out a dimensional reduction on a D-dimensional torus of gravity coupled to an second rank anti-symmetric tensor gauge field and a scalar, that is, they are just the usual Kaluza–Klein states.

Vector massless states must have an oscillator excitation with a μ index and so a non-trivial N or $\bar{N}$. Hence as we examined $N = \bar{N}$ above, we next consider states with non-zero winding number of the forms

$$\alpha_{-1}^\mu |w_L, w_R\rangle \text{ and } \bar{\alpha}_{-1}^\mu |w_L, w_R\rangle. \tag{10.2.33}$$

These states have $N = 1, \bar{N} = 0$ or $N = 0, \bar{N} = 1$, respectively. Examining equations (10.2.29) and (10.2.30) we find that these states will be massless if $w_L^2 = 0, w_R^2 = 2$ and $w_L^2 = 2, w_R^2 = 0$, respectively. Hence we find twice as many massless vectors as there are vectors of length squared 2 in Λ^D. The only other possible massless vectors could arise when $N = 2, \bar{N} = 0$ or $N = 0, \bar{N} = 2$, but then there are no momenta in the compactified directions and one cannot satisfy equation (10.2.29).

It is instuctive to examine a particular example. Let us consider the lattice Λ^D given by $e_i^I = \sqrt{\alpha'}\delta_i^I$. As a result, $g_{ij} = \alpha'\delta_{ij}$, $g_{ij}^* = (\alpha')^{-1}\delta_{ij}$ and $e_i^{*I} = (1/(\sqrt{\alpha'}))\delta_i^I$. If we consider the states of equation (10.2.33) with $N = 1, \bar{N} = 0$, then $2D$ massless vectors occur when $m_i = \pm\delta_{ik}, n_i = \mp\delta_{ik}$ for $k = 1, \ldots, D$. For the states with $N = 0, \bar{N} = 1$, we find additional $2D$ massless vectors if $m_i = \pm\delta_{ik}, n_i = \pm\delta_{ik}$ for $k = 1, \ldots, D$. These, together with those of equation (10.2.31) provide us with $3D + 3D$ massless vectors which, it turns out, generate the gauge group $(SU(2)_L)^D \times (SU(2)_R)^D$. The precise non-Abelian nature of this symmetry only emerges from the construction of the interacting terms. In fact, one can show, in general, that this is largest gauge group that one can generate by any a torus compactification of the type considered in this section. As such, one can at best find can factors of $SU(2)$ and not any larger non-Abelian group.

The T duality symmetry of the string on a circle compactification generalises to the torus. One easily verifies that $e_i \to \alpha' e_i^*$, or equivalently $\Lambda^D \to \alpha'\Lambda^{D*}$, and $e_i^* \to (\alpha')^{-1}e_i$, or equivalently $\Lambda^{D*} \to \alpha^{-1}\Lambda^D$, and $m_i \leftrightarrow n_i$ at the same time, takes $L^I \to \alpha' p^I$, $p^I \to \frac{1}{\alpha'}L^I$ and so $w_L \to w_L$ and $w_R \to -w_R$. As before, we extend this to all the oscillators by taking $\alpha_n^I \to \alpha_n^I$ and $\bar{\alpha}_n^I \to -\bar{\alpha}_n^I$. One can easily verify that this preserves the mass-shell conditions of equations (10.2.29)–(10.2.30) and so it preserves the spectrum of the string. This transformation is equivalent to the simple transformation $x_L^I \to x_L^I$ and $x_R^I \to -x_R^I$, that is, $x = x_L + x_R \to x_L - x_R$. Again this is a symmetry of the string scattering amplitudes.

The vertex operator discussion for the string on a torus extends in a straightforward way from that on a circle and the generalisation of equation (10.2.35) is

$$: \exp p_L^I x_L^I \exp p_R^I x_R^I : . \tag{10.2.34}$$

10.3   Compactification in the presence of background fields [10.6]

In the above, we have assumed the closed string to be compactified on a torus T^D which has an associated lattice Λ^D. The closed string contains the graviton as one of its infinite number of space-time fields. However, the graviton determines the space-time structure and, in particular, the background values of the graviton determine the torus, and hence the associated lattice Λ_D, on which the string propagates. As such, it is natural to consider a formulation of string theory where the space-time metric has constant background values G_{IJ} in the compactified directions and the usual Minkowski metric in the uncompactified directions. Having chosen non-trivial background values for the graviton, it is natural to also consider non-trivial background values for the other massless fields of the closed string. In particular, we will take the massless anti-symmetric tensor gauge field that occurs in the closed string to have constant background values, denoted B_{IJ}, in the compactified directions.

We start from the usual action of equation (2.1.10) except that we replace the space-time Minkowski metric in the compactified directions by the values of the space-time background metric and repeat the discussion of chapter 2 to find the physical state constraints taking into account the quantisation conditions on the compactified momenta. What is not obvious is that the spectrum of states can be influenced by taking the massless anti-symmetric tensor gauge field to have non-trivial background values in the compactified directions. However, we will find this to be the case and indeed, we will find that it can lead to an enhanced gauge symmetry compared to that found in the previous section.

Consequently, we consider a string with action $A = A^U + A^C$, where A^U is the contribution from the uncompactified coordinates and is given by equation (2.1.10), while A^C is the contribution from the compactified coordinates and is given by

$$A^C = -\frac{1}{4\pi\alpha'} \int d^2\xi \sum_{I,J} \{\sqrt{-g}g^{\alpha\beta}\partial_\alpha x^I \partial_\beta x^J G_{IJ} - \epsilon^{\alpha\beta}B_{IJ}\partial_\alpha x^I \partial_\beta x^J\}, \tag{10.3.1}$$

where $\epsilon^{\alpha\beta} = -\epsilon^{\beta\alpha}$, $\epsilon^{01} = 1$. We take the string to live on a torus, and so the coordinates x^I are subject to the boundary condition

$$x^I(\tau, \sigma + 2\pi) = x^I(\tau, \sigma) + 2\pi n^I, \quad n^I \in \mathbf{Z}. \tag{10.3.2}$$

This equation differs from equation (10.2.6) since now the geometry of the torus is encoded in the values of the background fields G_{IJ} and B_{IJ}.

We now follow the discussion of chapter 2, but for the above action. The term containing B_{IJ} is a total derivative and, consequently, it does not contribute to the equations of motion. The boundary contribution that arises from this term when calculating the equation of motion is also automatically satisfied as a result of equation (10.3.2). We may therefore take x^I to be given by the sum of a left-handed and a right-handed piece as in equation (2.1.50). We can also adopt the gauge choice of equation (2.1.68). However, the B_{IJ} term does change the momentum, which becomes

$$P_I(\sigma) = \frac{\delta A^C}{\delta \partial_\tau x^I} = \frac{1}{2\pi\alpha'}(G_{IJ}\dot{x}^J + B_{IJ}x^{J'}). \tag{10.3.3}$$

As before, the constraints follow from the variation of $g_{\alpha\beta}$. Since the second term of equation (10.3.1) is independent of $g_{\alpha\beta}$, the constraints, when expressed in terms of

x^I and G_{IJ}, are independent of B_{IJ}. The contribution to the energy-momentum tensor $T_{\alpha\beta} = T^U_{\alpha\beta} + T^C_{\alpha\beta}$ from the compactified coordinates is given by

$$T^C_{++} = \partial_+ x^I \partial_+ x^J G_{IJ}, \quad T^C_{--} = \partial_- x^I \partial_- x^J G_{IJ}, \tag{10.3.4}$$

where $\partial_\pm x^I = \frac{1}{2}(\dot{x}^I \pm x'^I)$. We introduce the oscillators through the variables

$$\wp^k = \frac{1}{\sqrt{2\alpha'}} e^k_j (\dot{x}^J + x'^J) = \sum_n \alpha^k_n e^{-in\sigma} \tag{10.3.5}$$

and

$$\bar{\wp}^k = \frac{1}{\sqrt{2\alpha'}} e^k_j (\dot{x}^J - x'^J) = \sum_n \bar{\alpha}^k_n e^{in\sigma}. \tag{10.3.6}$$

In these equations, we have introduced the symbols e^k_j which are defined to satisfy $\sum_k e^k_I e^k_J = G_{IJ}$. The reader may observe that we used such a symbol in the last section and it will eventually turn out that the symbols represent the same quantities as there. The coordinates x^I and the momenta P_I of equation (10.3.3) obey the usual Poisson bracket or commutator relations for the classical and quantum theory respectively. Using equation (10.3.3) the above variables can be expressed as

$$\wp^k = \frac{1}{\sqrt{2\alpha'}} e^{*k}_I ((2\pi\alpha')P_I + (G_{IJ} - B_{IJ})x'^J), \tag{10.3.7}$$

$$\bar{\wp}^k = \frac{1}{\sqrt{2\alpha'}} e^{*k}_I ((2\pi\alpha')P_I - (G_{IJ} + B_{IJ})x'^J). \tag{10.3.8}$$

The objects e^{*k}_I are defined to be dual vectors to the e^{kI} in the sense that $\sum_I e^{*k}_I e^l_I = \delta^k_l$. It is straightforward to verify that $\wp^k$ and $\bar{\wp}^k$ obey the same commutation relations as in the absence of the background fields and as a result the oscillators obey their usual commutation relations, the only non-vanishing ones being

$$[\alpha^k_n, \alpha^l_m] = n\delta_{n+m,0}\delta^{kl}, \quad [\bar{\alpha}^k_n, \bar{\alpha}^l_m] = n\delta_{n+m,0}\delta^{kl}. \tag{10.3.9}$$

The contribution to the energy-momentum tensor in equation (10.3.4) from the compactified directions can be written in the form

$$T^C_{++} = \frac{\alpha'}{2} \sum_k \wp^k \wp^k, \quad T^C_{--} = \frac{\alpha'}{2} \sum_k \bar{\wp}^k \bar{\wp}^k. \tag{10.3.10}$$

As a result, the Virasoro operators which are given by equation (2.1.56) have a contribution L^C_n and $\bar{L}^C_n$'s from the compactified coordinates which take the usual form in terms of the oscillators of equation (10.3.5) and (10.3.6).

At first sight it may appear that the background fields B_{IJ} do not play any role in the physical state conditions. However, the wavefunctions we wish to consider contain factors of the form $e^{i\sum_I p_I x^I}$, where p_I are the zero modes of the P_I of equation (10.3.3), that is,

$$p_I = \int_{-\pi}^{\pi} d\sigma P_I. \tag{10.3.11}$$

Under the periodicity condition of equation (10.3.2) single valuedness of the wavefunctions implies that

$$\sum p_I r^I \in \mathbf{Z} \text{ for all } r^I \in \mathbf{Z}, \text{ or equivalently } p_I = m_I \in \mathbf{Z}. \tag{10.3.12}$$

Since it is the momenta which are quantised, we must express all quantities in terms of them and it is in this step that the background fields B_{IJ} enter the game. Adding and subtracting equations (10.3.7) and (10.3.8) we find that

$$\wp^k + \bar{\wp}^k = \frac{2e^{\star kI}}{\sqrt{2\alpha'}}(2\pi\alpha' P_I - B_{IJ}x^J) \tag{10.3.13}$$

and

$$\wp^k - \bar{\wp}^k = \frac{2e^{\star kI}}{\sqrt{2\alpha'}}(G_{IJ}x^J). \tag{10.3.14}$$

Taking the zero mode of equations (10.3.13) and (10.3.14) and the conditions of equations (10.3.2) and (10.3.12) we find that

$$w_L^k \equiv \alpha_0^k = e^{\star kI}\left(\sqrt{\frac{\alpha'}{2}}m_I - \frac{1}{\sqrt{2\alpha'}}B_{IJ}n^J + \frac{1}{\sqrt{2\alpha'}}G_{IJ}n^J\right) \tag{10.3.15}$$

and

$$w_R^k \equiv \bar{\alpha}_0^k = e^{\star kI}\left(\sqrt{\frac{\alpha'}{2}}m_I - \frac{1}{\sqrt{2\alpha'}}B_{IJ}n^J - \frac{1}{\sqrt{2\alpha'}}G_{IJ}n^J\right). \tag{10.3.16}$$

We can identify the $2D$ component vectors of the form $W^T = (w_L, w_R)$ as belonging to a $\Lambda^{D,D}$, which as we will now show is even and self-dual. We define the scalar product of the lattice to be such that for any two vectors W and W'

$$W \cdot W' = \sum_k \{w_L^k w_L^{k'} - w_R^k w_R^{k'}\} = \sum_I \{n^I m_I' + n^{I'} m_I\}. \tag{10.3.17}$$

Clearly, $W \cdot W' = 2\sum_I n^I m_I$, so the lattice is even. The basis vectors of $\Lambda^{D,D}$ are given by

$$F_1^{kI} = \sqrt{\frac{\alpha'}{2}}\begin{pmatrix} e^{\star kI} \\ e^{\star kI} \end{pmatrix}, \quad F_2^{kI} = \frac{1}{\sqrt{2\alpha'}}\begin{pmatrix} e^{\star kJ}(G_{JI} - B_{JI}) \\ e^{\star kJ}(-G_{JI} - B_{JI}) \end{pmatrix}, \tag{10.3.18}$$

where the index I labels the basis vectors and the index k their individual components. We note that $W = F_1^I m_I + F_2^I n_I$. We note for future purposes that the basis can be written as a $2D \times 2D$ matrix,

$$F = (F_1 \quad F_2). \tag{10.3.19}$$

The metric of the lattice $\Lambda^{D,D}$ can be computed from the basis vectors of equation (10.3.18) using the scalar product of equation (10.3.17); denoting the metric by J, we final it is given by

$$J = \begin{pmatrix} 0 & 1 \\ 1 & 0 \end{pmatrix}. \tag{10.3.20}$$

Since $|\det J| = 1$ the lattice is unimodular and so $\Lambda^{D,D}$ is an even self-dual lattice. Hence, for each set of values of the background fields G_{IJ}, B_{IJ} we have constructed an even self-dual lattice $\Lambda^{D,D}$.

We observe that

$$(w_L)^2 + (w_R)^2 = Z^T M Z, \tag{10.3.21}$$

where

$$Z = \begin{pmatrix} m \\ n \end{pmatrix} \tag{10.3.22}$$

and

$$M = \begin{pmatrix} \alpha' G^{-1} & -G^{-1}B \\ BG^{-1} & \dfrac{1}{\alpha'}(G - BG^{-1}B) \end{pmatrix}. \tag{10.3.23}$$

We can absorb the factors of α' by making the field redefinitions $G_{IJ} \to \alpha' G_{IJ}$, $B_{IJ} \to \alpha' B_{IJ}$.

The physical state conditions are given by equations (3.1.45) and (3.1.46). The latter condition becomes

$$(L_0 - \bar{L}_0)\psi = \left\{ \sum_I n^I m_I + N - \bar{N} \right\} |\psi\rangle = \left(\frac{1}{2} Z^T J Z + N - \bar{N} \right) |\psi\rangle = 0, \tag{10.3.24}$$

where

$$N = \sum_k \sum_{n=1}^{\infty} \alpha^k_{-n} \alpha^k_n + \sum_\mu \sum_{n=1}^{\infty} \alpha^\mu_{-n} \alpha^\mu_n \text{ and } \bar{N} = \sum_k \sum_{n=1}^{\infty} \bar{\alpha}^k_{-n} \bar{\alpha}^k_n + \sum_\mu \sum_{n=1}^{\infty} \bar{\alpha}^\mu_{-n} \bar{\alpha}^\mu_n.$$

The other condition involving L_0 and $\bar{L}_0$ is given by

$$(L_0 + \bar{L}_0 - 2)\psi = \left\{ \frac{\alpha'}{2} p^\mu p_\mu + \frac{1}{2} Z^T M Z + N + \bar{N} - 2 \right\} |\psi\rangle = 0 \tag{10.3.25}$$

and so

$$\alpha' \text{mass}^2 = Z^T M Z + 2(N + \bar{N} - 2). \tag{10.3.26}$$

Of course, one must also enforce the other physical state conditions $L_n|\psi\rangle = 0$, $\bar{L}_n|\psi\rangle = 0$, $n \geq 1$.

The spectrum of the resulting string which is determined by the above two conditions involves momenta and winding in the uncompactified directions. However, these conditions depend on the D^2 values of the background fields G_{IJ}, B_{IJ}, or, equivalently, on the possible values of w_L, w_R, which when combined into W belong to the even self-dual lattice $\Lambda^{(D,D)}$.

Examining the form of the vectors given in equations (10.3.15) and (10.3.16) when $B_{IJ} = 0$, and comparing them with those of equations (10.2.23) and (10.2.24) we can identify the background field G_{IJ} as the lattice metric we adopted in the previous section. Thus the lattice metric is given by the values of the space-time metric in the compactified directions.

In order to demonstrate the possibility of gaining a larger gauge group we will give a simple example based on the two-dimensional root lattice, $\Lambda_{SU(3)}$ of SU(3) which has the basis vectors e_1 and e_2 given earlier. Let us choose $\alpha' = 2$ for simplicity and we take $B_{IJ} = -3\epsilon_{IJ}$. Clearly, there are massless vector physical states which have $w_L = w_R = 0$ and are of the form of equation (10.2.31). These four vectors are of Kaluza–Klein origin. Examining the physical state condition of equation (10.3.25), we conclude that further massless vector states must have either $\bar{N} = 1$, $N = 0$ and as a result $w_L^2 = 2$, $w_R = 0$ or $\bar{N} = 0$, $N = 1$ and $w_L^2 = 2$, $w_R = 0$. Let us consider states of the first type. From equation (10.3.16) we find that $w_R = 0$ implies that $m_1 = n_1 - 2n_2$, $m_2 = n_1 + n_2$. From equation (10.3.24) we conclude that $\sum_I n^I m_I = 1$ and so $n_1^2 + n_2^2 - n_1 n_2 = 1$, or $(n_1 - n_2)^2 + n_1^2 + n_2^2 = 2$. The only solutions for (n_1, n_2) are $(\pm 1, 0)$, $(0, \pm 1)$ and $\pm(1, 1)$ and indeed $w_L^2 = 2$ for each of these. Thus we find six massless vectors from these states. From states of the latter type we find another six. Hence, in all we find 16 massless vectors and it can be shown that these have interactions corresponding to the gauge group $SU(3)_L \times SU(3)_R$. The six vector states with $w_L^2 = 2$, $w_R = 0$ correspond to the six roots of one of the $SU(3)$s.

10.4   Description of the moduli space

In this section, we will show that the space of possible string theories constructed in the previous section is parameterised by the coset space [10.5]

$$\frac{O(D, D; \mathbf{R})}{O_L(D; \mathbf{R}) \times O_R(D; \mathbf{R})}. \tag{10.4.1}$$

As we will discuss, in addition points in the coset are to be identified under $O(D, D; \mathbf{Z})$ transformations. This coset space has the dimension D^2, which is the expected dimension as it agrees with the number of possible background field values. Indeed, we will show that the basis vectors of the lattice $\Lambda^{D,D}$ can be regarded as coset representatives for this coset.

The string theories constructed in the previous section were specified by even self-dual lattices. Although Euclidean even self-dual lattices are not, in general, unique, even self-dual lattices with Lorentzian or indeed more general indefinite signatures are unique up to the orthogonal rotations. Clearly, such a rotation does not change the scalar products and so must lead to a even self-dual lattice. Let us consider one of the even self-dual lattices of the previous section. All the other even self-dual lattices $\Lambda^{D,D}$ considered in the previous section must be related by $O(D, D; \mathbf{R})$ transformations to this chossen lattice. As a result, the space of all lattices $\Lambda^{D,D}$ is transitive and so must be of the form of a coset involving the group $O(D, D; \mathbf{R})$ with respect to an appropriate sub-group or isotropy group. The isotropy group is just the one that leaves invariant the chosen lattice. If we were to carry out orthogonal rotations separately on the left and right momenta, w_L and w_R, then this would preserve their scalar products and so the physical state conditions. As such, it leads to the same string theory. This explains the presence of the isotropy group $O_L(D; \mathbf{R}) \times O_R(D; \mathbf{R})$. Thus we have established the form of the coset of equation (10.4.1). The lattice $\Lambda^{D,D}$ is also preserved by $O(D, D; \mathbf{Z})$ transformations as these simply relabel the basis vectors of the lattice and as such we must identify points in the coset that are related under this group. In what follows we will give a derivation of the coset which does not rely on the above uniqueness theorem. The advantage is that it gives a more concrete idea of how the coset arises.

We begin by writing any vector W of the lattice $\Lambda^{D,D}$ in the form

$$W = FZ, \tag{10.4.2}$$

where Z is the column vector composed of integers as in equation (10.3.22) and F are basis vectors of the lattice $\Lambda^{D,D}$ as given in equation (10.3.19). We recall that the latter constitute a $2D \times 2D$ matrix whose first indices are their component indices, that is, $k, l, \ldots$, and whose second indices, that is, $I, J, \ldots$, label the basis vectors themselves. The physical state conditions of equations (10.3.24) and (10.3.25) involve Z and M and J. In terms of the basis vectors F of the lattice, the quantities M and J given in equations (10.3.23) and (10.3.20) can be rewritten as

$$F^T F = M \tag{10.4.3}$$

and

$$F^T \eta F = J, \tag{10.4.4}$$

where as before

$$\eta = \begin{pmatrix} 1_{D \times D} & 0 \\ 0 & -1_{D \times D} \end{pmatrix}, \quad J = \begin{pmatrix} 0 & 1_{D \times D} \\ 1_{D \times D} & 0 \end{pmatrix}. \tag{10.4.5}$$

We recognise η as the usual flat metric of the underlying vector space $\mathbf{R}^{(D,D)}$ and J as the metric of the lattice $\Lambda^{(D,D)}$. The two metrics correspond to two different choices of coordinates on $\mathbf{R}^{(D,D)}$, the coordinates x^a, y_a with the scalar product $x^a x^a - y_a y_a$ and the coordinates z^a, $\bar{z}_a$ with the metric $z^a \bar{z}_a$. The relation between the two sets of coordinates being $z^a = \frac{1}{\sqrt{2}}(x^a + y_a)$ and $\bar{z}^a = \frac{1}{\sqrt{2}}(x^a - y_a)$.

We define elements $\hat{g}$ of the group $\hat{O}(D, D; \mathbf{R})$ to be all $2D \times 2D$ matrices which satisfy

$$\hat{g}^T J \hat{g} = J. \tag{10.4.6}$$

One can verify that if $\hat{g}$ is in $\hat{O}(D, D; \mathbf{R})$ so is $\hat{g}^T$. By taking the inverse of the above equation, using the relation $J^{-1} = J$, and then multiplying by $\hat{g}$ and $(\hat{g}^T)^{-1}$ on the left- and right-hand sides, respectively, one finds that $\hat{g} J \hat{g}^T = J$.

We can also define the more standard group $O(D, D; \mathbf{R})$ to be all $2D \times 2D$ matrices such that

$$g^T \eta g = \eta. \tag{10.4.7}$$

We can change from one group to another by making the change $\hat{g} = B^T g B$, where the matrix B is given by

$$B = \frac{1}{\sqrt{2}} \begin{pmatrix} 1 & 1 \\ -1 & 1 \end{pmatrix} \tag{10.4.8}$$

and satisfies

$$B^T \eta B = J. \tag{10.4.9}$$

Thus the two groups $O(D, D; \mathbf{R})$ and $\hat{O}(D, D; \mathbf{R})$ are isomorphic. In fact, the matrix B also obeys the equation $B^T B = 1$ and so is itself an element of $O(D, D; \mathbf{R})$. If the two groups $\hat{O}(D, D; \mathbf{R})$ and $O(D, D; \mathbf{R})$ act on the vectors $\hat{V}$ and V, respectively, then these vectors are related by $V = B \hat{V}$, $\hat{V} = B^T V$.

We now carry out a change of basis of the underlying space $\mathbf{R}^{(D,D)}$ such that the metric of $\mathbf{R}^{(D,D)}$ is changed from η to J. This induces a corresponding change in the lattice basis vectors and we denote the basis matrix after this change by $\hat{F}$. Using the discussion above, we find that this change is achieved by the matrix B and in particular $F = B\hat{F}$. Carrying out this change we find that equations (10.4.3) and (10.4.4) become

$$\hat{F}^T \hat{F} = M \tag{10.4.10}$$

and

$$\hat{F}^T J \hat{F} = J. \tag{10.4.11}$$

The advantage of carrying out this change of basis now becomes clear: the last equation just implies that $\hat{F}$ is an element of $\hat{O}(D, D; \mathbf{R})$. Hence, we conclude that each of the even self-dual lattices $\Lambda^{D,D}$, and so the string theories constructed in the previous section, corresponds to an element of $\hat{O}(D, D; \mathbf{R})$. However, as we will now see this correspondence is not unique.

Corresponding to the transformation to the hatted basis, the vectors of the lattice $\hat{W}$ are related to those before the change W by $\hat{W} = B^T W$ and equation (10.4.2) becomes $\hat{W} = \hat{F} Z$. Using equations (10.3.15) and (10.3.16) and carrying out the change of basis, we identify the matrix $\hat{F}$ to be given by

$$\hat{F} = \begin{pmatrix} e^T & 0 \\ -e^{-1}B & e^{-1} \end{pmatrix}, \tag{10.4.12}$$

where the $D \times D$ matrix e is given by e_{Ik} and the matrix e^{-1} by e^{*kl}. The relationship between e and the space-time metric G can be written in matrix form as $ee^T = G$.

The physical state conditions involve w_L^2 and w_R^2 and as such they are invariant under $O_L(D; \mathbf{R}) \times O_R(D; \mathbf{R})$ transformations, which act on the momenta w_L and w_R in the obvious way. These transformations act on the vectors of $\Lambda^{D,D}$ by $W \to W' = hW$, where the matrix h has the form

$$h = \begin{pmatrix} h_L & 0 \\ 0 & h_R \end{pmatrix}, \tag{10.4.13}$$

where h_L and h_R belong to $O_L(D, \mathbf{R})$ and $O_R(D, \mathbf{R})$, respectively. Indeed, this is the maximal Lie group that preserves the square of the left and right momenta separately and so the maximal Lie group that preserves the on-shell conditions. The matrices h are just those that belong to both $O(2D; \mathbf{R})$ and $O(D, D; \mathbf{R})$.

The physical state conditions of equations (10.3.24) and (10.3.25) can also be written in terms of F using equations (10.4.3) and (10.4.4). Under the transformation h, the basis matrix F changes as $F \to F' = hF$. One readily sees that this transformation preserves the physical state conditions if and only if h belongs to $O(2D; \mathbf{R})$ and $O(D, D; \mathbf{R})$. Alternatively, the physical state conditions of equations (10.3.24) and (10.3.25) can also be written in terms of $\hat{F}$ using equations (10.4.10) and (10.4.11). In this basis $\hat{h} = B^T hB$ and $\hat{F} \to \hat{F}' = \hat{h}\hat{F}$. Since h preserves the metric η, $\hat{h}$ will preserve the metric J and as a consequence $\hat{F}'$ will still be an element of $\hat{O}(D, D; \mathbf{R})$, that is, $(\hat{F}')^T J\hat{F}' = J$. Further, as B and h are elements of $O(2D; \mathbf{R})$, so is $\hat{h}$ and as such the left-hand side of equation (10.4.10) is preserved and so we again conclude that this transformation preserves the physical state conditions.

Hence, we conclude that basis matrices F that are related by $O_L(D; \mathbf{R}) \times O_R(D; \mathbf{R})$ lead to the same string theory. It follows that the string theories constructed in the previous section are parameterised by a space that has isotropy group $O_L(D; \mathbf{R}) \times O_R(D; \mathbf{R})$.

We can also consider transformations $g \in \hat{O}(D, D; \mathbf{R})$ which act on the basis vectors of the lattice $\Lambda^{(D,D)}$ by transforming the labels of the basis vectors rather than the component indices as above. Put another way these transformations do not change the basis of the underlying vector space $R^{D,D}$, but are a transformation of the basis vectors of the lattice. In terms of the matrix $\hat{F}$ these transformations act as

$$\hat{F} \rightarrow \hat{F}' = \hat{F}\hat{g}. \tag{10.4.14}$$

This transformation does preserve the expression of equation (10.4.11), that is, the metric of the lattice, and so the physical state condition of equation (10.4.24). It also leads to a lattice with basis vectors $\hat{F}'$ and lattice metric J and so to an even self-dual lattice. However, such a transformation does not preserve the expression of equation (10.4.10) and so does not preserve the physical state condition of equation (10.3.25). As such, assuming that we can interpret $\hat{F}'$ as the basis vectors of one of the strings constructed in the previous section, this new string theory will be different to the one constructed from $\hat{F}$.

We can summarise the transformations of $\hat{F} \in \hat{O}(D, D; \mathbf{R})$ by the equation

$$\hat{F} \rightarrow \hat{h}\hat{F}\hat{g}, \tag{10.4.15}$$

where $\hat{h} \in \hat{O}_L(D; \mathbf{R}) \times \hat{O}_R(D; \mathbf{R})$ and $g \in \hat{O}(D, D; \mathbf{R})$ is a transformation. Using an $\hat{O}_L(D; \mathbf{R}) \times \hat{O}_R(D; \mathbf{R})$ transformations on the left one can show that one can bring any $\hat{F} \in \hat{O}(D, D; \mathbf{R})$ to be of the form of of equation (10.4.12) for suitable values of the background fields B_{IJ} and G_{IJ}. Hence any $\hat{F} \in \hat{O}(D, D; \mathbf{R})$ describes the basis vectors of an even self-dual lattice and corresponds to one of the string theories constructed in the previous section. Furthermore, since the transformations by $\hat{O}_L(D; \mathbf{R}) \times \hat{O}_R(D; \mathbf{R})$ lead to equivalent string theories we have shown the desired conclusion, namely all such the string theories constructed in the previous section can be parameterised by the coset of equation (10.4.1).

There is one subtlety to this conclusion. We noted that carrying out a $\hat{g} \in \hat{O}(D, D; \mathbf{R})$ of the form $\hat{F} \rightarrow \hat{F}\hat{g}$ takes one generically to a new string theory. However, this is not the case if $\hat{g} \in \hat{O}(D, D; \mathbf{Z})$. Although such a transformation takes $\hat{F}$ to another element of $\hat{O}(D, D; \mathbf{R})$, it will transform the lattice $\Lambda^{D,D}$ to itself as it corresponds to taking $Z \rightarrow \hat{g}Z$. In effect, it just uses an alternative equivalent basis of the even self-dual lattice and so does not affect the spectrum of states of the string. Hence we must identify points in the coset of equation (10.4.1) which are related by $\hat{O}(D, D; \mathbf{Z})$ transformations in this way as mentioned below equation (10.4.1). In fact, this symmetry is just another example of the T duality discussed above. The reader may wish to consider the case of the circle of section 10.1 from this viewpoint.

It only remains to find the transformation of the background fields induced by the transformation $g \in \hat{O}(D, D; \mathbf{R})$. Let us combine the background fields into the $D \times D$ matrix $E \equiv G + B$. Such a transformation will not preserve the form of equation (10.4.12), but we may use a simultaneous compensating local transformation to preserve this form and so deduce the transformation of E. To avoid this step we will use an alternative, method of finding this result. We consider the mapping

$$\hat{g}(K) = (aK + b)(cK + d)^{-1}, \tag{10.4.16}$$

where K is any $D \times D$ matrix and a, b, c, d are also $D \times D$ matrices such that

$$\hat{g} = \begin{pmatrix} a & b \\ c & d \end{pmatrix} \tag{10.4.17}$$

is a member of $\hat{O}(D, D; \mathbf{R})$. We note that the $\hat{F}$ of equation (10.4.12) satisfies

$$\hat{F}^T(1) = E \tag{10.4.18}$$

and as a result

$$E' = (\hat{F}')^T(1) = (\hat{g}^T \hat{F})^T(1) = \hat{g}^T(\hat{F}^T)(1) = \hat{g}^T(E). \tag{10.4.19}$$

Hence, the transformation of the background fields is given by [10.7–10.9]

$$E' = (a^T E + c^T)(b^T E + d^T)^{-1}, \tag{10.4.20}$$

where we have used the result

$$(\hat{g})^T = \begin{pmatrix} a^T & c^T \\ b^T & d^T \end{pmatrix}. \tag{10.4.21}$$

These formulae can also be used to find the transformations of $\hat{O}(D, D; \mathbf{Z})$. This is a generalisation of the so-called T duality symmetry found for a the circle in section 10.1. It can be shown that $\hat{O}(D, D; \mathbf{Z})$ transformations are a symmetry of the scattering amplitudes of the compactified string [10.31]. This transformation is further discussed in section 17.1.

10.5   Heterotic compactification [10.4]

The emergence of only $(SU(2)_L)^D \times (SU(2)_R)^D$ from a torus compactification with $B_{ij} = 0$ is rather disappointing given the well-known relationship between string vertex operators, lattices and Lie groups, which is studied in chapter 16. However, there is one very important method of enlarging the gauge group, other than using a background anti-symmetric tensor gauge field and this leads to much larger gauge groups. At first sight, this method requires an unnatural step. For the torus compactification, the reader will recall that, although we started with the field x^I, it was useful to split it into left and right fields x_L^I and x_R^I even though their zero modes were identified, that is, $q_L^I = q_R^I$ and $p_L^I = p_R^I$. In this section we wish to regard x_L^I and x_R^I as distinct, that is, independent fields, which have the same decomposition as before, that is,

$$x_L^I = q_L^I + \frac{\sqrt{2\alpha'}}{2}\alpha_0^I(\tau + \sigma) + i\sqrt{\frac{\alpha'}{2}} \sum_{\substack{n=-\infty \\ n\neq 0}}^{\infty} \frac{1}{n}\alpha_n^I e^{-in(\tau+\sigma)}, \tag{10.5.1}$$

$$x_R^{I'} = q_R^{I'} + \frac{\sqrt{2\alpha'}}{2}\bar{\alpha}_0^{I'}(\tau - \sigma) + i\sqrt{\frac{\alpha'}{2}} \sum_{\substack{n=-\infty \\ n\neq 0}}^{\infty} \frac{1}{n}\bar{\alpha}_n^{I'} e^{-in(\tau-\sigma)}, \tag{10.5.2}$$

except that now $q_L^I \neq q_R^{I'}$. Indeed, even the index ranges of I and I' can be different as we may not have the same number of left and right moving fields. We will, in particular, no longer identify the zero modes. We also have the commutation relations

$$[\alpha_n^I, \alpha_m^J] = n\delta_{n+m,0}\delta^{IJ}, \quad [\bar{\alpha}_n^{I'}, \bar{\alpha}_m^{J'}] = n\delta_{n+m,0}\delta^{I'J'},$$

$$[q_L^I, p_L^J] = i\delta^{IJ}, \quad [q_R^{I'}, p_R^{J'}] = i\delta^{I'J'} \tag{10.5.3}$$

with all other commutators vanishing. As before $p_L^I = \sqrt{2/\alpha'}\alpha_0^I$ and $p_R^{I'} = \sqrt{2/\alpha'}\bar{\alpha}_0^{I'}$. This means, in particular, that all the commutators between left and right objects vanish. These relations are those corresponding to two independent fields: a x_L^I which is left moving, $((\partial/\partial\tau) - (\partial/\partial\sigma))x_L^I = 0$ and a $x_R^{I'}$ which is right moving, $((\partial/\partial\tau) - (\partial - \partial\sigma))x_R^{I'} = 0$. Such chiral scalar fields are discussed further in the next section.

Since x_L^I and $x_R^{I'}$ are independent we can regard them as living on different tori, T_L^D and $T_R^{D'}$. These tori, may even be of different dimensions. Let T_L^D be generated by lattices Λ_L in the sense that $T_L^D = R^D/2\pi\Lambda_L$ and $T_R^{D'}$ be generated by lattices Λ_R in the sense that $T_R^{D'} = R^{D'}/2\pi\Lambda_R$. If the left fields have winding numbers L^I, they satisfy $x_L^I(\tau + \sigma + 2\pi) = x_L^I(\tau + \sigma) + 2\pi L_L^I$ and similarly for the right fields $x_R^{I'}(\tau - \sigma - 2\pi) = x_R(\tau - \sigma) + 2\pi L_R^{I'}$. Explicitly carrying out these shifts on the x_L^I and x_R^I given in equations (10.5.1) and (10.5.2), we find that these conditions imply that $(\alpha'/2)p_L \in \Lambda_L$ and $(\alpha'/2)p_R \in \Lambda_R$.

However, we must also check for the single valued nature of the wavefunction. This implies that $\exp i\sum_I p_L^I x_L^I$ and $\exp i p_R^{I'} x_R^{I'}$ be separately single valued. This in turn implies that

$$\frac{\alpha'}{2}p_L \in (\Lambda_L)^*, \quad \frac{\alpha'}{2}p_R \in (\Lambda_R)^*. \tag{10.5.4}$$

Consequently, we have $(\alpha'/2)p_L \in \Lambda_L \cap (\Lambda_L)^*$ and similarly for p_R. Clearly, one way to solve these conditions is to use a self-dual lattice. The appearance of much more stringent conditions is a consequence of the fact that we have now demanded two equivalent conditions on the string field: one for its left part and one for its right part.

The on-shell conditions are the same as usual and the L_ns and $\bar{L}_n$s depend on the α_n^I and $\bar{\alpha}_n^I$ oscillators in the usual way. The mass-shell condition, $(L_0 + \bar{L}_0 - 2)|\psi\rangle = 0$, implies that the particles have masses given by

$$(\text{mass})^2 = \frac{1}{2}\sum_{I=1}^D (p_L^I)^2 + \frac{1}{2}\sum_{I'=1}^{D'} (p_R^{I'})^2 + \frac{2}{\alpha'}(N + \bar{N} - 2) \tag{10.5.5}$$

and the condition $(L_0 - \bar{L}_0)|\psi\rangle = 0$ implies the constraint

$$\frac{\alpha'}{4}\left[\sum_{I=1}^D (p_L^I)^2 - \sum_{I=1}^{D'} (p_R^{I'})^2\right] = -(N - \bar{N}). \tag{10.5.6}$$

Obviously, if $T_L^D \neq T_R^{D'}$ we cannot build an x^I which would belong to a torus compactification since even if we had the same number of left and right moving compactified coordinates $x^I = x_L^I + x_R^{I'}$ would not be subject to the equivalence relations for a torus which is the same for left and right moving fields. Indeed, we have lost the picture of a space-time compactification altogether and we cannot regard the string as living on any space-time as far as the compactified dimensions are concerned. Since we live in a four-dimensional world, we should construct string theories that have the fields x^μ, $\mu = 0, 1, 2, 3$ of Minkowski space-time, however, we can take the remaining fields to be left and right movers that arise from no space-time. An extreme, but important, example is to take no right moves at all other than x_R^μ of Minkowski space-time.

Determining the spectrum of such a string follows much the same pattern as in section 10.2. We find that it contains the graviton, an anti-symmetric tensor and a scalar which in the previous cases were of Kaluza–Klein origin, as well as the generalisation of the scalars of equation (10.2.32), which are now given by

$$\alpha_{-1}^I \bar{\alpha}_{-1}^{J'} |0, 0\rangle, \tag{10.5.7}$$

and a generalisation of the vector states of equation (10.2.31), which now are

$$\alpha_{-1}^I \bar{\alpha}_{-1}^\mu |0, 0\rangle, \alpha_{-1}^\mu \bar{\alpha}_{-1}^{I'} |0, 0\rangle. \tag{10.5.8}$$

The latter describe the gauge group $[U(1)_L]^D \times [U(1)]_R^{D'}$. We will find additional massless vectors by considering states of the form of equation (10.2.33). As there, these states are massless if either $w_L^2 = 2$, $w_R = 0$ or $w_R^2 = 2$, $w_L = 0$, where now $w_L^I = \sqrt{\alpha'/2} p_L^I = \alpha_0^I$ and $w_R^{I'} = \sqrt{\alpha'/2} p_R^{I'} = \bar{\alpha}_0^{I'}$. Thus the additional massless vector states are given by

$$\bar{\alpha}_{-1}^\mu |w_L^2 = 2, w_R = 0\rangle, \alpha_{-1}^\mu |w_L = 0, w_R^2 = 2\rangle. \tag{10.5.9}$$

In the case studied in section 10.2, w_L and w_R arose from one torus. However, in this section w_L and w_R arise from the different independent tori and massless states exist for every vector of length squared 2 in either $\sqrt{2/\alpha'} \Lambda_L$ or $\sqrt{2/\alpha'} \Lambda_R$.

In fact, all the above massless vectors have the interactions of the non-Abelian group $G_L \times G_R$, where the roots of G_L are the vectors w_L such that $w_L^2 = 2$ and similarly for w_R and G_R. The vectors of equation (10.5.8) correspond to the Cartan subalgebra of $G_L \times G_R$. Thus we find many more massless states as the lattices Λ^L and Λ^R are not related.

As in section 10.2, one can consider the non-Euclidean lattice, $\Lambda^{L,R}$, which consists of all the vectors $W = (w_L, w_R)$ whose scalar product is

$$w_1 \cdot w_2 = \sum_{I=1}^{D} w_{1L}^I w_{2L}^I - \sum_{I'=1}^{D'} w_{1R}^{I'} w_{2R}^{I'}. \tag{10.5.10}$$

It follows from equation (10.5.6) that this lattice $\Lambda^{L,R}$ is integral and even. However, it is just the direct sum of the left and right lattices and, unlike the lattices of section 10.2, it will not, in general, be self-dual. One can show that string constructed by the method of this section will be modular invariant if and only if the lattice $\Lambda^{L,R}$ is a self-dual lattice. In the event that we have only left movers in the compactified direction, this means that Λ^L is a Euclidean lattice which is even and self-dual.

Even self-dual Euclidean lattices are rather rare, at least for low dimensions. They can only occur in the dimensions $8n$ for n an integer. In eight dimensions there only exists one such lattice and it is the root lattice of E_8. There are two even self-dual Euclidean lattices of dimension 16, these being associated with the groups $E_8 \times E_8$ and $SO(32)$. There are 24 such lattices in dimension 24. Even Lorentzian, or indefinite signature, self-dual lattices only exist in dimensions $8n + 2p$, for n an integer, and there is only one such lattice in each of these dimensions. Here p is the number of minus signs in the diagonal metric of the lattice. For a Lorentzian lattice $p = 1$. While for the lattices of the section 10.3, that is, $\Lambda^{D,D}$, we have $p = D$ and there is such an even self-dual lattice if we take $n = 0$ in the above formula. These uniqueness statements are up to the appropriate orthogonal rotations which clearly take one even self-dual lattice to another.

10.6   The heterotic string

The construction of the heterotic string depends on a peculiarity of motion in $1 + 1$ space-time. Obviously, motion in this space-time takes place in only one dimension, that is, it is along a line, or segment of a line, Seen from an embedding space-time the motion is either to the left or the right. Unlike in higher dimensions, the direction of the motion has a Lorentz invariant meaning. The invariant nature of the direction of the motion in $1 + 1$ dimensions depends on the existence of the Lorentz invariant tensor $\epsilon^{\alpha\beta} = -\epsilon^{\beta\alpha}; \epsilon^{01} = +1$ which can be used to write the Lorentz invariant equation

$$\partial_\alpha \phi = \pm \epsilon_{\alpha\beta} \partial^\beta \phi. \tag{10.6.1}$$

These two equations are the same as the equations

$$(\partial_0 \mp \partial_1)\phi = 0. \tag{10.6.2}$$

In terms of light-cone coordinates $\xi^+ = \tau + \sigma$, $\xi^- = \tau - \sigma$, these equations take a particularly simple form

$$\partial_+ \phi \equiv \frac{1}{2}\left(\frac{\partial}{\partial \xi^0} + \frac{\partial}{\partial \xi^1}\right)\phi = 0, \, \partial_- \phi \equiv \frac{1}{2}\left(\frac{\partial}{\partial \xi^0} - \frac{\partial}{\partial \xi^1}\right)\phi = 0. \tag{10.6.3}$$

The solutions to these equations are $\phi = \phi_R(\xi^-)$ and $\phi = \phi_L(\xi^+)$, respectively, where ϕ_L and ϕ_R are arbitrary functions. As anticipated, they represent motion in a given direction and we will refer to them as left and right moving, respectively.

Such scalar fields do not possess actions of a simple form. The most obvious action would be the space-time integral of $\partial_\alpha \phi \partial^\alpha \phi$; however,

$$\partial_\alpha \phi \partial^\alpha \phi = \pm \epsilon_{\alpha\beta} \partial^\beta \phi \partial^\alpha \phi = \mp \partial^\beta \phi \epsilon_{\beta\alpha} \partial^\alpha \phi = -\partial^\beta \phi \partial_\beta \phi = 0 \tag{10.6.4}$$

and so the candidate action vanishes identically. This is an example of a more general phenomenon. Given a rank $d - 1$ gauge field in any Minkowski space-time of even dimension $2d$ we can impose the self-duality constraint on its field strength:

$$F_{\mu_1 \cdots \mu_d} = \pm \frac{1}{d!} \epsilon_{\mu_1 \cdots \mu_d \nu_1 \cdots \nu_d} F^{\nu_1 \cdots \nu_d}, \tag{10.6.5}$$

where $F_{\mu_1 \cdots \mu_d} = d\partial_{[\mu_1} A_{\cdots \mu_d]}$. The most obvious action involves the product of two field strengths with all their indices contracted. However,

$$F_{\mu_1 \cdots \mu_d} F^{\mu_1 \cdots \mu_d} = \pm \frac{1}{d!} \epsilon_{\mu_1 \cdots \mu_d \nu_1 \cdots \nu_d} F^{\nu_1 \cdots \nu_d} F^{\mu_1 \cdots \mu_d}$$

$$= (\pm)\frac{1}{d!}(-1)^{d^2} F^{\nu_1 \cdots \nu_d} \epsilon_{\nu_1 \cdots \nu_d \mu_1 \cdots \mu_d} F^{\mu_1 \cdots \mu_d}$$

$$= (-1)^{d^2} F^{\nu_1 \cdots \nu_d} F_{\nu_1 \cdots \nu_d} \tag{10.6.6}$$

and hence it vanishes if d is odd. Such a phenomenon can only occur in space-times of dimensions $4n + 2$, $n \in \mathbf{Z}$. The case of $d = 1$ fits into this pattern as we can regard ϕ as the 'gauge field,' but being only a scalar it has no gauge invariance. The most celebrated example is the rank 4 gauge field and its self-dual five form field strength in ten dimensions.

In $1 + 1$ space-time, an on-shell Weyl spinor is automatically either left or right moving. A Weyl spinor χ is one that obeys the condition $\gamma_5 \chi = \pm \chi$. Using the definition $\gamma_5 = \gamma^0 \gamma^1$, we can rewrite this condition as $(\gamma_0 \mp \gamma_1)\chi = 0$. The latter condition can be written as $\gamma_- \chi = 0$ and $\gamma_+ \chi = 0$ for the upper and lower signs, respectively, where the light-cone γ matrices were defined in equation (6.1.38) by $\gamma_+ \equiv \frac{1}{2}(\gamma_0 + \gamma_1)$ and $\gamma_- \equiv \frac{1}{2}(\gamma_0 - \gamma_1)$. In light-cone notation the free equation of motion $\not{\partial}\chi = 0$ becomes $\gamma_+ \partial_- \chi + \gamma_- \partial_+ \chi = 0$. Taking $\gamma_5 \chi = \chi$, or $\gamma_- \chi = 0$, we find the equation of motion becomes $\gamma_+ \partial_- \chi = 0$. However, this is equivalent to $\partial_- \chi = 0$, which just states that it is left moving. A spinor of the other chirality is found to be right moving by the same argument. Unlike left, or right, moving scalars we can write down a simple action to describe the action for such spinors. It is none other than the usual action with the spinor being subject to the Weyl condition. Indeed, for a chiral spinor for which $\gamma_5 \chi = \chi$, the standard action $-\frac{i}{2} \int d^2 x \bar{\chi} \not{\partial} \chi$ becomes $-i \int d^2 x \chi_+^* \partial_- \chi_+$, where χ_+ is the upper component of χ.

Let us now consider a closed string. Its world sheet is a $(1 + 1)$-dimensional space-time which we parametized by (τ, σ). The world-sheet spatial dimension is just topologically a circle. The string is built out of fields that live on this $(1 + 1)$-dimensional space-time and so we can consider some of them to be only left or right moving. This simple observation is the mechanism behind the construction of the heterotic string. Although, in retrospect this was a rather simple extension, it was a step that was only realised after a series of developments that strongly hinted at this possiblility. It is amusing to recall how this came about. It had been known [10.10] since 1978 that there existed only one supergravity theory in eleven dimensions and two supergravity theories in ten dimensions which were maximal, that is, possessed a supersymmetry parameter with 32 components, and also one other supergravity theory in ten dimensions invariant under a supersymmetry with only 16 components. The two supergravity theories with maximal supersymmetry in ten dimensions were called IIA and IIB supergravity, while the one with less supersymmetry was called type I supergravity. Motivated by the realisation that the first theories were the low energy effective actions for the IIA and IIB superstring theories, they were constructed in [10.11–10.13] for IIA and [10.14–10.16] for IIB. The type I supergravity theory in ten dimensions was found in [10.17, 10.18]. The supersymmetric Yang–Mills theory that also existed in ten dimensions had been previously constructed in [10.19, 5.1] and the coupling between this theory and type I supergravity was given in [10.20]. This coupling was the low energy effective action for the open and closed type I superstring theories. Although the supergravity coupling to the supersymmetric Yang–Mills theory was possible for all gauge groups, type I string theory was already known to be only possible for the gauge groups $U(n)$, $USp(2n)$ and $SO(n)$ [10.21].

After the construction of these theories, it was realised [10.22] that quantum field theories could suffer from gravitational anomalies. These were anomalies in translation symmetries and they could only arise if the theory contained chiral spinors or gauge fields whose field strengths were self-dual or both. Since IIA supergravity has no fields of this type it must be anomaly free. However, it was shown [10.22] that IIB supergravity is also anomaly free due to a subtle cancellation between the anomaly contributions from its chiral spinors and self-dual gauge field. It was then shown that the anomalies also cancelled for two other ten-dimensional theories: these were type I supergravity coupled to supersymmetric Yang–Mills theory with the gauge groups SO(32) [10.23] and that with $E_8 \times E_8$ [10.23, 10.24].

This result was puzzling because although there existed a type I string with SO(32) gauge group there was, as already metioned, no known type I string that possessed the gauge group $E_8 \times E_8$. A striking earlier observation [10.25, 10.26] was that the string vertex operators, whose momenta were restricted to belong to certain classes of lattices, generated the Lie group whose roots were the points of length squared 2 in the lattice. Further, the Euclidean self-dual lattice only existed in dimensions that were multiples of 8 and there were only two such 16-dimensional self-dual Euclidean lattices, which were associated with the groups SO(32) and $E_8 \times E_8$, a fact that was stressed in [10.27]. It was clear to the few people working in this field that the existence of the two self-dual lattices in 16 dimensions with these same groups was not a coincidence and that there should exist a string theory with gauge group $E_8 \times E_8$ whose low energy effective action was type I supergravity coupled to supersymmetric Yang–Mills theory with gauge group $E_8 \times E_8$. It was suggested [10.28] that there should exist a string theory which had a 16-dimensional torodial internal space whose momenta belonged to the $E_8 \times E_8$ root lattice. This was the historical background that lead to the heterotic string [10.29] which we now discuss. The story illustrates well that some of the most important discoveries even if simple are only obvious with hindsight.

As indicated above, the heterotic string is constructed from fields that are left and right movers. The right moving sector consists of the scalar fields $x_R^\mu(\tau - \sigma)$ and the fermionic fields $\psi_R^\mu(\tau - \sigma)$, $\mu = 0, 1, \ldots, 9$, and as such can admit a right moving local world-sheet supersymmetry. The left sector consists of the fields $x_L^\mu(\tau + \sigma)$, $\mu = 0, 1, \ldots, 9$, and $x_L^I(\tau + \sigma)$, $I = 1, \ldots, 32$. Since the left sector possesses no world-sheet fermions it cannot carry any world-sheet supersymmetry. The fields $x_L^\mu(\tau + \sigma)$ and $x_R^\mu(\tau - \sigma)$, $\mu = 0, 1, \ldots, 9$, have their zero modes identified and these zero modes are taken to be the coordinates of ten-dimensional Minkowski space. We therefore expand these fields as

$$x_L^\mu(\tau + \sigma) = q_L^\mu + \frac{\sqrt{2\alpha'}}{2} \alpha_0^\mu(\tau + \sigma) + i\sqrt{\frac{\alpha'}{2}} \sum_{\substack{n=-\infty \\ n \neq 0}}^{\infty} \frac{1}{n} \alpha_n^\mu e^{-in(\tau+\sigma)} \qquad (10.6.7)$$

and

$$x_R^\mu(\tau - \sigma) = q_R^\mu + \frac{\sqrt{2\alpha'}}{2} \bar{\alpha}_0^\mu(\tau - \sigma) + i\sqrt{\frac{\alpha'}{2}} \sum_{\substack{n=-\infty \\ n \neq 0}}^{\infty} \frac{1}{n} \bar{\alpha}_n^\mu e^{-in(\tau-\sigma)}, \qquad (10.6.8)$$

where $q_L^\mu = q_R^\mu \equiv x_0^\mu(0)$, $\alpha_0^\mu = \bar{\alpha}_0^\mu$. Hence, $x^\mu \equiv x_L^\mu + x_R^\mu$, $\mu = 0, \ldots, 9$, behaves like the bosonic coordinate for any of the closed strings we have considered previously.

We take the right moving ψ_R^μ to be Majorana–Weyl spinors. If we choose the Weyl condition $\gamma_5 \psi_R^\mu = -\psi_R^\mu$, then, following our discussion at the beginning of this section, we find that $\partial_+ \psi_R^\mu = 0$ and so ψ_R^μ is automatically right moving. The Weyl condition implies that only the lower component of ψ_R^μ, denoted ψ_{R-}^μ, is non-zero and the Majorana condition implies that this is real.

The left moving scalars $x_L^I(\tau + \sigma)$, $I = 1, \ldots, 16$, are defined to lie on a torus associated with an even self-dual Euclidean lattices of dimension 16. As mentioned above, there are only two such lattices and they are associated with the groups $E_8 \times E_8$ or SO(32) and we will denote these lattices by $\Lambda_{E_8 \times E_8}$ and $\Lambda_{SO(32)}/\mathbf{Z}_2$, respectively. As discussed in earlier sections, the $x_L^I(\tau + \sigma)$ are identified under $\sigma \rightarrow \sigma + 2\pi$ as

$$x_L^I(\tau + \sigma + 2\pi) = x_L^I(\tau + \sigma) + 2\pi L_L^I, \tag{10.6.9}$$

where L_L^I is a vector that belongs to one of the above self-dual lattices. For these coordinates we now find ourselves in precisely the situation covered in section 10.5 with no x_R^I. We can take x_L^I to have the expansion

$$x_L^I = q_L^I + \frac{\sqrt{2\alpha'}}{2}\alpha_0^I(\tau + \sigma) + i\sqrt{\frac{\alpha'}{2}} \sum_{\substack{n=-\infty \\ n \neq 0}}^{\infty} \frac{1}{n}\alpha_n^I e^{-in(\tau+\sigma)}, \tag{10.6.10}$$

where $p_L^I = \sqrt{\frac{2}{\alpha'}}\alpha_0^I$. We recall that the single valued nature of the wavefunction implies that the zero mode momentum p_L must lie on the dual lattice, which is, in this case, the same lattice as it is self-dual. Hence, we have that

$$p_L^I = \sum_{i=1}^{16} n_i e_i^I, \qquad n_i \in \mathbf{Z}, \tag{10.6.11}$$

where e_i are the basis vectors of the lattice.

For the more interesting case of $E_8 \times E_8$, the self-dual lattice is just the direct sum of two root lattices of E_8, that is, $\Lambda_{E_8 \times E_8} = \Lambda_{E_8} \oplus \Lambda_{E_8}$. The basis of Λ_{E_8} can be chosen to be the simple roots of Λ_{E_8} and then the lattice metric $g_{ij} = e_i.e_j$ is just the Cartan matrix of E_8, which is given by

$$g_{ij} = \begin{pmatrix} 2 & -1 & & & & & & \\ -1 & 2 & -1 & & & & & \\ & -1 & 2 & -1 & & & & \\ & & -1 & 2 & -1 & & & \\ & & & -1 & 2 & -1 & & -1 \\ & & & & -1 & 2 & -1 & \\ & & & & & -1 & 2 & \\ & & & & -1 & & & 2 \end{pmatrix}, \tag{10.6.12}$$

where the blank entries represent zeros. We can verify that Λ_{E_8}, and so $\Lambda_{E_8} \oplus \Lambda_{E_8}$, is self-dual by checking that $\det g_{ij} = 1$.

In fact, the root lattice of SO(32) is not self-dual as is most easily verified by computing that the determinant of its Cartan matrix is 4. However, within the weight lattice of SO(32), which is just the dual of $\Lambda_{SO(32)}$, there is a self-dual lattice. The coset space $\Lambda_{SO(32)}^*/\Lambda_{SO(32)}$ is isomorphic to $\mathbf{Z}_2 \times \mathbf{Z}_2$. The lattice which is spanned by one of the spinor weights of SO(32) and the roots of SO(32) is the lattice $\Lambda_{SO(32)}/\mathbf{Z}_2$ and this is a self-dual lattice. A more complete discussion of lattices and this point in particular is given in chapter 16.

In effect, we can think of the left moving sector of the heterotic string as a left moving bosonic string in 26 dimensions, 16 of which are compactified on the $\Lambda_{E_8 \times E_8}$ or $\Lambda_{SO(32)}^*/\mathbf{Z}_2$ lattice. The right moving sector is just that for an $N = 1$ superstring discussed in chapters 6 and 7.

The world-sheet action for the heterotic string is composed out of x^μ, ψ_R^μ, $\mu = 0, 1, \ldots, 9$, and x_L^I and is given by $A = A_1 + A_2$, where

$$A_1 = \frac{1}{2\pi\alpha'} \int d^2\xi \left\{ -\frac{1}{2}\partial_\alpha x^\mu \partial^\alpha x^\nu - \frac{i}{2}\overline{\psi}^\mu \slashed{\partial} \psi^\nu \right\} \eta_{\mu\nu} + \cdots \tag{10.6.13}$$

and

$$A_2 = \frac{1}{2\pi\alpha'} \int d^2\xi \left\{ \sum_I \left(-\frac{1}{2}\partial_\alpha x^I \partial^\alpha x^I \right) + \cdots \right\}, \tag{10.6.14}$$

where $+\cdots$ is the coupling to the appropriate right moving supergravity for the right moving fields and only gravity for left moving fields. As we noted earlier, there is no simple action for left moving scalars. We can use the above A_2 action to derive the equations of motion provided we supplement it by the constraint

$$\partial_- X^I + \cdots = 0. \tag{10.6.15}$$

However, this is, in fact, the equation of motion as it is left moving. One can also represent the 16 scalars X^I by 32 Majorana–Weyl fermions ψ^i, $i = 1, \ldots, 32$ ($\gamma_5\psi^i = -\psi^i$), which are automatically left moving. For these we can write down an action without a constraint and so replace A_2 by

$$A_2 = \frac{1}{2\pi\alpha'} \int \left(-\frac{i}{2}\overline{\psi}^i \slashed{\partial} \psi^i + \cdots \right) d^2\xi. \tag{10.6.16}$$

The supersymmetry transformations and the coupling to $N = 1$ supergravity and gravity in the right and left sectors is reviewed in chapter 21 of [1.11].

The variation of the action A by the graviton and gravitino leads to the constraints. These set to zero the energy-momentum tensor and the supersymmetry current, respectively, once we make an appropriate gauge choice for the graviton and gravitino. This follows from the fact that the graviton and gravitino couple, at lowest order, to the energy-momentum tensor and supersymmetry current, respectively, and the higher order terms do not occur in the gauge in which the gravitino vanishes and the graviton is of the form $g_{\alpha\beta} = \eta_{\alpha\beta}e^\phi$. These constraints are given by

$$T_{++} \equiv \partial_+ x^\mu \partial_+ x^\nu \eta_{\mu\nu} + \partial_+ x^I \partial_+ x^I = 0 \tag{10.6.17}$$

in the left moving sector and

$$T_{--} \equiv \left\{ \partial_- x^\mu \partial_- x^\nu - \frac{i}{2}\overline{\psi}^\mu \partial_- \psi^\nu \right\} \eta_{\mu\nu}, \quad J_{--} \equiv 2\partial_- x^\mu \psi_\mu = 0 \tag{10.6.18}$$

in the right moving sector.

The quantisation proceeds much as before: the oscillators α_n^μ, $\overline{\alpha}_n^\mu$ and α_n^I of equations (10.6.7), (10.6.8) and (10.6.10) obey their usual relations,

$$[\alpha_n^I, \alpha_m^J] = n\delta_{n+m,0}\delta^{IJ}, \quad [\alpha_n^\mu, \alpha_m^\nu] = n\delta_{n+m,0}\eta^{\mu\nu}[\overline{\alpha}_n^\mu, \overline{\alpha}_m^\nu] = n\delta_{n+m,0}\eta^{\mu\nu}. \tag{10.6.19}$$

The quantisation of the spinor ψ_R^μ follows that given in chapters 6 and 7 and in particular equation (6.1.79) for ψ_μ now becomes

$$\psi_{R-}^\mu \delta\psi_{R-}^\mu \big|_{-\pi}^\pi = 0. \tag{10.6.20}$$

We can then adopt the periodic R boundary conditions

$$\psi_{R-}^{\mu}(\sigma + \pi) = \psi_{R-}^{\mu}(\sigma - \pi) \qquad \text{(R)} \tag{10.6.21}$$

or anti-periodic NS boundary conditions

$$\psi_{R-}^{\mu}(\sigma + \pi) = -\psi_{R-}^{\mu}(\sigma - \pi) \qquad \text{(NS)}. \tag{10.6.22}$$

Corresponding to these boundary conditions we expand ψ_{R-}^{μ} as

$$\psi_{R-}^{\mu} = \sqrt{\alpha'} \sum_{n \in \mathbf{Z}} \overline{d}_n^{\mu} e^{-in(\tau - \sigma)} \qquad \text{(R)}, \tag{10.6.23}$$

$$\psi_{R-}^{\mu} = \sqrt{\alpha'} \sum_{r \in \mathbf{Z} + \frac{1}{2}} \overline{b}_r^{\mu} e^{-ir(\tau - \sigma)} \qquad \text{(NS)}. \tag{10.6.24}$$

The Fourier transform of the right moving constraints of equation (10.6.17) is defined, as in equation (6.1.98) and (6.1.99), and we find in the left moving sector the operators

$$L_n = \sum_{\mu} \sum_{m} :\alpha_{n-m}^{\mu} \alpha_{m\mu}: + \sum_{I} \sum_{m} :\alpha_{n-m}^{I} \alpha_{m}^{I}:. \tag{10.6.25}$$

In the right moving sector, we find that the Fourier modes of the constraints of equation (10.6.18) are given by

$$\overline{L}_n = \sum_{m} :\overline{\alpha}_{n-m} \cdot \overline{\alpha}_m: + \frac{1}{2} \sum_{r} \left(r + \frac{n}{2} \right) :\overline{b}_{-r} \cdot \overline{b}_{n+r}:, \tag{10.6.26}$$

$$\overline{G}_r = \sum_{m} \overline{\alpha}_m \cdot \overline{b}_{r-m} \tag{10.6.27}$$

in the NS sector and

$$\overline{L}_n = \sum_{m} :\overline{\alpha}_{n-m} \cdot \overline{\alpha}_m: + \frac{1}{2} \sum_{m} \left(\frac{n}{2} + m \right) :\overline{d}_{-m} \cdot \overline{d}_{m+n}:, \tag{10.6.28}$$

$$F_n = \sum_{m} \alpha_m \cdot d_{n-m} \tag{10.6.29}$$

in the R sector. In these equations

$$\alpha_0^{\mu} = \overline{\alpha}_0^{\mu} = \sqrt{\frac{\alpha'}{2}} p^{\mu}, \qquad \alpha_L^{I} = \sqrt{\frac{\alpha'}{2}} p_L^{I}. \tag{10.6.30}$$

We can now verify that the heterotic string does not possess any conformal anomalies. As discussed in chapter 9, this cancellation must happen separately for the left and right moving sectors. In the right moving sector, we have ten scalars x_R^{μ} and ten fermions ψ_R^{μ}. These lead to a conformal anomaly of $c_R = 10 \times 1 + 10 \times \frac{1}{2} = 15$. The ghost action is independent of the details of the action for x_R^{μ} and ψ_R^{μ} and depends only on the local symmetries. In the right moving sector the local symmetries are a local right moving supersymmetry and the uusal reparameterisations. Hence, in this sector we have b, c ghosts for the reparameterisations and β, γ ghosts for the local supersymmetry. These contribute $c = -26$ and $c = +11$, respectively, to the conformal anomaly. Consequently, the conformal anomaly in the right moving sector is $c = 15 - 26 + 11 = 0$ as required. Had we taken the heterotic string to live in D dimensions, we would have found the dimension $D = 10$ by demanding the cancellation of the anomaly.

In the right moving sector we have only 26 scalar fields and the only local symmetry is reparameterisation invariance. As a result, the only ghosts are the b, c pair which give $c = -26$, so cancelling the anomaly due to the 26 scalar fields.

The physical states of the heterotic string are subject to the physical state conditions which, in the NS sector, are given by

$$(\bar{L}_0 - \tfrac{1}{2})|\Phi\rangle = 0 = (L_0 - 1)|\Phi\rangle, \tag{10.6.31}$$

$$\bar{G}_r|\Phi\rangle = 0, \qquad r \geq \tfrac{1}{2}, \tag{10.6.32}$$

and, in the R sector, by

$$\bar{L}_0|\Phi\rangle = 0 = (L_0 - 1)|\Phi\rangle, \tag{10.6.33}$$

$$\bar{F}_n|\Phi\rangle, \qquad n \geq 1, \tag{10.6.34}$$

as well as

$$L_n|\Phi\rangle = \bar{L}_n|\Phi\rangle = 0 \qquad n \geq 1 \tag{10.6.35}$$

in both sectors.

It is straightforward to analyse the lowest mass physical states. The heterotic string does not contain a tachyon. Examining the physical state condition of the form $(L_0 - 1)|\Phi\rangle = 0$ we conclude that such a state would have to have a momentum p^μ in Minkowski space x^μ that obeyed $p^2 = (2/\alpha')\alpha_0 \cdot \alpha_0 = (2/\alpha')$. However, for a momentum associated with ten-dimensional Minkowski space-time $\alpha_0^\mu = \bar{\alpha}_0^\mu$ and as a result $\bar{L}_0 \geq 1$, which is incompatible with the physical state condition $(\bar{L}_0 - \tfrac{1}{2})|\Phi\rangle = 0$.

The same type of argument does not hold if the momentum is associated with the left-handed internal space x^I as this is not related to any right-handed momentum. The physical states can be written as a tensor product of left moving and right states, that is, $|\Phi\rangle = |\phi\rangle_L \otimes |\phi'\rangle_R$ apart from the space-time zero modes, that is the space-time momenta p^μ which are in common to left and right states. In fact, we will suppress the dependence on this momentum and only display the momentum of the compactified left movers. In the NS sector we find the massless states, that is, $p^\mu p_\mu = 0$, to be given by

$$s_{\mu\nu}\alpha_{-1}^\mu|0\rangle_L \otimes \bar{b}_{-\frac{1}{2}}^\nu|0\rangle_R \tag{10.6.36}$$

as well as

$$A_\mu \left|0, p_L^2 = \frac{2}{\alpha'}\right\rangle_L \otimes \bar{b}_{-\frac{1}{2}}^\mu|0\rangle_R, \; A_\mu^I \alpha_{-1}^I|0\rangle_L \otimes \bar{b}_{-\frac{1}{2}}^\mu|0\rangle_R. \tag{10.6.37}$$

The states $|0\rangle$ means it is annihilated by all the postively moded oscillators in the corresponding sector. The first states in equation (10.6.36) correspond to the graviton, the anti-symmetric second rank tensor and the scalar, or dilaton, field. Those of equation (10.6.37) correspond to vector fields; there are 480 states in the first and 16 in the second making 496 vectors in all. The 480 states arise due to the 480 vectors of length squared 2 that correspond to the simple roots of the Lie algebras $E_8 \times E_8$, or $SO(32)$, depending upon which lattice we choose for the construction. The 16 correspond to the Cartan subalgebra elements of these Lie algebras. It is no surprise to learn, given the discussion in the previous sections and chapter 16, that these vectors are the gauge particles for the gauge groups $E_8 \times E_8$, or SO(32) depending on the lattice taken.

In the R sector we have following massless states:

$$\psi_{\mu\alpha} \alpha^{\mu}_{-1} |0\rangle_L \otimes |0\rangle_{R\alpha} \tag{10.6.38}$$

and

$$\lambda_{\mu\alpha} \left|0, \; p_L^2 = \frac{2}{\alpha'}\right\rangle_L \otimes |0\rangle_{R\alpha}, \; \lambda^I_{\mu\alpha} \alpha^I_{-1} |0\rangle_L \otimes |0\rangle_{R\alpha}. \tag{10.6.39}$$

Here the symbol $|0\rangle_{R\alpha}$ means the element in the right sector which is a space-time spinor but is annihilated by all the positively moded d_m oscillators. The relevent discussion is similar to that in chapter 7. The states of equation (10.6.38) correspond to the gravitino and the dilatino. The latter arises as the γ trace of the states which is present in a similar way to the presence of the trace of the states of equation (10.6.36) leads to the dilaton. The states of equation (10.6.39) are the 496 gauginos which arise in the same way as the 496 vectors. They belong to the adjoint representation of the gauge group.

As we will see in chapter 13, these particles can be collected into supersymmetry multiplets carrying 16 supersymemtries. These theories are, in fact, type I supergravity $(e_\mu{}^a, \phi, A_{\mu\nu}, ; \psi_{\mu\alpha}, \chi_\alpha)$ and $N = 1$ Yang–Mills theory (A_μ, λ_α) with gauge group $E_8 \times E_8$, or SO(32). As we mentioned before, the choice of a self-dual lattice in the construction of the left moving sector is required by demanding that the resulting string be modular invariant.

11 The physical states and the no-ghost theorem

I spent a week in Portadown one Sunday

Irish saying

Upon quantisation, the vibrational modes of the classical open string become an infinite number of quantum point particles, each of which corresponds to a state carrying an irreducible representation of the Poincaré group. At the free level these physical states are described in the old covariant approach of chapter 3 by the Virasoro conditions

$$L_n|\psi\rangle = 0, \quad n \geq 1, \quad (L_0 - 1)|\psi\rangle = 0. \tag{11.0.1}$$

The norm of a quantum state can be interpreted as a probability density and so for a string theory to be consistent the norms of all the solutions to equation (11.0.1) must be positive. Given the infinite number of solutions and the very large number of negative states in a general string state, due to the presence of the α_n^0 oscillators, this is a very non-trivial requirement. The purpose of this chapter is to prove that the physical states do have positive norm. This result was taken by the early workers in string theory to be a sign of the miraculous properties of string theory. In this process we will evaluate a number of commutators. In the old days one simply pushed the operators past each other using the oscillator relation $[\alpha_n^\mu, \alpha_m^\nu] = g^{\mu\nu} n \delta_{n+m,0}$. We refer the reader to the old reviews on string theory for such derivations. Here, however, we will make use of the perhaps slightly quicker operator product expansion techniques discussed in chapters 8 and 9. We will begin by considering the open string and then extend the result to the closed string.

In the gauge covariant approach discussed in chapter 12, the physical states will be taken to be those that satisfy

$$Q|\chi\rangle = 0; \quad b_0|\chi\rangle = 0. \tag{11.0.2}$$

We will also show in this chapter that the solutions to these conditions are essentially the same as those of equation (11.0.1).

11.1 The no-ghost theorem

We now prove the no-ghost theorem [3.13, 3.14] for the open bosonic string. Use will be made of the simplifications given in reference [11.1] This theorem is essential for the consistency of string theory, for it guarantees that physical states, that is, those which satisfy the Virasoro conditions $(L_n - \delta_{n,0})|\psi\rangle = 0$, $n \geq 0$, have positive norm. Put another way, it states that the string has sufficient gauge invariances to remove the infinite number of negative norm states that occur in the Hilbert space H generated by α^{μ}_{-n}, $\mu = 0, \ldots, D - 1$, $n = 1, 2, \ldots$ acting on $|0, p\rangle$, where $\alpha^{\mu}_n|0, p\rangle = 0$, $n \geq 1$, $\alpha^{\mu}_0|0, p\rangle = \sqrt{2\alpha'}p^{\mu}|0, p\rangle$. The method of the proof is to explicitly find all the solutions of the Virasoro conditions and then to show that these have positive norm. It is far from clear what these solutions are in terms of the α^{μ}_n. However, we can introduce the so-called DDF operators [11.2].

$$A^{(s)}_n = \oint \frac{dz}{2\pi i} \epsilon^{(s)}_{\mu}: \wp^{\mu}(z)e^{ink \cdot Q(z)}: \quad s = 1, \ldots, D - 1, \tag{11.1.1}$$

where the 4-momentum k^{μ} is light-like, $k_{\mu}k^{\mu} = 0$ and $\epsilon^{(s)}_{\mu}$ are any vectors such that $k^{\mu}\epsilon^{(s)}_{\mu} = 0$. The field Q^{μ} is defined by

$$Q^{\mu}(z) \equiv q^{\mu} - ip^{\mu}\ln z + i\sum_{n=-\infty}^{\infty} \frac{\alpha^{\mu}_n}{n}z^{-n} \tag{11.1.2}$$

and $\wp^{\mu} \equiv i(\partial/\partial z)Q^{\mu}$. The operator in the above equation is discussed in sections 9.1.1 and 9.3.1. The condition $k^{\mu}\epsilon^{(s)}_{\mu} = 0$ ensures that there are no normal ordering problems between $\wp^{\mu}$ and the exponential factor. The reader comparing this with the old literature may like to know that the pioneers often used $\wp^{\mu}(z)$ to mean the $z\wp^{\mu}(z)$ given here.

The field Q^{μ} is related to the coordinate $x^{\mu}(\tau, \sigma)$ of the open bosonic string whose form was found in terms of oscillators in equation (2.1.80). The precise relationship is given by

$$Q^{\mu}(e^{i\tau}) = x^{\mu}(\tau, \sigma = 0) = q^{\mu} + \sqrt{2\alpha'}\left(\alpha^{\mu}_0\tau + i\sum_{n=-\infty,n\neq 0}^{\infty} \frac{1}{n}\alpha^{\mu}_n e^{in\tau}\right). \tag{11.1.3}$$

Taking $\alpha^{\mu}_0 = \sqrt{2\alpha'}p^{\mu}$, $z = e^{i\tau}$ and setting $2\alpha' = 1$, as we will do henceforth in this chapter, we indeed recover the expression of Q^{μ} of equation (11.1.2). We also note that $z\wp^{\mu} = (d/d\tau)x^{\mu}(\tau, \sigma = 0)$. This field ws discussed in section 9.3.

When normal ordering operators we put all the destruction operators α^{μ}_n, $n = 1, 2, \ldots$, to the right of the creation operators α^{μ}_{-m}, $m = 1, 2, \ldots$, and factors of q^{μ} to the left of factors of p^{μ}. Carrying out the normal ordering, we find that even for $k^{\mu}k_{\mu} \neq 0$

$$z^{k^2/2}:e^{ik \cdot Q(z)}: = z^{k^2/2}e^{ik \cdot Q_<(z)}e^{ik \cdot q}z^{k \cdot P}e^{ik \cdot Q_>(z)} = e^{ik \cdot Q_<(z)}e^{ik \cdot Q_0(z)}e^{ik \cdot Q_>(z)}, \tag{11.1.4}$$

where

$$Q^{\mu}(z) \equiv Q^{\mu}_<(z) + Q^{\mu}_0(z) + Q^{\mu}_>(z) \tag{11.1.5}$$

and

$$Q^{\mu}_0 \equiv q^{\mu} - ip^{\mu}\ln z, \quad Q^{\mu}_>(z) \equiv i\sum_{n=1}^{\infty} \frac{\alpha^{\mu}_n}{n}z^{-n}, \quad Q^{\mu}_<(z) \equiv i\sum_{n=-\infty}^{-1} \frac{\alpha^{\mu}_n}{n}z^{-n}. \tag{11.1.6}$$

We note that $\wp^{\mu}(z) = \sum_{n=-\infty}^{\infty} \alpha^{\mu}_n z^{-n-1}$.

The operator $A^{(s)}_n$ is only well defined acting on states with momentum p^{μ} satisfying $p \cdot k \in Z$ as only then is the integrand a single valued function of z. The momentum of the state changes from p^{μ} to $p^{\mu} - nk^{\mu}$ under the action of $A^{(s)}_{-n}$. The advantage of the DDF operators is that they commute with the L_n, as we will now demonstrate. For later work it will be advantageous to define the operator

$$A_n^\mu = \oint \frac{dz}{2\pi i} \frac{1}{2} :[\wp^\mu(z)e^{ink\cdot Q(z)} + e^{ink\cdot Q(z)}\wp^\mu(z)]:.$$ (11.1.7)

The symmetrisation in equation (11.1.7) is required in order to adopt a normal ordered expression for A_n^μ, which is self-adjoint. It should be understood to be a short hand for

$$\tfrac{1}{2}(:\wp^\mu(z)e^{ik\cdot x(z)} + e^{ik\cdot Q(z)}\wp^\mu(z):) \equiv e^{ik\cdot Q_<(z)}\left[\wp_<^\mu(z)e^{ik\cdot q}z^{p\cdot q} + \frac{1}{2z}(p^\mu e^{ik\cdot q}z^{k\cdot q}\right.$$

$$\left. + e^{ik\cdot q}z^{k\cdot q}p^\mu) + e^{ik\cdot q}z^{k\cdot q}\wp_>^\mu(z)\right]e^{ik\cdot Q_>(z)},$$ (11.1.8)

where

$$\wp_>^\mu(z) = \sum_{n=1}^{\infty}\alpha_n^\mu z^{-n-1}, \quad \wp_<^\mu(z) = \sum_{n=1}^{\infty}\alpha_{-n}^\mu z^{n-1}.$$ (11.1.9)

Although the expression of equation (11.1.8) is obviously self-adjoint, the normal ordering prescription of equation (11.1.9) is not the same as that used in equation (11.1.4), that is, it is not given by putting α_{-n}^μ, $n = 1, 2, \ldots$, to the left of α_m^μ, $m = 1, 2, \ldots$, and q^μ to the left of p^μ. Indeed, it is also not the normal ordering prescription that applies in conformal field theory. In the latter, normal ordered operators are defined by taking the finite parts of the operator product expansion and this is indeed the same as the α_{-n}^μ, $n = 1, 2, \ldots$, to the left of α_m^μ, $m = 1, 2, \ldots$, and q^μ to the left of p^μ prescription. Hence, we cannot apply in a straightforward way the operator product techniques of chapter 8 to the expression of equation (11.1.8). However, we observe that we may rewrite A_n^μ as

$$A_n^\mu = \oint \frac{dz}{2\pi i} :\wp^\mu(z)e^{ink\cdot Q(z)}: + \frac{n}{2}k^\mu\phi_n,$$ (11.1.10)

where

$$\phi_n = \oint \frac{dz}{2\pi i} :e^{ink\cdot Q(z)}:.$$ (11.1.11)

Written in this way we can use the standard normal ordering prescription on both expressions and as this is the normal ordering implied by conformal field theory we can use the operator product expansion and other conformal field theory techniques on these quantities. Clearly, for the operator of equation (11.1.1) either normal ordering prescription yields the same expression due to the constraint $k^\mu\epsilon_\mu^{(s)} = 0$ and so the operator of equation (11.1.1) is actually self-adjoint.

Using operator product techniques it was shown in chapter 8 that

$$[L_n, \phi_m] = -n \oint \frac{dz}{2\pi i} z^{n-1}:e^{imk\cdot Q(z)}:$$ (11.1.12)

and that

$$\left[L_n, \oint \frac{dz}{2\pi i} :\wp^\mu(z)e^{imk\cdot Q(z)}:\right] = \frac{mk^\mu}{2}n(n+1)\oint \frac{dz}{2\pi i} z^{n-1}:e^{imk\cdot Q(z)}:.$$ (11.1.13)

Consequently, we make the crucial observation that

$$[L_n, A_m^\mu] = \frac{n^2}{2} mk^\mu \oint \frac{dz}{2\pi i z} z^{n-1} :e^{imk \cdot Q(z)}:.$$

(11.1.14)

As a result of the constraint $k \cdot \epsilon^{(s)} = 0$,

$$[L_n, A_m^{(s)}] = 0.$$

(11.1.15)

Consequently, $A_n^{(s)}$ acting on any physical state must create another physical state or give zero.

In fact, $A_n^{(s)}$ only has two $D-2$ independent degrees of freedom. It obviously only has $D-1$ degrees of freedom due to the restriction $k \cdot \epsilon^{(s)} = 0$ and we lose another one since, for $\epsilon^\mu \sim k^\mu$, we have the relations

$$k_\mu A_n^\mu = \oint \frac{dz}{2\pi i} :k \cdot \wp^\mu e^{ink \cdot Q}: = \begin{cases} \frac{1}{n} \oint \frac{dz}{2\pi i} \frac{d}{dz} e^{ink \cdot Q} = 0 & \text{if } n \neq 0, \\ k \cdot p & \text{if } n = 0. \end{cases}$$

(11.1.16)

When analysing physical states, a considerable simplification can be achieved by working in a fixed Lorentz frame. We choose the tachyon to have momentum $p_0^+ = p_0^- = 1$, $p_0^i = 0$, where we use the light-cone notation of chapter 4, which, for a vector V^μ, was to take $V^\pm = \frac{1}{\sqrt{2}}(V^{D-1} \pm V^0)$ with the other spatial components being represented by themselves. We note that this is not the same as the light-cone conventions we used for the two-dimensional world sheet as explained in appendix B. We also select the light-like momentum k^μ to be $k^- = 1$, $k^+ = k^i = 0$. These choices satisfy $p_0^2 = 2$, $p_0 \cdot k = 1$ and $k^2 = 0$. The $D-2$ independent DDF operators can be chosen to be the A_n^i, $i = 1, \ldots, D-2$, and equation (11.1.16) becomes $A_n^+ = 0$ for $n \neq 0$. On the tachyonic state $|0, p_0\rangle$, the A_{-1}^i act as

$$A_{-1}^i |p_0\rangle = \oint \frac{dz}{2\pi i} \wp_{<}^i(z) e^{-ik \cdot Q_{<}(z)} |0, p_0 - k\rangle = \alpha_{-1}^i |0, p_0 - k\rangle,$$

(11.1.17)

where

$$\alpha_n^\mu |0, p_0\rangle = 0, n \geq 1, \alpha_0^\mu |0, p_0\rangle = p_0^\mu |0, p_0\rangle.$$

(11.1.18)

The A_n^i, $n > 0$, are destruction operators in the sense that

$$A_n^i |0, p_0\rangle = 0, \quad n \geq 1.$$

(11.1.19)

Hence A_{-1}^i create $D-2$ transverse photon states, which we recognise as all the level 1 physical states with positive definite norm. As we will see, what is more remarkable is that for $D = 26$ arbitrary polynomials of A_{-n}^i acting on the tachyon state create all the physical states with positive definite norm. The A_n^i obey the algebra

$$[A_n^i, A_m^j] = np \cdot k\delta^{ij}\delta_{n+m,0}.$$

(11.1.20)

We note that the states of interest to us have $p \cdot k = (p_0 - nk) \cdot k = 1$ and hence on these states they obey the same relation as α_n^i. We establish this result using the operator product expansion of equations (9.1.38) and (9.1.40), which imply that

$$:\wp^i(z)e^{ink \cdot Q(z)}::\wp^j(\zeta)e^{imk \cdot Q(\zeta)}: = \frac{\delta^{ij}}{(z-\zeta)^2} :e^{i(nk \cdot Q(z)+mk \cdot Q(\zeta))}: + \text{analytic},$$

(11.1.21)

$i, j = 1, 2, \ldots, D - 2$, since $k^i = k^j = 0$. Substituting in the commutator, we find that

$$[A_n^i, A_m^j] = \oint_{C_0} \frac{d\zeta}{2\pi i} \oint_{C_s} \frac{dz}{2\pi i} \frac{\delta^{ij}}{(z - \zeta)^2} :e^{i(nk \cdot Q(z) + mk \cdot Q(\zeta))}:$$

$$= \delta^{ij} \oint_{C_0} \frac{d\zeta}{2\pi i} :e^{imk \cdot Q(\zeta)} \frac{d}{d\zeta} e^{ink \cdot Q(\zeta)}:$$

$$= \begin{cases} \delta^{ij} \oint_{C_0} \frac{d\zeta}{2\pi i} \frac{n}{n+m} \frac{d}{d\zeta} :e^{i(mk \cdot Q(\zeta) + nk \cdot Q(\zeta))}: = 0 \text{ if } n + m \neq 0, \\ \delta^{ij} \oint_{C_0} \frac{d\zeta}{2\pi i} \, nk \cdot \wp(\zeta) = \delta^{ij} nk \cdot p \text{ if } n + m = 0, \end{cases} \tag{11.1.22}$$

where C_0 is a contour around $\zeta = 0$ and C_1 is a contour around ζ.

We wish to supplement the A_n^i by other operators such that they generate the same Hilbert space H as the α_n^μ acting on the oscillator vacuum. Towards this end we consider the L_n and the operator

$$K_n = k \cdot \alpha_n = \oint \frac{dz}{2\pi i} z^n k \cdot \wp(z). \tag{11.1.23}$$

The set A_n^i, L_n, K_n obeys the algebra of equation (11.1.20) as well as

$$[A_n^i, K_m] = 0, \quad [A_n^i, L_m] = 0,$$

$$[L_n, L_m] = (n - m)L_{n+m} + \frac{D}{12} n(n^2 - 1)\delta_{n+m,0}, \tag{11.1.24}$$

$$[L_n, K_m] = -mK_{n+m}, \quad [K_n, K_m] = 0.$$

We observe that

$$K_n|0, \, p_0\rangle = 0, \quad n > 0. \tag{11.1.25}$$

Let us consider the states

$$|D\rangle = A_{-n_1}^{i_1} \cdots A_{-n_p}^{i_p} |0, \, p_0\rangle, \tag{11.1.26}$$

which we will refer to as *DDF* states. These have positive norm by virtue of equation (11.1.20) and satisfy the Virasoro conditions

$$(L_n - \delta_{n,0})|D\rangle = 0 \quad n \geq 0, \tag{11.1.27}$$

as a result of equation (11.1.24). They also obey the condition

$$K_n|D\rangle = 0, \quad n \geq 1; \quad K_0|D\rangle = p \cdot k|D\rangle = |D\rangle. \tag{11.1.28}$$

Let us consider all states of the form

$$|\{\lambda, \mu\}, D\rangle = L_{-1}^{\lambda_1} L_{-2}^{\lambda_2} \cdots L_{-n}^{\lambda_n} K_{-1}^{\mu_1} \cdots K_{-m}^{\mu_m} |D\rangle, \tag{11.1.29}$$

where $|D\rangle$ is a DDF state with the *exception* that it is built on a vacuum state $|0, p\rangle$ with $p \cdot k = 1$ but p^2 arbitrary. In other words, it satisfies $L_n|D\rangle = 0$, $n \geq 1$, but not necessarily $(L_0 - 1)|D\rangle = 0$. We will continue to use the symbol $|D\rangle$ for convenience, but it will now mean states of the form of equation (11.1.26) with an appropriately adjusted momentum To do otherwise, that is, take $p^2 = 0$, would imply that $|\{\lambda, \mu\}, D\rangle$ can never be a physical state unless it contains no L_{-n} or K_{-n}. The actual value of p^2 is given by imposing that $L_0 - 1$ vanishes on the state of equation (11.1.29).

We define the index l by

$$l = \sum_{r=1}^{n} r\lambda_r + \sum_{s=1}^{m} s\mu_s. \tag{11.1.30}$$

For a given value of l the above states will be linearly independent provided the matrix

$$M^{(l)}_{\{\lambda,\mu\};\{\lambda',\mu'\}} = \langle\{\lambda,\mu\},D|\{\lambda',\mu'\},D\rangle \tag{11.1.31}$$

has non-zero determinant. This is indeed the case; to see why let us examine the case of $l = 1$ for which the set of states $|\{\lambda,\mu\},D\rangle$ consists of the states

$$L_{-1}|D\rangle \quad \text{and} \quad K_{-1}|D\rangle. \tag{11.1.32}$$

Their norms are given by

$$\langle D|L_1 L_{-1}|D\rangle = 2\langle D|L_0|D\rangle, \ \langle D|L_1 K_{-1}|D\rangle = \langle D|K_{-1}L_1|D\rangle = \langle D|K_0|D\rangle, \tag{11.1.33}$$

and $\langle D|K_1 K_{-1}|D\rangle = 0$. Hence $M^{(1)}_{\{\lambda,\mu\};\{\lambda',\mu'\}}$ is the matrix

$$M^{(1)} = \begin{pmatrix} 2L_0 & K_0 \\ K_0 & 0 \end{pmatrix} = \begin{pmatrix} 2L_0 & 1 \\ 1 & 0 \end{pmatrix}. \tag{11.1.34}$$

Here we used that $K_0|D\rangle = |D\rangle \neq 0$ and so $M^{(1)}$ has determinant $-K_0^2 = -1$. In equation (11.1.34) we have for simplicity not given the expectation values, that is, L_0 in the matrix really denotes $L_0 = \langle D|L_0|D\rangle$. In general, $M^{(l)}_{\{\lambda,\mu\};\{\lambda',\mu'\}}$ is evaluated by pushing the K_n and L_n to the right, where they annihilate using the algebra of equation (11.1.20). Consequently, $M^{(p)}$ can be evaluated in terms of $\langle D|L_0|D\rangle = 0$ and $\langle D|K_0|D\rangle = \langle D|p\cdot k|D\rangle$ alone. At the next level $l = 2$ we adopt the ordering of our states to be

$$L^2_{-1}|D\rangle, \ L_{-2}|D\rangle, \ L_{-1}K_{-1}|D\rangle, \ K_{-2}|D\rangle, \ K^2_{-1}|D\rangle. \tag{11.1.35}$$

It is straightforward to show that

$$M^{(2)} = \begin{pmatrix} 12.L_0^2 & 6L_0 & 4L_0K_0 + 2K_0 & 2K_0 & 2K_0^2 \\ 6L_0 & 4L_0 + D/2 & 3K_0 & 2K_0 & 0 \\ 4L_0K_0 + 2K_0 & 3K_0 & K_0^2 & 0 & 0 \\ 2K_0 & 2K_0 & 0 & 0 & 0 \\ 2K_0^2 & 0 & 0 & 0 & 0 \end{pmatrix}. \tag{11.1.36}$$

We observe that the entries below the diagonal from lower left to upper right vanish and that all the entries on this diagonal are non-zero. The determinant is up to a sign the product of these diagonal entries and so is non-zero. The vanishing entries occur as we can never destroy a K_n and

$$\langle D|K_{n_1} \cdots K_{n_p}|D\rangle = 0 \tag{11.1.37}$$

unless $n_1 = 0 = n_2 = \cdots = n_p$.

In fact, this pattern of zeros is the general case, provided we adopt a certain ordering of the basis. First, let us define an ordering of two states which have two strings of L_n alone. Given two sets of L_n, $L^{\lambda_1}_{-1}, \ldots, L^{\lambda_n}_{-n}$ and $L^{\lambda'_1}_{-1}, \ldots, L^{\lambda'_m}_{-m}$, we denote generically them by $\{\lambda\}$

and $\{\lambda'\}$. We take the state with $\{\lambda\}$, to be ordered before the state with $\{\lambda'\}$ if $\{\lambda\} > \{\lambda'\}$, where this latter symbol means $\sum_r r\lambda_r > \sum_r r\lambda'_r$ or if $\sum_r r\lambda_r = \sum_r r\lambda'_r$ and $\lambda_1 > \lambda'_1$ or if $\sum_r r\lambda_r > \sum_r r\lambda'_r$ and $\lambda_1 = \lambda'_1$ and $\lambda_2 > \lambda'_2$, etc. Given two states with strings of Ks alone we adopt the same notation, that is, $\{\mu\} > \{\mu'\}$, but we order them in opposite ordering.

Let us consider states with a string of Ls and Ks such as occurs in equation (11.1.29) which we denote by $\{\lambda, \mu\}$. We take the state with $\{\lambda, \mu\}$ to be ordered before the state $\{\lambda', \mu'\}$ if $\{\lambda, \mu\} > \{\lambda', \mu'\}$, which means that if $\{\lambda\} > \{\lambda'\}$ or if $\{\lambda\} = \{\lambda'\}$ and $\{\mu\} < \{\mu'\}$. Put in words, we first order them according to their string of Ls, but if these are the same we then order them in the opposite order to the Ks order rule. The reader may easily verify that this is indeed the ordering adopted for the $l = 1$, and 2 cases given above. A little thought shows that with this ordering $M^{(l)}$ has the pattern of zeros required, that is, the same as in equations (11.1.34) and (11.1.36), and has non-zero terms along its diagonal from lower left to upper right. Consequently, $\det M^{(l)}$ is non-zero and all states of the form of equation (11.1.29) for a given DDF state $|D\rangle$ are linearly independent.

We can build states of the form of equation (11.1.29) on any DDF states. A little thought shows that two such states $|D\rangle$ and $|D'\rangle$, which are distinct, obey $\langle D|D'\rangle = 0$. Furthermore, the states $|\{\lambda, \mu\}, D\rangle$ and $|\{\lambda', \mu'\}, D'\rangle$ have a scalar product, which after carrying out the L_n and K_m algebra is proportional to $\langle D|D'\rangle$ and so also vanishes. Thus it follows that all states of the form of equation (11.1.29) built on all DDF states are linearly independent.

The Hilbert space H was taken to be all states of the form

$$\prod_{\mu=0}^{D-1} (\alpha^{\mu_1}_{-1})^{\lambda_1, \mu_1} (\alpha^{\mu_2}_{-2})^{\lambda_2, \mu_2} \cdots (\alpha^{\mu_n}_{-n})^{\lambda_n, \mu_n} |0, p\rangle, \tag{11.1.38}$$

where $\alpha^{\mu}_n |0, p\rangle = 0$; $n \geq 1$. As we discussed in chapter 3 each of these states is an eigenstate of $N = L_0 - 1 = \sum_{n=1}^{\infty} \alpha^{\mu}_{-n}\alpha_{\mu n}$ and at level n (that is, eigenvalue $N = n$) there are $T^D(n)$ such states, where

$$\sum_{n=0}^{\infty} T^D(n)x^n = \prod_{n=1}^{\infty} \frac{1}{(1 - x^n)^D}. \tag{11.1.39}$$

The states of equation (11.1.29) are also linearly independent and also eigenstates of N. The number of such states at a given level is a question of combinatorics and is determined only by the type of oscillators in $|D\rangle$ and the nature of the operators L_{-n} and K_{-n}. For these states, we therefore have at level n, $T^D(n)$ states which is the same number of states as in the Hilbert space H generated by the α^{μ}_n. Consequently, the states of equation (11.1.29) are not only linearly independent, but must span H, that is, form a basis for H. Put another way, the mapping from α^{μ}_n to A^i_n, K_n or L_n given by the definitions of equations (11.1.7) and (11.1.23) and the standard definition of L_n in terms of α^{μ}_n must be invertible.

We can now establish the no-ghost theorem. Given any element of $|\chi\rangle \in H$ we can obviously write it in the form

$$|\chi\rangle = |k\rangle + |s\rangle, \tag{11.1.40}$$

where $|k\rangle \in K$ is the subspace of states of the form

$$K^{\mu_1}_{-1} \cdots K^{\mu_n}_{-n} |D\rangle \tag{11.1.41}$$

including the states with no K_{-n}, and $|s\rangle \in S$, where S is defined to be all states in H which contain at least one factor of L_{-n}, $n \geq 1$. Due to the linear independence, proved above, this decomposition is unique. Using the relations $[L_{-1}, L_{-2}] = L_{-3}$, $[L_{-1}, L_{-3}] = 2L_{-4}$ etc. we can generate any L_{-n} from L_{-1} and L_{-2} and so any state in S can be written in the form

$$|s\rangle = L_{-1}|\psi_1\rangle' + L_{-2}|\psi_2\rangle, \tag{11.1.42}$$

where $|\psi_1\rangle'$, $|\psi_2\rangle \in H$. We may also write it as

$$|s\rangle = L_{-1}|\psi_1\rangle + \tilde{L}_{-2}|\psi_2\rangle, \tag{11.1.43}$$

where $\tilde{L}_{-2} = L_{-2} + \frac{3}{2}L_{-1}^2$ is an operator which we recognise from our discussion of zero-norm physical states of chapter 3.

Let us now consider a state $|\chi\rangle \in H$ which is a physical state, that is, $(L_n - \delta_{n,0})|\chi\rangle = 0$, $n \geq 0$. The operator L_0 takes the subspaces K to K and S to S and, as they are linearly independent, we may conclude that $(L_0 - 1)|\chi\rangle = 0$ implies that

$$(L_0 - 1)|k\rangle = 0, \tag{11.1.44}$$

$$(L_0 - 1)|s\rangle = 0. \tag{11.1.45}$$

The latter constraint translates into

$$L_0|\psi_1\rangle = 0, \quad (L_0 + 1)|\psi_2\rangle = 0. \tag{11.1.46}$$

The operator L_1 maps K to K, since $L_1|D\rangle = 0$ and $[L_1, K_{-n}] = nK_{-n+1}$, and it also takes S to S since

$$L_1 L_{-1}|\psi_1\rangle + L_1 \tilde{L}_{-2}|\psi_2\rangle = 2L_0|\psi_1\rangle + L_{-1}L_1|\psi_1\rangle + 6L_{-1}|\psi_2\rangle + L_{-2}L_1|\psi_2\rangle$$

$$= L_1(L_{-1}|\psi_1\rangle + \tfrac{9}{2}|\psi_2\rangle) + \tilde{L}_{-2}(L_1|\psi_2\rangle) \in S. \tag{11.1.47}$$

Due to the linear independence of states in K and S, $L_1|\chi\rangle = 0$ implies

$$L_1|k\rangle = 0, \quad L_1|s\rangle = 0. \tag{11.1.48}$$

Similarly, $\tilde{L}_2 = \tilde{L}_{-2}^{\dagger}$ maps K to K for the same reason as above, and as a consequence of the relation

$$\tilde{L}_2 (L_{-1}|\psi_1\rangle + L_{-2}|\psi_2\rangle) = L_{-1}\tilde{L}_2|\psi_1\rangle + (-9L_{-1}L_1 + \tfrac{1}{2}D - 13)|\psi_2\rangle$$

$$+ \tilde{L}_{-2}L_2|\psi_2\rangle, \tag{11.1.49}$$

$\tilde{L}_2$ maps S to S provided $D = 26$. Thus the condition $\tilde{L}_2|\chi\rangle = 0$ implies that

$$\tilde{L}_2|s\rangle = 0, \quad \tilde{L}_2|k\rangle = 0, \tag{11.1.50}$$

Since any L_n can be generated by commutators of L_1 and L_2, the above discussion can be summarised as: if $|\chi\rangle$ is a physical state (that is, $L_n|\chi\rangle = \delta_{n,0}|\chi\rangle$, $n \geq 0$), then, in terms of the decomposition of equation (11.1.40), $|s\rangle$ and $|k\rangle$ are also physical states:

$$(L_n - \delta_{n,0})|k\rangle = 0, \quad (L_n - \delta_{n,0})|s\rangle = 0 \; n \geq 0, \tag{11.1.51}$$

provided $D = 26$.

The physical nature of $|k\rangle$ considerably restricts its possible form. Such a state can be written as

$$|k\rangle = (K_{-m})^{\lambda_m} \cdots (K_{-2})^{\lambda_2} (K_{-1})^{\lambda_1} |D\rangle, \qquad (11.1.52)$$

since the K_{-n} commute. As such, we can order the K_{-m} with the highest m first. Acting with

$$(L_1)^{\lambda_1} (L_2)^{\lambda_2} \cdots (L_m)^{\lambda_m} \qquad (11.1.53)$$

we must annihilate $|s\rangle$; however, using the relations of equation (11.1.24), $K_n|0, p\rangle = 0$, $n \geq 1$, and $K_0|D\rangle = |D\rangle$, we find that the result is proportional to

$$(K_0)^{\Sigma_r \lambda_r} |D\rangle \neq 0. \qquad (11.1.54)$$

Such a state cannot therefore be a physical state unless it contains no K_{-n}. Thus the only physical states in K are the DDF states.

Let us summarise our findings. For $D = 26$, any physical state $|\chi\rangle$ (that is, $(L_n - \delta_{n,0})|\chi\rangle = 0$, $n \geq 0$) can be written as the sum of a DDF state $|D\rangle$ given in equation (11.1.26) and a physical spurious state $|s\rangle$:

$$|\chi\rangle = |D\rangle + |s\rangle, \qquad (11.1.55)$$

where $(L_n - \delta_{n,0})|D\rangle = 0$, $(L_n - \delta_{n,0})|s\rangle = 0$, $n \geq 0$, and $|s\rangle = L_{-1}|\psi_1\rangle + \tilde{L}_{-2}|\psi_2\rangle$.

The one exception to this result is states which have $p^\mu = 0$, since for these states the above choice of momentum based on p_0^μ cannot apply. It is straightforward to solve the physical state conditions for these states. They can only be of the form $\alpha_{-1}^\mu|0, 0\rangle$, $\mu = 0, 1, \ldots, 25$. In what follows we will not always explicitly mention these states but they are to be understood to be included in the space of physical states.

We are now in a position to prove the central result.

The no-ghost theorem   The physical states of the open bosonic string have positive norm if $D \leq 26$.

Using the above results, the proof is obvious. The scalar product of a physical state and a physical spurious state vanishes since

$$\langle \chi|s\rangle^\dagger = \langle s|\chi\rangle = ((\langle\psi_1|L_1 + \langle\psi_2|\tilde{L}_2)|\chi\rangle = 0. \qquad (11.1.56)$$

This includes physical spurious states $|s\rangle$ themselves which therefore have zero norm. Consequently, if $|\chi\rangle$ is physical, then

$$\langle\chi|\chi\rangle = \langle D|D\rangle. \qquad (11.1.57)$$

However, the scalar product of any DDF states is positive definite since it only depends on the relation $[A_n^i, A_{-m}^j] = n\delta_{n,m}\delta^{ij}$, which involves the positive definite metric δ^{ij}. Hence, we have shown the result.

We note that for $D = 26$, the positive definite states are the DDF states and at level n there are $T^{24}(n)$ of these, where

$$\sum_{n=0}^{\infty} T^D(n)x^n = \prod_{n=1}^{\infty} \frac{1}{(1 - x^n)^D}. \qquad (11.1.58)$$

When the space-time dimension D is less than 26 the physical states also have positive norm since they form a subspace of the positive norm states in the 26-dimensional theory once we adjoin the oscillators α_n^μ, $\mu = D - 1, D, \ldots, 25$ to those of the D-dimensional string theory. For $D > 26$ the norms of the physical states are not all positive as can be seen by exhibiting some negative norm physical states. At level 2, the most general state is of the form $|\psi\rangle = (d_1 p \cdot \alpha_{-2} + d_2 (p \cdot \alpha_{-1})^2 + d_3 \alpha_{-1} \cdot \alpha_{-1})|0; p\rangle$, where d_1, d_2 and d_3 are constants. If $p^2 = -2$, then $(L_0 - 1)|\psi\rangle = 0$, while $L_1|\psi\rangle = L_2|\psi\rangle = 0$ imply $d_1 - 2d_2 + d_3 = 0$ and $-4d_1 - 2d_2 + Dd_3 = 0$, respectively. These in turn imply that $d_1 = d_3 (\frac{1}{5}(D - 1))$ and $d_2 = d_3 (\frac{1}{10}(D + 4))$. The norm of this physical state is then given by $\langle \psi | \psi \rangle = (2d_3^2/25)(-D + 26)(D - 1)$. Clearly, if $D > 26$ this is a physical state with negative norm. Although the no-ghost theorem holds if $D < 26$, it is not true that the operators A_{-n}^i, $i = 1, 2, \ldots, D - 2$, generate all the physical states of positive definite norm as they do when $D = 26$.

The closed string obeys the analogous theorem to the open string. All its physical states are positive if $D \leq 26$. The states of the closed string can be written as tensor products of open string states with the exception of the zero modes which require more care.

11.2   The zero-norm physical states

In this section, we wish to give a more explicit expression for the zero-norm physical states and then present a shorter proof of the no-ghost theorem. The key is to extend the DDF operators given in the previous section to consider all the A_n^μ of equation (11.1.7). The operator product expansion

$$:\wp^\mu(z)e^{ik \cdot Q(z)} :: \wp^\nu(\zeta)e^{ik' \cdot Q(\zeta)}:$$

$$= : \left[\frac{\eta_{\nu\mu}}{(z - \zeta)^2} + \frac{k'^\mu \wp^\nu(\zeta)}{(z - \zeta)} - \frac{k^\nu \wp^\mu(z)}{(z - \zeta)} - \frac{k'_\mu k^\nu}{(z - \zeta)^2} \right]$$

$$\times e^{ik \cdot Q(z) + ik' \cdot Q(\zeta)} : (z - \zeta)^{k \cdot k'} + \cdots \tag{11.2.1}$$

is found by expanding $e^{ik \cdot Q}$ and using Wick's theorem. Making the replacements $k^\mu \to nk^\mu$, $k'^\mu \to = mk'^\mu$, the above operator product expansion leads to the commutator

$$[A_n^\mu, A_m^\nu] = n\delta_{n+m,0} p \cdot k\eta^{\mu\nu} + mk^\mu A_{n+m}^\nu - nk^\nu A_{n+m}^\mu$$

$$+ n^3 \delta_{n+m,0} p \cdot kk^\mu k^\nu. \tag{11.2.2}$$

Examining equation (11.1.14) we find that not all components of A_n^μ commute with L_n and so they will not all produce physical states when acting on the tachyon state. This difficulty can be overcome by considering the operator

$$\mathcal{A}_m^\mu = A_m^\mu - \frac{m}{2} k^\mu F_m(k), \tag{11.2.3}$$

where

$$F_m(k) = \oint_{c_0} \frac{dz}{2\pi i z} \left(\frac{k \cdot (z\wp(z))'}{k \cdot \wp(z)} \right) e^{imk \cdot Q(z)}, \tag{11.2.4}$$

where $(z\wp(z))' = (d/dz)(z\wp(z))$. The operator $\mathcal{A}_m^\mu$ is to be distinguished from the operator $A_m^{(s)}$ used previously by the different character of its upper indices and from A_m^μ. At first sight $F_n(k)$ looks ill defined as the denominator $k \cdot \wp(z)$ might vanish. We have in mind that $F_n(k)$ will act on DDF states, on which $k \cdot \wp(z) = k \cdot p + \text{oscillators} = k \cdot (p_0 - mk) + \text{oscillators} = 1 + \text{oscillators}$. As such, we can expand the denominator in the usual way.

Using the product expansion techniques, we find that

$$
[L_n, F_m(k)] = \oint_{c_0} \frac{d\zeta}{2\pi i \zeta} \int_{C_\zeta} dz z^{n-1} : \left\{ \frac{m}{(z - \zeta)} \frac{k \cdot \wp(z)}{k \cdot \wp(\zeta)} k \cdot (z\wp(z))' e^{imk \cdot Q(\zeta)} \right.
$$

$$
- \frac{k \cdot \wp(z)}{(k \cdot \wp(\zeta))^2} \frac{k \cdot (\zeta \wp(\zeta))'}{(z - \zeta)^2} e^{imk \cdot Q(\zeta)} + 2k \cdot \wp(z) \frac{\zeta^2}{(z - \zeta)^3} e^{imk \cdot Q(\zeta)}
$$

$$
\left. + \frac{k \cdot \wp(z)}{(z - \zeta)^2} \frac{1}{k \cdot \wp(\zeta)} e^{imk \cdot Q(\zeta)} \right\} :
$$

$$
= \oint_{c_0} \frac{d\zeta}{2\pi i \zeta} n^2 \zeta^n e^{imk \cdot Q(\zeta)}.
\tag{11.2.5}
$$

This result, when taken with equation (11.1.14), implies that

$$
[L_n, \mathcal{A}_m^\mu] = 0.
\tag{11.2.6}
$$

Using equation (11.1.16) we have the relation $k \cdot \mathcal{A}_m = \delta_{m,0} k \cdot p$. However, the remaining $D - 1$ $\mathcal{A}_m^\mu$ may be used to create physical states. The reader may derive the equations

$$
[\mathcal{A}_m^\nu, F_n] = n k^\nu F_{n+m} + k^\nu m^2 \delta_{n+m,0} k \cdot p,
$$

$$
[F_n, F_m] = 0,
\tag{11.2.7}
$$

to find that the commutator of $\mathcal{A}_m^\mu$ with itself is given by

$$
[\mathcal{A}_n^\mu, \mathcal{A}_m^\nu] = \delta_{n+m,0} p \cdot k \eta^{\mu\nu} n + m k^\mu \mathcal{A}_{n+m}^\nu - n k^\nu \mathcal{A}_{n+m}^\mu + 2n^3 k \cdot p k^\mu k^\nu \delta_{m+n,0}.
\tag{11.2.8}
$$

To identify the zero-norm physical states, we will adopt a particular division of the $\mathcal{A}_n^\mu$. In addition to k^μ, we introduce, another light-like $\bar{k}^\mu$ such that $k \cdot \bar{k} = -1$, $\bar{k}^2 = 0$. One such choice is $\bar{k}^- = 0$, $\bar{k}^+ = -1$, $\bar{k}^i = 0$ provided we also adopt the previous choice $k^- = 1$, $k^+ = k^i = 0$. We now split $\mathcal{A}_n^\mu$ into $\tilde{B}_n = \bar{k}_\mu \mathcal{A}_n^\mu$ and those components, denoted B_n^i, $i = 1, \ldots, D - 2$, which are orthogonal to k^μ and $\bar{k}^\mu$. For the above choices of k^μ and $\bar{k}^\mu$, which we now adopt, $\tilde{B}_n = -\mathcal{A}_n^-$ and $B_n^i = \mathcal{A}_n^i = A_n^i$, for $i = 1, \ldots, D - 2$.

The algebra of these operators is given by

$$
[B_n^i, B_m^j] = n \delta_{n+m,0} \eta^{ij},
$$

$$
[\tilde{B}_n, \tilde{B}_m] = (n - m) \tilde{B}_{n+m} + 2n^3 \delta_{n+m,0},
\tag{11.2.9}
$$

$$
[B_n^i, \tilde{B}_m] = n B_{n+m}^i.
$$

The algebra of the $\tilde{B}_n$ is a Virasoro algebra for which the zero mode generator, L_0, has been redefined. Using the Sugawara construction on the B_n^i, we can create another Virasoro type of algebra with generators

$$
l_n = \frac{1}{2} \sum_{i=1}^{D-2} : B_{-p}^i B_{n+p}^i : .
\tag{11.2.10}
$$

The operator l_n is by definition normal ordered by placing B^i_{-p} to the left of B^i_p for $p \geq 1$. The operator l_n contains $D - 2$ oscillators and obeys the algebra

$$[l_n, l_m] = (n - m)l_{n+m} + \frac{D - 2}{11.}n(n^2 - 1)\delta_{n+m,0}, [l_n, B^i_m] = -mB^i_{n+m}.$$

$$(11.2.11)$$

If we define the operator

$$B_n = \tilde{B}_n - l_n + 1,$$

$$(11.2.12)$$

we find that it commutes with B^i_m:

$$[B_n, B^i_m] = 0,$$

$$(11.2.13)$$

which implies that it also commutes with l_n:

$$[B_n, l_n] = 0.$$

$$(11.2.14)$$

The operator B_n has the commutator

$$[B_n, B_m] = [\tilde{B}_n, \tilde{B}_m] - [l_n, l_m]$$

$$= (n - m)B_{n+m} + \left\{2 - \frac{(D - 2)}{11.}\right\}n(n^2 - 1)\delta_{n+m,0}.$$

$$(11.2.15)$$

In 26 dimensions, which we now take to be the case, the central term disappears and so we have the relation

$$[B_n, B_m] = (n - m)B_{n+m}.$$

$$(11.2.16)$$

Since the $B^i_n = A^i_n$ and B_n commute with L_n, they may be used to create physical states of the form

$$(B_{-1})^{b_1} \cdots (B_{-n})^{b_n} A^{i_1}_{-n_1} \cdots A^{i_p}_{-n_p}|0; p_0\rangle = (B_{-1})^{b_1} \cdots (B_{-n})^{b_n}|D\rangle,$$

$$(11.2.17)$$

where $|D\rangle$ is the state discussed in the previous section.

Acting on $|0, p_0\rangle$ and adopting the above choices of k^μ and $\bar{k}^\mu$, we find that

$$B^i_{-n}|0; p_0\rangle = (\alpha^i_{-n} + \cdots)|0; p_0 - nk\rangle,$$

$$B_{-n}|0; p_0\rangle = -\left(\alpha^-_{-n} - \left(1 - \frac{n}{2}\right)\alpha^+_{-n} + \frac{n}{2}\alpha^+_{-n} + \cdots\right)|0; p_0 - nk\rangle,$$

$$(11.2.18)$$

where $+ \cdots$ stands for terms with more than one oscillator. Taking $n = 1$ one easily verifies that they do indeed give the 24 positive definite norm physical states and the one zero-norm physical state discussed in chapter 3. The latter is the one created by B_{-1}. We also note that $B_0|0, p_0\rangle = (p^-_0 - 1)|0, p_0\rangle = 0$.

The norm of the physical state of equation (11.2.17) may be evaluated by first doing the $B^i_{-n} = A^i_n$ algebra, which gives a result proportional to

$$\langle 0, p_0|(B_n)^{b_n} \cdots (B_1)^{b_1}(B_{-1})^{b_1} \cdots (B_{-n})^{b_n}|0, p_0\rangle.$$

$$(11.2.19)$$

This expression may be evaluated by moving B_p, $p \geq 1$, to the right, using equation (11.2.16), where they annihilate. The scalar product must always vanish as $B_0|0, p_0\rangle = 0$. Consequently, all the physical states of equation (11.2.17) are of positive norm and are of zero norm if and only if they involve any factor of B_{-n}.

It remains to show that these are the only solutions to the Virasoro conditions. To demonstrate this, we introduce one further set of operators

$$\phi_{n,m} = \oint \frac{dz}{2\pi i z} z^n e^{imk \cdot Q(z)}, \tag{11.2.20}$$

where $\phi_{0,m} \equiv \phi_m$. One finds that

$$[L_p, \phi_{n,m}] = -(n+p)\phi_{n+p,m} \tag{11.2.21}$$

and

$$[B_n^i, \phi_{m,p}] = 0, \qquad [B_p, \phi_{n,m}] = m\phi_{n,m+p}, \qquad [\phi_{n,p}, \phi_{m,q}] = 0. \tag{11.2.22}$$

Acting on the tachyon state, we find the result

$$\phi_{n,m}|0; p_0\rangle = \oint \frac{dz}{2\pi i z} z^{n+m} e^{imk.Q_<(z)}|0; p_0 - mk\rangle$$

$$= (-mk \cdot \alpha_{n+m} + \cdots)|0; p_0 - mk\rangle, \tag{11.2.23}$$

which vanishes unless $n + m \leq 0$. In particular, we observe that

$$\phi_{-m}|0; p_0\rangle = (mk \cdot \alpha_{-m} + \cdots)|0; p_0 - mk\rangle, \qquad m \geq 1, \tag{11.2.24}$$

$$\phi_0|0; p_0\rangle = |0; p_0\rangle. \tag{11.2.25}$$

We recall that the only non-zero component of k^μ is $k^- = 1$ and so we find the oscillator α_{-1}^+ in the first term of equation (11.2.24). This, in conjunction with equation (11.2.18), suggests that the operators B_n^i, B_n and ϕ_n generate the same Hilbert space as α_n^μ. We therefore consider all the states of the form

$$|\{\mu, k\}, D\rangle \equiv (B_{-1})^{\mu_1} \cdots (B_{-n})^{\mu_n} (\phi_{-m})^{k_m} \cdots (\phi_{-1})^{k_1}|D\rangle, \tag{11.2.26}$$

where $|D\rangle$ is a DDF state.

The scalar product of all such states is encoded in the matrix

$$N^{(l)}_{\{\mu, k\}, \{\mu', k'\}} = \langle\{\mu, k\}, D | \{\mu', k'\}, D\rangle, \tag{11.2.27}$$

where $l = \sum_r r\mu_r + \sum_s s k_s$. The DDF states are annihilated by the positive modes of ϕ_n and B_m:

$$\phi_n|D\rangle = 0, \quad n \geq 1, \qquad B_m|D\rangle = 0, \quad m \geq 1, \tag{11.2.28}$$

but

$$\phi_0|D\rangle = |D\rangle, \qquad B_0|D\rangle = 0. \tag{11.2.29}$$

The matrix $N^{(l)}$ is evaluated using equations (11.2.28) and (11.2.29) and the commutation relations of B_n and ϕ_m which, using equation (11.2.22), are given by

$$[\phi_n, \phi_m] = 0, \qquad [B_n, \phi_m] = m\phi_{n+m}. \tag{11.2.30}$$

Clearly, $N^{(l)}$ can only depend on the expectation value of ϕ_0 as this is the only operator to give such a non-vanishing expectation value. The determinant of $N^{(l)}$ is non-zero, for similar reasons as for the analogous matrix $M^{(l)}$ considered in the previous section. The matrix element

$$\langle D| \phi_{n_1} \cdots \phi_{n_p}|D\rangle \tag{11.2.31}$$

is non-zero only if $n_1 = n_2 = \cdots = 0$. When evaluating $N^{(l)}$ we can never destroy a ϕ_m and so it will be non-zero only if all the ϕ_m present are converted to ϕ_0s by the B_n. Adopting the analogous ordering to that given in the previous section, the entries below the diagonal from the lower left to the upper right of $N^{(l)}$ all vanish and the diagonal entries are all non-zero.

Thus the determinant of $N^{(l)}$ is non-vanishing and all states of the form of equation (11.2.26) are linearly independent. At a given value of $N = \sum_{n=1}^{\infty} \alpha^{\mu}_{-n} \alpha^{\mu}_n$, say n, the number of such states is $T^{26}(n)$. This is the same as the number of states in H. Thus any state in H can be written in the form of equation (11.2.26).

If $|\{\mu, k\}, D\rangle$ is a physical state, it will be annihilated by $(L_1)^{k_1} \cdots (L_m)^{k_m}$. This factor commutes with the B_{-p} terms and falls on the ϕ_{-m} terms. Using equations (11.2.21) and (11.2.29) we conclude that

$$(L_1)^{k_1} \cdots (L_m)^{k_m} |\{\mu, k\}, D\rangle = e(B_{-1})^{\mu_1} \cdots (B_{-n})^{\mu_n} (\phi_0)^{\sum rk_r} |D\rangle$$
$$= e(B_{-1})^{\mu_1} \cdots (B_{-1})^{\mu_n} |D\rangle. \tag{11.2.32}$$

In the above, e is a non-vanishing constant and so the only physical states are those that involve no factors of ϕ_{-p}.

We may summarise as follows. For the open bosonic string in 26 dimensions all states of the form

$$(B_{-1})^{\mu_1} \cdots (B_{-n})^{\mu_n} (\phi_{-m})^{k_m} \cdots (\phi_{-1})^{k_1} |D\rangle \tag{11.2.33}$$

are linearly independent and form a basis for H which is the vector space spanned by polynomials α^{μ}_n acting on the oscillator vacuum state. A state is physical if and only if it has no ϕ_{-p} factors. A physical state is a zero-norm physical state if and only if it has one or more factors of B_{-p}. The remaining states, which involve no ϕ_{-p} or B_{-p}, are physical states with positive definite norm. Thus we have not only proved the no-ghost theorem, but we have also classified all the zero-norm physical states. There are $T^{25}(n)$ physical states at level n of which $T^{24}(n)$ have positive definite norm.

11.3   The physical state projector

It is useful to have a projection operator which projects onto the space of physical states with positive definite norm, and vanishes on physical states of zero norm and non-physical states. Such an operator was discovered by Brink and Olive [11.3]; it is constructed from the operator

$$E = N - N^{tr}, \tag{11.3.1}$$

where

$$N = L_0 - \frac{1}{2}(\alpha_0)^2 = \sum_{n=1}^{\infty} \alpha^{\mu}_{-n} \alpha_{n\mu}, \tag{11.3.2}$$

$$N^{tr} = \sum_{i=1}^{D-2} \sum_{n=1}^{\infty} A^i_{-n} A^i_n. \tag{11.3.3}$$

We begin by establishing the properties of E. Substituting $A_0^\mu = p^\mu$ in equation (11.2.2) yields

$$[p^\mu, A_m^i] = mk^\mu A_n^i. \tag{11.3.4}$$

This, together with the equation $[L_0, A_m^i] = 0$, implies that

$$[N, A_m^i] = -mA_m^i. \tag{11.3.5}$$

On the other hand, equation (11.2.11) states that

$$[N^{tr}, A_m^i] = -mA_m^i. \tag{11.3.6}$$

Taking the difference of these equations, we find that E commutes with A_m^i:

$$[E, A_m^i] = 0. \tag{11.3.7}$$

The commutation relations of E with L_n and K_m are found using equations in the previous sections to be

$$[E, L_n] = [L_0, L_n] = -nL_n \quad \text{and} \quad [E, K_n] = [L_0, K_n] = -nK_n. \tag{11.3.8}$$

In section 11.1 we showed that any state in H could be written in terms of A_n^i, L_n and K_n acting on the oscillator vacuum state as written in equation (11.1.29). The operator E acting on $|\{\lambda, \mu\}, D\rangle$ gives

$$E|\{\lambda, \mu\}, D\rangle = c|\{\lambda, \mu\}, D\rangle, \tag{11.3.9}$$

where

$$c = \sum_{r=1}^{n} r \lambda_r + \sum_{s=1}^{m} s \mu_s. \tag{11.3.10}$$

Thus E can only take positive integer values and is zero only on DDF states.

The Brink–Olive projector is given by

$$\tau = \oint_{C_0} \frac{dy}{2\pi i y} y^E. \tag{11.3.11}$$

On a DDF state $\tau|D\rangle = |D\rangle$, but on any other state it vanishes. It follows trivially from this observation that $\tau^2 = \tau$.

The operator E may be rewritten provided we introduce the operator

$$D_n = \oint \frac{dz}{2\pi i} z^{n-2} : \frac{1}{k \cdot \wp(z)} : . \tag{11.3.12}$$

For similar reasons to those spelt out in the context of the operator F_n in the previous section, D_n is well defined. Evaluating its commutator with L_n by operator product techniques yields

$$[L_n, D_m] = \oint_{C_0} \frac{d\zeta}{2\pi i} \zeta^{(m-2)} \oint_{C_\zeta} \frac{dz}{2\pi i} z^{(n+1)} \left\{ -\frac{1}{(z - \zeta)^2} : \frac{k \cdot \wp(z)}{(k \cdot \wp(\zeta))^2} : \right\}$$

$$= -\oint_{C_0} \frac{d\zeta}{2\pi i} \zeta^{(n+m-2)} \left\{ \frac{n}{k \cdot \wp(\zeta)} + \frac{1}{(k \cdot \wp(\zeta))^2} \frac{d}{d\zeta} (\zeta k \cdot \wp(\zeta)) \right\}$$

$$= -(2n + m)D_{n+m}. \tag{11.3.13}$$

As A_n^i is in a direction orthogonal to k^μ and $k^2 = 0$ we easily conclude that

$$[A_n^i, D_m] = 0,$$
(11.3.14)

and

$$[D_n, D_m] = 0, \quad [K_n, D_m] = 0.$$
(11.3.15)

Consider the expression

$$E' = -(D_0 - 1)(L_0 - 1) - \sum_{n=1}^{\infty}(D_{-n}L_n + L_{-n}D_n).$$
(11.3.16)

We will show that E' is equal to E if $D = 26$. Since D_n and L_n commute with A_n^i, so does E'. Using equations (11.3.13) and (11.3.15) we find, for arbitrary dimension D, that

$$[E', L_n] = -nL_n + \frac{(D-26)}{11}n^2(n-1)D_n, \quad [E', K_m] = -mK_m.$$
(11.3.17)

Consequently, E and E', for $D = 26$, have the same commutation relation with A_n^i, L_n and K_n and one easily sees that $E'|0; p\rangle = 0 = E|0; p\rangle$. However, any state in H can be written in terms of these operators acting on the oscillator vacuum state $|0; p\rangle$. Consequently, $E - E'$ vanishes on any state in H and so $E = E'$

When deriving the second relation in equation (11.3.17) we made use of the identity

$$\sum_{n=-\infty}^{\infty} K_{-n}D_{n+m} = \delta_{m,0}.$$
(11.3.18)

This follows from substituting the expression

$$zk \cdot \wp(z) = 1 + \sum_{n=-\infty,n\neq o}^{\infty} K_n z^{-n}$$
(11.3.19)

into the numerator of the trivial identity

$$\oint \frac{dz}{2\pi iz} z^m \frac{k \cdot \wp(z)}{k \cdot \wp(z)} = \delta_{m,0}$$
(11.3.20)

and using the definition of D_n.

11.4   The cohomology of Q

The BRST charge Q plays an important role in the gauge covariant string theory discussed in chapter 12, and in the related BRST approach to string theory given in chapter 3. Our principal task in this section is to find the cohomology classes of Q. We first consider the equation

$$Q|\chi\rangle = 0,$$
(11.4.1)

where in the coordinate representation $|\chi\rangle$ is an arbitrary functional of $x^\mu(\sigma)$ and the ghosts $c(\sigma)$ and $b(\sigma)$. It is given in terms of an oscillator basis in chapters 3 and 12. Next we define the equivalence relation $|\chi_1\rangle \sim |\chi_2\rangle$ if

$$|\chi_1\rangle = |\chi_2\rangle + Q|s\rangle \tag{11.4.2}$$

for any state $|s\rangle$. Given all solutions of $Q|\chi\rangle = 0$, we can divide them into equivalence classes according to the above equivalence relation. The set of such equivalence classes is called the cohomology classes of Q. Note that if $|\chi_1\rangle$ is a solution of $Q|\chi_1\rangle = 0$, then $Q^2 = 0$ implies that $Q|\chi_2\rangle = 0$.

The physical states of many systems (see appendix A) are often just the cohomology classes of Q. It is therefore natural to expect this to be the case for string theory. In particular one may expect that the cohomology of the string theory BRST charge Q is just the DDF states. Such expectations are also in line with the formulation of the free gauge covariant action of the string given in chapter 12. In this section we will find that this is indeed true with one important qualification. Our proof follows [11.4] closely.

An important property of Q is the equation

$$L_n^{tot} = \{Q, b_n\} = L_n^\alpha + L_n^{gh}, \tag{11.4.3}$$

where we recall that L_n^{tot} is the Virasoro operator, which includes α_n^μ and the ghost oscillators c_n and b_n and is given in equation (3.2.74). We write

$$L_0^{tot} = R^{tot} + \tfrac{1}{2}\alpha_0^2 - 1, \tag{11.4.4}$$

where R^{tot} counts the total mode number of any state including the contribution of the ghost oscillators.

Let us consider a state $|r^{tot}, p^\mu\rangle$ which is an eigenstate of R^{tot} and α_0^μ with the eigenvalues indicated, and satisfies $Q|r^{tot}, p^\mu\rangle = 0$. Equation (11.4.3) then implies that

$$
\begin{aligned}
L_0^{tot}|r^{tot}, p^\mu\rangle &= (r^{tot} + \tfrac{1}{2}p^2 - 1)|r^{tot}, p^\mu\rangle \\
&= Q\,b_0|r^{tot}, p^\mu\rangle. \tag{11.4.5}
\end{aligned}
$$

Clearly, if $r^{tot} + \tfrac{1}{2}p^2 - 1 \neq 0$, we may divide by this quantity to find that

$$|r^{tot}, p^\mu\rangle = Q\frac{b_0}{r^{tot} + \tfrac{1}{2}p^2 - 1}|R^{tot}, p^\mu\rangle. \tag{11.4.6}$$

Hence, states for which $r^{tot} + \tfrac{1}{2}p^2 - 1 \neq 0$ are in the same equivalence class as the zero state. In these steps we have used the relations

$$[Q, L_n^{tot}] = 0 = [Q, \alpha_0^\mu] \tag{11.4.7}$$

to interchange the order of the factors. Consequently, states in a non-trivial cohomology class must satisfy

$$\tfrac{1}{2}p^2 = 1 - r^{tot}. \tag{11.4.8}$$

The derivation begins with the observation that

$$\tilde{E} = \{Q, S\}, \tag{11.4.9}$$

where

$$S = - \sum_{n=-\infty}^{\infty} b_{-n}(D_n - \delta_{n,0}) \tag{11.4.10}$$

and

$$\tilde{E} = E - \sum_{n=1}^{\infty} n(c_{-n}b_n + b_{-n}c_n) = -(D_0 - 1)(L_0 - 1) - \sum_{n=1}^{\infty}(D_{-n}L_n + L_{-n}D_n)$$

$$+ \sum_{n=1}^{\infty} n(c_{-n}b_n + b_{-n}c_n). \tag{11.4.11}$$

We recognise E as the object used to construct the physical state projection operator given earlier in this chapter where it was shown to be equal to E' of equation (11.3.16). Equation (11.4.9) is proved by straightforward evaluation using equations (11.4.3) and (11.3.13), and the definition of L^{gh}. Indeed, we find that

$$-\{Q, S\} = \left\{ Q, \sum_{n=1}^{\infty} b_{-n}D_n + \sum_{n=0}^{\infty} D_{-n}b_n - b_0 \right\}$$

$$= \sum_{n=1}^{\infty} (\{Q, b_{-n}\}D_n - b_{-n}[Q, D_{-n}])$$

$$+ \sum_{n=0}^{\infty} ([Q, D_{-n}]b_n + D_n\{Q, b_n\} - L_0^{tot})$$

$$= \sum_{n=1}^{\infty} \left(L_{-n}^{tot}D_n - \sum_{m=-\infty}^{\infty} b_{-n}c_{-m}[L_m^{\alpha}, D_n] \right)$$

$$+ \sum_{n=0}^{\infty} \left(\sum_{m=-\infty}^{\infty} c_{-m}[L_m^{\alpha}, D_{-n}]b_n + D_{-n}L_n^{tot} \right) - L_0^{tot}$$

$$= \sum_{n=1}^{\infty} (L_{-n}^{\alpha}D_n + D_{-n}L_n^{\alpha}) - L_0^{\alpha} - L_0^{gh} + D_0L_0^{\alpha} + \sum_{n=-\infty}^{\infty} L_{-n}^{gh}D_n$$

$$+ \left[\sum_{m=-\infty}^{\infty} \left(\sum_{n=1}^{\infty} b_{-n}c_{-m}(2m + n)D_{n+m} - \sum_{n=0}^{\infty} c_{-m}(2m - n)D_{m-n}b_n \right) \right]. \tag{11.4.12}$$

Taking care of the normal ordering implicit in the definitions of L_n^{gh}, we recognise the last term, given in square brackets, as

$$- \sum_{m=-\infty}^{\infty} L_{-m}^{gh}D_m - D_0. \tag{11.4.13}$$

The first part in equation (11.4.13) cancels the next to last term in equation (11.4.12) leaving us with the desired result, that is, equation (11.4.9).

It follows from equation (11.3.15) that

$$S^2 = \sum_{n=-\infty}^{\infty} \sum_{m=-\infty}^{\infty} b_{-n}D_n b_{-m}D_m - \left\{ b_0, \sum_{m=-\infty}^{\infty} B_{-m}D_m \right\} + b_0^2$$

$$= \frac{1}{2} \sum_{n=-\infty}^{\infty} \sum_{m=-\infty}^{\infty} b_{-n}b_{-m}[D_n, D_m] = 0 \qquad (11.4.14)$$

and so

$$[S, \tilde{E}] = S\{Q, S\} - \{Q, S\}S = 0. \qquad (11.4.15)$$

Similarly, one has

$$[\tilde{E}, Q] = 0. \qquad (11.4.16)$$

From the definition of E in equation (11.3.1) and equation (11.4.11) we find that we may write $\tilde{E}$ as

$$\tilde{E} = -N^{tr} + R^{tot}. \qquad (11.4.17)$$

Consider now all solutions of $Q|\chi\rangle = 0$. Since Q commutes with N^{tr}, R^{tot} and α_0^μ, which also commute with themselves, we may label such solutions by their eigenstates n^{tr}, r^{tot} and p^μ, viz $|n^{tr}, r^{tot}, p^\mu\rangle$. Equation (11.4.9) then implies that

$$\tilde{E}|n^{tr}, r_0^{tot}, p^\mu\rangle = QS|n^{tr}, r^{tot}, p^\mu\rangle = (r^{tot} - n^{tr})|n^{tr}, r^{tot}, p^\mu\rangle. \qquad (11.4.18)$$

Thus $(R^{tot} - N^{tr})$ must vanish on a non-trivial cohomology class of Q.

Any state $|\chi\rangle$ may be written as

$$b_{-p_1} \cdots b_{-p_s} c_{-q_1} \cdots c_{-q_t} |\{\lambda, \mu\}, D\rangle|\pm\rangle$$

$$= b_{-p_1} \cdots b_{-p_s} c_{-q_1} \cdots c_{-q_t} (L_{-1}^\alpha)^{\lambda_1} \cdots (L_{-n}^\alpha)^{\lambda_n} (K_{-1})^{\mu_1} \cdots (K_{-m})^{\mu_m} |D\rangle|\pm\rangle,$$

$$(11.4.19)$$

where $|D\rangle$ is a DDF eigenstate and the vacua $|\pm\rangle$ are those given in chapter 3 and are defined by the conditions

$$b_0|-\rangle = 0, \quad c_0|+\rangle = 0. \qquad (11.4.20)$$

Examining equation (11.4.17) we note that $\tilde{E}$ is equal to the total number of modes of all the oscillators minus that due to the DDF oscillators A_n^i. As such, we find that the states of equation (11.4.19) are eigenstates of $\tilde{E}$, with eigenvalues which are positive integers. The only states with eigenvalue zero are those containing only DDF states. It follows, from the above considerations, that all states that contain a ghost oscillator, L_{-n} or K_{-m} must be Q acting on some state and so belong to the cohomology class of Q containing the state zero. This leaves the states containing only DDF oscillators as possible candidates for the non-trivial cohomology classes.

We therefore consider whether the DDF states

$$\Pi_i \Pi_n (A_{-n_i}^i)^{\tau_{n_i}} |0, p_0\rangle|\pm\rangle, \qquad (11.4.21)$$

which have states with momentum $p^\mu = p_0^\mu - N^{tr}k^\mu$, where for these states $N^{tr} = \sum n_i \tau_{n_i}$, are annihilated by Q. We have now been specific about the choice of ghost vacuum with respect to the zero modes b_0 and c_0 using the conventions of chapter 3. The condition of equation (11.4.8) can be written for these states as

$$\tfrac{1}{2}p^2 = -N^{tr} + 1,\tag{11.4.22}$$

since $\tilde{E} = R^{tot} - N^{tr}$ vanishes on non-trivial cohomology classes.

We first consider the states with no DDF operators $|0; p_0\rangle|\pm\rangle$. Acting with Q we find that

$$Q|0; p_0\rangle|+\rangle = 0,$$

$$Q|0; p_0\rangle|-\rangle = (\tfrac{1}{2}p_0^2 - 1)|0; p_0\rangle|+\rangle.\tag{11.4.23}$$

The latter state is annihilated by Q only if $p_0^2 = 2$. This is also the condition that this state be in a non-trivial cohomology class of Q. We must also impose $p_0^2 = 2$ on the state $|p_0\rangle|+\rangle$ in order that it too is in a non-trivial cohomology class. Since L_n^α commutes with A_m^i, so does Q; it then follows that all the states in equation (11.4.21) are annihilated by Q and belong to non-trivial cohomology classes of Q provided $\tfrac{1}{2}p^2 = 1 - N^{tr}$. It is important to observe that we have two copies of DDF states corresponding to the two possible ghost vacua. These choices of p^2 are just those obtained from the physical states condition that $(L_0 - 1)$ applied to these states vanishes.

An exception not covered by the above discussion is the case of states with zero momentum mentioned in section 11.1. In this case equation (11.4.6) implies that $r^{tot} = 1$ and it is straightforward to analyse these states explicitly using the form of the BRST charge of equation (3.2.84). One can readily verify that the states

$$\alpha_{-1}^\mu|0, 0\rangle|\pm\rangle, \quad \text{and} \quad b_{-1}|0, 0\rangle|-\rangle\tag{11.4.24}$$

also belong to non-zero cohomology classes of Q.

Hence, the cohomology classes of Q are just the states of equations (11.4.21) and (11.4.23). These are essentially two copies of the DDF states built on the two ghost vacua, $|\pm\rangle$. Therefore we have twice the number of states expected, that is, two copies of the physical states. Thus string theory does not quite follow the usual pattern, that is, the physical states are just the cohomology classes of Q without some additional restriction.

Although gauge covariant string theory is the subject of chapter 12 it will be beneficial to make some remarks about it here. The gauge covariant open bosonic string has the equation of motion

$$Q|\chi\rangle = 0.\tag{11.4.25}$$

We may fix the gauge symmetry

$$\delta|\chi\rangle = Q|\Lambda\rangle\tag{11.4.26}$$

by the condition

$$b_0|\chi\rangle = 0.\tag{11.4.27}$$

The physical states are therefore those which satisfy the equations (11.4.25) and (11.4.27). Further, if $|\chi_1\rangle \sim |\chi_2\rangle$, then $|\chi_1\rangle$ and $|\chi_2\rangle$ have the same scalar product with all physical states and so are physically indistinguishable. Thus, the distinct physical states are the cohomology class of Q which also satisfies equation (11.4.27). We note that equations

(11.4.25) and (11.4.27) imply $L_0^{tot}|\chi\rangle = 0$ on the state, which we found above is a condition that a state be in the non-trivial cohomology class of Q. However, the cohomology classes of Q are the states of of equations (11.4.21) and (11.4.23) but equation (11.4.27) requires that we take only the states with the $|-\rangle$ vacuum and discard the states with a $|+\rangle$ vacuum. Hence, the physical states are just those of equations (11.4.21) and (11.4.23), but taking only the $|-\rangle$ vacuum. Thus we indeed find only one copy of the DDF states and the state with zero momentum. In chapter 12 we will provide an alternative proof that $Q|\chi\rangle = 0 = b_0|\chi\rangle$ are the correct physical states.

To summarise, the physical states of the string are the cohomology classes of Q subject to $b_0|\chi\rangle = 0$. In many reviews the physical states were often incorrectly discussed as they omit mention of equation (11.4.27).

12 Gauge covariant string theory

First they told me it was wrong, then they told me it was obvious.

Murray Gell-Mann

In this chapter we will formulate the free bosonic string and superstrings as a field theories in their critical dimensions in a way that manifestly possesses Lorentz invariance. While it is to be expected that these theories possess the gauge symmetries of the massless modes we will find that such a formulation of string theory possesses an infinite number of local symmetries. The actions have considerable elegance and make a surprising use of the BRST formalism even though they are classical objects.

One of the most remarkable aspects of the evolution of modern physics has been the central role played by symmetry. The most important are local symmetries which place very strong constraints on the form of the theory that possesses them. This first discovery of a gauge symmetry was by Weyl in the electromagnetic equations of Maxwell and continued in the construction of the Standard Model which possesses the SU(3) $\otimes$ SU(2) $\otimes$ U(1) gauge symmetry. Another crucial tool which played an important role in modern physics is the Feynman path integral description of quantum field theory. This formulation possesses the advantage that it is constructed from an action which can be used to manifestly display the symmetries of interest and, as a result, one can more easily derive those properties of the theory which are a consequence of its symmetries.

An illustration of the importance of the interplay between local symmetries and the path integral formulation was the demonstration of the renormalisability of theories which contain spontaneously broken gauge symmetries. Essential use was made in this proof of the BRST symmetry of the gauge fixed action and the corresponding Ward identities which follow from the Feynman path integral. Another advantage of the path integral formulation is that it can be used to calculate certain non-perturbative effects, such as spontaneous symmetry breaking and instanton effects.

It is hoped that supersymmetry and string theory will provide us with further symmetries which will, in one way or another, be realised in nature. Indeed, as we saw in chapter 3, gauge transformations for the massless spin-1 particle of the open bosonic string were associated with the Virasoro operator L_{-1}. In particular the zero-norm physical states $L_{-1}|\Omega\rangle$ contained the gauge transformation of the spin-1 particle, while the corresponding physical state condition $L_1|\psi\rangle = 0$ was a gauge-fixing condition. However, there exist all the other Virasoro operators, which suggests that the string has an infinite number of gauge symmetries. As such, it is natural to seek an action for string theory which encodes its

293

symmetries and can be used in a Feynman path integral. We might expect such an action to be of the form

$$A = \int Dx^{\mu}(\sigma)L(\psi[x^{\mu}(\sigma)]) \qquad (12.0.1)$$

and the vacuum to vacuum amplitude, before gauge fixing and ghosts are added, to be of the form

$$\int D\psi \exp\left(\frac{i}{\hbar}A\right). \qquad (12.0.2)$$

Although the final result is generically of this form, we will find that this description will prove to be inadequate in a number of important ways.

An obvious Feynman path integral formulation does exist if we are prepared to work within the light-cone formulation of chapter 4 where it was straightforward to write down a free action. However, this formulation lacks manifest Lorentz invariance and all the local symmetries of string theory that the existence of the Virasoro constraints suggests are present. At the free level, the fact that these symmetries are not manifestly present in the formulation is not so important and indeed the light-cone formulation of string theory is easier to quantise than its covariant analogue. However, for the interacting theory the loss of these symmetries presents a very serious disadvantage. Although, in principle it is possible to discover all the properties of a theory by quantising it in a light-cone formulation, this can be difficult. As with any theory in which some gauge fixing or renormalisation procedure has obscured the symmetries, one must verify, at every order of perturbation theory, that the Ward identities corresponding to all the symmetries are indeed valid. In this process one may have to add correcting terms to the action. This has rarely been carried out for any theory. If one finds that one cannot add local terms to the action, then this is the signal that the theory has an anomaly. Indeed, in a light-cone formulation anomalies can appear in the Lorentz symmetry.

What we would like is an action for string theory that is manifestly Lorentz invariant and possesses all its local symmetries. It was hoped that this would provide a complete formulation of string theory from which all its properties could, at least in principle, be derived. This would include a formulation in which one could calculate non-perturbative phenomenona. To provide such an action for the free bosonic string is the task of this chapter.

12.1   The problem

The physical states of the open bosonic string are described in a Lorentz invariant manner by the Virasoro conditions described in chapter 2; they were $L_n|\psi\rangle = 0$, $n \geq 1$ and $(L_0 - 1)|\psi\rangle = 0$. We wish to find an action, or equivalently, an equation of motion from which these on-shell constraints follow. To get a feel for the problem let us first solve it for the lowest levels. At the lowest level we have the tachyon φ whose equation of motion is $(\alpha'\partial^2 + 1)\varphi = 0$. This follows at the lowest level from the equation $(L_0 - 1)|\psi\rangle = 0$ if we identify $|\psi\rangle = (\varphi(x) + \cdots)|0\rangle$. Hence at the lowest level there is nothing to do in the sense that the Virasoro constraint and the equation of motion coincide.

At the next level, we find the vector A_μ, which is subject to $\partial^2 A_\mu = 0$ as a result of the equation $(L_0 - 1)|\psi\rangle = 0$, and $\partial^\mu A_\mu = 0$, which is a consequence of $L_1|\psi\rangle = 0$. The equation from which these follow is well known to be

$$\partial^2 A_\mu - \partial_\mu \partial^\nu A_\nu = 0. \tag{12.1.1}$$

The constraint $\partial^\mu A_\mu = 0$ arises as a consequence of gauge fixing the gauge symmetry $\delta A_\mu = \partial_\mu \Lambda$.

We now wish to express the equation of motion of φ and A_μ in terms of the string field $|\psi\rangle$, which in the Schrödinger representation is a functional of $x^\mu(\sigma)$, and the Virasoro operators L_n working at level 1. Recalling the form of L_{-1} of equation (3.1.18) we recognise that

$$\delta|\psi\rangle = L_{-1}|\Lambda^1\rangle \tag{12.1.2}$$

gives $\delta A_\mu = \partial_\mu \Lambda$ at this level if we use the expansion of $|\psi\rangle$ of equation (3.1.20) and let $|\Lambda^1\rangle = (\Lambda(x^\mu) + \cdots)|0\rangle$, where $\alpha_n^\mu|0\rangle = 0$, $n \geq 1$. Since we are working only up to level 1 we release the constraints $(L_0 - 1)|\psi\rangle = 0$ and $L_1|\psi\rangle = 0$, but retain all higher order constraints:

$$L_2|\psi\rangle = 0 = L_1^2|\psi\rangle = L_3|\psi\rangle = L_2 L_1|\psi\rangle \text{ etc.} \tag{12.1.3}$$

To be consistent we must also impose on $|\Lambda^1\rangle$ the conditions

$$L_1|\Lambda^1\rangle = 0 = L_2|\Lambda^1\rangle = L_3|\Lambda^1\rangle \text{ etc.} \tag{12.1.4}$$

We must find an equation from which the above two equations, $(L_0 - 1)|\psi\rangle = 0$ and $L_1|\psi\rangle = 0$ follow, or equivalently, an equation that possesses the gauge invariance of equation (12.1.2). We now take $|\psi\rangle = (\varphi(x) + iA_\mu \alpha_{-1}^\mu + \cdots)|0\rangle$. Remembering to maintain the level, the most general such expression is

$$(L_0 - 1 + aL_{-1}L_1)|\psi\rangle = 0, \tag{12.1.5}$$

where a is a constant. Varying as in (12.1.5) we find it is gauge invariant if

$$(L_0 - 1 + aL_{-1}L_1)L_{-1}|\Lambda^1\rangle = (L_0 - 1 - 2a + 2aL_0)L_{-1}|\Lambda^1\rangle = 0, \tag{12.1.6}$$

where we have used equation (12.1.4). Clearly, we require $a = -\frac{1}{2}$. We recover the first of the Virasoro conditions $L_1|\psi\rangle = 0$ as a choice for the gauge symmetry of equation (12.1.2), whereupon we obtain equation (12.1.5) which implies $(L_0 - 1)|\psi\rangle = 0$.

The action from which the equation of motion of equation (12.1.5) follows is given by

$$A = \tfrac{1}{2}\langle\psi|(L_0 - 1 - \tfrac{1}{2}L_{-1}L_1)|\psi\rangle. \tag{12.1.7}$$

We may write equation (12.1.5) in a slightly different way by introducing another string field $|\phi^1\rangle$ which is subject to the same constraints as $|\Lambda^1\rangle$. The two equations are

$$(L_0 - 1)|\psi\rangle + L_{-1}|\phi^1\rangle = 0 \tag{12.1.8}$$

and

$$2|\phi^1\rangle = -L_1|\psi\rangle. \tag{12.1.9}$$

The reason for this rewriting will become apparent as we develop the theory further. In terms of component fields these equations are the trivial equation for the tachyon and that for the spin-1 particle, which are of the forms $(\alpha'\partial^2 - 1)A_\mu + \sqrt{2\alpha'}\partial_\mu\phi^1 = 0$ and $2\phi^1 = -\sqrt{2\alpha'}\partial^\mu A_\mu$.

One might hope that one could continue to higher levels in the same vein; however, as we now see there are substantial differences. So far we have considered spins 0 and 1, but the equations of motion for these spins as well as spin $\frac{1}{2}$ differs in an important way from those of all higher spins. For spins 0, $\frac{1}{2}$ and 1 the on-shell conditions and the equations of motion are constructed from the same number of fields. This is not true for any higher spin field, massive or massless: for example, take a massless spin-2 particle. It is described by the on-shell conditions

$$\partial^\mu h_{\mu\nu} = 0, \ \partial^2 h_{\mu\nu} = 0 \tag{12.1.10}$$

and

$$h_{\mu\nu} = h_{\nu\mu}, \ h_\mu{}^\mu = 0. \tag{12.1.11}$$

The equation of motion is given by the linearised Einstein equation (see equation (9.2) of reference [1.11] for example) and involves the field $h_{\mu\nu}$ subject to $h_{\mu\nu} = h_{\nu\mu}$, but includes the trace $h_\mu{}^\mu$. The important point is that the latter trace is not present in the on-shell conditions, but one cannot formulate a correct equation of motion without it. The spin-$\frac{3}{2}$ particle is rather similar, but in this case the spin-$\frac{1}{2}$ part, $\gamma^\mu\psi_{\mu\alpha}$, of the spinor is absent from the on-shell conditions, but present in the Rarita–Schwinger equation.

As another example, we consider a massive spin-2 particle whose on-shell conditions are

$$0 = \partial^\mu h_{\mu\nu} = h_\mu{}^\mu = (\partial^2 - m^2)h_{\mu\nu}, \tag{12.1.12}$$

where $h_{\mu\nu} = h_{\nu\mu}$. Readers may show for themselves that there exists no Lorentz covariant wave equation in terms of $h_{\mu\nu}$ with $h_\mu{}^\mu = 0$ which leads to equation (12.1.12). One writes down the most general Lorentz covariant equation for $h_{\mu\nu}$, which has at most two space-time derivatives, and then shows that for no choice of the arbitrary coefficients can one deduce (12.1.12). The correct equations require not only the field $h_{\mu\nu}$ subject to $h_{\mu\nu} = h_{\nu\mu}$ and $h_\mu{}^\mu = 0$, but also a further field $\phi(x^\mu)$; they are given by

$$(-\partial^2 + m^2)h_{\mu\nu} + \partial_\mu\partial^\rho h_{\rho\nu} + \partial_\nu\partial^\rho h_{\rho\mu} - \frac{2}{D}\eta_{\mu\nu}\partial\rho\partial^\lambda h_{\rho\lambda}$$
$$= \frac{D-2}{D-1}\left(\partial_\mu\partial_\nu\phi - \frac{\eta_{\mu\nu}}{D}\partial^2\phi\right) \tag{12.1.13}$$

and

$$\partial^\mu\partial^\nu h_{\mu\nu} = \left(\partial^2 - \frac{D}{D-2}m^2\right)\phi, \tag{12.1.14}$$

where D is the dimension of space-time. Readers may verify that they lead to the required conditions:

$$\phi = 0 = \partial^\mu h_{\mu\nu} = (-\partial^2 + m^2)h_{\mu\nu}. \tag{12.1.15}$$

The field ϕ can be though of as the trace of $h_{\mu\nu}$.

The on-shell conditions for a particle of a given spin arise from embedding the unitary irreducible representations of the Poincaré group into a reducible non-unitary representation of the Poincaré group that contains a tensor representation, for example, ϕ_μ, $\phi_{\mu\nu}$, of the Lorentz group. The on-shell conditions are none other than the projection conditions which ensure that the reducible representation contains only the original irreducible representation. For example, a massive spin-1 particle is described by the field $\varphi_i(p) = 1, 2, \ldots, D - 1$. We embed this in the representation $\varphi_\mu(p)$ which is subject to $(p^2 + m^2)\varphi_\mu = 0$ and $p^\mu \varphi_\mu = 0$. The latter condition, in the rest frame $p^\mu = (m, 0, 0 \ldots, 0)$, becomes $\varphi_0 = 0$ and we recover the original representation. These two conditions follow from the single equation $-(m^2/\partial^2)\varphi_\mu = P_\mu{}^\nu\varphi_\nu$, where $P_\mu{}^\nu = \delta_\mu^\nu - \partial_\mu\partial^\nu/\partial^2$. Multiplying by ∂^2 we recognise the field equation.

For certain large infinite classes of spins higher than 1 the equations of motion, including the required extra fields which were systematically introduced, have been found [12.1]. We call these extra fields supplementary fields. Their appearance is related to the structure of the projectors referred to above. As mentioned above for a spin-1 particle the projector when multiplied by ∂^2 becomes local, that is, contains no $(\partial^2)^{-1}$ and is the equation of motion. However, for higher spins the projector when multiplied by ∂^2 is still non-local and cannot be identified with the equation of motion. We cannot multiply by $(\partial^2)^2$ as then we have terms with too many space-time derivatives in the equation of motion. The supplementary fields are necessary to find an equation of motion that is local and has only at most two space-time derivatives. The string contains, above the first level, particles of spin greater than 1. One may therefore expect it to require supplementary fields. To find how many, we must find what the spins of the on-shell states are and compare these with the fields required for the corresponding field equations with the fields present in $\Psi(x^\mu(\sigma))$ at the corresponding level. The fastest way to establish the particle content is to examine the light-cone formulation, as discussed in chapter 4, and reconstruct the relevant representations of the Poincaré group.

We now give some examples of how this works. Let us consider the second level of the open bosonic string. The light-cone functional ψ contains at this level the terms

$$\{\cdots + i\alpha_2^{i\dagger}A_i^2 + h_{ij}\alpha_1^{i\dagger}\alpha_1^{j\dagger} + \cdots\}\psi_0, \tag{12.1.16}$$

where $i, j = 1, \ldots, D - 2$. We have therefore $\frac{1}{2}(D - 2)(D - 1) + (D - 2) = \frac{1}{2}(D - 2)(D - 1) - 1$ states which belong to the symmetric traceless representation of the little group $SO(D - 1)$. We have often referred to this state as 'massive spin 2', being the higher dimensional analogue of the massive graviton in four dimensions. In the covariant theory, the string functional contains the fields

$$\{\cdots + i\alpha_2^{\mu\dagger}A_\mu^2 + h_{\mu\nu}\alpha_1^{\mu\dagger}\alpha_1^{\nu\dagger} + \cdots\}\psi_0, \tag{12.1.17}$$

that is, the fields A_μ^2 and $h_{\mu\nu}$ at the second level. Examining the on-shell constraints of equation (3.1.24) we find that it possesses the on-shell gauge symmetry

$$\delta h_{\mu\nu} = \partial_\mu\xi_\nu - \partial_\nu\xi_\mu - \frac{10}{D+4}\eta_{\mu\nu}\partial^\rho\xi_\rho, \quad \delta A_\mu^2 = -\sqrt{2\alpha'}\left(\frac{\xi_\mu}{\alpha'} + \frac{(D-6)}{(D+4)}\partial_\mu\partial^\rho\xi_\rho\right),$$

where $(\alpha'\partial^2 - 1)\xi_\mu = 0$. Thus A_μ^2 is gauge away, implying $h_\mu^\mu = 0$ and leaving only $h_{\mu\nu}$. From our previous considerations following equation (12.1.12) we see that we require one further field ϕ in order to be able to construct satisfactory equations of motion. This field ϕ will be the lowest component of a supplementary string field denoted $|\chi\rangle$, which is of the form

$$|\chi\rangle = \{\phi(x^\mu) + \cdots\}\psi_0. \tag{12.1.18}$$

As another example, we consider the first level of the closed bosonic string which we take to be unoriented (that is, to have the symmetry $\alpha_n^\mu \leftrightarrow \bar\alpha_n^\mu$). In the light-cone formalism the states, at this level, are

$$\{\cdots + \alpha_1^{i\dagger}\bar\alpha_1^{j\dagger}h_{ij} + \cdots\}\psi_0. \tag{12.1.19}$$

The number of such states is $(D-2)(D-1)/2$ and they form the symmetric traceless second rank tensor and the singlet representations of the little group $SO(D-2)$. Consequently, at this level we have a graviton and a spin-0 particle corresponding to $h_i{}^i$. By contrast in the Lorentz covariant theory we have the one field $h_{\mu\nu} = h_{\nu\mu}$, which includes its trace, and the corresponding on-shell conditions are

$$\partial^\mu h_{\mu\nu} = 0 = \partial^2 h_{\mu\nu}. \tag{12.1.20}$$

Writing down the most general equation of the form $\partial\partial h = 0$, one soon discovers that such an equation can never lead to the conditions of equation (12.1.20).

The solution to this problem is well known, since we wish to describe Einstein's theory coupled to a spin-0 particle; the gravity is described by $h_{\mu\nu}$, including its trace, subject to the linearised Einstein equation and the spin-0 particle is described by an additional scalar φ subject to $\partial^2\varphi = 0$. This additional scalar φ does not appear in the string functional ψ and so must be the first component of a new string functional that we must introduce.

Since the open and closed strings contain massless spin 1 and spin 2 respectively, they must possess the corresponding gauge invariances when contained in a Lorentz covariant theory such as gauge covariant string theory. However, as all objects associated with the string are functionals of $x^\mu(\sigma)$, it is natural to expect the gauge parameter of the string to also be of this form. This would imply that the higher levels of the string also possess gauge symmetries despite the fact that they describe massive states. In fact, this is the case as is confirmed by the presence of the zero-norm physical states which exist at most higher levels and are described in Chapters 2 and 11.

Given a theory with a local gauge symmetry and which is at most second order in derivatives, it is often uniquely determined once the gauge symmetry is specified. This is the case both for Yang–Mills theory and Einstein's gravity at the linearised level. The Noether technique can then be used to determine the full theories. As such, if we could find the gauge symmetry of string theory we would expect it to determine the free string theory and by some analogue of the Noether technique the interacting theory. This programme has never been carried out, but it is useful to realise that the interactions could be found in this way.

We can now extend the gauge covariant open string theory, given to the first level in equation (12.1.7), to the second level. In doing this we must incorporate the supplementary field $|\chi\rangle$ of equation (12.1.18). For this construction we release some of the constraints of equation (12.1.3) and now impose on $|\psi\rangle$ and $|\chi\rangle$ the set of constraints

$$L_3|\psi\rangle = 0 = L_2 L_1|\psi\rangle = L_1^3|\psi\rangle = L_4|\psi\rangle = \cdots; L_1|\chi\rangle$$
$$= 0 = L_2|\chi\rangle = \cdots, \tag{12.1.21}$$

that is, impose all constraints at level 3 and above. The most general equations of motion for $|\psi\rangle$ and $|\chi\rangle$ which maintain the level and agree with our result above at level 1 are given by

$$\left(L_0 - 1 - \frac{1}{2}L_{-1}L_1 - \frac{\gamma}{4}L_{-2}L_2\right)|\psi\rangle + \left(L_{-1}^2 + \frac{3}{2}\beta L_{-2}\right)|\chi\rangle = 0, \tag{12.1.22}$$

$$\left(L_1^2 + \frac{3}{2}\beta L_2\right)|\psi\rangle = (aL_0 + b)|\chi\rangle. \tag{12.1.23}$$

The use of the same β in equations (12.1.22) and (12.1.23) is required by demanding that the equations of motion follow from an action. An alternative way of searching for a gauge invariance is to demand that L_1 in equation (12.1.22) should vanish when we use equation (12.1.23). This is a consequence of using the gauge transformation of equation (12.1.2) in the action. Carrying this out and using constraints of equation (12.1.21) we find that

$$-\frac{1}{2}L_{-1}\left(L_1^2 + \frac{3}{2}\gamma L_2\right)|\psi\rangle + L_{-1}(4L_0 + 2L_{-1} + \frac{9}{2}\beta)|\chi\rangle = 0. \tag{12.1.24}$$

To eliminate $|\psi\rangle$ by using equation (12.1.23) requires $\gamma = \beta$ and we then find the condition

$$-\frac{1}{2}L_{-1}(aL_0 + b)|\chi\rangle + L_{-1}(4L_0 + 2 + \frac{9}{2}\beta)|\chi\rangle = 0. \tag{12.1.25}$$

Hence, we conclude that $a = 8$ and $b = 4 + 9\beta$. Applying L_2 in a similar way fixes $\beta = 1$, and so we find the equations of motions to be

$$\left(L_0 - 1 - \frac{1}{2}\sum_{n=1}^{2}\frac{1}{n}L_{-n}L_n\right)|\psi\rangle + (L_{-1}^2 + \frac{3}{2}L_{-2})|\chi\rangle = 0,$$
$$(L_1^2 + \frac{3}{2}L_2)|\psi\rangle = (8L_0 + 13)|\chi\rangle. \tag{12.1.26}$$

This system of equations is, in fact, invariant under the transformations, with parameters $|\Lambda^1\rangle$ and $|\Lambda^2\rangle$:

$$\delta_1|\psi\rangle = L_{-1}|\Lambda^1\rangle, \quad \delta_1|\chi\rangle = \frac{1}{2}L_1|\Lambda^1\rangle, \quad \delta_2|\psi\rangle = L_{-2}|\Lambda^2\rangle,$$
$$\delta_2|\chi\rangle = \frac{3}{2}|\Lambda^2\rangle \tag{12.1.27}$$

with

$$L_2|\Lambda^1\rangle = L_1^2|\Lambda^1\rangle = 0 = \cdots, \tag{12.1.28}$$

$$L_1|\Lambda^2\rangle = L_2|\Lambda^2\rangle = 0 = \cdots. \tag{12.1.29}$$

We stress that equations (12.1.26) are $|\Lambda^2\rangle$ invariant only for $D = 26$. The corresponding action is given by

$$\frac{1}{2}\langle\psi|(L_0 - 1 - \frac{1}{2}L_{-1}L_1 - \frac{1}{4}L_{-2}L_2)|\psi\rangle + \langle\psi|(L_{-1}^2 + \frac{3}{2}L_{-2})|\chi\rangle$$
$$-\frac{1}{2}\langle\chi|(8L_0 + 13)|\chi\rangle. \tag{12.1.30}$$

This completes the action up to the second level.

It will be instructive to rewrite the second level result in a kind of first order formalism. Completing the squares in the $L_1|\psi\rangle$ and $L_2|\psi\rangle$ terms in the action, we may rewrite expression (12.1.30) as

$$\tfrac{1}{2}\langle\psi|(L_0 - 1)|\psi\rangle - \tfrac{1}{4}((\langle\psi|L_{-1} - 2\langle\chi|L_1)(L_1|\psi\rangle - 2L_{-1}|\chi\rangle))$$

$$- \tfrac{1}{8}((\langle\psi|L_{-2} - 6\langle\chi|)(L_2|\psi\rangle - 6|\chi\rangle)) - 2\langle\chi|(L_0 + 1)|\chi\rangle, \qquad (12.1.31)$$

where we have used the fact that at this level $L_1|\chi\rangle = 0$. If we now introduce the auxiliary fields $|\phi^{(1)}\rangle$ and $|\phi^{(2)}\rangle$ we may rewrite the action as

$$\tfrac{1}{2}\langle\psi|(L_0 - 1)|\psi\rangle + \langle\phi^{(1)}|(L_1|\psi\rangle + L_{-1}|\zeta^1{}_1\rangle) + \langle\phi^{(2)}|(L_2|\psi\rangle + 3|\zeta^1{}_1\rangle)$$

$$+ \langle\phi^{(1)}|\phi^{(1)}\rangle + 2\langle\phi^{(2)}|\phi^{(2)}\rangle - \tfrac{1}{2}\langle\zeta^1{}_1|(L_0 + 1)|\zeta^1{}_1\rangle, \qquad (12.1.32)$$

where $|\zeta^1{}_1\rangle = -\tfrac{1}{2}|\chi\rangle$. One of the first steps [12.2] in the discovery of gauge covariant string theory was the construction up to the sixth level involving the systematic introduction of supplementary fields along the lines given in this section.

12.2   The solution

The gauge covariant equation of motion of the open bosonic string is given by [12.3, 12.4]

$$\tfrac{1}{2}\langle\chi|Q|\chi\rangle, \qquad (12.2.1)$$

where Q is the BRST charge given in chapter 3. It is invariant in 26 dimensions under the gauge transformation

$$\delta|\chi\rangle = Q|\Lambda\rangle \qquad (12.2.2)$$

since $Q^2 = 0$ in this dimension. The field $|\chi\rangle$ is an arbitrary functional of $x^\mu(\sigma)$ and the ghosts are as given in equation (3.2.60) with the one constraint

$$N|\chi\rangle = -\tfrac{1}{2}|\chi\rangle, \qquad (12.2.3)$$

where N is the ghost number operator of equation (3.2.97). One cannot fail to be impressed by the elegance of expression (12.2.1). It has, however, a very unexpected feature, namely it is a gauge invariant theory which makes use of the BRST tools of the first quantised string theory for its construction. While one may expect the gauge fixed and ghost extended BRST theory which possesses a BRST symmetry to involve the BRST tools of the first quantised theory, one could not expect a gauge covariant second quantised theory, which contains no Grassmann odd space-time fields to use such tools. Indeed, gauge covariant theories have never been constructed in this way in the past. The prototype example is the Yang–Mills theory and the action $\int d^4x(-\tfrac{1}{4}(F_{\mu\nu})^2)$ possesses no obvious trace of the BRST machinery. The elegance of the free action of expression (12.2.1) has misled some researchers to believe that it is somehow obvious. It is possible that gauge covariant string theory, with its surprising use of the BRST formalism, contains a deep lesson which we have yet to understand.

We will shortly show that expression (12.2.1) really is the correct result, that is, it leads to the physical states of string theory as embodied by the Virasoro conditions (that is, $L_n\psi = 0$, $n \geq 1$, $(L_0 - 1)\psi = 0$). This, we will show later. A more elementary condition is that it should contain no Grassmann odd space-time fields. The most general functional of $x^\mu(\sigma)$ and the ghost coordinates $b(\sigma)$ and $c(\sigma)$ is given in equation (3.2.50) where we adopt the notation $|\chi\rangle = \psi|-\rangle + \phi|+\rangle$. We note that in this equation ψ and φ are functionals of all these coordinates except for the dependence on the ghost zero mode coordinates which are treated separately, so giving rise to the two vacua $|\pm\rangle$. Also, $\psi^{m_1\cdots m_a}_{n_1\cdots n_b}$ and $\varphi^{m_1\cdots m_c}_{n_1\cdots n_d}$ are functionals of only $x^\mu(\sigma)$. In this section we will continue with the use of this notation. Implementing the condition of equation (12.2.3), we find that field of equation (3.2.50) contains the fields

$$\psi^{m_1\cdots m_a}_{n_1\cdots n_b} \tag{12.2.4}$$

with $a = b \equiv k$ and

$$\phi^{m_1\cdots m_c}_{n_1\cdots n_d} \tag{12.2.5}$$

with $c = d + 1 = k + 1$. We will denote these fields in the generic form by ψ^k_k and ϕ^{k+1}_k and the discussion around equation (3.2.60) tells us that these are indeed Grassmann even.

Corresponding to the condition on $|\chi\rangle$ of equation (12.2.3) the gauge parameter $|\Lambda\rangle$ satisfies

$$N|\Lambda\rangle = -\tfrac{3}{2}|\Lambda\rangle \tag{12.2.6}$$

as a result of the equation $[N, Q] = Q$. Equation (12.2.6) implies that if we write

$$|\Lambda\rangle = \Lambda|-\rangle + \Omega|+\rangle,$$

then the field content of $|\Lambda\rangle$ is Λ^{k+1}_k and Ω^{k+2}_k.

Another elementary check on expression (12.2.1), or equivalently, its equation of motion

$$Q|\chi\rangle = 0, \tag{12.2.7}$$

is that it reduces, at the first level, to that found in equations (12.1.8) and (12.1.9). At this level, we may write

$$|\chi\rangle = \psi|-\rangle + \phi^1 b^\dagger_1|+\rangle, \tag{12.2.8}$$

where ψ and ϕ are functionals of $x^\mu(\sigma)$ and Q is given by

$$Q =: \sum_{n=0,1,-1} c_{-n}L^\alpha_n + : 2b_0 c_1 c_{-1} - b_1 c_{-1}c_0 + b_{-1}c_1 c_0 - c_0 : + \cdots \tag{12.2.9}$$

where $+$ denotes higher level terms and we have used equation (3.2.94). Acting with Q on $|\chi\rangle$ we find that

$$Q|\chi\rangle = ((L^\alpha_0 - 1)\psi + L^\alpha_{-1}\phi^1)|+\rangle + (L^\alpha_1 \psi + 2\phi^1)c_{-1}|-\rangle \tag{12.2.10}$$

taking into account the constraints of equation (12.2.4). Taking $Q\chi = 0$ we find our previous result, namely equations (12.1.8) and (12.1.9). Similarly, we find that the gauge symmetry of equation (12.2.2) is given at this level by

$$\delta\psi = L^\alpha_{-1}|\Lambda^1\rangle, \quad \delta\phi = L^\alpha_0|\Lambda^1\rangle \tag{12.2.11}$$

where $|\Lambda\rangle = \Lambda^1 b^\dagger_1|-\rangle$.

At the next level we must take

$$|\chi\rangle = (\psi + c^\dagger_1 b^\dagger_1 \zeta^1_1)|-\rangle + (b^\dagger_1 \phi_1 + b^\dagger_2 \phi_2)|+\rangle. \tag{12.2.12}$$

We note that equation (12.2.12) has the field content we previously found and we leave it as an exercise for the reader to recover the second level result of equation (12.1.26).

To proceed further we must express $Q|\chi\rangle = 0$ in terms of the component fields that is, functionals of $x^\mu(\sigma)$ alone. To this end, and recalling the role the zero modes b_0 and c_0 play in the vacuum structure, we express Q in the form

$$Q = c_0 K - 2b_0 M + d + D, \tag{12.2.13}$$

where K, M, d and D do not contain any zero modes, $d^\dagger = D$, and they are given by

$$K = L_0 - 1 + \sum_{n=1}^{\infty}(nc_n^\dagger b_n + nb_n^\dagger c_n),$$

$$M = -\sum_{n=1}^{\infty} nc_n^\dagger c_n,$$

$$d = \sum_{n=1}^{\infty} c^{n\dagger}\left(L_n + \sum_{m,p=1}^{\infty} f^p_{m,-n}b_p^\dagger c^m + \tfrac{1}{2}c^{m\dagger}f^p_{m,n}b_p\right), \tag{12.2.14}$$

$$D = \sum_{n=1}^{\infty} c^n\left(L_{-n} + \sum_{m,p=1}^{\infty} f^p_{m,-n}c^{m\dagger}b_p + \tfrac{1}{2}b_p^\dagger f^p_{n,m}c^m\right).$$

Here $f^p_{m,n} = (m-n)\delta_{m+n,p}$ are the structure constants of the Virasoro algebra. They satisfy the relations

$$d^2 = D^2 = 0, [K, d] = 0 = [K, D] = [K, M],$$
$$\tag{12.2.15}$$
$$[M, d] = [M, D] = 0,$$

$$\{d, D\} - 2MK = 0 \tag{12.2.16}$$

if $D = 26$. As a consequence we can again verify that

$$Q^2 = \{d, D\} - 2MK = 0. \tag{12.2.17}$$

Let us now study the action of Q on an arbitrary functional. If $|\chi\rangle$ is Grassmann odd we find that

$$Q|\chi\rangle = (2M\phi + (d+D)\psi)|-\rangle + (K\psi + (d+D)\phi)|+\rangle, \tag{12.2.18}$$

while if $|\chi\rangle$ is Grassmann even

$$Q|\chi\rangle = (-2M\phi + (d+D)\psi)|-\rangle + (-K\psi + (d+D)\phi)|+\rangle, \tag{12.2.19}$$

where K, M, d and D are now defined by acting on $\psi^{m_1\cdots m_a}_{n_1\cdots n_b}$ as

$$(K\psi)^{m_1\cdots m_a}_{n_1\cdots n_b} = (L_0 - 1 + m + n)\psi^{m_1\cdots m_a}_{n_1\cdots n_b},$$

$$(M\psi)^{m_1\cdots m_{a-1}}_{n_1\cdots n_{b+1}} = (-1)^b a\, n_{p[n_1}\psi^{pm_1\cdots m_{a-1}}_{n_2\cdots n_{b+1}]},$$

$$(d\psi)^{m_1\cdots m_a}_{n_1\cdots n_{b+1}} = L_{[n_1}\psi^{m_1\cdots m_a}_{n_2\cdots n_{b+1}]} + a f^{[m_1}_{p,[-n_1}\psi^{pm_2\cdots m_a]}_{n_2\cdots n_{b+1}]} - \tfrac{1}{2}bf^p_{[n_1,n_2}\psi^{m_1\cdots m_a}_{pn_3\ldots n_{b+1}]}, \tag{12.2.20}$$

$$D\psi^{m_1\cdots m_{a-1}}_{n_1\cdots n_b} = (-1)^b a\left[L_{-p}\psi^{pm_1\cdots m_{a-1}}_{n_1\cdots n_b} + bf^q_{[n_1,-p}\psi^{pm_1\cdots m_{a-1}}_{qn_2\cdots n_b]}\right]$$
$$- \tfrac{1}{2}(a-1)f^{m_1}_{k,l}\psi^{klm_2\cdots m_{a-1}}_{n_1\cdots n_b},$$

where $m = \sum_{i=1}^{a} m_i$, $n = \sum_{i=1}^{b} n_j$ and $\eta_{pn} = n\delta_{p,n}$. The difference in signs between $|\chi\rangle$ being Grassmann even and odd comes from pushing the zero modes through the oscillators and fields onto the vacuum.

For example, we find that if $\psi_0^0 = \psi$; $\phi_0^1 = \phi^n$, then

$$(d\psi_0^0) = L_n\psi, \quad (D\phi_0^1) = \sum_{n=-\infty}^{\infty} L_{-n}\phi^n, \quad (d\phi_0^1) = L_m\phi^n + (2m+n)\phi^{n+m},$$

$$(12.2.21)$$

$$(M\phi_0^1) = n\phi^n, \quad (K\phi_0^1) = (L_0 - 1 + n)\phi^n.$$

The equation $Q|\chi\rangle = 0$ can then be rewritten as

$$K\psi_a^a + D\phi_a^{a+1} + d\phi_{a-1}^a = 0, \quad d\psi_a^a + D\psi_{a+1}^{a+1} + 2M\phi_a^{a+1} = 0, \quad (12.2.22)$$

while the gauge transformation $\delta|\chi\rangle = Q|\Lambda\rangle$ is equivalent to

$$\delta\psi_a^a = d\Lambda_{a-1}^a + D\Lambda_a^{a+1} - 2M\Omega_{a-1}^{a+1},$$

$$\delta\phi_a^{a+1} = -K\Lambda_a^{a+1} + d\Omega_{a-1}^{a+1} + D\Omega_a^{a+2}. \quad (12.2.23)$$

The action $\frac{1}{2}\langle\chi|Q|\chi\rangle$ becomes

$$\frac{1}{2}\langle\psi|K|\psi\rangle + \langle\psi|d|\phi\rangle + \langle\psi|D|\phi\rangle + \langle\phi|M|\phi\rangle, \quad (12.2.24)$$

where the sum over indices is implied, that is, $\langle\psi|K|\psi\rangle = \sum_k\langle\psi_k^k|K|\psi_k^k\rangle$.

We now demonstrate that $\frac{1}{2}\langle\chi|Q|\chi\rangle$ has the physical states of the open bosonic string. We will show [12.3] this by fixing the gauge and adding the corresponding ghosts to the action, and we will then show that the bosonic minus fermionic degrees of freedom is the same as the number of physical degrees of freedom of the string, that is, the light-cone count, that is, $T^{24}(N)$ at level N, where

$$\sum_{N=0}^{\infty} T^D(N)x^N = \prod_{N=0}^{\infty} \frac{1}{(1-x^N)^D}. \quad (12.2.25)$$

We start with the classical action whose field content is ψ_k^k and ϕ_k^{k+1}. The BRST formulation is found as explained in appendix A. Examining equation (12.2.23) and observing the presence of K in $\delta\phi_k^{k+1}$ we can use Λ_k^{k+1} to set $\phi_k^{k+1} = 0$, that is, fix the gauge. The corresponding ghost term is given by

$$\bar{\Lambda}_{k-1}^k(-K\Lambda_{k-1}^{k-1} + D\Omega_k^{k-2} + d\Omega_{k-1}^{k+1}). \quad (12.2.26)$$

In the above, $\bar{\Lambda}_{k+1}^k$ are the anti-ghosts and we use, as an economy of notation, the same notation for ghost fields and the symmetries from which they arise, that is, Λ_k^{k+1} and Ω_k^{k+2} in equation (12.2.26) are the ghost fields which occur as symmetries in equation (12.2.23). We recall that in the BRST procedure every local symmetry leads to a ghost with the same index structure, but opposite Grassmann parity to the parameter of the symmetry. We observe that the action of equation (12.2.26) is of the form $\langle\bar{\Lambda}|Q|\Lambda\rangle$ for appropriate fields $\bar{\Lambda}$ and Λ.

The action of equation (12.2.26), however, possesses a so-called 'hidden' invariance $|\Lambda\rangle = Q|\Lambda'\rangle$. Since $[Q, N] = -N$ we find that $N|\chi\rangle = -\frac{1}{2}|\chi\rangle$ implies that $(N + \frac{5}{2})|\Lambda'\rangle =$

0, and so it has the content $\{\Lambda_{k-1}^{\prime k+1}, \Omega_{k-1}^{\prime k+2}\}$. We fix the invariance $\Lambda_{k-1}^{\prime k+1}$ by setting $\Omega_k^{k+1} = 0$, that is, we insert $\delta(\Omega_{k-1}^{k+1})$ in the functional integral, and add the ghost term

$$\bar{\Lambda}_{k+1}^{\prime k-1}(K\Lambda_{k-1}^{\prime k+1} + d\Lambda_{k-2}^{\prime k+1} + D\Lambda_{k-1}^{\prime k+2}). \tag{12.2.27}$$

This term again has an invariance corresponding to $|\Lambda'\rangle = Q|\Lambda''\rangle$ requiring ghost for ghost for ghosts. Clearly this process continues indefinitely.

The net result is a BRST action with the field content

$$\psi_k^k, \ \Lambda_k^{k+1}, \ \Lambda_{k-1}^{\prime k+1}, \ \Lambda_{k-1}^{\prime\prime k+2}, \dots$$

$$\bar{\Lambda}_{k+1}^k, \ \bar{\Lambda}_{k+1}^{\prime k-1}, \ \bar{\Lambda}_{k+2}^{\prime\prime k-1}, \dots \tag{12.2.28}$$

all of which occur in the action with the operator K.

We observe that in this set there is one and only one tensor of the type $\Lambda_{n_1 \cdots n_b}^{m_1 \cdots m_a}$ for each a and b, and consequently these fields may be neatly fitted into a functional $|\chi\rangle$ which is completely general except that it has no $|+\rangle$ vacuum, that is, it satisfies $b_0|\chi\rangle = 0$, and so is of the form

$$|\chi\rangle = \sum_{\{n\}\{m\}} c^{n_1\dagger} \cdots c^{n_b\dagger} b_{m_1}^\dagger \cdots b_{m_a}^\dagger \Lambda_{n_1 \cdots n_b}^{m_1 \cdots m_a} |-\rangle, \tag{12.2.29}$$

where $\Lambda_{n_1 \cdots n_b}^{m_1 \cdots m_a}$ can be either a ghost or an anti-ghost depending on its index type. The free BRST action is therefore given by

$$\tfrac{1}{2}\langle\chi|c_0 K|\chi\rangle = \tfrac{1}{2}\langle\chi|[c_0 b_0, Q]|\chi\rangle, \tag{12.2.30}$$

where now $|\chi\rangle$ only satisfies $b_0|\chi\rangle = 0$.

Of course, it is no longer gauge invariant, but is BRST invariant, namely

$$\delta|\chi\rangle = \lambda Q|\chi\rangle, \tag{12.2.31}$$

where λ is the rigid Grassmann odd BRST parameter. The gauge-fixed action of equation (12.2.30) was originally found by Siegel [12.5], before the development of gauge covariant string theory, by constructing the second quantised BRST free theory directly from the first quantised string theory.

We now demonstrate that the BRST action possesses the correct number of degrees of freedom [12.3]. Let us write the tensor Λ_{b+c}^{a+c} in the form

$$\Lambda_{n_1 \cdots n_b; p_1 \cdots p_c}^{m_1 \cdots m_a; p_1 \cdots p_c}, \tag{12.2.32}$$

indicating that it has c indices in common. This tensor first contributes at level $M = \sum_{i=1}^a m_i + \sum_{i=1}^b n_i + 2\sum_{i=1}^c p_i$ and is Grassmann even or odd according to whether $\sum m_i + \sum n_i$ is even or odd. The net (Bose–Fermi) number of tensors at level M is $c(M)$, where

$$\sum_{M=0}^{\infty} c(M)x^M = \prod_{n=1}^{\infty}(1 - 2x^n + x^{2n}) = \prod_{n=1}^{\infty}(1 - x^n)^2. \tag{12.2.33}$$

One way to think about the derivation of this equation is first to consider the count arising from just the ghost oscillators b_n and c_n. The corresponding expansion is $\phi + c_n^\dagger \phi_n + b_n^\dagger \phi^n + c_n^\dagger b_n^\dagger \phi^n{}_n$ as $c_n^2 = 0 = b_n^2$. The contribution to the right-hand side of equation (12.2.33) is $(1 - 2x^n + x^{2n})$.

Each tensor in equation (12.2.32) first contributes at level M, is a functional of $x^\mu(\sigma)$, and so can be expanded in terms of α_n^μ oscillators acting on the vacuum. Hence, it contributes $T^{26}(N-M)$ at level N. As a result, the net total number of degrees of freedom at level N in χ is given by

$$p(N) = \sum_{M=0}^{N} c(M) T^{26}(N-M),$$

and so

$$\sum_{N=0}^{\infty} p(N) x^N = \sum_{N=0}^{\infty} \sum_{M=0}^{\infty} c(M) T^{26}(N-M) x^{N-M} x^M = \prod_{n=1}^{\infty}(1-x^n)^2 \prod_{n=1}^{\infty} \frac{1}{(1-x^n)^{26}}$$

$$= \prod_{n=1}^{\infty}(1-x^n)^{-24} = \sum_{n=0}^{\infty} T^{24}(N) x^N. \tag{12.2.34}$$

Hence, we recover the light-cone count and so confirm that $\langle \chi|Q|\chi \rangle$ does describe the spectrum of string theory.

A much shorter proof is as follows:

$$\sum_{M=0}^{\infty}\{n_B(M) - n_F(M)\} x^{\hat{N}} = STrx^{[\sum_{n=1}^{\infty}(\alpha_n^\mu \alpha_{n\mu} + c_n^\dagger b_n + b_n^\dagger c_n)]}$$

$$= \prod_{n=1}^{\infty}(1-x^n)^2 \prod_{n=1}^{\infty} \frac{1}{(1-x^n)^{26}} = \prod_{n=1}^{\infty} \frac{1}{(1-x^n)^{24}}, \tag{12.2.35}$$

where we have made use of the fermionic partition function and $\hat{N} = N - \frac{1}{2}(c_0 b_0 - b_0 c_0)$.

It is instructive to examine the number of fields at low levels in the BRST theory. At the first level

$$|\chi\rangle = (l + i\alpha_1^{\mu\dagger} A_\mu)|-\rangle + b_1^\dagger e_1|-\rangle + c_1^\dagger \bar{e}_1|-\rangle + \cdots \tag{12.2.36}$$

and we recognise e_1 and $\bar{e}_1$ as the ghost and anti-ghost for the gauge field A_μ and l is the tachyon. This is the correct result and the count of bosonic minus fermionic states at level one is $D - 1 - 1 = D - 2$.

We now give an alternative demonstration that the gauge covariant action does describe the correct physical states of string theory making use of chapter 11. It is good first to recall why Maxwell's equation describes the correct $D - 2$ states of the photon. The wave equation (12.1.1) for the photon has the gauge invariance $\delta A_\mu = \partial_\mu \Lambda$, which we may use to set $\partial_\mu A^\mu = 0$, whereupon the wave equation becomes $\partial^2 A_\mu = 0$. The resulting system has an on-shell gauge symmetry $\delta A_\mu = \partial_\mu \Lambda$ with $\partial^2 \Lambda = 0$. The physical states are then given by $\partial^2 A_\mu = 0$, subject to the removal of the on-shell gauge degree of freedom, that is, $\delta A_\mu = \partial_\mu \Lambda$. That is, we consider fields A_μ and A'_μ to be the same if $A'_\mu = A_\mu + \partial_\mu \Lambda$.

To analyse the states we choose a particular frame of reference. Let k^μ, $k^2 = 0$, be the momentum of the photon. We now write A_μ in terms of a particular basis:

$$A_\mu = k_\mu \lambda + \epsilon_\mu^i \mu_i + \bar{k}_\mu \tau, \tag{12.2.37}$$

where λ, μ_i, $i = 1, 2, \ldots, D - 2$, and τ are arbitrary functions of k_μ subject only to $k^2 = 0$. Also we take the polarisations to obey $k^\mu \epsilon_\mu^i = 0$, and $\bar{k}^\mu \epsilon_\mu^i = 0$, where $k^\mu \bar{k}_\mu = 1$ as well as

$\bar{k}^2 = 0$. The gauge choice $\partial^\mu A_\mu = 0$ implies that $\tau = 0$, while the on-shell gauge symmetry allows us to set $\lambda = 0$. We are left with the $D - 2$ degrees of freedom $\epsilon^i_\mu \mu_i$, which are the correct physical states.

Let us follow the same procedure for the string. We begin with $|\chi\rangle$ subject to $(N + \frac{1}{2})|\chi\rangle = 0$ and the equation of motion $Q|\chi\rangle = 0$. We first choose the gauge $b_0|\chi\rangle = 0$, so eliminating the states based on the $|+\rangle$ vacuum. The on-shell gauge symmetry is $\delta|\chi\rangle = Q|\Lambda\rangle$. As such the physical states obey $Q|\chi\rangle = 0 = b_0|\chi\rangle$, which implies that $K|\chi\rangle = \{b_0, Q\}|\chi\rangle$, and are subject to the on-shell gauge symmetry which can be stated as taking $|\chi'\rangle$ and $|\chi\rangle$ to be equivalent if $|\chi'\rangle = |\chi\rangle + Q|\Lambda\rangle$. That is, we find the cohomology classes of Q as indeed we did in chapter 11. The result was just the physical states of the string. The choice of frame entered in the construction of the DDF operators. Having solved for the equivalence classes, we can impose that $(N + \frac{1}{2})|\chi\rangle = 0$, but as the solution to this constraint contains the non-zero equivalence classes of Q, subject to $b_0|\chi\rangle = 0$, the result is the same.

12.3   Derivation of the solution

In this section we discuss one of the paths that lead to the free gauge covariant open string action [12.3, 12.4]. These two references resulted from two distinct approaches to the problem. Reference [12.3] was a natural sequel to [12.2, 12.6, 12.7], which were motivated by the realisation that supplementary fields were required. They showed how to incorporate the supplementary fields at the low levels of the string as explained in section 12.1 and then how to systematically incoorporate them at all levels whilst maintaining a gauge invariant action. The alternative approach began with the BRST second quantised free theory given in [12.5]. This was constructed by making extensive use of the BRST machinery of the first quantised theory given in chapter 3. The authors of [12.8] and [12.9] then proceeded to work backwards from the BRST theory to the gauge covariant theory. This was a non-trivial task since the free BRST action of equation (12.2.30) has a particularly unilluminating structure, namely the kinetic operator is $\partial^2 +$ the number operator. In [12.4] the result of [12.8] and [12.9] was generalised to give the well-known result. We will not pursue this latter approach here.

In this section we will present the first derivation [12.2, 12.3, 12.6, 12.7]. This has the advantage that it gives some insight into what the gauge covariant free action really contains and it gives a systematic way of finding all the free gauge covariant actions which we will find in subsequent sections.

We now begin the process of incorporating supplementary fields at all levels of the string in conjunction with the gauge symmetries along the lines of [12.2, 12.3, 12.6, 12.7]. In view of the considerations of section 12.1 it is natural to adopt the gauge symmetry

$$\delta\psi = \sum_{n=1}^{\infty} L_{-n}\Lambda^n, \tag{12.3.1}$$

where Λ^n are gauge parameters which are functionals of $x^\mu(\sigma)$. The first two terms were found earlier in this chapter, see equations (12.1.27), and the extension to the other terms is very natural in view of the Virasoro conditions. An action which has the above gauge symmetry must satisfy

$$0 = \sum_{n=1}^{\infty} \left(\Lambda^n, L_n \frac{\delta A}{\delta \psi} \right) + \sum_i \left(\delta \Phi^i, \frac{\delta A}{\delta \Phi_i} \right), \tag{12.3.2}$$

where Φ^i represents all other string fields such as the supplementary fields and $\delta\Phi^i$ are their gauge variations corresponding to that of equation (12.3.1). If we enforce the Φ^i equations of motion, the last term vanishes and we conclude that in this case

$$L_n \frac{\delta A}{\delta \psi} = 0, \quad n = 1, 2, \ldots. \tag{12.3.3}$$

Applying L_n to $\delta A/\delta \psi$ we find the terms $L_n \psi$ which, since the result must vanish upon the use of the equations of motion, must be specified as the equations of motion of fields ϕ^n, $n = 1, 2, \ldots$. The introduction of such fields is consistent with the second level result of equation (12.1.32), which contains the fields ϕ^n, $n = 1, 2$. Such terms must occur in the action as $\sum_{n=1}^{\infty} (\phi^n, L_n \psi)$ and so in the ψ equation in the form

$$(L_0 - 1)\psi + \sum_{m=1}^{\infty} L_{-m}\phi^m + \cdots = 0. \tag{12.3.4}$$

However, the action of L_n on the ψ equation of motion now has new terms containing $L_n \phi^m$ which must, by the same logic, be specified as a result of the equation of motion of fields ζ_m^n, $n, m = 1, 2, \ldots$. These must occur in the action in the form $\sum_{n,m=1}^{\infty} (\zeta_m^n, L_n \phi^m)$ and so the ϕ^n equation of motion takes the form

$$L_n \psi + \sum_{p=1}^{\infty} L_{-p}\zeta_n^p + d_n \phi^n + \sum_{p,m} \delta(n - p - m)d_p^m \zeta_m^p = 0, \tag{12.3.5}$$

where we have added all possible additional terms of the correct level with arbitrary constants d_n and d_p^m. Similarly the ζ_m^n equation of motion is of the form

$$L_n \phi^m - (L_0 + e_n^m)\zeta_n^m + f_n^m \phi^{n+m} = 0, \tag{12.3.6}$$

where e_n^m and f_n^m are constants. The last terms of equations (12.3.5) and (12.3.6) must come from the same term in the action, which is of the form $\sum_{,m=1}^{\infty} (\zeta_m^n, f_n^m \phi^{n+m})$ implying that $d_p^m = f_p^m$.

Applying L_n to the ψ equation of motion results in the condition

$$(L_0 + n - 1)L_n \psi + \sum_{m=1}^{\infty} L_{-m}L_n \phi^m + \sum_{m=1}^{n-1}(n + m)L_{n-m}\phi^m$$

$$+ \sum_{m=n+1}^{\infty} (n + m)L_{n-m}\phi^m + (2nL_0 + \tfrac{1}{12}Dn(n^2 - 1))\phi^n = 0 \tag{12.3.7}$$

if we use all other equations of motion. Let us denote the five terms in this equation by T_i, $i = 1, 2, 3, 4, 5$, respectively. Upon using equation (2.3.5) the first term becomes

$$T_1 = -\sum_{p=1}^{\infty} L_{-p}(L_0 + n + p - 1)\zeta_n^p$$

$$- c_n(L_0 + n - 1)\phi^n - \sum_{p,m=1}^{\infty} \delta(n - p - m)f_p^m(L_0 + n - 1)\zeta_m^p. \qquad (12.3.8)$$

Using equation (12.3.6) the second term is given by

$$T_2 = -\sum_{m=1}^{\infty}\{L_{-m}(L_0 + e_n^m)\zeta_n^m + L_{-m}f_n^m\phi^{n+m}\}, \qquad (12.3.9)$$

while the third term is given by

$$T_3 = \sum_{m=1}^{n-1}((n + m)(L_0 + e_{n-m}^m)\zeta_{n-m}^m - f_{n-m}^m\phi^n\}. \qquad (12.3.10)$$

Examining the above terms, which must all cancel, we conclude that

$$e_n^m = n + m - 1, \quad c_n = 2n, \quad f_n^m = 2n + m. \qquad (12.3.11)$$

All the remaining terms vanish due to the identity

$$-\sum_{m=1}^{n-1}(n + m)(2n - m) + \tfrac{1}{2}Dn(n^2 - 1) - 2n(n - 1) = 0, \qquad (12.3.12)$$

provided $D = 26$.

We summarise the result:

$$(L_0 - 1)\psi + \sum_{n=1}^{\infty} L_{-n}\phi^n = 0,$$

$$L_n\psi + \sum_{m=1}^{\infty} L_{-m}\zeta_n^m + \sum_{p,m=1}^{\infty} \delta(n - p - m)(2p + m)\zeta_m^p = 0, \qquad (12.3.13)$$

$$L_n\phi^m - (L_0 + n + m - 1)\zeta_n^m + (2n + m)\phi^{n+m} = 0.$$

The action from which it follows is

$$\tfrac{1}{2}(\psi, (L_0 - 1)\psi) + \sum_{n=1}^{\infty}(\phi^n, L_n\psi) + \sum_{n,m=1}^{\infty}(L_n\phi^m, \zeta_m^n) + \sum_{n=1}^{\infty} n(\phi^n, \phi^n)$$

$$- \tfrac{1}{2}\sum_{n,m=1}^{\infty}(\zeta_m^n, (L_0 + n + m - 1)\zeta_n^m) + \sum_{,m=1}^{\infty}(2n + m)(\phi^{n+m}, \zeta_n^m). \qquad (12.3.14)$$

It is straightforward to extend the gauge symmetry of equation (12.3.1) and one finds that

$$\delta\psi = \sum_{n=1}^{\infty} L_{-n}\Lambda^n, \quad \delta\phi^n = -(L_0 + n - 1)\Lambda^n,$$

$$\delta\zeta_m^n = -L_m\Lambda^n - (2m+n)\Lambda^{n+m} \tag{12.3.15}$$

is a symmetry of the above system if $D = 26$.

In fact, one can set $\phi^n = 0, n = 3, 4, \ldots, \zeta_m^n = 0, n, m = 3, 4, \ldots$, and still have a system which is gauge invariant under $\Lambda^n, n = 1, 2$ [12.7]. This latter system was called the finite set and the one above the infinite set for obvious reasons.

Of course, one must not only find a gauge invariant action, but one which leads to the correct count of states. Carrying out the BRST quantisation as explained in the previous section and counting the bosonic minus fermionic degrees of freedom one finds that counting works at the third (sixth) levels for the infinite (finite) set, but, at first sight, fails at higher levels. In fact, examination of the systems shows the existence of additional symmetries, which when correctly taken into account enable one to show that the count of states is correct up to the sixth (tenth) level [12.3]. Unfortunately, the higher the level the more complicated the additional symmetries become and the more difficult it is to carry out the count of states. Although it is unknown whether the count is correct to all orders, in view of its gauge invariance to all orders and correct count at those levels where it has been carefully carried out, it would seem likely that these systems do indeed have the correct physical states.

To give the reader a feeling for the additional symmetries, we list the first such symmetry which is given by [12.3]

$$\delta\psi = 0, \delta\phi^2 = -L_{-2}\Omega_3 + \tfrac{1}{4}L_{-1}L_1\Omega_3, \quad \delta\phi^2 = L_{-1}\Omega_3, \delta\phi^3 = \Omega_3 - \tfrac{1}{4}L_{-1}L_1\Omega_3,$$

$$\delta\phi^4 = -\tfrac{1}{2}L_1\Omega_3, \delta\zeta_2^1 = -4\Omega_3, \delta\zeta_1^2 = 2\Omega_3, \quad \delta\zeta_1^3 = -\tfrac{1}{2}L_1\Omega_3, \delta\zeta_3^1 = \tfrac{3}{2}L_1\Omega_3. \tag{12.3.16}$$

The reader may verify that this is a good symmetry up to fourth level. The count of ghosts is changed because of the hidden symmetry which then develops.

The nature of the additional symmetries suggests that they have resulted as a consequence of gauge fixing certain symmetries in a simpler system that has yet more supplementary fields. The net effect of all this discussion is to suggest that there should exist a new symmetry, $\delta\phi^n = L_{-m}\Omega^{nm} + \cdots$, and so we should insist that essentially L_n on the ψ, ϕ and indeed all equations of motion should vanish. Following the same argument as before this would require us to introduce a ϕ_m^{np} to specify $L_n\zeta_p^m$ and at all levels to introduce the set of fields

$$\psi_{m_1\cdots m_p}^{n_1\cdots n_p}, \phi_{m_1\cdots m_p}^{n_1\cdots n_{p+1}}, \quad p = 0, 1, 2\cdots. \tag{12.3.17}$$

These may be written in our previous notation as

$$\psi_p^p, \phi_p^{p+1}. \tag{12.3.18}$$

Clearly, we are to identify $\psi_0^0 = \psi, \psi_1^1 = \{\zeta_n^m\}$ and $\phi_0^1 = \{\phi^n\}$.

To carry out this final step it will be useful to rewrite our results using the d, D, M and K of equations (12.2.20). The result of equations (12.2.13) can be written as

$$K\psi_0^0 + D\phi_0^1 = 0, \quad d\psi_0^0 + D\psi_1^1 + 2M\phi_0^1 = 0, \quad K\psi_1^1 + d\phi_0^1 = 0 \quad (12.3.19)$$

and the gauge invariances of equations (12.3.15) can be written as

$$\delta\psi_0^0 = D\Lambda_0^1, \quad \delta\phi_0^1 = -K\Lambda_0^1, \quad \delta\psi_1^1 = d\Lambda_0^1. \quad (12.3.20)$$

The gauge invariance of equations (12.3.19) is obvious using equations (12.2.15) and (12.2.16). We now consider the extension to include the fields of equation (12.3.18). Writing the most general terms which extend equations (12.3.18) and (12.3.19) and taking care to balance the number of indices in an equation, one inevitably arrives at the generalisations

$$K\psi_p^p + D\phi_p^{p+1} + d\phi_{p-1}^p = 0, \quad d\psi_p^p + D\psi_{p-1}^{p+1} + 2M\phi_p^{p+1} = 0 \quad (12.3.21)$$

with the gauge invariances

$$\delta\psi_p^p = d\Lambda_p^p + D\Lambda_p^{p+1} - 2M\Omega_{p-1}^{p+1}, \delta\phi_p^{p+1} = -K\Lambda_p^{p+1} + d\Omega_{p-1}^{p+1} + D\Omega_p^{p+2}. \quad (12.3.22)$$

We have indeed recovered the solution [12.3, 12.4] given in the previous section. The final step is to recognise the result as being none other than $Q|\chi\rangle = 0$ with $|\chi\rangle$ subject to $(N + \frac{1}{2})|\chi\rangle = 0$.

12.4   The gauge covariant closed string [12.10, 12.11]

The elegance of the $\langle\chi|Q|\chi\rangle$ for the open string might tempt one to postulate that the same action is true for the closed string. However, as we shall see this is not the case. In the case of the closed string we have four ghost zero modes $c_0, b_0, \bar{c}_0$ and $\bar{b}_0$ and as a consequence the vacuum has the four possible states

$$|\pm, \pm\rangle \text{ and } |\pm, \mp\rangle, \quad (12.4.1)$$

where

$$c_0|+, \pm\rangle = 0, \quad b_0|-, \pm\rangle = 0$$

and

$$\bar{c}_0|\pm, +\rangle = 0, \quad \bar{b}_0|\pm, -\rangle = 0. \quad (12.4.2)$$

All four vacua are annihilated by $c_n, b_n, \bar{c}_n$ and $\bar{b}_n$ for $n \geq 1$.

The BRST charge Q for the closed string was discussed in section 3.2 and was found to have the form

$$Q =: \sum_{=-\infty}^{\infty} c_n L_{-n} - \frac{1}{2}\sum_{n,m,p=-\infty}^{\infty} f_{n,m}^p b_p c_{-n} c_{-m} - c_0 :$$

$$+ (c_n \to \bar{c}_n, b_n \to \bar{b}_n), \quad (12.4.3)$$

where $f^p_{n,m} = (n - m)\delta_{n+m,p}$. Isolating the zero modes it can be written as

$$Q = c_0 K - 2b_0 M + d + D + \bar{c}_0 \bar{K} - 2\bar{b}_0 \bar{M} + \bar{d} + \bar{D}, \qquad (12.4.4)$$

where K, M, d, D are independent of the zero modes and are given by equations (12.2.14), their barred analogues being obtained by putting bars on all quantities. Of course, they still obey equations (12.2.15) and (12.2.16) and their barred versions. The most general functional contains terms of the form

$$c^\dagger_{n_1} \cdots c^\dagger_{n_a} b^\dagger_{m_1} \cdots b^\dagger_{m_b} \bar{c}^\dagger_{p_1} \cdots \bar{c}^\dagger_{p_c} \bar{b}^\dagger_{q_1} \cdots \bar{b}^\dagger_{q_c} \, \psi^{m_1 \cdots m_b \, q_1 \cdots q_c}_{n_1 \cdots n_a \, p_1 \cdots p_c} |-, -\rangle \qquad (12.4.5)$$

plus similar terms multiply all the other vacua of expressions (12.4.1). Given the expression for the operators $K, M, \dots, \bar{d}, \bar{D}$ in terms of oscillators we can find their action on the component fields $\psi^{m_1 \cdots m_b \, q_1 \cdots q_c}_{n_1 \cdots n_a \, p_1 \cdots p_c}$. Since the unbarred operators do not act on barred oscillators, the action of K, M, d, D is on the $m_1 \cdots m_b, n_1 \cdots n_a$ indices leaving the indices $q_1 \cdots q_c, p_1 \cdots p_c$ inert and vice versa for the barred quantities, for example,

$$(d\psi)^{m_1 \cdots m_a \, p_1 \cdots p_c}_{n_1 \cdots n_{b+1} \, q_1 \cdots q_d} = L_{[n_1} \psi^{m_1 \cdots m_a \, p_1 \cdots p_c}_{n_2 \cdots n_{b+1}] \, q_1 \cdots q_d} + a f^{[m_1}_{m}, {}_{[-n_1} \psi^{m_2 \cdots m_a] \, p_1 \cdots p_c}_{n_2 \cdots n_{b+1}] \, q_1 \cdots q_d}$$
$$- \tfrac{1}{2} b f^n_{[n_1 n_2} \psi^{m_1 \cdots m_a \, p_1 \cdots p_c}_{n n_3 \cdots n_{b+1}] \, q_1 \cdots q_d}. \qquad (12.4.6)$$

Recalling the discussion of equations (3.2.57) and (3.2.58) we conclude that the only non-zero scalar products between the vacua are

$$\langle \pm, \pm | \mp, \mp \rangle \text{ and } \langle \pm, \mp | \mp, \pm \rangle. \qquad (12.4.7)$$

Since Q contains either terms which are linear in the ghost zero modes or terms with no zero mode ghosts, the expression $\langle \chi | Q | \chi \rangle$, where χ is a functional of $x^\mu(\sigma)$, and the ghosts, cannot, unlike the open string, contain any terms bilinear in the same zero mode ghosts. This means it cannot be the correct result as it cannot contain the term $\langle \psi | (L_0 + \bar{L}_0 - 2) | \psi \rangle$. As a little further thought shows it vanishes, indeed $\langle \chi | Q | \chi \rangle$ is actually Grassmann odd!

The safest approach to finding the correct gauge covariant action for the closed bosonic string is to mimic the steps of section 12.3 for the open string. Let us denote the original string functional ψ of $x^\mu(\sigma)$ in terms of which the Virasoro constraints of the closed string are defined by ψ^0_0. Following the same logic as in the open case, we first construct the closed string analogue of the infinite set. As such, we must introduce fields whose equations of motion contain $L_n \psi, \bar{L}_n \psi, L_n \phi^m, \bar{L}_n \bar{\phi}^m, L_n \bar{\phi}^m$ and $\bar{L}_n \phi^m$, that is, $\phi^n, \bar{\phi}^n, -\zeta^n{}_m, \bar{\zeta}^n{}_m, -\bar{\chi}^n{}_m$ and $\chi^n{}_m$, respectively, all of which are functionals of $x^\mu(\sigma)$. We may write these fields as

$$\phi^{10}_{00}, \phi^{01}_{00}, \psi^{10}_{10}, \psi^{01}_{01}, \psi^{10}_{01}, \psi^{01}_{10},$$

where

$$\phi^{00}_{00} = \psi, \ \phi^{10}_{00} = \{\phi^n\}, \ \psi^{01}_{00} = \{\bar{\phi}^n\}, \ \psi^{10}_{10} = \{-\zeta^n{}_m\}, \ \psi^{01}_{01} = \{\bar{\zeta}^m_n\},$$
$$\psi^{10}_{01} = \{-\bar{\chi}^n{}_m\}, \ \psi^{01}_{10} = \{\chi^n{}_m\}.$$

The signs are so as to agree with the notation used in [12.10]. We note that the first index structure on the above fields corresponds to the unbarred sector of the closed string and the second index to the barred sector.

As was done for the open string in expression (12.3.17), we introduce the further fields

$$\psi^{k_1 k_2}_{k_3 k_4}, \text{ with } k_1 + k_2 - k_3 - k_4 = 0,$$
$$\phi^{k_1 k_2}_{k_3 k_4}, \text{ with } k_1 + k_2 - k_3 - k_4 = 1. \tag{12.4.8}$$

All fields are subject to the algebraic constraint

$$(K - \bar{K})\psi = (K - \bar{K})\phi = 0 \tag{12.4.9}$$

corresponding to the constraint in equation (3.1.46).

Following the same logic as for the open string the above, we find that the the equations of motion for the gauge covariant closed bosonic string are given by

$$(K + \bar{K})\psi^{k_1 k_2}_{k_3 k_4} + d\phi^{k_1 k_2}_{k_3 -1 k_4} + d\bar{\phi}^{k_1 k_2}_{k_3 k_4 -1}(-1)^{k_1+k_3} + D\phi^{k_1 +1 k_2}_{k_3 k_4}$$
$$+ \bar{D}\phi^{k_1 k_2 +1}_{k_3 k_4}(-1)^{k_1+k_3} = 0, \tag{12.4.10}$$

$$d\psi^{k_1 k_2}_{k_3 k_4} + \bar{d}\psi^{k_1 k_2}_{k_3 +1 k_4 -1}(-1)^{k_1+k_3+1} + D\psi^{k_1 +1 k_2}_{k_3 +1 k_4} + \bar{D}\psi^{k_1 k_2 +1}_{k_3 -1 k_4}(-1)^{k_1+k_3+1}$$
$$+ M\phi^{k_1 +1 k_2}_{k_3 k_4} + \bar{M}\phi^{k_1 k_2 +1}_{k_3 +1 k_4 -1} = 0 \tag{12.4.11}$$

and their invariance is given by the transformations

$$\delta\psi^{k_1 k_2}_{k_3 k_4} = D\Lambda^{k_1 +1 k_2}_{k_3 k_4} + \bar{D}\Lambda^{k_1 k_2 +1}_{k_3 k_4}(-1)^{k_1+k_3} + d\Lambda^{k_1 k_2}_{k_3 -1 k_4} + \bar{d}\Lambda^{k_1 k_2}_{k_3 k_4 -1}(-1)^{k_1+k_3},$$
$$\delta\phi^{k_1 k_2}_{k_3 k_4} = -(K + \bar{K})\Lambda^{k_1 k_2}_{k_3 k_4} + d\Lambda^{k_1 k_2}_{k_3 -1 k_4} + D\Lambda^{k_1 +1 k_2}_{k_3 k_4} + \bar{d}\Lambda^{k_1 k_2}_{k_3 k_4 -1}(-1)^{k_1+k_3}$$
$$+ \bar{D}\Lambda^{k_1 k_2 +1}_{k_3 k_4}(-1)^{k_1+k_3}. \tag{12.4.12}$$

Following the same argument as in section 12:2 the reader may verify that these equations do indeed lead to the correct count of physical states, namely those manifest in the closed string light-cone theory.

We now wish to cast the result for the gauge covariant string in terms of the general functional $|\chi\rangle$ of expression (12.4.5). However, we note that even if we impose a number constraint on such a general functional $|\chi\rangle$ we will have twice as many fields as those found just above as a consequence of the four vacua. To write the solution of equations (12.4.10) and (12.4.11) in terms of $|\chi\rangle$ we must impose additional restrictions using the zero modes. We now demonstrate two ways of solving this problem.

Rather than use Q, let us use the object

$$\hat{Q} = Q - \delta_0(K - \bar{K}) + 2\tilde{\delta}_0(M - \bar{M}) = \gamma_0(K + \bar{K}) - 2\tilde{\gamma}_0(M + \bar{M}) + Q^1, \tag{12.4.13}$$

where

$$\gamma_0 = \tfrac{1}{2}(c_0 + \tilde{c}_0), \quad \tilde{\gamma}_0 = \tfrac{1}{2}(b_0 + \tilde{b}_0), \quad \delta_0 = \tfrac{1}{2}(c_0 - \tilde{c}_0), \quad \tilde{\delta}_0 = \tfrac{1}{2}(b_0 - \tilde{b}_0)$$

and

$$Q^1 = d + D + \tilde{d} + \tilde{D}.$$

We note that

$$\hat{Q}^2 = -(M - \bar{M})(K - \bar{K}) \tag{12.4.14}$$

and so vanishes on all the fields of interest to us. The most straightforward procedure is to consider a space in which δ_0 and $\tilde{\delta}_0$ do not occur. The vacuum in this space is annihilated by $c_n, b_n, \tilde{c}_n, \tilde{b}_n$ for $n \geq 1$ and the $|+\rangle$ vacuum satisfies

$$\gamma_0|+\rangle = 0, \tag{12.4.15}$$

while

$$|-\rangle = 2\bar{\gamma}_0|+\rangle. \tag{12.4.16}$$

The most general functional of $x^\mu(\sigma)$ and the ghosts with the above vacua is given by

$$|\chi\rangle = \sum_{n,m,q} c^\dagger_{n_1} \cdots b^\dagger_{m_1} \cdots \tilde{c}^\dagger_{p_1} \cdots \tilde{b}^\dagger_{q_1} \cdots \psi^{m_1 \cdots m_b \, q_1 \cdots q_d}_{n_1 \cdots n_a \, p_1 \cdots p_c} |-\rangle$$

$$+ \sum_{n,m,p,q} c^\dagger_{n_1} \cdots b^\dagger_{m_1} \cdots \tilde{c}^\dagger_{p_1} \cdots \tilde{b}^\dagger_{q_1} \cdots \phi^{m_1 \cdots m_b \, q_1 \cdots q_d}_{n_1 \cdots n_a \, p_1 \cdots p_c} |+\rangle. \tag{12.4.17}$$

The gauge invariant theory should contain only Grassmann even fields of equation (12.4.8) and so we subject $|\chi\rangle$ to the constraint

$$\left[\sum_{n=1}^{\infty} [(c^\dagger_n b^n - b^\dagger_n c_n) + (b_n \Leftrightarrow \tilde{b}_n, c_n \Leftrightarrow \tilde{c}_n)] + \tfrac{1}{2}\gamma_0\bar{\gamma}_0 \right] |\chi\rangle = 0. \tag{12.4.18}$$

We are still enforcing the condition

$$(K - \tilde{K})|\chi\rangle = 0. \tag{12.4.19}$$

The action [12.10, 12.11] which leads to the correct equations of motion of equations (12.4.10) and (12.4.11) is given by

$$\langle \chi | \hat{Q} | \chi \rangle \tag{12.4.20}$$

and it is invariant under

$$\delta|\chi\rangle = \hat{Q}|\Lambda\rangle \tag{12.4.21}$$

provided one uses equation (12.4.14).

We may also formulate the free action in an alternative way by reintroducing δ_0 and $\tilde{\delta}_0$. The most general functional is then of the form

$$|\chi\rangle = \psi|-, +\rangle + \phi|+, +\rangle + \tilde{\psi}|+, -\rangle + \tilde{\phi}|-, -\rangle. \tag{12.4.22}$$

We take a $-$ vacuum to be Grassmann odd and a $+$ vacuum to be Grassmann even. We note that if $|\chi\rangle$ is odd and $\tilde{\phi} = \tilde{\psi} = 0$, and we recover the previous field content. The above vacua are defined by the equations

$$\gamma_0|+, \pm\rangle = 0 = \delta_0|\pm, +\rangle \tag{12.4.23}$$

and

$$|-, +\rangle = 2\gamma_0|+, +\rangle, \quad |+, -\rangle = 2\tilde{\delta}_0|+, +\rangle, \quad |-, -\rangle = 4\bar{\gamma}_0\tilde{\delta}_0|+, +\rangle. \tag{12.4.24}$$

This differs slightly from our previous definition of the vacuum, which used the zero modes c_0, b_0 and $\tilde{c}_0, \tilde{b}_0$, but we will, for economy, use the same symbols:

$$\langle -, -|+, +\rangle = \langle +, +|-, -\rangle^\dagger = 4\langle +, +|\tilde{\delta}_0\bar{\gamma}_0|+, +\rangle = -\langle +, +|-, -\rangle \tag{12.4.25}$$

and as a result we may choose

$$\langle -, -|+, +\rangle = i. \tag{12.4.26}$$

By further manipulations one finds that

$$\langle +, -|-, +\rangle = i = -\langle -, +|+, -\rangle. \tag{12.4.27}$$

We again impose the following algebraic constraints on $|\chi\rangle$:

$$(K - \tilde{K})|\chi\rangle = 0 \tag{12.4.28}$$

and

$$\left[\sum_{n=1}^{\infty} [(c_n^\dagger b_n - b_n^\dagger c_n) + (c_n \Leftrightarrow \tilde{c}_n, b_n \Leftrightarrow \tilde{b}_n)] - 2\bar{\delta}_0 \delta_0 + 2\gamma_0 \bar{\gamma}_0 \right] |\chi\rangle = 0. \tag{12.4.29}$$

This last equation ensures that we have the field content of expressions (12.4.8), but in addition the fields $\tilde{\phi}^n$ and $\tilde{\psi}$, which have the reversed index structure to ϕ and ψ, respectively. As noted above, the gauge covariant free action cannot be $\langle \chi|Q|\chi\rangle$, and the correct action [12.10] is

$$i\langle \chi|\bar{\delta}_0 Q|\chi\rangle. \tag{12.4.30}$$

We note that we can replace Q by $\hat{Q}$ since $\bar{\delta}_0(Q - \hat{Q})$ vanishes on $|\chi\rangle$. This action is Hermitian as a result of the fact that $i\langle \chi|\{\bar{\delta}_0, Q\}|\chi\rangle$ vanishes due to equation (12.4.27). Evaluating this action, we find we can recover an action which leads to the correct field equations of equations (12.4.10) and (12.4.11). The fields $\tilde{\phi}$ and $\tilde{\psi}$ drop out as a consequence of the factor $\bar{\delta}_0$. The action of equation (12.4.30) is invariant under

$$\delta|\chi\rangle = Q|\Lambda\rangle \tag{12.4.31}$$

and this reproduces the transformation laws of equations (12.4.12).

An interesting extension of the above is to release the constraint $(K - \bar{K})|\chi\rangle = 0$ and write an action such that this condition also occurs as one of the equations of motion. To achieve this end we introduce another string field $|\Phi\rangle$ in addition to $|\chi\rangle$ neither of these fields being now subject to equation (12.4.17). Let us then consider the action [12.12]

$$A = i\langle \chi|(\bar{\delta}_0 Q - Q\bar{\delta}_0)|\chi\rangle + 2i(\langle \chi|Q|\Phi\rangle - \langle \Phi|Q|\chi\rangle) \tag{12.4.32}$$

with the gauge transformations

$$\delta|\chi\rangle = Q|\Lambda\rangle, \quad \delta|\Phi\rangle = -(K - \bar{K})|\Lambda\rangle, \tag{12.4.33}$$

where $|\Lambda\rangle$ is also not subject to $(K - \bar{K})|\Lambda\rangle = 0$. Varying the above action we find that

$$\delta A = 2i\langle \Lambda|(Q\bar{\delta}_0 Q|\Lambda\rangle + (K - \bar{K})Q)|\chi\rangle + \text{h.c.} = 0 \tag{12.4.34}$$

since

$$\{\bar{\delta}_0, Q\} = -(K - \bar{K}). \tag{12.4.35}$$

12.5   The gauge covariant superstring

In this section we will construct the gauge covariant action for the open superstring. We begin by assembling the necessary machinery assuming some knowledge of chapters 6 and 7. The generators of the superconformal algebra are denoted L_A. In the R sector $L_A = \{L_n, F_m; n, m = 0, \pm 1, \ldots\}$ and in the NS sector they are $L_A = \{L_n, G_r, n = 0, \pm 1, \ldots; r = \pm\frac{1}{2}, \pm\frac{3}{2}, \ldots\}$. Their algebra is of the form

$$[L_A, L_B\} = f^C_{AB} L_C, \tag{12.5.1}$$

where $[\,,\}$ denotes a commutator or anti-commutator as appropriate.

In detail in the R sector these relations are given by

$$\{F_m, F_n\} = 2L_{n+m} + \frac{D}{2}m^2 \delta_{m+n,0}, \quad [L_m, F_n] = \left(\frac{m}{2} - n\right)F_{n+m},$$

$$[L_m, L_n] = (m - n)L_{n+m} + \frac{D}{8}m^3 \delta_{m+n,0} \tag{12.5.2}$$

and in the Neveu–Schwarz sector by

$$[G_r, G_s] = 2L_{r+s} + \frac{D}{2}\left(r^2 - \frac{1}{4}\right)\delta_{r+s,0}, \quad [L_m, G_r] = \left(\frac{m}{2} - r\right)G_{r+m},$$

$$[L_m, L_n] = (m - n)L_{m+n} + \frac{D}{8}m(m^2 - 1)\delta_{m+n,0}. \tag{12.5.3}$$

We now find the analogues of K, M, d and D of the bosonic model. To do this we examine the BRST charge. In the superstring we have ghost and anti-ghost fields corresponding to the world-sheet supersymmetry of the original action and so we find, in addition to the usual reparameterisation ghosts $b(\sigma)$ and $c(\sigma)$, the Grassmann even ghosts $g(\sigma)$ and $\bar{g}(\sigma)$. The reader may wish to consult appendix A for the procedure. The corresponding ghost modes are denoted by g_r and $\bar{g}_r$. They correspond to the classical constraints $G_r = 0$ (see chapter 3). They obey the relations

$$[g_r, \bar{g}_s] = \delta_{r+s,0}, \quad [g_r, g_s] = 0 = [\bar{g}_r, \bar{g}_s] \tag{12.5.4}$$

and have the Hermiticity properties

$$g_r^\dagger = g_{-r}, \quad \bar{g}_r^\dagger = \bar{g}_{-r}. \tag{12.5.5}$$

The most general functional of the $x^\mu(\sigma)$ and all the ghost $b(\sigma)$, $c(\sigma)$, $g(\sigma)$ $\bar{g}(\sigma)$ is of the form

$$|\chi_{NS}\rangle = \sum_{n,m,r,s} c^{n_1\dagger} \cdots b^\dagger_{m_1} \cdots g^{r_1\dagger} \cdots \bar{g}^\dagger_{s_1} \cdots \psi^{m_1 \cdots \, s_1 \cdots}_{n_1 \cdots \, r_1 \cdots} |-\rangle$$

$$+ \sum_{n,m,r,s} c^{n_1\dagger} \cdots b^\dagger_{m_1} \cdots g^{r_1\dagger} \cdots \bar{g}^\dagger_{s_1} \cdots \phi^{m_1 \cdots s_1 \cdots}_{n_1 \cdots r_1 \cdots} |+\rangle, \tag{12.5.6}$$

where the vacua obey the equations

$$g^r|\pm\rangle = c_n|\pm\rangle = b_m|\pm\rangle = \bar{g}_s|\pm\rangle = 0, \quad n, m, r, s > 0 \tag{12.5.7}$$

and

$$c_0|+\rangle = 0; \quad |-\rangle = b_0|+\rangle. \tag{12.5.8}$$

Following similar arguments to those for the open string in section 12.3, we find that the gauge covariant open string in the NS sector has a field content of the form

$$\psi_K^K, \; \phi_K^{K+1},$$

(12.5.9)

where K and $K+1$ stand this number of upper and lower indices and can be either n or r. In order to restrict the most general functional of equation (12.5.6) $|\chi\rangle_{NS}$ to have this field content we impose the constraint

$$\left(N - \sum_{r=\frac{1}{2}}^{\infty} (g^{r\dagger}\bar{g}^r - \bar{g}_r^\dagger g^r)\right)|\chi_{NS}\rangle = 0,$$

(12.5.10)

where N is the b, c ghost number operator of equation (3.2.97).

The BRST charge corresponding to the superconformal group is

$$Q^{N-S} =: \sum_A \beta_A L_{-A} - \frac{1}{2} \sum_{A,B,C} \bar{\beta}_c f_{AB}{}^C \beta_{-A}\beta_{-B} - \frac{1}{2}c_0 :,$$

(12.5.11)

where

$$\beta_A = (c_n, g_r), \quad \bar{\beta}_A = (b_n, \bar{g}_r).$$

(12.5.12)

One could find this by computing the Noether charge for the BRST symmetry or trust that the prescription given in equation (3.2.89) is correct. One can show that that $(Q^{NS})^2 = 0$ in ten dimensions [12.13,12.14]. We may rewrite this BRST charge so as to display its dependence on the ghost zero modes:

$$Q^{NS} = c_0 K - 2b_0 M + D + d,$$

(12.5.13)

where K, M, D and d do not involve c_0 and b_0, satisfy $D = d^\dagger$, and are defined by the same procedure as in the bosonic case. They obey the relations

$$d^2 = D^2 = 0, [d, K] = [D, K] = [M, d] = [M, D] = 0,$$

(12.5.14)

$$\{d, D\} = 2MK.$$

Following the same steps as we did for the open string, we find that gauge covariant open superstring in the NS sector has the equations of motion

$$K\psi + (D+d)\phi = 0,$$

(12.5.15)

$$d\psi + D\psi + 2M\phi = 0.$$

The field involved is just the functional of $x^\mu(\sigma)$ of equation (12.5.10). Equation (12.5.6) is invariant under the transformations

$$\delta\psi = d\Lambda + D\Lambda - 2M\Omega,$$

(12.5.16)

$$\delta\phi = -K\Lambda + d\Omega + D\Omega.$$

The NS sector is almost identical to the open bosonic string. The above equations of motion [12.13] follow from the action [12.11, 12.13]

$$\langle\chi_{NS}|Q^{NS}|\chi_{NS}\rangle$$

(12.5.17)

and this action is obviously invariant under

$$\delta|\chi_{NS}\rangle = Q^{NS}|\Lambda_{NS}\rangle. \tag{12.5.18}$$

The reader may verify, using arguments similar to those in section 12.2, that the count of physical states is that for the open superstring in the NS sector.

We now repeat the procedure in the R sector where life is a little more complicated. In this sector we denote the ghost and anti-ghost corresponding to the world-sheet supersymmetry $f(\sigma)$ and $\bar{f}(\sigma)$. They correspond to the classical constraints $F_n = 0$. As such, we introduce the oscillators f_n and $\bar{f}_n$ which obey the relations

$$[f_n, \bar{f}_m] = \delta_{n+m,0}, \quad [f_n, f_m] = 0 = [\bar{f}_n, \bar{f}_m]. \tag{12.5.19}$$

They obey the Hermiticity properties $f_n^\dagger = f_n$, $\bar{f}_n^\dagger = -\bar{f}_{-n}$.

The super-Virasoro charge is given by

$$Q^R =: \sum_A \beta_A L_{-A} - \frac{1}{2} \sum_{A,B,C} \bar{\beta}_C f_{AB}{}^C \beta_{-A} \beta_{-B}, \tag{12.5.20}$$

where $\beta_A = \{c_n, f_n\}$ and $\bar{\beta}_A = \{b_n, \bar{f}_n\}$. In ten dimensions one finds that $(Q^R)^2 = 0$ [12.13, 12.14]. We may rewrite Q^R so as to display its dependence on the ghost zero modes;

$$Q^R =: f_0 F + c_0 K - 2b_0 M + D + d + \bar{f}_0 J - b_0 f_0^2. \tag{12.5.21}$$

We note that F, D, d, M, K and J obey the relations

$$\{d, D\} = 2KM + FJ, \quad F^2 = K, \quad J = [M, F], \quad d^2 = D^2 = [M, K] = 0 \tag{12.5.22}$$

and all other commutators, and anti-commutators vanish.

Following the by now well trod path, we conclude that the gauge covariant string requires the following functionals of $x^\mu(\sigma)$:

$$\psi_K^K, \phi_K^{K+1}, \tag{12.5.23}$$

where K denotes the number of indices of either type and, following the arguments of section 12.3, the gauge covariant equations of motion in the R sector are given by [12.13]

$$F\psi + (D+d)\phi = 0, \quad -(MF + FM)\phi + (D+d)\psi = 0. \tag{12.5.24}$$

They are invariant under the gauge transformations

$$\delta\psi = (D+d)\Lambda - (MF + FM)\Omega, \quad \delta\phi = F\Lambda + (D+d)\Omega. \tag{12.5.25}$$

We now wish to cast the above results in a more compact form in terms of the general functional of the coordinates and ghosts and the BRST charge Q^R. However, we observe that f_0 is a Grassmann even and so no power of it vanishes. Clearly, the set of fields of expressions (12.5.23) does not include the infinite set of fields that would be contained in an arbitrary expansion in f_0. We therefore consider an expansion of the form [12.13–12.15]

$$|\chi_R\rangle = (\psi + f_0\phi)|-\rangle + F\phi|+\rangle, \tag{12.5.26}$$

where

$$\psi = c^{n_1\dagger} \cdots b_{m_1}^\dagger \cdots f^{p_1\dagger} \cdots \bar{f}_{q_1}^\dagger \cdots \psi_{n_1\cdots n_b p_1\cdots p_c}^{m_1\cdots m_a q_1\cdots q_d} \tag{12.5.27}$$

and similarly for ϕ.

The vacua obey the conditions

$$0 = c_n|\pm\rangle = b_n|\pm\rangle = f_n|\pm\rangle = \bar{f}_n|\pm\rangle, n \geq 1 \tag{12.5.28}$$

and

$$\bar{f}_0|\pm\rangle = 0, \; c_0|+\rangle = 0, \; b_0|+\rangle = |-\rangle. \tag{12.5.29}$$

In order that $|\chi_R\rangle$ only contain the fields of expressions (12.5.23) we must also impose the constraint

$$\left(N - \sum_{n=1}^{\infty}(f^{n\dagger}\bar{f}_n - \bar{f}_n^\dagger f^n) - f_0\bar{f}_0\right)|\chi_R\rangle = 0, \tag{12.5.30}$$

where N is the b, c ghost number operator of equation (3.2.97).

We can now examine the action of Q^R on $|\chi_R\rangle$. One finds

$$Q_R|\chi_R\rangle = (\tilde{\psi} + f_0\tilde{\phi})|-\rangle + F\tilde{\phi}|+\rangle, \tag{12.5.31}$$

where

$$\tilde{\psi} = (D+d)\psi - (MF + FM)\phi \tag{12.5.32}$$

and

$$\tilde{\phi} = F\psi + (D+d)\phi. \tag{12.5.33}$$

We observe that the action of Q maintains the form of the expansion of equation (12.5.26). The field equations (12.5.24) are given by

$$c_0 Q_R|\chi_R\rangle = 0. \tag{12.5.34}$$

We now wish to find an action from which this follows. There are a number of ways of doing this; however, one method, which is also useful in establishing the space-time supersymmetry of the open superstrings, is as follows. Let us introduce

$$\chi_R = \psi + f_0\phi + F\phi c_0. \tag{12.5.35}$$

We note, using equation (12.5.26), that

$$|\chi_R\rangle = \chi_R|-\rangle. \tag{12.5.36}$$

We now introduce a new type of vacuum $|\tilde{\pm}\rangle$ which satisfies all the conditions of equations (12.5.28) and (12.5.29) that $|\pm\rangle$ does with the exception of the condition $\bar{f}_0|\pm\rangle = 0$, which we now replace by

$$f_0|\tilde{\pm}\rangle = 0. \tag{12.5.37}$$

We take Hermitian conjugation to be such that

$$(|\pm\rangle)^\dagger = \langle\tilde{\pm}|, \;\; (|\tilde{\pm}\rangle)^\dagger = \langle\pm| \tag{12.5.38}$$

and consequently we have the equations

$$\langle\tilde{\pm}|f_0 = 0 \;\; \text{and} \;\; \langle\pm|\bar{f}_0 = 0. \tag{12.5.39}$$

The action that leads to equation (12.5.24) is given by [12.13, 12.15]

$$-\tfrac{1}{2}\langle-|\bar{f}_0\chi_R^\dagger c_0 Q_R\chi_R|-\rangle.\qquad(12.5.40)$$

The gauge covariant superstring action was also discussed in [12.16].

Adding the NS action of equation (12.5.1) to the R action of equations (12.5.24) and imposing the GSO projection discussed in chapter 7 we have the gauge covariant action for the open superstring. This combined action was shown to possess space-time supersymmetry in [12.14].

13 Supergravity theories in four, ten and eleven dimensions

> I was at first almost frightened when I saw such mathematical force made to bear upon the subject, and then wondered to see the subject stood it so well.
>
> Faraday

We take a supergravity theory to be one that has some supersymmetry and contains gravity, but no higher spin fields. In ten and eleven dimensions, there exist only four supergravity theories which are distinguished by their different underlying supersymmetry algebras. The types of spinor one can have in different dimensions was studied in chapter 5 and it is key to understanding the different possible supersymmetry algebras. In turn, the number of components of the supercharge determines the number of degrees of freedom and the spins of the particles in the supermultiplet. By analysing the irreducible representations of supersymmetry algebras in general dimensions it was shown [10.10] than there is only one supersymmetry multiplet in eleven dimensions containing no spin higher than gravity, no such multiplets above eleven dimensions and only four such supermultiplets in ten dimensions. A Majorana spinor in ten and eleven dimensions has $32 = 2^{10/2}$ components. In eleven dimensions there is only one supergravity theory [13.1] whose supersymmetry algebra contains a supercharge which is a Majorana spinor. In ten dimensions we can have Majorana–Weyl spinors, which have 16 components. In this dimension there are two maximal supergravity theories: the IIA supergravity theory [10.11–10.13] which has a supersymmetry algebra with one Majorana spinor and so two Majorana–Weyl spinors of opposite chirality, and the IIB supergravity theory [10.14–10.16], which contains a super-symmetry algebra with two Majorana–Weyl spinors of the same chirality. We can truncate these theories to find the type I supergravity theory [10.17–10.18] whose supersymmetry algebra contains a single Majorana–Weyl spinor, which has 16 components. This last theory couples [10.20] to the ten-dimensional super Yang–Mills theory [10.19, 5.1], which is based on the same supersymmetry algebra.

Unlike the other theories, the IIA supergravity theory possesses an extension to include a cosmological term. However, up to this ambiguity the supergravity theories in ten and eleven dimensions are uniquely determined by the supersymmetry they possess and the requirement that they possess terms with no more than two space-time derivatives. Their uniqueness has an extremely important consequence. The IIA and IIB string theories possess the same space-time supersymmetry algebra as the IIA and IIB supergravity theories and so their quantum corrections, whether perturbative, or non-perturbative, will also possess the corresponding space-time supersymmetry. As such, any perturbative, or non-perturbative, correction to IIA and IIB superstring theories which has terms with no more than two space-time derivatives in its bosonic terms must occur in the corresponding IIA and IIB supergravity theories. Consequently, these supergravity theories are the *complete* low energy effective actions for the IIA and IIB superstring theories. Given our incomplete understanding of string theory and, in particular, our lack of a systematic method of computing non-perturbative effects, the supergravity theories have played a crucial role in understanding the quantum effects of the superstrings, their symmetries and even what theory may eventually replace superstrings. In particular, they have led to the realisation that we must also include branes. Indeed, the IIA and IIB supergravity theories and the eleven-dimensional supergravity theory contain some of the most studied equations in string theory.

We begin in section 13.1 by explaining how to construct supergravity theories. In section 13.2 we explain what is a non-linear realisation as these play an important role in many supergravity theories. However, the reader who is keen to find what the ten- and eleven-dimensional supergravity theories are may want to jump sections 13.3, 13.4, 13.5 and 13.7.

In dimensions nine and less there exists a unique maximal supergravity theory that can be found by dimensionally reducing either of the maximal ten-dimensional supergravity theories on tori. These theories, as well as the IIB supergravity theory in ten dimensions, possess symmetries which have important consequences in string theory. In section 13.6 we discuss how these symmetries arise when supergravity theories are dimensionally reduced.

Finally, in section 13.8, we discuss solutions of generic supergravity theories and, in particular, the maximally supergravity theories in ten and eleven dimensions. Some of these solutions correspond to branes and we will find the commonly used brane solutions in ten and eleven dimensions.

13.1   Four ways to construct supergravity theories

In this section we explain how to construct a supergravity theory. There are four main ways. The method, which has been used most often is the Noether method, which requires only a knowledge of the on-shell states of the theory. It was used to construct the four-dimensional $N = 1$ supergravity in its on-shell [13.2, 13.3] and off-shell [13.4, 13.5] formulations. This method was also used in the construction of the supergravity theory in eleven dimensions [13.1] and a variant of it [13.6] was used to find many of the properties of the IIB supergravity theory. Although this method does not make use of any sophisticated mathematics and can be rather lengthy, it is very powerful. Starting from only the linearised theory for the relevant supermultiplet, it gives a systematic way of finding the final non-linear theory. To illustrate the method clearly and without undue technical complexity we explain, in section 13.1.1, how to construct the Yang–Mills theory from the linearised theory and then, after discussing the linearised supergravity theory in four dimensions, apply the Noether procedure to find the full non-linear theory. In fact, we will only carry out the first steps in this Noether procedure, but these clearly illustrate how to find the final result, most of whose features are already apparent at an early stage of the process. We then give the $N = 1$ supergravity theory in four dimensions.

The second method uses the superspace [1.10] description of supergravity theories. Supergravity theories in superspace share a number of similarities with the usual theory of general relativity. They are built from a supervielbein and a spin connection and are invariant under superdiffeomorphisms. Using the supervielbein and spin connection we can construct covariant derivatives and then define torsions and curvatures in the usual way. The superspace formulation differs from the usual formulation of general relativity in the nature of the tangent space group. The tangent space of the superspace formulation of supergravity contains Grassmann odd and Grassmann even sub-spaces; however, the tangent space group is only the Lorentz group which rotates the Grassmann odd sub-space into itself and the Grassmann even sub-space into itself. In general relativity the tangent space group is also the Lorentz group, but in this case it acts on all components of the tangent space in a different way, that is, the tangent space belongs to an irreducible representation of the Lorentz group.

The use of a restricted tangent space group in the superspace formulation allows us to take some of the torsions and curvatures to be zero since these constraints are invariant under the Lorentz tangent space group. In fact, the torsions and curvatures form a highly reducible representation of the Lorentz group and, as we will see, the imposition of such constraints is precisely what is required to find the correct theory of supergravity. Hence the problem of finding the superspace formulation of supergravity is to find which of the torsions and curvatures should be set to zero. This method was first successfully carried out for the $N = 1$ supergravity theory in four dimensions [13.7]. We require different sets of constraints for the on-shell and off-shell supergravity theory. Clearly one gets from the latter to the former by imposing more constraints. To find the constraints for the off-shell theory, when this exists, can be rather difficult; however it turns out that to find the constraints for the on-shell theory is very straightforward and remarkably requires only dimensional analysis and a knowledge of the on-shell states in x-space. The latter are determined in a straightforward way from the representations of the supersymmetry algebra. Introducing a linearised action that leads to the correct on-shell constraints one finds that it is invariant under certain Abelian gauge transformations. From these one can deduce the dimensions of all the gauge-invariant quantities. By introducing a notion of geometric dimension, which in effect absorbs all factors of Newton's constant κ into the fields, one finds that for sufficiently low dimensions and certain Lorentz character there are no gauge-invariant tensors in x-space. The superspace torsions and curvatures are gauge invariant and so the superspace tensors with these dimensions and Lorentz character must then vanish as there is no available x-space tensor that their lowest component could equal. In this, apparently rather trivial way one arrives at a set of superspace constraints.

We can substitute these superspace constraints into the Bianchi identities satisfied by the torsions and curvatures to find constraints on the torsions and curvatures of higher dimension. From this set of constraints one can find an x-space theory and it turns out that in all known cases this theory is none other than the full corresponding on-shell supergravity theory. In other words, the constraints deduced from dimensional analysis and the use of the Bianchi identities are sufficient to find the on-shell supergravity in superspace and hence also in x-space. This method does not apply to the theory of gravity alone as the torsions and curvatures found there have only vector indices and do not achieve the low dimensions that occur in superspace theories where spinor indices occur. In section 13.1.2 we explicitly carry out this programme for the four-dimensional $N = 1$ supergravity theory and recover the theory we found by the Noether method. This procedure was used to find the full IIB supergravity, discussed in section 13.5, in superspace and in x-space [111].

The third method of finding supergravity theories is by gauging certain space-time groups. We will see that by gauging the super Poincaré group one can find the $N = 1$ four-dimensional supergravity theory [13.8, 13.9]. Theories which contain gravity are not in general gauge theories and so one must implement certain constraints. It was this method which provided the first algebraic proof of the invariance of supergravity theories [13.8] and it was used to construct the theories of conformal supergravity [13.10].

The final method is dimensional reduction. Given a supergravity theory we may take some dimensions to be circles and suppress the dependence of the fields on the coordinates of these circles to find a supergravity theory of lower dimension. This technique is useful because supergravity theories in higher dimensions are generally much simpler than those in lower dimensions and so are much easier to construct using one of the above methods. The technique of dimensional reduction will be explained in section 13.1.4; it was used to construct the IIA supergravity theory in ten dimensions, as we explain in section 13.4.

13.1.1 The Noether method

Any theory whose non-linear form is determined by a gauge principle can be constructed by a Noether procedure which was first understood in the context of gravity [13.11]. We begin by illustrating the Noether technique in the simplest possible context, namely, the construction of the Yang–Mills theory from the linearised (free) theory. This follows rather closely the treatment given in chapter 7 of [1.11].

We begin with a free level, or linearised, theory of spin-1 fields that is invariant under two distinct transformations: rigid transformations and local Abelian transformations. We recall that a rigid symmetry is one whose parameters are space-time independent while a local symmetry has space-time dependent parameters. The rigid transformations belong to a group S with generators R_i which satisfy

$$[R_i, R_j] = s_{ij}{}^k R_k. \tag{13.1.1}$$

The $s_{ij}{}^k$ are the structure constants of the group S. Under these rigid transformations the vector fields A_a^i are taken to transform as

$$\delta A_a^i = s_{jk}{}^i T^j A_a^k, \tag{13.1.2}$$

where T_j are the constant infinitesimal group parameters. The other type of transformations are the local Abelian transformations which are given by

$$\delta A_a^i = \partial_a \Lambda^i. \tag{13.1.3}$$

Clearly both these transformations form a closed algebra and, in particular, one finds that $[\delta_\Lambda, \delta_T] A_a^i = \partial_a (s_{jk}{}^i T^j \Lambda^k)$. The linearised theory which is invariant under the transformations of equations (13.1.2) and (13.1.3) is given by

$$A^{(0)} = \int d^4x \{ -\tfrac{1}{4} f_{ab}^i f^{abj} \} s_{ij}, \tag{13.1.4}$$

where $f_{ab}^i = \partial_a A_b^i - \partial_b A_a^i$ and s_{ij} is the Cartan–Killing metric. The latter is given in terms of the structure constants by $s_{ij} = s_{ik}{}^l s_{jl}{}^k$. The reader may verify that it is invariant under adjoint transformations, that is, verify the relation $s_{ij}{}^k s_{kl} = s_{lj}{}^k s_{ik}$. The proof can be found around equation (16.1.6).

The non-linear theory is found in a series of steps, the first of which is to make the rigid transformations local, that is, $T^j = T^j(x)$. Now, $A^{(0)}$ is no longer invariant under

$$\delta A_a^i = s_{jk}{}^i T^j(x) A_a^k, \tag{13.1.5}$$

but, since it was invariant when T^j was a constant, its variation may be written in the form

$$\delta A^{(0)} = \int d^4x \{ (\partial_a T^i(x)) j^{aj} \} s_{ij}, \tag{13.1.6}$$

where

$$j^{ak} = -s_{ij}{}^k A_b^i f^{abj}. \tag{13.1.7}$$

In fact, j^{ak} is the Noether current corresponding to the symmetry of equation (13.1.2) of the linearised theory. The variation of any action subject to the equations of motion always vanishes as it contains terms which are the equation of motion times the variation of the field. Clearly, this holds for any variation and so when T^j is a function of x^μ. As such, the right-hand side of equation (13.1.6) vanishes for any $T^j(x^\mu)$ and so $\partial_a j^{ak} = 0$ when the equations of motion apply, that is, we have a conserved current. This is why it is called the Noether method.

Now consider the action A_1

$$A_1 = A^{(0)} - \tfrac{1}{2} g \int d^4x (A_a^i j^{aj}) s_{ij}, \tag{13.1.8}$$

where g is the gauge coupling constant; it is invariant *to order* g^0 *provided* we combine the local transformation $T^i(x)$ with the local transformation $\Lambda^i(x)$ with the identification $\Lambda^i(x) = (1/g)T^i(x)$. That is, the initially separate local and rigid transformations of the linearised theory become knitted together into a single local transformation given by

$$\delta A_a^i = \frac{1}{g} \partial_a T^i(x) + s_{jk}{}^i T^j(x) A_a^k(x). \tag{13.1.9}$$

The first term in the transformation of δA_a^i yields, in the variation of the last term in A_1, a term which cancels the unwanted variation of $A^{(0)}$.

We now continue with this process of amending the Lagrangian and transformations order by order in g until we obtain an invariant Lagrangian. The variation of A_1 under the second term of the transformation of equation (13.1.9) is of order g and is given by

$$\delta A_1 = \int d^4x \{ -g(A_a^i A_b^j s_{ij}{}^k)(A_b^l \partial_a T^m s_{lm}{}^n) \} s_{kn}. \tag{13.1.10}$$

An action invariant to order g is

$$A_2 = A_1 + \int d^4x \frac{g^2}{4} (A_a^i A_b^j s_{ij}{}^k)(A^{bl} A^{am} s_{lm}{}^n) s_{kn} = \frac{1}{4} \int d^4x F_{ab}^k F^{abn} s_{kn}, \tag{13.1.11}$$

where

$$F_{ab}^i = \partial_a A_b^i - \partial_b A_a^i - g s_{jk}{}^i A_a^j A_b^k. \tag{13.1.12}$$

In fact, the action A_2 is invariant under the transformations of equation (13.1.9) to all orders in g and so represents the final answer, and is, of course, the well-known action of Yang–Mills theory. The commutator of two transformations on A_a^i is

$$[\delta_{T_1}, \delta_{T_2}]A_a^i = s_{jk}^i T_2^j \left(\frac{1}{g}\partial_a T_1^k + s_{lm}^k T_1^l A_a^m\right) - (1 \leftrightarrow 2)$$

$$= \frac{1}{g}\partial_a T_{12}^i + s_{ij}^k T_{12}^j A_a^k, \qquad (13.1.13)$$

where

$$T_{12}^i = s_{jk}^i T_2^j T_1^k$$

and so the transformations form a closed algebra. In the last step we used the Jacobi identity satisfied by the structure constants.

For supergravity and other local theories the procedure is similar, although somewhat more complicated. The essential steps are first to identify the rigid and local transformations of the free theory and then find invariant Lagrangians order by order in the appropriate coupling constant. This is achieved in general not only by adding terms to the action, but also by adding terms to the transformation laws of the field. If the latter process occurs, one must also check the closure order by order in the gauge coupling constant.

Although one can use a Noether procedure which relies on the existence of an action, one can also use a Noether method which uses the transformation laws alone [13.6]. This works, in the Yang–Mills case, as follows: upon making the rigid transformation local as in equation (13.1.5) one finds that the algebra no longer closes, that is,

$$[\delta_\Lambda, \delta_T]A_a^i = s_{ij}^i(\partial_a T^j \Lambda^k) = \partial_a(s_{ij}^i(T^j \Lambda^k)) - s_{jk}^i(\partial_a T^j)\Lambda^k. \qquad (13.1.14)$$

The first term is an Abelian transformation of equation (13.1.3) but the second term is not a transformation allowed so far. The closure is achieved by identifying the two transformations by taking $\Lambda^i(x) = (1/g)T^i(x)$. The transformation of A_a^i is then given by equation (13.1.9) and carrying out the closure we find it does close to all orders of g^0 and indeed to all orders in g and the process stops here. In general, however, one must now close the algebra order by order in the coupling constant by adding terms to the transformation laws and so to the closure relations for the algebra. Having the full transformations, it is then easy to find the full equations of motion, or the action when that exists. The latter is particularly simple for supersymmetric theories as the on-shell algebra implies the equations of motion.

We will now construct $N = 1$ $D = 4$ supergravity using the Noether method. We will start with the on-shell states and construct the linearised theory, first without and then with auxiliary fields. The $N = 1$ irreducible representations of supersymmetry which include a spin-2 particle, that is, the graviton, contain either a spin-$\frac{3}{2}$ or a spin-$\frac{5}{2}$ fermion. We will choose the spin-$\frac{3}{2}$ particle, the theory with the spin-$\frac{5}{2}$ particle has not been constructed and may not have causal propagation. The linearised theory possesses rigid supersymmetry and local Abelian gauge invariances. The latter invariances are required, in order that the fields describe the correct massless on-shell states. In what follows we will use the conventions of appendix A of reference [1.11], for example, $\gamma_5 = -i\gamma^0\gamma^1\gamma^2\gamma^3$.

These on-shell states of spin-2 and spin-$\frac{3}{2}$ particles are represented by a symmetric second rank tensor field, $h_{\mu\nu}$, subject to $\partial^2 h_{\mu\nu} = 0 = \partial^\mu h_{\mu\nu}$ and $\eta^{\mu\nu} h_{\mu\nu} = 0$, and a Majorana vector spinor, $\psi_{\mu\alpha}$, subject to $\partial\!\!\!/\psi_\mu = 0 = \gamma^\mu \psi_\mu = \partial^\mu \psi_\mu$. As discussed in chapter 12 the unique free field equations that lead to these on-shell states are constructed from the fields $h_{\mu\nu}$ and ψ_μ, which are not traceless and γ-traceless. They are given by

$$E_{\mu\nu} = 0, \quad R^\mu = 0, \tag{13.1.15}$$

where $E_{\mu\nu} = R^L_{\ \mu\nu} - \frac{1}{2}\eta_{\mu\nu}R^L$, $R^L_{\ \mu\nu}{}^{ab}$ is the linearised Riemann tensor given by

$$R^L_{\mu\nu ab} = -\partial_a\partial_\mu h_{b\nu} + \partial_b\partial_\mu h_{a\nu} + \partial_a\partial_\nu h_{b\mu} - \partial_b\partial_\nu h_{a\mu},$$

$$R^{Lb}_\nu = R^L_{\ \mu\nu}{}^{ab}\delta^\mu_a, R^L = R^L_{\ \mu}{}^a\delta^\mu_a$$

and

$$R^\mu = \varepsilon^{\mu\nu\rho\kappa} i\gamma_5\gamma_\nu\partial_\rho\psi_\kappa. \tag{13.1.16}$$

They are invariant under the Abelian local transformations

$$\delta h_{\mu\nu} = \partial_\mu\xi_\nu(x) + \partial_\nu\xi_\mu(x), \quad \partial\psi_{\mu\alpha} = \partial_\mu\eta_\alpha(x). \tag{13.1.17}$$

These transformations are crucial in order to obtain the required on-shell conditions given above.

We must now search for the supersymmetry transformations. For the free theory, the supersymmetry transformations are linear transformations, that is, they are first order in the fields $h_{\mu\nu}$ and $\psi_{\mu\alpha}$. Using dimensional analysis and Lorentz symmetry the most general linearised supersymmetry transformations are of the form

$$\delta h_{\mu\nu} = \frac{1}{2}(\bar\varepsilon\gamma_\mu\psi_\nu + \bar\varepsilon\gamma_\nu\psi_\mu) + \delta_1\eta_{\mu\nu}\bar\varepsilon\gamma^\kappa\psi_\kappa,$$

$$\delta\psi_\mu = +\delta_2\sigma^{ab}\partial_a h_{b\mu}\varepsilon + \delta_3\partial_\nu h^\nu_{\ \mu}\varepsilon. \tag{13.1.18}$$

δ_1, δ_2 and δ_3 are constants which will be determined by the demanding that these supersymmetry transformations and the local transformations of equation (13.1.17) should form a closing algebra when the field equations of equation (13.1.15) hold.

Evaluating the commutator of a Rarita–Schwinger local transformation with parameter $\eta_\alpha(x)$ of equation (13.1.17) and a supersymmetry transformation with parameter ε_α of equation (13.1.18) on $h_{\mu\nu}$, we find that

$$[\delta_\eta, \delta_\varepsilon]h_{\mu\nu} = \frac{1}{2}(\bar\varepsilon\gamma_\mu\partial_\nu\eta + \bar\varepsilon\gamma_\nu\partial_\mu\eta) + \delta_1\eta_{\mu\nu}\bar\varepsilon\partial\!\!\!/\eta. \tag{13.1.19}$$

This is a local transformation of equation (13.1.17) with parameter $\frac{1}{2}\bar\varepsilon\gamma_\mu\eta$ on $h_{\mu\nu}$ provided $\delta_1 = 0$. Similarly, calculating the commutator of a local transformation of $h_{\mu\nu}$ and a supersymmetry transformation on $h_{\mu\nu}$ automatically yields the correct result, that is, zero. However, evaluating the commutator of a supersymmetry transformation and a local transformation with parameter ξ_μ on $\psi_{\mu\alpha}$ yields

$$[\delta_{\xi_\mu}, \delta_\varepsilon]\psi_\mu = +\delta_2\sigma^{ab}\partial_a(\partial_\mu\xi_b)\varepsilon + \delta_3\partial_\nu\partial^\nu\xi_\mu\varepsilon + \delta_3\partial_\nu\partial_\mu\xi^\mu\varepsilon, \tag{13.1.20}$$

which is a local transformation on ψ_μ, provided $\delta_3 = 0$. Hence, we take $\delta_1 = \delta_3 = 0$.

We must test the commutator of two supersymmetries. On $h_{\mu\nu}$ we find the commutator of two supersymmetries to give

$$[\delta_{\varepsilon_1}, \delta_{\varepsilon_2}]h_{\mu\nu} = +\tfrac{1}{2}\{\bar{\varepsilon}_2\gamma_\mu\delta_2\sigma^{ab}\partial_a h_{b\nu}\varepsilon_2 + (\mu \leftrightarrow \nu)\} - (1 \leftrightarrow 2)$$

$$= \delta_2\{\bar{\varepsilon}_2\gamma^b\varepsilon_1\partial_\mu h_{b\nu} - \bar{\varepsilon}_2\gamma^a\varepsilon_1\partial_a h_{\mu\nu} - (\mu \leftrightarrow \nu)\}. \qquad (13.1.21)$$

This is a local transformation on $h_{\mu\nu}$ with parameter $\delta_2\bar{\varepsilon}_2\gamma^b\varepsilon_1 h_{b\nu}$ as well as a space-time translation. The magnitude of this translation coincides with that dictated by the supersymmetry algebra we adopt, namely $\{Q_\alpha, Q_\beta\} = 2(\gamma^a C)_{\alpha\beta}P_a$, provided $\delta_2 = -1$, which is the value we now adopt.

It is rather more complicated to verify that the transformations of equation (13.1.18), with the above values of the parameters, together with the Abelian gauge transformations of equation (13.1.17) do indeed form a closed transformation on the gravitino. This is carried out in detail in chapter 9 of [1.11] along the lines discussed here. The reader can verify that the transformations of equation (13.1.18) with the values of $\delta_1 = \delta_3 = 0$, $\delta_2 = -1$ do indeed leave the linearised equations of motion of $h_{\mu\nu}$ and $\psi_{\mu\alpha}$ of equation (13.1.15) invariant. Thus we have found a representation of supersymmetry carried by the fields $h_{\mu\nu}$ and $\psi_{\mu\alpha}$ subject to their local Abelian transformations when their field equations hold.

The action from which the field equations of equation (13.1.16) follow is given by

$$A = \int d^4x\{-\tfrac{1}{2}h^{\mu\nu}E_{\mu\nu} - \tfrac{1}{2}\bar{\psi}_\mu R^\mu\}. \qquad (13.1.22)$$

It is invariant under the transformations of equation (13.1.18) provided we adopt the values for the parameters δ_1, δ_2 and δ_3 found above. This invariance holds without use of the field equations. This situation is typical of supersymmetric field theories without the auxiliary fields; we have a set of transformations which close only with the use of the equations of motion, but they leave an action invariant without using the the equations of motion. Of course, we can never use the equations of motion when proving the invariance of an action!

We now wish to find a linearised formulation which is built from fields which carry a representation of supersymmetry without imposing any restrictions (that is, equations of motion), namely, we find the auxiliary fields. As a guide to their number we can apply the Fermi–Bose counting rule. This states that there are equal number of fermions and bosons on a representation of supersymmetry for which P_a is a one-to-one operator (for a proof see, for example, [1.11]). Since the theory contains local transformations, the rule only applies to the gauge invariant states as only these states carry a representation of supersymmetry alone and not that of supersymmetry and local transformations. On-shell, $h_{\mu\nu}$ has two helicities and so does $\psi_{\mu\alpha}$; however, off-shell, $h_{\mu\nu}$ contributes $(5 \times 4)/2 = 10$ degrees of freedom minus 4 gauge degrees of freedom giving 6 bosonic degrees of freedom. On the other hand, off-shell $\psi_{\mu\alpha}$ contributes $4 \times 4 = 16$ degrees of freedom minus 4 gauge degrees of freedom, giving 12 fermionic degrees of freedom. Hence the auxiliary fields must contribute 6 bosonic degrees of freedom. If there are n auxiliary fermions there must be $4n + 6$ bosonic auxiliary fields.

Let us assume that a minimal formulation exists, that is, there are no auxiliary spinors. Let us also assume that the bosonic auxiliary fields occur in the Lagrangian as squares without derivatives (like F and G) and so are of dimension 2. Hence we have six bosonic auxiliary fields; it only remains to find their Lorentz character and transformations. We will assume that they consist of a scalar M, a pseudo-scalar N and a pseudo-vector b_μ. This is called the old minimal formulation [13.4, 13.5]. In fact, there is another possibility, we can use the two fields A_μ and $a_{\kappa\lambda}$ which possess the local transformations $\delta A_\mu = \partial_\mu \Lambda$; $\delta a_{\kappa\lambda} = \partial_\kappa \Lambda_\lambda - \partial_\lambda \Lambda_\kappa$. We note that a contribution $\varepsilon_{\mu\nu\rho\kappa} A^\mu \partial^\nu a^{\rho\kappa}$ to the action would not lead to propagating degrees of freedom. This is called the new minimal formulation [13.4, 13.5] and we will not discuss it further here.

The transformation of the fields $h_{\mu\nu}$, $\psi_{\mu\alpha}$, M, N and b_μ must reduce on-shell to the on-shell transformations found above. This restriction, dimensional arguments and the fact that if the auxiliary fields are to vanish on-shell they must vary into field equations requires the transformations to be of the forms [13.4, 13.5]

$$\delta h_{\mu\nu} = \tfrac{1}{2}(\bar{\varepsilon}\gamma_\mu \psi_\nu + \bar{\varepsilon}\gamma_\nu \psi_\mu),$$

$$\delta\psi_{\mu\alpha} = -\sigma^{ab}\partial_a h_{b\mu}\varepsilon - \tfrac{1}{3}\gamma_\mu(M + i\gamma_5 N)\varepsilon + b_\mu i\gamma_5\varepsilon + \delta_6 \gamma_\mu \rlap{/}b i\gamma_5\varepsilon,$$

$$\delta M = \delta_4 \bar{\varepsilon}\gamma \cdot R, \qquad\qquad\qquad (13.1.23)$$

$$\delta N = \delta_5 i\bar{\varepsilon}\gamma_5\gamma \cdot R,$$

$$\delta b_\mu = +\delta_7 i\bar{\varepsilon}\gamma_5 R_\mu + \delta_8 i\bar{\varepsilon}\gamma_5\gamma_\mu\gamma \cdot R.$$

The constants δ_4, δ_5, δ_6, δ_7 and δ_8 are determined by the restriction that the transformations of equations (13.1.23) and the local transformations of equations (13.1.17) should form a closed algebra. For example, the commutator of two supersymmetries on M gives

$$[\delta_{\varepsilon_1}, \delta_{\varepsilon_2}]M = \delta_4\{-\bar{\varepsilon}_2\gamma^\mu\varepsilon_1\partial_\mu M + 16\bar{\varepsilon}_2\sigma^{\mu\nu}i\gamma_5\varepsilon_1(1 + 3\delta_6)\partial_\mu b_\nu\}, \qquad (13.1.24)$$

which is the required result provided $\delta_4 = -\tfrac{1}{2}$ and $\delta_6 = -\tfrac{1}{3}$. Evaluating the commutator of two supersymmetries on all fields we find a closing algebra, without using any field equations, provided

$$\delta_4 = -\tfrac{1}{2}, \quad \delta_5 = -\tfrac{1}{2}, \quad \delta_6 = -\tfrac{1}{3}, \quad \delta_7 = \tfrac{3}{2} \quad \text{and} \quad \delta_8 = -\tfrac{1}{2}. \qquad (13.1.25)$$

We henceforth adopt these values for the parameters. An action which is constructed from the fields $h_{\mu\nu}$, $\psi_{\mu\alpha}$, M, N and b_μ and is invariant under the transformations of equations (13.1.23) with the above values of the parameters is

$$A_0 = \int d^4x\{-\tfrac{1}{2}h_{\mu\nu}E^{\mu\nu} - \tfrac{1}{2}\bar{\psi}_\mu R^\mu - \tfrac{1}{3}(M^2 + N^2 - b^\mu b_\mu)\}. \qquad (13.1.26)$$

Upon elimination of the auxiliary fields M, N and b_μ, using their equations of motion of equation (13.1.12), equation (13.1.26) reduces to the algebraically on-shell Lagrangian.

We can now construct the non-linear supergravity theory by applying the Noether technique discussed at the beginning of this section. Just as in the case of Yang–Mills, the linearised theory possesses the local Abelian invariances of equations (13.1.17) as well as the rigid (that is, constant parameter) supersymmetry transformations of equations (13.1.23). We first make the parameter of the rigid transformations space-time dependent, that is, set $\varepsilon = \varepsilon(x)$ in equations (13.1.23). The linearised action of equation (13.1.26) is then no longer invariant, but its variation must be of the form

$$\delta A_0 = \int d^4x \partial_\mu \bar{\varepsilon}^\alpha j^\mu_\alpha, \qquad\qquad (13.1.27)$$

since it is invariant when $\bar{\varepsilon}^\alpha$ is a constant. The object j^μ_α is linear in $\psi_{\mu\alpha}$ and the bosonic fields $h_{\mu\nu}$, M or N and b_μ. As such, on dimensional grounds, it must be of the generic form

$$j_{\mu\alpha} \propto \partial h \psi + M\psi + N\psi + b\psi.$$

Consider now the action, A_1 given by

$$A_1 = A_0 - \frac{\kappa}{4} \int d^4x \bar{\psi}^\mu j_\mu, \tag{13.1.28}$$

where κ is the gravitational constant. The action A_1 is invariant to order κ^0 provided we combine the now local supersymmetry transformation of equations (13.1.23) with the local transformation on ψ_μ of equations (13.1.17) with parameter $\eta(x) = (2/\kappa)\varepsilon(x)$. That is, the transformation of ψ_μ becomes

$$\delta\psi_\mu = \frac{2}{\kappa}\partial_\mu\varepsilon(x) - \partial_a h_{b\mu}\sigma^{ab}\varepsilon(x) - \frac{1}{3}\gamma_\mu(M + i\gamma_5 N)\varepsilon(x)$$

$$+ i\gamma_5\left(b_\mu - \frac{1}{3}\gamma_\mu\gamma^\nu b_\nu\right)\varepsilon(x). \tag{13.1.29}$$

The remaining fields transform as before except that ε is now space-time dependent.

As for the Yang–Mills case the two invariances of the linearised action have become knitted together to form one transformation, the role of gauge coupling now being played by the gravitational constant, κ. The addition of the term $(-\kappa/4)\int d^4x \bar{\psi}^\mu j_\mu$ to A_0 does the required job; its variation is

$$-\int d^4x \frac{\kappa}{4} \cdot 2 \cdot \left(\frac{2}{\kappa}\right)(\partial_\mu\bar{\varepsilon})j^\mu + \text{terms of order } \kappa^1. \tag{13.1.30}$$

The order κ^1 terms do not concern us at the moment. We note that $j_{\mu\alpha}$ is linear in $\psi_{\mu\alpha}$ and so we get a factor of 2, one from the $\delta\psi_{\mu\alpha}$ in $j_{\mu\alpha}$ and the other from the $\bar{\psi}_\mu$ that multiplies $j_{\mu\alpha}$.

We now proceed to find an invariant action order by order in κ by adding terms to the action and also to the transformations of the fields. For example, if we added a term to $\delta\psi_\mu$, say $\delta\bar{\psi}_\mu = \cdots + \bar{\varepsilon}X_\mu\kappa$, then in the linearised action we receive a contribution $-2\kappa\bar{\varepsilon}X_\mu R^\mu$ from the variation of ψ_μ. It is necessary at each step (order of κ) to check that the transformations of the fields form a closed algebra. In fact, any ambiguities that arise in the procedure are resolved by demanding that the algebra closes.

The final set of transformations is [13.4, 13.5]

$$\delta e_\mu^a = \kappa\bar{\varepsilon}\gamma^a\psi_\mu,$$

$$\delta\psi_\mu = 2\kappa^{-1}D_\mu(w(e,\psi))\varepsilon + i\gamma_5(b_\mu - \tfrac{1}{3}\gamma_\mu\slashed{b})\varepsilon - \tfrac{1}{3}\gamma_\mu(M + i\gamma_5 N)\varepsilon,$$

$$\delta M = -\frac{1}{2}e^{-1}\bar{\varepsilon}\gamma_\mu R^\mu - \frac{\kappa}{2}i\bar{\varepsilon}\gamma_5\psi_\nu b^\nu - \kappa\bar{\varepsilon}\gamma^\nu\psi_\nu M + \frac{\kappa}{2}\bar{\varepsilon}(M + i\gamma_5 N)\gamma^\mu\psi_\mu,$$

$$\delta N = -\frac{e^{-1}}{2}i\bar{\varepsilon}\gamma_5\gamma_\mu R^\mu + \frac{\kappa}{2}\bar{\varepsilon}\psi_\nu b^\nu - \kappa\bar{\varepsilon}\gamma^\nu\psi_\nu N - \frac{\kappa}{2}i\bar{\varepsilon}\gamma_5(M + i\gamma_5 N)\gamma^\mu\psi_\mu,$$

$$\delta b_\mu = \frac{3i}{2}e^{-1}\bar{\varepsilon}\gamma_5\left(g_{\mu\nu} - \frac{1}{3}\gamma_\mu\gamma_\nu\right)R^\nu + \kappa\bar{\varepsilon}\gamma^\nu b_\nu\psi_\mu - \frac{\kappa}{2}\bar{\varepsilon}\gamma^\nu\psi_\nu b_\mu$$

$$- \frac{\kappa}{2}i\bar{\psi}_\mu\gamma_5(M + i\gamma_5 N)\varepsilon - \frac{i\kappa}{4}\varepsilon_\mu^{\ bcd}b_b\bar{\varepsilon}\gamma_5\gamma_c\psi_d, \tag{13.1.31}$$

where

$$R^{\mu} = \varepsilon^{\mu\nu\rho\kappa} i\gamma_5\gamma_\nu D_\rho\big(w(e,\psi)\big)\psi_\kappa \quad \text{and} \quad D_\mu\big(w(e,\psi)\big)\psi_\nu = \left(\partial_\mu + w_{\mu ab}\frac{\sigma^{ab}}{4}\right)\psi_\nu$$

and

$$w_{\mu ab} = \tfrac{1}{2}e^{\nu}{}_a(\partial_\mu e_{b\nu} - \partial_\nu e_{b\mu}) - \tfrac{1}{2}e_b{}^{\nu}(\partial_\mu e_{a\nu} - \partial_\nu e_{a\mu}) - \tfrac{1}{2}e_a{}^{\rho}e_b{}^{\sigma}(\partial_\rho e_{\sigma c} - \partial_\sigma a_{\rho c})e_{\mu}^c$$
$$+ \tfrac{1}{4}\kappa^2(\bar\psi_\mu\gamma_a\psi_b + \bar\psi_a\gamma_\mu\psi_b - \bar\psi_\mu\gamma_b\psi_a). \tag{13.1.32}$$

They form a closed algebra, the commutator of two supersymmetries on any field being

$$[\delta_{\varepsilon_1}, \delta_{\varepsilon_2}] = \delta_{supersymmetry}(-\kappa\xi^\nu\psi_\nu) + \delta_{general\ coordinate}(2\xi_\mu)$$
$$+ \delta_{Local\ Lorentz}(-\tfrac{2}{3}\kappa\varepsilon_{ab\lambda\rho}b^\lambda\xi^\rho - \tfrac{2}{3}\kappa\bar\varepsilon_2\sigma_{ab}(M + i\gamma_5 N)\varepsilon_1 + 2\xi^d w_d{}^{ab}), \tag{13.1.33}$$

where $\xi_\mu = \bar\varepsilon_2\gamma_\mu\varepsilon_1$. The transformations of equations (13.1.31) leave invariant the action [13.4, 13.5]

$$A = \int d^4x \left\{ \frac{e}{2\kappa^2}R - \frac{1}{2}\bar\psi_\mu R^\mu - \frac{1}{3}e(M^2 + N^2 - b_\mu b^\mu) \right\}, \tag{13.1.34}$$

where

$$R = R_{\mu\nu}{}^{ab}e_a{}^\mu e_b{}^\nu, \quad R_{\mu\nu}{}^{ab}\frac{\sigma_{ab}}{4} = [D_\mu, D_\nu] \quad \text{and} \quad R^\mu = \epsilon^{\mu\nu\rho\kappa} i\gamma_5\gamma_\nu D_\rho\psi_\kappa.$$

The auxiliary fields M, N and b_μ may be eliminated using their algebraic equations of motion to obtain an action which was the form in which the $D = 4, N = 1$ supergravity was originally found in [13.2, 13.3]. Indeed, this was the first supergravity that was constructed.

As discussed at the beginning of this section one could also find the full theory by working with the algebra of field transformations alone.

The original formulation of supergravity in second order form [13.2] was shown to be invariant under supersymmetry using a computer. However, an algebraic proof of this invariance was given in [13.8, 13.12] for this theory and used for all subsequent supergravity theories. This proof uses the 1.5 order formalism which arose in the gauge approach to supergravity theories, which was discussed in section 9.3. Indeed, the proof of the invariance for $D = 4, N - 1$ supergavity to first order in fermions is given in this section. The reader will find the proof to all orders in chapter 10 of [1.11].

It may seem that the auxiliary fields are not very useful as their equations of motion just set them to zero. However, having a supersymmetry algebra that only closes using the equations of motion has considerable drawbacks. For example, if one couples the supergravity theory without auxiliary fields to supermatter, then the equations of motion will change and hence so will the supersymmetry algebra. Consequently, for every different matter coupling one requires a different supersymmetry algebra. With auxiliary fields the supersymmetry algebra is universal and this allows one to build the analogue of the tensor calculus of general relativity for supergravity [13.13]. This was the essential tool for the construction of the most general coupling of $D = 4, N = 1$ supergravity to $N = 1$ matter, that is, the Wess–Zumino model [13.14] and the $N = 1$ Yang–Mills theory [13.15], a result which underpins the quest to find a realistic model of supersymmetry.

We close this section by noting that one can carry out the Noether procedure to construct Einstein's theory of pure gravity. The linearised action is just that of equation (13.1.26) with all the other fields set to zero. This action possesses as symmetries the rigid translation

$$\delta h_{\mu\nu} = \zeta^\lambda \partial_\lambda h_{\mu\nu} \tag{13.1.35}$$

and the local transformation

$$\delta h_{\mu\nu} = \partial_\mu \xi_\nu + \partial_\nu \xi_\mu. \tag{13.1.36}$$

Letting the translation parameter ζ^λ become space-time dependent, that is, $\zeta^\lambda(x)$, the action is no longer invariant and its variation is of the form $\int d^4x \partial^\tau \zeta^\lambda T_{\tau\lambda}$. Here $T_{\tau\lambda}$ has dimension 4 as h and ζ have dimensions 1 and -1, respectively. As such, it must be of the form $\partial h \partial h$. It is just the energy momentum tensor for a spin two field. An action invariant to order κ^0 is given by

$$-\frac{\kappa}{6} \int d^4x h^{\tau\lambda} T_{\tau\lambda}, \tag{13.1.37}$$

provided we knitted together the above two transformations by identifying $\xi_\nu = (1/\kappa)\zeta_\nu$. This means that we now have only one transformation under which

$$\delta h_{\mu\nu} = \frac{1}{\kappa} \partial_\mu \zeta_\nu + \frac{1}{\kappa} \partial_\nu \zeta_\mu + \zeta^\lambda \partial_\lambda h_{\mu\nu}. \tag{13.1.38}$$

This variation of $h_{\mu\nu}$ contains the first few terms of an Einstein general coordinate transformation of the vielbein, which is given at lowest order in terms of $h_{\mu\nu}$ by

$$e_\mu{}^a = \eta_\mu{}^a + \kappa h_\mu{}^a. \tag{13.1.39}$$

We obtain an invariant action order by order in κ by adding terms to the action and to the transformations of the fields.

We close this section by giving an extension of $N = 1$ supergravity to include a cosmological constant. We can add to the action of equation (13.1.34) the terms

$$\int d^4x \det e \left(6m^2 + \frac{m\kappa}{2} \bar\psi_\mu \gamma^{\mu\nu} \psi_\nu\right). \tag{13.1.40}$$

The reader may verify that the action is still invariant if we add to the transformations of equations (13.1.31) the term $\delta\psi_\mu = 2m\gamma_\mu \epsilon$. For example, varying the cosmological term we find $\int d^4x \det e(m^2\kappa\bar\epsilon\gamma^\mu\psi_\mu)$, which is cancelled by the variation of ψ under the new addition in the second term.

13.1.2 The on-shell superspace method

Given the on-shell states of any supergravity theory and the linearised action, or equations of motion, from which they follow we can easily find the dimensions of the possible gauge invariant quantities of the theory. Comparing these with the dimensions of the superspace torsions and curvatures one finds that some of the latter have dimensions which are lower than any of the gauge invariant quantities in the space-time formulation of the theory. Since the lowest component of any superfield is just an expression in space-time we must conclude that the lowest components of these torsions and curvatures must vanish. However, if the lowest component of any superfield vanishes the entire superfield vanishes and so the corresponding torsions and curvatures for which there is no gauge invariant expression as its lowest component must vanish. Remarkably, it turns out that these constraints are sufficient to put the theory on shell, that is, they give the field equations of the theory.

Let us begin with a discussion of the geometry of superspace for a supergravity theory. The geometrical framework [13.7, 13.16] of superspace supergravity has many of the constructions of general relativity, but also has some essential differences. A more general account of this using the same conventions and containing more details, but a slightly different perspective, can be found in chapter 16 of [1.11].

For simplicity we will consider the case of $D = 4$, $N = 1$ superspace. We begin with an eight-dimensional manifold $z^\Pi = (x^\mu, \theta^{\underline{\alpha}})$, where x^μ is a Grassmann even coordinate and $\theta^{\underline{\alpha}}$ is a Grassmann odd coordinate. On this manifold a supergeneral coordinate reparameterisation has the form

$$z^\Pi \to z^{\Pi'} = z^\Pi + \xi^\Pi, \tag{13.1.41}$$

where $\xi^\Pi = (\xi^\mu, \xi^{\underline{\alpha}})$ are arbitrary functions of z^Π.

Just as in general relativity we can consider scalar superfields, that is, fields for which

$$\phi'(z') = \phi(z) \tag{13.1.42}$$

and superfields with superspace world indices φ_Λ; for example,

$$\varphi_\Lambda = \frac{\partial \phi}{\partial z^\Lambda}. \tag{13.1.43}$$

The latter transform as

$$\varphi'_\Lambda(z') = \frac{\partial z^\Pi}{\partial z^{\Lambda'}} \varphi_\Pi(z). \tag{13.1.44}$$

The transformation properties of higher order tensors are given by the obvious generalisation.

We must now specify the geometrical structure of the manifold. For reasons that will become apparent, the superspace formulation is essentially a vielbein, rather than a metric, formulation. We introduce supervielbeins $E_\Pi^{\ N}$ which transform under the supergeneral coordinate transformations as

$$\delta E_\Pi^{\ N} = \xi^\Lambda \partial_\Lambda E_\Pi^{\ N} + \partial_\Pi \xi^\Lambda E_\Lambda^{\ N}. \tag{13.1.45}$$

The $N, M, \ldots$ indices are tangent space indices, while $\Pi, \Lambda, \ldots$ are world indices. For vector tangent indices we use $m, n, \ldots$ and for vector world indices we use $\mu, \nu, \ldots$. For spinors we will use two-component notation: for tangent indices we use $A, B, \ldots$ and $\dot{A}, \dot{B}, \ldots$ and for world indices we underline the same indices. More details of the conventions are given in appendix A of [1.11].

We now take the tangent space group to be just the Lorentz group. It acts on the supervielbein by transforming its tangent space index as $\delta E_\Pi^{\ N} = E_\Pi^{\ M} \Lambda_M^{\ N}$, where

$$\Lambda_M^{\ N} = \begin{pmatrix} \lambda_m^{\ n} & 0 & 0 \\ 0 & -\frac{1}{4}(\sigma_{mn})_A^{\ B}\Lambda^{mn} & 0 \\ 0 & 0 & +\frac{1}{4}(\sigma_{mn})_{\dot{A}}^{\ \dot{B}}\Lambda^{mn} \end{pmatrix}. \tag{13.1.46}$$

The matrix $\Lambda_m{}^n$ is an arbitrary function on superspace and it governs not only the rotation of the vector index, but also the rotation of the spinorial indices. Since we are dealing with an eight-dimensional manifold one could choose a much larger tangent space group. For example, $\Lambda_M{}^N$ could be an arbitrary matrix that preserves a tangent space metric in this larger space. An obvious candidate is the metric for $OSp(4, 1)$. If this were the case one could introduce a metric $g_{\Pi\Lambda} = E_\Pi{}^N g_{NM} E_\Lambda{}^M$ and one would have a formulation which mimicked Einstein's general relativity [13.17]. However, such a formulation cannot lead to the x-space component $N = 1$ supergravity given earlier. One way to see this is to observe that the above tangent space group does not coincide with that of rigid superspace, which is just the coset space of the super Poincaré/Lorentz group. As such, it has the Lorentz group as the tangent space group and it acts as in equation (13.1.46). As linearised superspace supergravity must admit rigid superspace in the flat limit it follows that a formulation based on an $OSp(4, 1)$ tangent group will not contain linearised supergravity in this limit. In fact the $OSp(4, 1)$ formulation has a higher derivative action.

An important consequence of this restricted tangent space group is that tangent supervectors $V^N = V^\Pi E_\Pi{}^N$ belong to a reducible representation of the Lorentz group. This allows one to write down many more invariants. For example, the objects

$$V^m V_m, \quad V^A V^B \varepsilon_{AB}, \quad V^{\dot A} V^{\dot B} \varepsilon_{\dot B \dot A} \tag{13.1.47}$$

are all separately invariant.

We define a Lorentz valued spin connection:

$$\Omega_{\Lambda M}{}^N = \begin{pmatrix} \Omega_{\Lambda m}{}^n & 0 & 0 \\ 0 & -\frac{1}{4}\Omega_\Lambda{}^{mn}(\sigma_{mn})_A{}^B & 0 \\ 0 & 0 & \frac{1}{4}\Omega_\Lambda{}^{mn}(\bar\sigma_{mn})_{\dot A}{}^{\dot B} \end{pmatrix}. \tag{13.1.48}$$

This object transforms under supergeneral coordinate transformations as

$$\delta\Omega_{\Lambda M}{}^N = \xi^\Pi \partial_\Pi \Omega_{\Lambda M}{}^N + \partial_\Lambda \xi^\Pi \Omega_{\Pi M}{}^N \tag{13.1.49}$$

and under tangent space rotation as

$$\delta\Omega_{\Lambda M}{}^N = -\partial_\Lambda \Omega_M{}^N + \Omega_{\Lambda M}{}^S \Lambda_S{}^N + \Omega_{\Lambda R}{}^N \Lambda_M{}^R (-1)^{(M+R)(N+R)}. \tag{13.1.50}$$

The covariant derivatives are then defined by

$$D_\Lambda = \partial_\Lambda + \frac{1}{2}\Omega_\Lambda{}^{mn} J_{mn}, \tag{13.1.51}$$

where J_{mn} are the appropriate Lorentz generators. We adopt the convention that $(\frac{1}{2}J_{pq}w^{pq})V_n = w_n{}^m V_m$ and $(\frac{1}{2}J_{pq}w^{pq})\chi_\alpha = \frac{1}{4}(w_{nm}\gamma^{mn})_\alpha{}^\beta \chi_\beta$ (see appendix A of [1.11]). The covariant derivative with tangent indices is

$$D_N = E_N{}^\Lambda D_\Lambda, \tag{13.1.52}$$

where $E_N{}^\Lambda$ is the inverse vielbein defined by

$$E_N{}^\Lambda E_\Lambda{}^M = \delta_N{}^M \tag{13.1.53}$$

or

$$E_\Lambda{}^M E_M{}^\Pi = \delta_\Lambda{}^\Pi. \tag{13.1.54}$$

Equipped with supervielbein and spin connection we define the torsion and curvature tensors as usual as

$$[D_N, D_M\} = T_{NM}{}^R D_R + \tfrac{1}{2} R_{NM}{}^{mn} J_{mn}, \tag{13.1.55}$$

where $[\,,\}$ means take the commutator or anti-commutator appropriate to the Grassmann character of the objects involved. Using equations (13.1.51) and (13.1.52) we find that

$$T_{MN}{}^R = E_M{}^\Lambda \partial_\Lambda E_N{}^\Pi E_\Pi{}^R + \Omega_{MN}{}^R - (-1)^{MN}(M \leftrightarrow N), \tag{13.1.56}$$

$$R_{MN}{}^{rs} = E_M{}^\Lambda E_N{}^\Pi (-1)^{\Lambda(N+\Pi)}\{\partial_\Lambda \Omega_\Pi{}^{rs} + \Omega_\Lambda{}^{rk} \Omega_{\Pi k}{}^s - (-1)^{\Lambda\Pi}(\Lambda \leftrightarrow \Pi)\}. \tag{13.1.57}$$

We recall that $(-1)^{MN}$ is $+1$ except when M and N are fermionic indices when it is -1. For example, $[A_M, B_N] = A_M B_N - (-1)^{MN} B_N A_M$.

The supergeneral coordinate transformations can be rewritten using these tensors

$$\delta E_\Lambda{}^M = -E_\Lambda{}^R \xi^N T_{NR}{}^M + D_\Lambda \xi^M, \tag{13.1.58}$$

$$\delta \Omega_{\Lambda M}{}^N = E_\Lambda{}^R \xi^P R_{PRM}{}^N, \tag{13.1.59}$$

where $\xi^N = \xi^\Lambda E_\Lambda{}^N$ and we have discarded a Lorentz transformation.

The torsion and curvature tensors satisfy Bianchi identities which follow from the identity

$$\big[D_M, [D_N, D_R\}\} - \big[[D_M, D_N\}, D_R\} + (-1)^{RN}\big[[D_M, D_R\}, D_N\} = 0. \tag{13.1.60}$$

They read

$$\begin{aligned}
I^{(1)}_{RMN}{}^F &\equiv [-(-1)^{(M+N)R} D_R T_{MN}{}^F + T_{MN}{}^S T_{SR}{}^F + R_{MNR}{}^F] \\
&\quad + [(-1)^{MN} D_N T_{MR}{}^F - (-1)^{NR} T_{MR}{}^S T_{SN}{}^F - (-1)^{NR} R_{MRN}{}^F] \\
&\quad + [-D_M T_{NR}{}^F + (-1)^{(N+R)M} T_{NR}{}^S T_{SM}{}^F + (-1)^{(N+R)M} R_{NRM}{}^F] = 0
\end{aligned} \tag{13.1.61}$$

and

$$\begin{aligned}
I^{(2)}_{RMN}{}^{mn} &\equiv [(-1)^{(M+N)R} D_R R_{MN}{}^{mn} + T_{MN}{}^S R_{SR}{}^{mn}] \\
&\quad - (-1)^{NR}(R \to N,\ N \to R,\ M \to M \text{ in the first bracket}) \\
&\quad + (-1)^{(N+R)M}(M \to N,\ N \to R,\ R \to M \text{ in the first bracket}) = 0.
\end{aligned} \tag{13.1.62}$$

It can be shown that if $I^{(1)}_{MNR}{}^F$ holds, then $I^{(2)}_{RMN}{}^{mn}$ is automatically satisfied. This result holds in the presence of constraints on $T_{MN}{}^R$ and $R_{MN}{}^{mn}$ and is a consequence of the restricted tangent space choice [13.18]. For all Grassmann odd indices, that is, $R, M, N = A, B, C$, we find that

$$\begin{aligned}
I_{ABC}{}^N &= -D_A T_{BC}{}^N + T_{AB}{}^S T_{SC}{}^N + R_{ABC}{}^N - D_C T_{AB}{}^N + T_{CA}{}^S T_{SB}{}^N \\
&\quad + R_{CAB}{}^N - D_B T_{CA}{}^N + T_{BC}{}^S T_{SA}{}^N + R_{BCA}{}^N = 0
\end{aligned} \tag{13.1.63}$$

and for Grassmann odd indices with one Grassmann even index

$$\begin{aligned}
I_{ABr}{}^N &= -D_A T_{Br}{}^N + T_{AB}{}^S T_{Sr}{}^N + R_{ABr}{}^N - D_r T_{AB}{}^N + T_{rA}{}^S T_{SB}{}^N \\
&\quad + R_{rAB}{}^N + D_B T_{rA}{}^N - T_{Br}{}^S T_{SA}{}^N - R_{BrA}{}^N = 0,
\end{aligned} \tag{13.1.64}$$

while for two Grassmann even indices and one odd we have

$$I_{Anr}{}^N = -D_A T_{nr}{}^N + T_{An}{}^s T_{sr}{}^N + R_{Anr}{}^N - D_r T_{An}{}^N + T_{rA}{}^s T_{sn}{}^N$$
$$+ R_{rAn}{}^N - D_n T_{rA}{}^N + T_{nr}{}^s T_{sA}{}^N + R_{nrA}{}^N = 0. \tag{13.1.65}$$

Clearly, one can replace any undotted index by a dotted index and the signs remain the same. For rigid superspace all the torsions and curvatures vanish except for $T_{A\dot{B}}{}^n = -2i(\sigma^n)_{A\dot{B}}$.

The dimensions of the torsions and curvature can be deduced from the dimensions of D_N from equation (13.1.55). If F and B denote fermionic and bosonic indices, respectively, then the dimensions of the derivatives are given by

$$[D_F] = \tfrac{1}{2}; \quad [D_B] = 1 \tag{13.1.66}$$

and as a result

$$[T_{FF}{}^B] = 0; \quad [T_{FF}{}^F] = [T_{FB}{}^B] = \tfrac{1}{2}, [T_{FB}{}^F] = [T_{BB}{}^B] = 1;$$
$$[T_{BB}{}^F] = \tfrac{3}{2}, \tag{13.1.67}$$

while

$$[R_{FF}{}^{mn}] = 1; \quad [R_{FB}{}^{mn}] = \tfrac{3}{2}; \quad [R_{BB}{}^{mn}] = 2. \tag{13.1.68}$$

It is useful to consider the notion of the geometric dimension of fields. This is the dimension of the field as it appears in the torsions and curvature. Such expressions never involve κ as they are given by the geometry and not the dynamics of the theory. They are non-linear in certain bosonic fields, such as the vielbein $e_\mu{}^n$, and as a result these fields must have zero geometric dimensions. The dimensions of the other fields are determined in relation to $e_\mu{}^n$ to be given by

$$[e_\mu{}^n] = 0 \quad [\psi_\mu{}^\alpha] = \tfrac{1}{2} \quad [M] = [N] = [b_\mu] = 1. \tag{13.1.69}$$

These dimensions can, for example, be read off from the supersymmetry transformations. These dimensions differ from the canonical assignment of dimension by one unit. The difference comes about as we have absorbed factors of κ into the fields.

Having set up the appropriate geometry of superspace we are now in a position to derive the equations of motion of the $D = 4$, $N = 1$ supergravity theory using its superspace setting and solely from a knowledge of the on-shell states in the irreducible representation as described above. The on-shell states are represented by $h_{\mu\nu}$ and $\psi_\mu{}^\alpha$, which, as discussed previously, have the gauge transformations

$$\delta h_{\mu\nu} = \partial_\mu \xi_\nu + \partial_\nu \xi_\mu, \quad \delta \psi_{\mu\alpha} = \partial_\mu \eta_\alpha. \tag{13.1.70}$$

We have omitted the non-linear terms, as it is sufficient to consider only the linearised theory to determine the dimensions of the gauge invariant quantities. The geometric dimension of $h_{\mu\nu}$ is 0 while that of $\psi_\mu{}^\alpha$ is $\tfrac{1}{2}$. The lowest dimension gauge objects invariant under these gauge transformations are of the form

$$\partial \psi \quad \text{and} \quad \partial \partial h, \tag{13.1.71}$$

which have dimensions $\tfrac{3}{2}$ and 2, respectively.

Consider now the supertorsion and curvature; these objects at $\theta = 0$ must correspond to gauge invariant x-space objects. If there is no such object, then the corresponding tensor must vanish at $\theta = 0$ and hence to all orders in θ. The only dimension 0 tensors are $T_{AB}{}^n$ and $T_{A\dot{B}}{}^n$. There are no dimension 0 covariant objects except the numerically invariant tensor $(\sigma^n)_{A\dot{B}}$. Hence, we must conclude on grounds of Lorentz invariance that

$$T_{AB}{}^n = 0, \quad T_{A\dot{B}}{}^n = c(\sigma^n)_{A\dot{B}}, \tag{13.1.72}$$

where c is a constant. We choose $c \neq 0$ in order to agree with rigid superspace. The reality properties of $T_{A\dot{B}}{}^n$ imply that c is imaginary and we can normalize it to take the value $c = -2i$.

There are also no dimension $\frac{1}{2}$ covariant tensors in x-space and so

$$T_{A\dot{B}}{}^{\dot{C}} = T_{AB}{}^C = T_{Am}{}^n = 0. \tag{13.1.73}$$

There are no dimension 1 covariant objects in x-space. This would not be the case if one had an independent spin connection, $w_{\mu}{}^{rs}$, for $\partial e + w + \cdots$ would be a covariant quantity. When $w_{\mu}{}^{rs}$ is not an independent quantity, it must be given in terms of $e_{\mu}{}^n$ and $\psi_{\mu}{}^{\alpha}$ in such a way as to render the above dimension 1 covariant quantity 0. Hence, as we do not have an independent spin connection, that is, in second order formalism, we have the constraints

$$T_{nA}{}^{\dot{B}} = T_{nA}{}^B = R_{AB}{}^{mn} = 0 = R_{A\dot{B}}{}^{mn}. \tag{13.1.74}$$

In other words, every dimension 0, $\frac{1}{2}$, and 1 torsion and curvature vanishes with the exception of $T_{A\dot{B}}{}^n = -2i(\sigma^n)_{A\dot{B}}$.

The dimension $\frac{3}{2}$ tensors can involve at $\theta = 0$ the spin $\frac{3}{2}$ object $\partial\psi$ and so these will not all be zero. The only remaining non-zero tensors are $T_{mn}{}^A$, $R_{Ar}{}^{mn}$ and $R_{st}{}^{mn}$ and of course $T_{A\dot{B}}{}^n = -2i(\sigma^n)_{A\dot{B}}$. However, as we will now demonstrate, the constraints of equations (13.1.72)–(13.1.74) are sufficient to specify the entire on-shell theory. The first non-trivial Bianchi identity has dimension $\frac{3}{2}$ and is

$$I_{nB\dot{D}}{}^{\dot{C}} = -D_n T_{B\dot{D}}{}^{\dot{C}} + T_{nB}{}^F T_{F\dot{D}}{}^{\dot{C}} + R_{nB\dot{D}}{}^{\dot{C}} + D_{\dot{D}} T_{nB}{}^{\dot{C}} - T_{\dot{D}n}{}^F T_{FB}{}^{\dot{C}}$$
$$- R_{\dot{D}nB}{}^{\dot{C}} - D_B T_{\dot{D}n}{}^{\dot{C}} + T_{B\dot{D}}{}^F T_{Fn}{}^{\dot{C}} + R_{B\dot{D}n}{}^{\dot{C}} = 0. \tag{13.1.75}$$

Using the above constraints this reduces to

$$-2i(\sigma^m)_{B\dot{D}} T_{mn}{}^{\dot{C}} - R_{nB\dot{D}}{}^{\dot{C}} = 0. \tag{13.1.76}$$

Tracing on $\dot{D}$ and $\dot{C}$ then yields

$$(\sigma^m)_{B\dot{D}} T_{mn}{}^{\dot{D}} = 0. \tag{13.1.77}$$

This is the Rarita–Schwinger equation as we will demonstrate shortly.

The spin 2 equation must have dimension 2 and is contained in the $I_{Bmn}{}^A$ Bianchi identity.

$$I_{Bmn}{}^A = -D_B T_{mn}{}^A + T_{Bm}{}^F T_{Fn}{}^A + R_{Bmn}{}^A - D_n T_{Bm}{}^A + T_{nB}{}^F T_{Fm}{}^A$$
$$+ R_{nBm}{}^A - D_m T_{nB}{}^A + T_{mn}{}^F T_{FB}{}^A + R_{mnB}{}^A = 0. \tag{13.1.78}$$

Application of the constraints gives

$$-D_B T_{mn}{}^A + T_{mnB}{}^A = 0. \tag{13.1.79}$$

On contracting with $(\sigma^m)_{\dot{B}A}$ we find

$$(\sigma^m)_{\dot{B}A} D_B T_{mn}{}^A = 0 = R_{mnB}{}^A (\sigma^m)_{\dot{B}A}. \tag{13.1.80}$$

Using the fact that $R_{mnB}{}^A = -\frac{1}{4}R_{mn}{}^{pq}(\sigma_{pq})_B^A$ yields the result

$$R_{mn} - \frac{1}{2}\eta_{mn}R = 0, \quad \text{or} \quad R_{mn} = 0, \tag{13.1.81}$$

where $R_{mn} = R_{msn}{}^s$.

We now wish to demonstrate that the constraints of equation (13.1.80) are the spin $\frac{3}{2}$ and spin 2 equations. The $\theta = 0$ components of $E_\mu{}^n$ and $E_\mu{}^A$ are denoted as follows:

$$E_\mu{}^n(\theta = 0) = e_\mu{}^n, \quad E_\mu{}^A(\theta = 0) = \frac{1}{2}\psi_\mu{}^A. \tag{13.1.82}$$

At this stage the above equation is simply a definition of the x-space fields $e_\mu{}^n$ and $\psi_\mu{}^A$. The $\theta = 0$ components of $E_A{}^n$ may be gauged away by an appropriate supergeneral coordinate transformation. Under such a transformation

$$\delta E_A{}^n(\theta = 0) = \xi^\Pi \partial_\Pi E_A{}^n|_{\theta=0} + \partial_A \xi^\Pi E_\Pi{}^n|_{\theta=0} = \cdots + \partial_A \xi^\mu e_\mu{}^n + \cdots \tag{13.1.83}$$

and it is clear that we may choose $\partial_A \xi^\mu$ so that $E_A{}^n = 0$. Similarly we may choose

$$E_{\dot{A}}{}^{\dot{B}} = \delta_A^{\dot{B}}, \quad E_{\dot{A}}{}^B = \delta_A^B, \quad E_{\dot{A}}{}^{\dot{B}} = 0. \tag{13.1.84}$$

To summarise

$$E_\Pi{}^M(\theta = 0) = \begin{pmatrix} e_\mu{}^n & \frac{1}{2}\psi_\mu{}^A & \frac{1}{2}\psi_\mu{}^{\dot{A}} \\ 0 & \delta_B^A & 0 \\ 0 & 0 & \delta_{\dot{B}}^{\dot{A}} \end{pmatrix}. \tag{13.1.85}$$

For the spin connection Ω_Π^m we define

$$\Omega_\mu{}^{mn}(\theta = 0) = w_\mu{}^{mn}. \tag{13.1.86}$$

We can use a Lorentz transformation to set

$$\Omega_\alpha{}^{mn}(\theta = 0) = 0 \tag{13.1.87}$$

using equation (13.1.50).

We now evaluate the lowest components of the torsions and curvatures in terms of the above x-space fields. The best way to do this is first to find these objects with their lower indices converted to world indices using the inverse supervielbein. For the torsion we find, using equation (13.1.56), that

$$T_{\Pi\Lambda}^R = (-1)^{\Pi(N+\Lambda)}E_\Lambda{}^N E_\Pi{}^M T_{MN}^R = -\partial_\Pi E_\Lambda{}^R + \Omega_{\Pi\Lambda}{}^R - (-1)^{\Pi\Lambda}(\Pi \leftrightarrow \Lambda), \tag{13.1.88}$$

where $\Omega_{\Pi\Lambda}{}^R = (-1)^{\Pi(N+\Lambda)}E_\Lambda{}^N \Omega_{\Pi N}{}^R$. At $\theta = 0$ we then find using equation (13.1.85) that

$$T_{\mu\nu}{}^{\dot{A}} = -\frac{1}{2}\partial_\mu \psi_\nu{}^{\dot{A}} + \Omega_{\mu\nu}{}^{\dot{A}} - (\mu \leftrightarrow \nu) = -\frac{1}{2}\psi_{\mu\nu}{}^{\dot{A}}, \tag{13.1.89}$$

where

$$\psi_{\mu\nu}{}^{\dot{A}} \equiv D_\mu \psi_\nu{}^{\dot{A}} - (\mu \leftrightarrow \nu), \quad D_\mu \psi_\nu{}^{\dot{A}} = \partial_\mu \psi_\nu{}^{\dot{A}} - \psi_\nu{}^{\dot{B}}w_{\mu\dot{B}}{}^{\dot{A}} \tag{13.1.90}$$

and

$$\Omega_{\mu\nu}{}^{\dot{A}} = E_\nu{}^N w_{\mu N}{}^{\dot{A}} = \frac{1}{2}\psi_\nu{}^{\dot{B}}w_{\mu\dot{B}}{}^{\dot{A}}$$

at $\theta = 0$. We now find the torsion with all tangent indices which is given in terms of $T_{\mu\nu}{}^{\dot{A}}$ by the relation

$$T_{\mu\nu}{}^{\dot{A}}(\theta = 0) = E_\mu{}^N(\theta = 0)E_\nu{}^M(\theta = 0)T_{NM}{}^{\dot{A}}(\theta = 0)(-1)^{NM}$$

$$= e_\mu{}^n e_\nu{}^m T_{nm}{}^{\dot{A}}(\theta = 0), \tag{13.1.91}$$

where we have used the constraints $T_{Bn}{}^{\dot{A}} = T_{\dot{B}\dot{C}}{}^{\dot{A}} = 0$. Consequently,

$$0 = (\sigma^m)_{AB} T_{mn}{}^{\dot{B}}(\theta = 0) = -\tfrac{1}{2}(\sigma^m)_{A\dot{B}} e_m{}^\mu e_n{}^\nu \psi_{\mu\nu}{}^{\dot{B}} \tag{13.1.92}$$

and we recognise the Rarita–Schwinger equation on the right-hand side.

To be complete we must also show that $w_\mu{}^{mn}$ is the spin connection given in terms of $e_\mu{}^n$ and $\psi_\mu{}^A$. This follows from the constraint $T_{nm}{}^r = 0$. We note that

$$T_{\mu\nu}{}^r(\theta = 0) = -\partial_\mu e_\nu{}^r + w_{\mu\nu}{}^r - (\mu \leftrightarrow \nu). \tag{13.1.93}$$

However,

$$T_{\mu\nu}{}^r(\theta = 0) = E_\mu{}^N(\theta = 0)E_\nu{}^M(\theta = 0)T_{NM}{}^r(\theta = 0)(-1)^{MN}$$

$$-\tfrac{1}{4}\psi_\mu{}^A \psi_\nu{}^{\dot{B}} T_{A\dot{B}}{}^r(\theta = 0) - \psi_\nu{}^{\dot{B}}\psi_\nu{}^A T_{\dot{B}A}{}^r(\theta = 0)$$

$$= +\tfrac{1}{2} i \psi_\nu{}^{\dot{B}}(\sigma^r)_{A\dot{B}}\psi_\mu{}^A - (\mu \leftrightarrow \nu). \tag{13.1.94}$$

Consequently, we find that

$$w_{\mu n}{}^m e_\nu{}^n - \partial_\nu e_\mu{}^m - (\mu \leftrightarrow \nu) = +\frac{i}{2}\psi_\nu{}^{\dot{B}}(\sigma^m)_{A\dot{B}}\psi_\mu{}^A - (\mu \leftrightarrow \nu), \tag{13.1.95}$$

which can be solved in the usual way to yield the correct expression for $w_{\mu n}{}^m$, that is, the one of equation (13.1.32).

The derivation of the spin-2 equation follows a similar pattern:

$$R_{\mu\nu}{}^{mn}(\theta = 0) = \partial_\mu w_\nu{}^{mn} + w_\mu{}^{mr} w_{\nu r}{}^n - (\mu \leftrightarrow \nu). \tag{13.1.96}$$

However,

$$R_{\mu\nu}{}^{nm}(\theta = 0) = E_\mu{}^N(\theta = 0)E_\nu{}^M(\theta = 0)R_{NM}{}^{mn}(\theta = 0)(-1)^{MN}$$

$$= e_\mu{}^p e_\nu{}^q R_{pq}{}^{nm}(\theta = 0)$$

$$+ \tfrac{1}{2}\left(\psi_\mu{}^{\dot{A}} e_\nu{}^p R_{\dot{A}p}{}^{nm} + \psi_\mu{}^A e_\nu{}^p R_{Ap}{}^{nm}(\theta = 0) - (\mu \leftrightarrow \nu)\right). \tag{13.1.97}$$

The object $R_{Ap}{}^{nm}$ can be found from the Bianchi identity $I_{Anr}{}^s$

$$0 = I_{Anr}{}^s = -D_A T_{nr}{}^s + T_{An}{}^F T_{Fr}{}^s + R_{Anr}{}^s - D_r T_{An}{}^s + T_{rA}{}^F T_{Fn}{}^s$$

$$+ R_{rAn}{}^s - D_n T_{rA}{}^s + T_{nr}{}^F T_{FA}{}^s + R_{nrA}{}^s. \tag{13.1.98}$$

Using the constraints we find that

$$R_{Anr}{}^s + R_{rAn}{}^s = +2i T_{nr}{}^{\dot{B}}(\sigma^s)_{A\dot{B}}. \tag{13.1.99}$$

From equation (13.1.92) after stripping of the σ^m matrix from both sides, we find that

$$R_{Anr}{}^s + R_{rAn}{}^s = -i(\sigma^s)_{A\dot{B}} e_n{}^\mu e_r{}^\nu \psi_{\mu\nu}{}^{\dot{B}}. \tag{13.1.100}$$

Contracting equation (13.1.57) with $e^{\nu}{}_{m}$ we find

$$e_{m}^{\nu}R_{\mu\nu}{}^{nm}(\theta=0) \equiv R_{\mu}^{n} = e_{\mu}^{p}R_{pm}{}^{nm} + \tfrac{1}{2}(\psi_{\mu}^{A}R_{Am}{}^{nm} + \psi_{\mu}^{\dot{A}}R_{\dot{A}m}{}^{nm}). \qquad (13.1.101)$$

Equation (13.1.100) then gives

$$e_{\mu}^{p}R_{pm}{}^{nm} = R_{\mu}^{n} - \left(\frac{i}{2}\psi_{\mu}^{A}(\sigma^{m})_{A\dot{B}}e_{m}^{\lambda}e^{n\tau}\psi_{\lambda\tau}^{\dot{B}} + \text{h.c.}\right). \qquad (13.1.102)$$

Using equation (13.1.81) ($R_{mn} = 0$) on the right-hand side of this equation yields the spin-2 equation of $N = 1$ supergravity.

One should also analyse all the remaining Bianchi identities and show that they do not lead to any inconsistencies. However, it can be shown that the other Bianchi identities are now automatically satisfied [13.18, 13.19].

The superspace formulation of $D = 4$, $N = 1$ supergravity, including the on-shell and off-shell constraints, was given in [13.7, 13.16]. A discussion of the off-shell constraints, and so the theory containing the auxiliary fields, can be found in section 16.2 of [1.11].

The on-shell superspace method was used [10.15] to find the full equations of motion in superspace and so in x-space of the IIB supergravity.

13.1.3 Gauging of space-time groups

It has been known for many years that Einstein's theory of general relativity contains a local Lorentz symmetry. When the action is expressed in first order formalism the spin connection is the gauge field and the Riemann tensor is the field strength for a Lorentz group [13.19]. However, Einstein's theory is not of the form of field strength squared and so it is not a Yang–Mills theory. As such, it might not appear to be such a great idea to gauge the Poincaré group, but this was what was done in [13.8] and it has proved to be useful. In fact, [13.8] gauged the super Poincaré group and deduced the supergravity theory from this viewpoint. For simplicity, we will take the case of four dimensions and the $N = 1$ super Poincaré algebra, which is given by

$$[J_{ab}, P_{c}] = \eta_{bc}P_{a} - \eta_{ac}P_{b}, \quad [J_{ab}, J_{cd}] = \eta_{bc}J_{ad} + \eta_{ad}J_{bc} - \eta_{ac}J_{bd} - \eta_{bd}J_{ac},$$

$$[Q_{\alpha}, P_{a}] = 0, \quad \{Q_{\alpha}, Q_{\beta}\} = -2(\gamma^{a}C^{-1})_{\alpha\beta}P_{a},$$

$$[Q_{\alpha}, J_{cd}] = \tfrac{1}{2}(\gamma_{cd})_{\alpha}{}^{\beta}Q_{\beta}. \qquad (13.1.103)$$

We introduce the Lie algebra valued gauge field

$$A_{\mu} = e_{\mu}^{a}P_{a} - \tfrac{1}{2}w_{\mu}{}^{ab}J_{ab} + \tfrac{1}{2}\bar{\psi}_{\mu}^{\alpha}Q_{\alpha} \qquad (13.1.104)$$

and define the covariant derivative $\hat{D}_{\mu} = \partial_{\mu} - A_{\mu}$. We could introduce the coupling constant κ in front of A_{μ} but we will for simplicity set it to 1. The appearance of the vielbein, spin connection and gravitino is already apparent at this stage.

The field strengths, defined by $[\hat{D}_{\mu}, \hat{D}_{\nu}] = -R_{\mu\nu}{}^{a}P_{a} + \tfrac{1}{2}R_{\mu\nu}{}^{ab}J_{ab} - \tfrac{1}{2}\bar{\Psi}_{\mu\nu}Q$, are found to be

$$R_{\mu\nu}{}^{a} = \partial_{\mu}e_{\nu}^{a} - \partial_{\nu}e_{\mu}^{a} + \omega_{\mu}{}^{a}{}_{c}e_{\nu}^{c} - \omega_{\nu}{}^{a}{}_{c}e_{\mu}^{c} + \tfrac{1}{2}\bar{\psi}_{\mu}\gamma^{a}\psi_{\nu},$$

$$R_{\mu\nu}{}^{ab} = \partial_{\mu}\omega_{\nu}{}^{ab} + \omega_{\mu}{}^{ac}\omega_{\nu c}{}^{b} - (\mu \leftrightarrow \nu),$$

$$\Psi_{\mu\nu} = (\partial_{\mu} - \tfrac{1}{4}\gamma_{cd}\omega_{\mu}{}^{cd})\psi_{\nu} - (\mu \leftrightarrow \nu) \equiv D_{\mu}\psi_{\nu} - (\mu \leftrightarrow \nu). \qquad (13.1.105)$$

The variation of the fields under gauge transformations is given by $\delta A_\mu = \partial_\mu \Lambda - [A_\mu, \Lambda]$ and so with $\Lambda = \epsilon^a P_a - \frac{1}{2} w^{ab} J_{ab} + \frac{1}{2} \bar{\eta}^\alpha Q_\alpha$ we find that

$$\delta e_\mu^a = \partial_\mu \epsilon^a - \omega^a{}_c e_\mu^c + \omega_\mu{}^{ac} \epsilon_c + \frac{1}{2} \bar{\eta} \gamma^a \psi_\mu,$$

$$\delta \omega_\mu{}^{ab} = \partial_\mu \omega^{ab} - (\omega^{ac} \omega_{\mu c}{}^b - \omega^{bc} \omega_{\mu c}{}^a), \tag{13.1.106}$$

$$\delta \psi_\mu = 2(\partial_\mu - \tfrac{1}{4} \gamma_{cd} \omega_\mu{}^{cd}) \eta + \tfrac{1}{4} \gamma_{cd} \omega^{cd} \psi_\mu \equiv D_\mu \eta + \tfrac{1}{4} \gamma_{cd} \omega^{cd} \psi_\mu.$$

We notice part of $\hat{D}_\mu$ contains the covariant derivative for the Lorentz group $D_\mu = \partial_\mu + \frac{1}{2} \omega_\mu^{cd} J_{cd}$. Indeed, the field strength $R_{\mu\nu}{}^{ab}$ is just the field strength for the Lorentz group and the fields transform in the expected way under local Lorentz transformations, for example, $\delta e_\mu^a = -\omega^a{}_c e_\mu^c$.

We now consider the most general action which is linear in the field strengths. Such an action cannot in general be invariant under the gauge transformations as it is not quadratic in the curvature, but we will only demand invariance subject to the constraint

$$R_{\mu\nu}{}^a = 0. \tag{13.1.107}$$

We now pause to first consider the case of pure gravity, and so set $\psi_\mu = 0 = \eta$ in the above transformations. Thus we are considering the gauge theory of the Poincaré group. Taking account that it must be invariant under local Lorentz transformations and second order in space-time derivatives, the action must, up to factors of $\det e_\mu^a$, which we discuss later, be of the form

$$\frac{1}{8} \int d^4 x \, \epsilon^{\mu\nu\rho\lambda} \epsilon_{abcd} e_\mu^a e_\nu^b R_{\rho\lambda}{}^{cd}. \tag{13.1.108}$$

The variation of the action is given by

$$\frac{1}{4} \int d^4 x \, \epsilon^{\mu\nu\rho\lambda} \epsilon_{abcd} e_\mu^a D_\nu e^b R_{\rho\lambda}{}^{cd}$$

$$= -\frac{1}{8} \int d^4 x \, \epsilon^{\mu\nu\rho\lambda} \epsilon_{abcd} (C_{\nu\mu}{}^a e^b R_{\rho\lambda}{}^{cd} + 2 e_\mu^a e^b D_\nu R_{\rho\lambda}{}^{cd}), \tag{13.1.109}$$

where $D_\mu e_\nu^a = \partial_\mu e_\nu^a - \omega_\mu{}^a{}_c e_\nu^c$. The first term on the right-hand side of the above equation vanishes due to the constraint of equation (13.1.107) and the last term vanishes due to the Bianchi identity $[D_{[\mu}[D_\rho, D_{\lambda]}]] = 0 = \frac{1}{2}[D_{[\mu} R_{\rho\lambda]}{}^{cd} J_{cd}]$. Thus we arrive at an invariant action which is Einstein's. The action of equation (13.1.108) has ω_μ^{cd} as an independent field. However, its equation of motion algebraically expresses ω_μ^{cd} in terms of e_μ^a and this is just the same relation that follows from solving the constraint we have adopted. Thus, with hindsight, it was consistent to adopt the constraint of equation (13.1.107).

The method can be further understood by the realisation that a general coordinate transformation of the vielbein can be written as

$$\delta_\epsilon e_\mu^a = \xi^\nu \partial_\nu e_\mu^a + \partial_\mu \xi^\nu e_\nu^a + R_{\mu\nu}{}^a \xi^\nu + \xi^\nu \omega_\nu{}^a{}_c e_\mu^c, \tag{13.1.110}$$

where $\xi^\mu = \epsilon^a e_a{}^\mu$. Hence the constraint also converts the gauge transformation associated with translations into a general coordinate transformation and a local Lorentz rotation, which explains why proceeding in the above way leads to Einstein's theory. Of course, the constraint of equation (13.1.107) is not invariant under local Poincaré transformations, but nonetheless we have found Einstein's theory in a simple way. This connection with general coordinate transformations makes it clear that had we taken an action with factors of $\det e^a_\mu$ in various places, then, as we are considering second order in space-time derivatives, we would have found the same result, that is, just Einstein's action. Of course, we can add the usual cosmological constant.

Let us now return to the gauge theory of the super Poincaré group. We wish to find an action that is invariant subject to the constraint of equation (13.1.107). The most general action which is linear in the field strengths and is invariant under local Lorentz transformations is given by

$$-\frac{1}{8}\int d^4x\, \epsilon^{\mu\nu\rho\lambda}(\epsilon_{abcd}e^a_\mu e^b_\nu R_{\rho\lambda}{}^{cd} - 2if\bar{\psi}_\mu\gamma_5\gamma_\nu\Psi_{\rho\lambda}), \tag{13.1.111}$$

where f is a constant and $\gamma_\mu = e^a_\mu\gamma_a$. Varying the action under a supersymmetry gauge transformation η leads to

$$-\frac{1}{8}\int d^4x\, \epsilon^{\mu\nu\rho\lambda}(2\epsilon_{abcd}\bar{\eta}\gamma^a\psi_\mu e^b_\nu R_{\rho\lambda}{}^{cd} - 2if\bar{\eta}D^{\leftarrow}_\mu\gamma_5\gamma_\nu\Psi_{\rho\lambda}$$
$$- if\bar{\psi}_\mu\gamma_5\gamma_\nu R_{\rho\lambda}{}^{cd}\gamma_{cd}\eta - if\bar{\psi}_\mu\gamma_5\gamma^c\Psi_{\rho\lambda}\eta\gamma_c\psi_\nu). \tag{13.1.112}$$

Integrating the derivative in the second term by parts, and using the constraint of equation (13.1.107), we find a contribution that is given by

$$-\frac{1}{8}\int d^4x\, \epsilon^{\mu\nu\rho\lambda}(if\bar{\eta}\gamma_5\gamma_\nu R_{\mu\rho}{}^{cd}\gamma_{cd}\Psi_\lambda) \tag{13.1.113}$$

and a term that is cubic in the gravitino. The third term in expression (13.1.112) can be written as

$$-\frac{1}{8}\int d^4x\, \epsilon^{\mu\nu\rho\lambda}(if\bar{\eta}R_{\rho\lambda}{}^{cd}\gamma_{cd}\gamma_5\gamma_\nu\Psi_\mu). \tag{13.1.114}$$

We can now see that the terms linear in the gravitino cancel if $f = -1$. The terms cubic in the gravitino then cancel if one performs a Fierz reschuffle (see chapter 10 and the conventions of appendix A given in [1.11] for more details). We have used the identity $i\{\gamma_a\gamma_5, \gamma_{cd}\} = 2\epsilon_{acdb}\gamma^b$. Thus we find an invariant action subject to the constraint of equation (13.1.107).

As for the case of pure gravity one can solve the constraint of equation (13.1.107) for $w_\mu{}^{ab}$ in terms of e^a_μ. The result is precisely what one obtains by taking the equation of motion of $w_\mu{}^{ab}$ determined from the action with $f = -1$. Hence setting $R_{\mu\nu}{}^a = 0$ just takes the theory from first to second order formalism. We recognise the supersymmetry transformations of the gravitino and e^a_μ are those found in equation (13.1.31) with the auxiliary fields set to zero. The theory of supergravity with a cosmological constant obtained by gauging the super de Sitter group was subsequently found in [13.9]. The only real difference is that the gravity part of the action can be written as a term quadratic in field strengths.

The above method of deriving the supergravity action had one very substantial advantage: it led to a simple algebraic proof of its invariance under local supersymmetry transformations which previously used a computer. The trick is to notice that the variation of the action is just the variation of the fields multiplied by their field equations. In particular, in the variation of $w_\mu{}^{ab}$ one finds its own field equation, but this just algebraically expresses $w_\mu{}^{ab}$ in terms of the derivative of the vielbein and a term quadratic in the gravitino. To summarise, we can vary the action taking the spin connection to be varied separately and then vary the vielbein and gravitino that are contained within it; we then find this term vanishes as the variation multiplies the field equation for the spin connection which vanishes when we express it in terms of the vielbein and gravitino. Thus we need only vary the vielbein and the gravitino which are much simpler to deal with. This way of viewing the theory became known as 1.5 order formalism.

Similar to the case of pure gravity, the constraint of equation (13.1.107) implies that a translation gauge transformation leads to a general coordinate transformation plus a local Lorentz transformation and a supersymmetry transformation. Thus although setting $R_{\mu\nu}{}^a = 0$ does not preserve the gauge transformations of the theory, and in this respect adopting it is a bit of a sleight of hand, we do find a supergravity theory which is invariant under a set of transformations. It is, of course, the correct theory.

The construction of the theories of conformal supergravity was carried out using this gauging method [13.10]. The key to getting the gauge method to work is to guess the appropriate constraints. However, since these constraints break the original gauge transformations they are not always easy to find. Much effort has been devoted to developing the method discussed into a systematic procedure without any breaking of the underlying symmetries by hand. One such work was that of [13.20] where the full gauge symmetry was realised, but was spontaneously broken. The gauging of space-time groups has been useful in the study of higher spins and is the starting point for the Torino programme of constructing supergravity theories.

13.1.4 Dimensional reduction

Having constructed a supergravity theory we can find supergravity theories in lower dimensions by dimensionally reducing the theory on a torus. This preserves all the supersymmetries of the original theory. The resulting theory has many more fields and so is more complicated than the original theory. Given this complexity, it is often quicker to construct the lower dimensional theory using dimensional reduction rather than one of the methods given in the preceding sections. One surprise in carrying out this procedure is that the dimensionally reduced theories possess some unexpected symmetries which are associated with the scalar fields of the theory.

Let us first consider dimensional reduction on a one-dimensional torus, that is, a circle S^1. Let us take a theory in $d + 1$ dimensions and let us take the circle to have coordinate x_d and the remaining coordinates of the theory to be x^μ, $\mu = 0, 1, \ldots, d - 1$. We also adopt the convention that hatted indices run over all $d + 1$ indices, but unhatted indices only run over the d indices not associated with the circle, for example, $\hat{\mu}, \hat{\nu} = 0, 1, \ldots, d$, while $\mu, \nu = 0, 1, \ldots, d - 1$. We take the circle coordinate x_d to be such that $x'_d \sim x_d$ if $x'_d = x_d + 2\pi n R$, $n \in \mathbf{Z}$, where $2\pi R$ parameterises the range of x^d.

Given any of the fields in the $(d + 1)$-dimensional supergravity we can exploit its periodic dependence on the circle to take its Fourier transform on x_d. In particular, if ϕ represents any of these fields we find that

$$\phi(x^\mu, x_d) = \phi(x^\mu) + \sum_{n, n \neq 0} e^{in\theta} \phi_n(x^\mu), \qquad (13.1.115)$$

where $\theta = x_d/R$ for $0 \le \theta < 2\pi$. In this equation we have suppressed any possible vector or spinor indices that the field may possess. Thus from each particle in $d + 1$ dimensions we find an infinite number of particles in d dimensions. The non-zero modes (that is, $n \neq 0$), however, will lead to massive particles whose masses are given by their momentum in the circle direction and so are proportional to $1/R$. Such massive particles are called Kaluza–Klein particles. In the limit when the radius of the circle is very small these particles become infinitely massive and can be neglected, whereupon one is left with a finite set of massless particles, the zero modes. Discarding the massive particles can also be achieved by taking all the $(d+1)$-dimensional fields to be independent of x_d and in the rest of this section this is what we will do.

In this chapter we will largely be concerned with the dimensional reduction of the bosonic sector of supergravity theories whose actions contain gravity and usually gauge fields whose rank can be greater than 1. In what follows we will consider the dimensional reduction of such terms separately. Dimensional reduction has a long history [13.21].

We take the metric of the $(d+1)$-dimensional theory to be of the form

$$ds^2_{d+1} = e^{2\alpha_d\phi} ds^2_d + e^{-2(d-2)\alpha_d\phi}(dx_d + A_\mu dx^\mu)^2, \tag{13.1.116}$$

where

$$\alpha_d = \sqrt{\frac{1}{2(d-1)(d-2)}} \,. \tag{13.1.117}$$

This particular choice of constant α_d ensures that after performing the dimensional reduction we find Einstein's action and not the Riemann scalar multiplied by an exponential of ϕ and that the kinetic term for ϕ has the standard normalisation. The corresponding form of the metric is

$$g_{\hat\mu\hat\nu} = \begin{pmatrix} e^{2\alpha_d\phi} g_{\mu\nu} + e^{-2(d-2)\alpha_d\phi} A_\mu A_\nu & A_\mu e^{-2(d-2)\alpha_d\phi} \\ A_\mu e^{-2(d-2)\alpha_d\phi} & e^{-2(d-2)\alpha_d\phi} \end{pmatrix}, \tag{13.1.118}$$

and the vielbein and its inverse are

$$e_{\hat\mu}^{\hat m} = \begin{pmatrix} e^{\alpha_d\phi} e_\mu^m & e^{-(d-2)\alpha_d\phi} A_\mu \\ 0 & e^{-(d-2)\alpha_d\phi} \end{pmatrix}, \quad (e^{-1})_{\hat m}^{\hat\mu} = \begin{pmatrix} e^{-\alpha_d\phi} e_m^\mu & -A_\nu e^{-\alpha_d\phi}(e^{-1})_m^\nu \\ 0 & e^{(d-2)\alpha_d\phi} \end{pmatrix}. \tag{13.1.119}$$

We have used the local Lorentz symmetry to set $e_d{}^m = 0$.

The circle was parameterised by $0 \leq x_d \leq 2\pi R$. However, the parameter R has no physical meaning. The only physical distance is the diffeomorphism invariant distance around the circle, denoted $2\pi R^{phy}_{d+1}$, which is given by

$$2\pi R^{phy}_{d+1} = \int_0^{2\pi R} dx_d \sqrt{g_{dd}} = 2\pi R e^{-(d-2)\alpha_d\phi} \quad \text{or} \quad R^{phy}_{d+1} = e^{-(d-2)\alpha_d\phi} R. \tag{13.1.120}$$

We have used that ϕ is independent of x_d.

Performing the dimensional reduction on the Einstein term, we find that

$$\int d^{d+1}x e R = 2\pi R \int d^d x e \left(R - \tfrac{1}{2}\partial_\mu\phi\partial^\mu\phi - \tfrac{1}{4}e^{-2(d-1)\alpha_d\phi} F_{\mu\nu}F^{\mu\nu}\right), \tag{13.1.121}$$

where $F_{\mu\nu} = \partial_\mu A_\nu - \partial_\nu A_\mu$.

We now consider the dimensional reduction of a p-form field strength $F_{\hat{\mu}_1\cdots\hat{\mu}_p} = p\partial_{[\hat{\mu}_1}A_{\hat{\mu}_2\cdots\hat{\mu}_p]}$ in the $(d+1)$-dimensional theory. We take $A_{\hat{\mu}_1\cdots\hat{\mu}_{p-2}d} = A_{\mu_1\cdots\mu_{p-2}}$ and if none of the indices is in the circle direction we take the gauge field to be given by taking $\hat{\mu}_1 = \mu_1, \ldots, \hat{\mu}_{p-1} = \mu_{p-1}$ to find $A_{\mu_1\cdots\mu_{p-1}}$. Using this ansatz and equation (13.1.118) we find that

$$\int d^{d+1}x \frac{e}{2p!} F_{(p)}^2$$

$$= 2\pi R \int d^d x \left(\frac{e}{2p!} e^{-2(p-1)\alpha_d\phi} \hat{F}_{(m)}^2 + \frac{e}{2(p-1)!} e^{2(d-p)\alpha_d\phi} F_{(p-1)}^2 \right). \qquad (13.1.122)$$

It is to be understood in these equations that the fields are always those of the relevant dimension involved (for example, in equation (13.1.122) e refers to the $(d+1)$-dimensional volume element on the left-hand side and the d-dimensional volume element on the right-hand side). We also use the notation $F_{(p)}^2 = F_{m_1\cdots m_p}F_{n_1\cdots n_p}\eta^{m_1 n_1}\cdots\eta^{m_p n_p}$. In deriving these results we have used the above expressions for the vielbein to find that the field strengths in $d+1$ and d dimensions, referred to tangent space, are related by

$$F_{n_1\cdots n_{p-1}d} = e^{-\alpha_d(p-d+1)}F_{n_1\cdots n_{p-1}}$$

and

$$\hat{F}_{n_1\cdots n_p} \equiv (e^{-1})_{n_1}{}^{\hat{\mu}_1}\cdots(e^{-1})_{n_p}{}^{\hat{\mu}_p}F_{\hat{\mu}_1\cdots\hat{\mu}_p} = e^{-p\alpha_d}(F_{n_1\cdots n_p} - pF_{[n_1\cdots n_{p-1}}A_{n_p]}),$$

where the field strengths in d dimensions that appear on right-hand side of these equations are defined by $F_{\mu_1\cdots\mu_{p-1}} = (p-1)\partial_{[\mu_1}A_{\mu_2\cdots\mu_{p-1}]}$ and $F_{\mu_1\cdots\mu_p} = p\partial_{[\mu_1}A_{\mu_2\cdots\mu_p]}$.

We may repeat this procedure to obtain a dimensional reduction on an n-torus. From the metric $g_{\hat{\mu}\hat{\nu}}$ one finds the metric in the lower-dimensional theory $g_{\mu\nu}$, n vectors $g_{\mu i}$, $i = 1, \ldots, n$, and $n(n+1)/2$ scalars g_{ij}, $i, j = 1, \ldots, n$. The last includes n scalars ϕ_i, $i = 1, \ldots, n$ that are the diagonal components of the the metric which appear in the action as e^{ϕ_i} to some power. The vectors $g_{\mu i}$ are often called graviphotons. From a p-form gauge field $A_{\hat{\mu}_1\cdots\hat{\mu}_p}$ we get $A_{\mu_1\cdots\mu_p}$, $A_{\mu_1\cdots\mu_{p-1}i}$, $A_{\mu_1\cdots\mu_{p-2}ij}$, $\ldots, A_{1_1\cdots i_p}$, although some of the latter may vanish if $n < p$. Thus we find scalars from the gauge fields when all the indices are taken to lie on the torus.

The action for the field strengths in $m+1$ dimensions of the generic form

$$\int d^{m+1}x \frac{e}{2p!} e^{\vec{\beta}\cdot\vec{\phi}} F_{(p)}^2, \qquad (13.1.123)$$

where we have collected the ϕ_i up into a vector $\vec{\phi}$. However, once we reach $p+1$ dimensions, we can dualise any p-form field strengths into scalars. The action for such a field strength is of the form

$$\int d^{p+1}x \frac{e}{2p!} e^{\vec{\beta}\cdot\vec{\phi}} F_{(p)}^2. \qquad (13.1.124)$$

A completely equivalent action is given by

$$\int d^{p+1}x \frac{e}{2p!} e^{\vec{\beta}\cdot\vec{\phi}} F_{(p)}^2 + \frac{e}{p!} \varphi\partial_\mu F_{\nu_1\cdots\nu_p}\epsilon^{\mu\nu_1\cdots\nu_p}, \qquad (13.1.125)$$

where $F_{\mu_1 \cdots \mu_p}$ is to be thought of as an independent field. The Lagrange multiplier φ has an equation of motion that imposes the Bianchi identity on $F_{\mu_1 \cdots \mu_p}$, which in turn implies it is the curl of a $(p-1)$-form. Enforcing this we find the original action. However, we can also integrate by parts so that there are no derivatives acting on $F_{\mu_1 \cdots \mu_p}$ and eliminating $F_{\mu_1 \cdots \mu_p}$ by its algebraic equation of motion $F^{\nu_1 \cdots \nu_p} = e^{-\vec{\beta} \cdots \vec{\phi}} \epsilon^{\mu \nu_1 \cdots \nu_p} \partial_\mu \varphi$. Substituting this back into expression (13.1.125) leads to the equivalent action

$$\int d^{p+1}x \frac{e}{2} e^{-\vec{\beta} \cdot \vec{\phi}} \partial_\mu \varphi \partial^\mu \varphi, \tag{13.1.126}$$

which describes a scalar. Note the change in sign with which the vector $\vec{\beta}$ occurs and the fact that the new scalar field has a different symbol to those scalars arising from diagonal components of the metric, which are denoted by ϕ. The field strengths for which one carries out this duality change arise from field strengths in the higher dimensional theory. The one exception being the 2-form field strengths of the graviphoton which can be dualised in three dimensions and arise from the Riemann tensor. In three dimensions, as one can dualise all 2-form field strengths and so all vector gauge fields, we can express the result in three dimensions entirely in terms of scalar fields.

Thus if we compactify down to three dimensions we can write the action in the form

$$\int d^3x e \left[R - \tfrac{1}{2} \partial_\mu \vec{\phi} \cdot \partial^\mu \vec{\phi} - \tfrac{1}{2} \sum_{\vec{\alpha}} e^{\vec{\beta} \cdot \vec{\phi}} \partial_\mu \chi_{\vec{\beta}} \partial^\mu \chi_{\vec{\beta}} + \cdots \right]. \tag{13.1.127}$$

Here $\vec{\phi} = (\phi_1, \ldots, \phi_n)$, $\vec{\beta}$ are constant n-vectors and $\chi_{\vec{\beta}}$ are the other scalar fields which carry the index $\vec{\beta}$ to denote with which exponential factor of ϕ they occur. There are two type of scalars labelled by $\vec{\phi}$ and $\chi_{\vec{\beta}}$. The former arise from the diagonal components of the metric while the latter arise from off-diagonal components of the metric and components of the gauge fields. Some of the latter will have arisen from dualisation. The different origins of these two types of scalars result in them having different symmetries. The latter, that is, $\chi_{\vec{\beta}}$, possess remnants of general coordinate and gauge transformations that ensure that they can only have derivative interactions, that is, appear in the form $\partial_\mu \chi_{\vec{\beta}}$. On the other hand, the $\vec{\phi}$ scalars have non-derivative interactions, that is, apart from their kinetic term they appear as naked fields.

In fact, the above discussion is slightly over simplified as it neglects the effects of Chern–Simmons terms, which can mean that some field strengths do not satisfy the simple Bianchi identity $dF = 0$, but one with additional terms. This results in changes to expression (13.1.125). The net result of all this is to replace the kinetic terms $\partial_\mu \chi_{\vec{\alpha}}$ by $A_{\vec{\alpha}}^{\vec{\beta}} \partial_\mu \chi_{\vec{\beta}}$, where $A_{\vec{\alpha}}^{\vec{\beta}}$ is a non-degenerate matrix depending on the scalars $\chi_{\vec{\alpha}}$. Their presence is denoted by the $+ \cdots$ in expression (13.1.127), but these extra terms will play no role in subsequent discussions.

It is instructive to compute the number of scalars in a particular example. Let us consider eleven-dimensional supergravity, whose bosonic sector consists of gravity coupled to a 3-form gauge field, and dimensionally reduce this on an 8-torus. We find directly the scalar fields g_{ij} and A_{ijk} for $i, j, k = 1, \ldots, 8$, which makes $36 + 56$ scalars. However, in three dimensions we find the fields $g_{\mu i}$ and $A_{\mu ij}$ which dualise to give another $8 + 28$ scalars. Thus in all we have 128 scalars in three dimensions. As we will discuss in section 13.6 the resulting theory in three dimensions has scalars which belong to the non-linear realisation of E_8 with local subgroup SO(16). We note here that the dimension of this coset space is $248 - 120 = 128$ as it should be.

Rather than reduce on a n-dimensional torus in n steps as described above, one can perform the dimensional reduction in a single step. The reader can find an account of this as well as all the changes in the components of the Riemann tensor in, for example, [13.22].

13.2   Non-linear realisations

Given a quantum field theory with a rigid symmetry group G which is spontaneously broken to a subgroup H, then Goldstone's theorem [13.23] states that the theory possesses $D_G - D_H$ massless particles, called Goldstone particles, where D_G and D_H are the dimensions of the groups G and H, respectively. It is also true, in many circumstances, that the effective low energy theory is described by the non-linear realisation of G with local subgroup H which we are about to describe. The great advantage of this result is that if one is studying a phenomenon which one suspects results from a spontaneous symmetry breaking, then one can deduce the non-linear realisation from the low energy scattering data of the massless particles. Having deduced the non-linear realisation, one has discovered the symmetry G of the underlying theory. It is important to note that for this one did not need to have any understanding of the underlying dynamics. However, possessing the symmetry one may then use it to try to understand what the underlying dynamics could be. Non-linear realisations have played a very important role in our understanding of the symmetries in quantum field theories and in particular those of the strong interactions and particle physics in general.

Let G be a group and H one of its subgroups. In the group G we define the equivalence relation $g_1 \sim g_2$ if $g_2^{-1} g_1 \in H$. It is straightforward to show that this relation divides G into a set of equivalence classes which are disjoint and which all have the same number of elements. The set of equivalence classes is called the the coset space by G/H. Each equivalence class can be denoted by [g], where g is any element contained in the class under consideration and has the form $[g] = \{gh, \ \forall \ h \in H\}$. The formalism given below is quite general, but it is often applied to the case where G is a non-compact group and H is its maximally non-compact subgroups.

We first present the original formulation of non-linear realisations used in the classic papers [13.24]. Let us consider the coset space G/H with coset representatives $g(\phi)$, where ϕ parameterise the group element. If the generators of H are denoted by H_i and the remaining generators of G are T_a we may use the exponential parameterisation to choose $g(\phi) = \exp(\phi^a T_a)$ in a coordinate chart around the identity of the group. We take the group element to depend on space-time, that is, $\phi(x^\mu)$. These will turn out to be the fields in the resulting field theory. Under the action of the group we find that

$$g(\phi(x)) \rightarrow g_0 g(\phi(x)) = g(\phi'(x))h(\phi(x), g_0) \qquad (13.2.1)$$

or, in exponential parameterisation

$$\exp(\phi^a(x)T_a) \rightarrow g_0(\exp(\phi^a(x)T_a)) = \exp(\phi^{a'}(x)T_a)\exp(u^i(\phi(x), g_0)H_i).$$

$$(13.2.2)$$

The $h(\phi(x), g_0)$ part of the transformation is required to maintain our choice of coset representative.

The central problem in the theory of non-linear realisation is to construct an action, or equation of motion, that is invariant under the transformations of equation (13.2.1). The usual method is to consider the Cartan forms which belong to the Lie algebra and so can be written in the form

$$\Omega_\mu \equiv g^{-1}\partial_\mu g = f_\mu^a T_a + w_\mu^i H_i. \tag{13.2.3}$$

Under the transformations we find that Ω_μ transforms as

$$\Omega_\mu \to h^{-1}\partial_\mu h + h^{-1}\Omega_\mu h. \tag{13.2.4}$$

The theory simplifies if we restrict ourselves to reductive cosets which are those for which the commutator $[T_a, H_i] = c_{ai}{}^b T_b$, that is, the result of the commutator can be written in terms of only the coset generators T_a. Put another way the generators T_a form a representation of H. In this case the above transformation rule implies that

$$f_\mu \equiv f_\mu^a T_a \to h^{-1} f_\mu h, \quad w_\mu \equiv w_\mu^i H_i \to h^{-1} w_\mu h + h^{-1}\partial_\mu h. \tag{13.2.5}$$

We can think of f_μ^a as a vielbein on G/H defining a set of preferred frames and w_μ^i as the connection associated with H transformations.

Although the fields depend on space-time, this has played no role in the group theory and is essentially a dummy variable over which we will integrate. In some important applications, which we will consider later, space-time will arise due to some of the generators of the algebra used in the non-linear realisation, one example being the Poincaré group.

An invariant action is given by

$$\int d^D x\, \eta^{\mu\nu} Tr(f_\mu f_\nu). \tag{13.2.6}$$

In this action, and what follows, we take the generators to be in a suitable matrix representations and then the trace is just the usual matrix trace. In fact, up to a constant, the trace is independent of the representation used as the algebra has a unique, up to a constant, invariant scalar product, namely the Killing metric. The corresponding equation of motion is given by

$$D_\mu f_a^\mu = 0, \tag{13.2.7}$$

where we have introduced the covariant derivative $D_\mu f_\nu \equiv \partial_\mu f_\nu + [w_\mu, f_\nu]$. It is straightforward to see using equation (13.2.5) that the above Lagrangian and so the equation of motion are invariant under the symmetry of equation (13.2.1)

In fact, the above choice of action is not always unique. If the representation of H carried by T_a, or equivalently f_μ, is reducible, then we can also consider the trace of $(f_\mu)^2$ for each irreducible representation separately and so in the action we get one arbitrary constant for each irreducible representation.

Matter fields can be incorporated in the non-linear realisation. Consider a matter field E that transforms under some linear representation of G, that is, $E \to E' = U(g_0)E = D(g_0^{-1})E$ under $g_0 \in G$, where D acts as a matrix on the indices of E. From this representation we can construct one which transforms non-linearly by taking $C(\phi) = D(g(\phi(x))^{-1})E$.

Under the action of the group

$$U(g_0)C(\phi) = U(g_0)(D(g(\phi(x))^{-1})E) = D(g(\phi(x))^{-1})U(g_0)E$$
$$= D(g(\phi(x))^{-1})D(g_0^{-1})E = D((g_0g(\phi(x)))^{-1})E$$
$$= D(h(\phi, g_0))D(g(\phi'(x))^{-1})E = D(h(\phi, g_0))C(\phi').$$

In carrying out this step we have used the D as a representation, that is, $D(g_1)D(g_2) = D(g_1g_2)$. The reader may be wondering why we find $D(h)$ acting on C and not $D(h^{-1})$ as one might expect for the passive interpretation we are using. It is due to our use of the formula $g \to g_0gh$ and the resulting group composition rules for the two hs. In particular, the hs occur in the opposite order to what one might expect, that is, $h_2h_1 = h_3$, where the $g_0^1g = g'h_1^{-1}$ etc. One finds $D(h^{-1})$ if one adopts the convention $g \to g_0gh^{-1}$. Clearly, one can invert the procedure and go from the non-linear realisation to the linear realisation by taking $E = D(g(\phi(x)))C(\phi)$. The covariant derivative of C is given by $D_\mu C = \partial_\mu C - \omega_\mu C$, where ω_μ is taken in the representation of H with which C transforms. It transforms as $D_\mu C \to D(h)D_\mu C$.

There is another more general way to formulate non-linear realisations that is used more often, but is largely equivalent to the historical method given above. We consider the elements of the group G to depend on space-time, that is, $g(x^\mu)$, and will construct a theory built from $g(x^\mu)$ which is invariant under

$$g(x^\mu) \to g_0g(x^\mu)h(x^\mu), \tag{13.2.8}$$

where $h \in H$ is a local (that is, space-time dependent) transformation and $g_0 \in G$ is a rigid transformation (that is, does not depend on space-time). We note that the rigid and local transformations may be carried out completely independently. This defines what the non-linear realisation is and one only has to find what is dynamics that is invariant under this symmetry. The most frequently used approach is to build the dynamics out of the Cartan forms:

$$\Omega_\mu \equiv g^{-1}\partial_\mu g \equiv f_\mu^a T_a + w_\mu^i H_i. \tag{13.2.9}$$

Under the symmetry transformations of equation (13.2.8) the objects in equation (13.2.9) are invariable under the rigid g_0 transformations, but under the local h transformation $\Omega_\mu \to h^{-1}\partial_\mu h + h^{-1}\Omega_\mu h$. The sub-algebra and coset pieces transform as

$$f_\mu \equiv f_\mu^a T_a \to h^{-1}f_\mu h, \quad \omega_\mu \to h^{-1}\omega_\mu h + h^{-1}\partial_\mu h, \tag{13.2.10}$$

if we restrict ourselves to the reductive cosets mentioned above. In terms of an exponential parameterisation we may write $g = \exp(\phi^a(x)T_a)\exp(u^i(x)H_i)$, where the generators of H are denoted by H_i and the remaining generators of G are denoted by T_a. An invariant action is given by

$$\int d^D x \, \eta^{\mu\nu} Tr(f_\mu f_\nu). \tag{13.2.11}$$

One can use the local H transformation of equation (13.2.8) to bring the group element to the form $g = \exp(\phi^a(x)T_a)$, that is, choose a particular coset representative. The rigid g_0 transformation of equation (13.2.8) then requires a compensating local transformation to maintain this choice with the result that one finds the transformation law of equation (13.2.2).

In fact, having made this particular gauge choice the formulation of the non-linear realisation being discussed here just becomes that of the historical formulation given initially. It is important to note that once one has made this gauge choice the Cartan forms are not invariant under the rigid g_0 transformations as we are required to make a local compensating transformation to maintain the choice of coset and this does not leave the Cartan forms invariant.

One can further extend the last formulation to find a third example [13.25]. In the formulation of the non-linear realisation of equation (13.2.8) we introduce a gauge field A_μ which is in the Lie algebra of H, that is, $A_\mu = A_\mu^i H_i$, which is taken to transform under the local subgroup as

$$A_\mu \to h^{-1} A_\mu h - \partial_\mu h^{-1} h \tag{13.2.12}$$

and is inert under the rigid G transformations. Using the gauge field we can define a covariant derivative by

$$D_\mu g = \partial_\mu g - g A_\mu. \tag{13.2.13}$$

which transforms as

$$D_\mu g \to (D_\mu g) h. \tag{13.2.14}$$

Then the modified Cartan forms

$$g^{-1} D_\mu g \tag{13.2.15}$$

are invariant under the rigid transformation g_0 of equation (13.2.8), but transform under local H transformations as

$$g^{-1} D_\mu g \to h^{-1} (g^{-1} D_\mu g) h. \tag{13.2.16}$$

An invariant action is given by

$$\int d^D x \mathrm{Tr}\{(g^{-1} D_\mu g)(g^{-1} D_\nu g)\} \eta^{\mu\nu}. \tag{13.2.17}$$

We observe that A_μ occurs in the action without any space-time derivative and so can be eliminated by its equation of motion. Varying A_μ we find its equation of motion is given by

$$\mathrm{Tr}\{\delta A^\mu (g^{-1} \partial_\mu g - A_\mu)\} = 0. \tag{13.2.18}$$

If we write

$$g^{-1} \partial_\mu g = A_\mu + P_\mu, \tag{13.2.19}$$

then, since δA^μ is an arbitrary element of the Lie algebra of H, the equation of motion tells us that P_μ is in the orthogonal to the Lie algebra of H and so can be written in terms of the coset generators T_a. Using equation (13.2.19) we may eliminate A_μ algebraically from the action to find

$$\mathrm{Tr} \int d^D x P_\mu P_\nu \eta^{\mu\nu}. \tag{13.2.20}$$

This is the same action as expression (13.2.6), or expression (13.2.11), since comparing equations (13.2.9) and (13.2.19) we conclude that P_μ and $f_\mu^a T_a$ are the equal and hence so are $w_\mu^i H_i$ and A_μ once we use the equation of motion of the latter. Consequently, the three methods of constructing the non-linear realisation are equivalent. Their equations of motion, which for the action of expression (13.2.17) is given by $D_\mu P^\mu = \partial_\mu P^\mu + [A_\mu, P^\mu] = 0$, agree.

Incorporating matter fields into the above formulation of the non-linear realisation proceeds much like before. Given a linear realisation E we define $C(\phi) = D(g(\phi(x))^{-1})E$ which transforms as $C(\phi) \to D(h)C(\phi')$. The covariant derivative is given by $D_\mu C = \partial_\mu C - A_\mu C$ and it transforms as $D_\mu C \to D(h)D_\mu C$.

As the action of expression (13.2.20), or the previous equivalent ones, is invariant under the symmetry group G we expect to have corresponding conserved currents. The local transformations of equation (13.2.8) give rise to identically conserved current. However, the rigid transformations of equation (13.2.8) give rise to conserved currents that can be found by making the rigid transformations space-time dependent and identifying the coefficient of the space-time derivative of the now local parameter. Under $g \to g_0(x)g$ we note from equation (13.2.19) that

$$\delta A_\mu + \delta P_\mu \to g^{-1}(g_0^{-1}\partial_\mu g_0)g. \tag{13.2.21}$$

The variation of the action of expression (13.2.20) is then given by

$$2\text{Tr}\int d^D x P_\mu \delta P_\nu \eta^{\mu\nu} = 2\text{Tr}\int d^D x P_\mu (\delta A_\nu + \delta P_\nu)\eta^{\mu\nu}$$

$$= 2\text{Tr}\int d^D x P_\mu g^{-1}(g_0^{-1}\partial_\nu g_0)g\eta^{\mu\nu}$$

$$= 2\text{Tr}\int d^D x g P_\mu g^{-1}(g_0^{-1}\partial_\nu g_0)\eta^{\mu\nu}. \tag{13.2.22}$$

In the second line of this equation we used the fact that A_μ is orthogonal to P_μ. As $(g_0^{-1}\partial_\mu g_0)$ is an arbitrary element of the Lie algebra of G we can identify the current as

$$J_\mu = g P_\mu g^{-1}. \tag{13.2.23}$$

It is straightforward to verify that it is indeed conserved by virtue of the equations of motion.

A particular case of non-linear realisations that will be important for us is when G is a finite dimensional semi-simple Lie group and we make a particular choice of local subgroup that is related to the Cartan involution. This involution and the theory of Kac–Moody algebras is discussed in detail in chapter 16, but we give the basic equations here in order to carry out the non-linear realisation. The generators of the group G consist of those of the Cartan sub-algebra $\vec{H}$, which by definition commute, and the remaining generators $E_{\vec{\alpha}}$ whose commutators with $\vec{H}$ can be 'diagonalised' to be of the form

$$[\vec{H}, E_{\vec{\alpha}}] = \vec{\alpha}E_{\vec{\alpha}}, \tag{13.2.24}$$

where $\vec{\alpha}$ are the roots of the algebra. This way of writing the algebra is called the Cartan–Weyl basis. We may split the roots into positive and negative ones: a root is positive (negative) if its first non-vanishing element in a chosen basis is positive (negative). In addition the positive roots can be written as linear combinations, with non-negative integer coefficients, of the so-called simple roots. Essentially the simple roots are a basis for the positive roots with positive integers as coefficients. The number of simple roots is equal to the rank of the algebra and so the number of Cartan sub-algebra elements $\vec{H}$. The Cartan matrix is given in terms of the simple roots α_a by

$$A_{ab} = 2\frac{\vec{\alpha}_a \cdot \vec{\alpha}_b}{\vec{\alpha}_a \cdot \vec{\alpha}_a}. \qquad (13.2.25)$$

We will explain in chapter 16 that the commutators of the Cartan sub-algebra generators and the generators associated with the simple roots and the negative simple roots satisfy a number of relations, called the Serre relations, which are completely determined by the Cartan matrix. Furthermore, given these Serre relations one can deduce the complete Lie algebra. In other words, given the Serre relations the full Lie algebra is uniquely determined. Thus the full Lie algebra is encoded in the Cartan matrix. It turns out that the Cartan matrix for a finite dimensional semi-simple Lie algebra is integer valued with 2s on the diagonal and has off-diagonal entries which are negative integers or zero, with the zeros are symmetrically distributed and finally the Cartan matrix is positive definite, that is, $v^T A v > 0$ for any real vector v. Furthermore, every matrix of this kind leads to one of the finite dimensional semi-simple Lie algebras, that is, one of the algebras in the so-called Cartan list. Thus there is a one-to-one map between matrices of this kind and finite dimensional semi-simple Lie algebra

Finite dimensional semi-simple Lie algebras are invariant under the Cartan involution

$$I : (E_{\vec{\alpha}}, \vec{H}) \rightarrow -(E_{-\vec{\alpha}}, \vec{H}). \qquad (13.2.26)$$

It is an involution and so by definition $I^2 = 1$ and $I(AB) = I(A)I(B)$ for any two elements A and B of the algebra. It follows that this operation preserves the commutation relations of the algebra, in particular equation (13.2.24). The set of Cartan involution invariant generators is given by $E_{\vec{\alpha}} - E_{-\vec{\alpha}}$ and these generate a Cartan involution invariant sub-group. If we denote G to be the group, then $I(G)$ denotes the Cartan involutions sub-group and similarly for the algebra. The Cartan involution invariant sub-groups of SU(n), SO(n, n), E(6) E_7 and E_8 are SO(n), SO(n) $\otimes$ SO(n), USp(8), SU(8) and SO(16), respectively. The reader may check that the number of generators is the required number. For example, for SU(n) the number of generators in the Cartan invariant sub-algebra is the number in SU(n) minus the number of Cartan sub-algebra generators divided by 2. This is because there are equal numbers of positive and negative root generators and the total number is the number of generators minus those in the Cartan sub-algebra. This is $(n^2 - 1 - (n - 1))/2 = n(n - 1)/2$, which is the number of generators in SO(n), as it should be. We note that the Cartan involution invariant sub-algebra does not include the Cartan generators of the original algebra.

Let us now consider a non-linear realisation of a semi-simple finite dimensional group with the local sub-group being the Cartan involution invariant subgroup. Adopting the second of the two approaches given above we have $g \rightarrow g_0 g h$ of equation (13.2.8). The general group element may be written in the form

$$g = e^{\sum_{\vec{\alpha}>0} \chi^{\vec{\alpha}} \cdot E_{\vec{\alpha}}} e^{-\frac{1}{\sqrt{2}} \vec{\phi} \cdot \vec{H}} e^{\sum_{\vec{\alpha}<0} \chi^{\vec{\alpha}} \cdot E_{\vec{\alpha}}},$$

where $\alpha > 0$ means all positive roots and $\alpha > 0$ all negative roots. Using the local h transformation the group element can be written in the form

$$g = e^{\sum_{\vec{\alpha}>0} \chi^{\vec{\alpha}} \cdot E_{\vec{\alpha}}} e^{-\frac{1}{\sqrt{2}} \vec{\phi} \cdot \vec{H}}, \qquad (13.2.27)$$

where the sum is only over the positive roots.

We will now construct the dynamics using the Cartan forms. $\Omega_\mu = g^{-1}\partial_\mu g$ and we can write it as

$$\Omega_\mu = P_\mu + Q_\mu, \quad \text{where} \quad P_\mu = \tfrac{1}{2}(\Omega_\mu - I(\Omega_\mu)), \quad Q_\mu = \tfrac{1}{2}(\Omega_\mu + I(\Omega_\mu)). \quad (13.2.28)$$

These then transform under the transformations of equation (13.2.8) as

$$P_\mu \to h^{-1}P_\mu h, \quad Q_\mu \to h^{-1}Q_\mu h + h^{-1}\partial_\mu h. \quad (13.2.29)$$

It follows that $I(P_\mu) = -P_\mu$ and $I(Q_\mu) = Q_\mu$, so these objects are in the coset and local sub-algebra, respectively. Starting from the group element of equation (13.2.27) we find that

$$\Omega_\mu = -\frac{1}{\sqrt{2}}\partial_\mu\vec{\phi}\cdot\vec{H} + \sum_{\vec{\alpha}>0}\partial\chi^{\vec{\alpha}}\cdot E_{\vec{\alpha}}e^{\frac{1}{\sqrt{2}}\vec{\alpha}\cdot\vec{\phi}} + \cdots \quad (13.2.30)$$

where $+\cdots$ stands for terms of higher order in $\chi^{\vec{\alpha}}$. An invariant action is therefore given by $-\int d^d x\,\mathrm{Tr}(P_\mu P^\mu)$. We note that $D_\mu P_\nu \equiv \partial_\mu P_\nu + [Q_\mu, P_\nu]$ transforms covariantly, $D_\mu P_\nu \to h^{-1}D_\mu P_\nu h$ and that the equation of motion is $D_\mu P_\nu \eta^{\mu\nu} = 0$.

The generators of G can be taken to obey

$$\mathrm{Tr}(H_i H_j) = \delta_{ij}, \quad \mathrm{Tr}\left(E_{\vec{\alpha}}E_{-\vec{\beta}}\right) = \delta_{\vec{\alpha},\vec{\beta}}, \quad (13.2.31)$$

where $c_{\vec{\alpha}}$ are constants; for a more detailed discussion of this point see chapter 12. Taking into account that the trace is non-vanishing only for two generators whose root sum is zero we find that the action

$$-\int d^d x\,\mathrm{Tr}(P_\mu P^\mu) = \int d^d x\left(-\tfrac{1}{2}\partial_\mu\vec{\phi}\cdot\partial^\mu\vec{\phi}\right.$$

$$\left. -\tfrac{1}{2}\sum_{\vec{\alpha}>0}e^{+\sqrt{2}\vec{\phi}\cdot\vec{\alpha}}c_{\vec{\alpha}}\partial_\mu\chi^{\vec{\alpha}}\cdot\partial^\mu\chi^{\vec{\alpha}} + \cdots\right). \quad (13.2.32)$$

The $+\cdots$ means the addition of terms of higher order in $\chi^{\vec{\alpha}}$. Neglecting such terms results from evaluating only terms resulting from commutators of $\vec{H}$ and $E_{\vec{\alpha}}$ or $\vec{H}$ and $E_{-\vec{\alpha}}$, but not commutators of the non-zero root generators with themselves. We note that while the fields $\chi^{\vec{\alpha}}$ appear only with space-time derivatives those associated with the Cartan sub-algebra generators can appear naked and they multiply the kinetic term of the $\chi^{\vec{\alpha}}$ fields by an exponential containing the roots of the algebra. We will return to discuss the similarity between this last equation and equation (13.1.127) in section 13.1.

It is instructive to compute the action for the non-linear realisation with group SL(2, **R**) with local subgroup SO(2). The generators in Cartan–Weyl basis are E, F, H, which obey $[H, E] = \sqrt{2}E$, $[H, F] = -\sqrt{2}F$ and $[E, F] = \sqrt{2}H$. The general group element is given by $g = e^{Ex}e^{-(\phi/\sqrt{2})H}e^{uF}$. We may use the local sub-algebra, which is generated by $E - F$, to choose the group element to be of the form $g = e^{Ex}e^{-(\phi/\sqrt{2})H}$. The Cartan form is then given by $\Omega_\mu = -\frac{1}{\sqrt{2}}\phi H + \partial_\mu\chi e^\phi E$ and so

$$P_\mu = -\frac{1}{\sqrt{2}}\partial_\mu\phi H + \partial_\mu\chi e^\phi\frac{(E+F)}{2} \equiv P_{\mu 1}H + P_{\mu 2}\frac{(E+F)}{\sqrt{2}},$$

$$Q_\mu = \partial_\mu\chi e^\phi\frac{(E-F)}{2}. \quad (13.2.33)$$

Under a local transformation $h = e^{\theta(E-F)/\sqrt{2}}$, the Cartan forms transform as $P_\mu \to h^{-1}P_\mu h$ and so $P_{\mu 1} \to \cos\theta P_{\mu 1} + \sin\theta P_{\mu 2}$, $P_{\mu 2} \to \cos\theta P_{\mu 2} - \sin\theta P_{\mu 1}$. As a result we find that the complex quantity $P_\mu \equiv \frac{1}{\sqrt{2}}(P_{\mu 1} + iP_{\mu 2}) = -ie^\phi \partial_\mu(-\chi + ie^{-\phi})$ transforms as $P_\mu \to e^{-i\theta}P_\mu$. An invariant action is then given by

$$\int d^d x \frac{1}{2}\mathrm{Tr}P_\mu (P_\nu)^* \eta^{\mu\nu} = \int d^d x \left(-\frac{1}{2}\partial_\mu\phi\partial^\mu\phi - \frac{1}{2}\partial_\mu\chi\partial^\mu\chi e^{2\phi}\right)$$

$$= -\int d^d x \frac{1}{2}\frac{|\partial_\mu\tau|^2}{(\mathrm{Im}\tau)^2}, \qquad (13.2.34)$$

where in the last equality we have used that $P_\mu = -i\partial_\mu\tau/\mathrm{Im}\tau$, where $\tau = -\chi + ie^\phi$. The reader can verify that the general construction of the action of equation (13.2.32) gives the same result.

For the choice of sub-group we are considering there is another way to construct the dynamics. We first define the operation denoted #, which consists of the inverse followed by the Cartan involution. Hence, on a group element g it is given by $g^\# = I(g^{-1})$ and on an element of the algebra by $A^\# = -I(A)$, but for two elements A and B of the algebra $(AB)^\# = I(B)I(A)$ as the inverse it is given by inverts the order of elements. On the algebra it acts as $(E_{\vec{a}})^\# = E_{-\vec{a}}$ and $(H_i)^\# = H_i$. For orthogonal groups # coincides with the transpose. We note that $(\exp(E_{\vec{a}}))^\# = \exp(E_{-\vec{a}})$ and so $(\exp(E_{\vec{a}} - E_{-\vec{a}}))^\# = \exp(-(E_{\vec{a}} - E_{-\vec{a}}))$, or $h^\# = h^{-1}$ if h is in the Cartan involution invariant sub-group, that is, the local subgroup. If we define $\mathcal{M} = gg^\#$, then under equation (13.2.8) it transforms as

$$\mathcal{M} \to g_0 \mathcal{M} g_0^\# \qquad (13.2.35)$$

and in particular it is invariant under the local sub-group. In contrast to the method used above to construct the dynamics, here we first solve the problem of finding objects invariant under the local sub-algebra and from these construct objects invariant under the rigid G transformations. Clearly, the Lagrangian

$$\mathcal{L}_{G/H} = \tfrac{1}{4}\mathrm{Tr}\left(\partial_\mu\mathcal{M}\partial^\mu\mathcal{M}^{-1}\right) = -\tfrac{1}{4}\mathrm{Tr}\left(\mathcal{M}^{-1}\partial_\mu\mathcal{M}\mathcal{M}^{-1}\partial^\mu\mathcal{M}\right) \qquad (13.2.36)$$

is invariant under rigid G transformations and local H transformations.

Using the form of g of equation (13.2.27) we find that

$$\mathcal{M} = e^{\sum_{\vec{a}>0}\chi^{\vec{a}}\cdot E_{\vec{a}}} e^{-\sqrt{2}\vec{\phi}\cdot\vec{H}} e^{\sum_{\vec{a}>0}\chi^{\vec{a}}\cdot E_{-\vec{a}}}. \qquad (13.2.37)$$

The action of equation (13.2.36) is equal to the action of equation (13.2.32) found using the previous method, up to a numerical factor. This can be proved by straightforward evaluation of the two quantities in terms of $\vec{\phi}$ and χ, or by evaluating them in terms g and using that $\mathrm{Tr}(A^\# B^\#) = \mathrm{Tr}(AB)$ for any two elements A and B of the algebra. The latter identity follows from the fact that the trace, or equivalently the scalar product on the algebra, is invariant under the Cartan involution.

Let us work out one of the simplest examples, the non-linear realisation of $G = SU(1, 1)$ with local subgroup $H = U(1)$. The group $G = SU(1, 1)$ is the set of two by two matrices g of determinant 1 which acts on the column vector $\begin{pmatrix} z_1 \\ z_2 \end{pmatrix}$ by $\begin{pmatrix} z_1 \\ z_2 \end{pmatrix} \to g\begin{pmatrix} z_1 \\ z_2 \end{pmatrix}$ in such a way as to preserve $|z_1|^2 - |z_2|^2$. The most general element of SU(1, 1) can be written in the form

$$g = \begin{pmatrix} u & v \\ v^* & u^* \end{pmatrix},$$

(13.2.38)

subject to $uu^* - vv^* = 1$. The entries of this matrix, u, v and their complex conjugates, are taken to be functions of space-time. The U(1) sub-group is all elements of the form

$$h = \begin{pmatrix} e^{-2i\theta} & 0 \\ 0 & e^{+2i\theta} \end{pmatrix}.$$

(13.2.39)

An infinitesimal element of $G = SU(1, 1)$ can therefore be written in the form $g = I + A$, where A is given by

$$A = -2ia\sigma_3 + b_1\sigma_1 - b_2\sigma_2 = \begin{pmatrix} -2ia & b \\ b^* & +2ia \end{pmatrix},$$

(13.2.40)

where a, b_1 and b_2 are real, $b = b_1 + ib_2$ and σ_i, $i = 1, 2, 3$, are the Pauli matrices.

An alternative parameterisation of elements of $G = SU(1, 1)$ is given by exponentiating the above infinitesimal element;

$$e^A = \begin{pmatrix} \cosh\rho - 2ia\dfrac{\sinh\rho}{\rho} & b\dfrac{\sinh\rho}{\rho} \\ b^*\dfrac{\sinh\rho}{\rho} & \cosh\rho + 2ia\dfrac{\sinh\rho}{\rho} \end{pmatrix},$$

(13.2.41)

where $\rho^2 = b^*b - 4a^2$. The U(1) sub-group is generated by $i\sigma_3$ and its elements take the form

$$h = \begin{pmatrix} e^{-2ia} & 0 \\ 0 & e^{+2ia} \end{pmatrix}.$$

(13.2.42)

The transformation of g under the non-linear realisation is given by equation (13.2.8), namely $g \to g_0 g h$.

In this particular example we may resort to a trick to find the transformations of the fields under the rigid and local transformations. If we write $g = (C_+, C_-)$, where C_+ and C_- are two column vectors, then $C_\pm$ transforms as $C_\pm \to gC_\pm e^{\mp 2i\theta}$ under local U(1) transformations. Taking the ratio of the top and bottom components of the column vectors $C_\pm$ and denoting it by $c_\pm$, it follows that $c_\pm$ is inert under local U(1) transformations, but transforms as

$$c_\pm \to \frac{u_0 c_\pm + v_0}{v_0^* c_\pm + u_0^*}$$

(13.2.43)

under a rigid SU(1,1) transformation which we may write in the form of equation (13.2.38), but with all entries possessing a subscript 0.

We can use a local U(1) transformation to bring g to be of the form

$$g = \frac{1}{\sqrt{1 - \varphi\varphi^*}} \begin{pmatrix} 1 & \varphi \\ \varphi^* & 1 \end{pmatrix}.$$

(13.2.44)

This choice is most easily achieved using the form of SU(1, 1) elements given in equation (13.2.41). Examining the second column vector in g we find that $c_- = \varphi$ and so φ transforms as under G as

$$\varphi \to \frac{u_0 \varphi + v_0}{v_0^* \varphi + u_0^*}. \tag{13.2.45}$$

The Cartan forms $\Omega_\mu = g^{-1} \partial_\mu g = P_\mu + Q_\mu$ given in equation (13.2.9) belong to the Lie algebra of SU(1,1) with Q_μ in the U(1) local sub-algebra and P_μ in the coset. As a result we may write

$$\Omega_\mu = \begin{pmatrix} 2iQ_\mu & P_\mu \\ \bar{P}_\mu & -2iQ_\mu \end{pmatrix}, \tag{13.2.46}$$

where $\bar{P}_\mu = (P_\mu)^*$. Under an infinitesimal local U(1) transformation of equation (13.2.42) we find that they transform as

$$\delta Q_\mu = -\partial_\mu a, \quad \delta P_\mu = 4ia P_\mu. \tag{13.2.47}$$

For the choice of g of equation (13.2.44) we find that

$$Q_\mu = -\frac{i}{4} \frac{(-\varphi \partial_\mu \varphi^* + \varphi^* \partial_\mu \varphi)}{(1 - \varphi \varphi^*)}, \quad P_\mu = \frac{\partial_\mu \varphi}{(1 - \varphi \varphi^*)}. \tag{13.2.48}$$

The invariant action is then given by

$$-\frac{1}{2} \int d^D x P_\mu \bar{P}^\mu = -\frac{1}{2} \int d^D x \frac{\partial_\mu \varphi \partial^\mu \varphi^*}{(1 - \varphi \varphi^*)^2}. \tag{13.2.49}$$

The corresponding equation of motion is found by substituting the expressions of equations (13.2.48) into $D_\mu P^\mu \equiv \partial_\mu P^\mu + 4i Q_\mu P^\mu = 0$ or varying the above action.

As we will discuss in section 13.5 the two scalars that occur in the IIB supergravity theory belong to the coset SU(1,1) with local sub-algebra U(1). As such, they must have an equation of motion that is given by the above equation decorated by other terms containing the superpartners of the scalars.

So far we have considered non-linear realisation of groups that did not contain any generators that were associated with space-time. In other words we have considered internal groups and any symmetries that space-time might possess were imposed upon the non-linear realisation as an additional requirement when constructing the dynamics. However, one can also construct non-linear realisations when the group does contain space-time generators. Indeed, one of the most used constructions is when the group contains only generators associated with space-time. As before the group element transforms as in equation (13.2.1), or equation (13.2.8). One of the simplest and most well-known examples is to take the Poincaré group as the group of the non-linear realisation and the local sub-group to be the Lorentz group with generators J_{ab}. For concreteness we recall the commutation relations of the Poincaré group:

$$[J_{ab}, P_c] = \eta_{bc} P_a - \eta_{ac} P_b, \quad [P_a, P_b] = 0. \tag{13.2.50}$$

In this case, the coset generators are just the translation generators P_a and the group element can be written in the form

$$g = \exp(x^a P_a). \tag{13.2.51}$$

Carrying out the transformations induced by rigid group elements belonging to the Poincaré group we find that x^a transforms by a translation and a Lorentz transformation, that is, under a Poincaré transformation, and so we may interpret x^a as the coordinates of Minkowski space. Another celebrated example is to take the group to be that generated by a super-symmetry algebra. If one takes all the generators of the supersymmetry algebra to be in the local sub-algebra except those of the space-time translations and supercharges, then one finds the traditional superspace with coordinates x^a, θ^i_α. There are other choices of local sub-algebra and then one finds other interesting superspaces. There is a very large literature on such constructions and so we will not pursue this here. We note that in these two examples the parameterisations of the group elements are interpreted as the coordinates of the space-time.

One can also consider the non-linear realisations of groups that contain space-time generators [13.26] but have additional generators which are not in the local sub-algebra and lead instead to fields defined on space-time. Let us consider a non-linear realisation involving the conformal group [13.27]. The commutation relations of the conformal algebra are those of the Poincaré group plus

$$[J_{ab}, K_c] = -\eta_{ac}K_b + \eta_{bc}K_a, \quad [P_a, D] = P_a, \quad [K_a, D] = -K_a, \quad [J_{ab}, D] = 0,$$

$$[K_a, K_b] = 0, \quad [P_a, K_b] = +2\eta_{ab}D - 2J_{ab}, \tag{13.2.52}$$

together with the other commutators that vanish. The group element g is taken to transform as usual, that is, $g \to g_0 g h$, where h is a local transformation that belongs to the local sub-group. We take the local sub-algebra to be the Lorentz algebra and so our coset representative can be chosen as

$$g = e^{x^a P_a} e^{\phi^a(x)K_a} e^{\sigma(x)D}. \tag{13.2.53}$$

We note that we have not treated the coefficients of the coset generators in an equal way as those of P_a are by definition the space-time coordinates, but the coefficients of the remaining generators are functions of space-time.

It is straightforward to show that the Cartan forms are given by

$$\Omega = g^{-1}dg = dx^a(e^\sigma P_a + e^{-\sigma}(\partial_a\phi^b - \phi^c\phi_c\delta^b_a + 2\phi_a\phi^b)K_b$$

$$+ (\partial_a\sigma + 2\phi_a)D + (-\delta^c_a\phi^d + \delta^d_a\phi^c)J_{cd})$$

$$\equiv dx^\mu(e_\mu{}^a P_a + G_{\mu b}K^b + G_\mu D + \tfrac{1}{2}\omega^{cd}_\mu J_{cd}). \tag{13.2.54}$$

Under the transformation $g \to g_0 g h$ we find that $\Omega \to h^{-1}\Omega h + h^{-1}dh$ and is inert under the rigid g_0 transformations. As such, under a Lorentz transformation $e_\mu{}^a = e^\sigma \delta^a_\mu$ transforms as a vector on its a index and $\omega^{cd}_\mu = \tfrac{1}{2}(-\delta^c_a\phi^d + \delta^d_a\phi^c)$ transforms like a Lorentz spin connection. The μ indices are not transformed under local Lorentz transformations. We may then identify $e_\mu{}^a$ as a vielbein and ω^{cd}_μ as a spin connection. The remaining Cartan forms transform under local Lorentz transformation as their indices suggest, that is, $G_{\mu b}$ is a vector and G_μ a scalar.

Although $\Omega = dx^\mu \Omega_\mu$ is indeed invariant under the rigid transformation, up to the compensating local Lorentz transformations, Ω_μ is not since x^μ are not inert. Quantities that transform under just Lorentz transformation are given by $\Omega_a \equiv (e^{-1})_a{}^\mu \Omega_\mu = e^{-\sigma}\delta^\mu_a \Omega_\mu$

and we may use these to build our dynamics. The resulting quantities are

$$\nabla_a \sigma = e_a{}^\mu \nabla_\mu \sigma = e^{-\sigma}(\partial_a \sigma + 2\phi_a), \quad \nabla_a \phi_b = e^{-2\sigma}(\partial_a \phi^b - \phi^c \phi_c \delta_a^b + 2\phi_a \phi^b).$$

$$(13.2.55)$$

As $\nabla_a \sigma$ transforms covariantly under the Lorentz group and is inert under rigid transformations, it can be set to zero and still preserve all the symmetries of the non-linear realisation. As a result, we can eliminate ϕ_a in terms of $\partial_a \sigma$, by setting $\nabla_a \sigma = 0$, namely $2\phi_a = -\partial_a \sigma$. This leaves σ as the only remaining Goldstone field.

The covariant derivative of the non-linear realisation denoted Δ_a of a matter field B which transforms as $B \to D(h^{-1})B$, where D is a representation of the Lorentz group, is given by

$$\Delta_a B = e^{-\sigma}(\partial_a + \partial^b \sigma \Sigma_{ab})B.$$

$$(13.2.56)$$

In particular, for a vector field A_a we have

$$\Delta_a A_b = e^{-\sigma}(\partial_a A_b + \eta_{ab} \partial^c \sigma A_c - \partial_b \sigma A_a),$$

$$(13.2.57)$$

where the representation D of the Lorentz group is given by $D(1 - \frac{1}{2}w_{cd}J^{cd}) = 1 - \frac{1}{2}w_{ab}\Sigma_{ab}$ for infinitesimal w_{cd}.

Using $g \to g_0 g h_c$, where h_c is the compensating Lorentz transformation, it is straightforward to find the transformations under dilations and special conformal transformations of σ from the group element of equation (13.2.53). We find that

$$\delta\sigma = (2x \cdot \beta x \cdot \partial - x^2 \beta \cdot \partial)\sigma + 2\beta_\mu x^\mu + \lambda.$$

$$(13.2.58)$$

The matter field B is found to transform as

$$\delta B = (2x \cdot \beta x \cdot \partial - x^2 \beta \cdot \partial)B + (\beta^a x^b - \beta^b x^a)\Sigma_{ab}B.$$

$$(13.2.59)$$

No dilation of B occurs as dilations are not part of the local sub-group.

The equation of motion of σ should carry no Lorentz indices and an invariant equation is given by

$$\nabla_a \phi_b \eta^{ab} = 0 \quad \text{or} \quad 2\partial^2 \sigma + (D-2)(\partial_\mu \sigma \partial^\mu \sigma) = 0.$$

$$(13.2.60)$$

Another way to construct the equation of motion is to use the connection $w_\mu{}^{cd}$ which transforms under local Lorentz group as a spin connection. As such, the usual Einstein action is invariant and evaluating this we find that

$$\int d^D x e R = -3(D-1)(D-2) \int d^D x e^{(D-2)\sigma} \partial_\mu \sigma \partial^\mu \sigma.$$

$$(13.2.61)$$

Here we have used the vielbein to contract the indices on the Riemann tensor. The resulting equation of motion is just equation (13.2.59). This non-linear realisation can be thought of as just that for a theory in which only the trace part of the metric is left.

We now give an outline of the general method of constructing non-linear realisations that involve space-time coordinates and fields. We consider a group that contains the generators L_N and we denote the remaining generators by the generic symbol K^*. We will assume that the generators L_A from a representation of the K^*s, which themselves form a group denoted G. The general group element is of the form

$$g = e^{z^A L_A} e^{\phi(z) \cdot K} \equiv g_L g_K, \tag{13.2.62}$$

where $g_L = e^{z^A L_A}$. We recognise z as the coordinates and ϕ as the fields. The local sub-group can be used to set some of the fields ϕ to zero. The discussion below holds if one makes this choice, or one works with the general group element. The Cartan forms can be written as

$$\Omega = g^{-1} dg = dz^\Pi E_\Pi{}^A L_A + dz^\Pi G_{\Pi,*} K^*. \tag{13.2.63}$$

Since Ω is invariant under $g \to g_0 g$, it follows that each of the coefficients of the above generators is invariant, that is, $dz^\Pi E_\Pi{}^A$ and $dz^\Pi G_{\Pi,*}$ are invariant. However, dz^Π does transform under g_0 and so $E_\Pi{}^A$ and $G_{\Pi,*}$ are not invariant. To find quantities that only transform under the local sub-algebra we can rewrite Ω as

$$\Omega = g^{-1} dg = dz^\Pi E_\Pi{}^A (L_A + G_{A,*} K^*), \tag{13.2.64}$$

where we recognise that $G_{A,*} = (E^{-1})_A{}^\Pi G_{\Pi,*}$. It follows that $G_{A,*}$ are inert under g_0 transformations and just transform under local transformations. As such, they are useful quantities with which to construct the dynamics as one must now only solve the problem of finding objects which are invariant under the local symmetry. We may think of $G_{A,*}$ as covariant derivatives of the fields ϕ at least in the directions perpendicular to the local subalgebra as these transform covariantly. If we had chosen coset representatives, then a g_0 transformation would require a local compensating transformation, but as the final result is invariant under local transformations these compensaitions do not affect the dynamics.

Using the second form of the group element of equation (13.2.62) we find that the Cartan forms can be expressed as

$$\Omega = g_K^{-1} dz^A L_A g_K + g_K^{-1} dg_K, \tag{13.2.65}$$

where we have now assumed that the generators L_A commute and we have used that these generators belong to a representation of G. We recognise

$$dz^\Pi E_\Pi{}^A L_A = g_K^{-1} dz^A L_A g_K = dz^T EL, \tag{13.2.66}$$

where in the last equation we have used an obvious matrix notation in that the matrix E has the elements $E_\Pi{}^A$. We can think of this object as a generalised vielbein. The reader will have noticed that the generators L can carry either a Π or an A index depending on the context; this is not a change carried out with the vielbein and $L_A = L_\Pi \delta_A^\Pi$.

Under a rigid $g_0 \in G$ transformation $g \to g_0 g$, the different parts of the group element transform as $g_L \to g_0 g_L (g_0)^{-1}$ and $g_K \to g_0 g_K$. As a result the coordinates transform under G as

$$z^\Pi L_\Pi \to g_0 z^\Pi L_\Pi (g_0)^{-1}. \tag{13.2.67}$$

We can define the action of the representation of G to which the generators L_Π belong by $U(k)(L_\Pi) \equiv k^{-1} L_\Pi k = D(k)_\Pi{}^A L_A$, where $k \in G$ and $D(k)_\Pi{}^A$ is the matrix representation. As a result of equation (13.2.67), we find that in matrix notation $dz^I \to dz^{T'} = dz^T D(g_0^{-1})$, or putting in the indices $dz^\Pi \to dz^{\Pi'} = dz^\Lambda D(g_0^{-1})_\Lambda{}^\Pi$. The derivative $\partial_\Pi = \partial/\partial z^\Pi$ in the generalised space-time transforms as $\partial'_\Pi = D(g_0)_\Pi{}^\Lambda \partial_\Lambda$. Examining equation (13.2.66), we note that E is equal to the matrix $D(g_K)_\Pi{}^\Lambda$.

As the Cartan form is inert under rigid transformations, their action on the coordinates must be compensated by that on E and so $E_\Pi{}^{\Lambda'} = D(g_0)_\Pi{}^\Lambda E_\Lambda{}^A$. The generalised vielbein $E_\Pi{}^{\Lambda'}$ transforms on its upper index by a local transformation and so we can think of the upper index as a tangent index and the lower index as a world index.

As for the case of internal symmetries, one can construct invariant actions using the Cartan forms or by using $M \equiv g_K I_c(g_K^{-1})$, where I_c is the Cartan involution. It is easy to see that M is inert under local transformations but transforms as $M \to M' = g_0 M I_c(g_0^{-1})$ under rigid transformations. Using that $E = D(g_k)$ we find that M in the l representation is given by $D(M) = D(g_k)D(I_c(g_k^{-1})) = EE^T$, where E^T is defined by this equation. As the notation suggests, for many groups it is just the transpose. Writing out the indices explicitly $D(M)_{\Pi\Lambda} = (EE^T)_{\Pi\Lambda}$ and we can write its rigid transformation as $D(M)_{\Pi\Lambda} \to D(g_0^{-1})_\Lambda{}^\Gamma D(M)_{\Gamma\Theta} D(I_c(g_0^{-1}))_\Pi{}^\Theta$.

We close this section by formulating gravity in D dimensions as a non-linear realisation. We begin by considering the non-linear realisation of the algebra of IGL(D) whose algebra is given by

$$[K^a{}_b, K^c{}_d] = \delta^c_b K^a{}_d - \delta^a_d K^c{}_b, \quad [K^c{}_b, P_a] = -\delta^c_a P_b. \tag{13.2.68}$$

We take as the local sub-group the Lorentz group SO($1, D-1$) which is generated by $J_{ab} = K^c{}_{[b}\eta_{c|a]} - K^c{}_{[a}\eta_{c|b]}$. The general group element can be written as

$$g = e^{x^a P_a} e^{h_a{}^b(x) K^a{}_b}, \tag{13.2.69}$$

where we have not used the local sub-group to make any choice of representative.

The Cartan forms are given by

$$\mathcal{V} = g^{-1} dg = dx^\mu e_\mu{}^a (P_a + \Omega_{ab}{}^c K^b{}_c), \tag{13.2.70}$$

where $e_\mu{}^a \equiv (e^h)_\mu{}^a$ and $\Omega_{a,b}{}^c \equiv (e^{-1})_a{}^\mu (e^{-1}\partial_\mu e)_b{}^c$. The Cartan forms are inert under rigid IGL(D) but transform under local SO($1, D-1$) transformations as $\mathcal{V} \to h^{-1}\mathcal{V}h + h^{-1}dh$. As such, the object $e_\mu{}^a$ transforms under its a index with a local Lorentz transformation. It also transforms on its μ index so as to compensate the change in dx^μ in order to keep the Cartan form inert. As such, it is a likely candidate for the vielbein of general relativity.

The decomposition of the GL(D) Cartan forms into coset and sub-group generators, as was done in equation (13.2.28), corresponds to taking the indices on $K^a{}_b$ to be symmetric and anti-symmetric, respectively, and so we may write

$$\mathcal{V} = dx^\mu e_\mu{}^a (P_a + \tfrac{1}{2} S_{a,b}{}^c K^b{}_c + K_c{}^b \tfrac{1}{2} Q_{a,b}{}^c J^b{}_c), \tag{13.2.71}$$

where $Q_{a,b}{}^c = \Omega_{a,[b}{}^{c]}$, $S_{a,b}{}^c = \Omega_{a,(b}{}^{c)}$. The object in this last equation just transforms just under local Lorentz transformations on its indices, the first transforms covariantly and the second like a connection. The most generally invariant action which is second order in space-time derivatives is given by

$$A = \int d^D x \det e \{ d_1 \eta^{ab} D_a S_{b,c}{}^c + d_2 D_c S_b{}^{bc} + d_3 S_a{}^{ac} S_{d,c}{}^d + d_4 S_d{}^{ac} S_{a,c}{}^d$$

$$+ d_5 \eta^{ab} S_{a,d}{}^c S_{b,c}{}^d + d_6 S_a{}^{ac} S_{c,d}{}^d + d_7 \eta^{bc} S_{b,d}{}^d S_{c,a}{}^a \}, \tag{13.2.72}$$

where $d_1, \ldots, d_7$ are constants that are not determined by the non-linear realisation and the covariant derivative is given by $D_a S_{b,c}{}^d = \partial_a S_{b,c}{}^d + Q_{a,b}{}^e S_{e,c}{}^d + Q_{a,c}{}^e S_{b,e}{}^d - Q_{a,e}{}^d S_{a,c}{}^e$. The above action can be expressed in terms of the metric using the relation $S_{a,bc} = \frac{1}{2} e_b{}^\mu e_c{}^\nu e_a{}^\lambda \partial_\lambda g_{\mu\nu}$.

Constructing gravity from a non-linear realisation was first proposed in [13.81] using the metric formualtion and in four dimensions. The authors took the simultaneous non-linear realisation of IGL(D), as given in the last few equations, with the conformal group. This was a slightly complicated procedure, whose first step was to realise that the Goldstone field associated with dilation generator of the conformal group and the diagonal parts of the metric, arising from $K^a{}_a$ in IGL(D), should be identified as their corresponding generators have the same action on x^μ. One then constructs an invariant action only from objects which transformed covariantly under both non-linear realisations. The result was Einstein's theory of general relativity. The result was to some extent expected since it had previously been shown [13.82] that the closure of the coordinate transformations of these two groups was the infinite dimensional group of general coordinate transformations. Alternatively, one could start from the action above and demand conformal invariance, so fixing the above constants. An account of these methods in D dimensions and involving the vielbein was given in the early papers on E_{11}. An alternative approach is to take the action of equation (13.2.72) and fix the constants by demanding that it be invariant under diffeomorphisms. The result is that $d_1 = -\frac{1}{2} = -d_2 = d_3 = -d_4 = 2d_5 = -d_6 = 2d_7$.

We can also construct the action using $M = g I_c(g^{-1})$ as explained above. In this case the generators L_A are the generators P_a, G is the group GL(D) and $I_c(g^{-1}) = g^T$. As such, we find that $D(M) = e e^T$, or written with indices $g_{\mu\nu} \equiv D(M)_{\mu\nu} = e_\mu{}^a \eta_{ab} e_\nu{}^b$. We find that $g_{\mu\nu}$ is inert under local transformations and transforms under rigid GL(D) indices on its two indices. The space-time derivative $\partial_\mu = \partial/\partial x^\mu$ also transforms under rigid GL(D) transforms.

We can construct the most general invariant action that is bilinear in space-time derivatives out of M and ∂_μ; it is given by

$$A = \int d^D x \det e (c_1 \partial_\lambda g_{\mu\nu} \partial_\rho (g^{-1})^{\mu\nu} (g^{-1})^{\lambda\rho} + c_2 \partial_\lambda (g^{-1})^{\mu\nu} \partial_\mu g_{\rho\nu} (g^{-1})^{\lambda\rho}$$

$$+ c_3 \partial_\lambda (g^{-1})^{\lambda\tau} \partial_\mu (g^{-1})^{\mu\nu} g_{\tau\nu} + c_4 \partial_\lambda \det g \partial_\rho (\det g)^{-1} (g^{-1})^{\lambda\rho}$$

$$+ c_5 \det g \partial_\rho (\det g)^{-1} \partial_\lambda (g^{-1})^{\lambda\rho}), \tag{13.2.73}$$

where $c_1, \ldots, c_5$ are constants.

This action is invariant diffeomorphism, which is equivalent to conformal invariance if $2c_1 = \frac{1}{2} = -c_2 = -2c_4 = -c_5$ and $c_3 = 0$. Indeed one can verify that the result is Einstein's theory using the identity

$$\int d^D x \sqrt{-\det g} R = \int d^D x \sqrt{-\det g} \{ -\frac{1}{2} \partial^\tau g^{\nu\lambda} \partial_\nu g_{\tau\lambda} + \frac{1}{4} \partial_\nu g_{\rho\kappa} \partial^\nu g^{\rho\kappa}$$

$$+ \frac{1}{4} \partial_\nu (\ln \det g) \partial^\nu (\ln \det g) - \frac{1}{2} \partial^\nu (\ln \det g) \partial^\mu g_{\mu\nu} \}$$

$$= \int d^D x \det e \{ -\frac{1}{2} \partial^\tau g^{\nu\lambda} \partial_\nu g_{\tau\lambda} + \frac{1}{4} \partial_\nu g_{\rho\kappa} \partial^\nu g^{\rho\kappa}$$

$$+ \partial_\mu \ln \det e \partial^\mu \ln \det e - \partial^\mu \ln \det e \partial^\lambda g_{\mu\lambda} \}, \tag{13.2.74}$$

which is valid in D dimensions and using the definitions $\partial^\mu = g^{\mu\nu}\partial_\nu$, $\Gamma^\lambda_{\mu\nu} = \frac{1}{2}g^{\lambda\tau}(\partial_\nu g_{\tau\mu} + \partial_\mu g_{\tau\nu} - \partial_\tau g_{\mu\nu})$ and $R_{\mu\nu}{}^\rho{}_\lambda = \partial_\mu \Gamma^\rho_{\nu\lambda} + \Gamma^\rho_{\mu\kappa}\Gamma^\kappa_{\nu\lambda} - (\mu \to \nu)$.

13.3   Eleven-dimensional supergravity

As described in chapter 5 the only non-trivial representation of the Clifford algebra in eleven dimensions is inherited from that in ten dimensions and so has dimension $2^{10/2} = 32$. We also inherit the ten-dimensional properties $\epsilon = 1$ and so $B^T = B$ and $C = -C^T$. Eleven-dimensional supergravity is based on the eleven-dimensional supersymmetry algebra with Majorana spinor Q_α, which has 32 real components. As we discussed in chapter 5 the algebra takes the form [5.7]

$$\{Q_\alpha, Q_\beta\} = (\gamma^m C^{-1})_{\alpha\beta} P_m + (\gamma^{mn} C^{-1})_{\alpha\beta} Z_{mn} + (\gamma^{mnpqr} C^{-1})_{\alpha\beta} Z_{mnpqr}, \qquad (13.3.1)$$

where P_m is the usual translation operator and Z_{mn} and Z_{mnpqr} are central charges. Although these play little apparent role in the construction of the supergravity theory they are very important in M theory.

Eleven is the maximal number of dimension in which one can have a supergravity theory [10.10]. By a supergravity theory we mean a theory with spins 2 and less. This follows from the study of the massless irreducible representations of supersymmetry. We recall that the irreducible representations of the Poincaré group are specified (actually induced) by the irreducible representation of the little group which is just the group that leaves a fixed momentum invariant. For a theory in D dimensions, and for the massless case, this is in effect just the group $SO(D-2)$. Similarly, the irreducible representation of the supersymmetry algebra is specified by the little algebra that leaves invariant a given momentum. Now the little algebra, for the massless case, contains $SO(D-2)$ and the supercharges in terms of $SO(D-2)$ spinor representation. One finds [10.10] that in dimensions greater than eleven the irreducible representations of the super Poincaré group have spins greater than 2 and so any supersymmetric theory in these dimensions must have spins greater than 2. By spin 2 we mean a symmetric traceless representation of the little group, that is, $h_{ij} = h_{ji}$, $h_i{}^i = 0$ and by spins greater than 2 we mean a higher rank tensor representation of $SO(D-2)$. The maximal supergravity theory in four dimensions possesses $N = 8$ supersymmetry, which is an algebra with eight Majorana supercharges, each of which has four real components, making 32 supercharges in all. In fact, this theory was obtained by dimensional reduction of the eleven-dimensional supergravity theory.

That a supergravity theory can exist in at most eleven dimensions can be explained by considering the more familiar irreducible representations of supersymmetry in four dimensions (see, for example, [1.11] for a review). In four dimensions for a massless particle in a given Lorentz frame only two of the four components of the supercharges act non-trivially on the physical states. This can be traced to the fact that $\det \gamma^a p_a = 0$ when $p^2 = 0$ and so the right-hand side of the anti-commutator of the supercharges must have zeros when diagonalised. Taking the expectation value of the supercharges which lead to these zeros we find the norm vanishes; however, the norms of all physical states must be positive definite and so we conclude that these supercharges must vanish on physical states. The anti-commutation relations of the remaining non-trivial supercharges form a Clifford algebra: one raises the helicity by $\frac{1}{2}$ and the other lowers the helicity by $\frac{1}{2}$. Choosing the supercharges that raise the helicity to annihilate the vacuum, the physical states are given by the action of the remaining supercharges. If the supersymmetry algebra has N Majorana supercharges, the physical states are given by the action of N creation operators, each of which lowers the helicity by $\frac{1}{2}$. Consequently, if we take the vacuum to have helicity 2, the lowest helicity state in the representation will be $2 - N/2$. To have a supergravity theory we cannot have less than helicity -2 and hence the limit $N \leq 8$. Thus the maximally supersymmetric supergravity theory in four dimensions has $4 \times 8 = 32$ supercharges Given a maximal supergravity theory in a dimension greater than 4 we can, as explained in section 13.1.4, dimensionally reduce it on a torus by taking all the fields to be independent of the extra dimensions to find a supergravity theory in four dimensions. As supersymmetry is preserved in the torus reduction the number of supercharges is unchanged. As such, the maximal number of supercharges must be 32 as if it was more we would construct a supergravity theory in four dimensions with more supercharges than 32 contradicting the result just found. Hence, a supergravity theory in a higher dimension can only arise if that dimension possesses a spinor representation that has dimension 32 or less. Thus we find the desired result; eleven is the highest number of dimension in which a supergravity theory can exist.

The irreducible representation, or particle content, of eleven-dimensional supergravity was found in [10.10] by analysing the irreducible representations of the supersymmetry algebra of equation (13.4.1), without central charges. One could also deduce it by requiring that it dimensionally reduce to the irreducible representation of the four-dimensional $N = 8$ supergravity. We now give a more ad hoc argument for the particle content of eleven-dimensional supergravity.

Eleven-dimensional supergravity must be invariant under general coordinate transformations (that is, local translations) and local supersymmetry transformations. To achieve these symmetries it must possess the equivalent 'local gauge' fields, the vielbein e^a_μ and the gravitino $\psi_{\mu\alpha}$. The last transforms as $\delta\psi_{\mu\alpha} = \partial_\mu\eta_\alpha + \cdots$ and so must be the same type of spinor as the supersymmetry parameter, which in our case is a Majorana spinor.

For future use we now give the on-shell count of degrees of freedom of the graviton and gravitino in D dimensions. As these are massless fields, the relevant bosonic part of the little group which classifies the irreducible representation is $SO(D - 2)$. The graviton belongs to a second rank symmetric traceless tensor of $SO(D - 2)$, $h_{ij} = h_{ji}$, $h^i{}_i = 0$, and as such has $\frac{1}{2}(D - 2)(D - 1) - 1$ degrees of freedom on-shell. The gravitino has $(D - 3)cr$ real components. Here c is the dimension of the Clifford algebra in dimension $D - 2$ and so is given by $c = 2^{(D/2)-1}$ if D is even and $c = 2^{((D-1)/2)-1}$ if D is odd. The quantity r is 2, 1 or $\frac{1}{2}$ when $\psi_{\mu\alpha}$ is a Dirac, Majorana or Majorana–Weyl spinor, respectively. In terms of little group representations, the gravitino is a vector spinor ϕ_i, $i = 1, \ldots, D - 2$, which is γ-traceless $\gamma^i\phi_i = 0$. The result follows from the fact that the gravitino is the irreducible vector spinor representation of the little group in $D - 2$ dimensions and so has $(D - 2)cr$ components, but the γ-trace subtracts another spinor's worth of components (that is, cr). We also note that in D dimensions a rank p anti-symmetric gauge field belongs to the anti-symmetric rank p tensor representation of $SO(D - 2)$, $A_{i_1\ldots i_p} = A_{[i_1\ldots i_p]}$, and so has $((D - 2)\cdots(D - p - 1)/p!)$ degrees of freedom on-shell.

Using the above counts, for eleven dimensions we find that the graviton and gravitino have, respectively, 44 and 128 degrees of freedom on-shell. However, in any supermultiplet of on-shell physical states the fermionic and bosonic degrees of freedom must be equal.

Assuming that there are no further fermionic degrees of freedom we require another 84 bosonic on-shell degrees of freedom. If we take these to belong to an irreducible representation of SO(9), then the unique solution would be a third rank anti-symmetric tensor. This can only arise from a third rank gauge field $A_{\mu_1\mu_2\mu_3}$ whose fourth rank gauge field $F_{\mu_1\mu_2\mu_3\mu_4} \equiv 4\partial_{[\mu_1}A_{\mu_2\mu_3\mu_4]}$, the anti-symmetry being with weight 1.

The eleven-dimensional supergravity Lagrangian was constructed in [13.1] and is given by

$$L = +\frac{e}{4\kappa_{11}^2}R(\Omega(e,\psi)) - \frac{e}{48}F_{\mu_1\cdots\mu_4}F^{\mu_1\cdots\mu_4} - \frac{e}{2}\bar{\psi}_\mu\gamma^{\mu\nu\varrho}D_\nu\left(\frac{1}{2}(\Omega+\hat{\Omega})\right)\psi_\varrho$$

$$-\frac{1}{192}e\kappa_{11}(\bar{\psi}_{\mu_1}\gamma^{\mu_1\cdots\mu_6}\psi_{\mu_2} + 12\bar{\psi}^{\mu_3}\gamma^{\mu_4\mu_5}\psi^{\mu_6})(F_{\mu_3\cdots\mu_6} + \hat{F}_{\mu_3\cdots\mu_6})$$

$$+\frac{2\kappa_{11}}{(12)^4}\epsilon^{\mu_1\cdots\mu_{11}}F_{\mu_1\cdots\mu_4}F_{\mu_5\cdots\mu_8}A_{\mu_9\mu_{10}\mu_{11}}, \tag{13.3.2}$$

where

$$F_{\mu_1\cdots\mu_4} = 4\partial_{[\mu_1}A_{\mu_2\mu_3\mu_4]}, \quad \hat{F}_{\mu_1\cdots\mu_4} = F_{\mu_1\cdots\mu_4} + 3\kappa_{11}\bar{\psi}_{[\mu_1}\gamma_{\mu_2\mu_3}\psi_{\mu_4]} \tag{13.3.3}$$

and

$$\Omega_{\mu mn} = \hat{\Omega}_{\mu mn} - \frac{1}{4}\kappa_{11}^2\bar{\psi}_\nu\gamma_{\mu mn}{}^{\nu\lambda}\psi_\lambda,$$

$$\hat{\Omega}_{\mu mn} = \Omega_{\mu mn}^{(0)}(e) + \frac{1}{2}\kappa_{11}^2(\bar{\psi}_\nu\gamma_n\psi_m - \bar{\psi}_\nu\gamma_m\psi_n + \bar{\psi}_n\gamma_\mu\psi_m). \tag{13.3.4}$$

The symbol $\Omega_{\mu mn}^{(0)}(e)$ is the usual expression for the spin connection in terms of the vielbein e_μ^n. The reader can find this in equation (13.1.36) by setting $\psi_\mu = 0$. We also define $D_\mu(\Omega)\psi_\nu = (\partial_\mu - \frac{1}{4}\Omega_\mu{}^{mn}\gamma_{mn})\psi_\nu$ and $[D_\mu, D_\nu] = -\frac{1}{4}R_{\mu\nu m}{}^n\gamma^m{}_n$. The reader who wishes to have more conventional normalisation for the terms in the Lagrangian can achieve this by rescaling the action over all and then scaling all the fields except the vielbein.

The above action is invariant under the local supersymmetry transformations

$$\delta e_\mu{}^m = \kappa_{11}\bar{\epsilon}\gamma^m\psi_\mu, \quad \delta A_{\mu_1\mu_2\mu_3} = -\frac{3}{2}\bar{\epsilon}\gamma_{[\mu_1\mu_2}\psi_{\mu_3]},$$

$$\delta\psi_\mu = \frac{1}{\kappa_{11}}D_\mu(\hat{\Omega})\epsilon + \frac{1}{12^2}(\gamma_\mu{}^{\nu_1\cdots\nu_4} - 8\delta_\mu^{\nu_1}\gamma^{\nu_2\nu_3\nu_4})\hat{F}_{\nu_1\ldots\nu_4}\epsilon \tag{13.3.5}$$

as well as the obvious gauge symmetry $\delta A_{\mu_1\mu_2\mu_3} = 3\partial_{[\mu_1}\Lambda_{\mu_2\mu_3]}$.

Although the result may at first sight look complicated most of the terms can be easily found using the Noether method. As discussed in section 13.1, we start from the linearised theory for the graviton, the gravitino and the 3-form gauge field. The linearised action is bilinear in the fields and is invariant under a set of rigid supersymmetry transformations, which are linear in the fields as well as the local Abelian transformations $\delta h_{\mu\nu} = \partial_\mu\xi_\nu + \partial_\nu\xi_\mu$, $\delta\psi_{\mu\alpha} = \partial_\mu\eta_\alpha$, $\delta A_{\mu_1\mu_2\mu_3} = \partial_{[\mu_1}\Lambda_{\mu_2\mu_3]}$. The linearised supersymmetry transformations are found by using dimensional analysis and closure of the linearised supersymmetry algebra. We next let the rigid supersymmetry parameter ϵ_α depend on space-time and identify the parameter of the Abelian local transformation of ψ_μ with the now local supersymmetry parameter, that is, we identify $\eta_\alpha = (1/\kappa_{11})\epsilon_\alpha$. We know that the final result will be invariant under general coordinate transformations and so we may at each step in the Noether procedure insert the full vielbein instead of the linearised gravity field so as to ensure this invariance. Even at this stage in the Noether procedure we recover all the terms in the transformations in the fields given in equation (13.4.5) except for some of the terms in the spin connection $\hat{\Omega}$. For the action, we find all the terms except the last two terms and again some terms in the spin connection of the gravitino. The action at this stage is not invariant under the now local supersymmetry transformations and as explained in section 13.1 we can gain invariance at order κ_{11}^0 by adding a term of the form $\bar{\psi}^{\mu\alpha} j_{\mu\alpha}$, where $j_{\mu\alpha}$ is the Noether current for the supersymmetry of the linearised theory. This term is none other than the second to last term in the above action. To gain invariance to order κ_{11}^1 we must cancel the variations of this second to last term under the supersymmetry transformation. This is achieved if we add the last term to the action. Hence, even at this early stage in the procedure we have accounted for essentially all the terms in the action and transformation laws. While one can pursue the Noether procedure to the end, to find the final form of the action and transformations laws, it is perhaps best to guess the final form of the connection and verify that the action is invariant and the local supersymmetry transformations close. The former can readily be achieved by using the so-called 1.5 order formalism [13.8, 13.12] which is reviewed in section 13.1.3.

Carrying out the Noether method on the algebra the reader will soon understand that a hat on an object just means that it has been corrected at higher orders in κ_{11} so as to ensure that its variation under supersymmetry does not contain the derivative of the supersymmetry parameter.

One of the most interesting terms in the action is the last term in equation of (13.3.2), which has a naked gauge field, but is easily seen to be gauge invariant. This term is called a Chern–Simmons term and its presence in the action has a number of important implications.

The eleven-dimensional action contains only one coupling constant, the gravitational constant κ_{11}, which has the dimensions of (mass)$^{-9/2}$ and so defines a Planck mass m_p by $\kappa_{11} = m_p^{-9/2}$. We note that there are no scalars in the action whose expectation value could be used to define another coupling constant. If we scale the fields by $\psi_\mu \to \kappa_{11}^{-1}\psi_\mu$ and $A_{\mu_1\mu_2\mu_3} \to \kappa_{11}^{-1}A_{\mu_1\mu_2\mu_3}$ we find that all factors of κ_{11} drop out of the supersymmetry transformation laws and the action scales by an overall factor of κ_{11}^{-2}. As such, when expressed in terms of these variables, the classical field equations do not contain κ_{11}. As a result, the value of the coupling constant κ_{11} has no physical meaning at the classical level. Another way to see this fact is to observe that if, after carrying out the above redefinitions, we Weyl-rescale the fields by

$$e_\mu^m \to e^{-\alpha} e_\mu^m, \quad \psi_\mu \to e^{-\frac{\alpha}{2}}\psi_\mu \quad \text{and} \quad A_{\mu_1\mu_2\mu_3} \to e^{-3\alpha}A_{\mu_1\mu_2\mu_3} \tag{13.3.6}$$

as well as scale the coupling constant by $\kappa_{11} \to e^{-\frac{9\alpha}{2}}\kappa_{11}$, we then find that the action is invariant. In deriving this result we used the equation

$$R(e^\tau e_\mu^m) = e^{-2\tau}(R(e_\mu^m) + 2(D-1)D^\mu D_\mu \tau + (D-1)(D-2)g^{\mu\nu}\partial_\mu\tau\partial_\nu\tau),$$

$$\tag{13.3.7}$$

where R is the fully contracted Riemann tensor in dimension D. Above we only required the case τ is a constant. This particular result can be easily seen by noting that the spin connection is scale invariant, but the Ricci tensor is contracted using two inverse vielbeins. Of course, this is not a symmetry of the action in the usual sense as we have rescaled the coupling constant. However, as the coupling constant only occurs as a prefactor multiplying the entire action, it is a symmetry of the classical equations of motion. Hence, we can only specify the value of the constant κ_{11} with respect to a particular metric.

These transformations are not a symmetry of the quantum theory, where, in the path integral, the prefactor of κ_{11}^{-2} which multiplies the action occurs as $\kappa_{11}^{-2}\hbar$, where $\hbar$ is Planck's constant. In this case we can absorb the rescaling either in κ_{11} or $\hbar$. The above Weyl scaling of the vielbein implies that the proper distance d^2s scales as $d^2s \to e^{-2\alpha}d^2s$. Taking the scaling to be absorbed by $\hbar$ we find that $\hbar \to e^{-9\alpha}\hbar$ and so small proper distance corresponds to small $\hbar$. Put another way small κ_{11}, or equivalently small $\hbar$, is the same as working at small proper distance.

In this book we have used the signature $\eta_{nm} = \text{diag}(-1, 1, \ldots, 1)$. However, all the original papers on supergravities in eleven and ten dimensions used the signature $\eta_{nm} = \text{diag}(1, -1, \ldots, -1)$. Since these papers contain many more formulae than those given here it will be useful to give a systematic procedure for changing between the two signatures. To go to the latter metric we must take $\eta_{nm} \to -\eta_{nm}$, $\gamma^a \to i\gamma^a$, $e_\mu{}^n \to e_\mu{}^n$, $\psi_\mu \to \psi_\mu$ and $A_{\mu_1\ldots\mu_p} \to A_{\mu_1\ldots\mu_p}$ for any p. We must also change our definition of the Dirac bar of spinors. However, in supersymmetric theories, the spinors are often Majorana and the charge conjugation matrix is unchanged, since the transpose of the γ matrices is the same even if we multiply them by i. As such, when carrying out the change of signature we can write the Dirac conjugate of a spinor in terms of its Majorana conjugate, carry out the change and then write them back in terms of the Dirac conjugate.

Using these rules it is straightforward to carry out the change of signature. One finds, for example, that $g^{\mu\nu}\partial_\mu\sigma\partial_\nu\sigma \to -g^{\mu\nu}\partial_\mu\sigma\partial_\nu\sigma$ and $R \to -R$, since $R = R_{\mu\nu a}{}^b(e^{-1})_c{}^\mu(e^{-1})_b{}^\nu\eta^{ac}$. To give a more complicated example,

$$-\frac{i}{2}\bar{\psi}_\mu\gamma^{\mu\nu\rho}D_\nu\psi_\rho \to -\frac{1}{2}\bar{\psi}_\mu\gamma^{\mu\nu\rho}D_\nu\psi_\rho. \tag{13.3.8}$$

The factor of $-i$ is made up of i^3 from the three γ matrices and there are no η_{mn} factors as all the indices are in the correct positions. Another example is given by

$$\bar{\psi}^\mu\gamma^{\nu\rho}\psi^\tau \to -\bar{\psi}^\mu\gamma^{\nu\rho}\psi^\tau. \tag{13.3.9}$$

In this case the factor of -1 is composed of i^2 for the two γ matrices and $(-1)^2$ from the two factors of η_{mn} which occur due to the raising of the indices on the ψ_μ.

We close this section by commenting on the normalisation of fields in supergravity actions. Of course, one is free to normalise the fields as one likes but a canonical normalisation is often adopted for the bosonic fields which is specified by writing the supergravity action in D dimensions in the generic form

$$\frac{1}{2\kappa_D^2}\int d^Dx\, e\left(R - \frac{1}{2p!}F_{\mu_1\cdots\mu_p}F^{\mu_1\cdots\mu_p} - \frac{1}{2}\bar{\psi}_\mu\gamma^{\mu\nu\rho}D_\nu\psi_\rho - \frac{1}{2}\bar{\lambda}\gamma^\mu D_\mu\lambda + \cdots\right),$$

$$\tag{13.3.10}$$

where κ_D is the gravitational coupling in D dimensions. As noted above one can always rescale the gauge and fermionic fields by $1/\kappa_D$ to get the overall factor of κ_D^{-1} multiplying the entire action as displayed above. The gravitational coupling in D dimensions has the dimensions of $(\text{length})^{(D-2)/2}$ and so we can define a Plank length in D dimensions by $l_P^D \equiv (\kappa_D)^{2/(D-2)}$ and a Plank mass by $l_P^D \equiv 1/m_P^D$.

To find this normalisation for the eleven-dimensional supergravity of equation (13.3.2) we can multiply the action by 2 and then take $e_\mu{}^a \to e_\mu{}^a$, $A_{\mu_1\mu_2\mu_3} \to (1/2\kappa_{11})A_{\mu_1\mu_2\mu_3}$ and $\psi_\mu \to (1/2\kappa_{11})\psi_\mu$. This also changes the supersymmetry transformation, for example, $\delta e_\mu{}^a = \frac{1}{2}\bar{\epsilon}\gamma^a\psi_\mu$, etc. The Lagrangian is now given by

$$
L = +\frac{1}{2\kappa_{11}^2}\left(eR(\Omega(e,\psi)) \right) - \frac{e}{48}F_{\mu_1\cdots\mu_4}F^{\mu_1\cdots\mu_4} - \frac{e}{2}\bar{\psi}_\mu\gamma^{\mu\nu\varrho}D_\nu\left(\frac{1}{2}(\Omega+\hat{\Omega}) \right)\psi_\varrho
$$

$$
-\frac{1}{192.2}e(\bar{\psi}_{\mu_1}\gamma^{\mu_1\cdots\mu_6}\psi_{\mu_2} + 12\bar{\psi}^{\mu_3}\gamma^{\mu_4\mu_5}\psi^{\mu_6})(F_{\mu_3\cdots\mu_6} + \hat{F}_{\mu_3\cdots\mu_6})
$$

$$
+\frac{1}{(12)^4}\epsilon^{\mu_1\cdots\mu_{11}}F_{\mu_1\cdots\mu_4}F_{\mu_5\cdots\mu_8}A_{\mu_9\mu_{10}\mu_{11}}\Bigg) \tag{13.3.11}
$$

with suitable adjustments for the definitions of the field strengths etc. We began with the historical normalisation in equation (13.3.2) so that the reader can access the old literature to find aspects of the eleven-dimensional and IIA theory not discussed in this book.

13.4   The IIA supergravity theory

In this and the next section, we give the supergravity theories in ten dimensions which have the maximal supersymmetry. There are two such theories, called IIA and IIB.

In ten dimensions the non-trivial representation of the Clifford algebra has dimension $2^{10/2} = 32$. As we found in chapter 5, the matrices B and C associated with complex conjugation and transpose, respectively, of the γ matrices obey the properties $B^T = B$ and $C = -C^T$ (that is, $\epsilon = 1$). In ten dimensions a Majorana spinor has 32 real components; however, we can also have Majorana–Weyl spinors and these only have 16 real components.

In the discussion at the beginning of section 13.3 we found that a supergravity theory can only be based on a supersymmetry algebra with 32, or fewer, supercharges. If we consider the supersymmetry algebra with a 32-component Majorana spinor we find the IIA supergravity theory. Clearly, when we decompose the Majorana spinor into Majorana–Weyl spinors we get two such spinors: one of each chirality. The other ten-dimensional supersymmetry algebra with 32 supercharges has two Majorana–Weyl spinors of the same chirality and the corresponding supergravity is IIB supergravity.

The supersymmetry algebra for a single Majorana–Weyl supercharge, which has 16 real components, was given chapter 5. There are two supersymmetric theories that are based on this algebra, the $N = 1$ Yang–Mills theory [5.1, 10.19] and the $N = 1$ supergravity theory [10.17, 10.18], which is more often called type I supergravity. The coupling between the two theories was given in [10.20].

Upon dimensional reduction of eleven-dimensional supergravity to ten dimensions, by taking the eleventh dimension to be a circle, we will obtain a ten-dimensional theory that possesses a supersymmetry algebra based on a 32-component Majorana supercharge. The resulting theory can only be IIA supergravity. Indeed, this was how IIA supergravity was found [11.11–11.13].

The importance of the IIA and IIB supergravity theories, which was the main motivation for their construction, is that they are the low energy effective theories of the corresponding IIA and IIB closed string theories in ten dimensions. Type I supergravity coupled to $N = 1$

Yang–Mills theory is the effective action for the low energy limit of the $E_8 \otimes E_8$ or $SO(32)$ heterotic string.

As we have mentioned the IIA supergravity theory is based on a supersymmetry algebra which contains one Majorana spinor Q_α, $\alpha = 1, \ldots, 32$. Following the discussion of section 5.5, we conclude that the anti-commutator of two supersymmetry generators can have central charges of rank p, where $p = 1, 2 \bmod 4$ and so the corresponding anti-commutator is given by [5.7]

$$\{Q_\alpha, Q_\beta\} = (\gamma^m C^{-1})_{\alpha\beta} P_m + (\gamma^{mn} C^{-1})_{\alpha\beta} Z_{mn} + (\gamma^{m_1 \cdots m_5} C^{-1})_{\alpha\beta} Z_{m_1 \cdots m_5}$$
$$+ (\gamma^{m_1 \cdots m_4} \gamma_{11} C^{-1})_{\alpha\beta} Z_{m_1 \cdots m_4} + (\gamma^m \gamma_{11} C^{-1})_{\alpha\beta} Z_m + (\gamma_{11} C^{-1})_{\alpha\beta} Z,$$

$$(13.4.1)$$

where $\gamma_{11} = \gamma_0 \gamma_1 \cdots \gamma_9$. We need only take $p \le 5$, since we may use the equation

$$\gamma^{m_1 \cdots m_s} \gamma_{11} = \frac{1}{(10-s)!} (-1)^{s(s-1)/2} \epsilon^{m_1 \cdots m_s m_{s+1} \cdots m_{10}} \gamma_{m_{s+1} \cdots m_{10}},$$

$$(13.4.2)$$

to eliminate terms with $p \ge 6$. We recall that $\epsilon^{01 \cdots 9} = 1$.

We now derive the IIA supergravity theory from the eleven-dimensional supergravity given in the previous section using the method of dimensional reduction explained in section 13.1.4. We take the eleventh dimension to be a circle S^1 of radius R, that is, $x^{10'} = x^{10} + 2\pi nR$, $n \in \mathbf{Z}$, where $2\pi R$ parameterises the range of x^{10}. The index convention of section 13.1.4 is changed in this section so that now hatted indices run over all eleven-dimensional indices, but unhatted indices only run over the ten-dimensional indices, for example, $\hat{\mu}, \hat{\nu} = 0, 1, \ldots, 10$ while $\mu, \nu = 0, 1, \ldots, 9$. Given any of the fields in the eleven-dimensional supergravity we can take its Fourier transform in x^{10} as in equation (13.1.115), but as we are only interested in the massless modes, or zero modes, we take all the eleven-dimensional fields to be independent of x^{10}.

This reduction proceeds in the following generic manner

$$D = 11 \qquad e_{\hat\mu}^{\hat m} \qquad A_{\hat\mu_1 \hat\mu_2 \hat\mu_3} \qquad \psi_{\hat\mu\alpha}$$
$$\downarrow \qquad\qquad \downarrow \qquad\qquad \downarrow$$
$$D = 10 \qquad e_\mu{}^m, B_\mu, \sigma \quad A_{\mu_1\mu_2\mu_3}, A_{\mu_1\mu_2} \quad \psi_{\mu\alpha}, \lambda_\alpha$$

$$(13.4.3)$$

After having carried out the dimensional reduction we want to arrive at a supergravity theory that has its kinetic terms in canonical form, for example, we want to find the usual Einstein term without any factors of the scalar field. As explained in section 13.1.4, this places restrictions on the dimensional reduction ansatz. In particular, the vielbein is dimensionally reduced using equation (13.1.119) and so, with $\phi = -\sigma$, it takes the form

$$e_{\hat\mu}^{\hat m} = \begin{pmatrix} e^{-\frac{1}{12}\sigma} e_\mu{}^m & e^{\frac{2}{3}\sigma} A_\mu \\ 0 & e^{\frac{2}{3}\sigma} \end{pmatrix}, \quad (e^{-1})_{\hat m}^{\hat\mu} = \begin{pmatrix} e^{\frac{1}{12}\sigma} e_m{}^\mu & -e^{\frac{1}{12}\sigma} A_\nu (e^{-1})_m{}^\nu \\ 0 & e^{-\frac{2}{3}\sigma} \end{pmatrix}.$$

$$(13.4.4)$$

The 3-form gauge field reduces in an obvious way: $A_{\mu_1\mu_2\mu_3} = A_{\mu_1\mu_2\mu_3}$, $A_{\mu_1\mu_2} = A_{\mu_1\mu_2 10}$, where $\mu_1, \mu_2, \mu_3 = 0, 1, \ldots, 9$. The field A_μ is a gauge field whose gauge transformation has a parameter that arises from the general coordinate transformations with parameter ξ^{10} in the eleven-dimensional theory. The component $e_{\hat\mu=10}{}^m$ of the vielbein can be chosen to be zero as a result of a local Lorentz transformation with parameter $w^m{}_{10}$. We adopt the same definitions as in [10.11] except that $B_\mu = \frac{1}{2}A_\mu$, $\sigma^{there} = \frac{2}{3}\sigma^{here}$ and we must change the signature as explained in the last section. We can change the definition of γ^{11} so that now $\gamma^{11} = \gamma^1 \cdots \gamma^{10}$. These two steps can be combined by making the change $\gamma^{11} \to -i\gamma^{11}$ from the formulae of [10.11].

The gravitino appears in the eleven-dimensional action in the form

$$\psi_{\hat\mu} e_m{}^{\hat\mu} = e^{+\frac{1}{12}\sigma}(\psi_\mu e_m{}^\mu - A_\mu e_m{}^\mu \psi_{\hat{10}}) \equiv e^{+\frac{1}{12}\sigma}\psi'_m,$$

where the eleven-dimensional gravitino with a world index is denoted by $\psi_{\hat\mu} = (\psi_{\hat\mu}, \psi_{\hat{10}})$ and $\psi_{\hat\mu} e_{10}{}^{\hat\mu} = e^{-\frac{2}{3}\sigma}\psi_{\hat{10}}$. In order to find kinetic terms for the spinors in ten dimensions that have the standard form we must carry out Weyl rescalings and diagonalisation. Let us denote the ten-dimensional spinors by ψ_μ and λ, then the relations to the eleven-dimensional spinors with a world index that achieves this are given by

$$\psi_{\hat{10}} = \frac{2\sqrt{2}}{3}e^{\frac{17}{24}\sigma}\lambda, \quad \psi_{\hat\mu} - A_\mu \frac{2\sqrt{2}}{3}e^{\frac{17}{24}\sigma}\lambda = e^{-\frac{1}{24}\sigma}\left(\psi_\mu - \frac{1}{6\sqrt{2}}\gamma_\mu\gamma^{11}\lambda\right),$$

$$\mu = 0, 1, \ldots, 10. \tag{13.4.5}$$

The symbols $\psi_{\hat{10}}$ and $\psi_{\hat\mu}$ at the beginning of the first and second equations, respectively, are the eleven-dimensional gravitino with a world index, while the one after the equality, ψ_μ, is the gravitino in ten dimensions with a world index.

The resulting ten-dimensional IIA supergravity theory is given by

$$L = L^B + L^F, \tag{13.4.6}$$

where the first term contains all the terms which are independent of the fermions and the second term is the remainder. The bosonic part is found using the dimensional reduction formulae of section 13.1.4. In particular in the vielbein ansatz of equation (13.1.119) we must take $\alpha_d = \frac{1}{12}$ and use the equations (13.1.121) and (13.1.122) for the dimensional reduction of the Riemann tensor and the 4-form field strength terms. The dimensional reduction of the Chern–Simons term is easily performed. The result is given by [10.11–10.13]

$$L^B = \frac{1}{2\kappa_{10}^2}\left\{eR(w(e)) - \frac{1}{12}ee^{\frac{\sigma}{2}}\hat{F}_{\mu_1\ldots\mu_4}\hat{F}^{\mu_1\cdots\mu_4} - \frac{1}{3}ee^{-\sigma}F_{\mu_1\cdots\mu_3}F^{\mu_1\cdots\mu_3}\right.$$

$$-\frac{1}{4}ee^{\frac{3}{2}\sigma}F_{\mu_1\mu_2}F^{\mu_1\mu_2} - \frac{1}{2}e\partial_\mu\sigma\partial^\mu\sigma$$

$$\left. + \frac{1}{2\times(12)^2}\epsilon^{\mu_1\cdots\mu_{10}}F_{\mu_1\cdots\mu_4}F_{\mu_5\cdots\mu_8}A_{\mu_9\mu_{10}}\right\}, \tag{13.4.7}$$

where

$$F_{\mu_1\mu_2} = 2\partial_{[\mu_1}A_{\mu_2]}, \tag{13.4.8}$$

$$F_{\mu_1\mu_2\mu_3} = 3\partial_{[\mu_1}A_{\mu_2\mu_3]}, \tag{13.4.9}$$

$$\hat{F}_{\mu_1\cdots\mu_4} = 4(\partial_{[\mu_1}A_{\cdots\mu_4]} + A_{[\mu_1}F_{\mu_2\mu_3\mu_4]}). \tag{13.4.10}$$

We have taken the liberty of suitably scaling the fields by κ_{10}^{-1} as discussed above.

The fermionic part of the Lagrangian is much more complicated and is given by [10.11–10.13]

$$L^F = \frac{4e}{2\kappa_{10}^2} \left\{ -\frac{1}{2}\bar\psi_{\mu_1}\gamma^{\mu_1\mu_2\mu_3}D_{\mu_2}\psi_{\mu_3} - \frac{1}{2}\bar\lambda\gamma^\mu D_\mu\lambda + \frac{\sqrt2}{4}\bar\lambda\gamma^{11}\gamma^{\mu_1}\gamma^{\mu_2}\psi_{\mu_1}\partial_{\mu_2}\sigma \right.$$

$$+ \frac{1}{96}e^{\frac{\sigma}{4}}\left\{ -\bar\psi_{\mu_1}\gamma^{\mu_1\cdots\mu_6}\psi_{\mu_2} - 12\bar\psi^{\mu_3}\gamma^{\mu_4\mu_5}\psi^{\mu_6} + \frac{1}{\sqrt2}\bar\lambda\gamma^{11}\gamma^{\mu_1}\gamma^{\mu_3\cdots\mu_6}\psi_{\mu_1} \right.$$

$$\left. + \frac{3}{4}\bar\lambda\gamma^{11}\gamma^{\mu_3\cdots\mu_6}\lambda \right\}F_{\mu_3\cdots\mu_6} - \frac{1}{24}e^{-\frac{\sigma}{2}}\left\{ \bar\psi_{\mu_1}\gamma^{11}\gamma^{\mu_1\cdots\mu_5}\psi_{\mu_2} - 6\bar\psi^{\mu_3}\gamma^{11}\gamma^{\mu_4}\psi^{\mu_5} \right.$$

$$\left. - \sqrt2\bar\lambda\gamma^{\mu_1}\gamma^{\mu_3\mu_4\mu_5}\psi_{\mu_1} \right\}F_{\mu_3\mu_4\mu_5} - \frac{1}{16}e^{\frac{3\sigma}{4}}\left\{ \bar\psi_{\mu_1}\gamma^{11}\gamma^{\mu_1\cdots\mu_4}\psi_{\mu_2} + 2\bar\psi^{\mu_3}\gamma^{11}\psi^{\mu_4} \right.$$

$$\left. + \frac{3}{\sqrt2}\bar\lambda\gamma^{\mu_1}\gamma^{\mu_3\mu_4}\psi_{\mu_1} - \frac{5}{4}\bar\lambda\gamma^{11}\gamma^{\mu_3\mu_4}\lambda \right\}F_{\mu_3\mu_4} + \text{quartic} \right\}, \qquad (13.4.11)$$

where the covariant derivative acting on the spinors is just the usual covariant derivative with the spin connection, which is just a function of the vielbein.

The supersymmetry transformations of the IIA fields can be deduced by using the dimensional ansatz for the fields in the transformation of eleven-dimensional fields of equation (13.3.6). One finds that [10.11–10.13]:

$$\delta e_\mu{}^n = \epsilon\gamma^m\psi_\mu, \qquad \delta\sigma = \sqrt2\bar\lambda\gamma^{11}\epsilon,$$

$$\delta A_\mu = e^{-3\sigma/4}\left\{ \bar\psi_\mu\gamma^{11}\epsilon + \frac{3\sqrt2}{4}\bar\lambda\gamma_\mu\epsilon \right\},$$

$$\delta A_{\mu_1\mu_2} = e^{\sigma/2}\left\{ -\bar\psi_{[\mu_1}\gamma_{\mu_2]}\gamma^{11}\epsilon + \frac{1}{2\sqrt2}\bar\lambda\Gamma_{\mu_1\mu_2}\epsilon \right\},$$

$$\delta A_{\mu_1\mu_2\mu_3} = \frac{e^{-\sigma/4}}{2}\left\{ 3\bar\psi_{\mu_1}\gamma_{\mu_2\mu_3}\epsilon + \frac{1}{2\sqrt2}\bar\lambda\gamma^{11}\gamma_{\mu_1\mu_2\mu_3}\epsilon \right\}$$

$$+ 3e^{\sigma/2}\left\{ -A_{[\mu_1}\bar\psi_{\mu_2}\gamma_{\mu_3]}\gamma^{11}\epsilon + \frac{1}{2\sqrt2}A_{[\mu_1}\bar\lambda\gamma_{\mu_2\mu_3]}\epsilon \right\},$$

$$\delta\lambda = \frac{\sqrt2}{4}\hat D_\mu\sigma\gamma^\mu\gamma^{11}\epsilon - \frac{3}{16\sqrt2}e^{3\sigma/4}\gamma^{\nu_1\nu_2}\epsilon F_{\nu_1\nu_2} + \frac{1}{12\sqrt2}e^{-\sigma/2}\gamma^{\nu_1\nu_2\nu_3}\epsilon F_{\nu_1\nu_2\nu_3}$$

$$- \frac{1}{96\sqrt2}e^{\sigma/4}\gamma^{\nu_1\nu_2\nu_3\nu_4}\gamma^{11}\epsilon F_{\nu_1\nu_2\nu_3\nu_4} + \text{quadratic},$$

$$\delta\psi_\mu = D_\mu\epsilon - \frac{1}{64}e^{3\sigma/4}(\gamma_\mu{}^{\nu_1\nu_2} - 14\delta_\mu^{\nu_1}\gamma^{\nu_2})\gamma^{11}\epsilon F_{\nu_1\nu_2}$$

$$- \frac{1}{48}e^{-\sigma/2}(\gamma_\mu{}^{\nu_1\nu_2\nu_3} - 9\delta_\mu^{\nu_1}\gamma^{\nu_2\nu_3})\gamma^{11}\epsilon F_{\nu_1\nu_2\nu_3} + \frac{1}{128}e^{\sigma/4}$$

$$\times \left(\gamma_\mu{}^{\nu_1\nu_2\nu_3\nu_4} - \frac{20}{3}\delta_\mu^{\nu_1}\gamma^{\nu_2\nu_3\nu_4} \right)\gamma^{11}\epsilon F_{\nu_1\nu_2\nu_3\nu_4} + \text{quadratic}, \qquad (13.4.12)$$

where ϵ is the suitably defined parameter of local supersymmetry transformations. The reader will be able to extract from [10.11] the higher order terms in fermions in the action and transformations following the same procedure.

To put the action in the canonical form of equation (13.3.10) the reader should carry out the transformations

$$e_\mu{}^m \to e_\mu{}^m, \quad A_\mu \to A_\mu, \quad A_{\mu\nu} \to \frac{A_{\mu\nu}}{2}, \quad A_{\mu\nu\rho} \to \frac{A_{\mu\nu\rho}}{2},$$

$$\sigma \to \sigma, \quad \psi_\mu \to \frac{\psi_\mu}{2}, \lambda \to \frac{\lambda}{2}$$

on the action and transformation laws. We note that the Chern–Simmons term is then given by

$$+\frac{1}{2\kappa_{10}^2}\left\{\frac{3}{4 \times (12)^3}\epsilon^{\mu_1\cdots\mu_{10}}F_{\mu_1\cdots\mu_4}F_{\mu_5\cdots\mu_8}A_{\mu_9\mu_{10}}\right\}.$$

The IIA action has an SO(1, 1) invariance with parameter c that transforms the fields as

$$\sigma \to \sigma + c, \ A_\mu \to e^{-3c/4}A_\mu, \ A_{\mu\nu} \to e^{c/2}A_{\mu\nu}, \ A_{\mu\nu\rho} \to e^{-c/4}A_{\mu\nu\rho}, \quad (13.4.13)$$

while the vielbein in ten dimensions is inert. This symmetry has its origin in the eleven-dimensional theory and, in particular, the Weyl scalings of equation (13.3.6) given in the previous section. Indeed, if one carries out the transformations of equation (13.3.6) and a diffeomorphism on x^{10} of the form $x^{10} \to e^{-9\alpha}x^{10}$ one finds the transformations of equation (13.4.13) if we identify $\alpha = c/12$. For example, we find that $A_{\mu\nu} \to A'_{\mu\nu} = e^{-3\alpha}e^{9\alpha}A_{\mu\nu}$. The first factor of $e^{-3\alpha}$ is just that of equation (13.3.6) and the second factor of $e^{9\alpha}$ is a diffeomorphism once we recall that $A_{\mu\nu} = A_{\mu\nu 10}$.

Although the above diffeomorphism is a symmetry of the eleven-dimensional theory, the transformation of equation (13.3.6) is not, indeed it scales the Lagrangian L by a factor $L \to e^{-9\alpha}L$. We will now explain why the above combination is a symmetry of the ten-dimensional theory. Carrying out the diffeomorphism, from the active viewpoint, transforms $\int dx^{10} \to e^{-9\alpha}\int dx^{10}$ and so the Lagrangian L in eleven dimensions changes as $L \to e^{9\alpha}L$. Thus the combination of the two transformations leaves the Lagrangian invariant. However, in the dimensional reduction we neglect all the dependence of the fields on x^{10} in L which then becomes the ten-dimensional Lagrangian. Thus we have a symmetry in ten dimensions as the factor of $\int dx^{10}e_{10}{}^{10}$, which is an essential part of the eleven-dimensional theory, just leads to a factor of $2\pi R_{11}$ which multiples the Lagrangain of the ten dimensions. One can explain this from a slightly different viewpoint. The latter when combined with the factor κ_{11}^{-2} defines the ten-dimensional Newtonian coupling constant which is given by $\kappa_{10}^{-2} = 2\pi R_{11}\kappa_{11}^{-2}$. This has the dimensions of (mass)$^{-8}$, as it should. To keep the range of x^{10} the same under the diffeomorphism we must also scale R by $R \to e^{-9\alpha}R$. Under the transformation of equation (13.3.6) we find an invariant result if we also scale the eleven-dimensional Newtonian constant by $\kappa_{11}^{-2} \to e^{9\alpha}\kappa_{11}^{-2}$. Thus the ten-dimensional Newtonian constant is invariant under the two transformations and so what it multiplies, that is, the ten-dimensional action, must also be invariant. The phenomenon described here whereby a symmetry of the equations of motion becomes a symmetry of an action found by dimensional reduction also occurs much more generally than this example.

We now consider the sigma model approach to string theory and in particular how it gives rise to the IIA supergravity theory. This approach begins with the two-dimensional string action given by

$$-\frac{1}{4\pi\alpha'} \int d^2\xi \sqrt{-g} g^{\alpha\beta} (\partial_\alpha x^\mu \partial_\beta x^\nu g^s_{\mu\nu} - \epsilon^{\alpha\beta} \partial_\alpha x^\mu \partial_\beta x^\nu A_{\mu\nu})$$

$$+\frac{1}{4\pi} \int d^2\xi \sigma R^{(2)} + \cdots, \qquad (13.4.14)$$

where $R^{(2)}$ is the curvature scalar in two dimensions and $+\cdots$ stands for terms containing fermions. This is just the string action we have discussed in earlier chapters, such as chapter 2, except that it is coupled to the fields $g^s_{\mu\nu}$, $A_{\mu\nu}$ and σ, which are the metric, anti-symmetric tensor gauge field and dilaton which occur in the massless string spectrum. These background fields fields are taken to be functions of $x^\mu(\tau, \sigma)$. The meaning of the superscript s on the metric will become clear shortly. The constant α' has the dimensions of $(\text{mass})^{-2}$. The combination in front of the first term of the string action is called the string tension T_1 (that is, $T_1 \equiv 1/2\pi\alpha'$). The first term is just the usual string action discussed in chapter 2.

In this approach we demand that the two-dimensional action be conformally invariant as a quantum field theory [13.28]. Essentially one computes the beta functions associated with the couplings and sets them to zero. One finds that one recovers the tree-level string scattering equations and thus at lowest order in α' we obtain the terms of lowest order in space-time derivatives, that is, the supergravity equations of motion. However, the result is expressed in terms of the fields as defined in equation (13.4.14) and it turns out that these are not the same variables which we used above to construct the IIA supergravity theory. In fact, one finds that the Einstein term does not appear as eR above, but rather as $e^{-2\sigma} eR$. The reason for this precise form will be given shortly. Using equation (13.3.7) we conclude that the string vielbein $e_\mu^{s\,m}$ is related to the vielbein used in the above expression for the IIA supergravity theory by $e_\mu^{s\,m} = e^{1/4\sigma} e_\mu^{\,m}$. All the other fields are unchanged. We refer to the action which contains the usual Einstein term, that is, eR, as the action in the Einstein frame and the action derived from the sigma model approach as the action in the string frame. We note that the action of equation (13.4.14) does not include all the massless fields of the IIA string theory.

These considerations also apply to the bosonic strings. The only difference is the change to the string frame, which is carried out in all dimensions in section 17.4.

Making this change in the bosonic part of the IIA Lagrangian of equation (13.4.7) and dropping the 's' superscript on the string vielbein we find that the Lagrangian in the string frame becomes

$$L_B = \frac{1}{2\kappa_{10}^2} \left\{ ee^{-2\sigma} \left\{ R + 4\partial_\mu \sigma \partial^\mu \sigma - \frac{1}{3} F_{\mu_1\mu_2\mu_3} F^{\mu_1\mu_2\mu_3} \right\} \right.$$

$$+ \left\{ -\frac{1}{12} eF'_{\mu_1\cdots\mu_4} F'^{\mu_1\cdots\mu_4} - \frac{1}{4} eF_{\mu_1\mu_2} F^{\mu_1\mu_2} \right\}$$

$$\left. + \frac{1}{2\times 12^2} \epsilon^{\mu_1\cdots\mu_{10}} F_{\mu_1\cdots\mu_4} F_{\mu_5\cdots\mu_8} A_{\mu_9\mu_{10}} \right\}. \qquad (13.4.15)$$

As we have mentioned, the IIA supergravity theory is the lower energy limit of the IIA string theory, which is obtained as the string tension goes to zero (that is, $\alpha' \to \infty$). In this limit one is left with only the massless particles of the IIA supergravity theory. It will be very useful to know how these particles arise in the IIA string. This closed string theory in its formulation with manifest world surface supersymmetry, that is, the NS–R formulation studied in chapters 6 and 7, has four sectors: the $NS \otimes NS$, the $R \otimes R$, $R \otimes NS$ and the $NS \otimes R$ corresponding to the different boundary conditions that can be adopted for the two-dimensional spinor ψ_μ in the theory. The $NS \otimes R$ and $R \otimes NS$ sectors contain the fermions, while the $NS \otimes NS$ and the $R \otimes R$ sectors contain the bosons. It is straightforward to solve the physical state conditions in these sectors and, as explained in chapter 7, one finds that the bosonic fields of the IIA supergravity arise as

$$\underbrace{e_\mu^a, A_{\mu\nu}, \sigma,}_{NS \otimes NS} \quad \underbrace{A_{\mu\nu\varrho}, A_\mu,}_{R \otimes R} \quad \underbrace{\psi_{\mu\alpha}, \lambda_\alpha.}_{NS \otimes R \text{ and } R \otimes NS} \tag{13.4.16}$$

Looking at the IIA Lagrangian in the string frame of equation (13.4.15) we find that all the fields that arise in the $NS \otimes NS$ sector occur in a different way from those in the $R \otimes R$ sector: while the former have a factor of $e^{-2\sigma}$ the latter do not. This is related to the fact that the fields from the $R \otimes R$ sector do not appear in the two-dimensional string action of equation (13.4.14) and so one cannot compute their scattering using this formulation.

The IIA supergravity has two parameters. It contains the gravitational coupling constant κ_{10}, which appears as a factor of $1/\kappa_{10}^2$ multiplying the entire Lagrangian. It also contains the parameter $e^{\langle\sigma\rangle}$. The IIA string also has two parameters: the string tension T and the string coupling constant g_s. Since the low energy effective action of the string is IIA supergravity these two sets of parameters must be related to each other.

We first consider how the parameters arise in the string theory. In a second quantised formulation of string theory, one finds that the action can be written in a way where the string coupling only occurs as a prefactor of g_s^{-2}. Examples of such formulations are the light-cone gauge action and the gauge covariant action studied in chapters 4 and 18, and 13, respectively. Essentially the coupling g_s appears in front of the cubic vertex. However, just as in Yang–Mills theory one can can redefine the fields so as to absorb the g_s apart from an overall g_s^{-2} in front of the action. The parameter α' only occurs in these formulations through the masses of the the particles, or equivalently, the L_n operators that occur in these formulations. In the path integral formulation, the action becomes multiplied by $\hbar^{-1}$ and so we find that $\hbar$ and g_s only occur in the combination $\hbar g_s^2$. The power of Planck's constant measures the number of loops. Indeed, if we have any Feynman graph with n loops, I propagators and V vertices, the power of $\hbar$ associated with an n-loop graph is given by $\hbar^{(I-V)}$. Here we have used the topological relation $n = I - V + 1$, that each vertex carries a power of $\hbar$ and each propagator an inverse power of $\hbar$ and so we find the factor $\hbar^{(n-1)}$. Our previous discussion then implies that each n-loop diagram has a power $g_s^{(2n-2)}$ associated with it.

Now let us examine how the parameters arise in the effective action calculated from string theory. The first quantised string action in two dimensions of equation (13.4.14) contains for constant σ, the term $\langle\sigma\rangle\chi$, where χ is the Euler number which is given by $\chi = (1/4\pi) \int d^2\xi R^{(2)}$. For a closed Riemann surface of genus g it is given by $\chi = 2 - 2g$. The string sweeps out a Riemann surface and the number of holes in the surface is the genus and also the number of string loops, that is, $g = n$. Thus evaluating the path integral, which contains e^{-S} in Euclidean space, for n-loop string scattering we find a factor of $e^{(2n-2)\langle\sigma\rangle}$. Taken together with the considerations of the previous paragraph, we conclude that the IIA string coupling and the expectation value of the dilaton are related by $g_s = \langle e^\sigma \rangle$ [13.29]. At tree level, $g = 0 = n$ and so we have a factor of $e^{-2\langle\sigma\rangle}$ and since we will always obtain the gravity contribution eR at tree level we will always find a term $e^{-2\langle\sigma\rangle}eR$ in the effective action. This explains the result we assumed above for the appearance of this term. Since κ_{10} has the dimension of $(\text{mass})^{-4}$ it must be proportional to $(\alpha')^2$ and it is given by the relation

$$2\kappa_{10}^2 = (2\pi)^7 (\alpha')^4 e^{2\langle\sigma\rangle}.$$

The factor of $e^{\langle\sigma\rangle}$ follows from the remarks above, but the other factors require detailed computation.

We now finish our discussion of the IIA supergravity theory by noting some points that will be useful for our later discussions of string duality and brane dynamics. The IIA supergravity theory has the gauge fields A_μ, $A_{\mu\nu}$ and $A_{\mu\nu\varrho}$. This means the IIA theory has gauge fields of rank q, where $q = 1, 2, 3$, and these have corresponding field strengths of rank $q + 1 = 2, 3, 4$. Given a field strength $F_{\mu_1\cdots\mu_{q+1}}$ of rank $q + 1$ we can define the dual field strength by

$$F_{\mu_1\cdots\mu_{(D-1-q)}} = \frac{1}{(q+1)!}\epsilon_{\mu_1\cdots\mu_{10}} F^{\mu_D-q\cdots\mu_{10}}. \tag{13.4.17}$$

Hence, the duals of the above field strengths are forms of rank $q = 6, 7, 8$. When the original field strengths are on-shell we can, at least at the linearised level, write the dual field strengths in terms of dual gauge fields of ranks 5, 6 and 7. Hence the IIA theory has gauge fields of ranks $p = 1, 2, 3, 5, 6, 7$ if we include the dual gauge fields as well as the original ones. It is instructive to list the above gauge fields according to the string sector in which they arise. In the $NS \otimes NS$, we find gauge fields of ranks 2 and 6, while in the $R \otimes R$ sector we find gauge fields of ranks 1, 3, 5 and 7. We observe that classifying the gauge fields according to the different sectors splits them into fields of even and odd rank, respectively. The theory also contains a scalar σ whose 'field strength' has rank 1 and dual gauge field has rank 8. If we include these we find gauge fields of all ranks up to 8 with the exception of 4. The above discussion of duality relations can require modification at the non-linear level.

From equation (13.4.4) we read off the component of the vielbein associated with the circle to be $e_{10}^{10} = e^{(2\langle\sigma\rangle/3)}$ with corresponding metric $g_{10\,10} = e^{(4\langle\sigma\rangle/3)}$. The parameter R introduced into the defining range of the variable x^{10} has no physical meaning as it only parameterises the range of x^{10}. As discussed in equation (13.1.120) we can compute the physical radius of the circle in the eleventh dimension, which we denote by R_{11}. We find that $R_{11} = Re^{(2\langle\sigma\rangle/3)}$. As the string coupling is given by $g_s = e^{\langle\sigma\rangle}$, we find that

$$g_s = \left(\frac{R_{11}}{R}\right)^{3/2}. \tag{13.4.18}$$

The above relationship between the radius R_{11} of compactification of the eleven-dimensional theory and the IIA string coupling constant implies that as $R_{11} \to \infty$ we find that $g_s \to \infty$.

Thus in the strong coupling limit of the IIA string the radius of the circle of compactification becomes infinite suggesting that the theory decompactifies [13.30, 13.31].

We now consider some properties of the Kaluza–Klein modes which we have so far ignored in the reduction from eleven dimensions. Their masses are given by terms of the form $\{(e^{-1})_{11}^{\hat{\mu}}\partial_{\hat{\mu}}\bullet\}^2 = \{e^{-2\sigma/3}\partial_{11}\bullet\}^2$, where we have used equation (13.4.4) and $\bullet$ represents any field. Examining the expansion of equation (13.1.115) we find that the masses of Kaluza–Klein particles are given by $ne^{-2\langle\sigma\rangle/3}/R$ for integer n. However, using the relationship between R and R_{11} given just above we find that the masses of the Kaluza–Klein particles are given by n/R_{11}. The gauge field A_μ which originated from the eleven-dimensional metric couples to the Kaluza–Klein particles in a way which is governed by the derivative

$$(e^{-1})_m^{\hat{\mu}}\partial_{\hat{\mu}}\bullet = e^{\frac{1}{12}\sigma}(e^{-1})_m^{\mu}(\partial_\mu - A_\mu\partial_{11})\bullet,$$

where we have again used equation (13.4.4). As such, we find that the Kaluza–Klein particles have charges given by n/R for integer n. From the IIA string perspective, the A_μ field is in the $R \otimes R$ sector and so the Kaluza–Klein particles couple with these charges to the $R \otimes R$ sector. It turns out that IIA supergravity possesses solitonic particle solutions that have precisely the masses and charges of the Kaluza–Klein particles [13.30, 13.31]. Thus IIA supergravity, and so in effect the IIA string, knows about at least some of the particle content of the theory in eleven dimensions and not only the massless modes that arise after the compactification. It is this observation and that in the above paragraph that underlies the conjecture [13.30], [13.31] that the strong coupling limit of the IIA string theory is an eleven-dimensional theory, called M theory, whose low energy limit is eleven-dimensional supergravity.This idea is further discussed in section 17.4.

The IIA supergravity theory has an extension that also possesses maximal supersymmetry and has a cosmological constant. It can be thought of as a massive deformation, but up to the value of this parameter it is unique. In lower dimensions the maximal supergravity theories, whose only dimensional parameter is the gravitational constant, also admit massive deformations. These theories possess a cosmological constant, but are generally called gauged supergravities, because they generically contain vector fields that carry a Yang–Mills gauge symmetry. These theories will be discussed in section 17.5.

13.5   The IIB supergravity theory

The IIB supergravity theory has an underlying 32-component supersymmetry algebra that contains two Majorana–Weyl spinors of the same chirality. As we will see, it contains a fourth rank gauge field which has a self-dual rank 5 field strength. Such a field does not possess an action [13.32] and as a result the IIB theory also does not possess an action. Consequently, the usual Noether method cannot be used. Since the IIB supergravity theory cannot be obtained from eleven-dimensional supergravity by dimensional reduction it cannot be constructed using this method. The term cubic in fields of the IIB theory in the light-cone gauge Hamiltonian were constructed in [13.33]. The key to constructing the IIB theory was the realisation that one could use a variant of the Noether method [13.6] on the transformation laws alone. As explained earlier in section 13.1.1, this method begins with the rigid supersymmetry transformations and local Abelian transformations of the linearised theory. Letting the supersymmetry parameter become space-time dependent, the transformation laws no longer close; however we may still close the supersymmetry algebra order by order in κ by adding terms to the transformation laws provided we also identify the now local spinor parameter of the supersymmetry transformation with the spinor parameter that occurs in the local Abelian transformation of the gravitino. In this way the complete supersymmetry transformation laws of the IIB theory and the SU(1, 1) symmetry of the theory were found [10.14]. The fact that the two scalars of the theory belonged to the coset SU(1, 1)/U(1) considerably simplified the construction of the theory. The supersymmetry transformation laws only close subject to the field equations . and so in carrying out the closure one finds the field equations. This was carried out in [10.16], where the field equations up to terms quadratic in the fermions were found. At the same time the full field equations of IIB supergravity were found [10.15] in superspace and x-space using the on-shell superspace techniques of section 3.1.2.

The methods used in these calculations are explained in sections 13.1 and 13.2 and although the ideas are straightforward the calculations themselves are technically complicated.

13.5.1 The algebra and field content

The IIB supergravity is based on a supersymmetry algebra whose two supercharges Q^i_α, $i = 1, 2$, $\alpha = 1, \ldots, 32$ are Majorana–Weyl spinors of the same chirality. They therefore obey the conditions

$$(Q^i_\alpha)^* = Q^i_\alpha, \quad \gamma_{11} Q^i = Q^i. \tag{13.5.1}$$

As discussed in chapter 5 the Majorana relation involves the matrix B, but it turns out that in ten dimensions this can be chosen to be the identity matrix. The supersymmetry algebra was given in equation (5.5.8), which we rewrite here:

$$\{Q^i_\alpha, Q^j_\beta\} = (P_- \gamma_m C^{-1})_{\alpha\beta} \delta^{ij} P^m + (P_- \gamma_m C^{-1})_{\alpha\beta} Z^{ij}_m + (P_- \gamma_{m_1 m_2 m_3} C^{-1})_{\alpha\beta} \epsilon^{ij} Z^{m_1 m_2 m_3}$$

$$+ (P_- \gamma_{m_1 \cdots m_5} C^{-1})_{\alpha\beta} Z^{m_1 \cdots m_5 ij}, \tag{13.5.2}$$

where $P_- = \frac{1}{2}(I - \gamma_{11})$, and the central charge Z^{ij}_m is symmetric and traceless in its ij indices, $Z^{m_1 \cdots m_5 ij}$ is only symmetric in its ij indices, but self-dual, or anti-self-dual, in its $m_1 \cdots m_5$ indices. The central charges are important for the branes that exist in the IIB theory.

The IIB supersymmetry algebra includes an SO(2), or U(1), generator that rotates the $i, j, \ldots$ indices on the supercharges. However, it is often more useful to work instead with the complex Weyl supercharges $Q_\alpha = Q^1_\alpha + iQ^2_\alpha$, $\bar{Q}_\alpha = Q^1_\alpha - iQ^2_\alpha = (Q_\alpha)^*$. The corresponding U(1) generator denoted R $(R^\dagger = -R)$ acts on the supercharges as

$$[Q_\alpha, R] = iQ_\alpha, \quad [\bar{Q}_\alpha, R] = -i\bar{Q}_\alpha. \tag{13.5.3}$$

The field content of the IIB theory is the smallest irreducible representation of the above supersymmetry algebra and its corresponding off-shell field content is given by

$$e^m_\mu, \ A_{\mu\nu}, \ \varphi, \ B_{\mu\nu\rho\kappa}, \ \psi_{\mu\alpha}, \ \lambda_\alpha. \tag{13.5.4}$$

The fields $A_{\mu\nu}$ and φ are complex, while the gauge field $B_{\mu\nu\rho\kappa}$ is real. The spinors are complex Weyl spinors; the graviton $\psi_{\mu\alpha}$ is of the opposite chirality to λ_α, but has the same chirality as the supersymmetry parameter ϵ_α. Recalling our discussion above equation (4.1.2) we find that the fields of equation (13.5.4) lead to 35, 56, 2, 35, 112 and 16 on-shell degrees of freedom, respectively. In this count of states we have taken account of the fact that the gauge field $B_{\mu_1\mu_2\mu_3\mu_4}$ defines the linearised fifth-rank field strength $g_{\mu_1\mu_2\mu_3\mu_4\mu_5} \equiv 5\partial_{[\mu_1}B_{\mu_2\mu_3\mu_4\mu_5]}$, which satisfies the self-duality condition given by

$$g_{\mu_1\mu_2\mu_3\mu_4\mu_5} = \frac{1}{5!}\epsilon_{\mu_1\mu_2\mu_3\mu_4\mu_5\nu_1\nu_2\nu_3\nu_4\nu_5}g^{\nu_1\nu_2\nu_3\nu_4\nu_5} \equiv {}^*g_{\mu_1\mu_2\mu_3\mu_4\mu_5}. \tag{13.5.5}$$

Without the self-duality condition this gauge field corresponds to a particle that belongs to the fourth-rank totally anti-symmetric representation of the little group SO(8). The self-duality condition above corresponds to the constraint that this representation is self-dual and hence the 35 degrees of freedom given above. Thus the supermultiplet of IIB supergravity has the required 128 bosonic degrees of freedom and 128 fermionic degrees of freedom on-shell.

The fields of equation (13.5.4) transform under the U(1) transformations and their R weights are 0, 2, 4, 0, 1 and 3, respectively. Clearly, real fields must have R weight 0 and the gravitino must have the opposite R weight to the supercharge Q_α, since ψ_μ can be thought of as the gauge field of supersymmetry.

The standard action for a rank 4 gauge field would be of the form

$$\int d^{10}x \; g_{\nu_1\nu_2\nu_3\nu_4\nu_5}g^{\nu_1\nu_2\nu_3\nu_4\nu_5}, \tag{13.5.6}$$

where g is the 5-form associated with the gauge field. However, this action can be written in form language as $\int g \wedge {}^*g = \int g \wedge g = -\int g \wedge g = 0$. This discussion can be rephrased without using forms as follows: using the self-duality condition we can rewrite one of the field strengths in terms of $\star g$; swapping the indices on the ϵ symbol such that the last five indices are at the beginning, for which we incur a minus sign, and using the self-duality condition once more, we again find that the above action vanishes as it is equal to minus itself.

Clearly this result holds for any rank $2n + 1$ self-dual gauge field strength in a space-time of dimension $4n + 2$, $n \in \mathbf{Z}$ [13.32]. As there is no simple action for a fourth-rank gauge field, there is no action for the IIB theory itself. Although some actions have been constructed they are either not manifestly Lorentz invariant or involve some possibly unphysical fields and we will content ourselves with deriving the equations of motion. The reader will sometimes find in the literature an action written together with the self-duality constraint; while this is a convenient way of summarising the dynamics of the rest of the fields, it is not supersymmetric unless all the fields are on-shell.

The massless fields in the IIB string are just those that occur in the IIB supergravity and indeed the IIB supergravity is the theory that describes the effective action of the low energy limit of the IIB string. As explained in chapter 7 the massless modes of the the IIB string theory are contained in the physical state conditions which in the NS–R formulation has $NS \otimes NS$, $R \otimes R$, $R \otimes NS$ and $NS \otimes R$ sectors corresponding to the possible boundary conditions for the two-dimensional spinor ψ_μ in the theory. The last two sectors contain the fermions, while the $NS \otimes NS$ and the $R \otimes R$ sectors contain the bosons. The bosons arise in these two sectors as

$$\underbrace{e_\mu^a, \; A_{\mu\nu}^1, \; \phi,}_{NS \otimes NS} \quad \underbrace{A_{\mu\nu}^2, \; \chi, \; B_{\mu\nu\rho\tau},}_{R \otimes R} \tag{13.5.7}$$

where $A_{\mu\nu}^1$ and $A_{\mu\nu}^2$ are real fields which make up the complex fields $A_{\mu\nu}$, and we replace the previously discussed complex scalar φ by two real fields, χ and ϕ. The precise way in which these decompositions are defined will be specified later. The fact that the rank 2 gauge fields are split between the $NS \otimes NS$ and $R \otimes R$ sectors has important consequences for discussions of string duality. Since the physical state conditions in the $NS \otimes NS$ sector are exactly the same as in the IIA string we should not be surprised to find that this sector contains exactly the same bosonic field content.

The two scalars of the IIB theory belong to a non-linear realisation of the SU(1, 1) symmetry of the theory with local subgroup U(1). This non-linear realisation was explicitly worked out in section 13.2 and the above complex scalar φ is just that discussed there. The equation of motion of the IIB scalars must, in the absence of the other IIB fields, reduce to those of equation (13.2.49). Hence, the IIB equations of motion for the scalars must be of this form, but will also include other terms containing the superpartners of the scalars. We observe that the scalars only occur through P_μ or, for the derivatives of other fields with non-zero U(1) weight, through the U(1) connection Q_μ. Both these Cartan forms are given by equation (13.2.48). As these objects arise from the space-time derivative acting on the group element they have geometric dimension 1. This fact plays a crucial role in the way the scalars are encoded into the IIB theory as it allows the use of dimensional analysis to restrict the way the scalars can occur in the theory. This would not be the case if they could occur naked, that is, without a space-time derivative.

Let us now examine how the rank 2 tensor gauge field $A_{\mu\nu}$ can occur in the theory. The field strength of $A_{\mu\nu}$ is given by $\mathfrak{I}_{\mu_1\mu_2\mu_3} = 3\partial_{[\mu_1} A_{\mu_2\mu_3]}$. From the viewpoint of the SU(1, 1) non-linear realisation these, and all the other fields of the theory, must be matter representations, discussed in section 13.2. In fact, the gauge fields can only transform under rigid SU(1, 1) transform and must be inert under the local U(1) transformations, since if the gauge fields were to have a non-trivial U(1) transformation their field strengths would not transform covariantly due to the space-time derivative and the space-time dependence of U(1) transformations. One could attempt to avoid this conclusion by including the U(1) connection Q_μ in the definition of the field strength; however, then the corresponding field strength would not be invariant under the standard Abelian gauge transformations of the gauge field. These fields transform under a rigid $g_0 \in SU(1, 1)$ as

$$\begin{pmatrix} \bar{\mathfrak{I}}_{\mu\nu\rho} \\ \mathfrak{I}_{\mu\nu\rho} \end{pmatrix} \rightarrow D(g_0^{-1}) \begin{pmatrix} \bar{\mathfrak{I}}_{\mu\nu\rho} \\ \mathfrak{I}_{\mu\nu\rho} \end{pmatrix}. \tag{13.5.8}$$

As we are dealing with the doublet representation, $D(g_0^{-1}) = g_0$, where the latter element is that of equation (13.2.38) with 0 subscript added. As discussed in section 13.2, we can use the scalars to construct a non-linear realisation from the linear representation of the field strengths by defining

$$\begin{pmatrix} \bar{F}_{\mu\nu\rho} \\ F_{\mu\nu\rho} \end{pmatrix} \rightarrow D((g(\varphi))^{-1}) \begin{pmatrix} \bar{\mathfrak{I}}_{\mu\nu\rho} \\ \mathfrak{I}_{\mu\nu\rho} \end{pmatrix}, \tag{13.5.9}$$

where $D((g(\varphi))^{-1})$ is the inverse of the group element of equation (13.2.44). These new quantities are invariant under the rigid SU(1, 1) but transformation under local U(1) transformations as

$$\begin{pmatrix} \bar{F}_{\mu\nu\rho} \\ F_{\mu\nu\rho} \end{pmatrix} \rightarrow D(h) \begin{pmatrix} \bar{F}_{\mu\nu\rho} \\ F_{\mu\nu\rho} \end{pmatrix} \quad \text{or equivalently} \quad \bar{F}_{\mu\nu\rho} \rightarrow e^{-2ia}\bar{F}_{\mu\nu\rho},$$

$$F_{\mu\nu\rho} \rightarrow e^{2ia}\bar{F}_{\mu\nu\rho}, \tag{13.5.10}$$

where $D(h) = h$, and the latter group element is that of equation (13.2.42). To derive this result we have used that the scalars transform as $g \rightarrow g_0 g h$. Thus the $\bar{F}_{\mu\nu\rho}$ and $F_{\mu\nu\rho}$ have U(1) weights -2 and 2, respectively. We recall that the coset derivative of the scalars transforms as $P_\mu \rightarrow e^{4ia}P_\mu$, that is, P_μ has weight 4. Using the form of g given in equation (13.2.44) we readily find that

$$\begin{pmatrix} \bar{F}_{\mu\nu\rho} \\ F_{\mu\nu\rho} \end{pmatrix} = \frac{1}{\sqrt{1-\varphi\varphi^*}} \begin{pmatrix} 1 & -\varphi \\ -\varphi^* & 1 \end{pmatrix} \begin{pmatrix} \mathfrak{F}_{\mu\nu\rho} \\ \mathfrak{F}_{\mu\nu\rho} \end{pmatrix} = \begin{pmatrix} \mathfrak{F}_{\mu\nu\rho} - \varphi\mathfrak{F}_{\mu\nu\rho} \\ \mathfrak{F}_{\mu\nu\rho} - \varphi^*\mathfrak{F}_{\mu\nu\rho} \end{pmatrix}.$$

We note that this differs from that in reference [10.15], which uses the active approach.

The rank 4 gauge field is real and so must be a singlet under the SU(1, 1) and U(1) transformations. Its field strength, like that of the rank 2 fields, has geometric dimension 1.

13.5.2 The equations of motion

The derivation of the equations of motion is too complicated to reproduce here. However, many features of the equations can be deduced using the features of the IIB theory discussed above. The equations of motion must be invariant under general coordinate transformations and contain terms of the same U(1) weight and geometric dimension. Furthermore, given any one field its equation of motion must reduce to the required form in the absence of the other fields. In addition, the gauge fields can only occur in the objects $F^{\mu_1\mu_2\mu_3}$ and $\bar{F}^{\mu_1\mu_2\mu_3}$ of equation (13.5.9) as well as a rank 5 field strength $G_{\mu_1\cdots\mu_5}$. These field strengths all have geometric dimension 1. The kinetic term for the vielbein must occur through the usual Ricci tensor $R_{\mu\nu}$, which has geometric dimension 2 and as we discussed above the scalars belong to the non-linear realisation of SU(1, 1) with local subgroup U(1) and hence are only contained in the geometric dimension 1 objects P_μ or Q_μ. The latter can only occur as part of a covariant derivative of other fields. Finally, $F^{\mu_1\mu_2\mu_3}$, $\bar{F}^{\mu_1\mu_2\mu_3}$ and P_μ have the U(1) weights 2, -2 and 4, respectively, while $G_{\mu_1\cdots\mu_5}$ and $R_{\mu\nu}$ have U(1) weight 0 and as we are dealing with supergravity we only want bosonic terms which are second order in space-time derivatives.

Let us begin with the scalars; their equation of motion must generalise equation (13.2.49) and hence all the terms in the equation must have U(1) weight 4 and geometric dimension 2. The only possible candidate is $F_{\mu_1\mu_2\mu_3}F^{\mu_1\mu_2\mu_3}$. The equation for $A_{\mu\nu}$ must contain $D^{\mu_3}F_{\mu_1\mu_2\mu_3}$ and so it has geometric dimension 2 and U(1) weight 2. The only such terms we can add are $\bar{F}_{\mu_1\mu_2\mu_3}P^{\mu_3}$ and $G_{\mu_1\cdots\mu_5}F^{\mu_3\cdots\mu_5}$. The equation of motion for the fourth-rank gauge field is just the self-duality condition for a rank-five generalisation of the field strength which can only be the usual field strength as well as terms of the form

$i(A_{[\mu_1\mu_2}\Im^*_{\mu_3\cdots\mu_5]} - A^*_{[\mu_1\mu_2}\Im_{\mu_3\cdots\mu_5]})$ as this term is real, has geometric dimension 1 and U(1) weight 0.

The equations of motion of IIB supergravity in the absence of fermions are given by [10.15, 10.16]

$$D^\mu P_\mu = \tfrac{1}{6}F_{\mu_1\mu_2\mu_3}F^{\mu_1\mu_2\mu_3}, \tag{13.5.11}$$

$$D^{\mu_3}\bar{F}_{\mu_1\mu_2\mu_3} = -F_{\mu_1\mu_2\mu_3}\bar{P}^{\mu_3} - \tfrac{i}{6}G_{\mu_1\cdots\mu_5}F^{\mu_3\cdots\mu_5}, \tag{13.5.12}$$

$$R_{\mu\nu} = 2\bar{P}_{(\mu}P_{\nu)} + \bar{F}_{(\mu}{}^{\nu_1\nu_2}F_{\nu)\nu_1\nu_2} - \tfrac{1}{12}g_{\mu\nu}\bar{F}_{\mu_1\mu_2\mu_3}F^{\mu_1\mu_2\mu_3} + \tfrac{1}{96}G_\mu{}^{\mu_1\cdots\mu_4}G_{\nu\mu_1\cdots\mu_4}, \tag{13.5.13}$$

$$G_{\mu_1\cdots\mu_5} = {}^*G_{\mu_1\cdots\mu_5}, \tag{13.5.14}$$

where

$$G_{\mu_1\cdots\mu_5} = 5\partial_{[\mu_1}A_{\mu_2\cdots\mu_5]} + 10i(A_{[\mu_1\mu_2}\Im^*_{\mu_3\cdots\mu_5]} - A^*_{[\mu_1\mu_2}\Im_{\mu_3\cdots\mu_5]}). \tag{13.5.15}$$

The field transformations were originally found using the Noether method on the algebra [10.15], however, here we use the notations and conventions of [10.14]. They are given by

$$\delta e_\mu{}^m = \bar{\epsilon}\sigma^m\psi_\mu + \epsilon\sigma^m\bar{\psi}_\mu, \quad \delta v = -2\epsilon\lambda u, \quad \delta u = 2\bar{\epsilon}\bar{\lambda}v,$$

$$\delta(\bar{A}_{\mu\nu}, A_{\mu\nu})g = (\epsilon\sigma_{\mu\nu}\bar{\lambda} - 2\bar{\epsilon}\sigma_{[\mu}\bar{\psi}_{\nu]}, -\bar{\epsilon}\sigma_{\mu\nu}\lambda - 2\epsilon\sigma_{[\mu}\psi_{\nu]}) + \text{quadratic},$$

$$\delta A_{\mu\nu\rho\kappa} = 12i(A_{\mu\nu}\delta\bar{A}_{\rho\kappa} - \bar{A}_{\mu\nu}\delta A_{\rho\kappa}) + \text{quadratic},$$

$$\delta\lambda = \tfrac{1}{24}F_{\mu\nu\rho}\sigma^{\mu\nu\rho}\epsilon - \tfrac{1}{2}P_\mu\sigma^\mu\bar{\epsilon} + \text{quadratic}, \tag{13.5.16}$$

$$\delta\psi_\mu = \nabla_\mu\epsilon + \frac{3}{16}F_{\mu\nu\rho}\sigma^{\nu\rho}\epsilon - \frac{1}{48}F_{\nu\rho\kappa}\sigma_\mu{}^{\nu\rho\kappa}\bar{\epsilon}$$

$$- \frac{i}{192}G_{\mu\nu\rho\kappa\tau}\sigma^{\nu\rho\kappa\tau}\epsilon + \text{quadratic},$$

where + quadratic means the addition of terms quadratic in fermions. The u and v are the degrees of freedom in the SU(1, 1) non-linear realisation as given in equation (13.2.38). The ten-dimensional γ matrices are given by

$$\gamma^m = \begin{pmatrix} 0 & \sigma^m \\ \hat{\sigma}^m & 0 \end{pmatrix}, \tag{13.5.17}$$

where $\sigma^m = -i(I, \sigma^i)$, $\hat{\sigma}^m = -i(I, -\sigma^i)$ and σ^i, $i = 1, \ldots, 9$ obey $\{\sigma^i, \sigma^j\} = 2\delta^{ij}$ and a ten-dimensional spinor χ is written in terms of two Weyl spinors as

$$\chi = \begin{pmatrix} \phi \\ \bar{\phi} \end{pmatrix}. \tag{13.5.18}$$

More details can be found in [10.15]. Their form is also consistent with dimensional analysis, U(1) weight and the other features used above to restrict the form of the equations of motion. We note that [10.15] defined the F_{mnp} and $\bar{F}_{mnp}$ in terms of a row vector which can be found by taking the transpose of equation (13.5.9).

The equations of motion and field transformations to all orders in fermions are given in equations (10.4) and (10.6) of [10.15], provided one carries out the change of signature as explained at the end of section 13.3. The above formulae were also calculated in the subsequent reference [13.83] and found to agree.

13.5.3 The SL(2, **R**) version

The group $SU(1, 1)$ is isomorphic to the group $SL(2, \mathbf{R})$. Although this is only a cosmetic change, for some purposes it is better to formulate the theory in a manner where the $SL(2, \mathbf{R})$ form of the symmetry is manifest, As we explained in section 13.2, $g \in SU(1, 1)$ acts on the column vector $\begin{pmatrix} z_1 \\ z_2 \end{pmatrix}$ by $\begin{pmatrix} z_1 \\ z_2 \end{pmatrix} \to g \begin{pmatrix} z_1 \\ z_2 \end{pmatrix}$. If we denote the ratios of the column vector by $z = z_1/z_2$ then the action of $SU(1, 1)$ becomes

$$z \to \frac{uz + v}{v^*z + u^*}. \tag{13.5.19}$$

This action is such that it takes the unit disc $|z| \leq 1$ to itself. We can map the unit disc to the upper half-plane $H = \{w : \text{Im } w \geq 0\}$ by the transformation

$$z \to w = -i\left(\frac{1 - z}{1 + z}\right). \tag{13.5.20}$$

The action induced by the transformation of equation (13.5.19) on $w \in H$ is given by

$$w \to \frac{aw - b}{-cw + d}, \tag{13.5.21}$$

where $ad - bc = 1$ and a, b, c, d are real. In this last transformation we recognise the action of the group $SL(2, \mathbf{R})$, corresponding to the element

$$\hat{g}_0 = \begin{pmatrix} u & -b \\ -c & d \end{pmatrix} \in SL(2, \mathbf{R}). \tag{13.5.22}$$

It is well known that $SL(2, \mathbf{R})$ is the largest group which maps the upper half-plane to itself and so we should not be surprised that in mapping from the unit disc onto the upper half-plane the action of $SU(1, 1)$ becomes that of $SL(2, \mathbf{R})$. The precise relationship between the parameters of the two groups is given by

$$a = \frac{1}{2}(u + u^* - v - v^*), \quad b = \frac{i}{2}(-u + u^* - v + v^*),$$

$$c = -\frac{i}{2}(-u + u^* + v - v^*), \quad d = \frac{1}{2}(u + u^* + v + v^*). \tag{13.5.23}$$

For the $SU(1, 1)$ formulation of the IIB theory given above we found that the complex scalar field φ of equation (13.2.45) transformed under $SU(1, 1)$ in the same way as the variable z of equation (13.5.20). Hence, if we make the transformation from φ to the variable τ by

$$\varphi \to \tau = -i\left(\frac{1 - \varphi}{1 + \varphi}\right), \tag{13.5.24}$$

then τ transforms under $SL(2, \mathbf{R})$ just like w, that is,

$$\tau \to \frac{a\tau - b}{-c\tau + d}. \tag{13.5.25}$$

The reader may like to verify that the scalar SL$(2, \mathbf{R})$ action of equation (13.2.34) becomes the S$(1, 1)$ action of equation (13.2.49) using the above change of variable from τ to φ.

We could have constructed the IIB supergravity with an SL$(2, \mathbf{R})$ from the outset. The scalars would then belong to the non-linear realisation of SL$(2, \mathbf{R})$ with local subgroup SO(2) and we would consider a general element of SL$(2, \mathbf{R})$ given by

$$\hat{g}(x) = \begin{pmatrix} e(x) & f(x) \\ g(x) & h(x) \end{pmatrix},$$
(13.5.26)

which transforms as usual as $\hat{g}(x) \rightarrow \hat{g}_0 \hat{g}(x) \hat{k}(x)$, where $\hat{k}(x) \in$ SO(2) has the from

$$\hat{k}(x) = \begin{pmatrix} \cos\theta & \sin\theta \\ -\sin\theta & \cos\theta \end{pmatrix}$$
(13.5.27)

and $\hat{g}_0$ is given in equation (13.5.22). We can use the local subgroup to choose $\hat{g}(x)$ to be of the form

$$\hat{g}(x) = \begin{pmatrix} e^{-\phi/2} & \chi e^{\phi/2} \\ 0 & e^{\phi/2} \end{pmatrix}.$$
(13.5.28)

In carrying out a rigid $\hat{g}_0$ transformation on this coset representative we must carry out a compensating transformation before we can recognise the result on the fields ϕ and χ. However, a quicker method is to recognise that $\tau \equiv (ie - f)/(h - ig)$ transforms in a way that is independent of $\hat{k}$ transformations but that under a $\hat{g}_0$ transformation it changes as

$$\tau \rightarrow \frac{a\tau - b}{-c\tau + d}.$$
(13.5.29)

For our choice of representative above $\tau = -\chi + ie^{-\phi}$ and so we may identify this τ with the one that occurs in equation (13.5.24).

Given the above coset representative $\hat{g}(x)$ we can compute the Cartan form Ω_μ and then find P_μ and Q_μ. The results agree precisely with those above equation (13.2.33), where we computed these quantities using abstract group elements. To make contact between the two approaches one takes

$$E = \begin{pmatrix} 0 & 1 \\ 0 & 0 \end{pmatrix}, \quad F = \begin{pmatrix} 0 & 0 \\ 1 & 0 \end{pmatrix} \quad \text{and } H = \frac{1}{\sqrt{2}} \begin{pmatrix} 1 & 0 \\ 0 & -1 \end{pmatrix}$$
(13.5.30)

in the SL$(2, \mathbf{R})$ group element $g = e^{E\chi} e^{-(\phi/\sqrt{2})H}$ used in that section. One can also verify that the $\hat{k}$ above is given by $\hat{k} = e^{\theta(E-F)}$. Adding the vielbeins into the scalar action of equation (13.2.34) it becomes

$$\int d^d x \det e \operatorname{Tr} P_\mu P^\mu = \int d^d x \det e(-\tfrac{1}{2}\partial_\mu\phi\partial^\mu\phi - \tfrac{1}{2}\partial_\mu\chi\partial^\mu\chi e^{2\phi}).$$
(13.5.31)

It remains to find new variables for the rank 2 gauge field that transform in a recognisable way under SL$(2, \mathbf{R})$. Let us write the complex field strength $\Im_{\mu\nu\rho}$ that occurs in the SU$(1, 1)$ formulation as $\Im_{\mu\nu\rho} = \hat{\Im}_{2\mu\nu\rho} + i\hat{\Im}_{1\mu\nu\rho}$, then an explicit calculation, using (13.5.23), shows that the transformation law of equation (13.5.8) for $\Im_{\mu\nu\rho}$ becomes

$$\begin{pmatrix} \hat{\Im}_{1\mu\nu\rho} \\ \hat{\Im}_{2\mu\nu\rho} \end{pmatrix} \rightarrow \begin{pmatrix} d & b \\ c & a \end{pmatrix} \begin{pmatrix} \hat{\Im}_{1\mu\nu\rho} \\ \hat{\Im}_{2\mu\nu\rho} \end{pmatrix} = D(\hat{g}_0^{-1}) \begin{pmatrix} \hat{\Im}_{1\mu\nu\rho} \\ \hat{\Im}_{2\mu\nu\rho} \end{pmatrix},$$
(13.5.32)

where $\hat{g}_0$ is that given in equation (13.5.22).

Given the linearly transforming field strengths $\hat{\Im}_{i\mu\nu\rho}$, $i = 1, 2$ we can construct the non-linear transforming quantities using section 13.2:

$$\begin{pmatrix} \hat{F}_{1\mu\nu\rho} \\ \hat{F}_{2\mu\nu\rho} \end{pmatrix} = D((\hat{g}(x))^{-1}) \begin{pmatrix} \hat{\Im}_{1\mu\nu\rho} \\ \hat{\Im}_{2\mu\nu\rho} \end{pmatrix} = \begin{pmatrix} e^{\phi/2}(\hat{\Im}_{1\mu\nu\rho} - \chi\hat{\Im}_{2\mu\nu\rho}) \\ e^{-\frac{\phi}{2}}\hat{\Im}_{2\mu\nu\rho} \end{pmatrix}, \quad (13.5.33)$$

which now transform under just the local subgroup SO(2). Introducing the complex quantity $\hat{F}_{\mu\nu\rho}$ this transformation becomes

$$\hat{F}_{\mu\nu\rho} \equiv \hat{F}_{1\mu\nu\rho} + i\hat{F}_{2\mu\nu\rho} = e^{\phi/2}(\hat{\Im}_{1\mu\nu\rho} + \tau\hat{\Im}_{2\mu\nu\rho}) \rightarrow e^{-i\theta}\hat{F}_{\mu\nu\rho}. \quad (13.5.34)$$

Clearly, an SL(2, **R**) invariant action takes the form

$$\int d^{10}x \det e \hat{F}_{\mu_1\mu_2\mu_3}(\hat{F}_{\nu_1\nu_2\nu_3})^* = \int d^{10}x \det e(e^\phi(\hat{\Im}_1 - \chi\hat{\Im}_2)^2 + e^{-\phi}\hat{\Im}_2^2), \quad (13.5.35)$$

where $\hat{\Im}_1^2 = \hat{\Im}_{1\mu_1\mu_2\mu_3}\hat{\Im}_{1\mu_1\mu_2\mu_3}g^{\mu_1\nu_1}g^{\mu_2\nu_2}g^{\mu_3\nu_3}$, etc.

Making the transition to string frame $(e_E)_\mu{}^n = e^{-\phi/4}(e_s)_\mu{}^n$, the above actions for the scalars and the 3-form take the generic form

$$\int d^{10}x \det e_s \left(-\frac{1}{2}e^{-2\phi}\partial_\mu\phi\partial^\mu\phi - \frac{1}{2}\partial_\mu\chi\partial^\mu\chi \right.$$

$$\left. - \frac{1}{2.3!}(\hat{\Im}_1 - \chi\hat{\Im}_2)^2 - \frac{1}{2.3!}e^{-2\phi}\hat{\Im}_2^2 \right). \quad (13.5.36)$$

We note that the factors of e^ϕ in the first two terms are consistent with the fact that ϕ and χ are in the NS–NS sectors and R–R sectors, respectively, and we then conclude that $\hat{\Im}_{2\mu_1\mu_2\mu_3}$ and $\hat{\Im}_{1\mu_1\mu_2\mu_3}$ are in the NS–NS sectors and R–R sectors, respectively. This confirms our previous assignments.

Using the above equations we can substitute into the equations of motion (13.5.11)–(13.5.15) to find a formulation that is manifestly SL(2, **R**) invariant. One finds that the $NS \otimes NS$ fields (that is, $e_\mu{}^a$, $\hat{A}_{2\mu\nu}$ and ϕ) have identical equations of motion as those of the $NS \otimes NS$ sector of the IIA supergravity. In fact, since these fields do not include the rank 4 gauge field we can formulate the dynamics of the $NS \otimes NS$ sector of the theory in terms of an action and this action will have a Lagrangian which is the first term of equation (13.4.15), if we choose to work with the string metric.

We close with some comments on some of the features of the IIB theory that will be most relevant to our later discussions of string duality and brane dynamics. Like the IIA theory, the IIB theory has two coupling constants: the Newtonian constant κ, whose dependence we have suppressed, and the expectation value of $\langle e^\phi \rangle$. For the same reasons as we explained for the IIA string the latter plays the role of the IIB string coupling constant, that is, $g_s = \langle e^\phi \rangle$. In general, an SL(2, **R**) transformation changes from weak to strong string coupling. For example, the transformation $\tau = -\chi + ie^{-\phi} \rightarrow \tau' = -\chi' + ie^{-\phi'} = -\frac{1}{\tau}$ implies that

$$g'_s = \langle e^{\phi'} \rangle = \langle e^{-\phi} \rangle = \frac{1}{g_s} \quad (13.5.37)$$

if we take $\langle \chi \rangle = 0$. This obviously maps from weak coupling, g_s small, to the strong coupling, g_s large.

The SL$(2, \mathbf{R})$ transformation mixes the two real rank 3 field strengths $\hat{\mathfrak{F}}^i_{\mu\nu\rho}$, $i = 1, 2$, one of which arises from the $NS \otimes NS$ sector and the other from the $R \otimes R$ sector of the string theory. Here a generic SL$(2, \mathbf{R})$ transformation rotates the 3-form field strength in the $NS \otimes NS$ sector into that in the $R \otimes R$ sector and vice versa.

The IIB theory cannot be obtained from the eleven-dimensional supergravity by a straightforward dimensional reduction. However, if we were to reduce the IIB supergravity theory on a circle to nine dimensions, then the resulting supergravity theory would have an underlying supersymmetry algebra with 32 supercharges, which being an odd dimension has no notion of chirality, and so can only contain two 16-component Majorana spinors. This maximal 32-component supersymmetry algebra is unique and so it is precisely the same as the supersymmetry algebra that would result if we were to reduce the non-chiral IIA theory on a circle to nine dimensions. As a result the maximal supergravity theory in nine dimensions is also unique. That dimensionally reducing the IIA and IIB supergravity theories on a circle to nine dimensions leads to the same supergravity theory, see, for example, [10.11], will have important consequences for string theory.

13.6   Symmetries of the maximal supergravity theories in dimensions less than ten

All supergravity theories were found by using one of the four methods given in section 13.1. We found that the IIB supergravity theory possessed two scalars and these belonged to a non-linear realisation of SU$(1, 1)$ with local subgroup U(1) [10.14]. The eleven-dimensional supergravity theory does not possess any scalars, and the ten-dimensional IIA supergravity theory has only one scalar. However, all the other maximal supergravity theories can be obtained by dimensional reduction from either of these theories. As explained in section 13.1.4, as one reduces on a torus of larger and larger dimension one finds more and more scalars and in three dimensions one finds only scalars. It has been found that the scalars that belong to any supergravity multiplet occur in the action in a non-linear realisation. Although this is a consequence of supersymmetry, the symmetries that arise in the non-linear realisation are unexpected. This feature was first observed in the four-dimensional $N = 4$ supergravity theory, which possesses an SL$(2, \mathbf{R})$ symmetry with U(1) as local sub-group. Thus in this case the two scalars belong to a non-linear realisation of the group SL$(2, \mathbf{R})$ with local algebra U(1) [13.34]. One of the most intriguing such discoveries was that the maximal supergravity theory in four dimensions contains 70 scalars and has E_7 symmetry [13.35]. In particular, when the eleven-dimensional supergravity is dimensionally reduced on a k-torus to $11 - k$, $k = 1, \ldots, 8$ dimensions the resulting scalars can be expressed as a non-linear realisation [13.36, 13.37]. Table 13.6.1 gives the groups that arise and the corresponding local subgroups of the non-linear realisation. It follows from the observation at the end of the previous section that that one finds precisely the same theories if one dimensionally reduces the IIB supergravity theory on a $k - 1$ torus. We note that in this case the torus is one dimension less but that one begins with an SL$(2, \mathbf{R})$ symmetry.

The dimensional reduction of eleven-dimensional supergravity to three dimensions leads to a theory which is invariant under E_8 [13.37] and that to two dimensions leads to a theory

Table 13.6.1 Symmetries of the maximal supergravities

D	G	H
11	1	1
10, IIB	SL(2)	SO(2)
10, IIA	SO(1, 1)/Z_2	1
9	GL(2)	SO(2)
8	$E_3 \sim$ SL(3) × SL(2)	U(2)
7	$E_4 \sim$ SL(5)	USp(4)
6	$E_5 \sim$ SO(5, 5)	USp(4) × USp(4)
5	E_6	USp(8)
4	E_7	SU(8)
3	E_8	SO(16)

that is invariant under the affine extension of E_8 [13.38], which is also called E_9. As we have discussed, the two scalars in the IIB supergravity theory belong to a SU(1, 1), or SL(2, **R**)), non-linear realisation with local algebra U(1).

In this section we will use the constructions outlined in [13.39] and many of the technical steps required for the dimensional reductions carried out in this section can also be found in [13.40]. In the exposition we follow [13.41].

In section 13.2 we found the action which results from the dimensional reduction on a torus of a theory containing gravity and form gauge fields. The final result in three dimensions contains only scalar fields and was given in equation (13.1.127). We will now consider when such an action is a non-linear realisation. The latter possesses an action which was given in equation (13.2.32). The reader will immediately see that there are some similarities. Indeed, the notation has been deliberately set up so as to help make the comparison. The diagonal components of the metric in the dimensional reduction had the symbol ϕ. We also used the symbol ϕ for fields in the non-linear realisation which arise in the non-linear realisation from the Cartan sub-algebra. Comparing the two expressions it is clear that we should try to identify these two fields. The remaining scalar fields, that is, those that are the non-diagonal components of the metric and those arising from the gauge fields in the dimensional reduction, were called χ, while those arising from the positive roots of the algebra used in the non-linear realisation were also called χ and, as the notation suggests, we should identify these. Comparing the form of the kinetic term of the χ fields in the dimensional reduction in equation (13.1.127) and that in the non-linear realisation in equation (13.2.32) one realises that if the dimensional reduction leads to a non-linear realisation, then we should identify the vectors $\vec{\beta}$ in equation (13.1.127) as $\vec{\beta} = \sqrt{2}\vec{\alpha}$, where $\vec{\alpha}$ are the roots of the Lie algebra from which the non-linear realisation is constructed. In particular, we can see if the putative roots read off in this way have scalar products which form the Cartan matrix of Lie algebra. The Cartan matrix is formed from the scalar products of the simple roots. A brief discussion of such matters was given in section 13.2 and there is a more extensive discussion in chapter 17. This is clearly a necessary condition that we find in three dimensions a non-linear realisation, but it turns out to be in most cases also sufficient. We recall that the non-linear realisation we were examining in section 13.2 was constructed assuming that the local sub-group was the Cartan involution invariant sub-group.

Let us begin by considering pure gravity in D space-time dimensions with the usual action

$$S = \int d^D x e R. \tag{13.6.1}$$

We now carry out the dimensional reduction on an n-torus by taking one circle at a time. For the first circle we use equation (13.1.121) and find the Riemann scalar, a gravipho-ton (the $g_{D-1\mu}$, $\mu = 0, 1, \ldots, D-2$ components of the metric) and one scalar from the diagonal part of the metric on the circle (the g_{D-1D-1} component). At the next step we repeat the procedure on the Riemann scalar, using equation (13.1.121), but also on the graviphoton rank 2 field strength using equation (13.1.122) to produce a graviphoton $(g_{D-1\mu}\mu = 0, 1, \ldots, D-3)$, or rank 2 field strength, and a rank 1 field strength, that is, the derivative of one scalar. This scalar is the off-diagonal component of the metric g_{D-1D-2}. We also find one other scalar corresponding to the diagonal component of the circle on which we are dimensionally reducing.

The ϕ scalars are the diagonal components of the metric and, for this case, all the χ scalars must originate by creating second-rank field strengths from R and subsequently converting them to scalars. Let us first consider the class of scalars that arise from the curvature scalar in $D-k$ dimensions producing a 2-form field strength in $D-k-1$ dimensions and then turn this into a scalar $D-k-2$ dimensions. That is, for the first k circle reductions we just take the first term in equation (13.1.121), for the next reduction on a circle we take the third term in the same equation to arrive at a 2-form field strength and for the reduction on the next circle we take the last term in equation (13.1.122) applied to this field strength.

Reducing pure gravity in D dimensions on an n-torus results in $n-1$ such scalars with the putative roots

$$\vec{\beta}_k = \sqrt{2}\vec{\alpha}_k = (0, \ldots, 0, -2(D-k-2)\alpha_{D-k-1}, 2(D-k-4)\alpha_{D-k-2}, 0, \ldots, 0),$$

$$\tag{13.6.2}$$

where α_d is given in equation (13.1.117). The index k takes the range $k = 0, \ldots, n-2$ and there are k zeros on the left and $n-2-k$ zeros on the right. The first non-trivial entry arises from the scalar that appears in front of the 2-form in $D-k-1$ dimensions using equation (13.1.121) and the second non-trivial entry results from the conversion of the field strength to a scalar. This scalar arises from the metric component $g_{D-k-1D-k-2}$. It is not hard to see that the putative roots in the above equation satisfy

$$\vec{\alpha}_k \cdot \vec{\alpha}_l = \begin{cases} 2 & k = l, \\ -1 & |k-l| = 1, \\ 0 & |k-l| \geq 2. \end{cases} \tag{13.6.3}$$

We define a root to be positive if its first non-zero component from the *right* is positive and with this definition the $\vec{\alpha}_k$ above are positive roots. Note one usually starts from the left, but it is a matter of choice and the above is the choice which is useful here. With this definition all the above roots are positive.

We now find the roots associated with the rest of the χ scalars. As opposed to above, one can do this by not going directly from a 2-form to a scalar, that is, having created a second-rank field strength one does not convert it to a scalar in the reduction on the next circle. This results in an additional $\frac{1}{2}(n-1)(n-2)$ scalars which arise from g_{ij}, $i \geq j\, i, j = 1, \ldots, n$, where $i \neq j$ and $i \neq j-1$. Their associated roots take the form $\vec{\alpha}_k + \vec{\alpha}_{k+1} + \cdots + \vec{\alpha}_{k+l}$, where $k = 0, \ldots, n-2$ and $l = 1, \ldots, n-k-2$. As such, the roots of equation (13.3.6) are a basis of all are roots and so they are simple roots. From equations (13.6.3) and (13.2.26) we can read off the putative Cartan matrix. We recognise that it does have all the good properties to be a Cartan matrix, 2s on the diagonal, negative integers or 0s off the diagonal, the zeros are symmetrically distributed, and is a positive definite matrix. As such, it must correspond to one of the finite dimensional semi-simple Lie algebras listed by Cartan. In fact, the corresponding Dynkin diagram is that of A_{n-1}, or $SL(n)$,

$$\bullet \;-\; \bullet \;-\; \cdots \;-\; \bullet \;-\; \bullet$$

Diagram 13.6.1

for $n \leq D - 4$. Thus dimensional reduction of gravity in D dimensions on an n-torus leads to scalar fields which for $n \leq D - 4$ can belong to a non-linear realisation of $SL(n)$. The Cartan involution invariant sub-group is $SO(n)$. This implies there are $n^2 - 1 - \frac{1}{2}n(n-1)$ scalars which indeed agrees with the number of scalars we have found, that is, those in g_{ij} minus 1. The missing scalar is one of the n scalars arising from the diagonal components of the metric, $n-1$ of them are needed for the Cartan sub-algebra of $SL(n)$ and the final one results in an additional $U(1)$, which leads to $GL(n)$.

The above holds when the dimension of the final theory is 4 or above. However, when we arrive in three dimensions (that is, for $n = D - 3$) there is an additional effect: we can use equation (13.1.125) to dualise the graviphotons into scalars. Let us dualise the first such graviphton that arises (that is, after compactifaction to $D - 1$ dimensions on a single S^1). This is essentially dualising the $g_{D-1\mu}$, $\mu = 0, 1, 2$, component of the metric to find a scalar. This leads to the root

$$\sqrt{2}\vec{\delta} = (2(D-2)\alpha_{D-1}, 2\alpha_{D-2}, 2\alpha_{D-3}, \ldots, 2\alpha_3). \tag{13.6.4}$$

This root is positive. As its first entry is positive, it cannot be written as a sum of the roots of equation (13.6.3) and so it is also simple. One can show that

$$\vec{\delta} \cdot \vec{\alpha}_k = \begin{cases} 0 & k \neq 0, \\ -1 & k = 0, \end{cases} \qquad \vec{\delta} \cdot \vec{\delta} = 2. \tag{13.6.5}$$

Dualising the remaining vectors in three dimensions leads to $D - 4$ non-simple positive roots of the form $\vec{\delta} + \vec{\alpha}_0 + \vec{\alpha}_1 + \cdots + \vec{\alpha}_l$, $l = 0, \ldots, D - 3$. Thus we find that the action of gravity dimensionally reduced to three dimensions leads to scalars which belong to a non-linear realisation with algebra A_{D-3} or $SL(D-2)$ whose associated Dynkin diagram is given by

$$\bullet \;-\; \bullet \;-\; \bullet \;-\; \cdots \;-\; \bullet \;-\; \bullet \;-\; \bullet$$

Diagram 13.6.2

We note that two simple roots, the first and last in the diagram, appear upon compactification to three dimensions. Thus the dualisation of the graviphotons leads to symmetry enhancement in three dimensions above what one might normally expect. In particular, if we go from four to three dimensions one finds two scalars that belong to a non-linear realisation of $SL(2)$. This symmetry was discovered long ago [13.42] in the reduction of four-dimensional gravity and it has played an important role in the generation of solutions. We note that dualising the graviphoton was crucial to obtaining the enhancement.

Let us now consider the dimensional reduction of gravity coupled to a 3-form field strength $H_{\mu\nu\rho}$ and a dilaton ϕ in D dimensions. In particular, we will consider the action

$$S = \int d^D x e (R - \tfrac{1}{2}\partial_\mu \phi \partial^\mu \phi - \tfrac{1}{12} e^{\beta\phi} H_{\mu\nu\lambda} H^{\mu\nu\lambda}), \qquad (13.6.6)$$

where β is a constant. For $\beta = \sqrt{8/D-2}$, equation (13.6.6) is the effective action of the closed bosonic string (with the tachyon set to zero) or the $NS \otimes NS$ sector of the IIA, or IIB, supergravity theories. The reduction of equation (13.6.6) was discussed in [13.43].

Clearly, if we dimensionally reduce this system on an n torus we find the same scalars and roots that we did for pure gravity, coming from the eR term in equation (13.6.6). However, for this action we also begin with an additional scalar ϕ already in D dimensions. The existence of ϕ implies that the putative root vectors are now $(n+1)$-dimensional. In particular the roots $\vec{\alpha}_k$ discussed above for gravity gain an extra column on their left containing a zero. The reader will recall that we use the symbol ϕ for all scalars that will eventually belong to the Cartan sub-algebra and it is their position on the vector which records their origin. In particular, the scalars not in the first column arise from the diagonal components of the metric.

Let us now consider the scalars and their roots that originate from the 3-form. The first scalar from $H_{\mu\nu\rho}$ arises from the reduction on the first two circles, or a 2-torus. It results from converting the field strength $H_{(3)}$ to a $H_{(2)}$ and then to a $H_{(1)}$ using equation (13.1.122). We recognise $H_{(1)}$ as the derivative of a scalar. The scalar is given by $A_{D-1 D-2}$. The resulting putative root is given by

$$\sqrt{2}\vec{\rho} = (\beta, 2(D-4)\alpha_{D-1}, 2(D-4)\alpha_{D-2}, 0, \ldots, 0), \qquad (13.6.7)$$

where there are $n-2$ zeros on the right. A little work shows that

$$\vec{\rho} \cdot \vec{\alpha}_k = \begin{cases} 0 & k \neq 1, \\ -1, & k = 1, \end{cases} \qquad \vec{\rho} \cdot \vec{\rho} = \frac{\beta^2}{2} + 2\frac{D-4}{D-2}. \qquad (13.6.8)$$

Unlike the case of gravity this new root leads to scalar products which do not necessarily belong to the Cartan matrix of a Kac–Moody algebra. In particular, we must ensure that $\vec{\rho} \cdot \vec{\rho}$ is an integer. For $D > 6$ the only possibility is to fix $\vec{\rho} \cdot \vec{\rho} = 2$, that is, $\vec{\rho}$ is the same length as the $\vec{\alpha}_k$ roots. This, in turn, determines the coupling to the 3-form to be

$$\beta = \sqrt{\frac{8}{D-2}}. \qquad (13.6.9)$$

This dilaton coupling β is precisely that which is found in the the bosonic string effective action in D dimensions and the IIA supergravity theory.

Continuing with the compactification on an n-torus, with $n \leq D-5$, we find additional scalars and their (positive) roots. However, they are not simple, but rather of the form $\vec{\rho} + \sum_k n_k \vec{\alpha}_k$, where n_k are positive integers. The simple roots are summarised by the Dynkin diagram D_n:

Diagram 13.6.3

whose maximally non-compact form is $O(n, n)$. Thus the scalars should belong to the non-linear realisation $O(n, n)$ with local sub-algebra $O(n) \otimes (n)$, which is indeed the Cartan involution invariant subgroup.

Upon compactification to four dimensions we may dualise $H_{\mu\nu\lambda}$ to obtain a scalar and its associated root

$$\sqrt{2}\vec{\gamma} = (-\beta, 4\alpha_{D-1}, 4\alpha_{D-2}, \ldots, 4\alpha_4). \tag{13.6.10}$$

One can explicitly check that $\vec{\gamma}$ is simple and satisfies

$$\vec{\gamma} \cdot \vec{\alpha}_k = 0, \quad \vec{\gamma} \cdot \vec{\rho} = 0, \quad \vec{\gamma} \cdot \vec{\gamma} = \frac{\beta^2}{2} + 2\frac{D-4}{D-2}, \tag{13.6.11}$$

where $k = 0, \ldots, D - 6$. Note that $\vec{\gamma}$ automatically has the same length as $\vec{\rho}$.

Continuing to three dimensions we obtain more putative roots by dualising the vectors. However, in contrast to the gravity case we find that none of these give simple roots. In particular, we now find that $\vec{\delta} = \vec{\rho} + \vec{\gamma} + \vec{\alpha}_0 + \cdots + \vec{\alpha}_{D-5}$ and hence $\vec{\delta}$ is no longer simple. In addition, we find that $\vec{\gamma} \cdot \vec{\alpha}_{D-5} = -1$. Thus the simple roots are those of the Dynkin diagram D_{D-2} and corresponding maximally non-compact group is $O(D - 2, D - 2)$:

Diagram 13.6.4

Here the roots on the bottom row are the roots $\vec{\alpha}_k$ from the gravity sector and appear from right to left as we compactify, with the first (right most) simple root arising in $D - 2$ dimensions and the last (left most) simple root arising in three dimensions. In the top row the right root $\vec{\rho}$ appears after compactification to $D - 2$ dimensions and the left root $\vec{\gamma}$ appears upon compactification to four dimensions.

We conclude by noting that for $D \leq 6$ there is another possible choice for β that leads to an integral Cartan matrix. Namely, we could set $\beta = \sqrt{12 - 2D/D - 2}$, so that the length of $\vec{\rho}$ and $\vec{\gamma}$ is $\sqrt{2}$. However, in this case we find that $\vec{\rho} \cdot \vec{\gamma} = 2$, which is positive and hence cannot be identified with a Kac–Moody algebra.

Finally we can consider gravity coupled to a 4-form $G_{\mu\nu\lambda\rho}$ with no scalar in D dimensions. In eleven dimensions this is the bosonic sector of eleven-dimensional supergravity if we neglect the Chern–Simons term as we are doing here. In this case we start with the action

$$S = \int d^D x e (R - \tfrac{1}{48} G_{\mu\nu\lambda\rho} G^{\mu\nu\lambda\rho}). \tag{13.6.12}$$

Clearly when we dimensionally reduce this action on an n-torus we obtain all of the roots we found from the gravity sector alone. However, in addition we find scalars and their roots from compactifying the 4-form. Let us first consider the scalar that arises by taking the indices of the rank 3 gauge field to be on the first tree circles on a 3-torus. That is, we use equation (13.1.122) to change from $G^2_{(4)}$ to $G^2_{(3)}$, then to $G^2_{(2)}$ and finally to $G^2_{(1)}$ which we recognise as the derivative of a scalar which is essentially the $A_{D-1D-2D-3}$ component of the gauge field. In this process one finds exponentials of the diagonal components of the metric resulting in the putative root

$$\sqrt{2}\vec{\epsilon} = (2(D-5)\alpha_{D-1}, 2(D-5)\alpha_{D-2}, 2(D-5)\alpha_{D-3}, 0, \ldots, 0), \tag{13.6.13}$$

where there are $n - 3$ zeros on the right. It is straightforward to check that

$$\vec{\epsilon} \cdot \vec{\alpha}_k = \begin{cases} 0 & k \neq 2 \\ -1 & k = 2 \end{cases}, \quad \vec{\epsilon} \cdot \vec{\epsilon} = 3\frac{D-5}{D-2}. \tag{13.6.14}$$

Further compactification to lower dimensions introduces additional scalars and their roots, but one can check that none of these are simple.

In addition, in five dimensions, we can dualise $G_{\mu\nu\lambda\rho}$ to obtain a new root $\vec{\eta}$, with the same length as $\vec{\epsilon}$. While $\vec{\eta}$ is always a positive linear combination of $\vec{\epsilon}$ and $\vec{\alpha}_k$, the coefficients are integers only if $D = 3a + 5$ for a non-negative integer a. Similarly in three dimensions $\vec{\delta}$ is a positive linear combination of the other roots but only with integer coefficients if $D = 3b + 2$ for a non-negative integer b.

Thus only eleven-dimensional supergravity M theory has $\vec{\epsilon} \cdot \vec{\epsilon} = 2$ and we find that the dimensional reduction to three dimensions results in a theory which is a non-linear realisation with algebra E_8 and local sub-algebra SO(16), which is the Cartan involution invariant subgroup. The reader can verify that one finds 128 scalars in this non-linear realisation which is the number we have just found from dimensional reduction. The Dynkin diagram of E_8 is

Diagram 13.6.5

As above, the bottom row of simple roots $\vec{\alpha}_k$ are those of the gravity sector and appear from right to left as we compactify, starting with the reduction to nine dimensions on the right and ending in three dimensions on the left. The root $\vec{\epsilon}$ on the top row appears in eight dimensions. For any other dimension the Cartan metric does not have integer entries. Hence only eleven-dimensional M theory possesses a coset symmetry.

If we dimensionally reduce to seven dimensions using only a 7-torus, then we find scalars that belong to a non-linear realisation with algebra E_7 and local sub-algebra SU(8) as well as vector gauge fields and, of course, gravity. The algebras in higher dimensions are obtained by deleting dots from the right of the E_8 Dynkin diagram.

In all the above we have tested a necessary condition for a theory when dimensionally reduced on a torus to possess scalars which belong to a non-linear realisation. However, this does not by itself guarantee that the theory really is a non-linear realisation. In fact for all the theories considered above this is the case. Although we only considered the terms bilinear in the χ fields, the complete non-linear realisation is found by considering all terms in the original theory before dimensional reduction. In the case of eleven-dimensional supergravity this includes the Chern–Simons term. In three dimensions we have just scalars, but in higher dimensions we generally have other fields which are either inert under the symmetry or belong to matter representations of the non-linear realisation, as discussed in equation (13.2.24). Clearly, for the theory to have the symmetry that occurs in the non-linear realisation of the scalars, the other fields must couple to the scalars and themselves in such a way as to respect the symmetry. In all the theories considered above this is indeed the case no matter what the dimension of the torus used to do the reduction.

One can also consider a further reduction to two dimensions of pure gravity, the effective action of the bosonic string and eleven-dimensional supergravity. However, in three dimensions and less, gravity has no degrees of freedom and so further compactification will not lead to new scalars in the reduced Lagrangian. In fact, the symmetry of the system is increased upon compactification to two dimensions. In particular, as described in [13.38], the rigid G symmetry of the non-linear realisation found in three dimensions is enlarged to be the corresponding affine symmetry when the theory is reduced to two dimensions. Thus the total symmetry group is the affine version of the above groups G whose Dynkin diagram is obtained by adding a single root to the Dynkin diagrams above. These algebras are discussed in chapter 16.

For the wide class of theories we considered, we see that only a very small set of actions lead upon dimensional reduction to coset symmetries. In particular, we found that type actions consisting of gravity coupled to a 3-form gauge field only lead to a non-linear realisation in the familiar case of the bosonic sector of eleven-dimensional supergravity. On the other hand we found that actions consisting of gravity coupled to a 2-form gauge field and a dilaton in any dimension lead to a non-linear realisation, but only for a precise coupling of the dilaton to the 2-form. These two cases are associated with M theory and string theory, respectively. From the point of view of dimensional reduction the emergence of a coset structure is somewhat miraculous. The discussion [13.41] given here is complementary to that of [13.40], in that we have start with a family of higher dimensional actions and ask which ones lead to a coset structure in three dimensions, whereas [13.40] started with a coset structure in three dimensions and then lifted it up to a higher dimensional theory. This latter process is called oxidation and one can find theories which upon dimensional reduction to three dimensions lead to non-linear realisations for all the possible finite dimensional semi-simple Lie algebras listed by Cartan [13.40].

A more efficient, but more sophisticated in terms of group theory, way of deriving the results of this section was given in [13.22].

13.7   Type I supergravity and supersymmetric Yang–Mills theories in ten dimensions

Type I supergravity was the first supergravity theory in ten dimensions to be constructed [10.17, 10.18]. Its underlying algebra contains a Majorana–Weyl spinor supercharge whose anti-commutator was given in equation (5.5.4). The field content is given by

$$\underbrace{e_\mu^a, \phi,}_{NS\otimes NS} \quad \underbrace{A_{\mu\nu},}_{R\otimes R} \quad \text{plus} \quad \underbrace{\psi_{\mu\alpha}, \lambda_\alpha,}_{NS\otimes R} \tag{13.7.1}$$

where we have also indicated the sectors in the type I closed string from which they come. The gravitino $\psi_{\mu\alpha}$ and the spinor λ_α are Majorana–Weyl spinors of opposite chirality and all the bosonic fields are real.

As explained at the end of chapter 7, this theory can be obtained from the IIB theory by a projection which corresponds in the string theory to world-sheet parity, that is, it exchanges left and right moving modes. In the IIB theory the projection operator Ω acts on the fields so as to change the signs of $\Im_{2\mu\nu\rho}$, χ and $B_{\mu\nu\rho\sigma}$, which are in the $NS \otimes NS$, $R \otimes R$ and $R \otimes R$

sectors, respectively, but leaves inert e_μ^a, $\Im_{1\mu\nu\rho}$ and ϕ which are in $NS \otimes NS$, $R \otimes R$ and $NS \otimes NS$ sectors, respectively. To recover type I supergravity we keep only fields which are left inert by Ω and so we find only the latter fields. On the spinors we impose a Majorana condition which leads to a gravitino and another fermion which are both Majorana–Weyl although of opposite chirality.

We can also obtain type II supergravity from IIA supergravity theory. We impose the obvious Weyl conditions on the gravitino and spinor in the IIA theory, and to be consistent with supersymmetry, we must also set $A_{\mu\nu\rho} = 0 = A_\mu$. It is straightforward to find the Lagrangian for the bosonic fields by truncating the corresponding action for IIA theory of equation (13.4.7).

The $N = 1$ Yang–Mills theory is based on the supersymmetry algebra with one Majorana–Weyl supercharge, that is, the same supersymmetry algebra as type I supergravity. It consists of a gauge field A_μ and one Majorana–Weyl spinor λ_α. Since it does not contain a gravitino it is only invariant under a rigid supersymmetry. This theory [5.1, 10.19] is easily derived. We first write down the linearised transformation laws that are determined up to two constants by dimensional analysis. The constants are then fixed by demanding that the supersymmetry transformations and the linearised gauge transformations form a closed algebra. The full theory is uniquely found by demanding that the algebra closes and that the action be invariant under the usual non-Abelian gauge transformation of the gauge field. The result is given by

$$\int d^{10}x \operatorname{Tr}(-\tfrac{1}{4}F_{\mu\nu}F^{\mu\nu} - \tfrac{1}{2}\bar\lambda\gamma^\mu D_\mu\lambda), \tag{13.7.2}$$

which is invariant under

$$\delta A_\mu^i = \bar\epsilon\gamma_\mu\lambda^i, \quad \delta\lambda^i = -\tfrac{1}{2}F_{\mu\nu}^i\gamma^{\mu\nu}\epsilon, \tag{13.7.3}$$

where

$$F_{\mu\nu} = \partial_\mu A_\nu - \partial_\nu A_\mu + g[A_\mu, A_\nu] \tag{13.7.4}$$

is the Yang–Mills field strength and

$$D_\mu\lambda = \partial_\mu\lambda + g[A_\mu, \lambda]. \tag{13.7.5}$$

Varying the action of equation (13.7.2) we find that

$$\delta A = \int d^{10}x\{-F^{\mu\nu}\bar\epsilon\gamma_\nu D_\mu\lambda - \tfrac{1}{2}\gamma^{\mu\nu}F_{\mu\nu}\gamma^\rho D_\rho\lambda\}, \tag{13.7.6}$$

where we have used the result $\delta F_{\mu\nu} = \bar\epsilon\gamma_\nu D_\mu\lambda - \bar\epsilon\gamma_\mu D_\nu\lambda$. Using the identity $\gamma^{\mu\nu}\gamma^\rho = \gamma^\mu\eta^{\nu\rho} - \gamma^\nu\eta^{\mu\rho} + \gamma^{\mu\nu\rho}$, we find the variation of equation (13.7.6) vanishes at first order in the fermions, provided we use the Bianchi identity, $D_{[\mu}F_{\nu\rho]} = 0$. The cubic terms in the fermions are more complicated and to see they vanish one must use the Fierz identity explained in chapter 6.

We can dimensionally reduce this theory from ten dimensions to d dimensions on a $10 - d$ torus keeping only the massless modes. This preserves the number of supersymmetries and so we find supersymmetric theories with 16 supersymmetries. We take the gauge field to be given by

$$
A_{\hat{\mu}} = \begin{cases} A_\mu, & \text{for} \quad \hat{\mu} = \mu = 0, 1, \ldots, d-1, \\ \phi_i & \text{for} \quad \hat{\mu} = i = d, \ldots, 9. \end{cases} \tag{13.7.7}
$$

The resulting bosonic part of the action is given by

$$
\int d^d x \text{Tr} \left(-\tfrac{1}{4} F_{\mu\nu} F^{\mu\nu} - \tfrac{1}{2} \sum_i D_\mu \phi_i D^\mu \phi_i - \tfrac{1}{4} \left(\sum_{i,j} [\phi_i, \phi_j]^2 \right) \right), \tag{13.7.8}
$$

where $D_\mu \phi_i = \partial_\mu \phi_i + g[A_\mu, \phi_i]$. In four dimensions we find the well-known $N = 4$ supersymmetric Yang–Mills theory [5.1], indeed this was how it was found.

As the $N = 1$ Yang–Mills theory in ten dimensions and the type I supergravity are based on the same supersymmetry algebra, we can construct a coupling between them. This was found in [10.20]. If we take the gauge group to be SO(32), or $E_8 \otimes E_8$, we find the theory that results from the low energy limit of the corresponding heterotic string theory, or in the former case the type I string theory, with these gauge groups.

13.8   Solutions of the supergravity theories

As explained earlier in this chapter, the IIA and IIB supergravity theories are the complete low energy effective actions of the IIA and IIB superstrings; by complete we mean that they contain all perturbative and non-perturbative effects. Our understanding of superstring theory is rather limited in that there is, even as a matter of principle, no really effective way of systematically computing non-perturbative effects and indeed there is no underlying theory of what string theory really is. As a result the IIA and IIB supergravity theories have played a crucial role in the development of superstring theory. One important aspect of this has been the discovery of solutions corresponding to branes in these theories and one consequence of this has been that a fundamental theory must include branes as well as strings. A similar story applies to the eleven-dimensional supergravity theory, which is thought to be one of the limits of the ill-understood M theory. In this section we will find the most important brane solutions of the eleven-dimensional and IIA and IIB supergravity theories.

13.8.1 Solutions in a generic theory

Supergravity theories, in general, consist of gravity coupled to scalars and gauge fields of various ranks plus fermions. As we are considering solutions we can forget about the fermionic terms in the action in this section. In order to be as general as possible we will consider solutions to a generic theory in D dimensions of the form

$$
A = \frac{1}{2\kappa_D^2} \int d^D x \sqrt{-g} \left(R - \frac{1}{2} \partial^\mu \phi \partial_\mu \phi - \sum_i \frac{1}{2 \times n_i!} e^{a_i \phi} F_{\mu_1 \ldots \mu_{n_i}} F^{\mu_1 \ldots \mu_{n_i}} \right), \tag{13.8.1}
$$

where $F_{\mu_1\cdots\mu_{n_i}} = n_i\partial_{[\mu_1}A_{\cdots\mu_{n_i}]}$ and $R_{\mu\nu}{}^\mu{}_\rho = R_{\nu\rho}$, $R = g^{\mu\nu}R_{\mu\nu}$. The constant κ_D is related to the usual gravitational constant by $2\kappa_D^2 = 16\pi G_D$. The equations of motion determined by varying with respect to the metric, $g_{\mu\nu}$, are the Einstein equations and we also have equations of motion arising from varying the gauge fields and the variation of the field ϕ, called the dilaton. For the generic action of equation (13.8.3) these are given by

$$R^\mu{}_\nu = \frac{1}{2}\partial^\mu\phi\partial_\nu\phi + \sum_i \frac{1}{2n_i!}e^{a_i\phi}$$

$$\times \left(n_i F^{\mu\lambda_2\cdots\lambda_{n_i}}F_{\nu\lambda_2\cdots\lambda_{n_i}} - \frac{n_i-1}{D-2}\delta^\mu{}_\nu F_{\lambda_1\cdots\lambda_{n_i}}F^{\lambda_1\cdots\lambda_{n_i}}\right). \tag{13.8.2}$$

$$\partial_\mu(\sqrt{-g}e^{a_i\phi}F^{\mu\lambda_2\cdots\lambda_{n_i}}) = 0, \tag{13.8.3}$$

$$\frac{1}{\sqrt{-g}}\partial_\mu(\sqrt{-g}\partial^\mu\phi) - \sum_i \frac{a_i}{2.n_i!}e^{a_i\phi}F_{\lambda_1\lambda_2\cdots\lambda_{n_i}}F^{\lambda_1\lambda_2\cdots\lambda_{n_i}} = 0. \tag{13.8.4}$$

We also have the Bianchi identity

$$\partial_{[\lambda_1}F_{\lambda_2\cdots\lambda_{n_i+1}]} = 0. \tag{13.8.5}$$

In the action of equation (13.8.1) we can have field strengths of rank 1 corresponding to the presence of more scalars. The sum is over all possible field strengths although in a specific supergravity theory only a few of these will actually be present, and usually only one of these will be active for a particular solution. The above action does not include the Chern–Simmons term that supergravity theories can possess. While this term plays an important role in brane dynamics we will explicitly check that its contribution to the field equations vanishes for the solutions we are considering and so we have omitted it. Modulo this point the generic action of equation (13.8.1) encompasses all the supergravity theories. Solutions to such a generic theory were considered in references [13.44, 13.45] and then in [13.46–13.52].

We may also write the field equations in terms of the dual field strength which we may define by

$$\sqrt{-g}e^{a_i\phi}F^{\mu_1\cdots\mu_{n_i}} = \frac{1}{(D-n_i)!}\epsilon^{\mu_1\cdots\mu_{n_i}\nu_1\cdots\nu_{D-n_i}}\tilde{F}_{\nu_1\cdots\nu_{D-n_i}}. \tag{13.8.6}$$

Substituting this into the above field equations we find that they retain their form except that they now involve $\tilde{F}_{\nu_1\cdots\nu_{D-n_i}}$ and a_i is replaced by $-a_i$. Indeed, substituting for the dual field strength in the gauge equation of motion of equation (13.8.3) we obtain the Bianchi identity for the dual gauge field strength, while substituting in the Bianchi identity of equation (13.8.5) we find the equation of motion for the dual gauge field. This relies on the precise dilaton contribution, given above, to the above definition of the dual field strength. For the other equations we use the relation $\epsilon^{\mu_1\cdots\mu_n\tau_1\cdots\tau_m}\epsilon_{\nu_1\cdots\nu_n\tau_1\cdots\tau_m} = -n!m!\delta^{\mu_1}_{[\nu_1}\cdots\delta^{\mu_n}_{\nu_n]}$.

We now look for solutions that correspond to branes and in particular a p-brane which has a $(p+1)$-dimensional world volume. We take the brane to live in the background space-time such that in this space-times the coordinates of the brane world volume are $t, x_j, j = 1,\ldots, p$ and those transverse to the brane are $y_b, b = 1,\ldots, d$, where $d = D - p - 1$ is the number of dimensions transverse to the brane. In doing so we have assumed that the brane is static in the background space-time. Indeed, we will assume that all the fields are invariant under $t \to t + a$, where a is a constant. We will also assume that the fields are invariant under constant translations in x_j, that is, $x_j \to x_j + a_j$. This implies that the metric components are independent of t and x_j. We also assume that the world volume has an SO(p) symmetry acting on its spatial coordinates and transverse to the brane we have a SO($D - p - 1$) symmetry. Furthermore, we assume that the line element is invariant under $t \to -t$ ruling out terms linear in dt. The line element is then of the form

$$ds^2 = -B^2 dt^2 + C^2 \sum_{j=1}^{j=p} dx_j^2 + E^2 \sum_{a=1}^{d} (dy_a)^2 + F^2 \left(\sum_a y_a dy_a \right)^2, \qquad (13.8.7)$$

where B, C, E, F depend only on $r = \sqrt{\sum_a y_a y_a}$.

Given the special role of the radial coordinate r, it will often prove useful to use spherical coordinates $r, \theta_1, \ldots, \theta_{d-1}$ instead of the transverse coordinates y_a. The relationship between Cartesian and spherical coordinates for the transverse coordinates is given by

$$y_a = r \sin \theta_1 \cdots \sin \theta_{a-1} \cos \theta_a, \ a = 1, \ldots, d - 1,$$

$$y_d = r \sin \theta_1 \cdots \sin \theta_{d-1}. \qquad (13.8.8)$$

The reader will recognise the familiar expressions for the change for three-dimensional spatial coordinates. The two relevant terms in the line element are given by

$$\sum_{a=1}^{d} dy_a dy_a = dr^2 + r^2 d\Omega_{d-1}^2,$$

$$d\Omega_{d-1}^2 = d\theta_1^2 + \sin^2 \theta_1 d\theta_2^2 + \cdots + \sin^2 \theta_1 \cdots \sin^2 \theta_{d-2} d\theta_{d-1}^2, \qquad (13.8.9)$$

$$\sum_{a=1}^{d} y_a dy_a = r dr.$$

In terms of spherical coordinates the above line element is given by

$$ds^2 = -B^2 dt^2 + C^2 \sum_{j=1}^{j=p} dx_j^2 + E^2 (dr^2 + r^2 d\Omega_{d-1}^2) + F^2 r^2 dr^2. \qquad (13.8.10)$$

By carrying out a change of coordinates $r \to r'$ one can easily show that one can set $F = 0$ and so, dropping the prime on r, we are left with the metric of equation (13.8.7) or (13.8.10) with $F = 0$, that is,

$$ds^2 = -B^2 dt^2 + C^2 \sum_{j=1}^{j=p} dx_j^2 + E^2 \sum_{a=1}^{d} (dy_a)^2, \qquad (13.8.11)$$

or equivalently

$$ds^2 = -B^2 dt^2 + C^2 \sum_{j=1}^{j=p} dx_j^2 + E^2 (dr^2 + r^2 d\Omega_{d-1}^2). \qquad (13.8.12)$$

We will find in chapter 14 that a p-brane couples to a rank $p+1$ gauge field. The ansatz for a rank $p+1$ gauge field must also obey the above symmetry requirements and one possibility is to take

$$F_{ti_1\cdots i_p a} = \epsilon_{i_1\cdots i_p}\partial_a A(r) \text{ or equivalently } A_{ti_1\cdots i_p} = (-1)^{p+1}\epsilon_{i_1\cdots i_p}A(r). \qquad (13.8.13)$$

In terms of radial components the field strength in spherical coordinates is given by

$$F_{ti_1\cdots i_p r} = \epsilon_{i_1\cdots i_p}\frac{dA(r)}{dr}. \qquad (13.8.14)$$

We refer to this as the *electric* ansatz. The other ansatz compatible with the above symmetry requirements is given by

$$F^{a_1\cdots a_{d-1}} = \frac{e^{-a_i\phi}\epsilon^{a_1\cdots a_d}}{\sqrt{-g}}\partial_{a_d}\tilde{A}(r). \qquad (13.8.15)$$

This is referred to as the *magnetic* ansatz. If we calculate the dual field strength given by equation (13.8.6), then we find that it takes the form of the ansatz that occurs in equation (13.8.13). The above agrees with our usual notions of electric and magnetic fields for the case of four space-time dimensions and for Maxwell's theory in the absence of gravity. In this case we have a point particle, that is, a $(p=0)$-brane which couples to a rank 1 gauge field with a rank 2 field strength. The electric ansatz of equation (13.8.13) is $F_{0a} = \partial_a A(r)$, while the magnetic ansatz of equation (13.8.14) is $F_{ab} = \epsilon_{abc}\partial_c\tilde{A}(r)$.

We now assume that only the rank $p+2$ field strength is non-zero and simplify the notation by taking $a_i = a$. The other field in the theory, that is, the dilaton, also depends only on r.

The first step in finding solutions of the above type is to compute the components of the Ricci tensor in terms of the ansatz of equation (13.8.11). The non-zero curvature components referred to *Cartesian* coordinates are given by

$$R_t{}^t = E^{-2}\left\{-\sum_a \partial_{y_a}\partial_{y_a}\ln B - \sum_a \partial_{y_a}\ln B\partial_{y_a}\Psi\right\}, \qquad (13.8.16)$$

$$R_{x_i}{}^{x_j} = E^{-2}\delta_i^j\left\{-\sum_a \partial_{y_a}\partial_{y_a}\ln C - \sum_a \partial_{y_a}\ln C\partial_{y_a}\Psi\right\}, \qquad (13.8.17)$$

$$R_{y_a}{}^{y_b} = E^{-2}\left\{-\sum_a \partial_{y_a}\partial_{y_a}\Psi + \partial_{y_a}\ln E\partial_{y_b}\Psi + \partial_{y_b}\ln E\partial_{y_a}\Psi - \partial_{y_a}\ln B\partial_{y_b}\ln B\right.$$

$$- p\partial_{y_a}\ln C\partial_{y_b}\ln C - (d-2)\partial_{y_a}\ln E\partial_{y_b}\ln E - \delta_b^a\sum_c \partial_{y_c}\partial_{y_c}\ln E$$

$$\left.- \delta_b^a\sum_c \partial_{y_c}\ln E\partial_{y_c}\Psi\right\}, \qquad (13.8.18)$$

where $\Psi = \ln B + p\ln C + (d-2)\ln E$, We observe that if $\Psi = 0$, that is, if

$$BC^p E^{d-2} = 1, \qquad (13.8.19)$$

the Ricci tensor becomes particularly simple. This condition is related to preservation of some supersymmetry by the solution for those generic theories which are the bosonic sectors of supergravity theories. As we will be primarily interested in such theories, and in addition for solutions that preserve some supersymmetry, we will later adopt this condition. Indeed, we will find that solutions of the maximal supergravity theories in ten and eleven dimensions that obey the condition of equation (13.8.19) possess the above symmetries and preserve half of the 32 supersymmetries. However, the meaning of the condition of equation (13.8.19) is not obvious for non-supersymmetric theories such as pure gravity.

We will need the components of the Ricci tensor when referred to radial coordinates. One can just start from the ansatz of equation (13.8.12), but it is simpler to make the coordinate change from the curvature referred to Cartesian coordinates given in the above equation to spherical coordinates. We first change the derivatives with respect to y_a to those with respect to r. We note that as B is a function of r only

$$\partial_a B = \frac{x_a}{r} B', \quad \partial_a \partial_b B = \delta_b^a \frac{1}{r} B' + \frac{y_a y_b}{r^2} \left(B'' - \frac{1}{r} B' \right),$$

$$\sum_a \partial_a \partial_a B = B'' + \frac{d-1}{r} B', \tag{13.8.20}$$

where $B' = dB/dr$, $B'' = d^2 B/dr^2$ and similarly for C and E. Substituting these into equations (13.8.16), (13.8.17) and (13.8.18) we find that the Ricci tensor referred to Cartesian coordinates is given by

$$R_t{}^t = \frac{1}{E^2} \left\{ -(\ln B)'' - \frac{d-1}{r} (\ln B)' - (\ln B)' \Psi' \right\}, \tag{13.8.21}$$

$$R_{x_i}{}^{x_j} = \frac{1}{E^2} \left\{ -(\ln C_i)'' - \frac{d-1}{r} (\ln C)' - (\ln C)' \Psi' \right\} \delta_i^j, \tag{13.8.22}$$

$$R_{y_a}{}^{y_b} = \delta_a^b \frac{1}{E^2} \left\{ -(\ln E)'' - \frac{d-1}{r} (\ln E)' - (\ln E)' \Psi' - \frac{1}{r} \Psi' \right\}$$

$$+ \frac{y_a y_b}{r^2} \frac{1}{E^2} \left\{ - \Psi'' + \frac{1}{r} \Psi' + 2(\ln E)' \Psi' - (\ln B)'^2 \right.$$

$$\left. - p(\ln C)'^2 - (d-2)(\ln E)'^2 \right\}. \tag{13.8.23}$$

We observe that the last components of the Ricci tensor have the form $R_{y_a}{}^{y_b} = \delta_a^b R_1 + (y_a y_b/r^2) R_2$. Carrying out the coordinate change from Cartesian coordinates to spherical coordinates we find that

$$R_r{}^r = \frac{\partial y_a}{\partial r} \frac{\partial r}{y_b} R_a{}^b = R_1 + R_2, \quad R_{\theta_\alpha}{}^{\theta_\beta} = R_1 \delta_\alpha^\beta, \quad \alpha, \beta = 1, \ldots, d-1. \tag{13.8.24}$$

The other components of the Ricci tensor in the t and x_i directions are equal in the two coordinate systems. Thus the Ricci tensor referred to *spherical* coordinates for the above ansatz for the metric is given by

$$R_t{}^t = \frac{1}{E^2}\left\{-(\ln B)'' - \frac{d-1}{r}(\ln B)' - (\ln B)'\Psi'\right\}, \tag{13.8.25}$$

$$R_{x_i}{}^{x_j} = \frac{1}{E^2}\left\{-(\ln C)'' - \frac{d-1}{r}(\ln C)' - (\ln C)'\Psi'\right\}\delta^i_j, \tag{13.8.26}$$

$$R_r{}^r = \frac{1}{E^2}\left\{-\Psi'' + (\ln E)'\Psi' - (\ln E)'' - \frac{d-1}{r}(\ln E)'\right.$$

$$\left. - (\ln B)'^2 - p(\ln C)'^2 - (d-2)(\ln E)'^2\right\}, \tag{13.8.27}$$

$$R_\alpha{}^\beta = \frac{1}{E^2}\left\{-(\ln E)'' - \frac{d-1}{r}(\ln E)' - (\ln E)'\Psi' - \frac{1}{r}\Psi'\right\}\delta^\beta_\alpha. \tag{13.8.28}$$

The equation that is easiest to solve is that for the gauge field. Using the metric of equation (13.8.12) and the gauge ansatz of equation (13.8.13) we find that equation (13.8.3) becomes

$$\frac{d}{dr}(e^{a\phi}(BC^pE)^{-1}(Er)^{d-1}A(r)') = 0 \tag{13.8.29}$$

and so we find that

$$F_{ti_1\cdots i_p r} = \epsilon_{i_1\cdots i_p}BC^pEe^{-a\phi}\frac{Q}{(Er)^{d-1}}, \tag{13.8.30}$$

where Q is an integration constant which is proportional to the electric charge. The reader will recognise equation (13.8.30) for the case of four dimensions and a rank 2 field strength, that is, Maxwell's equations. Having solved the gauge field equation of motion, the gauge field strength, or equivalently the quantity $A(r)$, can be found once we know B, C, E and ϕ. Hence we must now solve the metric equation to determine these quantities.

In the equations of motion of the other fields we find the quantities

$$F_{\mu_1\cdots\mu_{p+2}}F^{\mu_1\cdots\mu_{p+2}} = (p+2)!F_{ti_1\cdots i_p r}F^{ti_1\cdots i_p r} = -(p+2)!e^{-2a\phi}\frac{Q^2}{(Er)^{2(d-1)}}, \tag{13.8.31}$$

as well as

$$F^{\mu\lambda_2\cdots\lambda_{p+2}}F_{\nu\lambda_2\cdots\lambda_{p+2}} = (\delta^\mu_t\delta^t_\nu + \delta^\mu_i\delta^i_\nu + \delta^\mu_r\delta^r_\nu)(p+1)!F_{ti_1\cdots i_p r}F^{ti_1\cdots i_p r}$$

$$= -(\delta^\mu_t\delta^t_\nu + \delta^\mu_i\delta^i_\nu + \delta^\mu_r\delta^r_\nu)(p+1)!e^{-2a\phi}\frac{Q^2}{(Er)^{2(d-1)}}. \tag{13.8.32}$$

All other components vanish.

For the above brane ansatz and using the solution for the gauge field of (13.8.30), the metric equations of motion become

$$R_t{}^t = \frac{1}{E^2} \left\{ -(\ln B)'' - (\ln B)' \left[\Psi' + \frac{d-1}{r} \right] \right\}$$

$$= -\frac{d-2}{2(D-2)} e^{-a\phi} \frac{Q^2}{(Er)^{2(d-1)}}, \tag{13.8.33}$$

$$R_{x_i}{}^{x_i} = \frac{1}{E^2} \left\{ -(\ln C)'' - (\ln C)' \left[\Psi' + \frac{d-1}{r} \right] \right\}$$

$$= -\frac{d-2}{2(D-2)} e^{-a\phi} \frac{Q^2}{(Er)^{2(d-1)}}, \tag{13.8.34}$$

$$R_r{}^r = \frac{1}{E^2} \left\{ -(\Psi)'' + (\ln E)' \Psi' - (\ln E)'' - (d-2)((\ln E)')^2 \right.$$

$$\left. - ((\ln B)')^2 - p((\ln C)')^2 - \frac{(d-1)}{r} (\ln E)' \right\}$$

$$= \frac{1}{2} \frac{1}{E^2} \phi'^2 - \frac{d-2}{2(D-2)} e^{-a\phi} \frac{Q^2}{(Er)^{2(d-1)}}, \tag{13.8.35}$$

$$R_{\theta_a}{}^{\theta_a} = \frac{1}{E^2} \left\{ -(\ln E)'' - (\ln E)' \frac{(d-1)}{r} - (\ln E)' \Psi' - \frac{\Psi'}{r} \right\}$$

$$= \frac{p+1}{2(D-2)} e^{-a\phi} \frac{Q^2}{(Er)^{2(d-1)}}. \tag{13.8.36}$$

The other metric equations of motion are automatically satisfied. Finally, we have the equation of motion of the dilaton which depends only on r as a consequence of the above assumed symmetries. We note that equation (13.8.4) becomes

$$\frac{1}{\sqrt{-g}} \partial_r \left(\sqrt{-g} g^{rr} \partial_r \phi \right) = \frac{1}{BC^p E (Er)^{d-1}} \left[BC^p E (Er)^{d-1} \frac{1}{E^2} \phi' \right]'$$

$$= \frac{1}{E^2} \left\{ \phi'' + \frac{d-1}{r} \phi' + \phi' \left[(\ln B)' + p(\ln C)' \right. \right.$$

$$\left. \left. - (\ln E)' + (d-1)(\ln E)' \right] \right\}. \tag{13.8.37}$$

Reorganising this equation we find that the dilaton equation of motion becomes

$$\frac{1}{E^2} \left\{ \phi'' + \phi' \left[\Psi' + \frac{d-1}{r} \right] \right\} = -\frac{a}{2} e^{-a\phi} \frac{Q^2}{(Er)^{2(d-1)}}. \tag{13.8.38}$$

We will now solve the equations of motion assuming that $\Psi = 0$, or equivalently equation (13.8.19) holds. In this case the equations of motion simplify dramatically and equations (13.8.33)–(13.8.36) and (13.8.38) are given by

$$\frac{d}{dr} \left\{ r^{(d-1)} (\ln B)' \right\} = \frac{d-2}{2(D-2)} e^{-a\phi} E^{-2(d-2)} \frac{Q^2}{(r)^{d-1}}, \tag{13.8.39}$$

$$\frac{d}{dr} \left\{ r^{(d-1)} (\ln C)' \right\} = \frac{d-2}{2(D-2)} e^{-a\phi} E^{-2(d-2)} \frac{Q^2}{(r)^{d-1}}, \tag{13.8.40}$$

$$-\frac{1}{r^{d-1}}\frac{d}{dr}(r^{d-1}(\ln E)') - (d-2)((\ln E)')^2 - ((\ln B)')^2 - p((\ln C)')^2$$

$$= \frac{1}{2}\phi'^2 - \frac{d-2}{2(D-2)}e^{-a\phi}E^{-2(d-2)}\frac{Q^2}{(r)^{2(d-1)}}, \tag{13.8.41}$$

$$\frac{d}{dr}\left\{(r^{(d-1)}(\ln E)'\right\} = -\frac{p+1}{2(D-2)}E^{-2(d-2)}e^{-a\phi}\frac{Q^2}{(r)^{d-1}}, \tag{13.8.42}$$

$$\frac{d}{dr}\left\{r^{(d-1)}\phi'\right\} = -\frac{a}{2}E^{-2(d-2)}e^{-a\phi}\frac{Q^2}{(r)^{d-1}}. \tag{13.8.43}$$

We now impose a further condition on the solution, namely that the brane has an SO(1, p) symmetry enlarging the previously adopted SO(p) symmetry in the spatial world volume of the brane. The metric then takes the form

$$ds^2 = B^2\left(-dt^2 + \sum_{j=1}^{j=p}dx_j^2\right) + E^2\sum_{a=1}^{d}(dy_a)^2, \tag{13.8.44}$$

or, equivalently,

$$ds^2 = B^2\left(-dt^2 + \sum_{j=1}^{j=p}dx_j^2\right) + E^2(dr^2 + r^2d\Omega_{d-1}^2). \tag{13.8.45}$$

Comparing this with equation (13.8.11) or (13.8.11) we find that we must set $B = C$, whereupon equation (13.8.39) becomes identical to equation (13.8.40). The condition $\Psi = 0$ of equation (13.8.19) becomes

$$B^{p+1}E^{d-2} = 1. \tag{13.8.46}$$

Using this condition equations (13.8.39) and (13.8.42) become identical and so we are left with the latter:

$$\frac{d}{dr}\left\{(r^{(d-1)}(\ln E)'\right\} = -\frac{p+1}{2(D-2)}E^{-2(d-2)}e^{-a\phi}\frac{Q^2}{(r)^{d-1}}. \tag{13.8.47}$$

Furthermore, using this equation to eliminate the first term in equation (13.8.41), setting $B = C$ and eliminating B using equation (13.8.46) we find that equation (13.8.41) becomes

$$-\frac{(D-2)(d-2)}{(p+1)}((\ln E)')^2 = \frac{1}{2}(\phi')^2 - \frac{1}{2}e^{-a\phi}E^{-2(d-2)}\frac{Q^2}{(r)^{2(d-1)}}. \tag{13.8.48}$$

The final equation left to consider is the dilaton equation (13.8.43) which we repeat

$$\frac{d}{dr}\left\{r^{(d-1)}\phi'\right\} = -\frac{a}{2}E^{-2(d-2)}e^{-a\phi}\frac{Q^2}{(r)^{d-1}}. \tag{13.8.43}$$

Hence, if we assume that the brane world volume has SO(1, p) symmetry, SO(d) symmetry in the transverse space to the brane, translation symmetry of the world volume coordinates, or equivalently stated ISO(1, p) $\otimes$ SO(d) symmetry, and if equation (13.8.46) holds (that is, $\Psi = 0$), then we only need to solve equations (13.8.43), (13.8.47) and (13.8.48).

We now solve the above equations for a *theory with no dilaton*. Then we need only solve equations (13.8.47) and (13.8.48) with $\phi = 0$. The reader may verify that in deriving these latter two equations we have not used the dilaton equation. Taking the square root of equation (13.8.48) we find that

$$E^{d-2}(\ln E)' = \sqrt{\frac{(p+1)}{2(D-2)(d-2)} \frac{|Q|}{(r)^{d-1}}}. \tag{13.8.49}$$

The solution to this equation is

$$E = \left(1 + \sqrt{\frac{(p+1)}{2(D-2)(d-2)} \frac{|Q|}{(r)^{d-2}}}\right)^{1/(d-2)},$$

$$B = \left(1 + \sqrt{\frac{(p+1)}{2(D-2)(d-2)} \frac{|Q|}{(r)^{d-2}}}\right)^{-1/(p+1)}. \tag{13.8.50}$$

We have taken the integration constant to be 1 in order to have a solution that is asymptotically (that is, $r \to \infty$) like Minkowski space-time and we have used equation (13.8.46) to determine B. Equation (13.8.47) is then identically satisfied

For a *theory with a dilaton* we replace the right-hand side of equation (13.8.47) using equation (13.8.43) to find that

$$\frac{d}{dr}\{(r^{(d-1)}(\ln E)'\} = \frac{p+1}{a(D-2)}\frac{d}{dr}\{r^{(d-1)}\phi'\}, \tag{13.8.51}$$

which implies that

$$\ln E = \frac{p+1}{a(D-2)}\phi. \tag{13.8.52}$$

We have set the two integrations constants to zero as we insist on having a solutions with vanishing dilaton and Minkowski metric at infinity. Substituting equation (13.8.52) into equation (13.8.48) we find that

$$\frac{\Delta}{a^2(D-2)}(\phi')^2 = \frac{1}{2}e^{-2\Delta\phi/a(D-2)}\frac{Q^2}{(r)^{2(d-1)}}, \tag{13.8.53}$$

where

$$\Delta = (p+1)(d-2) + \tfrac{1}{2}a^2(D-2). \tag{13.8.54}$$

Taking the square root of equation (13.8.53) leads to

$$\frac{d}{dr}\left(e^{(\Delta/a(D-2))\phi}\right) = \sqrt{\frac{\Delta}{2(D-2)}}\frac{|Q|}{r^{d-1}}, \tag{13.8.55}$$

which can readily be integrated to give

$$e^{(\Delta/a(D-2))\phi} = 1 + \frac{1}{d-2}\sqrt{\frac{\Delta}{2(D-2)}}\frac{|Q|}{r^{d-2}}. \tag{13.8.56}$$

We have chosen the plus sign in equation (13.8.55) so that the resulting dilaton in equation (13.8.56) does not possess a singularity for $r > 0$ and we have chosen the asymptotic value of the dilaton to be zero. Using equation (13.8.52) and (13.8.46) we find that the complete solution if we assume $\mathrm{ISO}(1, p) \otimes \mathrm{SO}(d)$ symmetry and $\Psi = 0$ is given by

$$B = N^{-(d-2)/\Delta}, \quad E = N^{(p+1)/\Delta}, \quad e^\phi = N^{a(D-2)/\Delta},$$

$$F_{i_1 \cdots i_p r} = (\pm)\sqrt{\frac{2(D-2)}{\Delta}} \frac{d}{dr}\left(N^{-1}\right), \tag{13.8.57}$$

with

$$N = 1 + \frac{1}{d-2}\sqrt{\frac{\Delta}{2(D-2)}} \frac{|Q|}{r^{d-2}}, \tag{13.8.58}$$

where Δ was given in equation (13.8.54). Since the equations of motion involve Q^2 we can have two signs and the sign in the field strength is the same as that in $(\pm) = Q/|Q|$. We find that the remaining equation (13.8.43) is automatically satisfied. The corresponding metric for this p-brane solution is

$$ds^2 = N^{-2(d-2)/\Delta}\left(-dt^2 + dx_1^2 + \cdots + dx_p^2\right) + N^{2(p+1)/\Delta}\left(dr^2 + r^2 d\Omega_{d-1}^2\right). \tag{13.8.59}$$

We notice that the fall off in the direction transverse to the brane is like $1/r$ to the power of the number of transverse directions minus 2.

We observe that the solution of equation (13.8.51) for a theory with no dilaton is also given by equation (13.8.59) if we set $a = 0$ and so these equations provide the spherically symmetric solutions for all generic theories that also satisfy equation (13.8.46), that is, $\Psi = 0$. Hence, if the theory we are studying contains a rank $p + 1$ gauge field, and so a rank $p + 2$ field strength, it will possess the spherically symmetric p-brane solution given here.

In deriving this we have used the electric ansatz of equation (13.8.13). However, we can also reformulate the equations of motion in terms of the dual field strength $\tilde{F}_{\mu-1 \cdots \mu_{D-p-2}}$, which is of rank $D - p - 2$. The field equations are then those for the dual field strength except with the change $a \to -a$. Hence, we will also find a solution for the dual field strength which corresponds to a $(D - p - 4)$-brane. This solution will be given by equations (13.8.57) and (13.8.58) provided we put tildes on all quantities and take $\tilde{p} = D - p - 4, \tilde{d} = p + 3$ and $\tilde{a} = -a$. This is equivalent to taking the magnetic ansatz of equation (13.8.15) in the original theory; indeed as we have previously noted taking the electric ansatz for the dual field strength is the same, using equation (13.8.6), as taking the magnetic ansatz for the original field strength.

We have found p-brane solutions in the so-called isotropic form of the metric of equation (13.8.11). However, it is often useful to have the solution in the form

$$ds^2 = B^2\left(-dt^2 + \sum_{j=1}^{j=p} dx_j^2\right) + G^2 d\hat{r}^2 + \hat{r}^2 d\Omega_{d-1}^2, \tag{13.8.60}$$

where the coordinates of this metric are $t, x_j, \hat{r}, \theta_\alpha$. We can think of this as adopting another choice of coordinates when going from equation (13.8.10) to equation (13.8.12). Comparing the metric of equation (13.8.12) and that of equation (13.8.60) we conclude that

$$\hat{r} = Er, \quad \frac{d\hat{r}}{dr} = \frac{E}{G}. \tag{13.8.61}$$

For the solution we have in equation (13.8.57) this implies that

$$\hat{r} = N^{(p+1)/\Delta}(r)r, \quad G = \frac{1 + \dfrac{1}{d-2}\sqrt{\dfrac{\Delta}{2(D-2)}\dfrac{|Q|}{r^{d-2}}}}{1 + \dfrac{a^2}{2(d-2)}\sqrt{\dfrac{(D-2)}{2\Delta}\dfrac{|Q|}{r^{d-2}}}}. \tag{13.8.62}$$

This change simplifies considerably for the case of a theory for which the dilaton coupling vanishes (that is, $a = 0$) and we find that the metric in the new coordinates is given by

$$ds^2 = \left(1 - \frac{b}{\hat{r}^{d-2}}\right)^{2/(p+1)}\left(-dt^2 + \sum_{j=1}^{j=p} dx_j^2\right) + \left(1 - \frac{b}{\hat{r}^{d-2}}\right)^{-2} d\hat{r}^2 + \hat{r}^2 d\Omega_{d-1}^2, \tag{13.8.63}$$

where

$$b = \frac{1}{d-2}\sqrt{\frac{\Delta}{2(D-2)}}|Q|.$$

We must also carry out a corresponding coordinate change on the gauge field strength.

In the above we adopted the choice $\Psi = 0$. *However, one can make progress even when $\Psi \neq 0$ if we still assume* $\mathrm{ISO}(1, p) \otimes \mathrm{SO}(d)$ *symmetry.* Let us consider the ansatz of equation (13.8.12) with the form field given by equation (13.8.14). The relevant equations are (13.8.33)–(13.8.36) and (13.8.38) with $C = B$ and $e^\Psi = B^{p+2}E^{d-2}$. We note that

$$R_t{}^t + pR_{x_i}{}^{x_i} + (d-2)R_{\theta_\alpha}{}^{\theta_\alpha} = -\frac{d^2\Psi}{dr^2} - \left(\frac{d\Psi}{dr}\right)^2 - \frac{(2d-3)}{r}\frac{d\Psi}{dr} = 0. \tag{13.8.64}$$

We may express this equation as

$$\frac{d^2\Psi}{du^2} + \left(\frac{d\Psi}{du}\right)^2 - \frac{1}{u}\frac{d\Psi}{du} = 0, \quad \text{where } u = -\frac{1}{(d-2)r^{d-2}}. \tag{13.8.65}$$

Provided $\Psi \neq 0$, we may then write this equation as

$$\frac{d}{du}\left(\ln\frac{de^\Psi}{du}\right) = \frac{1}{u}. \tag{13.8.66}$$

Integrating this equation implies that $\Psi = \ln(e + cu^2)$, where e and c are constants. Demanding a solution that is asymptotically flat implies that $e = 1$ and so

$$\Psi = \ln\left(1 - \left(\frac{u}{u_0}\right)^2\right) = \ln\left(1 - \left(\frac{r_0}{r}\right)^{2(d-2)}\right). \tag{13.8.67}$$

We note that to recover $\Psi = 0$ we must take $u_0 \to \infty$ or $r_0 \to 0$.

Using equation (13.8.38) to eliminate the right-hand side of equation (13.8.33) we find that it can be written as

$$\frac{1}{r^{d-1}\left((\ln B)' + \dfrac{(d-2)}{a(D-2)}\phi'\right)}\frac{d}{dr}\left(\ln\left(r^{d-1}(\ln B)' + \frac{(d-2)}{a(D-2)}\phi'\right)\right) = -\Psi'. \tag{13.8.68}$$

Substituting Ψ from equation (13.8.67) we find that

$$(\ln B)' + \frac{(d-2)}{a(D-2)}\phi' = (d-2)r_0^{(d-2)}\frac{1}{r^{d-1}(1-(r_0/r)^{2(d-2)})} \tag{13.8.69}$$

and integrating again we have

$$\ln B = -\frac{(d-2)}{a(D-2)}\phi + \ln\frac{(1-(r_0/r)^{(d-2)})}{(1+(r_0/r)^{(d-2)})}, \tag{13.8.70}$$

where we have adjusted the integration constant in order to obtain an asymptotically flat solution. Recalling that $\psi = (p+1)\ln B + (d-2)\ln E$, we can trade B for E to find that

$$\ln E = -\frac{p}{(d-2)}\ln\left(1-\left(\frac{r_0}{r}\right)^{(d-2)}\right) + \frac{(p+2)}{(d-2)}\ln\left(1+\left(\frac{r_0}{r}\right)^{(d-2)}\right)$$

$$+ \frac{(p+1)}{a(D-2)}\phi. \tag{13.8.71}$$

Thus from equations (13.8.70) and (13.8.71) we can eliminate B and E to leave us only with ϕ.

To solve the equations it is advantageous to change the variable from r to u. Equation (13.8.33) becomes

$$\frac{d^2\ln B}{du^2} + \frac{d}{du}(\ln B)\frac{d\Psi}{du} = \frac{(d-2)Q^2}{2(D-2)}B^{2(p+1)}e^{-2\Psi}e^{-a\phi}, \tag{13.8.72}$$

while equation (13.8.35) minus equation (13.8.36) is given by

$$\frac{d^2\Psi}{du^2} + \frac{d}{(d-2)u}\frac{d\Psi}{du} - 2\frac{d}{du}(\ln E)\frac{d\Psi}{du} + (d-2)\left(\frac{d\ln E}{du}\right)^2$$

$$+ (p+1)\left(\frac{d\ln B}{du}\right)^2 = -\frac{1}{2}\left(\frac{d\phi}{du}\right)^2 + \frac{Q^2}{2}E^{-2(d-2)}e^{-a\phi}. \tag{13.8.73}$$

Equation (13.8.36) is given by

$$\frac{d^2\ln E}{du^2} + \frac{d}{du}(\ln E)\frac{d\Psi}{du} - \frac{1}{(d-2)u}\frac{d\Psi}{du} = -\frac{(p+1)Q^2}{2(D-2)}E^{-2(d-2)}e^{-a\phi} \tag{13.8.74}$$

and finally equation (13.8.38) becomes

$$\frac{d^2\phi}{du^2} + \frac{d\phi}{du}\frac{d\Psi}{du} = -\frac{aQ^2}{2}E^{-2(d-2)}e^{-a\phi}. \tag{13.8.75}$$

We note that equation (13.8.72) can be rewritten as

$$e^\Psi\frac{d}{du}\left(e^\Psi\frac{d\ln B}{du}\right) = \frac{(d-2)Q^2}{2(D-2)}B^{2(p+1)}e^{-a\phi}. \tag{13.8.76}$$

We now introduce yet another variable, v, such that

$$\frac{d}{dv} = e^\Psi\frac{d}{du} \quad \text{or} \quad v = \frac{u_0}{2}\ln\left(\frac{1+(u/u_0)}{1-(u/u_0)}\right). \tag{13.8.77}$$

Equation (13.8.76) now reads

$$\frac{d^2 \ln B}{dv^2} = \frac{(d-2)Q^2}{2(D-2)} B^{2(p+1)} e^{-a\phi}.$$

(13.8.78)

Before finding the most general solution it will be useful to solve the equations in the absence of the dilaton, that is, $\phi = 0 = a$, whereupon equation (13.8.78) takes the form

$$\frac{d^2 T}{dv^2} = ce^T,$$

(13.8.79)

which multiplying by dT/dv implies that

$$\frac{d}{dv}\left(\frac{dT}{dv}\right)^2 = 2c\frac{d}{dv}e^T,$$

(13.8.80)

which, in turn, implies that

$$\left(\frac{dT}{dv}\right)^2 = 2ce^T + \text{constant}.$$

(13.8.81)

Let us denote the solution by $T(v, c)$. In fact, two solutions are

$$T(v, c) = -2 \ln\left(e_1 + \sqrt{\frac{c}{2}}v\right) \quad \text{and} \quad T(v, c) = -2 \ln \sinh\left(\sqrt{\frac{c}{2}}v + e_2\right), \quad (13.8.82)$$

where e_1 and e_2 are constants.

Hence if $\phi = a = 0$ the solution to equation (13.8.78) is given by

$$B = e^{T(v, c_B)/2(p+1)}, \quad E = e^{-T(v, c_B)/2(d-2)}\left(1 - \left(\frac{u}{u_0}\right)^2\right)^{1/(d-2)},$$

(13.8.83)

where we identify $c_B = (d-2)(p+1)Q^2/(D-2)$ and we have used equation (13.8.71) with $\phi = 0$.

We now return to find the most general solution. We cast the other equations in terms of the variable v rather than u. Beginning with equation (13.8.73) we solve for E in terms of B and Ψ and then eliminating the latter using equation (13.8.67) we find that this equation becomes

$$\frac{(p+1)(D-2)}{(d-2)}\left(\frac{d \ln B}{dv}\right)^2 - \frac{4}{u_0^2}\frac{(d-1)}{(d-2)} = -\frac{1}{2}\left(\frac{d\phi}{dv}\right)^2 + \frac{Q^2}{2}e^{-a\phi}B^{2(p+1)},$$

(13.8.84)

while equation (13.8.74), when E is eliminated, is identical to equation (13.8.78). Finally, starting with equation (13.8.75), we eliminate E using equation (13.8.71) and changing variables from u to v we find that it can be written as

$$\frac{d^2\hat{\phi}}{dv^2} = -\frac{a^2}{2}Q^2 e^{-2\Delta\hat{\phi}/a(D-2)},$$

(13.8.85)

where

$$\hat{\phi} = \phi + \frac{2(p+1)a(D-2)}{\Delta}\frac{v}{u_0}.$$

Equation (13.8.85) is of the form of equation (13.8.79) provided we identify $K = -2\Delta\hat{\phi}/a(D-2)$ with $c_\phi = \Delta Q^2/(D-2)$ and so the general solutions is

$$e^\phi = e^{-\frac{a(D-2)}{2\Delta}T(v,c_\phi)}e^{-\frac{2(p+1)a(D-2)}{\Delta}\frac{v}{u_0}}. \tag{13.8.86}$$

From equations (13.8.70) and (13.8.71) we find that the remaining fields are given by

$$B = e^{\frac{(d-2)}{2\Delta}T(v,c_\phi)}e^{-\frac{a^2(D-2)}{\Delta}\frac{v}{u_0}}, \quad E = e^{-\frac{(p+1)}{2\Delta}T(v,c_\phi)}e^{\frac{a^2(D-2)(p+1)}{(d-2)\Delta}\frac{v}{u_0}}\left(1-\left(\frac{u}{u_0}\right)^2\right)^{\frac{1}{(d-2)}}. \tag{13.8.87}$$

Thus equations (13.8.86) and (13.8.87) give the most general p-brane solution to the generic theory of equation (13.8.1) with $\text{ISO}(1, p)\otimes \text{SO}(D-p-1)$ symmetry. A general solution with such symmetry was found in [13.53]. The field strength for the rank $p+2$ field in found from equation (13.8.30) with $B = C$ and the above values of B and E.

In order to admit an asymptotically flat space-time we require $T \to 0$ as $v \to 0$, which is equivalent to $u \to 0$ and to $r \to \infty$. This reduces the number of parameters in T to one. For example, in the first solution of equation (13.8.82) we must take $e_1 = 1$ leaving the constant c. This parameter, together with u_0 means that we have a two-parameter solution, which corresponds to having independent mass and charge. One can recover the solutions of equation (13.8.57) and (13.8.58) by taking $u_0 \to \infty$, so setting $\Psi = 0$ and $v = u = -1/(d-2)r^{d-2}$, then taking the first solution of equation (13.8.82) and identifying $N = e^{-T/2}$.

In the absence of a dilaton, that is, $\phi = 0$, and an electric charge $Q = 0$ we find that equation (13.8.78) has the solution $\ln B = ev$ for constant e. Putting this into equation (13.8.84) we find that it is satisfied provided

$$e = \pm\frac{2}{u_0}\sqrt{\frac{d-1}{(p+1)(D-2)}}.$$

Substituting for v using equations (13.8.77) and (13.8.65) we find the solution

$$B = \left(\frac{\left(1-\left(\frac{r_0}{r}\right)^{(d-2)}\right)}{\left(1+\left(\frac{R_0}{r}\right)^{(d-2)}\right)}\right)^{\sqrt{(d-1)/(p+1)(D-2)}},$$

$$E = B^{-(p+1)/(d-2)}\left(1-\left(\frac{r_0}{r}\right)^{2(d-2)}\right)^{1/(d-2)}. \tag{13.8.88}$$

For the case of a point particle, that is, $p = 0$ we find the solution for a spherically symmetric black hole in D dimensions. It is consoling to recover the solution for a four-dimensional spherically symmetric black hole in isometric form which is given by

$$ds^2 = -\left(\frac{1-\frac{m}{2r}}{1+\frac{m}{2r}}\right)^2 dt^2 + \left(1+\frac{m}{2r}\right)^4(dr^2 + r^2(d\theta^2 + \sin^2\theta d\varphi^2)), \tag{13.8.89}$$

where $m = r_0/2$. Using equation (13.8.61) we may change to the more familiar form of the solution in Schwarzschild coordinates

$$ds^2 = -\left(1 - \frac{m}{2\hat{r}}\right) dt^2 + \frac{d\hat{r}^2}{\left(1 - \frac{m}{2\hat{r}}\right)} + \hat{r}^2(d\theta^2 + \sin^2\theta d\varphi^2). \qquad (13.8.90)$$

The relation between the two coordinates being that $\hat{r} = r(1 + m/2r)^2$.

In the above we first have found the most general p-brane solutions that possess ISO(1, p) $\otimes$ SO(d) symmetry, which are also subject to the condition of equation (13.8.19), that is, $\psi = 0$. We will see in the next section that these solutions provide all the well-known solutions associated with a single brane or a set of parallel branes. These solutions have only one parameter and so their mass and charge must be related. We then found the most general p-brane solution with ISO(1, p) $\otimes$ SO(d) symmetry. These were two-parameter solutions and so they have an independent mass and charge. Clearly, one can modify the spherically symmetric ansatz to find two other interesting solutions. In particular, we can take only SO(p) symmetry in the world volume. This means taking $B \neq C$. Subtracting equation (13.8.34) from equation (13.8.33), we find that

$$\frac{d}{dr}\left\{\ln\frac{d}{dr}\ln\frac{B}{C}\right\} = -\left(\Psi + \frac{d-1}{r}\right). \qquad (13.8.91)$$

Using the general solution for Ψ of equation (13.8.67) we may solve for $\ln B/C$. Proceeding in this way we can find the most general solution with SO(p) $\otimes$ SO(d) symmetry. This would take us beyond the scope of this book, but we note the interesting solution of this type often referred to as a black brane solution, which we give in Schwarzschild coordinates [13.45, 13.47, 13.54–13.59]:

$$ds^2 = -\left(1 - \left(\frac{\hat{r}_+}{\hat{r}}\right)^{d-2}\right)\left(1 - \left(\frac{\hat{r}_-}{\hat{r}}\right)^{d-2}\right)^{\left(\frac{2(d-2)}{\Delta} - 1\right)} dt^2$$

$$+ \left(1 - \left(\frac{\hat{r}_-}{\hat{r}}\right)^{d-2}\right)^{\frac{2(d-2)}{\Delta}} \sum_{j=1}^{p} dx_j^2$$

$$+ \left(1 - \left(\frac{\hat{r}_-}{\hat{r}}\right)^{d-2}\right)^{\frac{a^2(D-2)}{\Delta(d-2)} - 1}\left(1 - \left(\frac{\hat{r}_+}{\hat{r}}\right)\right)^{-1} d\hat{r}^2 \qquad (13.8.92)$$

$$+ \hat{r}^2\left(1 - \left(\frac{\hat{r}_-}{\hat{r}}\right)^{d-2}\right)^{\frac{a^2(D-2)}{\Delta(d-2)}} d\Omega_{d-1}^2,$$

$$e^{\frac{\Delta}{(D-2)a}\phi} = \left(1 - \left(\frac{\hat{r}_-}{\hat{r}}\right)^{(d-2)}\right)^{-1},$$

$$F_{ty_1...y_p r} = \sqrt{\frac{2(D-2)}{\Delta}}\left(\frac{\hat{r}_+}{\hat{r}_-}\right)^{\frac{d-2}{2}}\left(1 - \left(\frac{\hat{r}_-}{\hat{r}}\right)^{(d-2)}\right)'$$

The field strength was evaluated from equation (13.8.30). The hat on the r coordinate indicates that this is not the radial coordinate of polar coordinates but an additional coordinate transformation has been carried out. Using equation (13.8.61) one can convert this to isotropic coordinates if one so wishes.

One can also take even less symmetry in the world volume and find solutions corresponding to intersecting branes. The first intersecting brane solution was found in [13.47], but was not interpreted as such. The earliest fully recognised intersecting solutions were found in [13.60] and [13.61]. In next section we will give such solutions in the context of the eleven-dimensional supergravity theory.

It is interesting to compute the mass and charge and examine the global characteristics of the above solutions. In particular, what the structure and nature of their horizons are and how one can extend the space-times. That is, what is the analogue of Kruskal coordinates which is the maximal extension of space-time for the Schwarzschild solution? These interesting topics are beyond the scope of this book.

In the above we have focussed on brane solutions, but there are many other interesting solutions. One such class is found by taking the field strength to be proportional to the alternating symbol. Let us take the first $p + 2$ coordinates of space-time to be x^μ, $\mu = 0, 1, \ldots, p + 1$ and the rest of the coordinates to be y^a, $a = 1, \ldots, D - p - 2$. We then set

$$F_{\mu_1 \ldots \mu_{p+2}} = c\epsilon_{\mu_1 \ldots \mu_{p+2}}, \tag{13.8.93}$$

where c is a constant and all other components of the field strength are zero. We set the dilaton $\phi = 0$. The gauge field equation (13.8.3) is then automatically satisfied and the metric equation becomes

$$R^\mu{}_\nu = -\frac{(D - p - 3)c^2}{2(D - 2)}\delta^\mu_\nu, \quad R^a{}_b = \frac{(p + 1)c^2}{2(D - 2)}\delta^a_b, \tag{13.8.94}$$

all other components being zero. The simplest solutions are found by taking a product space-time and in each of these taking the maximally symmetric solution. This means taking a $D - p - 2$ sphere, S^{D-p-2} in the y coordinate part and anti-de Sitter space AdS_{p+2} in the x coordinate part. The choice of spaces corresponds to the negative and positive cosmological constants for the two spaces, which are given by $\Lambda_x = -(D - p - 3)c^2/2(D - 2)$ and $\Lambda_y = (p + 1)c^2/2(D - 2)$.

Anti-de Sitter space AdS_n with a cosmological constant $-3d$ is almost like a sphere but with a different signature in the metric. If we introduce coordinates u^n, $n = 1, \ldots, n + 1$ in flat $n + 1$ space, then it is the surface defined by

$$u^n \eta_{nm} u^m = -1/d^2. \tag{13.8.95}$$

where $\eta_{nm} = \text{diag}(-1, 1, \ldots, 1, -1)$. One can deduce the metric in spherical coordinates on this surface induced from the flat metric of the background space-time to be

$$ds_n^2 = -(1 + d^2 r^2)dt^2 + (1 + d^2 r^2)^{-1}dr^2 + r^2 d\Omega_{n-2}^2, \tag{13.8.96}$$

where $d\Omega_{n-2}^2$ is the metric on the sphere S^{n-2} given in equation (13.8.9).

Clearly, AdS_n space-time has the isometry group $SO(2, p + 1)$. Hence the total isometry group of the above solution which is the space $AdS_{p+2} \otimes S^{D-p-2}$ is $SO(2, p + 1) \otimes SO(D - p - 1)$. Clearly, one can find a similar solution for the dual field strength and the corresponding space is $AdS_{D-p-2} \otimes S^{p+2}$ with isometry group $SO(2, D - p - 3) \otimes SO(p + 3)$. as it corresponds to taking $p \rightarrow D - p - 4$.

13.8.2 Brane solutions in eleven-dimensional supergravity

The bosonic sector of eleven-dimensional supergravity, discussed in section 13.3 consists of gravity coupled to a 3-form gauge field, hence it is of the form of the generic action of equation (13.8.1), with the important exception that it also possesses a Chern–Simmons term. The latter contributes only to the gauge field equation of motion by a term proportional to

$$\epsilon^{\mu_1 \cdots \mu_{11}} F_{\mu_4 \cdots \mu_7} F_{\mu_8 \cdots \mu_{11}}. \tag{13.8.97}$$

We find that this term vanishes for the field strength ansatz of equation (13.8.13) and so the previous section provides us with solutions to the full supergravity theory. Hence this supergravity theory possesses a 2-brane solution, called the M2-brane. Substituting $D = 11, d = 8, p = 2$ and $a = 0$, we find that $\Delta = (p + 1)(d - 2) + \frac{1}{2}a^2(D - 2) = 18$ for which the solution of equation (13.8.57) becomes line element given by [13.46]

$$ds^2 = N_2^{-2/3}(-(dt)^2 + (dx_1)^2 + (dx_2)^2) + N_2^{1/3}((dy_1)^2 + \cdots + (dy_8)^2), \tag{13.8.98}$$

together with a 4-form field strength

$$F_{t12a} = \pm \frac{\partial}{\partial y_a} N_2^{-1}, \ i = a, \ldots, 8. \tag{13.8.99}$$

In these equations N_2 is a harmonic function of the form $N_2 = 1 + |Q|/6r^6$, where $r^2 = (y_1)^2 + \cdots + (y_8)^2$ For brevity we will often omit the two possible signs for the field strength.

However, we can also consider the dual field strength, or equivalently, the magnetic ansatz of equation (13.8.15) and find a 5-brane solution called the M5-brane. Now $D = 11, \tilde{d} = 5, \tilde{p} = 5$ and $\tilde{a} = 0$ and so $\tilde{\Delta} = 18$ This solution [13.47] has the line element

$$ds^2 = N_5^{-\frac{1}{3}}(-(dt)^2 + (dx_1)^2 + \cdots + (dx_5)^2) + N_5^{\frac{2}{3}}((dy_1)^2 + \cdots + (dy_5)^2), \tag{13.8.100}$$

where $N_5 = 1 + k/r^3$, $k = |Q|/3$ and $r^2 = (y_1)^2 + \cdots + (y_5)^2$. The corresponding value of the dual field strength is

$$\tilde{F}_{1 \cdots 6a} = \frac{\partial}{\partial y_a} N_5^{-1}. \tag{13.8.101}$$

Using the duality relation we find that the 4-form field strength is given by

$$F_{a_1 \cdots a_4} = \pm \epsilon_{a_1 \cdots a_4 b} \frac{\partial}{\partial y_b} N_5^{-1}, \ a_1, \ldots, a_4, b = 1, \ldots, 5. \tag{13.8.102}$$

For completeness we also record some other solutions of eleven-dimensional supergravity that play an important part in discussions on M theory. Gravity is well known to possess gravity waves encoded in the Brinkmann solution [13.62]. Since eleven-dimensional supergravity possesses gravity it will also have such a solution which is called the pp wave, or M wave, solution [13.63]. The solution in a D-dimensional space-time containing gravity is given by

$$ds^2 = -(1 - K)(dt)^2 + (1 + K)(dx_1)^2 - 2Kdtdx_1 + ((dx_2)^2$$
$$+ \cdots + (dx_{D-1})^2), \tag{13.8.103}$$

where $1 + K = 1 + k/r^{D-4}$ and $r^2 = (x_2)^2 + \cdots + (x_{D-1})^2$. Hence, taking $D = 11$ we find the pp wave solution in eleven-dimensional supergravity. This is a purely gravitational solution, meaning that the 3- and 6-form gauge fields vanish.

Another purely gravitational solution is the so-called Kaluza–Klein (KK) monopole, which exists in any D-dimensional theory containing gravity [13.64]. It has a metric given by

$$ds^2 = (-(dt)^2 + (dx_1)^2 + \cdots + (dx_{D-5})^2 + H^{-1} \left(dz + \sum_{i=a}^{3} A_a dx_a \right)^2$$
$$+ H \sum_{i=a}^{3} (dx_a)^2, \tag{13.8.104}$$

where $\partial_a A_b - \partial_b A_a = -\epsilon_{abc} \partial_c H$ with $H = 1 + Q/r$. The line element in the z, y_a directions is that of Euclidean Taub–NUT space for which the z directions must be compactified, that is, be an S^1 with period $4\pi Q$. For eleven-dimensional supergravity we must take $D = 11$ of course. In fact, all the above solutions preserve half of the 32 supersymmetries of the theory.

Another very important solution is that discussed at the end of the last section which has the space-time $AdS_4 \otimes S^7$, with isometry group $SO(2, 3) \otimes SO(8)$, for which we may take the 4-form field to be proportional to alternating symbol. A similar solution has space-time, and $AdS_7 \otimes S^4$, with isometry group $SO(2, 3) \otimes SO(8)$ and in this case we take the dual 7-form field strength to be proportional to the alternating symbol. These are called the Freund–Rubin solutions [13.65]. They have the remarkable property that they preserve all of the 32 supersymmetries. Another of the very limited number of solutions that preserve all the supersupersymmetries is a modification of the plane wave solution above to include a non-trivial contribution from the 4-form field strength [13.66].

We now present, without their derivation, some solutions for intersecting branes. The solution for two intersecting M2-branes whose world volumes are in the 012 and 034 directions is given by [13.61, 13.67]

$$ds^2 = (H_1 H_2)^{1/3} [- (H_1 H_2)^{-1} dt^2 + H_1^{-1} \left(dx_1^2 + dx_2^2 \right) + H_2^{-1} \left(dx_3^2 + dx_4^2 \right)$$
$$+ \left(dy_1^2 + \cdots + dy_6^2 \right)], \tag{13.8.105}$$
$$F_{t12a} = \frac{c_1}{2} \partial_a (H_1)^{-1}, \qquad F_{t34a} = \frac{c_2}{2} \partial_a (H_2)^{-1}, \qquad a = 1, \ldots, 6.$$

where $c_i = \pm 1, i = 1, 2$. The harmonic functions $H_i, i = 1, 2$ depend on $y_a, a = 1, \ldots, 6$ and are of the form $H_i = 1 + a_i/r_i^4$, where $r_i^2 = \sum_a (y_a - y_a^i)^2$ in which y_a^i are the positions of the 2-branes in the transverse space. It turns out that this solution preserves eight of the original 32 supersymmetries. We note that the brane in the world volume in the 012 directions does not depend on the 34 directions and so is delocalised in this sense and similarly for the other M2-brane. It is sometime useful to write the M2-brane configuration using the notation

$$
\begin{array}{ccccccc}
0 & 1 & 2 & . & . & \cdots \\
0 & . & . & 3 & 4 & \cdots
\end{array}
\tag{13.8.106}
$$

The entries indicate which directions the branes live in.

We now consider a solution for an M2-brane intersecting an M5-brane [13.61, 13.67]

$$ds^2 = H_1^{2/3}H_2^{1/3}\big[H_1^{-1}H_2^{-1}(-dt^2 + dx_1^2) + H_1^{-1}(dx_2^2 + dx_3^2 + dx_4^2 + dx_5^2)$$
$$+ H_2^{-1}(dx_6^2) + (dy_1^2 + dy_2^2 + dy_3^2 + dy_4^2)\big], \tag{13.8.107}$$

$$F_{6abc} = \frac{c_1}{2}\epsilon_{abcd}\partial_d H_1, \qquad F_{t16a} = \frac{c_2}{2}\frac{\partial_a H_2}{H_2^2},$$

where $c_i = \pm 1, i = 1, 2$. and the functions $H_i, i = 1, 2$ are harmonic functions that depend on $y_a, a = 1, \ldots, 4$. Using the above notation we denote the configuration by

$$
\begin{array}{ccccccc}
0 & 1 & 2 & 3 & 4 & 5 & . & \cdots \\
0 & 1 & . & . & . & . & 6 & \cdots
\end{array}
\tag{13.8.108}
$$

This makes it clear that the world volumes of the M2- and M5-branes intersect in the directions 01 and so on a 1-brane.

Finally we give a solution for two intersecting M5-branes [13.60, 13.61, 13.67];

$$ds^2 = (H_1 H_2)^{2/3}[(H_1 H_2)^{-1}(-dt^2 + dx_1^2 + dx_2^2 + dx_3^2) + H_1^{-1}(dx_4^2 + dx_5^2)$$
$$+ H_2^{-1}(dx_6^2 + dx_7^2) + (dy_1^2 + dy_2^2 + dy_3^2)], \tag{13.8.109}$$

$$F_{67ab} = \frac{c_1}{2}\epsilon_{abc}\partial_c H_1, \qquad F_{45ab} = \frac{c_2}{2}\epsilon_{abc}\partial_c H_2,$$

where $c_i = \pm 1, i = 1, 2$ and the functions $H_i, i = 1, 2$ are harmonic functions that depend on $y_a, a = 1, \ldots, 3$. We may write their configurations as

$$
\begin{array}{cccccccc}
0 & 1 & 2 & 3 & 4 & 5 & . & . & \cdots \\
0 & 1 & 2 & 3 & . & . & 6 & 7 & \cdots
\end{array}
\tag{13.8.110}
$$

The last two solutions preserve only eight of the original 32 supersymmetries and are delocalised in the sense discussed above. Clearly, these are only a few of the many intersecting solutions that one can have.

Given any solution the variation of the gravitino must vanish, $\delta\psi_\mu = 0$, as classically all the fermions vanish. This variation was given in section 10.3 and it contains the derivative of the spinor parameter as well as connection and field strength terms. The number of independent spinor solutions to this equation is the number of supersymmetry parameters preserved. This point is further discussed in section 13.8.4. The global properties of the above solutions and, in particular, the nature of their horizons are discussed in [13.68, 13.69].

13.8.3 Brane solutions in the ten-dimensional maximal supergravity theories

Let us first consider the *IIA supergravity theory* whose action is given in equation (13.4.6). It consists of gravity coupled to a dilaton and field strengths of ranks 2, 3 and 4. Hence, apart from the Chern–Simmons term it is of the form of our generic action of equation (13.8.1). We recall that the field strengths of ranks 2 and 4 are associated with the R–R sector and that rank 3 is in the NS–NS sector. The dilaton and gravity also arise in the latter sector. If we also consider the dual field strengths, then we have in addition field strengths of ranks 6, 8 in the R–R sector and 7 in the NS–NS sector. Examining equation (13.4.7) we identify the dilaton coupling to be given by

$$a_{p+2} = \eta \frac{(3-p)}{2}, \quad \text{where } \eta = \begin{cases} 1, & R\text{–}R, \\ -1 & NS\text{–}NS, \end{cases} \tag{13.8.111}$$

for $p = 1, 2, 3$ if we identify $\sigma = \phi$; however, as one easily verifies, it also gives the correct dilaton coupling for the dual fields as under $p \to 6 - p$ it reverses sign. One can verify that the Chern–Simmons term vanishes for all the solutions we will discuss below. For the electric ansatz the field strength is non-zero only when it carries the world volume indices and one transverse index; however, the indices on the ε present in the Chern–Simmons term, which is contracted with the two field strengths must have all indices different. A similar argument applies to the magnetic ansatz.

Hence in this theory we will find spherically symmetric p-brane solutions for $p = 0, 1, 2, 4, 5, 6$. Evaluating Δ of equation (13.8.54) we find using equation (13.8.111) that for all the branes $\Delta = 16$. From equation (13.8.57) we read off the p-brane solutions to be

$$ds^2 = N_p^{-(7-p)/8}(-(dt)^2 + (dx_1)^2 + \cdots + (dx_p)^2)$$
$$+ N_p^{(p+1)/8}((dy_1)^2 + \cdots + (dy_d)^2), \tag{13.8.112}$$

where $N_p = 1 + |Q|/(7-p)r^{7-p}$. The dilaton given by

$$e^\phi = (N_p)^{a_{p+2}/2}, \tag{13.8.113}$$

and a gauge field, with world indices, is given by

$$A_{1\cdots p+1} = \pm(N_p^{-1} - 1). \tag{13.8.114}$$

The extra term of -1 is optional as it vanishes when we compute the field strength, but it has the advantage that the gauge potential has no constant term.

Consequently in the NS–NS sector we find a 1-brane and a 5-brane solution which couple to the gauge fields of ranks 2 and 6, the latter being the dual gauge field of the former. The 1-brane solution is a string solution that couples to the 2-form gauge field such as we associated with the string itself and is called the F1-brane. The second solution is called the NS5-brane. In the R–R sector we have p-branes solutions [13.48–13.52] for $p = 0, 2, 4, 6$ which couple to gauge fields of ranks 1, 3, 5, 7. These are called Dp-brane solutions.

We note that in the R–R sector we have branes for all even p except $p = 8$. This would couple to a rank 9 anti-symmetric tensor gauge field $A_{\mu_1 \cdots \mu_9}$. Such a field can be consistently included in the IIA theory. It has a rank 10 field strength and so can appear in the action as

$$\int d^{10}x \det e (F_{\mu_1 \cdots \mu_{10}} F^{\mu_1 \cdots \mu_{10}}). \tag{13.8.115}$$

The equation of motion of $A_{\mu_1 \cdots \mu_9}$ sets

$$F_{\mu_1 \cdots \mu_{10}} = \epsilon_{\mu_1 \cdots \mu_{10}} c, \tag{13.8.116}$$

where c is a constant. Substituting this back in the above action we find a cosmological constant. In fact, using this mechanism one can recover [13.70] the massive IIA supergravity theory of [13.71]. This latter theory is often called the Romans theory and it can be thought of as a deformation of the familiar IIA supergravity theory by a new dimensional parameter. There is only one such deformation allowed that preserves the 32 supersymmetries. The corresponding 8-brane solution is given by substituting $p = 8$ in equation (13.8.112) and has a metric of the form [13.70, 13.72]

$$ds^2 = (N_8)^{\frac{1}{8}}((-dt)^2 + (dx_1)^2 + \cdots + (dx_8)^2) + (N_8)^{\frac{9}{8}}(dy_1)^2,$$

$$A_{1\ldots9} = (N_8)^{-1} - 1 \tag{13.8.117}$$

and the dilaton and gauge field given by

$$e^\phi = (N_8)^{-\frac{5}{4}}, A_{1\ldots9} = (N_8)^{-1} - 1. \tag{13.8.118}$$

We also have the purely gravitational solutions: the pp wave and KK monopole solutions given by taking $D = 10$ in equations (13.8.103) and (13.8.104). All the above solutions preserve half of the 32 supersymmetries of the theory and are the most usually discussed half BPS solutions of the IIA supergravity theory.

Let us now consider the *IIB supergravity theory* discussed in section 13.5. This theory consists of gravity, a dilaton and one rank 3 field strength in the NS–NS sector and another rank 3 field strength, a scalar and a self-dual rank 6 field strength in the R–R sector. Taking into account the possibility of dual field strengths we find that this theory possesses gauge fields of ranks 2 and 6 in the NS–NS sector and ranks 0, 2, 4, 6 and 8 in the R–R sector. As such, we will find p-branes for $p = -1, 1, 3, 5, 7$. The IIB theory does not have an action, but we can read off the dilaton coupling from the equations of motion, and in particular the metric field equation, to be

$$a_{p+2} = \eta \frac{(3 - p)}{2}, \tag{13.8.119}$$

where $\eta = 1$ in the R–R sector and $\eta = -1$ in the NS–NS sector. As in the IIA case this formula also works for the dual fields. We find that for this theory $\Delta = 16$ for all branes and from equation (13.8.57) we find the p-brane solution to have the metric [13.48–13.52]

$$ds^2 = (N_p)^{-(7-p)/8}(-(dt)^2 + (dx_1)^2 + \cdots + (dx_p)^2)$$

$$+ (N_p)^{(p+1)/8}((dy_1)^2 + \cdots + (dy_d)^2), \tag{13.8.120}$$

where $N_p = 1 + |Q|/(7 - p)r^{7-p}$, the dilaton is given by

$$e^\phi = (N_p)^{-\eta(p-3)/4} \tag{13.8.121}$$

and the gauge field is given by

$$A_{1 \cdots p+1} = \pm (N_p^s)^{-1} - 1. \tag{13.8.122}$$

Thus we find $(p = 1, 5, 7)$-brane solutions in the NS–NS sector. The 1-brane is just a string which couples to the rank 2 gauge field in this sector and it is corresponds to the fundamental string itself. In the R–R sector we have $(p = -1, 1, 3, 5, 7)$-branes. Thus in this sector we find another 1-brane, but this is a D-brane as it occurs in the R–R sector and is called the D1 string. These two 1-branes transform as a doublet under SU(1, 1) or equivalently SL(2, **R**), of the IIB theory as is obvious from the fact that the rank 2 gauge fields to which they couple transform in this way. We also find an instanton, that is, the $p = -1$, which lives at a point of space-time [13.73].

The R–R sector contains branes for all odd p, except $p = 9$. This is called a space filling brane as there are no transverse coordinates and so the world volume of the brane is space-time itself. Substituting $p = 9$ into equation (13.8.115) leads to the solution

$$ds^2 = N(-(dt)^2 + (dx_1)^2 + \cdots + (dx_9)^2), \ e^\phi = N^{-\frac{3}{2}}, \tag{13.8.123}$$

where now N is a constant as there are no transverse coordinates. There are world-sheet arguments associated with D-branes which suggest such a 9-brane should exist in the IIB theory [13.74]. We also have the purely gravitational solutions: the pp wave and KK monopole solutions given by taking $D = 10$ in equations (13.8.103) and (13.8.104). All the above solutions preserve half of the 32 supersymmetries of the theory.

We also have the solutions considered at the end of the last section. In particular, if we take the 5-form field strength to be proportional to alternating symbol, namely

$$F_{\mu_1 \cdots \mu_5} = d\epsilon_{\mu_1 \cdots \mu_5}, \ F_{a_1 \cdots a_5} = d\epsilon_{a_1 \cdots a_5}, \tag{13.8.124}$$

all other components and fields, other than the metric, vanishing. The presence of non-vanishing field strengths in both spaces is required as the field strength is self-dual. For the Riemann tensor we find from equation (13.8.2) that

$$R^\mu{}_\nu = -\frac{c^2}{2}\delta^\mu_\nu, \ R^a{}_b = \frac{c^2}{2}\delta^a_b, \tag{13.8.94}$$

all other components being zero. Hence we have a solution $AdS^5 \otimes S^5$ with isometry group SO(2, 4) ⊗ SO(6) [13.75]. This preserves all of the 32 supersymmetries of the theory. Another solution that has this property is a modification of the pp wave solution to include a non-trivial 5-form field strength [13.76]. These two solutions play an important role in the AdS–CFT correspondence.

13.8.4 Brane charges and the preservation of supersymmetry

The p-brane solutions that we have been discussing possess energy density, generally mass and charge. These can be measured by computing the surface integrals at infinity of the corresponding long range gravitational and gauge fields. Given a p-brane we can define its electric charge to be

$$Q_e = \frac{1}{2\kappa_D^2} \int_{S^{D-p-2}} \star F, \tag{13.8.125}$$

where the integration is over the $(D - p - 2)$-sphere and F is the field strength of the gauge field A to which the brane couples. The integral is to be taken over a sphere at infinity that is in an asymptotic region of space at large sphere radius. More generally one can integrate over any surface which is the boundary ∂M_{D-p-1}, which is in the asymptotic region, of a spatial subspace M_{D-p-1} of dimension $D - p - 1$. The integral will, of course, vanish unless the boundary actually contains the brane whose charge is to be measured. As specified the above charge is known as the electric charge. In the expression of equation (13.8.125) we are to think of $\star F$ as a form which is generically given by $\star F_c d\mu$ where $\star F_c$ are the components of the dual of the field strength and $d\mu$ is the measure over which we are going to integrate. As the surface of integration is specified, the measure $d\mu$ is also specified and so therefore are components of the field strength that occur in the integration. For the case of the sphere, and using spherical coordinates as given in equation (13.8.9), we conclude that $d\mu = r^{D-p-2} d\theta_1 \wedge (\sin \theta_1 d\theta_2) \wedge \cdots \wedge (\sin \theta_1 \cdots \sin \theta_{D-p-3} d\theta_{D-p-2})$ and so it is the $(\star F)_{\theta_1 \cdots \theta_{D-p-2}}$ component of the dual of the field strength occurs in the integral. As such we see that the component of the field strength that occurs lives in the time and other world volume directions as well as the radial direction, that is, $F_{ti_1 \cdots i_p r}$ whose expression is given in equation (13.8.30). We note that asymptotically $F_{ti_1 \cdots i_p r} = \epsilon_{i_1 \cdots i_p} Q / r^{D-p-2}$, where Q was the integration constant in the solution. The factor of $r^{-(D-p-2)}$ in the field strength cancels with that in the measure and so the magnitude of the electric charge is found to be $Q_e = \Omega_{D-p-2} Q$, where Ω_{D-p-2} is the volume of the unit $(D - p - 2)$-sphere.

The magnetic charge is given by

$$Q_m = \frac{1}{2\kappa_D^2} \int_{S^{p+2}} F, \qquad\qquad (13.8.126)$$

where analogous definitions apply and one can replace the sphere by the boundary ∂M_{p+3} of a suitable spatial M_{p+3} surface at infinity.

For an infinite planar brane, as indeed are many of the above solutions, one can consider the flat space in which they live to be a very large torus. In effect, the brane is wrapped once on this torus. The definition of which is electric and which is magnetic is referred to the field strength traditionally used to describe the supergravity actions. For the case of eleven dimensions this is a four-dimensional field strength and then the 2-brane and the 5-brane have electric and magnetic charges, respectively. This follows, for example, for the 2-brane as we need a 7-sphere to enclose the 2-brane and so we require a 7-form in the integral which must be the dual of the 4-form field strength. A similar argument implies that the 5-brane must carry a magnetic charge.

We can calculate the electric charge using two different surfaces. The difference in the charge is the integral of $\star F$ over the boundary of one surface minus that over the other surface. However, by Stokes's theorem, this can be written as $\int d \star F$ over the volume enclosed by the two surfaces, but by the equation of motion this vanishes provided the deformation did not include any new charged sources. In fact, this last statement requires some further clarification. The equation of motion of the gauge field is not generally given by $d \star F = 0$, indeed for the 2-brane of M theory it is of the generic form $d \star F_4 + F_4 \wedge F_4 = 0$. However, for the solutions we are considering the fields drop off sufficiently fast that the additional non-linear terms in the equations of motion do not contribute to the integral at the boundary at infinity and can be dropped. As such the charges we calculate are independent of the precise surface at infinity chosen. For the magnetic charge $dF = 0$ as a result of the

Bianchi identity and so the same argument applies. That the magnetic charge is the integral of a quantity that is a space-time derivative only serves to emphasise that magnetic charges arise due to non-trivial topological effects.

One can also calculate the charges using surfaces that are not at infinity, but then one must use the charge which includes the non-linear terms such that the object under the integral is annihilated by d by virtue of its equation of motion. For the case of the 2-brane in eleven dimensions this means one takes the integral of $\star F + A \wedge F$.

To illustrate the above discussion we first consider the familiar case of a point particle in four dimensions. The particle can be taken to be at rest in the spatial 3-space M_3 and the electric charge is the integral over a boundary at infinity of this 3-space, for example, the 2-sphere S^2. The above charge then reduces to

$$Q_e = \frac{1}{2\kappa_D^2} \int_{S^2} \star F = \frac{1}{2\kappa_D^2} \int r^2 d\Omega \underline{E} \cdot \hat{r}, \qquad (13.8.127)$$

where $\star F = (\star F)_{\theta\phi} d\theta \wedge \sin\theta d\phi$ in which r, θ and ϕ are the usual spherical coordinates. As such, it is the $E_r = F_{0r}$ component of the field strength that occurs in the integral. We have denoted $\hat{r}$ to be the unit vector in the radial direction. Taking $\underline{E} = (e/r^2)\hat{r}$ we arrive at the usual relationship between the asymptotic form of the field strength and the electric charge the particle carries, that is, $Q_e = e$.

As a further example we consider the 2-brane in eleven dimensions. The M2-brane occupies a space with two spatial dimensions leaving an eight-dimensional transverse space M_8 embedded in the ten-dimensional spatial part of space-time. The electric charge is given by the integral

$$Q_e = \frac{1}{2\kappa_D^2} \int_{S^7} \star F_4, \qquad (13.8.128)$$

where $F_4 = dA_3$ and S^7 is a sphere at infinity. For the case of the planar M2 solution given in equation (13.8.99) the ten-dimensional spatial part of space-time at infinity can be thought of as $R^2 \times S^7$. The R^2 correspond to the spatial directions in which the 2-brane lies, while the transverse space is the S^7 over which we integrate. In particular, the spatial world volume of the brane lies in the x^1, x^2 directions of the background space and the transverse spatial volume is in the eight directions $x^3, \ldots, x^{10}$, which are bounded by a 7-sphere at infinity. As we are integrating over a 7-sphere we must be evaluating the electric charge. Up to a numerical factor we find that the the constant in the solution equals the charge of the brane. We note that we find a factor of r^7 in the integral over the surface but this is cancelled by the factor of r^{-7} coming from the field strength. As noted above, this cancellation occurs for all branes as the gauge field drops off as r^{-1} to the power of the spatial dimensions of the brane minus 2.

Let us also consider the 5-brane in M theory, which, if we assume it is static and planar, can be taken to lie in the $x^1, \ldots, x^5$ directions of the background space. The five-dimensional transverse spatial volume is in the $x^6, \ldots, x^{10}$ directions. This is bounded at infinity by a 4-sphere over which we will integrate to find the charge. As such, the 5-brane carries a magnetic charge and again the constant in the solution becomes proportional to the charge.

Above we denoted the charges by objects without indices. However, to calculate the charges one must choose, for the electric case, an M_{D-p-1} spatial sub-volume of the $D-1$ spatial volume in which the brane lives and it is the boundary ∂M_{D-p-1} over which we integrate the field strength. We can think of the embedding of M_{D-p-1} as being specified by a p-form. As a result the electric charge is really a rank p form. Indeed we found when evaluating the electric charge that it was proportional to $\epsilon_{i_1 \cdots i_p}$ and so lies in the spatial world volume directions. Similarly the magnetic charge is actually a rank $D - p - 4$ form. For a given brane and choice of surface the integral will vanish unless it is correctly oriented to capture the flux of the brane. For the planar 2-brane considered above, the only surface one can take that results in a non-zero answer takes M_8 to be the spatial volume transverse to the 2-brane. The mass is given by computing the ADM mass which is a suitable integral over a derivative of the metric at spatial infinity.

In this section, we have solved the field equations in the region of space-time in which there are no sources, as is evident by the absence of any term on the right-hand side of equation (13.8.3). However, in order to have a charge there must be a source some where in space-time. We will find in the next section that a p-brane moves according to an action that consists of two terms, the volume it sweeps out plus a coupling to a rank $p + 1$ gauge field of the background supergravity theory. The latter term is of the form $\int d^{p+1}\xi A_{\mu_1 \cdots \mu_{p+1}}(X^\mu) j^{\mu_1 \cdots \mu_{p+1}}$, where $j^{\mu_1 \cdots \mu_{p+1}}$ is the corresponding current which is toplogical and has the generic form $j^{\mu_1 \cdots \mu_{p+1}} = q_p dX^{\mu_1} \wedge \cdots \wedge dX^{\mu_{p+1}}$. Here $X^\mu(\xi)$ describes the embedding of the p-brane in the background space-time and q_p is the charge. See equation (14.2.2). The term that contains the volume swept out also has a constant multiplying it and this is called the p-brane tension denoted T_p. In general, the charge and tension of the p-brane are independent constants but for the p-branes we are studying they are equal as a consequence of the large amount of supersymmetry the p-branes possesses.

As such, the equation of motion of the gauge field will contain this topological current times a δ function, ensuring that the term is evaluated at the position of the p-brane, that is, $J^{\mu_1 \cdots \mu_{p+1}} = \int d^{p+1}\xi \delta(x^\mu - X^\mu(\xi)) j^{\mu_1 \cdots \mu_{p+1}}$. The integration is over the world volume of the brane. Indeed, this term appears on the right-hand side of equation (13.8.3). In form language, the equation of motion as it comes from the action can be written as $\star d \star F = J$. Taking the dual it becomes $d \star F = \star J$. We can use Gauss' theorem to evaluating the charge of equation (13.8.125) by expressing it as

$$Q_e = \frac{1}{2\kappa_D^2} \int_{\partial M_{D-p-1}} \star F = \frac{1}{2\kappa_D^2} \int_{M_{D-p-1}} d \star F = \frac{1}{2\kappa_D^2} \int_{M_{D-p-1}} \star J. \qquad (13.8.129)$$

As such, we find that the charge of the p-brane as computed by equation (13.8.125) is proportional to the charge q_p as it appears in the p-brane action and so proportional to the tension T_p of the p-brane.

To see if a particular solution preserves some supersymmetry we just take the explicit solution and vary it under the supersymmetry transformations of the corresponding super-gravity theory. As the fermions of the theory vanish for the classical solution the supersymmetry variations of the bosonic field vanish automatically; however, we must also set the variations of the fermion fields to zero. For example, for the case of eleven-dimensional supergravity there is only one fermion, the gravitino, and examining equation (13.3.5) we find that any preserved supersymmetry must satisfy the equation

$$\delta\psi_\mu = 0 = \frac{1}{\kappa} D_\mu(\hat{\Omega})\epsilon + \frac{1}{12^2}(\Gamma_\mu{}^{\nu_1 \cdots \nu_4} - 8\delta_\mu^{\nu_1}\Gamma^{\nu_2\nu_3\nu_4})\hat{F}_{\nu_1 \cdots \nu_4}\epsilon. \qquad (13.8.130)$$

Substituting the expression for the field strengths of the solution under consideration one can solve such equation to find the non-vanishing supersymmetry parameters and so which supersymmetries are preserved by the solution. Carrying out this calculation for the 2-brane in eleven dimensions one finds that the preserved supersymmetry parameters are of the form $\epsilon = N_2^{-1/6}\epsilon_0$, where ϵ_0 is a constant parameter and satisfies the condition $\Gamma^{012}\epsilon = \pm\epsilon$. As N_2 tends to 1 at spatial infinity the preserved supersymmetries are the constants ϵ_0 at infinity. For the 5-brane in eleven dimensions one finds that the preserved supersymmetry parameters are of the form $\epsilon = N_5^{-1/12}\epsilon_0$, where ϵ_0 is a constant parameter and it satisfies the condition $\Gamma^{012345}\epsilon = \pm\epsilon$. To find the preserved supersymmetry parameters it is useful to to rewrite the supersymmetry parameter, which is a spinor of $\mathrm{Spin}(1, D-1)$, in terms of the Spin group preserved by the solutions which for the planar p-brane solution considered above is $\mathrm{Spin}(1, p) \otimes \mathrm{Spin}(D-p-1)$. One then also uses a corresponding decomposition of the γ matrices. In other words we write $\epsilon = \eta_1 \otimes \eta_2$, where η_1 and η_2 are spinors of $\mathrm{Spin}(1, p)$ and $\mathrm{Spin}(D-p-1)$, respectively.

Whether a solution preserves some supersymmetry can also be answered by examining the supersymmetry algebra. For the case of eleven dimensions, which we now consider, this involves the algebra of equation (13.3.1), which on the right-hand side involves the momentum but also the central charges. Clearly, if a supersymmetry transformation with parameter ϵ_P preserves the solution its action on all the fields of the theory will vanish and so therefore will $\{Q, Q\}_{\alpha\beta}\bar{\epsilon}_P^\beta = 0$ on all the fields as $Q_\beta\bar{\epsilon}_P^\beta$ generates such a variation. As we will see it is relatively easy to look for zero modes of the anti-commutator $\{Q, Q\}$.

Let us first consider the case when none of the central charges is active and so $\{Q_\alpha, Q_\beta\} = (\Gamma^m C^{-1})_{\alpha\beta} P_m$. If the right-had side acting on a spinor has a non-trivial solution we require $\det(\Gamma^m P_m) = 0 = (P^2)^{16}$ and so we conclude that the momentum P^μ must be null. In this case we can adopt a Lorentz frame, where $P^\mu = m(1, \pm 1, \dots, 0)$, and then we find that

$$\{Q_\alpha, Q_\beta\} = m((\Gamma_0(1 \pm \Gamma^{01})C^{-1}))_{\alpha\beta}. \tag{13.8.131}$$

The parameter of the preserved supersymmetry must then obey $(1 \pm \Gamma^{01})\epsilon_P = 0$. Since $(\Gamma^{01})^2 = I$ we find the matrix multiplying ϵ_P is a projector and so for a given solution we have a given sign and we find that half of the supersymmetries are preserved. Thus we may conclude that any solution that has only the momentum active, that is, none of the central charges active, must preserve half of the supersymmetries, in which case it is null, or none of the supersymmetries. As one might immediately guess this corresponds to the pp wave solution of equation (13.8.103) as this solution only involves the gravitational field. Indeed, one can substitute the solution into $\delta\psi_\mu = 0$ and verify that the preserved supersymmetry parameters do indeed satisfy the above equation. One can also compute P^μ using an ADM mass formula and verify that the momentum of the pp wave is indeed null.

Now let is consider the case when the momentum and the 2-form central charge are active. Assuming we are dealing with a massive object we can take $P^\mu = (m, 0, \dots, 0)$. We also assume that the 2-form charge has only one non-zero component, which we take to be Z^{12}. As a result the anti-commutator of two supersymmetries takes the form

$$\{Q_\alpha, Q_\beta\} = (m\Gamma_0(1 + b\Gamma^{012})(C^{-1}))_{\alpha\beta}, \tag{13.8.132}$$

where $b = Z^{12}m^{-1}$. We can evaluate the determinant as follows:

$$\det(1 + b\Gamma^{012}) = e^{\text{Tr}\ln(1+b\Gamma^{012})} = \exp\left(\text{Tr}\left(b\Gamma^{012} - \frac{b^2 I}{2} + \frac{b^3\Gamma^{012}}{3} + \cdots\right)\right)$$

$$= e^{16\ln(1-b^2)} = (1 - b^2)^{16}. \tag{13.8.133}$$

In evaluating this expression we have used the facts that $(\Gamma^{012})^2 = I$ and $\text{Tr}\Gamma^{012} = 0$. Hence we will preserve some supersymmetry if and only if $b = \pm 1$ and in this case the preserved supersymmetry parameter will satisfy $(1 \pm \Gamma^{012})\epsilon_P = 0$. As this condition is of the form of a projector acting on ϵ_P we conclude, as for the previous case, that for this configuration either half the supersymmetries are preserved if $Z^{12} = \pm m$ or otherwise all supersymmetries are broken.

As the reader will have guessed this possibility corresponds to the 2-brane solution of equation (13.8.98) and one can verify by computing $\delta\psi_\mu$ that this does preserve half of the possible 32 supersymmetries and the preserved supersymmetries satisfy the above constraint.

As we will explain in chapter 14, a p-brane couples to a rank $p + 1$ gauge field and this automatically possesses a rank p conserved charge, that is, the charge J mentioned above, which becomes a central charge in the supersymmetry algebra [13.77]. By calculating this charge one can verify that the above identifications of solutions with supersymmetries preserved by the supersymmetry algebra are indeed correct.

The general picture for all such brane solutions of the eleven-dimensional supergravity theories with only the momentum and one component of a rank p-form central charge non-zero is rather similar. One finds that either half the supersymmetries are preserved provided the mass and central charge are equal up to a numerical factor or otherwise no supersymmetries are preserved. The preserved supersymmetries satisfy the condition $\Gamma^{01\cdots p}\epsilon = \pm\epsilon$ and this is also the condition that one finds by examining $\delta\psi_\mu = 0$. This result also holds for the IIA and IIB theories. The IIA and IIB supersymmetry algebras are given in sections 13.4 and 13.5, respectively. The arguments are similar to those given above. The reader may wish to carry out the calculation for the 5-form central charge in eleven dimensions which corresponds to the 5-brane solution.

The supersymmetry algebra allows one to derive a bound involving the mass and central charges of the branes. Multiplying the anti-commutator of two supersymmetry charges by two Grassmann *even* Majorana spinors η it becomes $2(\bar\eta Q)^2$. We note, using the definition of Dirac conjugate of chapter 4, that is, $\bar\eta = \eta^\dagger\Gamma^0$, that

$$(\bar\eta Q)^\dagger = (\eta^\dagger\Gamma^0 Q)^\dagger = Q^\dagger(\Gamma^0)^\dagger\eta = -Q^\dagger\Gamma^0\eta = -\bar Q\eta = \bar\eta Q. \tag{13.8.134}$$

In the last equality we have used the Majorana properties of the spinors, as described in chapter 4, but have taken into account the fact that η is Grassmann even. As a result we conclude that $(\bar\eta Q)$ is real and so $(\bar\eta Q)^2$ is positive. Doing this for one of the supersymmetry algebras above, that is, as given in equations (13.3.1) or (13.4.1) or (13.5.2), we find that the left-hand side, that is, $\bar\eta(\Gamma^\mu P_\mu + \cdots)\eta$, where $+\cdots$ stands for the central charge contributions, must therefore be positive. For the case of the 2-brane of eleven dimensions considered above we find the bound $P^0 \geq |Z^{12}|$. The simplest way to see this is to realise that as the eigenvalues of Γ^{012} are ± 1, then in a suitable basis the matrix $P^0 + Z^{12}\Gamma^{012}$ has the eigenvalues $P^0 \pm Z^{12}$. The 2-brane solution we considered earlier saturates this bound and only if this is the case do we find that some supersymmetries are preserved. Objects of this type are called Bogomol'nyi–Prasad–Sommerfield (BPS) branes. The bound is analogous to the Olive–Witten bound in supersymmetric Yang–Mills theory [13.78]. Clearly, for other branes where one has more than one component of the central charge active, or indeed several different central charges active, one can derive similar bounds. For the above solutions one can calculate their charge and ADM mass and verify that they really do saturate the bound.

We close this chapter by giving a brief account of the charge quantisation conditions for branes. As we saw above the 2-brane can only carry an electric charge and the 5-brane a magnetic charge. By using arguments similar to those used to derive the famous Dirac quantisation condition for electric and magnetic charge in four dimensions, one can show that a p-brane electric charge Q_e and its magnetic dual Q_m obey a quantisation condition [13.79, 13.80], sometimes referred to as the Dirac–Nepomechie–Teitelboim (DNT) quantisation condition, that we may generically write as

$$Q_e Q_m = 2\pi n, \quad n \in Z \tag{13.8.135}$$

for suitably normalised charges. Since the brane tensions are proportional to the brane charges it follows that the brane tensions will also obey a similar quantisation condition.

A generic brane cannot carry both electric and magnetic charge. However, they can do so when the field strength of the gauge field to which the brane couples has rank $D/2$. This is the case for the point particle in four dimensions, the field strength just being a 2-form. Maxwell's theory with the addition of magnetic monopoles is just such an example and the Dirac quantisation condition is

$$Q_e^1 Q_m^2 - Q_e^2 Q_m^1 = 2\pi n \quad n \in Z. \tag{13.8.136}$$

The upper label on the charges indicates with which of the two particles it is associated. Objects which carry both magnetic and electric charge are called dyons. Other examples are strings in six dimensions which have rank 3 field strengths and 2-branes in eight dimensions. In such cases the Dirac quantisation condition has two terms rather than the one found in equation (13.8.134). As the charges arise from solutions in the supergravity theories they belong to multiplets of the symmetries of these theories and in particular the symmetries discussed in section 13.6. However, these are continuous groups and these cannot leave the quantisation conditions invariant. As such, these symmetries must be broken by the solutions to be over some discrete field in such a way that they preserve the quantisation conditions. We will discuss this further in chapter 17. This discrete symmetry can be used to determine the form of the quantisation conditions. If $D = 4n$, then one finds a minus sign between the two terms, but if $D = 4n + 2$ one finds a plus sign. Of course, if the field strength involved is self-dual, as is the case for the rank 5 field strength of the IIB supergravity theory, then electric and magnetic charges are identical.

14   Brane dynamics

The raised nail gets hammered down.

Japanese saying

Quantum field theory has traditionally concerned the quantisation of point particles and so far this book has largely been about the classical, and then quantum, behaviour of strings. In this chapter we will consider objects that sweep out, as they move through space-time, spatial volumes of dimension greater than 1. We refer to such an object as being a p-brane if its world volume contains time and p spatial coordinates. A 0-brane is just a particle and a 1-brane is a string. In fact, p-branes for $p \geq 2$ occur in string theory as solitons in the corresponding low energy effective actions of string theories as well as 0-branes and 1-branes. Indeed, we found in section 13.8 that the IIA and IIB supergravities, which are the unique low energy effective actions for the IIA and IIB superstring theories, each have a whole family of p-brane solutions. Being solitons, p-branes are non-perturbative objects as seen from the perspective of the low energy effective actions, and, except for the fundamental strings, are non-perturbative from the viewpoint of string perturbation theory.

In this chapter, we will study p-branes as objects in their own right and, using supersymmetry, we will find what possible superbranes can exist and give their dynamics. We will see later in this book that p-branes are related by duality symmetries to the perturbative strings that we studied earlier. As such, string duality symmetries imply that string theory should be extended to a theory that includes branes on an equal footing with strings.

We will close this chapter by finding solutions to the M5-brane equations of motion and use these to find the low energy effective action for the $N = 2$ supersymmetric Yang–Mills quantum field theory to all orders in the coupling constant from the dynamics of the M5-brane.

14.1   Bosonic branes

Bosonic p-branes are extended objects that sweep out a $(p + 1)$-dimensional space-time surface, often called the world volume, in a background space-time. The former and latter are also sometimes refered to as the world surface and target space-time, respectively. Such objects are not supersymmetric as their dynamics is described by only bosonic variables. A p-brane sweeps out a $(p + 1)$-dimensional world surface M, with coordinates ξ^m, $m = 0, 1, \ldots, p$ in a D-dimensional background space-time $\underline{M}$ with coordinates $X^{\underline{n}}$, $\underline{n} = 0, 1, \ldots, D - 1$. As the symbols imply, for the world, or coordinate, indices we

420

use $m, n, p, \ldots$ for the embedded p-brane world volume and $\underline{m}, \underline{n}, \underline{p}, \ldots$ for background, or target, space-time. The corresponding tangent space indices are $a, b, c, \ldots$ for world volume indices and $\underline{a}, \underline{b}, \underline{c}, \ldots$ for background space-time indices. This notation is used extensively in the literature on this subject. The reader should have no difficulty making the transition from the $\mu, \nu, \ldots$ and $\alpha, \beta, \gamma, \ldots$ used in the previous chapters for the background space-time and world volume indices, respectively, of the string. As we considered the string in a flat background we had no need to distinguish world and tangent indices for the background space-time and we used only world indices for the world volume or background space-time. The surface M swept out by the p-brane in the background space-time $\underline{M}$ is specified by the functions $X^{\underline{n}}(\xi^n)$ which extremise the action

$$-T \int d^{p+1}\xi \sqrt{-\det g_{mn}}, \tag{14.1.1}$$

where

$$g_{mn} = \partial_n X^{\underline{n}} \partial_m X^{\underline{m}} g_{\underline{nm}} \tag{14.1.2}$$

and $g_{\underline{nm}}$ is the metric of the background space-time often referred to as the background metric. The constant T is the brane tension and has the dimensions of $(\text{mass})^{p+1}$. The metric g_{mn} is the metric induced on the world surface M by the background metric of the background space-time. As such, we recognise the action in equation (14.1.1) as the volume swept out by the p-brane world surface. Hence, like the string and point particle, a p-brane moves so as to extremise the volume of the surface it sweeps out. If the background space-time is flat the background metric is just the Minkowski metric $g_{\underline{mm}} = \eta_{\underline{mm}}$. A 0-brane is just a point particle and $T = m$, where m is the mass of the particle. A 1-brane is just the usual bosonic string and the action of equation (14.1.1) is the Nambu action for the string if we take the background metric to be flat. In this case we often write $T = 1/2\pi\alpha'$, where α' is the string Regge slope parameter. The reader will have noticed that we are using capital X for brane coordinates in this chapter. The above action for the 2-brane was found in [2.3].

The action of equation (14.1.1) is invariant under reparameterisations of both the background space-time $\underline{M}$ and the world volume space-time M. Under the former $X^{\underline{m}} \to X^{\underline{m}'}(X^{\underline{n}}), \xi^n \to \xi^n$ and so $\partial_n X^{\underline{m}} \to \partial_n X^{\underline{m}'}(X) = \partial_n X^{\underline{p}} \partial_{\underline{p}} X^{\underline{m}'}$. The background metric $g_{\underline{nm}}$ changes under a background reparameterisation in the usual way:

$$g'_{\underline{nm}}(X^{\underline{s}'}) = \frac{\partial X^{\underline{p}}}{\partial X^{\underline{n}'}} \frac{\partial X^{\underline{q}}}{\partial X^{\underline{m}'}} g_{\underline{pq}}(X^{\underline{s}}).$$

This change compensates that for the derivatives of the coordinates in such a way that g_{mn}, and so the action, is invariant.

Under world volume reparameterisations $\xi^n \to \xi^{n'}(\xi), X^{\underline{m}'}(\xi') = X^{\underline{m}}(\xi), g'_{\underline{nm}}(X^{\underline{s}'}) = g_{\underline{nm}}(X^{\underline{s}})$ and so $\partial_n X^{\underline{m}} \to \partial X^{\underline{m}'}/\partial \xi^{n'} = (\partial \xi^p/\partial \xi^{n'})\partial_p X^{\underline{m}}$. As a result, $\det g'_{mn} = J^2 \det g_{mn}$ where $J = \det(\partial \xi^p/\partial \xi^{n'})$ and the action is invariant as $d^{p+1}\xi' = J^{-1}d^{p+1}\xi$. The argument is essentially identical to that for the special case of the string in chapter 2, that is, equation (2.1.6) and what follows.

The bosonic brane does not have enough symmetry to determine its couplings to the fields in the background space-time other than gravity. However, a p-brane naturally couples to an anti-symmetric rank $p+1$ gauge field $A_{\underline{m_1}\cdots\underline{m}_{p+1}}(X^{\underline{n}})$ living on the background space-time by a term of the form

$$q \int d^{p+1}\xi \, \epsilon^{n_1\cdots n_{p+1}} \partial_{n_1} X^{\underline{m_1}} \cdots \partial_{n_{p+1}} X^{\underline{m}_{p+1}} A_{\underline{m_1}\cdots\underline{m}_{p+1}}(X^{\underline{n}}), \tag{14.1.3}$$

where q is the coupling constant which is just the charge of the brane. This term is also invariant under reparameterisation of the background space-time and also of the p-brane world volume. This is straightforward to show for the former by using the standard formula for the general coordinate transformation of the tensor $A_{\underline{m_1}\cdots\underline{m}_{p+1}}$. For reparameterisations of the world volume one uses the identity

$$\epsilon^{n_1\cdots n_{p+1}} \frac{\partial \xi^{m_1}}{\partial \xi^{n_1\prime}} \cdots \frac{\partial \xi^{m_{p+1}}}{\partial \xi^{n_{p+1}\prime}} = J \epsilon^{m_1\cdots m_{p+1}}.$$

For example, the motion of a 0-brane, that is, a point particle, is described by the functions $X^{\underline{n}}(\tau)$, where $\xi^0 = \tau$, and it naturally couples to a vector field $A_{\underline{n}}$ in the form

$$e \int d\tau \frac{dX^{\underline{n}}}{d\tau} A_{\underline{n}}(X^{\underline{m}}). \tag{14.1.4}$$

If we couple this expression to that in equation (14.1.1) for a flat background space-time, then the equations of motion for $X^{\underline{n}}$ are nothing but those describing the motion of a charged point particle in an electromagnetic field according to the Lorentz force law. Indeed, taking the action of equation (14.1.1) with $T = m$ in Minkowski space-time, that is, the action of equation (1.1.1), and varying with respect to $X^{\underline{p}}$ we find its equation of motion to be

$$m \frac{d}{d\tau} \left(\frac{\dot{X}_{\underline{p}}}{\sqrt{-\dot{X}^2}} \right) + e \frac{d\dot{X}^{\underline{q}}}{d\tau} F_{\underline{qp}} = 0,$$

where $F_{\underline{qp}} = 2\partial_{[\underline{q}} A_{\underline{p}]}$. For the gauge choice $\tau = t \equiv X^0$, taking $E_i = F_{i0}$ and $F_{ij} = \epsilon_{ijk} B_k$, for $i, j = 1, 2, \ldots, D-1$, and in the non-relativistic limit where $-\dot{X}^2 = 1$, the equation of motion becomes $md^2\underline{X}/dt^2 = e(\underline{E} + d\underline{X}/dt \wedge \underline{B})$, which is the Lorentz force law. The reader may find it educational to reinstate the speed of light c in this calculation rather than take it to be 1 as we have done.

A 1-brane, that is, a string, couples to a 2-form A_{nm} in the form

$$q \int d^2\xi \, \epsilon^{mn} \partial_m X^{\underline{m_1}} \partial_n X^{\underline{m_2}} A_{\underline{m_1 m_2}}. \tag{14.1.5}$$

Since we are coupling a gauge field to the brane the new term in the action should preserve gauge invariance. We find that under the gauge transformation $\delta A_{\underline{m_1}\cdots\underline{m}_{p+1}} = (p+1)\partial_{[\underline{m_1}} \Lambda_{\underline{m_2}\cdots\underline{m}_{p+1}]}$ the variation of the contribution to the action of equation (14.1.3) becomes

$$q(p+1) \int d^{p+1}\xi \, \epsilon^{n_1\cdots n_{p+1}} \partial_{n_1} X^{\underline{m_1}} \cdots \partial_{n_{p+1}} X^{\underline{m}_{p+1}} \partial_{\underline{m_1}} \Lambda_{\underline{m_2}\cdots\underline{m}_{p+1}}$$

$$= q(p+1) \int d^{p+1}\xi \, \epsilon^{n_1\cdots n_{p+1}} \partial_{n_2} X^{\underline{m_2}} \cdots \partial_{n_{p+1}} X^{\underline{m}_{p+1}} \partial_{n_1} \Lambda_{\underline{m_2}\cdots\underline{m}_{p+1}}$$

$$= q(p+1) \int d^{p+1}\xi \, \partial_{n_1} (\epsilon^{n_1\cdots n_{p+1}} \partial_{n_2} X^{\underline{m_2}} \cdots \partial_{n_{p+1}} X^{\underline{m}_{p+1}} \Lambda_{\underline{m_2}\cdots\underline{m}_{p+1}}) = 0,$$

provided we have a gauge transformation whose parameter has the appropriate properties at any boundary of the brane.

We can split the background space-time indices $\underline{n}$, $\underline{m}$, into those associated with the directions longitudinal to the brane, that is, those in the world volume of the brane, and those directions which are transverse to the brane. We denote the former by $n, m, \ldots = 0, \ldots, p$, and the latter by $n', m', \ldots = p+1, \ldots, D-1$. A useful gauge choice is the so-called static gauge in which we use the reparameterisation transformations of the world surface to identify the $p+1$ longitudinal coordinates $X^n(\xi)$, $n = 0, 1, \ldots, p$, with the coordinates ξ^n, $n = 0, 1, \ldots, p$ of the p-brane world volume; in other words

$$X^n(\xi) = \xi^n, \ n = 0, 1, \ldots, p. \tag{14.1.6}$$

This leaves the transverse coordinates $X^{n'}(\xi), n' = p+1, \ldots, D-1$, to describe the dynamics of the brane. We can think of the $D - p - 1$ transverse coordinates as the Goldstone bosons, or zero modes, of the broken translations due to the presence of the p-brane. This discussion assumes that the brane lies along $p+1$ the coordinate axes, however, clearly an analogous result holds even if this is not the case.

As we have explained in the previous chapter, the low energy effective action of a superstring theory which possesses space-time supersymmetry must be a supergravity theory and we found in section 13.8 that these theories possess p-branes solutions for values of p that are in one-to-one correspondence with the presence of gauge fields in these theories consistent with the above coupling. For such solutions the supergravity fields do not depend on the world volume coordinates of the brane, that is, the time and the p spatial coordinates of the p-brane world surface, but they do depend on the $d = D - p - 1$ coordinates transverse to the brane through a harmonic function which we called N. The transverse coordinates just tell us the position of the brane in the background space-time.

It will be instructive first to consider a simpler, and better understood, example of a soliton, namely the monopole (that is, a 0-brane) that occurs in four dimensions. Monopoles arise as static solutions in the $N = 2$ supersymmetric Yang–Mills field theory. The space of parameters required to specify the solution is called the moduli space. If there are K monopoles, then the moduli space has dimension $4K$. For one monopole this four-dimensional space is $\mathbf{R}^3 \otimes S^1$. It is made up of the three spatial coordinates $\mathbf{R}^3$ which specify the position of the monopole and a further parameter associated with gauge transformations which have a non-trivial behaviour at infinity. For $K \geq 2$ the moduli space is more complicated, but an explicit metric is known for the case of $K = 2$ [14.1]. Since there exist static solutions containing several monopoles, it follows that monopoles do not experience any forces when at rest. However, when they are set in motion they do experience velocity dependent forces. In general, to find the behaviour of the monopoles requires a very complicated calculation in the field theory in which they arise as solitons. However, if the monopoles have only a small amount of energy above their rest masses, then we can approximate their motion in terms of the coordinates of their moduli space [14.2]. The motion is then described by an action whose dynamical variables are the coordinates of the moduli space, which are given a time dependence corresponding to the motion of the monopoles.

For only one monopole the motion is described by $x^i(\tau), i = 1, 2, 3, \eta(\tau)$, where we denote $\xi^0 = \tau$. This makes contact with the theory of non-linear realisations in that we originally have a system with translational invariance which is broken by the presence of the monopole and the theory is described by a non-linear realisation which involves the

Goldstone bosons corresponding to the broken translations which are none other than three of the moduli X^i, $i = 1, 2, 3$. A related interpretation exists for the fourth moduli η which is related to the existence of non-trivial gauge transformations at infinity. A more detailed account of these ideas can be found in [14.3].

In a similar spirit we can consider the moduli space and action which describe the low energy behaviour of p-brane solitons. However, in this chapter we consider only a single p-brane. The moduli space of this solution is in many cases just the position of the p-brane which is just the coordinates transverse to the brane. We could let these moduli depend on time in the effective action and then deduce the action for the low energy motion of the p-brane. In deriving the brane dynamics in this chapter we will not follow this approach, but it is good to bear it in mind. Indeed, we can think of the action of equation (14.1.1) as the effective action which describes the low energy behaviour of a single p-brane soliton. For certain important branes the moduli space is larger due to the existence of large gauge transformations of the background field and in this case the dynamics of the branes is more complicated and has the coordinates which describe the position of the brane, but in addition fields living on its world volume.

There is an alternative interpretation of the action of equation (14.1.1): we might believe that a certain p-brane was a fundamental object, then we might take the action of equation (14.1.1) to describe its dynamics completely. This was precisely the viewpoint adopted for the string and is the starting point for much of the discussion in this book!

14.2   Types of superbranes

A super-p-brane can be viewed as a $(p + 1)$-dimensional *bosonic* sub-manifold M, with coordinates ξ^n, $n = 0, 1, \ldots, p$, that moves through a background superspace $\underline{M}$ with coordinates

$$Z^{\underline{N}} = (X^{\underline{n}}, \Theta^{\underline{\alpha}}), \tag{14.2.1}$$

where $\Theta^{\underline{\alpha}}$ is a spinor and so is Grassmann odd. We use the superspace index convention that $\underline{N}, \underline{M}, \ldots$ and $\underline{A}, \underline{B}, \ldots$ represent the world and tangent space indices of the background space-time. Later, we will use the same symbols without the underlining to represent the corresponding indices of the world surface, or volume, superspace. We can write the motion of the brane in the background superspace as $Z^{\underline{N}}(\xi^n)$.

In this section we wish to find which types of superbranes can exist in which space-time dimensions with particular emphasis on ten and eleven dimensions. We will also discuss the features of brane dynamics which are generic to all branes, leaving to the next three sections a more detailed discussion of the dynamics of the specific types of branes.

Superbranes come in various types. The simplest are those whose dynamics can be described entirely by specifying the superworld surface $Z^{\underline{N}}(\xi^n)$ that the p-brane sweeps out in the background superspace. We refer to these branes as *simple superbranes*. They are also called type I branes in some of the literature and they should not to be confused with type I strings. As we shall see, there are other types of branes that have fields living in their world volumes, which transform non-trivially under the brane Lorentz group $SO(1, p)$. In particular, we will discuss branes that have vectors and second-rank anti-symmetric tensor gauge fields on their world volumes.

Any super-p-brane has an action of the form

$$A = A_1 + A_2. \tag{14.2.2}$$

The first term is given by

$$A_1 = -T \int d^{p+1}\xi \sqrt{-\det g_{mn}} + \cdots, \tag{14.2.3}$$

where

$$g_{mn} = \partial_n Z^{\underline{N}} \partial_m Z^{\underline{M}} g_{\underline{NM}}, \tag{14.2.4}$$

$$g_{\underline{NM}} = E_{\underline{N}}{}^{\underline{a}} E_{\underline{M}}{}^{\underline{b}} \eta_{\underline{ab}} \tag{14.2.5}$$

and $E_{\underline{N}}{}^{\underline{a}}$ is the supervielbein on the background superspace. Its expression for a flat background superspace is given in equation (14.2.7). We use the same symbol for the world surface metric as for the bosonic case; the reader will be able to distinguish between the two as a result of the context. The $+\cdots$ denotes terms involving fields living in the brane world volume and background fields. The constant T is the p-brane tension and has the dimensions of (mass)$^{p+1}$ since the action is dimensionless and $X^{\underline{n}}$ has the dimension of (mass)$^{-1}$. If we consider a brane that is static we can think of the tension T as the mass per unit spatial volume of the brane.

The symbol $g_{\underline{NM}}$ is not really a background metric in the usual sense since when it is expressed in terms of the supervielbien the sum on the tangent space indices is restricted to be only over the Grassmann even subspace. Such a restricted summation is possible as the superspace tangent space group is just the Lorentz group. This is a universal feature of superspaces used to formulate supergravities which are not just straightforward generalisations of Einstein's general relativity. As we discussed in chapter 13, the theory of local superspace is formulated in terms of the supervielbein since the metric is not uniquely defined.

The second part of the action of equation (14.2.2), namely the term A_2, includes the quantity

$$A_2 = q \int d^{p+1}\xi \, \epsilon^{n_1 \cdots n_{p+1}} \partial_{n_1} X^{\underline{m}_1} \cdots \partial_{n_{p+1}} X^{\underline{m}_{p+1}} A_{\underline{m}_1 \cdots \underline{m}_{p+1}} + \cdots, \tag{14.2.6}$$

where $A_{\underline{m}_1 \cdots \underline{m}_{p+1}}$ is a background gauge field.

The superbranes we study in this section possess space-time supersymmetry and so the background space-time fields must belong to supermultiplets that exist in the background superspace. In this review, we will take these supermultiplets to be the supergravity theories that have the background supersymmetry algebra possessed by the p-brane. The field content of the possible supergravity theories can be deduced from the supersymmetry algebra using the method of induced representation in the same way as for the Poincaré group. As explained in the previous chapter, the supergravity theory is essentially unique if the supersymmetry algebra has 32 supercharges and the type of spinors contained in the supersymmetry algebra are specified. Thus, unlike bosonic branes, the possible background fields are specified by the background supersymmetry of the brane, if that supersymmetry has 32 supercharges. If the brane has 16 supercharges, then there is a larger choice of background supergravity multiplets; however, the possible supergravity theories are still very limited. The general coupling of a p-brane to the background supergravity fields is complicated, but it always contains a coupling to a $(p+1)$-gauge field in the form of equation (14.1.3). However, since this $(p+1)$-gauge field must be one of the background fields of the supergravity theory, this places an important restriction on which superbranes can arise for a given dimension and background space-time supersymmetry algebra.

The action of equation (14.2.2), whose precise forms will be given explicitly later in this chapter, is clearly invariant under superreparameterisations of the background superspace $\underline{M}$, but only under bosonic reparameterisations of the world surface M since the embedded manifold M is a bosonic manifold. It will also turn out to be invariant under a Fermi–Bose symmetry called κ-supersymmetry which is a generalisation of the κ-supersymmetry found for the point particle in section 1.2.2 in whose context it was first found [1.12]. This is a complicated symmetry that ensures that the fermions have the correct number of degrees of freedom on-shell and we will discuss it in more detail in the next section. This symmetry relates the terms in A_1 to those in A_2 and vice versa, and in fact fixes uniquely the form of A_2 given A_1.

Even if the background space-time is flat superspace, the supervielbein has a non-trivial dependence on the superspace coordinates given by

$$\partial_m Z^{\underline{N}} E_{\underline{N}}{}^{\underline{a}} = \partial_m X^{\underline{a}} - \frac{i}{2} \bar{\Theta} \gamma^{\underline{a}} \partial_m \Theta, \quad \partial_m Z^{\underline{N}} E_{\underline{N}}{}^{\underline{\alpha}} = \partial_m \Theta^{\underline{\alpha}}. \tag{14.2.7}$$

In this case, the background superreparameterisation invariance reduces to just rigid supersymmetry;

$$\delta X^{\underline{a}} = \frac{i}{2} \bar{\epsilon} \gamma^{\underline{a}} \Theta, \quad \delta \theta^{\underline{\alpha}} = \epsilon^{\underline{\alpha}}. \tag{14.2.8}$$

We note that the action of equation (14.2.3) involves only $\Pi_m{}^{\underline{a}} \equiv \partial_m Z^{\underline{N}} E_{\underline{N}}{}^{\underline{a}}$ and does not depend on $\partial_m Z^{\underline{N}} E_{\underline{N}}{}^{\underline{\alpha}}$.

If we consider the case of a super-1-brane, the action of equation (14.2.2) for flat superspace is just the Green–Schwarz action [6.5] for the superstring as discussed in chapter 6. The case $p = 2$ in eleven-dimensional background superspace is often referred to as the membrane. The action for the supermembrane in eleven dimensions was found in [14.4, 14.5].

Note that the action of equation (14.2.2) does not appear to possess world volume supersymmetry. We recall that for the case of a super-1-brane (that is, the superstring discussed in chapter 6) there are two formulations, the Green–Schwarz formulation, the one given here, and the original NS–R formulation together with the GSO projection. The latter is formulated in terms of $x^{\underline{n}}$ and a field $\psi^{\underline{n}}$ which, in contrast to above, possesses a background space vector index and is a spinor with respect to the two-dimensional world sheet. This formulation is manifestly invariant under world-surface reparameterisations, but not under background superreparameterisations. However, if one goes to light-cone gauge, then one finds that the Green–Schwarz formulation has a hidden world-sheet supersymmetry, while the NS–R formulation has a hidden background space-time supersymmetry provided we carry out the GSO projection. Indeed, if we make this projection, then both formulations describe the same theory. We note the corresponding story did not hold for the Brink–Schwarz particle discussed in section 1.2.2 and the spinning particle discussed in section 1.2.1 whose quantisations lead to different theories.

It is thought [14.6] that all branes which possess κ-supersymmetry in their Green–Schwarz formulation actually have a hidden world surface supersymmetry. Although the analogue of a NS–R formulation is not known for p-branes when $p > 1$, there does exist a superembedding formalism [14.7–14.13]. In this formulation the p-brane sweeps out a *supermanifold* which is embedded in the background superspace. Although this approach leads to equations of motions and not an action, it has the advantage that it possesses super-reparameterisation invariance in both the world volume and the background superspaces. The κ-symmetry is then just part of the superreparameterisations of the world volume. Its complicated form is a result of the gauge fixing required to go from the superembedding formalism to the formulation of branes considered in this section. This natural explanation of κ-symmetry was first found in [1.14] and it is explained for the point particle in section 1.3. We will comment further on the superembedding formalism in section 14.5.

The superbranes also possess a super static gauge for which the bosonic coordinates take the form of equation (14.1.6), that is, $X^n = \xi^n$, $n = 0, 1, \ldots, p$, while κ-supersymmetry can be used to set half of the fermions $\Theta^{\underline{\alpha}}(\xi) = (\Theta^\alpha(\xi), \; \Theta^{\alpha'}(\xi))$ to vanish, that is, $\Theta^\alpha = 0$. The indices α and α' go over complementary halves of the original range. While the remaining $D - p - 1$ bosonic coordinates $X^{n'}$, $n' = p + 1, \ldots, D - 1$, correspond to the Goldstone bosons associated with the breaking of translations by the super-p-brane, the remaining $\Theta^{\alpha'}$ are the Goldstone fermions corresponding to the breaking of half of the supersymmetries by the superbrane.

As discussed above the superbranes we are interested in possess world-volume supersymmetry, albeit that it is hidden in the formulations we consider, and as such the fields in the p-brane world volume must belong to a supermultiplet of the world-volume space-time. In this case the world-volume fields must have equal numbers of fermionic and bosonic degrees of freedom on-shell. Let us first consider a p-brane that arises in a theory that has maximal background supersymmetry. This would occur if the brane moved in a background superspace corresponding to a maximal supergravity theory. In this case, if the brane breaks half of the 32 supersymmetries of the background space-time, it will, in super static gauge, have only sixteen $\Theta^{\alpha'}$ which will lead to eight fermionic degrees of freedom on-shell. If we are dealing with a simple superbrane these must be matched by the coordinates $X^{n'}$, which are non-trivial in static gauge. As a result the brane must have eight transverse coordinates $X^{n'}$ and so $D - 8 = p + 1$ world-volume coordinates X^n. Thus if we are in eleven dimensions, the only simple superbrane is a 2-brane while if we are in ten dimensions the only simple superbrane is a 1-brane. Hence, if we have a simple superbrane that lives in a background superspace with maximal supersymmetry and the brane breaks half the supersymmetries, then the only possibilities are a 2-brane that lives in eleven dimensions, a 1-brane that lives in ten dimensions and a particle that lives in nine dimensions. Of course, that if one has lower brane supersymmetry one can find 2- and 1-branes in lower dimensions.

There also exist super-D-branes, usually called just D-branes, whose dynamics requires a vector field A_n, $n = 0, 1, \ldots, p$, living in the brane world volume in addition to the superspace coordinates $X^{\underline{n}}$, $\Theta^{\underline{\alpha}}$ which describe the embedding of the brane in the background superspace. In this case, if we have a brane that arises in the background of a maximal supergravity theory and which breaks half of this supersymmetry, then we again have eight fermionic degrees of freedom on-shell. This must be balanced by the vector which has $p - 1$ degrees of freedom on-shell and the remaining $D - p - 1$ transverse coordinates from which we deduce that $D = 10$. Hence, such super-D-branes can only exist in ten dimensions. For super-D-branes this simple counting argument allows branes for all p; however, as we will see shortly not all values of p actually occur in a given background. We will also discuss branes with higher rank gauge fields living on their world surface.

To find further restrictions on which branes actually exist we can use the argument mentioned above. A super-p-brane must couple to a $p + 1$ gauge field which must belong to the background supergravity theory. Hence to see which super-p-branes can exist we need only find the ranks of the gauge fields which are present in the corresponding supergravity theory. When doing this we must bear in mind that if we have a rank $p + 1$ gauge field which has a rank $p + 2$ field strength F_{p+2} we can take its dual to produce a rank $D - p - 2$ field strength G_{D-p-2}, which has a corresponding rank $D - p - 3$ dual gauge field B_{D-p-3}. The original field strength and the dual field strength, at lowest order in all fields, are related by

$$G_{\underline{n}_1 \cdots \underline{n}_{(D-p-2)}} = \frac{1}{(p+2)!} \epsilon_{\underline{n}_1 \cdots \underline{n}_{(D-p-2)} \underline{n}_{(D-p-1)} \cdots \underline{n}_D} F^{\underline{n}_{(D-p-1)} \cdots \underline{n}_D}. \tag{14.2.9}$$

A more precise formula, in the absence of Chern–Simons terms, is that of equation (13.8.6). Hence, if the original rank $p + 1$ gauge field couples to a p-brane, the dual gauge field has rank $D - p - 3$ and couples to a $(D - p - 4)$-brane. We note that if the F_{p+2} field strength satisfies its equation of motion this implies the Bianchi identity for F_{D-p-2} and vice versa.

Let us begin with branes in eleven dimensions. The eleven-dimensional supergravity theory was described in section 13.3. It possesses a third-rank gauge field $A_{n_1 n_2 n_3}$ which should couple to a 2-brane. The dual potential has rank 6, that is, $B_{\underline{n}_1 \cdots \underline{n}_6}$, and so this couples to a 5-brane. Thus in eleven dimensions we expect only a 2-brane and a 5-brane. The 2-brane is just the simple superbrane we discussed above. As explained above, the 5-brane has sixteen Goldstone fermions and these lead to eight degrees of freedom on-shell. However, a 5-brane in an eleven-dimensional space-time has only five bosonic transverse coordinates leading to only five degrees of freedom on-shell. Clearly, we require another three degrees of freedom on-shell. These must form a representation of the little group $SO(4)$ of the six-dimensional brane world-volume Lorentz group. If we assume that they belong to an irreducible representation of $SO(4)$, they can only be a second-rank self-dual anti-symmetric tensor. On-shell this corresponds to a second-rank anti-symmetric gauge field whose field strength obeys a self-duality condition. The dynamics of the 5-brane is significantly more complicated than that of simple branes and will be given in section 14.4.

Let us now consider the superbranes in ten dimensions that possess a background IIA supersymmetry algebra. In section 13.4 we found that the IIA supergravity theory has gauge fields of ranks 1, 2, 3 of which the rank 2 arises in the $NS \otimes NS$ sector, while ranks 1 and 3 arise in the $R \otimes R$ sector of the IIA string. Including their duals which always belong to the same sector as the original gauge field from which they arise, we find that the IIA theory contains gauge fields of ranks 1, 2, 3, 5, 6, 7. Of these ranks 2 and 6 arise in the $NS \otimes NS$ sector of the string while ranks 1, 3, 5 and 7 arise in the $R \otimes R$ sector of the string. This suggests the existence of p-branes for $p = 1, 5$ that can couple to the $NS \otimes NS$ sector of the IIA string and p-branes for $p = 0, 2, 4, 6$ that can couple to the $R \otimes R$ sector of the IIA string. Of these only the 1-brane can be a simple brane and it is this brane that couples to the $A_{\mu\nu}$

Table 14.2.1 The possible superbranes with $p \leq 8$

Theory			p					
M theory			2_S			5		
IIA	0_D	1_S	2_D		4_D	5	6_D	8_D
IIB		$1_S + 1_D$		3_D		$5 + 5_D$		7_D

gauge field of the IIA string. It is, in fact, the IIA string itself and it is sometimes therefore called the fundamental string. The 5-brane is the straightforward dimensional reduction of the eleven-dimensional 5-brane that will be discussed later in this chapter. The remaining branes with $p = 0, 2, 4, 6$ are the D-branes discussed above that have a vector field living in their world volume. In the literature an 8-brane is also discussed. The associated gauge field has rank 9 with a rank 10 field strength $F_{\underline{n}_1 \cdots \underline{n}_{(10)}}$. The corresponding kinetic term in the action is of the form $-\frac{1}{10!} \int d^{10}x \sqrt{-\det g_{nm}} F^{\underline{n}_1 \cdots \underline{n}_{(10)}} F_{\underline{n}_1 \cdots \underline{n}_{(10)}}$. However, as the field strength has rank 10, the equation of motion of the gauge field implies that $F_{\underline{n}_1 \cdots \underline{n}_{(10)}} = c\epsilon_{\underline{n}_1 \cdots \underline{n}_{(10)}}$, where c is a constant. Substituting this back into the action we find it becomes $\int d^{10}x\, c^2 \sqrt{-\det g_{nm}}$, which is a cosmological constant. In fact, there does exist a unique supergravity theory in ten dimensions with an underlying 32 component IIA supersymmetry algebra [13.71]. It can be viewed as a deformation of the IIA supergravity theory discussed above to include a massive parameter that is associated with the cosmological constant. This theory can be formulated to include a 9-form gauge field which indeed plays the role just described [13.70]. Indeed, this theory does possess an 8-brane that couples to this 9-form gauge field and is associated with the cosmological constant.

Let us now turn to the IIB theory and its branes. The original gauge fields in the IIB supergravity theory discussed in section 13.5 are of rank 2 in the $NS \otimes NS$ sector and ranks 0, 2, 4 in the $R \otimes R$ sector. If we include their dual gauge fields we have gauge fields of ranks 2, 6 in the $NS \otimes NS$ sector and ranks 0, 2, 4, 6, 8 in the $R \otimes R$ sector. We do not include two rank 4 gauge fields as their field strength is the rank 5 self-dual field strength of the IIB theory. This suggests that there exist 1- and 5-branes which couple to the $NS \otimes NS$ sector of the IIB theory and -1-, 1-, 3-, 5-, 7-branes which couple to the $R \otimes R$ sector. The latter are the D-branes. The 1-brane which couples to the rank 2 gauge field in the $NS \otimes NS$ sector is the IIB string itself. The 5-brane which couples to the $NS \otimes NS$ sector is a more complicated object. Note that it is the 3-brane that couples to the rank 4 gauge field whose field strength satisfies a self-duality property. As one might expect this D-brane possesses a self-duality symmetry [14.14]. The ($p = -1$)-brane which couples to the $R \otimes R$ sector occupies just a point in space-time and so is an instanton.

The branes discussed above in eleven and ten dimensions are listed in table 14.2.1. In this table a 'D' or 'S' subscript denotes the superbrane to be a D-brane or simple brane, respectively. We note that the D-branes only couple to the $R \otimes R$ sector of the corresponding string theory and have p even and odd for the IIA and IIB theories, respectively. In fact the IIA and IIB theories possess higher p-branes, which we will discuss in chapter 17. We have not included the D-instanton of the IIB theory.

In section 13.8 we found that there are solutions in the respective supergravity theories corresponding to all the above branes. The p-brane actions can be thought of as the low-energy motions of these solitons.

We now discuss one further guide to determining which superbranes can occur. Given a p-brane one can construct the following current

$$j^{n\underline{m}_1 \cdots \underline{m}_p} = \epsilon^{nn_1 \cdots n_p} \partial_{n_1} X^{\underline{m}_1} \cdots \partial_{n_p} X^{\underline{m}_p},$$
(14.2.10)

which is automatically conserved. The rank p charge associated with this current is given by

$$Z^{\underline{m}_1 \cdots \underline{m}_p} = q \int d^p \xi \, j^{0\underline{m}_1 \cdots \underline{m}_p},$$
(14.2.11)

where the integral is over the p spatial coordinates of the brane world volume. Our previous discussion on the coupling of the p-brane to a $(p+1)$-form gauge background gauge field can be restated as the $(p+1)$-form background gauge field couples to the current $\int d^{p+1}\xi \, j^{n\underline{m}_1 \cdots \underline{m}_p} \partial_n X^{\underline{p}} A_{\underline{p}\underline{m}_1 \cdots \underline{m}_p}$. The current of equation (14.2.11) can be written in the form $\partial_{n_1} (\epsilon^{nn_1 \cdots n_p} X^{\underline{m}_1} \partial_{n_2} X^{\underline{m}_2} \cdots \partial_{n_p} X^{\underline{m}_p})$ and is itself a divergence. If, for example, the brane is wrapped on some non-trivial p cycle in the background space-time, then the charge $q^{-1} Z^{\underline{m}_1 \cdots \underline{m}_p}$ measures the number of times the p-brane wraps around the p cycle. We recall that it is just this current that appears on right-hand side of the equation of motion of the gauge field and so is used to calculate the charge of the brane.

Computing the background supersymmetry algebra which is a symmetry of the above action for the p-brane it turns out [13.77] that the above charge occurs as a central charge in the supersymmetry algebra. Hence, if a p-brane arises in a particular theory we expect to find its p-form central charge in the corresponding supersymmetry algebra. The existence of such charges is required by the considerations at the end of chapter 13 on the amount of supersymmetry branes preserve. However, we can turn this argument around and find out which central charges can occur in the appropriate supersymmetry algebra and then postulate the existence of a p-brane for each corresponding p-form central charge. Which central charges can occur for a specified type of supercharge was discussed in chapter 5. For example, in equation (5.5.5), we found that the eleven-dimensional supersymmetry algebra with one Majorana spinor, which is the supersymmetry algebra appropriate to eleven-dimensional supergravity, had a 2-form and a 5-form central charge. Hence in M theory we expect from the above argument to find a 2-brane and a 5-brane. This agrees with our previous considerations. The reader can also calculate the same considerations for the IIA and IIB supersymmetry algebras, given in earlier chapters, and find agreement with the above proposed branes in each theory.

14.3   Simple superbranes

In this section we give the complete dynamics of simple superbranes [14.4, 14.5, 14.6]. By definition these are branes whose dynamics is described completely by only the embedding coordinates $Z^{\underline{N}} = (X^{\underline{n}}, \Theta^{\underline{\alpha}})$ which describe the surface of the brane in the background superspace. Such objects include the fundamental superstring in ten dimensions and the supermembrane in eleven dimensions. The action for a simple super-p-brane is given by

$$A = A_1 + A_2.$$
(14.3.1)

The first term is given by

$$A_1 = -T \int d^{p+1}\xi \sqrt{-\det g_{mn}}, \tag{14.3.2}$$

where

$$g_{mn} = \partial_n Z^{\underline{N}} \partial_m Z^{\underline{M}} g_{\underline{NM}}, \tag{14.3.3}$$

$$g_{\underline{NM}} = E_{\underline{N}}{}^a E_{\underline{M}}{}^b \eta_{\underline{ab}}, \tag{14.3.4}$$

where $E_{\underline{N}}{}^{\underline{A}}$ is the supervielbein of the background superspace. The second term in the action of equation (14.3.1) is given by

$$A_2 = -\frac{T}{(p+1)!} \int d^{p+1}\xi \, \epsilon^{n_1 \cdots n_{p+1}} \partial_{n_1} Z^{\underline{M_1}} E_{\underline{M_1}}{}^{\underline{A_1}} \cdots \partial_{n_{p+1}} Z^{\underline{M_{p+1}}} E_{\underline{M_{p+1}}}{}^{\underline{A_{p+1}}} B_{\underline{A_1} \cdots \underline{A_{p+1}}}.$$

$$\tag{14.3.5}$$

In this expression, $B_{\underline{A_1} \cdots \underline{A_{p+1}}}$ is a background superspace $p+1$ gauge field referred to superspace tangent indices. Its corresponding superspace $(p+1)$-form is

$$B = \frac{1}{(p+1)!} E^{\underline{A_1}} \wedge \cdots \wedge E^{\underline{A_{p+1}}} B_{\underline{A_1} \cdots \underline{A_{p+1}}} = dZ^{\underline{N_{p+1}}} \wedge \cdots \wedge dZ^{\underline{N_1}} B_{\underline{N_1} \cdots \underline{N_{p+1}}},$$

$$\tag{14.3.6}$$

where $E^{\underline{A}} = dZ^{\underline{N}} E_{\underline{N}}{}^{\underline{A}}$.

The reader will notice that, up to a combinatoric factor, the coefficient in front of this second term is equal to that for the first term. The coefficient in front of the second term is the coupling of the p-brane to the background gauge field and so can be interpreted as the charge of the p-brane. In principle, one could take this coefficient to have a different value, but one finds that for the type of supersymmetric branes we are considering, it is fixed to take the value given above. Technically this arises as the brane action must possess κ-supersymmetry, described below, for consistency. This is a reflection of the fact that the brane action can be interpreted as describing the low energy motion of the brane as that of a soliton of the background supergravity theory and we are interested in such solutions that preserve some fraction of the supersymmetry, generally half, of the background supergravity. The equality of the two coefficients can be thought of as a BPS condition as the second term gives rise to the charge of the brane which must be balanced against the mass density of the brane, encoded in the first term in the action, in order that the brane preserve some supersymmetry.

The geometry of the background superspace is described as in section 13.1.2 using supervielbeins and connections, but now with the addition of the gauge superfield B. The covariant objects are the torsions and curvatures which are defined as in chapter 13 and have the same expressions in terms of the supervielbeins and connections as before, but now we have in addition the superfield strength H of the gauge superfield B given by the exterior derivative, in superspace of course, acting on the superspace gauge field, that is, $H = dB$. The superspace torsions and curvatures satisfy Bianchi identities as in section 13.1.2, but in addition we have the easily derived identity

$$D_{\underline{A_1}} H_{\underline{A_2} \cdots \underline{A_{p+2}}} + T_{\underline{A_1 A_2}}{}^{\underline{B}} H_{\underline{B A_3} \cdots \underline{A_{p+1}}} + \text{supercyclic permutations} = 0. \tag{14.3.7}$$

In this expression the superfield strength is refered to the superspace tangent space using the supervielbein as was done for the gauge in equation (14.3.6).

It is obvious that both terms of the actions of equations (14.3.2) and (14.3.5) are invariant under background superspace reparameterisations and so the action possesses background supersymmetry.

In the case that the background superspace is flat, then the supervielbein is given as in equation (14.2.6). However, the gauge field B also has a non-trivial form. The resulting torsions all vanish except for

$$T_{\underline{\alpha\beta}}{}^{\underline{a}} = i(\gamma^{\underline{a}}C^{-1})_{\underline{\alpha\beta}} \tag{14.3.8}$$

and

$$H_{\underline{\alpha\beta a_1\cdots a_p}} = -i(-1)^{\frac{1}{4}p(p-1)}(\gamma_{\underline{a_1\cdots a_p}}C^{-1})_{\underline{\alpha\beta}}. \tag{14.3.9}$$

In terms of forms we may express these results as

$$T^{\underline{a}} = \tfrac{1}{2}id\bar\Theta\gamma^{\underline{a}}d\Theta \tag{14.3.10}$$

and

$$H = \frac{i}{2p!}E^{\underline{a_p}}\cdots E^{\underline{a_1}}d\bar\Theta\gamma_{\underline{a_1\cdots a_p}}d\Theta, \tag{14.3.11}$$

where we have omitted the wedge symbol for simplicity.

Since H is an exact form it follows that $dH = 0$. Using this on equation (14.3.11), and the form of $E^{\underline{a}}$ of equation (14.3.6), we find the equivalent condition [14.6]

$$d\bar\Theta\gamma_{\underline{a}}d\Theta d\bar\Theta\gamma^{\underline{ab_1\cdots b_{(p-1)}}}d\Theta = 0. \tag{14.3.12}$$

Since Θ is Grassmann odd, $d\Theta$ is Grassmann even and hence the above condition must hold for a Grassmann even spinor of the appropriate type that is complex, Weyl, Majorana or Majorana–Weyl, but is otherwise arbitrary. This means that we can remove the $d\Theta$s from the above equation provided we include a projector if the spinor is Weyl, and symmetrise on all the free spinor indices with the inclusion of appropriate matrices C if it is Majorana and on the first and third and second and fourth indices separately if it is complex. The matrix C arises from expressing $d\bar\Theta$ in terms of $d\Theta$ using the definition of a Majorana spinor. For example, if Θ is Majorana then we find the identity

$$(\gamma_{\underline{a}}C^{-1})_{\underline{(\alpha\beta}}(\gamma^{\underline{ab_1\cdots b_{p-1}}}C^{-1})_{\underline{\delta\epsilon)}} = 0. \tag{14.3.13}$$

Since we require that the simple brane can exist in a flat background superspace, which must contain the superfield B whose field strength H is closed, simple branes will only exist if equation (14.3.13), or the analogous conditions for other spinors, holds. This condition just depends on the properties of the γ matrices in the dimensions of interest and it can be shown by using the appropriate Fierz identity. It holds for Majorana spinors in eleven dimensions if $p = 2$, that is, for the membrane of M theory. It also holds for $p = 1$ in ten dimensions for Majorana spinors and Majorana–Weyl spinors of opposite chirality corresponding to the presence of the fundamental strings of the IIA and IIB theory, respectively. Of course, for the case of the string we have already found the action and we know that it is invariant as explained in chapter 6. In fact a membrane also exists in seven dimensions and strings also in dimensions 3, 4, 6 for suitable types of spinors. For all cases for which the identity works one finds that the corresponding simple brane has equal numbers of fermionic and bosonic degrees of freedom, corresponding to the fact that it possesses a hidden world volume supersymmetry. In fact, we earlier used this condition to find the possible degrees of freedom of simple branes in ten and eleven dimensions.

The constraints of equations (14.3.8) and (14.3.9), as well as others, hold for any background superspace that describes an on-shell supergravity theory. This can be seen using the same arguments, based on dimensional analysis, employed to find the constraints on superspace torsions and curvatures in chapter 13.

Finally, it remains to discuss the κ-symmetry of the action of equation (14.3.1). The variation of the coordinates $Z^{\underline{N}}$ under this symmetry when referred to the tangent basis is given by

$$\delta Z^{\underline{N}} E_{\underline{N}}{}^{\underline{a}} = 0, \quad \delta Z^{\underline{N}} E_{\underline{N}}{}^{\underline{\alpha}} = (1+\Gamma)^{\underline{\alpha}}{}_{\underline{\beta}} \kappa^{\underline{\beta}}, \tag{14.3.14}$$

where the local parameter $\kappa^{\underline{\beta}}(\xi)$ is a world-volume scalar, but a background space-time spinor. The matrix Γ is given by

$$\Gamma = \frac{(-1)^{\frac{1}{4}(p-2)(p-1)}}{(p+1)!\sqrt{-\det g_{nm}}} \epsilon^{n_1 \cdots n_{p+1}} \partial_{n_1} Z^{\underline{N}_1} E_{\underline{N}_1}^{\underline{a}_1} \cdots \partial_{n_{p+1}} Z^{\underline{N}_{p+1}} E_{\underline{N}_{p+1}}^{\underline{a}_{p+1}} \gamma_{a_1 \cdots a_{p+1}}. \tag{14.3.15}$$

The matrix Γ obeys the remarkably simple property $\Gamma^2 = 1$ and as a result $\frac{1}{2}(1 \pm \Gamma)$ are projectors. Clearly a κ of the form $\kappa = \frac{1}{2}(1 - \Gamma)\kappa^1$ leads to no contribution to the symmetry and as a result only half of the components of κ actually contribute. The remaining half allow us to gauge away the corresponding half of Θ. This property of κ-symmetry is essential for getting the correct number of on-shell fermion degrees of freedom, namely eight. In fact, the action is only invariant under κ-symmetry if the background fields obey their equations of motion. In particular, the action is only invariant if the background superspace torsions and curvatures obey the constraints that imply the equations of motion for the component supergravity fields.

For the case when the background superspace is flat the above κ-symmetry reduces to

$$\delta X^{\underline{n}} = \frac{i}{2}\bar{\Theta}\gamma^{\underline{n}}\delta\Theta, \quad \delta\Theta = (1+\Gamma)\kappa. \tag{14.3.16}$$

The reader will recognise these for the special case of the superstring formulated by the Green–Schwarz action considered in chapter 6.

The form of the redundancy discussed above in the κ invariance leads to very difficult problems when gauge fixing the κ-symmetry and so quantising the theory. Imposing a gauge condition on the spinor Θ to fix the κ symmetry can only fix part of the symmetry and leaves unfixed that part of κ which is given by κ^1. A little thought shows that as a result the ghost action itself will have a local symmetry which must be itself fixed. In fact, this process continues and results in an infinite number of ghosts for ghosts corresponding to the infinite set of invariances. This would not in itself be a problem, but it has proved impossible to find a Lorentz-covariant gauge-fixed formulation. The problem of quantising can also be seen from the viewpoint of the Hamiltonian approach where, like for the case of the analogous point particle, discussed in chapter 1, one finds that the system possesses first- and second-class constraints that cannot be separated in a Lorentz-covariant manner.

14.4   D-branes

As we discussed, D-branes have a vector field A_n living on their world surface in addition to the embedding coordinates $X^{\underline{n}}$, $\Theta^{\underline{\alpha}}$. The action for the D-branes is a generalisation of that of the simple brane of equation (14.3.1) to include this vector field A_n. In the absence of background fields other than the background metric and when $\Theta^{\underline{\alpha}} = 0$ the action is just the Born–Infeld action

$$-T \int d^{p+1}\xi \sqrt{-\det(g_{nm} + cF_{nm})}, \tag{14.4.1}$$

where $F_{mn} = 2\partial_{[m}A_{n]}$ and c is a constant. This action has the merit that it is general coordinate invariant. We may evaluate it using the expansion

$$
\begin{aligned}
&\sqrt{-\det(g_{nm} + cF_{nm})} \\
&= \sqrt{-\det(g_{nm})}\sqrt{\det(\delta^n_m + cF^n{}_m)} = \sqrt{-\det(g_{nm})} \exp \frac{1}{2}\operatorname{Tr}\ln(\delta^n_m + cF^n{}_m) \\
&= \sqrt{-\det(g_{nm})} \exp \frac{1}{2}\operatorname{Tr}\left(-\frac{1}{2}c^2 F^n{}_p F^p{}_m - \frac{c^4}{4}F^n{}_p F^p{}_q F^q{}_r F^r{}_m + \cdots\right) \\
&= \sqrt{-\det(g_{nm})}\left(1 + \frac{c^2}{4}F_{nm}F^{nm} - \frac{c^4}{8}F^n{}_p F^p{}_q F^q{}_r F^r{}_n + \frac{c^4}{32}(F_{nm}F^{nm})^2 + \cdots\right).
\end{aligned}
\tag{14.4.2}
$$

Thus the Born–Infeld action contains the usual Maxwell terms but also terms of higher order in the field strength squared. It was proposed in the hope that it might cure the infinite self-energy of a charged point particle.

The full action was found [14.16–14.18] by using κ-symmetry. The bosonic part is given by

$$S_1 + S_2. \tag{14.4.3}$$

The first part is a generalisation of the volume swept out to include the world-volume gauge field and background 2-form field:

$$S_1 = -T_p \int d^{p+1}\xi e^{-\phi}\sqrt{-\det(g_{nm} + \mathcal{F}_{nm})}, \tag{14.4.4}$$

where ϕ is the dilaton field, $\mathcal{F}_{nm} = 2\pi\alpha' F_{nm} + B_{nm}$ and B_{nm} is the pull back to the brane world volume of the massless background 2-form, that is, $B_{nm} = (\partial X^{\underline{n}}/\partial\xi^n)(\partial X^{\underline{m}}/\partial\xi^m)B_{\underline{nm}}$. We see that S_1 involves all the massless fields in the NS–NS sector of the superstring, namely the dilaton ϕ, the graviton g_{nm} and the 2-form $B_{\underline{nm}}$. Their action is that of the IIA or IIB supergravity theories of chapter 13. This part of the action is the same for the IIA and IIB theories as their NS–NS sectors are the same. The vector field possesses the usual Abelian gauge transformation $\delta A_n = \partial_n\Lambda$. However, $\mathcal{F}_{nm}$, and so the action, is invariant under the transformations $\delta B_{nm} = 2\partial_{[n}\Lambda_{m]}$ and $\delta A_n = -(1/2\alpha')\Lambda_n$, where $\Lambda_n = \partial_n X^{\underline{m}}\Lambda_{\underline{m}}$.

The second term is given by

$$S_2 = q \int \{C \wedge e^{\mathcal{F}}\}_{p+1} \wedge \mathcal{G}. \tag{14.4.5}$$

The terms in this equation require quite some explanation. The object C is constructed from the R–R massless gauge fields pulled back to the brane world volume. We include the dual gauge fields as discussed in sections 13.4 and 13.5 for the IIA and IIB theories, respectively. To agree with common usage we will denote these fields by C rather than the A used in these sections. It is just the formal sum of all the R–R forms in the theory so the typical term is given by

$$C = \cdots + C_{n_1 \cdots n_q} d\xi^{n_1} \wedge \cdots \wedge d\xi^{n_q} + \cdots$$

$$= \cdots + C_{\underline{n}_1 \cdots \underline{n}_q} \frac{dX^{\underline{n}_1}}{d\xi^{n_1}} \cdots \frac{dX^{\underline{n}_q}}{d\xi^{n_q}} d\xi^{n_1} \wedge \cdots \wedge d\xi^{n_q} + \cdots . \tag{14.4.6}$$

We recall that the R–R forms are odd in the IIA theory and even in the IIB theory. The quantity $e^{\mathcal{F}} = e^{\mathcal{F}_{nm} d\xi^n \wedge d\xi^m} = 1 + \mathcal{F}_{nm} d\xi^n \wedge d\xi^m + \cdots$, that is, we formally expand the exponential. The symbol $\{\bullet\}_{p+1}$ means take only the $(p+1)$-form from the result of all the expansions and it is this that is integrated over. The quantity q is the charge of the brane; κ-symmetry requires that it is related to the tension T_p. Finally, $\mathcal{G} = 1 + \cdots$, where $+\cdots$ stands for forms constructed from the Riemann curvature. These terms are related to those that appear in gravitational anomalies. For the case of the D1 string in the IIB theory this term has the form

$$S_2 = q \int (C\mathcal{F}_{nm} + C_{nm}) d\xi^n \wedge d\xi^m. \tag{14.4.7}$$

As we will discuss in the next section, N coincident branes possess vectors that belong to a U(N) gauge theory. The corresponding action at lowest order in the fields is just that for the ten-dimensional Yang–Mills action dimensionally reduced to $p + 1$ dimensions. This action is given in section 13.7. We note that it contains the potential $\frac{1}{4}\text{Tr}(\sum_{I \leq J} [\phi_I, \phi_J]^2)$. The minimum is given when ϕ_I belongs to the Cartan subalgebra and these values can be thought of as the positions of the N D-branes. The full action is much more complicated.

As we discussed at the beginning of this section the brane action can be viewed as the Goldstone action for the breaking of symmetries by the presence of the brane. The transverse coordinates $X^{n'}, \Theta^{\alpha'}$ correspond to the breaking of translations in the background space-time and space-time supersymmetry, respectively. The world volume gauge fields arise from the breaking of the rank 1 central charge symmetry found in the IIA and IIB supersymmetry algebras [14.19].

14.5   Branes in M theory

As we have discussed above, there exist a 2-brane and a 5-brane in eleven dimensions. The 2-brane is a simple brane and so its dynamics was given in section 6.3 if we take $p = 2$. The dynamics of the 5-brane is the subject of this section. As we discussed in section 14.2 the 5-brane contains five scalar fields and a 16-component spinor corresponding to the breaking of translations and supersymmetry by the 5-brane. It also contains, living on its world volume, an anti-symmetric second-rank tensor gauge field whose field strength obeys a self-duality condition. This latter condition is required to obtain the correct number of degrees of freedom. While the form of this condition is obvious at the linearised level, it is an important feature of the 5-brane dynamics that the self-duality condition in the full theory is a very non-linear condition on the third-rank field strength.

As for other branes, the bosonic indices of the fields of the 5-brane can be decomposed into longitudinal and transverse indices (that is, for the world background space-time indices $\underline{n} = (n, n')$) according to the decomposition of the eleven-dimensional Lorentz group $SO(1, 10)$ into $SO(1, 5) \times SO(5)$ corresponding to the presence of the 5-brane. The corresponding decomposition of the eleven-dimensional spin group Spin $(1, 10)$ is into $Spin(1, 5) \times$ Spin (5). These spin groups moded by Z_2 are isomorphic to their corresponding Lorentz groups in the usual way. Spin(5) is isomorphic to the group USp(4) which is defined below equation (5.3.18). Thus the 5-brane possesses $Spin(1, 5) \times Spin(5)$ or $Spin(1, 5) \times USp(4)$ symmetry.

The dynamics of the 5-brane is described by the embedding coordinates $X^{\underline{n}}$, $\Theta^{\underline{\alpha}}$ and the 2-form B_{mn} which lives in the world volume. If we work in static gauge, then $X^n = \xi^n$, $n = 0, 1, \dots, 5$ and we are left with the transverse coordinates $X^{n'}$, $n' = 6, \dots, 10$. Clearly, these are five real scalars which are $Spin(1, 5)$, or $SO(1, 5)$, singlets, but belong to the vector representation of $SO(5)$. This representation corresponds to the second-rank anti-symmetric tensor representation ϕ_{ij}, $i, j = 1, \dots, 4$, of the $USp(4)$ group which is traceless with respect to the anti-symmetric metric Ω^{ij} of this group. The gauge field B_{mn} is a singlet under $SO(5)$, or equivalently $USp(4)$, as it lives in the world volume.

We now discuss the decomposition of the spinor which first occurs as an eleven-dimensional spinor $\Theta^{\underline{\alpha}}$. Although we began with spinor indices $\underline{\alpha}$ that take 32 values we can, as above for all branes, split these indices into two pairs of indices each taking 16 values $\underline{\alpha} = (\alpha, \alpha')$. Corresponding to the decomposition of $Spin(1, 10)$ into $Spin(1, 5) \times$ Spin(5) these spinor indices can be written as $\alpha \rightarrow \alpha i$ and $\alpha' \rightarrow {}^i_\alpha$ when appearing as superscripts and $\alpha \rightarrow \alpha i$ and $\alpha' \rightarrow {}^\alpha_i$ when appearing as subscripts. Here the above indices of the groups $Spin(1, 5)$ and $USp(4)$ are denoted by $\alpha, \beta, \dots = 1, \dots, 4$ and $i, j, \dots = 1, \dots, 4$, respectively. It should be clear whether we mean α to be sixteen- or four-dimensional depending on the absence or presence of $i, j, \dots$ indices respectively. For example, we will write $\Theta^{\alpha'} \rightarrow \Theta^i_\alpha$, $\Theta^\alpha \rightarrow \Theta^{\alpha i}$, $\Theta_\alpha \rightarrow \Theta_{\alpha i}$ and $\Theta_{\alpha'} \rightarrow \Theta^\alpha_i$.

We may use the expected κ-symmetry to set $\Theta^\alpha = 0$ leaving us with spinors to carry the indices Θ^i_α, $\alpha = 1, \dots 4$, $i = 1, \dots, 4$. The spinors have the required 16 real components. We recall from chapter 5 that in six dimensions one cannot have Majorana spinors, but one can have symplectic Majorana spinors and even symplectic Majorana–Weyl spinors. The fermions of the 5-brane, Θ^i_α, belong to the four-dimensional vector representation of $USp(4)$ and are $USp(4)$ symplectic Majorana–Weyl spinors and so obey equation (5.3.17). Since they are Weyl we can work with their Weyl projected components which take only four values as opposed to the usual $8 = 2^3$ components for a non-chiral six-dimensional spinor.

The fields of the 5-brane belong to the so-called $(2, 0)$ tensor multiplet which, by definition, transforms under $(2, 0)$ six-dimensional supersymmetry. The $(2, 0)$ notation means that the supersymmetry parameter is a $USp(4)$ symplectic Majorana–Weyl spinor. By contrast, we note that $(1, 0)$ supersymmetry means that the supersymmetry parameter is a $USp(2)$ symplectic Majorana–Weyl spinor, while if the parameter is just a $USp(4)$ symplectic Majorana spinor the supersymmetry would be denoted by $(2, 2)$.

The classical equations of motion of the 5-brane in superstatic gauge and in the absence of fermions and background fields, including gravity, are [14.11]

$$G^{mn}\nabla_m\nabla_nX^{a'} = 0 \tag{14.5.1}$$

and

$$G^{mn}\nabla_mH_{npq} = 0, \tag{14.5.2}$$

where the world-surface indices are $m, n, p = 0, 1, \ldots, 5$ and $a, b, c = 0, 1, \ldots, 5$ for world and tangent indices, respectively. The transverse tangent indices are $a', b' = 6, 7, 8, 9, 10$. We now define the symbols that occur in these equations of motion. The usual induced metric for a p-brane is given, in static gauge and Minkowski background space-time, by

$$g_{mn} = \eta_{mn} + \partial_mX^{a'}\partial_nX^{b'}\delta_{a'b'}. \tag{14.5.3}$$

The covariant derivative in the equations of motion is defined with the Levi–Civita connection with respect to the metric g_{mn}. Its action on a vector field T_n is given by

$$\nabla_mT_n = \partial_mT_n - \Gamma_{mn}{}^pT_p, \tag{14.5.4}$$

where

$$\Gamma_{mn}{}^p = \partial_m\partial_nX^{a'}\partial_rX^{b'}g^{rs}\delta_{a'b'}. \tag{14.5.5}$$

We define the world-volume vielbein associated with the above metric in the usual way: $g_{mn} = e_m{}^a\eta_{ab}e_n{}^b$. It is not the inverse of this metric that appears in the above equations of motion, but another inverse metric G^{mn} which is related to the usual induced metric given above by the equation:

$$G^{mn} = (e^{-1})_c{}^m\eta^{ca}m_a{}^dm_d{}^b(e^{-1})_b{}^n, \tag{14.5.6}$$

where the matrix m is given by

$$m_a{}^b = \delta_a{}^b - 2h_{acd}h^{bcd}. \tag{14.5.7}$$

The field h_{abc} is an anti-symmetric 3-form which is self-dual:

$$h_{abc} = \frac{1}{3!}\varepsilon_{abcdef}h^{def}, \tag{14.5.8}$$

but it is not the curl of a 3-form gauge field. It is related to the field $H_{mnp} = 3\partial_{[m}B_{np]}$ which appears in the equations of motion and is the curl of a gauge field, but H_{mnp} is not self-dual. The relationship between the two fields is given by

$$H_{mnp} = e_m{}^ae_n{}^be_p{}^c(m^{-1})_c{}^dh_{abd}. \tag{14.5.9}$$

Clearly, the self-duality condition on h_{abd} transforms into a condition on H_{mnp} and vice versa for the Bianchi identity $dH = 0$.

The above 5-brane equations of motion were found using the superembedding formalism. This formalism, including the superspace embedding condition of equation (14.5.12), was first given within the context of the superparticle in [1.14]. Further work on the superparticle was carried out in [14.20] and [14.21]. The superembedding formalism was first applied to p-branes in [14.7–14.9] and to the M theory 5-brane in [14.10]. The 5-brane equations of motion to all orders in the background space-time fields but to first order in fermions were derived in [14.11]. Reference [14.13] contains a review of the embedding formalism. The superembedding approach was also used to find p-brane actions and equations of motion in [14.23, 14.24].

We now explain the simple idea that underlies this formalism and use it to derive the 5-brane equations of motion [14.10, 14.11]. We consider a background superspace $\underline{M}$ in which a brane sweeps out a superspace M. We note that in this set up the embedded manifold is itself a superspace unlike for the formulation of superbrane dynamics discussed in earlier sections. The world volume has the coordinates $z^M = (x^n, \theta^\mu)$, where $\mu, \nu, \ldots$ denote curved world-volume spinor indices. The background superspace has the coordinates $Z^{\underline{M}} = (X^{\underline{m}}, \Theta^{\underline{\alpha}})$ and the brane sweeps out a world volume $Z^{\underline{M}}(z^N)$. On the two superspaces $\underline{M}$ and M we have a set of preferred frames or supervielbeins. The frame vector fields on the background manifold $\underline{M}$ and the 5-brane world volume M are denoted by $E_{\underline{A}} = E_{\underline{A}}^{\underline{M}} \partial_{\underline{M}}$ and $E_A = E_A^M \partial_M$, respectively. We recall that we use the superspace index convention that $\underline{N}, \underline{M}, \ldots$ and $\underline{A}, \underline{B}, \ldots$ represent the world and tangent indices of the background superspace $\underline{M}$, while $N, M, \ldots$ and $A, B, \ldots$ represent the world and tangent space indices of the embedded superspace M.

Since the supermanifold M is embedded in the supermanifold $\underline{M}$, the frame vector fields of M must point somewhere in $\underline{M}$. Exactly where they point is encoded in the coefficients $E_A^{\underline{A}}$ which relate the vector fields E_A and $E_{\underline{A}}$, that is,

$$E_A = E_A^{\underline{A}} E_{\underline{A}}. \tag{14.5.10}$$

Applying this relationship to the coordinate $Z^{\underline{M}}$ we find the equation

$$E_A^{\underline{A}} = E_A^{\,N} \partial_N Z^{\underline{M}} E_{\underline{M}}^{\underline{A}}. \tag{14.5.11}$$

It is now straightforward to express the torsion and curvature tensors of M in terms of those of $\underline{M}$ plus terms involving a suitable covariant derivative of $E_A^{\underline{A}}$; one finds that

$$\nabla_A E_B^{\underline{C}} - (-1)^{AB} \nabla_B E_A^{\underline{C}} + T_{AB}^{\,C} E_C^{\underline{C}} = (-1)^{A(B+\underline{B})} E_B^{\underline{B}} E_A^{\underline{A}} T_{\underline{AB}}^{\underline{C}}, \tag{14.5.12}$$

where the derivative ∇_A is covariant with respect to both embedded and background super-spaces, that is, it has connections which act on both underlined and non-underlined indices. The simplest way to derive this equation is to rewrite the covariant derivative of M using equation (14.5.10) and the notation of chapter 13, that is, we can write

$$\nabla_A = E_A^M D_M = E_A + \Omega_A = E_A^{\underline{C}} E_{\underline{C}} + \Omega_A = E_A^{\underline{C}} \nabla_{\underline{C}} + \Omega_A - E_A^{\underline{C}} \Omega_{\underline{C}} \tag{14.5.13}$$

and then compute the commutator, or anti-commutator, of ∇_A and ∇_B as appropriate to find the superspace torsion and curvatures of M in terms of those of $\underline{M}$ and covariant derivatives of $E_A^{\underline{C}}$.

The tangent space of the superspaces $\underline{M}$ and M can be divided into their Grassmann odd and even sectors, that is, the odd and even pieces of M are spanned by the vector fields E_α and E_a, respectively, and similarly for $\underline{M}$. The superembedding formalism has only one assumption: the odd tangent space of M should lie in the odd tangent space of $\underline{M}$. This means that

$$E_\alpha^{\underline{a}} = 0. \tag{14.5.14}$$

To proceed one substitutes this condition into the relationships of equation (14.5.12) between the torsions and curvatures of M in terms of $\underline{M}$ and analyses the resulting equations in order of increasing dimension. For example, at dimension 0 one finds the equation

$$E_\alpha{}^{\underline\alpha} E_\beta{}^{\underline\beta} T_{\underline{\alpha\beta}}{}^{\underline{c}} = T_{\alpha\beta}{}^c E_c{}^{\underline{c}}. \tag{14.5.15}$$

The eleven-dimensional background superspace can be constructed from the usual super-vielbeins and connections for the Lorentz group. However, it is useful, particularly when studying the 5-brane or the 2-brane, to introduce a background rank 3 gauge superfield A_{MNP} with a corresponding superfield strength H_{MNPQ}. Following the dimensional arguments given in chapter 13, which include the notion of geometric dimension, we conclude that

$$T_{\underline{\alpha\beta}}{}^{\underline{c}} = -i(\Gamma^{\underline{c}} C^{-1})_{\underline{\alpha\beta}}, \tag{14.5.16}$$

where $\Gamma^{\underline{c}}$ denote the elements of the Clifford algebra in the eleven dimensions. Later we will use γ^c to denote the elements of the chirally projected six-dimensional Clifford algebra. Substituting this into equation (14.3.14) we conclude that

$$E_\alpha{}^{\underline\alpha} E_\beta{}^{\underline\beta} (\Gamma^{\underline{c}} C^{-1})_{\underline{\alpha\beta}} = i T_{\alpha\beta}{}^c E_c{}^{\underline{c}}. \tag{14.5.17}$$

The solution to this equation is given by

$$E_\alpha{}^{\underline\alpha} = u_\alpha{}^{\underline\alpha} + h_\alpha{}^{\beta'} u_{\beta'}{}^{\underline\alpha} \tag{14.5.18}$$

and

$$E_a{}^{\underline{a}} = m_a{}^b u_b{}^{\underline{a}} \tag{14.5.19}$$

as well as

$$T_{\alpha\beta}{}^c = -i(\Gamma^c C^{-1})_{\alpha\beta} \rightarrow -i\Omega_{ij}(\gamma^c)_{\alpha\beta}, \tag{14.5.20}$$

where Ω_{ij} is the USp(4) inverse of the invariant metric mentioned above. The variables u with spinor indices are elements of the group Spin(1, 10) and in particular the pair $(u_\alpha{}^{\underline\alpha}, u_{\alpha'}{}^{\underline\alpha})$ together making up an element of the group Spin(1, 10). Similarly, $u_{a'}{}^{\underline{a}}$ is part of the pair $(u_a{}^{\underline{a}}, u_{a'}{}^{\underline{a}})$ which is a group element of SO(1, 10). We recall that the group elements of Spin(1, 10) and SO(1, 10) are related by the equation

$$u_\gamma{}^{\underline{a}} u_\delta{}^{\underline\beta} (\Gamma^{\underline{c}} C^{-1})_{\underline{\alpha\beta}} = i(\Gamma^{\underline{d}} C^{-1})_{\gamma\delta} u_{\underline{d}}{}^{\underline{c}}. \tag{14.5.21}$$

That the indices in equation (14.5.17) are not over the full range allows the possibility of having the tensor $h_\alpha{}^{\beta'}$ in the solution. In fact, this tensor is given by

$$h_\alpha{}^{\beta'} \rightarrow h_{\alpha i\beta}{}^j = \tfrac16 \delta_i{}^j (\gamma^{abc})_{\alpha\beta} h_{abc}, \tag{14.5.22}$$

where h_{abc} is self-dual as the chirally projected gamma matrix that multiplies it is self-dual, and

$$m_a{}^b = \delta_a{}^b - 2h_{acd} h^{bcd}. \tag{14.5.23}$$

We note that there is, in fact, no geometric gauge invariant dimension 0 object that $T_{\alpha\beta}^c$ in equation (14.5.20) can equal other than the γ matrix of this equation.

It is useful to introduce a normal basis $E_{A'} = E_{A'}{}^{\underline{A}}E_{\underline{A}}$ of vectors at each point on the world volume. The odd–odd and even–even components of the normal matrix $E_{A'}{}^{\underline{A}}$ can be chosen to be

$$E_{\alpha'}{}^{\underline{\alpha}} = u_{\alpha'}{}^{\underline{\alpha}}, \quad E_{a'}{}^{\underline{a}} = u_{a'}{}^{\underline{a}}. \tag{14.5.24}$$

Evaluating equation (14.5.15) at higher dimensions, using equation (14.5.14), we find the further equations; this procedure is much the same as for the usual analysis of the superspace Bianchi identities for super Yang–Mills or supergravity theories. Indeed, at dimension $\frac{1}{2}$ we find, after some work, that

$$E_a{}^{\underline{\alpha}}E_{\underline{\alpha}}{}^{\beta'}(\Gamma^a)_{\beta'}{}^{\alpha} = 0, \tag{14.5.25}$$

while at dimension 1 we find that

$$\eta^{ab}\nabla_a E_b{}^{\underline{\alpha}}E_{\underline{\alpha}}{}^{b'} = -\tfrac{1}{8}(\Gamma^{b'a})_{\gamma'}{}^{\beta} Z_{a\beta}{}^{\gamma'} \tag{14.5.26}$$

and

$$\hat{\nabla}^c h_{abc} = -\tfrac{1}{32}(\Gamma^c\Gamma_{ab})_{\gamma'}{}^{\beta}Z_{c\beta}{}^{\gamma'}, \tag{14.5.27}$$

where

$$Z_{a\beta}{}^{\gamma'} = E_{\beta}{}^{\underline{\beta}}\left(T_{a\underline{\beta}}{}^{\underline{\gamma}} - K_{a\underline{\beta}}{}^{\underline{\gamma}}\right) E_{\underline{\gamma}}{}^{\gamma'}. \tag{14.5.28}$$

The tensors $T_{a\underline{\beta}}{}^{\underline{\gamma}}$ and $K_{a\underline{\beta}}{}^{\underline{\gamma}}$ are defined as

$$T_{a\underline{\beta}}{}^{\underline{\gamma}} = E_a{}^{\underline{\alpha}}T_{\underline{\alpha\beta}}{}^{\underline{\gamma}}, \tag{14.5.29}$$

$$K_{a\underline{\beta}}{}^{\underline{\gamma}} = E_a{}^{\underline{\delta}}E_{\underline{\beta}}{}^{\gamma}(\nabla_\gamma E_{\underline{\delta}}{}^{\delta'})E_{\delta'}{}^{\underline{\gamma}}. \tag{14.5.30}$$

while

$$\hat{\nabla}^a h_{bcd} = \nabla^a h_{bcd} - 3X_{a,[b}{}^e h_{cd]e} \tag{14.5.31}$$

with

$$X_{a,b}{}^c = (\nabla^a u_b{}^{\underline{c}})u_{\underline{c}}{}^c. \tag{14.5.32}$$

Although one can also derive the on-shell constraints on the background superspace torsions, curvatures and gauge superfield strengths, that is, equations of motion, from the superembedding approach, it is simpler to use the known values for these background constraints. In addition to equation (14.5.16) these are given by [14.25, 14.26]

$$T_{\underline{a\beta}}{}^{\underline{\gamma}} = -\tfrac{1}{36}(\Gamma^{\underline{bcd}})_{\underline{\beta}}{}^{\underline{\gamma}}H_{\underline{abcd}} - \tfrac{1}{288}(\Gamma_{\underline{abcde}})_{\underline{\beta}}{}^{\underline{\gamma}}H^{\underline{bcde}}. \tag{14.5.33}$$

In this equation $H_{\underline{abcd}}$ is totally anti-symmetric and is the dimension 1 component of the closed superspace 4-form H converted to tangent superspace indices using the inverse supervielbein. Its only other non-vanishing component is

$$H_{\underline{aby\delta}} = -i(\Gamma_{\underline{ab}}C^{-1})_{\underline{\gamma\delta}}. \tag{14.5.34}$$

To find the equations of motion in terms of x-space fields we evaluate equations (14.5.25)–(14.5.27) at $\theta^\mu = 0$, where we recall that the superspace of the brane world volume M has the coordinates x^n and θ^μ, which are Grassmann even and odd, respectively. We will find the spinor, scalar and 3-form equations, respectively. Using world-volume superreparameterisation symmetry we can, as explained in chapter 13, choose the world-volume supervielbein to have the components

$$E_m{}^a(x, \theta) = E_m{}^a(x) + O(\theta), \quad E_m{}^\alpha(x, \theta) = E_m{}^\alpha(x) + O(\theta),$$

$$E_\mu{}^a(x, \theta) = \dot{0} + O(\theta), \quad E_\mu{}^\alpha(x, \theta) = \delta_\mu{}^\alpha + O(\theta), \tag{14.5.35}$$

and the inverse takes the form

$$E_a{}^m(x, \theta) = E_a{}^m(x) + O(\theta), \quad E_a{}^\mu(x, \theta) = E_a{}^\mu(x) + O(\theta),$$

$$E_\alpha{}^m(x, \theta) = 0 + O(\theta), \quad E_\alpha{}^\mu(x, \theta) = \delta_\alpha{}^\mu + O(\theta), \tag{14.5.36}$$

where $E_a{}^m(x)$ is the inverse of $E_m{}^a(x)$. The component field $E_m{}^\alpha(x)$ is the world-volume gravitino, which, like the world-volume metric, is determined by the embedding approach. The field $E_a{}^\mu(x)$ is linearly related to the gravitino.

Using the embedding condition of equation (14.5.14) in equation (14.5.11) and the choice $E_\alpha{}^m(x, \theta) = 0$ of equation (14.5.36) we conclude that

$$\partial_\mu Z^{\underline{M}} E_{\underline{M}}{}^\alpha = 0 \qquad \text{at } \theta^\mu = 0. \tag{14.5.37}$$

As a result we find that

$$E_a{}^{\underline{a}} = E_a{}^m \varepsilon_m{}^{\underline{a}}, \quad E_a{}^{\underline{\alpha}} = E_a{}^m \varepsilon_m{}^{\underline{\alpha}} \qquad \text{at } \theta^\mu = 0, \tag{14.5.38}$$

where we have used the definitions

$$\varepsilon_m{}^{\underline{a}}(x) = \partial_m Z^{\underline{M}} E_{\underline{M}}{}^{\underline{a}}, \quad \varepsilon_m{}^{\underline{\alpha}}(x) = \partial_m Z^{\underline{M}} E_{\underline{M}}{}^{\underline{\alpha}} \qquad \text{at } \theta^\mu = 0. \tag{14.5.39}$$

The coordinates $Z^{\underline{M}}$ describing the motion of the superbrane in $\underline{M}$ are functions of the superspace M. However, at $\theta^\mu = 0$ we can identify them with the usual coordinates of a superbrane, that is, we take $Z^{\underline{m}} = X^{\underline{M}}(x)$ and $Z^{\underline{\alpha}} = \Theta^{\underline{\alpha}}(x)$ at $\theta^\mu = 0$. We recognise $\varepsilon_m{}^{\underline{a}}(x)$ as the generalisation to the p-brane of the symbol $\Pi_m{}^{\underline{a}}$ that occurs in the Green–Schwarz formulation of the superstring of chapter 6 once we work in the flat superspace using equation (14.2.7).

The usual world-volume metric for the superbrane is defined by

$$g_{mn}(x) = \varepsilon_m{}^{\underline{a}} \varepsilon_n{}^{\underline{b}} \eta_{\underline{ab}}, \tag{14.5.40}$$

whose corresponding vielbein is defined as usual by $g_{mn}(x) = e_m{}^a \eta_{ab} e_n{}^b$. However, we note from equation (14.5.19) that

$$E_a{}^{\underline{a}} E_b{}^{\underline{b}} \eta_{\underline{ab}} = m_a{}^c m_b{}^d \eta_{cd}. \tag{14.5.41}$$

It is to be understood that these equations are to be taken at $\theta^\mu = 0$. Hence we have in effect two world-volume vielbeins $e_n{}^a$, one associated with the usual brane metric and another, $E_n{}^a(x)$, which arises in superspace. They are related by

$$e_a{}^m = (m^{-1})_a{}^b E_b{}^m. \tag{14.5.42}$$

From the latter we can define an alternative world-volume metric by

$$G^{mn} = E_a{}^m(x) E_b{}^n(x) \eta^{ab} = ((m^2)^{ab} e_a{}^m e_b{}^n)(x), \tag{14.5.43}$$

which will make its appearance later. We also note the relation

$$u_a{}^{\underline{a}} = e_a{}^m \varepsilon_m{}^{\underline{a}}, \tag{14.5.44}$$

which follows from equation (14.5.38).

It remains to identify how the 2-form B_{mn} living in the world volume arises. To construct the superspace geometry of the world-volume superspace M we introduced supervielbeins and connections, but it is advantageous to also introduce a world-volume, second-rank gauge superfield. The corresponding torsions and curvatures were encountered above, however, we now also have for the gauge superfield a world-volume superfield strength H_{MNP} which is referred to tangent frame as

$$H_{MNP} = E_P{}^C E_N{}^B E_M{}^A H_{ABC}(-1)^{((B+N)M+(P+C)(M+N))}. \tag{14.5.45}$$

It turns out that the only non-vanishing component when referred to tangent space is H_{abc}, which is related to h_{abc} by

$$H_{abc} = m_a{}^d m_b{}^e h_{cde}. \tag{14.5.46}$$

Evaluating this at $\theta^\mu = 0$ one finds

$$H_{mnp}(x) = (E_m{}^a E_n{}^b E_p{}^c H_{abc})(x), \tag{14.5.47}$$

so that, using equation (14.5.42), we find

$$h_{abc}(x) = m_a{}^d e_d{}^m e_b{}^n e_c{}^P H_{mnp}(x). \tag{14.5.48}$$

Although the field h_{abc} is a self-dual anti-symmetric tensor it is not the curl of a 2-form potential. In fact, it turns out that an analysis of the world-volume superspace geometry implies that the superfield H_{MNP} obeys the equation

$$dH_3 = -\tfrac{1}{4}H_4, \tag{14.5.49}$$

where H_4 is the pull back of the background space 4-form, that is,

$$H_{MNPQ} = \partial_M Z^{\underline{M}} \cdots \partial_Q Z^{\underline{Q}} H_{\underline{MNPQ}}. \tag{14.5.50}$$

This equation implies the x-space equation $H_{mnp} = 3\partial_{[n}B_{mp]} - \tfrac{1}{4}A_{nmp}$ and A_{nmp} is the pull back of the background space-time 3-form gauge superfield.

We are now in a position to write the equations of motion in terms of $\varepsilon_m{}^{\underline{a}}$, $\varepsilon_m{}^{\underline{\alpha}}$ and $H_{mnp}(x)$, that is, in terms of the world-volume x-space fields $X^{\underline{m}}(x)$, $\Theta^{\underline{\alpha}}(x)$ and $B_{mn}(x)$. After some work one finds the equations of motion of the 5-brane to all orders in the bosonic background and world-volume fields, but only to first order in fermions [14.11].

The result at dimension $\tfrac{1}{2}$ is the spinor equation and it is given by

$$\varepsilon_a(1 - \Gamma)\Gamma^b m_b{}^a = 0, \tag{14.5.51}$$

where $\Gamma_m = \varepsilon_m{}^{\underline{a}}\Gamma_{\underline{a}}$, $\Gamma_b = e_b{}^m \Gamma_m$, $\varepsilon_b = e_b{}^m \varepsilon_m$ and the background space-time spinor indices are suppressed. In this equation the matrix Γ is given by

$$\Gamma = (-1 + \tfrac{1}{3}\Gamma^{mnp}h_{mnp})\Gamma_{(0)}, \tag{14.5.52}$$

where

$$\Gamma_{(0)} = \frac{1}{6!\sqrt{-g}}\epsilon^{m_1 \cdots m_6}\Gamma_{m_1 \cdots m_6}, \qquad h_{mnp} = e_m{}^a e_n{}^b e_p{}^c h_{abc}. \tag{14.5.53}$$

In carrying out this calculation the matrix Γ is defined by $\frac{1}{2}(1+\Gamma)_{\underline{\alpha}}{}^{\underline{\beta}} = E_{\underline{\alpha}}{}^{\alpha} E_{\alpha}{}^{\underline{\beta}}$ and is then evaluated to give the above expression.

The dimension 1 scalar equation is found to be

$$G^{mn}\nabla_m \varepsilon_n{}^{\underline{c}} = \frac{1}{\sqrt{-g}}\left(1 - \frac{2}{3}\text{Tr}k^2\right)\epsilon^{m_1\cdots m_6}$$

$$\times \left(\frac{1}{6^2 \times 4 \times 5}H^{\underline{a}}{}_{m_1\cdots m_6} + \frac{2}{3}H^{\underline{a}}{}_{m_1 m_2 m_3}H_{m_4 m_5 m_6}\right)(\delta_{\underline{a}}{}^{\underline{c}} - \varepsilon_{\underline{a}}{}^m \varepsilon_m{}^{\underline{c}}),$$

$$(14.5.54)$$

where $k_a{}^b = h_{acd}h^{bcd}$, the background space-time indices on H_4 and H_7 have been converted to world volume indices with factors of $\varepsilon_m{}^{\underline{a}}$ and

$$H_{\underline{d_1}\cdots\underline{d_4}} = \frac{1}{7!}\epsilon_{\underline{d_1}\cdots\underline{d_4}\underline{e_1}\cdots\underline{e_7}}H^{\underline{e_1}\cdots\underline{e_7}}.$$

$$(14.5.55)$$

The 3-form equation is

$$G^{mn}\nabla_m H_{npq} = \frac{1}{(1 - \frac{2}{3}\text{Tr}k^2)}e_p^a e_q^b (4Y + 4mY + mmY)_{ab},$$

$$(14.5.56)$$

where Y_{ab} is given by

$$Y_{ab} = (\tilde{K} + m\tilde{K} + \frac{1}{4}mm\tilde{K})_{ab}$$

$$(14.5.57)$$

with $\tilde{K}_{ab} = -(1/36 \times 4!)\epsilon_{abcdef}H^{cdef}$, $(m\tilde{K})_{ab} = m_{[a}{}^c \tilde{K}_{b]c}$, $(mm\tilde{K})_{ab} = m_a{}^c m_b{}^d \tilde{K}_{cd}$ and with mY mmY defined in a similar way to the $m\tilde{K}$ and $mm\tilde{K}$.

We note from equation (14.5.39) that to first order in fermions

$$\varepsilon_m{}^{\underline{a}}(x) = \partial_m X^{\underline{m}} E_{\underline{m}}{}^{\underline{a}} + \partial_m \Theta^{\underline{\alpha}} E_{\underline{\alpha}}{}^{\underline{a}} = \partial_m X^{\underline{m}} E_{\underline{m}}{}^{\underline{a}}.$$

$$(14.5.58)$$

Furthermore, in a flat superspace background $\varepsilon_m{}^{\underline{a}}(x) = \partial_m X^{\underline{a}}$ and setting the background 4-form field to zero we recover equations (14.5.1) and (14.5.2) from equations (14.5.53) and (14.5.55), respectively.

Finally we show how the κ-symmetry transformations emerge automatically from the embedding formalism. They are related to the Grassmann odd world volume diffeomorphisms, which are an obvious symmetry of the embedding formalism. The world volume has the coordinates $z^M = (x^n, \theta^\mu)$ and we consider an infinitesimal world surface diffeomorphism $\delta z^M = v^M$. This induces the change of coordinates $Z^{\underline{N}} \to Z^{\underline{N}} + v^M \partial_M Z^{\underline{N}}$ describing the embedding of the brane in the background superspace. When refered to the background superspace tangent frame we find that

$$\delta z^{\underline{A}} \equiv \delta z^M E_M{}^{\underline{A}} = v^A E_A{}^{\underline{A}}.$$

$$(14.5.59)$$

In carrying out the last step we have used equation (14.5.11) and defined v^A by $v^M = v^A E_A{}^M$. For a Grassmann odd transformation ($v^a = 0$) one has

$$\delta z^{\underline{a}} = 0, \qquad \delta z^{\underline{\alpha}} = v^\alpha E_\alpha{}^{\underline{\alpha}}.$$

$$(14.5.60)$$

The vanishing of the Grassmann even variation $\delta z^{\underline{a}}$ is the characteristic feature of κ-symmetry and it follows from the basic embedding condition equation (14.5.14).

The relation between the parameter v^α and the familiar κ transformation parameter $\kappa^{\underline{\alpha}}$ can be expressed as

$$v^\alpha = \kappa^{\underline{\gamma}} E_{\underline{\gamma}}{}^\alpha,$$

$$(14.5.61)$$

whereupon the κ transformation rule takes the form

$$\delta z^{\underline{\alpha}} = \kappa^{\underline{\gamma}}(1 + \Gamma)_{\underline{\gamma}}{}^{\underline{\alpha}}. \tag{14.5.62}$$

In this equation we have absorbed a factor of 2 into the definition of κ and the matrix Γ was defined from $\frac{1}{2}(1 + \Gamma)_{\alpha}{}^{\underline{\beta}} = E_{\underline{\alpha}}{}^{\alpha} E_{\alpha}{}^{\underline{\beta}}$ as before. The above equations are valid for any θ^{μ}, but we are taking them to be evaluated at $\theta^{\mu} = 0$, so that they are component results.

There remains the determination of the κ-symmetry transformation of the anti-symmetric tensor field B_{mn}. However, it is more convenient to compute the κ transformations rule for the field $h_{abc}(x)$. Carrying out the corresponding reparameterisation one finds, after several steps, that the final result is given by

$$\delta h_{abc} = \frac{i}{16} m_{[a|}{}^{d} \, \varepsilon_d (1 - \Gamma) \Gamma_{|bc]} \kappa, \tag{14.5.63}$$

where $\Gamma_a = \Gamma^m e_{ma}$ and the background space-time spinor indices are suppressed. One can check that this transformation preserves the self-dual property of h_{abc} provided one uses the spinor equation.

In the above we have missed out a number of the steps in the derivation which the reader can find by studying the original reference [14.11]. However, we have explained a number of conceptual issues that are omitted in this paper. Between the two readers can hopefully derive completely the dynamics of the 5-brane if they so wish.

In fact, the superembedding formalism for the 5-brane leads to the equations of motion for the fields of the brane and the equations of motion for the background fields (that is, the on-shell superspace constraints on superspace torsions curvatures and field strengths). Although above, for simplicity, we assumed the latter, they do come out of the formalism as must be the case as one has incorporated κ symmetry. The same story holds for the 2-brane.

Although the embedding condition of equation (14.5.14) is very natural in that all the geometry of superspace is contained in the odd sectors of the tangent space of supermanifolds, its deeper geometrical significance is unclear. However, the power of this approach became evident once it was shown to lead to the correct dynamics for the most sophisticated brane, the M theory 5-brane. Although the above results hold for many branes, they do not hold for all branes unless the embedding condition is in general supplemented by a further condition.

There is another formulation of the dynamics of the 5-brane given in [14.23] and [14.27]. Although this formulation involves an action, this is not necessarily an advantage as has been pointed out in [14.28]. Reference [14.23] used the superembedding formalism, but in a different way to that considered in this section.

14.6   Solutions of the 5-brane of M theory

In this section we will construct some solutions of the 5-brane equations of motion given in the last section. In particular we will find 3-brane [5.8] and string [14.29] solutions. From the perspective of eleven dimensions these can be interpreted as the 5-brane self-intersecting and a 2-brane ending on a 5-brane, respectively. To find which brane solutions are likely to be present in the 5-brane we can follow the same type of arguments that we gave in section 14.2 when we looked for brane solutions of the maximal supergravity theories. The 5-brane is constructed from a self-dual 3-tensor h_{mnp}, $(m, n, p = 0, \ldots, 5)$, five scalars $X^{b'}$ $(b' = 6', \ldots, 10')$, describing the position of the 5-brane in the space transverse to its world volume, and 16 fermions Θ_α^i, $(\alpha = 1, \ldots, 4, i = 1, \ldots, 4)$ and is invariant under $(2, 0)$ supersymmetry, which has the supercharges Q_α^i. This is the 16-component world-volume supersymmetry as opposed to the 32-component supersymmetry of the background superspace in eleven dimensions which is also a symmetry of the brane dynamics. We are using world-volume reparameterisation invariance to set the embedding coordinates $X^a = \xi^a$, $a = 0, 1, \ldots, 5$ and κ-supersymmetry to set $\Theta^{\alpha i} = 0$.

The most general form for the $(2, 0)$ supersymmetry algebra in six dimensions was given in equation (5.5.9), which we repeat here using the Weyl projected matrices we have been using above

$$\{Q_\alpha^i, Q_\beta^j\} = \Omega^{ij}(\gamma^m)_{\alpha\beta} P_m + (\gamma^m)_{\alpha\beta} Z_m^{ij} + (\gamma^{mnp})_{\alpha\beta} Z_{mnp}^{ij}. \tag{14.6.1}$$

Here Ω^{ij} is the invariant tensor of $USp(4) \cong Spin(5)$, P_m is the momentum, Z_m^{ij} is in the symmetric **5** of $Spin(5)$ and Z_{mnp}^{ij} is self-dual and in the anti-symmetric **10** of $Spin(5)$. In fact, the left-hand side can be thought of as the most general 16×16 symmetric matrix and therefore has $16 \times 17/2 = 136$ degrees of freedom, which is the same number that appears on the right-hand side: $6 + 5 \times 6 + 10 \times 10 = 136$.

Supersymmetric p-brane solutions possess a rank p topological charge that will appear in the supersymmetry algebra on the right-hand side of the anti-commutators of two super-symmetries. We recall that this has to happen in order for the solutions to preserve some supersymmetry. Looking at the above supersymmetry algebra we may expect to find 1-brane and 3-brane solitons and, recalling section 13.8.4, as only one central charge is active these should preserve half of the world-volume supersymmetry, that is, eight supersymmetries.

We can also look for sources for other branes. The 5-brane has only scalars and a 2-form gauge field. The 2-form has a rank 3 field strength which is self-dual, but the scalars have a rank 1 'field strength' which is dual to a rank 5 field strength that we can express as coming from a rank 4 potential. As a p-brane couples to a rank $p + 1$ gauge field we find the same result, namely a 1-brane and a 3-brane. From the scalar we might get a -1-brane, or instanton, but it turns out that in this particular theory it is not present. We will now explicitly construct the 1- and 3-brane solutions.

14.6.1 The 3-brane

Since we will have a 3-brane living within the 5-brane we will denote the six-dimensional coordinates of the 5-brane world volume by xs with hatted variables $\hat{m}, \hat{n} = 0, 1, 2, \ldots, 5$, that is, $x^{\hat{n}}$, instead of the variables ξ^n used earlier to denote the world-volume coordinates. We take the 5-brane to lie in the directions $X^0, X^1, X^2, X^3, X^4, X^5$ of the eleven-dimensional space in which it is embedded and we seek a static planar 3-brane which we can take to lie in the plane x^0, x^1, x^2, x^3 with the 5-brane world volume. The coordinates transverse to the 3-brane still in the 5-brane world volume are x^4, x^5. We will denote these latter coordinates in particular by x^m, that is, using unhatted indices. Thus when we see an $\hat{m}$ index it can take the range $0, 1, \ldots, 5$, but a unhatted index m can only take the values $m = 4, 5$. Following a similar discussion to that above for branes in supergravity, we will assume that all the

5-brane fields only depend on the transverse coordinates, that is, x^4 and x^5. This implies that the solution has an $ISO(1, 3)$ symmetry.

In static gauge we take $X^0 = x^0, \ldots, X^5 = x^5$. The coordinates transverse to the 5-brane, that is, the scalar fields, are X^6, X^7, X^8, X^9 and X^{10}, as before we denote these by $X^{a'}$. To find a 3-brane we take X^7, X^8 and X^9 to be constants leaving two active scalars X^6 and X^{10}. One can, of course, take either of the two scalars to be non-trivial. Let us dualise one of the remaining scalars X^6, say to a 5-form

$$G_{\hat{m}\hat{n}\hat{p}\hat{q}\hat{r}} = \epsilon_{\hat{m}\hat{n}\hat{p}\hat{q}\hat{r}\hat{s}} \partial^{\hat{s}} X^6. \tag{14.6.2}$$

We now assume that

$$G_{0123m} = \epsilon_{mn} \partial^n X^6 \equiv v_m, \quad X^{10} = \phi, \quad h_{mnp} = 0, \tag{14.6.3}$$

where ϵ_{mn} is the volume element on the space transverse to the 3-brane but still within the 5-brane. The metric of the 5-brane induced from the flat background space-time is $g_{\hat{m}\hat{n}} = \partial_{\hat{m}} X^{\underline{a}} \eta_{\underline{ab}} \partial_{\hat{n}} X^{\underline{b}}$. Taking into account that we are in static gauge, that is, $X^0 = x^0, \ldots, X^5 = x^5$, and that X^7, X^8, X^9 are constants, it takes the form

$$g_{\hat{m}\hat{n}} = \begin{pmatrix} -1 & 0 & \\ 0 & \mathbf{1}_{3\times3} & \\ & & \delta_{mn} + \partial_m X^6 \partial_n X^6 + \partial_m X^{10} \partial_n X^{10} \end{pmatrix}, \tag{14.6.4}$$

where $\mathbf{1}_{3\times3}$ is the unit matrix in three dimensions. Since the 3-form is zero, the equations of motion of equation (14.5.54) take the form

$$g^{mn} \nabla_m \nabla_n X^6 = g^{mn} \nabla_m \nabla_n X^{10} = 0, \tag{14.6.5}$$

where ∇ is the covariant derivative using the Levi–Civita connection of $g_{\hat{m}\hat{n}}$. For the metric given just above, this connection takes the form of equation (14.5.5). To arrive at the final form of the ansatz it is helpful now to determine the condition that half of the supersymmetry is preserved by the soliton. The linearised supersymmetry variation of the 5-brane fermion is given by [14.29]

$$\delta\Theta_\beta^{\,j} = \epsilon^{\alpha i}\left(\tfrac{1}{2}(\gamma^m)_{\alpha\beta}(\gamma_{b'})_i^{\,j}\partial_m X^{b'} - \tfrac{1}{6}(\gamma^{mnp})_{\alpha\beta}\delta_i^{\,j}h_{mnp}\right). \tag{14.6.6}$$

Taking $h_{mnp} = 0$, this becomes

$$\delta\Theta_\beta^{\,j} = \tfrac{1}{2}\epsilon^{\alpha i}(\gamma^m)_{\alpha\beta}(\gamma_{b'})_i^{\,j}\partial_m X^{b'}. \tag{14.6.7}$$

Inserting our ansatz of equation (14.6.3) into $\delta\Theta_\beta^{\,j} = 0$ one finds

$$0 = \tfrac{1}{2}\epsilon^{\alpha i}(\gamma^4)_{\alpha\gamma}(\gamma_6)_i^{\,k} \times \left\{\left[\delta_k^{\,j}\delta_\beta^\gamma v_5 + (\gamma_{610})_k^{\,j}(\gamma^{45})_\beta^\gamma \partial_5\phi\right]\right.$$
$$\left. + (\gamma_{610})_k^{\,l}\left[\delta_l^{\,j}\delta_\beta^\gamma \partial_4\phi + (\gamma_{610})_l^{\,j}(\gamma^{45})_\beta^\gamma v_4\right]\right\}. \tag{14.6.8}$$

Setting

$$v_m = \pm\partial_m\phi \quad \text{or equivalently} \quad -\partial_4 X^6 = \pm\partial_5 X^{10}, \ \partial_5 X^6 = \pm\partial_4 X^{10}, \tag{14.6.9}$$

we find that the solution will be invariant under supersymmetries which satisfy

$$\epsilon_0^{\alpha i}(\gamma_{610})_i^{\,j}(\gamma^{45})_\alpha^{\,\beta} = \mp\epsilon_0^{\beta j}. \tag{14.6.10}$$

We note that the condition of equation (14.6.9) is equivalent to the statement that $X^6 + iX^{10} \equiv s$ is a (anti-)holomorphic function of $x^4 + ix^5 \equiv z$ for the minus (plus) sign in equation (14.6.9). Put more simply, s is a (anti-)holomorphic function of z. The solution then preserves half the supersymmetries as one can verify that the matrix operator that occurs in equation (14.6.10) is a projector. One can verify that these conditions are true to all orders [5.8]. As the original 5-brane is invariant under sixteen supersymmetries this implies that the 3-brane solutions preserve eight of them.

If we substitute the condition of equation (14.6.9) into the metric of equation (14.6.4) we find that it becomes diagonal. Indeed, by explicitly writing out the non-diagonal components we find that they vanish and that $g_{mn} = \delta_{mn}(1 + (\partial\phi)^2)$, $m, n = 4, 5$ and so the inverse metric is given by $g^{mn} = \delta_{mn}/(1 + (\partial\phi)^2)$, $m, n = 4, 5$. Here we have defined $(\partial\phi)^2 = \delta^{mn}\partial_m\phi\partial_n\phi$. Using the expression for the Christoffel symbol of equation (14.5.5) we find that the equations of motion for the only non-trivial fields X^6, X^{10} are satisfied if

$$\frac{1}{(1 + (\partial\phi)^2)^2}\partial^2\phi = 0, \tag{14.6.11}$$

where $\partial^2\phi = \delta^{mn}\partial_m\partial_n\phi$. In fact, the equation of motion for X^6 is trivially satisfied, but that for X^{10} gives the above result.

Thus the solution corresponding to N 3-branes located at y_I ($I = 0, \ldots, N - 1$) has the general form

$$G_{0123m} = \pm\partial_m\phi, \quad \phi = \phi_0 + \sum_{I=0}^{N-1} Q_I \ln|z - y_I|, \tag{14.6.12}$$

where ϕ_0 and Q_I are constants. Clearly this solution has bad asymptotic behaviour unless $\sum Q_I = 0$; however, one can still define the charge of a single 3-brane as

$$Q = \int_{S_\infty^1} \star G_5, \tag{14.6.13}$$

where S_∞^1 is the transverse circle at infinity and $\star$ is the flat six-dimensional Hodge star. The presence of the conformal factor in equation (14.6.11) indicates that the equations of motion are satisfied even at the points where the solution is ill behaved and hence no sources are needed.

The 3-brane has a four-dimensional world volume and we can wonder what is the dynamics of such a brane. As explained at the beginning of this chapter, one can find the dynamics for a slowly moving brane using techniques from the theory of monopoles. We take the solution for the 3-branes and we let the moduli depend on the world-volume coordinates of the 3-brane; the dynamics is then found by substituting this into the equations of motion of the 5-brane.

Clearly there are two bosonic moduli describing the location of the 3-brane in the transverse space of the 5-brane and so we will find two scalar fields in the dynamics. The fermionic zero modes, or fermionic moduli, can be thought of as Goldstone fields for the broken supersymmetries due to the presence of the 3-brane. We recall that the solution breaks half the supersymmetries and as we began with sixteen there are eight broken and eight preserved supersymmetries. As such there are eight fermionic degrees of freedom and these can only be two Majorana spinors as we are dealing with a $N = 2$

supersymmetric theory in four dimensions. On-shell we have two bosonic degrees of freedom and four fermionic degrees of freedom and so we are missing two bosonic degrees of freedom. The simplest possibility is a vector and indeed the 2-form gauge fields of the 5-brane do contributes a vector field to the dynamics. This is more subtle and is due to the large gauge transformations of the 2-form gauge field, which are non-vanishing at infinity and lead to moduli. We recall that theories are only invariant under small gauge transformations. Thus we find a theory with four fermionic and four bosonic degrees of freedom on-shell and so we find that the dynamics of the 3-brane is built from a $D = 4$, $N = 2$ vector multiplet. This confirms the presence of the vector which contributes two degrees of freedom on-shell.

Since the scalars transform under Spin(5), by choosing an arbitrary pair of scalars $(X^{a'}, X^{b'})$ with $a' \neq b'$ for our solution, we obtain a multiplet of 3-branes transforming as a $(\frac{5 \times 4}{2} = 10)$-dimensional representation of Spin(5), in agreement with the algebra. In the M theory interpretation, which we will discuss in section 14.7, the 3-brane represents the intersection of two 5-branes; these scalars point along the two directions of the external 5-brane which are transverse to the world volume of the 5-brane we are considering.

14.6.2 The self-dual string

We will consider a static string whose world sheet lies in the x^0, x^1 directions of the world volume of the 5-brane, which is in the directions $x^0, \ldots, x^5$. We take all the 5-brane fields to be independent of x^0 and x^1. In this section, and as before, we denote the six-dimensional world-volume indices which take the full range with a hat, that is, $\hat{a}, \hat{b}, \ldots, \hat{m}, \hat{n} \ldots = 0, 1, \ldots, 5$, and denote in particular the four coordinates transverse to x^0, x^1, that is, the coordinates x^2, x^3, x^4, x^5, as carrying the indices $a, b, \ldots, m, n \ldots = 2, 3, 4, 5$. We take only one of the scalar fields $X^{a'}$ transverse to the 5-brane to be active thus breaking the SO(5) symmetry to SO(4). We choose this scalar field to be X^{10}. The other scalar fields are assumed to be constants. We now assume that

$$X^{10} \equiv \phi \quad h_{01a} \equiv v_a \quad h_{abc} = \epsilon_{abcd} v^d \tag{14.6.14}$$

with the other components of h_{abc} vanishing. The reader may verify that the ansatz respects the self-duality of h_{abc}.

We must now work out the quantities that appear in the equations of motion. The metric induced on the brane from the flat background world volume of the 5-brane is given by

$$g_{\hat{m}\hat{n}} = \begin{pmatrix} -1 & 0 & \\ 0 & 1 & \\ & & \delta_{mn} + \partial_m \phi \partial_n \phi \end{pmatrix}. \tag{14.6.15}$$

The vielbein $e_{\hat{m}}^{\hat{a}}$ associated with $g_{\hat{m}\hat{n}}$ can be found to have the form

$$e_{\hat{m}}^{\hat{a}} = \begin{pmatrix} 1 & 0 & \\ 0 & 1 & \\ & & \delta_m^a + c \phi_m \phi^a \end{pmatrix}, \tag{14.6.16}$$

where $\phi_n \equiv \partial_n \phi$, $c \equiv (|\phi_n|^2)^{-1}(-1 \pm \sqrt{1 + |\phi_n|^2})$, $|\phi_n|^2 \equiv \phi_n \phi_m \delta^{mn}$ and we adopt the convention that the derivative of ϕ only ever carries a lower world index (that is, $\phi_a = \delta_a^n \phi_n$). The inverse is given by

$$(e^{-1})_{\hat{a}}{}^{\hat{m}} = \begin{pmatrix} 1 & 0 \\ 0 & 1 \\ & \delta_a^m \mp \frac{c}{\sqrt{1+\phi^2}}\phi_a\phi^m \end{pmatrix}. \tag{14.6.17}$$

We find that the determinants are given by

$$\det g_{mn} = e^{\mathrm{tr}\ln g_{mn}} = 1 + |\phi_n|^2, \quad \det e_m^a = e^{\mathrm{tr}\ln e_m^a} = \sqrt{1 + |\phi_n|^2}. \tag{14.6.18}$$

The matrix m which plays an important role in relating the self-dual 3-form to the 3-form that is the derivative of the gauge field is readily found to take the form

$$m_{\hat{a}}{}^{\hat{b}} \equiv \delta_{\hat{a}}{}^{\hat{b}} - 2h_{\hat{a}\hat{c}\hat{d}}h^{\hat{b}\hat{c}\hat{d}}$$

$$= \begin{pmatrix} 1 + 4v^2 & 0 \\ 0 & 1 + 4v^2 \\ & & (1 - 4v^2)\delta_a{}^b + 8v_a v^b \end{pmatrix}, \tag{14.6.19}$$

where $v^2 = v^a v_a$. Its inverse is given by

$$(m^{-1})_{\hat{a}}{}^{\hat{b}} = \frac{1}{1 - 16(v^2)^2} \begin{pmatrix} 1 - 4v^2 & 0 \\ 0 & 1 - 4v^2 \\ & & (1 + 4v^2)\delta_a{}^b - 8v_a v^b \end{pmatrix}. \tag{14.6.20}$$

At this point it is desirable to refine our ansatz to ensure that the solution preserves half of the supersymmetries. Substituting equation (14.6.14) into the linearised supersymmetry transformation of equation (14.6.6) we find that, at this level, the preserved supersymmetries must satisfy

$$\delta_0 \Theta_\alpha^j = 0 = \epsilon^{\alpha i} (\tfrac{1}{4} \delta_\alpha{}^\gamma \delta_i{}^k \partial_a \phi - (\gamma^{01})_\alpha{}^\gamma (\gamma_{10})_i{}^k v_a). \tag{14.6.21}$$

Therefore if we take

$$V_n \equiv H_{01n} = \pm \tfrac{1}{4} \partial_n \phi, \quad n = 2, \dots, 5, \tag{14.6.22}$$

the solution will be invariant under linearised supersymmetries which satisfy

$$\epsilon_0^{\beta j} = \pm (\gamma^{01})_\alpha{}^\beta (\gamma_{10})_i{}^j \epsilon_0^{\alpha i}. \tag{14.6.23}$$

As the operator that appears here is a projector we find that we preserve half of the supersymmetries, that is, eight. At the linearised level h_{abc} and $H_{mnp} \delta_a^m \delta_b^n \delta_c^p$ are equal; however, we have written equation (14.6.22) in such a way that it is valid in the full theory. For a discussion of the preservation to all orders see reference [14.29].

We now calculate the components of $H_{\hat{m}\hat{n}\hat{p}}$, which we recall are given in terms of $h_{\hat{a}\hat{d}\hat{e}}$ by $H_{\hat{m}\hat{n}\hat{p}} = e_{\hat{m}}{}^{\hat{a}} e_{\hat{n}}{}^{\hat{b}} e_{\hat{m}}{}^{\hat{c}} (m^{-1})_{\hat{c}}{}^{\hat{e}} h_{\hat{a}\hat{b}\hat{e}}$. The only non-vanishing components are

$$H_{01m} \equiv V_m = \frac{1}{(1 + 4v^2)} e_m{}^a v_a,$$

$$\tag{14.6.24}$$

$$H_{mnp} = \frac{\det e}{(1 - 4v^2)} \epsilon_{mnpq} e_a{}^q v^a.$$

Using the first of these equations, and the supersymmetry preserving condition of equation (14.6.22), we find that

$$v_a = \frac{1}{2} \frac{\phi_a}{1 + \sqrt{1 + |\phi_n|^2}},$$ (14.6.25)

where we recall that $\phi_a = \delta_a^n \phi_n$. A short calculation using this result in the second of equations (14.6.24) shows that

$$H_{mnp} = \pm \tfrac{1}{4} \epsilon_{mnpq} \phi_r \delta^{rq}.$$ (14.6.26)

Finally we calculate the inverse metric $G^{\hat{m}\hat{n}}$ that occurs in the 5-brane equation of motion:

$$
\begin{aligned}
G^{\hat{m}\hat{n}} &= (m^2)^{\hat{a}\hat{b}} e_{\hat{a}}^{\hat{m}} e_{\hat{b}}^{\hat{n}} \\
&= \begin{pmatrix} -(1+4v^2)^2 & 0 & \\ 0 & (1+4v^2)^2 & \\ & & (1-4v^2)^2 g^{mn} + 16 v^a v^b e_a^m e_b^n \end{pmatrix}.
\end{aligned}
$$ (14.6.27)

Expressing this in terms of ϕ alone we find that

$$G^{mn} = \delta^{mn} \frac{4}{(|\phi_n|^2)^2} \left(1 \pm \sqrt{1 + |\phi_n|^2} \right)^2.$$ (14.6.28)

We note that we could write $G^{\hat{m}\hat{n}} = \eta^{\hat{a}\hat{b}} E_{\hat{a}}^{\hat{m}} E_{\hat{b}}^{\hat{n}}$, introducing the new vielbein $E_{\hat{a}}^{\hat{m}}$ which is related to our usual vielbein by $e_{\hat{a}}^{\hat{m}} = (m^{-1})_{\hat{a}}^{\hat{b}} E_{\hat{b}}^{\hat{m}}$.

Let us now solve the equations of motion. For the above field values the scalar equation becomes

$$G^{mn} \nabla_m \phi_n = 0,$$ (14.6.29)

while the equation for the gauge field leads to the two equations

$$G^{mn} \nabla_m V_n = 0$$ (14.6.30)

and

$$G^{mn} \nabla_m H_{mpq} = 0,$$ (14.6.31)

depending on the values we take. The equation of motion for $H_{01m} \equiv V_n$ now reduces to that of ϕ if we use equation (14.6.22). Using equation (14.6.28) we find that the scalar equation becomes

$$\delta^{mn} \partial_m \partial_n \phi = 0.$$ (14.6.32)

The equation for H_{mnp} becomes identically satisfied. We must also ensure that H_{mnp} is closed. We find that the equation

$$\epsilon^{mnpq} \partial_m H_{npq} = 0$$ (14.6.33)

reduces to the scalar equation.

Therefore we find that the solutions are given by harmonic functions on the flat transverse space $\mathbf{R}^4$. For N strings located at y_I^m, $(I = 0, 1, \ldots, N - 1)$ the solution reduces to

$$H_{01m} = \pm \tfrac{1}{4}\partial_m\phi, \quad H_{mnp} = \pm \tfrac{1}{4}\epsilon_{mnpq}\delta^{qr}\partial_r\phi, \quad \phi = \phi_0 + \sum_{I=0}^{N-1} \frac{2Q_i}{|x - y_I|^2}, \quad (14.6.34)$$

where ϕ_0 and Q_I are constants. Note that the solution is smooth everywhere except at the centres of the strings. Furthermore, the presence of the conformal factor in equation (14.6.28), which is present in the scalar equation of motion, implies that the equations of motion are satisfied even at the poles of ϕ, so that no source terms are required and the solution is truly solitonic. Clearly, the string soliton given above is self-dual even though in general the tensor H_{mnp} need not be. If we consider a single string, then we find it has the same electric and magnetic charges

$$Q_E = \int_{S^3_\infty} \star H = \int_{S^3_\infty} H = Q_M, \quad (14.6.35)$$

where S^3_∞ is the transverse sphere at infinity, S^3_1 the unit sphere and $\star$ is the flat six-dimensional Hodge star. To calculate the mass per unit length of this string one could wrap it around a circle of radius R and reinterpret the string as a 0-brane in five dimensions. The mass per unit length times $2\pi R$ is then identified with the mass of the 0-brane.

We now consider what the action is that describes the motion of the self-dual string. As before the dynamics is constructed from the moduli of the solution, which are given a dependence on the world-volume coordinates of the string. We therefore consider the moduli of a single string soliton. Clearly, there are four bosonic zero modes y_0^m coming from the location of the the string in the four transverse dimensions. The fermionic moduli are generated by the broken supersymmetries $\epsilon_\alpha{}^i$, which satisfy

$$\epsilon^{\beta j} = (\gamma^{01})_\alpha{}^\beta (\gamma_{10})_i{}^j \epsilon^{\alpha i}. \quad (14.6.36)$$

Here we have chosen a given sign for the supersymmetry preservation condition corresponding to a choice of sign in equation (14.6.22). If we take the spinors to satisfy $\gamma_{10}\epsilon = \pm\epsilon$, then they will satisfy $(\gamma^{01}\epsilon = \pm\epsilon)$, which are left and right moving for the upper and lower signs, respectively. Therefore we have four left and four right moving fermions on the string world sheet. Thus we have equal numbers of fermionic and bosonic moduli. This confirms that there are no other bosonic moduli which might, for example, have arisen from large gauge transformations. Hence the string has a $(4, 4)$ supermultiplet of fields.

From the M theory point of view the strings obtained here should be interpreted as the end of an infinitely extended 2-brane ending on a 5-brane [14.30]. The scalar $X^{10} = \phi$ then corresponds to the direction of the membrane transverse to the 5-brane. Clearly the SO(5) symmetry rotates the choice of this direction. The infinite mass per unit length can be seen as arising from the infinite length of this membrane. In terms of the notation of equation (13.8.106) the 2-brane ending on the 5-brane has the configuration

$$\begin{array}{ccccccccccc} 0 & 1 & . & . & . & . & . & . & . & . & 10 \\ 0 & 1 & 2 & 3 & 4 & 5 & . & . & . & . & . \end{array} \quad (14.6.37)$$

By dimensionally reducing the 5-brane to ten dimensions and using T duality as explained in chapter 17 one can find solutions to the D-brane equations of motion in ten dimensions [14.29]. Such solutions were also found by a different method in reference [14.31].

14.7   Five-brane dynamics and the low energy effective action of the $N = 2$ Yang–Mills theory

One of the most unexpected developments in string theory was that brane dynamics could be used to find properties of quantum field theories. We noted in section 4.4 chapter that D-branes have vector gauge fields living on their world volume and their dynamics is just part of the dynamics of the brane. As such brane dynamics contains the field theories of interest, but as we will see it can account for sophisticated properties of quantum field theory. In this section we will use the classical dynamics of the 5-brane to derive the complete low energy effective action of the four-dimensional $N = 2$ SU(2) Yang–Mills quantum field theory spontaneously broken to U(1) [14.32]. This effective action had been previously derived and is called the Seiberg–Witten action [14.33]. It was found using essentially two inputs: the first was the complete perturbative result for this action which had been known for many years [14.34, 14.35] and the second was an application of electromagnetic duality [14.36]. One of the most mysterious features of the Seiberg–Witten effective action is the way that it is related to an associated Riemann surface. In particular, it has been shown how the effective action can be constructed given the Riemann surface using a specific recipe. However, how this surface arises naturally in the theory is not apparent. The work of [14.33] was generalised to the spontaneously broken gauge groups SU(N) with corresponding Riemann surfaces [14.37].

The calculation given in this section can be thought of as a prototype derivation of quantum field theory using brane dynamics. We will also see in particular how the Riemann surface arises naturally in this context. In this section we will use material in chapters 15 and 17 and the reader may wish to consult these before reading this section.

Let us begin by considering the IIA theory with a system of NS5-branes and 4-branes [14.38]. The configuration is similar to that discussed around equation (15.3.19), we place two parallel NS5-branes in the $X^0, X^2, X^1, X^3, \ldots, X^5$ plane separated along the X^6 direction by a fixed distance. These two NS5-branes will preserve 16 of the 32 space-time supersymmetries. Next we introduce N parallel D4-branes in the X^0, X^1, X^2, X^3, X^6 plane. These D4-branes stretch between the two NS5-branes and reduce the number of preserved supersymmetries to eight. In terms of the notation of equation (13.8.106) we show only the longitudinal directions of the branes in the form

$$
\begin{array}{llllllll}
\text{NS5} & 0 & 1 & 2 & 3 & 4 & 5 & \quad \cdots \\
\text{D4} & 0 & 1 & 2 & 3 & & 6 & \cdots
\end{array}
\tag{14.7.1}
$$

The configuration is shown in figure 14.7.1.

The world volume of the N parallel D4-branes will carry vector gauge fields that belong to the gauge group U(N) if they are coincident and U(N) spontaneously broken to U(1)s if they are separated. However, the presence of the NS5-branes restricts the world volume of the D4-branes in the x^6 directions to a finite interval and at low energy, that is, Taylor expanding in the x^6 momentum and keeping only the constant term, we find that the 4-branes have in effect only a four-dimensional world volume in the directions x^0, x^1, x^2, x^3. In fact, we will take a system that excludes the overall U(1) factor of the gauge group U(N) which corresponds to the centre of mass motion of the D4-branes. Thus the non-trivial brane dynamics of this system is that of an $N = 2$ supersymmetric Yang–Mills theory with gauge group U(N) spontaneously broken to U(1)s. We note that in the limit of small string

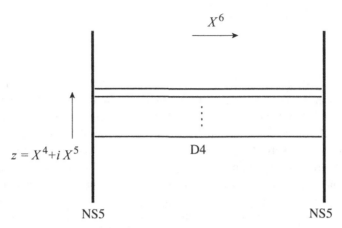

Figure 14.7.1   Two NS5s and N D4s.

coupling g_s the mass density of the NS5-branes is large compared with the D4-branes as they go as $1/g_s^2$ and $1/g_s$, respectively, and so at low energy the NS5-branes are essentially static, or fixed in position.

Now let us consider the limit of large string coupling. As discussed in chapter 17, M theory, that is, eleven-dimensional supergravity in the low energy limit, dimensionally reduced on a circle is the IIA theory. However, in the limit of large string coupling the physical radius of the circle becomes large and the IIA theory decompacitifies to the eleven-dimensional theory. Thus in this limit we must add a new coordinate x^{10} belonging to the circle which is periodic with period $2\pi R$. The NS5-branes simply lift to M5-branes in the directions X^μ, $\mu = 0, 1, \ldots, 5$. The D4-branes also lift to M5-branes, only wrapped on the X^{10} dimension and so in the directions X^0, X^1, X^2, X^3, X^6 and x^{10}. Thus it was realised [14.38] that in eleven dimensions the entire configuration appears as intersecting M5-branes and indeed is a single M5-brane wrapped on a two-dimensional manifold which was a Riemann surface embedded in the four-dimensional space with coordinates x^4, x^5, X^6, X^{10}; see figure 14.7.2. Indeed, the Riemann surfaces that arise in this way are those known to be associated with the low energy effective action of the $N = 2$ supersymmetric Yang–Mills for the corresponding gauge groups.

We will now derive the complete low energy effective action of $N = 2$ supersymmetric Yang–Mills theory from the dynamics of the 5-brane [14.32], that is, from an eleven-dimensional perspective with a coordinate X^{10} which we assume describes a circle of radius R. The dynamics of the 5-brane was given in section 14.5 and a central role will be played by the 3-brane solutions that they possess which were discussed in section 14.6.1. We will see that they correspond to the wrapping of the 5-brane around the Riemann surface. We now consider a 3-brane solution, given in the section 14.6.1, whose world volume lies in the (x^0, x^1, x^2, x^3), the directions with x^4 and x^5 are the directions transverse to the 3-brane but in the world volume of the 5-brane. We take the directions X^7, X^8, X^9 to be constants leaving the directions X^6 and X^{10} to be active, that is, we take them to depend on the coordinates x^4 and x^5 transverse to the 3-brane. This corresponds to the configuration above for which we are trying to account. We recall that we are in static gauge and so $X^0 = x^0, \ldots, X^5 = x^5$ and so these directions, which are those in the world volume of the 5-brane, can be denoted by small or large symbols.

In section 14.6.1 we found that this preserves half of the 5-brane supersymmetries, that is, it leaves eight supersymmetries, if

$$\partial_4 X^6 = \partial_5 X^{10}, \qquad \partial_5 X^6 = -\partial_4 X^{10}, \tag{14.7.2}$$

and that the resulting theory is built out of the $N = 2$ vector multiplet on the four-dimensional world volume, (x^0, x^1, x^2, x^3), of the 3-branes. Adopting the complex notation $s = (X^6 + iX^{10})/R$ and $z = \Lambda^2 (x^4 + ix^5)$, where Λ is a mass scale, we recognise this last equation as the Cauchy–Riemann equation. Thus s is a function of z only. Although any choice of this function will solve the field equations, when we choose a specific solution we have a function which we can think of satisfying one complex equation in a space of two complex dimensions, that is, it defines a one complex dimensional surface, that is, a Riemann surface, in a two complex dimensional space.

We can think of the M5-brane as moving in the remaining eight-dimensional space with coordinates $x^0, \ldots, x^5, X^6, X^{10}$, in which $x^0, \ldots, x^3$ are flat $\mathbf{R}^4$ and x^4, x^5, X^6, X^{10} form a non-trivial four-dimensional space Q. The M5-brane field equations in the presence of these 3-branes then imply that the M5-brane is wrapped on a Riemann surface Σ which is embedded in the four-dimensional space Q. The volume of this Riemann surface is set by the scale R^2. Given the geometrical construction of the Riemann surface presented here, it is perhaps more natural to assign z the dimensions of length. However, we have assigned z the dimensions of mass in order to make contact with the literature on the low energy effective action of the $N = 2$ theory.

As such, we define $t = e^{-s}$ and consider a 3-brane configuration defined by

$$F(t, z) = 0, \tag{14.7.3}$$

where F is a complex polynomial.

Let us consider the IIA picture in ten dimensions obtained by taking the small R limit. In this case the parts of the 5-brane which lie on the circle in the X^{10} direction give rise to D4-branes in the $(x^0, x^1, x^2, x^3, X^6)$ plane, while the parts of the 5-brane that do not lie in the X^{10} direction lead to NS5-branes in the $(x^0, x^1, x^2, x^3, x^4, x^5)$ directions. If we think of $F(t, z) = 0$ as a polynomial in t for fixed z its zeros correspond to the positions of the NS5-branes (see fig 14.7.2) and to gain the IIA configuration originally discussed we take the polynomial to be quadratic in t. Alternatively if we think of the polynomial $F(t, z) = 0$ as a function of z for fixed t, then the zeros correspond to the positions of the 4-branes and so a polynomial of degree N in z corresponding to the presence of N 4-branes and the gauge group $SU(N)$. Thus our polynomial equation takes the form

$$F = t^2 - 2B(z)t + \Lambda^{2N} = 0, \tag{14.7.4}$$

where

$$B(z) = z^N + u_{N-1} z^{N-2} + \cdots + u_1 = \prod_{i=1}^{N} (z - e_i) \tag{14.7.5}$$

and the e_i are functions of the u_i and correspond to the positions of the N 4-branes. The e_i, or u_i, determine the Riemann surface and we can think of them as its moduli. The e_i are subject to $\sum_i^N e_i = 0$ as we have not taken a z^{N-1} term in B. This can be scaled away by a shift in z and so it corresponds to the centre of mass of the 4-branes which we can freeze to be a

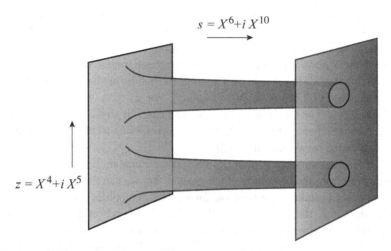

$$s = X^6 + i X^{10}$$

$$z = X^4 + i X^5$$

Figure 14.7.2   A single 5-brane wrapped on a Riemann sphere.

constant; it corresponds to the absence of the U(1) factor in the unusual gauge group U(N). We have rescaled t by Λ in order to keep equation (14.7.4) dimensionally correct. It is the Riemann surface considered in [14.38]. The choice of F of equation (14.7.4) corresponds to the scalar function

$$s = -\ln\left(B + \sqrt{Q}\right), \tag{14.7.6}$$

where we have introduced the polynomial

$$Q \equiv B^2(z) - \Lambda^{2N}. \tag{14.7.7}$$

We can also write the above equation in the more familiar form

$$y^2 = Q = B(z)^2 - \Lambda^{2N}, \tag{14.7.8}$$

where $y = t - B$. It is apparent from this equation that we are not taking any Riemann surface, but a hyperelliptic surface which can be defined in this way. It can be thought of as a sphere CP^1, with a set of branch cuts which begin and end at the zeros of the polynomial Q.

It is instructive to consider the 3-brane solution of equation (14.6.24) which is a collection of N 3-branes at the positions d_i, $i = 1, \dots, N$, with charges q_i

$$s = s_0 - \sum_i q_i \ln(z - d_i), \tag{14.7.9}$$

where s_0 is an arbitrary constant which we set to zero. The corresponding surface is

$$F(t, z) \equiv t^2 - \prod_i (z - d_i)^{2q_i} = 0, \tag{14.7.10}$$

where t has been suitably rescaled by Λ. For $q_i = \frac{1}{2}$ it is non-singular since we do not have any repeated roots.

Let us further describe the choice of F of equation (14.7.4) and its relation to 3-branes. If $\Lambda = 0$, then this solution is described by equation (14.7.10) with all the $q_i = 1$ and the positions d_i of the 3-branes are the N roots of B. As we have mentioned a z^{N-1} term in equation (14.7.5) corresponds to the centre of mass coordinate for the 3-branes (its coefficient u_N is the sum of the roots). In fact, including this term would produce an infinite contribution to the action we will construct and so has been frozen out. For a non-zero value of Λ the Riemann surface is generically non-singular, but the picture in terms of the simple 3-branes of equation (14.7.8) is slightly obscured. For $|z| \gg \Lambda$ we again see N distinct 3-branes located at d_i but for finite z this is not the case. In terms of the ten-dimensional type IIA picture the choice of equation (14.7.4) corresponds to two NS5-branes with N D4-branes suspended between them [14.38].

We now summarise the strategy that lies behind the calculation that we are about to do [14.32]. We recall from the beginning of the chapter that we can find the motion of slowly moving monopoles by taking the solution for several monopoles and letting the moduli of the solution, that is, the parameters of the solution, depend on time. We can think of this time as the world time of the particle that is in this case the monopole. Monopoles are solutions to the $N = 2$ supersymmetric SU(2) Yang–Mills theory that preserve half the supersymmetry of the theory. We can find the action for the motion of the monopoles by taking the action for this theory and substituting the solutions letting the moduli depend on time. For a review see [14.3]. In our case, we are interested in the low energy behaviour of the $N = 2$ Yang–Mills theory, which is the same as the low energy dynamics of the brane configuration described above. This is determined at low energy by the behaviour of the slowly moving 3-branes. Important for the application of this technique is that the solution for the 3-branes found previously is a static solution and so the 3-branes at rest experience no forces between them, just like the monopoles. As such we should take the solution above and let the parameters, in this case the moduli of the Riemann surface, depend on the world-volume coordinates $x^0, \ldots, x^3$ of the 3-brane. We put this correspondence in the table below:

theory	$N = 2$ super Yang–Mills	5-brane
solution	monopoles	3-branes
parameters	monopole moduli space	moduli of Riemann surface
coordinates	t	$x^0, \ldots, x^3$

The positions of the 3-branes are encoded in the variables u_i and so letting the positions become dependent on $x^0, \ldots, x^3$ is the same as letting the u_i depend on these coordinates.

We now wish to evaluate the dynamics [14.32] according to the above strategy. For simplicity we take only the scalar fields to be non-trivial and set the 3-form field of the M5-brane to zero. This has the consequence that we have set the vector fields in the four-dimensional effective theory to zero. For such a configuration the equation of motion is $g^{\hat{m}\hat{n}} \nabla_{\hat{m}} \nabla_{\hat{n}} s = 0$, where $\hat{m}, \hat{n} = 0, 1, \ldots, 5$. This equation of motion follows from the usual world-volume swept out action

$$I_5 = \int d^6 x \sqrt{-\det g_{\hat{m}\hat{n}}}, \qquad (14.7.11)$$

where, for a 3-brane configuration,

$$g_{\hat{m}\hat{n}} = \eta_{\hat{m}\hat{n}} + \tfrac{1}{2}(\partial_{\hat{m}} s \partial_{\hat{n}} \bar{s} + \partial_{\hat{m}} \bar{s} \partial_{\hat{n}} s). \qquad (14.7.12)$$

One can then evaluate the determinant in equation (14.7.11) to find

$$-\det g_{\hat{m}\hat{n}} = (1 + \tfrac{1}{2}\partial_{\hat{m}}s\partial^{\hat{m}}\bar{s})^2 - \tfrac{1}{4}|\partial_{\hat{m}\hat{n}}s\partial^{\hat{m}}s|^2. \tag{14.7.13}$$

This result is most easily found by considering the equation

$$-\det g_{\hat{m}\hat{n}} = \frac{1}{6!}\epsilon^{\hat{m}_1\cdots\hat{m}_6} g_{\hat{m}_1\hat{n}_1} \cdots g_{\hat{m}_6\hat{n}_6} \epsilon^{\hat{n}_1\cdots\hat{n}_6}, \tag{14.7.14}$$

substituting in the expression of equation (14.7.12) and expanding in powers of s. In the above expressions and below we have suppressed factors of R which appear with s. To obtain the low energy effective action for the 3-brane we will keep only terms of second order in the derivatives ∂_μ. Using the equation of motion of s one readily finds

$$I_5 = \tfrac{1}{2}\int d^6x\,\partial_\mu s\partial^\mu\bar{s}. \tag{14.7.15}$$

Letting the moduli depend on the world-volume coordinates of the 3-brane, that is, $u_i = u_i(x)$, the above action becomes

$$I_5 = \frac{1}{4i}\int d^4x\,\partial_\mu u_i\partial^\mu\bar{u}_j\int_\Sigma \omega^i\wedge\bar{\omega}^j, \tag{14.7.16}$$

where $\omega^i = -\partial s/\partial u_i\,dz$. We have used the change of coordinates $d\xi^1\wedge d\xi^2 = (i/2)dz\wedge d\bar{z}$ for the coordinates $x^4 \equiv \xi^1$ and $x^5 \equiv \xi^2$.

We now evaluate the effective action of equation (14.7.16). It is a straightforward calculation to see, using equation (14.7.6) that

$$\omega^i = -\frac{\partial s}{\partial u_i}dz = \frac{z^{i-1}}{\sqrt{Q}}dz \equiv \lambda^i, \tag{14.7.17}$$

where we recognise λ^i is the ith holomorphic form of the Riemann surface. Using the Riemann bilinear relation we may express the effective action as

$$I_5 = \frac{1}{4i}\int d^4x\,\partial_\mu u_i\partial^\mu\bar{u}_j\sum_{k=1}^{N-1}\left(\int_{A_k}\lambda^i\int_{B^k}\bar{\lambda}^j - \int_{A_k}\bar{\lambda}^j\int_{B^k}\lambda^i\right), \tag{14.7.18}$$

where A_k and B^k are a basis for the a and b cycles on the Riemann surface.

Next we note that $\partial_\mu u_i\lambda^i = \partial_\mu\lambda_{SW}$, where we have introduced the Seiberg–Witten differential λ_{SW} which satisfies [14.33]

$$\frac{\partial\lambda_{SW}}{\partial u_i} = \lambda^i. \tag{14.7.19}$$

Therefore if we define the new scalar modes

$$a_i = \int_{A_i}\lambda_{SW}, \qquad a_D^i = \int_{B^i}\lambda_{SW}, \tag{14.7.20}$$

we obtain the effective action

$$I_5 = \frac{1}{4i}\int d^4x\,\left(\partial_\mu a_k\partial^\mu\bar{a}_D^k - \partial_\mu\bar{a}_k\partial^\mu a_D^k\right) = -\frac{1}{2}\mathrm{Im}\left(\int d^4x\,\partial_\mu\bar{a}_i\partial^\mu a_j\tau^{ij}\right), \tag{14.7.21}$$

where $\tau^{ij} = \partial a_D^i / \partial a_j$. This is the Seiberg–Witten low energy effective action [14.33] for $N = 2$ supersymmetry SU(N) gauge theory spontaneously broken to U(1)$^{N-1}$. The Riemann surface of equation (14.7.4) is indeed that associated with this gauge group and hence the correct Seiberg–Witten differential for $N = 2$ SU(N) Yang–Mills theory. We note that

$$\tau^{ij} = \frac{\partial a_D^i}{\partial a_j} = \sum_k \frac{\partial a_D^i}{\partial u_k} \frac{\partial u_k}{\partial a_j} = \frac{\int_{B_i} \lambda^k}{\int_{A_k} \lambda^j} \tag{14.7.22}$$

using equations (14.7.19) and (14.7.20). As such we recognise τ^{ij} as the period matrix of the Riemann surface which is defined by

$$\int_{B_i} \lambda^k = \tau^{ik}, \qquad \int_{A_i} \lambda^j = \delta_{jk}. \tag{14.7.23}$$

The holomorphic 1-forms are chosen such that the latter equation holds.

Although we have only derived the scalar part of the full Seiberg–Witten effective action for $N = 2$ SU(N) Yang–Mills theory, the vector and spinor terms are determined from this term by $N = 2$ supersymmetry. The vector terms have been determined [14.39] using an extension of the above calculation but using a non-trivial 3-form gauge field strength on the 5-brane. This calculation makes use of some of the more subtle features of the 5-brane dynamics and in particular the complications of the self-duality condition on the tree form field strength.

We note that we have derived the complete *quantum* non-perturbative Seiberg–Witten effective action from the *classical* dynamics of the M theory 5-brane. Given our conventional understanding of the relationship between classical and quantum theories it is surprising to see such a detailed and complete connection. This can be seen as a consequence of the relation between the IIA theory and the eleven-dimensional, or M theory. Although M theory has only one scale M_{Planck}, the solution for the 3-branes introduces two more scales R and Λ, which occur as integration constants. Reintroducing $\hbar$ we would find that it occurs in combination with these integration constants. Since these constants, and so $\hbar$, do not play the role of a perturbative parameter in the derivation we have given, from the M theory perspective it is perhaps not so surprising to have found the full non-perturbative Seiberg–Witten effective action.

We close this section by giving some details of the Seiberg–Witten effective action so that the connection becomes clearer. For simplicity we consider only the case of gauge group SU(2) spontaneously broken to U(1) and as a result we have only one moduli u, one a and one a_D. The action of the $N = 2$ supersymmetric U(1) theory which has only two space-time derivatives is given in $N = 2$ superspace by

$$\int d^4x d^4\theta F(A) + \text{c.c.}, \tag{14.7.24}$$

where $x^\mu, \theta^{\alpha i}, i = 1, 2$ are the coordinates of $N = 2$ superspace and A is a $N = 2$ chiral superfield (see [1.11] for a review). Due to the latter condition the integral is over only half of the $N = 2$ superspace. The complete perturbative contribution has

$$F = \frac{\tau_{cl}}{2} A^2 + \frac{i}{2\pi} k A^2 \ln \frac{a^2}{\Lambda^2}, \tag{14.7.25}$$

where $\tau_{cl} = \theta + i(4\pi/g^2)$ where θ is the theta angle, g is the gauge coupling, Λ is a renormalisation scale and k is a constant which is related to the beta function which has only a one-loop contribution [14.34, 14.35].

The $N = 2$ superfield A contains two $N = 1$ superfields, the chiral superfield φ and the, also chiral, field strength W_α. These contain as x-space fields two scalars, denoted a and $\bar{a}$, and the spinor of the Wess–Zumino multiplet and the vector and spinor of the gauge multiplet, respectively. The action of equation (14.7.24) is given in terms of $N = 1$ superfields by

$$\frac{1}{4\pi}\text{Im}\left\{\int d^4x d^4\theta \frac{dF}{d\varphi}\bar{\varphi} + \int d^4x d^2\theta \frac{1}{2}\frac{d^2F}{d\varphi^2}W_\alpha W^\alpha\right\}. \qquad (14.7.26)$$

The scalar contribution comes from the first term, which, neglecting the spinor and auxiliary field contributions, can be written (see [1.11]) as

$$\frac{1}{16 \times 4\pi}\text{Im}\left\{\int d^4x D^2\bar{D}^2\left(\frac{dF}{d\varphi}\bar{\varphi}\right)\right\} = \frac{1}{16 \times 4\pi}\text{Im}\left\{\int d^4x \frac{dF}{d\varphi}D^2\bar{D}^2\bar{\varphi}\right\}$$

$$= \frac{1}{4\pi}\text{Im}\left\{\int d^4x \frac{dF}{da}\partial^2\bar{a}\right\} = -\frac{1}{4\pi}\int d^4x \partial_\mu a \partial^\mu \bar{a}\left(\frac{d^2F}{da^2} - \frac{d^2\bar{F}}{d\bar{a}^2}\right). \qquad (14.7.27)$$

Comparing with equation (14.7.21) we recognise that

$$\tau = \frac{d^2F}{da^2}, \qquad (14.7.28)$$

which specifies the function F in terms of the period of the Riemann surface. The Seiberg–Witten action contains an infinite number of instanton corrections.

15 D-branes

You do not know the meaning of novelty until you hear me speak.
Elizabeth Gaskell from her book *Cranford* as serialised by the BBC

We will begin by showing that string theory allows a more general kind of open string than those considered in chapters 3 and 4 and their supersymmetric extensions in chapters 6 and 7. The main difference is that they obey different boundary conditions at the ends of the string. The open strings we studied earlier had the boundary condition $(\partial x^\mu/\partial\sigma)(\tau, \sigma) = 0$ at their end points for all μ; however, one can have open strings that have ends points with the boundary condition $(\partial x^\mu/\partial\tau)(\tau, \rho) = 0$ in the some, or all, directions. The above boundary conditions correspond to setting the derivative of $x^\mu(\tau, \sigma)$ perpendicular and normal to the boundary to zero. Traditionally these types of boundary conditions are called Neumann (N) or Dirichlet (D) respectively. Corresponding to the different boundary conditions we find different types of open strings. At a given end, and in a given spatial direction, we can impose either N or D boundary conditions. As such, we can have open strings with D boundary conditions at both ends in certain spatial directions and N boundary conditions at each end for the remainder directions. However, we will also consider strings, which for a given direction obey N boundary conditions at one end and D boundary conditions at the other end.

One obvious consequence of the D boundary condition is that the ends of the open string are fixed in time for those directions in which we impose the D boundary conditions. Given a sub-manifold M of the background space-time $\underline{M}$ through which the open strings move we may choose our boundary conditions such that one or both of the ends of the open string are fixed to be on M. Put another way the end points of the open string can be thought as defining the sub-manifold M. The importance of these new types of open strings stems from the interpretation of M as one of the p-branes discussed in chapter 14. In this case, it turns out that the interactions of the open strings describe the dynamics of the p-brane. The p-branes are non-perturbative objects, indeed they arise as solitonic solutions of the low energy effective actions of the corresponding string theories which for superstring theories are the corresponding supergravity theories. However, the interactions of these open strings can be calculated using standard open string perturbation theory. As such, these open strings allow the computation of brane dynamics using open string perturbation theory. Branes for which we can use this description are called D-branes and their low energy degrees of freedom include rank 1 gauge fields in their world volume, as well as the usual coordinates describing their embedding in space-time $\underline{M}$. It turns out that supersymmetric D-branes only couple to gauge fields the belong to the R–R sector of the closed string background. We found in chapter 14 that the p-branes that couple to the R–R sector have p even in the

460

IIA string and p odd in the IIB string and that they all had rank 1 gauge fields living on their world volumes.

Open strings with D boundary conditions at one end and N boundary conditions at the other were first formulated in [15.1] in the context of off-shell string theory. However, open strings and their connection with D-branes, as discussed above, were first formulated in the context of the bosonic string [15.2, 15.3] and then in the superstring [15.4]. It was in [15.4] that the relation to R–R fields was observed and D-branes became the subject of extensive study.

The interplay between open strings and the D-branes to which they are attached has had deep implications for our understanding of strings, Yang–Mills theories and quantum field theories in general. One of the most interesting effects is that N coincided parallel D-branes have a world volume theory that possesses a U(N) Yang–Mills symmetry. This is central to the conjectured equivalence, the AdS/CFT conjecture, [15.5–15.7] between IIB string theory on $AdS_5 \otimes S^5$ and the $N = 4$ Yang–Mills theory for which there is now very considerable evidence.

15.1 Bosonic D-branes [15.2, 15.3]

Let us consider an open string whose action is given by equation (2.1.10), that is,

$$A = -\frac{1}{4\pi\alpha'} \int d^2\xi \sqrt{-g} g^{\alpha\beta} \partial_\alpha x^\mu \partial_\beta x^\nu \eta_{\mu\nu}. \tag{15.1.1}$$

This string moves between two configurations which are fixed at an initial and a final time in such a way as to extremise this action. Following the same discussion up to equation (2.1.15) we find its variation is given by

$$\delta A = \int d^2\xi \left\{ \frac{\delta A}{\delta g_{\alpha\beta}} \delta g_{\alpha\beta} + \frac{\delta A}{\delta \partial_\alpha x^\mu} \delta(\partial_\alpha x^\mu) \right\} = 0. \tag{15.1.2}$$

As in chapter 2, the first term leads to equation (2.1.20), that is, $T_{\alpha\beta} = 0$, and it plays no further part in equation (15.1.2). To treat the second term we must integrate by parts and the equation becomes

$$0 = \int d^2\xi \left\{ -\delta x^\mu \partial_\alpha \frac{\delta A}{\delta(\partial_\alpha x^\mu)} + \partial_\alpha \left(\delta x^\mu \frac{\delta A}{\delta(\partial_\alpha x^\mu)} \right) \right\}. \tag{15.1.3}$$

The first term in this equation leads to the equation of motion,

$$\delta A / \delta(\partial_\alpha x^\mu) = 0, \tag{15.1.4}$$

of equation (2.1.21) and we are left with the last term which is a boundary term. The boundary is just the two curves swept out by the two ends of the string which in previous chapters we have taken to be at $\sigma = 0$ and $\sigma = \pi$. The boundary is then parameterised by τ. It will prove advantageous to treat the boundary in a more general setting: we take Σ to be the surface swept out by the string and $\partial\Sigma$ to be its boundary. We take ξ^α to be the coordinates of Σ and the boundary to be parameterised by s, that is, $\xi^\alpha(s)$. The tangent to the boundary is given by $d\xi^\alpha/ds$.

Then the last term in equation (15.1.3) can be written as

$$0 = \int_{\partial\Sigma} v = \int_{\partial\Sigma} ds \frac{d\xi^\alpha}{ds} \delta x^\mu \varepsilon_{\alpha\beta} \frac{\delta A}{\delta(\partial_\beta x^\mu)}, \tag{15.1.5}$$

where v is the 1-form $v = d\xi^\alpha \delta x^\mu \varepsilon_{\alpha\gamma} (\delta A/\delta \partial_\gamma x^\mu)$. In deriving this result we have used Green's theorem

$$\int_\Sigma dv = \int_{\partial\Sigma} v \tag{15.1.6}$$

and

$$dv = \partial_\beta \left\{ \varepsilon_{\alpha\gamma} \delta x^\mu \frac{\delta A}{\delta \partial_\gamma x^\mu} \right\} d\xi^\beta \wedge d\xi^\alpha = d^2\xi \partial_\gamma \left(\delta x^\mu \frac{\delta A}{\partial \partial_\gamma x^\mu} \right), \tag{15.1.7}$$

where $d^2\xi = d\xi^0 \wedge d\xi^1$ and therefore $d\xi^\alpha \wedge d\xi^\beta = \varepsilon^{\alpha\beta} d^2\xi$.

As in chapter 2, we choose the gauge $g_{\alpha\beta} = e^{\hat{\phi}} \eta_{\alpha\beta}$ and then the equations of motion are those of equations (2.1.26) and (2.1.27) and

$$\frac{\delta A}{\delta(\partial_\beta x^\mu)} = -\frac{1}{2\pi\alpha'} \partial^\beta x^\mu. \tag{15.1.8}$$

As such, the boundary condition of equation (15.1.5) becomes

$$0 = \int_{\partial\Sigma} ds \frac{d\xi^\alpha}{ds} \delta x^\mu \varepsilon_{\alpha\beta} \partial^\beta x_\mu, \tag{15.1.9}$$

which we must take to vanish separately at each the end of the string as δx^μ is independent at both ends. One way to satisfy this boundary condition is to take the derivative of x^μ normal to the boundary to vanish:

$$\frac{d\xi^\alpha}{ds} \varepsilon_{\alpha\beta} \partial^\beta x^\mu \Big|_{\partial\Sigma} = 0. \tag{15.1.10}$$

Such a boundary condition is called an N boundary condition.

Taking the usual parameterisation of the open string (that is, $0 \le \sigma \le \pi$; $-\infty < \tau < \infty$), the boundary $\partial\Sigma$ is just the positions of the two ends of the open string at $\sigma = 0$ and $\sigma = \pi$ and the parameter s can be taken to be τ. In this case, at $\sigma = 0$, the above equation becomes

$$\frac{\partial x^\mu}{\partial\sigma}(\tau, 0) = 0. \tag{15.1.11}$$

If we impose it at the end of the string with $\sigma = 0$ and at the end with $\sigma = \pi$ and in all directions, then we have the boundary conditions of equations (2.1.23), that is, the open strings we studied in earlier chapters.

Let us consider x^μ to have its derivative along the boundary and in the direction μ to vanish:

$$\frac{d\xi^\alpha}{ds} \partial_\alpha x^\mu \Big|_{\partial\Sigma} = 0. \tag{15.1.12}$$

This is called a D boundary condition. In terms of the usual parameterisation it becomes at $\sigma = 0$ say,

$$\frac{\partial x^\mu}{\partial\tau}(\tau, 0) = 0. \tag{15.1.13}$$

This does not at first sight appear to satisfy the boundary condition of equation (15.1.9). However, we should take the variation δx^μ to satisfy the same boundary condition, that is, $(\partial/\partial\tau)\delta x^\mu(\tau, \sigma) = 0$ at $\sigma = 0$. Hence $\delta x^\mu(\tau, 0)$ is a constant. However, the variation of the action is carried out with $\delta x^\mu(\tau, \sigma) = 0$ at $\tau = \tau_i$ and $\tau = \tau_f$, where τ_i and τ_f are the initial and final times for the motion of the string. Imposing this latter requirement at the ends of the string and the above condition on δx^μ, we find that

$$\delta x^\mu(\tau, 0) = 0 = \delta x^\mu(\tau, \pi) \qquad (15.1.14)$$

and so the above boundary term of equation (15.1.9) also vanishes for the D boundary condition. The analogous discussion at the other end of the string, that is, $\sigma = \pi$, is identical.

One of the most important applications of the above open string is to describe fluctuations of D-branes. We recall from chapter 14 that a p-brane sweeps out a $(p + 1)$-dimensional world volume M in a background space-time $\underline{M}$. Let us consider to begin with a static p-brane whose world volume lies in the directions $\mu = 0, 1, \ldots, p$, which we call the longitudinal directions. The transverse directions are therefore $\mu = p + 1, \ldots, D - 1$. We then consider an open string that has both ends fixed on the above p-brane. In this case, the ends of the open string are fixed in the transverse directions, but their positions in the longitudinal directions are unspecified. As such, we adopt D type boundary conditions in the transverse directions and N type boundary conditions in the longitudinal directions:

$$\frac{\partial x^\mu}{\partial \sigma}(\tau, 0) = 0 = \frac{\partial x^\mu}{\partial \sigma}(\tau, \pi); \quad \mu = 0, 1, \ldots, p \qquad (15.1.15)$$

and

$$\frac{\partial x^\mu}{\partial \tau}(\tau, 0) = 0 = \frac{\partial x^\mu}{\partial \tau}(\tau, \pi); \quad \mu = p + 1, \ldots, D - 1. \qquad (15.1.16)$$

The D boundary conditions just state that the ends are fixed in these directions. However, for this particular application both ends of the string are fixed to the same surface and so we may replace the last condition by

$$x^\mu(\tau, 0) = a^\mu = x^\mu(\tau, \pi), \quad \mu = p + 1, \ldots, D - 1, \qquad (15.1.17)$$

where a^μ are the transverse coordinates that specify the position of the p-brane. The motion of such an open string is depicted in figure 15.1.1, where the world volume of the p-brane is drawn as a plane for simplicity. We note that the presence of the p-brane, or equivalently the different boundary conditions, break the $SO(1, D - 1)$ Lorentz symmetry of the background space to $SO(1, p) \times SO(D - p - 1)$.

We can also consider two p-branes which are parallel and have the same longitudinal and transverse directions as above. In this case, in addition to the open strings which begin and end on the same brane we can have, as shown in figure 15.1.2, open strings that begin on one brane and end on the second brane. In this case the open strings obey the same boundary conditions of equations (15.1.15) and (15.1.16). However, if one brane is at a_1^μ, $\mu = p + 1, \ldots, D - 1$ and the other at a_2^μ, $\mu = p + 1, \ldots, D - 1$, then the D boundary conditions may be written as

$$x^\mu(\tau, 0) = a_1^\mu, \; x^\mu(\tau, \pi) = a_2^\mu, \; \mu = p + 1, \ldots, D - 1, \qquad (15.1.18)$$

while the N boundary conditions are still given by equation (15.1.15).

Figure 15.1.1   An open string and a single D-brane.

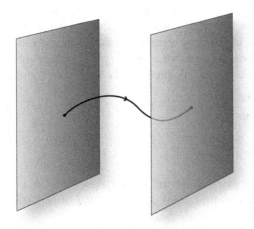

Figure 15.1.2   An open string stretched between two parallel D-branes.

In the above examples both ends of the open string have the same type of boundary conditions. However, this is not the case if we consider open strings that end on p-branes that are not parallel. To be concrete let us consider two 4-branes D_1 and D_2 which have their world-volume coordinates in the 01236 and 01456 directions, respectively. As shown in figure 15.1.3, let us consider open strings that have their end at $\sigma = 0$ on D_1 and their end at $\sigma = \pi$ on D_2. In this case the string obeys the boundary conditions

$$\frac{\partial x^\mu}{\partial \sigma}(\tau, 0) = 0, \quad \mu = 0, 1, 2, 3, 6, \quad \frac{\partial x^\mu}{\partial \tau}(\tau, \pi) = 0; \quad \mu = 4, 5, 7, \ldots, D-1$$

$$(15.1.19)$$

and

$$\frac{\partial x^\mu}{\partial \sigma}(\tau, \pi) = 0, \quad \mu = 0, 1, 4, 5, 6, \quad \frac{\partial x^\mu}{\partial \tau}(\tau, \pi) = 0; \quad \mu = 2, 3, 7, \ldots, D-1.$$

$$(15.1.20)$$

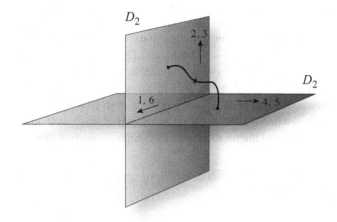

Figure 15.1.3   An open string between two non-parallel D-branes.

We now analyse the behaviour of the above open strings much as in chapter 2, but taking into account the different boundary conditions. We can impose on a given end, and for each direction, either D or N boundary conditions; however, as the above illustrates, the choice of boundary condition corresponds to the particular branes on which the open string begin and end.

Let us begin by giving the most general mode expansion regardless of the particular boundary conditions. In general, the strings will not end and begin on the same brane and we denote the difference in positions of the end points of the string by $x^\mu(\tau, \pi) - x^\mu(\tau, 0) \equiv -f^\mu(\tau)\pi$. It then follows that the variable $\hat{x}^\mu(\tau, \sigma) = x^\mu(\tau, \sigma) + f^\mu(\tau)\sigma$ satisfies the condition $\hat{x}^\mu(\tau, 0) = \hat{x}^\mu(\tau, \pi)$. As a result, we may apply Fourier's theorem to $\hat{x}^\mu(\tau, \sigma)$ and so write $x^\mu(\tau, \sigma)$ as

$$x^\mu(\tau, \sigma) = x_0^\mu(\tau) + \sum_{\substack{n \in Z \\ n \neq 0}} x_n^\mu(\tau)e^{in\sigma} + \sum_{r \in Z + \frac{1}{2}} y_r^\mu(\tau)e^{ir\sigma} + f^\mu(\tau)\sigma. \quad (15.1.21)$$

The presence of the half-integer modes results from applying Fourier's theorem to the range $0 \leq \sigma \leq \pi$ while this theorem is usually stated for the range $0 \leq \sigma < 2\pi$.

As we discussed above $x^\mu(\tau, \sigma)$ satisfies the wave equation, which implies that the Fourier modes obey the equations $\ddot{x}_n + n^2 x_n = 0 = \ddot{y}_r + r^2 y_r$. Taking both solutions, introducing new variables and suitable constants we find that we may write

$$x^\mu(\tau, \sigma) = a^\mu(0) + k^\mu(0)z + l^\mu(0)\bar{z} + i\sqrt{\frac{\alpha}{2}} \sum_{\substack{n \in Z \\ n \neq 0}} (\alpha_n^\mu(0)e^{-inz} + \hat{\alpha}_n^\mu(0)e^{-in\bar{z}})\frac{1}{n}$$

$$+ i\sqrt{\frac{\alpha'}{2}} \sum_{r \in Z + \frac{1}{2}} (c_r^\mu(0)e^{-irz} + \hat{c}_r^\mu(0)e^{-ir\bar{z}})\frac{1}{r}, \quad (15.1.22)$$

where

$$z = \tau + \sigma, \quad \bar{z} = \tau - \sigma. \quad (15.1.23)$$

We note that because this is a chapter that contains fermions we use z and $\bar{z}$ rather than ξ^+ and ξ^-.

The ends of the string obey N or D boundary conditions. Let us denote the directions that obey N and D boundary conditions at $\sigma = 0$ by N_0 and D_0, respectively, and similarly at the other end at $\sigma = \pi$ by N_π and D_π. To give an example, for the boundary conditions of equation (15.1.19) we have $\mu = 0, 1, 2, 3, 6 \in N_0$ while $\mu = 4, 5, 7, \ldots \in D_0$. If we impose the Neumann boundary conditions in the μth direction at both ends of the string, that is, $\mu \in N_0$ and $\mu \in N_\pi$, then we find that $k^\mu = l^\mu$, $\alpha_n^\mu = \hat{\alpha}_n^\mu$ and $c_r^\mu = \hat{c}_r^\mu = 0$. Identifying the centre of mass coordinate q^μ and the momentum as in chapter 2 we find that

$$x^\mu(\tau, \sigma) = q^\mu(0) + \frac{1}{2}\sqrt{2\alpha'}\alpha_0^\mu(0)(z + \bar{z}) + i\sqrt{\frac{\alpha'}{2}} \sum_{\substack{n \in Z \\ n \neq 0}} \frac{\alpha_n^\mu(0)}{n}(e^{-inz} + e^{-in\bar{z}}).$$

(15.1.24)

The oscillators are defined by equation (2.1.61). If we adopt N boundary conditions in all directions we would discover the usual open bosonic string and agreement with equation (2.1.79).

Let us now examine the effect of imposing the D boundary conditions at both ends of the string in the μth direction, that is, $\mu \in D_0$ and $\mu \in D_\pi$. In this case, we find that $\alpha_n^\mu = -\hat{\alpha}_n^\mu$, $c_r^\mu = \hat{c}_r^\mu = 0$, $k^\mu = -l^\mu$ and so

$$x^\mu(\tau, \sigma) = a_1^\mu + \frac{1}{2\pi}(a_2^\mu - a_1^\mu)(z - \bar{z}) + i\sqrt{\frac{\alpha'}{2}} \sum_{\substack{n \in Z \\ n \neq 0}} \frac{1}{n}\alpha_n^\mu(0)(e^{-inz} - e^{-in\bar{z}}).$$

(15.1.25)

Since $z - \bar{z} = 2\sigma$ we find that $x^\mu(\pi) = a_2^\mu$, as required by equation (15.1.18). The above expression is consistent with the fact that $x^\mu(\tau, \sigma) - (a_1^\mu + \frac{1}{2}(a_2^\mu - a_1^\mu)(z - \bar{z}))$ is a periodic function due to the boundary conditions of equation (15.1.18). We note that there is no dependence linear in τ and so no centre of mass momentum. There is also no centre of mass coordinate in the μth direction as a_1^μ and a_2^μ are constants. This fact will play an important role in what follows. Defining the oscillators by equation (2.1.61) we recognise that $\alpha_0^\mu = (1/\pi\sqrt{2\alpha'})(a_2^\mu - a_1^\mu)$.

There are two further possibilities corresponding to imposing mixed boundary conditions in a given direction. We can either impose in the μth direction N boundary conditions at $\sigma = 0$ and D boundary conditions at $\sigma = \pi$, that is, $\mu \in N_0$ and $\mu \in D_\pi$, or D boundary conditions at $\sigma = 0$ and N boundary conditions at $\sigma = \pi$, that is, $\mu \in D_0$ and $\mu \in N_\pi$. In the former case, we find that $k^\mu = l^\mu = \alpha_n^\mu = \hat{\alpha}_n^\mu = 0$ leading to

$$x^\mu(\tau, \sigma) = a_2^\mu + i\sqrt{\frac{\alpha'}{2}} \sum_{r \in Z + \frac{1}{2}} \frac{1}{r}c_r(0)(e^{-irz} + e^{-ir\bar{z}}),$$

(15.1.26)

while in the latter case

$$x^\mu(\tau, \sigma) = a_1^\mu + i\sqrt{\frac{\alpha'}{2}} \sum_{r \in Z + \frac{1}{2}} \frac{1}{r}c_r(0)(e^{-irz} - e^{-ir\bar{z}}).$$

(15.1.27)

Thus for mixed boundary conditions we find that $x^\mu(\tau, \sigma)$ contains a new type of oscillator first introduced into string theory in [15.1]. We note that for mixed boundary conditions we have no centre of momenta or coordinates.

By computing the momentum $P^\mu = -(1/2\pi\alpha')\dot{x}^\mu$ and imposing that the only non-vanishing fundamental Poisson bracket takes the form

$$\{x^\mu(\tau, \sigma), P^\nu(\tau, \sigma')\} = \eta^{\mu\nu}\delta(\sigma - \sigma'), \tag{15.1.28}$$

we conclude that the oscillators of all the above expansions of equations of $x^\mu(\tau, \sigma)$ obey the Poisson brackets

$$\{\alpha_n^\mu, \alpha_m^\nu\} = -in\delta_{n+m,0}\eta^{\mu\nu}, \quad \{c_r^\mu, c_s^\nu\} = -ir\delta_{r+s,0}\eta^{\mu\nu}, \tag{15.1.29}$$

with all other Poisson brackets vanishing. One has to carry out this calculation separately for each type of boundary condition.

The quantisation of the above open strings proceeds much as for the usual open bosonic string discussed in chapter 3. Turning the Poisson brackets of equation (15.1.29) into commutators one finds that α_n^μ and c_r^μ oscillators obey the commutator relations

$$[\alpha_n^\mu, \alpha_m^\nu] = n\eta^{\mu\nu}\delta_{n+m,0}, \quad [c_r^\mu, c_s^\nu] = r\eta^{\mu\nu}\delta_{r+s,0}, \tag{15.1.30}$$

with all other commutators between them vanishing. Classically, the open strings are subject to the constraints $T_{\alpha\beta} = 0$ resulting from the $g_{\alpha\beta}$ equation of motion. However, as in chapter 3, in the quantum theory we implemented only some of them on the wavefunction. We therefore adopt the physical state conditions,

$$L_n\psi = 0 \quad n \geq 1, \quad (L_0 - a)\psi = 0, \tag{15.1.31}$$

where the intercept $a = (D-2)/24 - e/16$ and e is the total number of N–D and D–N directions. This particular value will be justified in the next section. The L_n are defined in the same way in terms of x^μ as in chapter 3, that is, by equation (2.1.77), and are given in terms of oscillators by

$$L_n = \tfrac{1}{2} \sum_{\mu,\nu \,\in\, N-Nor D-D} \sum_{m\in\mathbf{Z}} : \alpha_m^\mu \alpha_{n-m}^\nu \eta_{\mu\nu} :$$

$$+ \tfrac{1}{2} \sum_{\mu,\nu \,\in\, N-Dor D-N} \sum_{s\in\mathbf{Z}+\frac{1}{2}} : c_s^\mu c_{n-s}^\nu \eta_{\mu\nu} :, \tag{15.1.32}$$

where the sum over μ depends as indicated on the boundary conditions at each end. We find that

$$L_0 = \frac{1}{4\pi^2\alpha'} \sum_{\mu,\nu \,\in\, D-D} (a_2^\mu - a_1^\mu)(a_2^\nu - a_1^\nu)\eta_{\mu\nu} + \frac{1}{2} \sum_{\mu,\nu \,\in\, N-N} \alpha_0^\mu \alpha_0^\nu \eta_{\mu\nu}$$

$$+ \sum_{\mu,\nu \,\in\, N-Nor D-D} \sum_{m\in\mathbf{Z}, m\neq 0} : \alpha_{-m}^\mu \alpha_m^\nu \eta_{\mu\nu} :$$

$$+ \sum_{\mu,\nu \,\in\, N-Dor D-N} \sum_{s\in\mathbf{Z}+\frac{1}{2}} : c_{-s}^\mu c_s^\nu \eta_{\mu\nu} : . \tag{15.1.33}$$

The first term arises due to the energy in the string stretched between the two D-branes at positions a_1^μ and a_2^μ, the second term contains the centre of mass momentum for directions in which there are N–N boundary conditions and the rest of the expression just counts the number of oscillators.

As for the more usual open bosonic string we may write any functional Ψ of $x^\mu(\sigma)$ in terms of the oscillators acting on a vacuum, but for the branes considered here we must include the c_r^μ oscillators as well as the usual oscillators if the string has directions with mixed boundary conditions. For an open string with only N–N, or D–D boundary conditions, that is, an open string that has both ends on the same p-brane, or has its ends on parallel branes, we may expand the wavefunction Ψ as follows;

$$\psi = \left\{ \phi(x^\mu) + \sum_{\nu \in N-N} \alpha^\nu_{-1} A_\nu(x^\mu) + \sum_{\nu \in D-D} \alpha^\nu_{-1} \varphi_\nu(x^\mu) + \cdots \right\} \psi_0, \quad (15.1.34)$$

where ψ_0 is annihilated by all the oscillators α^μ_n $n \geq 1$. The component fields $\phi, A_\mu, \varphi_\mu, \ldots$ just depend on the centre of mass coordinates $x^\mu_0 \equiv x^\mu$ in the N–N directions, that is, for $\mu \in N-N$, since, as we discovered above, there are no centre of mass coordinates in the D–D directions. We recall from chapter 14 that for a p-brane we can choose static gauge. In this case, this choice identifies $x^\mu = \xi^\mu$, $\mu \in N-N$. It follows that the component fields ϕ, φ_μ, and A_μ can therefore be thought of as living on the p-brane on which the open strings end. The decomposition of the sums into two sums over $\mu \in N-N$ and $\mu \in D-D$ corresponds to the breaking of the Lorentz group by the presence of the p-brane. The Lorentz group SO(1, $D-1$) of the background space-time is broken into SO($D-p-1$) $\times$ SO(1, p), where SO(1, p) is the Lorentz group of the p-brane. The above fields live in the $(p+1)$-dimensional space-time of the brane and so have Lorentz group SO(1, p) and have SO($D-p-1$) as an internal group. The fields A_μ, $\mu \in N-N$ transform as a vector of the brane Lorentz group and are inert under the internal group. However, the fields φ_μ, $\mu \in D-D$ are Lorentz scalars under the brane Lorentz group, but transform as vectors under the internal group SO($D-p-1$). The latter can be thought of as the transverse coordinates which describe the embedding of the brane in its background space-time. That is, if $x^\mu(\xi)$, $\mu = 0, \ldots, D-1$ describe the embedding of the brane in the background space-time, and we choose the N–N directions to be $\mu = 0, \ldots, p$, then the static gauge allows us to set $x^\mu = \xi^\mu$, $\mu = 0, \ldots, p$ and we identify $x^\mu = \varphi^\mu(\xi)$, $\mu = p+1, \ldots, D-1$.

Solving the physical state conditions of equation (15.1.31), in 26 dimensions, we find that for a single D-brane on which the open strings begin. or end, ϕ is a tachyon, A_μ, $\mu \in N-N$ is a world-surface massless vector and φ_μ, $\mu \in D-D$ are $D-p-1$ massless scalars on the world surface. The $L_1\psi = 0$ condition implies that $\partial^\mu A_\mu = 0$.

Now let us consider the case of N parallel D-branes which we label $i, j = 1, \ldots N$. There are now N^2 possible open strings connecting the N D-branes labelled by $[i, j]$ if the open string starts at brane i at $\sigma = 0$ and ends at brane j at $\sigma = \pi$. As we are dealing with unoriented strings, the open string that begins on brane i at $\sigma = 0$ and ends on brane j at $\sigma = \pi$ is distinguished from the open string $[j, i]$ which begins and ends in the opposite order. As such we have N^2 copies of the states that we have above; see figure 15.1.4. If the branes are coincident, then $a_1^\mu = a_2^\mu$ the additional term in L_0 in equation (15.1.33) vanishes and the states are still massless. In fact, they belong to the U(n) Yang–Mills theory coupled to $26 - p - 1$ scalars [15.8].

In the old days it was usual to introduce the so-called Chan–Patton factors to label the ends of the open string, but with the introduction of D-branes the labelling comes naturally due to the presence of the different D-branes on which the open strings can end. Indeed, for

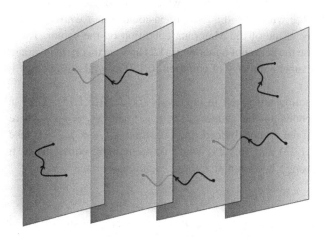

Figure 15.1.4   Open strings between N parallel D-branes.

the Yang–Mills theory in 26 dimensions one can consider them to end on space-time filling D-branes. It was known from the early days of string theory that scattering open strings with Chan–Patton factors leads at low energy to the Yang–Mills theory [3.1].

If the N parallel D-branes are not coincident, then L_0 in equation (15.1.33) has a first term which is non-zero and the masses are shifted upwards to become

$$(\text{mass})^2 = -p^2 = -\frac{1}{2\alpha'}(\alpha_0^\mu)^2$$

$$= \frac{1}{4\pi^2\alpha'^2} \sum_{\mu,\nu \,\in\, D-D} (a_2^\mu - a_1^\mu)(a_2^\nu - a_1^\nu)\eta_{\mu\nu} + \frac{1}{\alpha'}(\text{number operator} - 1).$$

$$(15.1.35)$$

Hence although we have the same number of states their masses are not the same as for the usual open bosonic string studied in chapter 3. The above states shown in equation (15.1.34) now have mass squared given by

$$\frac{1}{4\pi^2\alpha'^2} \sum_{\mu,\nu \,\in\, D-D} (a_2^\mu - a_1^\mu)(a_2^\nu - a_1^\nu)\eta_{\mu\nu}. \qquad (15.1.36)$$

In fact they make up a massive vector with p on-shell states and $26 - p - 2$ scalars. We recall that the states of a massive vector belong to the vector representation of the little group, which in this case is $SO(p)$ corresponding to the brane Lorentz group $SO(p+1)$. Hence, one of the scalars has become part of the vector to allow for the increase by one of the on-shell states of a vector when it becomes massive. This is similar to the mechanism of spontaneous symmetry breaking.

15.2   Super D-branes in the NS–R formulation

The above discussion of D-branes can be carried over to the case of open superstrings provided we extend the discussion of the boundary conditions on the bosons x^μ to the fermions. In this section, we use the Neveu–Schwarz–Ramond formulation given in chapter 6. Consequently, we work with x^μ and a two-dimensional spinor whose two components, χ^μ_+ and χ^μ_-, are real by virtue of the Majorana condition on the spinor. The superstring moves so as to extremise the action of equation (6.1.2) and the theory is invariant under the two-dimensional local supersymmetry of equation (6.1.7) as well as the other symmetries discussed in that chapter.

The equations of motion are given in equations (6.1.12)–(6.1.15). They are considerably simplified by choosing the gauge $g_{\alpha\beta} = \eta_{\alpha\beta}e^\phi$ and $\psi_\alpha = \gamma_\alpha\zeta$, where ζ is any two-dimensional Majorana spinor. Indeed, the constraints $T_{\alpha\beta} = j_{\alpha A} = 0$ take the form given in equations (6.1.23) and (6.1.24) while the remaining equations of motion are $\partial\!\!\!/\chi^\mu = \partial_\alpha\partial^\alpha x^\mu = 0$. Using z and $\bar z$ notation these equations become

$$\bar\partial\chi^\mu_+ = 0 = \partial\chi^\mu_-, \quad \partial\bar\partial x^\mu = 0. \tag{15.2.1}$$

When we choose the above gauge choice, the local supersymmetry of equation (6.1.7) becomes reduced to the superconformal algebra whose transformations are given by

$$\delta x^\mu = i(\varepsilon_-\chi^\mu_+ - \varepsilon_+\chi^\mu_-),$$
$$\delta\chi^\mu_+ = 2\partial x^\mu\varepsilon_-, \quad \delta\chi^\mu_- = -2\bar\partial x^\mu\varepsilon_+, \tag{15.2.2}$$

where

$$\bar\partial\varepsilon_- = 0 = \partial\varepsilon_+. \tag{15.2.3}$$

As a result, ε_- and ε_+ are arbitrary functions of z and $\bar z$, respectively.

When varying the action of equation (6.1.1) for the open superstring, as for the open bosonic string, we find the above equations of motion as well as some boundary conditions. The boundary conditions for the bosons x^μ are just the same as for the case of no fermions and were discussed in the previous section. It will prove useful to rewrite the results in terms of the z and $\bar z$ coordinates. The N and D boundary conditions at $\sigma = 0$, for example, are given by

$$\partial x^\mu = \bar\partial x^\mu \quad \mu \in N_0; \quad \partial x^\mu = -\bar\partial x^\mu \quad \mu \in D_0, \tag{15.2.4}$$

with identical equations at $\sigma = \pi$. We may write the above boundary condition as

$$\partial x^\mu = M^{(0)\mu}{}_\nu\bar\partial x^\nu, \tag{15.2.5}$$

where

$$M^{(0)\mu}{}_\nu = \begin{cases} \delta^\mu_\nu & \mu \in N_0, \\ -\delta^\mu_\nu & \mu \in D_0. \end{cases} \tag{15.2.6}$$

The boundary term arising from the variation of the spinors is discussed in chapter 6 and is given by equation (6.1.31), which can be written as

$$\chi^\mu_-\delta\chi_{-\mu} = \chi^\mu_+\delta\chi_{+\mu} \quad \text{at } \sigma = 0, \pi. \tag{15.2.7}$$

We impose the boundary condition separately at both ends of the string. Following the discussion in chapter 6 we still have equation (6.1.32) and we can always choose the spinor

χ^{μ} such that equation (6.1.33) holds, that is, $\chi_+^{\mu}(\tau, 0) = \chi_-^{\mu}(\tau, 0)$. Thus we are left with the choices of equations (6.1.34) and (6.1.35), which were

$$\chi_+^{\mu} = \chi_-^{\mu} \qquad R \tag{15.2.8}$$

or

$$\chi_+^{\mu} = -\chi_-^{\mu} \quad NS \tag{15.2.9}$$

at $\sigma = \pi$. As before, if we adopt the choice of equation (15.2.8) we refer to the string as being in the R sector, while if we adopt the choice of equation (15.2.9) we say it is in the NS sector. This discussion is identical to that for the ordinary open string studied in chapters 6 and 7 and as there it leads to integer moded supercharges in the R sector and half-integer moded ones in the NS sector.

As $\chi_{\pm}^{\mu}$ satisfies the equation of motion $\bar{\partial}\chi_+^{\mu} = 0 = \partial\chi_-^{\mu}$, the above boundary condition implies the mode expansions,

$$\chi_+^{\mu} = \alpha' \sum_n d_n^{\mu} e^{-inz}, \quad \chi_-^{\mu} = \alpha' \sum_n d_n^{\mu} e^{-inz} \qquad R \tag{15.2.10}$$

corresponding to equation (15.2.8) and

$$\chi_+^{\mu} = \alpha' \sum_r b_r^{\mu} e^{-inrz}, \quad \chi_- = \alpha' \sum_r b_r^{\mu} e^{-inrz} \qquad NS \tag{15.2.11}$$

corresponding to equation (15.2.9). So far the discussion of fermion modes is unchanged to that we had in chapter 6.

It is instructive to compute the supersymmetry transformations that correspond to the above boundary conditions. As we shall see the boundary conditions reduce the amount of supersymmetry by a half as compared to that for the closed string as they will identify ϵ_+ with ϵ_-. However, the precise identification will depend on the boundary conditions adopted. As such, we will begin by considering a generalisation of the supersymmetry transformations of equation (15.2.2) and in particular take a different supersymmetry variation with parameter ϵ^{μ} in each direction μ, that is, $\delta x_{\pm}^{\mu} = 2\partial x^{\mu} \epsilon_{\pm}^{\mu}$, where no sum is implied in this last equation. These generalised supersymmetry transformations will still be a symmetry of the equations of motion as these do not mix the different directions, but they will not in general preserve the constraints over which the μ indices are summed. Later we will relate the supersymmetry parameters in the different directions; however, this identification will depend on the type of boundary conditions we have.

Varying the boundary condition $\chi_+^{\mu} = \chi_-^{\mu}$ at $\sigma = 0$ we find that

$$\partial x^{\mu} \varepsilon_-^{\mu} = -\bar{\partial} x^{\mu} \varepsilon_+^{\mu}. \tag{15.2.12}$$

Using, respectively the D, and N boundary conditions on x^{μ} in equation (15.2.12) we find that at $\sigma = 0$

$$\varepsilon_-^{\mu} = -\varepsilon_+^{\mu}, \ \mu \in N_0; \quad \varepsilon_-^{\mu} = \varepsilon_+^{\mu}, \ \mu \in D_0. \tag{15.2.13}$$

At $\sigma = \pi$ we have the two possible boundary conditions of equations (15.2.8) and (15.2.9). Taking their supersymmetry variation, we find, depending whether x^{μ} satisfies N or D boundary conditions, that at $\sigma = \pi$

$$\varepsilon_-^{\mu} = -\varepsilon_+^{\mu}, \ \mu, \in N_{\pi}; \quad \varepsilon_-^{\mu} = \varepsilon_+^{\mu}, \ \mu \in D_{\pi} \tag{15.2.14}$$

for the boundary condition of equation (15.2.8), that is, for the R sector, while

$$\varepsilon_-^\mu = +\varepsilon_+^\mu, \quad \mu \in N_\pi; \quad \varepsilon_-^\mu = -\varepsilon_+^\mu, \quad \mu \in D_\pi \tag{15.2.15}$$

for the boundary condition of equation (15.2.9), that is, for the NS sector.

It is now straightforward to calculate the moding of $\varepsilon_\pm^\mu$. Using the conditions $\bar\partial\varepsilon_-^\mu = 0 = \partial\varepsilon_+^\mu$ we find that

$$\varepsilon_\mp^\mu = \sum_{r\in Z+\frac{1}{2}} \varepsilon_{\mp,r}^\mu e^{-ir(\tau\pm\sigma)} + \sum_{n\in Z} \varepsilon_{\mp,n}^\mu e^{-in(\tau\pm\sigma)}. \tag{15.2.16}$$

Imposing the boundary conditions at $\sigma = 0$ of equation (15.2.13) we find

$$\varepsilon_{-,r}^\mu = -\varepsilon_{+,r}^\mu, \quad \varepsilon_{-,n}^\mu = -\varepsilon_{+,n}^\mu, \quad \mu \in N_0 \tag{15.2.17}$$

and

$$\varepsilon_{-,r}^\mu = \varepsilon_{+,r}^\mu, \quad \varepsilon_{-,n}^\mu = \varepsilon_{+,n}^\mu, \quad \mu \in D_0. \tag{15.2.18}$$

However, at $\sigma = \pi$ we find that for the choice of boundary condition of equation (15.2.8), that is, for the R sector, that

$$\varepsilon_{-,r}^\mu = \varepsilon_{+,r}^\mu, \quad \varepsilon_{-,n}^\mu = -\varepsilon_{+,n}^\mu, \quad \mu \in N_0 \tag{15.2.19}$$

and

$$\varepsilon_{-,r}^\mu = -\varepsilon_{+,r}^\mu, \quad \varepsilon_{-,n}^\mu = \varepsilon_{+,n}^\mu, \quad \mu \in D_0, \tag{15.2.20}$$

while for the choice of boundary condition of equation (15.2.9), that is, for the NS sector, that

$$\varepsilon_{-,r}^\mu = -\varepsilon_{+,r}^\mu, \quad \varepsilon_{-,n}^\mu = \varepsilon_{+,n}^\mu, \quad \mu \in N_0 \tag{15.2.21}$$

and

$$\varepsilon_{-,r}^\mu = \varepsilon_{+,r}^\mu, \quad \varepsilon_{-,n}^\mu = -\varepsilon_{+,n}^\mu, \quad \mu \in D_0. \tag{15.2.22}$$

These equations determine which modes of the supersymmetry transformation are non-zero. For example, if in the direction μ we have N–N boundary conditions and we are in the R sector, that is, the boundary condition of equation (15.2.8), we have equations (15.2.17), (15.2.19), then $\varepsilon_{\pm,r}^\mu = 0$, $\varepsilon_{-,n}^\mu = -\varepsilon_{+n}^\mu$; alternatively for the boundary condition of equation (15.2.9) we find that $\varepsilon_{-,r}^\mu = \varepsilon_{+,r}^\mu$, $\varepsilon_{\pm,n}^\mu = 0$. Similarly, if we impose D–D boundary conditions in the μ direction, then we find that $\varepsilon_{\pm,r}^\mu = 0$, $\varepsilon_{-,n}^\mu = \varepsilon_{+n}^\mu$ for the boundary condition of equation (15.2.8), while $\varepsilon_{-,r}^\mu = \varepsilon_{+,r}^\mu$ $\varepsilon_{\pm,n}^\mu = 0$, for the boundary condition of equation (15.2.9). However, if we have an N–D boundary condition, then from equations (15.2.17), (15.2.20) and (15.2.22) we find that $\varepsilon_{-,r}^\mu = -\varepsilon_{+,r}^\mu$, $\varepsilon_{\pm,n}^\mu = 0$ for the choice of boundary condition of equation (15.2.8), while $\varepsilon_{-,n}^\mu = -\varepsilon_{+,n}^\mu$, $\varepsilon_{\pm,r}^\mu = 0$ for the boundary condition of equation (15.2.9). The same modes survive in the D–N case. Thus for N–N and D–D directions in the NS sector (that is, equation (15.2.8)) we have only a half-integer moded supersymmetry parameter, while in the R sector (that is, equation (15.2.9)) we have only a integer moded supersymmetry parameter. The surviving moding in the directions with mixed boundary conditions is the opposite.

We must now find a consistent identification for all the supersymmetry parameters ϵ^μ in the different directions. The string begins and ends on branes that always have a time direction in their world volume and so the strings always have at least one N–N direction which we can identify with the one supersymmetry parameter. In open strings with directions that have N–N boundary conditions we can identify $\varepsilon_n \equiv \varepsilon^\mu_{-,n} = -\varepsilon^\mu_{+,n}$ in the R sector and $\varepsilon_r \equiv \varepsilon^\mu_{-,r} = -\varepsilon^\mu_{+,r}$ in the NS sector, while in directions that have D–D boundary conditions we can take $\varepsilon_n \equiv \varepsilon^\mu_{-,n} = \varepsilon^\mu_{+,n}$ in the R sector and $\varepsilon_r \equiv \varepsilon^\mu_{-,r} = \varepsilon^\mu_{+,r}$ in the NS sector. It is clear that this identification is compatible with all the above boundary conditions. However, for open strings that have D–N or N–D boundary conditions as well as N–N or D–D boundary conditions the surviving moding in the different directions is opposite and so the only identification that is possible with the demand to have just one supersymmetry parameter is to set the supersymmetry parameters in these directions to zero, that is, $\delta x^\mu = 0 = \delta \chi^\mu$, $\mu \in D - N$ or $N - D$.

Taking the Fourier transform of the supersymmetry variations of equation (15.2.2) and using the modings found above, one can find the supersymmetry variation of the individual modes. For a μ direction that has N–N boundary conditions we find that

$$\delta\alpha^\mu_n = \sqrt{2\alpha'}n \sum_m \varepsilon_m d^\mu_{n-m}, \quad \delta q^\mu = i\alpha' \sum_n \varepsilon_n d^\mu_{-n}\delta, \quad d^\mu_n = \sqrt{\frac{2}{\alpha'}} \sum_m \alpha^\mu_{n-m}\varepsilon_m$$

$$(15.2.23)$$

and

$$\delta\alpha^\mu_n = \sqrt{2\alpha'}n \sum_s \varepsilon_s b^\mu_{n-s}, \quad \delta q^\mu = i\alpha' \sum_r \varepsilon_r b^\mu_{-r}, \quad \delta b^\mu_r = \sqrt{\frac{2}{\alpha'}} \sum_s \alpha^\mu_{r-s}\varepsilon_s$$

$$(15.2.24)$$

in the R and NS sectors, respectively.

In the D–D directions there is no centre of mass or momentum, that is, zero modes. The analogous formulae for these directions are the same except that the equations which concern the variation of the center of mass and momentum are, of course, absent and these variables are also absent on the right-hand side of the variations of the remaining variables. The quantisation of the fermionic sector of the open string discussed above follows closely the steps given in chapter 6 for the usual open string. The b_r and d_n oscillators obey the same anti-commutation as before as given in relations of equations (6.1.71) and (6.1.72). Introducing the wavefunction Ψ we demand that it obey the physical state conditions

$$L_n\Psi = 0 = F_n\Psi = 0 \quad n \geq 0 \tag{15.2.25}$$

in the R sector and

$$L_n\psi = 0, \quad n \geq 1, \quad G_r\psi = 0, \quad r \geq \frac{1}{2}, \quad \left(L_0 - \frac{e}{8} - \frac{1}{2}\right)\psi = 0 \tag{15.2.26}$$

in the NS sector. In this last equation e is the number of mixed directions, that is, N–D or D–N directions. The L_n, F_n and G_r are as in equations (6.1.77) and (6.1.73)–(6.1.74) for the R and NS sectors, respectively. The expression for L_0 for the bosonic oscillators is given by equation (15.1.33) and so depends on the separation of the D-branes.

We now give an argument [15.9] that justifies the intercept of equation (15.2.26) for the superstring and equation (15.1.31) for the bosonic string. The string consists of an infinite number of oscillators. We recall that the standard bosonic harmonic oscillator has the Hamiltonian,

$$H = p^2 + q^2 = a^\dagger a + \tfrac{1}{2} \tag{15.2.27}$$

where $a = p + iq$, $a^\dagger = p - iq$ and so $[a, a^\dagger] = i$. While the standard fermionic oscillator has Hamiltonian

$$H = b^\dagger b - \tfrac{1}{2} \tag{15.2.28}$$

with $\{b, b^\dagger\} = i$. We observe that the zero point energies are $+\tfrac{1}{2}$ and $-\tfrac{1}{2}$, respectively. Strings can contain bosonic oscillators which obey $[a_n^\mu, a_n^{\nu\dagger}] = n\eta^{\mu\nu}$ $[c_r^\mu, c_r^{\mu\dagger}] = r\eta^{\mu\nu}$ and fermionic oscillators that obey $\{b_r^\mu, b_r^{\mu\dagger}\} = r\eta^{\mu\nu}$ or $\{d_n^\mu, d_n^{\nu\dagger}\} = n\eta^{\mu\nu}$. Taking into account the rescaling of the Hamiltonians by the appropriate factors, the corresponding zero point energies are given by $n/2$ or $r/2$ and $-r/2$ or $-n/2$, respectively. Hence if a string has n_α, n_c, n_d and n_b species of α_n^μ, c_s^μ, d_n^μ and b_r^μ oscillators, respectively, its zero point energy ε_0 is given by

$$\varepsilon_0 = +\frac{n_\alpha}{2}\sum_{n=0}^\infty n + \frac{n_c}{2}\sum_{r=\frac{1}{2}}^\infty r - \frac{1}{2}n_d\sum_{n=0}^\infty n - \frac{1}{2}n_b\sum_{r=\frac{1}{2}}^\infty r. \tag{15.2.29}$$

This expression is, of course, divergent; however, it can be regulated using the ζ function which is defined by

$$\sum_{n=0}^\infty \frac{1}{(n+b)^s} = \zeta(s,b). \tag{15.2.30}$$

It can be shown [15.10] that

$$\zeta(-1,b) = -\tfrac{1}{12}(6b^2 - 6b + 1). \tag{15.2.31}$$

Writing ε_0 in terms of $\zeta(-1,0)$ and $\zeta(-1,\tfrac{1}{2})$, and using the above result we find that [15.9]

$$\varepsilon_0 = -\tfrac{1}{48}\{2(n_\alpha - n_d) - (n_c - n_b)\}. \tag{15.2.32}$$

The intercept a that occurs in the equation $(L_0 - a)\psi = 0$ as one of the physical state conditions is the negative of the zero point energy (that is, $a = -\varepsilon_0$) as L_0 is essentially the Hamiltonian. Hence the intercept is given by

$$a = \tfrac{1}{48}\{2(n_\alpha - n_d) - (n_c - n_b)\}. \tag{15.2.33}$$

Let us apply the above formula to various strings we have encountered. Let us first consider a bosonic string in a D-dimensional space-time, e directions of which obey N–D or D–N boundary conditions, the remainder being N–N or D–D. Taking into account that two of the oscillators are removed in the light-cone gauge, we have $D - e - 2\,\alpha_n^\mu$ oscillators and $e\,c_r^\mu$ oscillators and hence

$$a = \frac{D-2}{24} - \frac{e}{16}. \tag{15.2.34}$$

This is the intercept adopted in equation (15.1.31). Consider now the NS–R formulations of the superstring in D dimensions with e D–N or N–D directions. In the NS sector we have $(D - 2 - e)\,\alpha_n^\mu$ oscillators, $e\,c_r^\mu$ oscillators, $(D - 2 - e)\,b_r$ oscillators and $e\,b_n$ oscillators giving

$$a = \frac{D-2}{16} - \frac{e}{8}. \tag{15.2.35}$$

In contrast, in the R sector we have $(D - 2 - e)\,\alpha_n^\mu$ oscillators, $e\,c_r^\mu$ oscillators $(D - 2 - e)\,d_n$ oscillators and $e\,d_r$ oscillators and hence $a = 0$. This must be the case as the L_0 condition is the square of the F_0 condition which being Grassmann odd can have no intercept.

Let us apply the above theory to the case of a single D-brane whose world volume is in the directions $\mu = 0, 1, \ldots, p$. We consider open strings that begin and end on this one D-brane. As we discussed for the bosonic string, both ends of the string obey N boundary conditions in $\mu = 0, 1, \ldots, p$ directions and D boundary conditions in the remaining transverse $\mu = p + 1, \ldots, 9$ directions. The string functional Ψ can be written in terms of the oscillators α_n^μ, $n \geq 1$, as well as d_n^μ, $n \geq 1$, in the R sector and b_r^μ, $r \geq \frac{1}{2}$, in the NS sector acting on the vacuum. The component fields depend on the centre of mass coordinates in the N–N directions since there are no such coordinates in the D–D directions. We can choose static gauge to identify the world volume coordinates with those in the background space-time which lie in the brane world volume, that is, $\xi^\mu = x^\mu$, $\mu = 0, 1, \ldots, p$. The vacuum is annihilated by all the remaining oscillators. Thus Ψ is given by

$$\Psi_\alpha = (\lambda_\alpha(\xi^\nu) + \cdots) \tag{15.2.36}$$

and

$$\Psi = \left(\phi(\xi^\nu) + \sum_{\mu=0,\ldots p} b_{-\frac{1}{2}}^\mu A_\mu(\xi^\nu) + \sum_{\mu'=p+1,\ldots,9} b_{-\frac{1}{2}}^{\mu'} \varphi_{\mu'}(\xi^\nu) + \cdots \right) \Psi_0, \tag{15.2.37}$$

in the R and NS sectors, respectively.

The intercept is zero in both the NS and R sectors. Hence if we have just one D-brane on which the strings begin or end the massless states, classified according to the $SO(1, p)$ Lorentz group of the brane, are the spin $\frac{1}{2}\,\lambda_\alpha$, one spin 1, A_μ, $\mu = 0, \ldots, p$ and $10 - p - 1$ spin 0s, $\varphi_{\mu'}$, $\mu' = p + 1, \ldots, D - 1$. There is also a tachyon ϕ. However, imposing the usual GSO projector of chapter 7, we find that of these states, the spinor λ_α is now Weyl, the tachyon is eliminated, but A_μ and $\varphi_{\mu'}$ survive. The photon A_μ is a singlet under $SO(9 - p)$ but the scalars $\varphi_{\mu'}$ transform as a vector under this group. The Majorana–Weyl spinor λ_α transforms as a spinor under $\mathrm{Spin}(1, p) \times \mathrm{Spin}(10 - p - 1)$. Thus, we find the field content of $D = 10$, $N = 1$ Yang–Mills theory dimensionally reduced to $p + 1$ dimensions corresponding to the fact that the fields are living on the $(p + 1)$-dimensional D-brane.

15.3   D-branes in the Green–Schwarz formulation

We now give an account [15.11] of supersymmetric open strings and their corresponding D-branes using the Green–Schwarz formalism discussed in chapter 6. We recall that the Green–Schwarz formulation of the superstring consists of the fields $x^\mu(\tau, \sigma)$; $\mu = 0, 1, \ldots, D - 1$

and the Grassmann odd fields $\theta^i(\tau, \sigma)$, which are space-time spinors with an internal index $i = 1, 2$. We will mainly be interested in ten dimensions where the spinors are Majorana–Weyl and with either the same chirality (IIB) or opposite chirality (IIA).

The first step is to find the boundary conditions that arise when deriving the equations of motion of the Green–Schwarz formulation of equation (6.2.48). When carrying out this derivation in chapter 6 we omitted mention of the boundary terms that must arise when we vary the action of equation (6.2.4). As explained for the case of the bosonic string in chapter 2, we must also adopt boundary conditions on the fields such that the boundary terms vanish separately. Following arguments similar to those leading to equation (15.1.5), we find that the boundary term that arises from varying the Green–Schwarz action is given by

$$
\int_{\partial\Sigma} \frac{d\xi^\alpha}{ds} ds \varepsilon_{\alpha\beta} \left(\delta x^\mu \frac{\delta A}{\delta(\partial_\beta x^\mu)} + \delta\theta^i \frac{\delta A}{\delta \partial_\beta \theta^i} \right). \tag{15.3.1}
$$

Choosing the gauge choice $g_{\alpha\beta} = \eta_{\alpha\beta} e^\phi$ we find that the above boundary term vanishes if

$$
\delta x^\mu (\Pi_{1\mu} + i(\theta^1 \gamma_\mu \partial_0 \theta^1 - \theta^2 \gamma_\mu \partial_0 \theta^2))
$$
$$
+ \{ -\Pi_{1\mu}(-i\bar{\theta}^j \gamma^\mu \delta\theta^j) - i\partial_0 x^\mu (\bar{\theta}^1 \gamma_\mu \delta\theta^1 - \bar{\theta}^2 \gamma_\mu \delta\theta^2)
$$
$$
+ \bar{\theta}^1 \gamma^\mu \partial_0 \theta^1 \bar{\theta}^2 \gamma^\nu \delta\theta^2 - \bar{\theta}^1 \gamma^\mu \delta\theta^1 \bar{\theta}^2 \gamma_\mu \partial_0 \theta^2 \} = 0 \tag{15.3.2}
$$

at the boundary at $\sigma = 0$ and $\sigma = \pi$ separately. We recall that $\Pi_\alpha^\mu = \partial_\alpha x^\mu - i\bar{\theta}^j \gamma^\mu \partial_\alpha \theta^j$. The analysis of the first term in this equation, which is essentially only a function of x^μ, is the same as for the bosonic open string of section 15.1. Consequently, we conclude that ends of the string obey N or D boundary conditions. Since δx^μ vanishes in the directions that have D boundary conditions, we find no additional constraint in these directions, In the N directions the boundary terms containing δx^μ vanish at $\sigma = 0$ if

$$
(\theta^1 \gamma^\mu \partial_0 \theta^1 - \theta^2 \gamma^\mu \partial_0 \theta^2) - (\theta^1 \gamma^\mu \partial_1 \theta^1 + \theta^2 \gamma^\mu \partial_1 \theta^2) = 0, \quad \mu \in N_0, \tag{15.3.3}
$$

with a similar condition at the other end of the string at $\sigma = \pi$. The second term comes from the θs contained in Π_1^μ. We will return to this term shortly.

Examining the $\delta\theta^i$ variations in equation (15.3.2), we find that at $\sigma = 0$

$$
\bar{\theta}^1 \gamma^\mu \delta\theta^1 + \bar{\theta}^2 \gamma^\mu \delta\theta^2 = 0, \mu \in D_0,
$$
$$
\bar{\theta}^1 \gamma^\mu \delta\theta^1 - \bar{\theta}^1 \gamma^\mu \delta\theta^2 = 0, \mu \in N_0. \tag{15.3.4}
$$

The difference in sign corresponds to fact that $\partial_1 x^\mu$ and $\partial_0 x^\mu$ are undetermined in the D and N directions, respectively. We will also require that the higher order terms in $(\theta)^3$ must vanish separately. We now place a condition on θ^i of the form $\theta^2 = P_0 \theta^1$. Since θ^1 and θ^2 are Majorana spinors, that is, $\bar{\theta}^i = (\theta^i)^T C$, this condition implies that $\bar{\theta}^2 = \bar{\theta}^1 \hat{P}_0$, where $\hat{P}_0 = C^{-1} P_0^T C$. Since $\delta\theta^i$ satisfies the same conditions as θ^i equation (15.3.4) implies that

$$
\gamma^\mu + \hat{P}_0 \gamma^\mu P_0 = 0, \mu \in D_0,
$$
$$
\gamma^\mu - \hat{P}_0 \gamma^\mu P_0 = 0, \mu \in N_0. \tag{15.3.5}
$$

Let us choose the N_0 boundary conditions to be in $\mu = 0, 1, 2, \ldots, p$. If we take $P_0 = \gamma^{01\cdots p}$ and using the properties

$$\hat{P}_0 = C^{-1}P_0^T C = (-1)^{p+1}(-1)^{\frac{1}{2}p(p+1)}P_0 \quad \text{and} \quad P_0^2 = -(-1)^{\frac{1}{2}p(p+1)},$$

(15.3.6)

given in chapter 5, we find that the above boundary conditions are automatically satisfied.

We now return to the terms resulting from δx^μ terms of equation (15.3.3). As already noted such terms vanish automatically for D boundary conditions. For N boundary conditions the terms in the first bracket vanish as the condition $\theta^2 = P_0\theta^1$ implies that at $\sigma = 0$ $\partial_0\theta^2 = P_0\partial_0\theta^1$ and similarly at $\sigma = \pi$. The terms in the second bracket vanish if we require $\partial_\sigma\theta^2 = -P_0\partial_\sigma\theta^1$ and $\partial_\sigma\theta^2 = -P_\pi\partial_\sigma\theta^1$ at the end points $\sigma = 0$ and $\sigma = \pi$, respectively.

One can verify that the terms cubic in θ in equation (15.3.2) also now vanish if we use ' the above conditions and so we have found that the complete boundary term vanishes.

In IIA theory string theory, we choose θ^1 and θ^2 to have opposite chiralities. Since $P_0\gamma^{11} = (-1)^{(p+1)(D-p-1)}\gamma^{11}P_0$ we find that the condition $\theta^2 = P_0\theta^1$ requires D-branes to have even p. In type IIB string theory, on the other hand, θ^1 and θ^2 have the same chirality and we therefore find that D-branes have odd p. As a result, we find that the open strings considered here end on branes, that is, by definition D-branes, that couple to the R–R sector of the massless supergravity theory.

To summarise, if we have an open string which is attached to one D-brane and ends on another D-brane, the θ^i are subject to the constraints

$$\theta^2 = P_0\theta^1, \quad \theta^2 = P_\pi\theta^1$$

(15.3.7)

at $\sigma = 0$ and $\sigma = \pi$ respectively. Here P_0 is given by the product of all the γ-matrices in the longitudinal directions of the D-brane at $\sigma = 0$. The matrix P_π is formed in the same way, but involving the longitudinal directions of the D-brane at $\sigma = \pi$.

We now consider the consequences of the boundary conditions of equation (15.3.4) on the supersymmetry transformations. As $\delta\theta^i = \epsilon^i$, for constant ϵ the only supersymmetry variations $\delta\theta^i$ that preserve the boundary conditions on θ^i are those for which

$$\epsilon^2 = P_0\epsilon^1, \quad \epsilon^2 = P_\pi\epsilon^1, \quad \text{which implies} \quad \epsilon^1 = P_0^{-1}P_\pi\epsilon^1.$$

(15.3.8)

As ϵ^1 is a Majorana–Weyl spinor, which has 16 components, this condition implies that we have only eight supersymmetries. An exception is when the two D-branes are parallel for which $P_0 = P_\pi$ and the above condition is automatically filled; this system has 16 supersymmetries [15.2].

The number of supersymmetries will be less than eight if we have more projector conditions as we will encounter shortly.

Let us now consider the supersymmetry variation of x^μ, namely $\delta x^\mu = i\bar{\epsilon}^j\gamma^\mu\theta^j$, and use the above equations. We find that

$$\delta x^\mu \mid_{\sigma=0} = i\bar{\epsilon}^1 \left[\gamma^\mu + \hat{P}_0\gamma^\mu P_0\right]\theta^1 \mid_{\sigma=0},$$

$$\delta\partial_\sigma x^\mu \mid_{\sigma=0} = i\bar{\epsilon}^1 \left[\gamma^\mu - \hat{P}_0\gamma^\mu P_0\right]\partial_\sigma\theta^1 \mid_{\sigma=0}.$$

(15.3.9)

Demanding that $\delta x^{\mu} = 0$ for $\mu \in D_0$ and $\delta \partial_1 x^{\mu} = 0$ for $\mu \in N_0$, we find no new conditions as these are automatically satisfied as a result of equation (15.3.5). Analogous conditions arise at $\sigma = \pi$. Hence we learn that boundary conditions on x^{μ} are compatible with the supersymmetry preserved in equation (15.3.8). To make clear this interplay between the supersymmetry that is preserved and the D-brane configuration, let us consider the example where the ϵs obey a condition that involves $P_0 = \gamma^{012\cdots p}$, then to preserve supersymmetry we must impose N boundary conditions on x^{μ} for $\mu = 0, \ldots, p$ and D boundary conditions for $\mu = p+1, \ldots, 9$. If we take the projector $P_0 = -\gamma^{012\cdots p}$, we must still impose N boundary conditions in the directions $\mu = 0, \ldots, p$ and D boundary conditions for $\mu = p+1, \ldots, 9$, but this brane has the opposite orientation and charge and is called an anti-brane. Thus requiring a particular space-time supersymmetry, that is, equation (15.3.8), is preserved we learn that the $\sigma = 0$ end point of the string is fixed to lie on a surface, that is, the D-brane, whose world-volume directions are those that occur in the γ matrix that constitutes P_0. For example, if $P_0 = \gamma^{01\cdots p}$, then the D-brane lies in the $(x^0, x^1, x^2, \ldots, x^p)$ plane. Similarly one can determine on which plane the $\sigma = \pi$ end point of the open string must end from a knowledge of P_{π}.

We now study the consequences of the condition $\theta^2 = P_0 \theta^1$ on the κ symmetry of the string which we found in chapter 6 to be given by $\delta x^{\mu} = i \bar{\theta}^j \gamma^{\mu} \delta \theta^j$ and $\delta \theta^j = 2 i \gamma^{\mu} \Pi_{\mu \alpha} \kappa^{j \alpha}$. The above condition implies that $P_0 \gamma^{\mu} \Pi_{\mu \alpha} \kappa^{1 \alpha} = \gamma^{\mu} \Pi_{\mu \alpha} \kappa^{2 \alpha}$. This is automatically satisfied if we take $\kappa^{10} = e P_0 \kappa^{20}$ and $\kappa^{11} = -e P_0 \kappa^{21}$, where $\hat{P}_0 = e P_0 = (-1)^{\frac{1}{2} p(p+1) + p + 1} P_0$. We are using the fact that $\Pi_{\mu \alpha}$ vanishes if $\alpha = 0$ and $\mu \in D$ and if $\alpha = 1$ and $\mu \in N$ and equation (15.3.5). As a result we find that the κ transformations are given by

$$\delta \theta^1 = 2 i \gamma^{\mu} \Pi_{\mu \alpha} \kappa^{1 \alpha} \tag{15.3.10}$$

and

$$\delta x^{\mu} = \begin{cases} 0 & \mu \in D, \\ 2 i \bar{\theta}^1 \gamma^{\mu} \delta \theta^1 & \mu \in N. \end{cases} \tag{15.3.11}$$

Computing the states of the open strings is most easily done in light-cone gauge, which was discussed for the bosonic string in chapter 4. As discussed there we can use the invariance under under world-volume reparameterisations to choose the conformal gauge $g_{\alpha \beta} = \eta_{\alpha \beta} e^{\phi}$. We may also use the residual conformal symmetry to choose

$$x^+(\tau, \sigma) \equiv x^0(\tau, \sigma) + x^1(\tau, \sigma) = x^+ + 2 \alpha' p^+ \tau, \tag{15.3.12}$$

where x^+ in the last equation is independent of τ and σ and the normalisation of p^+ is appropriate for an open string. Using the κ symmetry we may make one further gauge choice, $\gamma^{01} \theta^i = \theta^i$, and reduce our spinor to

$$S^i = \tfrac{1}{2}(1 + \gamma^{01}) \theta^i. \tag{15.3.13}$$

The light-cone projector involves a choice of direction (here we assume it is x^1) and sign (here we assumed a plus sign). However these cannot be chosen arbitrarily since the parameters of the κ symmetry are also subject to constraints, discussed above, in particular those that follow from the constraints of equation (15.3.7). Indeed, we must ensure that the direction chosen is such that the surviving κ symmetry is not eliminated by the matrix projector, that is, $\tfrac{1}{2}(1 + \gamma^{01})$ for the above choice of direction. This means in practice that γ^{01} must commute with the projectors $P_0^{-1} P_{\pi}$ and any other projectors which we will encounter later. In addition, in some two-dimensional cases these projectors may imply $\gamma^{01}\theta^i = \pm\theta^i$. This means that the θ^i are chiral fermions on the world sheet. However, we must choose the light-cone projector to act non-trivially, otherwise the unphysical field components will not be projected out. Thus in these cases we must chose the light-cone condition $\gamma^{01}\theta^i = \mp\theta^i$. We will assume these conditions are satisfied in what follows.

As discussed in chapter 4, given the above choices we may solve for x^- and P^-. The equations of motion for the remaining variables take on the very simple form

$$(\partial_0^2 - \partial_1^2)x^\mu = 0,\ \mu \neq 0, 1,\quad (\partial_0 + \partial_1)\theta^1 = 0,\ (\partial_0 - \partial_1)\theta^2 = 0. \tag{15.3.14}$$

The oscillator expressions for the fields x^μ were given in equations (15.1.24)–(16.1.27). We recall that only for the directions with N–N boundary conditions did we find zero modes, that is, a centre of mass momentum and coordinate. As a result of equation (15.3.14) the spinor fields S^i take the form

$$S^1 = \sum_n S_n^1 e^{-in(\tau-\sigma)} + \sum_r S_r^1 e^{-in(\tau-\sigma)},$$
$$S^2 = \sum_n S_n^2 e^{-in(\tau+\sigma)} + \sum_r S_r^2 e^{-in(\tau+\sigma)}. \tag{15.3.15}$$

Next we must impose the boundary conditions of equation (15.3.7) to find that $S^2 = P_0 S^1$ at $\sigma = 0$ and $S^2 = P_\pi S^1$ at $\sigma = \pi$. This leads to

$$\left.\begin{array}{l} S_n^2 = P_0 S_n^1 \\ S_r^2 = P_0 S_r^1 \end{array}\right\} \text{ at } \sigma = 0,\qquad \left.\begin{array}{l} S_n^2 = P_\pi S_n^1 \\ S_r^2 = -P_\pi S_r^1 \end{array}\right\} \text{ at } \sigma = \pi. \tag{15.3.16}$$

These equations determine S_n^2 and S_r^2 in terms of S_n^1 and S_r^1 and also place the constraints

$$S_n^1 = P_0^{-1} P_\pi S_n^1,\quad S_r^1 = -P_0^{-1} P_\pi S_r^1. \tag{15.3.17}$$

To summarise, the spinors S_n^2 and S_r^2 have been eliminated in terms of S_n^1 and S_r^1, which are subject to equation (15.3.17). Since $(P_0^{-1} P_\pi)^2 = 1$ and as $P_0^{-1} P_\pi$ is just a γ matrix $\text{Tr}(P_0^{-1} P_\pi) = 0$ unless $P_0^{-1} P_\pi$ is a multiple of the identity. As a result, there are only three cases for $P_0^{-1} P_\pi$: (1) all eight eigenvalues equal 1, (2) all eight eigenvalues equal -1, (3) four eigenvalues are $+1$ and four eigenvalues are -1. In cases (1) and (2) we have $P_0 = P_\pi$ and $P_0 = -P_\pi$, respectively, and so we have in effect two D-branes that are parallel, while for case 3 we have D-branes which are not parallel and so have some directions with D–N or N–D boundary conditions.

We now outline the calculation of the quantum states. In the quantum theory we have the above oscillators which obey the usual commutation relations and anti-commutations relations. We introduce a vacuum state $|0\rangle$ which is annihilated by all the oscillators that have positive modes, that is, $n, r > 0$. The action of a non-negatively moded oscillator on the ground state then leads to fields with increased mass compared to the ground state. The exception, apart from the usual momentum operator α_0^μ are the operators S_0^1.

Let us suppose that we have N S_0^{1A} oscillators labelled by $A = 1, \ldots, N$. As these obey the anti-commutation relation $\{S_0^{1A}, S_0^{1B}\} = \delta^{AB}$, the S_0^{1A} form a Clifford algebra in N dimensions. As discussed in chapter 5 there is a unique representation of this Clifford algebra which has dimension $2^{N/2}$, where we assume N is even. We may construct this representation by defining $N/2$ operators that are taken to annihilate the vacuum and $N/2$

helicity raising operators that act on the vacuum creating new states of increasing 'helicity'. The $2^{N/2}$ states constitute a supersymmetric multiplet containing equal numbers of bosonic and fermionic states. In general, the states obtained by this method do not contain all the positive and negative helicity states required to form a particle and are therefore not CPT self-conjugate. To rectify this we must add another supermultiplet with the reversed 'helicities'. In a few cases the multiplet is CPT self-conjugate and this last step is not necessary. This constitutes a brief outline of this counting of states and the reader may study [1.11] for an account of supersymmetric state construction using the supersymmetry algebra in four dimensions. Helicity is a four-dimensional concept arising from the SO(2) little group for massless particles; in higher dimensions one must use the Casimirs of the larger little group. In the above we have used 'helicity' rather loosely but it is meant to denote the appropriate variables.

Let us apply this construction of the states to several cases. First we consider case (1), corresponding to Dp-branes with $P_0 = P_\pi$. This is the case if the open strings end on the same D-brane, or go between two parallel D-branes. This system has 16 space-time supersymmetries. We have only N–N, or D–D, boundary conditions and so the fields x^μ lead, in light-come gauge, to eight independent integer moded oscillators α_n^μ. We note that the constraint of equation (15.3.17) is satisfied if $S_r^1 = S_r^2 = 0$, while the S_n^1 are unconstrained. As θ^1 is a Majorana–Weyl spinor it generically has sixteen components; however, the constraint $\gamma^{01}\theta^1 = \theta^1$ reduces the number of S_0^1 to eight. Thus we have eight independent integer moded fermionic oscillators. Remembering the discussion of the intercept above equation (15.2.33), we have intercept $a = 0$. As we have eight surviving S_0^1 operators we find a Clifford algebra which is represented by $2^4 = 8 + 8$ massless states. This is, in fact, a CPT self-conjugate multiplet. It can only be that of ten-dimensional Abelian theory dimensionally reduced to the dimension of the world volume of the D-brane.

We can also consider N parallel coincident branes as we did for the bosonic string. The discussion proceeds along similar lines. If they are N coincident branes, then we find the states of the string belong to the adjoint representation of U(N) and it follows that the massless states form the maximal super-Yang–Mills vector multiplet in $p + 1$ dimensions [15.8]. If there is a space between the parallel D-branes, then the masses are shifted as for the bosonic string.

Next we consider case (2), corresponding to an open string with one end on a Dp-brane and the other on an anti-Dp-brane. Here $P_0 = -P_\pi$ and so by equation (15.3.8) no space-time supersymmetries are preserved. The corresponding x^μ fields are the same as for case (1) and so there are eight integer moded scalars. However, by equation (15.3.17) there are eight half integer moded fermions S_r^1 and so no spinorial zero modes. Thus we have only one state with intercept $+1/2$, which is a tachyon [15.12].

In case (3), the open strings go between D-branes that are not parallel and then we have N–N, D–D, and a mixture of N–D and D–N boundary conditions. The bosonic oscillators in the directions with N–N and D–D boundary conditions are the α_n^μ, while in the directions with N–D or D–N boundary conditions we have the c_r^μ oscillators. For the fermionic oscillators we have a mixture of the S_n^1 and S_r^1 which obey equation (15.3.17). Indeed, we will have four S_n and four S_r. This reduces the number of S_0^1 oscillators to four, that is, by a factor of 2 compared to case (1). As such the lowest states forms a four-dimensional representation of Clifford algebra in four dimensions which carries a representation of supersymmetry with eight parameters. If we take the vacuum to have helicity $-\frac{1}{2}$ acting with the two helicity raising operators we find states with helicity $-\frac{1}{2}$, 0 and $\frac{1}{2}$. Although at first sight this looks like a CPT invariant multiplet, careful thought about the action of the U(2) R symmetry carried by the supersymmetry generators leads to the conclusion that one must add the conjugate multiplet. Essentially the helicity 0 states carry the doublet representation of SU(2) and this is not a real representation. Adding this multiplet we arrive at the $N = 2$ hypermultiplet. The references to helicity in this paragraph really refer to four dimensions, but the appropriate statements in other dimensions are easy to substitute. In general, one finds the hypermultiplet, dimensionally reduced to the intersection of the D-branes [15.13].

If ν is the number of N–D, or D–N, boundary conditions we have $8 - \nu$ α oscillators, ν c oscillators, and four integer moded and four half-integer moded S oscillators. As a result the intercept is $a = \frac{1}{12}(4 - \nu)$. We note that if $\nu = 4$ we have intercept 0 and so the above states are massless if the branes are coincident.

If there are N_1 D-branes at the $\sigma = 0$ end of the string and N_2 D-branes at the other end, then we have $N_1 N_2$ possible open strings and these form the bifundamental representation of $U(N_1) \times U(N_2)$. We note that this is not the the adjoint representation confirming our choice to put the states in the hypermultiplet by selecting the vacuum to have helicity $-\frac{1}{2}$. Adding the CPT conjugates occurs naturally in that the open strings are unoriented and so we must also include strings with P_π and P_0 applied at the end points $\sigma = 0$ and $\sigma = \pi$, respectively, which automatically adds the CPT conjugates set of states.

Let us give an example of two D4-branes that are configured as

$$
\begin{array}{cccccccc}
0 & 1 & 2 & 3 & . & . & 6 & \cdots \\
0 & 1 & . & . & 4 & 5 & 6 & \cdots
\end{array}
\tag{15.3.18}
$$

A direction at a given end of the open string is N or D depending if that direction is in, or transverse to, the world volume of the brane at that end, respectively. Hence the directions 0, 1, 6 are N–N, directions 2, 3, 4, 5 are either N–D or D–N, while directions 7, 8, 9 are D–D. For this case $P_0 = \gamma^{01236}$ and $P_\pi = \gamma^{01456}$ and so $(P_0)^{-1} P_\pi = -\gamma^{2345}$. As such, from equation (15.3.17) we find that $S_n^1 = -\gamma^{2345} S_n^1$ and $S_r^1 = \gamma^{2345} S_r^1$ and we have four S_n^1 and four S_r^1. Consequently, we have four spinorial zero modes S_0^1 and so $2^2 = 4$ states. These have eight supersymmetries. We recall that we only have centre of mass coordinates in N–N directions and so the fields only depend on these three coordinates, which in this case are in the 0, 1, 6 directions. As such, we find a field theory living on the brane intersection three-dimensional world volume. These states can be thought of as the dimensional reduction of the hypermultiplet of four dimensions to three dimensions.

A degenerate situation occurs for case (3) when $P_0^{-1} P_\pi = -\gamma^{23456789}$. As $\gamma^{01} S^1 = S^1$, we find that $P_0^{-1} P_\pi S^1 = -S^1$ and hence there are no S_n^1 and eight S_r^1. This corresponds to eight x^μ with D–N or N–D boundary conditions (that is, an open string stretched between a Dp-brane and a D($p + 8$)-brane). The corresponding intercept is 0. As we have no fermionic massless modes we find only one state which is massless which must be a singlet under space-time supersymmetry [15.14].

Many interesting phenomena can be found in systems that involve D-branes suspended between other branes. The prototype example is to suspend a D-brane configuration between two NS5-branes. In the first such example two D3-branes were placed between NS5-branes

[15.15] in the type IIB theory. To be concrete we will study this system and in particular the configuration

$$
\begin{array}{llllllll}
0 & 1 & 2 & 3 & 4 & 5 & & \cdots \\
0 & 1 & 2 & . & . & . & 6 & \cdots \\
0 & 1 & 2 & . & . & . & 6 & \cdots
\end{array}
\qquad (15.3.19)
$$

The first entry is the NS5-brane and the second and third are two parallel D3-branes. Corresponding to the presence of the 5-brane we have additional boundary conditions [15.15]. We are to think of the 5-branes as static; at weak coupling they have a larger energy density and they are infinite in two more dimensions than the 3-branes. The presence of the N5-branes breaks half of the remaining supersymmetries, leaving those which satisfy

$$
\epsilon^1 = \gamma^{012345}\epsilon^1 \equiv P_{NS}\epsilon^1. \qquad (15.3.20)
$$

The additional boundary conditions we must impose should preserve these supersymmetries. The D3-branes are confined to lie between the NS5-branes and so their x_6 dependence is confined to an interval. As such, the ends of the open strings, which end on the D3-branes, are also confined to this interval One can think of this as a kind of compactification. As such, we can Fourier expand the fields, but we will keep only the first, that is, constant x_6, mode and so there is no centre of mass momentum in the x^6 direction and as such it is a constant. Therefore the string end points must be fixed at a definite value of x^6 and under a supersymmetry variation we have at the $\sigma = 0$ end of the string

$$
\delta x^\mu \mid_{\sigma=0} = \bar{\epsilon}^1 [\gamma^\mu + \hat{P}_0 \gamma^\mu P_0]\theta^1 = 0, \qquad (15.3.21)
$$

where we have used equation (15.3.9) The analogous condition at $\sigma = \pi$ has an identical matrix structure as $P_0\theta^1 = P_\pi\theta^1$ and $P_0\epsilon^1 = P_\pi\epsilon^1$. For directions in which we have D–D, D–N or N–D boundary conditions the supersymmetry variation of equation (15.3.21) vanishes automatically due to equation (15.3.5). However, this is not the case for the directions with N–N boundary conditions of which the direction 6 is one and, in this case, demanding we preserve the symmetries implies the condition

$$
\bar{\epsilon}^1 \gamma^6 \theta^1 \mid_{\sigma=0,\pi} = 0, \qquad (15.3.22)
$$

for all ϵ^1 such that $P_{NS}\epsilon^1 = P_0^{-1}P_\pi\epsilon^1 = \epsilon^1$. Next we write $\epsilon^1 = \frac{1}{4}(1 + P_0^{-1}P_\pi)(1 + P_{NS})\eta$ for an arbitrary η and substitute this for ϵ^1 in (15.3.22). Since (15.3.22) is now true for all η we find that the boundary conditions introduced by this dimensional reduction must project out spinors

$$
(1 + P_\pi^{-1}P_0)(1 - P_{NS})\theta^1 \mid_{\sigma=0,\pi} = 0. \qquad (15.3.23)
$$

We have used that $\hat{P}_{NS} = -P_{NS}$ which was found by substituting $p = 5$ in equation (15.3.6). We will consider configurations where that $(P_\pi)^{-1}P_0$ and P_{NS} commute and then this conditions becomes

$$
(1 - P_{NS})\theta^1 \mid_{\sigma=0,\pi} = 0, \qquad (15.3.24)
$$

where we have used the fact that $P_\pi^{-1}P_0\theta^1 = \theta^1$ at $\sigma = 0, \pi$.

In fact, the above consideration is not quite correct. One should only impose equation (15.3.24) on the zero mode part of θ^1, or equivalently S_0^1. This can be seen by demanding that the states that survive are a subset of those before one introduces the NS5-branes. This is a reasonable request as one can take the NS5-branes to be very far apart, but then the conditions is the same. This will not be the case if we use the constraint on all of θ^1 as this will lead to a reduction in the number of the S oscillators and, by the considerations above, change the intercept. Thus what we really should impose is

$$(1 - P_{NS})S_0^1 = 0. \tag{15.3.25}$$

The requirement that the states in the presence of the NS5-brane are a subset of those in its absence will prove to be a useful criterion in what follows.

There are other ways to compactify a brane configuration which reduce the massless spectrum. For example, one could place the configuration between two D5-branes [15.15] or compactify x^6 on the orbifold $S^1/\mathbf{Z}_2$. The analysis of these situations follows in an analogous manner.

If we suspend the two D-branes between two NS5-branes as in equation (15.3.19), then we must now take account of the constraints of equations (15.3.20) and (15.3.25). The first removes half of the sixteen space-time supersymmetries to leave eight. While the second leaves only four independent S_0^1 acting on the vacuum. Two of the S_0^1 annihilate the vacuum leaving us with two helicity raising operators. As a result we obtain only four states. If we take the case of D3-branes, and so have a four-dimensional world volume, and take the vacuum to have helicity -1, these states carry the helicities $-1, -\frac{1}{2}, 0$. However, we cannot have a theory with a spin 1 which has just helicity 1 as this violates the CPT theorem, and so we must add the supermultiplet with the opposite helicity states. For the case of coincident D3-branes we obtain four copies of the $N = 2$ vector multiplet which belong to the U(2) $N = 2$ Yang–Mills theory, which consists of one spin 1, two spin $\frac{1}{2}$s and two scalars [14.15].

Let us suspend this D-brane configuration corresponding to case (3) between two NS5-branes. Corresponding to equation (15.3.20) we find half the number of supersymmetries that occur in the same situation but with the NS5-branes absent, so leaving us with four. Similarly, equation (15.3.25) now leads to only two non-vanishing zero modes, one of which defines the vacuum and one acts on it to raise its helicity. After including the CPT conjugate states we therefore obtain two bosons and two fermions and this leads to a Wess–Zumino multiplet a dimensionally reduced to the intersecting brane world volume. An example of such a configuration is to choose $P_0 = \gamma^{0123456}$, $P_\pi = \gamma^{0123678}$ and rather than adding NS5-branes we compactify x^6 on a the orbifold $S^1/\mathbf{Z}_2 \cong [0, 1]$. This corresponds to two D6-branes intersecting over a 4-brane preserving $N = 1$ supersymmetry in four dimensions:

$$
\begin{array}{lcccccccc}
\text{D6}: & 1 & 2 & 3 & 4 & 5 & 6 & & \\
\text{D6}: & 1 & 2 & 3 & & & 6 & 7 & 8
\end{array}
\tag{15.3.26}
$$

where $x^6 \in [0, 1]$. After compactification the surviving supersymmetries satisfy $\gamma^6\epsilon = \epsilon$. In terms of the chiral components this gives $\epsilon^1 = \gamma^6\epsilon^2 = \gamma^{012345}\epsilon^1$. This is the same projector as in equation (15.3.20) and therefore we find the same constraint (15.3.25) on S_0^1. The ground state forms a chiral four-dimensional $N = 1$ Wess–Zumino multiplet in the bifundamental of the two D6-brane gauge groups.

Finally, an interesting degenerate case occurs if $P_0^{-1}P_\pi S_0^1 = -P_{NS}S_0^1$, since then $P_{NS}S_0^1 = S_0^1 = -P_0^{-1}P_\pi S_0^1 = -S_0^1$ and so we have no fermionic zero modes. Thus there is no space-time supersymmetry in such a string and the only surviving massless modes are the lowest helicity components of the ground state. Since the ground state is a hypermultiplet this gives two helicity $-\frac{1}{2}$ fermions, taking into account the unoriented nature of the open strings. To be specific one could take $P_0 = \gamma^{01236}$, $P_\pi = \gamma^{01346}$ and $P_{NS} = \gamma^{012345}$ which can be pictured as

$$
\begin{array}{llllllll}
\text{NS5}: & 0 & 1 & 2 & 3 & 4 & 5 & \\
\text{D4}: & 0 & 1 & 2 & 3 & & & 6 \\
\text{D4}: & 0 & 1 & & & 4 & 5 & 6
\end{array}
\tag{15.3.27}
$$

that is, two D4-branes intersecting over a 2-brane and suspended between two NS5-branes. After performing the dimensional reduction induced by the NS5-branes one finds two chiral fermions in the resulting two-dimensional theory. Clearly there are many other possibilities. For example, one could add more NS5-branes, dimensionally reducing additional directions of a brane configuration.

An interesting application of these constructions is to see if one can find the Standard Model. Branes tend to lead to $U(N)$ gauge groups, supersymmetry and gauge symmetry breaking, chiral fermions and confinement in some cases. However, this is what one observes in the Standard Model, that is, the gauge group $SU(3) \otimes U(2)$, chiral fermions, no supersymmetry and confinement for the $SU(3)$ gauge group. If we were to take a configuration with three parallel NS5-branes labelled 1, 2 and 3 and put three parallel D3-branes between the first two and two parallel D3-branes between the second two, then one would find the gauge group $SU(3) \otimes SU(2) \otimes U(1)$, as well as some matter which does have some of the correct Y charges. Of course, this model is completely unrealistic in that it has $N = 2$ supersymmetry and so no chiral fermions. Much more sophisticated models with increasingly realistic features, along these lines, have been explored. One might aim to find a model that had no final supersymmetry, as does the Standard Model, but most efforts are aimed at finding a minimal extensions of the Standard Model which includes $N = 1$ supersymmetry.

16   String theory and Lie algebras

Everybody attributes the theory of Lie algebras to Cartan, but most of it was discovered by
Killing, except that is the Killing form, which was found by Cartan.
This statement reflects the neglect of Killing's work that occurred despite the clear
credit given to it by Cartan

Lie algebras have played a crucial role in the formulation of the electromagnetic, nuclear weak and strong forces as well as many other areas of theoretical physics. As string theory aims to provide a single theory of physics it is not unlikely that Lie algebras will play a central role. With this in mind we first give, in this chapter, a review of finite-dimensional semi-simple Lie algebras, that is, those in the list of Cartan. A proper understanding of these Lie algebras then allows us to define Kac–Moody algebras and discuss their properties, in particular a subclass of these algebras called Lorentzian algebras. We illustrate the general theory in the context of a Kac–Moody algebra called E_{11}. Finally, we show how string theory and its vertex operators lead very naturally to Lie algebras.

16.1   Finite dimensional and affine Lie algebras

16.1.1   A review of finite-dimensional Lie algebras and lattices

It is beyond the scope of this book to give a complete account of Lie algebras; however, in this and next section we will give an account that contains many of the main results. Although someone who is unfamiliar with Lie algebras could read this chapter it might be desirable to gain some familiarity with this subject first; some useful accounts are [5.5, 16.1]. Some textbooks covering the same material on finite dimensional Lie algebras are [5.4, 16.2, 16.3]. We will begin with an exposition which is familiar to physicists and then develop a point of view which is more often encountered in the mathematics literature, namely the Serre presentation of semi-simple Lie algebras. This starts from a very concise formulation of Lie algebras and deduces their structure and all their properties in a very elegant and efficient way. This viewpoint is essential for understanding the latter sections on Kac–Moody algebras. The main aim of these first two sections is to enable physicists to bridge the gap between the physics account of Lie algebras and that found in the mathematics literature.

485

A *Lie algebra* G over a field k is a vector space equipped with a map $G \times G \to G$ denoted by $(X, Y) \to [X, Y]$ which obeys the following three properties:

$$[X, aY + bZ] = a[X, Y] + b[X, Z],$$

$$[X, Y] = -[Y, X], \tag{16.1.1}$$

$$[[X, Y], Z] + [[Y, Z], X] + [[Z, X], Y] = 0$$

for all $X, Y, Z \in G$ and $a, b \in k$. The last condition is called the Jacobi identity. The second condition can also be written as $[Z, Z] = 0$ for all $Z \in G$ as can be seen by taking $Z = X + Y$. For us the field k is either the complex or the real numbers, and we then call the Lie algebra a complex or a real Lie algebra, respectively.

If T_a are a basis for G we can write

$$[T_a, T_b] = \sum_c f_{ab}{}^c T_c, \tag{16.1.2}$$

where the $f_{ab}{}^c$ are called the *structure constants*. They must satisfy $f_{ab}{}^c = -f_{ba}{}^c$ and the Jacobi identity; $3 \sum_e f_{[ab}{}^e f_{|e|c]}{}^d = \sum_e (f_{ab}{}^e f_{|e|c}{}^d + f_{ca}{}^e f_{|e|b}{}^d + f_{bc}{}^e f_{|e|a}{}^d) = 0$. In the physics literature T_a are often referred to as the generators.

A sub-algebra of a Lie algebra is a sub-set of the elements of the Lie algebra that is a Lie algebra in its own right. An Abelian Lie algebra is one for which all elements commute, that is, $[X, Y] = 0$ for elements X and Y in the Lie algebra.

Given any algebra A with a multiplication rule $(X, Y) \to X \cdot Y \in A$ which is associative, that is, $((X, Y), Z) = (X, (Y, Z))$, we can construct a Lie algebra by defining $[X, Y] = X \cdot Y - Y \cdot X$; this is often called the commutator. The reader may easily verify that the above properties of the Lie algebra hold. The commonest way to construct a Lie algebra is to consider a set of matrices, which being naturally equipped with the associative operation of matrix multiplication, allows us to take [,] to be the commutator, that is, $[X, Y] = XY - YX$.

The theory of Lie algebras is most efficiently developed using the complex numbers. Given a complex Lie algebra, if one finds a basis in which the structure constants are real, one can then work with this basis over the real numbers and so one has found a real Lie algebra. There are usually several ways to do this and so several real forms of a given complex Lie algebra. Given any real form of a Lie algebra we can revert to its complex form by working over the complex numbers. We will develop the general theory using the complex numbers. Physicists usually work with a particular real form of a Lie algebra. Hence to make contact with the physics literature we will often mention some of the real forms of the Lie algebras we are discussing.

Let us consider the Lie algebra, denoted SU(n), of $n \times n$ matrices A with complex entries that are anti-Hermitian ($A = -A^\dagger$), but taken over the real numbers **R**. We can use matrix multiplication to construct the Lie algebra. For the case of SU(2) a basis is given by

$$J_1 \equiv -\frac{i}{2}\sigma_1 = -\frac{i}{2}\begin{pmatrix} 0 & 1 \\ 1 & 0 \end{pmatrix}, \quad J_2 \equiv -\frac{i}{2}\sigma_2 = -\frac{i}{2}\begin{pmatrix} 0 & -i \\ i & 0 \end{pmatrix}, \quad J_3 \equiv -\frac{i}{2}\sigma_3$$

$$= -\frac{i}{2}\begin{pmatrix} 1 & 0 \\ 0 & -1 \end{pmatrix}. \tag{16.1.3}$$

The matrices σ_k are the three Pauli matrices. We find that they obey the relations $[J_i, J_j] = \epsilon_{ijk} J_k$, where ϵ_{ijk} is completely anti-symmetric in its three indices with $\epsilon_{123} = 1$. The anti-Hermitian character of the Lie algebra elements implies that the structure constants are real as they must be. If we take this algebra to be over the complex numbers $\mathbf{C}$ it is called the Lie algebra A_1.

Another commonly used real Lie algebra is SO(n), the set of $n \times n$ matrices R with real entries that obey $R^T = -R$. The reader may verify that for a suitable choice of basis of the generators the Lie algebras of SO(3) and SU(2) are the same.

Yet another popular real Lie algebra is that of SL(n, $\mathbf{R}$), the set of $n \times n$ matrices with real entries and trace zero. A basis for the SL(2, $\mathbf{R}$) Lie algebra is given by

$$S = \frac{1}{2}\begin{pmatrix} 0 & 1 \\ 1 & 0 \end{pmatrix}, \quad T = \frac{1}{2}\begin{pmatrix} 0 & 1 \\ -1 & 0 \end{pmatrix}, \quad U = \frac{1}{2}\begin{pmatrix} 1 & 0 \\ 0 & -1 \end{pmatrix}. \tag{16.1.4}$$

The Lie algebra is readily found to be

$$[U, T] = S, \quad [U, S] = T, \quad [S, T] = -U. \tag{16.1.5}$$

Comparing this with the algebra of SU(2) above we see that it is different. However, if we identify $J_1 = T, J_2 = iS$ and $J_3 = iU$, the algebras are the same. Thus, although the two real algebras are different, once taken over the complex numbers, that is, viewed as complex Lie algebras, they are the same algebra. They are different real forms of the complex Lie algebra SL(2, $\mathbf{C}$), also denoted A_1.

We can define a linear transformation on the space of the Lie algebra for any element X of the Lie algebra by

$$Y \rightarrow \mathrm{adj}(X)Y \equiv [X, Y], \quad \forall \, Y \in G. \tag{16.1.6}$$

It obeys the equation $\mathrm{adj}([X, Y]) = [\mathrm{adj}(X), \mathrm{adj}(Y)]$ for any elements X, Y of G. This is obvious as if we apply it to any element of the Lie algebra Z it automatically satisfies $[[X, Y], Z] = [X, [Y, Z]] - [Y, [X, Z]]$. As a result, $\mathrm{adj}(X)$ is a representation of the Lie algebra; it is called the *adjoint representation*. The reader can find more on representations in the next section, but the above is all that is required in this section.

To be concrete, given the algebra of equation (16.1.2) we can consider the transformation of any element T_b acting on any other element T_a by

$$T_a \rightarrow [T_b, T_a] = \sum_c f_{ba}{}^c T_c, \tag{16.1.7}$$

that is, the action of the generator T_b is represented by the matrix $f_{ba}{}^c$.

The adjoint representation allows us to define a map from $G \otimes G$ to the complex numbers, for any finite dimensional Lie algebra G, by

$$B(X, Y) = \frac{1}{c_{adj}}\mathrm{tr}(\mathrm{adj}(X)\mathrm{adj}(Y)). \tag{16.1.8}$$

In this equation $\mathrm{adj}(X)$ means the matrix corresponding to the linear transformation when it is carried out in a particular basis of the Lie algebra. However, the result $B(X, Y)$ is independent of the basis used. Here c_{adj} is a constant whose value can be chosen to give more elegant expressions. It is equivalent to an overall scaling of the elements of the Lie algebra which one might equally well choose from the outset. Clearly

$$B(X, Y) = B(Y, X) \quad \text{and} \quad B([X, Y], Z) = B(X, [Y, Z]) \tag{16.1.9}$$

using the standard properties of the trace. Hence B provides a candidate scalar product on G whose metric has components $G_{ab} = B(T_a, T_b)$ in the basis we are using. It is called the *Killing form*, or *Killing metric*, and it can be expressed in terms of the structure constants by $G_{ab} = \sum_{c,d} f_{ac}{}^d f_{bd}{}^c$. Taking $X = T_a$, $Y = T_b$ and $Z = T_c$ the above invariance property of the Killing form leads to the equation

$$\sum_d (f_{ba}{}^d G_{dc} + f_{bc}{}^d G_{ad}) = 0 \quad \text{or equivalently} \quad f_{abc} \equiv \sum_d f_{ab}{}^d G_{dc} = f_{[abc]}.$$

(16.1.10)

An *ideal*, I, is a sub-vector space of G which consists of elements X such that $[X, Y] \in I \ \forall \ Y \in G$. If I_1 and I_2 are two ideals of G, then $[I_1, I_2]$, the commutator of all elements of I_1 with I_2, is an ideal. This follows trivially by applying the Jacobi identity: $[X, [I_1, I_2]] = [[X, I_1], I_2] - [I_1, [I_2, X]]$ for any element $X \in G$. Given a Lie algebra G we can construct from the sets $G^{(n+1)} = [G^{(n)}, G^{(n)}]$, where $G^{(0)} = G$. A Lie algebra G is *soluble* if $G^{(n)} = 0$ for some n. Alternatively, given a Lie algebra we can construct the algebras $G_{(n+1)} = [G_{(n)}, G]$. We say a Lie algebra G is *nilpotent* if $G_{(n+1)} = 0$ for some $n \geq 1$. It is straightforward to see that $G^{(n)}$ and $G_{(n)}$ are ideals of G using that the commutator of two ideals is also an ideal. Then it follows that $G^{(n+1)} = [G^{(n)}, G^{(n)}] \subset G^{(n)}$ as $G^{(n)}$ is an ideal.

A Lie algebra is said to be *simple* if it is not an Abelian algebra and it possesses no ideal other than G itself. A Lie algebra is *semi-simple* if it does not possess a non-zero soluble ideal. Hence a semi-simple algebra can possess an ideal, in which case it is not simple, but this ideal may not be soluble as then it is not semi-simple. It turns out that an equivalent definition of a semi-simple Lie algebra is that is should not possess an Abelian ideal. Also one can show that any semi-simple Lie algebra is the direct sum of simple Lie algebras. It is not difficult to show that if the Killing form of a Lie algebra vanishes identically, then the algebra is soluble. A necessary and sufficient condition for a Lie algebra to be semi-simple is that the scalar product is non-degenerate, that is, if $B(X, Y) = 0$, $\forall \ X \in G$, then $Y = 0$. This is equivalent to $\det G_{ab} \neq 0$. We will now show this result, although the reader who wants to get ahead can skip the next three paragraphs on a first reading.

Let us take a semi-simple algebra to be one that has no Abelian ideal and consider a Lie algebra that has an Abelian ideal, whose generators we denote by putting a bar on the index, that is, $T_{\bar{a}}$. We use the notation that those elements in the ideal carry the indices $\bar{a}, \bar{b}, \ldots$, those not belonging to the ideal carry the indices $\hat{a}, \hat{b}, \ldots$ and generic elements carry the indices $a, b, \ldots$. Since the algebra is Abelian $f_{\bar{a}\bar{b}}{}^c = 0$ and as it is an ideal $f_{\bar{b}d}{}^{\hat{c}} = 0$. Then we find that $G_{a\bar{b}} = \sum_{c,d} f_{ac}{}^d f_{\bar{b}d}{}^c = \sum_{\bar{c},d} f_{a\bar{c}}{}^d f_{\bar{b}d}{}^{\bar{c}} = \sum_{\bar{c},d} f_{a\bar{c}}{}^d f_{\bar{b}d}{}^{\bar{c}} = 0$. Clearly in this case $\det G_{ab} = 0$ as it has a vanishing column.

Conversely, if $\det G_{ab} = 0$, then there exist solutions r^a such that $\sum_b G_{ab} r^b = 0$ and all the $\sum_a r^a T_a$ form an ideal. It is an ideal as $[r^a T_a, T_b] = \sum_c s^c T_c$, where $s^c = \sum_a r^a f_{ab}{}^c$; however, $\sum_c G_{ac} s^c = \sum_{c,d} G_{ac} f_{db}{}^c r^d = \sum_{c,d} f_{dbc} r^d = \sum_d f_{bcd} r^d = \sum_{c,d} f_{bc}{}^e G_{ed} r^d = 0$ using equation (16.1.10). Furthermore, the Killing form vanishes, on all generators of the form $r^a T_a$ as $(\sum_a r_1^a T_a, \sum_b r_2^b T_b) = \sum_{a,b} r_1^a r_2^b G_{ab} = 0$. Hence, we have found an ideal restricted to which the Killing form vanishes. Since the necessary and sufficient condition for a Lie algebra to be soluble is that its Killing form vanishes, it follows that the ideal is soluble and so the algebra is not semi-simple.

The adjoint representation is faithful if $\text{adj}(X) = \text{adj}(Y)$ implies that $X = Y$. If this is the case, then $[X, Z] = [Y, Z]$ for all $Z \in G$ and so $[X - Y, Z] = 0$. Hence $X - Y$ belongs to an ideal, which is also soluble as it is spanned by just one element whose commutator with itself vanishes, unless $X = Y$. Consequently, if a Lie algebra is semi-simple, then the adjoint representation is faithful. Furthermore, if the Lie algebra is simple, then the adjoint representation is irreducible. For if the adjoint representation is not irreducible, then there exists a sub-set $\hat{G}$ of G such that $\text{adj}(X)(Y) = [X, Y] \in \hat{G}$ for all $Y \in \hat{G}$ and for all $X \in G$. In this case $\hat{G}$ is an ideal. This contradicts the definition of simple. Since this ideal need not be soluble, this last theorem does not hold for semi-simple Lie algebras.

From now on in this and the next section, we will restrict our attention to finite dimensional semi-simple Lie algebras. In this case $\det G_{ab} \neq 0$ and so the inverse G^{ab} exists and we can raise and lower indices using it. In particular we can as above define $f_{abc} = f_{ab}{}^d G_{dc}$. It is clear from equation (16.1.10) that f_{abc} is totally anti-symmetric in its indices.

For a semi-simple Lie algebra we can find a *Cartan sub-algebra*, denoted H. This is a maximal set of commuting generators, H_i,

$$[H_i, H_j] = 0, \tag{16.1.11}$$

which is also adj diagonalisable. The latter means we can find a basis of the Lie algebra such that each element of it is an eigenvector of the adjoint representation restricted to the Cartan sub-algebra. This is equivalent to demanding that the adjoint representation restricted to the Cartan sub-algebra be completely decomposable. The two results are equivalent as all irreducible representations of an Abelian Lie algebra are one-dimensional; but H is an Abelian Lie algebra, so its representation adj H can be written as a sum of one-dimensional representations. Hence given a Cartan sub-algebra we have a basis of the rest of the Lie algebra E_α, such that

$$[H_i, E_\alpha] = \alpha_i E_\alpha. \tag{16.1.12}$$

Thus the structure constant that occurs in this commutator is given by $f_{i\alpha}{}^\beta = \alpha_i \delta_\alpha^\beta$. Using this form of the structure constants we see that the Killing form restricted to the Cartan sub-algebra is given by $G_{ij} = \sum_{\alpha,\beta} f_{i\alpha}{}^\beta f_{j\beta}{}^\alpha = \sum_\alpha \alpha_i \alpha_j$. The number of commuting elements H_i in the Cartan sub-algebra is called the *rank* of the algebra and we will often denoted it by r. It can be shown that there always exists a Cartan sub-algebra and that any two Cartan sub-algebras are related by an inner automorphism of the algebra and so are essentially identical. This means that if H_i is a basis for one Cartan sub-algebra, then $e^{-\text{adj}(X)}H_i, i = 1, \ldots, r$, form a basis of the other Cartan sub-algebra.

Equation (16.1.12) implies that with any generator E_α, not in the Cartan sub-algebra, we can associate a set of numbers α_i, one for each H_i. Taken together these α_i can be thought of as a vector, denoted α in an r-dimensional space, the i index labelling its components. The set of all αs is called the roots and is denoted by Δ. Thus the Lie algebra G consists of the elements

$$G = \{H_i, E_\alpha, \alpha \in \Delta\}. \tag{16.1.13}$$

It can be shown that there is a unique element of the Lie algebra for every root in the Lie algebra and that there are no roots which are identically zero.

Let us return to our previous example, the Lie algebra SU(2). We found that the basis adopted previously, J_i, $i = 1, 2, 3$, obeyed $[J_i, J_j] = \epsilon_{ijk}J_k$. Defining $H = i\sqrt{2}J_3$ and $J_{\pm} = (iJ_1 \mp J_2)$ we find that the Lie algebra becomes

$$[H, J_{\pm}] = \pm\sqrt{2}J_{\pm}, \quad [J_+, J_-] = \sqrt{2}H. \tag{16.1.14}$$

Clearly, H is a Cartan sub-algebra and the algebra has rank 1. We have only two roots $\pm\sqrt{2}$, that is, $\Delta = \{\sqrt{2}, -\sqrt{2}\}$. We note that the definition of the above generators required the complex numbers and so is outside the framework of the original SU(2) algebra, in fact we have the algebra A_1. The adjoint representation of A_1 is given by

$$\text{adj } H = \begin{pmatrix} 0 & 0 & 0 \\ 0 & \sqrt{2} & 0 \\ 0 & 0 & -\sqrt{2} \end{pmatrix}, \quad \text{adj } J_+ = \begin{pmatrix} 0 & -\sqrt{2} & 0 \\ 0 & 0 & 0 \\ \sqrt{2} & 0 & 0 \end{pmatrix},$$

$$\text{adj } J_- = \begin{pmatrix} 0 & 0 & \sqrt{2} \\ -\sqrt{2} & 0 & 0 \\ 0 & 0 & 0 \end{pmatrix}, \tag{16.1.15}$$

where we take the basis of the Lie algebra to be H, J_+, J_-. The Killing form is then easily found to have the only non-zero components given by $B(H, H) = 4/c_{adj}$, $B(J_+, J_-) = 4/c_{adj}$.

One can carry out a similar treatment for the algebra SL(2, **R**) and one finds the above A_1 algebra if one takes the generators $E = T + S$, $F = S - T$ and $H = \sqrt{2}U$.

The scalar product B on G satisfies $B(E_\alpha, E_\beta) = 0$ unless $\alpha + \beta = 0$ as is obvious from equation (16.1.9) applied with $X = E_\alpha, Y = H_i$ and $Z = E_\beta$ as this equation then implies that $(\alpha_i + \beta_i)B(E_\alpha, E_\beta) = 0$. Taking $X = H_i$, $Y = H_j$ and $Z = E_\alpha$ implies that $B(E_\alpha, H_i) = 0$. It follows that $B(E_\alpha, *)$, where $*$ is any generator, is zero except for $B(E_\alpha, E_{-\alpha}) = G_{\alpha-\alpha}$ and this must be non-zero as otherwise $\det G_{ij} = 0$. Consequently, if α is a root, so is $-\alpha$. Equation (16.1.12) does not set the scale of E_α and we may choose this scale to set $G_{\alpha-\alpha} = 1$. Thus the only non-zero components of the Killing metric are $G_{\alpha-\alpha} = 1 = G_{-\alpha\alpha}$ and those with all indices within the Cartan sub-algebra, G_{ij}. As a result, by thinking about the form of $B(X, Y)$, we see that if $\det G_{ab} \neq 0$ it follows that $\det G_{ij} \neq 0$. The last step can also be arrived at as follows. If there exists an h in the Cartan sub-algebra such that $B(h, k) = 0$ for all k in the Cartan sub-algebra, then, as $B(h, E_\alpha) = 0$, we find that $B(h, X) = 0$ for all $X \in G$ and so the scalar product B is degenerate on G. However, this conflicts with the semi-simple nature of G and so can not be the case. As a result, the scalar product B restricted to the Cartan sub-algebra is non-degenerate, which is equivalent to $\det G_{ij} \neq 0$.

Using the choice $G_{\alpha-\alpha} = 1$ we find that

$$f_{\alpha-\alpha}{}^i = \sum_k G^{ik}f_{\alpha-\alpha k} = \sum_k G^{ik}f_{k\alpha-\alpha} = \sum_k G^{ik}f_{k\alpha}{}^\alpha = \sum_k G^{ik}\alpha_k \equiv \alpha^i,$$

$$\tag{16.1.16}$$

where G_{ij} is the restriction of the Killing metric to the Cartan sub-algebra and G^{ij} is its inverse. This implies that

$$[E_\alpha, E_{-\alpha}] = \sum_i \alpha^i H_i. \tag{16.1.17}$$

We may write equation (16.1.12) in the form $[h, E_\alpha] = \alpha(h)E_\alpha$, where $h = \sum_i c^i H_i$ is any member of the Cartan sub-algebra and $\alpha(h) = \sum_i c^i \alpha_i$. Thus, using equation (16.1.12), any root α determines a complex number $\alpha(h)$ for any element of the Cartan sub-algebra h. Thus any root α can be thought of as a linear map from the Cartan sub-algebra H to the complex numbers $\mathbf{C}$, that is, $\alpha : H \to \mathbf{C}$ or $h \mapsto \alpha(h)$. Thus the roots belong to the vector space H^* dual to the Cartan sub-algebra H. Indeed from this viewpoint we can define $G^\alpha = \{X \in G \text{ such that } [h, X] = \alpha(h)X, \ \forall h \in H\}$. If $G^\alpha \neq 0$, we call α a root. As we have seen if α is a root, so is $-\alpha$ and one can show that this is the only multiple of α that is a root. As we have already mentioned, for finite dimensional semi-simple Lie algebras, there is only one element of the Lie algebra with the same root, that is, $\dim G^\alpha = 1$.

Since Cartan sub-algebra H possesses a non-degenerate scalar product, H^* is isomorphic to H. Indeed, using that the Killing form is non-degenerate on the Cartan sub-algebra, we can identify a unique element of H from any element of H^*; in particular by setting $\alpha(h) = B(h_\alpha, h)$ we find for each root α a unique member of the Cartan sub-algebra, h_α. Writing $h_\alpha = d^i H_i$ this identification implies that $\alpha(h) = \sum_i c^i \alpha_i = \sum_{ij} d^i c^j B(H_i, H_j) = \sum_{ij} d^i c^j G_{ij} = \sum_j c^j d_j$ for all c^i and so $h_\alpha = \sum_{ij} \alpha_i G^{ij} H_j = \sum_j \alpha^j H_j$. Thus the correspondence between the roots in H^* and the Cartan sub-algebra H is given by

$$\alpha \leftrightarrow h_\alpha = \sum_j \alpha^j H_j. \tag{16.1.18}$$

The great advantage of this way of viewing things is that it provides us with a scalar product on the space of roots by taking

$$(\alpha, \beta) \equiv B(h_\alpha, h_\beta), \text{ or equivalently, } (\alpha, \beta) = \sum_{i,j} \alpha^i G_{ij} \beta^j = \sum_i \alpha^i \beta_i. \tag{16.1.19}$$

Defining the roots as linear functionals on the Cartan sub-algebra also provides us with a definition of the roots that does not rely on a particular choice of basis of the Cartan sub-algebra. To become used to this way of thinking it is instructive to give an alternative derivation of equation (16.1.17). Using the invariance of the Killing form, we find that

$$B([E_\alpha, E_{-\alpha}], h) = B(E_\alpha, [E_{-\alpha}, h]) = \alpha(h)B(E_\alpha, E_{-\alpha}) = B(h_\alpha, h)B(E_\alpha, E_{-\alpha}), \tag{16.1.20}$$

which implies that

$$B([E_\alpha, E_{-\alpha}] - B(E_\alpha, E_{-\alpha})h_\alpha, h) = 0 \ \forall \ h \in H. \tag{16.1.21}$$

However, as the Killing form on the Cartan sub-algebra is non-degenerate we conclude that $[E_\alpha, E_{-\alpha}] = B(E_\alpha, E_{-\alpha})h_\alpha$, which with our above choice of $B(E_\alpha, E_{-\alpha}) = 1$ gives equation (16.1.10).

The Killing metric on the Cartan sub-algebra is given by

$$B(h_\alpha, h_\beta) = \sum_{\gamma \in \Delta} \gamma(h_\alpha)\gamma(h_\beta) = \sum_{\gamma \in \Delta} (\gamma, \alpha)(\gamma, \beta) = (\alpha, \beta) \tag{16.1.22}$$

for any two roots α and β. This result follows by simply using the definition of the Killing metric, evaluating the adjoint representation of h_α and taking the trace using that

$B(E_\alpha, E_\beta) = \delta_{\alpha,-\beta}$. In particular, we note that $[h_\alpha, E_\gamma] = \gamma(h_\alpha)E_\gamma$ and $[h_\alpha, H_i] = 0$ from which we find the first step. The second and final steps follow by using $\gamma(h_\alpha) = (\gamma, \alpha)$ and $(\alpha, \beta) = (h_\alpha, h_\beta)$, respectively. Taking $\beta = \alpha$, we find that $\sum_{\gamma\in\Delta}(\gamma, \alpha)(\gamma, \alpha) = (\alpha, \alpha)$, which implies that (α, α) is non-vanishing as the sum over γ is over all roots. Furthermore, dividing by $(\alpha, \alpha)^2$ it follows that

$$\frac{4}{(\alpha, \alpha)} = \sum_{\gamma\in\Delta}\left(2\frac{(\gamma, \alpha)}{(\alpha, \alpha)}\right)^2. \tag{16.1.23}$$

We will show in the next section (see equation (16.1.84)) that $2(\gamma, \alpha)/(\alpha, \alpha)$ is an integer and so we conclude that (α, α) is real and positive even though we are working over the complex numbers. It also follows that $(\beta, \alpha) = (h_\beta, h_\alpha)$ is real for any two roots α and β. One can show that the elements h_α span the Cartan sub-algebra.

Although we are analysing complex Lie algebras, it will prove very advantageous to consider the elements of the Cartan sub-algebra over the real numbers. Let us consider an arbitrary element of the Cartan sub-algebra $u = \sum_i c^i h_{\alpha_i}$ where h_{α_i} is an any basis of the Cartan sub-algebra constructed from the roots α_i and $c^i \in \mathbf{R}$. Following similar steps to those just above we find that

$$B(u, u) = \sum_{i,j} c^i c^j B(h_{\alpha_i}, h_{\alpha_j}) = \sum_{i,j}\sum_{\gamma\in\Delta} c^i c^j \gamma(h_{\alpha_i})\gamma(h_{\alpha_j})$$

$$= \sum_{\gamma\in\Delta}\left(\sum_i c^i\gamma(h_{\alpha_i})\right)^2 \geq 0, \tag{16.1.24}$$

since $\gamma(h_{\alpha_i}) = (h_\gamma, h_{\alpha_i})$ is real. Furthermore, if $(u, u) = 0$, then $\sum_i c^i\gamma(h_{\alpha_i}) = 0$ for all roots γ, and choosing γ corresponding to h_{α_k} we conclude that $\sum_i c^i B(h_{\alpha_k}, h_{\alpha_i}) = 0$. However, since the Killing metric restricted to the Cartan subalgebra is non-degenerate we conclude that $u = 0$. As such we have shown that the Killing form is a symmetric positive definite metric when restricted to the Cartan sub-algebra and similarly for the induced scalar product on the dual space of roots. Thus we have found a scalar product with the standard properties and so we are working in a Euclidean space. We note that the basis with respect to which we are using real numbers has its origins in the roots of the algebra.

Let us consider two distinct roots α and β. In this two-dimensional space we can write $(\alpha, \beta) = \sqrt{(\alpha, \alpha)(\beta, \beta)}\cos\theta$, since we are in a Euclidean space. As $2(\alpha, \beta)/(\alpha, \alpha) \in \mathbf{Z}$ we may write

$$2\frac{(\alpha, \beta)}{(\alpha, \alpha)}2\frac{(\beta, \alpha)}{(\beta, \beta)} = 4\cos^2\theta. \tag{16.1.25}$$

As such $4\cos^2\theta = 0, 1, 2, 3$. The reader will have no difficulty analysing the various possible integer values of $p = 2(\alpha, \beta)/(\alpha, \alpha)$, $q = 2(\alpha, \beta)/(\beta, \beta)$ and the resulting ratio of the lengths of the two roots. One finds that either $4\cos^2\theta = 0$ and $(\alpha, \beta) = 0$ or that

$$4\cos^2\theta = \begin{cases} 1, & \text{then } (p, q) = (\pm1, \pm1), \\ 2, & \text{then } (p, q) = (\pm1, \pm2) \text{ or } (p, q) = (\pm2, \pm1), \\ 3, & \text{then } (p, q) = (\pm1, \pm3) \text{ or } (p, q) = (\pm3, \pm1) \end{cases} \tag{16.1.26}$$

and that $(\alpha, \alpha) = d(\beta, \beta)$ for $d = 1, 2, 3$, respectively, for the last three cases and for a suitable choice of the order of the roots α and β. We note that if all the roots have the same length, then only the values $0, \pm 1$ are allowed, and that if $p = \pm 2$ or ± 3, then $q = \pm 1$ and vice versa.

Given the above facts, the Killing form restricted to the Cartan sub-algebra is a real symmetric matrix which we can diagonalise by an orthogonal transformation. Since it is also a positive definite matrix the eigenvalues must be positive and we can scale the generators to set them to be 1. Equivalently we can always choose a reordering and rescaling of the basis such that

$$B(H_i, H_j) = \delta_{ij}. \tag{16.1.27}$$

We will not, in general, use this result in what follows.

As we now explain the commutators between E_α and E_β are of the form

$$[E_\alpha, E_\beta] = \begin{cases} 0, & \text{if } \alpha + \beta \notin \Delta, \\ \sum_i \alpha^i H_i, & \text{if } \alpha + \beta = 0, \\ \epsilon(\alpha, \beta) E_{\alpha+\beta}, & \text{if } \alpha + \beta \in \Delta. \end{cases} \tag{16.1.28}$$

By considering the Jacobi identity involving H_i, E_α and E_β it is clear that the commutator $[E_\alpha, E_\beta]$, for $\alpha + \beta \neq 0$, is proportional to $E_{\alpha+\beta}$ when $\alpha + \beta \in \Delta$ and zero if $\alpha + \beta \notin \Delta$ and when $\alpha + \beta = 0$ is equal to $\alpha^i H_i$. For the latter case we have just used equation (16.1.17). It can be shown that there exist a choice of generators such that $\epsilon(\alpha, \beta) = \pm(p + 1)$. The integer p is specified as follows: if α and β are roots and $\alpha + \beta$ is a root and we consider the root string $\alpha, \alpha - \beta, \ldots \alpha - (p + 1)\beta$, where the final root is the last in the sequence, that is, $\alpha - (p + 1)\beta$ is not a root. In fact this choice of generators is just the Chevalley generators to be introduced shortly. When the algebra is written using equations (16.1.11), (16.1.12) and (16.1.28) it is said to be in the *Cartan–Weyl form* of the Lie algebra. We will soon give another more elegant way of writing the algebra.

The Lie algebra SL(n, **R**) consists of $n \times n$ real traceless matrices. Let us consider in detail the case of SL(3, **R**), a basis of which is given by

$$H_1 = \frac{1}{\sqrt{2}} \begin{pmatrix} 1 & 0 & 0 \\ 0 & -1 & 0 \\ 0 & 0 & 0 \end{pmatrix}, \quad H_2 = \frac{1}{\sqrt{6}} \begin{pmatrix} 1 & 0 & 0 \\ 0 & 1 & 0 \\ 0 & 0 & -2 \end{pmatrix}$$

and

$$E_{\alpha_1} = \begin{pmatrix} 0 & 1 & 0 \\ 0 & 0 & 0 \\ 0 & 0 & 0 \end{pmatrix}, \quad E_{\alpha_2} = \begin{pmatrix} 0 & 0 & 0 \\ 0 & 0 & 1 \\ 0 & 0 & 0 \end{pmatrix}, \quad E_{\alpha_3} = \begin{pmatrix} 0 & 0 & 1 \\ 0 & 0 & 0 \\ 0 & 0 & 0 \end{pmatrix},$$

$$\tag{16.1.29}$$

$$E_{-\alpha_1} = \begin{pmatrix} 0 & 0 & 0 \\ 1 & 0 & 0 \\ 0 & 0 & 0 \end{pmatrix}, \quad E_{-\alpha_2} = \begin{pmatrix} 0 & 0 & 0 \\ 0 & 0 & 0 \\ 0 & 1 & 0 \end{pmatrix}, \quad E_{-\alpha_3} = \begin{pmatrix} 0 & 0 & 0 \\ 0 & 0 & 0 \\ 1 & 0 & 0 \end{pmatrix}.$$

Calculating the commutators we find that

$$[H_i, H_j] = 0, \quad [H_i, E_{\pm\alpha_j}] = \pm(\alpha_j)_i E_{\pm\alpha_j}, \tag{16.1.30}$$

where

$$\alpha_1 = (\sqrt{2}, 0), \quad \alpha_2 = \left(-\frac{1}{\sqrt{2}}, \sqrt{\frac{3}{2}}\right), \quad \alpha_3 = \left(\frac{1}{\sqrt{2}}, \sqrt{\frac{3}{2}}\right). \tag{16.1.31}$$

We note that $\alpha^3 = \alpha^1 + \alpha^2$. The remaining commutators are given by

$$[E_{\alpha_i}, E_{-\alpha_i}] = \sum_j (\alpha_i)_j H_j, \quad [E_{\alpha_1}, E_{\alpha_2}] = E_{\alpha_3}, \quad [E_{-\alpha_1}, E_{-\alpha_2}] = -E_{-\alpha_3},$$

$$[E_{\alpha_1}, E_{-\alpha_3}] = -E_{-\alpha_2}, \quad [E_{\alpha_2}, E_{-\alpha_3}] = E_{-\alpha_1}, \quad [E_{-\alpha_1}, E_{\alpha_3}] = E_{\alpha_2}, \quad [E_{-\alpha_2}, E_{\alpha_3}] = -E_{-\alpha_1},$$

$$(16.1.32)$$

where $(\alpha_j)_i$ denotes the ith component of the vector α_j. The remaining commutators vanish. Cearly, H_1, H_2 are a Cartan sub-algebra and so the algebra has rank 2. In fact, the algebra is already in Cartan–Weyl basis and we recognise that it does have the correct features as derived in equation (16.1.28). The root space is given by $\Delta = \{\alpha_1, \alpha_2, \alpha_1 + \alpha_2; -\alpha_1, -\alpha_2, -\alpha_1 - \alpha_2\}$. If we take the Lie algebra to be over the complex numbers it is denoted by the algebra A_2, another real form of which is SU(3).

Given Δ we can divide it into two spaces denoted Δ_+ and Δ_-, which are called *the positive and negative root* spaces. We first choose a particular ordered basis of the vector space spanned by Δ. A root α is positive when written in terms of this chosen basis if its first component is positive and if this is zero its second component is positive etc. If a root is not positive, it is defined to be negative. We note that the actual division is basis dependent, but the general results are independent of the basis. Among the set of positive roots Δ_+ is a set of simple roots, α_a. A simple root is a positive root that cannot be written as a sum of two positive roots. The difference of two simple roots is not a simple root. To see this consider $\gamma = \alpha_a - \alpha_b$, where α_a and α_b are simple roots, then if γ is a positive root we can write $\alpha_a = \gamma + \alpha_b$, which is a contradiction, while if γ is a negative root we can write a similar contradictory equation for α_b. It follows from the definition of the simple roots that all roots in Δ_+ can be written in terms of simple roots, that is, for any root α in Δ_+, $\alpha = \sum_a n_a \alpha_a$ and it turns out that the n_a are positive integers or zero. As one might expect the number of simple roots is equal to the rank of the algebra r. If α_a is a simple root, $-\alpha_a \in \Delta_-$ and any $\beta \in \Delta_-$ can be written in terms of the elements $\{-\alpha_a, a = 1, \ldots, r\}$ of Δ_- with positive integer coefficients. As a result the elements of the Lie algebra can be written as

$$G = G_- + G_0 + G_+, \quad (16.1.33)$$

where $G_\pm = \{E_\alpha; \alpha \in \Delta_\pm\}$ and $G_0 = H$. An algebra is *simply laced* if all its simple roots have the same length squared which we may, and will, take to be 2.

For the algebra SL(3, **R**) discussed above if we take our chosen ordered basis to be $(0, 1)$, $(1, 0)$, then $\Delta_+ = \{\alpha_1, \alpha_2, \alpha_1 + \alpha_2\}$ and $\Delta_- = \{-\alpha_1, -\alpha_2, -\alpha_1 - \alpha_2\}$.

Given any root we may write it in terms of the simple roots, $\alpha = \sum_a n_a \alpha_a$ and its height h is defined to be $h = \sum_a n_a$. Since we are dealing with a finite dimensional algebra there must be a root with a maximum height and this is called the *highest root* and denoted by θ. Although there could be two or more roots with the highest length it turns out that it is unique. It can be written in terms of the simple roots $\theta = \sum_a m_a \alpha_a$ and we refer to the m_a as the *Kac labels*. The *Coxeter number*, $h(G)$, of the Lie algebra G is related to the height of the highest root by $h(G) = 1 + \sum_{a=1} n_a$. Clearly, $\theta = \alpha_1 + \alpha_2$ for A_2 and $h(A_2) = 3$.

Given the simple roots we can form the *Cartan matrix* A_{ab} of G by

$$A_{ab} = 2 \frac{(\alpha_a, \alpha_b)}{(\alpha_a, \alpha_a)}. \quad (16.1.34)$$

We note that to evaluate this the scalar product is constructed using the Killing metric restricted to the Cartan sub-algebra as discussed above. For A_2, one of whose real forms is SL(3, **R**), we find that

$$A_{ab} = \begin{pmatrix} 2 & -1 \\ -1 & 2 \end{pmatrix}. \tag{16.1.35}$$

We note that the Cartan matrix A_{ab} is symmetric if and only if the algebra is simple.

The Cartan matrix must have the properties

$$A_{aa} = 2, \tag{16.1.36}$$

$$\text{if } A_{ab} = 0 \text{ then } A_{ba} = 0, \tag{16.1.37}$$

A_{ab} can only take the values $0, -2, -3$ and -1

with the corresponding $A_{ba} = 0, -1, -1,$ and $-1, -2, -3,$

respectively, $\tag{16.1.38}$

$$\frac{A_{ab}}{(\alpha_b, \alpha_b)} = 2 \frac{(\alpha_a, \alpha_b)}{(\alpha_a, \alpha_a)(\alpha_b, \alpha_b)} \text{ is a positive definite matrix.} \tag{16.1.39}$$

The first two properties are obvious. The third property follows from equation (16.1.26) and the fact proved in the next section that $(\alpha_a, \alpha_b) \leq 0$ for any two simple roots α_a and α_b. The final property is proved as follows,

$$\sum_{a,b} c^a c^b \frac{A_{ab}}{(\alpha_b, \alpha_b)} = \left(\sum_a \frac{c^a \alpha_a}{(\alpha_a, \alpha_a)}, \sum_b \frac{c^b \alpha_b}{(\alpha_b, \alpha_b)} \right) \geq 0. \tag{16.1.40}$$

Here $c^a \in \mathbf{R}$ and the result follows as we showed in equation (16.1.23) that (γ, γ) is real and positive for any root γ.

Let us consider a particular c^a, that which has non-zero entries for only c^a and C^b, whereupon the positivity condition reduces to that for a 2×2 matrix with indices a and b. This matrix has 2s on the diagonals and taking the determinant to be positive we find that $A_{ab}A_{ba} < 4$ if $a \neq b$. For the simply laced case the bound becomes $(A_{ab})^2 \leq 3$ and so the off-diagonal elements can only take the values $\pm 0, \pm 1$. Thus we recover part of the result of equation (16.1.38) from equation (17.1.1.39).

Given any Cartan matrix we can draw a Dynkin diagram, which is a pictorial way of writing the Cartan matrix. We draw r dots, or nodes, labelled $a = 1, \ldots, r$ and connect node a with node b by $A_{ab}A_{ba}$ lines. If two nodes have associated simple roots of different length, we also draw an arrow pointing to the shorter of the two roots on any lines that may connect them. If all the simple roots have the same length, that is, the algebra is simply laced, we do not need any arrows. From the Dynkin diagram one can uniquely reconstruct the Cartan matrix for the finite dimensional semi-simple Lie algebras being studied in this section up to permutations of the labelling of the simple roots. This relies on the properties of the Cartan matrix contained in equations (16.1.36)–(16.1.39). The former value only occurs when there is a line in the Dynkin diagram between nodes a and b. For A_1 the Dynkin diagram is just a single node, that is, •, and for A_2 the Dynkin diagram is just two nodes joined by a single line, that is, • — •. Later when dealing with Kac–Moody algebras we will have to take a modified definition of the Dynkin diagram.

As a finite dimensional semi-simple Lie algebra must have a Cartan matrix which obeys the properties of equations (16.1.36)–(16.1.39), there are only a very limited number of possibilities for the Cartan matrix and so for the possible semi-simple Lie algebras. It is instructive to find all such algebras of rank 2. The Cartan matrix must be of the form

$$A = \begin{pmatrix} 2 & -n \\ -m & 2 \end{pmatrix},$$ (16.1.41)

where n and m are positive integers. Now as the Cartan matrix is positive definite it has $\det A > 0$ and so $\det A = 4 - nm > 0$. The possibilities are $n = m = 1$, which is A_2, $n = 1$, $m = 2$, which is C_2, of which SO(5) is a real form, and $n = 1$, $m = 3$, which is the exceptional algebra G_2.

The list of all possible semi-simple Lie algebras was found by Killing and Cartan and the result is the Lie algebras

$$A_n, B_n, C_n, D_n, ; G_2, F_4, \quad \text{and} \quad E_n, n = 6, 7, 8.$$ (16.1.42)

Particular real forms of the first few are SL(n, **R**), SO(n, $n + 1$), Sp($2n$), SO(n, n) respectively. Only the algebras A_n, D_n, E_n, $n = 6, 7, 8$ are simply laced. The Dynkin diagrams, simple roots and some other properties of most of these algebras are given in appendix D. As we will see shortly, for every Cartan matrix which satisfies the properties in equations (16.1.36)–(16.1.39) we do indeed find a finite dimensional semi-simple Lie algebra.

It will prove very useful to introduce the *Chevalley generators*. These consist of $E_a = \sqrt{(2/(\alpha_a, \alpha_a))} E_{\alpha_a}, a = 1, \ldots, r$, that is, the suitably scaled generators corresponding to the simple roots, the generators corresponding to the negative simple roots $F_a = \sqrt{(2/(\alpha_a, \alpha_a))} E_{-\alpha_a}$ and finally the generators of the Cartan sub-algebra which we take in the form $H_a = 2\alpha_a^i H_i / (\alpha_a, \alpha_a), a = 1, \ldots, r$. This makes $3r$ generators which using equations (16.1.11) (16.1.12) and (16.1.28) must obey the relations

$$[H_a, H_b] = 0,$$

$$[H_a, E_b] = A_{ab} E_b, \quad [H_a, F_b] = -A_{ab} F_b,$$ (16.1.43)

$$[E_a, F_b] = \delta_{a,b} H_a.$$

We note that these relations are completely specified by the Cartan matrix A_{ab} and that E_a, F_a, H_a for a given a obey the commutators of A_1. For the case of A_1 these commutators are those of the full algebra.

Equation (16.1.43) contains, in general, only a few of the commutators of the Lie algebra. However, Serre proved the following theorem.

Theorem   Given a finite dimensional semi-simple Lie algebra G with $3r$ Chevalley generators E_a, F_a, H_a which obey equation (16.1.43) and a Cartan matrix, that obeys the properties of equations (16.1.36)–(16.1.39), we may uniquely reconstruct the complete Lie algebra G if we take the Chevalley generators to obey the further relation

$$[E_a, [E_a, \ldots E_b] \ldots] = 0$$ (16.1.44)

where we have $1 - A_{ab}$ factors of E_a. There is a similar relation for the F_as.

This relation is called the Serre relation and it can also be written as

$$(adjE_a)^{1-A_{ab}}E_b = 0. \tag{16.1.45}$$

We note that this last relation is also completely specified by the Cartan matrix.

Given a Cartan matrix A_{ab} we can introduce the $3r$ Chevalley generators that obey equations (16.1.43) and (16.1.44). The Lie algebra is just given by the multiple commutators of E_a,

$$[E_{a_1}, [E_{a_2}, \dots [E_{a_{n-1}}, E_{a_n}] \dots], \tag{16.1.46}$$

and a similar set of commutators for the F_a. The algebra is then just all these commutators subject to equations (16.1.43) and (16.1.44). That is, it is the multiple commutators $[E_{\alpha_1}, [E_{\alpha_2}, [E_{\alpha_3} \dots [E_{\alpha_{n-1}}, E_{\alpha_n}] \dots]]$ subject to equations (16.1.43) and (16.1.44) with a similar commutators for the F_a. We need not write commutators involving E_a and F_a as using equation (16.1.43) and the Jacobi identity we may eliminate the E_a or F_a whichever are the fewer.

This is a remarkable result since, as we have already observed, equations (16.1.43) and (16.1.44) are completely specified by the Cartan matrix A_{ab} and so any finite dimensional semi-simple Lie algebra is completely specified by its Cartan matrix A_{ab}, or equivalently, its Dynkin diagram. This result ensures that for any acceptable Cartan matrix, that is, one that obeys equations (16.1.36)–(16.1.39), there is a corresponding unique finite dimensional semi-simple Lie algebra.

Let us examine how this goes for A_2 with Cartan matrix A_{ab} given in equation (16.1.35). This has rank 2 so we have the Chevalley generators H_a, E_a, F_a, $a = 1, 2$. Let us denote the roots associated with E_a by α_a, $a = 1, 2$. Considering the E_a commutators we have $[E_1, E_2]$ with root $\alpha_1 + \alpha_2$. However, we do not have $[E_1, [E_1, E_2]]$ as $1 - A_{12} = 2$ and so this multiple commutator vanishes due to the Serre relation. Similarly $[E_2, [E_2, E_1]] = 0$. Hence the positive root generators are E_1, E_2, $[E_1, E_2]$ with roots α_1, α_2 and $\alpha_1 + \alpha_2$. We have analogous expressions for the negative root generators, that is, F_1, F_2, $[F_1, F_2]$, and so the negative root roots are $-\alpha_1$, $-\alpha_2$ and $-(\alpha_1 + \alpha_2)$. Hence we have three positive root generators, three negative root generators and two Cartan sub-algebra generators. Furthermore, A_2 is simply laced as the Cartan matrix is symmetric and choosing the length squared of all the fundamental roots to be 2 we find that $(\alpha_1, \alpha_2) = -1$. The reader will recognise the root space of SL(3) and using Jacobi identities and equations (16.1.43) and (16.1.44) the A_2, or SL(3), Lie algebra.

For a simply laced algebra the Killing form on the Chevalley generators is given by

$$B(E_a, F_b) = \delta_{a,b}, \quad B(H_a, H_b) = A_{ab} \tag{16.1.47}$$

with all other scalar products being 0. These relations follow from our previous discussion, the first from $B(E_\alpha, E_{-\alpha}) = 1/c_{adj}$ for a suitable choice of c_{adj} and the second from the relation between the Chevalley generators and those of the Cartan–Weyl basis given above equation (16.1.43). We note that in the simply laced case the Killing metric evaluated on the the Cartan sub-algebra using the basis of the Chevalley generators H_a is just the Cartan matrix. For non-simply laced Lie algebras these equations cannot be correct as the Killing form is symmetric, but the Cartan matrix, unlike for the simply laced case, is not. For non-simply laced Lie algebras the Killing form is given by $B(E_a, F_b) = \delta_{a,b}d_a^{-1}$, $B(H_a, H_b) = A_{ab}d_b^{-1}$, where $d_a = (\alpha_a, \alpha_a)/2$. We observe that the insertion of the factors d_a do indeed lead to a symmetric Killing form. Inserting the Cartan matrix of equation (16.1.34). We find that $B(H_a, H_b) = 4(\alpha_a, \alpha_b)/(\alpha_a, \alpha_a)(\alpha_b, \alpha_b)$.

One can ask if the Killing metric given in equation (16.1.47) is the unique metric which has the invariance property of equation (16.1.9). In fact, the invariant metric is only unique up to an overall constant of proportionality for simple algebras. However, it turns out that every semi-simple algebra can be written in terms of a direct sum of simple algebras and so we find that the invariant metric of a semi-simple Lie algebra has as many constants as there are simple factors.

Using the notion of the roots as linear functions on the Cartan sub-algebra, the second of the equations (16.1.43) can be written as $[H_a, E_b] = \alpha_b(H_a)E_b = A_{ab}E_b$. The Cartan sub-algebra element h' corresponding to the root α_a is such that $\alpha_b(h) = B(h', h)$ for all elements h in the Cartan sub-algebra. We note that for the simply laced case $\alpha_b(H_a) = B(H_b, H_a) = A_{ab}$ using equation (16.1.47) together with the fact that H_a span the Cartan subalgebra implies that the root $\alpha_b \in H^*$ corresponds to $H_b \in H$. For the non-simply laced case α_a corresponds to $H_a(\alpha_a, \alpha_a)/2$ since $\alpha_a(H_b) = B(H_a(\alpha_a, \alpha_a)/2, H_b) = A_{ba}$.

Let us summarise the above development, which largely follows the historical path. A Lie algebra always possesses an adjoint representation. This enabled us to define a scalar product, the Killing form, on the Lie algebra. We then introduced the concept of a Cartan sub-algebra and diagonalising the adjoint action of these generators we defined a set of roots. The roots can be thought of as linear functionals on the Cartan sub-algebra. We then restricted our study to semi-simple Lie algebras, which is equivalent to demanding that the Killing form is non-degenerate. This allowed us to associate with each root a unique element of the the Cartan sub-algebra and so to define a scalar product on the space of roots. The roots were divided into positive and negative spaces and in the former we found a set of simple roots which are essentially a basis for the positive roots. The scalar product of these simple roots defined the Cartan matrix A_{ab} whose entries were found to take only a very restricted set of values. These restrictions allowed us to deduce all possible finite dimensional semi-simple Lie algebras.

The above did not require us to introduce a particular basis for Cartan sub-algebra, however, as in most of the physics literature we began with the algebra in so-called Cartan–Weyl form, given in equations (16.1.11), (16.1.12) and (16.1.28). We then introduced the Chevalley generators which obeyed the so-called Serre relations of equations (16.1.43)–(16.1.44), which only depend on the Cartan matrix. Finally, we found that any semi-simple Lie algebra could be uniquely reconstructed from its Cartan matrix, or equivalently its Dynkin diagram, using the Chevalley generators when subject to equations (16.1.43) and (16.1.44). The full algebra is formed from multiple commutators of the Chevalley generators.

We now summarise some basic facts about lattices. A lattice is given by

$$\Lambda = \left\{ \sum_{i=1}^{N} n_i e_i : n_i \in \mathbf{Z} \right\}, \tag{16.1.48}$$

where e_i, $i = 1, \ldots, N$ form a basis of a vector space V equipped with a scalar product. The corresponding lattices have a scalar product which they inherit from the underlying vector space. The metric is given by $g_{ij} \equiv e_i \cdot e_j$. We call the lattice Λ *Euclidean* if the underlying vector space has a metric of Euclidean signature (that is, has signature N) and

Lorentzian if the metric has signature $N - 2$, that is, the metric, when diagonalised, has only one negative eigenvalue. A lattice is said to be *integral* if $x \cdot y \in \mathbf{Z}$ for all $x, y \in \Lambda$. An *even lattice* is an integral lattice for which $x \cdot x \in 2\mathbf{Z}$ for all $x \in \Lambda$. If an integral lattice is not even it is called *odd*.

The unit cell of a lattice is the set of points $\{x = \sum_i a_i e_i : 0 \leq a_i < 1\}$; clearly it only contains one lattice point, namely $x = 0$. The volume of the unit cell denoted $\mathrm{vol}(\Lambda)$ is then given by $\mathrm{vol}(\Lambda) = \sqrt{|\det g_{ij}|}$. A lattice is *unimodular* if $\mathrm{vol}(\Lambda) = 1$ in which case there is only one point of Λ per unit volume.

The dual lattice Λ^* is the set of points

$$\Lambda^* = \{y : y \cdot x \in \mathbf{Z} \ \forall \, x \in \Lambda\}. \tag{16.1.49}$$

If the scalar product is non-degenerate, Λ^* is a lattice with a basis consisting of the vectors e_i^*, $i = 1, \ldots, N$, where $e_i^* \cdot e_j = \delta_{ij}$. The metric on Λ^* is $g_{ij}^* = e_i^* \cdot e_j^*$ and the volume of the unit cell $\mathrm{vol}(\Lambda^*) = \sqrt{|\det g_{ij}^*|}$. The definition of e_i^* implies that $\sum_{i=1}^N e_i(e_i^* \cdot e_j) = e_j$, which, in turn, implies that

$$\sum_{i=1}^N e_i^I e_i^{J*} = \delta^{IJ}. \tag{16.1.50}$$

In this equation the indices $I, J, \ldots$ denote the components of the basis vectors. Consequently, multiplying equation (16.1.50) by $e_{kI} e_{lJ}^*$ we find that the metrics of the lattices Λ and Λ^* are related by

$$g_{ij}^* = (g_{ij})^{-1} \tag{16.1.51}$$

and so $\mathrm{vol}(\Lambda^*) = (\mathrm{vol}(\Lambda))^{-1}$. A lattice is *self-dual* if $\Lambda^* = \Lambda$. Clearly, Λ is integral if and only if $\Lambda \subset \Lambda^*$. However, if $\mathrm{vol}(\Lambda) = 1$, then Λ and Λ^* have the same number of points per unit volume and so $\Lambda = \Lambda^*$ if and only if Λ is unimodular and integral.

Self-dual lattices are relatively rare. Indeed, Euclidean even self-dual lattices only exist in dimensions $8n$, $n = 1, 2, \ldots$. There is one such lattice with dimension 8 and two with dimension 16. We will give explicit constructions of these lattices later. Self-dual even Lorentzian lattices only exist in dimensions $8n + 2$, $n = 1, 2, \ldots$; however, in each such dimension the lattice is unique and is denoted by $\Pi^{8n+1,1}$. The simplest such lattice $\Pi^{1,1}$ is in dimension 2. These lattices can be defined as follows. Let $x = (x^1, \ldots, x^{D-1}; x^0) \in \mathbf{R}^{D-1,1}$, and let us denote by $r \in \mathbf{R}^{D-1,1}$ the vector $r = (\frac{1}{2}, \ldots, \frac{1}{2}; \frac{1}{2})$. Then $x \in \Pi^{D-1,1}$ provided that

$$x.r \in \mathbf{Z} \tag{16.1.52}$$

and, in addition, either

$$\text{all } x^\mu \in \mathbf{Z} \quad \text{or} \quad \text{all } x^\mu - r^\mu \in \mathbf{Z}. \tag{16.1.53}$$

In the above, the scalar product is the usual scalar product of Minkowski space, that is,

$$x.y = \sum_{i=1}^{D-1} x^i y^i - x^0 y^0. \tag{16.1.54}$$

Let us now consider in detail the lattice $\Pi^{1,1}$. Using equations (16.1.52) and (16.1.53) it is straightforward to verify that $\Pi^{1,1}$ consists of the vectors

$$(m, 2p+m) \quad \text{and} \quad (n+\tfrac{1}{2}, 2q+n+\tfrac{1}{2}) \quad \forall \, m, n, p, q \in \mathbf{Z}. \tag{16.1.55}$$

It is sometimes useful to use a description of the lattice $\Pi^{1,1}$ in terms of vectors $z = (z^+, z^-)$ that are related to the vectors x given above by

$$z^+ = x^0 + x^1, \qquad z^- = \tfrac{1}{2}(x^0 - x^1). \tag{16.1.56}$$

In terms of these vectors the scalar product becomes $x \cdot y = -z^+ w^- - z^- w^+$, where $w^{\pm}$ are also defined in terms of y^μ as in equations (16.1.56). In the basis described by (z^+, z^-), the vectors of the lattice $\Pi^{1,1}$ have the simple form

$$(n, m) \quad \forall \, n, m \in \mathbf{Z}. \tag{16.1.57}$$

The vector r is now simply $r = (1, 0)$.

The null vectors of $\Pi^{1,1}$ are clearly of the form $(n, 0)$ and $(0, m)$ and so the primitive null vectors can be taken to be given by $k \equiv (1, 0)$ and $\bar{k} \equiv (0, -1)$. We have chosen these vectors such that $k \cdot \bar{k} = 1$. Clearly, all vectors of the lattice $\Pi^{1,1}$ are of the form $pk + q\bar{k}$, where $p, q \in \mathbf{Z}$. There are only two points of length squared 2 in $\Pi^{1,1}$, namely $\pm(k + \bar{k})$.

To every finite dimensional semi-simple Lie algebra we can construct a lattice out of its roots. Indeed, given the simple roots α_a of a Lie algebra G we can define the lattice

$$\Lambda_G = \{n_a \alpha_a, \forall \; n_a \in \mathbf{Z}\}. \tag{16.1.58}$$

For a simply laced algebra all the roots have the same length squared, which we take to be 2, the metric on the root lattice is the Cartan matrix as $(\alpha_a, \alpha_b) = A_{ab}$. Since $A_{ab} = (\alpha_a, \alpha_b)$ is either 2, if $a = b$, or negative integers including 0 otherwise, Λ_G is an integral even lattice. Since the Cartan matrix is a positive definite symmetric form for this case it follows that Λ_G is a Euclidean lattice.

A useful construction is the notion of the coroot, $\check{\alpha} = 2\alpha/(\alpha, \alpha)$. The coroot lattice of a Lie algebra G is given by

$$\check{\Lambda}_G = \{n_a \check{\alpha}_a, n_a \in \mathbf{Z}\}. \tag{16.1.59}$$

For simply laced algebras the simple coroots and the simple roots are the same and so the root lattice and the coroot lattice are equal.

To give a simple example, let us consider the two-dimensional lattice for A_2, one of whose real forms is SL(3, $\mathbf{R}$). The lattice Λ_{A_2} is constructed from the simple roots of A_2. In the last section we found that, in a chosen basis, the simple roots are given by $\alpha_1 = e_1 = (\sqrt{2}, 0)$, $\alpha_2 = e_2 = (-1/\sqrt{2}, \sqrt{3/2})$. The dual lattice is generated by $e_1^* = (1/\sqrt{2}, \sqrt{1/6})$, $e_2^* = (0, \sqrt{2/3})$. Since

$$g_{ab} = A_{ab} = \begin{pmatrix} 2 & -1 \\ -1 & 2 \end{pmatrix},$$

we conclude that Λ_{A_2} is an integral and even lattice. However, $\mathrm{vol}\Lambda_{A_2} = \sqrt{\det A_{ab}} = \sqrt{3}$, implying that Λ_{A_2} is not unimodular.

The first non-trivial example of a self-dual Euclidean lattice occurs in eight dimensions where there is only one such lattice. In fact, it is the root lattice of E_8. This follows from the following observation. The determinant of the Cartan matrix of E_8 is 1 and, as the Cartan matrix is the metric on the E_8 root lattice, it follows that the E_8 root lattice is even and unimodular and so it must be the unique self-dual lattice of dimension 8. This lattice consists of the points

$$\Lambda_{E_8} = \{(x_1, \ldots, x_8); \; \forall \; x_i \in \mathbf{Z} \text{ or } \forall \; x_i \in \mathbf{Z} + \tfrac{1}{2}, \text{ provided for both cases}$$

$$\sum_i x_i = 0, \text{mod } 2\}. \tag{16.1.60}$$

The points of length squared 2 are $(\pm 1, \pm 1, 0^6)$, $(\pm 1, 0, \pm 1, 0^5)$ etc., meaning we take two entries with ± 1 with any combination of signs and zeros in all the other entries and $(\pm\tfrac{1}{2}, \ldots, \pm\tfrac{1}{2})$, all signs in each entry being allowed subject to the number of plus signs being even. We can choose a basis, that is, the simple roots, for the root lattice Λ_{E_8}, to be

$$\alpha_1 = (0, 0, 0, 0, 0, 1, -1, 0),$$
$$\alpha_2 = (0, 0, 0, 0, 1, -1, 0, 0),$$
$$\alpha_3 = (0, 0, 0, 1, -1, 0, 0, 0),$$
$$\alpha_4 = (0, 0, 1, -1, 0, 0, 0, 0),$$
$$\alpha_5 = (0, 1, -1, 0, 0, 0, 0, 0), \tag{16.1.61}$$
$$\alpha_6 = (-1, -1, 0, 0, 0, 0, 0, 0),$$
$$\alpha_7 = \left(\tfrac{1}{2}, \tfrac{1}{2}, \tfrac{1}{2}, \tfrac{1}{2}, \tfrac{1}{2}, \tfrac{1}{2}, \tfrac{1}{2}, \tfrac{1}{2}\right),$$
$$\alpha_8 = (1, -1, 0, 0, 0, 0, 0, 0).$$

The reader may verify that their scalar products lead to the Cartan matrix of E_8 whose corresponding Dynkin diagram is given in appendix D. The numbering of the simple roots is the same as the numbering of the nodes of this Dynkin diagram. The simple roots given here differ slightly from the choice in appendix D.

Given two lattices Λ_1 and Λ_2, it follows from the definition of the dual lattice that

$$(\Lambda_1 \oplus \Lambda_2)^* = \Lambda_1^* \oplus \Lambda_2^*. \tag{16.1.62}$$

Clearly $\Lambda_{E_8} \oplus \Pi^{1,1}$ is a self-dual lattice as each of the component lattices is self-dual. However, since it is also an even Lorentzian lattice it must be the unique even Lorentzian self-dual lattice in ten dimensions, namely $\Pi^{9,1}$.

For completeness we very briefly discuss the real forms of Lie algebras. In the above we have considered Lie algebras over the complex field and so for a generic choice of basis the structure constants will be complex. A real form of a complex Lie algebra is found by constructing a basis for the complex Lie algebra for which the structure constants are real. The real Lie algebra arises by taking this basis, but now over the real numbers. There is, in general, more than one real form for a given complex Lie algebra. For example, for A_1 we can have the real Lie algebras $SU(2)$ and $SL(2, \mathbf{R})$.

Given any semi-simple complex Lie algebra, we can take as a basis for the algebra the elements $T_\alpha = E_\alpha - E_{-\alpha} \; S_\alpha = i(E_\alpha + E_{-\alpha})$, for all positive roots α, and $H'_a = iH_a =$

$2i \sum_i \alpha_a^i H_i/(\alpha_a \alpha_a)$. The algebra is then given by

$$[H_a', T_\alpha] = 2\frac{(\alpha_a, \alpha)}{(\alpha_a \alpha_a)} S_\alpha, \quad [H_a', S_\alpha] = -2\frac{(\alpha_a, \alpha)}{(\alpha_a \alpha_a)} T_\alpha,$$

$$[T_\alpha, T_\beta] = N_{\alpha,\beta} T_{\alpha+\beta} \text{ if } \alpha + \beta \in \Delta,$$

$$[T_\alpha, S_\beta] = \begin{cases} N_{\alpha,\beta} S_{\alpha+\beta} & \text{if } \alpha + \beta \in \Delta \\ \sum_{a,b}(\alpha, \alpha_a) A_{ab}^{-1} H_b' & \text{if } \alpha = \beta \end{cases}, \quad (16.1.63)$$

$$[S_\alpha, S_\beta] = -N_{\alpha,\beta} T_{\alpha+\beta} \text{ if } \alpha + \beta \in \Delta$$

as well as commutators that vanish. We have used that $N_{\alpha,\beta} = -N_{-\alpha,-\beta}$. Since (α, β) is real for any two roots α and β and, for a suitable choice of the underlying E_α elements, $N_{\alpha,\beta}$ is also real, we find that the structure constants are indeed real. Hence the algebra constructed from T_α, S_α and H_a', when taken over the real numbers leads to a real Lie algebra.

In this basis the Killing metric takes the form

$$(H_a', H_b') = -\frac{4}{c_{adj}}\frac{(\alpha_a, \alpha_b)}{(\alpha_a, \alpha_a)(\alpha_b, \alpha_b)},$$

$$(T_\alpha \cdot T_\beta) = -\frac{2}{c_{adj}}\delta_{\alpha-\beta,0}, \quad (S_\alpha \cdot S_\beta) = -\frac{2}{c_{adj}}\delta_{\alpha-\beta,0}, \quad (S_\alpha \cdot T_\beta) = 0 \quad (16.1.64)$$

with all other scalar product vanishing. As $4(\alpha_a, \alpha_b)/(\alpha_a, \alpha_a)(\alpha_b, \alpha_b)$ is a positive definite matrix we find that the Killing form for the full algebra evaluated in this basis is negative definite.

A real Lie algebra is said to *compact* if the Killing form is either positive definite or negative definite. Clearly, in the latter case a positive definite metric is found by just taking it to be minus the Killing form. As such, the real form just given is compact. Indeed, we have shown that given a complex semi-simple Lie algebra we can always find a real form that is compact.

The existence of the above real form is related to the Cartan involution discussed further in section 16.7.3. This takes $E_\alpha \to -E_{-\alpha}$ and $H_a \to -H_a$. This involution preserves the Lie algebra, that is, it preserves the commutators. We see that under this transformation the above generators transform as

$$T_\alpha \to T_\alpha, S_\alpha \to -S_\alpha, H_a' \to -H_a'. \quad (16.1.65)$$

This explains why the elements T_α form a sub-algebra.

It turns out that for every involution of the real Lie algebra we have just constructed one can find another real form and, furthermore, one can find all possible real forms of a given semi-simple complex Lie algebra in this way; see [5.5] for a discussion.

16.1.2 Representations of finite dimensional semi-simple Lie algebras

As for the above discussion of the Lie algebras we analyse their representations over the complex numbers. A *representation* of a Lie algebra G is a map from G to the set of linear operators A of a vector space V, that is, $X \to A(x)$ such that $A([X, Y] = [A(X), A(Y)]$ for all $X, Y \in G$. Given a finite dimensional representation of a finite dimensional semi-simple

Lie algebra G we can choose a basis such that the states of the representation are eigenstates of the Cartan sub-algebra H of G. For the Cartan–Weyl basis of G with elements of H being H_i, this means that

$$H_i|\mu\rangle = \mu_i|\mu\rangle. \tag{16.1.66}$$

We call the eigenvectors μ_i the *weights* of the representation. We can think of the weights as being linear maps on the Cartan sub-algebra to the complex numbers by writing the above equation as $h|\mu\rangle = \mu(h)|\mu\rangle$, where h is any element of the Cartan algebra. If we write $h = \sum_i c^i H_i$, then $\mu(h) = \sum_i c^i \mu_i$. Using the Killing form we can then identify a unique element h_μ of the Cartan sub-algebra for every weight μ by setting $\mu(h) = B(h_\mu, h)$. Just as we did for the roots, this then defines a scalar product on the weights by $(\mu, \nu) = B(h_\mu, h_\nu)$.

In terms of the Chevalley generators in the Cartan sub-algebra, that is, $H_a = 2(\sum_i \alpha_a^i H_i)/(\alpha_a, \alpha_a)$, equation (16.1.66) becomes

$$H_a|\mu\rangle = 2\frac{(\alpha_a, \mu)}{(\alpha_a, \alpha_a)}|\mu\rangle. \tag{16.1.67}$$

The Chevalley generators E_b act on the state $|\mu\rangle$ so as to change its weight. Indeed, using equation (16.1.43), $E_b|\mu\rangle$ has weight $\mu + \alpha_b$ since

$$H_i(E_b|\mu\rangle) = (\mu_i + (\alpha_b)_i)(E_b|\mu\rangle),$$

or equivalently,

$$H_a(E_b|\mu\rangle) = \left(2\frac{(\alpha_a, \mu)}{(\alpha_a, \alpha_a)} + A_{ab}\right) E_b|\mu\rangle. \tag{16.1.68}$$

We get a similar result with a minus sign in front of α_b in the first equation and A_{ab} in the second equation if we consider $F_b|\mu\rangle$.

Let us assume that we are dealing with a representation for which there exists a state $|\Lambda\rangle$ with weight Λ such that $\Lambda + \alpha_b$ for $b = 1, \ldots, r$ is not a weight and for which every other state in the representation can be obtained by elements of the Lie algebra acting on this state. We call such a representation a *highest weight representation* and Λ the *highest weight* of the representation. In this case

$$E_b|\Lambda\rangle = 0; \quad b = 1, \ldots, r \tag{16.1.69}$$

and the states in the representation are given by

$$\prod_a (F_a)^{n_a}|\Lambda\rangle \tag{16.1.70}$$

with weights $\Lambda - \sum_a n_a \alpha_a$. The representation is entirely determined by its highest weight Λ and we may use it to denote the entire representation. We call the states obtained by the action of the F_as on the highest weight state $|\Lambda\rangle$ the descendants. Clearly, any irreducible finite dimensional representation is a highest weight representation and the set of states in equation (16.2.70) must terminate and so the descendants contain a state that is annihilated by the F_a. It can also happen that amongst the states of equation (16.1.70) there is a state other than the state $|\Lambda\rangle$ which is itself a highest weight state of the algebra G. Under the action of the algebra this state will just transform into polynominals of F_a acting on it. Hence this new highest weight state and all its descendants transform amongst themselves and the representation will only be irreducible if we set all these states to zero.

Theorem Given any finite dimensional highest weight representation Λ, the quantities q_a given by

$$q_a \equiv 2\frac{(\alpha_a, \Lambda)}{(\alpha_a, \alpha_a)}, \quad a = 1, \ldots, r \tag{16.1.71}$$

are positive integers or 0.

The proof uses the fact that E_a, F_a and H_a, for fixed a, form the sub-algebra A_1 and so we can use the familiar construction of the states of a particle with spin. Indeed, let us consider the highest weight state $|\Lambda\rangle$ such that $E_a|\Lambda\rangle = 0$, the states $(F_a)^n|\Lambda\rangle$ have weight $\Lambda - n\alpha_a$ and we may normalise these states such that $(F_a)^n|\Lambda\rangle = |\Lambda - n\alpha_a\rangle$. On the other hand, acting with E_a on the latter state leads to a state with weight $\Lambda - (n-1)\alpha_a$ and we may write $E_a|\Lambda - n\alpha_a\rangle = r_n|\Lambda - (n-1)\alpha_a\rangle$, where r_n is a constant which we will determine below. However,

$$\begin{aligned} E_a|\Lambda - n\alpha_a\rangle = r_n|\Lambda - (n-1)\alpha_a\rangle &= E_a F_a|\Lambda - (n-1)\alpha_a\rangle \\ &= ([E_a, F_a] + F_a E_a)|\Lambda - (n-1)\alpha_a\rangle \tag{16.1.72} \\ &= (H_a + r_{n-1})|\Lambda - (n-1)\alpha_a\rangle \end{aligned}$$

and so, using equation (16.1.67), we find the recurrence relation

$$r_n = r_{n-1} + \frac{2(\alpha_a, \Lambda)}{(\alpha_a, \alpha_a)} - 2(n-1). \tag{16.1.73}$$

Given that $r_0 = 0$, the solution to this relation is given by

$$r_n = 2n\frac{(\alpha_a, \Lambda)}{(\alpha_a, \alpha_a)} - n(n-1). \tag{16.1.74}$$

Since the representation is finite dimensional there exits a positive integer q_a such that $(F_a)^{q_a+1}|\Lambda\rangle = 0$. Consequently, $r_{q_a+1} = 0$ and so the above equation implies that

$$q_a = 2\frac{(\alpha_a, \Lambda)}{(\alpha_a, \alpha_a)} \tag{16.1.75}$$

and hence the desired result. The integers q_a, $a = 1, \ldots, r$ are called the *Dynkin indices* and they completely determine Λ. Given a highest weight Λ we can calculate the q_a from the above equation and find the number of steps before the process terminates.

Given a weight μ in the above procedure, we may write the sequence

$$\Lambda = \mu + p\alpha_a, \ldots, \mu, \ldots, \mu - m\alpha_a = \Lambda - q_a\alpha_a. \tag{16.1.76}$$

Such a set of weights is called the α_a root string through μ. We recognise $q_a = m + p$, using equation (16.1.75), and we find by adding $-p$ to each side that

$$m = p + 2\frac{(\alpha_a, \mu)}{(\alpha_a, \alpha_a)}. \tag{16.1.77}$$

If one knows the number of steps p one has taken in the root string, from this last equation knowing μ we can calculate m. If m is positive, then the sequence continues and $F_a|\mu\rangle \neq 0$, but if it is zero, then the string stops and $F_a|\mu\rangle = 0$. We note any weight in the root string is of the form $\mu = \Lambda - p\alpha_a$ and so

$$2\frac{(\alpha_a, \mu)}{(\alpha_a, \alpha_a)} = q_a - m \in \mathbf{Z}. \tag{16.1.78}$$

We started from the highest weight state $|\Lambda\rangle$ of the representation and found states in the root string of α_a using F_a. However, since $[E_c, F_a] = 0$ for $a \neq c$ all these states will be highest weight states for the other Chevalley generators E_c, $a \neq c$, that is, $E_c|\gamma\rangle = 0$ if $|\gamma\rangle$ denotes any one of these states. As such, on each of these states we may construct a root string of α_c using F_c. The number of states in the root string is given by $2(\alpha_c, \gamma)/(\alpha_c, \alpha_c)$. Continuing in this way one finds the entire representation. Since we are dealing with finite dimensional representations every representation will possess a lowest weight μ for which $\mu - \alpha_a$ is not a weight in the representation for any simple root α_a.

Above we have considered a root string using the simple roots or Chevalley generators. However, as E_α, $E_{-\alpha}$ and $H_\alpha = \alpha^i H_i$ form an A_1 algebra ($[E_\alpha, E_{-\alpha}] = \alpha^i H_i$, etc.) for any root α, we can use these to lower with $E_{-\alpha}$ on any state $|\Lambda\rangle$ that is annihilated by E_α. That is to construct the root string of α through Λ;

$$\Lambda + p\alpha, \ldots, \Lambda, \ldots \Lambda - m\alpha. \tag{16.1.79}$$

Here, by definition, $\Lambda + (p+1)\alpha$ and $\Lambda - (m+1)\alpha$ are not weights. Applying the same arguments as above we find that

$$m - p = 2\frac{(\alpha, \Lambda)}{(\alpha, \alpha)} \in \mathbf{Z} \tag{16.1.80}$$

and that the root string contains $p + m + 1$ terms. We again note that in deriving this result we have used the standard representations theory of A_1, one of whose real forms in SU(2).

Given the state $|\mu\rangle$ in the α_a string of the highest weight state $|\Lambda\rangle$ we can consider the transformation

$$S_{\alpha_a}(\mu) = \mu - 2\frac{(\alpha_a, \mu)}{(\alpha_a, \alpha_a)}\alpha_a. \tag{16.1.81}$$

This is a reflection in the plane through the origin perpendicular to the root α_a; it is called a *Weyl reflection*. We note from equation (16.1.77) that we can write $S_{\alpha_a}(\mu) = \mu + p\alpha_a - m\alpha_a = \Lambda - m\alpha_a$, which is a weight, corresponding to a non-vanishing state, in the α_a root string. Indeed, given any state in the representation we can construct the root string for any simple root that contains it and then apply the above argument to show that the Weyl reflection in the simple root leads to a weight, corresponding to a non-vanishing state, in the representation. More generally, we can consider the root string based on any root α, that is, the one based on the A_1 algebra consisting of E_α, $E_{-\alpha}$ and $\alpha^i H_i$ and find that the Weyl reflection in the root α, namely

$$S_\alpha(\mu) = \mu - 2\frac{(\alpha, \mu)}{(\alpha, \alpha)}\alpha, \tag{16.1.82}$$

corresponds to a non-vanishing state in the representation.

The *Weyl group* of a Lie algebra G is the set of all transformations generated by reflections in the roots of the algebra. It turns out that it is generated by a product of all Weyl reflections in the simple roots, S_{α_a}. We have found that the Weyl group acts on the weights belonging to any finite dimensional irreducible representation to give another weight in the representation.

We can apply the above to the adjoint representation. The highest weight of the adjoint representation is just the highest root θ as adding any root to β is not a root by definition. The corresponding highest weight state $|\theta\rangle$ satisfies $E_a|\theta\rangle = 0$, $a = 1, \ldots, r$. The weights in the adjoint representation are just the set of roots Δ in the algebra. Equation (16.1.78) implies that any two simple roots α_a and α_b in the algebra obey

$$2\frac{(\alpha_a, \alpha_b)}{(\alpha_a, \alpha_a)} \in \mathbf{Z} \tag{16.1.83}$$

and by considering the root string of any root α through any root β we find that

$$2\frac{(\alpha, \beta)}{(\alpha, \alpha)} \in \mathbf{Z}. \tag{16.1.84}$$

These two results proved crucial in our discussions of the last section.

It follows from the above discussion that the set of roots is invariant under Weyl reflections and so if α and β are any two roots so is $S_\alpha(\beta) = \beta - 2((\alpha, \beta)/(\alpha, \alpha))\alpha$. If we apply this to the distinct simple roots α_a and α_b, we conclude that $(\alpha_a, \alpha_b) \leq 0$ as otherwise we would have a root which was composed of two simple roots of different signs and this contradicts the decomposition of roots into positive and negative spaces. Another way to see this is to look at equation (16.1.77) taking $\mu = \alpha_b$; as $\alpha_b - \alpha_a$ is not a root then $m = 0$ and as p is a positive integer, or 0, we find the desired result.

We can use the above argument to prove the Serre relation of equation (16.1.45). The multiple commutator of n E_a with E_b has root $\alpha_b + n\alpha_a$ and if the action of another E_a vanishes this will be a highest weight $\Lambda = \alpha_b + n\alpha_a$. Lowering with F_a reverses the action of the E_a and then ends with a vanishing state. Applying equation (16.1.75) with $q_a = n$ we find that $n = 2(\alpha_a, \Lambda)/(\alpha_a, \alpha_a) = A_{ab} + 2n$ and so we find that $n = -A_{ab}$, thus deriving the Serre relation.

An important set of representations are those with highest weights Λ^b such that

$$2\frac{(\alpha_a, \Lambda^b)}{(\alpha_a, \alpha_a)} = \delta_a{}^b. \tag{16.1.85}$$

There are r such representations and we can associate the representation with highest weight Λ^b with a node labelled b of the Dynkin diagram of G. These representations are called the *fundamental representations* of G.

As an example let us consider the Lie algebra A_1 which has only one positive root and so only one simple root α with a length squared which we take to be 2. The three Chevalley generators are denoted by E, F and H. We have only one fundamental weight Λ which obeys $2(\alpha, \Lambda)/(\alpha, \alpha) = 1$. It is given by $\Lambda = \frac{1}{\sqrt{2}}$. The highest weight state is $|\Lambda\rangle$ and is subject to $E|\Lambda\rangle = 0$. The states in the representation are $|\Lambda\rangle$ and $F|\Lambda\rangle$, since as $q = 1$ we have only one step in the root string. As such $F^2|\Lambda\rangle = 0$. The state $F|\Lambda\rangle$ has weight $\Lambda - \alpha = -\frac{1}{\sqrt{2}}$ and we recognise the doublet representation of A_1.

We next consider the case of A_2. Using the values of the roots given in the previous section we find that the highest weights of the fundamental representations are given by $\Lambda^1 = (\frac{1}{\sqrt{2}}, \frac{1}{\sqrt{6}})$ and $\Lambda^2 = (0, \frac{2}{\sqrt{6}})$. Let us construct the fundamental representation with highest weight Λ^1. The highest weight state satisfies by definition $E_a|\Lambda^1\rangle = 0, a = 1, 2$. Since $q_2 = (\Lambda^1, \alpha_2) = 0$, we find that $F_2|\Lambda^1\rangle = 0$. As $q_1 = (\Lambda^1, \alpha_1) = 1$, acting with F_1 we find the state $F_1|\lambda^1\rangle = |\Lambda^1 - \alpha_1\rangle$, but this latter state is then annihilated by F_1. It is also annihilated by E_2 as $E_2F_1|\Lambda^1\rangle = F_1E_2|\Lambda^1\rangle = 0$, where we have used equation (16.1.43). The state $|\Lambda^1 - \alpha_1\rangle$ acts as the highest state in a new root string in which F_2 acts to give the state $F_2|\Lambda^1 - \alpha_1\rangle = |\Lambda^1 - \alpha_1 - \alpha_2\rangle$, but this state is then annihilated by F_2 since there is only one step in this root string as $(\Lambda^1 - \alpha_1, \alpha_2) = 1$. Furthermore, as $(\Lambda^1 - \alpha_1 - \alpha_2, \alpha_1) = 0$ it is also annihilated by F_1 and so we have no more states. Hence, we find a representation with three states that form the well-known **3**, or triplet, representation of A_2. We can represent these three states as a vector T_i, $i = 1, 2, 3$, that is, three objects with three lower indices. The representation with fundamental weight Λ^2 is denoted by $\bar{\mathbf{3}}$ and contains the three states S^i, $i = 1, 2, 3$ which carry upper indices.

It is instructive to find the fundamental representation of A_{n-1}, one real form of which is $\mathrm{SL}(n, \mathbf{R})$, with highest weight Λ^1. This algebra is discussed in appendix D.1 which includes the Dynkin diagram. By calculating the scalar products of the simple roots with the weights in the representation one find that it contains the weights

$$\Lambda^1, \Lambda^1 - \alpha_1, \Lambda^1 - \alpha_1 - \alpha_2, \ldots, \Lambda^1 - \alpha_1 - \cdots - \alpha_{n-1}. \tag{16.1.86}$$

These are the **n** representations of $\mathrm{Sl}(n, \mathbf{R})$ which are carried by the states T_i, $i = 1, \ldots, n$. The representation with highest weight Λ^{n-k} is realised on a tensor with k totally anti-symmetrised superscript indices, that is, $T^{i_1 \cdots i_k} = T^{[i_1 \cdots i_k]}$. Using the group invariant epsilon symbol $\epsilon^{i_1 \cdots i_n}$, this representation is equivalent to taking a tensor with $n - k$ lowered indices.

Although the precise decomposition of a given element of the Weyl group is not unique its length is defined to be the smallest number of simple root reflections required. However, there does exist a unique Weyl reflection, denoted W_0, that has the longest length. It turns out that this element obeys $W_0^2 = 1$, takes the positive simple roots to negative simple roots and its length is the same as the number of positive roots. As a result, $-W_0$ exchanges the positive simple roots with each other and, as Weyl transformations preserve the scalar product, it must also preserve the Cartan matrix. Consequently, it must lead to an automorphism of the Dynkin diagram and so the Lie algebra. Given any representation of G the highest and lowest weights are related by

$$\mu = W_0 \lambda. \tag{16.1.87}$$

Given the definition of the fundamental weights and carrying out a Weyl transformation W_0, we may conclude that, as $-W_0$ is an automorphism, the negatives of the highest and lowest weights of a given fundamental representation are the lowest and highest representations of one of the other fundamental representations. It is always the case that the two representations have the same dimension. However, it can, and often does, happen that a fundamental representation is taken to itself under the action of $-W_0$.

For $\mathrm{SL}(n)$ $W_0 = (S_{\alpha_1} \cdots S_{\alpha_{n-1}})(S_{\alpha_1} \cdots S_{\alpha_{n-2}}) \cdots (S_{\alpha_1} S_{\alpha_2}) S_{\alpha_1}$ and one finds that the highest and lowest weights of the fundamental representations are related by

$$W_0 \Lambda_{n-k} = \mu_{n-k} = -\Lambda_k \iff W_0 \mu_{n-k} = \Lambda_{n-k} = -\mu_k. \tag{16.1.88}$$

This result follows from the above remarks on W_0 as it must take a fundamental representation to a fundamental representation and correspond to an automorphism of the Dynkin diagram, which in this case is just takes the nodes k to $n - k$. The reader may verify this equation for the Λ_1 and Λ_{n-1} fundamental representations A_{n-1} by explicitly constructing the highest and lowest weights and then checking that minus the lowest is the highest weight of the other representation.

The weights of all the finite dimensional representations of G belong to a lattice that is called the *weight lattice* and is denoted Λ_G^W. Since any weight μ of a representation obeys equation (16.1.78), this implies that the weight lattice of G is contained in the dual lattice to the coroot lattice of G, namely $(\check{\Lambda}_G)^*$. Clearly, the root lattice is a sub-lattice of the weight lattice. The weight lattice must contain the fundamental weights of equation (16.1.85), but as these are nothing but a basis of the dual lattice to the coroot lattice, the weight lattice is just the dual of the coroot lattice; indeed

$$\Lambda_G^W = \left\{ \sum_b n_b \Lambda^b, n_b \in Z \right\}. \tag{16.1.89}$$

Let us define the fundamental Weyl chamber $\bar{C}$ of the weight lattice to be

$$\bar{C} = \{ x \in \Lambda_G^W; \ (x, \alpha_a) \geq 0, \ a = 1, \ldots, r \}. \tag{16.1.90}$$

Due to equation (16.1.71), any highest weight is contained in $\bar{C}$, but one can show that there is a one-to-one correspondence between highest weight finite dimensional irreducible representations of a finite dimensional semi-simple Lie algebra and elements in $\bar{C}$. Every weight x in $\bar{C}$ can be written in the form $x = \sum_a m_a \Lambda_a$ for m_a a positive integer or 0. Of course, not every such representation of the Lie algebra will extend to all Lie groups which have a particular real form of the complex Lie algebra, as is well known for the case of the spinor representation of A_1 and the group SO(3).

In the rest of this section we will assume, for simplicity, that we are dealing with simply laced Lie algebras, that is, all the simple roots have the same length squared which we choose to be 2. In this case the coroot lattice and the root lattice are the same and so $\Lambda_G^W = (\Lambda_G)^*$. We can define an equivalence relation on $(\Lambda_G)^*$ by $x_1 \sim x_2$ if $x_1 - x_2 \in \Lambda_G$. Under addition we can then define the Abelian group $(\Lambda_G)^*/\Lambda_G$. The centre of the group is given by $Z_G = \{a; ga = ag \ \forall \ g \in G\}$. The centre of the unique simply connected Lie group has a compact real form as its Lie algebra has a centre which is isomorphic to the group

$$Z_G = \frac{(\Lambda_G)^*}{\Lambda_G}. \tag{16.1.91}$$

Furthermore, $\det A_{ab} = |Z_G|$, where the symbol on the right-hand side is the number of elements in Z_G. For example, the centre of SU(2) is made up of the matrices $\pm I$, where I is the 2×2 identity matrix and so $|Z(SU(2)| = 2$, which does indeed equal the determinant of the Cartan matrix which is the 1×1 matrix which has value 2. The reader may like to check equation (16.1.91) for SU(n) for which $|Z(SU(n)| = n$. Since the determinant of the Cartan matrix of E_8 is 1 we must conclude that the centre of the group E_8, which is simply connected and has a Lie algebra with a compact real form, has only one element, namely the identity matrix. Given any sub-group Z of Z_G we can form the group G/Z to find a group that is not simply connected and has the first fundamental group $\pi_1(G/Z) = Z$.

It is instructive to spell out the above for the case of D_n, one of whose real forms is SO($2n$), whose Dynkin diagram in given in appendix D. The two spinoral representations, denoted s and $\bar{s}$, of SO($2n$) are the fundamental representations associated with the two nodes at the end of the branched fork and the vector representation, denoted v, is associated with the node at the end of the long tail. In the labelling of the Dynkin diagram of D_n in appendix D.2 these are the fundamental representations with highest weights $\Lambda_s = \Lambda_n, \Lambda_{\bar{s}} = \Lambda_{n-1}$ and $\Lambda_v = \Lambda_1$, respectively. The elements of the group $(\Lambda_G)^*/\Lambda_G$ are the cosets

$$\equiv [0], \ [\lambda_s], \ [\Lambda_{\bar{s}}], \ [\Lambda_v], \tag{16.1.92}$$

where 0, Λ_s, $\Lambda_{\bar{s}}$ and $[\Lambda_v]$ are the coset representatives respectively. In particular [0] is any root in the root lattice of G, that is, $[0] = \Lambda_G$. Using the fundamental weights in appendix D.2 one can show that

$$\frac{(\Lambda_{D_n})^*}{\Lambda_{D_n}} = \begin{cases} \mathbf{Z}_2 \otimes \mathbf{Z}_2 & \text{if } n \in 2\mathbf{Z}, \\ \mathbf{Z}_4 & \text{if } n \in 2\mathbf{Z}+1. \end{cases} \tag{16.1.93}$$

This is consistent with $\det A_{ab} = 4$ for D_n. We now prove this relation if n is even. In this case $\Lambda_s + \Lambda_s \in \Lambda_G$, $\Lambda_s + \Lambda_{\bar{s}} \in \Lambda_G + \Lambda_v$ and $\Lambda_v + \Lambda_s \in \Lambda_G + \Lambda_{\bar{s}}$ etc. and so, using addition as the group multiplication rule, we have the group multiplication rules $[\Lambda_s][\Lambda_s] = [0]$, $[\Lambda_s][\Lambda_{\bar{s}}] = [\Lambda_v]$ and $[\Lambda_v][\Lambda_s] = [\Lambda_{\bar{s}}]$ etc. Hence, if we write the elements of $\mathbf{Z}_2 \otimes \mathbf{Z}_2$ as $(e_1, e_2), (a_1, e_2), (e_1, a_2), (a_1, a_2)$ in the obvious notation, that is, $a_1^2 = e_1, a_2^2 = e_2$, these correspond to the cosets in equation (16.1.84).

As any group acts on its cosets in a natural way, so $\mathbf{Z}_2 \otimes \mathbf{Z}_2$ acts naturally on $(\Lambda_{D_n})^*/\Lambda_{D_n}$ by addition. A subgroup is $\mathbf{Z}_2^d = \{[0], [\Lambda_v]\}$ and so one can form the coset $\hat{\Lambda}_{D_n} \equiv ((\Lambda_{D_n})^*/\Lambda_{D_n})/\mathbf{Z}_2^d$. This consists of only two cosets which we may take to be $\hat{\Lambda}_{D_n} = \{[0], [\Lambda_s]\}$. In terms of the original lattice this corresponds to adding the spinor weight Λ_s to the root lattice Λ_{D_n}. Since the sum of two spinor weights of the same kind is in the root lattice, that is, $[\Lambda_s][\Lambda_s] = [0]$, it follows that a spinor weight and simple roots of D_n are not linearly independent. A basis for the new lattice $\hat{\Lambda}_{D_n}$ is to take the spinor weight Λ_s and the simple roots α_a, $a = 2, \ldots, n$ of D_n. In the discussion of D_n in appendix D.2 we find that $\Lambda_s = (\frac{1}{2}, \ldots, \frac{1}{2})$, which has length squared $n/4$, thus the new lattice is only integral if $n = 4p$ for integer p and even if $n = 8m$ for integer m.

For the case of D_8 we add the spinor weight $\Lambda_s = (\frac{1}{2}, \ldots, \frac{1}{2})$ of length squared 2. Looking at the roots of E_8 in equation (16.1.61) we recognise this as the root α_7 and so one finds the root lattice of E_8, that is, $\hat{\Lambda}_{D_8} = \Lambda_{E_8}$. Hence, we have passed from D_8 to E_8 by taking the simple roots to be those of D_8 but replacing the simple root α_1 with the spinor weight Λ_s of D_8. For D_{16} we add the spinor weight to the root lattice, but now this is $\Lambda_s = (\frac{1}{2}, \ldots, \frac{1}{2})$, which has length squared 4 and so the resulting lattice cannot be the root lattice of any finite dimensional semi-simple Lie algebra.

By computing the metric of the lattice, one can show that if $n = 8m$, $m \in \mathbf{Z}$, $\hat{\Lambda}_{D_n} = (\Lambda_{D_n})^*/\mathbf{Z}_2^d$ is an even self-dual lattice. Since there is only one self-dual even lattice of dimension 8 we conclude, as above, that $\Lambda_{E_8} = (\Lambda_{D_8})^*/\mathbf{Z}_2^d$. For dimension 16 we now have two self-dual lattices: $(\Lambda_{D_{16}})^*/\mathbf{Z}_2^d$ and $\Lambda_{E_8} \oplus \Lambda_{E_8}$, which is self-dual by virtue of equation (16.1.55). These lattices are not the same as the former has a basis element of length squared greater than 2. Thus we have found both even self-dual lattices with dimension 16. At dimension 24 there are 24 even self-dual lattices, the so-called Niemeier lattices. One of these is $(\Lambda_{D_{24}})^*/\mathbf{Z}_2^d$.

16.1.3 Affine Lie algebras

Given an finite dimensional semi-simple Lie algebra G with elements T_a in equation (16.1.2) we can define an enlarged Lie algebra G^+ by considering elements $T_n{}^a$, $n \in \mathbf{Z}$, and k which obey the algebra

$$[T_{a,n}, T_{b,m}] = f_{ab}{}^c T_{c,n+m} + n G_{ab} k \delta_{n+m,0}, \quad [T_{a,n}, k] = 0, \tag{16.1.94}$$

where G_{ab} is the Killing metric of the algebra G. The new algebra contains the original algebra, which consists of the $T_{a,0}$, as a sub-algebra G.

Utilising the Cartan–Weyl form of the algebra G given in equations (16.1.11)–(16.1.12) and (16.1.15), the above algebra takes the form

$$[H_{i,n}, H_{j,m}] = G_{ij}nk\delta_{n+m,0}, \quad [H_{i,n}, E_{\alpha,m}] = \alpha_i E_{\alpha,n+m} \tag{16.1.95}$$

and

$$[E_{\alpha,n}, E_{\beta,m}] = \begin{cases} 0 & \text{if } \alpha + \beta \notin \Delta, \\ \sum_i \alpha^i H_{i,n+m} + \dfrac{2n}{(\alpha,\alpha)}k\delta_{n+m,0} & \text{if } \alpha + \beta = 0, \\ \epsilon(\alpha,\beta)E_{\alpha+\beta,n+m} & \text{if } \alpha + \beta \in \Delta, \end{cases} \tag{16.1.96}$$

where Δ refers to the roots of G and $[k, \bullet] = 0$ with $\bullet$ any generator.

We must now select a Cartan sub-algebra in the extended algebra G^+. This should include $H_{i,0}$ and k; however, even with this set $E_{\alpha,n}$ has the same commutators, and so the same root, for all n. To resolve this degeneracy we introduce a new generator d with commutator relations

$$[d, T_{a,n}] = nT_{a,n}, \ [d, k] = 0 \text{ or equivalently } [d, E_{\alpha,n}] = nE_{\alpha,n},$$

$$[d, H_{i,n}] = nH_{i,n}, \ [d, k] = 0. \tag{16.1.97}$$

The element d just measures the level n. We now choose as our Cartan sub-algebra the elements

$$H_{i,0}, k, d. \tag{16.1.98}$$

The roots then have the form (α, n, m), where α are the roots of G and n and m are the values of the commutators of k and d in the generator in question. As such $E_{\alpha,n}$ and $H_{n,i}$ have the roots $(\alpha, 0, n)$ and $(0, 0, n)$, respectively. We note that the middle component of the root is always 0 as the generator k commutes with all elements. To select the positive roots we choose a basis whose first element is $(0, 0, 1)$ and then use the same basis for the remaining components that was used to choose the positive roots of G with two 0s as the last two components, that is, $\alpha \in \Delta_+$. As such, the positive roots are $(\alpha, 0, 0)$ with α a positive root, $(\beta, 0, n)$, $n > 0$ with β any root in G and $(0, 0, n)$, $n > 0$. The set of simple roots is given by

$$\hat{\alpha}_a = (\alpha_a, 0, 0), \ a = 1, 2, \ldots r, \quad \hat{\alpha}_{r+1} = (-\theta, 0, 1), \tag{16.1.99}$$

where as before α_a are the simple roots of G and θ is the highest root of G. It is straightforward to verify that the above simple roots are indeed simple roots. For example, they span the space of all positive roots using only positive integers due to the observation that $-\theta + \sum_a n_a\alpha_a$ leads to all roots of G for $n_a \geq 0$. We also note that none of the roots in equation (16.1.99) can be written in terms of a sum of the others.

As the algebra is infinite dimensional we cannot define a Killing metric, or scalar product on the algebra, using the trace of two adjoint representation matrices as we did for the finite dimensional semi-simple Lie algebras. However, an affine algebra possesses an invariant metric in the sense of equation (16.1.9). Indeed, the metric is almost completely determined by its invariance properties. As before the metric is always 0 unless it involves two elements whose roots sum to 0. One can show that

$$B(T_{a,n}, T_{b,m}) = \delta_{n+m,0} G_{ab}, \ B(T_{a,n}, k) = 0,$$

$$B(d, T_{a,m}) = 0, B(k, d) = 1, \quad B(k, k) = 0, \ B(d, d) = 0 \tag{16.1.100}$$

with similar relations for the algebra elements in the Cartan–Weyl basis. The last relation is not determined by symmetry and can be made 0 by a choice of d, that is, $d \to d' = d - \frac{1}{2}kB(d, d)$ if $B(d, d)$ is non-vanishing. Consequently, given two roots of the forms (α, n, m) and (β, p, q) their scalar product is given by $(\alpha, \beta) + nq + mp$. We note that, in contrast to the semi-simple Lie algebras studied previously, the affine algebra contains roots that have length squared 0, namely $(0, 0, n)$. The above metric is of block diagonal form, the first r components, associated with the roots of G are Euclidean, but the last two components have signature 0 as eigenvalues are of opposite sign. However, since the middle component of any root is always zero the roots can never have negative length squared.

The Cartan matrix of G^+ is given by $A_{ab} = 2(\hat{\alpha}_a, \hat{\alpha}_b)/(\hat{\alpha}_a, \hat{\alpha}_a)$ and is a $(r + 1) \times (r + 1)$ matrix. The Dynkin diagram has $r + 1$ nodes and is given by using the prescription in the last section. It is the Dynkin diagram for G, but with one extra node, called the affine node, which attaches itself to the Dynkin diagram of G in a way that depends on the highest root θ of G. For all simply laced algebras G, it turns out that $\theta^2 = 2$, consequently all the simple roots of the affine algebra also have length squared 2 and so G^+ too is simply laced. Then the number of lines between the node 1 a in the Dynkin diagram of G and the affine node labelled $r + 1$ is $(\alpha_a, \theta)^2$. In some cases the affine node can attach itself to more than one node of the Dynkin diagram of G. This is the case for A_{n-1}, where the affine node attaches itself by a single line to node 1 and by another single line to node $n - 1$.

For any affine algebra the determinant of the Cartan matrix is 0. We note that the highest root has the form $\theta = \sum_{a=1}^{r} m_a \alpha_a$ and so $\sum_{a=1}^{r+1} m_a \hat{\alpha}_a = (0, 0, 1)$ if $m_{r+1} = 1$. However, this vector has zero scalar product with all the simple roots, and as a result $\sum_{b=1}^{r+1} A_{ab} m_b = 0$. The desired result follows as the Cartan matrix has an eigenvector with eigenvalue zero.

Like for finite dimensional semi-simple Lie algebras, we can define Chevalley generators $\hat{E}_a$, $\hat{F}_a$ and $\hat{H}_a$ for $a = 1, \ldots, r + 1$ for affine Lie algebras. In particular, we take

$$\hat{H}_a = \frac{2\hat{\alpha}_a \cdot H}{(\hat{\alpha}_a, \hat{\alpha}_a)} = \begin{cases} H_a & \text{if } a = 1, \ldots, r, \\ -\theta^i H_i + d & \text{if } a = r + 1 \end{cases} \tag{16.1.101}$$

and

$$\hat{E}_a = \begin{cases} E_a, & \text{if } a = 1, \ldots, r, \\ E_{-\theta,1} & \text{if } a = r + 1 \end{cases} \tag{16.1.102}$$

with a similar equation for $\hat{F}_a$, $a = 1, \ldots, r + 1$. One can easily verify that these obey equations (16.1.43) and, in fact, they also obey the Serre relation of equation (16.1.44). Given only the Cartan matrix of the affine algebra one can uniquely reconstruct the affine algebra from the multiple commutators of the $\hat{E}_a$ and separately the $\hat{F}_a$ provided one satisfies equations (16.1.43) and (16.1.4).

It is well known that affine Lie groups arise by considering the compositions of maps of a circle S^1 into the Lie group G. This construction gave the affine algebra without a central term, but this can also be introduced in this way; however this is non-trivial [16.28].

It is very instructive to see how the central element k emerges from the presentation of the affine algebra in terms of Chevalley generators. For simplicity we will consider the case of A_1^+ whose Cartan matrix and Dynkin diagram are given by

$$A_{ab} = \begin{pmatrix} 2 & -2 \\ -2 & 2 \end{pmatrix} \qquad \text{and} \qquad \bullet = \bullet. \tag{16.1.103}$$

The Chevalley generators are $\hat{E}_1 = E$, $\hat{F}_1 = F$ and $\hat{H}_1 = \alpha H$, where E, F and H are the elements of A_1 in Cartan–Weyl basis and $\alpha = \sqrt{2} = \theta$ is the positive root. In terms of the generators of equation (16.1.68)–(16.1.69) we can identify $\hat{E}_2 = E_{-\theta,1}$, $\hat{F}_2 = F_{\theta,-1}$. We note that

$$[\hat{E}_1, \hat{F}_1] = \hat{H}_1 = \alpha H, \quad \text{and} \quad [\hat{E}_2, \hat{F}_2] = \hat{H}_2 = -\alpha H + 2k. \tag{16.1.104}$$

Consequently, $\frac{1}{2}(\hat{H}_1 + \hat{H}_2) = k$. It is straightforward to show that this k commutes with all the generators. Clearly, it commutes with $\hat{H}_1$ and $\hat{H}_2$ and

$$[\tfrac{1}{2}(\hat{H}_1 + \hat{H}_2), \hat{E}_1] = \tfrac{1}{2}(A_{11}\hat{E}_1 + A_{21}\hat{E}_1) = 0 \tag{16.1.105}$$

using the defining relations of the Serre generators. Similarly one finds that it commutes with $\hat{E}_2$ and the $\hat{F}_i$, $i = 1, 2$.

We note that $[\hat{E}_1, \hat{E}_2] = [E_{\alpha,0}, E_{-\alpha,1}] = \alpha H_1$, where H_1 corresponds to the root $(0, 0, 1)$. This root is null and the commutator of H_1 with any generator not in the Cartan subalgebra is non-vanishing and has its level increased by 1.

16.2   Kac–Moody algebras

Kac–Moody algebras were found in 1968 [16.4] and [16.5] and for reviews of Kac–Moody algebras the reader may try the exhaustive accounts in [16.3] and also [16.6, 16.7]. The key to understanding Kac–Moody algebras is to adopt the lesson of the first section in this chapter. Namely, any finite dimensional semi-simple Lie algebra is completely specified by its Cartan matrix once we adopt a set of Chevalley generators and the relations of equations (16.1.43) and (16.1.44)) that they must satisfy. For a Kac–Moody algebra, we begin with a Cartan matrix, which we require to have certain properties, and define the algebra using a set of Chevalley generators and the relations they must satisfy. Note that even though we will generally define infinite dimensional algebras we will formulate them in terms of a finite dimensional matrix.

A Kac–Moody algebra of rank r is defined by $r \times r$ matrix A_{ab} which by definition satisfies the following properties:

$$A_{aa} = 2, \tag{16.2.1}$$

$$A_{ab} \quad \text{for} \quad a \neq b \quad \text{are negative integers or zero} \tag{16.2.2}$$

and

$$A_{ab} = 0 \quad \text{implies} \quad A_{ba} = 0. \tag{16.2.3}$$

Strictly speaking we should only consider Cartan matrices A_{ab} which by swapping rows and columns cannot be put in block diagonal form in which case we would have two commuting algebras. Also the Cartan matrix A_{ab} should be symmetrisable, that is, there should exist rational numbers d_a such that $d_a A_{ab}$ is symmetric. Clearly, if A_{ab} is already symmetric, then we do not need the rational numbers d_a. The reader who is studying this subject for the first time may wish to just consider only the symmetric case so as to avoid the troublesome factors of d_a.

At first sight there does not seem to be any difference between the Cartan matrices specified in equations (16.2.1)–(16.2.3) and those we encountered in our study of finite dimensional semi-simple Lie algebras in section 16.1.1. In fact, the only difference is that we are not demanding that the Cartan matrix be positive definite.

Given the Cartan matrix A_{ab} the associated Kac–Moody algebra is formulated in terms of its Chevalley generators which consist of the commuting generators H_a, which are those of the Cartan sub-algebra, as well as the generators E_a and F_a which turn out to be those of the positive and negative simple roots, respectively. The Chevalley generators are required to obey

$$[H_a, H_b] = 0, \tag{16.2.4}$$

$$[H_a, E_b] = A_{ab} E_b, \quad [H_a, F_b] = -A_{ab} F_b, \tag{16.2.5}$$

$$[E_a, F_b] = \delta_{a,b} H_a, \tag{16.2.6}$$

as well as the Serre relation

$$(adjE_a)^{1-A_{ab}}(E_b) = [E_a, \dots [E_a, E_b] \dots] = 0, \ [F_a, \dots [F_a, F_b] \dots] = 0. \tag{16.2.7}$$

In equation (16.2.7) there are $1 - A_{ab}\ E_a$ in the first equation and the same number of F_a in the second equation.

Given the generalised Cartan matrix A_{ab}, one can uniquely reconstruct the entire Kac–Moody algebra by taking the multiple commutators of the simple root generators subject to the above Serre relations. In particular, one can find, at least as a matter of principle, the generators and roots of the Kac–Moody algebra as well as all their commutators.

Associated with any Cartan matrix we may draw a Dynkin diagram from which one can recover the Cartan matrix and so the entire Kac–Moody algebra. When the Cartan matrix is symmetric this is particularly simple. For a rank r Kac–Moody algebra we draw r nodes labelled from 1 to r. Between nodes labelled a and b, for $a \neq b$, we draw $-A_{ab}$ lines or links. We have considered a slightly different definition to that we used for the finite dimensional semi-simple Lie algebras as in the latter case it is true that $A_{ab}A_{ba} \leq 4$ and this allows us to uniquely reconstruct the Dynkin diagram from the earlier definition. As Kac–Moody algebras do not, in general, satisfy this condition we require a more general definition in order that the Dynkin diagram uniquely determines the Cartan matrix. In fact, for symmetric Cartan matrices the two definitions coincide as $|A_{ab}| \leq 1$, if $a \neq b$, for finite dimensional semi-simple Lie algebras. For a non-symmetric Cartan matrix one needs a more general definition of the Dynkin diagram.

Like all Lie algebras Kac–Moody algebras possess an adjoint representation which is defined in the same way as for the finite dimensional case, namely the generator Y has the action $X \rightarrow [Y, X]$, $\forall X \in G$. As we generally have an infinite number of generators, the adjoint representation is infinite dimensional; as such the trace is not well defined, and it cannot be used to define a scalar product on the Kac–Moody algebra. However, taking into account the commutators of equations (16.2.4)–(16.2.6), it can be shown that there is a scalar product, which is symmetric and invariant, that is, $B(X, Y) = B(Y, X)$ and $B([X, Y], Z) = B(X, [Y, Z])$, and on the Chevalley generators it is given by

$$B(E_a, F_b) = \delta_{a,b}, \quad B(H_a, H_b) = A_{ab} \tag{16.2.8}$$

with all other scalar products being 0. For a non-symmetric Cartan matrix, which by definition must be symmetrisable, we have $B(E_a, F_b) = \delta_{a,b} d_a^{-1}$, $B(H_a, H_b) = A_{ab} d_b^{-1}$. This is not the unique invariant scalar product but it is the one that agrees with the one we adopted for the finite semi-simple Lie algebras. Since Kac–Moody algebras include affine algebras, the scalar product restricted to the Cartan subalgebra cannot always be non-degenerate, but we can as in that case always introduce additional generators, such as d, which make it non-degenerate on this enlarged space. We will assume that this enlargement is carried out in the cases that require it in what follows.

Using the same arguments as we did the finite dimensional semi-simple Lie algebras we can view the roots as linear functionals on the Cartan sub-algebra H, namely $[h, E_\alpha] = \alpha(h) E_\alpha$ for any h in the Cartan sub-algebra and E_α any other generator in the Kac–Moody algebra defines the linear functional α; $h \to \alpha(h)$ for all h in the Cartan sub-algebra H. We can define the space $G^\alpha = \{X \text{ such that } [h, X] = \alpha(h)X \ \forall h \in H\}$. Then α is a root if $G^\alpha \neq 0$. Unlike in the finite dimensional semi-simple case, the dimension of G^α is not 1, indeed, it can be very large.

Using the above scalar product we can identify with every root α a unique element of the Cartan sub-algebra h_α by $\alpha(h) = B(h_\alpha, h)$. As such, we can define a scalar product on the roots by $(\alpha, \beta) = B(h_\alpha, h_\beta)$. The Chevalley generators E_a correspond to the simple roots α_a and we can, like before, consider the space of positive roots $\Delta_+ = \{\sum_a n_a \alpha_a; n_a \geq 0, n_a \in \mathbf{Z}\}$.

The Cartan matrix is then given in terms of the simple roots by

$$A_{ab} = 2\frac{(\alpha_a, \alpha_b)}{(\alpha_a, \alpha_a)}, \tag{16.2.9}$$

following the same argument as given below equation (16.1.47). Clearly, the Cartan matrix is symmetric if and only if all the simple roots have the same length squared which we may take to be 2. We call the algebra simply laced in this case.

If the symmetrised Cartan matrix is positive definite,

$$\sum_{a,b} v_a d_a A_{ab} v_b > 0 \tag{16.2.10}$$

for any real vector v_a, then we find the finite dimensional semi-simple Lie algebras considered in section 16.1, that is, the list of Killing and Cartan algebras of equation (16.1.42). We refer to this case as the finite case. In this case, $\det A > 0$, as the determinant is the product of the eigenvalues once one has diagonalised the Cartan matrix.

If A_{ab} is positive semi-definite, that is, $\sum_{a,b} v_a d_a A_{ab} v_b \geq 0$, for all real vectors v_a with equality attained for one, but only one, non-trivial vector, then we have an affine Kac–Moody algebra as studied in section 16.2. In this case $\det A = 0$. The simplest example is to take

$$A = \begin{pmatrix} 2 & -2 \\ -2 & 2 \end{pmatrix} \quad \text{or equivalently } 0 = 0, \tag{16.2.11}$$

which is affine A_1 or A_1^+. Affine A_2 or A_2^+ has the Cartan matrix

$$A = \begin{pmatrix} 2 & -1 & -1 \\ -1 & 2 & -1 \\ -1 & -1 & 2 \end{pmatrix} \qquad \text{or equivalently} \qquad \begin{matrix} \bullet \\ / \ \backslash \\ \bullet \!\!-\!\! \bullet \end{matrix} \, . \tag{16.2.12}$$

Finally, if A_{ab} is indefinite, that is, $\sum_{a,b} v_a d_a A_{ab} v_b$ is not positive semi-definite or positive, then we have a general Kac–Moody algebra. Although the above definition of a Kac–Moody algebra is rather simple and determines the algebra completely as a matter of principle, very little is known about these algebras, which contain an infinite number of generators. Indeed, in no case, except for some special algebras associated with string theory, is even a listing of the generators known. We will in later sections develop techniques to find at least some of the generators.

As for the finite dimensional semi-simple Lie algebras we may introduce the root lattice Λ_G which is the integer span of the simple roots: $\Lambda_G = \{\sum_a n_a \alpha_a, n_a \in \mathbf{Z}\}$. This is an even integral lattice for the simply laced case.

A special case of Kac–Moody algebras are *hyperbolic Kac–Moody* algebras which have an indefinite A_{ab}; but we also demand that deleting *any* node in its Dynkin diagram leads to a, possibly disconnected, set of connected Dynkin diagrams, each of which is of finite type except for at most one of affine type. We note that deleting a node in the Dynkin diagram corresponds to deleting a row or column of the Cartan matrix. Hyperbolic Kac–Moody algebras have a maximal rank of 10. For hyperbolic algebras it turns out that $(A^{-1})_{ab}$ are negative for all a and b in contrast to the finite case where this quantity is always positive.

The theory of representation of Kac–Moody algebras is a bit more complicated than that for finite dimensional semi-simple algebras of section 16.1.2. However, it is rather similar when we are dealing with highest weight representations, that is, those for which there exists a highest weight state $|\Lambda\rangle$ which satisfies $E_a |\Lambda\rangle = 0, a = 1, \ldots, r$. The states in the representation are then given by $\prod_{n_a} (F_a)^{n_a} |\Lambda\rangle$. However, for the Kac–Moody case we have an infinite set of states in the representation, but the representation is still completely specified by its highest weight Λ.

The r fundamental representations have the highest weights $\Lambda_1, \Lambda_2, \ldots, \Lambda_r$, which are defined by

$$\frac{2(\alpha_a, \Lambda_b)}{(\alpha_a, \alpha_a)} = \delta_{ab}, \quad \text{or equivalently,} \quad \Lambda_a = \sum_{c=1}^{r} \frac{(\alpha_a, \alpha_a)}{(\alpha_c, \alpha_c)} (A^{-1})_{ac} \alpha_c. \tag{16.2.13}$$

The second equation is only valid when A_{ab}^{-1} exists, that is, for all cases except the affine case. Taking the scalar product with Λ_c we find that

$$(A^{-1})_{ac} = \frac{2(\Lambda_a, \Lambda_c)}{(\alpha_a, \alpha_a)}. \tag{16.2.14}$$

The weight lattice Λ_G^W is the integer span of the fundamental weights and is the dual of the root lattice for the case of a symmetric Cartan matrix, that is, in the simply laced case $\Lambda_G^W = (\Lambda_G)^*$. As for the finite-dimensional semi-simple Lie algebras we can consider the quotient

$$\frac{(\Lambda_G)^*}{\Lambda_G} \equiv Z(G). \tag{16.2.15}$$

As before the number of elements $|(\Lambda_G)^*/\Lambda_G|$ in $(\Lambda_G)^*/\Lambda_G$ is given in terms of the Cartan matrix by

$$|\det A| = \left| \frac{(\Lambda_G)^*}{\Lambda_G} \right|. \tag{16.2.16}$$

As we noted in section 16.1, if A_{ab} is of finite type it is associated with a finite-dimensional semi-simple Lie algebra G, which can be exponentiated uniquely to construct a simply connected Lie group G whose centre is the finite Abelian group $Z(G)$, which is isomorphic to $(\Lambda_G)^*/\Lambda_G$.

Given the fundamental weights we can define the Weyl vector ρ by

$$\rho = \sum_{a=1}^{r} \frac{2}{(\alpha_a, \alpha_a)} \Lambda_a = \sum_{a,b=1}^{r} (A^{-1})_{ab} \frac{2\alpha_b}{(\alpha_b, \alpha_b)}. \tag{16.2.17}$$

It satisfies the fundamental property

$$\rho \cdot \alpha_a = 1, \quad a = 1, 2, \ldots, r. \tag{16.2.18}$$

In the finite case the Weyl vector is equal to the half the sum of the positive roots and the coefficients $\sum_{a=1}^{r} (2/(\alpha_a, \alpha_a))(A^{-1})_{ab}$ are all positive, while they are all negative in the hyperbolic case and in the Lorentzian case mixed signs are possible.

16.3   Lorentzian algebras

As noted above, very little is known about Kac–Moody algebras that are not finite dimensional or affine, indeed for no such Kac–Moody algebra is even a listing of the generators known, never mind the commutators they obey. Even though many of the finite dimensional semi-simple and affine Lie algebras have found applications in physical systems it is possible that the class of Kac–Moody algebras is too large and they may not, in general, have physical applications. As such, it may be desirable to try to find a sub-class of Kac–Moody algebras that are hopefully more tractable. In this hope we now define the sub-class of Lorentzian algebras [16.9]. This includes algebras of finite, affine and hyperbolic types and it will turn out that their generalised Cartan matrix can possess at most one negative eigenvalue. Like all Kac–Moody algebras, Lorentzian Kac–Moody algebras are specified by their Dynkin diagrams. However, by definition a *Lorentzian Kac–Moody algebra* must possess at least *one* node whose deletion yields a Dynkin diagram whose connected components are of finite type except for at most one of affine type. For simplicity we will restrict our attention to the case when the Kac–Moody algebra has a symmetric Cartan matrix which implies that all the simple roots have the same length squared which we take to be 2. At first sight Lorentzian Kac–Moody algebras are the same as hyperbolic Kac–Moody algebras, the difference is that in the former there exists one node for which one can find the desired decomposition while it the latter case one can delete any node and one must find the required decomposition. Clearly, hyperbolic Kac–Moody algebras are a sub-class of Lorentzian Kac–Moody algebras which form a much larger and more interesting class.

We shall call the overall Dynkin diagram C, and the selected node, whose deletion yields the reduced diagram C_R, the 'central' node. For many examples the central node is not uniquely determined by the property that C_R has only connected components of finite and affine type, and what we shall do in the following will apply to every admissible choice for the central node. Unlike the overall Dynkin diagram C, the reduced diagram C_R need not be connected. If it is disconnected we denote the n connected components $C_1, C_2, \ldots, C_n$. The Cartan matrix of C_R is obtained from that for C simply by deleting the row and column corresponding to the central node. Then the overall Dynkin diagram can be reconstructed from the reduced Dynkin diagram and the central node, denoted c, by adding those edges linking the latter to each node of C_R. The number of links of the central node c to the ith node is

$$\eta_i = -A_{ci}, \tag{16.3.1}$$

namely the entries in the row and column of the Cartan matrix of C whose deletion was just mentioned.

Note that if the connected components of C_R are all of finite or affine type, the off-diagonal elements in its Cartan matrix only take the values 0 or -1 as discussed in sections 16.1.1 and 16.2 for a symmetric Cartan matrix. For convenience affine A_1 is excluded as it has a -2 entry in its Cartan matrix. This limitation does not apply to the values of η_i, which could be any integer $0, 1, 2 \ldots$.

Let us consider an example which we will be interested in later, the algebra called E_{11} which has the Dynkin diagram

Clearly we can delete the node labeled 11 to find the algebra A_{10} and so E_{11} is a Lorentzian algebra. However, this is not the only node one can delete to find a finite algebra. Deleting the node labeled 10 leaves D_{10} as the remaining algebra. While deleting the node labeled d for $d \leq 7$ leads to the algebra $A_{d-1} \otimes E_{11-d}$. Note we define $E_4 = A_4$ and $E_5 = D_5$.

Let us now consider the case when *the reduced diagram is of finite type* and construct linearly independent simple roots and fundamental weights for the overall Dynkin diagram C in terms of those for the reduced diagram C_R. Those for the general case are given in [16.9]. The $r - 1$ simple roots for C_R, $\alpha_1, \alpha_2, \ldots, \alpha_{r-1}$ are linearly independent and span a Euclidean space of dimension $r - 1$ since the reduced diagram C_R is of finite type. They will suffice for the corresponding nodes of the overall Dynkin diagram C once they are augmented by the simple root for the central node which is given by

$$\alpha_c = -v + x, \quad \text{where} \quad v = \sum_{i=1}^{r-1} \eta_i \lambda_i = -\sum_{i=1}^{r-1} A_{ci} \lambda_i. \tag{16.3.2}$$

Here λ_i are the fundamental weights for the reduced diagram; they lie in the space spanned by the simple roots, while x is a vector in a one-dimensional vector space that is orthogonal to the roots space of C_R. Evidently this guarantees $\alpha_i \cdot \alpha_c = A_{ic}$ leaving only the condition

$$2 = A_{cc} = v^2 + x^2, \tag{16.3.3}$$

which determines the sign of x^2. Thus the simple roots of C span an r-dimensional vector space every element of which can be expressed in terms of the simple roots of C_R and x.

In terms of the $(r-1)$-dimensional vector space and the one-dimensional vector space associated with x we may write the simple roots of C as

$$\alpha_a = (0, \alpha_a), \quad a \neq c, \quad \alpha_c = (x, -\nu),$$

where the α_a in the brackets are those for C_R. This r-dimensional vector space is Euclidean, or Lorentzian, according to whether x^2 is positive or negative. Likewise if x^2 vanishes the simple roots span a space with a positive semi-definite metric and so constitute an affine root system whose Dynkin diagram will be recognisable as one of these fully classified algebras. This proves our earlier assertion that the Cartan matrix has at most one negative eigenvalue since it is formed from the scalar products of the simple roots.

The r fundamental weights for the original Lorentzian algebra, with Dynkin diagram C, will be denoted $\ell_c, \ell_1, \ell_2, \ldots \ell_{r-1}$ in order to distinguish them from the $r-1$ fundamental weights $\lambda_1, \ldots, \lambda_{r-1}$ for the reduced diagram. They are related to each other by

$$\ell_i = \lambda_i + \frac{\nu . \lambda_i}{x^2} x, \quad \ell_c = \frac{1}{x^2} x, \tag{16.3.4}$$

provided x^2 does not vanish, that is, the overall diagram C is not affine. The reader may easily check that these expressions do indeed satisfy the definition of equation (16.2.13) for the fundamental weights. The Weyl vector of the Kac–Moody algebra is then given by

$$R \equiv \sum_{j=c}^{r-1} \ell_j = \rho + \frac{(1 + \nu \cdot \rho)}{x^2} x, \tag{16.3.5}$$

where $\rho = \sum_{j=1}^{r-1} \lambda_j$ is the Weyl vector for the reduced diagram, (the same as the sum of the Weyl vectors for each connected component of C_R if it is disconnected). Notice that

$$R . \ell_c = \frac{(1 + \nu \cdot \rho)}{x^2}, \tag{16.3.6}$$

and hence has the same sign as x^2 given that $\nu \cdot \rho$ is positive when C_R is of finite type (as the quantities $\lambda_i \cdot \lambda_j$ are positive). Thus at least one of the coefficients in the expansion of the Weyl vector R in terms of simple roots is negative when C is Lorentzian.

As we will shortly see, the determinant of the Cartan matrix A for the overall diagram C is related to the Cartan matrix A_R of the reduced diagram C_R (obtained from A by deleting the row and column corresponding to the central node) by

$$\det A = x^2 \det A_R = (2 - \nu^2) \det A_R. \tag{16.3.7}$$

Equation (16.3.7) can be proven directly by using the expression for the Cartan matrix in terms of the scalar products of the roots to factorise $\det A$ into products of determinants of matrices made of the components of the simple roots of C. In particular, we write $(\alpha_a, \alpha_b) = \alpha_a^i \alpha_b^j G_{ij}$ and so $\det A = (\det \alpha_a^i)^2 \det G_{ij}$. However, the simple roots for nodes of C_R have no component in the direction of x and so evaluating the determinant we find that $\det \alpha_a^i = x$ times the square root of the determinant of C_R.

Let us now use (16.3.7) to determine a few determinants explicitly. We first consider the case of A_{n-1}, one of whose real forms is SU(n). Its Dynkin diagram is given in appendix D.1. Deleting the node labelled $n-1$ we find the algebra A_{n-2}, or SU($n-1$), and the ν of equation (16.3.2) is given by $\nu = \lambda_{n-2}$, where λ_{n-2} is the last fundamental weight S($n-1$). Using the fact, given in appendix D.1, that the scalar products of the fundamental weights of SU(n) are given by $(\lambda_i, \lambda_j) = i(n - j)/n$ for $i \leq j$ we find that $x^2 = n/(n - 1)$ and as a result

$$\det A(SU(n)) = \frac{n}{n - 1} \det A(SU(n - 1)) = n. \tag{16.3.8}$$

It is straightforward to write down the simple roots and fundamental weights of A_n and the algebras below using the deletions given here.

A similar argument applies to the Lie algebra D_n, one of whose real forms is SO($2n$). The Dynkin diagram is given in appendix D.2. We delete one of the spinor nodes, namely node n, leaving the algebra A_{n-1}, or SL(n). The corresponding $\nu = \lambda_{n-2}$ of the A_{n-1} algebra and as a result $x^2 = 4/n$. Substituting the last expressions in equation (16.3.7) we conclude that

$$\det A(SO(2n)) = \frac{4}{n} \det A(SU(n)) = 4. \tag{16.3.9}$$

Finally we consider E_n whose Dynkin diagram is given in appendices E.3–E.5. We delete node n to find A_{n-1} or SL(n). Then $\nu = \lambda_{n-3}$ and $x^2 = (9 - n)/n$ and we conclude that

$$\det A(E_n) = \frac{(9 - n)}{n} \det A(SU(n)) = 9 - n. \tag{16.3.10}$$

This shows that E_n is Lorentzian if $n \geq 10$, as indeed it must be as it is a Lorentzian algebra. Notice also that E_{10}, which is hyperbolic, has a Cartan matrix with determinant -1 so that its root lattice $\Lambda_R(E_{10})$ is an even, unimodular Lorentzian lattice. We note that E_{11} has a Cartan matrix with determinant -2.

16.4   Very extended and over-extended Lie algebras

We now discuss some particular classes of Lorentzian Kac–Moody algebras that are derived from finite dimensional semi-simple Lie algebras in a systematic way. In this section we shall *not* assume that G has a symmetric Cartan matrix, at least until we consider their weight lattices.

Let us begin by considering a finite dimensional semi-simple Lie algebra G of rank r whose simple roots α_i, $i = 1, \dots, r$, span the lattice Λ_G. Let us denote the highest root of G by θ; we will always normalise the simple roots of G such that $\theta^2 = 2$. This can always be done except for the rank 2 exceptional Lie algebra G_2, which our analysis does not cover. For the simply laced case this means that we take the simple roots all to have length squared 2.

We will proceed in a number of steps. We first enlarge the root lattice of Λ_G to be part of

$$\Lambda_G \oplus \Pi^{1,1}, \tag{16.4.1}$$

where $\Pi^{1,1}$ is the unique self-dual even Lorentzian lattice of dimension 2 discussed in section 16.1.1. In particular, we add to the simple roots of G the extended root

$$\alpha_0 = k - \theta, \tag{16.4.2}$$

where $k \in \Pi^{1,1} \subset \Lambda_G \oplus \Pi^{1,1}$ is the vector $k = (1, 0)$ introduced below equation (16.1.57). We note that $k^2 = 0$. The corresponding Lie algebra now has $r + 1$ simple roots, and, we note from section 16.1.3 that we have just added the additional simple root of equation

(16.1.99) and so we have obtained the affine Lie algebra of G, which we now denote by G^+. By construction we have $(\alpha_0, \alpha_0) = 2$. Let us denote the scalar products involving the new simple root as $(\alpha_0, \alpha_i) \equiv q_i'$ and so $2(\alpha_i, \alpha_0)/(\alpha_i, \alpha_i) \equiv q_i$. The corresponding Cartan matrix then has the form

$$A_{G^+} = \begin{pmatrix} & & & q_1 \\ & A_G & & \vdots \\ & & & q_r \\ q_1' & \cdots & q_r' & 2 \end{pmatrix}. \tag{16.4.3}$$

As explained in section 16.1.3, the determinant of the Cartan matrix of an affine algebra A_{G^+} vanishes, that is, $\det A_{G^+} = 0$.

Clearly, the roots of the affine algebra do not span the whole lattice $\Lambda_G \oplus \Pi^{1,1}$. Rather, the roots of the affine algebra can be characterised as the vectors y in this lattice which are orthogonal to k, that is, $y \cdot k = 0$.

We may further extend the affine Lie algebra by adding to the above simple roots yet another simple root to formulate a new algebra, namely by adding [10.27]

$$\alpha_{-1} = -(k + \bar{k}) \in \Lambda_G \oplus \Pi^{1,1}, \tag{16.4.4}$$

where $\bar{k} = (0, -1) \in \Pi^{1,1}$. We note that $\alpha_{-1}^2 = 2$, as well as $(\alpha_{-1}, \alpha_0) = -1$ and $(\alpha_{-1}, \alpha_i) = 0$, $i = 1, \ldots, r$. The Lie algebra so obtained is called the *over-extended Lie algebra*, and we shall denote it as G^{++}. The Cartan matrix associated to the over-extended Lie algebra has the structure

$$A_{G^{++}} = \begin{pmatrix} & & & q_1 & 0 \\ & A_G & & \vdots & \vdots \\ & & & q_r & 0 \\ q_1' & \cdots & q_r' & 2 & -1 \\ 0 & \cdots & 0 & -1 & 2 \end{pmatrix}. \tag{16.4.5}$$

Examining the determinant of this Cartan matrix we conclude that

$$\det A_{G^{++}} = 2 \det A_{G^+} - \det A_G = -\det A_G. \tag{16.4.6}$$

Clearly, the root lattice of G^+ is $\Lambda_{G^{++}} = \Lambda_G \oplus \Pi^{1,1}$. The algebra G^{++} is Lorentzian as we can take as our central node the last node we have added and deleting it yields an affine Lie algebra. The above sign of the determinant means that its Cartan matrix does have a negative eigenvalue.

We now enlarge the Lie algebra even further to find what is called the *very extended algebra* [16.9] by considering the lattice

$$\Lambda_G \oplus \Pi^{1,1} \oplus \Pi^{1,1} = \Lambda_{G^{++}} \oplus \Pi^{1,1}. \tag{16.4.7}$$

We denote the analogues of k and $\bar{k}$ in the second $\Pi^{1,1}$ lattice by l and $\bar{l}$, respectively. We now add the new simple root

$$\alpha_{-2} = k - (l + \bar{l}). \tag{16.4.8}$$

We then have that $(\alpha_{-2}, \alpha_{-2}) = 2$, $(\alpha_{-2}, \alpha_{-1}) = -1$, while all other scalar products involving α_{-2} vanish. Let us denote the resulting Kac–Moody algebra by G^{+++}. We will refer to it as a very extended algebra. The corresponding Cartan matrix is then of the form

$$
A_{G^{+++}} = \begin{pmatrix} & & & q_1 & 0 & 0 \\ & A_G & & \vdots & \vdots & \vdots \\ & & & q_r & 0 & 0 \\ q_1' & \cdots & q_r' & 2 & -1 & 0 \\ 0 & \cdots & 0 & -1 & 2 & -1 \\ 0 & \cdots & 0 & 0 & -1 & 2 \end{pmatrix}.
$$

(16.4.9)

Examining the determinant of this Cartan matrix we conclude that

$$
\det A_{G^{+++}} = 2 \det A_{G^{++}} - \det A_G = 2 \det A_{G^{++}} = -2 \det A_G. \qquad (16.4.10)
$$

The root lattice of G^{+++} consists of all vectors x in $\Lambda_G \oplus \Pi^{1,1} \oplus \Pi^{1,1}$ which are orthogonal to the time-like vector

$$
s = l - \bar{l} = (1, 1). \qquad (16.4.11)
$$

We note that G^{+++} is a Lorentzian algebra as we may take the affine node as the central node. As the determinant of the Cartan matrix of G^{+++} is negative its Cartan matrix does have one negative eigenvalue. We will refer to the three nodes we have added to form the G^{+++} algebra as the affine, over-extended and very extended nodes.

It is straightforward to calculate the fundamental weights of the over-extended and very extended algebras using equation (16.3.4). Essentially we just find expressions that satisfy the defining conditions of equation (16.2.13) as one can verify by using the above expressions for the simple roots. In the over-extended case the fundamental weights are given as

$$
\Lambda_i = \lambda_i^{\mathrm{f}} - (\lambda_i^{\mathrm{f}}, \theta)(k - \bar{k}), \quad i = 1, \ldots, r,
$$
$$
\Lambda_0 = -(k - \bar{k}), \qquad (16.4.12)
$$
$$
\Lambda_{-1} = -k,
$$

where λ_i^{f} are the fundamental weights of G. On the other hand, the fundamental weights of the very extended algebra are

$$
\Lambda_i = \lambda_i^{\mathrm{f}} - (\lambda_i^{\mathrm{f}}, \theta)\left(k - \bar{k} - \tfrac{1}{2}(l + \bar{l})\right), \qquad i = 1, \ldots, r,
$$
$$
\Lambda_0 = -\left(k - \bar{k} - \tfrac{1}{2}(l + \bar{l})\right),
$$
$$
\Lambda_{-1} = -k, \qquad (16.4.13)
$$
$$
\Lambda_{-2} = -\tfrac{1}{2}(l + \bar{l}).
$$

Similarly given the defining property of the Weyl vector of equation (16.2.18) we find that the Weyl vector of an over-extended algebra is given by [10.27]

$$
\rho = \rho_{\mathrm{f}} + h\bar{k} - (h + 1)k, \qquad (16.4.14)
$$

where ρ_{f} is the Weyl vector of the underlying finite dimensional Lie algebra G, and h is its Coxeter number. The Weyl vector of the very extended Kac–Moody algebras is given by [16.9]

$$\rho = \rho_f + h\bar{k} - (h+1)k - \tfrac{1}{2}(1-h)(l+\bar{l}). \tag{16.4.15}$$

Very extended algebras first arose in the context of finding fundamental symmetries. As we will discuss later, it is conjectured [16.10] that E_8^{+++}, often also called E_{11}, is a symmetry of the theory underlying string and branes. There are also less developed conjectures, namely D_{24}^{+++} is a symmetry of a theory underlying the bosonic string [16.10], A_{D-3}^{+++} a symmetry of D-dimensional gravity [13.41] and similar conjectures for any other very extended algebra [16.11].

It is interesting to consider the weight lattices of the over-extended and very extended algebras discussed above. For simplicity we shall only consider the simply laced case for which the weight lattice Λ_W is just the dual of the root lattice Λ_R, that is, $\Lambda_W = (\Lambda_R)^*$. These lattices can be easily found, using the fact that for any two lattices Λ_1 and Λ_2 we have equation (16.1.62), that is, $(\Lambda_1 \oplus \Lambda_2)^* = \Lambda_1^* \oplus \Lambda_2^*$. Now the lattice $\Pi^{1,1}$ is self-dual, and thus the weight lattice of G^{++} is simply given by

$$(\Lambda_{G^{++}})^* = \Lambda_G^* \oplus \Pi^{1,1}. \tag{16.4.16}$$

In particular, it follows that the coset

$$\frac{(\Lambda_{G^{++}})^*}{(\Lambda_{G^{++}})} = Z_G \tag{16.4.17}$$

as the $\Pi^{1,1}$ part factors out and we are left with the same result as for the finite algebras of equation (16.1.91). This result is consistent with the relation between the determinants of equation (16.4.6) and equation (16.1.90).

The weight lattice for G^{+++} is given by

$$(\Lambda_{G^{+++}})^* = \Lambda_G^* \oplus \Pi^{1,1} \oplus (r, -r) : \ 2r \in \mathbf{Z}. \tag{16.4.18}$$

In deriving (16.4.18) we have noted that the root lattice of G^{+++} is

$$\Lambda_{G^{+++}} = \Lambda_G \oplus \Pi^{1,1} \oplus \{(t, -t) : t \in \mathbf{Z}\} \tag{16.4.19}$$

since $l + \bar{l} = (1, -1)$. The last lattice in equation (16.4.18) is generated by $f = (\tfrac{1}{2}, -\tfrac{1}{2})$ which is not in $\Pi^{1,1}$, but for which $2f \in \Pi^{1,1}$. Thus we conclude that

$$\frac{(\Lambda_{G^{+++}})^*}{(\Lambda_{G^{+++}})} = \mathbf{Z}_G \times \mathbf{Z}_2. \tag{16.4.20}$$

Here the $\mathbf{Z}_2$ results from the fact that the dual lattice contains the vector f in the last factor of $\Pi^{1,1}$. This is consistent with the factor of 2 between the two determinants of equation (16.4.10).

We now discuss in more detail the over-extended and very extended algebras that arise when Λ_G is an even self-dual lattice of dimension r, or a sub-lattice of such a lattice. We recall that even self-dual Euclidean lattices only exist in dimensions $D = 8n$, $n = 1, 2, \ldots$.

As we have discussed in section 16.1 the first non-trivial example of such a lattice occurs in eight dimensions where there is only one such lattice, the root lattice of E_8. Let us denote the corresponding affine, over-extended and very extended algebras by $E_9 = E_8^+$, $E_{10} = E_8^{++}$ and $E_{11} = E_8^{+++}$, respectively. We choose the basis for the root lattice of E_8,

Λ_{E_8}, to be that of equation (16.1.61). In order to describe the affine extension and over-extension of E_8 we consider the lattice $\Lambda_{E_8} \oplus \Pi^{1,1}$. The affine root that gives E_9 is now given by

$$\alpha_0 = k - \theta = (-\theta; (1,0)), \tag{16.4.21}$$

where $\theta \in \Lambda_{E_8}$ is the highest root of E_8. In terms of the simple roots the highest root of E_8 is given in appendix D.5. and using equation(16.1.61) we find that it is explicitly given by the eight-dimensional vector

$$\theta = (0, 0, 0, 0, 0, 0, 1, -1). \tag{16.4.22}$$

Finally, the over-extended root that enhances this to E_{10} is given by

$$\alpha_{-1} = -(k + \bar{k}) = (\mathbf{0}; (-1, 1)). \tag{16.4.23}$$

The lattice $\Lambda_{E_8} \oplus \Pi^{1,1}$ is clearly self-dual and is of Lorentzian signature and even. Such lattices only occur in dimensions $D = 8n + 2$, $n = 0, 1, 2 \ldots$, and the lattice in each dimension is unique and is denoted by $\Pi^{8n+1,1}$ and is discussed in section 16.1. It follows that the root lattice of E_{10} is precisely this lattice for $n = 1$, that is, $\Lambda_{E_{10}} = \Pi^{9,1}$. Indeed, one can see that the above simple roots form a basis for this lattice which is given in equation (16.1.52).

To find the very extended algebra we consider the lattice

$$\Pi^{9,1} \oplus \Pi^{1,1} = \Lambda_{E_8} \oplus \Pi^{1,1} \oplus \Pi^{1,1} = \Pi^{10,2}, \tag{16.4.24}$$

where this lattice is the unique even self-dual lattice of signature $(10, 2)$. The very extended root is given by

$$\alpha_{-2} = k - (l + \bar{l}) = (\mathbf{0}; (1, 0); (-1, 1)). \tag{16.4.25}$$

From equation (16.4.20) it follows that

$$\frac{\Lambda^\star_{E_{11}}}{\Lambda_{E_{11}}} = \mathbf{Z}_2. \tag{16.4.26}$$

We next consider the over-extended and very extended Lie algebras associated to the rank $8m$ algebra D_{8m} for integer m. As explained in section 16.1 the corresponding root lattice is not self-dual since $\det A = 4$ in this case and indeed equation (16.2.28) holds. However, as also discussed there $\hat{\Lambda}_{D_{8m}} = \Lambda^\star_{D_{8m}}/\mathbf{Z}_2$ is an even self-dual lattice. The root lattice of D_{8m} is spanned by its simple roots which can be taken to be the vectors in $\mathbf{Z}^{24}$ of the form

$$\alpha_i = \left(0^{i-1}, 1, -1, 0^{8m-1-i}\right), \quad i = 1, \ldots, 8m - 1, \quad \alpha_{8m} = \left(0^{8m-2}, 1, 1\right). \tag{16.4.27}$$

As explained in section 16.1.2, $\hat{\Lambda}_{D_{8m}}$ is obtained from the root lattice $\Lambda_{D_{8m}}$ by taking a basis of the simple roots, but with α_1 replaced by the spinor weight

$$[(\tfrac{1}{2})^{8m}], \tag{16.4.28}$$

which has length squared $2m$.

The root lattice of D_{8m}^{++} is

$$\Lambda_{D_{8m}^{++}} = \Lambda_{D_{16}} \oplus \Pi^{1,1}. \tag{16.4.29}$$

It is spanned by the roots of equation (16.4.27), together with

$$\begin{aligned}
\alpha_0 &= \big((-1, -1, 0^{8m-2}); (1, 0)\big), \\
\alpha_{-1} &= \big((0^{8m}); (-1, 1)\big).
\end{aligned} \tag{16.4.30}$$

From equation (16.4.29) it follows that

$$\frac{\Lambda^*_{D_{8m}^{++}}}{\mathbf{Z}_2} = \frac{\Lambda^*_{D_{8m}}}{\mathbf{Z}_2} \oplus \Pi^{1,1} = \hat{\Lambda}_{D_{8m}} \oplus \Pi^{1,1} = \Pi^{8m+1,1}, \tag{16.4.31}$$

and thus $\Lambda_{D_{8m}^{++}}$ is contained in $\Pi^{8m+1,1}$.

The very extended algebra derived from D_{8m}, denoted D_{8m}^{+++}, is constructed by considering the lattice

$$\Lambda_{D_{8m}} \oplus \Pi^{1,1} \oplus \Pi^{1,1} \tag{16.4.32}$$

and adding the simple root

$$\alpha_{-2} = \big((0^{8m}); (1, 0); (-1, 1)\big). \tag{16.4.33}$$

As we discussed in section 16.1, for $m = 1$ we have $\hat{\Lambda}_{D_8} = E_8$, for $m = 2$ we have $\hat{\Lambda}_{D_{16}}$ which is one of the two even self-dual Euclidean lattices in 16 dimensions and for $m = 3$ we have $\hat{\Lambda}_{D_{24}}$ which is one of the 24 even self-dual Euclidean lattices in 24 dimensions, the so-called Niemeier lattices. For a review of such matters see [16.12].

16.5 Weights and inverse Cartan matrix of E_n

As we have seen the techniques introduced in section 16.3 for analysing algebras in terms of their sub-algebras can be useful even in the context of the finite-dimensional semi-simple Lie algebras, that is, those in the Cartan list. In this section we will use these techniques to find algebraic formulae for the weights and inverse Cartan matrix of E_n for $n \le 8$ [16.13]. By E_5 we mean D_5, while for $n = 9, 10, 11$ we recognise the algebras E_8^+, E_8^{++} and E_8^{+++}, respectively, considered in the last section. For higher n we mean the Lie algebra whose Dynkin diagram is that for E_8, but with $n - 8$ nodes added, each by a single line, on to the longest leg of the E_8 Dynkin diagram. The Dynkin diagram of E_n is given by

If we consider the case $n \le 8$ we have one of the finite dimensional semi-simple Lie algebras, $n = 9$ is an affine algebra and for $n \ge 10$ we have a more general Kac–Moody algebra. All these algebras are Lorentzian as we can delete the node n, the exceptional node, to find the remaining algebra is A_{n-1}. Indeed, we will take this to be our central node. As such, we are studying the decomposition of E_n to A_{n-1}. Let α_i and λ_i for $i = 1, \ldots, n - 1$

be the simple roots and fundamental weights, respectively, of A_{n-1}. Using equation (16.3.2) the roots of E_n can be written as

$$\alpha_i, \ i = 1, \ldots, n-1, \quad \alpha_n = x - \lambda_{n-3}, \tag{16.5.1}$$

where x is an element of a vector space that is orthogonal to the roots of A_{n-1} and $x^2 = (9-n)/n$ in order that $\alpha_n^2 = 2$. The fundamental weights of E_n are given by

$$l_i = \begin{cases} \lambda_i + \dfrac{3i}{(9-n)}x, & i = 1, \ldots, n-3, \\[2mm] \lambda_i + \dfrac{(n-3)(n-i)}{(9-n)}x, & i = n-3, \ldots, n-1 \end{cases} \tag{16.5.2}$$

and

$$l_n = \frac{n}{(9-n)}x \tag{16.5.3}$$

by using equation (16.3.4). In deriving this result we used the scalar products of the fundamental weights of A_{n-1} given in appendix D.1:

$$(\lambda_i, \lambda_j) = (A^{A_{n-1}})^{-1}_{ij} = \frac{i(n-j)}{n}, \ \text{for } i \le j. \tag{16.5.4}$$

The inverse Cartan matrix of E_n is given by

$$((A^{E_n})^{-1})_{ab} = l_a . l_b. \tag{16.5.5}$$

Using equation (16.5.4), we find the following algebraic formulae for the inverse Cartan matrix of E_n:

$$((A^{E_n})^{-1})_{ij} = \begin{cases} \dfrac{i(9-n+j)}{(9-n)}, & i, j = 1, \ldots, n-3, i \le j, \\[3mm] \dfrac{(n-j)((n-3)^2 - i(n-5))}{(9-n)}, & i, j = n-3, \ldots, n-1, i \le j, \\[3mm] 2\dfrac{i(n-j)}{(9-n)}, & i = 1, \ldots, n-3, j = n-3, \ldots, n-1, \end{cases} \tag{16.5.6}$$

$$((A^{E_n})^{-1})_{in} = \begin{cases} \dfrac{3i}{(9-n)}, & i = 1, \ldots, n-3, \\[3mm] \dfrac{(n-3)(n-i)}{(9-n)}, & i = n-3, \ldots, n-1 \end{cases} \tag{16.5.7}$$

and

$$((A^{E_n})^{-1})_{nn} = \frac{n}{(9-n)}. \tag{16.5.8}$$

It is easy to check that for $n = 8, 10$ the inverse Cartan matrix has integer valued entries as indeed should be the case for a self-dual root lattice. If $n \le 8$ they are all positive as is always the case for finite dimensional semi-simple Lie algebras. For $n = 10$ they are all negative as is also the case for every hyperbolic algebra.

16.6   Low level analysis of Lorentzian Kac–Moody algebras

16.6.1 The adjoint representation

As we have discussed, a Kac–Moody algebra is just the multiple commutators of the Chevalley generators subject to the Serre relations of equations (16.2.4)–(16.2.7). While this procedure specifies the algebra completely, it is difficult to carry out for generalised Kac–Moody algebras and indeed has so far proved impossible to carry out in practice with any real success. In this section we will give an alternative, more efficient, way of finding what are the generators of a Lorentzian algebra at least for low levels. For a Lorentzian algebra one can always find at least one node which when deleted leaves a finite dimensional semi-simple Lie algebra, and, or, at most one affine algebra. As we have stressed above, one can gain some understanding of the Lorentzian algebra by decomposing it in terms of this remaining well-understood algebra. The level of a generator is the number of times the Chevalley generator associated with the deleted node occurs in the multiple commutator of Chevalley generators that leads to the generator in question. We will explain this definition more explicitly below.

We begin by implementing the strategy [16.9] advocated for in section 16.3 for analysing Lorentzian algebras. For simplicity, we will consider the case when deleting the central node leaves only one connected Dynkin diagram, denoted C_R, which corresponds to one of the finite dimensional semi-simple Lie algebras. We also assume that the Cartan matrix of the Lorentzian Kac–Moody algebra is symmetric. As we choose the simple roots to have length squared 2, then $2 - v^2 = x^2$.

A positive (negative) root α of the Kac–Moody algebra C can be written using equation (16.3.2) as

$$\alpha = \pm \left(l\alpha_c + \sum_i m_i \alpha_i \right) = \pm \left(lx - lv + \sum_j m_j \alpha_j \right) = \pm lx + W, \qquad (16.6.1)$$

where

$$W = \mp lv \pm \sum_j m_j \alpha_j \in (\Lambda_{C_R})^* \qquad (16.6.2)$$

and l and m_j are positive integers. The upper sign is for a positive root and the lower sign for a negative root. We define the *level*, denoted l, of the roots of C to be the number of times the root α_c occurs. For positive roots this is the number of times the generator E_{α_c} occurs in the multiple commutator of E_α. There is an analogous definition for negative roots counting the number of times F_{α_c} occurs. As the above equation says, W is a weight of the reduced algebra C_R. For a fixed l, the root space of C can be described by its representation content with respect to C_R. We carry out the analysis of the algebra C in terms of representations of the algebra C_R level by level.

Acting with the generators of C_R on the adjoint representation of C takes one from one root of C to a new root of C. The original and the new roots of C are of the form of equation (16.6.1). However, under the action of such operators the first term of this equation, that is, $\pm lx$, is unchanged, and so the level of the new root is the same. As the positive or negative roots have all their coefficients of a given sign, we find that the action of these raising or lowering operators does not take one out of the positive root space, or negative root space, of C if one begins there. As a result, we may conclude that if a given representation of C_R occurs in C, then it must be contained entirely in the positive root space with a copy in the negative root space of C, since for every positive root α there is contained in the negative root space the root $-\alpha$. As such, to find the representations of C_R that occur in the decomposition of C it suffices to find the representations that occur in the negative, or positive, root space of C. The exception being when $l = 0$, which is for the adjoint representation of C_R.

We now discuss restrictions that one can place on the possible representations that can occur at a given level. Given an irreducible representation with highest weight Λ, it is completely specified by its Dynkin indices, which are defined by $p_j = (\Lambda, \alpha_j) \geq 0$ or equivalently $\Lambda = \sum_i p_i \lambda_i$. If a representation of C_R occurs at level l, then its highest weight Λ must occur among the possible weights W and so we may set $W = \sum p_i \lambda_i$ in equation (16.6.2) for level l and see if this can occur or not. Taking the scalar product of equation (16.6.2) with λ_j we find that a representation of C_R, with Dynkin indices p_j, is contained in the adjoint representation of C at level l in the positive roots if there exist positive integers m_k such that

$$\sum_j p_j \lambda_j = -l\nu + \sum_j m_j \alpha_j \quad \text{or equivalently}$$

$$\sum_j p_j (\lambda_j, \lambda_k) = -l(\nu, \lambda_k) + m_k \tag{16.6.3}$$

and in the negative root space if there exist positive integers m_k such that

$$\sum_j p_j (\lambda_j, \lambda_k) = l(\nu, \lambda_k) - m_k. \tag{16.6.4}$$

We recall that

$$(A^f_{ik})^{-1} = (\lambda_i, \lambda_k), \tag{16.6.5}$$

where $(A^f)^{-1}_{jk}$ is the inverse Cartan matrix of C_R, which for any finite dimensional algebra is positive definite for all i and k. Using equation (16.3.2) the first term on the right-hand side of equations (16.6.3) can be written as $+l \sum_j A_{cj} (A^f_{jk})^{-1}$, which is negative definite and fixed for a given l. For the negative roots the right-hand side of equations (16.6.4) can be written as $-l \sum_j A_{cj} (A^f_{jk})^{-1}$, which is positive definite and fixed for a given l. However, the integers p_j and m_k must take the values $0, 1, 2 \ldots$ in both cases. Equation (16.6.3) allows for some arbitrarily large values of p_k even for a fixed level l since we may compensate these by taking large positive values of m_k. In contrast, equation (16.6.4) does place strong constraints, for a given level l, on the possible integers p_j that can occur. Since these are the Dynkin indices of the representations that can occur we find a restriction on the possible representations of C_R contained in the adjoint representation of C. Given a solution to equation (16.6.4) we find not only the Dynkin indices p_i, but also the positive integers m_k that allow us to read off the root of the Lorentzian algebra, using equation (16.6.1) corresponding to the highest weight state. This method of proceeding was given for the hyperbolic Lie algebra E_{10} in [16.14] and for a general Lorentzian algebra in [16.16]. The case of a decomposition into several Lie algebras is given in [16.14].

Any Kac–Moody algebra with symmetric Cartan matrix has its root length squared bounded by $\alpha^2 \leq 2, 0, -2, \ldots$ [16.6]. Consequently, using equations (16.6.1) and (16.6.5) and the observation that $(2 - \nu^2) \det A_{C_R} = \det A_C$ we find that

$$\alpha^2 = l^2 \frac{\det A_C}{\det A_{C_R}} + \sum_{ij} p_i (A_{ij}^f)^{-1} p_j \leq 2, 0, -2, \ldots. \tag{16.6.6}$$

For the non-symmetric case we would include the odd integers below 2. Since the first term is fixed for a given l and the second term is positive definite this also places constraints on the possible values of p_i.

A solution to the constraints of equations (16.6.4) and (16.6.6) implies that the corresponding representation of C_R can occur in the root space of C, but it does not imply that it actually does occur. The above constraints find the vectors in the lattice spanned by the positive or negative roots of C that have length squared $2, 0, -2, \ldots$ and contain highest weight vectors of representations of the reduced subalgebra C_R. This is not the same as starting from the simple roots, taking their multiple commutators and imposing the Serre relations. By definition, the latter calculation leads to all the roots of the Kac–Moody algebra and no more. However, given the list of possible representations that satisfy the constraints of equations (16.6.4) and (16.6.6) it is easier to consider the set of associated generators and construct the Kac–Moody algebra C up to the level investigated by insisting that they satisfy the Jacobi identities. Of course, one should then make sure that the Serre relations hold. In the example studied below we will see that the Jacobi identity is sufficient to show that one of the potential representations of the reduced algebra allowed by the above constraints does not actually belong to the Kac–Moody algebra. Indeed, the criterion given above does not distinguish between the Kac–Moody algebra C and the algebra derived from the physical state vertex operators associated with the root lattice, which is defined in section 16.8. As the latter contains many more states than the Kac–Moody algebra, one expects to find these additional states by using the criterion above.

16.6.2 All representations

By using a trick one can apply the above techniques to analyse any representations of a Lorentzian algebra L [16.16]. Let us begin with the fundamental representation associated with node e of the Lorentzian algebra L. We first construct an enlarged Dynkin diagram by adding a new node, called $\star$, connected by a single line to the node labelled e of L. We can decompose the adjoint representation of the enlarged algebra $L^\star$ in terms of the original algebra L by deleting the added node $\star$ and using the techniques above but now applied to the algebra $L^\star$. That is, we can analyse it level by level in the level $m_\star$ associated with node $\star$. At level 0 we just find the adjoint representation of the original algebra L, but at level 1, that is, $m_\star = 1$, we find a representation of L. This follows as the level is preserved by the commutation relations and so the commutator of level 0 with level 1 gives level 1. In other words the level 1 states carry a representation of the algebra L.

The simple roots of $L^\star$ are those of L, that is, α_a, together with

$$\alpha_\star = y - \lambda_e^L, \tag{16.6.7}$$

where y is a vector that is orthogonal to the roots of L and λ_e^L is the fundamental weight of L associated with node e. Since $(\alpha_*, \alpha_*) = 2$, we find that $y^2 = 2 - (\lambda_e^L, \lambda_e^L)$. The reader may verify that these simple roots reproduce the Cartan matrix of L^*, in particular $(\alpha_*, \alpha_e) = -1$.

The roots of L^* may be written as

$$\beta = m_* \alpha_* + \sum_a n_a \alpha_a = m_* y - \Lambda^L, \tag{16.6.8}$$

where $\Lambda^L = m_* \lambda_e^L - \sum_a n_a \alpha_a$. For a positive root m_* and n_a are positive integers. To find the representation of L that occurs we set $\Lambda^L = \sum_a p_a \lambda_a^L$. Clearly a solution for $m_* = 1$ is $\Lambda = \lambda_e^L$ for which $\beta^2 = 2$ confirming that the level 1 states are the λ_e^L representation.

To find what is in the λ_e^L representation in terms of an algebra we are familiar with we can delete a node in the L algebra to find such an algebra and analyse the result at level $m_* = 1$ and level by level in the deleted node in L.

We can study any representation of L using very similar techniques. It we are interested in the representation of L with highest weight $\sum_a p_a \lambda_a^L$, we add a new node labelled by $\star$ to the Dynkin diagram of L attached to the node a by p_a lines for each a. The representation of interest is just the level, that is, $m_\star = 1$, and we can proceed as above.

To illustrate the procedure we will calculate the content of the fundamental representation of G^{+++} associated with the very extended node which we label by 1. This is sometimes called the l_1 representation of $G^{+++} = L$. The algebra $L^* = G^{++++}$ has the Dynkin diagram of G^{+++} to which a new node, labelled by $\star$, is attached by single line to the very extended node.

We now analyse the adjoint representation of G^{++++} at level 1 in $m_\star = 1$ and level by level in a node, labelled c of G^{+++} which we also delete. In general, the Dynkin diagram that results from deleting this node of G^{+++} will contain several finite-dimensional semi-simple Lie algebras which we label by $p, q = 1, 2, \ldots$, their Dynkin diagrams being labelled by $C^{(p)}$ and the corresponding sub-algebras being $G^{(p)}$. The nodes of $C^{(p)}$ are labelled by the indices $i, j, \ldots = 1, 2, \ldots$. The index range will be different for each $C^{(p)}$ and the ambiguities of labelling of the indices on a given object are resolved by the knowledge of Dynkin diagram to which the object is associated.

We label the simple roots and weights of $G^{(p)}$ by $\alpha_i^{(p)}$ and $\lambda_i^{(p)}$, the range of the index i being apparent from the algebra to which it belongs. The Dynkin diagram C of G^{+++} is then labelled in a way which is appropriate to the decomposition we wish to perform; we label the node which is to be deleted by c, while the remaining nodes, which belong to the sub-diagrams $C^{(p)}$, are labelled as for the each sub-diagram. In the cases of most interest to us the remaining Dynkin diagram C_R will contain two pieces. The very extended node will be contained in the $C^{(1)}$ Dynkin diagram.

The roots of G^{++++} may be written as in equation (16.6.8) and equation (16.6.7) becomes

$$\alpha_* = y - \lambda_1^{G^{+++}}, \tag{16.6.9}$$

where y is a vector that is orthogonal to the roots of G^{+++}. Since $\alpha_* \cdot \alpha_* = 2$ and $\lambda_1^{G^{+++}} \cdot \lambda_1^{G^{+++}} = \frac{1}{2}$, we conclude that $y^2 = \frac{3}{2}$.

Finding the decomposition resulting from deleting the node c of G^{+++} proceeds as explained in the last section. We may write the simple roots of G^{+++} as

$$\alpha_c = x - \nu, \quad \text{and} \quad \alpha_i^{(p)}, \tag{16.6.10}$$

where $\alpha_i^{(p)}$ are the roots of the algebras $G^{(p)}$ and

$$\nu = -\sum_{i,p} A_{ci(p)}^{G^{+++}} \lambda_i^{(p)} \frac{(\alpha_c, \alpha_c)}{(\alpha_i^{(p)}, \alpha_i^{(p)})}, \tag{16.6.11}$$

where $A_{ci(p)}^{G^{+++}}$ is the Cartan matrix of G^{+++} labelled as explained above. This formula takes account of the possibility of the algebra G being non-simply laced. The vector x is orthogonal to all the simple roots of the sub-algebras $G^{(p)}$ and its length is fixed by demanding that the length of α_c be as required. The fundamental weights of G^{+++} are then given in terms of x and $\lambda_i^{(p)}$ by

$$l_i^{(p)} = \lambda_i^{(p)} + \frac{\nu \cdot \lambda_i^{(p)}}{x^2} x, \quad l_c = \frac{x}{x^2}. \tag{16.6.12}$$

We note that $l_1^{(1)} = \lambda_1^{G^{+++}}$.

Using the above expressions, we may write any root β of G^{+++} as given in equation (16.6.8) in the form

$$\beta = m_* y + x \left(m_c - m_* \frac{\nu^{(1)} \cdot \lambda_1^{(1)}}{x^2} \right) - \sum_p \Lambda^{(p)}, \tag{16.6.13}$$

where

$$\Lambda^{(p)} = -\sum_i m_i^{(p)} \alpha_i^{(p)} + m_c \nu^{(p)} + \delta_{(1,p)} m_* \lambda_1^{(1)}, \tag{16.6.14}$$

where $\nu^{(p)}$ is the component of ν in the sub-algebra $G^{(p)}$, that is, $\nu^{(p)} = -\sum_i A_{ci(p)}^{G^{+++}} \lambda_i^{(p)}$.

We are decomposing the representation of G^{++++} in terms of $G^1 \otimes G^2 \otimes \cdots$ and if a representation of $G^{(p)}$ occurs in the decomposition of the l_1, or adjoint, representations of G^{+++}, then its highest weight must occur in $\Lambda^{(p)}$ in both the positive and negative root spaces of G^{+++}. Applying this to the negative roots of G^{++++}, we must find that there exists a root of G^{+++} such that

$$\Lambda^{(p)} = \sum_i p_i^{(p)} \lambda_i^{(p)}, \tag{16.6.15}$$

where $p_i^{(p)}$ are positive integers. Taking the scalar product with the fundamental weights of $G^{(p)}$ we then find the conditions

$$\sum_i p_i^{(p)} \lambda_i^{(p)} \cdot \lambda_j^{(p)} - m_c \nu^{(p)} \cdot \lambda_j^{(p)} - \delta_{(1,p)} m_* \lambda_1^{(1)} \cdot \lambda_j^{(1)} = -m_j^{(p)} \frac{2}{(\alpha_i^{(p)}, \alpha_i^{(p)})}, \tag{16.6.16}$$

where $p_i^{(p)}, m_i^{(p)}; m_*$ and m_c are all positive integers. The scalar products of the fundamental weights are related to the inverse Cartan matrices of the sub-algebras by

$$(A^{(p)})_{ij}^{-1} = \frac{2}{(\alpha_i^{(p)}, \alpha_i^{(p)})} (\lambda_i^{(p)}, \lambda_j^{(p)}) \quad \text{and} \quad (\alpha_i^{(p)}, \lambda_j^{(q)}) = \frac{2\delta_{p,q}\delta_{ij}}{(\alpha_i^{(p)}, \alpha_i^{(p)})}. \tag{16.6.17}$$

For finite dimensional semi-simple algebras, the inverse Cartan matrices are positive definite and as a result the above equation tightly constrains the possible Dynkin indices $p_i^{(p)}$ or representations of the sub-algebras $G^{(p)}$ that can arise.

The above decomposition does not apply when one deletes the over-extended node of G^{+++} as the fundamental weight associated with this node has length 0 and so cannot be written as x/x^2.

The length squared of the roots of G^{++++} which contain the highest weights of the sub-algebras are given by

$$\beta^2 = \frac{3}{2}m_*^2 + x^2 \left(m_c - m_* \frac{\nu^{(1)} \cdot \lambda_1^{(1)}}{x^2} \right)^2 + \sum_q \sum_{i,j} p_i^{(q)} (\lambda_j^{(q)}, \lambda_j^{(q)}) p_j^{(q)}. \quad (16.6.18)$$

This quantity is an integer and is bounded from above for all Kac–Moody algebras and for simply laced Kac–Moody algebras it can only take the values $2, 0, -2, \dots$. Hence, we find a further constraint on the possible Dynkin labels $p_j^{(q)}$ and so representations that can arise.

A hyperbolic Kac–Moody algebra possesses roots at all these values of β^2, but this is not so for more general Kac–Moody algebras. For a general Kac–Moody algebra the roots are either real or imaginary. By definition a real root not only has $\beta^2 = 2$, but must also be conjugate under the Weyl group to a simple root. A root is imaginary if it is not real. An imaginary root has $\beta^2 \le 0$, but must also be conjugate under the Weyl group to a root that is in the fundamental Weyl chamber and also has connected support on the Dynkin diagram of the Kac–Moody algebra, see [17.3, p377] for a discussion. Consider the integer coefficients of a root when expressed in terms of simple roots and for those integer coefficients that are zero let us delete the corresponding nodes of the Dynkin diagram. A connected root is one for which the resulting is a connected diagram [16.6]. We will see that not every solution we find will respect this condition so we will discard such solutions. As explained above, equations (16.6.16) and (16.6.18) are necessary, but not sufficient conditions, for representations to occur in the l_1 representation; although for almost all cases the solutions found do occur.

The most common case, and the one of most interest to us, is when the deletion of the node c of G^{+++} leads to only two sub-algebras. It is useful in this case to introduce a more easily understood notation. We denote the roots, weights and Dynkin indices of the sub-algebras by

$$\alpha_i^{(1)} = \beta_i, \ \lambda_i^{(1)} = \mu_i, \ p_i^{(1)} = q_i, \ m_i^{(1)} = m_i; \ \alpha_i^{(2)} = \alpha_i, \ \lambda_i^{(2)} = \lambda_i, \ p_i^{(2)}$$
$$= p_i, \ m_i^{(2)} = n_i. \quad (16.6.19)$$

We will now assume that G is simply laced and so all the simple roots of G^{+++} have length squared 2. In this notation equation (16.6.16) becomes

$$\sum_i q_i \mu_i \cdot \mu_j - m_c \nu^{(1)} \cdot \mu_j - m_* \mu_1 \cdot \mu_j = -m_j \quad (16.6.20)$$

for the first sub-algebra and

$$\sum_i p_i \lambda_i \cdot \lambda_j - m_c \nu^{(2)} \cdot \lambda_j = -n_j \quad (16.6.21)$$

for the second sub-algebra. It is straightforward to rewrite equation (16.6.18) for the new notation.

16.7   The Kac–Moody algebra E_{11}

16.7.1 E_{11} at low levels

We now apply the analysis of the previous sections to find the Lie algebra of very extended E_8, that is, E_8^{+++}, also called E_{11}, at low levels in terms of representations of A_{10} [16.10, 16.15]. The Dynkin diagram of E_8^{+++} is given in section 16.3. The central node, c is chosen to be the node labelled 11 and deleting it gives A_{10}. Clearly, at level 0 we have the adjoint representation of A_{10}. We recall from section 16.3 that $\alpha_c = \alpha_{11} = x - \lambda_8$ and $x^2 = -\frac{2}{11}$.

Using the inverse Cartan matrix of A_{D-1}, which is given in equation (D.1.6), we find that the constraint of equation (16.6.4) becomes

$$\sum_{j \leq k} j(11-k)p_j + \sum_{j > k} k(11-j)p_j = -11m_k + l \begin{cases} 8(11-k), & k \geq 8, \\ 3k, & k \leq 8, \end{cases} \quad (16.7.1)$$

where $m_k = 0, 1, 2 \ldots$ and $p_k = 0, 1, 2, \ldots$.

We first analyse this equation for level 1, that is, $l = 1$. For $k = 1$ and $k = 10$ we find it gives

$$\sum_j (11-j)p_j = -11m_1 + 3 \quad \text{and} \quad \sum_j jp_j = -11m_{10} + 8, \quad (16.7.2)$$

respectively. Adding these two equations we have

$$\sum_j p_j = -(m_1 + m_{10}) + 1, \quad (16.7.3)$$

from which it is clear that only one of the p_j is non-zero and it must be equal to 1. In this case $m_1 = 0 = m_{10}$ and it is easy to see from equation (16.7.2) that the only possibility is $p_8 = 1$. It it straightforward to verify that for this value all of other equations (16.7.1) are satisfied. In analysing this latter equation we have used equation (D.1.6) and the relations $\det A_{E_n} = 9 - n$ and $\det A_{A_n} = n + 1$. The corresponding $m_k = 0$, $k = 1, \ldots, 10$, and so the corresponding E_8^{+++} root is just α_{11}. We may write the root by giving its m_k values, that is, $(m_1, \ldots, m_{11}) = (0^{10}, 1)$. The 1 in the eleventh place corresponds to the fact that we are at level 1. Equation (16.6.6) is also satisfied and the root has length squared 2.

It is straightforward, if tedious, to verify that the constraints of equations (16.6.4) and (16.6.6) allow the following representations of A_{10} [16.10, 16.15]:

$$l = 1, \ p_8 = 1; \ l = 2, \ p_5 = 1; \ l = 3, \ p_{10} = 1, p_3 = 1 \text{ and } p_2 = 1;$$
$$l = 4, \ p_1 = 1, p_{10} = 2 \text{ and } p_1 = 1, p_9 = 1 \text{ and } p_2 = 1, \ p_8 = 1 \text{ and } p_{10} = 1, \quad (16.7.4)$$

all other p_j being zero.

Having found the Dynkin indices of the SL(11) representations present we now give these representations in more concrete from. The Dynkin diagram of SL(n) is given in appendix D.1. As we mentioned in section 16.1.2, the fundamental representation with fundamental weight λ_{n-k} is carried by the rank k totally anti-symmetrised representation, that is, $T^{i_1 \ldots i_k}$, $i_1, \ldots, i_k = 1, \ldots, n$. This corresponds to having the representation with Dynkin index $p_{n-k} = 1$. Its Young tableau consists of just a column of $n - k$ boxes. The highest weight component is $T^{n-j \ldots n}$. More generally, an A_D representation with Dynkin indices p_j has a Young tableau which consists of p_{10} boxes in a row next above a row of p_9 two-column boxes etc. The corresponding tensors in general require certain symmetrisation and anti-symmetrisation conditions.

The positive root elements of the E_{11} algebra corresponding to the solutions in equation (16.7.4) are given by the SL(11) representations [16.10, 16.15]

$$R^{a_1 a_2 a_3}, R^{a_1 a_2 \dots a_6}, R^{a_1 a_2 \dots a_8, b} \tag{16.7.5}$$

at levels 1, 2 and 3 respectively. The last generator has the constraint $R^{[a_1 a_2 \cdots a_8, b]} = 0$, which can be read off from the Young tableau. We have not listed the generator $R^{a_1 a_2 \cdots a_9}$ as we will find out later that this is ruled out by the Jacobi identities. As discussed above, we find a copy in the negative root space and so we have the negative root elements

$$R_{a_1 a_2 a_3}, R_{a_1 a_2 \dots a_6}, R_{a_1 a_2 \dots a_8, b} \tag{16.7.6}$$

at levels $-1, -2, -3$ with the second to last generator satisfying an analogous constraint. We have not listed the generator $R_{a_1 a_2 \dots a_9}$ for the same reason as just mentioned.

We wish to identify the Chevalley generators of E_{11} amongst those above. To begin this process we start by considering the level 0 sub-algebra SL(11). However, to illustrate the procedure in general we consider SL(n) whose generators $\hat{K}^a{}_b, a, b = 1, \dots, n$, obey the commutator relations

$$[\hat{K}^a{}_b, \hat{K}^c{}_d] = \delta^c_b \hat{K}^a{}_d - \delta^a_d \hat{K}^c{}_b, \tag{16.7.7}$$

The Chevalley generators of SL(n), or A_{n-1}, are

$$E_a = \hat{K}^a{}_{a+1}, \ F_a = \hat{K}^{a+1}{}_a, \ \hat{H}_a = K^a{}_a - \hat{K}^{a+1}{}_{a+1}, a = 1, \dots, n. \tag{16.7.8}$$

The reader can verify that they do indeed satisfy equations (16.1.43) and (16.1.44) for the Cartan matrix of A_{n-1} given in appendix D.1. Also their multiple commutators lead to all the generators of A_{n-1}. For example, we find that $[\hat{K}^a{}_{a+1}, \hat{K}^{a+1}{}_{a+2}] = \hat{K}^a{}_{a+2}$ etc. These checks confirm that the above choice of Chevalley generators is indeed the correct one. Alternatively, one can show this result by realising that the Lie algebra of SL(n) is given by the matrices $(\hat{K}^a{}_b)_c{}^d = \delta^a_c \delta^d_b$ for $a \neq b$ together with suitable matrices of $\hat{K}^a{}_a$. Having found the roots, one can choose the positive roots such that they correspond to the generators with $\hat{K}^a{}_b, a < b$ and one then finds that the simple positive roots correspond to $K^a{}_{a+1}$.

For reasons that will become clear shortly we will now consider GL(11) rather than SL(11) as the extra commuting generator will fit naturally into the E_{11} algebra. We denote the generators of GL(11) by $K^a{}_b, a, b = 1, \dots, 11$. The SL(11) generators are related to those of GL(11) by

$$\hat{K}^a{}_b = K^a{}_b - \frac{1}{11} \delta^a_b \sum_c K^c{}_c. \tag{16.7.9}$$

The trace does not appear in equation (16.7.7) and so they also obey the same relations, that is,

$$[K^a{}_b, K^c{}_d] = \delta^c_b K^a{}_d - \delta^a_d K^c{}_b. \tag{16.7.10}$$

The generators of equation (16.7.5) belong to representations of SL(11) and in particular the 3-form $R^{c_1 \cdots c_3}$ has the commutator

$$[K^a{}_b, R^{c_1 \cdots c_3}] = \delta^{c_1}_b R^{a c_2 c_3} + \delta^{c_2}_b R^{a c_3 c_1} + \delta^{c_3}_b R^{a c_1 c_2} = 3\delta^{[c_1}_b R^{a c_2 c_3]}. \tag{16.7.11}$$

In this relation we have used the $K^a{}_b$ which belongs to GL(11), but as this is the first place that the extra GL(1) generator in GL(11) occurs we can normalise its action as it is given in this relation.

The positive root Chevalley generators E_a, $a = 1, \ldots, 11$, of E_{11} correspond to those of the simple roots. These are the level 0 simple roots of SL(11) and the level 1 simple root α_{11}. The former are those given in equation (16.7.8) once we remove the hats and the latter is the level 1 highest weight state of equation (16.7.4), that is, the $p_8 = 1$ solution found above. The highest weight state is the generator R^{91011} and as a result we conclude that $E_{11} = R^{91011}$. To summarise the Chevalley generators of E_{11} not including those in the Cartan subalgebra are given by [16.10]

$$E_a = K^a{}_{a+1}, a = 1, \ldots, 10, E_{11} = R^{91011}; F_a = K^{a+1}{}_a, a = 1, \ldots, 10,$$

$$F_{11} = R_{91011}, \tag{16.7.12}$$

where $R_{c_1 c_2 c_3}$ are the generators at level -1.

The Chevalley generators H_a of E_{11} are those of SL(11) of equation (16.7.8) and one other generator. They are given by [16.10]

$$H_a = K^a{}_a - K^{a+1}{}_{a+1}, a = 1, \ldots, 10,$$

$$H_{11} = \tfrac{2}{3}(K^9{}_9 + K^{10}{}_{10} + K^{11}{}_{11}) - \tfrac{1}{3}(K^1{}_1 + \cdots + K^8{}_8). \tag{16.7.13}$$

It is straightforward to verify that the Chevalley generators of equations (16.7.12) and (16.7.13) do indeed satisfy the defining relations $[H_m, E_n] = A_{mn}E_n$ with the Cartan matrix for E_{11} by using equations (16.7.11) and (16.7.12) as well as the other relations in equations (16.1.43) and (16.1.44). This confirms the above identification of the Chevalley generators.

Working with GL(11), rather than SL(11), did not introduce any additional assumption as we require an additional element in the Cartan sub-algebra of E_{11} whose action we are free to normalise as in equation (16.7.12). The defining conditions of the Chevalley generators then fix its relation to the Cartan generators of E_{11} to be as given in equation (16.7.13).

Rather than construct the E_{11} algebra by considering the multiple commutators of the Chevalley generators subject to the Serre relations, we can construct the algebra from the list of potential elements of the E_{11} algebra given in equation (16.7.5). We just write down all possible terms on the right-hand side of any commutator that have the correct level and A_{10} properties, then enforce the Jacobi identity. We now give the result of this calculation. We recall that the commutators of the generators of A_{10} with themselves are given up to, and including, level 3 by [16.10]

$$[K^a{}_b, K^c{}_d] = \delta^c_b K^a{}_d - \delta^a_d K^c{}_b, \tag{16.7.14}$$

$$[K^a{}_b, R^{c_1 \cdots c_6}] = \delta^{c_1}_b R^{ac_2 \cdots c_6} + \cdots, \quad [K^a{}_b, R^{c_1 \cdots c_3}] = \delta^{c_1}_b R^{ac_2 c_3} + \cdots, \tag{16.7.15}$$

$$[K^a{}_b, R^{c_1 \cdots c_8,d}] = (\delta^{c_1}_b R^{ac_2 \cdots c_8,d} + \cdots) + \delta^d_b R^{c_1 \cdots c_8,a}. \tag{16.7.16}$$

where $+ \cdots$ means the appropriate anti-symmetrisation, for example there are six and three terms respectively in the first and second equations in (16.7.15).

The commutators of the other positive root generators are given by

$$[R^{c_1 \cdots c_3}, R^{c_4 \cdots c_6}] = 2R^{c_1 \cdots c_6}, \quad [R^{a_1 \cdots a_6}, R^{b_1 \cdots b_3}] = 3R^{a_1 \cdots a_6[b_1 b_2, b_3]}. \tag{16.7.17}$$

One might have expected to see the generator $R^{a_1 \cdots a_9}$ on the right-hand side of the commutator in the last expression of equation (16.7.17). However, it is ruled out by the Jacobi identities and as such we must conclude that this generator does not actually occur in E_{11}.

The level 0 and negative level generators obey the relations

$$[K^a{}_b, R_{c_1 \cdots c_3}] = -\delta^a_{c_1} R_{bc_2 c_3} - \cdots, \quad [K^a{}_b, R_{c_1 \cdots c_6}] = -\delta^a_{c_1} R_{bc_2 \cdots c_6} - \cdots, \quad (16.7.18)$$

$$[K^a{}_b, R_{c_1 \cdots c_8, d}] = -(\delta^a_{c_1} R_{bc_2 \cdots c_8, d} + \cdots) - \delta^a_d R_{c_1 \cdots c_8, b}, \quad (16.7.19)$$

$$[R_{c_1 \cdots c_3}, R_{c_4 \cdots c_6}] = 2R_{c_1 \cdots c_6}, \quad [R_{a_1 \cdots a_6}, R_{b_1 \cdots b_3}] = 3R_{a_1 \cdots a_6[b_1 b_2, b_3]}. \quad (16.7.20)$$

Finally, the commutation relations between the positive and negative generators of up to level 3 are given by [16.17]

$$[R^{a_1 \cdots a_3}, R_{b_1 \cdots b_3}] = 18\delta^{[a_1 a_2}_{[b_1 b_2} K^{a_3]}{}_{b_3]} - 2\delta^{a_1 a_2 a_3}_{b_1 b_2 b_3} D,$$

$$[R_{b_1 \cdots b_3}, R^{a_1 \cdots a_6}] = \frac{5!}{2} \delta^{[a_1 a_2 a_3}_{b_1 b_2 b_3} R^{a_4 a_5 a_6]},$$

$$[R^{a_1 \cdots a_6}, R_{b_1 \cdots b_6}] = -5! \times 3 \times 3\delta^{[a_1 \cdots a_5}_{[b_1 \cdots b_5} K^{a_6]}{}_{b_6]} + 5!\delta^{a_1 \cdots a_6}_{b_1 \cdots b_6} D,$$

$$[R_{a_1 \cdots a_3}, R^{b_1 \cdots b_8, c}] = 8 \times 7 \times 2 \left(\delta^{[b_1 b_2 b_3}_{[a_1 a_2 a_3} R^{b_4 \cdots b_8]c} - \delta^{[b_1 b_2|c|}_{a_1 a_2 a_3} R^{b_3 \cdots b_8]} \right), \quad (16.7.21)$$

$$[R_{a_1 \cdots a_6}, R^{b_1 \cdots b_8, c}] = \frac{7! \times 2}{3} \left(\delta^{[b_1 \cdots b_6}_{[a_1 \cdots a_6} R^{b_7 b_8]c} - \delta^{c[b_1 \cdots b_5}_{a_1 \cdots a_6} R^{b_6 b_7 b_8]} \right),$$

where $D = \sum_b K^b{}_b$, $\delta^{a_1 a_2}_{b_1 b_2} = \frac{1}{2}(\delta^{a_1}_{b_1} \delta^{a_2}_{b_2} - \delta^{a_2}_{b_1} \delta^{a_1}_{b_2}) = \delta^{[a_1}_{b_1} \delta^{a_2]}_{b_2}$ with similar formulae when more indices are involved.

The computation of the first commutator of equations (16.7.21) is found by writing down the most general expression of level 0 and then fixing the coefficients using the fact that $[E_{11}, F_{11}] = H_{11}$, where these generators are given in equation (16.7.13). In particular,

$$[E_{11}, F_{11}] = [R^{91011}, R_{91011}] = 18\delta^{[910}_{[910} K^{11]}{}_{11]} - 2\delta^{91011}_{91011} D$$

$$= 2(\delta^{910}_{910} K^{11}{}_{11} + \delta^{119}_{119} K^{10}{}_{10} + \delta^{1011}_{1011} K^9{}_9) - \frac{1}{3} D = H_{11} \quad (16.7.22)$$

as required. The subsequent commutators are found by using the Jacobi identity. For example, for the case of $[R_{b_1 \cdots b_3}, R^{a_1 \cdots a_6}]$ we write the generator $R^{a_1 \cdots a_6} = \frac{1}{2}[R^{a_1 \cdots a_3}, R^{a_4 \cdots a_6}]$ and use the Jacobi identity and the first commutator in equations (16.7.21).

The generators that appear on the right-hand side of a commutator can be checked by seeing if the roots add up as required. For example, $[R^{678}, R^{91011}]$ could equal $R^{6 \cdots 11}$ as R^{91011} corresponds to the root α_{11} and R^{678} to the root $\alpha_{11} + \alpha_6 + 2\alpha_7 + 3\alpha_8 + 2\alpha_9 + \alpha_{10}$. The latter follows as R^{678} is just the multiple commutator of the $K^a{}_b$ corresponding to the above root with K^{91011}. One can check that this is precisely the root that corresponds to the level 2 generator $R^{6 \cdots 11}$ that appears in the solution of equation (16.7.4).

It is clear from the above that the positive and Cartan–Chevalley generators of E_{11} are contained in SL(11), that is, the generators $K^a{}_b$, and the generators R^{abc}. As such, all positive root generators of E_{11} can be expressed as multiple commutators of these generators. It is then obvious that the number of SL(11) indices on a generator is just three times the number of R^{abc} generators in the multiple commutator, but this is just $3l$, where l is the level. However, given the Dynkin indices of the SL(11) representation to which the generator belongs we can also compute the number of indices. The Dynkin index $p_{11-j} \equiv q_j$ leads to q_j blocks of j upper indices. Taking into account that we can also have blocks of 11 anti-symmetrised indices that are invariant under SL(11) transformations and so have no corresponding Dynkin index, we find the identity

$$11n + \sum_j jq_j = 3l. \tag{16.7.23}$$

The reader can readily see that this result works on the low level generators of E_{11}.

Using the above relation leads to a quicker method of finding the low level content of the adjoint representation. One can show that substituting for l in equation (16.7.1), or in the general analysis in equation (16.6.4), allows one to solve this equation and find a consistent set of integers m_j. However, one still has to solve equation (16.6.6) applied to E_{11} for which one can also substitute for l. Thus to find solutions one must only solve equation (16.7.23), which is very simple, and equation (16.6.6) applied to E_{11}. One can then find the corresponding roots from equation (16.7.1) for each set of allowable Dynkin indices. In fact, this method applies much more generally and in particular to other Lorentzian algebras.

16.7.2 The l_1 representation of E_{11}

In this section we will construct, at low levels, the content of the representation with highest weight Λ_1 which by definition obeys $(\Lambda_1, \alpha_a) = \delta_{a,1}$ in terms of representations of A_{10} [16.17]. This is often called the l_1 representation. We can do this either by following the analysis of section 17.1.2, applied to E_{11}, or by simply constructing the root string as discussed in section 16.1.2. In the first method we consider the Dynkin diagram found by adding a new node to the E_{11} Dynkin diagram which is attached to node 1 by a single line. We delete this new node keeping only level 1 states and then delete the node labelled 11 of the E_{11} Dynkin diagram to leave the A_{10} algebra. We analyse the results in terms of this algebra. At level 0 we have the A_{10} algebra and the fundamental representation associated with node 1 has A_{10} highest weight λ_1. This is the representation which is a tensor with a single lower index, that is, a vector. This is equivalent to a tensor with ten upper anti-symmetrised indices using the invariant epsilon symbol.

The fundamental weights of E_{11} can be read off from equation (16.3.4) to be

$$\Lambda_i = \lambda_i + \lambda_8 \cdot \lambda_i \frac{x}{x^2}, \ i = 1, \ldots, 10 \quad \Lambda_{11} = \frac{x}{x^2}. \tag{16.7.24}$$

In this section we use Λ_a rather than l_a for the highest weights of the fundamental representations associated with node a. Using equation (D.1.6) we find $\Lambda_1 = \lambda_1 - \frac{3}{2}x$ which obviously contains the states in the A_{10} representation λ_1 as we already found. The highest weight state $|\Lambda_1\rangle$ can be thought of as being at level $-\frac{3}{2}$.

We now consider the the second method. It is straightforward to construct the root string associated with the addition of the simple negative roots F_a on this highest weight state as explained in section 16.1.2. One finds

$$|\Lambda_1\rangle, |\Lambda_1 - \alpha_1\rangle, |\Lambda_1 - \alpha_1 - \alpha_2\rangle, \ldots, |\Lambda_1 - \alpha_1 - \cdots - \alpha_8 - \alpha_9\rangle,$$

$$|\Lambda_1 - \alpha_1 - \cdots - \alpha_8 - \alpha_{11}\rangle, \ldots. \tag{16.7.25}$$

The set of weights begins like that for A_{10} until one subtracts the weight α_8 and at this point one can subtract α_9 or α_{11}. This follows by calculating the scalar products with the simple roots and seeing if they are positive integers. Clearly, the weights which contain no α_{11}s collect up into the tensor $P^{a_1\cdots a_{10}}$. By calculating the scalar products with the simple roots one can see that we may rewrite the first weight with an α_{11} as

$$\Lambda_1 - \alpha_1 - \cdots - \alpha_8 - \alpha_{11} = \lambda_9 - \tfrac{5}{2}x. \tag{16.7.26}$$

This makes it clear that it contains the highest weight λ_9 for the A_{10} representation whose only non-vanishing Dynkin index is $p_9 = 1$. This is a second rank anti-symmetric tensor. It is in this way that new A_{10} representations in the weights appear.

Continuing in this manner we find that the representation l_1 contains the following A_{10} representations:

$$p_1 = 1,\ (-\tfrac{3}{2});\ p_9 = 1,\ (-\tfrac{5}{2});\ p_6 = 1,\ (-\tfrac{7}{2});\ ,\ p_4 = 1,\ p_{10} = 1,\ (-\tfrac{9}{2});\ \text{and}$$

$$p_3 = 1,\ (-\tfrac{9}{2});\ \ldots, \tag{16.7.27}$$

all other p_i vanishing. The number in brackets is the corresponding level. We may associate these representations with the tensors [16.17]

$$P^{a_1\ldots a_{10}}, Z^{a_1 a_2}, Z^{a_1\ldots a_5}, Z^{a_1\ldots a_7,b}, \ldots. \tag{16.7.28}$$

It is straightforward to identify the tensors associated with the root string of equation (16.7.25). The entries explicitly given correspond to $P^{2\cdots 11}$, $P^{13\cdots 11}$, $P^{124\cdots 11}$, $P^{12\cdots 81011}$ and Z^{1011}, respectively.

Given any Lie algebra G, a representation $u(A)$ of G is by definition is a set of linear operators $u(A)$, for each element $A \in G$, which obey the relation $u([A_1, A_2]) = [u(A_1), u(A_2)]$. If the representation is carried by the states $|X_s\rangle$ it defines the matrices $u(A)|X_s\rangle = -D(A)_s{}^t|\chi_t\rangle$. The minus sign is required in order that we do have a representation. We can enlarge the algebra G, by associating with each state in the representation $|X_s\rangle$ an element X_s of the enlarged algebra and adopting the commutation relations

$$[X_s, A] = -D(A)_s{}^r X_r, A \in g. \tag{16.7.29}$$

The reader may check that the Jacobi identities involving A_1, A_2 and X_s are automatically satisfied as $D([A_1, A_2]) = [D(A_1), D(A_2)]$. If we now specify commutation relations for the new elements X_s with themselves in such way that it is consistent with the Jacobi identities we will have defined a new algebra. This is called the semi-direct product algebra. One consistent choice is to take $[X_s, X_t] = 0$.

We will now construct, at low levels, the semi-direct product of E_{11} with its l_1 representation denoted by $E_{11} \oplus_s L_1$. At the lowest level we find the element $P^{a_1\cdots a_{10}}$, but it is simpler to work with the more familiar generator $P_a = \frac{1}{10!}\epsilon_{ba_1\cdots a_{10}}P^{a_1\cdots a_{10}}$. The highest weight Λ_1 of the l_1 representation by definition obeys $(\Lambda_1, \alpha_a) = \delta_{a,1}$. As such, the highest weight state of this representation $|\Lambda_1\rangle$ satisfies

$$E_a|\Lambda_1\rangle = 0 \quad \text{and} \quad H_a|\Lambda_1\rangle = (\Lambda_1, \alpha_a)|\Lambda_1\rangle = \delta_{a,1}|\Lambda_1\rangle,\ a = 1, \ldots, 11. \tag{16.7.30}$$

The generators corresponding to the highest weight state in the l_1 representation must have vanishing commutator with $E_a, a = 1, \ldots, 11$ and $H_a, a = 2, \ldots, 11$. As $K^a{}_b$ is of level 0 its commutator with P_a, which is in the vector representation of the SL(11) sub-algebra, can only be of the form $[K^a{}_b, P_c] = -\delta^a_c P_b + e\delta^a_b P_c$, where e is a constant. Using equation (16.7.13), and in particular $E_a = K^a{}_{a+1}, a = 1, \ldots, 10$ we conclude that the element corresponding to the highest weight state is P_1 and using the expression for H_{11} given in that equation we find that $[H_{11}, P_1] = 0$, as required by equation (16.7.30), only if $e = \frac{1}{2}$.

The reader may have noticed that P_1, or equivalently P^{11}, is not actually a highest weight as just indicated, for example, $[K^1{}_2, P_1] = -P_2$. They are, in fact, lowest weights of SL(11). It turns out that the subject has consistently used the name highest weight in the context of SL(11) representations that arise in the decompositions of E_{11} to SL(11) when what was actually meant was lowest weight. Here we have persisted with this unfortunate, but widely spread habit.

The commutators for the low level generators of the l_1 representation with $R^{a_1 a_2 a_3}$ are determined up to constants by demanding that the levels match and so we can take

$$[R^{a_1 a_2 a_3}, P_b] = 3\delta_b^{[a_1} Z^{a_2 a_3]}, \quad [R^{a_1 a_2 a_3}, Z^{b_1 b_2}] = Z^{a_1 a_2 a_3 b_1 b_2},$$

$$[R^{a_1 a_2 a_3}, Z^{b_1 \cdots b_5}] = Z^{b_1 \cdots b_5 [a_1 a_2, a_3]} + Z^{b_1 \cdots b_5 a_1 a_2 a_3}.$$

(16.7.31)

These equations define the normalisation of these generators of the l_1 representation charges. Using this result, the Jacobi identity and the fact that $e = \frac{1}{2}$ we conclude that [16.17]

$$[K^a{}_b, P_c] = -\delta^a_c P_b + \frac{1}{2}\delta^a_b P_c, \quad [K^a{}_b, Z^{c_1 c_2}] = 2\delta_b^{[c_1} Z^{|a|c_2]} + \frac{1}{2}\delta^a_b Z^{c_1 c_2},$$

$$[K^a{}_b, Z^{c_1 \cdots c_5}] = 5\delta_b^{[c_1} Z^{|a|c_2 \ldots c_5]} + \frac{1}{2}\delta^a_b Z^{c_1 \ldots c_5}.$$

(16.7.32)

We also find using the Jacobi identities and equations (16.7.14)–(16.7.20) and (16.7.29) that

$$[R^{a_1 \cdots a_6}, P_b] = -3\delta_b^{[a_1} Z^{\cdots a_6]}, \quad [R^{a_1 \cdots a_6}, Z^{b_1 b_2}] = -Z^{b_1 b_2 [a_1 \cdots a_5, a_6]} - Z^{b_1 b_2 a_1 \cdots a_6}.$$

(16.7.33)

The commutators with the negative root generators are given by

$$[R_{a_1 a_2 a_3}, P_b] = 0, \quad [R_{a_1 a_2 a_3}, Z^{b_1 b_2}] = 6\delta_{[a_1 a_2}^{b_1 b_2} P_{a_3]},$$

$$[R_{a_1 a_2 a_3}, Z^{b_1 \cdots b_5}] = \frac{5!}{2}\delta_{a_1 a_2 a_3}^{[b_1 b_2 b_3} Z^{b_4 b_5]},$$

(16.7.34)

The first equation, just follows from the fact that there is no level 0 object it can equal and the subsequent equations follow by using the equation (16.7.21) and the Jacobi identities.

Using the techniques of section 16.6.2 one can find the members of the l_1 representation at higher levels [16.13, 16.16, 16.17]:

$$P_a \ (0, 2), Z^{a_1 a_2} \ (1, 2), Z^{a_1 \cdots a_5} \ (2, 2), Z^{a_1 \cdots a_7, b} \ (3, 2), Z^{a_1 \cdots a_8} \ (3, 0),$$

$$Z^{b_1 b_2 b_3, a_1 \cdots a_8} \ (4, 2), Z^{(cd), a_1 \cdots a_9} \ (4, 2), Z^{cd, a_1 \cdots a_9} \ (4, 0),$$

$$Z^{c, a_1 \cdots a_{10}} \ (4, -2), \tilde{Z}^{c, a_1 \cdots a_{10}} \ (4, -2), Z \ (4, -4), Z^{c, d_1 \cdots d_4, a_1 \cdots a_9} \ (5, 2), \quad (16.7.35)$$

$$Z^{c_1 \cdots c_6, a_1 \cdots a_8} \ (5, 2), Z^{c_1 \cdots c_5, a_1 \cdots a_9} \ (5, 0), Z^{d_1, c_1 c_2 c_3, a_1 \cdots a_{10}} \ (5, 0),$$

$$Z^{c_1 \cdots c_4, a_1 \cdots a_{10}} \ (5, -2), Z^{(c_1 c_2, c_3)} \ (5, 2), Z^{c, a_1 a_2} \ (5, -2), Z^{c_1 \cdots c_3} \ (5, -4).$$

In the above, the first figure in the brackets refers to the level n_8, while the second figure is the length squared β^2 of the root in E_8^{++++} to which the highest weight of A_{10} representation belongs. The index range is $a, b, \ldots = 1, \ldots, 11$. We have not shown the blocks of 11 indices, but they are easily put in once one remembers that at level l we have $3l - 1$ indices. For example, the element at level 4 which is shown above with no indices really has a single block of 11 indices.

It is interesting to examine what adding the generators corresponding to the l_1 representation means in terms of the weight lattice. The weight Λ_1 equals $-\frac{1}{2}(l + \bar{l})$ in the notation of equation (16.4.7). The root lattice of E_{11} is given by

$$\Lambda_{E_8} \oplus \Pi^{(1,1)} \oplus \{(n, n),\ n \in \mathbf{Z}\}. \tag{16.7.36}$$

Adding the l_1 representation corresponds to adding the vector $-\frac{1}{2}(1, -1)$ to the last factor in the above decomposition and so one is working on the full weight lattice.

16.7.3 The Cartan involution invariant subalgebra of a Kac–Moody algebra

Given any Kac–Moody algebra G we can define the Cartan involution τ on the Chevalley generators by

$$\tau(E_a) = -F_a, \quad \tau(F_a) = -E_a, \quad \tau(H_a) = -H_a. \tag{16.7.37}$$

Indeed $\tau^2 = I$ and as it is an automorphism we take $\tau(AB) = \tau(A)\tau(B)$ for any two elements A and B in the algebra. One can verify that this preserves the Serre conditions of equations (16.1.43) and (16.1.44). Since the algebra is found from the multiple commutators of the E_a and the F_a it is defined on the entire Kac–Moody algebra. Using the Cartan involutions we can construct a sub-algebra of G, called $I(G)$, which is invariant under the Cartan involution. Indeed, $I(G)$ is just the multiple commutators of

$$E_a - F_a. \tag{16.7.38}$$

We note that $I(G)$ does not contain the Cartan sub-algebra of G and this makes it more difficult to construct. The above discussion applies to any Kac–Moody algebra including the finite-dimensional semi-simple Lie algebras. For A_1 we have only the generators E, F and H and $I(A_1) = \{E - F\}$.

Before considering E_{11} we find the Cartan involution sub-algebra of the more familiar A_{n-1} algebra. The generators of A_{n-1} are denoted by $K^a{}_b, a, b = 1, 2, \ldots, n$ and obey equation (16.7.7) with the Chevalley generators given in equation (16.7.8). The Cartan involution defined above induces an action $\tau(E_a) = \tau(K^a{}_{a+1}) = -F_a = -K^{a+1}{}_a$. Applying this to a commutator such as $[K^a{}_{a+1}, K^{a+1}{}_{a+2}] = K^a{}_{a+2}$ we find that the action of the Cartan involution on all the generators of A_n is given by

$$\tau(K^a{}_b) = -K^b{}_a. \tag{16.7.39}$$

We are raising and lowering indices with δ_{ab}. As such the Cartan involution invariant sub-algebra has the generators

$$J_{ab} = K^a{}_b - K^b{}_a. \tag{16.7.40}$$

It is straightforward to verify using equation (16.7.7) that J_{ab} obey the commutation relations of SO(n). As such, we recover the known fact that the Cartan involution invariant sub-algebra of SL(n) is SO(n).

A generator in the $\bar{n}$ representation of A_{n-1}, that is, with non-vanishing Dynkin index $p_n = 1$ transforms under A_{n-1} as

$$[K^a{}_b, R^c] = \delta^c_b R^a - \frac{\delta^a_b}{n} R^c, \tag{16.7.41}$$

with $K^a{}_b$ acting in a similar way for more complicated tensors. It is easy to verify that

$$[J_{ab}, R^c] = \delta^c_b R^a - \delta^c_a R^b \tag{16.7.42}$$

with a similar action on more complicated tensors. Hence, a tensor under SL(n) transforms under its Cartan involution invariant sub-algebra, SO(n), as the indices suggest. However, as δ_{ab} is an SO(n) invariant tensor, the representation of SL(n) is not always an irreducible representation of SO(n). For example, the tensor $T^{(ab)}$ of SL(n) with non-vanishing Dynkin index $p_n = 2$ does not transform irreducibly under SO(n), but as a singlet and a symmetric traceless tensor, $T^{(ab)} - (\delta^{ab}/n) \sum_c T^c{}_c$.

We recall from the end of section 16.1.1 that given any complex semi-simple Lie algebra we can construct a real form, which is compact, using the Cartan involution. The other possible real forms are then found from this form by using all possible involutions of this real algebra.

In fact, there are other Cartan involutions than the one considered above. We simply introduce a suitable number of minus signs into the Cartan involution of equation (16.7.37). These lead to different invariant sub-algebras, that is, for A_{n-1} we can find the invariant sub-algebra SO($p, n-p$). In fact, the Cartan involution one takes is related to the real form of the complex algebra one is considering and so one is really considering different real forms and their invariant sub-algebras.

We now consider the Cartan involution of the algebra E_{11}. By first considering its action on the Chevalley generators, that is, $\tau(R^{9\,10\,11}) = -R_{9\,10\,11}$, we find that it acts on the generators of GL(11) and those of equations (16.7.5) as

$$K^a{}_b \rightarrow -K^b{}_a, \quad R^{a_1a_2a_3} \rightarrow -R_{a_1a_2a_3}, \quad R^{a_1\cdots a_6} \rightarrow R_{a_1\cdots a_6}, \quad R^{a_1\cdots a_8,b} \rightarrow -R_{a_1\cdots a_8,b}. \tag{16.7.43}$$

The sub-algebra invariant under the Cartan involution, $I(E_{11})$, is generated by $E_a - F_a$ and at low levels it is given by

$$J_{ab} = K^c{}_b \eta_{ac} - K^c{}_a \eta_{bc}, \quad S_{a_1a_2a_3} = R^{b_1b_2b_3} \eta_{b_1a_1} \eta_{b_2a_2} \eta_{b_3a_3} - R_{a_1a_2a_3}, \tag{16.7.44}$$

$$S_{a_1\cdots a_6} = R^{b_1\cdots b_6} \eta_{b_1a_1} \cdots \eta_{b_6a_6} + R_{a_1\cdots a_6}, \tag{16.7.45}$$

$$S_{a_1\cdots a_8,c} = R^{b_1\cdots b_8,b} \eta_{b_1a_1} \cdots \eta_{b_8a_8} \eta_{bc} - R_{a_1\cdots a_8,c}. \tag{16.7.46}$$

In these equations η_{ab} is a diagonal matrix with $+1$ on all its diagonal elements if we adopt the Cartan involution of equation (16.7.43) in which case the generators J_{ab} are those of the algebra SO(11). However, we can adopt the modified Cartan involution $E_a \rightarrow -\eta_{aa}F_a$, $F_a \rightarrow -\eta_{aa}E_a$ and $H_a \rightarrow -H_a$, where η_{ab} is the Minkowski metric; in this case the Cartan involution invariant generators are given by equation (16.7.46) but with η_{ab} being the Minkowski metric and the generators J_{ab} are now those of the Lorentz algebra SO(1,10).

The $S_{a_1 a_2 a_3}$ and $S_{a_1 \cdots a_6}$ generators obey the commutators [16.17]

$$[S^{a_1 a_2 a_3}, S_{b_1 b_2 b_3}] = -18\delta^{[a_1 a_2}_{[b_1 b_2} J^{a_3]}{}_{b_3]} + 2S^{a_1 a_2 a_3}{}_{b_1 b_2 b_3}, \tag{16.7.47}$$

$$[S_{a_1 a_2 a_3}, S^{b_1 \cdots b_6}] = -\tfrac{5!}{2}\delta^{[b_1 b_2 b_3}_{a_1 a_2 a_3} S^{b_4 b_5 b_6]} - 3S^{b_1 \cdots b_6}{}_{[a_1 a_2, a_3]}. \tag{16.7.48}$$

The commutation relations of the Cartan involution invariant sub-algebra, whose elements are given in equations (16.7.44)–(16.7.46), and the elements of the the the l_1 representation are given by [16.17]

$$[S^{a_1 a_2 a_3}, P_b] = 3\delta^{[a_1}_b Z^{a_1 a_3]}, \quad [S_{a_1 a_2 a_3}, Z^{b_1 b_2}] = Z_{a_1 a_2 a_3}{}^{b_1 b_2} - 6\delta^{b_1 b_2}_{[a_1 a_2} P_{a_3]},$$

$$\tag{16.7.49}$$

$$[S_{a_1 a_2 a_3}, Z^{b_1 \cdots b_5}] = Z^{b_1 \cdots b_5}{}_{[a_1 a_2, a_3]} + Z^{b_1 \cdots b_5}{}_{a_1 a_2 a_3} - \tfrac{5!}{2}\delta^{[b_1 \cdots b_3}_{a_1 a_2 a_3} Z^{b_4 b_5]}.$$

In deriving these results we have used the last few equations of section 16.7.2.

16.8   String vertex operators and Lie algebras

One of the most remarkable developments in string theory has been the discovery of a close relationship between vertex operators, the object describing string interactions and Lie algebras. The representation of simply laced finite dimensional semi-simple Lie algebras and affine algebras using vertex operators was found in [16.18, 16.19], while the non-simply laced case was treated in [16.20]. An excellent account of this subject is [10.27].

We begin by recalling the contents of some previous chapters where we encountered the operator Q^μ, which was defined to be

$$Q^\mu(z) \equiv q^\mu - ip^\mu \ln z + i \sum_{n=-\infty, n\neq 0}^{\infty} \frac{\alpha_n^\mu}{n} z^{-n}. \tag{16.8.1}$$

The non-trivial commutators of the operators in Q^μ are given by

$$[q^\mu, p^\nu] = ig^{\mu\nu}, \quad [\alpha_n^\mu, \alpha_m^\nu] = ng^{\mu\nu}\delta_{n+m,0}, \tag{16.8.2}$$

where $g^{\mu\nu}$ is the constant metric of the underlying space-time, which for simplicity we may take to be that of Minkowski space. The oscillators α_n^μ are creation and annihilation operators that act on a vacuum $|0, s\rangle$ as $\alpha_n^\mu|0, s\rangle = 0$, $p^\mu|0, s\rangle = s^\mu|0, s\rangle$ for $n \geq 1$. As usual the operator $e^{is \cdot q}$ acting on a state with momentum r^μ leads to a state with momentum $r^\mu + s^\mu$: that is,

$$e^{is \cdot q}|0, r\rangle = |0, s+r\rangle. \tag{16.8.3}$$

The operator Q^μ arose in the open bosonic string whose embedding coordinate is $x^\mu(\tau, \sigma)$. Indeed, as explained in equation (11.1.3,) $Q^\mu(z) = x^\mu(\tau = -i\ln z, \sigma = 0)$ if we set $2\alpha' = 1$. Taking the derivative of Q^μ we find

$$\partial Q^\mu \equiv \frac{\partial}{\partial z} Q^\mu = -i \sum_{n=-\infty}^{\infty} \alpha_n^\mu z^{-n-1}. \tag{16.8.4}$$

The derivative can be written in terms of a derivative of the open string field $x^\mu(\tau, \sigma)$ with respect to τ at $\sigma = 0$. As open strings interact by splitting and joining at their end points it is natural that the vertex operator contains the operator Q^μ with its identification with x^μ

at the end of the string, that is, $\sigma = 0$. There is also the relation $\partial Q^\mu = (\partial/\partial z) x^\mu(\tau, \sigma)$, where on the right-hand side we have now taken the derivative of $x^\mu(\tau, \sigma)$ with respect to $z = e^{i(\tau+\sigma)}$ so eliminating the dependence on $\bar{z} = e^{-i(\tau+\sigma)}$.

The operator Q^μ also arises when the closed bosonic string lives on a torus as studied in chapter 10, where we recognise in equation (10.1.9) that $Q^\mu(z) = x_L^\mu(z)$, where the argument of x_L^μ is $z = e^{i(\tau+\sigma)}$ and we should take $\alpha' = 2$. It emerges most cleanly when we compactify on unrelated tori for left and right movers and then q_L^μ, p_L^μ and α_n^μ in torus reduction are identified with q^μ, p^μ and α_n^μ above. Comparing equations (10.1.34) and (16.8.2) we see that they obey the same commutators. As the closed string has been reduced on unrelated tori for left and right movers, the left and right momenta take values in the corresponding tori. In the construction of this chapter we will also demand that the momenta take values in a lattice.

The Virasoro operators operators associated with the operator $Q^\mu(z)$ were given in equation (3.1.13) by

$$L_n = \tfrac{1}{2} \sum_p : \alpha_{-p+n}^\mu \alpha_p^\nu g_{\mu\nu} := \oint dz : T(z) : z^{n+1}, \qquad (16.8.5)$$

where T is the required component of the energy momentum tensor

$$T(z) = -\tfrac{1}{2} : \partial Q^\mu \partial Q^\nu g_{\mu\nu} : . \qquad (16.8.6)$$

These agree with the Virasoro operators for the open string found in chapter 3 and those for a left moving scalar discussed, for example, in chapter 9.

The operator $Q^\mu(z)$ has conformal weight 0 with respect to T, that is, it obeys equation (8.6.8), with $h = 0$. We recall that $Q^\mu(z)$ is not an analytic function as it contains a $\ln z$ term. However, out of Q^μ we may construct operators with non-zero conformal weight as we did in equations (9.1.36) and (11.1.4):

$$U_T(k, z) \equiv :e^{ik \cdot Q(z)}: = e^{ik \cdot Q_<(z)} e^{ik \cdot q} z^{k \cdot p} e^{ik \cdot Q_>(z)} = z^{-k^2 - 2} e^{ik \cdot Q_<(z)} e^{ik \cdot Q_0(z)} e^{ik \cdot Q_>(z)},$$
$$(16.8.7)$$

where

$$Q^\mu(z) \equiv Q_<^\mu(z) + Q_0^\mu(z) + Q_>^\mu(z) \qquad (16.8.8)$$

and

$$Q_0^\mu(z) \equiv q^\mu - ip^\mu \ln z, \quad Q_>^\mu(z) \equiv i \sum_{n=1}^{\infty} \frac{\alpha_n^\mu}{n} z^{-n}, \quad Q_<(z) \equiv i \sum_{n=-\infty}^{-1} \frac{\alpha_n^\mu}{n} z^{-n}. \quad (16.8.9)$$

We recall that when normal ordering operators we put all the destruction operators α_n^μ, $n = 1, 2, \ldots$ to the right of the creation operators α_{-m}^μ, $m = 1, 2, \ldots$ and factors of q^μ to the left of factors of p^μ. The operator $U_T(k, z)$ has conformal weight $k^2/2$ and one can verify that it obeys equation (8.6.8) with $h = k^2/2$. If $k^2 = 2$ it is just the open string tachyon vertex operator and this is the case which will concern us most here, however, it will be advantageous to derive some formulae for this operator in the general case.

Having assembled all the relevant equations we may begin. We take the momenta to belong to an integral even lattice Λ. The physical states on the lattice are given by $L_n|\psi\rangle = 0$, $n \geq 1$ and $(L_0 - 1)|\psi\rangle = 0$. We can think of these as open string states and they are just those discussed in chapter 3, but now the momenta lie on the lattice Λ. As explained in chapter 8 and also in chapter 9, around equation (9.3.15), we may associate with every physical state a conformal operator of conformal weight 1. In particular, if $|\psi\rangle$ is the physical state corresponding to the conformal operator $U_\psi(r, z)$, where r^μ is the momentum of the state $|\psi\rangle$, then $\lim_{z\to 0} U_\psi(r, z)|0, 0\rangle = |\psi\rangle$ (see equation (8.8.1)). There is a one-to-one relationship between conformal operators of weight 1 and physical states as given the former we may use the above equation to construct a physical state. If the lattice vector r has length squared 2, then the only physical state is the tachyon state, $|0, r\rangle$, and the corresponding conformal operator of weight 1 is just the tachyon vertex operator $U_T(r, z)$ of equation (16.8.7). If $r^2 = 0$, then we have the physical states $\alpha^\mu_{-1}\epsilon_\mu|0, r\rangle$ with $r \cdot \epsilon = 0$ whose corresponding weight 1 conformal operator is given in equation (9.3.17) to be $U_P(r, z) = \epsilon_\mu:i\partial Q^\mu e^{ir\cdot Q(z)}:$. It is just the open string vertex operator for photon emission. A special case of the above is when $r^\mu = 0$ as then $\alpha^\mu_{-1}|0, 0\rangle$ are physical states which need no additional condition. The corresponding weight 1 conformal operator is $U_P(r, z) = :i\partial Q^\mu:$. These latter physical states are usually of little importance in string theory, but they will play a special role in what follows.

In general, the conformal operator of conformal weight 1 corresponding to the physical state $|\psi\rangle$ has the form

$$U_\psi(r, z) = z^{(r^2/2)-1}:e^{ir\cdot x(z)}f(\partial Q^\mu, \partial^2 Q^\mu, \ldots):, \tag{16.8.10}$$

where the function f is polynomial in its arguments and the sum of the number of ∂s that occurs is just the level and is given by $1 - r^2/2$. The operator U includes the polarisation tensors which are subject to constraints in order that the operator be of conformal weight 1.

Let us consider the contour integrals of the above vertex operators:

$$V_\psi(r) = \oint dz U_\psi(r, z). \tag{16.8.11}$$

As we will see, with one minor modification, these objects, which are associated with the lattice Λ, obey a Lie algebra. Since $U_\psi(r, z)$ is of conformal weight 1, using equation (8.6.8), the commutator with the Virasoro operators is given by

$$[L_n, U_\psi(r, z)] = \left[z^{n+1}\frac{d}{dz} + z^n(n+1)\right]U_\psi(r, z) = \frac{d}{dz}(z^{(n+1)}U_\psi(r, z)) \tag{16.8.12}$$

and as a result

$$[L_n, V_\psi(r)] = 0. \tag{16.8.13}$$

We can evaluate the commutators using the techniques of section 8.6. The normal ordering of two tachyon vertex operators was given in equation (9.1.40) and can be expressed as

$$U_T(r, z)U_T(s, w) = :e^{ir\cdot Q(z)}::e^{is\cdot Q(w)}: = (z - w)^{r\cdot s}:e^{ir\cdot Q(z)+is\cdot Q(w)}:. \tag{16.8.14}$$

The above expression was found for $|z| > |w|$, but it only has poles at $z = 0$, $w = 0$ and $z = w$ and so we may define it for all z and w outside these points by analytic continuation. We note that the operator product in the other order, that is, $U_T(s, w)U_T(r, z)$, is valid for $|w| \geq |z|$ and is obtained from the above expression by swapping z and w and r and s. However, the analytic continuation of the operators as written in equation (16.8.14) and their operator product expansion when written in the other order are not the same function. The two functions differ only by a factor of $(-1)^{r \cdot s}$.

The corresponding operator product relation is then read off to be

$$U_T(r, z)U_T(s, w)$$
$$= (z - w)^{r \cdot s} :e^{i(r+s) \cdot Q(w)}: + i(z - w)^{r \cdot s+1} :\partial(s \cdot Q(w))e^{i(r+s) \cdot Q(w)}: + \cdots), \quad (16.8.15)$$

where $+ \cdots$ means higher order terms in $(z - w)$. We recall that an operator product assumes that the operators are radially ordered, which corresponds to $|z| > |w|$ for equation (16.8.15). Equations (16.8.14) and (16.8.15) are valid for any r^2 and s^2.

In general, the operator product expansion of $V_{\psi_1}(r)$ with $V_{\psi_2}(s)$, where ψ_1 and ψ_1 are any physical states, is more complicated. The vertex operators contain $e^{ir \cdot Q(z)}$ and $e^{is \cdot Q(z)}$ factors, respectively, as well as polynomials in derivatives of Q and in their operator product expansion we find similar $(z - w)^{r \cdot s}$ factors as well as $(z - w)^{-p}$ factors for integer p coming from Wick contractions of $\partial^n Q^\mu$ terms in one of the operators with the exponentials or derivatives of Q terms in the other operator.

The general operator product expansions of $U_{\psi_1}(r, z)U_{\psi_2}(s, \zeta)$ and $U_{\psi_2}(s, \zeta)U_{\psi_1}(r, z)$, when analytically extended into the same regions, do not lead to the same functions. However, they do agree up to a factor of $(-1)^{r \cdot s}$ resulting from the $(z - w)^{r \cdot s}$ part. We note that $((-1)^{r \cdot s})^2 = 1$. For the case of two tachyon operators $\eta(r \cdot s) = (-1)^{r \cdot s}$. Such a factor did not arise for the operators involved in the evaluation of the commutators in section 8.6 as we tacitly assumed that we were only considering operators whose operator product expansions did agree. Taking this feature into account, and following the argument leading to equation (8.6.4), we conclude that

$$V_{\psi_1}(r)V_{\psi_2}(s) - (-1)^{r \cdot s}V_{\psi_2}(s)V_{\psi_1}(r) = \oint_{\Gamma_2} d\zeta \oint_{\Gamma} dz \overline{U_{\psi_1}(r, z)U_{\psi_2}(s, \zeta)}; \quad (16.8.16)$$

here the contour Γ is around the point $z = \zeta$ and the final integration Γ_2 is around the origin. The reader may wish to derive this equation from first principles for two tachyon operators using only equation (16.8.14).

Since $V_\psi(r)$ commutes with L_n so must the right-hand side of the above equation. As such, one suspects that it is the integral of a weight 1 conformal operator. The right-hand side of equation (16.8.16) is obviously the integral of

$$\oint_{\Gamma} dz \overline{U_{\psi_1}(r, z)U_{\psi_2}(s, \zeta)} = V_{\psi_1}(r)U_{\psi_2}(s, \zeta) - (-1)^{r \cdot s}U_{\psi_2}(s, \zeta)V_{\psi_1}(r), \quad (16.8.17)$$

where we have evaluated the commutator on the right-hand side as in chapter 8. However, looking at the right-hand side of the above equation we note that L_n commutes with $V_{\psi_1}(r)$ and $U_{\psi_2}(s, \zeta)$ is a conformal operator of weight 1 and as a result we can conclude that the right-hand side of equation (16.8.17) is a conformal operator of weight 1 and so is the vertex operator associated with a physical state. Hence the algebra of integrated vertex operators of equation (16.8.11) for all physical states forms a closed algebra. However, it is not a Lie algebra due to the troublesome factors $(-1)^{r \cdot s}$ found between the two terms.

To gain a Lie algebra, that is, turn the above equations into commutators, we must introduce some cocycle factors c_r which are functions of the momentum operator p^μ. They should obey the relations

$$c_r c_s = (-1)^{r \cdot s} c_s c_r, \quad c_r c_s = \epsilon(r, s) c_{r+s}, \quad r, s \in \Lambda, \tag{16.8.18}$$

where $\epsilon(r, s) = \pm 1$ and $[c_r, p^\nu] = 0$. The form of these factors is a little complicated and we refer the reader to [10.27, 16.6] for more details. Hence if we define the modified operators

$$G_\psi(r) = c_r V_\psi(r), \tag{16.8.19}$$

they obey the commutators

$$[G_{\psi_1}(r), G_{\psi_2}(s)] = \epsilon(r, s) c_{r+s} \oint_{\Gamma_2} d\zeta \oint_\Gamma dz \overline{U_{\psi_1}(r, z) U_{\psi_2}(s, \zeta)}, \tag{16.8.20}$$

where on the right-hand side we have a physical state vertex operator times its cocyle factor and so one of the new generators of the form of equation (16.8.19). We define *the algebra of physical states associated with the lattice* Λ, denoted G_Λ, to consist of the operators of equation (16.8.19) for all physical states ψ on the lattice Λ. As we have just seen their commutators form a Lie algebra. In general we have an infinite number of different physical states and so an infinite dimensional Lie algebra. As discussed above, the physical state operators include the momentum operators p^μ, which obviously commute amongst themselves and whose commutators with the other physical state operators are easily found to be

$$[p^\mu, G_\psi(r)] = r^\mu G_\psi(r), \tag{16.8.21}$$

which is consistent with the fact that $G_\psi(r)$ injects momentum r^μ into any state on which it acts. The momentum operators will turn out to be the Cartan sub-algebra of the Lie algebra G_Λ.

Let us evaluate some of these commutators for the simplest case of the operators associated with the tachyon using equation (16.8.14). Clearly, if $r \cdot s \geq 0$ we have no pole in equation (16.8.14) and so the commutator vanishes. However, if $r \cdot s = -1$, or $(r + s)^2 = 2$ as $r^2 = 2 = s^2$, we have a simple pole and so we find anther tachyon vertex operator, but with momentum $r + s$; thus

$$[G_T(r), G_T(s)] = \begin{cases} 0 & \text{if } r \cdot s \geq 0, \\ \epsilon(r, s) G_T(r + s) & \text{if } r \cdot s = -1. \end{cases} \tag{16.8.22}$$

Now let us consider the case of $r \cdot s = -2$. Using equation (16.8.14), we have that

$$[G_T(r), G_T(s)] = \epsilon(r, s) c_{r+s} \oint_{\Gamma_2} d\zeta :(ir \cdot \partial Q(\zeta)) e^{i(r+s) \cdot Q(\zeta)}:$$

$$= \epsilon(r, s) c_{r+s} \oint_{\Gamma_2} d\zeta :i\frac{(r - s)}{2} \cdot \partial Q(\zeta) e^{i(r+s) \cdot Q(\zeta)}:$$

$$= \epsilon(r, s) G_P(r + s), \tag{16.8.23}$$

where in the middle equality we have used the fact that the total derivative vanishes under the integral. This is just the generator associated with the photon physical state. Indeed, it has null momentum as $(r + s)^2 = 0$ and the polarisation $\epsilon^\mu = (s - r)^\mu/2$ satisfies the required physical state condition, namely, $\epsilon^\mu (s + r)_\mu = 0$. Clearly, as $r \cdot s$ becomes more negative one finds generators corresponding to higher and higher level physical states if indeed such states exist for the lattice being considered.

Physical states can occur for momentum $r^2 = 2, 0, -2, \ldots$, but only for the case of the tachyon, that is, $r^2 = 2$, does one find that there is just one physical state for a given momentum. Consequently, we will find, except when $r^2 = 2$, several physical state operators for a given momentum. One great advantage of the algebra of physical states is that one knows precisely how many generators at a given level it contains, as this is just the number of physical string states at that level which we have computed in previous chapters. For example, if the lattice has dimension 26 it is just the number of physical states in the bosonic string and is given in equation (11.1.58). However, if the dimension is less than 26, but greater than 2 the number of physical states is also known and is discussed in chapter 11. In fact, one even knows what the physical sates are explicitly using the DDF operators constructed in chapter 11 and so one has an almost explicit construction of the physical state symmetry generators $G_\psi (r)$. We note that this situation contrasts sharply with that for Kac–Moody algebras, where the number and nature of the generators are unknown except if they are finite dimensional semi-simple or affine.

We now consider the special case when the lattice Λ is the root lattice of a simply laced finite dimensional semi-simple Lie algebra, that is, $\Lambda = \Lambda_G$ in the notation of section 16.1.1. As discussed in that section such a lattice is integral and Euclidean and as the algebra is simply laced the lattice will also be even. We recall it consists of $\Lambda_G = \{\sum n_a \alpha_a\}$, where α_a are the simple roots of the Lie algebra G and the n_a are integers. The roots of the algebra are all points of length squared 2.

We now find the algebra of physical states G_Λ associated with this lattice. The first step is to find the physical states. Clearly, for the lattice points $r^2 > 2$ we have no physical states, for $r^2 = 2$ we have the tachyons $|0, r\rangle$ and if $r^2 = 0$ then $r^\mu = 0$, but we have the physical states $\alpha^\mu_{-1} |0, 0\rangle$. Hence, the algebra of physical states in this case consists of the operators

$$G_\Lambda = \{G_T (r), r \in \Lambda_G, r^2 = 2, \text{ and } p^\mu\}. \tag{16.8.24}$$

For any r^μ and s^μ such that $r^2 = 2 = s^2$, we have $(r + s)^2 = r^2 + s^2 + 2r \cdot s = 2(2 + r \cdot s) \geq 0$, as we have a Euclidean lattice, and so $r \cdot s = -2, -1, 0, \ldots$. However, as we noted above, if $r \cdot s \geq 0$, then the commutator of the physical state operators vanishes. If $r \cdot s = -1$, then $(r + s)^2 = 2$ and we find a simple pole and the commutator results in a vertex operator corresponding to another tachyon. However, if $r \cdot s = -2$, then $(r + s)^2 = 0$ in which case $r^\mu = -s^\mu$, as we have a Euclidean metric on the space of roots, and we find the momentum operators. We may read off the algebra of physical states from equations (16.8.22) and (16.8.23) to find

$$[G_T (r), G_T (s)] = \begin{cases} 0 & \text{if } r \cdot s \geq 0, \\ \epsilon(r, s) G_T (r + s) & \text{if } r \cdot s = -1, \\ r \cdot p & \text{if } r^\mu = -s^\mu, \end{cases} \tag{16.8.25}$$

as well as

$$[p^\mu, p^\nu] = 0, \quad [p^\mu, G_T (r)] = r^\mu G_T (r). \tag{16.8.26}$$

The only difference with the above discussion around equation (16.8.23) is that, in this case, for the last result in equation (16.8.25) $r \cdot s = -2$ implies that $r^\mu = -s^\mu$ and so the term in the exponential in equation (16.8.23) vanishes. The resulting algebra is precisely the Cartan–Weyl form of the Lie algebra corresponding to the root lattice we have taken, see equations (16.1.11), (16.1.12) and (16.1.28). The precise correspondence being $G_T(r) \leftrightarrow E_\alpha$, $p^\mu \leftrightarrow H_i$ with $r^\mu \leftrightarrow \alpha_i$. We note that if $r^2 = 2 = s^2$, the condition $r \cdot s = -1$ is that for a simple pole and so that for the commutator to lead to a vertex operator of momentum $r + s$. We note from the Lie algebra side that if α and β are two roots, then $\alpha + \beta$ is a root if and only if $(\alpha, \beta) = -1$. To see this point one can consider the root string of α through the state $|-\beta\rangle$ as was done in section 16.1.2. Since $(\alpha, -\beta) = 1$ we must conclude that $\alpha + \beta$ is the next state in the string and so is a root. Also any root has length squared 2 and this is the case only if $(\alpha, \beta) = -1$. That these two conditions, whose origins in the two approaches are very different, coincide is required in order for the construction to work. The role of the Cartan sub-algebra is played by the momentum operators and as there is only a tachyon operator for a given momentum of length squared 2 we have a unique generator for each such momentum. The appearance of the factor of $\epsilon(r, s)$ in the second line of equation (16.8.25) is consistent with the fact, noted below equation (16.1.28), that for simply laced algebras one finds just such a factor in just this place. The construction for the root lattices for non-simply laced algebras was given in [16.20].

This is a remarkable construction; by starting with the root lattice we have uniquely reconstructed the Lie algebra and it can be thought of as an alternative to the Serre construction given in section 16.1. One reason why it works is that the Serre relation of equation (16.1.44) is automatically encoded in the vertex algebra computation. Looking at that equation we see that it corresponds to a momentum of $\alpha_a(1 - A_{ab}) + \alpha_b$, whose length squared is $4 - 2A_{ab}$, but A_{ab} is a negative integer, or 0, and so the length squared is greater than 4. However, there exists no physical state corresponding to such a momentum and so the multiple commutator that appears in the Serre relation must vanish. We note that while this argument applies to the case here, that is, for finite dimensional semi-simple simply laced Lie algebras, we did not use anything that does not also hold for any Kac–Moody algebra and so the argument also applies in this general context. Essentially, string vertex operators automatically encode the Lie algebra of semi-simple Lie algebras if one takes the appropriate lattices.

We now consider the algebra of physical states that is associated with the root lattice of an affine algebra, G^+. We discussed this algebra in sections 16.1.3 and 16.4. In particular, in equation (16.4.1) we considered the lattice $\Lambda_G \oplus \Pi^{1,1}$ which contains the root lattice of G^+ with the roots being of the form $\alpha = \sum_{a=1}^{r} n_a \alpha_a + nk$, where α_a, $a = 1, \ldots, r$, are the simple roots of G and $k \in \Pi^{1,1} \subset \Lambda_G \oplus \Pi^{1,1}$ is the vector $k = (1, 0) \in \Pi^{1,1}$, which has $k^2 = 0$. Unlike that for Λ_G this lattice contains points of length squared 0, namely nk, but it has no points with negative length squared. As such, the physical states only occur for points of length squared 2 and 0. The points of length squared 2 are of the form $r + nk$, for any r a root in G. The corresponding physical states are just the tachyon states, $|0, r + nk\rangle$, with corresponding physical state operator $G_T(r + nk)$. The points of length squared 0 are of the form $r^\mu = nk^\mu$; their physical states are the photon physical states given by $\alpha_{-1}^\mu |0, nk\rangle$ and the corresponding physical state operator

$$G_P^\mu(nk) = \oint dz : i\partial Q^\mu e^{ink \cdot Q(z)} : . \tag{16.8.27}$$

We have indicated the normal ordering, but it is not required as $k^2 = 0$ and the polarisation tensor, which we have not written but which is assumed to be present, only has components in the μ directions which are orthogonal to the direction of the momentum k^μ.

Using the results above it is straightforward to calculate the algebra of physical states to find that

$$[G_P^\mu(nk), G_T(r+mk)] = r^\mu G_T(r+nk+mk), \tag{16.8.28}$$

$$[G_T(r+nk), G_T(s+mk)] = \begin{cases} 0 & \text{if } r \cdot s \geq 0, \\ \epsilon(r,s)G_T(r+s+nk+mk) & \text{if } r \cdot s = -1, \\ r \cdot G_P(nk+mk) + nk \cdot p\delta_{n+m,0} & \text{if } r^\mu = -s^\mu \end{cases} \tag{16.8.29}$$

and

$$[G_P^\mu(nk), G_P^\nu(mk)] = ng^{\mu\nu}k \cdot p\delta_{n+m,0}. \tag{16.8.30}$$

Identifying $H_{i,n} \leftrightarrow G_P^\mu(nk)$, $E_{\alpha,n} \leftrightarrow G_T(r+nk)$ and $k \leftrightarrow k \cdot p \leftrightarrow k$, with $\alpha_i \leftrightarrow r^\mu$, we find the affine algebra of equations (16.1.95) and (16.1.96)[10.27]. Here k is the central term of the affine algebra which is not to be confused with the momentum k^μ. The derivation operator is just given by $d = \bar{k} \cdot p$ with $\bar{k} = (0,1)$ so that $\bar{k}^2 = 0$ and $k \cdot \bar{k} = 1$.

There is another way to recover an affine algebra [10.27] without extending the lattice but working with the root lattice Λ_G. We just take the moments of the physical state vertex operators:

$$G_n(r) = \oint dz z^n U_T(r,z) \tag{16.8.31}$$

and the operators α_n^μ. Identifying these with $E_{\alpha,n}$ and $H_{i,n}$ we find the affine algebra of equations (16.1.95) and (16.1.96) with central term 1.

It is very interesting to consider the vertex algebra associated with Lorentzian lattices. The reader will be tempted to calculate the algebra of physical states for the over-extended and very extended algebras of section 16.4. In particular, one could consider the case of E_8^{++} whose root lattice is just $\Pi^{1,9}$, the unique self-dual even Lorentzian lattice in ten dimensions. The number of physical states on the Lorentzian lattice is just that for the bosonic string in ten dimensions, as discussed in chapter 11. However, this is not the same as the number of generators at the corresponding level in the over-extended algebra $E_8^{++} = E_{10}$ about which little is known except at low levels [16.21, 16.22]. There are always more generators in the algebra of physical states, that is, the vertex algebra associated with the root lattice of a Kac–Moody algebra, than in the Kac–Moody algebra itself; indeed the Kac–Moody algebra can be embedded in the vertex algebra [16.23]. However, the projection to find the elements of E_8^{++} from those of $G_{\Pi^{1,9}}$ is unknown.

It is certainly striking that the algebras we understand well, namely, the finite dimensional simply laced algebras and their affine extensions, emerge in a very simple way from the vertex algebras of physical states.

Another interesting algebra to construct is the algebra of physical states for the unique self-dual Lorentzian lattice of dimension 26, $\Pi^{1,25}$, that is, $G_{\Pi^{1,25}}$. The number of elements in the algebra is the same as the number of the bosonic string states in 26 dimensions. This is called the fake monster algebra [16.24]. It is a symmetry of bosonic string theory compactified on this lattice and this might be taken as some evidence of a vast symmetry underlying string theory [16.25, 16.26].

An important further development is the discovery of Lie algebras that are more general than Kac–Moody algebras and, in particular, Borcherds algebras. These are much more closely related to vertex operator algebras than Kac–Moody algebras. However, one can define them in an analogous way by further relaxing the Serre relations. For a review suitable for physicists see [16.27].

17 Symmetries of string theory

"We wanted the best, but it turned out as always."
"Here is what can happen when somebody's starting to reason".
"Wine we need for health, and the health we need to drink vodka".

Some sayings of Victor Chernomyrdin

In this chapter we begin by deriving the T duality perturbative symmetry of string theory from the world-sheet viewpoint in section 17.1. Although there is no complete formulation of string theory, the maximal supergravity theories in ten dimensions, studied in chapter 13, are the low energy effective actions for the type II strings. They are complete in that they contain all perturbative and all non-perturbative corrections. We will explain how these theories allow us to understand aspects of string theory that go beyond the well-understood perturbative regime. In particular, we will explain how the surprising exceptional group symmetries that the maximal supergravity theories possess are likely to be symmetries of string theory when suitably discretised. These symmetries have many features in common with the old electromagnetic duality symmetry of Maxwell's equations, whose generalisation are symmetries of the $N = 4$ supersymmetric Yang–Mills theory in four dimensions. This result is discussed in section 17.2.

We then discuss the connections between the maximal supergravity theories which suggest that there are corresponding connections between the string theories and an eleven-dimensional theory, usually called M theory. Finally we give an account of the conjectured E_{11} symmetry of an underlying theory of strings and branes.

17.1 T duality

In chapter 10 we considered the closed bosonic string compactified on a torus. We found that such a string possessed some unexpected symmetries called T duality which we now discuss in more detail. Let us first consider the compactification on a circle as given in section 10.1. The resulting mass spectrum was given in equation (10.1.19), which we repeat here for convenience:

$$\text{mass}^2 = \frac{1}{2}(p_L^2 + p_R^2) + \frac{2}{\alpha'}(N + \bar{N} - 2) = \frac{h^2}{R^2} + \frac{1}{(\alpha')^2}R^2 l^2 + \frac{2}{\alpha'}(N + \bar{N} - 2), \quad (17.1.1)$$

where the (Kaluza–Klein) momenta on the circle are h/R, l is the number of times the string wraps around the circle and N and $\bar{N}$ count n for each of the oscillators α_n^μ and $\bar{\alpha}_n^\mu$

550

in the state being considered, while the $(L_0 - \bar{L}_0)|\psi\rangle = 0$ condition of equation (11.1.20) was given by

$$hl = -N + \bar{N}. \tag{17.1.2}$$

We observed that the above two conditions and indeed all the physical state conditions are invariant under

$$h \leftrightarrow l, \ \alpha_n^d \leftrightarrow \alpha_n^d, \ \bar{\alpha}_n^d \leftrightarrow -\bar{\alpha}_n^d, \tag{17.1.3}$$

provided we also change

$$R \leftrightarrow \frac{\alpha'}{R} \quad \text{or} \quad \frac{R}{\sqrt{\alpha'}} \leftrightarrow \frac{\sqrt{\alpha'}}{R}. \tag{17.1.4}$$

We have taken the circle to be in the dth direction, hence the labels on the oscillators. These transformations are equivalent to

$$x_L^d(\tau + \sigma) \leftrightarrow x_L^d(\tau + \sigma), \ x_R^d(\tau - \sigma) \leftrightarrow -x_R^d(\tau - \sigma) \tag{17.1.5}$$

on the left and right moving modes, provided we also make a suitable change to the zero mode term. This is equivalent to the change $x^d(\tau, \sigma) = x_L^d(\tau + \sigma) + x_R^d(\tau - \sigma) \leftrightarrow y^d(\tau, \sigma) \equiv x_L^d(\tau + \sigma) - x_R^d(\tau - \sigma)$.

We observed in section 10.2 that the compactified string on a circle of radius R and one on a circle of radius α'/R are the same. Put another way the range of radii that are measurable by the string are from $\sqrt{\alpha'}$ to infinity instead of 0 to infinity as would be the case for a point particle and so the string cannot probe circles of radius less than $\sqrt{\alpha'}$. We also recall that at the self-dual radius $R = \sqrt{\alpha'}$ there were additional massless states and the gauge group increased from $U(1) \times U(1)$ to $SU(2) \times SU(2)$.

We next discussed the compactification on a general torus $T^D = R^D/2\pi \Lambda^D$ in section 10.2. The basis for Λ^D was denoted by e^I. We found that the left and right momenta, more specifically (w_L, w_R), belong to a lattice $\Lambda^{(D,D)}$ of dimension $2D$ whose expressions in terms of the basis vectors are given in equations (10.2.23) and (10.2.24). As we noted this is an even self-dual lattice. The string compactifed on the torus associated with the lattice Λ^D is equivalent to the string compactified on the dual lattice $(\Lambda^D)^*$ provided we also exchange the Kaluza–Klein momenta and the winding modes by $m_i \leftrightarrow n_i$ at the same time as these changes leave the physical state condition unchanged. As we noted already this is equivalent to

$$x_L^d(\tau + \sigma) \leftrightarrow x_L^d(\tau + \sigma), \ x_R^d(\tau - \sigma) \leftrightarrow -x_R^d(\tau - \sigma), \tag{17.1.6}$$

which, as before, is equivalent to the change $x^d(\tau, \sigma) = x_L^d(\tau + \sigma) + x_R^d(\tau - \sigma) \leftrightarrow y^d(\tau, \sigma) \equiv x_L^d(\tau + \sigma) - x_R^d(\tau - \sigma)$.

In sections 10.3 and 10.4 we carried out a string compactification in which the metric and anti-symmetric gauge field in the compactified directions have background values which correspond to the torus reduction. We found that the vectors (w_L, w_R) belong to a lattice $\Lambda^{(D,D)}$ of dimension $2D$ that was even and self-dual, but now this lattice is determined by the values of the background fields. The set of all possible string theories constructed in this way is labelled by the points in the coset

$$\frac{O(D, D; \mathbf{R})}{O_L(D; \mathbf{R}) \times O_R(D; \mathbf{R})}. \tag{17.1.7}$$

This coset has dimension D^2 consistent with the number of background fields which can be written as $E_{IJ} = G_{IJ} + B_{IJ}$ whose values label the coset.

Under the action of $O(D, D)$ we found that the matrix $E = G + B$ transforms as follows:

$$E' = (aE + b)(cE + d)^{-1}, \qquad (17.1.8)$$

where

$$\hat{g} = \begin{pmatrix} a & b \\ c & d \end{pmatrix} \qquad (17.1.9)$$

is the element of $O(D, D)$. This has a form that is a bit different to that of equation (10.4.21). However, we note that if $\hat{g} \in O(D, D)$, that is $\hat{g}$ obeys the defining condition $\hat{g}^T J \hat{g} = J$, then $\hat{g} J \hat{g}^T J \hat{g} = \hat{g}$ and so $(\hat{g}^T)^T J \hat{g}^T = J$, in other words if $\hat{g} \in O(D, D)$, then $\hat{g}^T \in O(D, D)$.

The above statement requires one modification. Given any lattice with a choice of basis it is equally well described by an alternative choice of basis that is found by the action of an integer valued matrix change acting on the original basis, provided its inverse is also integer valued. Thus the action of an integer valued matrix belonging to $O(D, D; \mathbf{Z})$, that is an $O(D, D; \mathbf{Z})$ matrix, will not change the lattice and so will not lead to a new string theory. These are the T duality transformations as seen from the background field way of viewing string compactification and its action is just that of equation (17.1.9) but now $\hat{g} \in O(D, D; \mathbf{Z})$.

To see more concretely what T duality symmetries are we will consider the decomposition of an $O(D, D)$ matrix. This will apply to any $O(D, D)$ matrix, but as we are considering T duality we have in mind for our appliaction that it will be integer valued. Substituting the g of equation (17.1.9) into the defining condition $g^T J g = J$ (see section 10.4) we find that

$$a^T d + c^T b = 1, \ a^T c + c^T a = 0, \ b^T d + d^T b = 0. \ d^T a + b^T c = 1. \qquad (17.1.10)$$

We note that the last conditions follows from the first. If we consider infinitesimal group elements, $a = 1 + \delta a$, $b = \delta b$ etc., then these conditions become

$$\delta a^T = -\delta d, \ \delta c = -\delta c^T, \ \delta b = -\delta b^T. \qquad (17.1.11)$$

Let us consider a transformation for which $b = c = 0$. In this case a is an arbitrary matrix and so belongs to $GL(n)$. The corresponding finite group element is of the form

$$\hat{g} = \begin{pmatrix} a & 0 \\ 0 & (a^T)^{-1} \end{pmatrix}, \qquad (17.1.12)$$

which is a T duality if $a \in GL(n, \mathbf{Z})$. Using equation (17.1.8) we see it takes $E' = aEa^T$. This is induced by a change of basis of the coordinates x^I. In fact, these are the large diffeomorphisms which are preserved by the torus.

Now let us consider a transformation for which $a = I = d$ and $c = 0$. Then we are left with $b = -b^T$ and the finite form of the group element is given by

$$\hat{g} = \begin{pmatrix} I & b \\ 0 & I \end{pmatrix}, \qquad (17.1.13)$$

which is a T duality if $b_{IJ} \in \mathbf{Z}$. In this equation I is the $n \times n$ identity matrix. This transformation has the effect $E' = E + b$, which implies that $B'_{IJ} = B_{IJ} + b_{IJ}$, that is, just a shift in the background B_{IJ} by integers.

Finally, we can consider transformations that have only one component of the matrices b and c non-zero. In particular, we take $b_{IJ} = C_{IJ} = \delta_{IJ}\delta_{IK} \equiv D_K$, where K is a fixed value, that is, they are diagonal matrices with only one non-zero entry down the diagonal. From equation (17.1.10) we find that $a = d = I - D_K$ and so the corresponding finite element of the T duality is given by

$$\hat{g} = \begin{pmatrix} I - D_K & D_K \\ D_K & I - D_K \end{pmatrix}. \tag{17.1.14}$$

This is the $O(D, D; \mathbf{Z})$ analogue of the $R \to \alpha'/R$ transformation of equation (17.1.4) since if we take a background with $B_{IJ} = 0$ and a diagonal metric $G_{IJ} = G_{JJ}\delta_{IJ}$ it leads, using equation (17.1.9), to the change

$$G'_{JJ} = G_{JJ}, \; I \neq K, \; G'_{KK} = \frac{1}{G_{KK}}. \tag{17.1.15}$$

Given the infinitesimal form of the algebra it is clear that any transformation of $O(D, D; R)$ can be made up of those of equations (17.1.12)–(17.1.14) and the same holds for the T duality group $O(D, D; \mathbf{Z})$. One can verify explicitly that the above transformations do indeed leave the physical state conditions invariant. For example, consider the change of equation (17.1.13). From equations (10.3.21)–(10.3.23) we see that $Z^T M Z$ changes by $-2m^T G^{-1}bn - n^T bG^{-1}bn - 2n^T BG^{-1}bn$, but this is just equivalent to replacing the sum over the integers m and n by $m \to m' = m - bn$ and $n \to n' = n$ and so this transformation leaves the physical state conditions inert. The reader can verify explicitly that the other two T duality transformations of equations (17.1.12) and (17.1.14) also leave invariant the physical state conditions. An interesting T duality transformation is found by taking $a = 0 = d$ and $b = I = c$ to find that

$$E' = \frac{1}{E}. \tag{17.1.16}$$

The reader may like to verify explicitly that this also leaves invariant the physical state conditions as we have already shown.

Having come this far it is useful to give the $O(D, D; \mathbf{R})$ Lie algebra. This can easily found by taking infinitesimial form of the group element, subject to the conditions of equation (17.1.11), and acting it on the $2D$-dimensional represenation in the obvious way. One then calculates the commutators of two such transformations and removing the parameters one find the Lie algebra in the standard manner:

$$[K^a{}_b, K^c{}_d] = \delta^c_b K^a{}_d - \delta^a_d K^c{}_b, \; [K^a{}_b, R^{cd}] = \delta^c_b R^{ad} - \delta^d_b R^{ac},$$

$$[K^a{}_b, \tilde{R}_{cd}] = -\delta^a_c \tilde{R}_{bd} + \delta^a_d \tilde{R}_{bc}, \tag{17.1.17}$$

$$[R^{ab}, \tilde{R}_{cd}] = \delta^{[a}_{[c} K^{b]}{}_{d]}, \quad [R^{ab}, R^{cd}] = 0 = [\tilde{R}_{ab}, \tilde{R}_{cd}].$$

As one sees the generators $K^a{}_b$ obey the GL(D) algebra and arise from transformations with $b = 0 = c$.

It is instructive to view T duality transformations from the perspective of the world-sheet approach to string theory in the presence of the background fields. We begin the action of equation (10.4.14) but consider only the part in the compactified directions, which in the notation of this section is given by

$$A = -\frac{1}{4\pi\alpha'} \int d^2\xi \sqrt{-g} g^{\alpha\beta} (\partial_\alpha x^I \partial_\beta x^J G_{IJ} - \epsilon^{\alpha\beta} \partial_\alpha x^I \partial_\beta x^J B_{IJ}) + \frac{1}{4\pi} \int d^2\xi \sigma R^{(2)}.$$

$$(17.1.18)$$

We now suppose that there is a U(1) isometry and as a result we can take the background metric anti-symmetric tensor and dilaton to be independent of x^d. The simplest case is when we have a circle in the x^d direction. We now consider the action

$$A = -\frac{1}{4\pi\alpha'} \int d^2\xi \sqrt{-g} g^{\alpha\beta} (G_{dd} Y_\alpha Y_\beta + 2G_{dI} Y_\alpha \partial_\beta x^I + \partial_\alpha x^I \partial_\beta x^J G_{IJ})$$

$$- \epsilon^{\alpha\beta} (2B_{dI} Y_\alpha \partial_\beta x^I + \partial_\alpha x^I \partial_\beta x^J B_{IJ} - 2y \partial_\alpha Y_\beta) + \frac{1}{4\pi} \int d^2\xi \sigma R^{(2)}. \quad (17.1.19)$$

In this action the indices I, J do not take the value d. This new action is equivalent to the previous action, since the y equation of motion implies that $\epsilon^{\alpha\beta} \partial_\alpha Y_\beta = 0$ and so, in the absence of any non-trivial topology, $Y_\beta = -\partial_\beta x^d$. Substituting for Y_β into the action we find the first action of equation (17.1.18). However, the equation of motion for the field Y_α is given by

$$Y_\alpha = -\frac{1}{G_{dd}} \left(G_{dI} \partial_\alpha x^I - \frac{\epsilon_\alpha{}^\beta}{\sqrt{-g}} (B_{dI} \partial_\beta x^I - \partial_\beta y) \right). \quad (17.1.20)$$

Since this equation is algebraic we can substitute it into the above action, which is equivalent in the path integral to integrating it out. The result is the action

$$A' = -\frac{1}{4\pi\alpha'} \int d^2\xi (\sqrt{-g} g^{\alpha\beta} (G'_{dd} \partial_\alpha y \partial_\beta y + 2G'_{dI} \partial_\alpha y \partial_\beta x^I + \partial_\alpha x^I \partial_\beta x^J G'_{IJ})$$

$$+ \epsilon^{\alpha\beta} (2B'_{dI} \partial_\alpha y \partial_\beta x^I + \partial_\alpha x^I \partial_\beta x^J B'_{IJ})) + \frac{1}{4\pi} \int d^2\xi \sigma R^{(2)}. \quad (17.1.21)$$

We observe that this is the same as the action of equation (17.1.18), but one of the coordinates of the string is y rather than x^d and the dependence on the background metric and anti-symmetric gauge field is changed. The transformed values of the fields are given by [17.1]

$$G'_{dd} = \frac{1}{G_{dd}}, \quad G'_{dI} = -\frac{B_{dI}}{G_{dd}}, \quad G'_{dI} = -\frac{G_{dI}}{G_{dd}},$$

$$G'_{IJ} = G_{IJ} - \frac{G_{dI} G_{dJ} - B_{dI} B_{dJ}}{G_{dd}}, \quad B'_{IJ} = B_{IJ} - \frac{G_{dI} B_{dJ} - G_{dI} B_{dJ}}{G_{dd}}. \quad (17.1.22)$$

We recognise this as a T duality transformation as given by equation (17.1.9) if we take the $n \times n$ matrices to be given by

$$a = \begin{pmatrix} 0 & 0 \\ 0 & I \end{pmatrix}, \quad b = \begin{pmatrix} 1 & 0 \\ 0 & 0 \end{pmatrix}, \quad c = \begin{pmatrix} 1 & 0 \\ 0 & 0 \end{pmatrix}, \quad d = \begin{pmatrix} 0 & 0 \\ 0 & I \end{pmatrix}. \quad (17.1.23)$$

In these equations I is the $n - 1 \times n - 1$ identity matrix and we have chosen the top component to correspond to the dth component. In fact, it is of the type of transformations given in equation (17.1.14).

We note that the dilaton σ is unchanged. However, it turns out that the integration of Y_β in the path integral leads to a change in the measure that is equivalent to the change $\sigma' = \sigma - \frac{1}{2}\ln G_{dd}$. [17.1]

The appearance of the T duality transformations can be seen from a different perspective by considering the toy model of a string with a flat world sheet with the action

$$-\int d^2\xi \, (Y_\alpha Y^\alpha - 2\epsilon^{\alpha\beta}y\partial_\alpha Y_\beta). \tag{17.1.24}$$

The y equation of motion implies that $\epsilon^{\alpha\beta}\partial_\alpha Y_\beta = 0$ and so we have $Y_\alpha = -\partial_\alpha x$ and putting this back in the action we have the usual action for a free scalar x. On the other hand, eliminating Y_α we have the action for the free scalar y. The equations of motion are the same as

$$\epsilon^{\alpha\beta}\partial_\beta y = -\partial^\alpha x \text{ or equivalently } \partial_1 y = \partial_0 x, \ \partial_0 y = \partial_1 x. \tag{17.1.25}$$

These imply the standard equations of motion for x and y, namely $\partial_\alpha\partial^\alpha x = 0$ and $\partial_\alpha\partial^\alpha y = 0$. If we take the solution of the equation of motion to be written as $x(\tau, \sigma) = x_L(\tau + \sigma) + x_R(\tau - \sigma)$ and $y(\tau, \sigma) = y_L(\tau + \sigma) + y_R(\tau - \sigma)$, then equation (17.1.25) implies, up to constants, that $x_L = y_L$ and $x_R = -y_R$ or if $x = x_L + x_R$, then $y = x_L - x_R$. Here we recognise the transformation corresponding to T duality found above.

The above discussion applies to the bosonic string and the bosonic sector of the superstrings. We will now extend it to the fermionic sector of the type II superstrings. Let us suppose that the ninth direction is a circle and we have performed a T duality, then equation (17.1.5) holds with $d = 9$. This is equivalent to carrying out a parity operation only on the right-handed coordinate. The usual parity operator acts on spinor space as Γ_{11}. The parity operator restricted to the one direction, acting on spinor space and only on right movers is given by $\Gamma_9\Gamma_{11}$ acting on the right-handed spinor indices with the identity operator on the left movers. The reader may also verify that it transforms vectors formed by sandwiching two spinors on Γ^μ in the correct way as $\Gamma_{11}\Gamma_9\Gamma^\mu\Gamma_9\Gamma_{11}$ equals Γ^μ if $\mu \neq 9$ and $-\Gamma^\mu$ if $\mu = 9$. The massless states in the $R \otimes R$ sector are contained in the bispinor given in equation (7.3.17) and as a result T duality corresponds to the change

$$S \to S' = S\Gamma_9\Gamma_{11}. \tag{17.1.26}$$

The physical state conditions impose chirality conditions on both indices of the bispinor. It is immediately clear that T duality reverses the chiralty of the right spinor index and so maps the IIA theory to the IIB theory and vice versa. Using equations (7.3.17)–(7.3.21) we find that T duality changes the field strengths as

$$F'_{\mu_1...\mu_n} = \mp F_{\mu_1\cdots\mu_n 9}, \quad F'_{\mu_1\cdots\mu_n 9} = \pm F_{\mu_1\cdots\mu_n}, \quad \mu_r \neq 9. \tag{17.1.27}$$

The two choices of sign arise from the action of Γ_{11} and the different chiralities of the right movers. This result is consistent with the change from IIA to IIB as the field strengths are of even rank in IIA and odd rank in IIB.

Since the gauge fields in the $R \otimes R$ sector couple to D-branes the T duality operation must have a corresponding effect on D-branes. Let us consider a D-brane on which there are open strings that must obey N and D boundary conditions in the longitudinal and transverse directions respectively. Examining boundary conditions that are written in the form of equation (15.2.4) it is obvious that the T duality change $\bar{\partial} x_L^9 \rightarrow -\bar{\partial} x_L^9$ will change N boundary conditions to D boundary conditions and vice versa.

Let us consider a D($p+1$)-brane wrapped on a circle in the ninth direction. The open strings living on the brane have the mode expansion of equation (10.1.8). Taking our coordinate to be the ninth rather than the twenty-fifth coordinate we find that a T duality transformation in the ninth direction implies that

$$\bar{\alpha}_0^{9\prime} = -\bar{\alpha}_0^9, \quad \bar{\alpha}_n^{9\prime} = -\bar{\alpha}_n^9. \tag{17.1.28}$$

Using equation (10.1.16) we see that the first of these equations implies

$$R_9' = \frac{\alpha'}{R_9}, \quad h' = l, \quad l' = h \tag{17.1.29}$$

as we have previously found. Imposing an N boundary condition we find that $\bar{\alpha}_0^9 = \bar{\alpha}_0^9$ and so there are no winding modes $l = 0$, but we do have momenta. After a T duality transformation we have D boundary conditions in the ninth direction and these imply that $\bar{\alpha}_0^{9\prime} = -\bar{\alpha}_0^{9\prime}$ and so now there is no momentum, that is, $h' = 0$, but we do have winding modes. In addition, since we now have D boundary conditions the end points of the string are fixed in the ninth direction and as such this direction is now transverse to the transformed brane. The open strings can then leave the brane and wrap around the circle in the ninth direction before returning to the brane. Before the T duality the open strings can move freely in the ninth direction but after the T duality they are frozen so as to occupy a fixed value of the ninth coordinate. This implies that after the transformation we have a Dp-brane which is no longer wrapped in the ninth direction. Conversely, given a Dp-brane which is not wrapped on the ninth direction it transforms to a D($p+1$)-brane which now is wrapped around the ninth direction [15.2, 17.2].

17.2   Electromagnetic duality

Around 1785 it was discovered by Coulomb that electric charges lead to a force that varies as $1/(\text{distance})^2$. Somewhat later, it was found (Ampere 1819) that electric currents produced magnetic fields. These seemed to be unrelated phenomena until it was realised (Faraday 1831) that a variable magnetic field could produce an electic current. This long series of developments culminated in Maxwell's equations (1863), which unified electric and magnetic fields in a set of equations. In order to find a consistent set of equations Maxwell was forced to introduce a new term in one of the previously known formulations and this in turn lead to the spectacular prediction of electromagnetic waves. Maxwell's equations in the absence of sources and in 3-vector notation are given in terms of the electric field $\underline{E}$ and magnetic field $\underline{B}$ by

$$\underline{\nabla} \cdot \underline{E} = 0, \quad \underline{\nabla} \cdot \underline{B} = 0,$$

$$\underline{\nabla} \times \underline{E} = -\frac{1}{c}\frac{\partial \underline{B}}{\partial t}, \quad \underline{\nabla} \times \underline{B} = \frac{1}{c}\frac{\partial \underline{E}}{\partial t}, \tag{17.2.1}$$

where,

$$\underline{\nabla} = \underline{i}\frac{\partial}{\partial x} + \underline{j}\frac{\partial}{\partial y} + \underline{k}\frac{\partial}{\partial z},$$

where $\underline{i}$, $\underline{j}$, $\underline{k}$ are unit orthogonal 3-vectors in space, $\underline{x} = \underline{i}x + \underline{j}y + \underline{k}z$ and c is the speed of light, which we will shortly set to 1 again.

Before proceeding with the main subject it is instructive to recall the role Maxwell's equations played in the discovery of symmetries. It was well known at the time of Maxwell, but not often discussed, that his, and indeed all physical laws, were invariant under space and separately time translations as well as space rotations. Indeed, it is easy to show that they are invariant under $\underline{x}' = \underline{x} + \underline{a}$, $t' = t$, $\underline{E}' = \underline{E}$ and $\underline{B}' = \underline{B}$ as well as $\underline{x}' = \underline{x}$, $t' = t + b$, $\underline{E}' = \underline{E}$ and $\underline{B}' = \underline{B}$, where $\underline{a}$ and b are constants. The role of 3-vector notation is just to encode the invariance of equations under spatial rotations and any equations that can be written in terms of this notation are invariant. The spatial rotations are linear transformations of the coordinates $\underline{x}$ which preserve $\underline{x} \cdot \underline{x}$ with similar transformations of the electric and magnetic fields.

Looking at Maxwell's equations we can see that space and time enter in a rather symmetric manner and it was Lorentz in 1904 who realised that these equations were invariant under the transformations that now carry his name and preserve $x^\mu \eta_{\mu\nu} x^\nu = c^2 t^2 - \underline{x} \cdot \underline{x}$. The corresponding-vector notation encodes Lorentz symmetry into equations and in particular we can write Maxwell's equations as

$$\partial^\mu f_{\mu\nu} = 0 \tag{17.2.2}$$

and

$$\partial_{[\mu} f_{\nu\rho]} = 0, \tag{17.2.3}$$

where $\partial_\mu = (\partial/c\partial t, \partial/\partial x, \partial/\partial y, \partial/\partial z)$. The field strength $f_{\mu\nu}$ is related to the electric and magnetic fields by $E_i = f_{0i}$ and $B_i = \frac{1}{2}\epsilon_{ijk} f_{jk}$.

Equation (17.2.3) is called the Bianchi identity and it can be solved by introducing the gauge potential A_μ and writing $f_{\mu\nu} = \partial_\mu A_\nu - \partial_\nu A_\mu$. It was eventually realised (Weyl 1920) that Maxwell's equations were invariant under gauge transformations $A'_\mu = A_\mu + \partial_\mu \Lambda(x)$. This symmetry is different to the ones discussed above in two respects: it does not change space or time, and so is internal in some sense, and it is not a rigid transformation in that the parameter Λ is not a constant but an arbitrary function of x^μ. The electromagnetic field couples to a point scalar particle with electric charge q with wavefunction ψ as $(i\hbar\partial_\mu + qA_\mu)\psi(x)$, and so to be invariant the wavefunction has the gauge transformation

$$\psi'(x) = e^{iq\Lambda(x)/\hbar}\psi(x). \tag{17.2.4}$$

We have introduced Planck's constant $\hbar$ to make it clear that this is a discussion involving quantum mechanics.

Yet another symmetry that turned up in Maxwell's equations was conformal symmetry (Cunningham and Bateman 1909). This preserves the line element $dx^\mu dx^\nu \eta_{\mu\nu}$ up to scale, that is, $dx^{\mu'} dx^{\nu'} \eta_{\mu\nu} = e^{\rho(x)} dx^\mu dx^\nu \eta_{\mu\nu}$, where ρ is any function of x^μ. These were studied at the beginning of chapter 8.

There is one further symmetry, which is the one of interest to us in this chapter, whose deeper significance has only become apparent in recent times. We note that Maxwell's equations, say in the form of equation (17.2.1), are invariant under

$$\underline{E}' + i\underline{B}' = e^{ia}(\underline{E} + i\underline{B}), \tag{17.2.5}$$

where a is a constant. This can be seen by writing the equations in terms of $\underline{\mathcal{E}} = \underline{E} + i\underline{B}$: they are $\underline{\nabla} \cdot \underline{\mathcal{E}} = 0$ and $\underline{\nabla} \times \underline{\mathcal{E}} = (i/c)\partial\underline{\mathcal{E}}/\partial t$. We note that the energy and momentum densities $\frac{1}{2}(\underline{E}+i\underline{B})^\star \cdot (\underline{E}+i\underline{B}) = \frac{1}{2}(\underline{E}^2 + \underline{B})^2$ and $\frac{1}{2i}(\underline{E}+i\underline{B})^\star \times (\underline{E}+i\underline{B}) = \underline{E} \times \underline{B}$, respectively, are invariant under this transformation. However, the Lagrangian, which is proportional to $\underline{E}^2 - \underline{B}^2$, is not invariant. This is an aspect of duality symmetries that will be important later.

The problem with this symmetry arises from the fact that Maxwell's equations possess a source only for electric charges which occurs on the right-hand side of the very first equation and to be a symmetry we would need to introduce [17.3] a magnetic source in the second equation whose charges q and g respectively transform as

$$q' + ig' = e^{ia}(q + ig). \tag{17.2.6}$$

Of course, no one has seen such a magnetic charge in isolation. We will return to this electromagnetic duality symmetry shortly.

Thus apart from supersymmetry and the diffeomorphisms of general relativity, many symmetries of modern physics had their origins in Maxwell's equations. Indeed Maxwell's equations are unique equations which are second order in space-time derivatives, invariant under special relativity and gauge transformations and constructed from A_μ.

We now derive the Dirac quantisation condition [17.3] using the argument of Wu and Yang [17.4]. Let us consider a source of magnetic field around which we take a 2-sphere S^2. The magnetic charge g is given by the magnetic flux leaving the 2-sphere by the formula

$$g = \int_{S^2} \underline{B} \cdot d\underline{S}. \tag{17.2.7}$$

As explained in appendix C we require two patches, S^2_N and S^2_S, to cover S^2 which can be taken to be around the north and south poles and overlapping around the equator. The magnetic charge is in the interior of the 2-sphere $\underline{\nabla} \cdot \underline{B} = 0$ on the two patches and so we may express $\underline{B}$ in terms of a vector gauge potential on each patch. Let the $\underline{A}^N$ and $\underline{A}^S$ be the vector gauge potentials on the two patches, respectively. The magnetic field is globally defined on the 2-sphere and so on the overlap the two vector gauge potentials must differ by a gauge transformation with parameter χ. In this case,

$$\int_{S^2} \underline{B} \cdot d\underline{S} = \int_{S^2_N} \underline{B} \cdot d\underline{S} + \int_{S^2_S} \underline{B} \cdot d\underline{S} = \int_{S^1} (\underline{A}^N - \underline{A}^S) \cdot d\underline{x}$$

$$= \int_{S^1} \underline{\nabla}\chi \cdot d\underline{x} = \chi(\phi = 2\pi) - \chi(\phi = 0), \tag{17.2.8}$$

where in the third step the second line integral is over the circle around the equator with azimuthal angle ϕ. If there was a globally defined vector gauge potential on S^2, then $\underline{A}^N = \underline{A}^S$ and we would have no magnetic charge as the integral would vanish. Let us assume that we have a magnetic charge and as a consequence the vector potential is not globally defined on S^2 and indeed its value jumps as we go around the equator.

The wavefunction must be single-valued on each patch and so around the equator, but under a gauge transformation $\psi \to e^{iq\chi/\hbar}\psi$ and so this will only be the case if $e^{iq\chi(\phi=0)/\hbar} = e^{iq\chi(\phi=2\pi)/\hbar}$ and as a result $\chi(\phi = 2\pi) - \chi(\phi = 0) = 2\pi m\hbar/q$ for integer m. Substituting this last relation into equation (17.2.8), and using equation (17.2.7), we find the Dirac quantisation condition

$$qg = 2\pi m\hbar, \ m \in \mathbf{Z}. \tag{17.2.9}$$

Clearly this quantisation condition breaks the electromagnetic duality symmetry of equation (17.2.6). However, as we will see electromagnetic duality lives to fight another day.

We can consider particles that carry electric and magnetic charges (called dyons). Let two such particles have electric and magnetic charges q_i and g_i with $i = 1, 2$ respectively, then the quantisation condition takes the form [17.5, 17.6]

$$q_1 g_2 - q_2 g_1 = 2\pi m\hbar, \ m \in \mathbf{Z}. \tag{17.2.10}$$

The quantisation condition for branes [13.79, 13.80] discussed in chapter 14 can be proved in a similar manner.

Let us now consider an SO(3) gauge theory which couples to scalar fields which are in the adjoint representation. The Lagrangian has the form

$$L = -\tfrac{1}{4}\text{Tr}F_{\mu\nu}F^{\mu\nu} - \tfrac{1}{2}\text{Tr}D_\mu \Phi D^\mu \Phi - V(\Phi), \tag{17.2.11}$$

where $F_{\mu\nu} = \partial_\mu A_\nu - \partial_\nu A_\mu + e[A_\mu, A_\nu]$, $D_\mu \Phi = \partial_\mu \Phi + e[A_\mu, \Phi]$, $A_\mu = A_\mu^a T_a$ and $\Phi = \Phi^a T_a$, T_a are the generators of the SO(3) and e is the coupling constant. The generators of SO(3) are normalised so as to obey $[T_a, T_b] = \epsilon_{abc}T_c$. Denoting the internal components of the gauge potential by $\underline{A}_\mu = (A_{\mu 1}, A_{\mu 2}, A_{\mu 3})$ and $\underline{\Phi} = (\Phi_1, \Phi_2, \Phi_3)$, the Lagrangian can be written as

$$L = -\tfrac{1}{4}\underline{F}_{\mu\nu}\underline{F}^{\mu\nu} - \tfrac{1}{2}D_\mu\underline{\Phi}D^\mu\underline{\Phi} - V(\underline{\Phi}) \tag{17.2.12}$$

where $\underline{F}_{\mu\nu} = \partial_\mu \underline{A}_\nu - \partial_\nu \underline{A}_\mu + e\underline{A}_\mu \times \underline{A}_\nu$, $D_\mu\underline{\Phi} = \partial_\mu \underline{\Phi} + e\underline{A}_\mu \times \underline{\Phi}$. The potential is assumed to have a minimum for which the scalar field has a non-trivial vacuum expectation value $\langle\underline{\Phi}\rangle$ whose length squared is taken to be $\langle\underline{\Phi}\rangle \cdot \langle\underline{\Phi}\rangle = v^2$. The SO(3) gauge symmetry is spontaneously broken to U(1) = SO(2). We also assume that the potential is positive definite. We could take $V = \lambda(\underline{\Phi} \cdot \underline{\Phi} - v^2)^2$, where λ is a constant.

Shifting the scalar field to its minimum we find the particle spectrum of the theory and their charges which can be read off from the coupling to the unbroken U(1) gauge vector. One finds a spin-1 massless and neutral gauge boson A_μ, two massive spin-1 particles $W_\mu^\pm$ of mass $e\hbar v$ and charge $\pm e\hbar$ and one massive spin-0 particle, the Higgs particle. Two of the scalar degrees of freedom are eaten by the two spin-1 to make them massive. We note that the masses of bosonic particles in a quantum field theory occur in their field equation in the operator $(\partial^2 - m^2/\hbar^2 + \cdots)$ so explaining the factors of $\hbar$ above, with similar considerations for the charges.

The energy of the above theory can be written as

$$\tfrac{1}{2}\int d^3x\{\underline{E}_i \cdot \underline{E}_i + \underline{B}_i \cdot \underline{B}_i + \underline{\Pi} \cdot \underline{\Pi} + D_i\underline{\Phi} \cdot D_i\underline{\Phi} + 2V(\underline{\Phi})\}, \tag{17.2.13}$$

where $\underline{\Pi} = D_0\underline{\Phi}$ and the index i denotes the space components and is summed over when repeated. We note that every term in the energy is positive.

The electric and magnetic charges corresponding to the preserved U(1) are given by

$$Q_e = \frac{1}{v}\int dS^i \underline{E}_i \cdot \underline{\Phi}, \quad Q_m = \frac{1}{v}\int dS^i \underline{B}_i \cdot \underline{\Phi}. \tag{17.2.14}$$

We notice that the electric and magnetic fields are projected in the internal space to lie in the direction of the Higgs field which takes the value $\langle \Phi \rangle$ at spatial infinity. We note that $\langle \Phi \rangle / v$ is a unit vector.

What brought the subject of magnetic monopoles to life was the discovery [17.7, 17.8] that the equations of motion of the Lagrangians of equation (17.2.11), or equivalently (17.2.12), possess solutions which carry magnetic charge, are non-singular and are of finite energy. At spatial infinity the Higgs field must take on their vacuum values in order to have finite energy, that is, $\langle \Phi \rangle^2 = v^2$, which defines a 2-sphere in the internal space. Thus the Higgs field is a map between the 2-sphere at infinity and the 2-sphere in the internal space just mentioned. It turns out that this map between 2-spheres can have non-trivial topology characterised by the winding number $m \in \mathbf{Z}$ and that the corresponding magnetic charge is

$$g = -\frac{4\pi}{e} m = -\frac{4\pi\hbar}{q_w} m, \qquad (17.2.15)$$

where $q_w = e\hbar$ is the charge of the W bosons discussed above. From now on we will set $\hbar = 1$. We notice that these are not the minimal charges allowed by the quantisation condition of equation (17.2.10). It turns out that the spinor representation of the gauge group SO(3) has particles of electric charge one half of that of the spin-1 particles and so the magnetic charge of equation (17.2.15) is indeed the minimal one allowed by the quantisation condition.

For a static solution the energy is the mass of the configuration and so equation (17.2.13) can be written as

$$M = \tfrac{1}{2} \int d^3x \{ \underline{E}_i \cdot \underline{E}_i + \underline{B}_i \cdot \underline{B}_i + D_i \underline{\Phi} \cdot D_i \underline{\Phi} + 2V(\underline{\Phi}) \}$$

$$\geq \tfrac{1}{2} \int d^3x \{ \underline{E}_i \cdot \underline{E}_i + \underline{B}_i \cdot \underline{B}_i + D_i \underline{\Phi} \cdot D_i \underline{\Phi} \} \qquad (17.2.16)$$

as the potential was assumed to be positive definite and, for a static solution, in the gauge $A_0 = 0$, $\underline{\Pi} = 0$. We may rewrite the energy as

$$M \geq \tfrac{1}{2} \int d^3x \{ (\underline{E}_i - \sin\theta D_i \underline{\Phi}) \cdot (\underline{E}_i - \sin\theta D_i \underline{\Phi})$$

$$+ (\underline{B}_i - \cos\theta D_i \underline{\Phi}) \cdot (\underline{B}_i - \cos\theta D_i \underline{\Phi}))$$

$$+ \sin\theta \int d^3x D_i \underline{\Phi} \cdot \underline{E}_i + \cos\theta \int d^3x D_i \underline{\Phi} \cdot \underline{B}_i$$

$$\geq \sin\theta \int d^3x D_i \underline{\Phi} \cdot \underline{E}_i + \cos\theta \int d^3x D_i \underline{\Phi} \cdot \underline{B}_i, \qquad (17.2.17)$$

valid for any value of θ. The last terms can be further evaluated by integration by parts:

$$\int d^3x D_i \underline{\Phi} \cdot \underline{B}_i = \int d^3x \partial_i (\underline{\Phi} \cdot \underline{B}_i) = vg, \qquad (17.2.18)$$

where we have used the Bianchi identity $D_i \underline{B}_i = 0$, the expression for the magnetic charge of equation (17.2.14) and that $D_i(\underline{\Phi} \cdot \underline{B}_i) = D_i(\underline{\Phi}) \cdot \underline{B}_i + \underline{\Phi} \cdot D_i(\underline{B}_i) = \partial_i(\underline{\Phi} \cdot \underline{B}_i)$. Thus we find that

$$M \geq ev\sin\theta + gv\cos\theta. \qquad (17.2.19)$$

As this is valid for any θ we can minimise with respect to θ to find that best bound is given by $\tan \theta = e/g$ and is given by [17.9, 17.10]

$$M \geq v\sqrt{q^2 + g^2}. \tag{17.2.20}$$

This bound will be achieved if the terms we have dropped in its derivation vanish and then the masses take the form [17.11]

$$M = v\sqrt{q^2 + g^2}, \tag{17.2.21}$$

where q is the electric charge of the solution. Solutions corresponding to particles that carry magnetic and electric charges were also found in [17.12]. We note that the derivation of the above equations is purely classical and so one would not expect it hold in a generic quantum field theory.

Let us return to considering a static solution with only a magnetic field. In this case $\sin \theta = 0$ and the bound is best achieved with $\cos \theta = 1$. To do this we must set the potential to zero which can be achieved by taking $\lambda = 0$ in the potential given above, but still taking the Higgs field to have a non-zero vacuum expectation value $\langle \Phi \rangle$. This is called the BPS limit. We also require [17.9]

$$\underline{B}_i = \pm D_i \Phi, \text{ in addition to the already assumed } \underline{E}_i = 0, \ D_0\Phi = 0. \tag{17.2.22}$$

One can show that any solution to equation (17.2.22) satisfies the full field equations of the theory. The great advantage is that the monopole solution considered earlier is very complicated, but equation (17.2.22) is a first order differential equation and so it is easier to study, indeed there is a large literature on its solutions.

As we have seen the solutions are classified by a topological invariant whose different sectors are labelled by the integer m, which determines the amount of magnetic charge the solution carries. The solutions for a given integer m and fixed energy are parameterised by a set of parameters called moduli. This space has been much studied and it can be endowed with the structure of a manifold with dimension $4m$ [17.13]. The metric of the moduli space is only known for $m = 1$ and $m = 2$. In chapter 14 we briefly discussed the case of $m = 1$ which corresponds to one particle. The case of $m = -1$ also describes one particle, but with the opposite magnetic charge and it can be thought of as an anti-monopole. As one might expect from the dimension of the moduli space, the sector with topological number m corresponds to m particles, each with the magnetic charge of equation (17.2.15). Since we are dealing with static solutions it follows that there can be no force between the monopoles when at rest. However, they do have velocity dependent forces and their motion at low energy can be found by a procedure which exploits the structure of the moduli space and is described in section 14.7. Indeed, the motion is described by the moduli that become functions of the world-line time of the monopole. Monopoles and anti-monopoles do have a force between them.

We now describe the original electromagnetic duality conjecture of Montonen and Olive [17.14]. These authors were struck by several similarities between the behaviour of electric and magnetic particles. The massive spin-1 $W^{\pm}$ particles which arise as excitations of the fundamental fields of the Lagrangian of equation (17.2.12) and the magnetic monopoles have masses that were of the form of vacuum expectation values of the Higgs field times their charges. The two are unified in equation (17.2.20), which is invariant under the electromagnetic duality transformations of equation (17.2.6). Furthermore, they found that the force between two W^+ particles when at rest vanished just like it did for two monopoles. However, also like monopoles, they have velocity dependent forces.

The magnetic solitons give rise to a U(1) magnetic field which falls as 1/(distance)^2 and one might imagine that this magnetic field can arise in a spontaneously broken gauge theory in which the monopole is the elementary quantum excitation rather than a solitonic solution and that in this theory the $W^\pm$ particles arise as solitonic excitations. Montonen and Olive conjectured that there is a symmetry transformation between these two theories such that one can compute the physical effects of all particles in either theory and the results will be the same. Under this transformation the charges of the elementary excitation in the two theories would have to transform into each other as

$$q \to g = -\frac{4\pi}{q}. \qquad (17.2.23)$$

We notice that if the original theory has a weak coupling, that is, $e \leq 1$, then the final theory has a strong coupling, that is, $g \geq 1$. It is this change from weak to strong coupling that is so interesting as it takes us from a perturbative to a non-perturbative regime, that is, from a regime where we can compute physical quantities quite well using perturbation theory to one where we generally find it quite difficult to find reliable results except in certain circumstances. We note that the real content of the Montonen–Olive conjecture is the existence of the two theories that have a transformation between them that interchanges electrically charged and magnetic particles and vice versa.

At first it was not clear how this duality could be true as, at least classically, the one monopole solution was spherically symmetric and so had no spin, while the $W^\pm$ have spin 1. Also the mass formula of equation (17.2.21) was derived using classical considerations and in a generic theory it is likely to be corrected by quantum effects in a way that does not respect duality symmetries.

Before proceeding further we need to introduce a new term in the Lagrangian of equation (17.2.12), namely [17.15]

$$-\frac{e^2\theta}{32\pi^2} \underline{F}_{\mu\nu} \star \underline{F}^{\mu\nu}, \qquad (17.2.24)$$

where $\star\underline{F}_{\mu\nu} = \frac{1}{2}\epsilon_{\mu\nu\rho\kappa}\underline{F}^{\rho\kappa}$ and θ is a parameter taking values $0 \leq \theta \leq 2\pi$. It is a total space-time derivative and so it does not affect the equations of motion. In the action such a term is integrated over and it is, in fact, a topological quantity, which takes integer values in the presence of instantons times 2θ.

However, this term does affect the Noether charge of the unbroken U(1) gauge transformations and so the allowed electric charge. Indeed, one can show that the electric charge q is modified by the presence of the θ term and is given by [17.15]

$$q = n_e e + \frac{e\theta}{2\pi}n_m, \quad n_e, n_m \in \mathbf{Z}, \quad g = n_m\frac{4\pi}{e}. \qquad (17.2.25)$$

In the second equation we have recorded the magnetic charge which is unchanged. We note that this does satisfy the quantisation condition of equation (17.2.10). Let us define

$$\tau = \frac{\theta}{2\pi} + i\frac{4\pi}{e^2}. \qquad (17.2.26)$$

We observe that the imaginary part of τ is positive definite. The part of the new Lagrangian which is the sum of that of equation (17.2.12) and the new term of equation (17.2.24) which contains the gauge fields can be rewritten as

$$-\frac{1}{32\pi}\text{Im}(\tau \underline{\mathcal{F}}_{\mu\nu} \cdot \underline{\mathcal{F}}^{\mu\nu}), \tag{17.2.27}$$

where $\underline{\mathcal{F}}_{\mu\nu} = \underline{F}_{\mu\nu} + i \star \underline{F}_{\mu\nu}$ and after making the transformation $\underline{A}_{\mu} \to (1/e)\underline{A}_{\mu}$ and $\underline{\Phi} \to (1/e)\underline{\Phi}$. Thus the two parameters e and θ can be combined into the one object τ.

The path integral is invariant under $\tau \to \tau + 1$ since θ multiplies $2\pi i$ times an integer in the exponential. As such, physics is invariant under $\tau \to \tau + 1$. We note that the transformation $\tau \to \tau' = -1/\tau$ leads to $e' = -4\pi/e$, if we set $\theta = 0$, which we recognise as the conjectured electromagnetic duality transformations of equation (17.2.22). However, *if* this was a symmetry for all θ we would have a symmetry group generated by

$$\tau \to \tau' = \tau + 1 \text{ and } \tau \to \tau' = -\frac{1}{\tau}. \tag{17.2.28}$$

In fact, these two transformations generate a group of transformations of the form

$$\tau' = \frac{a\tau + b}{c\tau + d}, \quad a, b, c, d \in \mathbf{Z}, \quad ad - cb = 1. \tag{17.2.29}$$

The set of matrices

$$g_0 = \begin{pmatrix} a & b \\ c & d \end{pmatrix} a, b, c, d \in \mathbf{Z}, \quad ad - cb = 1, \tag{17.2.30}$$

form the group SL(2, **Z**) with the group operation being the usual matrix composition. They act on τ by the action of equation (17.2.29) so as to respect the group composition law. It is easy to check that these transformations preserve the sign of Imτ and so we can take them to act on the upper half-plane. Thus SL(2, **Z**) acts as a transformation group on the upper half-plane. We note that $\pm I_{2\times 2}$, where $I_{2\times 2}$ is the unit matrix in two dimensions, have the same action on the upper half-plane.

Thus, modulo one small assumption, that is, the one above equation (17.2.28), Montonen–Olive duality implies that there is an SL(2, **Z**) symmetry. Thus incorporating the θ parameter allowed us to formulate electromagnetic duality as an SL(2, **Z**) symmetry.

Let us now examine the masses of the particles to see if they are invariant under SL(2, **Z**). Using equation (17.2.25), we have particles with magnetic charges $g = n_m 4\pi/e$ and electric charges $q = n_e e + n_m e\theta/2\pi$. In the BPS limit the mass squared formula of equation (17.2.21) takes the form

$$M_n^2 = v^2|q + ig|^2 = v^2 e^2|n_e + n_m\tau|^2 = 4\pi v^2 n^T A(\tau)n, \tag{17.2.31}$$

where n is the column charge number vector $n = \begin{pmatrix} n_m \\ n_e \end{pmatrix}$ and the matrix $A(\tau)$ is defined by

$$A(\tau) = \frac{1}{\text{Im}\tau}\begin{pmatrix} |\tau|^2 & \text{Re}\tau \\ \text{Re}\tau & 1 \end{pmatrix}. \tag{17.2.32}$$

The matrix $A(\tau)$ has an SL(2, **Z**) group interpretation. In terms of the notation of chapter 13, and in particular equation (13.5.30), let us consider the group element $g = e^{-\chi E}e^{-\frac{1}{\sqrt{2}}H}$ which we can think of as belonging to the coset SL(2, **R**) with local subgroup SO(2).

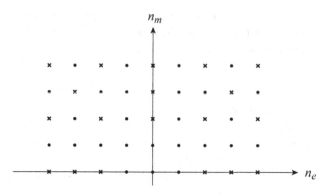

Figure 17.2.1   Allowed positive Dyon charges.

Then the usual coset action $g \rightarrow g_0 g h$ with $g_0 \in \mathrm{SL}(2, \mathbf{R})$ and $h \in \mathrm{SO}(2)$ leads to the transformation of τ of equation (17.2.29), provided we take $b \rightarrow -b$ and $c \rightarrow -c$, which we could have taken in the beginning as it is also an allowed $\mathrm{SL}(2, \mathbf{R})$ transformation. The object gg^T is invariant under local transformations but transforms under $gg^T \rightarrow g_0(gg^T)g_0^T$. Using the matrices of equation (13.5.30), we find that $-gg^T$ is equal to the $A(\tau)$ just above if we take $\tau = \chi + ie^{-\phi} = (\theta/2\pi) + i(4\pi/e^2)$. Thus we find that $A(\tau)$ transforms as $A(\tau) \rightarrow g_0 A(\tau)g_0^T$ and so the masses are indeed $\mathrm{SL}(2, \mathbf{Z})$ invariant provided we also transform the charges, that is, they are invariant under

$$\tau' = \frac{a\tau + b}{c\tau + d}, \quad n' = (g^{-1})^T n = \begin{pmatrix} d & -c \\ -b & a \end{pmatrix} n, \tag{17.2.33}$$

where g is the matrix of equation (17.2.30) and n the column vector of equation (17.2.31).

The original W^+ has the charge number vector $\begin{pmatrix} 0 \\ 1 \end{pmatrix}$ but under the transformation of equation (17.2.33) it becomes $n_3 \equiv \begin{pmatrix} -c \\ a \end{pmatrix}$. However, the integers a and c can have no common integer factor, that is, they are said to be coprime, as if they did then $\det g = ad - cb$ would have the same common factor and so could not be 1 as required by the definition of $\mathrm{SL}(2, \mathbf{Z})$. Thus a $\mathrm{SL}(2, \mathbf{Z})$ symmetry predicts the existence of dyonic particles, but only with charge numbers n_e and n_m which are coprime. If they are not coprime, then there exists an integer valued charge vector which lies on a line from the origin to n_3 and then n_3 can be written as the sum of two collinear charge vectors. In figure 17.2.1 we show all the charge vectors of the dyonic states predicted by $\mathrm{SL}(2, \mathbf{Z})$, that is, all dyons with a coprime charge number. The solid dots are the predicted dyons, that is, the coprime points, while the crosses are the points which are not coprime. We see that the only solely electrically charged particles are the $W^{\pm}$ with charge vectors $\begin{pmatrix} 0 \\ \pm 1 \end{pmatrix}$, while the dyons with two units of magnetic charge can only carry an electric charge that has an odd integer charge number n_e.

In fact if the charge number of a dyon is not coprime, then the dyon can decay into two dyons of smaller mass, that is, the dyon is unstable. Let us consider three dyons with charge vectors n_i, $i = 1, 2, 3$ such that the third can decay into the other two. Let us denote their electric and magnetic charges by q_i and g_i, respectively. From charge conservation

$q_3 = q_1 + q_2$ and $g_3 = g_1 + g_2$, but from equation (17.2.25) the charge vectors also add up: $n_3 = n_1 + n_2$. Looking at the first expression of equation (17.2.31) we see that the mass is given by the vacuum expectation value times the length of a 2-vector with components (q, g). It is then obvious from two-dimensional geometry that $M_{n_3} \leq M_{n_1} + M_{n_2}$ with equality if and only if the charges, and so the charge vectors, are collinear. The dyon can only decay if we have equality, that is, we have collinear charge vectors and so the charge vector n_3 is not coprime. The set of allowed charge vectors form a lattice. One can take the elementary electric and magnetic charge vectors, $\begin{pmatrix} 0 \\ 1 \end{pmatrix}$ and $\begin{pmatrix} 1 \\ 0 \end{pmatrix}$, respectively, to be a basis of the lattice; however, one can also choose any basis vectors that are related to these by an SL(2, **Z**) transformation. Thus from the viewpoint of the Montonen–Olive conjecture there are an infinite number of equivalent theories.

However, we have yet to resolve the objections raised in the paragraph after equation (17.2.23). We recall that the monopole solutions we discussed were spherically symmetric, while the electically charged $W^{\pm 1}$ particles carry spin 1; also the above derivations were classical and so were unlikely to survive in a generic quantum field theory. These problems are solved if we embed the above theory into one with supersymmetry. We would like the scalar and the spin-1 particles to belong to the same supermultiplet as then the scalars would automatically belong to the adjoint representation of the gauge group. We would also like a supermultiplet that has spins 1 or less. This is not possible for $N = 1$ supersymmetry, but there are two possibilities, one with $N = 2$ and one with $N = 4$ supersymmetry. For the case of $N = 2$, there are two multiplets with spins 1 or less: the vector multiplet, which has two scalars, one spin 1 and two Majorana spinors, and the hypermultiplet which has two Majorana fermions and four scalars. Clearly, it is the former which could contain the monopoles. For $N = 4$ supersymmetry there is only one supermultiplet that has spins 1 or less and it contains one spin 1, six scalars and four Majorana spinors.

We now focus on the $N = 4$ theory whose Lagrangian (see equation (13.7.8)) without the fermions, after the scaling of the fields by $1/e$ as discussed above, is given by

$$-\frac{1}{32\pi}\text{Im}(\tau \underline{\mathcal{F}}_{\mu\nu}\underline{\mathcal{F}}^{\mu\nu}) - \frac{1}{2e^2}D_\mu\underline{\Phi}^I D^\mu\underline{\Phi}^I - \frac{1}{4e^2}\sum_{I,J}(\underline{\Phi}^I \times \underline{\Phi}^J)^2, \qquad (17.2.34)$$

where we have included a θ term. We take SO(3) as our gauge group and underline the fields as before. The six scalars $\underline{\Phi}^I$ carry the indices $I, J = 1, \ldots, 6$. This theory has a manifest internal SO(6) symmetry which arises from the $N = 4$ supersymmetry algebra, which possesses an SO(6) symmetry that rotates the supercharges. The scalars can take on an expectation value $\langle\underline{\Phi}^I\rangle$ such that $\langle\underline{\Phi}^I\rangle \times \langle\underline{\Phi}^J\rangle = 0$ while still keeping the potential vanishing. The SO(3) gauge symmetry is then spontaneously broken to U(1). The generalisation to any gauge group of rank r is obvious except that now the symmetry is broken to $U(1)^r$. This theory obviously possesses the monopole solutions we discussed above, we just set all of the scalars to zero, but leave one combination of the six scalars, as we will discuss below, to be non-zero. We note that flat directions are a generic feature of supersymmetric theories and that if supersymmetry is preserved, they will remain flat to all orders of perturbation theory as the supersymmetric potential receives no quantum corrections [17.16]. Thus we see that supersymmetry begins to work its magic.

This theory has a vanishing β function and so is conformally invariant [17.17]. This is a desirable requirement as in this case the β function is automatically invariant under the change of coupling of equation (17.2.23).

When a gauge symmetry is spontaneously broken in any supersymmetric theory it is obvious that the resulting massive particles must also belong to supermultiplets, provided supersymmetry is also not broken. The 'usual' massive multiplets of $N = 4$ have 128 bosonic states and the one with the lowest spins contains all spins from 2 to 0 inclusive. This presents us with a puzzle as spontaneous symmetry breaking does not introduce any spins greater than 1. The resolution comes from the fact that by the so-called 'usual' massive representations we mean representations for which the only active operators on the right-hand side of the anti-commutator of two supersymmetry charges are the space-time translations, However, if the central charges on the right-hand side are active, then one can find further irreducible representations provided the central charges take very special values with respect to the mass of the particle. In this case one finds more zero eigenvalues for the matrix on the right-hand side and so some of the supercharges have vanishing anti-commutator. As the physical states belong to a Hilbert space with a positive definite norm these supercharges must vanish on the physical states and we find fewer states than in the generic massive case. These are sometimes called shortened, or BPS, multiplets, even though the BPS limit was invented long before this phenomenon was recognised. The irreducible representations of supersymmetry in four dimensions are discussed in chapter 8 of [1.11]; there the reader will find all the points discussed here spelt out in detail. In particular, the reader may like to examine tables 8.3 and 8.4 of that reference to see lists of some of these representations. In the case of $N = 4$ one can find a unique massive multiplet with spins 1 or less and it contains one spin 1 and five scalars and four Majorana spinors. Clearly, one of the scalars has been eaten by the spin 1 to make it massive as we might have hoped. Thus when the $N = 4$ gauge theory is spontaneously broken the massive states of a given charge must belong to this shortened supermultiplet; this applies to the electric $W^{\pm}$ particles, the monopoles and the dyons. However, for this to occur some of the central charges must be non-zero. Indeed the $N = 4$ supersymmetric Yang–Mills theory is the natural setting for the Montonen–Olive conjecture [17.18] and its $SL(2, \mathbf{Z})$ symmetry was proposed in reference [17.19].

If we have a solution of a supersymmetric theory, such as the monopole solution considered above, one can find from the supersymmetry field variations of the field how much supersymmetry the solution preserves. Since the supervariation of the bosonic fields gives fermions, but these are Grassmann odd and so are assumed to have zero expectation value, one only has to study the variations of the fermion fields. If the particles corresponding to the solutions lead to BPS supermultiplets, then we will find more supersymmetry is preserved than if they belong to the usual massive multiplets. Such solutions are often referred to as r-BPS solutions meaning they preserve a fraction r of the supersymmetries of the original theory. We will see below that the monopole solutions are half BPS.

As the supercharges are conserved, the central charges must also be conserved and, in the absence of some exceptional new conserved charges, they must be the charges we already know about, that is, the electric and magnetic charges. However, the central charges must obey a precise relation with the masses of the particles in order to get a shortened, or BPS, multiplet and so we find that we have a relationship that expresses the masses of the particles in terms of their electric and magnetic charges. This is just what we have in equation (17.2.21) and it turns out this is precisely what happens in supersymmetric theories [13.78]. What this relationship is can be found entirely from the supersymmetry algebra once one has identified the central charges in terms of the electric and magnetic charges. We carry out this step for the case of $N = 2$ supersymmetry below.

In fact, the $N = 4$ supersymmetry algebra can admit at most 12 central charges in the $6 + 6$ representation of SO(6) (see [1.11]) and they are the electric and magnetic charges in the direction of the six scalar fields:

$$Q_e^I = \frac{1}{ev} \int \underline{E}_i \cdot \underline{\Phi}^I dS_i, \quad Q_m^I = \frac{1}{ev} \int \underline{B}_i \cdot \underline{\Phi}^I dS_i. \tag{17.2.35}$$

Just to recall the notation, the dot product is in the internal space, while dS_i is the infinitesimal surface element in the direction normal to the surface. It turns out that to get shortened, or BPS, supermultiplets we must have

$$M^2 = v^2 \sum_I ((Q_e^I)^2 + (Q_m^I)^2), \tag{17.2.36}$$

which generalises the expression of equation (17.2.21) to be SO(6) invariant as well as SL(2, **Z**) invariant.

The derivation of the mass–charge relation from the supersymmetry algebra has one considerable advantage. As we increase the coupling constant of the quantum theory we do not expect to create new particles and certainly not particles of spin 2. However, as we have explained, the absence of such massive states implies that we have a shortened supermultiplet for which the mass and the charges must be related by equation (17.2.36). Thus we can expect this relation to hold at all values of the coupling constant in the quantum theory and so also in the non-perturbative regime of the theory. As such it should be independent of quantum effects. This overcomes one of the main objections raised after equation (17.2.23).

The other objection was that the monopole should carry spin 1. This arises in the supersymmetric theory due to the zero modes of the fermions which act much like the supercharges so as to create the supermultiplet of particles, which of course includes the particles of spin 1. This has been well studied in the low energy approximation using the description of the motion in terms of the moduli. The reader might like to consult the reviews referenced at the end of this section for further details.

It is believed that the $N = 4$ Yang–Mills theory is invariant under the SL(2, **Z**) transformations of equation (17.2.33). One striking piece of evidence for this comes from the realisation that the existence of dyons with two units of magnetic charge implies the existence of a unique harmonic 2-form on the moduli space with rather precise properties and this was indeed found to be the case [17.19].

We close this section by illustrating the properties of the BPS states in the context of the $N = 2$ Yang–Mills theory. The calculations are essentially the same as in the $N = 4$ Yang–Mills theory, but have the advantage that they are a little bit less complicated. As explained above, if we have a classical solution the amount of supersymmetry that is preserved can be found by demanding that the supersymmetric variation of the fermion field vanish:

$$\delta \underline{\lambda}_I = 0 = (-\tfrac{1}{2} \underline{F}_{\mu\nu} \gamma^{\mu\nu} - i\gamma^\mu (D_\mu \underline{A} - i\gamma_5 D_\mu \underline{B}) + ig\underline{A} \times \underline{B}\gamma_5)\epsilon_I, \tag{17.2.37}$$

where we have set to zero the auxiliary field in equation (12.17) of [1.11] and ϵ_I, $I = 1, 2$ are the supersymmetry parameters. Let us restrict the solution to be static and have only a magnetic field. We also set $\underline{A} = \alpha\underline{\Phi}$ and $\underline{B} = \beta\underline{\Phi}$, where α and β are constants. The variation then takes the form

$$0 = -(\tfrac{1}{2}\epsilon_{ijk}\underline{B}_k\gamma^{ij} + i\gamma^j D_j\underline{\Phi}(\alpha - i\gamma_5\beta))\epsilon_I, \tag{17.2.38}$$

where $i, j, k = 1, 2, 3$. Using the identity $\tfrac{1}{2}\epsilon_{\mu\nu\rho\kappa}\gamma^{\rho\kappa} = i\gamma_5\gamma_{\mu\nu}$, which implies that $-\tfrac{1}{2}\epsilon_{ijk}\gamma^{jk} = i\gamma_5\gamma_0\gamma_i$, we may rewrite this as

$$0 = -i\gamma_5\gamma_0\gamma_i D_i\underline{\Phi}(1 + P)\epsilon_I, \tag{17.2.39}$$

where we have set $\underline{B}_i = D_i\underline{\Phi}$, that is, our monopole solution, and $P = \gamma_0\gamma_5(\alpha - i\gamma_5\beta)$. It is easy to see using the conventions of appendix A of [1.11] that $P^2 = 1$, provided $\alpha^2 + \beta^2 = 1$, and it is also Hermitian, so it can have eigenvalues ± 1. Since the $\mathrm{Tr}P = 0$ we conclude that it has equal numbers of $+1$ and -1 eigenvalues. Hence we can preserve half of the supersymmetries, that is, those that obey $(1 + P)\epsilon_I = 0$. Thus when embedded in $N = 2$ Yang–Mills theory the monopole is a $\tfrac{1}{2}$-BPS solution. In fact, the same result holds for the embedding in $N = 4$ Yang–Mills theory.

We now derive the BPS conditions from the $N = 2$ supersymmetry algebra with central charges which is given by (see [1.11])

$$\{Q_\alpha^I, Q_\beta^J\} = 2((\gamma^\mu C)_{\alpha\beta}\delta^{IJ}P_\mu + i\epsilon^{IJ}(Z_1 C_{\alpha\beta} + Z_2(\gamma_5 C)_{\alpha\beta})). \tag{17.2.40}$$

Taking the particles to be in the rest frame, that is, $P_\mu = m\delta_{\mu,0}$, we may write the matrix on the right-hand side as

$$(\delta^{IJ}I_{4\times 4}m + i\epsilon^{IJ}A)2\gamma^0 C, \tag{17.2.41}$$

where we have suppressed the spinor indices and $A = (Z_1 + \gamma_5 Z_2)\gamma_0$. This matrix has zero eigenvalues, and so leads to shortened multiplets if

$$0 = \det(\delta^{IJ}I_{4\times 4}m + i\epsilon^{IJ}A) = \exp\{\mathrm{Tr}\ln(m\delta^{IJ}I_{4\times 4} + i\epsilon^{IJ}A)\}$$

$$= \exp\left\{\mathrm{Tr}\ \ln(\delta^{IJ}I_{4\times 4}m) + \tfrac{1}{2}\mathrm{Tr}\ \ln\delta^{IJ}I_{4\times 4}\left(1 - \frac{Z_1^2 + Z_2^2}{m^2}\right)\right\}$$

$$= \delta^{IJ}(m^2 - (Z_1^2 + Z_2^2))^2, \tag{17.2.42}$$

where $I_{4\times 4}$ is the 4×4 unit matrix arising from the spinor indices. In deriving this result we have used that $\mathrm{Tr}A = 0$, $A^2 = -(Z_1^2 + Z_2^2)$ and that $i\epsilon^{IJ}$ squares to the unit matrix. Hence we conclude that the dyonic states have masses such that

$$m = \sqrt{Z_1^2 + Z_2^2}. \tag{17.2.43}$$

Given that the central charges are the electric and magnetic charges times the field vaccum expectation value, we recover equation (17.2.21).

The reader may like to consult the reviews on electromagnetic duality of [17.19, 17.20, 17.21]. Indeed the author has found them useful when writing this section.

17.3   S and U duality

In the first section of this chapter we studied T duality. This symmetry transforms the dilaton, which is related to the string coupling by $g_s = e^\phi$, in such a way that if the coupling is small before the transformation it is small afterwards. Thus it transforms the perturbative regime of the theory to itself and indeed we showed that it is a symmetry at every order of string perturbation theory.

This is in contrast to the electromagnetic duality of $N = 4$ Yang–Mills theory which we studied in the previous section which maps the weak coupling regime to the strong coupling regime and as a result it takes perturbative to non-perturbative effects. It is usually very difficult to show the existence of symmetries like electromagnetic duality as it is difficult to compute the non-perturbative effects of theory in sufficient detail. However, as we saw supersymmetry can sometimes provide mechanisms that allow one to find relations which hold for all coupling constant regimes and these can be used to check if such symmetries are valid.

In this section we wish to study a symmetry of string theory that also takes the weak coupling regime to the strong coupling regime and vice versa. As we explained in chapter 13 the type II supergravities are the complete low energy effective actions for the corresponding type II string. That is, they contain all the perturbative and non-perturbative effects. As such, if there were such a symmetry it would have to be a symmetry of the type II supergravity theories. In effect, these theories provide an example of how supersymmetry can provide checks of such symmetries. A similar statement holds for the type I string theories and their low energy effective actions. When constructing the supergravity theories in chapter 13 we indeed found that there were some unexpected symmetries. In particular if any supergravity theory has scalar fields, then these belong to a non-linear realisation of some group G and this symmetry extends to be a symmetry of the full supergravity theory.

The simplest example of this was the SL(2, **R**) [10.14] of the IIB supergravity theory whose local subgroup was SO(2). This symmetry was discussed in section 13.5 on the IIB theory. We recall that there are two scalars, the dilaton ϕ and the axion χ, which can be combined into the complex quantity $\tau = -\chi + ie^{-\phi}$ which transforms under the non-linear realisation as in equation (13.5.29). By studying the sigma model approach to the IIB string perturbative theory we found that the IIB string coupling constant was given in terms of the dilaton by $g_s = e^{\langle\phi\rangle}$. The particular SL(2, **R**) transformation $\tau' = -1/\tau$ implies that

$$ g_s' = \frac{1}{g_s} \tag{17.3.1} $$

if we take $\chi = 0$. This makes it very clear that the SL(2, **R**) maps from weak to strong coupling and vice versa, like the electromagnetic duality symmetry of $N = 4$ Yang–Mills theory. It is called an S duality.

In addition to the two scalars, the bosonic fields of the IIB supergravity theory consist of a 4-form gauge field and the metric, both of which are inert under SL(2, **R**) provided we take the metric to be in Einstein frame, as well as two second rank gauge fields that transform as a doublet under SL(2, **R**) as given in equation (13.5.8). From the viewpoint of the IIB string, see chapter 7, one of the 2-form fields belongs to the $NS \otimes NS$ sector while the other belongs to the $R \otimes R$ sector. As discussed in chapter 14 they must couple to 1-branes: the 2-form in the $NS \otimes NS$ sector will couple to the IIB string itself, which is sometimes called the fundamental string, while the 2-form in the $R \otimes R$ sector will couple to a D 1-brane. We also have their dual fields, that is, two 6-form gauge fields. These couple to two 5-branes, one of which is in the $NS \otimes NS$ sector, and is referred to as the NS5-brane, and one in the $R \otimes R$ sector which is a D-brane. As the SL(2, $\mathbf{R}$) symmetry rotates the two rank 2 gauge fields it must also rotate the IIB string into the D string and vice versa. While the fundamental string, by definition, provides the mechanism to compute perturbative corrections of the IIB string, the D1-brane is a non-perturbative object. Its presence shows up in the IIB supergravity theory as a solitonic solution for which the 2-form in the $R \otimes R$ sector is active. Indeed in section 13.8.3 we found just such a solution. As discussed in the next section, one can compute its tension, which is the analogue for an extended object of the mass of a point particle and one finds that it has a factor $1/g_s$. Thus at weak IIB coupling it is very heavy, but at strong coupling it becomes light. All this is consistent with the non-perturbative nature of the SL(2, $\mathbf{R}$) symmetry.

One simple way to see in detail that the fundamental and D strings rotate into each other under SL(2, $\mathbf{R}$) symmetry is to examine their corresponding solutions of IIB supergravity given in equation (13.8.120). Under the change of equations (13.5.29) and (13.5.32) with $a = 0 = d$ and $b = 1 = -1$ that $\phi \to -\phi$, the metric remains the same and the gauge field changes, or preserves, its sign depending on whether it is $NS \otimes NS$ or $R \otimes R$. Examining the solution for the fundamental string say, that is, $\eta = -1$, we find that we do indeed get the solution for the D string. The metric remains the same as it should, but e^ϕ in equation (13.8.121) is given by the opposite power of N_p which corresponds to the change in the sign of η that takes us from the $NS \otimes NS$ to the $R \otimes R$ sector. The gauge field also changes as it should. It is very easy to see that the two 6-branes rotate into each other under SL(2, $\mathbf{R}$), but that the D3-brane transforms into itself.

Since the charge of a brane is given by an appropriate integral of the field strength of the gauge field to which the brane couples, the charges of the fundamental and D strings must also rotate under SL(2, $\mathbf{R}$) similarly to equation (13.5.8). However, the charges are subject to the quantisation condition of equation (13.8.136), which relates the electric charges of the 1-branes to the magnetic charges of the respective 5-branes. As a result the fundamental and D strings must have charges which are integer multiples of two fixed quantities. We will denote this set of integers by a column vector and refer to it as the charge number vector. The integer nature of these quantities will not be preserved by a general SL(2, $\mathbf{R}$) rotation. The situation here is very similar to that of the previous section for the electromagnetic duality symmetry and the quantisation condition of equation (17.2.10). As in that case we find that the quantisation condition is only respected by the sub-group SL(2, $\mathbf{Z}$) meaning 2×2 matrices with integer entries and determinant 1. This is easily verified by considering SL(2, $\mathbf{R}$) transformations of the charge vectors $\begin{pmatrix} 0 \\ 1 \end{pmatrix}$ and $\begin{pmatrix} 1 \\ 0 \end{pmatrix}$ which correspond to the fundamental string and D string. Thus if one includes solitonic solutions one can at best have an SL(2, $\mathbf{Z}$) symmetry and the same applies to string theory as this includes the D-brane. This symmetry includes the transformation of equation (17.3.1) and so takes the weak coupling regime theory to the strong coupling regime of the IIB string.

We found in section 13.5.3 that the field strengths $\hat{\mathfrak{S}}_{1\mu\nu\rho}$ and $\hat{\mathfrak{S}}_{2\mu\nu\rho}$ were in the $R \otimes R$ and $NS \otimes NS$ sectors of the superstring, respectively, and so an elementary charge of the fundamental string has charge number vector $\begin{pmatrix} 0 \\ 1 \end{pmatrix}$. Acting with an SL(2, **Z**) transformation we find, using equation (13.5.32), the charge number vector $\begin{pmatrix} b \\ a \end{pmatrix}$, where b and a are integers. Thus we find a state that is a combination of a fundamental strings and b D strings. We note that these integers are coprime, that is, have no common integer factor, since if they did $ad - bc$ would contain a common factor and so could not equal 1 as required by the definition of SL(2, **Z**). Hence, an SL(2, **Z**) symmetry implies that the theory must possess an infinite number of branes which have coprime charge number vectors. In particular, the SL(2, **Z**) transformation with $d = 0 = a$ and $b = 1 = -c$ rotates the charge vector of the fundamental string into the charge vector of the D1-brane.

The fundamental string and D string solutions given in section 13.8.3 preserved half of the 32 supersymmetries of the IIB supergravity theory. As this is a shortened, or BPS, multiplet some of the central charges in the IIB supersymmetry algebra of equation (13.5.2) must be active. Just as in the case of the $N = 4$ Yang–Mills theory this allows one to derive a relation between the charges of the solutions and their tensions (mass per unit length). This relation was already apparent in the construction of the dynamics of the fundamental string, and the D string in that the coefficient of the second term in the brane actions (often called the Wess–Zumino term) of chapter 14, which contains the coupling to the gauge field, is the charge, while the coefficient in front of the first term, which is the world volume swept out, is the tension; However κ symmetry ensures that these two coefficients are the same up to some innocuous factors. As the action of SL(2, **Z**) preserves supersymmetry the same conclusion applies to all the strings with coprime charge vector numbers. It also follows that the resulting relation between the charges and the tensions must be SL(2, **Z**) invariant. These have become known as (p, q) strings and their corresponding solutions of IIB supergravity are known. We note that if the charge numbers are greater than 1 they can correspond to parallel strings which carry a non-Abelian symmetry. Further evidence for the existence of (p, q) strings can be found in [16.8].

We note that if we believe in an SL(2, **Z**) symmetry, then we expect there to exist another description of the IIB theory in which the D string is fundamental and the fundamental string arises as a solitonic excitation. As such, the fundamental and D strings are really on an equal footing, or put another way, if one believes in fundamental strings and SL(2, **Z**) symmetry one must also believe in D strings and indeed in all branes. That the fundamental strings and D branes are treated in very different ways should really be seen as an artefact of our inadequate techniques for describing quantum field theories. Hopefully there exists a formulation of the theory that really does have all particles on an equal footing. We note that like the case of the $N = 4$ Yang–Mills theory this is a self-duality in that it maps the theory to itself. However, there are dualities in string theory that relate theories that arise in different contexts to each other. An appropriate example is the duality between the two-dimensional sine-Gordon model and the Thirring model in which the soliton of the former is equivalent to the quantum excitation of the field used to formulate the latter as found by Coleman and Skyrme in 1975.

A very similar story applies to the 5-branes, which also belong to a doublet and are dual to the strings.

The self-dual 4-form gauge field is in the $R \otimes R$ sector and it couples to a D3-brane. Since the 4-form is invariant under SL(2, **Z**) transformations we can expect that the brane action of this D3-brane, which is a non-linear extension of the $N = 4$ Yang–Mills theory in four dimensions, has an $SL(2, \mathbf{Z})$ symmetry.

In fact, the above S duality of the IIB theory was not the first time that it was suggested that string theory could possess a duality symmetry that transformed weak coupling regimes to strong ones. In particular, it was known that the low energy effective action for the heterotic string theory compactified on T^6 possessed an $SL(2, \mathbf{R})$ symmetry and the authors of [17.22, 17.23] suggested that the $SL(2, \mathbf{Z})$ sub-group is a symmetry of the heterotic string theory compactified on T^6. By considering the action of this symmetry on BPS states, in [17.22–17.25] evidence for this conjecture was given.

All supergravity theories that possess scalar fields are invariant under a symmetry under which the scalar fields belong to its non-linear realisation (see section 13.2). This phenomenon was first found for the case of $N = 4$ supergravity where the symmetry was $SL(2, \mathbf{R})$ [13.34], but the discovery of an E_7 symmetry of the four-dimensional supergravity theory was very surprising to those working in the field [13.35]. In fact, dimensionally reducing the eleven-dimensional supergravity theory on an n-dimensional torus leads to the maximal supergravity theory in $11 - n$ dimensions and this possesses an E_n symmetry. One can also dimensionally reduce IIB supergravity on an $n - 1$ torus and one finds the same symmetry; although the dimension of the torus is 1 less, one has in addition the $SL(2, \mathbf{R})$ symmetry [10.14] of the IIB theory. However, these maximal supergravity theories possess solitonic solutions whose charges must obey the quantisation condition and as a consequence only some discrete version of this symmetry can survive. As we have stressed, the maximal supergravity theories are the complete low energy effective actions for the type II string theories and it has been conjectured [17.26] that discrete versions of the symmetries of table 13.6.1 found in the supergravity theories were symmetries of string theory; they were renamed U dualities and are often denoted by $E_n(\mathbf{Z})$.

Dimensionally reducing any string theory on an $n - 1$ torus we expect to find a perturbative $O(n - 1, n - 1, \mathbf{Z})$ T duality. Dimensionally reducing any diffeomorphism invariant theory on an n torus leads to an $SL(n, \mathbf{Z})$ symmetry. These are the large diffeomorphisms that preserve the n torus. One way to see this is to think of the torus as a R^n subject to a set of equivalence relations which in turn define a lattice. The choice of basis vectors of the lattice is not unique, indeed given one choice of basis elements any other choice related by an $SL(n, \mathbf{Z})$ transformation is also an allowed basis. Thus dimensionally reducing on a n torus from the eleven-dimensional viewpoint, which includes the IIA theory, we expect to find as symmetries an $O(n - 1, n - 1, \mathbf{Z})$, T duality and the $SL(n, \mathbf{Z})$ remnant of diffeomorphism symmetries. While starting from the IIB viewpoint in ten dimensions we have the original non-perturbative $SL(2, \mathbf{Z})$ symmetry of the IIB theory and the same $O(n - 1, n - 1, \mathbf{Z})$ T duality and the $SL(n - 1, \mathbf{Z})$ diffeomorphisms which preserve the torus.

These are all part of the $E_n(\mathbf{Z})$ symmetry from which it is clear that the $E_n(\mathbf{Z})$ symmetry has perturbative and non-perturbative parts. In fact, one can think of the $E_n(\mathbf{Z})$ symmetry as the group generated by the $SL(2, \mathbf{Z})$ symmetry of the IIB theory and the $O(n - 1, n - 1, \mathbf{Z})$ T duality symmetry of the IIA theory. However, $E_n(\mathbf{Z})$ is also generated by the $O(n - 1, n - 1, \mathbf{Z})$ T duality symmetry and the $SL(n, \mathbf{Z})$ remnant of diffeomorphism symmetry from eleven dimensions. The way these symmetries all fit together is made clear in the E theory approach given in section 17.5.

17.4   M theory

As we will explain in chapter 18 string theory was found by proposing an S matrix for four scalar particles that satisfied a previously formulated wish list of properties. In a remarkable development the early pioneers were able to find string theory hidden in this scattering amplitude. Thus string theory was not proposed as a result of some general principles, but rather emerged and it was only after some time that it was clear that it was a theory that applied to the fundamental laws. Initially, perturbative string amplitudes were computed using the method of sewing, as explained in section 18.1, which was an extension of the methods by which string theory was first discovered. As time went by a number of other approaches to string theory were found. The first was the sum over world surfaces approach, described in section 18.2. This gave a very efficient method of computing perturbative string scattering amplitudes but only for the massless particles. Quite a bit later the string field theory approach was developed; it was discussed in chapter 12. Although even the simplest perturbative string scattering amplitude was very painful to compute using this method, it did allow the computation of some non-perturbative effects in more recent times leading to some interesting results. Even though closed string field theory did contain gravity, it involved operators like the Virasoro generators that contained contractions between space-time indices with a Minkowski metric; as a result closed string field theory is not formulated in a way that is background independent. Thus it has been clear for many years that there is no underlying theory that incorporates all aspects of strings and, in particular, their non-perturbative effects.

However, there was one area of certainty, that is, string theory at low energy. As explained in chapter 13 a supergravity theory can be constructed from its underlying supersymmetry algebra. However, any supersymmetry algebra with more than 32 supercharges will lead to a supergravity theory with spins higher than 2 and so will not, by definition, be a supergravity theory. There are only two supersymmetry algebras with 32 supercharges in ten dimensions and so only two maximal supergravity theories in ten dimensions, the IIA and IIB supergravity theories whose equations of motion were given in sections 13.4 and 13.5, respectively. These two supergravity theories are essentially unique as a result of the high degree of supersymmetry they possess. The IIA and IIB superstring theories are invariant under a space-time supersymmetry with 32 supercharges which are just the IIA and IIB supersymmetry algebras, respectively. Their low energy effective actions, which contain gravity and other fields, have terms with only two space-time derivatives. It follows that the two low energy effective actions of the IIA and IIB superstring theories are the IIA and IIB supergravity theories, respectively. However, as we have mentioned these supergravity theories are essentially unique and so they must be the complete low energy effective actions of the two superstrings; 'complete' means they contain all perturbative and non-perturbative effects. Thus at low energy we know string theory completely. As such, these theories have played, and will continue to play, a major role in the development of string theory as we will now see.

In chapter 13 we discussed the way the IIA, IIB, nine and eleven-dimensional supergravity theories are related. This is summarised in figure 17.4.1. Let us begin with the relation between the IIA and IIB supergravity theories. As we explained in chapter 13 there is only one maximal supergravity theory in nine dimensions and hence the dimensional reduction of the IIA and IIB supergravity theories on a circle leads to the same supergravity theory,

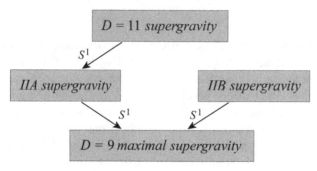

Figure 17.4.1   Relations between the IIA, IIB, $D = 9$ and $D = 11$ supergravity theories.

although the variables in which it is formulated as a result of the two different dimensional reductions are different. Thus at low energy the IIA and IIB string theories compactified on a circle are the same theory as their low energy effective actions agree. Assuming this holds at all energies we would conclude that the IIA and IIB string theories dimensionally reduced on circles of radius R_A and R_B, respectively, are the same theory. In fact this is the case the two theories are related by a T duality transformation such that $R_A \rightarrow \alpha'/R_B$. [15.2, 17.27, 17.28]. In the limit $R_A \rightarrow \infty$ the nine-dimensional theory decompactifies and one recovers the IIA string in ten dimensions. Similarly, in the limit $R_B \rightarrow \infty$, which is also the limit $R_A \rightarrow 0$, we recover the IIB string theory in ten dimensions. Just as for the reduction of the eleven-dimensional supergravity theory to find the IIA theory, see above equation (13.4.18), the radius of the circle is related to the expectation value of a scalar field in the theory in nine dimensions that arises from the diagonal component of the metric in ten dimensions and as a result each value of the radius corresponds to an expectation value for this scalar or, put in modern language, a point in the moduli space of the theory in nine dimensions. Thus the limit in which $R_A \rightarrow \infty$ is just a limit to a point in moduli space of the nine-dimensional theory and similarly for $R_B \rightarrow \infty$. Hence, two different limits in the moduli space of the nine-dimensional theory lead to the two different theories in ten dimensions. Starting with say the IIA theory in ten dimensions, one can compactify on a circle, carry out a T duality transformation and then take the appropriate limit to recover the IIB theory in ten dimensions. However, one cannot simply carry out the T duality transformation directly in the ten-dimensional theories and so this argument does not state that the IIA and IIB theories in ten dimensions are directly related. Clearly, the two string theories are also related if one dimensionally reduces on more than one circle.

We now turn to the relation between the IIA and eleven-dimensional supergravity theories. We recall that the former was first constructed from the latter by dimensional reduction on a circle. Indeed, we found in equation (13.4.18) that the IIA string coupling, $g_s = e^{\langle \sigma \rangle}$, and the radius of compactification, R_{11}, are related by $g_s \propto R_{11}^{\frac{3}{2}}$. Clearly, in the strong coupling limit of the IIA string, that is, as $g_s \rightarrow \infty$, the radius $R_{11} \rightarrow \infty$. However, in this limit the circle of compactification becomes flat and one expects to recover an eleven-dimensional theory whose low energy limit is eleven-dimensional supergravity. This realisation has lead to the conjecture [13.30, 13.31] that the strong coupling limit of IIA theory defines a consistent theory called M theory which possesses eleven-dimensional diffeomorphism

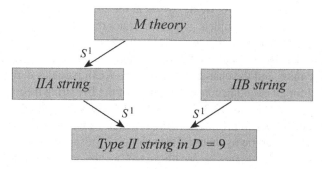

Figure 17.4.2   Relations between the IIA, IIB and $D = 9$ string theories and M theory.

invariance and has eleven-dimensional supergravity as its low energy effective action. The problem is that there is no known way to determine such a strong coupling limit and so very little is actually known about M theory.

The above relations between the IIA, IIB and nine-dimensional string theories and M theory are given in figure 17.4.2. They are essentially the same as the relation between the IIA, IIB and eleven-dimensional supergravity theories of figure 17.2.1 which has been known for many years, for example, they were discussed in reference [10.11]. This, of course, is not a coincidence as any relations between the superstrings would have to hold for their low energy effective actions and so for the IIA, IIB and eleven-dimensional supergravity theories. Turning this around the supergravity theories provide the best evidence for the relations between the IIA, IIB string theories and M theory.

The above conjecture raises a puzzle. When compactifying the eleven-dimensional supergravity on a circle one finds, in addition to the massless modes, an infinite set of massive modes called Kaluza–Klein modes. However, the IIA supergravity theory was obtained by throwing out all these Kaluza–Klein modes and retaining only the massless modes. The IIA string has an infinite number of massive modes, but none of these are the Kaluza–Klein modes. To see this we note that the perturbative states of the IIA string are not charged with respect to the gauge field A_μ since it belongs to the $R \otimes R$ sector of the IIA string and the massless fields in this sector enter only through their field strengths. As explained at the end of section 13.4, the Kaluza–Klein modes have a mass $m^{KK} \equiv n/R_{11}$ and a charge $q^{KK} \equiv n/R$ with respect to the U(1) gauge field A_μ in the IIA theory that arises from the graviton in eleven dimensions. However, the low energy effective action of M theory is by definition the eleven-dimensional supergravity and, as one can always compactify on a circle, the Kaluza–Klein modes must also be present in M theory. The puzzle is that if we are to believe that the strong coupling limit of IIA string theory is M theory, then these Kaluza–Klein modes should be also be present in the IIA string.

The charges of the Kaluza–Klein particles are derived from the momentum in the eleventh direction and this occurs in the supersymmetry anti-commutators of the eleven-dimensional supersymmetry algebra in the usual way. The ten-dimensonal supersymmetry algebra is just the dimensional reduction of that in eleven dimensions and so now the eleventh component of the momentum appears as a central charge. For the D0-brane this central charge is just its electric charge and the mass of the D0-brane and the electric charge are related in just such a way that one finds more zero eigenvalues on the right-hand side of the anti-commutator of the two supersymmetry generators. As we discussed in the section 17.2, and at the end of chapter 13, this leads to massive multiplets with fewer states than those of generic mass. Thus, the Kaluza–Klein states belong to massive supermultiplets that are an example of the shortened, or BPS, multiplets we encountered before. As argued previously, in such a case the relation between the mass and the charge of the Kaluza–Klein particles derived from the above values is valid for all coupling constant regimes of the IIA string theory.

The resolution of the puzzle [13.30, 13.31] comes from the existence of D0-branes in the IIA theory. As we found in section 13.8.3 and in particular equation (13.8.112), these arise as solutions in the IIA supergravity theory. This solution corresponds to the D0-brane of the IIA theory, which like all D-branes, as we discussed in chapter 15, do couple to the gauge fields in the $R \otimes R$ sector of the IIA string. The D0-branes are non-perturbatitve in nature, but this identification is in accord with fact that the masses of the Kaluza–Klein states are given by n/R_{11}, which as we shortly see can be written as $n/\sqrt{\alpha'} g_s$ when expressed in terms of the string coupling constant. These states become very massive as the coupling constant becomes small, but in the limit of large coupling they become massless. The appearance of an infinite number of massless modes indicates the transition to a theory in which a dimension has become decompactified.

The dynamics of a single D0-brane was given in section 14.4. Such a single D0-brane has indeed the correct mass and charge to be identified with the lowest mass Kaluza–Klein state. It is thought that the higher mass Kaluza–Klein states can be identified with bound states of D0-branes at threshold, that is, states consisting of multiple D0-branes that have the same energy as the energy state of two well-separated D0-branes. The dynamics of several parallel D0-branes is described by a supersymmetric generalisation of Born–Infeld theory, which possesses a U(N) gauge group if N is the number of parallel D0-branes. At low energy, this theory is just the dimensional reduction of $D = 10$ Yang–Mills theory to one dimension. Just as one quantises the point particle to discover that it corresponds to the Klein–Gordon equation and quantises a 1-brane (that is, string) to discover its particle spectrum, one must quantise this supersymmetric generalisation of non-Abelian Born–Infeld theory to discover the states in the IIA string arising from the presence of the D0-branes. It is thought that this quantum mechanical system does indeed have bound states that have the correct mass and charge to be identified with the Kaluza–Klein states [17.29].

There are also very interesting connections between the type I theories and indeed type I and type II string theories and M theory; we will not discuss these in this book and we refer the reader to [17.30].

As explained in section 13.8, the IIA, IIB and eleven-dimensional supergravity theories possess solutions that correspond to branes; however, for the case of the IIA and IIB supergravity theories only one of these corresponds to the fundamental strings of these theories. The role of branes was clarified by the symmetries, discussed in the previous section, that supergravity theories possess and are thought to be symmetries of the corresponding IIA and IIB string theories and their dimensional reductions. As discussed above the SL(2, **Z**) symmetry of the IIB theory rotates the fundamental string into the D string and if we believe in this symmetry we should believe that these two objects play equivalent roles in the underlying theory. As such, we should believe that branes are just as important as strings. Thus the underlying theory should be a theory of strings and branes. Unfortunately, our ability to compute almost anything with branes, as opposed to strings, is very limited, the exception being the use of the open strings associated with D-branes. This makes it even clearer that the underlying theory is quite different to the theory of strings as proposed for most of its history and explained in the opening chapters of this book. Clearly the principles of the underlying theory must incorporate strings and branes on an equal footing.

Above we have studied the relations between the ten-dimensional maximally super-symmetric string theories and M theory derived largely from the corresponding relations between the supergravity theories. As the branes are solutions of the supergravity theories it is inevitable that they will fit in with the connections we have found. Let us focus on the connections between the branes of M theory and those of the IIA string theory. M theory possesses the 2-brane, or membrane, and a 5-brane, while the IIA theory has the fundamental string and p-branes for p even. The latter theory was derived from the former theory by compactifying on a circle. If none of the world-volume directions of the 2-brane lie on this circle, that is, the 2-brane motion is outside the circle, then in ten dimensions we will find a 2-brane which must be the D2-brane. The gauge field of the D2-brane is accounted for by the x^{10} coordinate of the eleven-dimensional theory. We recall that in three dimensions a vector is equivalent to scalar as can easily be seen by introducing an independent field strength and a Lagrange multiplier which imposes that the field strength is a curl of the gauge field. However, if one of the world-volume directions lies in the circle direction, then in ten dimensions we find an extended object with one spatial direction, that is, a string. In effect the 2-brane has wrapped around the circle. It is clear that such an object will inherit the reparameterisation invariance and supersymmetry of the background and world volume of the eleven-dimensional 2-brane.

We now explicitly carry out this latter dimensional reduction for the 2-brane keeping only the bosonic terms for simplicity [17.31]. As in section 13.4 we take the compactification to be on the circle with coordinate x^{10} which we take to be parameterised by $0 \leq x^{10} \leq 2\pi R$. The 2-brane has in general a motion in eleven-dimensional space-time described by $x^{\hat{\mu}}(\xi^{\hat{\alpha}})$, $\hat{\mu} = 0, 1, \ldots, 10$, $\hat{\alpha} = 0, 1, 2$. We now take the 2-brane to wrap the circle n times and so take the reduction ansatz

$$
x^{\hat{\mu}}(\xi^{\hat{\alpha}}) = \begin{cases} x^{\mu}(\xi^{\alpha}, \sigma) = x^{\mu}(\xi^{\alpha}) & \mu = 0, 1, \ldots 9, \quad \alpha = 0, 1, \\ x^{10}(\xi^{\alpha}, \sigma) = n\sigma R. \end{cases} \tag{17.4.1}
$$

In this equation we have denoted the last coordinate on the 2-brane world volume by σ, that is, $\xi^2 = \sigma$ with the range $0 \leq \sigma \leq 2\pi$.

The 2-brane action was given in equation (14.3.1) and the part involving the world volume is of the form

$$
- T_2^M \int \Pi_{\hat{\alpha}} d\xi^{\hat{\alpha}} \sqrt{-\det g_{\hat{\alpha}\hat{\beta}}}, \tag{17.4.2}
$$

where

$$
g_{\hat{\alpha}\hat{\beta}} = \partial_{\hat{\alpha}} x^{\hat{\mu}} \partial_{\hat{\beta}} x^{\hat{\nu}} g_{\hat{\mu}\hat{\nu}} \tag{17.4.3}
$$

and we have added a superscript M to the symbol for the tension. Using the dimensional reduction ansatz of equation (13.4.4) we note that

$$
\partial_{\hat{\alpha}} x^{\hat{\mu}} e_{\hat{\mu}}{}^{m} = e^{-\frac{1}{12}\sigma} \partial_{\hat{\alpha}} x^{\mu} e_{\mu}{}^{m}, \quad \partial_{\hat{\alpha}} x^{\hat{\mu}} e_{\hat{\mu}}{}^{10} = e^{\frac{2}{3}\sigma} \partial_{\hat{\alpha}} x^{\mu} A_{\mu} + e^{\frac{2}{3}\sigma} \partial_{\hat{\alpha}} x^{10} \tag{17.4.4}
$$

and as a result

$$g_{\hat{\alpha}\hat{\beta}} = \begin{pmatrix} e^{-\frac{1}{6}\sigma}g_{\alpha\beta} + e^{\frac{4}{3}\sigma}A_\alpha A_\beta & nRe^{\frac{4}{3}\sigma}A_\alpha \\ nRe^{\frac{4}{3}\sigma}A_\beta & n^2R^2e^{\frac{4}{3}\sigma} \end{pmatrix},$$ (17.4.5)

where $A_\alpha = \partial_\alpha x^\mu A_\mu$. We can express the metric on the world surface in terms of a world-volume vielbein $\hat{e}$, which is given in matrix notation by

$$\hat{e} = \begin{pmatrix} e^{-\frac{1}{12}\sigma}e & e^{\frac{2}{3}\sigma}A \\ 0 & nRe^{\frac{2}{3}\sigma} \end{pmatrix},$$ (17.4.6)

where e is the vielbein for $g_{\alpha\beta}$ and A is the column vector with components A_α. It is then obvious that

$$\sqrt{-\det g_{\hat{\alpha}\hat{\beta}}} = \det \hat{e} = e^{\frac{1}{2}\sigma}R\det e = e^{\frac{1}{2}\sigma}R\sqrt{-\det g_{\alpha\beta}} = \sqrt{-\det g^s_{\hat{\alpha}\hat{\beta}}},$$ (17.4.7)

where in the equality we have changed the background metric from the metric in the Einstein frame to that in string frame. We recall that the σ model approach to the string is formulated in terms of the latter metric.

Substituting equation (17.4.7) into the 2-brane action of equation (17.4.2) we find, taking $n = 1$, that it becomes

$$2\pi R_{11}T_2^M \int d\xi^\alpha \sqrt{-\det g_{\alpha\beta}},$$ (17.4.8)

which is the action for the fundamental string in ten dimensions. We note that the result is independent of the gravi-photon, which is as it should be as the latter is in the $R \otimes R$ sector. Perhaps the reader can account for the obviously required factor of R_{11}, rather than R, in the above formula. We recognise that the tension of the string is related to the tension of the eleven-dimensional 2-brane by

$$2\pi R_{11}T_2^M = T_1 = \frac{1}{2\pi\alpha'}, \quad T_2^M = T_2^D.$$ (17.4.9)

In the latter equation we have included the relation between the tension of the 2-brane and the tension T_2^D of the D2-brane. This holds as the world-volume parts of the actions are unchanged by the dimensional reduction.

The reader will have no trouble carrying out the analogous dimensional reduction of the Wess–Zumino term of the 2-brane of equation (14.3.6) to find the Wess–Zumino term of the IIA string.

We now take the opportunity to give a simple derivation of the relation between the vielbein in the Einstein and the string frames in D dimensions. We must transform the term $\det eR$ in the Einstein frame to $e^{-2\sigma}\det eR$ in the string frame. Taking into account that the Riemann curvature involves two powers of the inverse vielbein we find that this is achieved by

$$e_\mu{}^m \rightarrow e^{-(2/D-2)\sigma}e_\mu^s{}^m.$$ (17.4.10)

The story for the 5-brane in eleven dimensions is rather similar: if it wraps on the circle we find the D4-brane of the IIA theory, while if it moves in the space-time outside the circle we find the NS5-brane of IIA theory. The relations between the tensions follow the same argument and are given by

$$2\pi RT_5^M = T_4^D, \quad T_5^M = T_5^{NS}.$$ (17.4.11)

Reference [14.11] gives the derivation of the dynamics of the D4-brane from the dynamics of the 5-brane, which was given in section 14.5.

The D6-brane arises from the Kaluza–Klein (KK) monopole in eleven dimensions whose solution was given in section 13.8.2. In fact, this solution has a compact, that is, circular, coordinate which is used to dimensionally reduce to IIA theory. As with all the branes discussed above, one can also carry out the dimensional reduction on the corresponding eleven-dimensional solutions to find the ten-dimensional solution.

We end this section by considering the relations between the constants and string tensions that occur in the IIA and IIB string theories and M theory. String theory has two constants: the Regge slope parameter α' and the string coupling constant g_s. We recall that the tension T_1 of the string is given by

$$T_1 = \frac{1}{2\pi\alpha'} = \frac{1}{2\pi l_s^2} = \frac{m_s^2}{2\pi} \tag{17.4.12}$$

as is evident from equation (2.1.1). In the last equation we have introduced a string scale l_s and mass m_s by $l_s = \sqrt{\alpha'} = 1/m_s$.

The maximal supergravity theories in D dimensions also have two constants, the gravitational constant κ_D, which we now take to occur in front of the Einstein term as $(1/2\kappa_D^2) \int d^D x \det e R$, and the expectation value of the dilaton field $\langle\sigma\rangle$. We note that the gravitational constant κ_D in D dimensions has the dimension in mass units of $(D-2)/2$(mass). Since the IIA supergravity theory is just the low energy effective action of the IIA string, it can be found by a string computation and this must also include the constants. The precise relationship was found in section 13.4 and is given by

$$\kappa_{10}^2 = (2\pi)^7 \alpha'^4 e^{2\langle\sigma\rangle} = (2\pi)^7 l_s^8 g_s^2, \quad g_s = e^{\langle\sigma\rangle}. \tag{17.4.13}$$

A similar statement holds for IIB theory.

On the other hand, M theory, whose low energy effective action is eleven-dimensional supergravity, has the gravitation constant κ_{11} and if we take the eleventh dimension to be a circle we also have the radius R_{11} of this circle. We note that R_{11} is the physical radius of the circle but the range of x^{10} is from 0 to $2\pi R$, where R is a parameter. Dimensionally reducing the Einstein term of the eleven-dimensional action on the circle we find that

$$2\pi R_{11}\kappa_{10}^2 = \kappa_{11}^2, \quad R_{11} = (g_s)^{\frac{2}{3}}R, \tag{17.4.14}$$

where the last equation is just equation (13.4.18).

By computing the force between two Dp-branes as an open string effect and a closed string transition and comparing the result one can find the tension of the Dp-branes to be [15.4]

$$T_p^D = \frac{\sqrt{\pi}}{\kappa_{10}}(2\pi l_s)^{(3-p)} = \frac{1}{(2\pi)^p}\frac{1}{l_s^{1+p}g_s}. \tag{17.4.15}$$

We recognise the g_s^{-1} typical of D-branes which also underlines their non-perturbative character. This computation assumes that the p-form gauge fields have a kinetic term with the normalisation $-1/(2(p+1)!2\kappa_{10}^2) \int d^{10}x \det e F_{\mu_1\dots\mu_{p+1}} F^{\mu_1\dots\mu_{p+1}}$. We note that the tension T_p^D of an electric brane and that of its dual magnetic brane T_{6-p}^D obey the equation

$$T_p^D T_{6-p}^D = \frac{2\pi}{2\kappa_{10}^2}. \tag{17.4.16}$$

We recognise this as the now correctly normalised quantisation condition [13.79, 13.80] with the minimal integer value on the right-hand side.

We found that the Kaluza–Klein modes have a mass equal to n/R_{11} but the lightest of these, that is, $n=1$, was to be identified with the D0-brane. The tension of a D0-brane is just the mass of this particle and from equation (17.4.13) we find this is given by $T_0^D = 1/l_s g_s$. As a result we can identify

$$R_{11} = g_s l_s. \tag{17.4.17}$$

We note that equation (17.4.12) then implies that $R = (g_s)^{\frac{1}{3}} l_s$. Using these formulae we can express the ten- and eleven-dimensional gravitational constants in terms of string parameters as

$$2\kappa_{10}^2 = (2\pi)^7 l_s^8 g_s^2, \quad 2\kappa_{11}^2 = (2\pi)^8 l_s^9 g_s^3. \tag{17.4.18}$$

Given the tension of the fundamental string of equation (17.4.12) we can compute the tension of the D2-brane using equation (17.4.11). We find that

$$T_2^D = \frac{1}{(2\pi)^2} \frac{1}{l_s^3 g_s}$$

in agreement with equation (17.4.15). Equation (17.4.11) also implies that the tension of the M2-brane is equal to the tension of the D2-brane and so

$$T_2^M = \frac{1}{(2\pi)^2} \frac{1}{l_s^3 g_s}.$$

In the same vein, the tension of the D4-brane is given by equation (17.4.15) to be

$$T_4^D = \frac{1}{(2\pi)^4} \frac{1}{l_s^5 g_s}$$

and from equation (17.4.11) we find the tension of the M5-brane. We summarise the results [17.32]

$$T_2^M = \frac{1}{(2\pi)^2} \frac{1}{l_s^3 g_s}, \quad T_5^M = \frac{1}{(2\pi)^5} \frac{1}{l_s^6 g_s^2}. \tag{17.4.19}$$

We can verify that these two tensions obey the equation

$$2\kappa_{11}^2 T_2^M T_5^M = 2\pi. \tag{17.4.20}$$

We recognise this as the quantisation condition between the electric and magnetic branes of M theory with the integer on the right-hand side taken to be 1. We note that they possess the minimal charges allowed by the quantisation condition.

17.5   E theory

As theoretical physics has developed it has become much more unified and this has been achieved by incorporating more and more symmetry. As we explained at the beginning of this chapter the unification of electric and magnetic forces led to Maxwell's equations which are invariant under Lorentz, conformal and gauge transformations as well as, in the absence of sources, electromagnetic duality transformations. The unification of the electromagnetic and weak forces was also achieved by the incorporation of an SU(2) $\otimes$ U(1) gauge symmetry which was spontaneously broken. Originally the weak interactions were formulated using a four-Fermi term, but this was not renormalisable. Then intermediate massive vector bosons were introduced with simple mass terms in the Lagrangian, but this violated unitarity and was also not renormalisable. A consistent theory was only found with the introduction of the Higgs boson and the spontaneously broken SU(2) $\otimes$ U(1) gauge symmetry.

String theory contains gauge bosons and gravity and is a consistent theory, meaning it possesses a divergence-free perturbative expansion. It must possess the corresponding symmetries, that is, gauge and diffeomorphism symmetries. As such, string theory has the potential to provide a consistent unified theory of gravity and the three other forces. However, as we found out in the previous sections, the underlying theory should include branes.

Gauge covariant string theory, discussed in chapters 12 and 18, is a theory with an infinite number of local symmetries with a gauge algebra in which the 3-string scattering vertex acts as a structure constant. However, as explained in the last section this theory suffers from a number of difficulties and cannot be regarded as a fundamental theory.

It would seem almost inevitable that an underlying theory of strings and branes would possess a very large symmetry algebra which unifies the symmetries of gravity and gauge theories in a single algebra. In 2001, the author proposed [16.10] that a particular Kac–Moody symmetry, called E_{11}, was a symmetry of an underlying theory of strings and branes. Unlike most of the material in this book, this is a subject which is still being developed and its precise place in the grand scheme is not completely clear. Nonetheless, it has had a number of successes and while we will not give an extensive review of this subject we will indicate the general thrust of this work. A subsequent proposal based on the Kac–Moody algebra E_{10}, which encodes space-time in a different way, was given in reference [16.14].

17.5.1 The eleven-dimensional theory

Kac–Moody algebras were explained in section 16.2. Section 16.7 is devoted to the E_{11} algebra and the decomposition into representations of the sub-algebra SL(11) was found at low levels in section 16.7.1. We recall that the Dynkin diagram of E_{11} is given by

Diagram 17.5.1   The E_{11} Dynkin diagram from eleven-dimensional viewpoint.

We wish to consider the non-linear realisation of E_{11} taking the local sub-algebra to be the Cartan involution invariant sub-algebra, denoted $I(E_{11})$ [16.10]. Constructions of this type are explained in section 13.2. We consider a group element $g \in E_{11}$ and construct a theory that is invariant under $g \to g_0 g$, where $g_0 \in E_{11}$ is a rigid transformation, that is, the parameters are constant, and also $g \to gh$, where $h \in I(E_{11})$ and is local meaning that, like g, it depends on space-time. We recall that $I(E_{11})$ was computed at low levels in section 16.7.3 and that at the lowest level it was found to be just SO(1,10). How the algebra E_{11} first arose in the context of supergravity, and why this led to the contents of this section is explained at the end of the section.

For a semi-simple finite dimensional Lie algebra one can write any group element as the product of a group element in the Borel sub-algebra, that is, the part generated by the Cartan sub-algebra and the positive root generators, multiplied by an element of the Cartan involution invariant subgroup. This is called the Igasawa decomposition. For example, for the case of SL(2, **R**) we have the Chevalley generators E, F and H; the Cartan involution invariant sub-algebra is generated by $E - F$, the Borel sub-algebra by H and E and we may write the general group element as $e^{a(E-F)} e^{cH} e^{bE}$. The analogous statement for E_{11} is identical and we will write the group elements in this way. In this case one can use the local sub-algebra to set the part of the group element that is in $I(E_{11})$ to be the identity element, leaving just the part generated by the Borel sub-algebra. As such, there is a one-to-one correspondence between the generators of the Borel sub-algebra and the fields of the theory. This holds regardless of which decomposition of E_{11} we consider and so it will be true for all the ten-dimensional and low dimensional theories we will construct later as non-linear realisations of E_{11}.

In table 17.5.1 we list the generators of the Borel sub-algebra decomposed into representations of SL(11). The first column gives the level, the second the Dynkin indices, denoted p_j in chapter 16, of the SL(11) representations, the third column gives integers, denoted m_k, which express the E_{11} root α corresponding to the highest state in the SL(11) representation in terms of simple roots α_k, that is, $\alpha = \sum_k m_k \alpha_k$, the fourth column is the length of the root squared, the fifth column is the outer multiplicity, which is the number of times that representations occur in the E_{11} algebra and the sixth column is the corresponding field in the non-linear realisation. Rather than study all these points now the reader may find it easier to consider the entries in the table as we develop the theory.

The analysis of section 16.7.1 gave the SL(11) representations of the generators and the fields are in the contragredient representation. We note that the last entry in column 4 is equal to the level as it must be as it is the number of times the root α_{11} occurs. The first three levels [16.10] were found analytically in section 16.7.1, see equation (16.7.5), and the same applies to the fourth level [16.15], but the higher levels were found using a computer programme [17.33]. If the outer multiplicity is 0, then the Serre relations forbid the occurrence of this representation and so it does not occur in E_{11}.

To construct the non-linear realisation we begin with a general group element of E_{11}, and having used the local sub-algebra to remove the part that is in $I(E_{11})$, we find, by examining equation (16.7.5) or table 17.5.1, that it takes the form

$$g_E = e^{h_a{}^b K^a{}_b} e^{\frac{1}{3!} A_{a_1 a_2 a_3} R^{a_1 a_2 a_3}} e^{\frac{1}{6!} A_{a_1 \cdots a_6} R^{a_1 \cdots a_6}} \cdots . \tag{17.5.1}$$

We take $h_a{}^b$, $A_{a_1 a_2 a_3}$ and $A_{a_1 \cdots a_6} \cdots$ to depend on space-time x^μ. By space-time we mean an eleven-dimensional manifold. Simply letting the fields depend on space-time is what

Table 17.5.1 The first few levels of E_8^{+++} decomposed with respect to its regular A_{10} sub-algebra

l	A_{10} Dynkin index	E_{11} root α	α^2	μ	Field
0	[1, 0, 0, 0, 0, 0, 0, 0, 0, 1]	(1, 1, 1, 1, 1, 1, 1, 1, 1, 1, 0)	2	1	$h_a{}^b$
1	[0, 0, 0, 0, 0, 0, 0, 1, 0, 0]	(0, 0, 0, 0, 0, 0, 0, 0, 0, 0, 1)	2	1	$A_{(3)}$
2	[0, 0, 0, 0, 1, 0, 0, 0, 0, 0]	(0, 0, 0, 0, 0, 1, 2, 3, 2, 1, 2)	2	1	$A_{(6)}$
3	[0, 0, 1, 0, 0, 0, 0, 0, 0, 1]	(0, 0, 0, 1, 2, 3, 4, 5, 3, 1, 3)	2	1	$h_{(8,1)}$
3	[0, 1, 0, 0, 0, 0, 0, 0, 0, 0]	(0, 0, 1, 2, 3, 4, 5, 6, 4, 2, 3)	0	0	$A_{(9)}$
4	[0, 1, 0, 0, 0, 0, 0, 1, 0, 0]	(0, 0, 1, 2, 3, 4, 5, 6, 4, 2, 4)	2	1	$A_{9,3}$
4	[1, 0, 0, 0, 0, 0, 0, 0, 0, 2]	(0, 1, 2, 3, 4, 5, 6, 7, 4, 1, 4)	2	1	$A_{10,1,1}$
4	[1, 0, 0, 0, 0, 0, 0, 0, 1, 0]	(0, 1, 2, 3, 4, 5, 6, 7, 4, 2, 4)	0	0	$A_{10,2}$
4	[0, 0, 0, 0, 0, 0, 0, 0, 0, 1]	(1, 2, 3, 4, 5, 6, 7, 8, 5, 2, 4)	−2	1	A_1
5	[1, 0, 0, 0, 0, 0, 1, 0, 0, 1]	(0, 1, 2, 3, 4, 5, 6, 8, 5, 2, 5)	2	1	$A_{10,4,1}$
5	[0, 1, 0, 0, 1, 0, 0, 0, 0, 0]	(0, 0, 1, 2, 3, 5, 7, 9, 6, 3, 5)	2	1	$A_{9,6}$
5	[1, 0, 0, 0, 0, 1, 0, 0, 0, 0]	(0, 1, 2, 3, 4, 5, 7, 9, 6, 3, 5)	0	0	$A_{10,5}$
5	[0, 0, 0, 0, 0, 0, 0, 1, 0, 1]	(1, 2, 3, 4, 5, 6, 7, 8, 5, 2, 5)	0	1	$A_{3,1}$
5	[0, 0, 0, 0, 0, 0, 1, 0, 0, 0]	(1, 2, 3, 4, 5, 6, 7, 9, 6, 3, 5)	−2	1	A_4
6	[1, 0, 0, 0, 1, 0, 0, 0, 1, 0]	(0, 1, 2, 3, 4, 6, 8, 10, 6, 3, 6)	2	1	$A_{10,6,2}$
6	[0, 1, 1, 0, 0, 0, 0, 0, 0, 1]	(0, 0, 1, 3, 5, 7, 9, 11, 7, 3, 6)	2	1	$A_{9,8,1}$
6	[0, 0, 0, 0, 0, 0, 1, 1, 0, 0]	(1, 2, 3, 4, 5, 6, 7, 9, 6, 3, 6)	2	1	$A_{4,1}$
6	[0, 0, 0, 0, 0, 1, 0, 0, 0, 2]	(1, 2, 3, 4, 5, 6, 8, 10, 6, 2, 6)	2	1	$A_{5,1,1}$
6	[1, 0, 0, 1, 0, 0, 0, 0, 0, 1]	(0, 1, 2, 3, 5, 7, 9, 11, 7, 3, 6)	0	1	$A_{10,7,1}$
6	[0, 0, 0, 0, 0, 1, 0, 0, 1, 0]	(1, 2, 3, 4, 5, 6, 8, 10, 6, 3, 6)	0	0	$A_{5,2}$
6	[0, 0, 0, 0, 1, 0, 0, 0, 0, 1]	(1, 2, 3, 4, 5, 7, 9, 11, 7, 3, 6)	−2	2	$A_{6,1}$
6	[0, 2, 0, 0, 0, 0, 0, 0, 0, 0]	(0, 0, 2, 4, 6, 8, 10, 12, 8, 4, 6)	0	0	$A_{9,9}$
6	[1, 0, 1, 0, 0, 0, 0, 0, 0, 0]	(0, 1, 2, 4, 6, 8, 10, 12, 8, 4, 6)	−2	1	$A_{10,8}$
6	[0, 0, 0, 1, 0, 0, 0, 0, 0, 0]	(1, 2, 3, 4, 6, 8, 10, 12, 8, 4, 6)	−4	1	A_7
7	[1, 0, 0, 1, 0, 0, 1, 0, 0, 0]	(0, 1, 2, 3, 5, 7, 9, 12, 8, 4, 7)	2	1	$A_{10,7,4}$
7	[0, 0, 0, 0, 1, 0, 0, 1, 0, 1]	(1, 2, 3, 4, 5, 7, 9, 11, 7, 3, 7)	2	1	$A_{6,3,1}$
7	[1, 0, 1, 0, 0, 0, 0, 0, 1, 1]	(0, 1, 2, 4, 6, 8, 10, 12, 7, 3, 7)	2	1	$A_{10,8,2,1}$
7	[0, 2, 0, 0, 0, 0, 0, 1, 0, 0]	(0, 0, 2, 4, 6, 8, 10, 12, 8, 4, 7)	2	1	$A_{9,9,3}$
7	[1, 0, 1, 0, 0, 0, 0, 1, 0, 0]	(0, 1, 2, 4, 6, 8, 10, 12, 8, 4, 7)	0	1	$A_{10,8,3}$
7	[0, 0, 0, 0, 1, 0, 1, 0, 0, 0]	(1, 2, 3, 4, 5, 7, 9, 12, 8, 4, 7)	0	1	$A_{6,4}$
7	[0, 0, 0, 1, 0, 0, 0, 0, 1, 1]	(1, 2, 3, 4, 6, 8, 10, 12, 7, 3, 7)	0	1	$A_{7,2,1}$
7	[0, 0, 0, 1, 0, 0, 0, 1, 0, 0]	(1, 2, 3, 4, 6, 8, 10, 12, 8, 4, 7)	−2	2	$A_{7,3}$
7	[1, 1, 0, 0, 0, 0, 0, 0, 0, 2]	(0, 1, 3, 5, 7, 9, 11, 13, 8, 3, 7)	0	1	$A_{10,9,1,1}$
7	[1, 1, 0, 0, 0, 0, 0, 0, 1, 0]	(0, 1, 3, 5, 7, 9, 11, 13, 8, 4, 7)	−2	2	$A_{10,9,2}$
7	[0, 0, 1, 0, 0, 0, 0, 0, 0, 2]	(1, 2, 3, 5, 7, 9, 11, 13, 8, 3, 7)	−2	3	$A_{8,1,1}$
7	[0, 0, 1, 0, 0, 0, 0, 0, 1, 0]	(1, 2, 3, 5, 7, 9, 11, 13, 8, 4, 7)	−4	3	$A_{8,2}$
7	[2, 0, 0, 0, 0, 0, 0, 0, 0, 1]	(0, 2, 4, 6, 8, 10, 12, 14, 9, 4, 7)	−4	1	$A_{10,10,1}$
7	[0, 1, 0, 0, 0, 0, 0, 0, 0, 1]	(1, 2, 4, 6, 8, 10, 12, 14, 9, 4, 7)	−6	4	$A_{9,1}$
7	[1, 0, 0, 0, 0, 0, 0, 0, 0, 0]	(1, 3, 5, 7, 9, 11, 13, 15, 10, 5, 7)	−8	1	A_{10}
8	[1, 0, 1, 0, 0, 1, 0, 0, 0, 1]	(0, 1, 2, 4, 6, 8, 11, 14, 9, 4, 8)	2	1	$A_{10,8,5,1}$
8	[0, 0, 0, 1, 0, 0, 1, 0, 1, 0]	(1, 2, 3, 4, 6, 8, 10, 13, 8, 4, 8)	2	1	$A_{7,4,2}$
8	[1, 1, 0, 0, 0, 0, 0, 0, 1, 1, 0]	(0, 1, 3, 5, 7, 9, 11, 13, 8, 4, 8)	2	1	$A_{10,9,3,2}$
8	[0, 0, 0, 0, 2, 0, 0, 0, 0, 1]	(1, 2, 3, 4, 5, 8, 11, 14, 9, 4, 8)	2	1	$A_{6,6,1}$
8	[0, 0, 1, 0, 0, 0, 0, 1, 0, 2]	(1, 2, 3, 5, 7, 9, 11, 13, 8, 3, 8)	2	1	$A_{8,3,1,1}$

(cont.)

Table 17.5.1 (*cont.*)

l	A_{10} Dynkin index	E_{11} root α	α^2	μ	Field
8	[0, 0, 0, 1, 0, 1, 0, 0, 0, 1]	(1, 2, 3, 4, 6, 8, 11, 14, 9, 4, 8)	0	1	$A_{7,5,1}$
8	[1, 1, 0, 0, 0, 0, 1, 0, 0, 1]	(0, 1, 3, 5, 7, 9, 11, 14, 9, 4, 8)	0	2	$A_{10,9,4,1}$
8	[1, 0, 0, 2, 0, 0, 0, 0, 0, 0]	(0, 1, 2, 3, 6, 9, 12, 15, 10, 5, 8)	2	1	$A_{10,7,7}$
8	[0, 2, 0, 0, 1, 0, 0, 0, 0, 0]	(0, 0, 2, 4, 6, 9, 12, 15, 10, 5, 8)	2	1	$A_{9,9,6}$
8	[0, 0, 1, 0, 0, 0, 0, 1, 1, 0]	(1, 2, 3, 5, 7, 9, 11, 13, 8, 4, 8)	0	1	$A_{8,3,2}$
8	[1, 0, 1, 0, 1, 0, 0, 0, 0, 0]	(0, 1, 2, 4, 6, 9, 12, 15, 10, 5, 8)	0	1	$A_{10,8,6}$
8	[0, 0, 1, 0, 0, 0, 1, 0, 0, 1]	(1, 2, 3, 5, 7, 9, 11, 14, 9, 4, 8)	-2	4	$A_{8,4,1}$
8	[2, 0, 0, 0, 0, 0, 0, 0, 1, 2]	(0, 2, 4, 6, 8, 10, 12, 14, 8, 3, 8)	2	1	$A_{10,10,3,1}$
8	[1, 1, 0, 0, 0, 1, 0, 0, 0, 0]	(0, 1, 3, 5, 7, 9, 12, 15, 10, 5, 8)	-2	2	$A_{10,9,5}$
8	[0, 0, 0, 1, 1, 0, 0, 0, 0, 0]	(1, 2, 3, 4, 6, 9, 12, 15, 10, 5, 8)	-2	2	$A_{7,6}$
8	[2, 0, 0, 0, 0, 0, 0, 0, 2, 0]	(0, 2, 4, 6, 8, 10, 12, 14, 8, 4, 8)	0	0	$A_{10,10,2,2}$
8	[0, 1, 0, 0, 0, 0, 0, 0, 1, 2]	(1, 2, 4, 6, 8, 10, 12, 14, 8, 3, 8)	0	1	$A_{9,2,1,1}$
8	[2, 0, 0, 0, 0, 0, 0, 1, 0, 1]	(0, 2, 4, 6, 8, 10, 12, 14, 9, 4, 8)	-2	2	$A_{10,10,3,1}$
8	[0, 1, 0, 0, 0, 0, 0, 0, 2, 0]	(1, 2, 4, 6, 8, 10, 12, 14, 8, 4, 8)	-2	2	$A_{9,2,2}$
8	[0, 0, 1, 0, 0, 1, 0, 0, 0, 0]	(1, 2, 3, 5, 7, 9, 12, 15, 10, 5, 8)	-4	3	$A_{8,5}$
8	[0, 1, 0, 0, 0, 0, 0, 1, 0, 1]	(1, 2, 4, 6, 8, 10, 12, 14, 9, 4, 8)	-4	6	$A_{9,3,1}$
8	[2, 0, 0, 0, 0, 0, 1, 0, 0, 0]	(0, 2, 4, 6, 8, 10, 12, 15, 10, 5, 8)	-4	2	$A_{10,10,4}$
8	[0, 1, 0, 0, 0, 0, 1, 0, 0, 0]	(1, 2, 4, 6, 8, 10, 12, 15, 10, 5, 8)	-6	7	$A_{9,4}$
8	[1, 0, 0, 0, 0, 0, 0, 0, 0, 3]	(1, 3, 5, 7, 9, 11, 13, 15, 9, 3, 8)	-2	2	$A_{10,1,1,1}$
8	[1, 0, 0, 0, 0, 0, 0, 0, 1, 1]	(1, 3, 5, 7, 9, 11, 13, 15, 9, 4, 8)	-6	7	$A_{10,2,1}$
8	[1, 0, 0, 0, 0, 0, 0, 1, 0, 0]	(1, 3, 5, 7, 9, 11, 13, 15, 10, 5, 8)	-8	6	$A_{10,3}$
8	[0, 0, 0, 0, 0, 0, 0, 0, 0, 2]	(2, 4, 6, 8, 10, 12, 14, 16, 10, 4, 8)	-8	3	$A_{1,1}$
8	[0, 0, 0, 0, 0, 0, 0, 0, 1, 0]	(2, 4, 6, 8, 10, 12, 14, 16, 10, 5, 8)	-10	5	A_2

one would do if one was dealing with an internal symmetry, but as we have seen E_{11} is a symmetry that also rotates space-time, for example, it contains the Lorentz group, and a more sophisticated approach is needed. Such an approach is given in section 17.5.6, however, at this stage we will also include in the symmetry algebra the space-time translation generators P_a and consider the group element $g = e^{x^a P_a} g_E$. We take P_a to have non-trivial commutation relations only with $K^a{}_b$; $[K^a{}_b, P_c] = -\delta^a_c P_b$. We will see later that this ad hoc step is the lowest level approximation in the more general procedure of section 17.5.6. How to carry out the appropriate non-linear realisation, and in particular the way space-time should be incorporated, has not been definitively settled; but we will find that carrying out the non-linear realisation, albeit in an incomplete way, will lead to the expected results as well as make predictions which have been verified in other ways. Thus although we do not possess the complete picture we will find that there is substantial evidence for the relevance of an underlying E_{11} symmetry.

At level 0 we have only the field $h_a{}^b$ which corresponds to generators $K^a{}_b$ of GL(11) and the local sub-algebra is just the Lorentz algebra SO(1,10). As explained at the end of section 13.2 this non-linear realisation, once we introduce the space-time translations, leads to Einstein gravity. As such, the field $h_a{}^b$ is the graviton, the precise identification being that the vielbein is given by $e_\mu{}^a = (e^h)_\mu{}^a$. The SL(11) part of the algebra corresponds to nodes 1–10 inclusive of the E_{11} Dynkin diagram and we will refer to it as the gravity line.

The field at the next level is the 3-form field $A_{a_1a_2a_3}$, which can be identified with the other bosonic dynamical field in the usual formulation of eleven-dimensional supergravity. At level 2 we find a 6-form field which is a good candidate to be the dual of the 3-form field in the sense that their field strengths are dual. The 1-field at level 3 can be thought of as the dual of the grativon, but at higher levels one has an infinite number of fields whose meaning is not immediately obvious. Thus we see that the E_{11} formulation encodes the fields expected in eleven-dimensional supergravity, that is, the graviton and the 3-form as well as their duals [16.10]. We will see that the existence of dual fields is a general characteristic of E_{11}.

To get a feel for how the non-linear realisation works without the complications of the gravity sector, which is just that given in section 13.2, let us consider just the 3- and 6-form gauge fields. That is, we take the group element of equation (17.5.1) with $h_a{}^b = 0$ and none of the fields indicated by the dots in equation (17.5.1). In this case flat and curved indices are the same as the vielbein $e_\mu{}^a = \delta_\mu^a$. Correspondingly, we will take the local sub-algebra to be just the Lorentz group. The corresponding E_{11} algebra just has the 3- and 6-form generators and is given by

$$[R^{a_1a_2a_3}, R^{a_4a_5a_6}] = 2R^{a_1a_2a_3a_4a_5a_6}, \tag{17.5.2}$$

all other commutators being zero. The Cartan form is given by

$$\mathcal{V} = g_E^{-1}\partial_a g_E = G_{a,a_1a_2a_3}R^{a_1a_2a_3} + G_{a,a_1\cdots a_6}R^{a_1\cdots a_6}, \tag{17.5.3}$$

where $G_{a,a_1a_2a_3} = \partial_a A_{a_1a_2a_3}$ and $G_{a,a_1\cdots a_6} = \partial_a A_{a_1\cdots a_6} - A_{[a_1a_2a_3}\partial_{|a|}A_{a_4a_5a_6]}$. We now only consider the totally anti-symmetrised part of the Cartan forms, namely the field strengths $F_{a_1a_2a_3a_4} \equiv 4G_{[a_1,a_2a_3a_4]}$ and $F_{a_1\cdots a_7} \equiv 7G_{[a_1,a_2\cdots a_7]}$. One way to motivate this restriction is to demand simultaneous invariance under the conformal group as one can do for gravity. One finds that the rigid E_{11} transformations that shift the 3- and 6-form fields become gauge transformations. Demanding that the equation be first order in space-time derivatives, the only possibility is given by

$$F_{a_1a_2a_3a_4} = \frac{1}{7!}\epsilon_{a_1a_2a_3a_4b_1\cdots b_7}F^{b_1\cdots b_7}. \tag{17.5.4}$$

Taking the divergence of this equation we find that the left-hand side does not vanish and we get the equation

$$\partial^{a_1}F_{a_1a_2a_3a_4} = -\frac{7}{2\times 8!}\epsilon_{a_2a_3a_4b_1\cdots b_8}F^{b_1\cdots b_4}F^{b_5\cdots b_8}. \tag{17.5.5}$$

This is indeed the equation of motion for the 3-form gauge field of eleven-dimensional supergravity including its Chern–Simmons term in the absence of gravity.

It is instructive to compute the rigid transformations of the non-linear realisation that shift the 3-form and 6-form generators. Again setting $h_a{}^b = 0$, for simplicity, and taking $g_0 = e^{c_{a_1a_2a_3}R^{a_1a_2a_3}/3!}$ and $g_0 = e^{c_{a_1\cdots a_6}R^{a_1\cdots a_6}/6!}$ in the transformation $g \to g_0 g$ we find the transformations

$$\delta A_{a_1a_2a_3} = c_{a_1a_2a_3}, \quad \delta A_{a_1\cdots a_6} = c_{a_1\cdots a_6} + 20c_{[a_1a_2a_3}A_{a_4a_5a_6]}. \tag{17.5.6}$$

It is straightforward to check that the Cartan forms and so the field strength are invariant under these transformations. We see they mix the 3-form and 6-form and so one can think of these as extensions of the electromagnetic duality transformations studied in section 17.2.

The part of the non-linear realisation that includes gravity, the 3- and 6-form fields, which introduces P_a and takes only the local Lorentz group as the local sub-group can be found in [17.34]. Although this paper was before the E_{11} conjecture of [16.10], and so E_{11} is not mentioned, the algebra used is that given later in the E_{11} paper [16.10] at low levels, that is, the one of equations (16.7.14)–(16.7.17). However, one does find, up to one constant which is not determined, the bosonic equations of motion of eleven-dimensional supergravity.

As we have mentioned the way this non-linear realisation was constructed suffers from a number of drawbacks. Much of the power of E_{11} has been lost as the local sub-group adopted is the same as that at the lowest level, only part of the Cartan forms, that is, the parts covariant under the conformal group, and the space-time generators P_a were introduced in an ad hoc way. Nonetheless, it did result in the correct fields and we recovered almost all the dynamics of eleven-dimensional supergravity. A more systematic approach will be discussed in section 17.5.6. In particular, as the reader will no doubt have guessed, given section 16.7.2, the space-time generator can be regarded as the first component of an E_{11} multiplet of generators.

The E_{11} conjecture grew out of the insight that, unlike most theories, eleven-dimensional supergravity could be formulated as a non-linear realisation. This involved an algebra, with the generators $K^a{}_b$, $R^{a_1 a_2 a_3}$, $R^{a_1 \cdots a_6}$ and P_a, which was also deduced from the construction [17.34]. However, unlike eleven-dimensional supergravity this algebra was not very aesthetic and it was proposed that one should consider instead a Kac–Moody algebra that contained this algebra, excluding the generator P_a. The smallest such Kac–Moody algebra was E_{11}. It was conjectured that a non-linear realisation of this algebra should be an extension of the eleven-dimensional supergravity theory [16.10]. In doing this it was realised that E_{11} included a field corresponding to the dual graviton and a formulation of linearised gravity involving the usual vielbein and this new dual gravity field was found [16.10].

17.5.2 The IIA and IIB theories

We now show how the IIA and IIB theories are contained in the E_{11} non-linear realisation. If we want to find a ten-dimensional theory that involves gravity we must find an SL(10) algebra that is part of the GL(D) algebra whose non-linear realisation describes the gravity sector. The Dynkin diagram, for SL(10), or A_9, consists of nine nodes connected by a single line. Assuming we must find this algebra by deleting nodes from the E_{11} Dynkin diagram, then we must find nine nodes in a row that we will call the gravity line for obvious reasons. Also assuming that we must begin with node 1 there are only two possibilities resulting from the bifurcation at node 8 of the E_{11} Dynkin diagram. We can take nodes 1–9 inclusive or we can take nodes 1–8 and node 11. These two possibilities correspond to decomposing E_{11} into representations of different A_9 sub-algebras and as a result the field content will be different. [16.10, 17.35]

As we will now show in detail that these decompositions correspond to the IIA and IIB theories, respectively. We draw the two Dynkin diagrams in diagrams 17.5.2 and 17.5.3, respectively, displaying the gravity line on the horizontal and all other nodes above the horizontal.

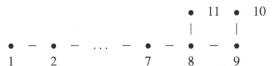

Diagram 17.5.2   The E_{11} Dynkin diagram from the IIA viewpoint.

Diagram 17.5.3   The E_{11} Dynkin diagram from the IIB viewpoint.

We observe in the Dynkin diagram of figure 17.5.3 that node 10 is not connected to the gravity line and so corresponds to an internal symmetry; as we will see it is the SL(2, **R**) symmetry of the IIB theory. From the decomposition analysis of section 16.6 it is easy to see that a node connected to the gravity line s dots from its end leads to a rank s field in the non-linear realisation. Thus for the 1 decomposition of diagram 17.5.2 we see that the non-linear realisation must contain rank 1 and rank 2 fields. While for the decomposition of diagram 17.5.2 we see that node 9 implies that it must contain a rank 2 field and in view of the attachment of this node to node 10 there must be an SL(2, **R**) doublet of these.

17.5.2.1   IIA

We can first delete node 10 and find a residual D_{10}, or SO(10,10) algebra, however, the representations are easier to interpret if we then delete node 11 and decompose with respect to the remaining SL(10) algebra. As such, the resulting SL(10) representations are labelled by two levels given by a pair of non-negative integers (l_1, l_2), where l_1 corresponds to the level associated with the node denoted 11 in the Dynkin diagram and l_2 to the node denoted 10. The first few levels are displayed in table 17.5.2 [16.10, 17.36].

Looking at the fields in table 17.5.2, we find at low levels precisely those corresponding to the bosonic part of IIA supergravity. We have 2-form $A_{a_1 a_2}$ in the NS–NS sector, the 1-form A_a and the 3-form $A_{a_1 a_2 a_3}$ in the R–R sector, along with their duals $A_{a_1 \cdots a_6}$, $A_{a_1 \cdots a_7}$ and $A_{a_1 \cdots a_5}$, respectively. At level $(0, 0)$ we find the GL(10) algebra, whose fields are those of gravity, and an additional scalar that is the dilaton, denoted A in this section. We note that the fields corresponding to Cartan sub-algebra of E_{11} are the nine fields arising from the Cartan sub-algebra generators of SL(10), the one field in the GL(1) factor in GL(10), all of which are related to the diagonal components of the metric, and the generator corresponding to the dilaton. At level $(3, 2)$ we find an 8-form $A_{a_1 \cdots a_8}$ which is the dual of the scalar and the dual gravity field $h_{a_1 \cdots a_7, b}$, while at level $(4,1)$ we find a 9-form. There then follow an infinite number of fields whose meaning is not immediately obvious.

We just noticed that there is a rank-9 field amongst the low level fields. This will have a field strength that has ten indices $F_{a_1 \cdots a_{10}}$ and its action will be of the form

$$\int d^{10}x \det e F_{a_1 \cdots a_{10}} F^{a_1 \cdots a_{10}}. \tag{17.5.7}$$

Table 17.5.2 The low level decomposition of E_{11} suitable for IIA

(l_1, l_2)	A_9 Dynkin index	E_{11} root α	α^2	μ	Field
(0, 0)	[1, 0, 0, 0, 0, 0, 0, 0, 1]	(1, 1, 1, 1, 1, 1, 1, 1, 0, 0)	2	1	$h_a{}^b$
(0, 0)	[0, 0, 0, 0, 0, 0, 0, 0, 0]	(0, 0, 0, 0, 0, 0, 0, 0, 0, 0, 0)	0	1	A
(1, 0)	[0, 0, 0, 0, 0, 0, 0, 1, 0]	(0, 0, 0, 0, 0, 0, 0, 0, 0, 0, 1)	2	1	A_2
(0, 1)	[0, 0, 0, 0, 0, 0, 0, 0, 1]	(0, 0, 0, 0, 0, 0, 0, 0, 0, 1, 0)	2	1	A_1
(1, 1)	[0, 0, 0, 0, 0, 0, 1, 0, 0]	(0, 0, 0, 0, 0, 0, 0, 1, 1, 1, 1)	2	1	A_3
(2, 1)	[0, 0, 0, 0, 1, 0, 0, 0, 0]	(0, 0, 0, 0, 0, 1, 2, 3, 2, 1, 2)	2	1	A_5
(3, 1)	[0, 0, 1, 0, 0, 0, 0, 0, 0]	(0, 0, 0, 1, 2, 3, 4, 5, 3, 1, 3)	2	1	A_7
(4, 1)	[1, 0, 0, 0, 0, 0, 0, 0, 0]	(0, 1, 2, 3, 4, 5, 6, 7, 4, 1, 4)	2	1	A_9
(2, 2)	[0, 0, 0, 1, 0, 0, 0, 0, 0]	(0, 0, 0, 0, 1, 2, 3, 4, 3, 2, 2)	2	1	A_6
(3, 2)	[0, 0, 1, 0, 0, 0, 0, 0, 1]	(0, 0, 0, 1, 2, 3, 4, 5, 3, 2, 3)	2	1	$h_{7,1}$
(3, 2)	[0, 1, 0, 0, 0, 0, 0, 0, 0]	(0, 0, 1, 2, 3, 4, 5, 6, 4, 2, 3)	0	1	A_8
(4, 2)	[0, 1, 0, 0, 0, 0, 0, 1, 0]	(0, 0, 1, 2, 3, 4, 5, 6, 4, 2, 4)	2	1	$A_{8,2}$
(4, 2)	[1, 0, 0, 0, 0, 0, 0, 0, 1]	(0, 1, 2, 3, 4, 5, 6, 7, 4, 2, 4)	0	1	$A_{9,1}$
(4, 2)	[0, 0, 0, 0, 0, 0, 0, 0, 0]	(1, 2, 3, 4, 5, 6, 7, 8, 5, 2, 4)	−2	2	A_{10}
(5, 2)	[1, 0, 0, 0, 0, 0, 1, 0, 0]	(0, 1, 2, 3, 4, 5, 6, 8, 5, 2, 5)	2	1	$A_{9,3}$
(5, 2)	[0, 0, 0, 0, 0, 0, 0, 1, 0]	(1, 2, 3, 4, 5, 6, 7, 8, 5, 2, 5)	0	1	A_2
(6, 2)	[0, 0, 0, 0, 0, 1, 0, 0, 0]	(1, 2, 3, 4, 5, 6, 8, 10, 6, 2, 6)	2	1	A_4
(3, 3)	[0, 1, 0, 0, 0, 0, 0, 0, 1]	(0, 0, 1, 2, 3, 4, 5, 6, 4, 3, 3)	2	1	$A_{8,1}$
(3, 3)	[1, 0, 0, 0, 0, 0, 0, 0, 0]	(0, 1, 2, 3, 4, 5, 6, 7, 5, 3, 3)	0	0	A_9
(4, 3)	[0, 1, 0, 0, 0, 0, 1, 0, 0]	(0, 0, 1, 2, 3, 4, 5, 7, 5, 3, 4)	2	1	$A_{8,3}$
(4, 3)	[1, 0, 0, 0, 0, 0, 0, 0, 2]	(0, 1, 2, 3, 4, 5, 6, 7, 4, 3, 4)	2	1	$A_{9,1,1}$
(4, 3)	[1, 0, 0, 0, 0, 0, 0, 1, 0]	(0, 1, 2, 3, 4, 5, 6, 7, 5, 3, 4)	0	1	$A_{9,2}$
(4, 3)	[0, 0, 0, 0, 0, 0, 0, 0, 1]	(1, 2, 3, 4, 5, 6, 7, 8, 5, 3, 4)	−2	2	A_1
(5, 3)	[0, 1, 0, 0, 1, 0, 0, 0, 0]	(0, 0, 1, 2, 3, 5, 7, 9, 6, 3, 5)	2	1	$A_{8,5}$
(5, 3)	[1, 0, 0, 0, 0, 0, 1, 0, 1]	(0, 1, 2, 3, 4, 5, 6, 8, 5, 3, 5)	2	1	$A_{9,3,2}$
(5, 3)	[1, 0, 0, 0, 0, 1, 0, 0, 0]	(0, 1, 2, 3, 4, 5, 7, 9, 6, 3, 5)	0	1	$A_{9,4}$
(5, 3)	[0, 0, 0, 0, 0, 0, 0, 1, 1]	(1, 2, 3, 4, 5, 6, 7, 8, 5, 3, 5)	0	1	$A_{2,1}$
(5, 3)	[0, 0, 0, 0, 0, 0, 1, 0, 0]	(1, 2, 3, 4, 5, 6, 7, 9, 6, 3, 5)	−2	3	A_3
(4, 4)	[1, 0, 0, 0, 0, 0, 1, 0, 0]	(0, 1, 2, 3, 4, 5, 6, 8, 6, 4, 4)	2	1	$A_{9,3}$
(4, 4)	[0, 0, 0, 0, 0, 0, 0, 0, 2]	(1, 2, 3, 4, 5, 6, 7, 8, 5, 4, 4)	2	1	$A_{1,1}$
(4, 4)	[0, 0, 0, 0, 0, 0, 0, 1, 0]	(1, 2, 3, 4, 5, 6, 7, 8, 6, 4, 4)	0	0	A_2
(5, 4)	[1, 0, 0, 0, 0, 1, 0, 0, 1]	(0, 1, 2, 3, 4, 5, 7, 9, 6, 4, 5)	2	1	$A_{9,4,1}$
(5, 4)	[0, 1, 0, 1, 0, 0, 0, 0, 0]	(0, 0, 1, 2, 4, 6, 8, 10, 7, 4, 5)	2	1	$A_{8,6}$
(5, 4)	[1, 0, 0, 0, 1, 0, 0, 0, 0]	(0, 1, 2, 3, 4, 6, 8, 10, 7, 4, 5)	0	1	$A_{9,5}$
(5, 4)	[0, 0, 0, 0, 0, 0, 1, 0, 1]	(1, 2, 3, 4, 5, 6, 7, 9, 6, 4, 5)	0	2	$A_{3,1}$
(5, 4)	[0, 0, 0, 0, 0, 1, 0, 0, 0]	(1, 2, 3, 4, 5, 6, 8, 10, 7, 4, 5)	−2	2	A_4
(5, 5)	[1, 0, 0, 1, 0, 0, 0, 0, 0]	(0, 1, 2, 3, 5, 7, 9, 11, 8, 5, 5)	2	1	$A_{9,6}$
(5, 5)	[0, 0, 0, 0, 0, 1, 0, 0, 1]	(1, 2, 3, 4, 5, 6, 8, 10, 7, 5, 5)	2	1	$A_{4,1}$
(5, 5)	[0, 0, 0, 0, 1, 0, 0, 0, 0]	(1, 2, 3, 4, 5, 7, 9, 11, 8, 5, 5)	0	0	A_5

The equation of motion sets $F_{a_1 \cdots a_{10}} = m \epsilon_{a_1 \cdots a_{10}}$, where m is a constant. Substituting this back into the action we find a cosmological constant. It has been known for many years that there is deformation of the IIA supergravity theory which is unique up to the value of the single massive deformation parameter. This theory was constructed in [13.71] and it is also invariant under the maximal number of 32 supersymmetries. Thus E_{11} automatically encodes the existence of this deformation and so the theory with a cosmological constant [17.36].

We will now construct a non-linear realisation which includes the above fields but leaves out, for simplicity, the 9-form and dual graviton, that is, using the generators [16.10, 17.37]

$$\tilde{K}^a{}_b, \ \tilde{R}, \ \tilde{R}^c, \ \tilde{R}^{c_1 c_2}, \ \tilde{R}^{c_1 c_2 c_3}, \ \tilde{R}^{c_1 \cdots c_5}, \ \tilde{R}^{c_1 \cdots c_6}, \ \tilde{R}^{c_1 \cdots c_7}, \ \tilde{R}^{c_1 \cdots c_8} \qquad (17.5.8)$$

together with their negative root analogues. We have placed a tilde on the generators to distinguish them from those in eleven dimensions. One can deduce their commutators directly from table 17.5.2 by writing down the most general terms consistent with preserving the levels, the Sl(10) character and Jacobi identities. The result is

$$[\tilde{K}^a{}_b, \tilde{K}^c{}_d] = \delta^c_b \tilde{K}^a{}_d - \delta^a_d \tilde{K}^c{}_b, \quad [\tilde{K}^a{}_b, P_c] = -\delta^a_c P_b,$$

$$[\tilde{K}^a{}_b, \tilde{R}^{c_1 \cdots c_p}] = \delta^{c_1}_b \tilde{R}^{ac_2 \cdots c_p} + \cdots, \qquad (17.5.9)$$

$$[\tilde{R}, \tilde{R}^{c_1 \cdots c_p}] = c_p \tilde{R}^{c_1 \cdots c_p}, \quad [\tilde{R}^{c_1 \cdots c_p}, \tilde{R}^{c_1 \cdots c_q}] = c_{p,q} \tilde{R}^{c_1 \cdots c_{p+q}},$$

where $+ \cdots$ means the appropriate anti-symmetrisations and the constants are given by

$$c_3 = -c_5 = -\tfrac{1}{4}, \quad c_2 = -c_6 = \tfrac{1}{2}, \quad c_1 = -c_7 = -\tfrac{3}{4}, \quad c_9 = \tfrac{5}{4},$$

$$c_{1,2} = -c_{2,3} = -c_{3,3} = c_{1,5} = c_{2,5} = 2, \quad c_{3,5} = 1, \qquad (17.5.10)$$

$$c_{2,6} = 2, \quad c_{1,7} = 3.$$

All not listed coefficients are equal to 0. The reader may wish to check the Jacobi identities. One must, in principle, also verify that the Serre relations hold. The first set of constants can be written as $c_{p+1} = \eta(p-3)/4$ where $\eta = 1$ or -1 if the generator is from the R–R sector or the NS–NS sector respectively.

We could also have derived the above algebra directly from the E_{11} algebra of equations (16.7.14)–(16.7.21). Treating the eleventh index, denoted by 11 as special and letting $a, b = 1, \ldots, 10$ we can define a set of generators in SL(10) representations:

$$\tilde{K}^a{}_b = K^a{}_b, \ \tilde{R} = \tfrac{1}{12}\left(-\sum_{a=1}^{10} K^a{}_a + 8K^{11}{}_{11}\right), \ \tilde{R}^a = 2K^a{}_{11}, \ \tilde{R}^{a_1 a_2} = R^{a_1 a_2 11},$$

$$\tilde{R}^{a_1 a_2 a_3} = R^{a_1 a_2 a_3} \ \tilde{R}^{a_1 \cdots a_5} = R^{a_1 \cdots a_5 11}, \ \tilde{R}^{a_1 \cdots a_6} = -R^{a_1 \cdots a_6}, \qquad (17.5.11)$$

$$\tilde{R}^{a_1 \cdots a_7} = \tfrac{1}{2} R^{a_1 \cdots a_7 11, 11}, \ \tilde{R}^{a_1 \cdots a_8} = \tfrac{3}{8} R^{a_1 \cdots a_8, 11}.$$

Computing their commutators we find the algebra of equation (17.5.9). In deriving this algebra, we have required that the generators $R^{a_1 \cdots a_8, b}$ are to be set to 0 and the relation $-8R^{11 a_1 \cdots a_7, a_8} = R^{a_1 \cdots a_7 a_8, 11}$. The latter follows from the fact that in eleven dimensions $R^{[a_1 \cdots a_8, b]} = 0$. Starting from the eleven-dimensional generators, one is free to take any definition of generators which belong to the SL(10) representation, but only the above choice obeys the algebra of equation (17.5.9).

The E_a and H_a Chevalley generators of E_{11} were given in terms of the eleven-dimensional generators in equations (16.7.12)–(16.7.13) and so using the above correspondence we can find them in terms of the IIA generators [16.10]:

$$E_a = \tilde{K}^a{}_{a+1}, \ a = 1, \ldots, 9, \ E_{10} = \tilde{R}^{10}, \ E_{11} = \tilde{R}^{910},$$

$$H_a = \tilde{K}^a{}_a - \tilde{K}^{a+1}{}_{a+1}, a = 1, \ldots, 9,$$

$$H_{10} = -\tfrac{1}{8}(\tilde{K}^1{}_1 + \cdots + \tilde{K}^9{}_9) + \tfrac{7}{8}\tilde{K}^{10}{}_{10} - \tfrac{3}{2}\tilde{R}, \qquad (17.5.12)$$

$$H_{11} = -\tfrac{1}{4}(\tilde{K}^1{}_1 + \cdots + \tilde{K}^8{}_8) + \tfrac{3}{4}(\tilde{K}^9{}_9 + \tilde{K}^{10}{}_{10}) + \tilde{R}.$$

Since the Chevalley generators of E_{11} outside GL(10) are contained in $\tilde{R}^{ab}$ and $\tilde{R}^a$, which are associated with nodes 11 and 10, respectively, an E_{11} generator will have $2l_1 + l_2$ SL(10). In table 17.5.2 blocks of ten indices, which do not transform under SL(10), are generally not shown.

The group element in the non-linear realisation of E_{11} up to the level considered is given by

$$g_E = \exp(h_a{}^b \tilde{K}^a{}_b)g_A \equiv g_h g_A,$$

where

$$g_A = e^{(1/8!)A_{a_1 \cdots a_8}\tilde{R}^{a_1 \cdots a_8}} e^{(1/7!)A_{a_1 \cdots a_7}\tilde{R}^{a_1 \cdots a_7}} e^{(1/6!)A_{a_1 \cdots a_6}\tilde{R}^{a_1 \cdots a_6}} e^{(1/5!)A_{a_1 \cdots a_5}\tilde{R}^{a_1 \cdots a_5}}$$

$$\times e^{(1/3!)A_{a_1 a_2 a_3}\tilde{R}^{a_1 a_2 a_3}} e^{(1/2!)A_{a_1 a_2}\tilde{R}^{a_1 a_2}} e^{A_{a_1}\tilde{R}^{a_1}} e^{AR}. \tag{17.5.13}$$

To construct the non-linear realisation we proceed as in the eleven-dimensional case. We take the fields $h_a{}^b, A, A_a, \ldots$ to depend on space-time and include the space-time generators $\tilde{P}_a$ in the symmetry algebra. We will take $\tilde{P}_a$ to have non-trivial relations with all the generators except those of SL(11) with which it has the commutator $[\tilde{K}^a{}_b, \tilde{P}_c] = -\delta^a_c \tilde{P}_a$. We also take the local sub-group to be just the Lorentz group. The non-linear realisation is then constructed as discussed at the end of section 13.2. As such, we take as our group element $g = g_x g_E$, where $g_x = e^{x^a \tilde{P}_a}$. A more systematic approach is discussed at the end of section 17.5.6.

The Cartan forms $\mathcal{V} = g^{-1}dg$ can be written as

$$\mathcal{V} \equiv dx^\mu \left(\hat{e}_\mu{}^a \tilde{P}_a + \Omega_{\mu a}{}^b \tilde{K}^a{}_b \right) + dx^\mu \left(\sum_{p=1,\ p\neq 4}^{9} \frac{1}{p!} e^{-c_p A} G_{\mu, a_1 \cdots a_p} \tilde{R}^{a_1 \cdots a_p} \right), \tag{17.5.14}$$

where

$$\Omega_{ab}{}^c \equiv (e^{-1})_a{}^\mu (e^{-1}\partial_\mu e)_b{}^c, \quad e_\mu{}^a = (e^h)_\mu{}^a. \tag{17.5.15}$$

The gravity part of the non-linear realisation can be constructed as in section 13.2. For simplicity, we will focus on the Cartan forms not belonging to GL(10); the full calculation can be found in [17.37] provided one sets the mass associated with the cosmological constant to 0. We will only take those combinations of the remaining Cartan forms that are totally anti-symmetric on all their indices. Like the case of gravity, this can be motivated by taking those Cartan forms which transform covariantly under the conformal group. This requirement then turns the rigid transformations of the non-linear realisation that shift the form fields into gauge transformations. Thus we define

$$\tilde{F}_{a_1 a_2 \cdots a_p} = p e^{-c_{p-1} A} e_{[a_1}{}^\mu G_{|\mu|, a_2 \cdots a_p]}. \tag{17.5.16}$$

Using the relations

$$e^{-A}de^A = dA - \tfrac{1}{2}[A, dA] + \tfrac{1}{6}[A, [A, dA]] - \tfrac{1}{24}[A, [A, [A, dA]]] + \cdots$$

$$= \frac{1 - e^{-A}}{A} \wedge dA, \tag{17.5.17}$$

$$e^{-A}Be^A = B - [A, B] + \tfrac{1}{2}[A, [A, B]] + \cdots = e^{-A} \wedge B,$$

where $1 \wedge B = B, A \wedge B = [A, B], A^2 \wedge B = [A, [A, B]]$ etc., We find that

$$\tilde{F}_a = D_a A, \quad \tilde{F}_{a_1 a_2} = 2e^{(3/4)A} D_{[a_1} A_{a_2]}, \quad \tilde{F}_{a_1 a_2 a_3} = 3e^{-(1/2)A} D_{[a_1} A_{a_2 a_3]},$$

$$\tilde{F}_{a_1 \cdots a_4} = 4e^{(1/4)A} (D_{[a_1} A_{a_2 \cdots a_4]} + 6A_{[a_1} D_{a_2} A_{a_3 a_4]}),$$

$$\tilde{F}_{a_1 \cdots a_6} = 6e^{-(1/4)A} \left(D_{[a_1} A_{a_2 \cdots a_6]} + 20 D_{[a_1} A_{a_2 a_3 a_4} A_{a_5 a_6]} \right),$$

$$\tilde{F}_{a_1 \cdots a_7} = 7e^{(1/2)A} \left(D_{[a_1} A_{a_2 \cdots a_7]} - 20 A_{[a_1 a_2 a_3} D_{a_4} A_{a_5 a_6 a_7]} \right.$$
$$\left. + 12 A_{[a_1} (D_{[a_2} A_{a_3 \cdots a_7]} + 20 A_{a_2 a_3} D_{[a_4} A_{a_5 a_6 a_7]}) \right), \tag{17.5.18}$$

$$\tilde{F}_{a_1 \cdots a_8} = 8e^{-(3/4)A} (D_{[a_1} A_{a_2 \cdots a_8]} - 42 A_{[a_1 a_2} (D_{a_3} A_{a_4 \cdots a_8]} + 10 D_{a_3} A_{a_4 a_5 a_6} A_{a_7 a_8]}),$$

$$\tilde{F}_{a_1 \cdots a_9} = 9 (D_{[a_1} A_{a_2 \cdots a_9]} - 24 A_{[a_1} D_{a_2} A_{a_3 \cdots a_9]} + 56 A_{[a_1 a_2} D_{a_3} A_{a_4 \cdots a_9]}$$
$$- 56 A_{[a_1 a_2 a_3} D_{a_4} A_{a_5 \cdots a_9]} + 1008 A_{[a_1} A_{a_2 a_3} D_{a_4} A_{a_5 \cdots a_9]}$$
$$+ \frac{8!}{4} A_{[a_1} A_{a_2 a_3} A_{a_4 a_5} D_{a_6} A_{a_7 a_8 a_9]} - 1120 A_{[a_1 a_2} A_{a_3 a_4 a_5} D_{a_6} A_{a_7 a_8 a_9]}),$$

where D_a is the covariant derivative, which is given by

$$D_a A_{a_1 \cdots a_p} = e_a{}^\mu (\partial_\mu A_{a_1 \ldots a_p} + (e^{-1} \partial_\mu e)_{a_1}{}^c A_{c a_2 \ldots a_p} + \cdots) \tag{17.5.19}$$

and $+ \cdots$ indicates the terms where $(e^{-1} \partial_\mu e)$ acts on the other indices of the gauge field.

We have taken the local sub-group to be only the Lorentz group under which the Cartan forms, and so the field strengths, transform as their indices suggest. If we demand that the equations of motion be only first order in derivatives, then the only possibility is given by

$$\tilde{F}^{a_1 \cdots a_p} = \frac{1}{(10-p)!} \epsilon^{a_1 \cdots a_{10}} \tilde{F}_{a_{p+1} \cdots a_{10}}, \qquad p = 1, 2, 3, 4. \tag{17.5.20}$$

Taking suitable space-time derivatives of these equations we find a set of second order equations which, up to field redefinition, are the same as those that follow from the action of equation (13.4.10). The construction of the gravity sector follows closely that given in section 13.2. To include the gauge fields in the Einstein equation is straightforward, but one cannot fix the coefficients of the terms that occur in the stress tensor on the right-hand side of this equation at the level of approximation used in the non-linear realisation.

17.5.2.2 IIB

Looking at the Dynkin diagram of E_{11} of diagram 17.5.3 we see that we can delete node 9 and decompose the adjoint representation in terms of $A_1 \otimes A_9$, that is, SL(2) $\otimes$ SL(10). However, we can also delete node 10 and decompose to representations of just SL(10) which are labelled by two levels: nodes 10 and 9 corresponding to the levels l_1 and l_2, respectively. The result can be found using the techniques given in chapter 16 and it is given at low levels in table 17.5.3 [17.36, 17.37]. Looking at the table one can readily recognise those that belong to the same SL(2) multiplet as they have the same highest root except for the tenth integer which changes as E_{10} and F_{10} act on the state. Thus it is relatively straightforward to reinstate the SL(2) factors and so in effect undelete node 10.

At low levels we recognise the field content of the IIB supergravity theory. At level $l_1 = 0 = l_2$ we have the graviton $h_a{}^b$ and a scalar, the dilaton σ. We see from the corresponding roots that the dilaton is part of the Cartan sub-algebra of E_{11}. This is also another scalar χ, the axion, whose corresponding generator is the Chevalley generator E_{10}. These two scalar fields belong to the SL(2, $\mathbf{R}$) coset associated with node 10. Further down the table we find a doublet of 2-forms $A^\alpha_{a_1 a_2}$, $\alpha = 1$, 2 and their 6-form duals $A^\alpha_{a_1 \cdots a_6}$, only one 4-form potential $A_{a_1 \cdots a_4}$, which indicates that its field strength will be self-dual, the dual graviton $h_{a_1 \cdots a_7, b}$ and a triplet of 8-forms $A^{\alpha\beta}_{a_1 \cdots a_8}$, where the latter is symmetric in its α and β indices. The 8-forms are the duals of the two scalars which belong to the SL(2, $\mathbf{R}$) non-linear realisation. For reasons we will explain later it turns out that the field strengths of the 8-forms obey a constraint which accounts for the mismatch in the degrees of freedom. We observe that there is no 9-form in the table, or indeed at any level, and so there is no massive deformation of the IIB theory.

Examining the table we find at level $l_2 = 5$ an SL(2, $\mathbf{R}$) quadruplet and doublet of 10-forms [17.37]. Ten-forms transform in the same way as scalars under SL(10) and so all their Dynkin indices vanish. We will explain below that generators can be thought of as carrying $2l_2$ SL(10) indices some of which are blocks of ten anti-symmetric indices and in this case this means one such block. The 10-forms have generators with the E_{11} roots $(1, 2, 3, 4, 5, 6, 7, 8, 5, i, 4)$ with $i = 1, 2, 3, 4$. However, for $i = 2, 3$ they have multiplicity $\mu = 2$, implying that there are two generators with these roots. The SL(2, $\mathbf{R}$) multiplet they belong to is apparent from studying the tenth component of the roots given, that is, the value i, which is the value of the Cartan sub-algebra element of SL(2, $\mathbf{R}$). Ten-forms couple to space-filling branes and they are the lead term in the Wess–Zumino part of the corresponding brane action. Thus although they do not lead to new propagating degrees of freedom in the background theory they play an important role in brane dynamics. These 10-forms were first predicted by E_{11}. Their existence was also found in a new formulation of the IIB supergravity theory. We recall from chapter 13 that one way the IIB theory was found was by closing the supersymmetry algebra. More recently it was found that if one added the dual 6- and 8-forms, then one could add a unique set of 10-forms which comprises a quadruplet and doublet [17.38]. Remarkably, the gauge transformation of these fields was only fixed by the closure at the non-linear level and this was in complete agreement, including precise coefficients, with that found from E_{11} [17.39]. A similar story applies to the 10-form in the IIA theory.

Examining table 17.5.3 we find, at low levels, the following generators

$$\hat{K}^a{}_b, \ \hat{R}_s, \ \hat{R}_s^{c_1 c_2}, \ \hat{R}_2^{c_1 \cdots c_4}, \ \hat{R}_s^{c_1 \cdots c_6}, \ \hat{R}_s^{c_1 \cdots c_8}, \qquad s = 1, 2. \tag{17.5.21}$$

These are the generators of the Borel sub-algebra to which we must add the negative root generators. The generators have labels 1 and 2 if their l_1 level is even and odd respectively, see table 17.5.3. When we have constructed the non-linear realisation, we find that they correspond to the fields in the NS–NS and R–R sectors if $s = 1$ and $s = 2$, respectively, for example, R_2 corresponds to the axion χ which is in the R–R sector and R_1 to the dilaton. The generators carry hats ($\hat{\ }$) to distinguish them from those in the IIA and eleven-dimensional theories. We have for simplicity set one of the 8-form generators to 0 and similarly the dual gravition generator. As for the IIA case, we can construct the corresponding commutators by writing down the most general possibilities consistent with preservation of the level, SL(10) character and Jacobi identities to find that [17.37]

$$[\hat{K}^a{}_b, \hat{K}^c{}_d] = \delta^c_b \hat{K}^a{}_d - \delta^a_d \hat{K}^c{}_b,$$

$$[\hat{K}^a{}_b, \hat{R}_s^{c_1 \cdots c_p}] = \delta^{c_1}_b \hat{R}_s^{ac_2 \cdots c_p} + \cdots, \qquad [\hat{R}_{s_1}^{c_1 \cdots c_p}, \hat{R}_{s_2}^{c_1 \cdots c_q}] = c_{p,q}^{s_1,s_2} \hat{R}_{s(s_1,s_2)}^{c_1 \cdots c_{p+q}}, \qquad (17.5.22)$$

$$[\hat{R}_1, \hat{R}_s^{c_1 \cdots c_p}] = d_p^s \hat{R}_s^{c_1 \cdots c_p}, \qquad [\hat{R}_2, \hat{R}_{s_1}^{c_1 \cdots c_p}] = \tilde{d}_p^s \hat{R}_{s(2,s_1)}^{c_1 \cdots c_p},$$

where $+ \cdots$ means the appropriate anti-symmetrisations. The generators $K^a{}_b$ satisfy the commutation relations of GL(10, **R**). The function $s = s(s_1, s_2)$ satisfies the properties $s(1, 1) = s(2, 2) = 1, s(1, 2) = s(2, 1) = 2$. In the last line of equations (17.5.22) we have split the commutators with the scalar generators into those for the dilaton (superscript 1) with coefficients d_p^s, and the axion (subscript 2) with coefficients $\tilde{d}_p^s$. One can see that the commutator with $\hat{R}_1$ is sector preserving, while the commutator with R_2 changes the sector. The Jacobi identity implies the following relations among the constants:

$$c_{q,r}^{s_2,s_3} c_{p,q+r}^{s_1,s(s_2,s_3)} = c_{p,q}^{s_1,s_2} c_{p+q,r}^{s(s_1,s_2),s_3} + c_{p,r}^{s_1,s_3} c_{q,p+r}^{s_2,s(s_1,s_3)}, \qquad (17.5.23)$$

where $c_{0,q}^{2,s_2} \equiv \tilde{d}_p^{s_2}$ and

$$(d_q^{s_3} + d_p^{s_1} - d_{p+q}^{s(s_1,s_3)}) c_{p,q}^{s_1,s_3} = 0. \qquad (17.5.24)$$

The constants can be chosen to be given by

$$d_2^1 = d_6^2 = -d_2^2 = -d_6^1 = \tfrac{1}{2}, \ d_0^2 = -d_8^2 = -1,$$

$$c_{2,2}^{1,2} = -c_{2,2}^{2,1} = -1, \ c_{2,4}^{2,2} = -c_{2,4}^{1,2} = 4, \ c_{2,6}^{1,2} = 1, \ c_{2,6}^{1,1} = -c_{2,6}^{2,2} = \tfrac{1}{2} \qquad (17.5.25)$$

$$\tilde{d}_2^1 = -\tilde{d}_6^2 = -\tilde{d}_8^2 = 1, \ \tilde{d}_2^2 = \tilde{d}_6^1 = \tilde{d}_8^1 = 0.$$

All not mentioned coefficients are 0.

We will now construct the non-linear realisation for the generators given in equation (17.5.21) [17.37]. As for the eleven-dimensional and IIA theories, we will introduce space-time by including the space-time translation generators into the algebra and take its only non-trivial commutator to be $[\hat{K}^a{}_b, \hat{P}_c] = -\delta^a_c \hat{P}_b$. We will also take the local sub-algebra to be just the Lorentz algebra.

The group element takes the form

$$g = \exp(x^\mu \hat{P}_\mu) \exp(h_a{}^b \hat{K}^a{}_b) g_A \equiv \exp(x^\mu \hat{P}_\mu) \exp(h_a{}^b \hat{K}^a{}_b) g_A, \qquad (17.5.26)$$

where

$$g_A = e^{(1/8!)A^2_{a_1 \cdots a_8} \hat{R}_2^{a_1 \cdots a_8}} e^{(1/8!)A^1_{a_1 \cdots a_8} \hat{R}_1^{a_1 \cdots a_8}} e^{(1/6!)(A^2_{a_1 \cdots a_6} \hat{R}_2^{a_1 \cdots a_6} + A^1_{a_1 \cdots a_6} \hat{R}_1^{a_1 \cdots a_6})}$$

$$\times e^{(1/4!)A^2_{a_1 \cdots a_4} \hat{R}_2^{a_1 \cdots a_4}} e^{(1/2!)(A^2_{a_1 a_2} \hat{R}_2^{a_1 a_2} + A^1_{a_1 a_2} \hat{R}_1^{a_1 a_2})} e^{\chi \hat{R}_2} e^{\sigma \hat{R}_1}. \qquad (17.5.27)$$

The Cartan form $\mathcal{V} = g^{-1} dg$ can be written as

$$\mathcal{V} \equiv dx^\mu (e_\mu{}^a \hat{P}_a + dx^\mu \Omega_{\mu a}{}^b \hat{K}^a{}_b) + dx^\mu \left(\sum_{p=1}^{8} \frac{1}{p!} e^{-d_{p-1}^s \sigma} G^s_{\mu, a_1 \cdots a_p} \hat{R}_s^{a_1 \cdots a_p} \right),$$

$$(17.5.28)$$

where $e_\mu{}^a = (e^h)_\mu{}^a$ and $\Omega_{ab}{}^c \equiv (e^{-1})_a{}^\mu (e^{-1} \partial_\mu e)_b{}^c$.

Proceeding as for the IIA case we will only consider the Cartan forms for the gauge fields of the theory and only use Cartan forms that are totally anti-symmetric in all their indices. As a result we define

$$\tilde{F}^s_{a_1 \cdots a_p} = p e^{-d^s_{p-1}\sigma} e_{[a_1}{}^\mu G^s_{|\mu|, a_2 \cdots a_p]}. \tag{17.5.29}$$

Using equation (17.5.17) we find that

$$\tilde{F}^1_a = D_a \sigma, \quad \tilde{F}^2_a = e^\sigma D_a \chi, \quad \tilde{F}^1_{a_1 a_2 a_3} = 3 e^{-\frac{1}{2}\sigma} \tilde{D}_{[a_1} A^1_{a_2 a_3]},$$

$$\tilde{F}^2_{a_1 a_2 a_3} = 3 e^{\frac{1}{2}\sigma} (D_{[a_1} A^2_{a_2 a_3]} - \chi D_{[a_1} A^1_{a_2 a_3]}),$$

$$\tilde{F}^1_{a_1 \cdots a_5} = 5(D_{[a_1} A_{a_2 \cdots a_5]} + 3 A^1_{[a_1 a_2} D_{a_3} A^2_{a_4 a_5]} - 3 A^2_{[a_1 a_2} D_{a_3} A^1_{a_4 a_5]}),$$

$$\tilde{F}^2_{a_1 \cdots a_7} = 7 e^{-\frac{1}{2}\sigma} \left(D_{[a_1} A^2_{a_2 \cdots a_7]} + 60 A^1_{[a_1 a_2} \left(D_{a_3} A^2_{a_4 \cdots a_7} \right. \right.$$
$$\left. \left. + A^1_{a_3 a_4} D_{a_5} A^2_{a_6 a_7]} - A^2_{a_3 a_4} D_{a_5} A^1_{a_6 a_7]} \right) \right),$$

$$\tilde{F}^1_{a_1 \cdots a_7} = 7 e^{\frac{1}{2}\sigma} \left(D_{[a_1} A^1_{a_2 \cdots a_7]} - 60 A^2_{[a_1 a_2} \left(D_{a_3} A^1_{a_4 \cdots a_7} + A^1_{a_3 a_4} D_{a_5} A^2_{a_6 a_7]} \right. \right.$$
$$\left. - A^2_{a_3 a_4} D_{a_5} A^1_{a_6 a_7]} \right) \right) + 7 e^{\frac{1}{2}\sigma} \chi \left(D_{[a_1} A^2_{a_2 \cdots a_7]} + 60 A^1_{[a_1 a_2} \left(D_{a_3} A^2_{a_4 \cdots a_7} \right. \right.$$
$$\left. \left. + A^1_{a_3 a_4} D_{a_5} A^2_{a_6 a_7]} - A^2_{a_3 a_4} D_{a_5} A^1_{a_6 a_7]} \right) \right),$$

$$\tilde{F}^1_{a_1 \cdots a_9} = 9 \left(D_{[a_1} A^1_{a_2 \cdots a_9]} - 7 \cdot 2 A^1_{[a_1 a_2} \left(D_{a_3} A^1_{a_4 \cdots a_9} - 6 \cdot 5 A^2_{a_3 a_4} \left(D_{a_5} A^2_{a_6 \cdots a_9]} \right. \right. \right.$$
$$\left. + \tfrac{1}{2} A^1_{a_5 a_6} D_{a_7} A^2_{a_8 a_9]} - \tfrac{1}{2} A^2_{a_5 a_6} D_{a_7} A^1_{a_8 a_9]} \right) \right)$$
$$+ 7 \cdot 2 A^2_{[a_1 a_2} \left(D_{a_3} A^2_{a_4 \cdots a_9} + 6 \cdot 5 A^1_{a_3 a_4} \left(D_{a_5} A^2_{a_6 \cdots a_9} \right. \right.$$
$$\left. \left. + \tfrac{1}{2} A^1_{a_5 a_6} D_{a_7} A^2_{a_8 a_9]} - \tfrac{1}{2} A^2_{a_5 a_6} D_{a_7} A^1_{a_8 a_9]} \right) \right) \right) + 9 \chi \left(D_{[a_1} A^2_{a_2 \cdots a_9} \right.$$
$$- 7 \cdot 4 A^1_{[a_1 a_2} \left(D_{a_3} A^2_{a_4 \cdots a_9} + 6 \cdot 5 A^1_{a_3 a_4} \left(D_{a_5} A^2_{a_6 \cdots a_9} \right. \right.$$
$$\left. \left. \left. + \tfrac{1}{2} A^1_{a_5 a_6} D_{a_7} A^2_{a_8 a_9]} - \tfrac{1}{2} A^2_{a_5 a_6} D_{a_7} A^1_{a_8 a_9]} \right) \right) \right),$$

$$\tilde{F}^2_{a_1 \cdots a_9} = 9 e^{-\sigma} \left(D_{[a_1} A^2_{a_2 \cdots a_9]} - 7 \cdot 4 A^1_{[a_1 a_2} \left(D_{a_3} A^2_{a_4 \cdots a_9} + 6 \cdot 5 A^1_{a_3 a_4} \left(D_{a_5} A^2_{a_6 \cdots a_9} \right. \right. \right.$$
$$\left. \left. \left. + \tfrac{1}{2} A^1_{a_5 a_6} D_{a_7} A^2_{a_8 a_9]} - \tfrac{1}{2} A^2_{a_5 a_6} D_{a_7} A^1_{a_8 a_9]} \right) \right) \right), \tag{17.5.30}$$

where D is as given in equation (17.5.19).

If we assume the equations of motion are only first order in space-time derivatives, then they can only be

$$\tilde{F}^s_{a_1 a_2 a_3} = \frac{1}{7!} \epsilon_{a_1 a_2 a_3 b_1 \cdots b_7} \tilde{F}^{s b_1 \cdots b_7}, \quad \tilde{F}_{a_1 \ldots a_5} = \frac{1}{5!} \epsilon_{a_1 \ldots a_5 b_1 \ldots b_5} \tilde{F}^{b_1 \ldots b_5}$$

$$\tilde{F}^s_a = \frac{1}{9!} \epsilon_{a b_1 \cdots b_9} \tilde{F}^{s b_1 \cdots b_9}, \quad s = 1, 2. \tag{17.5.31}$$

The non-linear realisation including the SL(2, **R**) triplet of 8-forms and 10-forms was carried out in [17.39]. This calculation was also done in a manifestly SL(2, **R**) way. Since the Cartan forms transform under the local sub-algebra, which is the Cartan involution invariant sub-algebra in this case, they transform under only the SO(2) part of the the SL(2, **R**) sub-group of E_{11}. This implies that the Cartan forms formed from the SL(2, **R**) triplet of 8-forms transform as a doublet and a singlet under local SO(2). As such, one can find an invariant constraint that sets one of the 9-form field strengths to 0. The two remaining field strengths act as the duals of the two scalar fields, further details can be found in [17.39].

We can obtain the more standard second order equations of IIB supergravity in terms of the original fields without their duals by differentiating equation (17.5.31). The result agrees with those of section 13.5, provided one makes suitable field redefinitions.

17.5.3 The common origin of the eleven-dimensional, IIA and IIB theories

To find the eleven-dimensional, IIA and IIB theories as non-linear realisations of E_{11} we have chosen different gravity sub-algebras. This resulted in taking different decompositions of E_{11} and so finding the different theories. However, like any Kac–Moody algebra, E_{11} has an essentially unique set of Chevalley generators in terms of which any generator can be expressed as a multiple commutator. When the root of the corresponding generator is expressed in terms of the simple roots the integer coefficients that arise are the number of times the corresponding Chevalley generators occur in the commutator. Given any generator in any of the three theories we know its E_{11} root and so we can find the generator with the same root in either of the other two theories. Thus we can map the generators, and so the fields, in any of the theories into those of either of the other two theories [17.40]. The list of generators and their corresponding E_{11} roots, at low levels, can be found using the analysis of section 16.6 and the results are given in tables 17.5.1–17.5.3.

The IIA theory is just the dimensional reduction of the eleven-dimensional theory and so the correspondence of the generators of the two theories is easy to find, at least at low levels. It is given at low levels in equation (17.5.11), which was checked by requiring that the generators so identified gave the IIA algebra of equation (17.5.9). Compared to the IIA case the relations for the IIB theory are more complicated as the gravity line of the IIB theory and the eleven-dimensional theory only agree on the first eight nodes and so they only have indices $a, b, \ldots = 1, 2, \ldots, 9$ in common. We will now find the correspondence between the generators of the IIB theory and the eleven-dimensional theory, up to scale, by identifying the generators with the same E_{11} roots.

The Chevalley generators were expressed in terms of the generators used in the eleven-dimensional theory in equations (16.7.14)–(16.7.21). The first step is to identify the E_a Chevalley generators in the IIB theory. The Chevalley generator E_{10} has by definition the root (0,0,0,0,0,0,0,0,0,1,0) and from table 17.5.3 we see that this corresponds to the axion χ and so to the generator R_2. Repeating this identification for all simple roots one finds

$$E_a = \hat{K}^a{}_{a+1}, a = 1, \ldots, 8, \quad E_9 = \hat{R}_1^{910}, \quad E_{10} = \hat{R}_2, \quad E_{11} = \hat{K}^9{}_{10}. \qquad (17.5.32)$$

The F_a Chevalley generators are readily found by using the Cartan involution. The Cartan sub-algebra Chevalley generators H_a are discovered by finding what linear combinations of the Cartan sub-algebra generators, that is, the generators $K^a{}_a$ and R_1, obey the defining equation $[E_a, H_b] = A_{ab}E_a$. The result is

$$H_a = \hat{K}^a{}_a - \hat{K}^{a+1}{}_{a+1}, a = 1, \ldots, 8, \quad H_9 = \hat{K}^9{}_9 + \hat{K}^{10}{}_{10} + \hat{R}_1 - \frac{1}{4}\sum_{a=1}^{10} \hat{K}^a{}_a,$$

$$H_{10} = -2\hat{R}_1, \quad H_{11} = \hat{K}^9{}_9 - \hat{K}^{10}{}_{10}. \qquad (17.5.33)$$

Table 17.5.3 The low-level decomposition of E_{11} suitable for IIB

(l_1, l_2)	A_9 Dynkin index	E_{11} root α	α^2	μ	Field
(0, 0)	[1, 0, 0, 0, 0, 0, 0, 0, 1]	(1, 1, 1, 1, 1, 1, 1, 0, 0, 1, 0)	2	1	$h_a{}^b$
(0, 0)	[0, 0, 0, 0, 0, 0, 0, 0, 0]	(0, 0, 0, 0, 0, 0, 0, 0, 0, 0, 0)	0	1	σ
(1, 0)	[0, 0, 0, 0, 0, 0, 0, 0, 0]	(0, 0, 0, 0, 0, 0, 0, 0, 0, 1, 0)	2	1	χ
(0, 1)	[0, 0, 0, 0, 0, 0, 0, 1, 0]	(0, 0, 0, 0, 0, 0, 0, 0, 1, 0, 0)	2	1	A_2
(1, 1)	[0, 0, 0, 0, 0, 0, 0, 1, 0]	(0, 0, 0, 0, 0, 0, 0, 0, 1, 1, 0)	2	1	A_2
(1, 2)	[0, 0, 0, 0, 0, 1, 0, 0, 0]	(0, 0, 0, 0, 0, 0, 1, 2, 2, 1, 1)	2	1	A_4
(1, 3)	[0, 0, 0, 1, 0, 0, 0, 0, 0]	(0, 0, 0, 0, 1, 2, 3, 4, 3, 1, 2)	2	1	A_6
(2, 3)	[0, 0, 0, 1, 0, 0, 0, 0, 0]	(0, 0, 0, 0, 1, 2, 3, 4, 3, 2, 2)	2	1	A_6
(1, 4)	[0, 1, 0, 0, 0, 0, 0, 0, 0]	(0, 0, 1, 2, 3, 4, 5, 6, 4, 1, 3)	2	1	A_8
(2, 4)	[0, 0, 1, 0, 0, 0, 0, 0, 1]	(0, 0, 0, 1, 2, 3, 4, 5, 4, 2, 2)	2	1	$h_{7,1}$
(2, 4)	[0, 1, 0, 0, 0, 0, 0, 0, 0]	(0, 0, 1, 2, 3, 4, 5, 6, 4, 2, 3)	0	1	A_8
(3, 4)	[0, 1, 0, 0, 0, 0, 0, 0, 0]	(0, 0, 1, 2, 3, 4, 5, 6, 4, 3, 3)	2	1	A_8
(1, 5)	[0, 0, 0, 0, 0, 0, 0, 0, 0]	(1, 2, 3, 4, 5, 6, 7, 8, 5, 1, 4)	2	1	A_{10}
(2, 5)	[0, 1, 0, 0, 0, 0, 0, 1, 0]	(0, 0, 1, 2, 3, 4, 5, 6, 5, 2, 3)	2	1	$A_{8,2}$
(2, 5)	[1, 0, 0, 0, 0, 0, 0, 0, 1]	(0, 1, 2, 3, 4, 5, 6, 7, 5, 2, 3)	0	1	$A_{9,1}$
(2, 5)	[0, 0, 0, 0, 0, 0, 0, 0, 0]	(1, 2, 3, 4, 5, 6, 7, 8, 5, 2, 4)	−2	2	A_{10}
(3, 5)	[0, 1, 0, 0, 0, 0, 0, 1, 0]	(0, 0, 1, 2, 3, 4, 5, 6, 5, 3, 3)	2	1	$A_{8,2}$
(3, 5)	[1, 0, 0, 0, 0, 0, 0, 0, 1]	(0, 1, 2, 3, 4, 5, 6, 7, 5, 3, 3)	0	1	$A_{9,1}$
(3, 5)	[0, 0, 0, 0, 0, 0, 0, 0, 0]	(1, 2, 3, 4, 5, 6, 7, 8, 5, 3, 4)	−2	2	A_{10}
(4, 5)	[0, 0, 0, 0, 0, 0, 0, 0, 0]	(1, 2, 3, 4, 5, 6, 7, 8, 5, 4, 4)	2	1	A_{10}
(2, 6)	[1, 0, 0, 0, 0, 0, 1, 0, 0]	(0, 1, 2, 3, 4, 5, 6, 8, 6, 2, 4)	2	1	$A_{9,3}$
(2, 6)	[0, 0, 0, 0, 0, 0, 0, 1, 0]	(1, 2, 3, 4, 5, 6, 7, 8, 6, 2, 4)	0	1	A_2
(3, 6)	[0, 1, 0, 0, 0, 1, 0, 0, 0]	(0, 0, 1, 2, 3, 4, 6, 8, 6, 3, 4)	2	1	$A_{8,4}$
(3, 6)	[1, 0, 0, 0, 0, 0, 0, 1, 1]	(0, 1, 2, 3, 4, 5, 6, 7, 6, 3, 3)	2	1	$A_{9,2,1}$
(3, 6)	[1, 0, 0, 0, 0, 0, 1, 0, 0]	(0, 1, 2, 3, 4, 5, 6, 8, 6, 3, 4)	0	1	$A_{9,3}$
(3, 6)	[0, 0, 0, 0, 0, 0, 0, 0, 2]	(1, 2, 3, 4, 5, 6, 7, 8, 6, 3, 3)	0	0	$A_{1,1}$
(3, 6)	[0, 0, 0, 0, 0, 0, 0, 1, 0]	(1, 2, 3, 4, 5, 6, 7, 8, 6, 3, 4)	−2	3	A_2
(4, 6)	[1, 0, 0, 0, 0, 0, 1, 0, 0]	(0, 1, 2, 3, 4, 5, 6, 8, 6, 4, 4)	2	1	$A_{9,3}$
(4, 6)	[0, 0, 0, 0, 0, 0, 0, 1, 0]	(1, 2, 3, 4, 5, 6, 7, 8, 6, 4, 4)	0	1	A_2
(2, 7)	[0, 0, 0, 0, 0, 1, 0, 0, 0]	(1, 2, 3, 4, 5, 6, 8, 10, 7, 2, 5)	2	1	A_4
(3, 7)	[1, 0, 0, 0, 0, 1, 0, 0, 1]	(0, 1, 2, 3, 4, 5, 7, 9, 7, 3, 4)	2	1	$A_{9,4,1}$
(3, 7)	[0, 1, 0, 1, 0, 0, 0, 0, 0]	(0, 0, 1, 2, 4, 6, 8, 10, 7, 3, 5)	2	1	$A_{8,6}$
(3, 7)	[1, 0, 0, 0, 1, 0, 0, 0, 0]	(0, 1, 2, 3, 4, 6, 8, 10, 7, 3, 5)	0	1	$A_{9,5}$
(3, 7)	[0, 0, 0, 0, 0, 0, 0, 2, 0]	(1, 2, 3, 4, 5, 6, 7, 8, 7, 3, 4)	2	1	$A_{2,2}$
(3, 7)	[0, 0, 0, 0, 0, 0, 1, 0, 1]	(1, 2, 3, 4, 5, 6, 7, 9, 7, 3, 4)	0	1	$A_{3,1}$
(3, 7)	[0, 0, 0, 0, 0, 1, 0, 0, 0]	(1, 2, 3, 4, 5, 6, 8, 10, 7, 3, 5)	−2	3	A_4
(4, 7)	[1, 0, 0, 0, 0, 1, 0, 0, 1]	(0, 1, 2, 3, 4, 5, 7, 9, 7, 4, 4)	2	1	$A_{9,4,1}$
(4, 7)	[0, 1, 0, 1, 0, 0, 0, 0, 0]	(0, 0, 1, 2, 4, 6, 8, 10, 7, 4, 5)	2	1	$A_{8,6}$
(4, 7)	[1, 0, 0, 0, 1, 0, 0, 0, 0]	(0, 1, 2, 3, 4, 6, 8, 10, 7, 4, 5)	0	1	$A_{9,5}$
(4, 7)	[0, 0, 0, 0, 0, 0, 0, 2, 0]	(1, 2, 3, 4, 5, 6, 7, 8, 7, 4, 4)	2	1	$A_{2,2}$
(4, 7)	[0, 0, 0, 0, 0, 0, 1, 0, 1]	(1, 2, 3, 4, 5, 6, 7, 9, 7, 4, 4)	0	1	$A_{3,1}$
(4, 7)	[0, 0, 0, 0, 0, 1, 0, 0, 0]	(1, 2, 3, 4, 5, 6, 8, 10, 7, 4, 5)	−2	3	A_4
(5, 7)	[0, 0, 0, 0, 0, 1, 0, 0, 0]	(1, 2, 3, 4, 5, 6, 8, 10, 7, 5, 5)	2	1	A_4
(4, 8)	[0, 1, 1, 0, 0, 0, 0, 0, 1]	(0, 0, 1, 3, 5, 7, 9, 11, 8, 4, 5)	2	1	$A_{8,7,1}$

We note that the E_a Chevalley generators outside GL(10) are contained in $\hat{R}^{ab}$ and $\hat{R}$ associated with nodes 9 and 10 and so a general generator will have $2l_2$ SL(10) indices. As for the IIA case generators can have blocks of anti-symmetrised ten indices that do not transform under SL(10).

Given the expressions of the Chevalley generators in the IIB formulation above and in eleven dimensions in equations (16.7.12)–(16.7.13) we conclude, by taking their commutators in each theory, that is, [17.37]

$$\hat{K}^a{}_b = K^a{}_b, \quad a, b = 1, \ldots, 9, \quad \hat{K}^{10}{}_{10} = \tfrac{1}{3}\sum_{a=1}^{9} K^a{}_a - \tfrac{2}{3}(K^{10}{}_{10} + K^{11}{}_{11}),$$

$$\hat{R}_1 = -\tfrac{1}{2}(K^{10}{}_{10} - K^{11}{}_{11}) \quad a = 1, \ldots, 8, \quad \hat{K}^9{}_{10} = R^{91011}, \tag{17.5.34}$$

$$\hat{R}_2 = K^{10}{}_{11}, \quad \hat{R}^{910}_1 = K^9{}_{10}.$$

These relations together with their respective algebras of equations (17.5.22) and (16.7.14)–(16.7.21) allow us to work out the correspondence between the eleven-dimensional and IIB generators at all levels one finds that [17.40]

$$\hat{R}^{a10}_1 = K^a{}_{10}, \quad \hat{R}^{ab}_1 = R^{ab11}, \quad \hat{R}^{a10}_2 = -K^a{}_{11}, \quad \hat{R}^{ab}_2 = -R^{ab10},$$

$$\hat{R}^{a_1\cdots a_3 10}_2 = -R^{a_1\cdots a_3}, \quad \hat{R}^{a_1\cdots a_4}_2 = 2R^{a_1\cdots a_4 1011},$$

$$\hat{R}^{a_1\cdots a_5 10}_2 = -\tfrac{1}{2}R^{a_1\cdots a_5 11}, \quad \hat{R}^{a_1\cdots a_6}_2 = \tfrac{1}{2}R^{a_1\cdots a_6 1011,11},$$

$$\hat{R}^{a_1\cdots a_5 10}_1 = \tfrac{1}{2}R^{a_1\cdots a_5 10}, \quad \hat{R}^{a_1\cdots a_6}_1 = -\tfrac{1}{2}R^{a_1\cdots a_6 1011,10},$$

$$\hat{R}^{a_1\cdots a_7 10}_2 = \tfrac{1}{2}R^{a_1\cdots a_7 11,11}, \quad \hat{R}^{a_1\cdots a_7 10}_1 = -\tfrac{1}{2}(R^{a_1\cdots a_7 10,11} + R^{a_1\cdots a_7 11,10}),$$

$$\hat{R}^{a_1\cdots a_7 10,10}_1 = R^{a_1\cdots a_6}, \quad \hat{S}^{a_1\cdots a_7 10}_2 = -\tfrac{1}{2}\hat{R}^{a_1\cdots a_6 10,10},$$

$$\hat{R}^{a_1\cdots a_7 10,a} = -\tfrac{3}{4}(R^{a_1\cdots a_7 10,11} - R^{a_1\cdots a_7 11,10}),$$

$$\hat{R}^{a_1\cdots a_6 10,b} = \tfrac{1}{4}(R^{a_1\cdots a_6 b11,10} - R^{a_1\cdots a_6 b10,11}) - R^{a_1\cdots a_6 1011,b}.$$

(17.5.35)

We close this section by giving a higher level example of the correspondence between the IIA and eleven-dimensional theory. We found that the former possessed a 9-form generator that led to the IIA theory with a cosmological constant. In particular, examining table 17.5.2 we find that the 9-form generator $R^{a_1\cdots a_9}$ of E_{11} arises from the root $(0, 1, 2, 3, 4, 5, 6, 7, 4, 1, 4)$, which corresponds to the highest weight component $R^{3\cdots11}$. On the other hand, in the non-linear realisation of E_{11} with the A_{10} gravity sub-algebra that leads to the eleven-dimensional theory the root $(0, 1, 2, 3, 4, 5, 6, 7, 4, 1, 4)$ is the highest weight component of the level 4 generator $R^{(ab)}_c$, that is, $R^{(1111)}{}_1$, or raising the lowered index with ϵ, $R^{(1111)2\cdots11}$. Since the A_9 sub-algebra of A_{10} is in common we can identify all the components of the IIA 9-form generator from its highest weight components in the obvious way to find that $\tilde{R}^{a_1\cdots a_9} = R^{(1111)a_1\cdots a_9 11}$.

Thus the IIA generator $R^{a_1\cdots a_9}$, which is associated with the massive IIA theory and the 8-brane, corresponds in the eleven-dimensional theory to the level 4 generators $R^{(ab)}_c$, or equivalently the field $A_{a_1\cdots a_9}$ of the IIA theory corresponds to the level 4 eleven-dimensional field $A^c_{(ab)}$. This strongly suggests that the eleven-dimensional E_{11} non-linear realisation contains not only the ten-dimensional IIA supergravity, but also the massive IIA supergravity theory. However, to achieve this one must include in the non-linear realisation the fields corresponding to the generators of level 4. This would be an extension of eleven-dimensional supergravity that has so far not been constructed.

17.5.4 Theories in less than ten dimensions

To find a theory in D dimensions from E_{11} we must identify an A_{D-1}, or SL(D), sub-algebra corresponding to D-dimensional gravity [16.13]. Proceeding as above we see that now there is only one choice: the first $D - 1$ dots and so the gravity line contains nodes 1 to $D - 1$ inclusive. To express the E_{11} fields in terms of those we recognise in D dimensions we delete the Dth dot and decompose the algebra in terms of the remaining algebra which is $A_{D-1} \otimes E_{11-D}$. We take $E_5 = D_5 = SO(5, 5)$ $E_4 = A_4 = SL(5)$ etc., that is, the symmetry groups found in table 13.6.1. The E_{11} Dynkin diagram with the gravity line drawn on the horizontal is given in diagram 17.5.4

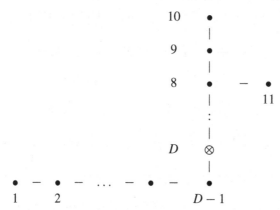

Diagram 17.5.4   The E_{11} Dynkin diagram from the viewpoint of the D-dimensional theory.

To find the nine-dimensional theory from the IIA theory of diagram 17.5.2 one deletes node 9, while to find it from the IIB theory of diagram 17.5.3 one deletes node 11. However, the gravity line for the two theories is the same as it consists of nodes 1–8 inclusive and so the theories are the same. This is consistent with the fact that there is a unique supergravity theory in nine dimensions.

The decomposition into representations of $A_{D-1} \otimes E_{11-D}$ can be carried out using the techniques of section 16.6 and table 17.5.4 contains the results for the form fields, that is, fields whose indices are totally anti-symmetric [17.41, 17.42]. We note that the generators belong to the contragredient representations to the fields and it is the former which were computed in chapter 16.

Let us denote the representation of E_{11-D} of the form field of rank n by R_n. The forms can be divided into three classes. The first is those of ranks less than $D - 2$ which correspond to propagating degrees of freedom. One obvious feature of these is that every form field $A_{a_1 \cdots a_n}$ of rank n appears with its dual field $A_{a_1 \cdots a_{D-n-2}}$ and in this sense it is a democratic formulation. The duality relations that the fields strength satisfy will be discussed in the next section. The other two classes are the forms of rank $D - 1$, which we will discuss in

Table 17.5.4 The A_{D-1} representations of the forms fields in any dimension

D	E_{11-D}	A_a	$A_{a_1 a_2}$	$A_{a_1 a_2 a_3}$	$A_{a_1 \cdots a_4}$	$A_{a_1 \cdots a_5}$	$A_{a_1 \cdots a_6}$	$A_{a_1 \cdots a_7}$	$A_{a_1 \cdots a_8}$
8	$SL(3) \otimes SL(2)$	$(\bar{3}, 2)$	$(3, 1)$	$(1, 2)$	$(\bar{3}, 1)$	$(3, 2)$	$(1, 3)$ $(8, 1)$	$(\bar{3}, 2)$ $(6, 2)$	$(3, 1)$ $(15, 1)$ $(3, 1)$ $(3, 3)$
7	$SL(5)$	$\overline{10}$	5	$\bar{5}$	10	24	$\overline{40}$ $\overline{15}$	70 45 5	
6	$SO(5, 5)$	16	10	$\overline{16}$	45	144	320 $\overline{126}$ 10		
5	E_6	27	$\overline{27}$	78	351	$\overline{1728}$ $\overline{27}$			
4	E_7	56	133	912	8645 133				
3	E_8	248	3875 1	147250 3875 248					

the next paragraph, and rank D that correspond to space-filling branes which we will not discuss further.

We recall that in the IIA theory we found a 9-form and realised that this corresponds to a theory with a cosmological constant and so a massive deformation of the IIA theory. Looking at table 17.5.4 we find, in a given dimension, that there are whole E_{11-D} multiplets worth of $D-1$ forms and so a corresponding number of theories with a cosmological constant. Such theories have been studied from almost the beginning of supergravity, the first being the massive deformation of the $N=1$ four-dimensional supergravity theory studied at the end of section 13.1.1 [17.43]. An important example of such theories was the deformation of the $N=8$ four-dimensional supergravity theory [17.44]. This illustrated a rather generic feature not present in the $N=1$ four-dimensional supergravity case, or the massive IIA theory discussed in the previous section. The massive deformations not only turn on a cosmological constant, but also in almost all cases a non-Abelian, or in some simpler cases an Abelian, gauge symmetry under which the other fields of the theory are charged. We note that in a supergravity theory before the deformation the gauge fields carry Abelian gauge symmetry but none of the other fields transform under this symmetry. As a result these theories, and indeed all such theories with a cosmological constant as a class, are often referred to as gauged supergravities. One can find these theories by adding a cosmological term and then using supersymmetry to find what other terms one must add. Although simply said, it is very complicated to carry out and the practitioners often relied on a number of short cuts. Even before all such theories were found the maximally supersymmetric gauged supergravity theories were classified [17.45–17.49]. The result was that these theories were classified by something called the embedding tensor and this belonged to a representation of E_{11-D}.

In table 17.5.4 the $D - 1$ forms predicted by E_{11}, and so the gauged supergravities, are given [17.41, 17.42]. For example, in five dimensions they occur as the $\mathbf{351}$ representation of E_6. These representations are in complete agreement with those found from the classification based on supersymmetry. However, in contrast to this latter technique, the results of table 17.5.4 are rather easy to derive. Indeed, the reader may like to derive the multiplet of $D - 1$ forms in D dimensions by starting with the fields in eleven dimension given in table 17.5.1 and its extensions and carry out by hand the dimensional reduction. For the higher dimensions this is very simple to carry out, but it becomes more lengthy for the cases of four and three dimensions. One finds for the case of three dimensions that one must use fields up to level 14 in the eleven-dimensional decomposition whose associated roots have rather negative length squared. Thus one is probing deep into the E_{11} algebra and finding results that are correct. This is one of a number of examples when E_{11} makes predictions that are usually the preserve of supersymmetry.

For example, to find the results for the 6-forms in seven dimensions we can dimensionally reduce the eleven-dimensional theory on a 4-torus, whose directions are rotated under an $SL(4, \mathbf{R})$ symmetry and are labelled by $i, j = 1, \ldots, 4$. The fields that occur in the non-linear realisation in eleven dimensions carry blocks of indices that are, for the most part, totally anti-symmetrised in each block. They also carry $3l$ indices if they occur at level l. Given a block of eleven indices we can take six of them to belong to the now seven-dimensional space-time leaving five totally anti-symmetrised indices, which we can only take to be internal indices. However, we only have four internal indices and so any field with a block of eleven indices will not lead to a 6-form. In table 17.5.1 we have not shown the blocks of eleven indices, but they are easily reinstated; for example, the field A_1 at level 4 has $3 \times 4 = 12$ indices and so one block of eleven indices. Given a block of ten indices we can take six to belong to space-time and then four to the internal space; the latter lead to a scalar of $SL(4)$. Clearly, if we have more than one block of ten indices we cannot get a 6-form. By the same argument, a block of nine indices will lead to six anti-symmetised indices in space-time and then three totally anti-symmetrised internal indices, indeed, $\phi_{a_1 \cdots a_6 ijk} \sim \epsilon_{ijkl} \phi_{a_1 \cdots a_6}{}^l$. By the same argument a block of eight indices leads to a field $\phi_{a_1 \cdots a_6 ij}$.

Looking at table 17.5.1, or even better computing the generators from first principles using chapter 16, we see that the only fields that can lead to 6-forms in seven dimensions are

$$A_{a_1 \cdots a_6}, h_{a_1 \cdots a_8, b}, A_{a_1 \cdots a_9, b_1 b_2 b_3}, A_{a_1 \cdots a_{10}, (bc)}, A_{a_1 \cdots a_{10}, b_1 \cdots b_4, c}$$

and upon dimensional reduction we find the 6-forms

$$A_{a_1 \cdots a_6}; A_{a_1 \cdots a_6, ij, k}, A_{a_1 \cdots a_6}{}^k; A_{a_1 \cdots a_6}{}^{[ij]}, A_{a_1 \cdots a_6}{}^{(ij)}; A_{a_1 \cdots a_6, (ij)}; A_{a_1 \cdots a_6, i},$$

where we have reexpressed the results in terms of irreducible representations of $SL(4)$, for example, $A_{a_1 \cdots a_6, [ij,k]} = 0$ and so we also get the second term between the semi-colons.

The U duality group in seven dimensions is $SL(5)$ and so it only remains to express the result in terms of representations of this group. The $\overline{\mathbf{15}}$ representation of $SL(5)$ is the representation $\phi^{IJ} = \phi^{JI}$, $I, J = 1, \ldots, 5$ and it arises from the $SL(4)$ representations $A_{a_1 \cdots a_6}$, $A_{a_1 \cdots a_6}{}^{(ij)}$ and $A_{a_1 \cdots a_6}{}^k$. The reader can verify that the remaining fields belong to the $\mathbf{40}$ representation of $SL(5)$ which is carried by $\phi_{IJ, K} = -\phi_{JI, K}$ with $\phi_{[IJ, K]} = 0$. In carrying out this computation it is useful to realise that the latter condition implies that

$\phi_{4j,k} + \phi_{k4,j} + \phi_{jk,4} = 0$ and so we only get one second rank anti-symmetric SL(4) tensor. Thus by a back-of-the-envelope calculation we have classified all the gauged maximal supergravity theories in seven dimensions. We note it involved a field at level 5 and that the field $A_{a_1 \cdots a_{10},[bc]}$ had multiplicity 0. The reader may like to recover this result by adding a cosmological term to the undeformed seven-dimensional maximal supergravity theory and finding all the possible theories that preserve all the supersymmetries; only joking!

The classification method based on supersymmetry used to find the gauged supergravity theories is quite different to the discussions used to relate the different parts of M theory, which usually just consider the undeformed type II and type I theories. However, it must surely be that the gauged supergravities are also part of the larger picture. The advantage of the E_{11} viewpoint is that it places all the gauged maximal supergravity theories in a single unified structure together with the usually discussed maximal undeformed theories. Indeed, we have seen that E_{11} provides a unified framework for all maximal supergravity theories.

Many of the gauged supergravity theories have within the framework of usual supergravity, or indeed string theory, no higher dimensional origin. However, in E_{11} one can relate the effects in any dimension to those in any other dimension, including eleven dimensions, in a way similar to the identification of the 9-form in the IIA theory with the corresponding level 4 field in eleven dimensions, as we discussed earlier.

One can also carry out the computation of the dynamics of the gauged supergravity theories using the E_{11} non-linear realisation [17.50–17.52]. Indeed, a systematic construction method for all maximal gauged supergravity theories in all dimensions was given in the last of these references. An important part in this calculation is played by the existence of the hierarchy of fields given in table 17.5.4: by the hierarchy we mean the set of form fields of the table of all ranks in a given dimension. Indeed, using this hierarchy provided by E_{11} has become the standard method of constructing gauged supergravity theories.

The introduction of dual fields for all form fields in the context of maximal supergravity theories was first given in [17.53]. However, this was done by fiat rather than as a result of an underlying algebra.

We end this section by commenting on the way T and S dualities appear in E_{11-D}, that is, U duality, transformations, as this is particularly simple to see from the E_{11} viewpoint. To find the theory in D dimensions resulting from compactification of the eleven-dimensional theory on a $11 - n$ torus we must delete node D leading to the GL(D) algebra of gravity and the E_{11-D}, that is, U duality, symmetry, see diagram 17.5.4. If we delete node 10 in the IIA theory, see diagram 17.5.2, we find the algebra O(10,10). However, if we are in D dimensions only O($n - 1$, $n - 1$) of this algebra is contained in E_{11-D}. These are the T duality transformations as is verified in section 17.5.7. However, as seen from the viewpoint of the IIB theory node 10 corresponds to the SL(2) symmetry of the IIB theory. Hence it is obvious that U duality is just T duality combined with the SL(2) symmetry of the IIB theory. As the former is known to be true, U duality is true if the SL(2) symmetry of the IIB theory holds. As discussed previously, the symmetries are suitably discretised.

17.5.5 Duality symmetries and conditions

We noted above that every form field in D dimensions appears with its dual. We recall that by a form field we mean one whose SL(D) indices are totally antisymmetric. The presence of dual fields suggests that the form fields should obey duality conditions on the field strengths; as we found for the 3-form and 6-form of eleven-dimensional supergravity and for the forms of the IIA and IIB theories. Such equations imply the more usual equations of motion which are second order in space-time derivatives. Just as in eleven dimensions, the E_{11} transformations rotate the form fields into each other generalising electromagnetic duality transformations. We recall from section 17.2 that the standard electromagnetic duality transformations did not leave the action invariant, but did preserve the equations of motion and so when incorporating electromagnetic duality transformations it is desirable to look for equations of motion rather than an action. It is the purpose of this section to give the equations of motion for the form fields as first order duality conditions using the representations and Cartan forms specified by E_{11} in table 17.5.4. This discussion is taken from [17.52].

We begin by considering form fields whose field strengths have a rank that is not half that of the dimensions of space-time, that is, those that do not obey a self-duality condition. Examining table 17.5.4 we find that for every form field of rank n for $n < \frac{1}{2}D$, with a field strength F_{n+1} of rank $n + 1$, that belongs to a representation $\mathbf{R}_n$ there is a dual form field of rank $D - n - 2$ with a field strength F_{D-n-1} of rank $D - n - 1$ which is in that representation $\mathbf{R}_{D-n-2}$ which is the conjugate representation, that is, $\mathbf{R}_{D-n-2} = \overline{\mathbf{R}}_n$.

The field strengths are just the Cartan forms of E_{11} associated with the form generators provided we take them to be totally anti-symmetric in all their space-time indices, as we did for the cases of ten and eleven dimensions. This has the effect of turning the E_{11} transformations that lead to shifts in the form fields into gauge transformations and so the form fields become gauge fields. This computation is very straightforward and can be found in [17.52] for the maximal supergravities in all dimensions. The Cartan forms are inert under the rigid action of the non-linear realisation and only transform under the local subgroup transformations, that is, only under the Cartan involution invariant sub-group $I(E_{11-D})$ rather than the rigid E_{11-D} transformations. It is instructive to consider how this comes about in the context of the IIB supergravity theory. In chapter 13, we found that the field strengths of the second rank form fields transformed as a doublet under rigid SL(2, $\mathbf{R}$), but by using the scalars, as they appear in the SL(2, $\mathbf{R}$) non-linear realisation, we could convert the field strengths to objects that are inert under the rigid transformations but do transform under the local sub-group SO(2), see equation (13.5.33). From the general theory of non-linear realisations this can be viewed as the conversion of a linear realisation of a matter field into the non-linear realisation. Here, and in the non-linear realisation of the IIB theory in ten dimensions of section 17.5.2.2, all the fields, including the scalars, are part of the same non-linear realisation and the E_{11} non-linear realisation automatically arranges for the Cartan forms to contain field strengths for the form fields but decorated by the scalars in just such a way that they transform just under the local sub-group.

Examining the representations of E_{11-D} of the form fields for $D \leq 7$, given in table 17.5.4, we find that under the decomposition to representations of $I(E_{11-D})$ we find only one irreducible real representation of $I(E_{11-D})$. Indeed, we find that the form fields and their duals belong to the same representation of $I(E_{11-D})$. For example, in seven dimensions the 2-forms belong to the $\mathbf{5}$ of SL(5, $\mathbf{R}$), while their dual form fields, the 3-forms, belong to the $\overline{\mathbf{5}}$ of SL(5, $\mathbf{R}$). The Cartan involution invariant sub-group is I(SL(5, $\mathbf{R}$)) = SO(5) and these two form fields both belong to the real $\mathbf{5}$ representation of this group, that is, the vector representation. As such the duality conditions for all such form fields can only be of the generic form

$$F_{n+1} = \star F_{D-n-1},$$
(17.5.36)

where $\star$ is the usual space-time dual using the ϵ symbol.

For dimensions $D \geq 8$ the form fields belong to representations of E_{11-D} that decompose into at most two distinct irreducible real representations of $I(E_{11-D})$ and their dual form fields belong to precisely the same representations of $I(E_{11-D})$. Then there is one duality condition of the above type for each representation of $I(E_{11-D})$ that occurs and it relates the form field to its dual in the same representation.

Traditionally, scalars in the non-linear realisation of a group E_{11-D} obey a second order differential equation. As discussed in section 13.2 the Cartan forms for the scalars, which belong to the adjoint representation of E_{11-D}, can be divided into two parts, one of which is invariant under the Cartan involution and by definition is in the adjoint representation of $I(E_{11-D})$, while the other part acquires a minus sign under the action of the Cartan involution. These two parts transform inhomogeneously and homogeneously under the local symmetry $I(E_{11-D})$; see equation (13.2.28) and the surrounding discussion. The equation of motion is usually formulated as the covariant derivative of the second part. Here the scalars are a non-linear realisation of E_{11-D}, but obey duality relations together with the rank $D - 2$ forms which are also in the adjoint representation of E_{11-D}. As such, can be derived they can be decomposed into the same representations of $I(E_{11-D})$ and so can be divided into two corresponding parts, namely the adjoint representation of $I(E_{11-D})$ and a remainder. The duality relation equates the space-time dual of the second part of the Cartan forms of the scalar to the analogous part of the Cartan forms for the $D - 2$ forms. The usual second order equation involving just the scalars results from the derivative of this duality relation.

For odd dimensional space-times there are clearly no generalised self-duality conditions. We split the analysis for even dimensions into the cases $D = 4m$ and $D = 4m + 2$ for integer m, which are different due to the fact that $\star\star = -1$ and $\star\star = +1$, respectively, when acting on a $(D/2)$-form. Let us begin with the latter case, that is, dimensions 10 and 6. As is well known in ten dimensions we have a 4-form field that is a singlet of $SL(2, \mathbf{R})$ and obeys a self-duality condition of the form $F_5 = \star F_5$. In six dimensions the forms belong to the **10** of $SO(5, 5)$, which decomposes into the reducible representation $(\mathbf{5}, \mathbf{1}) \oplus (\mathbf{1}, \mathbf{5})$ of $SO(5) \otimes SO(5) = I(SO(5, 5))$. The duality condition which is invariant under the $SO(5) \otimes SO(5)$ transformations of the field strength can only be of the form

$$\begin{pmatrix} F_3 \\ F_3' \end{pmatrix} = \star \begin{pmatrix} F_3 \\ -F_3' \end{pmatrix},$$
(17.5.37)

where F_3 and F_3' belong to the $(5, 1)$ and $(1, 5)$ representations of $SO(5) \otimes SO(5)$, respectively. The minus sign is required as the final theory can be written in terms of field strengths that satisfy no self-duality conditions with equations of motion that are second order in space-time derivatives. As such we must have equal numbers of self-dual and anti-self-dual field strengths.

Let us now consider the case of $D = 4m + 2$, that is, dimensions 8 and 4. In this case the forms belong to an irreducible representation of E_{11-D} that breaks into two irreducible representations of $I(E_{11-D})$ which are related by complex conjugation. In eight dimensions the 3-forms belong to the $(\mathbf{1}, \mathbf{2})$ representation of $SL(3, \mathbf{R}) \otimes SL(2, \mathbf{R})$, which breaks into $(\mathbf{1}, \mathbf{1}^+)$ and $(\mathbf{1}, \mathbf{1}^-)$ representations of $I(SL(3, \mathbf{R}) \otimes SL(2, \mathbf{R})) = SO(3) \otimes SO(2)$. As such, the unique invariant field equation is of the form

$$\begin{pmatrix} F_4 \\ F_4^* \end{pmatrix} = i \star \begin{pmatrix} F_4 \\ -F_4^* \end{pmatrix}, \tag{17.5.38}$$

where F_4 and F_4^* belong to the $(\mathbf{1}, \mathbf{1}^+)$ and $(\mathbf{1}, \mathbf{1}^-)$ representations of $SO(3) \otimes SO(2)$. The i found in this equation is due to the fact that $\star\star = -1$ in this dimension and the minus sign then results from the consistency with respect to complex conjugation. In four dimensions the 1-forms belong to the $\mathbf{56}$-dimensional representation of E_7 which decompose into the $\mathbf{28} \oplus \overline{\mathbf{28}}$ of representations $I(E_7) = SU(8)$. The self-duality condition can only be of the form

$$\begin{pmatrix} F_2 \\ F_2^* \end{pmatrix} = i \star \begin{pmatrix} F_2 \\ -F_2^* \end{pmatrix}, \tag{17.5.39}$$

where F_2 and F_2^* are the $\mathbf{28} \oplus \overline{\mathbf{28}}$ representations of $SU(8)$.

It is consoling to see the way the representations of the form fields, dictated by E_{11}, cooperate with the demand that the form field equations be duality conditions. As explained, it is straightforward to derive these equations from the underlying E_{11} algebra. One can rescale the fields so as to adjust certain constants in the duality relations, however, these constants are fixed once one also writes down the field equation for gravity as this involves the stress tensor and so the field strength squared.

We have noted that E_{11} always encodes the dual of gravity and so gravity itself should also be described by a first order equation duality equation.

As we will now show the E_{11} non-linear realisation must contain many more duality relations involving fields at all higher levels [17.54]. We will work in eleven dimensions but similar results exist in all dimensions. We observed at the end of section 16.7.1 that the number of A_{10}, or $SL(11)$, indices on a generator satisfied equation (16.7.23), that is, the equation $11n + \sum_j j q_j = 3l$, where the Dynkin indices are specified by $p_{11-j} = q_j$, l is the level and n the number of blocks of 11 totally anti-symmetrised indices. Let us search for all the generators of E_{11} with no blocks of 10 or 11 totally antisymmetric indices; this means taking $n = q_{10} = 0$. Using the fast method of finding the generators given at the end of section 16.7.1 we can substitute for l into equation (16.6.6) applied to E_{11} and we find that

$$\alpha^2 = \frac{1}{9} \left[\sum_{i=1}^{9} i(9-i) q_i^2 + 2 \sum_{i<j} i(9-j) q_i q_j \right] = 2, 0, -2, \dots. \tag{17.5.40}$$

We observe that, unlike in the general case, the right-hand side is positive definite and it is straightforward to find all the possible solutions to this equation. They are given by [17.54]

$$p_8 = q_3 = 1 \quad p_2 = q_9 = m; \quad p_5 = q_6 = 1 \quad p_2 = q_9 = m;$$
$$p_{10} = q_1 = 1, \quad p_3 = q_8 = 1 \quad q_9 = m \tag{17.5.41}$$

for any non-negative integer m. The additional solution $q_9 = m$ with all the other Dynkin indices 0 turns out not to exist in E_{11} due to the Serre relations, in fact, one can rule it out using the Jacobi identities. The corresponding fields are

$$A_{3,9,9,\ldots}, \quad A_{6,9,9,\ldots}, \quad h_{8,1,9,9,\ldots}, \tag{17.5.42}$$

where $A_3 = A_{a_1 a_2 a_3}, A_{3,9} = A_{a_1 a_2 a_3, b_1 \cdots b_9}$ etc.

If we take no blocks of nine, then we find the fields of E_{11} at levels 1, 2 and 3, that is, the 3-form, its dual the 6-form and the dual graviton. The physical states of a theory are found by decomposing into representations of the little group, which here is SO(9), as we are dealing with a massless eleven-dimensional theory. In this case, the blocks of nine indices transform as a singlet and the field $A_{3,9}$ transforms in the same way as the original field A_3. As such, the field $A_{3,9}$ can be used to describe the propagation of the same degrees of freedom as A_3, just as does the field A_6. A similar result holds for the other fields which have a set of three or six indices and any number of blocks of nine indices. The same story also applies to the gravity field $h_a{}^b$ and the fields $h_{8,1,9,9,\ldots}$. Thus we find that the non-linear realisation of E_{11} contains fields corresponding to the infinite possible dual descriptions of the bosonic degrees of freedom of eleven-dimensional supergravity which are usually described by the 3-form and the graviton. One expects that the E_{11} dynamics contains an infinite sequence of duality relations and it would be good to find them. It would seem highly likely that the well-known infinite set of duality relations for a scalar field in two dimensions is lifted to eleven dimensions by the E_{11} non-linear realisation.

Although we have now found a meaning for an infinite number of the higher level fields in E_{11} there are still an infinite number of fields whose meaning is as yet unclear.

The reader may wonder why it is desirable to consider the equation of motion as duality conditions between dual gauge fields when one can eliminate one of the gauge fields to find equations of motion that are second order in space-time derivatives. The reason is that duality symmetries rotate one gauge field into another in a local manner and if one has only one of the gauge fields, then a symmetry of this kind would have to be non-local and so much more difficult to handle.

17.5.6 Brane charges, the l_1 representation and generalised space-time

In section 16.7.2 we constructed the fundamental representation of E_{11} associated with node 1, denoted the l_1 representation [16.17]. In equation (16.7.35) we decomposed this representation into representations of SL(11); we summarise the result for the first few levels here for convenience:

$$P_a; Z^{a_1 a_2}; Z^{a_1 \cdots a_5}; Z^{a_1 \cdots a_7, b}, Z^{a_1 \cdots a_8}; Z^{a_1 \cdots a_8, b_1 b_2 b_3}, Z^{a_1 \cdots a_9, (bc)}, Z^{a_1 \cdots a_9, b_1 b_2},$$

$$Z^{a_1 \cdots a_{10}, b}, Z^{a_1 \cdots a_{11}}; Z^{a_1 \cdots a_9, b_1 \cdots b_4, c}, Z^{a_1 \cdots a_8, b_1 \cdots b_6}, Z^{a_1 \cdots a_9, b_1 \cdots b_5}, \ldots \tag{17.5.43}$$

where $a = 1, \ldots, 11$. This decomposition is associated with the deletion of node 11 to leave the SL(11) algebra and it is the one appropriate to the eleven-dimensional theory.

The appearance of the space-time translations P_a as the first component is not surprising as at the lowest level the l_1 representation is just the first fundamental representation of SL(11) and this is the vector representation. The first three representations of the l_1

representation in eleven dimensions have the correct index structure to be interpreted as the charges associated with the point particle, the 2-brane and the 5-brane. We observe that these are the central charges of the eleven-dimensional supersymmetry algebras.

It is natural to conjecture that the l_1 representation is the multiplet of brane charges [16.16, 16.17]. The Wess–Zumino term in the brane action is the product of the background field, to which the brane couples, contracted with a topological current, the integral of whose time component is the brane charge. Thus there is a correspondence between the brane charges and the background fields to which the brane couples. For example, the 2-brane and 5-brane of eleven dimensions couple to the fields $A_{a_1 a_2 a_3}$ and $A_{a_1 \cdots a_6}$, respectively, and the corresponding brane charges are Z^{ab}, $Z^{a_1 \cdots a_5}$ and for the pp wave, the corresponding field is the graviton $h^a{}_b$ and the charge is the space-time translations P_a. Thus at low levels the fields, or equivalently the generators, in the adjoint representation of E_{11} and the elements of the l_1 representation are indeed in the required correspondence. At the next level we find that the dual graviton corresponds to the charge $Z^{a_1 \cdots a_7, b}$ which is the charge associated with the Kaluza–Klein monopole. One can show that this correspondence holds at all levels [16.16], so providing evidence for the conjecture.

Just as we did for the adjoint representation we can study the l_1 representation in D dimensions. The most direct approach is to take the charges in the l_1 representation in eleven dimensions, carry out the dimensional reduction by hand and collect the result into representations of E_{11-D}. As explained in section 16.6.2, a more powerful method is to consider the enlarged Dynkin diagram by adding a new node connected to the node labelled 1 by a single line and realise that the l_1 representation is the generators at level 1 with respect to the new node added. One can then find the representations in any dimension just as we did for the adjoint representation, that is, by deleting node D to leave the algebra $SL(D) \otimes E_{11-d}$, after we have also deleted the new node. The results [16.13, 17.55, 17.56] are given in table 17.5.5.

Table 17.5.5 gives the l_1 representation itself, rather than in the case of the adjoint representation where the tables listed the fields corresponding to the generators. It is instructive to derive one of the entries of the table and we now derive the scalar charges in seven dimensions from the charges in eleven-dimensional theory on a 4-torus. The directions of the torus have the indices $i, j = 1, \ldots, 4$ and they rotate under an SL(4) symmetry. Examining the eleven-dimensional charges of equation (17.5.43) we find that the only charges that lead to scalar charges are P_i and Z^{ij}. There are no other conributions as a charge at level l carries $3l - 1$ SL(11) indices, although we have not shown the blocks of 11 totally anti-symmetric indices. These charges collect into the **10** representations of the U duality group SL(5) in seven dimensions; this represenation is carried by $Z_{IJ} = -Z_{JI}$, $i, j = 1, \ldots, 5$. Deriving the result in this way makes clear the eleven-dimensional origin of the charges in seven dimensions. The reader may like to verify some of the other entries in a similar way.

Brane charges below ten dimensions are discussed in [17.57–17.59]. In these references the authors took a particular charge, found, for example, by dimensional reduction, and applied U duality, that is, E_{11-d} transformations, in D dimensions to find the brane charges in the same E_{11-D} multiplet. They carried this out for the point particle and some other branes. For example, for the point particle we can take the starting charge to be one of the Kaluza–Klein particles, that is, take the charge n/R_i, where R_i is one of the radii of the torus used in the dimensional reduction. The results predicted by E_{11} in table 17.5.5 agree

Table 17.5.5 The l_1 representation of the form charges of E_{11} in D dimensions

D	E_{11-D}	Z	Z^a	$Z^{a_1a_2}$	$Z^{a_1\cdots a_3}$	$Z^{a_1\cdots a_4}$	$Z^{a_1\cdots a_5}$	$Z^{a_1\cdots a_6}$	$Z^{a_1\cdots a_7}$
8	$SL(3)\otimes SL(2)$	$(3,2)$	$(\bar{3},1)$	$(1,2)$	$(3,1)$	$(\bar{3},2)$	$(1,3)$	$(3,2)$	$(6,1)$
							$(8,1)$	$(6,2)$	$(18,1)$
							$(1,1)$		$(3,1)$
									$(6,1)$
									$(3,3)$
7	$SL(5)$	10	$\bar{5}$	5	$\bar{1}0$	24	40	70	
						1	15	50	
							10	45	
								5	
6	$SO(5,5)$	$\bar{1}6$	10	16	45	$1\bar{4}4$	320		
					1	16	126		
							120		
5	E_6	$\bar{2}7$	27	78	$3\bar{5}1$	1728			
				1	$\bar{2}7$	351			
						27			
4	E_7	56	133	912	8645				
			1	56	1539				
					133				
					1				
3	E_8	248	3875	147250					
		1	248	30380					
			1	3875					
				248					
				1					

precisely with these multiplets, that is, the first two columns of table 17.5.5 agree with the point particle and string charge multiplets found earlier [17.57–17.59].

While it is obvious that the decomposition of the l_1 multiplet would lead to representations of E_{11-D}, it did not have to lead to the correct representations where they were known. Put another way, the charge representations in D dimensions did not have to assemble into a single representation of E_{11}, so giving a unified origin for all brane charges. Although the charges at low levels have a simple SL(D) index structure, carrying just a single block of totally anti-symmetrised SL(D) indices, at higher levels one finds charges with very exotic index structures which are not usually discussed. Knowing that they have a common origin with more conventional charges gives one confidence that these exotic charges are also important. It is fair to say that there is considerable evidence that the l_1 multiplet does contain all the brane charges. We note that the l_1 representation predicts an infinite number of charges whose interpretation at all levels is as yet unknown.

When computing the non-linear realisations of the eleven-dimensional, IIA and IIB theories above we took a number of short cuts: we took the local sub-group to be that at lowest level, that is, the Lorentz group, we introduced the space-time translations P_a in an ad hoc manner and we also used only parts of the Cartan forms to construct the dynamics. Despite this we found the correct dynamics up a few undetermined constants. We will now present a more systematic way of carrying out the non-linear realisation.

The most substantial drawback of the previous considerations was the introduction of the space-time translations P_a in an ad hoc manner as a way of introducing space-time into the non-linear realisation. We recall that E_{11} is not an internal symmetry as it contains the SL(11) transformations of gravity which also contain Lorentz transformations. As such, a rigid SL(11) transformation must change not only the fields, but also the space-time coordinates. This was achieved by inserting $e^{x^a P_a}$ into the group element, however, we then must consider how E_{11} acts on the space-time translations. We saw that the space-time translations were just the first component in the infinite dimensional l_1 representation and so it is natural to introduce this representation into the non-linear realisation. Indeed, it has been proposed [16.17] that one should consider the non-linear realisation of the semi-direct product group $E_{11} \otimes l_1$. The structure of such a group was explained in section 16.7.2 and non-linear realisations of this type were discussed in section 13.2.

The group element $g \equiv g_L g_E$ was given in equation (13.2.62) and

$$g_L = e^{x^a P_a} e^{x_{ab} Z^{ab}} e^{x_{a_1 \cdots a_5} Z^{a_1 \cdots a_5}} \tag{17.5.44}$$

and

$$g_E = e^{h_a{}^b K^a{}_b} e^{\frac{1}{3!} A_{a_1 a_2 a_3} R^{a_1 a_2 a_3}} e^{\frac{1}{6!} A_{a_1 \cdots a_6} R^{a_1 \cdots a_6}} \cdots . \tag{17.5.45}$$

The fields $h_a{}^b$, $A_{a_1 a_2 a_3}$ and $A_{a_1 \cdots a_6}$, ... are now taken to depend on x^a, x_{ab} and $x_{a_1 \cdots a_5}$, ...

Thus we now have a generalised space-time that contains an infinite number of coordinates. We have coordinates associated to all the branes. Instead of just the x^μ, which are associated with the translation generators and so in a sense with the pp wave, or point particle, we also have coordinates x_{ab} which are associated with the charges Z^{ab}, and so the M2-brane. This is rather natural in the sense that events can be measured by all brane probes. Another advantage of the generalised space-time is that if we are in less than ten dimensions, the coordinates with fixed type of space-time indices will form U duality multiplets. For example, examining table 17.5.5 we find that there are a $\mathbf{\overline{27}}$ representation of E_6 scalar coordinates in six dimensions. This contains a $\mathbf{5}$ of SL(5), which are the components of the coordinates x^i which no longer belong to the space-time of the theory after the dimensional reduction; however, these coordinates do not carry a representation of the E_6 U duality transformations. The adjoint representation of E_{11} encoded an infinite number of fields that carry the infinite number of possible duality transformations; the l_1 representation achieves the same goal, but for the coordinates.

The coordinates with totally anti-symmetrised space-time indices in the non-linear realisation appropriate for theories in less that ten dimensions can be read off from the charges in table 17.5.5. For example, in seven dimensions we have ten scalar charges that belong to the $\mathbf{10}$-dimensional representation of SL(5). As one can easily trace the relationship between the eleven-dimensional brane charges and the charges in seven dimensions, see, for example, the case of scalar charges in seven dimensions discussed above, one can, using the obvious connection between brane charges and coordinates, relate the coordinates in the dimensionally reduced theory to the brane charges in eleven dimensions. Thus one extends the relation between the usual coordinates x^i and the internal space-time translations P_i to a relation involving U duality multiplets of corrdinates and charges.

In the generalised space-time we have for every field type a corresponding set of coordinates and so a new geometry generalising that of Riemann; the latter possesses only the metric and its corresponding coordinates x^a. Non-linear realisations of the type we are adopting here are explained in section 13.2 and in particular from equation (13.2.62) onwards. Looking at equation (13.2.63) we find an expression for the generalised vielbein on the generalised space-time. Indeed, at low levels, it is easy to compute.

The non-linear realisations we constructed above adopted a simplification of the procedure advocated here in that we only took the coordinates x^a. However, E_{11}, except for the level 0 sub-algebra SL(11), rotates this coordinate into the higher coordinates, nonetheless we also kept some of higher level fields in the adjoint representation of E_{11}. Of course, in the non-linear realisation of $E_{11} \otimes l_1$ one should include the higher coordinates and this has been done in [17.60, 17.61] at low levels; one finds a field theory on the generalised space-time and one can wonder how this is related to the field theories to which we are used. Some thoughts on this point can be found in [17.63]. Thus we arrive at the unsolved problem: although there is much evidence for the adjoint representation of E_{11} and the l_1 representation as the multiplet of brane charges, there is still a problem with how to introduce a space-time in a way that we understand.

17.5.7 Weyl transformations of E_{11} and the non-linear realisation of its Cartan sub-algebra

As we have seen so little is known about Kac–Moody algebras that it is difficult to systematically calculate the general properties of a non-linear realisation based upon them. However, by setting to zero all the fields of the non-linear realisation, except those associated with the Cartan sub-algebra, the group element takes the very simple form

$$g = \exp\left(\sum_a q_a H_a\right). \tag{17.5.46}$$

The fields q_a associated with the Cartan sub-algebra are then the only fields of the theory. Provided one only carries out operations that preserve the Cartan sub-algebra, it is then almost trivial to examine the consequences. Such is the case for Weyl transformations [16.11].

As we found in section 16.7.1, the Cartan sub-algebra of E_{11} consists of the generators $K^a{}_a$, $a = 1, \ldots, 11$ and so one can also write the group element of equation (17.5.46) in the form

$$g = \exp\left(\sum_{a=1}^{11} h_a{}^a K^a{}_a\right) = \exp(h^T K). \tag{17.5.47}$$

In the second equation we have used matrix notation, for example, the symbol K is a column vector whose ath component is $K^a{}_a$. The relationship between the Chevalley generators H_a and the 'physical' generators $K^a{}_a$ was found in section 16.7.1 to be given by

$$H_a = K^a{}_a - K^{a+1}{}_{a+1}, a = 1, \ldots, 10,$$

$$H_{11} = -\tfrac{1}{3}(K^1{}_1 + \ldots + K^8{}_8) + \tfrac{2}{3}(K^9{}_9 + K^{10}{}_{10} + K^{11}{}_{11}). \tag{17.5.48}$$

We may write this in matrix form as

$$K = \rho H \quad \text{and so} \quad q = \rho^T h, \qquad \bullet \tag{17.5.49}$$

the last equation follows from equating the group elements in equations (17.5.46) and (17.5.48). We recall that the vielbein is given by $e_\mu{}^a = (e^h)_\mu{}^a$ and so we find that keeping only fields associated with the Cartan sub-algebra implies keeping only the diagonal parts of the vielbein and so the metric. We introduce the notation

$$e_a{}^a = (e^h)_a{}^a \equiv e^{p^a}. \tag{17.5.50}$$

In terms of the metric this implies that $g_{aa} = e^{2p^a}\eta_{aa}$.

The Weyl reflection S_a corresponding to the simple root α_a on any weight β is given (see equation (16.1.82)) by

$$S_a\beta = \beta - 2\frac{(\beta, \alpha_a)}{(\alpha_a, \alpha_a)}\alpha_a.$$

For the simple roots this becomes

$$S_a\alpha_b = \alpha_b - 2\frac{(\alpha_b, \alpha_a)}{(\alpha_a, \alpha_a)}\alpha_a \equiv (s_a)_b{}^c\alpha_c. \tag{17.5.51}$$

Given the correspondence between the roots and the elements of the Cartan sub-algebra explained in chapter 16 we conclude that the action of the Weyl transformation S_a on the Cartan sub-algebra of a Kac–Moody algebra is given by

$$H'_b = S_aH_b = (s_a)_b{}^c H_c \quad \text{or in matrix notation} \quad H' = s_aH. \tag{17.5.52}$$

Since the Weyl group acts on Cartan sub-algebra generators to give Cartan sub-algebra generators, it makes sense to consider their action on elements restricted to be of the form of equation (17.5.46). Writing the group element of equation (17.5.46) in matrix form $g = \exp(q^T H)$, we conclude that the Weyl group acts on the fields q,

$$q'^T = q^T s_a, \text{ or } q' = s_a^T q, \tag{17.5.53}$$

as $s_a^2 = I$.

To find the physical effects of the Weyl transformations we need to find their action on the physical variables $h_a{}^a$. Using the matrix notation and, in particular, $H = \rho K$, the effect of the Weyl transformation is $S_aK = K' = \rho^{-1}s_a\rho K \equiv r_aK$ and so the physical fields h transform as

$$h' = r_a^T h. \tag{17.5.54}$$

Let us first consider the Weyl transformation generated by the simple root α_c, $c \neq 1, 8, 10, 11$. Its action on the Chevalley generators H_a is easily read off using equations (17.5.51) and (17.5.52) to be

$$S_cH_a = H_a \quad a \neq c, a \neq c \pm 1, \ S_cH_c = -H_c \quad S_cH_{c\pm1} = H_c + H_{c\pm1}. \tag{17.5.55}$$

It is then straightforward to see from equation (17.5.48) that the corresponding transformation on $K^a{}_a$ is given by

$$K^a{}_a \leftrightarrow K^{a+1}{}_{a+1}. \tag{17.5.56}$$

This relation is actually also valid for the Weyl reflections in all the other simple roots except that in α_{11}, that is, for the values $c = 1, \ldots, 10$, as is easily verified separately for $n = 1, 8, 10$.

Hence the Weyl transformation associated to a simple root α_c, $c = 1, \ldots, 10$ induces the transformation

$$p^a \leftrightarrow p^{a+1} \quad a = n, n \neq 11, \tag{17.5.57}$$

all other p^a being unchanged. In other words, these Weyl transformations just interchange two of the diagonal components of the metric.

It remains to find the action of the Weyl transformation generated by the simple root α_{11}. Following the same arguments we find that

$$S_{11}H_a = H_a, a \neq 8, a \neq 11, S_{11}H_{11} = -H_{11}, S_{11}H_8 = H_8 + H_{11}. \tag{17.5.58}$$

Examining the H_a that are unchanged we conclude that

$$K^a{}_a \rightarrow K^a{}_a + y \quad a = 1, \ldots, 8, \quad K^a{}_a \rightarrow K^a{}_a + x \quad a = 9, 10, 11. \tag{17.5.59}$$

The remaining conditions imply that $y = 0$ and $x = -H_{11}$. As a result, we find that

$$S_{11}K^a{}_a = K^a{}_a, \quad a = 1, \ldots, 8,$$
$$S_{11}K^a{}_a = K^a{}_a + \tfrac{1}{3}(K_1^1 + K_2^2 + \cdots + K_8^8) - \tfrac{2}{3}(K_9^9 + K_{10}^{10} + K_{11}^{11}), \tag{17.5.60}$$
$$a = 9, 10, 11.$$

Translated in terms of the variables p^a of equation (17.5.50) we conclude that

$$p'^a = p^a + \tfrac{1}{3}(p^9 + p^{10} + p^{11}) \quad a = 1, \ldots, 8 \quad p'^a = p^a - \tfrac{2}{3}(p^9 + p^{10} + p^{11}),$$
$$a = 9, 10, 11. \tag{17.5.61}$$

The above relations hold in the eleven-dimensional theory, however, if we dimensionally reduce the theory the effects are particularly easy to interpret. We recall from equation (13.1.46) that the physical radii of a rectangular torus, that is, the metric is diagonal in the compactified directions, is given by $R_a = l_p \int e_a{}^a dx_a$ and so we find that

$$h_a{}^a = ln\frac{R_a}{l_p} \text{ or } p_a = \frac{R_a}{l_p}. \tag{17.5.62}$$

The Weyl transformations S_a, $a = 1, \ldots, 10$ imply that

$$R_a \leftrightarrow R_{a+1}, \ l_p \rightarrow l_p. \tag{17.5.63}$$

However, S_{11} induces the transformations of equation (17.5.61), which in turn implies that

$$R_9' = \frac{l_p^3}{R_{10}R_{11}}, \ R_{10}' = \frac{l_p^3}{R_{11}R_9}, \ R_{11}' = \frac{l_p^3}{R_9R_{10}}, \ (l_p')^3 = \frac{l_p^6}{R_9R_{10}R_{11}}. \tag{17.5.64}$$

In deriving these equations we have assumed that there is at least one dimension that is not compactified and so whose radius is unchanged. From the viewpoint of the IIA, or the IIB, string, we recognise this as a doubled T duality [17.58, 17.59]. However, we have in addition the change of the Planck scale. In a similar way one can derive the effects of the E_{11} Weyl transformations in the IIA and IIB theories [17.62]. These include the T duality transformations of the IIA string and the non-perturbative S duality of the IIB theory.

18 String interactions

"The pole, yes but under very different circumstances from those expected. Great God this is an awful place and terrible enough for us to have laboured to it without the reward of priority. "

Scott on reaching the South Pole to find the flag of Amundsen

The discovery of string theory is a remarkable story that began as an attempt to understand hadronic dynamics. It was believed that quantum field theory could not account for the dynamics of hadrons and that one might be able to simply write down the S matrix which was to satisfy a set of properties one of which was called duality. An S matrix for initially four, and then any number, of spin-0 particles of the same mass which did satisfy all the desired properties was eventually found. Although it might seem that these S matrix elements could not be of much value, using physical and mathematical consistency, the early pioneers deduced from them the S matrix for the scattering of an infinite number of particles that were identified as those exchanged in the scattering of the spin-0 particles. Eventually these particles were identified with the states of a string. In telling this story in section 18.1 we will also get a good feeling for what string scattering amplitudes look like, what are their properties and, indeed, for string theory itself.

There is no known complete theory of strings that can be used to compute, even as a matter of principle, all their properties. However, there are a number of approaches that can be used to compute the perturbative scattering of strings. One of the earliest and most used approaches involves a sum over the world surfaces swept out by the string and this is the subject of section 18.2.

In section 18.3 we give an alternative, less well-known approach, which goes under the name of the group theoretic approach. One of the unusual aspects of string theory is that the theory used to calculate string scattering is often more complicated than the result itself, namely the string scattering amplitudes. The group theoretic approach gives an elegant set of equations which the scattering amplitudes obey and these are then used to determine the amplitudes directly.

Finally in section 18.4 we discuss interacting string field theory.

18.1   Duality, factorisation and the origins of string theory

String theory arose from an attempt to understand hadronic interactions in the late 1960s. At that time it was felt that the complexity of hadronic phenomena which included many states of spin 2 and above could not be described by a quantum field theory. This was

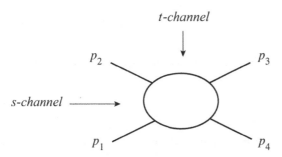

Figure 18.1.1   Four-particle scattering.

understandable since quantum field theory was plagued with infinities which were only removed by a complicated renormalisation process in certain theories such as the spin-0 ϕ^4 theory and QED which contains spin-1 and spin-$\frac{1}{2}$ particles. Also it did not seem likely that quantum field theories for higher spin particles could be renormalisable. Furthermore, the significance of non-Abelian gauge symmetry had not yet become apparent and so it was not at all clear what Lagrangian one should take as one's starting point. In an alternative approach, which grew out of the ideas of Heisenberg, it was assumed that one should only discuss physical quantities and so deal directly with the S matrix. It was to be deduced through a set of physical properties that it should satisfy. These included some well-known properties, such as unitary, causality and Lorentz invariance, which are present in any quantum field theory, and some less familiar requirements such as factorisation and duality. It was this last property, which we will explain as we go along, that was the most crucial. Duality arose historically in a particular approximation which enjoyed some success in accounting for the observed hadronic scattering data [18.1]. There began a search [18.2–18.5] to find such an S matrix and the first S matrix which actually realised some of these desirable properties was the Veneziano formula [18.6] for the scattering of four spin-0 states. We take the spin-0 particles to have momenta p_i, $i = 1, \ldots, 4$ and define

$$s = -(p_1 + p_2)^2, \qquad t = -(p_2 + p_3)^2, \qquad u = -(p_1 + p_3)^2, \qquad (18.1.1)$$

which are the rest frame (energies)^2 in the various channels also called s, t and u (see figure 18.1.1).

We take the four spin-0 particles to have the same mass, which we denote by μ. Since $-p_i^2 = \mu^2$ one finds that

$$s + t + u = 4\mu^2 = -\sum_{i=1}^{4} p_i^2. \qquad (18.1.2)$$

Neglecting the trivial term corresponding to no scattering, the S-matrix for the four spin-0 particles was postulated to be a sum of three terms:

$$T^4(p_i) = F(s, t) + F(t, u) + F(u, s), \qquad (18.1.3)$$

where

$$F(s, t) = g^2 \frac{\Gamma(-\alpha(s))\Gamma(-\alpha(t))}{\Gamma(-\alpha(s) - \alpha(t))} \qquad (18.1.4)$$

and

$$\alpha(s) = \alpha' s + \alpha_0. \tag{18.1.5}$$

The constants α' and α_0 were called the slope and intercept, respectively. The Γ function has its usual integral representation

$$\Gamma(u) = \int_0^\infty dt e^{-t} t^{u-1}, \tag{18.1.6}$$

which is valid for Re $u > 0$. In this region, one can show by integration by parts that

$$\Gamma(u+1) = u\Gamma(u). \tag{18.1.7}$$

One observes from equation (18.1.6) that $\Gamma(1) = 1$, while $\Gamma(n+1) = n!$ for any positive integer n.

We would like to define Γ for all u. This is achieved by demanding that Γ is given by equation (18.1.6) for Re $u > 0$ and also satisfies equation (18.1.7), which was derived for Re $u > 0$, but is now taken to hold for all u. From equation (18.1.7) we then find, for $u \neq 0, -1, -2, \dots$, that

$$\Gamma(u) = \frac{1}{(u)} \frac{1}{(u+1)} \cdots \frac{1}{(u+n)} \Gamma(u+n+1). \tag{18.1.8}$$

As $\Gamma(u+n+1)$ is defined by equation (18.1.6) for Re $u > -n - 1$, one finds a definition of $\Gamma(u)$ in this region. Since n can be as large as one likes we have defined $\Gamma(u)$ in the whole complex plane. At the points $u = 0, -1, -2, \dots$, $\Gamma(u)$ has a simple pole. Hence $\Gamma(u)$ is analytic at all points in the complex plane except at the points $u = 0, -1, -2$ at which it has simple poles and the point $|u| = \infty$, where it has an essential singularity.

Actually $F(s, t)$ is recognizable in terms of the well-known beta function which is defined by

$$B(x, y) = \frac{\Gamma(x)\Gamma(y)}{\Gamma(x+y)} \tag{18.1.9}$$

and has the integral representation

$$B(x, y) = \int_0^1 t^{x-1}(1-t)^{y-1} dt. \tag{18.1.10}$$

This result follows from the integral representation of $\Gamma(u)$ of equation (18.1.6) after the appropriate change of variables. Starting from the product of two Γ functions and changing the two integration variables to ξ^2 and η^2 and then changing to r, θ we find that

$$\Gamma(x)\Gamma(y) = 4 \int_0^\infty \int_0^\infty d\xi d\eta \; \eta^{2x-1} \xi^{2y-1} e^{-\eta^2 - \xi^2}$$

$$= 4 \int_0^{\frac{\pi}{2}} \int_0^\infty r^{2x+2y-1} (\cos\theta)^{2x-1} (\sin\theta)^{2y-1} e^{-r^2} dv d\theta$$

$$= 2\Gamma(x+y) \int_0^{\frac{\pi}{2}} (\cos\theta)^{2x-1} (\sin\theta)^{2y-1} d\theta. \tag{18.1.11}$$

After the further substitution $\cos\theta = u^{\frac{1}{2}}$ we recognise the second term in the last line as the integral representation of equation (18.1.10). Although $B(x, y)$ is valid for Re $x > 0$,

Re $y > 0$, we can define its value on the whole complex plane from its expression in terms of Γ functions with the above definition, or equivalently, by working with the equivalent recursion relations for $B(x, y)$. A good reference giving more details on the Γ and many other functions is reference [18.7].

As a result $B(x, y)$, which is a symmetric function of x and y, for y fixed, has poles at $x = 0, -1, -2, \ldots$; an essential singularity at $|x|$ is infinite, but is otherwise analytic. Note it has no poles in both x and y as then also $\Gamma(x + y)$ has a simple pole. Expanding the second bracket in equation (18.1.10), we find, for Re $y > 0$, that

$$B(x, y) = \sum_{n=0}^{\infty} \frac{(-1)^n}{n!} \frac{1}{(x+n)} \frac{\Gamma(y)}{\Gamma(-n+y)}.$$

(18.1.12)

This expression includes all the poles in $B(x, y)$ implied by the definition of $\Gamma(x)$ for Re $x > 0$ given above. Using equation (18.1.9), we may write

$$F(s, t) = g^2 \int_0^1 dx (x)^{-\alpha(s)-1} (1-x)^{-\alpha(t)-1}.$$

(18.1.13)

Using equation (18.1.12), we find that

$$F(s, t) = g^2 \sum_{n=0}^{\infty} \frac{1}{n!} \frac{(\alpha(t)+1)(\alpha(t)+2)\cdots(\alpha(t)+n)}{(n-\alpha(s))}.$$

(18.1.14)

Hence we find poles, that is, resonances, in the s channel whenever

$$\alpha(s) = \alpha' s + \alpha_0 = 0, 1, 2, \ldots.$$

(18.1.15)

This corresponds to particles being exchanged in these channels of masses given by

$$m^2 = \frac{1}{\alpha'}(n - \alpha_0)$$

(18.1.16)

for n a positive integer. The fact that the residue is a polynomial of degree n indicates that one of the particles of this mass being exchanged is of spin $J = n$. We recall that a particle of spin J has J Lorentz indices and so couples to two scalars as $\phi \partial_{\mu_1}, \ldots, \partial_{\mu_J} \phi$. Thus from two such vertices one finds J factors of momentum squared. Hence, spin J is the highest particle exchanged at this mass. Thus for the highest spin particles of a given mass $m_J^2 = (J - \alpha_0)/\alpha'$, and so the theory contains particles of ever-increasing spin whose (mass)^2 is linearly rising with their spin. Since we have a residue which is a polynomial in t, we also have other particles. If we were to identify the mass μ of the external spin-0 particles with the lowest mass particle in the spectrum which is also a scalar we would find that

$$\mu^2 = -\frac{\alpha_0}{\alpha'}.$$

(18.1.17)

The only reason to do this so far is the desire to identify the spin-0 particles being scattered with those being exchanged and so not introduce more scalar particles. However, later we will find more convincing reasons. We also note that unless $\alpha' > 0$, we will have an infinity of tachyons. However, even if $\alpha' > 0$ and $\alpha_0 > 0$ we will have one or more tachyons according to equation (18.1.17).

F is symmetric in s and t and, using equation (18.1.12), we may therefore write F as

$$F(s,t) = \sum_{n=0}^{\infty} \frac{P_n(t)}{n - \alpha(s)} = \sum_{n=0}^{\infty} \frac{P_n(s)}{n - \alpha(t)}, \tag{18.1.18}$$

where $P_n(x)$ is a polynomial of degree n in x which is given by

$$P_n(x) = g^2 \frac{1}{n!}(\alpha(x) + 1)(\alpha(x) + 2) \cdots (\alpha(x) + n). \tag{18.1.19}$$

This is the expression of the remarkable property of *duality* [18.1], which was the most important motivating property mentioned earlier that was crucial for the discovery of the Veneziano formula, namely $F(s,t)$ can be expressed in terms of poles in s for fixed t, but *also* expressed in terms of poles in t for fixed s. Clearly, to get duality, one requires an infinite number of poles as otherwise the first term in equation (18.1.18) would be a finite polynomial in t and so analytic in t.

Another way of expressing duality is to denote $F(s,t) \equiv A(1,2,3,4)$. Cycling the variables, that is, $p_i \to p_{i+1}$, we find $s \to t$ and $t \to s$ and so

$$A(2,3,4,1) = F(t,s) = F(s,t) = A(1,2,3,4). \tag{18.1.20}$$

That is, the symmetry of $F(s,t)$ under exchange of s and t, which ensures duality, is the same as $A(1,2,3,4)$, being the same if we cycle its indices. We also note that under $1 \leftrightarrow 2$, $3 \to 3$, $4 \to 4$, then $s \to s$ and $t \leftrightarrow u$ and so $F(s,u) = A(2,1,3,4)$. As such, the three terms in the non-trivial S matrix can be written as

$$T^4(p_i) = A(1,2,3,4) + A(2,1,3,4) + A(1,3,2,4). \tag{18.1.21}$$

We note that the poles in the various channels arise from the end points of the integration regions. For example, in equation (18.1.13) the s-channel and t-channel poles arise from the $x = 0$ and 1 end points, respectively.

The three terms for the S matrix are given by

$$g^2 \int_0^1 dx \; x^{-\alpha(s)-1}(1-x)^{-\alpha(t)-1} + g^2 \int_0^1 dx \; x^{-\alpha(t)-1}(1-x)^{-\alpha(u)-1}$$

$$+ g^2 \int_0^1 dx \; x^{-\alpha(u)-1}(1-x)^{-\alpha(s)-1}. \tag{18.1.22}$$

We note that they can be combined into one term, namely

$$g^2 \int_{-\infty}^{\infty} |x|^{-\alpha(s)-1}|1-x|^{-\alpha(t)-1}, \tag{18.1.23}$$

by making use of the substitutions

$$\frac{1}{1-x} = \bar{x} \quad \text{and} \quad \bar{x} = \frac{x-1}{x} \tag{18.1.24}$$

in the second and third terms, respectively, using $s + t + u = 4\mu^2$ and dropping the bars, provided we *assume* the relation

$$4\alpha'\mu^2 + 3\alpha_0 + 1 = 0. \tag{18.1.25}$$

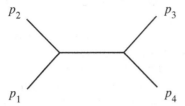

Figure 18.1.2   Lowest order scattering in φ^3.

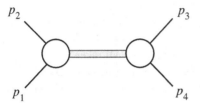

Figure 18.1.3   Factorisation in the s-channel of four-particle scattering.

This in conjunction with equation (18.1.17) implies that $\alpha_0 = 1$ and $\mu^2 = -1/\alpha'$. In fact, we will see later the consistency of the theory requires $\alpha_0 = 1$ and $\mu^2 = -1/\alpha'$ and so the above conditions will be valid.

The property of duality is not present in the individual Feynman graphs that arise in quantum field theory. There the S-matrix is made up of Feynman graphs. For example, in ϕ^3 theory the tree-level graph of figure 18.1.2 is equal to $g^2 s^{-1}$ which is not expressible in terms of poles in t, the poles in t arising from a different Feynman graph.

It is instructive to examine the S matrix in a given channel, that is, to examine its dependence on the momentum squared of two chosen particles. For particles 1 and 2 this means the momentum $-(p_1 + p_2)^2 = s$ and we refer to this as the s-channel. *Factorisation* for the four-point function means that if one picks a given channel, say the s-channel, then the residues of the poles in s can be written as a product of two factors: one factor corresponds to the amplitude for states 1 and 2 to interact with the particle being exchanged, and the other corresponds to the amplitude for states 3 and 4 interacting with the particle being exchanged; see figure 18.1.3 in which the grey line corresponds to all the possible particles being exchanged. In equations, this would mean

$$F(s,t) = \sum_{n=0}^{\infty} \sum_{d} \frac{g(1,2;n,d)g(3,4,n,d)}{(-(\alpha's^2 + \alpha_0) + n)}, \tag{18.1.26}$$

where the sum over d denotes the sum over all the states with the same mass which are exchanged. As we have stated, factorisation means that the first factor depends only on the properties (for example, momenta) of particles 1 and 2 and the particle being exchanged, and does not depend on the properties of particles 3 and 4. An equivalent restriction also holds for the second factor. Using partial wave analysis, one can identify what are the spins of the states that are being exchanged and also read off the on-shell scattering amplitudes between the two spin-0 states and all the particles being exchanged. The only problem occurs when more than two particles of the same spin and mass are exchanged in which case one cannot identify the amplitude for each spin separately. In an S matrix theory, of which any quantum field theory is an example, it can be shown from unitarity that for an arbitrary amplitude the residues of any poles will factorise, see reference [18.8] for a review. In the beginning of string theory, one was in effect simply handed an S matrix and one did not know in advance whether it was unitary or that it would factorise. That factorisation did take place was one of the features that greatly encouraged the early workers in string theory.

We now give a concrete account of the above discussion and carry out the factorisation explicitly for the first two levels. Let us take the two particles that are incoming to be labeled by 1 and 2 and the outgoing particles by 3 and 4. We will work in the centre of mass frame of the collision and let us denote the total energy by $2E$. Then the incoming particles have 4-momentum $p_1^\mu = (E, \underline{p})$ and $p_1^\mu = (E, -\underline{p})$. The outgoing particles have 4-momentum $p_3^\mu = (-E, -\underline{p}')$ and $p_4^\mu = (-E, \underline{p}')$. The 4-moment a are all taken to be incoming as is illustrated by the conservation law $\sum_i p_i^\mu = 0$ which we have used to fix the energy of the outgoing particles in terms of the incoming particles. Since $p_i^2 = -\mu^2$, where μ^2 is the mass squared of the external spin-0 particle, we have that $E^2 = \underline{p}^2 + \mu^2 = \underline{p}'^2 + \mu^2$ and so $\underline{p}^2 = \underline{p}'^2$. We take the scattering angle to be θ and it is oriented such that $\underline{p} \cdot \underline{p}' = \underline{p}^2 \cos\theta$. Using these formulae it is now straightforward to find that

$$s = 4E^2, \qquad t = -2(\underline{p})^2(1 - \cos\theta) = -\tfrac{1}{2}(s - 4\mu^2)(1 - \cos\theta). \qquad (18.1.27)$$

The exchange of a resonance of spin J and mass m in the s-channel gives the contribution (see reference [18.8] for a review)

$$\frac{g^2}{s - m^2} P_J(\cos\theta), \qquad (18.1.28)$$

where P_J is the Legendre polynomial, to the scattering amplitude. The symbol g will turn out to denote the coupling constant. In fact $P_0(\cos\theta) = 1$, $P_0(\cos\theta) = \cos\theta$ and $P_2(\cos\theta) = \tfrac{1}{2}(3(\cos\theta)^2 - 1)$.

The lowest mass particle exchanged has a mass squared, m^2, given by $\alpha'm^2 = -\alpha_0$ and its residue is a constant, implying that only a spin-0 particle is exchanged at this mass. The next pole occurs when $s = m^2 = (1 - \alpha_0)(1/\alpha')$ and the residue at this pole is given by

$$g^2(\alpha't + \alpha_0 + 1) = g^2\left[\left(-\frac{\alpha's}{2} + 2\alpha'\mu^2\right)(1 - \cos\theta) + \alpha_0 + 1\right]$$

$$= g^2\left[\left(2\alpha'\mu^2 + \frac{3}{2}\alpha_0 + \frac{1}{2}\right)\right.$$

$$\left. - \left(2\alpha'\mu^2 - \frac{(1 - \alpha_0)}{2}\right)\cos\theta\right]. \qquad (18.1.29)$$

Hence, in general, we have one spin-0 (the constant term) particle and one spin-1 (the $\cos\theta$ term) particle. However, if $4\alpha'\mu^2 + 3\alpha_0 + 1 = 0$, which will turn out to be the case, we only have one spin-1 particle.

Using the above explicit expressions for the 4-momenta and $s = (1 - \alpha_0)(1/\alpha')$ we find that

$$(p_2 - p_1).(p_3 - p_4) = -\cos\theta\left[4\mu^2 - \frac{(1 - \alpha_0)}{\alpha'}\right] \qquad (18.1.30)$$

and hence the spin-1 residue can be written as

$$\frac{\alpha' g^2}{2}(p_2 - p_1) \cdot (p_3 - p_4), \qquad (18.1.31)$$

which does indeed factorise into two pieces. In particular, associated with particles 1 and 2 we find the factor $\left(\sqrt{\alpha'}g/\sqrt{2}\right)(p_2 - p_1)^\mu$ with a similar factor for particles 3 and 4. In terms of Lagrange field theory, we recognise the spin-1–two spin-0 vertex to be one of the form $g\phi\partial^\mu\phi A^\mu \sqrt{\alpha'}$, which is consistent with this result. As the reader may verify at the next level one finds that there is in general one spin 2, one spin 1 and one spin-0 particle. If we adjust α_0 (that is, $\alpha_0 = 1$) as above, then the spin 1 particle disappears. Having done this in 26 dimensions, one finds only a spin 2 particle but in all other dimensions there is a spin-2 and a spin-0 particle. However, for dimensions greater than 26 one finds that the spin-0 particle gives a negative contribution to the residue and so unitarity is violated. Thus, we learn even at this stage that the theory which has the four-spin-0 scattering of equation (18.1.3) must live in 26 or less dimensions with the theory in 26 dimensions having a distinctly different spectrum to that in less than 26.

The next step taken by the early pioneers was to guess the scattering of N spin-0 particles, all of the same mass squared μ^2 [18.9–18.14]. The key was to generalise the parametric form of equation (18.1.13) for the four-point scattering. Let us label the N external legs in cyclic order by $i = 1, \cdots, N$ beginning with any leg and associate a real variable z_i, called the Koba–Nielsen variable [18.14], with leg i such that $z_{i+1} > z_i$. We next define the quantity

$$A(1, 2, \ldots, N) = g^{N-2} \int_{z_1 < z_2} \cdots \int_{z_{N-1} < z_N} dz_1 \cdots (dz_a) \cdots (dz_b) \cdots (dz_c) \cdots dz_N$$

$$\times |z_b^0 - z_c^0||z_a^0 - z_b^0||z_c^0 - z_a^0| \cdot \prod_{\substack{i,j \\ 1 \leq i < j \leq N}} |z_i - z_j|^{2\alpha' p_i \cdot p_j} \qquad (18.1.32)$$

from which we will construct the S-matrix. The notation in this formula means we omit the integrations over z_a, z_b and z_c, which are assigned the arbitrary values z_a^0, z_b^0 and z_c^0, respectively, provided $z_a^0 < z_b^0 < z_c^0$. At first sight, the formula is rather strange in that it singles out three zs, that is, z_a, z_b, z_c for special treatment and also gives them preferred values. In fact, this is not really the case; we will now demonstrate that $A(1, 2, \cdots, N)$ does not depend on which three variables one singles out not be integrated over and it also does not depend on the three values we choose them to take. To demonstrate this requires the use of projective transformations, which are transformations of the form

$$z \to \frac{az + b}{cz + d}, \qquad (18.1.33)$$

where z is any point on the half-plane. Since any such transformation can be written as a product of an inversion, a dilation and translations, to prove the projective invariance of A it suffices to demonstrate it for each of these particular transformations (see section 8.1). It is instructive to rewrite A as a measure and an integrand

$$g^{n-2} \int_{z_1 < z_2} \cdots \int_{z_{N-1} < z_N} d\mu \; I,$$ (18.1.34)

where the split is as follows:

$$d\mu = dz_1 \cdots (dz_a) \cdots (dz_b) \cdots (dz_c) \cdots dz_N \; |z_b^0 - z_c^0||z_c^0 - z_a^0||z_c^0$$
$$- z_a^0| \prod_{i=1}^{N} |z_{i+1} - z_i|^{-1}$$ (18.1.35)

and

$$I = \prod_{\substack{i,j \\ 1 \le i < j \le N}} |z_i - z_j|^{2\alpha' p_i \cdot p_j} \prod_{i=1}^{N} |z_{i+1} - z_i|,$$ (18.1.36)

where $z_{N+1} = z_1$.

For some purposes, it is more useful to write $A(1, 2, \ldots, N)$ as

$$A(1, 2, \ldots, N) = \int \cdots \int \prod_{i=1}^{N} dz_i \; \delta(z_a - z_a^0)\delta(z_b - z_b^0)\delta(z_c - z_c^0)$$
$$\times \prod_{i=1}^{N} |z_{i+1} - z_i|^{-1} \prod_{i=1}^{N} \Theta(z_{i+1} - z_i)||z_b - z_c||z_c - z_a||z_c - z_a| \; I,$$ (18.1.37)

where z_a^0, z_b^0 and z_c^0 are the chosen values of these three coordinates. Clearly, $d\mu$ and I are invariant separately under translations and $d\mu$ is manifestly dilation invariant. Under a dilation, we find that, upon using momentum conservation,

$$I \to I \left\{ \prod_{\substack{i,j \\ 1 \le i < j \le N}} |\lambda|^{2\alpha' p_i \cdot p_j} \right\} |\lambda|^N = I|\lambda|^N \; \exp \sum_{\substack{1 \le i < j \le N \\ i,j}} 2(p_i \cdot p_j)\alpha' \ln |\lambda|$$

$$= |\lambda|^N \exp \sum_{i=1}^{N} \sum_{\substack{j=1 \\ j \ne i}}^{N} \alpha' p_i \cdot p_j \ln |\lambda| I = I|\lambda|^{N - \alpha' \sum_{i=1}^{N} p_i^2} = |\lambda|^{N(1 + \alpha' \mu^2)} I.$$ (18.1.38)

Consequently, I is dilaton invariant, provided $p_i^2 \equiv -\mu^2 = 1/\alpha'$. This implies that the mass of spin-0, which appears in the original amplitude, is a tachyon, but all the other particles exchanged have non-tachyonic masses. Adopting this condition and identifying the mass of spin 0 with the lowest mass spin particle that is exchanged in the theory, that is, equation

(18.1.17), implies $\alpha_0 = 1$. In fact, one can slightly generalise equation (18.1.27) for A in such a way that the restriction is not required; one puts in a factor $\prod_{i=1}^{N} |z_{i+1} - z_i|^{(-1+\alpha'\mu^2)}$. However, we will not make this addition.

Under an inversion (that is, $z_i \to -1/z_i$), the measure $d\mu$ is easily found to be invariant while the integrand I transforms as

$$
I \to I \prod_{\substack{i,j \\ 1 \leq i < j \leq N}} \left\{ |\frac{1}{z_i}||\frac{1}{z_j}|^{2\alpha' p_i \cdot p_j} \prod_{i=1}^{N} |\frac{1}{z_i}||\frac{1}{z_{i+1}}| \right\}
$$

$$
= I \exp \left\{ -\left[\sum_{i=1}^{N} \ln|z_i| + \ln|z_{i+1}| \right] - \left\{ \sum_{\substack{i,j \\ 1 \leq i < j \leq N}} 2\alpha'(p_i \cdot p_j)(\ln|z_i| + \ln|z_j|) \right\} \right\}
$$

$$
= I \exp 2 \sum_{i=1}^{N} \ln|z_i|\{-1 + \alpha' p_i^2\}. \tag{18.1.39}
$$

It is invariant when $p_i^2 \equiv -\mu^2 = 1/\alpha'$, which was the same condition as for dilation invariance. Thus A is projectively invariant when $p_i^2 \equiv -\mu^2 = 1/\alpha'$. We recall that given three distinct points, there always exists a projective transformation which can take them to any other three distinct points (see section 8.1). Since both the measure and the integrand are invariant, we may make a projective transformation to make z_a, z_b and z_c take on any desired values and the value of A is unchanged. Furthermore, as the result is projective invariant we may move all the z_i to be on any contour which is related by a projective transformation to the real axis. One such example is the unit circle. We recall that the cyclic order of the z_i is always preserved under a projective transformation.

A little more subtle is the demonstration that we may choose any three points not to be integrated over and the result is the same. In order to show the general case, it suffices to show that we can choose z_{a+1} rather than z_a to be the variable not integrated over. The idea is that starting from equation (18.1.32), where z_a is fixed, we may always perform the projective transformation T that takes z_{a+1} to a constant denoted z_{a+1}^0 and z_b and z_c to be as before, that is, to take the values z_b^0 and z_c^0 respectively. Thus we consider the transformation

$$
T \quad \text{maps} \quad \begin{pmatrix} z_{a+1} & z_b^0 & z_c^0 \\ z_{a+1}^0 & z_b^0 & z_c^0 \end{pmatrix}. \tag{18.1.40}
$$

Let us write this transformation as $z' = T(z) = (az+b)/(cz+d)$ where a, b, c and d depend on z_{a+1}, z_{a+1}^0, z_b^0 and z_c^0. We note that $dT(z)/dz = (ad - bc)/(cz+d)^2$.

The transformation T has two fixed points z_b^0 and z_c^0 and so may also be written in the form

$$
\frac{T(z) - z_b^0}{T(z) - z_c^0} = K \frac{(z - z_b^0)}{(z - z_c^0)}, \tag{18.1.41}
$$

where K is independent of z, but depends on z_{a+1}, $z_{a+1}^0 z_b^0$ and z_c^0. Taking $z = z_{a+1}$ we find the precise dependence is given by

$$K = \frac{(z_{a+1}^0 - z_b^0)}{(z_{a+1}^0 - z_c^0)} \frac{(z_{a+1} - z_c^0)}{(z_{a+1} - z_b^0)}. \qquad (18.1.42)$$

Although $T(z_{a+1})$ is fixed to be z_{a+1}^0, a constant, z_a is now no longer fixed to be z_a^0 and is now a function of z_{a+1}. Differentiating equation (18.1.41), remembering that K is a function of z_{a+1} and then using equations (18.1.42) and (18.1.41), we find that

$$\frac{dT(z)}{dz_{a+1}} = \frac{(T(z) - z_b^0)}{(z_{a+1} - z_b^0)} \frac{(T(z) - z_c^0)}{(z_{a+1} - z_c^0)}. \qquad (18.1.43)$$

Let us write the expression for $A(1, 2, \ldots, N)$ of equation (18.1.32) in the form

$$A(1, 2, \ldots, N) = \int \cdots \int \prod_{\substack{i=1 \\ i \neq a,b,c}}^{N} dz_i |z_a^0 - z_b^0| |z_a^0 - z_c^0| |z_b^0 - z_c^0| \ f(z_1, \ldots, z_N), \qquad (18.1.44)$$

where z_a^0, z_b^0 and z_c^0 are the fixed values of z_a, z_b and z_c. Since A is projective invariant under $z_i' = \tilde{T}(z_i) = (\tilde{a} z_i + \tilde{b})/(\tilde{c} z_i + \tilde{d})$ and as

$$\prod_{\substack{i=1 \\ i \neq a,b,c}}^{N} dz_i' |z_a' - z_b'| |z_c' - z_b'| |z_a' - z_c'|$$

$$= \left\{ \prod_{i=1}^{N} \frac{d\tilde{T}(z_i)}{dz_i} \right\} \left\{ \prod_{\substack{i=1 \\ i \neq a,b,c}}^{N} dz_i |z_b - z_a| |z_c - z_b| |z_a - z_c| \right\}, \qquad (18.1.45)$$

we must conclude that

$$f(z_1', \ldots, z_N') = \left\{ \prod_{i=1}^{N} \frac{d\tilde{T}(z_i)}{dz_i} \right\}^{-1} f(z_1, \ldots, z_N)$$

$$= \prod_{i=1}^{N} \left\{ \frac{(\tilde{a}\tilde{d} - \tilde{b}\tilde{c})}{|(\tilde{c} z_i + \tilde{d})|^2} \right\}^{-1} f(z_1, \ldots, z_N). \qquad (18.1.46)$$

One can, of course, also find this result from the explicit form of f.

We now make the change of variables $z_i' = T(z_i, z_{a+1})$, $i \neq a+1$ and $z_{a+1} \to z_{a+1}$. Since the Jacobian for this transformation is

$$\prod_{\substack{i=1 \\ i \neq a,b,c,a+1}}^{N} \left(\frac{dT}{dz_i} \right)^{-1}, \qquad (18.1.47)$$

we find that

$$A = \int \cdots \int \prod_{\substack{i=1 \\ i \neq a,b,c,a+1}}^{N} dz_i' |z_a - z_b||z_c - z_b||z_c - z_a| dz_{a+1} \prod_{\substack{i=1 \\ i \neq a,b,c,a+1}}^{N} \left(\frac{dT}{dz_i}\right)^{-1}$$

$$\times f(z_1, \ldots, z_N)$$

$$= \int \cdots \int \prod_{\substack{i=1 \\ i \neq a,b,c,a+1}}^{N} dz_i' \; dz_{a+1} |z_a' - z_b'||z_c' - z_b'||z_c' - z_a'| \prod_{\substack{i=1 \\ i \neq a+1}}^{N} \left(\frac{dT}{dz_i}\right)^{-1}$$

$$\times f(z_1, \ldots, z_a, z_{a+1}, \ldots z_N)$$

$$= \int \cdots \int \prod_{\substack{i=1 \\ i \neq a,b,c,a+1}}^{N} dz_i' \; dz_{a+1} |z_a' - z_b^0||z_c^0 - z_b^0||z_c^0 - z_a'| \frac{(ad - bc)}{(cz_{a+1} + d)^2}$$

$$\times f(z_1', \ldots, z_a', z_{a+1}^0, \ldots, z_b^0, \ldots, z_c^0, \ldots, z_n'). \tag{18.1.48}$$

In deriving this result we have used that

$$z_a' - z_b' = \frac{(ad - bc)(z_a - z_b)}{(cz_a + d)(cz_b + d)}.$$

Finally, we must make the change of variables $z_{a+1} \rightarrow w = T(z_a)$. From equation (18.1.43), we note that

$$\frac{dw}{dz_{a+1}} = \frac{(z_a' - z_b^0)}{(z_{a+1} - z_b^0)} \frac{(z_a' - z_c^0)}{(z_{a+1} - z_c^0)} \tag{18.1.49}$$

and by explicit computation we find that

$$(z_{a+1}' - z_b^0) \; (z_{a+1}' - z_c^0) = \frac{(ad - bc)^2}{(cz_{a+1} + d)^2} (z_{a+1} - z_b^0) \frac{(z_{a+1} - z_c^0)}{(cz_b^0 + d)(cz_c^0 + d)}. \tag{18.1.50}$$

From equation (18.1.41) we note that

$$\frac{dT}{dz} = K \frac{(T(z) - z_c)^2}{(z - z_c)^2} = \frac{1}{K} \frac{(T(z) - z_b)^2}{((z) - z_b)^2} \tag{18.1.51}$$

and consequently substituting z_b and z_c as appropriate we have that

$$\frac{(ad - bc)}{(cz_b + d)^2} = \frac{dT}{dz}\bigg|_{z=z_b} = K, \qquad \frac{ad - bc}{(cz_c + d)^2} = \frac{dT}{dz}\bigg|_{z=z_c} = \frac{1}{K}. \tag{18.1.52}$$

Thus

$$\frac{(ad - bc)^2}{(cz_b + d)^2(cz_c + d)^2} = 1.$$

Putting all this together, equation (18.1.32) becomes

$$A(1, 2, \ldots, N) = \int \cdots \int \prod_{\substack{i=1 \\ i \neq a,b,c,a+1}}^{N} dz_i' d\omega |z_{a+1}^0 - z_b^0||z_{a+1}^0 - z_c^0||z_b^0 - z_c^0|$$

$$\times f(z_1', \ldots, \omega, z_{a+1}^0, \ldots, z_b^0, \ldots z_c^0, \ldots z_n'). \tag{18.1.53}$$

However, ω is a dummy variable and we can call it z_a which gives the required result. Readers may satisfy themselves that this change from z_a to z_{a+1} is consistent with the cyclic ordering of the points. By repeating the process we find that we could have taken the point chosen to be a constant to be any point z_c between z_a and z_b and the discussion is unchanged. More generally, we find we may choose any three points to be the ones chosen to be constants and given then any values consistent with the cyclic ordering of the limits on their values.

The problem encountered here is rather similar to that encountered in gauge theories where one has an action which is invariant under a local symmetry and one wants to integrate in the functional integral, not over all gauge field configurations, but only over this space when moded out by gauge equivalent field configurations. In the case here we could have started with the expression

$$\int \prod_{i=1}^{N} dz_i \prod_{\substack{i,j \\ 1 \leq i < j \leq N}} |z_i - z_j|^{2\alpha' p_i \cdot p_j}. \tag{18.1.54}$$

This is, however, projectively invariant and so contains a factor corresponding to the volume of the non-compact projective group $SL(2, \mathbf{R})$, which is infinite. To eliminate this degeneracy we can proceed as for gauge theories: we fix the projective invariance by fixing the values of three of the zs and then introduce the corresponding Faddeev–Popov determinant. This determinant is precisely the factor $|z_A^0 - z_b^0||z_b^0 - z_c^0||z_c^0 - z_a^0|$ if z_a, z_b, z_c are the three variables which have been fixed. This way of thinking about the matter arises when we consider the sum of world surfaces approach to string theory in the next section.

As a result of the above considerations we conclude that the formula for $A(1, \ldots, N)$ is cyclically symmetric:

$$A(1, 2, .., N) = A(N, 1, 2, \ldots, N - 1), \tag{18.1.55}$$

Examining equation (18.1.32) we see that if $z_i \rightarrow z_{i+1}$, $p_i \rightarrow p_{i+1}$, the result is the same but the three fixed zs move up by one. However, as we have just seen, the integral is the same no matter which three zs are fixed and which value is chosen for them

The expression for the non-trivial part of the S matrix for the scattering of N spin-0 particles is given by

$$T^{(N)}(p_i) = \sum_{\substack{\text{non-cyclic and} \\ \text{non-reversed perms}}} A(1, 2, \ldots, N). \tag{18.1.56}$$

As indicated, we have not included terms in which the order of all the legs is reversed. For example, if we have $V(1, 2, \ldots, N)$ we do not include $V(N, N - 1, \ldots, 2, 1)$ as well, since this is just a question of which way around the graph to label the external legs. There are $\frac{1}{2}(N - 1)!$ such terms in the sum in equation (18.1.56).

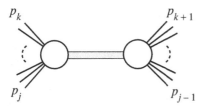

Figure 18.1.4   Factorisation of N-particle scattering.

To recover the four-point function we can take $z_1^0 = 0$, $z_2 = x$ $(0 < x < 1)$, $z_3^0 = 1$ and $z_4^0 = \infty$ in equation (18.1.32). One finds

$$A(1, 2, 3, 4) = g^2 \int_0^1 dx |1 - 0||\infty - 0||\infty - 1||x|^{2\alpha' p_1 \cdot p_2} |1|^{2\alpha' p_1 \cdot p_3}| - \infty|^{2\alpha' p_1 \cdot p_4}$$

$$\times |x - 1|^{2\alpha' p_2 \cdot p_3} |x - \infty|^{2\alpha' p_2 \cdot p_4} |1 - \infty|^{2\alpha' p_3 \cdot p_4}$$

$$= g^2 \int_0^1 dx |x|^{-\alpha(s)-1} |1 - x|^{-\alpha(t)-1} \tag{18.1.57}$$

which is indeed the four-point function we had at the beginning of this chapter. The factor of ∞ comes to the power zero and we have used $\alpha' p_i^2 = -1$.

Duality for an N-point function means the following. We sub-divide the external legs labelled $1, \ldots, N$ into two non-overlapping sets, each containing consecutive legs, say $\{j, \ldots, N, 1, \ldots, k\}$ and $\{k+1, \ldots, j-1\}$. The corresponding channel variable is given by

$$s_{jk} = -\left(\sum_{i=j}^{k} p_i^{\mu}\right)^2 = -\left(\sum_{i=k+1}^{j-1} p_i^{\mu}\right)^2. \tag{18.1.58}$$

Duality states that the amplitude $A(1, 2, \ldots, N)$ should be expressible in terms of a sum over poles in s_{jk} for *any* choice of j and k. Clearly, the cyclic symmetry of $A(1, 2, \ldots, N)$ implies that if A is a sum of poles in a given channel, it must be expressible as a sum of poles in all other channels related by cyclic symmetry.

Factorisation then means that for a given channel, say s_{jk},

$$A(1, 2, \ldots, N) = \sum_{\text{poles } c_{jk}} \sum_{d} \frac{g(p; c, d) \, g(q; c, d)}{-s_{jk}^2 + c_{jk}}. \tag{18.1.59}$$

The sum over d means a sum over all the intermediate states with the same mass squared, c_{jk}, and the labels p and q denote the dependence on the momenta on legs $j, \ldots, k$ and on legs $k+1, \ldots, j-1$, respectively. As the formula implies, the first factor $g(p; c, d)$ depends only on the properties of the particles $j, \ldots, k$ and the intermediate state, and not on the properties of the particles $k+1, \ldots, j-1$ and vice versa for the second factor $g(q; c, d)$. The factorisation discussed above is illustrated in figure 18.1.4.

One can also apply factorisation to the amplitudes that are themselves the result of factorising the four-point spin-0 scattering. In this way one arrives at an amplitude with two or more simultaneous poles (see figure 18.1.5). In particular, we may divide the external

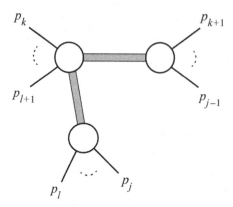

Figure 18.1.5   Double factorisation.

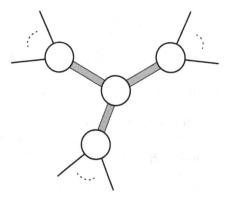

Figure 18.1.6   Higher factorisation.

legs into m non-overlapping sets each containing consecutive legs, that is, $\{j, j+1, j+2\}$, $\{j+3, j+4\}$, etc. The channel variables are then denoted by

$$s_{j,j+2}, \quad s_{j+3,j+4}, \quad \cdots . \tag{18.1.60}$$

Then factorisation means that the amplitude A is expressible as a sum of simultaneous poles in $s_{j,j+2}$, $s_{j+3,j+4}, \ldots$, that is, can be written in the generic form

$$\sum_{\text{poles } c_n} \left\{ \prod_{n=1}^{m} \frac{1}{1(-s_n + c_n)} g(p_n, c_n) \quad H_{c_1 \cdots c_m} \right\}, \tag{18.1.61}$$

where the degeneracy at a given mass level is to be understood. This particular factorisation is shown in figure 18.1.6 for $m = 3$.

The early pioneers of string theory showed that not only did the N spin-0 amplitude $A(1, 2, \ldots, N)$ of equation (18.1.32) satisfy duality, but it also factorised. As mentioned before, while the factorisation is guaranteed by unitarity in S matrix theory, or quantum field theory, it will not, in general, occur in any would-be amplitude.

Thus, given the amplitude for the scattering of any number of spin-0 particles we may deduce, by examining the appropriate residues, not only the mass spectrum, but also the scattering amplitudes between any of the particles found in the theory. However, having deduced the arbitrary scattering amplitudes, one must verify that one gets the same result for a given amplitude, no matter what factorisation one uses to obtain it. All the scattering amplitudes deduced from equation (18.1.32) pass this very stringent requirement, called the bootstrap condition.

We will now give an account of the argument that was originally used to show that the N spin-0 amplitude factorised and resulted a classification of the particles that were contained in the theory. We will follow the original path and although this is not the quickest one it will allow the reader to see how string theory began and also gain some appreciation of these amazing calculations that are rather forgotten these days.

While it is straightforward to deduce the scattering amplitudes for low level mass states using factorisation, as one looks at higher and higher masses one finds more and more particles being exchanged. To deduce the scattering amplitudes for any particles was a very formidable calculation. Looking at the $A(1, 2, \ldots, N)$ of equation (18.1.32) we see that it contains factors of $\exp(\sum_{i,j} 2\alpha' p_i p_j \ln |z_i - z_j|)$ and if the amplitude is to factorise we must find a way to rewrite this expression such that, for the selected channel, the momenta that are incoming and momenta that are outgoing occur in separate factors. The solution of this problem was very non-trivial and it involved the surprising step of introducing an infinite set of 'harmonic' oscillators α_n^μ which obeyed [3.5–3.9]

$$[\alpha_n^\mu, \alpha_m^\nu] = \eta^{\mu\nu} n \delta_{n+m,0} \tag{18.1.62}$$

with the Hermeticity property $(\alpha_n^\mu)^\dagger = \alpha_{-n}^\mu$. The corresponding vacuum is defined by $\alpha_n^\mu |0\rangle = 0, n > 1$. The vacuum at this stage just concerns the above oscillators and does not consider the momentum. They next introduced the operator [3.5–3.9]

$$\hat{V}(p) = \exp\left(\sqrt{2\alpha'} p_\mu \sum_{n=1}^{\infty} \frac{(\alpha_n^\mu)^\dagger}{n}\right) \exp\left(-\sqrt{2\alpha'} p_\mu \sum_{n=1}^{\infty} \frac{\alpha_n^\mu}{n}\right) \tag{18.1.63}$$

and the propagator

$$D(s) = \int_0^1 dz z^{R-\alpha's-2} = \frac{1}{R - \alpha's - 1}, \tag{18.1.64}$$

where $R = \sum_{n=1}^{\infty} \alpha_n^\dagger \cdot \alpha_n$. When reading the pioneering papers, such as [3.5]–[3.9], it is useful to know that in the original papers the authors used the more conventional normalised harmonic oscillator $a_n^\mu = \alpha_n^\mu / \sqrt{|n|}$. In what follows we will take $\alpha_0 = 1$ and $\mu^2 = -1/\alpha'$.

Let us consider the expression

$$\hat{A}(1, 2, \ldots, N) = \langle 0 | \hat{V}(p_{N-1}) D(s_{N-2}) \cdots D(s_3) \hat{V}(p_3) D(s_2) \hat{V}(p_2) | 0 \rangle$$

$$= \int_0^1 dx_2 \cdots \int_0^1 dx_{N-2} \langle 0 | \hat{V}(p_{N-1}) x_{N-2}^{(R-\alpha's_{N-2}-2)} \cdots$$

$$\hat{V}(p_3) x_2^{(R-\alpha's_2-2)} \hat{V}(p_2) | 0 \rangle, \tag{18.1.65}$$

where $s_k = -(\sum_{j=1}^k p_j)^2$. Using the equation

$$x^R \alpha_n^\mu x^{-R} = \alpha_n^\mu x^{-n}, \quad x^R (\alpha_n^\mu)^\dagger x^{-R} = (\alpha_n^\mu)^\dagger x^n \tag{18.1.66}$$

we can move all the factors of x to the right to find that

$$\hat{A} = \int dz_2 \cdots \int dz_{N-2} z_2^{-\alpha' s_2 - 2} z_3^{-\alpha' s_3 + \alpha' s_2 - 1} \cdots z_{N-2}^{-\alpha' s_{N-2} + \alpha' s_{N-1} - 1}$$

$$\times \langle 0 | \hat{V}(p_{N-1}, z_{N-1}) \cdots \hat{V}(p_3, z_3) \hat{V}(p_2, z_2) | 0 \rangle, \qquad (18.1.67)$$

where $z_i = \prod_{k=i}^{N-2} x_k$, $z_{N-1} = 1$ and

$$\hat{V}(p, z) = \exp\left(\sqrt{2\alpha'} p_\mu \sum_{n=1}^{\infty} \frac{(\alpha_n^\mu)^\dagger}{n} z^n \right) \exp\left(-\sqrt{2\alpha'} p_\mu \sum_{n=1}^{\infty} \frac{\alpha_n^\mu}{n} z^{-n} \right). \qquad (18.1.68)$$

The new variables z_i obey the constraints $z_2 < z_3 < \cdots < z_{N-1} = 1$, which is implicit in the above integration. When changing to the new variables of integration we have used that

$$\frac{dx_2}{x_2} \frac{dx_3}{x_3} \cdots \frac{dx_{N-2}}{x_{N-2}} = \frac{dz_2}{z_2} \frac{dz_3}{z_3} \cdots \frac{dz_{N-2}}{z_{N-2}}. \qquad (18.1.69)$$

To evaluate the above expression we push all the destruction operators to the right using the formula

$$e^A e^B = e^B e^A e^{[A,B]}, \qquad (18.1.70)$$

provided $[A, B]$ is a constant. In particular, we note, as discussed in section 9.1.1, that

$$\exp\left(-\sqrt{2\alpha'} p_{i\mu} \sum_{n=1}^{\infty} \frac{\alpha_n^\mu}{n} z_i^{-n} \right) \exp\left(\sqrt{2\alpha'} p_{j\mu} \sum_{n=1}^{\infty} \frac{(\alpha_n^\mu)^\dagger}{n} z_j^n \right)$$

$$= \exp\left(2\alpha' p_i \cdot p_j \ln\left(1 - \frac{z_j}{z_i}\right) \right) \exp\left(\sqrt{2\alpha'} p_{j\mu} \sum_{n=1}^{\infty} \frac{(\alpha_n^\mu)^\dagger}{n} z_j^n \right)$$

$$\times \exp\left(-\sqrt{2\alpha'} p_{i\mu} \sum_{n=1}^{\infty} \frac{\alpha_n^\mu}{n} z_i^{-n} \right), \qquad (18.1.71)$$

consequently we find that

$$\hat{A} = \int dz_2 \cdots \int dz_{N-2} z_2^{-\alpha' s_2 - 2} z_3^{-\alpha' s_3 + \alpha' s_2 - 1} \cdots z_{N-2}^{-\alpha' s_{N-2} + \alpha' s_{N-1} - 1}$$

$$\times \exp\left(2\alpha' \sum_{i=j=2, i>j}^{N-1} p_i \cdot p_j \ln\left(1 - \frac{z_j}{z_i}\right) \right)$$

$$= \int dz_2 \cdots dz_{N-2} \prod_{i=j=1, i>j}^{N-1} (z_i - z_j)^{2\alpha' p_i \cdot p_j}. \qquad (18.1.72)$$

We note that the product in the last equality contains an additional factor, that is, $i = 1$, compared to the sum in previous equality. We recognise this as the amplitude $A(1, 2, \ldots, N)$ of equation (18.1.32) where the coordinates z_1, z_{N-1} and z_{N-2} are not integrated over and chosen to be 0, 1 and ∞ respectively.

The advantage of writing $\hat{A}$ in this way is that it is clearly factorisable. Indeed, we can insert a complete set of states to rewrite $\hat{A}$ as

$$\hat{A} = \sum \langle 0 | \hat{V}(p_{N-1}) D(s_{N-2}) \cdots \hat{V}(p_{i+1}) | \phi \rangle \langle \phi | D(s_i) | \phi \rangle \langle \phi | \hat{V}(p_i) \cdots$$
$$\times \hat{V}(p_3) D(s_2) \hat{V}(p_2) | 0 \rangle \equiv \sum_\phi \langle \Psi_1 | \phi \rangle \langle \phi | D(s_i) | \phi \rangle \langle \phi | \Psi_2 \rangle. \qquad (18.1.73)$$

As required for factorisation, we have a factor on the left depending on momenta that are on the left times a factor with poles times a factor on the right that depends on the momenta that flow to the right away from the pole. We do not need to insert two sets of complete sets as the propagator is essentially diagonal.

The next step is to take the operators above and incorporate into them the position and momentum operators q^μ and p^μ, which as usual obey $[q^\mu, p^\nu] = i\eta^{\mu\nu}$. We define the new operator [3.5–3.9]

$$Q^\mu(z) \equiv q^\mu - i2\alpha' p^\mu \ln z + i\sqrt{2\alpha'} \sum_{n=-\infty}^{\infty} \frac{\alpha_n^\mu}{n} z^{-n} \qquad (18.1.74)$$

and define the new vertex operator

$$V(k, z) \equiv :e^{ik \cdot x(z)}: = e^{ik \cdot Q_<(z)} e^{ik \cdot q} z^{k \cdot p} e^{ik \cdot Q_>(z)} = z^{-2\alpha' k^2/2} e^{ik \cdot Q_<(z)} e^{ik \cdot Q_0(z)} e^{ik \cdot Q_>(z)}, \qquad (18.1.75)$$

where

$$Q^\mu(z) \equiv Q_<^\mu(z) + Q_0^\mu(z) + Q_>^\mu(z) \qquad (18.1.76)$$

and

$$Q_0^\mu(z) \equiv q^\mu - i2\alpha' p^\mu \ln z, \quad Q_>^\mu(z) \equiv i\sqrt{2\alpha'} \sum_{n=1}^{\infty} \frac{\alpha_n^\mu}{n} z^{-n},$$

$$Q_<(z) \equiv i\sqrt{2\alpha'} \sum_{n=-\infty}^{-1} \frac{\alpha_n^\mu}{n} z^{-n} . \qquad (18.1.77)$$

It was conventional to set $2\alpha' = 1$ in the old literature. We recall that when normal ordering operators we put all the destruction operators α_n^μ, $n = 1, 2, \ldots$, to the right of the creation operators α_{-m}^μ, $m = 1, 2, \ldots$, and factors of q^μ to the left of factors of p^μ. Having introduced the momentum into the operators we require an appropriate vacuum $|0, p\rangle$ which obeys $\alpha_n^\mu |0, p\rangle = 0$, $n \geq 1$ and carries momentum p^μ. We are interested in tachyons and so for us $k^2 = 1/\alpha'$.

Then the amplitude $\hat{A}(1, 2, \ldots, N)$ can be written as

$$\hat{A}(1, 2, \ldots, N) = \int dz_2 \cdots \int dz_{N-2}$$
$$\times \langle 0, 0 | V(p_N, z_N) V(p_{N-1}, z_{N-1}) \cdots V(p_2, z_2) V(p_1, z_1) | 0, 0 \rangle. \qquad (18.1.78)$$

We have inserted two new vertex operators at the beginning and end of the expression and we note that

$$|0, p\rangle = \lim_{z \to 0} V(p, z)|0, 0\rangle, \quad \langle 0, -p| = \lim_{z \to \infty} \langle 0, 0|V(p, z). \tag{18.1.79}$$

To understand the projective invariance of $\hat{A}(1, 2, \ldots, N)$ when written in its above operator formulation the early workers introduced the operators [3.10, 3.11]

$$L_n = \tfrac{1}{2} \sum_m : \alpha_{n-m} \cdot \alpha_m :, \tag{18.1.80}$$

where $\alpha_0^\mu = \sqrt{2\alpha'} p^\mu$. Actually one needs only $n = 1, \pm 1$ for projective invariance, but the role of the other operators soon became clear. The propagator of equation (18.1.64) can then be written as

$$D(p) = (L_0 - 1)^{-1} = \int_0^1 dx\, x^{L_0-2}. \tag{18.1.81}$$

An operator $C(z)$ of conformal weight h is defined by

$$[L_n, C(z)] = \left[z^{n+1} \frac{d}{dz} + z^n (n+1)h \right] C(z). \tag{18.1.82}$$

Actually the original workers had a slightly different definition of conformal weight which considered the operator $z^h C(z)$ and so instead of the factor $(n+1)$ on the right-hand side of the above equations one just had n. When reading the old literature it is important to keep track of this difference. One finds that Q^μ is an operator of conformal weight 1 and $V(p, z)$ an operator of conformal weight $\sqrt{2\alpha'} k^2/2$. Using this fact it is straightforward to show that the states Ψ that arise in the factorisation, that is, in equation (18.1.73), for the value of the momenta for which the pole occurs obey the constraints [3.4, 3.10]

$$L_n|\Psi\rangle = 0, \quad n > 1, \quad (L_0 - 1)|\Psi\rangle = 0. \tag{18.1.83}$$

The reader will by now have the strong feeling of *deja vu*. The above are just the physical state conditions of the open bosonic string of chapter 3.

Using the analogy of electrostatic charges on a sheet, it was realised that the scattering amplitudes above arose from the scattering of strings [18.15–18.17]. From a different perspective two-dimensional relativistic string actions were proposed and investigated in references [2.1, 2.2]. For a review of this interesting development see [18.18]. However, the reader will have noted that even before this realisation many aspects of string theory were already known.

Of course, it would have been quicker to consider from the outset the operators with the momentum included, in which case we can consider the expression

$$\hat{A}(1, 2, \ldots, N) = \langle 0, -p_N|V(p_{N-1}, 1)D \cdots V(p_3, 1)DV(p_2, 1)0, |p_1\rangle$$

$$= \int_0^1 dx_2 \cdots \int_0^1 dx_{N-2} \langle 0, -p_N|V(p_{N-1}, 1)x_{N-2}^{(L_0-1)} \cdots$$

$$\times V(p_3, 1)x_2^{(L_0-1)}V(p_2, 1)|0, p_1\rangle$$

$$= \int dz_2 \cdots \int dz_{N-2} \langle 0, -p_N|V(p_{N-1}, z_{N-1}) \cdots V(p_2, z_2)$$

$$V(p_1, z_1)|0, p_1\rangle, \tag{18.1.84}$$

which is equal to the expression of equation (18.1.78) and can, as above, be shown to be equal to the expression $A(1, 2, \ldots, N)$ that occurs in the amplitude for N spin-0 scattering. In the first line we use the propagator of equation (18.1.81) and we have used the fact that $V(p, z)$ is an operator of conformal weight 1 to move the $x^{(L_0 - 1)}$ factors to the right as well as that $(L_0 - 1)|0, p_1\rangle = 0$.

As we have mentioned, by repeated factorisation one can find the scattering amplitude for any particles in the theory. To actually carry this out in practice the oscillator formalism developed above was crucial. The scattering of three arbitrary particles was found by Scuito [18.19] and a physically equivalent cyclically symmetric vertex was found by Caneschi, Schwimmer and Veneziano in [18.20]. The latter was given by

$$
V^{csv} = \int dq_1 \int dq_2 \int dq_3 \delta(q_1 + q_2 + q_3) {}_1\langle 0, q_1 |{}_2\langle 0, q_2 |{}_3\langle 0, q_3 |
$$

$$
\times \exp\left\{ -\sum_{m=0}^{\infty} \sum_{n=1}^{\infty} \sum_{r=1}^{3} \alpha_m^{(r)\mu} \frac{(-1)^m (n-1)!}{m!(n-m)!} \alpha_{n\mu}^{(r+1)} \right\}. \tag{18.1.85}
$$

where $\alpha_{m\mu}^{(4)} = \alpha_{m\mu}^{(1)}$. The scattering of three physical states $|\psi_i\rangle$, $i = 1, \ldots, 3$ was given by

$$
V^{csv} |\psi_1\rangle_1 |\psi_2\rangle_2 |\psi_3\rangle_3. \tag{18.1.86}
$$

To evaluate this amplitude one takes the positively moded oscillators in $|\psi_i\rangle$, $i = 1, \ldots, 3$, and moves them onto vacuum in the vertex on the left where they annihilate; on the way they will, in general, bring down factors from the exponential which will result in momentum factors. Taking three tachyons as external states, that is, $|\psi_i\rangle = |0, p_i\rangle$ with $p_i^2 = 1/\alpha'$ we find that the result is a constant. However, if we take three massless spin-1 states, that is,

$$
|\psi_i\rangle = \epsilon_{i\mu}(k)\alpha_{-1}^{(i)\mu} |0, p_i\rangle \tag{18.1.87}
$$

with the on-shell conditions $k^\mu \epsilon_{i\mu}(k) = 0$ and $k^2 = 0$ which result from equation (18.1.83), we find the scattering is given by

$$
-\sqrt{2\alpha'}\{(k_1 \cdot \epsilon_2)(\epsilon_1 \cdot \epsilon_3) + (k_2 \cdot \epsilon_3)(\epsilon_1 \cdot \epsilon_2) + (k_3 \cdot \epsilon_1)(\epsilon_2 \cdot \epsilon_3)
$$

$$
+ 2\alpha'(k_1 \cdot \epsilon_2)(k_2 \cdot \epsilon_3)(k_3 \cdot \epsilon_1)\}. \tag{18.1.88}
$$

This agrees with the result for Yang–Mills theory after one drops the last terms in the limit of $\alpha' \to 0$.

We may write the above three-string vertex in the form

$$
V^{csv} = \int dq_1 \int dq_2 \int dq_3 \delta(q_1 + q_2 + q_3) ({}_1\langle 0, q_1 |{}_2\langle 0, q_2 |{}_3\langle 0, q_3 |)
$$

$$
\times \exp\left\{ -\sum_{r=1}^{3} \oint_0 dz \partial Q^{\mu(r)}(z) Q_{\mu>}^{(r+1)} \left(\frac{1}{1-z} \right) \right\}. \tag{18.1.89}
$$

If we take one of the legs, say leg 3, to be sandwiched on the tachyon state, then we would expect to recover the above tachyon vertex operator up to factors that are 1 on physical states. Let us define a vertex with two external legs by

$$
V(1, 2) = V^{csv} |0, p\rangle \Omega^{(2)}, \tag{18.1.90}
$$

where Ω is called the twist operator and is given by

$$\Omega = e^{-L_1}(-1)^{(L_0 - (2\alpha'/2)p^2)} \tag{18.1.91}$$

where $p^2 = 1/\alpha'$. Given any two states $|\chi_1\rangle$ and $|\chi_2\rangle$ we can define from $V(1, 2)$ a vertex operator V which acts on kets to give kets by

$$V(1, 2)|\chi_1\rangle_1|\chi_2\rangle_2 \equiv \langle\chi_1|V|\chi_2\rangle \tag{18.1.92}$$

for all states $|\chi_1\rangle_1$ and $|\chi_2\rangle_2$. One can show that the V which appears in this equation is just the tachyon emission vertex $V(p, 1)$ discussed above. If one takes the photon physical states instead of the tachyon on leg 2, then one finds the photon emission vertex, etc. The same applies to any other physical state. We note that $\Omega = 1$ on physical states and so one could, in principle, drop this factor; however, its inclusion leads to simpler results that we recognise.

While one could deduce the scattering amplitudes of an arbitrary number of particles by factorisation, it is easier to effectively reverse this process. Knowing that the amplitudes factorise one can build them out of the arbitrary three-string vertex V^{csv} of equation (18.1.85) and the propagator $D = (L_0 - 1)^{-1}$. This process is referred to as sewing. This has graphs which look like the ϕ^3 quantum field theory. However, there the resemblance stops; the combinatoric rule is not that of the ϕ^3 quantum field theory. By construction each graph found in the sewing procedure is cyclically symmetric and instead one sums over all inequivalent cyclic permutations and does not include terms for which the external labels are reversed. This is the same sum as we found in our formula for the S matrix of equation (18.1.56) and earlier versions of this formula. Indeed, the graph found by sewing can be thought of as the first term in the sum and the remaining terms arise due to the sum. As such, we find that we have only $(N - 1)!/2$ terms which all come with weight 1. The weights between the contributions were deduced using unitarity.

One could also construct loop contributions by sewing. Such graphs will contain loops and so three-string vertices for which two of the legs are not external legs and so not on-shell. In this case unphysical sates can propagate around the loops. This is to be expected as the theory involves massless spin-1 particles which are known to be associated with such problems, but in string theory these are only one of an infinite tower of such states. In the early days of string theory it was known [11.3] how to insert physical state projectors into one-loop graphs so as to eliminate the unwanted states and find the correct results. At higher loops the early results did contain the unwanted states, nonetheless, the early workers did compute essentially all open bosonic string multi-loop graphs. For example, the reader may enjoy reading the all loop and any external state calculations in [18.21].

For reviews of the material in this section including more details of the factorisation process and more references the reader can consult [18.22–18.29]. Reading these papers one wonders if such a sustained, highly technical, largely unsupported and ignored development would be possible today.

The discovery of the Faddeev–Poppov ghosts and the BRST formalism [1.2] made it clear how to eliminate the unphysical states and find the correct results. The three-string vertex incorporating ghost oscillators was found [12.10] using the techniques given in section 18.3. The discovery of this object opened the way to repeat the sewing procedure with this vertex and find the correct results for the scattering of bosonic strings, and modulo

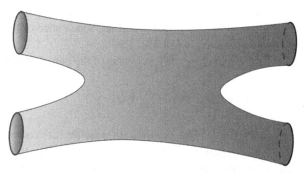

Figure 18.2.1   Four-string scattering at tree level.

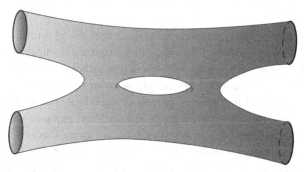

Figure 18.2.2   Four-string scattering at one loop.

technical difficulties also for the superstrings, for all physical states and loops. The results are not so very different to those found in the old days.

18.2   The path integral approach

The closed oriented bosonic string sweeps out a two-dimensional surface; two such surfaces are illustrated in figures 18.2.1 and 18.2.2. The first displays the scattering of four strings at tree level, that is, no loops, and the second shows the one-loop scattering of four closed strings. We have assumed that we have Wick rotated from the original Minkowski space world sheet to a Euclidean world sheet with positive definite metric as explained in section (8.1). The topology of two-dimensional surfaces is particularly simple in that they are classified by a single number, that is, the number of handles g they possess. This is measured by the Euler number $\chi = (1/4\pi) \int d^2\xi R = 2 - 2g$ for a surface with no boundaries and $\chi = (1/4)\pi \int d^2\xi R = 2 - 2g - k$ if we also have k boundaries. Each emitted string corresponds to a disc cut out of the surface and so one boundary. In the approach explained in this section the scattering of the string is given by a sum over the world surface with certain interaction terms inserted. Riemann surfaces and many of their properties appeared in the earliest multi-loop calculations of the dual model and, in particular, in [18.21] a sum over surfaces to describe string scattering was a motivation in the realisation that the old dual model was a string theory by, [18.15–18.16]. The systematic approach began in [18.30, 18.31]. The modern approach, which uses the metric on the world sheet and takes account of the conformal anomaly, was given in [18.32]. Many aspects of the calculation including the evaluation of determinants and the appearance of the moduli space of the Riemann surface were discussed in [18.33].

At first sight, the amplitude for the scattering of N strings is of the generic form

$$N \int \mathcal{D}x^\mu \mathcal{D}g_{\alpha\beta} \int d^2z_1 \cdots d^2z_N e^{-S[x,g]} V(z_1 \bar{z}_1) \cdots V(z_N, \bar{z}_N), \qquad (18.2.1)$$

where S is the action of equation (2.1.10), N is the normalisation factor, z are coordinates on the world surface and $V(z, \bar{z})$ is an insertion in the path integral that describes the emission of a string at position $z, \bar{z}$. In this formalism one does not describe the emission of all strings but only the tachyon and the strings associated with the the graviton and the antisymmetric tensor. For each of these strings one needs a different $V(z, \bar{z})$ whose form we will discuss later.

The path integral of equation (18.2.1) is undefined as the action $S[x, g]$ is invariant under two-dimensional diffeomorphisms and Weyl rescalings as discussed in chapter 2, but the integration is over all fields. In this section we will give a first principles derivation of the correct path integral which takes account of the points glossed over when this was discussed in chapter 3. The normalisation will be just so as to remove the infinite factor which is the volume of the two-dimensional diffeomorphism group and the Weyl group so we formally take $N = 1/\text{vol(diff)vol(weyl)}$. The situation is just like that for gauge theories where the integral over gauge related configurations leads to a divergence. For gauge theories the solution is described in appendix A, we work with a gauge fixed action and introduce the Faddeev–Popov ghosts, or equivalently find a BRST invariant action. In chapter 3 we carried out this method for string theory and so we should expect to find a gauge fixed action and to have to introduce the b and c ghosts resulting in a path integral based on the BRST invariant action of equation (3.2.16).

Thus the path integral we should consider is not over the two-dimensional metric but only over metrics that are not related by Weyl rescalings or two-dimensional diffeomorphisms. We can regard two metrics, $g_{\alpha\beta}$ and $g'_{\alpha\beta}$, as equivalent, denoted by $g'_{\alpha\beta} \sim^D g_{\alpha\beta}$, if they are related by a diffeomorphism, which if it is small implies that $g'_{\alpha\beta} = \nabla_\alpha \zeta_\beta + \nabla_\beta \zeta_\alpha$. This way of writing a diffeomorphism of the metric is shown below equation (3.2.2). We can alternatively regard two metrics as equivalent, denoted $g'_{\alpha\beta} \sim^W g_{\alpha\beta}$, if they are related by Weyl rescalings, that is, $g'_{\alpha\beta} = e^\phi g_{\alpha\beta}$. Finally, we can also consider that two metrics as equivalent, denoted $g'_{\alpha\beta} \sim g_{\alpha\beta}$, if they are related by a Weyl and a general coordinate transformation: taking the latter to be small this means that $g'_{\alpha\beta} = e^\phi (\nabla_\alpha \zeta_\beta + \nabla_\beta)$. The coset space of all such metrics subject to the last equivalence relations, denoted $\sim$, is called the *moduli space* $\mathcal{M}$ of the surface. With this equivalence relation we may write $\mathcal{M}$ as

$$\mathcal{M} = \{g_{\alpha\beta} : g'_{\alpha\beta} \sim g_{\alpha\beta}\} = \frac{\text{two-dimensional metrics}}{\text{diffeomorphisms and Weyl rescalings}}. \qquad (18.2.2)$$

Once we have used Weyl rescaling and reparameterisations to fix the metric in the path integral we are left with the remaining components of the metric which are the moduli. As such, the space of moduli is of central importance to the evaluation of string scattering amplitudes. To gain a better understanding of this space we will now discuss some features of two-dimensional surfaces.

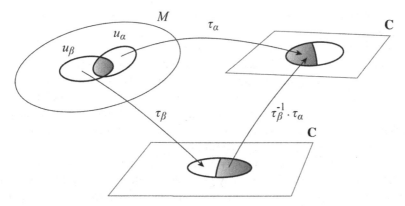

Figure 18.2.3   Charts for a complex manifold.

A *complex manifold* is like a differential manifold except that we use complex coordinates instead of real coordinates and analytic functions instead of differential functions. To be precise, we cover the manifold M by a set of charts U_α which are one to one onto maps φ_α from an open set of M to an open set of the complex numbers $\mathbf{C}$, such that on the overlap region between two charts the transition functions $\varphi_\beta^{-1} \cdot \varphi_\alpha$ is an analytic function, see figure 18.2.3. For a review suited to physicists, see [18.34]. Some books use the word holomorphic rather than analytic; the meaning is the same. The prototype example is the Riemann sphere of appendix D, where we have two charts z and ζ which are related by $z = 1/\zeta$. We say that such a collections of charts and transitions functions defines a *complex structure* on the manifold.

A map f between two complex manifolds M and M' is analytic if it is analytic when referred to the charts of the two manifolds, that is, $\varphi_\alpha^{-1} \cdot f \cdot \varphi_\alpha'$ is analytic. A map from M to M which is analytic and one to one onto is called an automorphism of M.

An *orientable manifold* is a manifold which, when viewed as a manifold with real coordinates, possesses a set of charts such that the determinant of the transition functions between any overlapping charts is positive, that is, if x_α^n and x_β^n are the real coordinates in two such charts, then $\det(\partial x_\alpha^n / x_\beta^m)$ is positive.

In fact, any orientable two-dimensional manifold is a complex manifold. Let us consider it as a real manifold with a patch with coordinates x^1 and x^2. As we discussed in chapter 3 on a two-dimensional manifold we may use reparameterisation transformations to bring the metric on a given patch into so-called isothermal form, that is, $d^2s = e^\rho((dx^1)^2 + (dx^2)^2) = e^\rho dz d\bar{z}$, where $z = x^1 + ix^2$ and $\bar{z} = x^1 - ix^2$ in any local coordinate patch. In terms of the metric this means that $g_{z\bar{z}} = \frac{1}{2}e^\rho$ and $g_{zz} = 0 = g_{\bar{z}\bar{z}}$. We may do the same on an overlapping patch with the complex coordinate w, that is, $d^2s = e^{\rho'} dw d\bar{w}$. The two coordinates must be related in such a way as to give the same d^2s and so $e^\rho dz d\bar{z} = e^{\rho'} dw d\bar{w}$. The coordinate transformation between the two charts $w = f(z, \bar{z})$ implies that $dw = (\partial w/\partial z)dz + (\partial w/\partial \bar{z})d\bar{z}$, but this will only preserve the line element if $(\partial w/\partial \bar{z})/(\partial \bar{w}/\partial z) = 0$. Thus either $\partial w/\partial \bar{z} = 0$ or $\partial \bar{w}/\partial \bar{z} = 0$. However, only the former preserves the orientation of M and so the transformation must be an analytic transformation $w = f(z)$, hence the result. Indeed, under the analytic transformation, $w = f(z)$, the metric $d^2s = e^{\rho'} dw d\bar{w}$ takes the form $d^2s = e^{\rho'} dw d\bar{w} = e^\rho dz d\bar{z}$ where

$e^\rho = (df/dz)(d\bar{f}/d\bar{z})e^{\rho'}$. A *Riemann surface* is a two-dimensional surface equipped with a complex structure.

Metrics that are related by a conformal rescaling, that is, $g'_{\alpha\beta} \sim {}^W g_{\alpha\beta}$, are said to have the same *conformal structure* . In terms of isothermal coordinates the conformal structure in a given patch is just the $dzd\bar{z}$ part of the metric. As such, the Weyl independent part of the metric, that is, the conformal structure, is encoded in the complex coordinates, or charts, that define the complex manifold. Hence, specifying the conformal structure is equivalent to specifying the coordinate charts up to changes by analytic function and so it is the same as defining the manifold as a complex manifold.

Given a two-dimensional manifold M with a given complex structure, the metric can be chosen to be given by $d^2s = e^\rho dzd\bar{z}$ in any given chart with similar expressions in all other coordinate charts. Let us consider another two-dimensional manifold, M' that has the same topology as M, that is, the same number of handles, but has a different complex structure. If we view the two manifolds, as real manifolds then we can choose the coordinates x^1 and x^2 on M such that $d^2s = e^\rho dzd\bar{z}$, where $z = x^1 + ix^2$. The manifold M', viewed as a complex manifold, cannot have complex coordinates that are obtained from those of M by an analytic transformation as the two metrics will then be conformally equivalent and so the two manifolds will have the same complex structure. The metric of M' can be written in terms of the coordinates x^1 and x^2 and so in terms of z and $\bar{z}$. It must be of the form $d^2s = e^{\rho'}|dz + \mu^z_{\bar{z}}d\bar{z}|^2$, where $\mu^z_{\bar{z}}$ is a function of z and $\bar{z}$ and is called the Beltrami differential. In order for the metric to be positive we require $|\mu^z_{\bar{z}}| < 1$. Let M' be a complex manifold with complex coordinate w, carrying out the diffeomorphism $w = f(z, \bar{z})$ we have the line element

$$d^2s = e^{\hat\rho} dwd\bar{w} = e^{\hat\rho} \frac{\partial f}{\partial z}\frac{\partial \bar{f}}{\partial \bar{z}}\left|dz + \frac{\partial f}{\partial \bar{z}}\left(\frac{\partial f}{\partial z}\right)^{-1} d\bar{z}\right|^2 .$$

Hence we can obtain the metric of M' when written in terms of the set of complex coordinates associated with M from the metric when written in terms of its own preferred complex coordinates w by a general coordinate transformation if the Beltrami differential satisfies $(\partial f/\partial z)\mu^z_{\bar{z}} = \partial w/\partial \bar{z}$. It can be shown that one can always find such f locally and indeed in all charts provided one accepts that it can have an isolated singularities.

On the space of tensors on the manifold we can define a scalar product. Given two tensors R and S of the same rank we define their scalar product by

$$(S, R) = \int d^2\xi \, S_{\alpha\beta\ldots}R_{\delta\epsilon\ldots}\sqrt{-g}g^{\alpha\delta}g^{\beta\epsilon}\cdots. \tag{18.2.3}$$

Given we have a positive definite metric, this really does obey the properties of a scalar product and, in particular, if $S = R$ then it is positive and if the scalar product vanishes then so does S.

We will now investigate the moduli space. It is the part of the metric that cannot be fixed by a coordinate transformation or Weyl rescaling. Given an arbitrary metric $g_{\alpha\beta}$ it can be related to an infinitesimally close reference metric $\hat{g}_{\alpha\beta}$ by a general coordinate transformation ζ_α and Weyl rescaling ϕ, that is, $g_{\alpha\beta} \sim \hat{g}_{\alpha\beta}$, if

$$\delta g_{\alpha\beta} = g_{\alpha\beta} - \hat{g}_{\alpha\beta} = \phi g_{\alpha\beta} + \nabla_\alpha \zeta_\beta + \nabla_\beta \zeta_\alpha. \tag{18.2.4}$$

Removing the trace, that is, only considering the effect of general coordinate transformations, this condition implies that there exists a globally defined vector field ζ_α such that

$$\delta g_{\alpha\beta} - g_{\alpha\beta}\delta g_{\delta\gamma}g^{\delta\gamma} \equiv t_{\alpha\beta} \equiv (P\zeta)_{\alpha\beta}, \qquad (18.2.5)$$

where

$$(P\zeta)_{\alpha\beta} = \nabla_\alpha\zeta_\beta + \nabla_\beta\zeta_\alpha - g_{\alpha\beta}g^{\gamma\delta}\nabla_\gamma\zeta_\delta \qquad (18.2.6)$$

and $t_{\alpha\beta}$ is an arbitrary symmetric traceless tensor. The operator P is a map from the space of vectors ζ_α into the space of symmetric traceless tensors $t_{\alpha\beta}$.

Suppose two metrics, $g_{\alpha\beta}$ and $\hat{g}_{\alpha\beta}$, are not related by a general coordinate transformation and Weyl transformation, then $t_{\alpha\beta}$ is not of the form $(P\zeta)_{\alpha\beta}$ for any ζ_α. Let us consider the quantity

$$\int d^2\xi\sqrt{-g}\left(t_{\alpha\beta} - \tfrac{1}{2}(P\zeta)_{\alpha\beta}\right)g^{\alpha\gamma}g^{\beta\delta}\left(t_{\gamma\delta} - \tfrac{1}{2}(P\zeta)_{\gamma\delta}\right) = (t - P\zeta, t - P\zeta)$$

$$(18.2.7)$$

for any ζ_α. This is an integral over the Riemann surface of an integrand $\left(t - \tfrac{1}{2}(P\zeta)\right)^2$ which is positive and does not vanish for any ζ_α. Since the integral is finite, it must possess a maximum which we can find by minimising with respect to ζ_α to find the equation

$$P^\dagger(t - P\zeta) = 0, \qquad (18.2.8)$$

where $P^\dagger$ is the adjoint of P. It maps symmetric traceless 2-tensors $S_{\alpha\beta}$ to vectors according to the rule.

$$(P^\dagger S_{\alpha\beta})_\alpha = -2\nabla^\beta S_{\alpha\beta}. \qquad (18.2.9)$$

Since $t - P\zeta$ never vanishes, by assumption, we find that $P^\dagger$ has a zero mode, that is, $P^\dagger t^0 = 0$ for some non-trivial t^0. Conversely, given any non-trivial zero mode t_0 of $P^\dagger$, it cannot be written in the form $t^0 = P\zeta$. For if it could, then $(\zeta, P^\dagger t_0) = 0 = -(P\zeta, t_0) = -(P\zeta, P\zeta)$, which implies that $P\zeta = 0$ and hence $t^0 = 0$ in contradiction with our original assumption.

Consequently, there is a one-to-one correspondence between zero modes of $P^\dagger$ and symmetric traceless second rank tensors which are not of the form $P\zeta$ for some ζ. If we denote the zero modes of $P^\dagger$ by $t^i_{\alpha\beta}$, then we may write $t_{\alpha\beta} = \tfrac{1}{2}(P\zeta)_{\alpha\beta} + t^i_{\alpha\beta}\delta m_i$, where m_i parameterise moduli space. This, in turn, means there is a one-to-one correspondence between zero modes of $P^\dagger$ and metrics $g_{\alpha\beta}$ which cannot be brought by a reparameterisation and Weyl rescaling to a given infinitesimally close metric $\hat{g}^{\alpha\beta}$.

Let us now consider the significance of zero modes of P, that is, if there exists a ζ_0 such that $P\zeta_0 = 0$. Then the infinitesimal coordinate transformations ζ and $\zeta + \zeta_0$ lead, using equation (18.2.4), to $t_{\alpha\beta} = 0$ and so to the same $\sqrt{-g}g^{\alpha\beta}$. Hence, even if two metrics are related by a diffeomorphism, then specifying $t_{\alpha\beta}=0$, or equivalently $\sqrt{-g}g^{\alpha\beta} = \sqrt{-\hat{g}}\hat{g}^{\alpha\beta}$, does not imply that the general coordinate transformation is zero. As such, if we want to specify completely a general coordinate transformation we must find some further condition. The equation

$$(P\zeta)_{\alpha\beta} = 0 \qquad (18.2.10)$$

is known as a conformal Killing equation. It is the condition for a globally defined general coordinate transformation to preserve the metric up to a Weyl transformation. In fact, we studied just this condition in equation (8.1.6) and we found, in equation (8.1.25), that the solutions were analytic vector fields, that is, $\zeta^z = \zeta^z(z)$. We require that the diffeomorphism be defined at all points of the Riemann sphere and so they are globally defined analytic vectors. It follows that they correspond to automorphism of the Riemann surface. In section 8.1.2 we found that the solutions to equation (18.2.10) for the Riemann sphere which are globally defined. They are the three analytic vector fields ∂_z, $z\partial_z$, $z^2\partial_z$ and their anti-analytic companions which together generate SL(2, **C**). A genus 1 Riemann surface has only one globally defined analytic vector field, but higher genus Riemann surfaces have none.

The moduli space $\mathcal{M}$ was defined earlier to be the parts of $g_{\alpha\beta}$ which cannot be removed by a coordinate transform or Weyl rescaling, that is, the metric subject to the equivalence relation $g_{\alpha\beta} \sim \hat{g}_{\alpha\beta}$. The moduli space is in one-to-one correspondence with the space of zero modes of $P^\dagger$. Their number depends on the genus of the surface. The Riemann–Roch theorem states that the number of zero modes of $P^\dagger$ minus the number of zero modes of P is $3g - 3$ for any surface of genus g. The number of zero modes of P is the number of globally defined vector fields on the surface. At genus 0, that is, the Riemann sphere, there are three globally defined vector fields and so three zero modes of P and consequently a genus 0 Riemann surface has no moduli. At genus 1 we have only one globally defined vector field and so the number of moduli is one. At higher genus there are no globally defined vector fields and so there are $3g - 3$ zero modes of $P^\dagger$ and thus the dimension of the moduli space is $3g - 3$. We note that these numbers are the complex dimensions of the moduli space, the real dimension is twice this.

We can now return to the task in hand, namely to correctly define the path integral of equation (18.2.1). We can write metric $g_{\alpha\beta}$ which is close to some reference metric $\hat{g}_{\alpha\beta}$ as a Weyl transformation, a general coordinate transformation and a part that is in the moduli space:

$$g_{\alpha\beta} - \hat{g}_{\alpha\beta} = \nabla_\alpha \delta\zeta_\beta + \nabla_\beta \delta\zeta_\alpha + 2g_{\alpha\beta}\delta\sigma + t^i_{\alpha\beta}\delta m_i \equiv \delta v_{\alpha\beta} + 2g_{\alpha\beta}\delta v + t^i_{\alpha\beta}\delta m_i,$$
(18.2.11)

where as before m^i are parameters describing the moduli space. Taking the trace implies that

$$\delta v_{\alpha\beta} = \nabla_\alpha \delta\zeta_\beta + \nabla_\beta \delta\zeta_\alpha - g_{\alpha\beta}g^{\gamma\delta}\nabla_\gamma \delta\zeta_\delta = \left(P\delta\zeta\right)_{\alpha\beta}$$
(18.2.12)

and

$$\delta v = \delta\sigma + g^{\gamma\delta}\nabla_\gamma \delta\zeta_\delta.$$
(18.2.13)

The integration over metrics can be written as

$$\mathcal{D}g_{\alpha\beta} = \mathcal{D}v_{\alpha\beta}\mathcal{D}v\mathcal{D}m,$$
(18.2.14)

where $\mathcal{D}m$ is the integration over the moduli. We now change variables to integrate over coordinate transformations ζ_α, Weyl factor σ and moduli.

Let us work on the sphere where there are no moduli, that is, $\delta m_i = 0$ in equation (18.2.11). The change in integration variables from $v_{\alpha\beta}$ and v to ζ_α and σ is given by

$$\mathcal{D}v_{\alpha\beta}\mathcal{D}v = J\mathcal{D}\zeta_\beta'\mathcal{D}\sigma,$$
(18.2.15)

where J is the Jacobian

$$J = \left| \det \left(\frac{\partial(v, v_{\alpha\beta})}{\partial(\sigma, \zeta_\alpha)} \right) \right|.$$
(18.2.16)

As discussed in equation (3.2.20) we find that

$$J = \left| \det \begin{pmatrix} 1 & g^{\gamma\delta} \nabla_\gamma \\ 0 & P \end{pmatrix} \right|.$$
(18.2.17)

Consequently, the determinant is given by

$$J = |\det' P| = (\det' P^\dagger P)^{\frac{1}{2}}.$$
(18.2.18)

The prime in equation (18.2.15) corresponds to the fact that the value of the metric with its determinant removed, or traceless if infinitesimal, does not determine all the general coordinate transformations, in particular the globally defined analytic vector fields do not change this part of the metric. Therefore $\mathcal{D}'\zeta$ is an integration over all diffeomorphisms except those that are globally defined analytic vector fields. As such, this integration gives the volume of all the diffeomorphisms except those that are global analytic vector fields. The normalisation factor N defined at the beginning of this section divides the path integral by the volume of all diffeomorphisms and Weyl rescalings, thus the integration over $\mathcal{D}'\zeta\mathcal{D}\sigma$ almost cancels N but leaves the volume of the globally defined analytic vector fields which, in the case we are considering here, that is, the Riemann sphere, is the volume of SL(2, **C**), denoted vol(SL(2, **C**)). Since the integrand does not depend on the diffeomorphisms and Weyl rescalings as it is invariant (the vertex operators V do not depend on the metric), the result of carrying out the integration over the metric in the path integral is

$$\frac{1}{\text{vol(SL(2, }\mathbf{C}))} \int \mathcal{D}x^\mu \int d^2z_1 \ldots d^2z_N e^{-S[x,\hat{g}]} (\det' P^\dagger P)^{\frac{1}{2}} V(z_1\bar{z}_1)\ldots V(z_N, \bar{z}_N),$$
(18.2.19)

where $\hat{g}_{\alpha\beta}$ is the reference metric which we can choose to be given by $\hat{g}_{\alpha\beta} = e^\phi \eta_{\alpha\beta}$. We recall that the action does not depend on the Weyl factor ϕ. We note that the global analytic vector fields lead to Weyl rescaling so to integrate over them and all Weyl rescaling would be to double count.

We may use anti-commuting variables to rewrite the determinant factor as

$$\det'(P^\dagger P) = \int \mathcal{D}'c^\gamma \mathcal{D}b_{\alpha\beta} \exp \left\{ -\frac{i}{2\pi} \int d^2\xi \sqrt{-\hat{g}} \hat{g}^{\alpha\beta} b_{\beta\gamma} \widehat{\nabla}_\alpha c^\gamma \right\}.$$
(18.2.20)

Putting this in the path integral, we find the scattering amplitudes become

$$\frac{1}{\text{vol(SL(2, }\mathbf{C}))} \int \mathcal{D}x^\mu \int \mathcal{D}'c^\gamma \mathcal{D}b_{\alpha\beta} \int d^2z_1 \cdots d^2z_N e^{-S[x,\hat{g},b,c]} V(z_1\bar{z}_1) \cdots V(z_N, \bar{z}_N),$$
(18.2.21)

where $S[x, \hat{g}, b, c]$ is the BRST action of equation (3.2.16). Thus we recover the results of the BRST analysis of chapter 3.

We note that the ghost and anti-ghost equations of motion can be written as

$$(Pc)_{\alpha\beta} = 0 \quad \text{and} \quad P^\dagger b = 0.$$
(18.2.22)

Consequently, the zero modes of P and $P^\dagger$ correspond to zero modes of the c and b fields, respectively. Thus the zero modes of b correspond to the moduli of the surface and the zero modes of c to the globally defined analytic vector fields.

The prime that appears in the determinant of $P^\dagger P$ and the integral over the c ghost fields is due to the previously mentioned point that we have not integrated over globally defined analytic vector fields and so we exclude from the determinant the zero modes of P and the integration over c.

The vertex operators V are just the physical state vertex operators of conformal dimension $(1, 1)$ that we found in section 9.3.1. For the tachyon this was

$$V(z, \bar{z}) = e^{ip \cdot x(z, \bar{z})}, \quad p^2 = 2. \tag{18.2.23}$$

The vertex operator for the graviton was given in equation (9.3.18).

The integrand, including the vertex operators, is invariant under general coordinate transformations including those generated by the globally defined analytic vector fields. We will now show how to deal with the resulting infinity. As discussed in section (8.1.2) we can use this SL(2, $\mathbf{C}$) symmetry to fix any three points. In particular let us choose to fix three Koba–Nielsen points, which for simplicity we take to be z_1, z_2, z_3, to be z_1^0, z_2^0, z_3^0. The relation between the fixed coordinates and the original coordinates can be written in terms of an SL(2, $\mathbf{C}$) transformation as

$$z_i = z_i^0 + a_{-1} + a_0 z_i^0 + a_1 (z_i^0)^2, \quad i = 1, 2, 3 \tag{18.2.24}$$

when the two points are close by and a more complicated formula when they are far apart. We may then swap the integration over the z_1, z_2, z_3 to be over the parameters a_{-1}, a_0, a_1. This change of integration variable is as follows

$$d^2 z_1 d^2 z_2 d^2 z_3 = d^2 a_{-1} d^2 a_0 d^2 a_1 \left| \det \frac{\partial z_i}{\partial a_j} \right|^2$$

$$= d^2 a_{-1} d^2 a_0 d^2 a_1 |z_1^0 - z_2^0|^2 |z_2^0 - z_3^0|^2 |z_3^0 - z_1^0|^2. \tag{18.2.25}$$

The integral over the parameters a_{-1}, a_0, a_1 gives the volume of this group SL(2, $\mathbf{C}$), which just cancels that factor in equation (18.2.21). Dropping the superscript 0 on the $z_i, i = 1, 2, 3$, we find that the scattering amplitude becomes

$$|z_1 - z_2|^2 |z_2 - z_3|^2 |z_3 - z_1|^2 \int \mathcal{D}x^\mu \int \mathcal{D}'c^\gamma \mathcal{D}b_{\alpha\beta}$$

$$\times \int d^2 z_4 \cdots d^2 z_N e^{-S[x,\hat{g},b,c]} V(z_1 \bar{z}_1) \cdots V(z_N, \bar{z}_N)$$

$$= |z_1 - z_2|^2 |z_2 - z_3|^2 |z_3 - z_1|^2 \int d^2 z_4 \cdots d^2 z_N \langle V(z_1 \bar{z}_1) \ldots V(z_N, \bar{z}_N) \rangle \tag{18.2.26}$$

where we have written the expression in terms of its field theory expectation value as defined by the path integral.

Let us now evaluate this expression for tachyons by inserting the vertex operators of equation (18.2.23). The resulting expression is Gaussian in x^μ and so can readily be evaluated by completing the square in x^μ in the usual way. We can also just use equation (9.1.42) to find that the scattering of tachyons is given by

$$|z_1 - z_2|^2 |z_2 - z_3|^2 |z_3 - z_1|^2 \int d^2 z_4 \cdots d^2 z_N \prod_{r>s} |z_r - z_s|^{p_r \cdot p_s}. \qquad (18.2.27)$$

Let us comment on the vertex operators from the BRST perspective. If $V(z, \bar{z})$ is a conformal operator of conformal weight $(1, 1)$, then $\int d^2 z V(z, \bar{z})$ is BRST invariant. Using the form of Q given in chapter 3 and equation (8.2.31) we find that

$$[Q, V(z, \bar{z})] = \int_z dw c(w) T^x(w) V(z, \bar{z}) = \int_z dw c(w) \left(\frac{1}{(w-z)^2} + \frac{\partial V(z, \bar{z})}{(w-z)} \right)$$
$$= \partial(c(z) V(z, \bar{z})), \qquad (18.2.28)$$

which implies that Q commutes with $\int d^2 z V(z, \bar{z})$. In addition it is true that $c(z) \bar{c}(\bar{z}) V(z, \bar{z})$ is also BRST invariant. However, we can only use three factors of the latter as more would vanish due to the nature of the ghost vacuum. We could then take as our amplitude

$$\langle c(z_1) \bar{c}(\bar{z}_1) V(z_1, \bar{z}_1) c(z_2) \bar{c}(\bar{z}_2) V(z_2, \bar{z}_2) c(z_3) \bar{c}(\bar{z}_3) V(z_3, \bar{z}_3)$$

$$\times \int d^2 z_4 \cdots d^2 z_N V(z_4 \bar{z}_4) \ldots V(z_N, \bar{z}_N) \rangle \qquad (18.2.29)$$

as this is also BRST invariant. In doing this we have implicitly included the zero modes in the path integration over the ghost field c. In fact, the result is the same as

$$\langle c(z_1) \bar{c}(\bar{z}_1) c(z_2) \bar{c}(\bar{z}_2) c(z_3) \bar{c}(\bar{z}_3) \rangle = |z_1 - z_2|^2 |z_2 - z_3|^2 |z_3 - z_1|^2. \qquad (18.2.30)$$

Here we have used equation (8.3.13) and that the conformal weight of c is $(-1, 0)$.

In order to compute higher loop scattering one needs to consider how to represent higher genus Riemann surfaces. In appendix D, we conclude that a genus zero Riemann surface, that is, a sphere, is just the Riemann sphere. The uniformisation theorem due to Klein, Koebe and Poincaré showed, roughly speaking, that a genus 1 Riemann surface can be mapped analytically to a torus and genus $g \geq 2$ can be mapped to the upper half-plane H modulo some discrete subgroup K of SL(2, **R**), that is H/K.

By a torus we mean the points of the complex plane **C** modulo a lattice Λ. Taking the lattice to be given by $\Lambda = \{n_1 \omega_1 + n_2 \omega_2, n_1, n_2 \in \mathbf{Z}\}$ we define the equivalence relation $z' \sim z$ if $z' = z + n_1 \omega_1 + n_2 \omega_2$ for some $n_1, n_2 \in \mathbf{Z}$. The torus is the complex plane **C** modulo this equivalence relation. The fundamental region T is the region of **C** in which no two points are equivalent and every point in **C** is equivalent to some point in T. The points in T are then in one-to-one correspondence with those of the genus Riemann surface. We can take T to be the parallelogram with the corner points 0, ω_1, ω_2 and $\omega_1 + \omega_2$ as shown in figure 18.2.4. We can choose $\tau \equiv \omega_2/\omega_1 > 0$ as we are free to interchange ω_1 and ω_2; as such, the fundamental region belongs to the upper half-plane H. We can also carry out the analytic transformation $z \to z/\omega_1$ whereupon the fundamental regions become the parallelogram with the points $0, 1, \tau, \tau + 1$. The Riemann surface is found by identifying the boundaries of the parallelogram with the same marks as shown in figure 18.2.4.

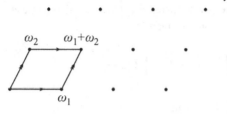

Figure 18.2.4   The fundamental region of the torus.

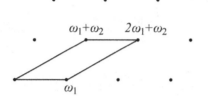

Figure 18.2.5   An alternative fundamental region of the torus.

However, this is not the only choice of the fundamental region: we may take it to have the corner points 0, ω_1', ω_2' and $\omega_1' + \omega_2'$ where

$$
\begin{pmatrix} \omega_1' \\ \omega_2' \end{pmatrix} = \begin{pmatrix} a & b \\ c & d \end{pmatrix} \begin{pmatrix} \omega_1 \\ \omega_2 \end{pmatrix},
\tag{18.2.31}
$$

with $a, b, c, d \in \mathbf{Z}$ and $ad - bc = \pm 1$. This follows as the points ω_1' and ω_2' are equivalent to ω_1 and ω_2. This condition is required in order that the inverse of this matrix should have integer entries as required by the demand that the corner points of the new fundamental region are equivalent to those of the previous choice. Only the choice $+1$ preserves the orientation. One such example is the transformation $\omega_1' = \omega_1$, $\omega_2' = \omega_2 + \omega_2$ which is drawn in figure 18.2.5. We note that that $\tau' = \omega_2'/\omega_1'$ is related to τ by

$$
\tau' = \frac{a\tau + b}{c\tau + d}.
\tag{18.2.32}
$$

It turns out that these transformations of τ arise from diffeomorphisms of the Riemann surface which cannot be deformed to the identity transformation. Such transformations did not arise for the Riemann sphere ($g = 0$). In fact, all the higher genus surfaces possess such diffeomorphisms. Let us denote the diffeomorphisms that are connected to the identity by Diff_0. We can define the equivalence relation between two diffeomorphisms f_1 and f_2 by $f_1 \sim f_2$ if $f_1^{-1} \cdot f_2 \in \text{Diff}_0$. The equivalence classes are just the different disconnected components of the diffeomorphisms. As such, all diffeomorphisms are given by the composition of Diff_0 with a discrete group Diff_{mcg} which is called the mapping class group. We may write Diff_{mcg} as the coset $\text{diff}_{mcg} = \text{Diff}/\text{Diff}_0$.

Let consider an analytic function f on a genus 1 Riemann surface. When studied from the viewpoint of its representation, the torus, it must be doubly periodic $f(z + n_1\omega_1 + n_2\omega_2) = f(z)$. However, this, in turn, defines an analytic function on $\mathbf{C}$ which must be a constant by Louville's theorem (the only analytic bounded function on $\mathbf{C}$ is a constant). We note that $\mathbf{C}$ does not contain the point at infinity. As such, the only analytic function on a genus 1

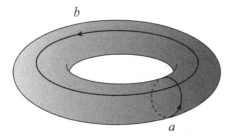

Figure 18.2.6   The homology of a genus 1 Riemann surface.

Riemann surface is a constant. If we consider an analytic vector field on a Riemann surface, by the same argument this implies the existence of a vector field $v(z)\partial/\partial z$ on the torus with v doubly periodic and so on $\mathbf{C}$. However, $\mathbf{C}$ has only one chart, namely a copy of itself and so there are no transition functions and we can use Louville's theorem to conclude that v is a constant. The same argument applies to any tensor including the symmetric traceless tensors. We also note that the only automorphism of $\mathbf{C}$ is of the form $az + b$ for constants a and b. This follows as it must be of the form $z \to f(z)$, where f is an analytic function and so can be written as a power series. Setting this to zero we always find a number of solutions, which is not allowed as it must be one to one unless it is of the above form.

It follows that the only automorphism of a genus Riemann surface is $b\partial/\partial z$, where b is a constant. It corresponds to the shift $z \to z + b$. As such, P has only one complex zero mode. Furthermore, the above discussion implies that the number of zero modes of P equals the number of zero modes of $P^\dagger$ for a genus 1 Riemann surface and so $P^\dagger$ also has only one zero mode. As such, the dimension of the moduli space of a genus 1 Riemann surface is 1. Since the only conformally invariant quantity associated with the torus is τ this must be related to the moduli space.

It will be useful to consider the homology of a manifold and we now give a very brief account. Let us consider a connected manifold M and denote the boundary of any manifold R by ∂R. The boundary of a boundary vanishes, that is, $\partial\partial R = \emptyset$. For example, the boundary of a disc is a circle but the boundary of a circle is the empty set. Let us consider any oriented sub-manifold S which is closed, that is, has no boundary, $\partial S = 0$. This is called a cycle. We say two such cycles S and S' are equivalent if they are not the boundary of some other manifold T, $\partial T = S' - S$. The collection of equivalence classes of such manifolds is called the *homology* of the manifold M. The homology of a genus 1 Riemann surface has two elements a, b, representatives of which are shown in figure 18.2.6. We note that they are chosen so as to only intersect once. We can think of the torus representation of the surface as that found by cutting along the a and b cycles and remembering the identification. A genus g Riemann surface has a homology with $2g$ elements, a_i, b_i, $i = 1, 2, \ldots, g$, representatives are shown in figure 18.2.7. Cutting along these cycles leads to the representation of the Riemann surface discussed above.

That the diffeomorphism corresponding to equation (18.2.31) is not connected to the identity can be seen by considering an element of the homology, say the a cycle, and seeing how it is mapped under the transformation. One finds that it is mapped to a different element of the homology and as two such cycles cannot be deformed into each other continuously we must conclude that neither can the transformation.

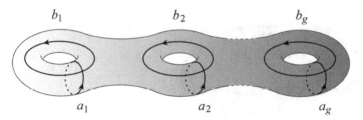

Figure 18.2.7   The homology of a genus g Riemann surface.

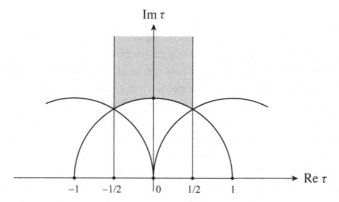

Figure 18.2.8   The moduli space at genus 1.

The moduli space as defined in equation (18.2.2) involved moding out by all diffeomorphisms and not just those connected to the origin. If we just did the latter we would find a space called Teichmuller space $\mathcal{T}$ which is given by

$$\mathcal{T} = \frac{\text{two-dimensional metrics}}{\text{Diff}_0 \text{ and Weyl rescalings}}, \tag{18.2.33}$$

the relation between the two spaces being

$$\mathcal{M} = \frac{\mathcal{T}}{\text{Diff}_{mcg}}. \tag{18.2.34}$$

At genus 1 $\text{Diff}_{mcg} = \text{SL}(2, \mathbf{Z})$, that is, the diffeomorphism corresponding to the transformations of equation (18.2.31), and $\mathcal{T}$ is just the space of τ, that is, the upper half-plane H. The moduli space is $H/\text{SL}(2, \mathbf{Z})$ and its fundamental region is given by $\{-\frac{1}{2} \le \text{Re } \tau \le 0, |\tau|^2 \ge 1\}$ as well as $\{0 \le \text{Re } \tau \le \frac{1}{2}, |\tau|^2 \ge 1\}$. It is shown in figure 18.2.8.

We now give a more concrete derivation of the moduli space of genus 1 Riemann surfaces. Given any Riemann surface we can represent it by the parallelogram in figure 18.2.4. In terms of the complex coordinates $z = x + iy, \bar{z} = x - iy$ of $\mathbf{C}$ associated with this surface the metric takes the form

$$d^2s = e^\phi (dx^2 + dy^2) = e^\phi dz d\bar{z}, \tag{18.2.35}$$

where $\phi(x + m + n\tau_1, y + n\tau_2) = \phi(x, y)$, $\tau = \tau_1 + \tau_2$ and, as discussed above, we have rescaled the parallelogram such that the corners are at $0, 1, \tau$ and $1 + \tau$. Let us carry out the coordinate change

$$x = \sigma_1 + \tau_1\sigma_2, \ y = \tau_2\sigma_2, \ \text{or equivalently} \ z = \frac{\xi}{2}(1 - i\tau) + \frac{\bar{\xi}}{2}(1 + i\tau),$$

$$\bar{z} = \frac{\bar{\xi}}{2}(1 + i\bar{\tau}) + \frac{\xi}{2}(1 - i\bar{\tau}), \tag{18.2.36}$$

where $\xi = \sigma_1 + i\sigma_2$. The above metric takes the form

$$d^2s = e^{\phi'} \left| d\xi + \frac{(1 + i\tau)}{(1 - i\tau)} d\bar{\xi} \right|^2 \tag{18.2.37}$$

where $e^{\phi'} = e^{\phi}|(1 - i\tau)|^2$ and $\phi'(\sigma_1 + m, \sigma_2 + n) = \phi'(x, y)$. We can identify $\mu = (1 + i\tau)/(1 - i\tau)$ as the Beltrami differential corresponding to the coordinate change to $\xi, \bar{\xi}$ coordinates from the complex coordinates $z, \bar{z}$ associated with the torus.

In terms of the $\xi, \bar{\xi}$ coordinates the corners of the parallelogram are at $0, 1, i$ and $1 + i$, which is a square. It is natural to refer all tori to this torus, the coordinate change is given by the Beltami differential which is parameterised by τ. Thus the moduli space is parameterised by τ modulo the transformation of equation (18.2.32).

Although we carried out the calculation of closed oriented scattering amplitudes at tree level, (genus 0) the calculation of the higher genus amplitudes proceeds along very similar lines. We can as before change the integration variable to be over the diffeomorphism, Weyl rescalings and now moduli. However, as the diffeomorphisms considered in this calculation are infinitesimal, for example, see equation (18.2.6), they must be the diffeomorphisms connected to the identity, Diff_0. As such, the integration in the path integral leads to the volume of Weyl rescalings and $\text{vol}(\text{Diff}_0)$, which does not completely cancel all of the normalisation factor N. Indeed, it leaves the integral divided by $\text{vol}(\text{Diff}_{mcg})$ and so we find an integral over the Teichmuller space, divided by $\text{vol}(\text{Diff}_{mcg})$. This is just the integration over the fundamental region of Teichmuller space, that is, moduli space itself. Thus one is left with an integral over x^μ and the moduli. The Jacobian of the change can again be written in terms of a path integral over the b and c ghosts However, there is a complication concerning the $3g - 3$ zero modes of the b ghost, which require $3g - 3$ factors involving the b ghost in the integrand in order for the integral not to vanish.

This discussion is incomplete for genus 1 as in that case we have one automorphism of the surface, that is, a zero mode of P. This is dealt with much as for the tree level: we find this part of Diff_0 is also not cancelled out, but is eliminated once we fix one of the Koba–Nielsen points. The result was found at one loop in [18.35] and in general in [18.36, 18.37]

We have assumed that there are no anomalies in the diffeomorphism and Weyl symmetries of this theory. In fact, this is only true, in general, for the former while for the latter it is only true if we work in 26 dimensions which we assume. In lower dimensions the anomaly has the effect of generating an additional scalar field, the so-called Liouville mode [18.32].

We noted above that the fundamental region of the genus 1 surface was not uniquely specified and that any two parallelograms related by equation (18.2.32) are physically equivalent. As such, we must get the same result no matter which of these infinite number of possible fundamental regions we take to integrate over and so the one-loop amplitude must be invariant under the transformations of equation (18.2.32). This is referred to as *modular* invariance and it places very strong constraints on the amplitudes. Indeed, one can apply it to the amplitude with no external strings and then it places constraints on the particles propagating around the loop and so the particle content of the string theory.

To be more precise; the result of a one-loop computation is an integral over the moduli of the surface which can be written in the form

$$\int \frac{d\tau d\bar{\tau}}{(\mathrm{Im}\tau)^2} f(\tau, \bar{\tau}), \tag{18.2.38}$$

suppressing all other dependences. The integration region is over one choice of fundamental region. The reader may easily check that $d\tau d\bar{\tau}/(\mathrm{Im}\tau)^2$ is invariant under the transformation of equation (18.2.32); however, in order to get the same result if we change our mind on which fundamental region to integrate over $f(\tau, \bar{\tau})$ must be invariant under the modular transformations of equation (18.2.32). In fact, for the closed string one-loop diagram with no external strings it can be shown that $f(\tau, \bar{\tau}) = (\mathrm{Im}\tau)^{-12}|\eta(\tau)|^{-48}$. Using [15.10] the reader may show that this latter quantity is indeed modular invariant. This is related to the number of states propagating around the loop as one may suspect from the the appearance of the Dedekind η function.

There are two commonly used ways to represent a Riemann surface; above we used the Fuchsian representation of a genus Riemann surface. In general, one can think of this method as cutting the g A and B cycles of the surface. We close this section by briefly discussing the other method, namely the Schottky method, which was used extensively in computing string amplitudes with loops, at least in the early days. This can be thought of as cutting only the B cycles and leaving the A cycles.

Let us consider g elements P_n of SL(2, $\mathbf{C}$) and their action on the Riemann sphere $\hat{C}$. We denote the group generated by the $P_n, n = 1, \dots, g$ by G. It has the elements $P_1, \dots, P_n, P_1^2, P_1P_2, P_1P_3, \dots, P_1^3, \dots$. A fixed point z_0 of a transformation P is one such that $P(z_0) = z_0$. A Schottky group is the group G, as above, but with the restrictions that the fixed points of all its elements form a discrete set. We can now ask what is the fundamental region F of G, that is, the region such that any point of $\hat{C}$ can be obtained from a point of F by the action of G and that no two points of F are related by a transformation of G. It has been shown that the fundamental region F is the region outside the $2n$ isometric circles of P_n and P_n^{-1}, $n = 1, \dots, g$. The isometric circle of a transformation $z \to (az + b)/(cz + d)$ is defined by the equation $|cz + d|^2 = |ad - bc|$. It is straightforward to show that the isometric circle C_n^{-1} of P_n^{-1} is mapped into the isometric circle of C_n of P_n by the action of P_n as illustrated in figure 18.2.9.

It can be shown that any Riemann surface can be represented by a suitable Schottky group and that this Schottky group is unique up to a single SL(2, $\mathbf{C}$) transformation S such that $P_n \to SP_nS^{-1}$ as well as up to modular transformations.

At one loop the torus is represented by the region outside the two isometric circles corresponding to a transformation P and its inverse P^{-1}. The transformation P identifies C_1^{-1} with C_1, in effect connecting the circles by a B cycle and so recovering the genus 1 Riemann surface, see figure 18.2.10.

For the closed string we find that each Riemann surface is characterised by P_n, $n = 1, \dots, g$ elements of SL(2, $\mathbf{C}$) minus the overall SL(2, $\mathbf{C}$) transformation S. These contain $3g - 3$ complex parameters. In terms of the $2g$ isometric circles these parameters correspond to the positions of their centres, $2g$ complex parameters, their radii, a further g real parameter

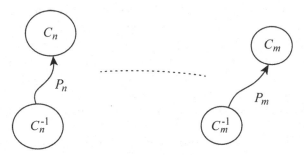

Figure 18.2.9   The Schottky representation of a Riemann surface.

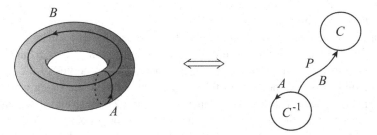

Figure 18.2.10   The Schottky representation of a genus 1 Riemann surface.

as P_n and P_n^{-1} have the same radius, namely $\sqrt{ad - bc}/c$, and g real parameters which arise from how one identifies a specific point on C_n^{-1} with a point on C_n, that is, g angles. The -3 is due to the fact that SP_nS^{-1} generates a group that represents the same Riemann surface. These $3g - 3$ parameters encode the $3g - 3$ possible complex structures of a Riemann surface of genus g. The case of $g = 1$ is slightly different and it has only one complex modulus.

For the open string, we first double the surface by a similar procedure as for the $g = 0$ case to construct a closed Riemann surface and then we take the P_n to be the elements of $SL(2, \mathbf{C})/Z_2$. Hence the surface has a moduli space with $3g - 3$ real parameters.

18.3   The group theoretic approach

This approach [18.38–18.44] took the attitude that the simplest thing in string theory was the result, that is, the S-matrix itself. It defined the string vertex through a set of very simple equations and the measure, that is, the function of moduli, that multiplied it by demanding the decoupling of physical zero-norm states. This allows for an efficient derivation of string scattering. In some sense this is an extreme example of the original philosophy of Heisenberg that led to the S-matrix programme.

The approach also had the advantage that it led to a systematic understanding of string vertices. The N string vertices found by sewing in the old days, as discussed in section 18.1, were rather complicated and depending on the precise sewing prescription could be different, although leading to the same physical result. These differences and the general properties of N string vertices become understood in the approach given in this section.

The group theoretic method applies to the open and closed bosonic strings and the superstrings for any external string states and at any loop order. However, in this section we will largely concentrate on the open bosonic string at genus 0, that is, tree level, as this illustrates the method with the least technical complications.

An alternative approach which shares some of the features of the group theoretic approach was given in [18.46–18.48].

As N strings scatter they sweep out a world sheet, which is a two-dimensional surface of genus g with some marked points, or punctures, where the outgoing strings leave the surface. We refer to these points as Koba–Nielsen points. As explained in section 9.3.1 the closed string at tree level sweeps out a cylinder, or strip, with coordinates $\varsigma = \tau + i\sigma$ after we have carried out the rotation to Euclidean space; however, we may conformally map this to the Riemann sphere with coordinate $z = e^\varsigma$. The open string, at tree level, sweeps out a strip which, after the rotation to Euclidean space, has coordinate $\varsigma = \tau + i\sigma$ with $0 \leq \sigma \leq \pi$. We extended the range of σ by reflection to $-\pi \leq \sigma \leq \pi$ and then as for the closed string we can map to the Riemann sphere. The properties of the Riemann sphere are the subject of appendix D. In this chapter we will take the string world sheet to be the Riemann sphere. The Riemann sphere is just the 2-sphere and it requires two coordinate patches to cover it associated with the north and south poles. Corresponding to the outgoing strings it also has a set of marked points, the Koba–Nielsen points. As will become clear we will also require local coordinate patches around each Koba–Nielsen point which we will denote generically by the coordinate ξ. This is not to be confused with our previous used of this symbol in chapter 2.

Any string theory possesses certain conformal operators, denoted generically by the symbol $R(\xi)$, which depend on the world surface, and act on the string Hilbert space $\mathcal{H}$. By definition, see chapter 8, a conformal operator of weight d behaves under a conformal transformation $\xi \to \xi'(\xi)$, where ξ is a local coordinate on the world surface, as

$$R(\xi) \to R'(\xi') = UR(\xi)U^{-1} = R(\xi)\left(\frac{d\xi'}{d\xi}\right)^d . \tag{18.3.1}$$

Here U is the operator on the string Hilbert space that corresponds to the conformal transformation $\xi \to \xi'(\xi)$, see chapters 8 and 9. The operator U can be written locally in the form

$$U = \exp\left(\sum_{n=-\infty}^{\infty} a_n L_n\right). \tag{18.3.2}$$

For the open string we only consider operators of this type, but for closed strings conformal operators have left and right weights $(d, \bar{d})$, but we will mainly only require operators with a weight which is either $(d, 0)$ and behaves as above, or $(0, \bar{d})$ in which case we will have the conformal transformation on the coordinate $\bar{\xi}$ which is implemented by $\bar{L}_n$.

The set of conformal operators for the open bosonic string, see chapter 9, includes ∂Q^μ, which has conformal weight 1. Equation (18.3.1) also holds for the embedding field itself Q^μ with weight 0. We recall that Q^μ includes a term with $\ln \xi$ and so it is not holomorphic. The energy momentum tensor $T(\xi)$ transforms with weight 2, but also has an anomalous term which involves the Schwarzian derivative of the conformal transformation, see section 8.4. In particular, it obeys,

$$T'(\xi') = T(\xi) \left(\frac{d\xi'}{d\xi} \right)^2 + \frac{c}{12} (S\xi')(\xi),$$

where

$$(Sf)(\xi) = \frac{f'''(\xi)}{f'(\xi)} - \frac{3}{2} \left(\frac{f''(\xi)}{f'(\xi)} \right)^2$$

and c is the central charge associated with T. We also have the ghosts b and c of conformal weights 2 and -1 respectively. As the reader will notice our use of the work conformal operator is a little loose as we have included some whose transformations are a little more complicated than that of equation (18.3.1).

For the closed bosonic string the set of conformal operators include x^μ and ∂x^μ and the energy-momentum tensor T provided one includes its anomalous term.

In the group theoretic approach, the scattering amplitude, at genus 0, of N arbitrary physical states $|\chi_i\rangle$ is given in terms of a vertex V, by the expression

$$W(|\chi_1\rangle, |\chi_2\rangle, \ldots, |\chi_N\rangle) \equiv \int_{\mathcal{M}_{z_k}} \Pi'_i dz_i f_m(z_i) V |\chi_1\rangle_{(1)} |\chi_2\rangle_{(2)} \cdots |\chi_N\rangle_{(N)}. \quad (18.3.3)$$

Here $\mathcal{M}_{z_k}$ is the moduli space of Riemann surfaces which, at genus 0, is simply the positions of the Koba–Nielsen points z_i. The function $f_m(z_i)$ is the *measure*, and V is called the *vertex*. The prime on the integral over the Koba–Nielsen points indicates that the integral is over only $N - 3$ points, It will turn out that the integral is independent of which three are not integrated over and what are the values they take. At genus greater than 0 the integration is also over the moduli of the surface and the vertex and measure f_m depend on the moduli.

The vertex is a multi-linear map from N copies of the Hilbert space of the string into the complex numbers:

$$V : \mathcal{H}_1 \otimes \cdots \otimes \mathcal{H}_N \to \mathbf{C}. \quad (18.3.4)$$

It depends on the Koba–Nielsen variables. The same vertex provides the scattering for any string states once it is saturated with the corresponding physical states as in equation (18.3.3). Therefore once we know the vertex and the measure, all scattering amplitudes are known. We note that it treats all string states on the same footing. We will define what is meant by the vertex very soon and later show how it and the measure are determined. Examples of the vertices we will find are those that arise in the sewing programme that led to string theory described in section 18.1.

The string vertex V is defined by demanding that it satisfy the equations

$$V R^{(i)}(\xi_i) = V R^{(j)}(\xi_j) \left(\frac{d\xi_j}{d\xi_i} \right)^d. \quad (18.3.5)$$

Here ξ_i and ξ_j denote coordinate patches on the string world surface that are valid in the neighbourhood of the Koba–Nielsen points z_i and z_j, respectively. The coordinate ξ_i associated with the Koba–Nielsen point z_i is required to vanish at this point, that is, $\xi_i(z_i) = 0$, and similarly $\xi_j(z_j) = 0$. Furthermore, the two coordinates are related in their overlap region by an analytic transformation which we denote by $\xi_j = \tau_{ji}(\xi_i)$. The situation

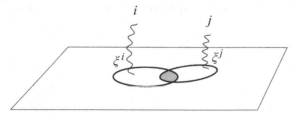

Figure 18.3.1   Strings leaving a Riemann surface and their overlapping coordinate patches.

is illustrated in figure 18.3.1. In equation (18.3.5) it is important to realise that the conformal operator $R^{(j)}$ of weight d acts on the jth leg, or string Hilbert space $\mathcal{H}^{(j)}$ of the vertex. This equation states that when the action of two conformal operators on the vertex is defined in an overlap region, then each operator gives the same result when acting on the corresponding Hilbert spaces and the appropriate coordinate transformation is made. Equation (18.3.5) holds for the set of conformal operators associated with the string under consideration. At very first sight equation (18.3.5) seems to be just the definition of the conformal operator and so trivially true; however, it is important to note that the operators act on different Hilbert spaces in a highly non-trivial manner. As we will see it does indeed determine the vertex. Some overlap identities can be derived as a consequence of some others. The obvious example being that ∂Q^μ follows from the identity for Q^μ by differentiation of the overlap equation.

One is free to choose whatever coordinate system on the string world sheet one likes provided it obeys the above criterion, and one finds correspondingly different string vertices. In fact, the vertex only depends on the transition functions τ_{ji} between the coordinate patches rather than the local coordinates themselves as only the former quantities actually occur in the overlap equations. To see this we write ξ_j in equation (18.3.5) as $\tau_{ji}(\xi_i)$ and we note that ξ_i plays the role of a dummy variable. It is often useful to refer the coordinates ξ_j to some underlying coordinate z, such as the coordinate usually used to describe the Riemann sphere of appendix C. Then we may write

$$\xi_i(z) = \sum_{n=1}^{\infty} a_n^i (z - z_i)^n \equiv (V_i)^{-1}(z) \quad , \qquad a_1^i \neq 0. \tag{18.3.6}$$

In terms of the coordinate z, the Koba–Nielsen points are denoted by $z = z_i$, $i = 1, \ldots, N$ and the transition of equation (18.3.6) does indeed satisfy $\xi_i(z_i) = 0$. We note that the transition functions are given by $\tau_{ji} = (V_j)^{-1} V_i$, since $\xi_j = (V_j)^{-1}(z) = \tau_{ji}((V_i)^{-1}z)$. The choice of different local coordinates ξ_i, subject to the above conditions, corresponds to taking different values of the constants a_n. However, as we will see different choices of local coordinates lead to the same physical scattering amplitudes.

The expression $V R^{(i)}(\xi_i)$ can be written as a Laurent series in ξ_i around $\xi_i = 0$, but when ξ_i approaches another Koba–Nielsen point, say z_j, this series diverges in general. To obtain a convergent expression as one approaches z_j one uses the identity of equation (18.3.5) in the overlap region where both $V R^{(i)}(\xi_i)$ and $V R^{(j)}(\xi_j)$ are convergent, and then approaches the point z_j using the latter expression. As such, the domain of validity of the coordinate ξ_i is defined in this sense of convergent series rather than in the sense of coordinate patches as used in the theory of differentiable manifolds.

As already discussed for scattering at genus 0 the world sheet is just the Riemann sphere with the marked Koba–Nielsen points and we can introduce a coordinate z which covers the whole sphere with the exception of the point at infinity where we may use the coordinate $\zeta = 1/z$. If none of the Koba–Nielsen points is at infinity, one can use the local coordinates $\xi_i = z - z_i$, but there are, of course, an infinite number of other choices. Should it happen that one has a Koba–Nielsen point at infinity, then one can use the local coordinate ζ for that string.

Hence, the vertex V by definition satisfies equation (18.3.5). What is far from obvious, but true, is that this equation determines the string vertex V up to a function of the Koba–Nielsen coordinates z_i. One may choose this function by taking $V \prod_i |0\rangle = 1$, but any other choice of the function is acceptable and will be taken into account when one computes the measure.

The amplitude of equation (18.3.3) also involves the measure f_m, which is determined from the requirement that the amplitude should vanish if any of the $|\chi_i\rangle$ is a zero-norm physical state. This means that

$$W(|\chi_1\rangle, |\chi_2\rangle, \ldots, |\chi_N\rangle) = 0 \tag{18.3.7}$$

if any of the $|\chi_i\rangle$ are physical zero-norm states. Examples of such states are given in chapter 3, equation (3.1.25) and later equations. This requirement is similar to the decoupling of such states in quantum electrodynamics and so we can think of equation (18.3.7) as a kind of Ward identity. In fact, this fixes the measure f_m up to a constant. The constant is determined as in the old days of sewing, or the sum over world surfaces approach, by requiring that the scattering amplitude is unitary.

To summarise the group theoretic approach, the string vertex V is by definition an object which obeys equations (18.3.5) for the conformal operators associated with the string, the string scattering amplitude is given by equation (18.3.3) and the measure is determined by demanding that zero norm physical states decouple. This procedure completely determines the scattering amplitudes of arbitrary string states.

Almost all of the vertex can be found more easily from some integrated identities which we now derive from the unintegrated identities above. Let us consider the expression

$$V\left(\oint_0 d\xi_i \varphi R^{(i)}(\xi_i) \right), \tag{18.3.8}$$

where φ is an arbitrary analytic, except for simple poles around the other Koba–Nielsen points, tensor of weight $1 - d$. That is, under a conformal transformation it transforms as

$$\varphi(\xi) \to \varphi'(\xi') = \varphi(\xi) \left(\frac{d\xi'}{d\xi} \right)^{1-d}. \tag{18.3.9}$$

As a result, $V\varphi R^{(i)}(\xi_i)d\xi_i$ is a tensor of weight 1, or simply, a 1-form. Let us now consider deforming the contour around z_i such that it encloses any other non-analytic behaviour of $V\varphi R^{(i)}(\xi_i)d\xi_i$. We find a set of contour integrals around the other Koba–Nielsen points. In carrying out this contour deformation as one approaches the neighbourhood of other Koba–Nielsen points, as explained above, one must use the equation (18.3.5) to maintain convergence. For the case of genus 0 the contour is pulled off to infinity where it vanishes, provided this is not a Koba–Nielsen point. Thus we find the integrated overlap equation

$$\sum_j V \left(\oint_0 d\xi_j \varphi R^{(j)}(\xi_j) \left(\frac{d\xi_j}{d\xi_i} \right)^{d-1} \right) = 0. \tag{18.3.10}$$

We now show how to dertermine the vertex explicitly so that the reader can see some more concrete expressions. We will return to the determination of the measure afterwards. In particular, we will compute open bosonic string scattering at the tree level. We adopt the particularly simple set of local coordinates

$$\xi_i(z) = (V_i)^{-1}(z) = z - z_i, \quad \text{or} \quad \xi_j = \xi_i + z_i - z_j. \tag{18.3.11}$$

We begin by deriving the N string vertex V. The overlap equation (18.3.5) for the conformal operator

$$Q^\mu(z) = -\frac{\overleftarrow{\partial}}{\partial \alpha_0^\mu} - \alpha_0^\mu \ln z + \sum_{\substack{n=-\infty \\ n \neq 0}}^{\infty} \frac{\alpha_n^\mu z^{-n}}{n} \tag{18.3.12}$$

of conformal weight 0 is given by

$$V Q^{\mu(i)}(z - z_i) = V Q^{\mu(j)}(z - z_j). \tag{18.3.13}$$

For $R(z) = P^\mu(z) \equiv \partial Q^\mu(z)$, the integrated overlap equation (18.3.10) becomes

$$V \left(\sum_{j=1}^N \oint d\xi_j \varphi_n P^{\mu(j)}(\xi_j) \right) = 0. \tag{18.3.14}$$

We can take φ_n to be of the form

$$\varphi_n = \frac{1}{(\xi_i)^n} = \frac{1}{(\xi_j + z_j - z_i)^n} = \sum_{p=0}^{\infty} \binom{-n}{p} \frac{(\xi_j)^p}{(z_j - z_i)^{n+p}} \qquad \forall j \neq i. \tag{18.3.15}$$

In fact, this set of φ form a complete set of all allowed φ. Substituting the φ into equation (18.3.14) and using the relation

$$P^\mu(z) = \sum_{n=-\infty}^{\infty} \alpha_{-n}^\mu z^{n-1} \tag{18.3.16}$$

we find the identity

$$0 = V \left(\alpha_{-n}^{\mu(i)} + \sum_{\substack{j=1 \\ j \neq i}}^N \sum_{p=0}^{\infty} \binom{-n}{p} \frac{\alpha_p^{\mu(j)}}{(z_j - z_i)^{n+p}} \right). \tag{18.3.17}$$

The solution to this equation is

$$V = \prod_{k=1}^N {}_{(k)}\langle 0| \exp \left(\sum_{\substack{i,j \\ i \neq j}} \left\{ -\frac{1}{2} \sum_{n,m=1}^{\infty} \alpha_n^{\mu(i)} \frac{(n+m-1)!(-1)^m}{m!n!(z_j - z_i)^{n+m}} \alpha_m^{\mu(j)} \right. \right.$$
$$\left. \left. - \sum_{n=1}^{\infty} \frac{\alpha_n^{\mu(i)} \alpha_0^{\mu(j)}}{n(z_j - z_i)^n} - \frac{1}{2} \alpha_0^{\mu(i)} N^{ij} \alpha_0^{\mu(j)} \right\} \right), \tag{18.3.18}$$

where N^{ij} is not determined by the integrated overlap and we have arbitrarily chosen the normalisation of V. To find N^{ij}, one uses the unintegrated Q^μ overlap of equation (18.3.13) and one finds the result to be

$$N^{ij} = - \ln |(z_j - z_i)|. \tag{18.3.19}$$

One could, from the beginning, have determined V completely from the Q^μ overlap; however, it is technically simpler to first carry out the partial determination given above with the integrated P^μ overlap. It is straightforward to evaluate the vertex for any choice of $(V_i)^{-1}$, or local coordinates.

We have used a somewhat short-hand notation in equation (18.3.18). By $\prod_{k=1}^{N}{}_{(k)}\langle 0|$ we actually mean

$$\prod_{k=1}^{N} \int d^D q_{k(k)} \delta \left(\sum_l q_l^\mu \right) \langle 0, q_k|, \tag{18.3.20}$$

where $|0, p\rangle$ obeys $\alpha_n^\mu |0, p\rangle = 0, n \geq 1$ and p^μ is its momentum.

We now continue with the formal development of the method. The vertex by definition satisfies the overlap equations (18.3.5) and to discover its properties it is often simpler to find how these equations behave under the desired transformation than to look into the details of the vertex itself. In particular we now consider the effect of conformal transformations on the vertex. Let us consider a conformal transformation $\xi \to \mathcal{N}(\xi)$ and denote by the same symbol $\mathcal{N}$ the operator of the realisation of the same conformal transformation on the string Fock space. We can carry out different conformal transformations on the different string Hilbert spaces and so on the different coordinate patches around the different Koba–Nielsen points. To be more precise, let us denote by $\mathcal{N}^{(i)}$ the operator that acts on the ith Hilbert space and by $\mathcal{N}_i$ the corresponding transformation $\xi_i \to \mathcal{N}_i(\xi_i)$ on the coordinates ξ_i around the ith Koba–Nielsen point. Given these conformal transformations we can define the conformally transformed vertex by

$$V(\hat{z}_i, \hat{\xi}_i) = V(z_i, \xi_i) \Pi_k \mathcal{N}^{(k)}. \tag{18.3.21}$$

The action of the conformal transformation may change the Koba–Nielsen points z_i and, as we have just mentioned, the local coordinates ξ_i used to define the vertex. We denote the new quantities with hats and show the explicit dependence of the vertex on them. In general, the operator $\mathcal{N}^{(k)}$ has the form

$$\mathcal{N}^{(k)} = \exp \left(\sum_{n=-\infty}^{\infty} c_n^k L_n^{(k)} \right), \tag{18.3.22}$$

where c_n^k are arbitrary constants. Under the above conformal transformation the vertex may also scale by a function of z_j which we have not explicitly indicated. As mentioned above, the behaviour of the vertex under a conformal transformation can be understood by studying its effect on the defining equations (18.3.5). Using the properties of conformal operators of equation (18.3.1) we find that equation (18.3.5) becomes

$$V(\hat{z}_i, \hat{\xi}_i) R^{(i)}(\mathcal{N}_i^{-1}(\xi_i)) = V(\hat{z}_i, \hat{\xi}_i) R^{(j)}(\mathcal{N}_j^{-1}(\xi_j)) \left(\frac{d(\mathcal{N}_j^{-1}(\xi_j))}{d(\mathcal{N}_i^{-1}(\xi_i))} \right)^d. \tag{18.3.23}$$

Writing $\xi_j = \tau_{ji}(\xi_i)$ and identifying the new coordinate as $\hat{\xi}_i = (\mathcal{N}_i)^{-1}\xi_i$, we find that the relation becomes

$$V(\hat{z}_i, \hat{\xi}_i)R^{(i)}(\hat{\xi}_i) = V(\hat{z}_i, \hat{\xi}_i)R^{(j)}(\mathcal{N}_j^{-1}\tau_{ji}\mathcal{N}_i(\hat{\xi}_i)) \left(\frac{d(\mathcal{N}_j^{-1}\tau_{ji}\mathcal{N}_i(\hat{\xi}_i))}{d\hat{\xi}_i} \right)^d .$$

$$(18.3.24)$$

Interpreting the argument of the operator R on the right-hand side of this equation as the new local coordinate, $\hat{\xi}_j = \hat{\tau}_{ji}(\hat{\xi}_i)$, we find that the transition functions for the new local coordinates of the conformal transformed vertex are given by the equation

$$\hat{\tau}_{ji} = (\mathcal{N}_j)^{-1}\tau_{ji}\mathcal{N}_i \qquad (18.3.25)$$

for all $i, j = 1, \ldots, N$. Since $\hat{\xi}_j = (\hat{V}_j)^{-1}(z) = (N_j)^{-1}\xi_j = (N_j)^{-1}(V_J)^{-1}(z)$ we conclude that $N_j = (V_j)^{-1}\hat{V}_j$. The new Koba–Nielsen points are also encoded as they are the points where $\hat{\xi}_j(\hat{z}_j) = 0$, or equivalently, when $(\hat{V}_j)^{-1}(\hat{z}_j) = 0$. The position of the ith Koba–Nielsen point is unchanged if $(N_i)^{-1}(0) = 0$.

Let us consider the effect on the vertex of a conformal transformation which changes the positions of the Koba–Nielsen points and the local coordinate systems $\xi_i \to \hat{\xi}_i = (\hat{V}_i)^{-1}z$ around each Koba–Nielsen point. This is implemented by the transformation $N_j = (V_j)^{-1}\hat{V}_j$ acting on the vertex:

$$V(v_r, \hat{z}_j, \hat{\xi}_j) = V(v_r, z_j, \xi_j) \prod_{i=1}^{N}(V_i)^{-1}\hat{V}_i. \qquad (18.3.26)$$

If $(\mathcal{N}_i)^{-1}(0) = 0$ for all i the transformations do not change the positions of the Koba–Nielsen points and just define new local coordinates around each Koba–Nielsen point. These are of the form of (18.3.22) with the sum for $n \geq 0$. Those that also shift the Koba–Nielsen points include L_{-1} in the sum. Since physical states are annihilated by $L_n - \delta_{n,0}, n \geq 0$, the former transformations do not affect physical amplitudes except for a scaling due to the L_0 term. Although, the L_0 term may lead to factors that multiply the vertex, this is compensated for when the measure is calculated.

Let us illustrate this discussion for the choice of coordinates used above to compute the vertex: $\xi_i = (V_i)^{-1}(z) = z - z_i$ or $\xi_i = \xi_k + z_k - z_i$. If we make a small change to the position of the jth Koba–Nielsen point, that is, $z_j \to \hat{z}_j = z_j + \epsilon$, $z_i \to \hat{z}_i = z_i$ for $i \neq j$, then the new coordinates are given by $\hat{\xi}_j = z - \hat{z}_j = z - z_j + \epsilon = (\hat{V}_j)^{-1}(z)$ and $\hat{\xi}_i = z - z_i = (\hat{V}_i)^{-1}(z)$ for $i \neq j$. As a result $(V_j)^{-1}\hat{V}_j(z) = z + \hat{z}_j - z_j = z + \epsilon$ and $(V_i)^{-1}\hat{V}_i(z) = z$ for $i \neq j$. The required conformal transformation corresponds to the operator $\mathcal{N}^{(j)} = (V_j)^{-1}\hat{V}_j = \exp(\epsilon L_{-1}^{(j)})$ acting on the $\mathcal{H}^j$ and $\mathcal{N}^{(i)} = (V_i)^{-1}\hat{V}_i$ is the identity transformation on $\mathcal{H}^i$ for $i \neq j$. As a result, equation (18.3.26) implies for this transformation that

$$\frac{\partial V}{\partial z^j} = VL_{-1}^{(j)}. \qquad (18.3.27)$$

As stated above, the measure is determined by ensuring that zero-norm physical states decouple. For the open bosonic string these states were found in chapter 3 to be given by

$$L_{-1}|\Omega\rangle \quad , \quad \left(L_{-2} + \tfrac{3}{2}L_{-1}^2\right)|\Omega'\rangle, \qquad (18.3.28)$$

where

$$L_n|\Omega\rangle = L_n|\Omega'\rangle = 0 \quad n > 0, \qquad L_0|\Omega\rangle = (L_0 + 1)|\Omega'\rangle = 0. \tag{18.3.29}$$

Taking one of the physical states in equation (18.3.3) to be one of these zero-norm states we may interpret it as an infinitesimal conformal transformation applied to the vertex times $|\Omega'\rangle$ or $|\Omega\rangle$. Such transformations will, in general, change the positions of the Koba–Nielsen points and the coordinate systems. Consequently, we find, generically, that

$$\int \prod_k{}' dz_k f_m \left(\sum_j g_j \frac{\partial V}{\partial z_j} - cV \right) |\chi_1\rangle_{(1)} \cdots |\Omega\rangle_{(i)} \cdots |\chi_N\rangle_{(N)} = 0, \tag{18.3.30}$$

where all the states $|\chi\rangle_{(i)}$ are physical and so obey $(L_n - \delta_{n,0})|\chi\rangle = 0, n \geq 0$ and the change induced in the Koba–Nielsen points is given by

$$z_j \rightarrow z_j + g_j(z_j). \tag{18.3.31}$$

Transformations that do not change the Koba–Nielsen coordinates, but lead to a change of coordinates around each Koba–Nielsen point involve $L_n, n \geq 0$ and so they give factor 1 on the physical states except for the L_0 contribution which leads to the constant c in the above equation. Thus integrating by parts we find that

$$0 = \sum_i \frac{\partial}{\partial z_i}(g_i f_m) + c f_m. \tag{18.3.32}$$

Thus the decoupling of the zero-norm physical states leads to a set of first order differential equations for the measure, f_m. In fact, one has as many differential equations as one has Koba–Nielsen variables which are integrated over and so the measure is determined by them.

We can now determine the measure for the vertex found in equation (18.3.18), which used the coordinates $\xi_j = z - z_j$. Taking the zero-norm physical state $L_{-1}^{(j)}|\Omega\rangle_j$ on the jth string Hilbert space and arbitrary physical states $|\chi_k\rangle$ on all the other Hilbert spaces to the integrated vertex of equation (18.3.4) and setting the result to 0 we find that

$$0 = \int \prod_{k=1}^{N-3} dz_k f_m \frac{\partial V}{\partial z^j}|\chi\rangle_1 \cdots |\Omega\rangle_j \cdots |\chi\rangle_N; \quad j = 1, \ldots, N-3, \tag{18.3.33}$$

which implies that $\partial f_m/\partial z^j = 0$, or that f_m is a constant. We have chosen the unintegrated Koba–Nielsen points to be $N-2$, $N-1$ and N.

This procedure applies to the legs for which the Koba–Nielsen points are integrated over. However, there are three for which this is not the case, and one must also show that the states $L_{-1}^{(j)}|\Omega\rangle_{(j)}$ decouple for $j = N-2$, $N-1$, N. To show this result we need the integrated identities of equation (18.3.10) for the energy-momentum tensor T which has $d = 2$ and so φ has $d = -1$, which is a vector field. The integrated relation has the form

$$V \left\{ \sum_{j=1}^N \oint_{\xi_j=0} d\xi_j \varphi \left(T^j(\xi_j) \frac{d\xi_j}{d\xi_i} + \frac{c}{12}(S\xi_j)(\xi_i)\frac{d\xi_i}{d\xi_j} \right) \right\} = 0. \tag{18.3.34}$$

The additional term is due to the Schwarzian derivative which appears in the transformation of the energy-momentum tensor. Let us take the vector field φ to be the three analytic vector fields that live on the Riemann sphere, namely $\varphi = 1, z = \xi_j + z_j, z^2 = (\xi_j + z_j)^2$. The corresponding identities are given by

$$V \sum_j L_{-1}^{(j)} = 0, \quad V \sum_j (L_0^{(j)} + z_j L_{-1}^{(j)}) = 0, \quad V \sum_j (L_1^{(j)} + 2z_j L_0^{(j)} + z_j^2 L_{-1}^{(j)}) = 0.$$

$$(18.3.35)$$

The last three equations express the $SU(1, 1)$ invariance of the tree level scattering amplitude which was discussed in section 18.3. One can use these three identities to convert any L_{-1} acting on the three legs whose Koba–Nielsen points are not integrated over in terms of L_{-1}s on the other legs as well as L_0s and L_1s. The last two give a multiplicative factor and zero, respectively, on the physical states and one can readily deduce the decoupling of the required zero-norm physical states.

To complete the discussion of decoupling, one must also show that states of the form $(L_{-2} + \frac{3}{2}L_{-1}^2)|\Omega'\rangle$ decouple. This is shown using the vertex preserving integrated T identities which have a φ with a simple pole at one Koba–Nielsen point. In particular, if we choose $\varphi = \xi_j^{-1}$, we find L_{-2}^j on leg j and $L_n, n \geq -1$ on the other legs. Using this identity one can show that this zero-norm physical state decouples.

Thus we have completely determined the open string tree level vertex V and the measure f_m and so the scattering amplitudes.

We note that in the older literature on the group theoretic approach (except the more recent [18.44]), the equations look a bit different because a different convention for a conformal operator was used. In this chapter we employ the commonly used definition of recent times given in equation (18.3.1). However, in the old literature and references in [18.38–18.43] which proposed the group theoretic approach, a $(\xi_i)^{-d}$ was factored out of the conformal operators, which led to the transformation rule

$$\frac{R(\xi_i)}{(\xi_i)^d} \rightarrow \frac{R(\xi_j)}{(\xi_j)^d} \left(\frac{d\xi_j}{d\xi_i} \right)^d .$$

$$(18.3.36)$$

The early papers on the group theoretic approach are rather difficult to read as the final formulation became apparent as it evolved. The reader can consult the review of [18.45] and the review contained as the second section of [18.44].

The reader can readily derive the scattering amplitudes for the closed bosonic string at genus 0, the only part requiring thought is the purely momentum part which is given in section 5 of [18.44]. The same group theoretic approach can be applied to superstrings provided one makes the appropriate generalisations. In particular, this involves a larger set of conformal operators, including those arising from the fermionic coordinates Ψ^μ. We refer the reader to [18.42, 18.43]. The ghosts oscillators were incorporated in [18.10, 18.40, 18.41].

A discussion of scattering amplitudes involving loops can be found in [18.38–18.41]. This subject is covered in the reviews of [18.45] and that contained in [18.44]. The discussion proceeds in much the same way except that there are additional identities corresponding to the representation of the Riemann surface. The high energy behaviour of string theory was derived in [18.44].

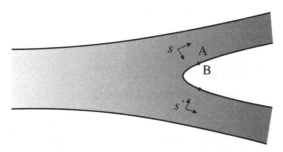

Figure 18.4.1   The observer dependent interaction points.

18.4   Interacting open string field theory

18.4.1 Light-cone string field theory

It has been known for a long time how to construct a field theory of strings in the light-cone gauge [4.1]. This is a field theory built out of the functionals of $x^i(\sigma)$, $i = 1, \ldots, 24$, as well as the centre of mass coordinates q^- and the universal time τ, which appears in x^+ once it is gauge fixed. The free action was given in chapter 4. In this formulation strings interacted in a very simple way by joining at their end points. The corresponding world sheet of the string is given in figure 18.4.1 We note that the point at which the interaction takes place is observer dependent and to illustrate this we have also shown the frames of reference of two observers in figure 18.4.1 The different space-time points at which the two observers find the interaction to have taken place are marked A and B.

At the moment of the interaction the incoming string can be identified, in the appropriate regions, with the two outgoing strings. By time reversal we must also allow two strings to become one by joining at their end points. To parameterise the interaction it is natural to assign the strings a length. Let the rth string have parameter σ_r and have length $\alpha_r\pi$, in which case

$$0 \leq \sigma_r \leq \alpha_r\pi. \tag{18.4.1}$$

The lengths are to be preserved in the interaction

$$\sum_{r=1}^{3} \alpha_r = 0. \tag{18.4.2}$$

The string lengths can be chosen to be given by the centre of momentum in the $+$ direction, specifically $2p^+ = \alpha$. They will then be automatically conserved. We take the length of incoming strings to be positive and that of outgoing strings to be negative. The string functionals will depend on p^+, once we have changed basis from q^- to p^+, and the action will have a factor $\int \prod_r (dp^{r+}/\sqrt{2p^{r+}})\delta(p_1^+ + p_2^+ + p_3^+)$, which we will not show explicitly in the vertex we derive below.

The string interaction can be made more mathematically precise. Let us consider the case of two strings joining to form a third and parameterise this process as a strip, see figure 18.4.2. Rather than use σ_r, it will prove useful to also parameterise the strings by using $\eta_r = \sigma_r/|\alpha_r|$, which is designed to take more standard values, that is, $0 \leq \eta_r \leq \pi$. At the

Figure 18.4.2   Two strings forming a third displayed as a strip.

interaction point string η_3^{int} on string 3, string 2 has length $\alpha_2\pi$ which then equals $\eta_3^{int}|\alpha_3|$ and so the interaction point is given by $\eta_3^{int} = \beta_2\pi$, where $\beta_2 = \alpha_2/\alpha_1 + \alpha_2$ and we define $\beta_1 = \alpha_1/\alpha_1 + \alpha_2$. It is then easy to see that the ηs on the three stings are related by

$$\eta_3 = \begin{cases} -\beta_2\eta_2 + \beta_2\pi & \text{for } 0 \le \eta_3 \le \beta_2\pi, \\ -\beta_1\eta_1 + \pi & \text{for } \beta_2\pi \le \eta_3 \le \pi. \end{cases} \qquad (18.4.3)$$

The interaction in coordinate space just encodes the identification of the strings and it is given by

$$V(x) \equiv V \mid x\rangle = N\delta\big(x_3^i(\eta_3) - \theta(-\eta_3 + \beta_2\pi)x_2^i(\eta_2) - \theta(\eta_3 - \beta_2\pi)x_1^i(\eta_1)\big),$$
$$(18.4.4)$$

where N is a function of the string lengths. At the moment they touch the δ function implies that they overlap, that is,

$$x^{(3)i}(\eta_3) = \begin{cases} x^{i(1)}(\eta_1) & \text{for } \beta_2\pi \le \eta_3 \le \pi, \\ x^{i(2)}(\eta_2) & \text{for } 0 \le \eta_3 \le \beta_2\pi. \end{cases} \qquad (18.4.5)$$

Hence, the cubic interaction term in the light-cone gauge field theory action is given by

$$A^{Int} = \frac{g}{3} \int \prod_r Dx^{i(r)}(\sigma)V(x)X\big(x^{i(1)}(\eta_1)\big)X\big(x^{i(2)}(\eta_2)\big)X\big(x^{i(3)}(\eta_3)\big). \qquad (18.4.6)$$

The reader may be wondering what is the relation between the δ function interaction of equation (18.4.4) and the result for three-string scattering as found in the dual model at the end of section 18.1. The latter is given in terms of oscillators and so we must change from the x-space basis, containing the δ function, to the oscillator basis. This is best carried out by first going to a momentum basis, then to a coherent state basis and finally recognizing this expression in terms of oscillators. Having the result in oscillator basis also has the advantage that it is well defined, unlike in the x-space basis where operations involving functional derivatives are generally not easily made into rigorous manoeuvres.

Let us now carry out this change of basis [18.49–18.51]. We consider the case of two strings joining to form a third and parameterise this process as a strip (see figure 18.4.2). As discussed in chapter 2 the Fourier expansion of $x^{ri}(\eta_r)$, which obeys the correct boundary conditions, is given by

$$x^{ri}(\eta_r) = x_0^{ri} + 2\sum_{n=1}^{\infty} \cos(n\eta_r)x_n^{ri}. \qquad (18.4.7)$$

Using the overlap condition of equation (18.4.5) we can match string 3 and string 1 in the region $0 \leq \eta_1 \leq \pi$ to find

$$\frac{1}{\pi} \int_0^\pi d\eta_1 \, x^{1i}(\eta_1)(\cos n\eta_1) = \frac{1}{\pi} \int_0^\pi d\eta_1 \, (\cos n\eta_1) x^{3i}(\eta_3), \qquad (18.4.8)$$

which implies that

$$x_0^{1i} = x_0^{3i} + \sum_{m=1}^\infty (-1)^m \frac{2}{m\beta_1} (\sin m\beta_1 \pi) \frac{x_m^{3i}}{\pi},$$

$$x_n^{1i} = \sum_{m=1}^\infty \frac{(-1)^n (-1)^m}{\pi} \left[\frac{-2m\beta_1}{n^2 - \beta_1^2 m^2} \right] (\sin n\beta_1 \pi) x_m^{3i}. \qquad (18.4.9)$$

Similarly for string 2 we find the equations

$$x_0^{2i} = x_0^{3i} + \sum_{m=1}^\infty \frac{2}{m\beta_2} (\sin m\beta_2 \pi) \frac{x_m^{3i}}{\pi}, \qquad x_n^{2i} = \sum_{m=1}^\infty \frac{2m\beta_2}{n^2 - \beta_2^2 m^2} (\sin m\beta_2 \pi) \frac{x_m^{3i}}{\pi}.$$

$$(18.4.10)$$

These results can be written in the form of column vectors and matrices:

$$x^1 = A^{1T} x_3, \quad x^2 = A^{2T} x^3, \quad \sum_{r=1}^3 \alpha^r x_0^r = 0, \quad x_0^1 - x_0^2 = \alpha_3 B^T x^3, \qquad (18.4.11)$$

where the matrices A_{nm}^1, A_{nm}^2 and the column vector B_n, $n, m = 1, 2, \ldots, \infty$, are read off from the preceding equations and T denotes the transpose. Consequently, the vertex of equation (18.4.4) can be rewritten as

$$V|x\rangle = N\delta \left(x_n^{1i} - \sum_{m=1}^\infty A^{1T}_{nm} x_m^{3i} \right) \delta \left(x_n^{2i} - \sum_{m=1}^\infty A^{2T}_{nm} x_m^{3i} \right)$$

$$\times \delta \left(\sum_{r=1}^3 \alpha^r x_0^{ri} \right) \delta \left(x_0^{1i} - x_0^{2i} - \alpha_3 \sum_{n=1}^\infty B_n x_n^{3i} \right). \qquad (18.4.12)$$

By a basis change we find the vertex in momentum space to be given by

$$V(p) \equiv V|p\rangle = \int \prod_{s,n} dx_n^{si} V|x_n^{si}\rangle \langle x_n^{si}|p_{in}^s\rangle = \int \prod_{s,n} dx_n^{si} \exp(ip_{si}^n x_n^{si}) V|x_n^{si}\rangle$$

$$= N\delta \left(p_0^{1i} + p_0^{2i} + p_0^{3i} \right) \delta \left(p^{3i} + A^2 p^{2i} + A^1 p^{1i} + B\mathcal{P}^i \right), \qquad (18.4.13)$$

where

$$\mathcal{P}^i = \alpha_1 p_0^{2i} - \alpha_2 p_0^{1i} = \alpha_2 p_0^{3i} - \alpha_3 p_0^{2i} = \alpha_3 p_0^{1i} - \alpha_1 p_0^{3i}.$$

To enable us to transform to oscillator basis we will discuss some elementary properties of coherent states. Consider a single one-dimensional harmonic oscillator; it is usual to change variables from x and p to the oscillators a and $a^\dagger$, which are given by $p = \frac{1}{\sqrt{2}}(a + a^+)$ and $x = (i/\sqrt{2})(a - a^+)$. The oscillators satisfy $[a, a^+] = 1$. We define the coherent state by

$|z\rangle = e^{za^\dagger}|\rangle$, where z is complex and the vacuum $|\rangle$ satisfies $a|\rangle = 0$. We observe that $|z\rangle$ satisfies the relations

$$a|z\rangle = z|z\rangle, \quad a^+|z\rangle = \frac{d}{dz}|z\rangle. \tag{18.4.14}$$

The overlap function between momentum and coherent states is given by

$$\langle p|z\rangle = \exp\left(-\frac{p^2}{2} - \frac{z^2}{2} + \sqrt{2}pz\right), \tag{18.4.15}$$

which is easily found by inserting $p = \frac{1}{\sqrt{2}}(a + a^+)$ and $x = (i/\sqrt{2})(a - a^+)$ and using the above relations. We also find that

$$\langle z'|z\rangle = \exp z^{*\prime}z \tag{18.4.16}$$

by using the relation $1 = \int dp|p\rangle\langle p|$. The completeness relation for the coherent states is given by

$$1 = \int dz dz^* \exp\left(-z^* z\right). \tag{18.4.17}$$

The generalisation to the string oscillators α_n^μ only requires factors of n in the appropriate places.

We next find the vertex in the coherent state basis

$$V(z) \equiv V|z\rangle = \int \prod_{r,n} dp_{rn} V|p_n^r\rangle\langle p_n^r|z\rangle$$

$$= \int \prod_{r,n} dp_{rn} \delta\left(\sum_{r=1}^{3} p_0^r\right) N\delta\left(p^3 + A^2 p^2 + A^1 p^1 + B\mathcal{P}\right)$$

$$\times \exp\left\{ - \tfrac{1}{2} p^{T^r} C^{-1} p^r - \tfrac{1}{2} z^{T^r} C^{-1} z^r + \sqrt{2} p^{T^r} C^{-1} z^r \right\}, \tag{18.4.18}$$

where $C_{mn} = n\delta_{mn}$. We do not integrate over p_0 and so the factor $\delta\left(\sum_{r=1}^{\infty} p_0^r\right)$ remains. To keep the expressions simple we will not show this factor but it is to be understood to be present. In the above and below we have suppressed the i indices until we find the final result. Rewriting the δ function as an integral we find

$$V|z\rangle = \int dp\, duN \exp\left\{-\tfrac{1}{2} p^{T^r} C^{-1} p^r - \tfrac{1}{2} z^{T^r} C^{-1} z^r + \sqrt{2} p^{T^r} C^{-1} z^r\right\}$$

$$\times \exp iu\left(\sum_{r=1}^{3} A^r p^r + B\mathcal{P}\right)$$

$$= \int dp\, duN \exp\left(-\tfrac{1}{2}(p^{Tr} - \sqrt{2}z^{Tr} - iu^T A^r C) C^{-1} \left(p^r - \sqrt{2}z^r - iCA^{Tr}u\right)\right)$$

$$\times \exp\left\{ +\tfrac{1}{2}\left(\sqrt{2}z^{Tr} + iu^T A^r C\right) C^{-1} \left(\sqrt{2}z^r + iCA^{Tr}u\right) \right.$$

$$\left. + iu^T B\mathcal{P} - \tfrac{1}{2} z^{T^r} C^{-1} z^r\right\}$$

$$= \int duN \exp\left\{\tfrac{1}{2} z^{T^r} C^{-1} z^r - \tfrac{1}{2} u^T \Gamma u + iu^T W\right\}, \tag{18.4.19}$$

where

$$\Gamma = \sum_{r=1}^{3} A^r C A^{Tr}, \quad W^T = \sqrt{2} z^{Tr} A^{Tr} + B^T \mathcal{P}, \quad A^3_{nm} = \delta_{n,m}. \tag{18.4.20}$$

Carrying out the final integration we obtain the result

$$V|z\rangle = N' \exp \left\{ \tfrac{1}{2} z^{Tr} C^{-1} z^r - \tfrac{1}{2} W^T \Gamma^{-1} W \right\}$$
$$= N' \exp \left\{ \tfrac{1}{2} z^r_n \bar{N}^{rs}_{nm} z^s_m + \bar{N}^r_m z^r_m \mathcal{P} + K \mathcal{P}^2 \right\}, \tag{18.4.21}$$

where $N' = N (\det \Gamma)^{(2-D)/2}$ and

$$\bar{N}^{rs}_{nm} = \delta^{rs} (C^{-1})_{nm} - 2 (A^{Tr} \Gamma^{-1} A^s)_{nm} \equiv N^{rs}_{nm} e^{n\tau_b/\alpha_r} e^{m\tau_b/\alpha_s},$$
$$\bar{N}^r_m = -\sqrt{2} (A^{Tr} \Gamma^{-1} B)_m \equiv N^r_m e^{m\tau_b/\alpha_r}, \quad K = -\tfrac{1}{2} B^T \Gamma^{-1} B. \tag{18.4.22}$$

Hence the vertex in terms of oscillators is given by the expression

$$V = {}_1\langle 0|{}_2\langle 0|{}_3\langle 0| N' \exp \left\{ \tfrac{1}{2} \alpha^{ir}_n e^{n\tau_b/\alpha_r} N^{rs}_{nm} e^{m\tau_b/\alpha_s} \alpha^{is}_m + N^r_m e^{m\tau_b/\alpha_r} \alpha^{ir}_m \mathcal{P}^i + K \mathcal{P}^i \mathcal{P}^i \right\}$$

$$\tag{18.4.23}$$

by virtue of equation (18.4.14).

It can be shown that [18.49–18.51]

$$N^{rs}_{nm} = -\frac{\alpha_1 \alpha_2 \alpha_3}{\alpha_r \alpha_s} \frac{mn}{n\alpha_s + m\alpha_r} f_n(\gamma_r) f_m(\gamma_s), \quad \gamma_r = -\frac{\alpha_{r+1}}{\alpha_r},$$
$$f_n(\gamma_r) = \frac{1}{n\gamma_r} \binom{n\gamma_r}{n} = \frac{(n\gamma_r - 1) \dots (n\gamma_r - n + 1)}{n!} \tag{18.4.24}$$

while

$$N^r_m = \frac{1}{\alpha_r} f_m(\gamma_r), \quad K = -\frac{\tau_b}{2\alpha_1 \alpha_2 \alpha_3}. \tag{18.4.25}$$

We note that the vertex is a complicated function of the string lengths. We may rewrite the vertex by reassigning the factors of $e^{n\tau_b/\alpha_r}$ as follows

$$V = {}_1\langle 0|{}_2\langle 0|{}_3\langle 0| \exp \left\{ \tfrac{1}{2} \alpha^{ir}_n N^{rs}_{nm} \alpha^{is}_m + N^r_m \alpha^{ir}_m \mathcal{P}^i \right\} \exp \left\{ \tau_b \sum_r \frac{1}{\alpha_r} \right\} \exp \left(\frac{\tau_b}{2} \sum_r P^{-(r)} \right).$$

$$\tag{18.4.26}$$

In this last equation P^- is proportional to the Hamiltonian (see chapter 4), which, taking account of the new length of the strings, is given by

$$P^- = \frac{2}{\alpha} \left(\frac{p^i p^i}{2} + \sum_{n=1}^{\infty} \alpha_{-n} \alpha_n - 1 \right).$$

We have used that $-\mathcal{P}^2 = \alpha_1 \alpha_2 \alpha_3 \sum_r p^{i(r)} p^{i(r)}/\alpha_r$. We have also chosen $N' = \exp(\tau_b \sum_r 1/\alpha_r)$, which is required in order that the theory be Lorentz invariant [18.52]. This vertex was first found in [18.52] using a sum over world-sheet approach and the last form given here is the form in which it appeared. We will not give this derivation but in the next section we will explain how to derive the key ingredient in this computation, namely

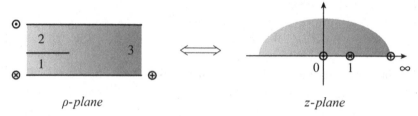

Figure 18.4.3   The map of three-string scattering to the upper half-plane.

the Green's function on the strip of figure 18.4.2, by using the map in figure 18.4.3. We note that the vertex is a rather complicated function of the string lengths. The three-string in the action is then given by $gV|\chi^1\rangle_1|\chi^2\rangle_2|\chi^3\rangle_3$.

Since the interaction is given by a δ function, it obeys, in the appropriate region, the relation

$$V\left(x^{3(\mu)}(\eta_3) - x^{\mu(1)}(\eta_1)\right) = 0, \tag{18.4.27}$$

which is none other than the identity $x\delta(x) = 0$. There are similar relations for the functional derivative of $x(\eta)$. We recognise these as similar to the relations which formed the basis for the group theoretic approach of section 18.3. Indeed, it was in the study of the covariantised light-cone approach, mentioned later, that the more general relations first arose.

In the light-cone string field theory there is also a four-string interaction [4.1] whose form can be deduced in a similar manner [18.50]. There is also a string field theory for closed strings [4.1] and one finds a very interesting open closed string vertex [4.1].

When comparing with the vertex V^{CSV} given at the end of section 18.1 it is important to realise that the vertices should only agree on-shell. Since physical states satisfy the Virasoro conditions $L_n|\chi\rangle = 0, n \geq 1$ and $(L_0 - 1)|\chi\rangle = 0$, the two vertices will lead to the same on-shell scattering if they are related by a transformation that involves $L_n, n \geq 0$ acting to the right.

18.4.2 Mapping the interacting string

The vertex in oscillator basis is well defined, but has a rather complicated form involving the f_m and N_{mn}^{rs} functions. From the point of view of the sum over the world surfaces of the strings these complications can be seen to arise as follows. In this approach, one starts with a two-dimensional field theory and one is required to find the Green's functions for point sources on the strip,

$$\left(\frac{\partial^2}{\partial\tau^2} - \frac{\partial^2}{\partial\sigma^2}\right) G(\sigma, \tau; \sigma', \tau') = 2\pi\delta(\sigma - \sigma')\delta(\tau - \tau'). \tag{18.4.28}$$

The Green's function is subject on the boundary to the condition

$$\partial G/\partial\eta(\sigma, \tau)(\sigma, \tau; \sigma', \tau') = f(\sigma, \tau),$$

where $\partial / \partial \eta$ is the normal derivative at the boundary and f is an arbitrary function corresponding to the emitted strings. This is identical to a problem in two-dimensional electrostatics where one must compute the potential on a strip due to some point charges on the boundary. Indeed, it was in thinking about this analogy that it was realised that the dual model was a theory of string scattering [18.15, 18.16]. It is most easily solved by transforming from the strip to the upper half-plane by a Schwartz–Christoffel map. The Green's function on the two-dimensional upper half-plane is rather simple: namely, for one charge it is $\ln |(z - z')|^2$. Upon transforming this simple function back to the strip, we will encounter the f_m and N_{mn}^{rs} functions.

We now discuss in detail the map from the strip to the upper half-plane [18.52]. Consider the three-string scattering shown in the figure 18.4.2. The complete strip is parameterised by $\rho = \tau + i\sigma$, $0 \leq \sigma \leq \alpha_3 \pi$, $-\infty < \tau < +\infty$ and the interaction is at $\tau = 0$. The individual strings have the coordinates

$$\rho_1 = \alpha_1 (\xi_1 + i\eta_1) \quad \text{for string 1,}$$

$$\rho_2 = \alpha_2 (\xi_2 + i\eta_2) + i\alpha_1 \pi \quad \text{for string 2,} \tag{18.4.29}$$

$$\rho_3 = \alpha_3 (\xi_3 + i\eta_3) - i\alpha_3 \pi \quad \text{for string 3,}$$

where $0 \leq \eta_r \leq \pi$; $\tau_r = \alpha_r \xi_r$ and $-\infty \leq \xi_r \leq 0$.

As an SL(2, **R**) transformation preserves the upper half-plane, the map from the strip to the latter is ambiguous up to such a transformation. The ambiguity is resolved if we select where on the real axis of the upper half-plane the three end points of the string are mapped. We will choose the end points of strings 1, 2 and 3 to be 1, 0 and ∞, respectively. The map from the strip with three strings to the upper half-plane is then given by the transformation [18.52]

$$\rho + \tau_b = \alpha_1 \ln(z - 1) + \alpha_2 \ln z. \tag{18.4.30}$$

This map is shown in figure 18.4.3. The coordinate of the strip is ρ and z is the coordinate on the upper half-plane.

One finds in particular that the real axis of the upper half-plane is mapped onto the boundary of the string. The three end points of the strings are mapped onto the three points of the upper half-plane as shown in figure 18.4.3. Starting at $z = +\infty$, (that is, at point $\oplus$) and moving along the real axis, we move along the strip boundary to the end of string 1 (that is, point $\otimes$ at $z = 1$). Here $\alpha_1 \ln(z - 1)$ changes argument by $\alpha_1 \pi$ and so we jump to the top of the first string and move towards the interaction point. When we arrive at this point we then travel along the bottom of string 2 towards the end of string 2 etc. Since the derivatives at the interaction point are discontinuous the map is not invertible at this point, that is,

$$\frac{d\rho}{dz} = 0 \quad \rightarrow \quad z_b = -\frac{\alpha_2}{\alpha_3} = \frac{\alpha_2}{|\alpha_3|} \tag{18.4.31}$$

and

$$\tau_b = \alpha_1 \ln \left(\frac{\alpha_1}{|\alpha_3|} \right) + \alpha_2 \ln \left(\frac{\alpha_2}{|\alpha_3|} \right). \tag{18.4.32}$$

The time evolution of the string on the upper half-plane is shown in the figure 18.4.4.

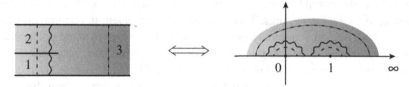

Figure 18.4.4   The time evolution of the interacting strings and upper half-plane.

To find the Green's function on the strip we will require the inverse of this map, that is, the map from the upper half-plane to the strip. To this end let us define the object $y(x)$ by

$$y \equiv \gamma \ln(1 + xe^y) = \gamma x + \frac{\gamma}{2}(2\gamma - 1)x^2 + \cdots = \gamma \sum_{n=1}^{\infty} f_n(\gamma)x^n. \qquad (18.4.33)$$

It can be shown that [18.52]

$$f_n(\gamma) = \frac{1}{n\gamma}\binom{n\gamma}{n} = \frac{1}{n!}(n\gamma - 1)(n\gamma - 2)\cdots(n\gamma - n + 1). \qquad (18.4.34)$$

The inverse map from the half-plane to the strip is given by [18.52]

$$z = \begin{cases} -\frac{1}{\upsilon_3}e^{-y_3(\upsilon_3)} & \text{for string 3,} \\ e^{-(\alpha_1/\alpha_2)y_1(\upsilon_1)} & \text{for string 1,} \\ \upsilon_2 e^{-(\alpha_1/\alpha_3)y_2(\upsilon_2)} & \text{for string 2,} \end{cases} \qquad (18.4.35)$$

where

$$\upsilon_r = e^{(\tau_b + \rho_r/\alpha_r)}; \quad y_r \equiv \gamma_r \ln(1 + \upsilon_r e^{y_r}), \quad \gamma_r = -\frac{\alpha_{r+1}}{\alpha_r}, \quad (\alpha_{3+1} \equiv \alpha_1). \qquad (18.4.36)$$

One can verify by substituting in equation (18.4.30) that this is the correct inverse map. We will not derive the vertex from this perspective [18.52], but the reader should be convinced that the functions in the vertex do arise in this method.

The generalisation to any number of interacting strings is straightforward. The map from the corresponding strip to the upper half-plane is given by

$$\rho = \sum_{r=1}^{N} \alpha_r \ln(z - z_r) + \text{constant}, \qquad (18.4.37)$$

where the strings have lengths α_r with $\sum_r \alpha_r = 0$. We can take the final string to be at infinity $z_N = \infty$ and then we omit the last terms, once we absorb an infinite constant in ρ. This then agrees with the map given above for three strings.

18.4.3 A brief discussion of interacting gauge covariant string field theory

The gauge covariant open string action has the form

$$\frac{1}{2}\langle \chi | Q | \chi \rangle + \frac{g}{3}V|\chi\rangle_1|\chi\rangle_2|\chi\rangle_3 + \cdots . \qquad (18.4.38)$$

The first term is the free action discussed in chapter 12, g is the string coupling and V is a three-string vertex of the generic form discussed in section 18.3. The indices 1, 2, 3 on $|\chi\rangle$ refer to the legs (Hilbert spaces) of the 3-vertex on which the string states act and $+\cdots$ means higher powers of g. The vertex will involve the oscillators not only for $x^\mu(\sigma)$ but also the $b(\sigma)$ and $c(\sigma)$ ghosts. The field $|\chi\rangle$ transforms as

$$\delta_3\langle\chi| = \frac{1}{g}{}_3\langle\Lambda|Q + V(|\chi\rangle_1|\Lambda\rangle_2 - |\Lambda\rangle_1|\chi\rangle_2) + \cdots. \tag{18.4.39}$$

We have assumed that the transformation and the action have the same 3-vertex, a result supported by low level computations.

Varying the action we find that it is invariant at order g if the 3-vertex is BRST invariant:

$$V\sum_{r=1}^{3}Q^{(r)} = 0. \tag{18.4.40}$$

The commutator of two transformations is given by

$$[\delta_{\Lambda'},\delta_\Lambda] = {}_3\langle\chi| = \frac{1}{g}V\{Q^{(1)}|\Lambda'\rangle_1|\Lambda\rangle_2 - Q^{(2)}|\Lambda\rangle_1|\Lambda'\rangle_2\} + \cdots$$

$$= -\frac{1}{g}V(|\Lambda'\rangle_1|\Lambda\rangle_2 - |\Lambda\rangle_1|\Lambda'\rangle_2)Q^{(3)} - (\Lambda\leftrightarrow\Lambda') + \cdots, \tag{18.4.41}$$

where we have used equation (18.4.40). Thus the algebra does close at lowest order. Indeed, we can read off from equation (18.4.41) the composite parameter and so the algebra of gauge transformations. The above steps to construct an open string gauge covariant theory and its explicit form at low levels are given in [12.10].

There are two proposals for an open string gauge covariant action. One mimics the light-cone string field theory, strings interact by joining at their end points, and one introduces, in an ad hoc way, string lengths [12.10, 18.53]. These are additional parameters that are not related to p^+, which now appear in a Lorentz covariant manner togther with the other components of the momentum, After these last papers appeared it was realised that if one adopted legitimate gauge conditions on the two-dimensional metric, as opposed to those usually chosen when first quantising the string, then the string does acquire additional parameters which can be interpreted as string lengths [3.17]. The incorporation of this result into the string field theory was not carried out.

In a very different proposal [12.4] the strings interact at their midpoints rather than their end points. This had the substantial advantage that it only requires a cubic vertex. However, as is apparent from the last section, the maps to the upper half-plane are much more complicated and the vertex when given in terms of oscillators reflects this complication. This proposal has been fully developed [18.54].

Appendix A   The Dirac and BRST methods of quantisation

In this appendix we outline the Dirac and BRST methods of quantisation which are used extensively in chapter two. We begin with the former.

A.1   The Dirac method [1.1]

Given a system with a Lagrangian L which is a function of the coordinates q_a and their time derivatives $\dot{q}_a$, we can compute the momenta, $p^a = \partial L / \partial \dot{q}_a$. It can happen that these momenta satisfy some constraints, $\phi_i = 0$. For many systems this is not the case and we can quantise as usual, but most of the theories used in modern physics do have such constraints and it is in these cases that one can use the Dirac method to quantise them. For consistency, we must demand that the time evolution of the constraints $\phi_i = 0$ vanish and in doing so we can generate new constraints. To be precise, since the time evolution is generated by the Hamiltonian, we take the Poisson bracket of the constraints with the Hamiltonian to vanish: $\{\phi_i, H^*\} = 0$. The Hamiltonian used for this is defined to be the usual Hamiltonian calculated from the Lagrangian, that is, $H = \sum_a p^a q_a - L$ plus the sum of the original constraints, each of which is multiplied by an arbitrary function of time v^i. In other words, we take $H^* = H + \sum_i v^i \phi_i$. Should one find new constraints in carrying out this process, we then demand that their time evolution vanishes and so we may find further constraints. We continue repeating this process until the time evolution of all the constraints vanishes.

We then calculate the Poisson brackets of all the constraints so obtained and identify the sub-set of the constraints whose Poisson brackets with themselves are only constraints and so vanish. The constraints in this subset we call the first class constraints and the remainder second class constraints. It can happen that in some systems one has only first class constraints and in some others only second class constraints, but in some systems of interest one has a mixture of first and second class constraints.

The usual definition of the Poisson bracket for any two functions f and g of the coordinates and momenta is

$$\{f, g\} = \sum_a \left(\frac{\partial f}{\partial q_a} \frac{\partial g}{\partial p^a} - \frac{\partial f}{\partial p^a} \frac{\partial g}{\partial x_a} \right). \tag{A.1.1}$$

It obeys the relations

$$\{f, g\} = -\{g, f\}, \ \{f, gk\} = g\{f, k\} + \{f, g\}k, \ \{f_1 + f_2, g\} = \{f_1, g\} + \{f_2, g\} \tag{A.1.2}$$

and the analogue of the Jacobi identity $\{\{f, g\}, k\} + \{\{k, f\}, g\} + \{\{g, k\}, f\} = 0$. We may use these relations to evaluate the Poisson bracket of any function of coordinates and momentum using the Poisson brackets involving just the coordinates and momenta, which are given by

$$\{q_a, q_b\} = 0, \quad \{p^a, p^b\} = 0, \quad \{q_a, p^b\} = \delta_a^b. \tag{A.1.3}$$

In the Dirac method, rather than use the original definition of the Poisson bracket of equation (A.1.1), we evaluate Poisson brackets using the relations of equation (A.1.2) and assume that the coordinates and momenta satisfy equation (A.1.3). The advantage is that we can handle Poisson brackets giving the time evolution of the constraints discussed above as these Poisson brackets involve the additional variables v^i which are not part of the original variables of the theory and so are not included in the usual definition of the Poisson bracket of equation (A.1.1). When carrying out this procedure, it is important to realise that although the constraints are taken to vanish, when we compute the Poisson brackets we only take the constraints to vanish *after* we have finished the computation. In this way a quantity that is taken to vanish at the end can have a non-vanishing Poisson bracket. For example, we find that $\{\phi_i, H^*\} = \{\phi_i, H + \sum_j v^j \phi_j\} = \{\phi_i, H\} + \sum_j v^j \{\phi_i, \phi_j\} + \sum_j \{\phi_i, v^j\} \phi_j = \{\phi_i, H\} + \sum_j v^j \{\phi_i, \phi_j\}$. In the intermediate line of this equation, the last term can be set to zero as it is proportional to the constraints and we have finished evaluating the Poisson bracket.

In calculating the time evolution of the constraints one may find that rather than imposing a further condition on the coordinates and momenta it leads to a condition which expresses some, or all, of the new variables v^i in terms of the coordinates and momenta. In this case we just substitute the most general such solution for v^i into the Hamiltonian.

To quantise a system with only first class constraints we replace the classical variables by operators and the Poisson brackets of the coordinates and momenta by commutators divided by $i\hbar$ if they are both Grassmann even or one is Grassmann even and one odd and by anti-commutators if they are both Grassmann odd. For example, for Grassmann even variables A, B, C if their Poisson bracket is

$$\{A, B\} = C, \tag{A.1.4}$$

these variables become operators which obey the commutator

$$[A, B] = i\hbar C. \tag{A.1.5}$$

In fact, one can often only apply this rule to a small set of the variables as one encounters operator ordering problems. However, one always implements it on the Poisson brackets of equation (A.1.3).

In the Schrödinger representation we represent the Grassmann even coordinates and momenta by

$$q_a \to q_a, \quad p^a \to -i\hbar \frac{\partial}{\partial q_a} \tag{A.1.6}$$

with analogous formulae for Grassmann odd quantities. We then introduce a wavefunction which depends only on the coordinates and the operators act on it. In this process the constraints become operators and we impose that they vanish when acting on the wavefunction. The time evolution is then given by the Schrödinger equation. Care must be taken to avoid any operator ordering problems. The quantisation of systems with second class constraints is much more complicated. Indeed, one has to change what one means by the Poisson bracket. The reader is referred to the excellent book by Dirac [1.1]. In systems with first and second class constraints one must separate the constraints into a set which is first class and a set which is second class in order to quantise the system.

The reader may feel that the Dirac quantisation procedure and the BRST method given below are a bit like cookery book recipes. However, it is good to remember that quantising a classical theory is not a logical process and it requires new physics. The justification for these quantisation procedures is that they lead to consistent quantisation procedures for those classical theories which we believe are required to describe nature. This does not mean that all these theories are consistent as quantum theories, the classic counterexample being Einstein's theory of gravity.

In chapter 1 we saw that the point particle has only one constraint whose time evolution is automatically zero. Hence, the Dirac procedure terminates very quickly and the quantisation of this system, which has one first class constraint, is relatively painless.

A.2   The BRST method

We now give the general procedure for constructing the BRST formulation of any theory with a local symmetry. We begin with a discussion of this procedure for Yang–Mills theory.

There are two reasons for fixing the gauge in Yang–Mills theories. The first is a practical reason: to have a perturbative expansion for small coupling we must be able to construct a propagator for the gauge field. This involves inverting the operator that occurs between the two gauge fields in the linearised theory. However, the invariance of the linearised action under lowest order gauge transformations implies that this operator has a non-zero kernel which is just these gauge transformations and so it cannot be inverted. The second reason is that the Feynman path integral integrates over all gauge field configurations including over all gauge equivalent configurations. The action has the same value for these equivalent configurations and one finds a resulting infinity arising from the integration. We can avoid the infinity coming from this integration and construct a propagator by introducing a gauge fixing term into the action. Originally, this term was introduced by hand, but it was discovered by Feynman that the resulting theory was non-unitary. He also found that unitarity could be restored, at least at one loop, if one introduced, by hand, particles of the the wrong statistics, the so-called ghosts, that circulated in the loops. It was found by Dewitt and independently by Faddeev and Popov that the introduction of the ghosts occurred naturally in a path integral formulation. In particular, they arose from a determinant which resulted from the gauge fixing process. Finally, it was realised by Becchi, Rouet, and Stora [1.2] and by Tyupin that the gauge fixed action including the ghost terms possessed a rigid symmetry with a Grassmann odd parameter. This BRST symmetry has played an increasingly important role in theoretical physics, but it can appear somewhat ad hoc and unmotivated and it is instructive to bear in mind the history given above.

Let us consider a group G whose anti-Hermitian generators T_a satisfy the Lie algebra

$$[T_a, T_b] = f_{ab}{}^c T_c. \qquad (A.2.1)$$

The gauge fields A_μ^a are in the adjoint representation of the group G and we define the matrix valued field $A_\mu = gA_\mu^a T_a$. It transforms under a gauge transformation as $A_\mu \to U^{-1}A_\mu U + U^{-1}\partial_\mu U$, where $U = \exp\left(w^a(x)T_a\right)$. The covariant derivative is defined by $D_\mu = \partial_\mu + A_\mu$ and the field strength is then given by $F_{\mu\nu} = gF_{\mu\nu}^a T_a = [D_\mu, D_\nu]$. We can normalise the T_a generators to satisfy a relation of the form $\text{Tr}(T_a T_b) = -c_F \delta_{ab}$, where c_F is a constant related to the value of the second order Casimir in the representation being considered.

The gauge invariant Yang–Mills action is given by

$$A^{YM} = -\frac{1}{4}\int d^4x F_{\mu\nu}^a F^{a\mu\nu} = \frac{1}{4g^2 c_F}\int d^4x \text{Tr}\{F_{\mu\nu}F^{\mu\nu}\} \tag{A.2.2}$$

We may rewrite the above gauge transformation of A_μ as $D_\mu \to U^{-1}D_\mu U$ and it is then obvious that the field strength transforms as $F_{\mu\nu} \to U^{-1}F_{\mu\nu}U$ and that the above action is gauge invariant.

Let us denote the gauge fixing condition by $G(A_\mu) = 0$. A popular choice is to take $G = \partial_\mu A^\mu = 0$. We implement this gauge fixing by adding to the Yang–Mills action the gauge fixing term

$$A^{gf} = -\int d^4x \text{Tr}\{BG(A_\mu)\}, \tag{A.2.3}$$

where $B = B^a T_a$ is a Lagrange multiplier.

Clearly, $A^{YM} + A^{gf}$ is not gauge invariant, but this symmetry is to be replaced by a rigid BRST symmetry with a rigid Grassmann odd parameter Λ. Under this symmetry the gauge field A_μ transforms as before, except that we must replace the local parameter $w^a T_a$ by $\Lambda c^a T_a$, where the fields c^a are Grassmann odd and are the famous ghosts. The transformation of the gauge field becomes

$$\delta A_\mu = \Lambda D_\mu c \equiv \Lambda(\partial_\mu c + [A_\mu, c]). \tag{A.2.4}$$

The ghosts are taken to transform as

$$\delta c = -\Lambda c \cdot c = -\frac{\Lambda}{2}\{c, c\} \qquad \text{or} \qquad \delta c^e = -\frac{\Lambda}{2}c^a c^b f_{ab}{}^e, \tag{A.2.5}$$

where $c \cdot c$ means matrix multiplication. We also introduce Grassmann odd anti-ghosts $b = b^a T_a$ which transform as

$$\delta b = \Lambda B, \quad \delta B = 0. \tag{A.2.6}$$

The most important characteristic of the BRST transformations of A_μ, c, b and B is that they are nilpotent, that is,

$$\delta_\Lambda \delta_\Pi(\text{any field}) = 0. \tag{A.2.7}$$

For example, for the ghosts we find that

$$\delta_\Lambda \delta_\Pi c = -\delta_\Lambda(\Pi c \cdot c) = \Pi(\Lambda c \cdot c \cdot c + c \cdot \Lambda c \cdot c) = 0, \tag{A.2.8}$$

since Λ and c are Grassmann odd. The reader may verify this nilpotency for the other fields.

The final task is to extend the original action plus the gauge fixing term of equation (A.2.3) by adding a third term in such a way that the total action is BRST invariant. This is achieved by taking the action

$$A^{BRST} = \frac{1}{4g^2 c_F} \int d^4x \mathrm{Tr}\{F_{\mu\nu}F^{\mu\nu}\} - \int d^4x \mathrm{Tr}BG + A^{gh}, \qquad (A.2.9)$$

where

$$A^{gh} = -\int d^4x \mathrm{Tr}\left\{b(x)\left\{\int d^4y \frac{\delta G(x)}{\delta A^\mu(y)} D^\mu c(y)\right\}\right\}.$$

For the case, $G = \partial^\mu A_\mu$, the above formula gives

$$A^{gh} = -\int d^4x \mathrm{Tr}(b\partial_\mu D^\mu c). \qquad (A.2.10)$$

The first term of the action of equation (A.2.9) is BRST invariant by itself as a consequence of its original gauge invariance. The BRST invariance of the second and third terms may be verified directly. However, we note that these terms arise as a BRST variation, namely

$$\Lambda(A^{gh} + A^{gf}) = -\int d^4x \delta_\Lambda \mathrm{Tr}\{b(x)G[A^\mu(x)]\} \qquad (A.2.11)$$

and since $\delta^2 = 0$, it follows that $\delta_\Pi(A^{gf} + A^{gh}) = 0$.

Let us now consider the Hermiticity properties of the ghost system. Let us choose $c = c^a T_a$ to be anti-Hermitian, that is, c^a to be Hermitian. Reality of the action implies that b is Hermitian as

$$\mathrm{Tr}\{(b\partial^\mu\partial_\mu c)^\dagger\} = \mathrm{Tr}\{(\partial^\mu\partial_\mu c)^\dagger b^\dagger\} = -\mathrm{Tr}\{(\partial^\mu\partial_\mu cb^\dagger)\} = \mathrm{Tr}\{(b^\dagger\partial^\mu\partial_\mu c)\}. \qquad (A.2.12)$$

Similarly, as Λc replaces $w = w^a T_a$ which is anti-Hermitian, Λ must be anti-Hermitian.

Corresponding to the rigid BRST symmetry of A^{BRST}, we may construct a BRST current and associated charge Q. The BRST action can be quantised in the usual way using the Dirac rule for replacing the fundamental Poisson brackets by commutators or anti-commutators depending on the Grassmann character of the fields involved. The BRST charge then becomes a differential operator and the physical states are those that satisfy

$$Q\Psi = 0, \qquad (A.2.13)$$

where Ψ is the wavefunction of the theory and so it depends on the coordinates of the original theory and the ghosts c^a. The momenta of the ghosts are usually the anti-ghosts. We regard two physical states Ψ_1 and Ψ_2 as equivalent, $\Psi_1 \sim \Psi_2$, if $\Psi_1 = \Psi_2 + Q\Omega$ for some Ω. To find the physical states of Yang–Mills theory we must find the Ψs which satisfy equation (A.2.13) and the physical states are then the equivalence classes as defined by the above the equivalence relation.

A more systematic derivation of the BRST theory starts from the path integral formulation of the quantum theory. This quantity is infinite as it contains an integral over an infinite number of gauge equivalent configurations. However, by manipulating this expression one can arrive at the action of equation (A.2.9). One then notices that this action has the above BRST symmetry. For an exposition of this viewpoint see [3.16].

The deduction of a quantum theory from its classical analogue cannot be achieved as a series of logical steps and some guess work is required. Indeed, the initial attempts to quantise systems which have very new features have often been incorrect. The BRST approach has been developed not so much as a general theory, but rather as a procedure they learnt from a number of examples of which Yang–Mills theory is the prototype. We now extract the general procedure whose justification is that the final result, namely a nilpotent set of transformations and an invariant action, defines a quantum theory which is unitary and whose physical observables are independent of how the gauge was fixed. Another point in its favour is that it allows one to demonstrate the renormalisability of spontaneously broken Yang–Mills theory for a general class of gauges.

We will replace the original local transformations $w^a(x^\mu)$ by rigid BRST transformations which have a parameter Λ that is of the opposite Grassmann parity to the transformation. We find a BRST invariant action as follows.

(i) We introduce a ghost c^a and anti-ghost b_a pair for each local symmetry w^a. The ghosts and anti-ghosts have opposite Grassmann parity to the transformation.
(ii) The original fields have BRST transformations given by the replacement $w^a \to \Lambda c^a$. The BRST transformations of the ghosts are given by $\delta c^a = \frac{1}{2}\tilde{f}^a_{12}$, where $\Lambda_1 \tilde{f}^a_{12} = f^a_{12}$ and f^a_{12} is the parameter resulting from the commutator of two of the original symmetry transformations, the first with parameter $w^a_1 = \Lambda_1 c^a$ and the second with parameter $w^a_2 = \Lambda c^a$. In other words, $[\delta_{\Lambda_1 c^a}, \delta_{\Lambda c^a}] = \delta_{f_{12}} = \delta_{\Lambda_1 \tilde{f}_{12}}$. These transformations are guaranteed to be nilpotent.
(iii) We next introduce the Lagrange multiplier B. The anti-ghost b and Lagrange multiplier B are taken to have the BRST transform $\delta b = \Lambda B$, $\delta B = 0$, respectively.
(iv) Finally, having chosen the gauge fixing function G, we add a corresponding term $A^{gf} = \int d^D x BG$, where D is the dimension of space-time to the original action. A BRST invariant action is given by

$$A^{BRST} = A^{orig} + A^{gf} + A^{gh}, \tag{A.2.14}$$

where $\Lambda(A^{gf} + A^{gh}) = \delta_\Lambda(\int d^D x b G)$ and A^{orig} is the action of the theory before the BRST procedure. It is BRST invariant since A^{orig} is invariant by virtue of its local symmetry and the second two terms are invariant by virtue of the nilpotency of the BRST transformations. The BRST action is then used in the path integral to define the quantum theory.

The above procedure works for many systems and in this book we apply the method to the point particle and string and their supersymmetric analogues. The physical states are defined in the same way as for the Yang–Mills theory.

It could happen that having carried out the above procedure, one finds that the ghost action has a local invariance. In this event, one must repeat the procedure adding ghosts for ghosts, etc., until one arrives at a set of nilpotent transformations and an invariant action. As we shall see this occurs for string field theory and the procedure requires an infinite number of steps.

Finally, we note that for the case of Yang–Mills theory, the prescription of point (ii) implies that $f_{12}^a = f_{bc}{}^a \Lambda_1 c^b \Lambda c^c$ and then $\tilde{f}_{12}^a = -f_{bc}{}^a \Lambda c^b c^c$. As a result, we recover, from the general procedure, the transformation given in equation (A.2.5).

The exposition given here is just an outline of the main quantisation techniques and it does not take into account the important developments of [1.3].

Appendix B   Two-dimensional light-cone and spinor conventions

In this appendix we give our conventions for light-cone coordinates and spinors in two dimensions.

B.1   Light-cone coordinates

We take light-cone coordinates to be defined by

$$\xi^+ \equiv \xi^0 + \xi^1 = \tau + \sigma, \quad \xi^- \equiv \xi^0 - \xi^1 = \tau - \sigma. \tag{B.1.1}$$

While we can work in any coordinate system, in terms of these coordinates conformal transformations take the particularly simple form

$$\xi^+ = f(\xi^+), \xi^- = g(\xi^-), \tag{B.1.2}$$

where f and g are independent functions. Conformal transformations play an important role in string theory and as a consequence using light-cone coordinates leads to particularly simple expressions for most quantities that occur in string theory. The derivatives in these coordinates become

$$\partial_+ \equiv \frac{\partial}{\partial\xi^+} = \frac{1}{2}\left(\frac{\partial}{\partial\xi^0} + \frac{\partial}{\partial\xi^1}\right), \quad \partial_- \equiv \frac{\partial}{\partial\xi^-} = \frac{1}{2}\left(\frac{\partial}{\partial\xi^0} - \frac{\partial}{\partial\xi^1}\right). \tag{B.1.3}$$

The components of a 1-form are related in the two-coordinate systems by

$$V_+ = \frac{\partial\xi^\alpha}{\partial\xi^+}V_\alpha = \tfrac{1}{2}(V_0 + V_1), \quad V_- = \tfrac{1}{2}(V_0 - V_1), \tag{B.1.4}$$

while for a second rank tensor,

$$N_{++} = \tfrac{1}{4}(N_{00} + N_{01} + N_{10} + N_{11}), N_{--} = \tfrac{1}{4}(N_{00} - N_{01} - N_{10} + N_{11}),$$
$$N_{+-} = \tfrac{1}{4}(N_{00} - N_{01} + N_{10} - N_{11}), \quad N_{-+} = \tfrac{1}{4}(N_{00} + N_{01} - N_{10} - N_{11}). \tag{B.1.5}$$

We have denoted the tensor components corresponding to ξ^+ and ξ^- by $+$ and $-$, respectively. For an energy-momentum tensor which is traceless and symmetric $T_{+-} = 0$ and the remaining two components are

$$T_{++} = \tfrac{1}{2}(T_{00} + T_{01}), \quad T_{--} = \tfrac{1}{2}(T_{00} - T_{01}). \tag{B.1.6}$$

The line element takes the form

$$ds^2 = -(d\xi^0)^2 + (d\xi^1)^2 = -d\xi^+ d\xi^-, \tag{B.1.7}$$

from which we read off the metric to be

$$\eta_{++} = 0, \ \eta_{--} = 0, \ \eta_{-+} = \eta_{+-} = -\tfrac{1}{2}, \tag{B.1.8}$$

and so its inverse is given by

$$\eta^{++} = 0, \ \eta^{--} = 0, \ \eta^{-+} = \eta^{+-} = -2. \tag{B.1.9}$$

Raising and lowering with this metric on a vector we have $V^+ = -2V_-$, $V^- = -2V_+$.

The integration measures of the two coordinates are related by

$$d^2\xi = -\tfrac{1}{2}d\xi^+ d\xi^-. \tag{B.1.10}$$

We will generally use ξ^+, ξ^- for the coordinates of the Minkowksi space-time world sheet, where they are equal to $\xi^+ = \tau + \sigma$, $\xi^- = \tau - \sigma$. However, we will use ς and $\bar{\varsigma}$ for the world sheet when described by a cylinder and rotated to Euclidean space. They are given by $\varsigma = \tau + i\sigma$ and $\bar{\varsigma} = \tau - i\sigma$. In contrast, we will use z and $\bar{z}$ for the Riemann sphere which is related by the mapping $z = e^\varsigma, \bar{z} = e^{\bar{\varsigma}}$.

B.2   Spinor conventions

We often denote the components of two-dimensional spinors by

$$\chi = \begin{pmatrix} \chi_+ \\ \chi_- \end{pmatrix}. \tag{B.2.1}$$

We take our two-dimensional γ-matrices as $\gamma^\mu = (i\sigma^2, \sigma^1)$, where σ^i are the Pauli matrices and $\gamma_5 = \gamma^0\gamma^1$. The charge conjugation matrix C is given by $C = -\gamma^0$ and one can verify that it satisfies its defining condition, namely $(\gamma^\mu)^T = -C\gamma^\mu C^{-1}$. The Dirac conjugate of a spinor is defined by $\bar{\chi}^D = -\chi^\dagger\gamma^0$ and the Majorana conjugate by $\bar{\chi}^C = \chi^T C$. We denote the components of the Majorana conjugate spinor by $\bar{\chi}^c = (\chi^+, \chi^-)$ and then we find that $\chi^\pm = \pm\chi_\mp$.

A Majorana spinor satisfies $\bar{\chi}^C = \bar{\chi}^D$ and so $\chi_\pm = (\chi_\pm)^*$. In two dimensions we can have Majorana–Weyl spinors, that is, $\bar{\chi}^C = \bar{\chi}^D$ and $\gamma_5\chi = \pm\chi$. As we noted the first condition implies that the spinor components are real and and the second implies that $\chi_\mp = 0$. As such, we are left with only one real component $\chi_\pm$, which carries a representation of spin(1,1), the covering group of the two-dimensional Lorentz group. The reader may apply chapter 5 to the two-dimensional case to recover the above discussion. In doing this the reader should note that the above definition of the adjoint differs slightly from the one advocated in chapter 5 where we took $\bar{\chi}^D = +\chi^\dagger\gamma^0$. Clearly, either choice is correct as a sign does not alter the way $\bar{\chi}^D$ transforms under the Lorentz group.

We note that $\bar{\chi}^C\gamma^\mu\chi$ is a vector and since $\gamma^+ = \gamma^0 + \gamma^1$ and $\gamma^- = \gamma^0 - \gamma^1$ we find that

$$\bar{\chi}^c\gamma^+\chi = 2\chi^+\chi^+, \ \text{and} \ \bar{\chi}^c\gamma^-\chi = 2\chi^-\chi^-. \tag{B.2.2}$$

As a result, we conclude that in two dimensions $\chi^+\chi^+$ and $\chi^-\chi^-$ transform like components of a two-dimensional vector under the Lorentz group, that is, like V^+ and V^-, respectively, under the Lorentz group.

This discussion illustrates the contrarywise nature of using $\pm$ to denote the components of both spinor and vectors. As long as one remembers which object one is dealing with one can understand what is the meaning of the $\pm$ index on a given object in a given place. The reader may like to look at the supercurrent in chapter 7, which carries spinor and vector indices to practise this.

In some quarters one exploits this observation to denote the V^+ and V^- on a vector by V^{++} and V^{--}, respectively, but the reader will be pleased to learn that we have resisted this temptation in this book, even though it would have led to a more uniform convention.

The above are the spinor conventions in Minkowski space.

Appendix C   The relationship between S^2 and the Riemann sphere

The Riemann sphere, $\mathbf{CP}^1$, is defined to be the complex 2-plane $\mathbf{C}^2$ with the origin $(0,0)$ removed (that is, $\mathbf{C}^2 - \{(0,0)\}$) which is subject to the equivalence relation

$$(\xi', \eta') \sim (\xi, \eta) \quad \text{if} \quad (\xi', \eta') = \lambda(\xi, \eta), \quad \lambda \in \mathbf{C}. \tag{C.0.1}$$

In terms of the components this last relation becomes $\xi' = \lambda\xi$, $\eta' = \lambda\eta$.

In the region of $\mathbf{CP}^1$ where $\eta \neq 0$, the equivalence relation states that $(\xi, \eta) \sim (\xi/\eta, 1)$ and so this region of $\mathbf{CP}^1$ has coordinate $z = \xi/\eta$. Alternatively, in the region of $\mathbf{CP}^1$ where $\xi \neq 0$, then $(\xi, \eta) \sim (1, \eta/\xi)$ and so this region has coordinate $z' = \eta/\xi$. Since the point $\xi = \eta = 0$ is excluded, all of $\mathbf{CP}^1$ is covered by these two coordinate patches. The relationship between the two coordinates on the overlap of the two patches being $z' = 1/z$.

In fact, the z coordinate patch covers all of $\mathbf{CP}^1$ except for the one point found by taking $z \to \infty$. This point corresponds to $z' = 0$ in the other coordinate patch. Hence, we can think of $\mathbf{CP}^1$ as the complex plane plus the latter point which is called the point at infinity.

The nature of the point at infinity is made clearer by the one-to-one onto map between the Riemann sphere and the 2-sphere S^2, which can be embedded in $\mathbf{R}^3$ in the usual way:

$$S^2 = \{(x_1, x_2, x_3) \ \in \mathbf{R}^3 : \ x_1^2 + x_2^2 + x_3^2 = \tfrac{1}{4}\}. \tag{C.0.2}$$

For convenience we have chosen the sphere to have radius $\frac{1}{2}$. The relation between S^2 and $\mathbf{CP}^1$ is given by the stereographic projection shown in figure C.0.1. The sphere S^2 has its centre at the point $(0,0,0)$ of the Cartesian coordinate system in $\mathbf{R}^3$ and we place a two-dimensional plane such that it is tangent to the sphere and touches at its south pole, that is, at the point $\left(0, 0, -\frac{1}{2}\right)$. The north pole of the sphere is at $\left(0, 0, \frac{1}{2}\right)$. The map between S^2 and the plane is given by the straight line drawn from the north pole through the point (x_1, x_2, x_3) on the sphere to a point on the plane which we denote by $\left(u, v, -\frac{1}{2}\right)$.

Consider now the triangle formed by the north pole, the south pole and the point $\left(u, v, -\frac{1}{2}\right)$ on the plane. This triangle can be projected onto the $y = 0$ plane and onto the $x = 0$ plane to yield two triangles whose respective angles θ and θ' at the north pole satisfy

$$\tan\theta = \frac{x_1}{\frac{1}{2} - x_3} = \frac{u}{1}, \quad \tan\theta' = \frac{x_2}{\frac{1}{2} - x_3} = \frac{v}{1}. \tag{C.0.3}$$

As a result, we find the relationship

$$\frac{u}{x_1} = \frac{v}{x_2} = \frac{1}{\frac{1}{2} - x_3}. \tag{C.0.4}$$

676

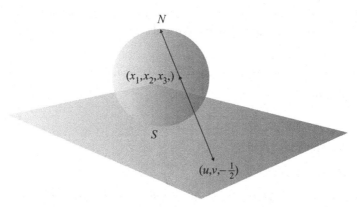

Figure C.0.1   The Riemann sphere.

If we denote $z = u + iv$, then it follows that $x_1 + ix_2 = z\left(\frac{1}{2} - x_3\right)$. Taking the modulus squared of this relation and using equation (C.0.2) we have that

$$x_1^2 + x_2^2 = |z|^2(\tfrac{1}{2} - x_3)^2 \text{ or } (x_3 + \tfrac{1}{2}) = |z|^2(\tfrac{1}{2} - x_3). \tag{C.0.5}$$

Solving this last equation for x_3 we find the points on the plane and 2-sphere are related by

$$x_3 = \frac{1}{2}\left(\frac{|z|^2 - 1}{|z|^2 + 1}\right) \quad x_1 + ix_2 = \frac{z}{|z|^2 + 1}. \tag{C.0.6}$$

We now identify the coordinate z just above with the coordinate also called z on the Riemann sphere. In terms of the corresponding coordinates $(\xi, \eta) \in \mathbf{C}^2$ of the complex 2-plane, the coordinate z takes the form $z = \xi/\eta$ and substituting this relation in equation (C.0.6) we find that the mapping from the Riemann sphere to S^2 is given by

$$x_1 + ix_2 = \frac{\xi\bar{\eta}}{|\xi|^2 + |\eta|^2}, \quad x_3 = \frac{1}{2}\frac{(|\xi|^2 - |\eta|^2)}{|\xi|^2 + |\eta|^2}. \tag{C.0.7}$$

From equation (C.0.7) we can calculate the inverse mapping from the S^2 to the Riemann sphere. In the z coordinate patch it is given by

$$z = \frac{(x_1 + ix_2)}{(\frac{1}{2} - x_3)}. \tag{C.0.8}$$

The other coordinate patch on $\mathbf{CP}^1$ has coordinate z' related by $z' = 1/z$ and so it is given in terms of the coordinates of the sphere by

$$z' = \frac{(x_1 - ix_2)}{(\frac{1}{2} + x_3)}. \tag{C.0.9}$$

We notice this map is defined everywhere except at the south pole. This coordinate patch can be found by drawing a straight line from the south pole through the point of the sphere to a plane which is tangent to the sphere touching at the north pole by repeating the analogous steps above. Taking these two coordinate patches into account it is clear that there is a one-to-one mapping from $\mathbf{CP}^1$ onto the 2-sphere S^2.

It is instructive to find the metric on $\mathbf{CP}^1$ induced from the standard metric on S^2, which is in turn that induced from its embedding in $\mathbf{R}^3$. If θ and φ are the usual spherical coordinates for the sphere, then

$$x_3 = \tfrac{1}{2}\cos\theta, \quad x_2 = \tfrac{1}{2}\sin\theta\sin\varphi, \quad x_1 = \tfrac{1}{2}\sin\theta\cos\varphi, \tag{C.0.10}$$

which can also be written as

$$x_1 + ix_2 = \tfrac{1}{2}e^{i\varphi}\sin\theta. \tag{C.0.11}$$

The metric on the 2-sphere is given by

$$ds^2 = \tfrac{1}{4}(d\theta^2 + \sin^2\theta\, d\varphi^2). \tag{C.0.12}$$

To find the corresponding metric on the Riemann sphere we must find the transformation between z and θ and φ. Using equations (C.0.6) and (C.0.9) we find that

$$2x_3 = \cos\theta = \frac{|z|^2 - 1}{|z|^2 + 1} \tag{C.0.13}$$

or that

$$|z| = \frac{1}{\tan(\theta/2)}. \tag{C.0.14}$$

This result together with equations (C.0.8) and (C.0.11) implies that

$$z = |z|e^{i\varphi} = \frac{e^{i\varphi}}{\tan(\theta/2)}. \tag{C.0.15}$$

As a result, we find that

$$dz = \frac{id\varphi e^{i\varphi}}{\tan(\theta/2)} + \frac{1}{2}e^{i\varphi}\frac{d\theta}{(\sin^2(\theta/2))} \tag{C.0.16}$$

and

$$(1 + |z|^2) = \frac{1}{(\sin^2(\theta/2))}. \tag{C.0.17}$$

Equations (C.0.16) and (C.0.17) then imply that the metric of equation (C.0.12) is given by

$$ds^2 = \frac{1}{4}\frac{dz d\bar{z}}{(1 + |z|^2)^2}. \tag{C.0.18}$$

Appendix D   Some properties of the classical Lie algebras

In this appendix we list the Dynkin diagrams, roots, weights, inverse Cartan matrices and some other properties of most of the classical finite dimensional simple Lie algebras.

D.1   The algebras A_{n-1}

Two real forms of A_{n-1} are SU(n) and SL(n). The Dynkin diagram is given by

$$
\begin{array}{ccccccccc}
\bullet & - & \bullet & - & \bullet & \cdots \quad \cdots \quad \cdots & \bullet & - & \bullet \\
1 & & 2 & & 3 & & n-2 & & n-1
\end{array}
$$

Let e_i, $i = 1, \ldots, n$, be a set of orthogonal unit vectors in $\mathbf{R}^n$, that is, $(e_i, e_j) = \delta_{ij}$. We can write the simple roots of A_{n-1} as

$$\alpha_i = e_i - e_{i+1}, \quad i = 1, \ldots, n. \tag{D.1.1}$$

The $n^2 - n$ roots of A_{n-1} are given by

$$\Delta = \{e_i - e_j, \quad i, j = 1, \ldots, n \; i \neq j\}, \tag{D.1.2}$$

while the positive roots are

$$\Delta_+ = \{e_i - e_j, \; j > i, \quad i, j = 1, \ldots, n\} \tag{D.1.3}$$

The highest root is $\theta = e_1 - e_n = \sum_{i=1}^{n-1} \alpha_i$. Hence, the Kac labels are given by $n_i = 1$ and the Coxeter number is $h = n$. The fundamental weights are given by

$$\lambda_i = \sum_{j=1}^{i} e_j - \frac{i}{n} \sum_{j=1}^{n} e_j. \tag{D.1.4}$$

In particular, we therefore have that the Weyl vector is given by

$$\rho = \sum_{i=1}^{n} \left(\frac{r + 2 - 2i}{2} \right) e_i. \tag{D.1.5}$$

Using the equation $A_{ij}^{-1} = (\lambda_i, \lambda_j)$, which is valid for simply laced algebras, we find that the inverse Cartan matrix is given by

$$A_{ij}^{-1} = (\lambda_i, \lambda_j) = i \frac{(n-j)}{n}, \tag{D.1.6}$$

where $j \geq i$. We note it is positive definite as it should be.

679

D.2   The algebras D_n

The most commonly used real form is $SO(2n)$. The Dynkin diagram is given by

Let e_i, $i = 1, \ldots, n$, be a set of orthonormal vectors in $\mathbf{R}^n$. The simple roots of D_n are given by

$$\alpha_i = e_i - e_{i+1}, \quad i = 1, \ldots, n-1, \quad \alpha_n = e_{n-1} + e_n. \tag{D.2.1}$$

The $2n(n-1)$ roots of D_n are given by

$$\Delta = \{\pm e_i \pm e_j, \quad i, j = 1, \ldots, n \, i \neq j\}, \tag{D.2.2}$$

while the positive roots are

$$\Delta_+ = \{e_i \pm e_j, \quad i < j \leq n\}. \tag{D.2.3}$$

The highest root is

$$\theta = e_1 + e_2 = \alpha_1 + 2\sum_{i=2}^{n-2}\alpha_i + \alpha_{n-1} + \alpha_n. \tag{D.2.4}$$

Hence, the Kac labels are given by $n_i = 2$, $i = 2, \ldots, n-2$, $n_1 = n_{n-1} = n_n = 1$, and the Coxeter number is $h = 2(n-1)$. The fundamental weights are given by

$$\lambda_i = \sum_{j=1}^{i} e_j, \quad i = 1, \ldots, n-2, \quad \lambda_{n-1} = \tfrac{1}{2}\sum_{j=1}^{n-1} e_j - \tfrac{1}{2}e_n,$$

$$\lambda_n = \tfrac{1}{2}\sum_{j=1}^{n-1} e_j + \tfrac{1}{2}e_n. \tag{D.2.5}$$

The Weyl vector is therefore of the form

$$\rho = \sum_{i=1}^{n-1}(n-i)\,e_i. \tag{D.2.6}$$

Using the equation $A_{ij}^{-1} = (\lambda_i, \lambda_j)$, we find that the inverse Cartan matrix is then

$$\begin{aligned}
A_{ij}^{-1} &= i, & &\text{for } i \leq j, \ i, j = 1, \ldots, n-2,\\
A_{i\,n-1}^{-1} &= A_{i\,n}^{-1} = \tfrac{i}{2}, & &\text{for } i = 1, \ldots, n-2,\\
A_{n-1\,n-1}^{-1} &= A_{n\,n}^{-1} = \tfrac{n}{4},\\
A_{n-1\,n}^{-1} &= \tfrac{n-2}{4}.
\end{aligned} \tag{D.2.7}$$

D.3   The algebra E_6

The Dynkin diagram of E_6 is given by

Let e_i, $i = 1, \ldots, 8$, be a set of orthonormal vectors in $\mathbf{R}^8$. The simple roots of E_7 are given by

$$\alpha_i = e_i - e_{i+1}, \ i = 1, \ldots, 3 \ \alpha_4 = e_4 + e_5, \alpha_5 = -\tfrac{1}{2}(e_1 + e_2 + \cdots e_7 + e_8),$$

$$\alpha_6 = e_4 - e_5. \tag{D.3.1}$$

The 72 roots are given by

$$\Delta = \{\pm e_i \pm e_j, \ i, j = 1, \ldots, 5, i \neq j\}$$

$$\oplus \left\{ \tfrac{1}{2} \sum_i \epsilon_i e_i, \ \epsilon_i = \pm 1, \prod_i \epsilon_i = 1, \ \epsilon_6 = \epsilon_7 = \epsilon_8 \right\}. \tag{D.3.2}$$

The highest root is given by

$$\Theta = \tfrac{1}{2}(e_1 + e_2 + e_3 + e_4 - e_5 - e_6 - e_7 - e_8). \tag{D.3.3}$$

The fundamental weights and the inverse Cartan matrix were given in section 16.6.

D.4   The algebra E_7

The Dynkin diagram of E_7 is given by

Let e_i, $i = 1, \ldots, 8$, be a set of orthonormal vectors in $\mathbf{R}^8$. The simple roots of E_7 are given by

$$\alpha_i = e_i - e_{i+1}, \ i = 1, \ldots, 4 \ \alpha_5 = e_5 + e_6, \alpha_6 = -\tfrac{1}{2}(e_1 + e_2 + \cdots e_7 + e_8),$$

$$\alpha_7 = e_5 - e_6. \tag{D.4.1}$$

The 126 roots are given by

$$\Delta = \{\pm e_i \pm e_j, \ i, j = 1, \ldots, 6, i \neq j\} \oplus \{\pm(e_7 + e_8)\}$$

$$\oplus \left\{ \tfrac{1}{2} \sum_i \epsilon_i e_i, \ \epsilon_i = \pm 1, \prod_i \epsilon_i = 1, \ \epsilon_7 = \epsilon_8 \right\}. \tag{D.4.2}$$

The highest root is given by

$$\Theta = -e_7 - e_8.$$ (D.4.3)

The fundamental weights and the inverse Cartan matrix were given in section (16.6).

D.5   The algebra E_8

The Dynkin diagram of E_8 is given by

Let e_i, $i = 1, \ldots, 8$, be a set of orthonormal vectors in $\mathbf{R}^8$. The simple roots of E_8 are given by

$$\alpha_i = e_i - e_{i+1}, \ i = 1, \ldots, 5 \ \alpha_6 = e_6 + e_7, \alpha_7 = -\tfrac{1}{2}(e_1 + e_2 + \cdots e_7 + e_8),$$

$$\alpha_8 = e_6 - e_7.$$ (D.5.1)

The 240 roots are given by

$$\Delta = \{\pm e_i \pm e_j, \ i, j = 1, \ldots, 8, i \neq j\} \oplus \left\{ \tfrac{1}{2} \sum_i \epsilon_i e_i, \ \epsilon_i = \pm 1, \prod_i \epsilon_i = 1 \right\}.$$

(D.5.2)

The highest root is given by

$$\Theta = e_1 - e_8.$$ (D.5.3)

The fundamental weights and the inverse Cartan matrix were given in section 17.6.

D.6   The algebras B_n

The most commonly used real form is $SO(2r + 1)$. The Dynkin diagram of B_n is given by

$$\bullet \ - \ \bullet \ - \ \bullet \ \ldots \ \ldots \ \ldots \ \bullet \ \Rightarrow \ \bullet$$
$$\ 1 \quad\quad 2 \quad\quad 3 \quad\quad\quad\quad\quad\quad\quad n-1 \quad\quad n$$

Let e_i, $i = 1, \ldots, n$, be a set of orthonormal vectors in $\mathbf{R}^n$. The simple roots of $SO(2r + 1)$ are given by

$$\alpha_i = e_i - e_{i+1}, \quad i = 1, \ldots, n - 1,$$ (D.6.1)

$$\alpha_n = e_n.$$

The roots of B_n are given by

$$\Delta = \{\pm e_i \pm e_j, \ \pm e_i, \ i, j = 1, \ldots, n, i \neq j\}.$$ (D.6.2)

The highest root is

$$\theta = e_1 + e_2 = \alpha_1 + 2\sum_{i=2}^{n}\alpha_i. \tag{D.6.3}$$

Hence, the Kac labels are given by $n_i = 2, i = 2, \ldots, n$ and $n_1 = 1$, and the Coxeter number is $h = 2n$. The fundamental weights are

$$\lambda_i = \sum_{j=1}^{i} e_j, \quad i = 1, \ldots, n-1; \quad \lambda_n = \tfrac{1}{2}\sum_{j=1}^{n} e_j. \tag{D.6.4}$$

Using the equation $A_{ij}^{-1} = 2(\lambda_i, \lambda_j)/(\alpha_i, \alpha_i)$ we find that the inverse Cartan matrix is given by

$$A_{ij}^{-1} = i, \quad \text{for } i \le j, i, j = 1, \ldots, n-1, \ A_{ni}^{-1} = i,$$
$$A_{in}^{-1} = \frac{i}{2}, \quad A_{nn}^{-1} = \frac{r}{2}. \tag{D.6.5}$$

D.7   The algebras C_n

The most commonly used real form is $\mathrm{Sp}(2n)$. The Dynkin diagram of C_n is given by

$$\overset{\bullet}{1} - \overset{\bullet}{2} - \overset{\bullet}{3} \quad \cdots \quad \cdots \quad \cdots \quad \overset{\bullet}{n-1} \ \Leftarrow \ \overset{\bullet}{n}$$

Let $e_i, i = 1, \ldots, n$ be a set of orthonormal vectors in $\mathbf{R}^n$. The simple roots of C_n are

$$\alpha_i = \frac{1}{\sqrt{2}}(e_i - e_{i+1}), \quad i = 1, \ldots, n-1,$$
$$\alpha_n = \sqrt{2}e_n. \tag{D.7.1}$$

The roots of C_n are given by

$$\Delta = \left\{ \frac{1}{\sqrt{2}}(\pm e_i \pm e_j), \ \pm\sqrt{2}e_i, \ i, j = 1, \ldots, n, i \ne j \right\}. \tag{D.7.2}$$

The highest root is

$$\theta = \sqrt{2}e_1 = 2\sum_{i=1}^{r-1}\alpha_i + \alpha_r. \tag{D.7.3}$$

Hence, the Kac labels are given by $n_i = 2, i = 1, \ldots, r-1$ and $n_n = 1$, and the Coxeter number is $h = 2n$. The fundamental weights are

$$\lambda_i = \frac{1}{\sqrt{2}}\sum_{j=1}^{i} e_j, \quad i = 1, \ldots, n. \tag{D.7.4}$$

We find that the inverse Cartan matrix is given by

$$A_{ij}^{-1} = i, \text{ for } j \ge i, i, j = 1, \ldots, r-1, \ A_{ir}^{-1} = i, A_{ri}^{-1} = \frac{i}{2}, A_{rr}^{-1} = \frac{r}{2}. \tag{D.7.5}$$

Chapter quote acknowledgements

Preface: Taken from the script of the Film Lawrence of Arabia, Horizon Pictures GB, rights now held by Columbia Pictures.

Chapter 1: From an opening speech at the Palestinian peace talks.

Chapter 2: Taken from the book Shackelton by Roland Huntford published by Hodder and Stoughton in 1986.

Chapter 4: The author wishes to thank the BBC, and especially the Editor of Horizon, for permission to publish this quote.

Chapter 5: Reproduced by permission of Penguin Books Ltd. From Piers the Ploughman by Langland, translation by J.F Goodridge, published in 1959.

Chapter 6: The author wishes to thank Professor Klaus Sibold for permission to publish this quote, which he collected.

Chapter 9: The author wishes to thank Professor David Farlie for permission to publish this quote.

Chapter 12: The author wishes to thank Professor Murray Gell-Mann for permission to publish this quote.

Chapter 13: The author wishes to thank Professor Leon Cooper for giving him this quote.

Chapter 15: The author wishes to thank the BBC drama for permission to use the quote from the script of Cranford by Heidi Thomas.

Chapter 17: Quotes attributed to Victor Chernomyrdin.

Chapter 18: From Scott's diary on reaching the South Pole to find the Flag of Amundsen; taken from the book *Race for the South Pole: the expedition diaries of Scott and Amundsen* by Roland Huntford, published by Hodder and Stoughton.

References

Chapter 1

[1.1] P. Dirac, *Lectures in Quantum Mechanics* (1964) Belfer Graduate School of Science, Yeshiva University, New York.

[1.2] C. Becchi, A. Rouet and R. Stora, *Renormalization of gauge theories*, Ann. Phys. NY., **98** (1976) 287.

[1.3] E. Fradkin, and G. Vilkovisky, *Quantization of relativistic systems with constraints*, Phys. Lett., **55B** (1975) 224; I. Batalin and G. Vilkovisky, *Relativistic S matrix of dynamical systems with boson and fermion constraints*, Phys. Lett., **69B** (1977) 309; E. Fradkin and T. Fradkina, *Quantization of relativistic systems with boson and fermion first and second class constraints*, Phys. Lett., **72B** (1978) 343.

[1.4] T. Kugo and I. Ojima, *Local covariant operator formalism of nonabelian gauge theories and quark confinement problem*, Suppl. Progr. Theoret. Phys., **66** (1979) 1.

[1.5] F. A. Berezin and M. S. Marinov, *Classical spin and Grassmann algebra*, Pisma Zh. Eksp. Teor. Fiz., **21** (1975) 678; *Particle spin dynamics as the Grassmann variant of classical mechanics*, Ann. Phys., **104** (1977) 336.

[1.6] R. Casalbuoni, *Relativity and supersymmetries*, Phys. Lett., **62B** (1976) 49.

[1.7] L. Brink, S. Deser, B. Zumino, P. Di Vecchia and P. Howe, *Local supersymmetry for spinning particles*, Phys. Lett., **64B** (1976) 435.

[1.8] L. Brink, P. Di Vecchia, and P. Howe, *A Lagrangian formulation of the classical and quantum dynamics of spinning particles*, Nucl. Phys., **B118** (1976) 76.

[1.9] L. Brink and J. Schwarz, *Quantum superspace*, Phys. Lett., **100B** (1981) 310.

[1.10] A. Salam and J. Strathdee, *On superfields and Fermi–Bose symmetry*, Phys. Rev., **Dll** (1975) 1521; *Feynman rules for superfields*, Nucl. Phys., **B86** (1975) 142.

[1.11] P. West, *Introduction to Supersymmetry and Supergravity* (1986, 1990) World Scientific. (There are two editions of this reference and some material can only be found in the second edition.)

[1.12] W. Siegel, *Hidden local supersymmetry in the supersymmetric particle action*, Phys. Lett., **128B** (1983) 397.

[1.13] I. Bengtsson and M. Cederwall, unpublished.

[1.14] D. Sorokin, V. Tkach and D. Volkov, *Superparticles, twistors and Siegel symmetry*, Mod. Phys. Lett., **A4** (1989) 901; D. Sorokin, V. Tkach, D. Volkov and A. Zheltuken, *From the superparticle Siegel symmetry to the spinning particle proper time supersymmetry*, Phys. Lett., **216B** (1989) 302.

[1.15] R. Penrose, *Twistor algebra*, J. Math. Phys., **8** (1967) 345; R. Penrose and M. MacCallum, *Twistor theory: an approach to the quantization of fields and space-time*, Phys. Rep., **6C** (1973) 243; R. Penrose and W. Rindler, *Spinors and Spacetime* (1986) Cambridge University Press.

685

[1.16] T. Shirafuji, *Lagrangian mechanics of massless particles with spin*, Progr. Theor. Phys., **70** (1983) 18.

[1.17] A. Ferber, *Supertwistors and conformal supersymmetry*, Nucl. Phys., **B132** (1977) 55.

[1.18] A. Bengtsson, I. Bengtsson, M Cederwall and N. Linden, *Particles, superparticles and twistors*, Phys. Rev., **D36** (1987) 1766; I. Bengtsson and M Cederwall, *Particles, twistors and the division algebras*, Nucl. Phys., **B302** (1988) 104.

[1.19] Y. Eisenberg and S. Solomon, *The twistor geometry of the covariantly quantized Brink–Schwarz superparticle*, Nucl. Phys., **B309** (1988) 709; *(Super)field theories from (super)twistors*, Phys. Lett., **B220** (1989) 562.

[1.20] A. Gumenchuk and D. Sorokin, *Relativistic superparticle dynamics and twistor correspondence*, Sov J. Nucl. Phys., **51** (1990) 350; D. Sorokin, *Double supersymmetric particle theories*, Fortshr. Phys., **38** (1990) 923.

[1.21] I. A. Bandos, *Relativistic superparticle dynamics and twistor correspondence*, Sov J. Nucl. Phys., **51** (1990) 906; *Multivalued action functionals, Lorentz harmonics, and spin*, JETP., Lett, **52** (1990) 205.

[1.22] F. Delduc, E. Ivanov and E. Sokatchev, *Twistor like superstrings with $D = 3, D = 4, D = 6$ target superspace and $N=(1,0)$, $N=(2,0)$, $N=(4,0)$ world sheet supersymmetry*, Nucl. Phys., **B384** (1992) 334.

[1.23] F. Delduc and E. Sokatchev, *Superparticle with extended worldline supersymmetry*, Class. Quantum Grav., **9** (1991) 361.

[1.24] A. Galperin and and E. Sokatchev, *A twistor like $D = 10$ superparticle action with manifest $N=8$ worldline supersymmetry*, Phys. Rev., **D46** (1992) 714.

[1.25] A. Galperin and and E. Sokatchev, *A twistor formulation of the nonheterotic superstring with manifest world sheet supersymmetry*, Phys. Rev., **D48** (1993) 4810.

[1.26] P. S. Howe and P. K. Townsend, *The massless superparticle as Chern–Simons mechanics*, Phys. Lett., **259B** (1991) 285.

[1.27] N. Berkovits, *A supertwistor description of the massless superparticle in ten-dimensional superspace*, Phys. Lett., **247B** (1990) 45.

[1.28] P. Howe and P. West, *The conformal group, point particles and twistors*, IJMP, **A7** (1992) 6639.

[1.29] N. Berkovits, *Super-poincaré invariant superstring field theory*, Nucl. Phys., **B450** (1995) 90.

[1.30] C. Teitelboim, *Quantum mechanics of the gravitational field*, Phys. Rev., **D25** (1982) 3159.

[1.31] S. Monaghan, *BRST Hamiltonian quantization of the supersymmetric particle in relativistic gauges*, Phys. Lett., **181B** (1986) 101.

[1.32] A. Neveu, and P. West, *Unification of the Poincaré group with BRST and Parisi–Sourlas supersymmetries*, Phys. Lett., **182B** (1986) 343.

Chapter 2

[2.1] Y. Nambu, *Proceedings of the Int. Conf. on Symmetries and Quark Modes*, Detroit 1969 (1970) Gordon and Breach.

[2.2] T. Goto, *Relativistic quantum mechanics of one-dimensional mechanical continuum and subsidiary condition of dual resonance model*, Progr. Theor. Phys., **46** (1971) 1560.

[2.3] P. Dirac, *An extensible model of the electron*, Proc. Roy. Soc. Lond., **A268** (1962) 57.

[2.4] P. Goddard, J. Goldstone, C. Rebbi and C. Thorn, *Quantum dynamics of a massless relativistic string*, Nucl. Phys., **B56** (1973) 109.

[2.5] L. Brink, P. Di Vecchia, and P. Howe, *A locally supersymmetric and reparametrization invariant action for the spinning string*, Phys. Lett., **65B** (1976) 471.

[2.6]   S. Deser and B. Zumino, *A complete action for the spinning string*, Phys. Lett., **65B** (1976) 369.

[2.7]   M. Schiffer and D. Spencer, *Functionals of Finite Riemann surfaces* (1954) Princeton University Press.

[2.8]   B. Zumino, *Relativistic strings and supergauges*, in *Renormalization and Invariance in Quantum Field Theory*, ed. E. Caianiello (1974) Plenum Press.

Chapter 3

[3.1]   A. Neveu and J. Scherk, *Gauge invariance and uniqueness of the renormalisation of dual models with unit intercept*, Nucl. Phys., **B36** (1972) 155.

[3.2]   T. Yoneya, *Connection of dual models to electrodynamics and gravidynamics*, Progr. Theor. Phys., **51** (1974) 1907; J. Scherk and J. Schwarz, *Dual models for nonhadrons*, Nucl. Phys., **B81** (1974) 118.

[3.3]   M. Kato and K. Ogawa, *Covariant quantization of string based on BRS invariance*, Nucl. Phys., **B212** (1983) 443; S. Hwang, *Covariant quantization of the string in dimensions $D = 26$ using a BRS formulation*, Phys. Rev., **D28** (1983) 2614.

[3.4]   E. Del Giudice and P. Di Vecchia, *Factorization and operator formalism in the generalized virasoro model*, Nuov. Cim., **5A** (1971) 90.

[3.5]   S. Fubini and G. Veneziano, *Level structure of dual-resonance models*, Nuov. Cim., **64A** (1969) 81.

[3.6]   S. Fubini, D. Gordon and G. Veneziano, *A general treatment of factorization in dual resonance models*, Phys. Lett., **29B** (1969) 679.

[3.7]   S. Fubini, and G. Veneziano, *Duality in operator formalism*, Nuov. Cim., **67A** (1970) 29.

[3.8]   S. Fubini and G. Veneziano, *Algebraic treatment of subsidiary conditions in dual resonance models*, Ann. Phys., **63** (1971) 12.

[3.9]   K. Bardakçi, and S. Mandelstam, *Analytic solution of the linear-trajectory bootstrap*, Phys. Rev., **184** (1969) 1640.

[3.10] F. Gliozzi, *Ward-like identities and the twisting operator in dual resonance models*, Nuovo Cim. Lett., **2** (1969) 846.

[3.11] G. Virasoro, *Subsidiary conditions and ghosts in dual resonance models*, Phys. Rev., **D1** (1970) 2933.

[3.12] J. Weis, unpublished.

[3.13] R. Brower, *Specturm-generating algebra and the no-ghost theorem for the dual model*, Phys. Rev., **D6** (1972) 1655. 2933.

[3.14] P. Goddard, P. and C. Thorn, *Compatibility of the dual pomeron with unitarity and the absence of ghosts in the dual resonance model*, Phys. Lett., **40B** (1972) 235.

[3.15] C. Lovelace, *Pomeron form-factors and dual Regge cuts*, Phys. Lett., **34B** (1971) 500.

[3.16] B. Lee, *Gauge theories*, *Methods in Field Theories, Les Houches Lectures*, 1975, eds. R. Balian and Z. Jinn–Justin (1975) North-Holland and World Scientific Publishing Company.

[3.17] A. Neveu, and P. West, *String lengths in covariant string field theory and OSp(26,2/2)*, Nucl. Phys., **B293** (1987) 266.

Chapter 4

[4.1]   M. Kaku and K. Kikkawa, *Field theory of relativistic strings, I Trees*, Phys. Rev., **D10**, no 4, (1974) 1100 (1974); *Field theory of relativistic strings, I Loops*, Phys. Rev., **D10**, no 6, (1974) 1823.

Chapter 5

[5.1] F. Gliozzi, D. Olive and J. Scherk, *supersymmetry, supergravity theories and the dual spinor model*, Nucl. Phys., **B122** (1977) 253.

[5.2] P. van Niewenhuizen, *Six lectures on supergravity*, in *Supergravity '81*, eds. by S. Ferrara and J. Taylor (1982) Cambridge University Press.

[5.3] T. Kugo and P. Townsend, *Supersymmetry and the division algebras*, Nucl. Phys., **B221**, (1983) 357.

[5.4] M. Naimark and A. Stern, *Theory of Group Representations* (1982) Springer-Verlag.

[5.5] A. Barut and R. Racka, *Theory of Group Representations and Applications* (1986) World Scientific Publishing.

[5.6] R. Haag, J. Lopuszanski and M. Sohnius, *All possible generators of supersymmetries of the S matrix*, Nucl. Phys., **B88** (1975) 81.

[5.7] J. Van Holten and A. Van Proeyen, $N = 1$ *supersymmetry algebras in* $D = 2$, $D=3$, $D = 4$ *mod 8*, J. Phys. Gen., **A15** (1982) 3763.

[5.8] P. Howe, N. Lambert and P. West, *The threebrane soliton of the M-fivebrane*, Phys. Lett., **419B** (1998) 79, arXiv:hep-th/9710033.

Chapter 6

[6.1] P. Ramond, *Dual theory for free fermions*, Phys. Rev., **D3** (1971) 2415.

[6.2] A. Neveu and J. H. Schwarz, *Factorizable dual model of pions*, Nucl. Phys., **B31** (1971) 86.

[6.3] A. Neveu and J. H. Schwarz, *Quark model of dual pions*, Phys. Rev., **D4** (1971) 1109.

[6.4] J. L. Gervais and B. Sakita, *Field theory interpretation of supergauges in dual models*, Nucl. Phys., **B34** (1971) 632.

[6.5] M. Green and J. Schwarz, *Covariant description of superstrings*, Phys. Lett., **136B** (1984) 367.

[6.6] L. Brink, P. Di Vecchia and P. Howe, *A locally supersymmetric and reparametrization invariant action for the spinning string*, Phys. Lett., **65B** (1976) 471.

[6.7] S. Desser and B. Zumino, *A complete action for the spinning string*, Phys. Lett., **65B** (1976) 369.

Chapter 7

[7.1] J. H. Schwarz, *Physical states and pomeron poles in the dual pion model*, Nucl. Phys., **B46** (1972) 62.

[7.2] R. C. Brower and K. A. Friedman, *Spectrum generating algebra and no ghost theorem for the Neveu–Schwarz model*, Phys. Rev., **D7** (1973) 535.

[7.3] P. Goddard and C. B. Thorn, *Compatibility of the dual pomeron with unitarity and the absence of ghosts in the dual resonance model*, Phys. Lett., **40B** (1972) 235; P. Goddard, C. Rebbi and C. B. Thorn, *Lorentz covariance and the physical states in dual resonance models*, Nuov. Cim., **12A** (1972) 425.

[7.4] E. Corrigan and P. Goddard, *The absence of ghosts in the dual fermion model*, Nucl. Phys., **B68** (1974) 189

[7.5] C. Thorn, *Embryonic dual model for pions and fermions*, Phys. Rev., **D4** (1971) 1112; J. Schwarz, *Dual quark-gluon model of hadrons*, Phys. Lett., **37B** (1971) 315.

[7.6] E. Corrigan and D. Olive, *Fermion meson vertices in dual theories*, Nuov. Cim., **11A** (1972) 749; E. Corrigan, *The scattering amplitude for four dual fermions*, Nucl. Phys., **B69** (1974) 325.

[7.7] L. Brink, D. Olive, C. Rebbi and J. Scherk, *The missing gauge conditions for the dual fermion emission vertex and their consequences*, Phys. Lett., **45B** (1973) 379.

[7.8] S. Mandelstam, *Manifestly dual formulation of the ramond model*, Phys. Lett., **46B** (1973) 447.

[7.9] E. Corrigan, P. Goddard R. A. Smith and D. Olive, *Evaluation of the scattering amplitude for four dual fermions*, Nucl. Phys., **B67** (1973) 477; J Schwarz and C. C. Wu, *Functions occurring in dual fermion amplitudes*, Nucl. Phys., **B73** (1974) 77; *Evaluation of dual fermion amplitudes*, Phys. Lett., **47B** (1973) 453.

[7.10] D. Bruce, E. Corrigan and D. Olive, *Group theoretical calculation of traces and determinants occurring in dual theories*, Nucl. Phys., **B95** (1975) 427.

[7.11] N. Berkovits, *Perturbative finiteness of superstring theory*, 4th International Winter Conference on Mathematical Methods in Physics (WC 2004), Rio de Janeiro, Brazil, published in PoS WC2004 (2004) 9.

[7.12] L. Alvarez-Gaume, P. Ginsparg, G. Moore and C. Vafa, *An $O(16) \times O(16)$ heterotic string*, Phys. Lett., **171B** (1986) 155; L. Dixon and J. Harvey, *String theories in ten dimensions without space-time supersymmetry*, Nucl. Phys., **B274** (1986) 93.

[7.13] M. Green and J. Schwarz, *Supersymmetrical dual string theory. 2. Vertices and trees*, Nucl. Phys., **B198** (1982) 252; *Supersymmetrical dual string theory. 3. Loops and renormalization*, Nucl. Phys., **B198** (1982) 441.

Chapter 8

[8.1] A. Belavin, A. Polyakov and A. Zamolodchikov, *Infinite conformal symmetry in two-dimensional quantum field theory*, Nucl. Phys., **B241** (1984) 333.

[8.2] K. Wilson, *Nonlagrangian models of current algebra*, Phys. Rev., **179** (1969) 1499; W. Zimmerman, *Lectures on Elementary Particles and Quantum Field Theory*, Brandeis Summer School (1970) MIT Press; A. Polyakov, *Conformal symmetry of critical fluctuations*, JETP Lett., **12** (1970) 381. For more references see I. Todorov, M. Minctchev and V. Petkova, *Conformal Invariance in Quantum Field Theory* (1978) Publicazone della classe di scienze della scuola normale superiore.

[8.3] H. Blote, J. Cardy and M. Nightingale, *Conformal Invariance, the Central charge, and universal finite size amplitudes at criticality*, Phys. Rev. Lett., **56** (1986) 742; I. Affleck, *Universal term in the free energy at a critical point and the conformal anomaly*, Phys. Rev. Lett., **56** (1986) 746; J. Cardy, *Conformal Invariance and Statistical Physics*, in *Les Houches Lectures 1988* eds. E. Brezin and J Zinn-Justin, (1989) Elsevier Publishers.

[8.4] C. Callan Jr, *Methods in Field Theory*, in *Methods in Field Theories, Les Houches Lectures*, 1975, eds. R. Balian and Z. Jinn-Justin (1975) North-Holland and World Scientific Publishing Company.

[8.5] V. Kac, *Highest weight representations of infinite dimensional Lie algebras*, Proc. ICM, Helsinki, (1978) 299; *Contravariant Form for the Infinite-Demensional Lie Algebras and Superalgebras* Springer lecture notes in Physics, **94** (1979) 441.

[8.6] V. Kac and A. Raina, *Bombay Lectures on Highest Weight Representations of Infinite Dimensional Lie Algebras*, Advanced Mathematics Series, Vol. 2 (1987) World Scientific Publishing.

[8.7] D. Friedan, E. Martinec and S. Shenker, *Conformal invariance, supersymmetry and string theory*, Nucl. Phys., **B271** (1986) 93.

[8.8]   C. Itzykson and J. Zuber, *Two-dimensional conformal invariant theories on a torus*, Nucl. Phys., **B275** (1986) 580; *The ADE classification of minimal and A1(1) conformal invariant theories*, Comm. Math. Phys., **113** (1987) 1; A. Kato, *Classification of modular invariant partition functions in two-dimensions*, Mod. Phys. Lett., **A2** (1987) 585. For a review see C. Itzykson and J. Drouffe, *Statistical Physics* Vols. 1 and 2, (1989) Cambridge University Press.

Chapter 9

[9.1]   C. Itzykson and J. Zuber, *Quantum Field Theory* (1985) McGraw Hill.
[9.2]   V. Dotsenko and V. Fateev, *Conformal algebra and multipoint correlation functions in two-dimensional statistical models*, Nucl. Phys., **B240** (1984) 312.
[9.3]   V. Dotsenko and V. Fateev, *Four point correlation functions and the operator algebra in the two-dimensional conformal invariant theories with the central charge $c < 1$*, Nucl. Phys., **B251** (1985) 691.
[9.4]   B. Feigen and D. Fuchs, *Topology, Proceedings of Leningrad Conference 1982*, eds. L. Faddeev and A. Mal'tsev, (1982) Springer.
[9.5]   G. Felder, *BRST approach to minimal models*, Nucl. Phys., **B317** (1989) 215.
[9.6]   A. Schellekens and N. Warner, *Anomalies, characters and strings*, Nucl. Phys., **B287** (1987) 317.

Chapter 10

[10.1]   E. Cremmer and J. Scherk, *Dual models in four-dimensions with internal symmetries*, Nucl. Phys., **B103** (1976) 399.
[10.2]   K. Kikkawa and M. Yamasaki, *Casimir effects in superstring theories*, Phys. Lett., **149B** (1984) 357.
[10.3]   N. Sakai and I. Senda, *Vacuum energies of string compactified on torus*, Prog. Theor. Phys., **75** (1986) 692.
[10.4]   F. Englert and A. Neveu, *Nonabelian compactification of the interacting bosonic string*, Phys. Lett., **163B** (1985) 349.
[10.5]   K. Narain, *New heterotic string theories in uncompactified dimensions < 10*, Phys. Lett., **169B** (1986) 41.
[10.6]   K. Narain, H. Sarmardi and E. Witten, *A note on toroidal compactification of heterotic string theory*, Nucl. Phys., **B279** (1987) 369.
[10.7]   A. Giveon, E. Rabinovici and G. Veneziano, *Duality in string background space*, Nucl. Phys., **B322** (1989) 167.
[10.8]   A. Shapere and F. Wilczek, *Selfdual models with theta terms*, Nucl. Phys., **B320** (1989) 669.
[10.9]   A. Giveon, E. Malkin and E. Rabinovici, *The Riemann surface in the target space and vice versa*, Phys. Lett., **B220** (1989) 551; *On discrete symmetries and fundamental domains of target space*, Phys. Lett., **B238** (1990) 57.
[10.10]   W. Nahm, *Supersymmetries and their representations*, Nucl. Phys., **B135** (1978) 149.
[10.11]   C. Campbell and P. West, *$N = 2$ $D = 10$ nonchiral supergravity and its spontaneous compactification.* Nucl. Phys., **B243** (1984) 112.
[10.12]   M. Huq and M. Namazie, *Kaluza–Klein supergravity in ten dimensions*, Class. Q. Grav., **2** (1985) 293.
[10.13]   F. Giani and M. Pernici, *$N = 2$ supergravity in ten dimensions*, Phys. Rev., **D30** (1984) 325.

[10.14] J, Schwarz and P. West, *Symmetries and transformation of chiral* $N = 2$ $D = 10$ *supergravity*, Phys. Lett., **126B** (1983) 301.

[10.15] P. Howe and P. West, *The complete* $N = 2$ $D = 10$ *supergravity*, Nucl. Phys., **B238** (1984) 181.

[10.16] J. Schwarz, *Covariant field equations of chiral* $N = 2$ $D = 10$ *supergravity*, Nucl. Phys., **B226** (1983) 269.

[10.17] A. H. Chamseddine, *Interacting supergravity in ten dimensions: the role of the six-index gauge field*, Phys. Rev., **D24** (1981) 3065.

[10.18] E. Bergshoeff, M. de Roo, B. de Wit and P. van Nieuwenhuizen, *Ten-dimensional Maxwell–Einstein supergravity, its currents, and the issue of its auxiliary fields*, Nucl. Phys., **B195** (1982) 97.

[10.19] L. Brink, J. Scherk and J. H. Schwarz, *Supersymmetric Yang–Mills theories*, Nucl. Phys., **B121** (1977) 77.

[10.20] G. Chapline and N. S. Manton, *Unification of Yang–Mills theory and supergravity in ten dimensions*, Phys. Lett., **120B** (1983) 105.

[10.21] N. Marcus and A. Sagnotti, *Tree level constraints on gauge groups for type I superstrings*, Phys. Lett., **119B** (1982) 97.

[10.22] L. Alvarez-Gaume and E. Witten, *Gravitational anomalies*, Nucl. Phys., **B234** (1984) 269.

[10.23] M. Green and J. Schwarz, *Anomaly cancellation in supersymmetric* $D = 10$ *gauge theory and superstring theory*, Phys. Lett., **149B** (1984) 117.

[10.24] J. Thierry-Mieg, *Remarks concerning the* $E_8 \times E_8$ *and* D_{16} *string theories*, Phys. Lett., **156B** (1985) 199.

[10.25] G. Segal, *Unitarity representations of some infinite dimensional groups*, Comm. Math. Phys., **80** (1981) 301.

[10.26] I. B. Frenkel and V. G. Kac, *Basic representations of affine Lie algebras and dual resonance models*, Invent. Math., **62** (1980) 23.

[10.27] P. Goddard and D. Olive, *Algebras, lattices and strings*, in *Vertex Operators in Mathematics and Physics*, eds J. Lepowsky, S. Mandelstam, I. Singer, (1985) Springer-Verlag.

[10.28] P. Freund, *Superstrings from twenty six-dimensions?*, Phys. Lett., **151B** (1985) 387.

[10.29] D. Gross, J. Harvey, E. Martinec and R. Rohm, *The heterotic string*, Phys Rev. Lett., **54** (1985) 502; *Heterotic string theory. 1. The free heterotic string*, Nucl. Phys., **B256**, 253 (1985); *Heterotic string theory. 2. The interacting heterotic string*, Nucl. Phys., **B267**, 75 (1986).

[10.30] A. Tseytlin, *Duality and dilaton*, Mod. Phys. Lett., **A6** (1991) 1721. M. Rocek and E. Verlinde, *Duality, quotients, and currents*, Nucl. Phys., **B373** (1992) 630; S. Schwarz and A. Tseytlin, *Dilaton shift under duality and torsion of elliptic complex*, Nucl. Phys., **B399** (1993) 691.

Chapter 11

[11.1] C. Thorn, *A detailed study of the physical state conditions in the covariantly quantised string*, Nucl. Phys., **B286** (1987) 61.

[11.2] E. Del Giudice, P. Di Vecchia and S. Fubini, *General properties of the dual resonance model*, Ann. Phys., **70** (1972) 378.

[11.3] L. Brink and D. Olive, *The physical state projection operator in dual resonance models for the critical dimension of space-time*, Nucl. Phys., **B56**, (1973) 253; *Recalculation of the the the unitary single planar dual loop in the critical dimension of space time*, Nucl. Phys., **B58** (1973) 237.

[11.4] M. Freedman and D. Olive, *The calculation of planar one loop diagrams in string theory using the BRS formalism*, Phys. Lett., **175B** (1986) 155.

Chapter 12

[12.1] L. Singh and C. Hagen, *Lagrangian formulation for arbitrary spin. 1. The boson case*, Phys. Rev., **D9** (1974) 898.

[12.2] A. Neveu, A. and P. West, *Gauge covariant local formulation of bosonic strings*, Nucl. Phys., **B268** (1986) 125.

[12.3] A. Neveu, H. Nicolai and P. West, *New symmetries and ghost structure of covariant string theories*, Phys. Lett., **167B** (1986) 307.

[12.4] E. Witten, *Noncommutative geometry and string field theory*, Nucl. Phys., **B268** (1986) 253.

[12.5] W. Siegel, *Covariantly second quantized string*, Phys. Lett., **142B** (1984) 276; *Covariantly second quantized string 2*, Phys. Lett., **149B** (1984) 157; *Covariantly Second Quantized String 3*, Phys. Lett., **149B** (1984) 162.

[12.6] A. Neveu, H. Nicholai and P. West, *Gauge covariant local formulation of free strings and superstrings*, Nucl. Phys., **B264** (1986) 573.

[12.7] A. Neveu, J. Schwarz and P. West, *Gauge symmetries of the free bosonic string field theory*, Phys. Lett., **164B** (1985) 51.

[12.8] T. Banks and M. Peskin, *Gauge invariance of string fields*, Nucl. Phys., **B264** (1986) 513.

[12.9] W. Siegel and B. Zweibach, *Gauge string fields*, Nucl. Phys., **B263** (1986) 105.

[12.10] A. Neveu and P. West, *The interacting gauge covariant bosonic string*, Phys. Lett., **B168** (1986) 192; *Symmetries of the interacting gauge covariant bosonic string*, Nucl. Phys., **B278** (1986) 601.

[12.11] T. Banks, M. Peskin, C. Preitschopf, D. Friedan and E. Martinec, *All free string theories are theories of forms*, Nucl. Phys., **B274** (1986) 71.

[12.12] A. Ballestrero and E. Maina, *Ramond–Ramond closed string field theory*, Phys. Lett., **182B** (1986) 317.

[12.13] Y. Kazama, A. Neveu, H. Nicolai, H. and P. West, *Symmetry structures of superstring field theories*, Nucl. Phys., **B276** (1986) 366.

[12.14] Y. Kazama, A. Neveu, H. Nicolai, and P. West, *Space-time supersymmetry of the covariant superstring*, Nucl. Phys., **B278** (1986) 833.

[12.15] J. Araytn and A. Zimerman, *Ghosts and the physical modes in the covariant free string field theories*, Int. J. Mod. Phys., **A1** (1986) 421; *On covariant formulation of the free Neveu–Schwarz and Ramond string models*, Phys. Lett., **166B** (1986) 130.

[12.16] H. Terao and S. Uehara, *Gauge invariant actions and gauge fixed actions of free superstring field theory*, Phys. Lett., **173** (1986) 134.

Chapter 13

[13.1] E. Cremmer, B. Julia and J. Scherk, *Supergravity theory in eleven-dimensions*, Phys. Lett., **76B** (1978) 409.

[13.2] D. Freedman, P. van Nieuwenhuizen and S. Ferrara, *Progress toward a theory of supergravity*, Phys. Rev., **D13** (1976) 3214; Phys. Rev., **D14** (1976) 912.

[13.3] S. Deser and B. Zumino, *Consistent supergravity*, Phys. Lett., **62B** (1976) 335.

[13.4] K. Stelle and P. West, *Minimal auxiliary fields for supergravity*, Phys. Lett., **B74** (1978) 330.

[13.5]   S. Ferrara and P. van Nieuwenhuizen, *The auxiliary fields of supergravity*, Phys. Lett., **B74** (1978) 333.

[13.6]   P. West, *Representations of Supersymmetry*, in *Supergravity 81*, eds. S. Ferrara and J. Taylor (1982) Cambridge University Press.

[13.7]   J. Wess and B. Zumino, Superspace formulation of supergravity, *Phys. Lett.*, **66B** (1977) 361; *Superfield Lagrangian for supergravity*, Phys. Lett., **74B** (1978) 51.

[13.8]   A. Chamseddine and P. West, *Supergravity as a gauge theory of supersymmetry*, Nucl. Phys., **B129** (1977) 39.

[13.9]   S. MacDowell and F. Mansouri, *Unified geometric theory of gravity and supergravity*, Phys. Rev. Lett., **38** (1977) 739.

[13.10] M. Kaku, P. van Niewenhuizen and P. K. Townsend, *Properties of conformal supergravity*, Phys. Rev., **D17** (1978) 3179.

[13.11] R. Kraichnan, *Special-relativistic derivation of generally covariant gravitation theory*, Phys. Rev., **98** (1955) 1118; A. Papapetrou, *Einstein's theory of gravitation and flat space*, Proc. Roy. Irish Acad., **52A** (1948) 11; S. Gupta, *Quantization of Einstein's gravitational field: general treatment*, Proc. Phys. Soc. London, **A65** (1952) 608; R. Feynman, contribution in Chapel Hill conference (1956); R. Feynman, F. Morinigo, W. Wagner and B. Hatfield, contribution in 'Feynman lectures on gravitation,' Addison-Wesley (1995) 232 (this article can be found at http://www.slac.stanford.edu/spires/find/hep/www?irn=3475700); W. Thirring, Forschritte der Physik, **7** (1959) 79; S. Desser, *Self-interaction and gauge invariance*, Gen. Rel. Grav., **1** (1970) 9.

[13.12] P. K. Townsend and P. van Nieuwenhuizen, *Geometrical interpretation of extended supergravity*, Phys. Lett., **67B** (1977) 439.

[13.13] S. Ferrara and P. van Nieuwenhuizen, *Tensor calculus for supergravity*, Phys. Lett., **76B** (1978) 404; K. S. Stelle and P. West, *Tensor calculus for the vector multiplet coupled to supergravity*, Phys. Lett., **77B** (1978) 376; S. Ferrara and P. van Nieuwenhuizen, *Structure of supergravity*, Phys. Lett., **78B** (1978) 573; K. S. Stelle and P. West, *Relation between vector and scalar multiplets and invariance in supergravity*, Nucl. Phys., **B145** (1978) 175.

[13.14] J. Wess and B. Zumino, *Supergauge transformations in four dimensions*, Nucl. Phys., **B70** (1974) 139; *A Lagrangian model invariant under supergauge transformations*, Phys. Lett., **49B** (1974) 52.

[13.15] J. Wess and B. Zumino, *Supergauge invariant extension of quantum electrodynamics*, Nucl. Phys., **B78** (1974) 1; S. Ferrara and B. Zumino, *Supergauge invariant Yang–Mills theories*, Nucl. Phys., **B79** (1974) 413; A. Salam and J. Strathdee, *Supersymmetry and nonabelian gauges*, Phys. Lett., **51B** (1974) 353.

[13.16] V. Akulov, D. Volkov and V. Soroka, *Gauge fields on superspaces with different holonomy groups*, JETP Lett., **22** (1975) 187.

[13.17] R. Arnowitt, P. Nath and B. Zumino, *Superfield densities and action principle in curved superspace*, Phys. Lett., **56B** (1975) 81; P. Nath and R. Arnowitt, *Generalized supergauge symmetry as a new framework for unified gauge theories*, Phys. Lett., **56B** (1975) 177; *Superspace formulation of supergravity*, Phys. Lett., **78B** (1978) 581.

[13.18] N. Dragon, *Torsion and curvature in extended supergravity*, Z. Phys., **C2** (1979) 29; E. Ivanov and A. Sorin, *Superfield formulation of Osp(1, 4) supersymmetry*, J. Phys. A. Math. Gen., **13** (1980) 1159.

[13.19] T. Kibble, *Lorentz invariance and the gravitational field*, J. Maths. Phys., **2** (1961) 212.

[13.20] K. S. Stelle and P. West, *Spontaneously broken de sitter symmetry and the gravitational holonomy group*, Phys. Rev., **D21**, (1980) 1466.

[13.21] Th. Kaluza, *On the problem of unity in physics*, Sitzungdber. Berl. Akad. (1921) 966; O. Klein, *Quantum theory and five-dimensional theory of relativity*, Z. Phys., **37** (1926) 895.

[13.22] N. Lambert and P. West, *Enhanced coset Symmetries and higher derivative corrections*, Phys. Rev., **D74** (2006) 065002, arXiv:hep-th/0603255.

[13.23] J. Goldstone, *Field theories with superconductor solutions*, Nuov. Cim., **19** (1961) 154. J. Goldstone, A. Salam and S. Weinberg, *Broken symmetries*, Phys. Rev., **127** (1962) 965.

[13.24] S. Coleman, J. Wess and B. Zumino, *Structure of phenomenological Lagrangians 1*, Phys. Rev., **177** (1969) 2239; C. Callan, S. Coleman, J. Wess and B. Zumino, *Structure of phenomenological Lagrangians 2*, Phys. Rev., **177** (1969) 2247.

[13.25] E. Ivanov and V. Ogievetsky, *Gauge theories as theories of spontaneous breakdown*, Lett. Math. Phys., **1** (1976) 309; *Gauge theories as theories of spontaneous breakdown*, JETP Lett., **23** (1976) 606.

[13.26] D. V. Volkov, *Phenomological Lagrangians*, Sov. J. Part. Nucl., **4** (1973) 3, D. V. Volkov and V. P. Akulov, *Possible universal neutrino interaction*, JETP Lett., **16** (1972) 438; *Is the Neutrino a Goldstone Particle?*, Phys. Lett., **46B** (1973) 109.

[13.27] A. Salam and J. Strathdee, *Nonlinear realizations. 1: The role of Goldstone bosons*, Phys. Rev., **184** (1969) 1750, C. Isham, A. Salam and J. Strathdee, *Spontaneous, breakdown of conformal symmetry*, Phys. Lett., **31B** (1970) 300; *Non-linear realisations of space-time symmetries. scalar and tensor gravity*, Ann. Phys., **62** no 1 (1971) 98.

[13.28] C. Lovelace, *Strings in curved space*, Phys. Lett., **135B** (1984) 75; K. Callan, D. Friedan, E. Martinec and M. Perry, *Strings in background fields*, Nucl. Phys., **B262** (1985) 593. For a review see A. Tseytlin, *Sigma model approach to string theory*, Int. J. Mod. Phys., **A4** (1989) 1257.

[13.29] E. Witten, *Some properties of O(32) superstrings*, Phys. Lett., **149B** (1984) 351; E. Fradkin and A. Tseytlin, *Effective field theory from quantized strings*, Phys. Lett., **158B** (1985) 316.

[13.30] P. Townsend, *The eleven dimensional supermembrane revisited*, Phys. Lett., **350B** (1995) 184, arXiv:hep-th/9501068.

[13.31] E. Witten, *String theory dynamics in various dimensions*, Nucl. Phys., **B443** (1995) 85, arXiv:hep-th/9503124.

[13.32] N. Marcus and J. Schwarz, *Field theories that have no manifestly Lorentz invariant formulation*, Phys. Lett., **115B** (1982) 111.

[13.33] M. Green and J. Schwarz, *Superstring interactions*, Phys. Lett., **122B** (1983) 143.

[13.34] S. Ferrara, J. Scherk and B. Zumino, *Algebraic properties of extended supersymmetry*, Nucl. Phys., **B121** (1977) 393; E. Cremmer, J. Scherk and S. Ferrara, *SU(4) invariant supergravity theory*, Phys. Lett., **74B** (1978) 61.

[13.35] E. Cremmer and B. Julia, *The N = 8 supergravity theory. I. The Lagrangian*, Phys. Lett., **80B** (1978) 48.

[13.36] B. Julia, in *Group disintegrations*, in *Superspace Supergravity*, eds. S.W. Hawking and M. Roček (1981) Cambridge University Press; E. Cremmer, *Dimensional Reduction In Field Theory And Hidden Symmetries In Extended Supergravity*, in Trieste Supergravity School 1981, p. 313; *Supergravities in 5 dimensions*, in *Superspace Supergravity*, eds. S.W. Hawking and M. Roček (1981) Cambridge University Press.

[13.37] N. Marcus and J. Schwarz, *Three-dimensional supergravity theories*, Nucl. Phys., **B228** (1983) 145.

[13.38] H. Nicolai, *The integrability of N = 16 supergravity*, Phys. Lett., **194B** (1987) 402; *On M-theory*, arXiv:hep-th/9801090; B. Julia nd H. Nicolai, *Conformal internal symmetry of 2-d sigma models coupled to gravity and a dilaton*, Nucl. Phys., **B482** (1996) 431, arXiv:hep-th/9608082.

[13.39] C. Pope unpublished notes, which can be found at faculty.physics.tamu.edu/pope/ihplec.pdf.

[13.40] E. Cremmer, B. Julia, H. Lu and C. Pope, *Higher dimensional origin of D = 3 coset symmetries*, arXiv:hep-th/9909099.

[13.41] N. Lambert and P. West, *Coset symmetries in dimensionally reduced bosonic string theory*, Nucl. Phys., **B615** (2001) 117, arXiv:hep-th/0107209.

[13.42] R. Geroch, *A method for generating solutions of Einstein's equations*, J. Math. Phys., **12** (1971) 918, **13** (1972) 394; J. Ehlers, Disertation, Hamburg University (1957).

[13.43] J. Maharana and J. Schwarz, *Noncompact symmetries in string theory*, Nucl. Phys., **B390** (1993) 3, arXiv:hep-th/9207016.

[13.44] G. Gibbons and K. Maeda, *Black holes and membranes in higher dimensional theories with dilaton fields*, Nucl. Phys., **B298** (1988) 741.

[13.45] G. Horowitz and A. Strominger, *Black strings and branes*, Nucl. Phys., **B360** (1991) 197.

[13.46] M. Duff and K. Stelle, *Multimembrane solutions of d = 11 supergravity*, Phys. Lett., **B253** (1991) 113.

[13.47] R. Guven, *Black p-brane solutions of 11-dimensional supergravity*, Phys. Lett., **B276** (1992) 49.

[13.48] M. Duff and J. Lu, *Black and super p-branes in diverse dimensions*, Nucl. Phys., **B416** (1994) 301.

[13.49] M. Duff and J. Lu, *The selfdual type IIB superthreebrane*, Phys. Lett., **273B** (1991) 409.

[13.50] M. Duff and J. Lu, *Elementary five-brane solutions of D = 10 supergravity*, Nucl. Phys., **B354** (1991) 141.

[13.51] C. Callan, J. Harvey and A. Strominger, *Supersymmetric string solitons*, hep-th/9112030.

[13.52] A. Dabholkar, G. Gibbons, J. Harvey and F. Ruiz-Ruiz, *Superstrings and solitons*, Nucl. Phys., **B340** (1990) 33.

[13.53] H. Lu and C. Pope, *p-brane taxonomy*, arXiv:hep-th/9702086.

[13.54] M. J. Duff, R. R. Khuri, R. Minasian and J. Rahmfeld, *New black hole, string and membrane solutions of the four dimensional heterotic string*, Nucl. Phys., B418 (1994) 195.

[13.55] M. Duff and J. Lu, *The black branes of M theory*, arXiv:hep-th/9604052.

[13.56] G. Gibbons, *Antigravitating black hole solitons with scalar hair in N = 4 supergravity*, Nucl. Phys., **B207** (1982) 337.

[13.57] P. Dobiasch and D. Maison, *Stationary, spherically symmetric solutions of Jordan's unified theory of gravity and electromagnetism*, Gen. Rel. Grav., **14** (1982) 231.

[13.58] A. Chodos and S. Detweiler, *Spherically symmetric solutions in five-dimensional general relativity*, Gen. Rel. Grav., **14** (1982) 879.

[13.59] D. Pollard, *Antigravity and classical solutions of five-dimensional Kaluza–Klein theory*, J. Phys., **A16** (1983) 565.

[13.60] G. Papadopoulos and P. Townsend, *Intersecting M branes*, Phys., **B380** (1996) 80.

[13.61] A. Tseytlin, *Composite BPS configurations of p-branes in 10 and 11 dimensions*, Class. Quant. Grav., **14** (1997) 2085, arXiv:hep-th/9702163.

[13.62] M. Brinkman, Proc. Natl. Acad. Sci., USA, **9** (1923) 1.

[13.63] C. Hull, *Exact pp wave solutions of eleven dimensional supergravity*, Phys. Lett., **B139** (1984) 39.

[13.64] S. Han and I. Koh, *N = 4 supersymmetry remaining in Kaluza–Klein monopole background in D = 11 supergravity*, Phys. Rev., **D31** (1985) 2503.

[13.65] P. Freund and M. Rubin, *Dynamics of dimensional reduction*, Phys. Lett., **B97** (1980) 233.

[13.66] J. Kowalski-Glikman, *Vacuum states in supersymmetric Kaluza–Klein theory*, Phys. Lett., **B134** (1984) 194.

[13.67] J. Gauntlett, D. Kastor and J. Traschen, *Overlapping branes in M theory*, Nucl. Phys., **B478** (1996) 544.

[13.68] M. Duff, G. Gibbons and P. Townsend, *Macroscopic superstrings as interpolating solitons*, Phys. Lett., **B332** (1994) 321.

[13.69] G. Gibbons, G. Horowitz and P. Townsend, *Higher dimensional resolution of dilatonic black hole singularities*, Class. Quant. Grav., **12** (1995) 297.

[13.70] E. Bergshoeff, M. de Roo, M. Green, G. Papadopoulos and P. Townsend, *Duality of type II 7-branes and 8-branes*, Nucl. Phys., **B470** (1996) 113, arXiv:hep-th/9601150.

[13.71] L. Romans, *Massive N = 2a supergravity in ten dimensions*, Phys. Lett., **B169** (1986) 374.

[13.72] J. Polchinski and E. Witten, *Evidence for heterotic type I string duality*, Nucl. Phys., **B460** (1996) 525, hep-th/9510169.

[13.73] G. Gibbons, M. Green and M. Perry, *Instantons and seven-branes in type IIB superstring theory*, hep-th/9511080. *Duality of type II 7-branes and 8-branes*, hep-th/9601150.

[13.74] J. Polchinski, *Dirichlet branes and Ramond–Ramond charges*, Phys. Rev Lett., **75** (1995) 4724.

[13.75] J. Schwarz, *Covariant field equations of chiral N = 2 D = 10 supergravity*, Nucl. Phys., **B226** (1983) 269.

[13.76] M. Blau, J. Figueroa-O'Farrill, C. Hull and G. Papadopoulos, *A new maximally supersymmetric background of IIB superstring theory*, JHEP, 0201(2002) 047, arXiv:hep-th/0110242.

[13.77] A de Azcarraga, J. Gauntlett, J. Izquierdo and P.K. Townsend, *Topological extensions of the supersymmetry algebra for extended objects*, Phys. Rev. Lett., **63** (1989) 2443.

[13.78] E. Witten and D. Olive, *Supersymmetry algebras that include topological charges*, Phys. Lett., **78B** (1978) 97.

[13.79] C. Teitelboim, *Monopoles of higher rank*, Phys. Lett., **167B** (1986) 69.

[13.80] R. Nepomechie, *Magnetic monopoles from antisymmetric tensor gauge fields*, Phys. Rev., **D31** (1984) 1921.

[13.81] A. Borisov and V. Ogievetsky, *Theory of dynamical affine and conformal symmetries as gravity theory of the gravitational field*, Theor. Math. Phys., **21** (1975) 1179.

[13.82] V. Ogievetsky, *Infinite-dimensional algebra of general covariance group as the closure of the finite dimensional algebras of conformal and linear groups*, Nuov. Cim., **8** (1973) 988.

[13.83] S. de Haro, A. Sinkovics and K. Skenderis, *On a supersymmetric completion of the R^4 term in the IIB supergravity*, Phys. Rev., **D67** (2003) 084010, arXiv:hep-th/0210080.

Chapter 14

[14.1] M. Atiyah and N. Hitchen, *The Geometry and Dynamics of Magnetic Monopoles*, (1988) Princeton University Press.

[14.2] N. Manton, *A remark on the scattering of BPS monopoles*, Phys. Lett., **110B** (1982) 54.

[14.3] N. Manton and P. Sutcliffe, *Topological Solitons*, (2007) Cambridge University Press.

[14.4] E. Bergshoeff, E. Sezgin and P. Townsend, *Properties of eleven dimensional supermembrane theory*, Ann. Phys., **185** (1988) 330.

[14.5] E. Bergshoeff, E. Sezgin and P. Townsend, *Supermembranes and eleven-dimensional supergravity*, Phys. Lett., **189** (1987) 75.

[14.6] A. Achucarro, J. Evans, P. Townsend and D. L. Wiltshire, *Super p-branes*, Phys. Lett., **198B** (1987) 441; J. Azcarraga and P. Townsend, *Topological extensions of the superymmetry algebra for extended objects*, Phys. Rev., **62** (1989) 1579.

[14.7] E. Bergshoff and E. Sezgin, *Twistor-like formulation of super p-branes*, Nucl. Phys., **B422** (1994) 329. Hep-th/9312168, E. Sezgin, *Space-time and worldvolume Supersymmetric Super p-brane actions*, hep-th/9312168.

[14.8] I. Bandos, D. Sorokin and and D. Volkov, *On the generalized action principle for superstrings and superbranes*, Phys. Lett., **352B** (1995) 269, arXiv:hep-th/9502141. I. Bandos, P. Pasti, D. Sorokin, M. Tonin and D. Volkov, *Superstrings and supermembranes in the doubly supersymmetric geometrical approach*, Nucl. Phys., **B446** (1995) 79, arXiv:hep-th/9501113.

[14.9] P. Howe and E. Sezgin, *Superbranes*, Phys. Lett., **390B** (1997) 133, arXiv:hep-th/9607227.

[14.10] P. Howe and E. Sezgin, $D = 11$, $p = 5$, Phys. Lett., **394B** (1997) 62, arXiv:hep-th/9611008.

[14.11] P. Howe, E. Sezgin and P. West, *Covariant fieldequations of the M theory five-brane*, Phys. Lett., **399B** (1997) 49, hep-th/9702008.

[14.12] P. Howe, E. Sezgin and P. West, *The six-dimensional self-dual tensor*, Phys. Lett., **B400** (1997) 255, hep-th/9702111.

[14.13] P. Howe, E. Sezgin, and P. West, *Aspects of superembeddings*, Contribution to the D. V. Volkov Memorial Volume, Lecture Notes in Physics, Vol. 509 (1998) Springer-Verlag; arXiv:hep-th/9705093

[14.14] A. Tseytlin, *Self-duallity of the Born–Infeld action and the Dirichlet 3-brane of type IIB superstring theory*, Nucl. Phys., **B469** (1996) 51; M. B. Green and M. Gutperle, *Comment on three-branes*, Phys. Lett., **377B** (1996) 28.

[14.15] M. Cederwall, A. Von Gussich, B. Nilsson and A. Westerberg, *The Dirichlet super-three-brane in ten-dimensional Type IIB supergravity*, Nucl Phys., **B490** (1997) 163, hep-th/9610148; E Bershoeff and P. Townsend, *Super branes*, Nucl Phys., **B490** (1997) 145, hep-th/9611173.

[14.16] M. Aganagic, C. Popescu and J. H. Schwarz, *D-brane actions with local kappa symmetry*, Phys. Lett., **393B** (1997) 311, hep-th/9610249.

[14.17] M. Cederwall, A. Von Gussich, B. Nilsson P. Sundell and A. Westerberg, *The Dirichlet super p-branes in ten-dimensional Type IIA and IIB supergravity*, Nucl. Phys., **B490** (1997) 179, hep-th/9611159.

[14.18] M. Aganagic, C. Popescu and J. Schwarz, *Gauge-invariant and gauge-fixed D-brane actions*, Nucl. Phys., **B495** (1997) 99, arXiv: hep-th/9612080.

[14.19] P. West, *Brane dynamics, central charges and* E_{11}, JHEP, **0503** (2005) 077, arXiv:hep-th/0412336.

[14.20] P. Howe and P. Townsend, *The massless superparticle as Chern–Simons mechanics*, Phys. Lett., **B259** (1991) 285; F. Delduc and E. Sokatchev, *Superparticle with extended worldline supersymmetry*, Class. Quant. Grav., **9** (1992) 361; A. Galperin, P. Howe and K. Stelle, *The superparticle and the Lorentz group*, Nucl. Phys., **B368** (1992) 248; A. Galperin and E. Sokatchev, *A twistor-like D = 10 superparticle action with manifest N = 8 world-line supersymmetry*, Phys. Rev., **D46** (1992) 714, arXiv:hep-th/9203051.

[14.21] F. Delduc, A. Galperin, P. S. Howe and E. Sokatchev, *A twistor formulation of the heterotic D = 10 superstring with manifest (8,0) worldsheet supersymmetry*, Phys. Rev., **D47** (1993) 578, hep-th/9207050; F. Delduc, E. Ivanov and E. Sokatchev, *Twistor-like superstrings with D = 3, 4, 6 target-superspace and N = (1, 0), (2, 0), (4, 0) world-sheet supersymmetry*, Nucl. Phys., **B384** (1992) 334, arXiv:hep-th/9204071.

[14.22] P. Pasti and M. Tonin, *Twistor-like formulation of the supermembrane in D = 11*, Nucl. Phys., **B418** (1994) 337, arXiv:hep-th/9303156.

[14.23] I. Bandos, K Lechner, A. Nurmagambetov, *et al.*, *Covariant action for the super fivebrane of M-theory*, Phys. Rev. Lett., **78** (1997) 4332, arXiv:hep-th/9701149.

[14.24] I. Bandos, *Generalized action principle and geometricalapproach for superstrings and super p-branes*, Mod. Phys. Lett., **A12** (1997) 799; arXiv:hep-th/9608094; I. Bandos, P. Pasti, D. Sorokin and M. Tonin, *Superbrane actions and geometrical approach*, in *Supersymmetry and Quantum Field Theory, Kharkar Conference* (1997); arXiv:hep-th/9705064; I. Bandos, D. Sorokin and M. Tonin, *Generalized action principle and superfield equations of motion for D = 10 Dp-branes*, Nucl. Phys., **B497** (1997) 275, arXiv:hep-th/9701127.

[14.25] E. Cremmer and S. Ferrara, *Formulation of eleven-dimensional supergravity in superspace*, Phys. Lett., **91B** (1980) 61.

[14.26] L. Brink and P. Howe, *Eleven-dimensional supergravity on the mass-shell in superspace*, Phys. Lett., **91B** (1980) 384.

[14.27] M. Perry and J. Schwarz, *Interacting chiral gauge fields in six-dimensions and Born–Infeld theory*, Nucl. Phys., **B489** (1997) 47, hep-th/9611065; M. Aganagic, J. Park, C. Popescu, and J. Schwarz, *Worldvolume action of the M-theory fivebrane*, Nucl. Phys., **B496**, 191; arXiv:hep-th/9701166.

[14.28] E. Witten, *The five-brane effective action in M theory*, J. Geom. Phys., **22** (1997) 103, arXiv:hep-th/9610234.

[14.29] P. Howe, N. Lambert and P. West, *The selfdual string soliton*, Nucl. Phys., **B515** (1998) 203, hep-th/9709014.

[14.30] A. Strominger, *Open P-branes*, Phys. Lett., **383B** (1996) 44, arXiv:hep-th/9512059.

[14.31] G. Gibbons, *Born–Infeld particles and Dirichlet p-branes*, Nucl. Phys., **B514** (1998) 603.

[14.32] P. Howe, N. Lambert and P. West, *Classical M-fivebrane dynamics and quantum $N = 2$ Yang–Mills*, Phys. Lett., **419** (1998) 79, arXiv:hep-th/9710034.

[14.33] N. Seiberg and E. Witten, *Electric–magnetic duality, monopole condensation, and confinement in $N = 2$ supersymmetric Yang–Mills theory*, Nucl. Phys., **B426** (1994) 19, hep-th/9407087; *Monopoles, Duality and Chiral Symmetry Breaking in $N = 2$ Supersymmetric QCD*, Nucl. Phys., **B431** (1994) 484, arXiv:hep-th/9408099.

[14.34] P. Howe, K. Stelle and P. West, *A class of finite four-dimensional supersymmetric field theories*, Phys. Lett., **124B** (1983) 55. For a review of perturbative results in rigid supersymmetric theories, see P. West, *Supersymmety and finiteness*, in *Proceedings of the 1983 Shelter Island II Conference on Qunatum Field Theory and Fundamental Problems of Physics*, eds. R. Jackiw, N. Kuri, S. Weinberg and E. Witten (1985) MIT Press.

[14.35] M. Grisaru and W. Siegel, *Supergraphity. 2. Manifestly Covariant Rules and Higher Loop Finiteness*, Nucl. Phys. **B201** (1982) 292; N. Seiberg, *Naturalness versus supersymmetric nonrenormalization theorems*, Phys. Lett., **318B** (1993) 469.

[14.36] C. Montonen and D. Olive, *Magnetic monopoles as gauge particles?*, Phys. Lett., **72B** (1977) 117.

[14.37] P. Argyres and A. Faraggi, *The vacuum structure and spectrum of $N = 2$ supersymmetric SU(N) gauge theory*, Phys. Rev. Lett., **74** (1995) 3931, hep-th/9411057; S. Klemm, W. Lerche, S. Theisen and S. Yankielowicz, *Simple singularities and $N = 2$ supersymmetric Yang–Mills theory*, Phys. Lett., **344B** (1995) 169, arXiv:hep-th/9411048

[14.38] E. Witten, *Solutions of four-dimensional field theories via M theory*, Nucl. Phys., **B500** (1997) 3, arXiv:hep-th/9703166.

[14.39] N. Lambert and P. West, *Gauge fields and M-fivebrane dynamics*, Nucl. Phys., **B524** (1998) 141, arXiv:hep-th/9712040.

Chapter 15

[15.1] E. Corrigan and D. Fairlie, *Off-shell states in dual resonance theory*, Nucl. Phys., **B91** (1975) 527.

[15.2] J. Dai, R. Leigh and J. Polchinski, *New connections between string theories*, Mod. Phys. Lett., **A4** (1989) 2073.

[15.3] R. Leigh, *Dirac–Born–Infeld action from Dirichlet sigma model*, Mod. Phys. Lett., **A4** (1989) 2767.

[15.4] J. Polchinski, *Dirichlet branes and Ramond–Ramond charges*, Phys. Rev. Lett., **75** (1995) 4724.

[15.5] J. Maldecena, *The large N limit of superconformal field theories*, Adv. Theor. Math. Phys., **2** (1997) 231, arXiv:hep-th/9711200.

[15.6] E. Witten, *Anti de Sitter sapce and holography*, Adv. Theor. Math. Phys., **2** (1998) 253, arXiv:hep-th/9802150.

[15.7] S. Gubser, I. Klebanov and A. Polyakov, *Gauge theory correlators from non-critical string theory*, Phys. Lett., **428B** (1998) 105, arXiv:hep-th/9802109

[15.8] E. Witten, *Bound states of strings and p-branes*, Nucl. Phys., **B460** (1996) 335, hep-th/9510135.

[15.9] L. Brink and H. Nielson, *A simple physical interpretation of the critical dimension of space-time in dual models*, Phys. Lett., **45B** (1973) 332.

[15.10] E. Whittaker and G. Watson, *A Course of Modern Analysis*, foruth edition (1927) page 267, Cambridge University Press.

[15.11] N. Lambert and P. West, *D-branes in the Green–Schwarz formalism*, Phys. Lett., **459B** (1999) 515, arXiv:hep-th/9905031

[15.12] M. Green, *Point-like states for type 2b superstrings*, Phys. Lett., **329B** (1994) 435, arXiv:hep-th/9403040.

[15.13] M. Bershadsky, V. Sadov and C. Vafa, *D-strings on D-manifolds*, Nucl. Phys., **463B** (1996) 398, arXiv:hep-th/9510225.

[15.14] J. Polchinski and E. Witten, *Evidence for heterotic type I string duality*, Nucl. Phys., **460B** (1996) 525, arXiv:hep-th/9510169.

[15.15] A. Hanany and E. Witten, *Type IIB superstrings, BPS monopoles, and three-dimensional gauge dynamics*, Nucl. Phys., **492B** (1997) 152, arXiv:hep-th/9611230.

Chapter 16

[16.1] B. Weybourne, *Classical Groups for Physicists* (1974) Wiley; R. Gilmore, *Lie Groups, Lie Algebras, and Some of Their Applications* (2006) Dover; H. Georgi, *Lie algebras in Particle Physics* (1982) Benjamin Cummings; R. Cahn, *Semi-simple Lie Algebras and Their Representations* (1984) Benjamin Cummings.

[16.2] J.-P. Serre, *Complex Semisimple Lie Algebras* (1987) Springer Verlag: J. Humphries, *Introduction to Lie Algebras and Representation Theory* (1972) Springer-Verlag.

[16.3] R. Carter, *Lie Algebras of Finite and Affine Type* (2005) Cambridge University Press.

[16.4] V. Kac, *Graded Lie algebras and symmetric spaces*, Funct. Anal. Appl., **2** (1968) 183.

[16.5] R. Moody, *A new class of lie algebras*, J. Algebra, **10** (1968) 211.

[16.6] V. Kac, *Infinite Dimensional Lie Algebras* (1983) Birkhauser.

[16.7] R. Moody and A. Pianzola, *Lie Algebras with Triangular Decompositions*, (1995) J. Wiley.

[16.8] S. Krass, R. Moody, J. Patera and S. Slansky, *Affine Lie Algebras, Weight Multiplicities and Branching Rules* (1990) University of California Press.

[16.9] M. Gaberdiel, D. Olive and P. West, *A class of Lorentzian Kac–Moody algebras*, Nucl. Phys., **B645** (2002) 403, arXiv:hep-th/0205068.

[16.10] P. West, E_{11} *and M Theory*, Class. Quant. Grav., **18** (2001) 4443, hep-th/0104081.

[16.11] F. Englert, L. Houart, A. Taormina and P. West, *The symmetry of M-theories*, JHEP, **0309** (2003) 020, arXiv:hep-th/0304206.

[16.12] J. H. Conway, N. J. A. Sloane, *Sphere Packings, Lattices and Groups* (1988) Springer-Verlag.

[16.13] P. West, E_{11} *origin of Brane charges and U-duality multiplets*, JHEP, **0408** (2004) 052, arXiv:hep-th/0406150.

[16.14] T. Damour, M. Henneaux and H. Nicolai, *E(10) and a 'small tension expansion' of M theory*, Phys. Rev. Lett., **89** (2002) 221601, arXiv:hep-th/0207267.

[16.15] P. West, *Very extended E_8 and A_8 at low levels, gravity and supergravity*, Class. Quant. Grav., **20** (2003) 2393, arXiv:hep-th/0212291.

[16.16] P. West and A. Kleinschmidt, *Representations of G^{+++} and the role of space-time*, JHEP, **0402** (2004) 033, arXiv:hep-th/0312247.

[16.17] P. West, E_{11}, *SL(32) and central charges*, Phys. Lett., **575B** (2003) 333, arXiv:hep-th/0307098.

[16.18] I. B. Frenkel and V. G. Kac, *Basic representations of affine Lie algebras and dual resonance models*, Invent. Math., **62** (1980) 23.

[16.19] G. Segal, *Unitarity representations of some infinite dimensional groups*, Comm. Math. Phys., **80** (1981) 301.

[16.20] P. Goddard, D. Olive and A. Schwimmer, *Vertex operators for non-simply-laced algebras*, Commun. Math. Phys., **107** (1986) 179.

[16.21] R. Gebert and H. Nicolai, *On E_{10} and the DDF construction*, Commun. Math. Phys., **172** (1995) 571, arXiv:hep-th/9406175

[16.22] R. Gebert, H. Nicolai and P. West, *Multistring vertices and hyperbolic Kac Moody Algebras*, Int. J. Mod. Phys., **A11** (1996) 429. arXiv:hep-th/9505106.

[16.23] O. Barwald, R. Gebert, M. Gunaydin and H. Nicolai, *Missing modules, the gnome Lie algebra, and E(10)*, Commun. Math. Phys., **195** (1998) 29, arXiv:hep-th/9703084

[16.24] J. Conway and J. Algebra, **80** (1983) 159; J. Conway and N. Sloane, Bull. Am. Math. Soc., **6** (2) (1982) 215.

[16.25] G. Moore, *Finite in all directions*, arXiv:hep-th/9305139.

[16.26] P. West, *Physical states and string symmetries*, Int. J. Mod. Phys., **A10** (1995) 761. arXiv:hep-th/9411029.

[16.27] R. Gilbert, *Introduction to vertex algebras, Borcherds algebras and the monster Lie algebra*, Int. J. Mod. Phys., **8** (1993) 5441.

[16.28] G. Segal and A. Pressley, *Loop Groups* (1986) Oxford University Press.

Chapter 17

[17.1] T. Buscher, *A symmetry of the string background field equations*, Phys. Lett., **B194** (1987) 59, **B201** (1988) 466.

[17.2] P. Horava, *Background duality of open string models*, Phys. Lett., **B231** (1989) 251.

[17.3] P. A. M Dirac, *Quantized singularities in the electromagnetic field*, Proc. R. Soc. Lond., **A133** (1931) 60.

[17.4] T. T. Wu and C. N. Yang, *Concept of nonintegrable phase factors and global formulation of gauge fields*, Phys. Rev., **D12** (1975) 3845.

[17.5] D. Zwanziger, *Quantum field theory of particles with both electric and magnetic charges*, Phys. Rev., **176** (1968) 1489.

[17.6] J. Schwinger, *A magnetic model of matter*, Science, **165** (1969) 757.

[17.7] G. 't Hooft, *Magnetic monopoles in unified gauge theories*, Nucl. Phys., **B79** (1974) 276.

[17.8] A. M. Polyakov, *Particle spectrum in the quantum field theory*, JETP Lett., **20** (1974) 194.

[17.9] E. Bogomol'nyi, *Stability of classical solutions*, Sov J. Nucl. Phys., **24** (1975) 449.

[17.10] M. Prasad and C. Sommerfeld, *An exact classical solution for the 't Hooft monopole and the Julia–Zee dyon*, Phys. Rev. Lett., **35** (1975) 760.

[17.11] S. Coleman, A. Neveu, S. Parke and C. M. Sommerfeld, *Can one dent a dyon?*, Phys. Rev., **D15** (1977) 544.

[17.12] B. Julia and A. Zee, *Poles with both magnetic and electric charges in nonabelian gauge theory*, Phys. Rev., **D11** (1975) 2227.

[17.13] E. Weinberg, *Parameter counting for multi-monopole solutions*, Phys. Rev., **D20** (1979) 936.

[17.14] C. Montonen and D. Olive, *Magnetic monopoles as gauge particles?*, Phys. Lett., **72B** (1977) 117.

[17.15] E. Witten, *Dyons of charge e theta/2 pi*, Phys. Lett., **86B** (1979) 283.

[17.16] P. West, *Supersymmetric effective potential*, Nucl. Phys., **B106** (1976) 219.

[17.17] M. Sohnius and P. West, *Conformal invariance in N = 4 supersymmetric Yang–Mills theory*, Phys. Lett., **100B** (1981) 245; S. Mandelstam, *Light cone superspace and the ultraviolet finiteness of the N = 4 model*, Nucl. Phys., **B213** (1983) 149.

[17.18] H. Osborn, *Topological charges for N = 4 supersymmetric gauge theories and monopoles of spin 1*, Phys. Lett., **83B** (1979) 321.

[17.19] A. Sen, *Dyon – monopole bound states, selfdual harmonic forms on the multi – monopole moduli space, and SL(2,Z) invariance in string theory*, Phys. Lett., **329B** (1994) 217.

[17.20] D. Olive, *Introduction to dualing* in *Duality and Supersymmetric Theories*, eds., by D. Olive and P. West (1999) Cambridge University Press.

[17.21] J. Gauntlett, *Supersymmetric monopoles and duality* in *Duality and Supersymmetric Theories*, eds. D. Olive and P. West (1999) Cambridge University Press.

[17.22] A. Font, L. Ibanez, D. Lust and F. D. Quevedo, *Strong–weak coupling duality and nonperturbative effects in string theory*, Phys. Lett., **249B** (1990) 35.

[17.23] S. J. Rey, *The confining phase of superstrings and axionic strings*, Phys. Rev., **D43** (1991) 526.

[17.24] A. Sen, *Quantization of dyon charge and electric magnetic duality in string theory*, Phys. Lett., **303B** (1993) 22; *Strong–weak coupling duality in four-dimensional string theory*, Int. J. Mod. Phys., **A9** (1994) 3707.

[17.25] J. Schwarz and A. Sen, *Duality symmetric actions*, Nucl. Phys., **B411** (1994) 35.

[17.26] C. M. Hull and P. K. Townsend, *Unity of superstring dualities*, Nucl. Phys., **B438** (1995) 109, hep-th/9410167.

[17.27] M. Dine, P. Huet and N. Seiberg, *Large and small radius in string theory*, Nucl. Phys., **B322** (1989) 301.

[17.28] E. Bergshoeff, C. Hull and T. Ortin, *Duality in the type-II superstring effective action*, Nucl. Phys., **B451** (1995) 547, hep-th/9504081.

[17.29] S. Sethi and M. Stern, *D-brane bound states redux*, Commun. Math. Phys., **194** (1998) 675, hep-th/9705046.

[17.30] A. Sen, *An Introduction to Non-Perturbative String Theory*, in *Duality and Supersymmetric Theories*, eds., D. Olive and P. West (1999) Cambridge University Press.

[17.31] M. J. Duff, P. S. Howe, T. Inami and K. S. Stelle, *Superstrings in D = 10 from supermembranes in D = 11.*, Phys. Lett., **191B** (1987) 70.

[17.32] M. Duff, J. Lu, and R. Minasian, *Eleven-dimensional origin of string-string duality: A one loop test*, Nucl. Phys., **B452** (1995) 261.

[17.33] H. Nicolai and T. Fischbacher, *Low level representations of E_{10} and E_{11}*, in *Kac-Moody Lie Algebras and Related Topics, the Proceedings of the Ramanujan International Symposium on Kac–Moody Algebras and Applications*, ISKMAA-2002, Chennai, India (2002); arXiv:hep-th/0301017.

[17.34] P. West, *Hidden superconformal symmetry in M theory*, JHEP, **08** (2000) 007, arXiv:hep-th/0005270.

[17.35] I. Schnakenburg and P. West, *Kac–Moody symmetries of IIB supergravity*, Phys. Lett., **517B** (2001) 421, arXiv:hep-th/0107181.

[17.36] A. Kleinschmidt, I. Schnakenburg and P. West, *Very-extended Kac–Moody algebras and their interpretation at low levels*, Class. Quant. Grav, **21** (2004) 2493; arXiv: hep-th/0309198.

[17.37] I. Schnakenburg and P. West, *Massive IIA supergravity as a non-linear realisation*, Phys. Lett., **540B** (2002) 137, arXiv:hep-th/0204207.

[17.38] E. A. Bergshoeff, M. de Roo, S. F. Kerstan, T. Ortin and F. Riccioni, *IIA ten-forms and the gauge algebras of maximal supergravity theories*, JHEP, **0607** (2006) 018, arXiv:hep-th/0602280.

[17.39] P. West, *E_{11}, ten forms and supergravity*, JHEP, **0603** (2006) 072, arXiv:hep-th/0511153.

[17.40] P. West, *The IIA, IIB and eleven dimensional theories and their common E_{11} origin*, Nucl. Phys., **B693** (2004) 76-102, arXiv:hep-th/0402140.

[17.41] F. Riccioni and P. West, *The E_{11} origin of all maximal supergravities*, JHEP, **0707** (2007) 063; arXiv:0705.0752.

[17.42] E. A. Bergshoeff, I. De Baetselier and T. A. Nutma, *E(11) and the embedding tensor*, JHEP, **0709** (2007) 047, arXiv: arXiv:hep-th/0705.1304.

[17.43] P. Townsend, *Cosmological constant in supergravity*, Phys. Rev., **D15** (1977) 2802.

[17.44] B. de Wit and H. Nicolai, *N = 8 supergravity with local $SO(8) \times SU(8)$ invariance*, Phys. Lett., **108B** (1982) 285; *N = 8 supergravity*, Nucl. Phys., **208B** (1982) 323.

[17.45] H. Samtleben and M. Weidner, *The maximal D = 7 supergravities*, Nucl. Phys., **725B** (2005) 383, arXiv:hep-th/0506237.

[17.46] B. de Wit, H. Samtleben and M. Trigiante, *On Lagrangians and gaugings of maximal supergravities*, Nucl. Phys., **655B** (2003) 93, arXiv:hep-th/0212239.

[17.47] B. de Wit, H. Samtleben and M. Trigiante, *Magnetic charges in local field theory*, JHEP, **0509** (2005) 016, arXiv:hep-th/0507289.

[17.48] H. Nicolai and H. Samtleben, *Maximal gauged supergravity in three dimensions*, Phys. Rev. Lett., **86** (2001) 1686, arXiv:hep-th/0010076.

[17.49] B. de Wit, H. Samtleben and M. Trigiante, *The maximal D = 5 supergravities*, Nucl. Phys., **716B** (2005) 215, arXiv:hep-th/0412173.

[17.50] F. Riccioni and P. West, *E(11)-extended spacetime and gauged supergravities*, JHEP, **0802** (2008) 039, arXiv:0712.1795

[17.51] F. Riccioni and P. West, *Local E(11)*, JHEP, **0904** (2009) 051. arXiv:0902.4678.

[17.52] F. Riccioni, D. Steele and P. West, *The E(11) origin of all maximal supergravities – the hierarchy of field-strengths*, JHEP, **0909** (2009) 095, arXiv:0906.1177.

[17.53] E. Cremmer, B. Julia, H. Lu and C. N. Pope, *Dualisation of dualities*, Nucl. Phys., **B523** (1998) 73, hep-th/9710119; *Dualisation of dualities, II: Twisted self-duality of doubled fields and superdualities*, Nucl. Phys., **B535** (1998) 242, arXiv:hep-th/9806106.

[17.54] F. Riccioni and P. West, *Dual fields and E_{11}*, Phys. Lett., **645B** (2007) 286-292, arXiv:hep-th/0612001.

[17.55] P. West, *Brane dynamics, central charges and E_{11}*, JHEP, **0503** (2005) 077, hep-th/0412336.

[17.56] P. Cook and P. West, *Charge multiplets and masses for E(11)*, JHEP, **11** (2008) 091, arXiv:0805.4451.

[17.57] S. Elitzur, A. Giveon, D. Kutasov and E. Rabinovici, *Algebraic aspects of matrix theory on T^d*, Nucl. Phys., **509B** (1998) 122, hep-th/9707217.

[17.58] N. Obers, B. Pioline and E. Rabinovici, *M-theory and U-duality on T^d with gauge backgrounds*, Nucl. Phys., **525B** (1998) 163, arXiv: hep-th/9712084.

[17.59] N. Obers and B. Pioline, *U-duality and M-theory, an algebraic approach*, hep-th/9812139; N. Obers and B. Pioline, *U-duality and M-theory*, Phys. Rep., **318** (1999) 113, arXiv: hep-th/9809039.

[17.60] F. Riccioni and P. West, *Local E(11)*, JHEP, **0904** (2009) 051, arXiv:hep-th/0902.4678.

[17.61] P. West, *E11, Generalised space-time and IIA string theory*, Phys. Lett. B to be published, arXiv:1009.2624

[17.62] P. West, *Generalised space-time and duality*, Phys. Lett., **693B** (2010) 373, arXiv:1006.0893.

[17.63] P. West, *Some simple predictions from E_{11} symmetry*, Phys. Lett., **603B** (2004) 63, arXiv:hep-th/0407088.

Chapter 18

[18.1] R. Dolan, D. Horn and C. Schmid, *Prediction of Regge parameters of rho poles from low-energy pi N data*, Phys. Rev. Lett., **19** (1967) 402; Phys. Rev., **166** (1968) 1968.

[18.2]   M. Ademollo, H. R. Rubinstein, G. Veneziano and M. A. Virasoro, *Bootstraplike conditions from superconvergence*, Phys. Rev. Lett., **14** (1967) 1402; Phys. Rev. 176 (1968) 1904.

[18.3]   S. Mandelstam, *Dynamics based on rising Regge trajectories*, Phys. Rev., **166** (1968) 1539.

[18.4]   C. Schmid, *Meson bootstrap with finite-energy sum rules*, Phys. Rev. Lett., **20** (1968) 628.

[18.5]   C. Schmid and J. Yellin, *Finite-energy sum rules and the process 0-minus + 0-minus —¿ 0-minus + 0-minus*, Phys. Rev., **182** (1969) 1449.

[18.6]   G. Veneziano, *Construction of a crossing – symmetric, Regge behaved amplitude for linearly rising trajectories*, Nuov. Cim., **57A** (1968)190.

[18.7]   A. Erdelyi, *Higher Transcendental Functions, The Bateman manuscript project*, vds I–III 1953 McGraw-Hill.

[18.8]   J. Eden, P. Landshoff, D. Olive and J. Polkinghorn, *Analytic S Matrix* (1966) Cambridge University Press.

[18.9]   K. Bardakci and H. Ruegg, *Reggeized resonance model for the production amplitude*, Phys. Lett., **28B** (1968) 342.

[18.10] M. A. Virasoro, *Generalization of veneziano's formula for the five-point function*, Phys. Rev. Lett., **22** (1969) 37.

[18.11] K. Bardakci and H. Ruegg, *Reggeized resonance model for arbitrary production processes*, Phys. Rev., **181** (1969) 1884.

[18.12] H. M. Chan and S. T. Tsou, *Explicit construction of the n-point function in the generalized veneziano model*, Phys. Lett., **28B** (1969) 485.

[18.13] C. G. Goebel and B. Sakita, *Extension of the Veneziano form to N-particle amplitudes*, Phys. Rev. Lett., **22** (1969) 257.

[18.14] Z. Koba and H. B. Nielsen, *Reaction amplitude for n mesons: A generalization of the Veneziano–Bardakci–Ruegg–Virasora model*, Nucl. Phys., **B10** (1969) 633.

[18.15] H. B. Nielsen, *An almost physical interpretation of the dual N point function, Proc. 15th International Conference on High Energy Physics, Kiev (1970).* Preprint (1969) Nordita.

[18.16] D. Farlie and H. B. Nielsen, *An analog model for ksv theory*, Nucl. Phys., **B20** (1969) 637.

[18.17] L. Susskind, *Dual-symmetric theory of hadrons. 1*, Nuov. Cim., **69A** (1970) 457; *Structure of hadrons implied by duality*, Phys. Rev., **D1** (1970) 1182.

[18.18] D. Fairlie, *The analogue model for string amplitudes*, to be published in *The Birth of String Theory*, eds. A. Cappelli, E. Castellani, F. Colomo and P. Di Vecchia, Cambridge University Press and other articles in this volume.

[18.19] S. Sciuto, *The general vertex function in dual resonance models*, Nuov. Cim., **2** (1969) 411.

[18.20] L. Caneschi, A. Schwimmer and G. Veneziano, *Twisted propagator in the operatorial duality formalism*, Phys. Lett., **30B** (1969) 351.

[18.21] V. Alessandrini, *A general approach to dual multiloop diagrams*, Nuov. Cim., **2A** (1971) 321; V. Allessendrini and D. Amati, *Properties of dual multiloop amplitudes*, Nuov. Cim., **4A** (1971) 793; C. Lovelace, *M-loop generalized veneziano formula*, Phys. Lett., **32** (1970) 703.

[18.22] V. Alessandrini, D. Amati, M. LeBellac and D. I. Olive, *The operator approach to dual multiparticle theory*, Phys. Rep., **1C** (1971) 170; as well as the supplement by D. Olive which appears immeadiately after this article.

[18.23] P. Frampton, *Dual Resonance Models*, (1974) Benjamin.

[18.24] S. Mandelstam, *Structural Analysis of Collision Amplitudes*, ed. Tran Than Van, Les Houches, June 1975, 593.

[18.25] C. Rebbi, *Dual models and relativistic quantum strings*, Phys. Rep., **12C** (1974) 1; J. Scherk, *An introduction to the theory of dual models and strings*, Rev. Mod. Phys., **47** (1975) 123.

[18.26] J. H. Schwarz, *Dual resonance theory*, Phys. Rep., **8C** (1973) 269.

[18.27] G. Veneziano, *An introduction to dual models of strong interactions and their physical motivations*, Phys. Rep., **9C** (1974) 199.

[18.28] J. Schwarz, *Superstring theory*, Phys. Rep., **89** (1982) 223.

[18.29] M. Green, *Supersymmetrical dual string theories and their field theory limits: A review*, Surv. High Energy Phys., **3** (1983) 127.

[18.30] C. Hsue, B. Sakita and M. Virasoro, *Formulation of dual theory in terms of functional integrations*, Phys. Rev., **D2** (1970) 2857.

[18.31] J. Gervais and B. Sakita, *Functional integral approach to dual resonance theory*, Phys. Rev., **D4** (1971) 2291.

[18.32] A. Polyakov, *Quantum geometry of fermionic strings*, Phys. Lett., **103B** (1981) 207, 211.

[18.33] O. Alvarez, *Theory of strings with boundaries: Fluctuations, topology, and quantum geometry*, Nucl. Phys., **B216** (1983) 125.

[18.34] P. Candelas, *Lectures on Complex Manifolds, Superstrings*, eds. L. Alvarez-Gaume, M. B Green, M. T. Grisaru, R. Iengo and E. Sezgin (1987) World Scientific.

[18.35] J. Polchinski, *Evaluation of the one loop string path integral*, Comm. Math. Phys., **104** (1986) 37.

[18.36] G. Moore and P. Nelson, *Absence of nonlocal anomalies in the polyakov string*, Nucl. Phys., **B266** (1986) 58.

[18.37] E. D'Hoker and D. Phong, *Multiloop amplitudes for the bosonic polyakov string*, Nucl. Phys., **B269** (1986) 205.

[18.38] A. Neveu and P. C. West, *Conformal mappings and the three string bosonic vertex*; Phys. Lett., **179B** (1986) 235; *Group theoretic approach to the perturbative string S matrix*, Phys. Lett., **193B** (1987) 187; *Cycling, twisting and sewing in the group theoretic approach to strings*, Commun. Math. Phys., **119** (1988) 585.

[18.39] A. Neveu and P. C. West, *A group theoretic method for string loop diagram*, Phys. Lett., **194B** (1987) 200; *Group theoretic approach to the open bosonic string multiloop S matrix*, Commun. Math. Phys., **114** (1988) 613; *String vertices and induced representations*; Nucl. Phys., **B320** (1989) 103.

[18.40] P. West and M. Freeman, *Ghost vertices for the bosonic string using the group theoretical approach to string theory*, Phys. Lett., **B205** (1988) 30.

[18.41] P. West, *Multiloop ghost vertices and the determination of the multiloop measure*, Phys. Lett., **205B** (1988) 38.

[18.42] A. Neveu and P. West, *The cyclic symmetric vertex for three arbitrary Neveu–Schwartz strings*, Phys. Lett., **180B** 34 (1986); *Neveu–Schwarz excited string scattering: A superconformal group computation*, Phys. Lett., **200B** (1988) 275; *Group theoretic approach to the superstring and its supermoduli*, Nucl. Phys., **311B** (1988) 79.

[18.43] M. D. Freeman and P. West, *Ramond string scattering in the group theoretic approach to string theory*, Phys. Lett., **217B** (1989) 259.

[18.44] N. Moeller and P. West, *Arbitrary four string scattering at high energy and fixed angle*, Nucl. Phys., **B729** (2005) 1, hep-th/0507152.

[18.45] P. West, *A brief review of the group theoretic approach to string theory*, in *Conformal Field Theories and Related Topics, Proceedings of Third Annecy Meeting on Theoretical Physics, LAPP, Annecy le Vieux, France*, eds. P. Binütruy, P. Sorba and R. Stora, Nucl. Phys. B (Proc. Suppl), **5B** (1988) 217.

[18.46] L. Alvarez-Gaume, C. Gomez and C. Reina, *Loop groups, Grassmanians and string theory*, Phys. Lett., **190B** (1987) 55.

[18.47] C. Vafa, *Operator formulation on Riemann surfaces*, Phys. Lett., **190B** (1987) 47.

[18.48] L. Alvarez-Gaume, C. Gomez, G. Moore and C. Vafa, *Strings in the operator formalism*, Nucl. Phys., **B303** (1988) 455.

[18.49] E. Cremmer and J. Gervais, *Combining and splitting relativistic strings*, Nucl. Phys., **B76** (1974) 209.

[18.50] E. Cremmer and J. Gervais, *Infinite field theory of interacting relativistic strings and dual theory*, Nucl. Phys., **B90** (1975) 410.

[18.51] J. Hopkinson, R. Tucker and P. Collins, *Quantum strings and the functional calculus*, Phys. Rev., **D12** (1975) 1653.

[18.52] S. Mandelstam, *Interacting-string picture of dual-resonance models*, Nucl. Phys., **B64** (1973) 205.

[18.53] H. Hata, K Itoh, T. Kugo, H. Kunitomo and K. Ogawa, *Covariant field theory of interacting string*, Phys. Lett., **172B** (1986) 186; *Manifestly covariant field theory of interacting string. 2*, Phys. Lett., **172B** (1986) 195.

[18.54] E. Cremmer, A. Schwimmer and C. Thorn, *The vertex function in Witten's formulation of string field theory*, Phys. Lett., **B179** (1986) 57; D. Gross and A. Jevicki, *Operator formulation of interacting string field theory*, Nucl. Phys., **B283** (1987) 1; *Operator formulation of interacting string field theory. 2*, Nucl. Phys., **B287** (1987) 225.

Index

Printed in the United States
by Baker & Taylor Publisher Services